International Congress of Mathematicians (ICM 2022)

Sections 1-4
Volume 3 of 7

Online
6-14 July 2022

Editors:

Dmitry Beliaev
Stanislav Smirnov

ISBN: 978-1-7138-9870-2

ICM

INTERNATIONAL CONGRESS OF MATHEMATICIANS
2022 JULY 6–14

SECTIONS 1–4

EDITED BY D. BELIAEV AND S. SMIRNOV

INTERNATIONAL MATHEMATICAL UNION
IMU

EMS PRESS

Editors

Dmitry Beliaev
Mathematical Institute
University of Oxford
Andrew Wiles Building
Radcliffe Observatory Quarter
Woodstock Road
Oxford OX2 6GG, UK

Email: belyaev@maths.ox.ac.uk

Stanislav Smirnov
Section de mathématiques
Université de Genève
rue du Conseil-Général 7–9
1205 Genève, Switzerland

Email: stanislav.smirnov@unige.ch

2020 Mathematics Subject Classification: 00B25

ISBN 978-3-98547-058-7, eISBN 978-3-98547-558-2, DOI 10.4171/ICM2022

Volume 1. Prize Lectures
ISBN 978-3-98547-059-4, eISBN 978-3-98547-559-9, DOI 10.4171/ICM2022-1
Volume 2. Plenary Lectures
ISBN 978-3-98547-060-0, eISBN 978-3-98547-560-5, DOI 10.4171/ICM2022-2
⟶ Volume 3. Sections 1–4
ISBN 978-3-98547-061-7, eISBN 978-3-98547-561-2, DOI 10.4171/ICM2022-3
Volume 4. Sections 5–8
ISBN 978-3-98547-062-4, eISBN 978-3-98547-562-9, DOI 10.4171/ICM2022-4
Volume 5. Sections 9–11
ISBN 978-3-98547-063-1, eISBN 978-3-98547-563-6, DOI 10.4171/ICM2022-5
Volume 6. Sections 12–14
ISBN 978-3-98547-064-8, eISBN 978-3-98547-564-3, DOI 10.4171/ICM2022-6
Volume 7. Sections 15–20
ISBN 978-3-98547-065-5, eISBN 978-3-98547-565-0, DOI 10.4171/ICM2022-7

This edition printed by Curran Associates, Inc.
ISBN 978-1-7138-9870-2

e-Book available: https://ems.press/books/standalone/275

No changes have been made to the content of these proceedings. There may be changes to pagination and minor adjustments for aesthetics.

Bibliographic information published by the Deutsche Nationalbibliothek
The Deutsche Nationalbibliothek lists this publication in the Deutsche Nationalbibliografie; detailed bibliographic data are available on the Internet at http://dnb.dnb.de.

Published by EMS Press, an imprint of the

European Mathematical Society – EMS – Publishing House GmbH
Institut für Mathematik
Technische Universität Berlin
Straße des 17. Juni 136
10623 Berlin, Germany

https://ems.press

CONTENTS

VOLUME 1

THE WORK OF THE FIELDS MEDALISTS AND THE IMU PRIZE WINNERS

PRIZE LECTURES

POPULAR SCIENTIFIC EXPOSITIONS

SUMMARIES OF PRIZE WINNERS' WORK

VOLUME 2

SPECIAL PLENARY LECTURES

PLENARY LECTURES

VOLUME 3

1. LOGIC

2. ALGEBRA

3. NUMBER THEORY – SPECIAL LECTURE

3. NUMBER THEORY

4. ALGEBRAIC AND COMPLEX GEOMETRY – SPECIAL LECTURE

4. ALGEBRAIC AND COMPLEX GEOMETRY

VOLUME 4

5. GEOMETRY – SPECIAL LECTURES

5. GEOMETRY

6. TOPOLOGY

7. LIE THEORY AND GENERALIZATIONS

8. ANALYSIS – SPECIAL LECTURE

8. ANALYSIS

VOLUME 5

9. DYNAMICS

10. PARTIAL DIFFERENTIAL EQUATIONS

11. MATHEMATICAL PHYSICS – SPECIAL LECTURE

11. MATHEMATICAL PHYSICS

VOLUME 6

12. PROBABILITY – SPECIAL LECTURE

12. PROBABILITY

13. COMBINATORICS – SPECIAL LECTURE

13. COMBINATORICS

14. MATHEMATICS OF COMPUTER SCIENCE – SPECIAL LECTURES

14. MATHEMATICS OF COMPUTER SCIENCE

VOLUME 7

15. NUMERICAL ANALYSIS AND SCIENTIFIC COMPUTING

16. CONTROL THEORY AND OPTIMIZATION – SPECIAL LECTURE

16. CONTROL THEORY AND OPTIMIZATION

17. STATISTICS AND DATA ANALYSIS

18. STOCHASTIC AND DIFFERENTIAL MODELLING

19. MATHEMATICAL EDUCATION AND POPULARIZATION OF MATHEMATICS

20. HISTORY OF MATHEMATICS

1. LOGIC

TAMENESS IN GEOMETRY AND ARITHMETIC: BEYOND O-MINIMALITY

GAL BINYAMINI AND DMITRY NOVIKOV

ABSTRACT

The theory of o-minimal structures provides a powerful framework for the study of geometrically tame structures. In the past couple of decades a deep link connecting o-minimality to algebraic and arithmetic geometry has been developing. It has been clear, however, that the axioms of o-minimality do not fully capture some algebro-arithmetic aspects of tameness that one may expect in structures arising from geometry. We propose a notion of *sharply o-minimal structures* refining the standard axioms of o-minimality, and outline through conjectures and various partial results the potential development of this theory in parallel to the standard one.

We illustrate some applications of this emerging theory in two main directions. First, we show how it can be used to deduce Galois orbit lower bounds—notably including in non-abelian contexts where the standard *transcendence methods* do not apply. Second, we show how it can be used to derive effectivity and (polynomial-time) computability results for various problems of unlikely intersection around the Manin–Mumford, André–Oort, and Zilber–Pink conjectures.

MATHEMATICS SUBJECT CLASSIFICATION 2020

Primary 03C64; Secondary 11U09, 32B20, 14G05

KEYWORDS

o-minimality, point counting, Pila–Wilkie theorem, André–Oort theorem, Yomdin–Gromov parameterization

INTERNATIONAL CONGRESS OF MATHEMATICIANS 2022 JULY 6–14

Proc. Int. Cong. Math. 2022, Vol. 3, pp. 1440–1461
DOI 10.4171/ICM2022/117

1. TAME GEOMETRY AND ARITHMETIC

1.1. O-minimal structures

The theory of o-minimal structures was introduced by van den Dries as an attempt to provide a framework of tame topology in the spirit of Grothendieck's "Esquisse d'un Programme" [42]. We refer the reader to this book for a general introduction to the subject and its history. For us, an o-minimal structure will always be an expansion of the ordered real field $\mathbb{R}_{\mathrm{alg}} := \{\mathbb{R}, +, \cdot, <\}$. Briefly, such an expansion is *o-minimal* if all definable subsets of $\mathbb{R}$ consist of finite unions of points and intervals.

Despite their apparent simplicity, it turns out that the axioms of o-minimality provide a broad framework of tame topology. In particular, one has good notions of dimension, smooth stratification, triangulation, and cell-decomposition for every definable set in an o-minimal structure. On the other hand, several natural and important structures turn out to be o-minimal. A few examples of particular importance for us in the present paper are $\mathbb{R}_{\mathrm{alg}}, \mathbb{R}_{\mathrm{an}}, \mathbb{R}_{\mathrm{an,exp}}$, and $\mathbb{R}_{\mathrm{Pfaff}}$. We will say a bit more on these in later sections.

1.2. Pila–Wilkie counting theorem

In [37], Pila and Wilkie discovered a "counting theorem" that would later find deep applications in arithmetic geometry. The theorem concerns the asymptotic density of rational (or algebraic) points in a definable set—as a function of height. We introduce this first, to motivate a broader discussion of the connection between tame geometry and arithmetic.

For $x \in \mathbb{Q}$, we denote by $H(x)$ the standard height of x. For a vector $x \in \mathbb{Q}^n$, we denote by $H(x)$ the maximum among the heights of the coordinates of x. For a set $A \subset \mathbb{R}^n$, we denote the set of $\mathbb{Q}$-points of A by $A(\mathbb{Q}) = A \cap \mathbb{Q}^n$ and denote

$$A(\mathbb{Q}, H) := \{x \in A(\mathbb{Q}) : H(x) \leqslant H\}. \tag{1.1}$$

For a set $A \subset \mathbb{R}^n$, we define the algebraic part A^{alg} of A to be the union of all connected semialgebraic subsets of A of positive dimension. We define the transcendental part A^{trans} of A to be $A \setminus A^{\mathrm{alg}}$.

Theorem 1 (Pila and Wilkie [37]). *Let $A \subset \mathbb{R}^m$ be a set definable in an o-minimal structure. Then for every $\epsilon > 0$ there exists a constant $C(A, \epsilon)$ such that for every $H \geqslant 1$,*

$$\#A^{\mathrm{trans}}(\mathbb{Q}, H) = C(A, \epsilon)H^{\epsilon}. \tag{1.2}$$

1.3. Transcendence methods, auxiliary polynomials

The use of transcendental (as opposed to algebraic) methods in the study of arithmetic questions has a long history. A common theme in these methods, running through the work of Schneider, Lang, Baker, Masser, and Wüstholz to name a few, is the use of *auxiliary polynomials*. We refer to [28] for a broad treatment of this subject.

The usefulness of polynomials in this context stems from their dual algebraic/analytic role. Suppose one is interested in the set $A(\mathbb{Q}, H)$ for some analytic set A. On the one hand, if a polynomial P, say, with integer coefficients, is evaluated at $x \in A(\mathbb{Q}, H)$ then $P(x)$ is again rational, and one can estimate its height in terms of H and the height of P. On the other hand, polynomials are extremely well-behaved analytic functions, and a variety of analytic methods may be used to prove upper bounds on the restriction of P to an analytic set A assuming it is appropriately constructed (say to vanish to high order at some points of A). One concludes from such an argument that P must vanish at every point in $A(\mathbb{Q}, H)$, for otherwise the height bound would contradict the upper bound.

The proof of the Pila–Wilkie counting theorem follows this classical line. However, it is fairly unique in the realm of transcendence methods in that the degrees of the auxiliary polynomials P are *independent* of the height, depending in fact only on ε. It is this unusual feature that makes it possible to prove the Pila–Wilkie theorem in the vast generality of o-minimal structures: polynomials of a given degree form a definable family, and the general machinery of o-minimality gives various finiteness statements uniformly for all such polynomials.

1.4. Beyond Pila–Wilkie theorem: the Wilkie conjecture

By contrast with the Pila–Wilkie theorem, most transcendence methods require the degrees of the auxiliary polynomials to depend on the height H of the points being considered—sometimes logarithmically and in some cases, such as the Schneider–Lang theorem, even linearly. A famous conjecture that seems to fall within this category is due to Wilkie.

Conjecture 2 (Wilkie [37]). *Let $A \subset \mathbb{R}^m$ be a set definable in $\mathbb{R}_{\exp}$. Then there exist constants $C(A), \kappa(A)$ such that for all $H \geqslant 3$,*

$$\#A^{\text{trans}}(\mathbb{Q}, H) = C(A)(\log H)^{\kappa(A)}. \tag{1.3}$$

The conclusion of the Wilkie conjecture is known to fail for general o-minimal structures, for instance, in $\mathbb{R}_{\text{an}}$ [40]. To achieve such asymptotics, it seems one would have to use auxiliary polynomials of degrees $d = (\log H)^q$, and o-minimality places no restrictions on the geometric complexity as a function of d.

In formulating his conjecture, Wilkie was probably influenced by Khovanskii's theory of fewnomials [25]. The latter implies fairly sharp bounds for the number of connected components of sets defined using algebraic and exponential functions (and more generally Pfaffian functions) as a function of the degrees of the equations involved. Below we attempt to axiomatize what it would mean for an arbitrary o-minimal structure to satisfy such sharp complexity bounds.

2. SHARPLY O-MINIMAL STRUCTURES

In this section we introduce *sharply o-minimal structures*, which are meant to endow a standard o-minimal structure with an appropriate notions comparable to dimension and degree in the algebraic case, and provide suitable control over these parameters under the basic logical operations. We first introduce the notion of a *format-degree* filtration (abbreviated *FD*-filtration) on a structure $\mathcal{S}$. This is a collection $\Omega = \{\Omega_{\mathcal{F},D}\}_{\mathcal{F},D\in\mathbb{N}}$ such that each

$\Omega_{\mathcal{F},D}$ is a collection of definable sets (possibly of different ambient dimensions), with

$$\Omega_{\mathcal{F},D} \subset \Omega_{\mathcal{F}+1,D} \cap \Omega_{\mathcal{F},D+1} \quad \forall \mathcal{F}, D \in \mathbb{N} \tag{2.1}$$

and $\bigcup_{\mathcal{F},D} \Omega$ is the collection of all definable sets in $\mathcal{S}$. We call the sets in $\Omega_{\mathcal{F},D}$ sets of *format* $\mathcal{F}$ and *degree* D. However, note that the format and degree of a set are not uniquely defined since Ω is a filtration rather than a partition.

We now come to the notion of a sharply o-minimal structure.

Definition 3 (Sharply o-minimal structure). A sharply o-minimal structure is a pair $\Sigma := (\mathcal{S}, \Omega)$ consisting of an o-minimal expansion of the real field $\mathcal{S}$ and an FD-filtration Ω; and for each $\mathcal{F} \in \mathbb{N}$, a polynomial $P_{\mathcal{F}}(\cdot)$ such that the following holds:

If $A \in \Omega_{\mathcal{F},D}$ then

(1) if $A \subset \mathbb{R}$, it has at most $P_{\mathcal{F}}(D)$ connected components,

(2) if $A \subset \mathbb{R}^{\ell}$ then $\mathcal{F} \geq \ell$,

(3) $A^c, \pi_{\ell-1}(A), A \times \mathbb{R}$, and $\mathbb{R} \times A$ lie in $\Omega_{\mathcal{F}+1,D}$.

Similarly if $A_1, \ldots, A_k \subset \mathbb{R}^{\ell}$ with $A_j \in \Omega_{\mathcal{F}_j,D_j}$ then

$$(4) \quad \bigcup_i A_i \in \Omega_{\mathcal{F},D}, \quad (5) \quad \bigcap_i A_i \in \Omega_{\mathcal{F}+1,D},$$

where $\mathcal{F} := \max_j \mathcal{F}_j$ and $D = \sum_j D_j$. Finally,

(6) if $P \in \mathbb{R}[x_1, \ldots, x_\ell]$ then $\{P = 0\} \in \Omega_{\ell,\deg P}$.

Given a collection $\{A_\alpha\}$ of sets generating a structure $\mathcal{S}$, and associated formats and degrees $\mathcal{F}_\alpha, D_\alpha$ one can consider the minimal FD-filtration Ω satisfying the axioms (2)–(6) above. We call this the FD-filtration generated by $\{(A_\alpha, \mathcal{F}_\alpha, D_\alpha)\}$. This will be sharply o-minimal if and only if axiom (1) is satisfied.

Definition 4 (Reduction of FD-filtrations). Let Ω, Ω' be two FD-filtrations on a structure $\mathcal{S}$. We say that Ω is *reducible* to Ω' and write $\Omega \leqslant \Omega'$ if there exist functions $a : \mathbb{N} \to \mathbb{N}$ and $b : \mathbb{N} \to \mathbb{N}[D]$ such that

$$\Omega_{\mathcal{F},D} \subset \Omega'_{a(\mathcal{F}),[b(\mathcal{F})](D)} \quad \forall \mathcal{F}, D \in \mathbb{N}. \tag{2.2}$$

We say that Ω, Ω' are equivalent if $\Omega \leqslant \Omega' \leqslant \Omega$.

We will usually try to prove that certain measures of complexity of definable sets depend polynomially on the degree, thinking of the format as constant. If one can prove such a statement for Ω'-degrees, and $\Omega \leqslant \Omega'$, then the same statement holds for Ω-degrees and in this sense Ω is reducible to Ω'.

Remark 5 (Effectivity). One can require further that a sharply o-minimal structure is *effective*, in the sense that the polynomial $P_{\mathcal{F}}(D)$ in Definition 3 is given by some explicit primitive recursive function of $\mathcal{F}$. Similarly, one may require a reduction $\Omega \leqslant \Omega'$ to be effective. Unless otherwise stated, all constructions in this paper are effective in this sense.

2.1. Examples and nonexamples

2.1.1. The semialgebraic structure

Consider the structure $\mathbb{R}_{\text{alg}}$ with the FD-filtration Ω generated by all algebraic hypersurfaces $\{P = 0\}$ with the format given by the ambient dimension and the degree given by $\deg P$. Then $(\mathbb{R}_{\text{alg}}, \Omega)$ is a sharply o-minimal structure. This is not an immediate statement: it follows from the results on effective cell decomposition, or elimination of quantifiers, in semialgebraic geometry [3].

Perhaps a more natural notion of format and degree in the semialgebraic category is as follows. Define $\Omega'_{\mathcal{F},D}$ to be the subsets of $\mathbb{R}^\ell$ with $\ell \leqslant \mathcal{F}$, that can be written as a union of basic sets

$$\{P_1 = \cdots = P_k = 0,\ Q_1 > 0, \ldots, Q_l > 0\} \tag{2.3}$$

with the sum of the degrees of the P_i and Q_j, over all basic sets, bounded by D. This is not sharply o-minimal according to our definition because it does not satisfy axiom (3), for instance. However, it is equivalent to Ω defined above.

2.1.2. The analytic structure $\mathbb{R}_{\text{an}}$

Not surprisingly, $\mathbb{R}_{\text{an}}$ is not sharply o-minimal with respect to any FD-filtration. Assume the contrary. Let $\omega_1 = 1$ and $\omega_{n+1} = 2^{\omega_n}$, and let $\Gamma = \{y = f(z)\} \subset \mathbb{C}^2$ denote the graph of the holomorphic function $f(z) = \sum_{j=1}^{\infty} z^{\omega_j}$ restricted to the disc of radius $1/2$ (which is definable in $\mathbb{R}_{\text{an}}$). Then by axioms (1), (5) and (6), the number of points in

$$\Gamma \cap \left\{ y = \varepsilon + \sum_{j=1}^{n} z^{\omega_j} \right\} \tag{2.4}$$

should be polynomial in ω_n, with the exact polynomial depending on the format and degree of Γ. But it is, in fact, $\omega_{n+1} = 2^{\omega_n}$ for $0 < \varepsilon \ll 1$, and we have a contradiction for $n \gg 1$.

2.1.3. Pfaffian structures

Let $B \subset \mathbb{R}^\ell$ be a domain, which for simplicity we take to be a product of (possibly infinite) intervals. A tuple $f_1, \ldots, f_m : B \to \mathbb{R}$ of analytic functions is called a *Pfaffian chain* if they satisfy a triangular system of algebraic differential equations of the form

$$\frac{\partial f_i}{\partial x_j} = P(x_1, \ldots, x_\ell, f_1, \ldots, f_i), \quad \forall i, j. \tag{2.5}$$

They are called *restricted* if B is bounded and $f_1, \ldots, f_m$ extend as real analytic functions to $\bar{B}$. A Pfaffian function is a polynomial $Q(x_1, \ldots, x_\ell, f_1, \ldots, f_m)$. We denote the structure generated by the Pfaffian functions by $\mathbb{R}_{\text{Pfaff}}$, and its restricted analog by $\mathbb{R}_{\text{rPfaff}}$.

Khovanskii [25] proved upper bounds for the number of connected components of systems of Pfaffian equations. This was later extended by Gabrielov and Vorobjov to sets defined using inequalities and quantifiers [21]. However, their results fall short of establishing the sharp o-minimality of $\mathbb{R}_{\text{rPfaff}}$. The problem is that for Gabrielov–Vorobjov's notion of format and degree, if $A \in \Omega_{\mathcal{F},D}$ then they are only able to show that $A^c \in \Omega_{P_{\mathcal{F}}(D), P_{\mathcal{F}}(D)}$

rather than $A^c \in \Omega_{\mathcal{F}+1, P_{\mathcal{F}}(D)}$ as required by our axioms. This is a fundamental difficulty, as it is essential in our setup that the format never becomes dependent on the degree.

In [13] the first author and Vorobjov introduce a modified notion of format and degree and prove the following.

Theorem 6. *There is an FD-filtration Ω on $\mathbb{R}_{\text{rPfaff}}$ that makes it into a sharply o-minimal structure. Moreover, Gabrielov–Vorobjov's standard filtration is reducible to Ω.*

We conjecture that this theorem extends to the structure $\mathbb{R}_{\text{Pfaff}}$, and this is the subject of work in progress by the first author and Vorobjov utilizing some additional ideas of Gabrielov [19].

2.2. Cell decomposition in sharply o-minimal structures

We recall the notion of a *cell* in an o-minimal structure. A cell $C \subset \mathbb{R}$ is either a point or an open interval (possibly infinite). A cell $C \subset \mathbb{R}^{\ell+1}$ is either the graph of a definable continuous function $f : C' \to \mathbb{R}$ where $C' \subset \mathbb{R}^{\ell}$ is a cell, or the area strictly between two graphs of such definable continuous functions $f, g : C' \to \mathbb{R}$ satisfying $f < g$ identically on C'. One can also take $f = -\infty$ and $g = \infty$ in this definition.

We say that a cell $C \subset \mathbb{R}^{\ell}$ is *compatible* with $X \subset \mathbb{R}^{\ell}$ if it is either strictly contained, or strictly disjoint from X. The following cell decomposition theorem can be viewed as the raison d'être of the axioms of o-minimality.

Theorem 7 (Cell decomposition). *Let $X_1, \dots, X_k \subset \mathbb{R}^{\ell}$ be definable sets. Then there is a decomposition of $\mathbb{R}^{\ell}$ into pairwise disjoint cells that are pairwise compatible with $X_1, \dots, X_k$.*

Given the importance of cell decomposition in the theory of o-minimality, it is natural to pose the following question.

Question 8. If $\mathcal{S}$ is sharply o-minimal and $X_1, \dots, X_k$ have format $\mathcal{F}$ and degree D, can one find a cell-decomposition where each cell has format $\operatorname{const}(\mathcal{F})$, and the number of cells and their degrees are bounded by $\operatorname{poly}_{\mathcal{S},\mathcal{F}}(k, D)$?

We suspect the answer to this question may be negative. Since cell decomposition is perhaps the most crucial construction in o-minimality, this is a fundamental problem. The following result rectifies the situation.

Theorem 9. *Let $(\mathcal{S}, \Omega)$ be sharply o-minimal. Then there exists another FD-filtration Ω' with $(\mathcal{S}, \Omega')$ sharply o-minimal such that $\Omega \leqslant \Omega'$, and in Ω' the following holds.*

Let $X_1, \dots, X_k \in \Omega'_{\mathcal{F},D}$, all subsets of $\mathbb{R}^{\ell}$. Then there exists a cell decomposition of $\mathbb{R}^{\ell}$ compatible with each X_j such that each cell has format $\operatorname{const}(\mathcal{F})$*, the number of cells is* $\operatorname{poly}_{\mathcal{F}}(k, D)$*, and the degree of each cell is* $\operatorname{poly}_{\mathcal{F}}(D)$.

In the structure $\mathbb{R}_{\text{rPfaff}}$, Theorem 9 is one of the main results of [13]. The general case is obtained by generalizing the proof to the general sharply o-minimal case, and is part of the PhD thesis of Binyamin Zack-Kutuzov.

2.3. Yomdin–Gromov algebraic lemma in sharply o-minimal structures

Let $I := (0,1)$. For $f : I^n \to \mathbb{R}^m$ a C^r-smooth map, we denote

$$\|f\|_r := \sup_{x \in I^n} \max_{|\alpha| \leqslant r} \left\| f^{(\alpha)}(x) \right\|. \tag{2.6}$$

The Yomdin–Gromov algebraic lemma is a result about C^r-smooth parametrizations of bounded norm for definable subsets of I^n. A sharply o-minimal version of this lemma is as follows.

Lemma 10. *Let $(\mathcal{S}, \Omega)$ be sharply o-minimal. Then there is a polynomial $P_{\mathcal{F},r}(\cdot)$ depending on the pair $(\mathcal{F}, r)$, such that for every $A \in \Omega_{\mathcal{F},D}$ the following holds. There exist a collection of maps $\{f_\alpha : I^{n_\alpha} \to A\}$ of size at most $P_{\mathcal{F},r}(D)$ such that $\bigcup_\alpha f_\alpha(I^{n_\alpha}) = A$; and $\|f_\alpha\|_r \leqslant 1$ and $n_\alpha \leqslant \dim A$ for every α.*

In the algebraic case, this result is due to Gromov [23], based on a similar but slightly more technically involved statement by Yomdin [44]. In the general o-minimal case, but without complexity bounds, the result is due to Pila and Wilkie [37]. In the restricted Pfaffian case, this result is due to the first author with Jones, Schmidt, and Thomas [6] using Theorem 9 in the $\mathbb{R}_{\mathrm{rPfaff}}$ case. The general case follows in the same way.

The following conjecture seems plausible, though we presently do not have an approach to proving it in this generality.

Conjecture 11. *In Lemma 10, one can replace $P_{\mathcal{F},r}(D)$ by a $P_{\mathcal{F}}(D, r)$, i.e., by a polynomial in both D and r, depending only on $\mathcal{F}$.*

In the structure $\mathbb{R}_{\mathrm{alg}}$ this was conjectured by Yomdin (unpublished) and by Burguet [14], in relation to a conjecture of Yomdin [45, CONJECTURE 6.1] concerning the rate of decay of the tail entropy for real-analytic mappings. The conjecture was proved in [9] by complex-analytic methods. We will say more about the possible generalization of these methods to more general sharply o-minimal structures in Section 3.

2.4. Pila–Wilkie theorem in sharply o-minimal structures

We now state a form of the Pila–Wilkie counting theorem, Theorem 1, with explicit control over the asymptotic constant.

Theorem 12. *Let $(\mathcal{S}, \Omega)$ be sharply o-minimal. Then for every $\epsilon > 0$ and $\mathcal{F}$ there is a polynomial $P_{\mathcal{F},\epsilon}(\cdot)$ depending on $(\mathcal{S}, \Omega)$, such that for every $A \in \Omega_{\mathcal{F},D}$ and $H \geqslant 2$,*

$$\#A^{\mathrm{trans}}(\mathbb{Q}, H) = P_{\mathcal{F},\epsilon}(D) \cdot H^\epsilon. \tag{2.7}$$

This result is based on Lemma 10, in the same way as the classical Pila–Wilkie theorem is based on the o-minimal reparametrization lemma. This reduction is carried out in [6] using Theorem 9 in the $\mathbb{R}_{\mathrm{rPfaff}}$ case. The general case follows in the same way.

2.5. Polylog counting in sharply o-minimal structures

We state a conjectural sharpening of the Pila–Wilkie theorem, in line with the Wilkie conjecture, in the context of sharply o-minimal structures. For $A \subset \mathbb{R}^\ell$, let

$$A(g,h) := \{x \in \mathbb{A} \cap \bar{\mathbb{Q}}^\ell : [\mathbb{Q}(x) : \mathbb{Q}] \leqslant g, h(x) \leqslant h\}, \tag{2.8}$$

where $h(\cdot)$ denotes the logarithmic Weil height.

Conjecture 13. *Let $(\mathcal{S}, \Omega)$ be sharply o-minimal. Then there is a polynomial $P_{\mathcal{F}}(\cdot,\cdot,\cdot)$ depending only on $(\mathcal{S}, \Omega)$ and $\mathcal{F}$, such that for every $A \in \Omega_{\mathcal{F},D}$ and $g, h \geqslant 2$,*

$$\# A^{\mathrm{trans}}(g,h) \leqslant P_{\mathcal{F}}(D,g,h). \tag{2.9}$$

The conjecture sharpens Pila–Wilkie in two ways. First, we replace the subpolynomial term H^ε by a polynomial in $h \sim \log H$. Second, we count algebraic points of arbitrary degree, and stipulate polynomial growth with respect to the degree as well.

Conjecture 13 is currently known only for the structure of *restricted elementary functions* $\mathbb{R}^{\mathrm{RE}} := (\mathbb{R}, +, \cdot, <, \exp|_{[0,1]}, \sin|_{[0,\pi]})$ where it is due to [8] (with a minor technical improvement in [5]).

Combining the various known techniques in the literature, it is not hard to see that Conjecture 11 implies Conjecture 13 in a general sharply o-minimal structure. In Section 5 we will see that Conjecture 13 has numerous applications in arithmetic geometry, going beyond the standard applications of the Pila–Wilkie theorem. We also discuss some partial results in the direction of Conjecture 13 in Section 4.4.

3. COMPLEX ANALYTIC THEORY

In this section we consider holomorphic analogs of the standard cell decomposition of o-minimality. We fix a sharply o-minimal structure $(\mathcal{S}, \Omega)$ throughout. We also assume that $\mathcal{S}$ admits cell-decomposition in the sense of Theorem 9, as we may always reduce to this case.

3.1. Complex cells

We start be defining the notion of a *complex cell*. This is a complex analog of the cells used in o-minimal geometry.

3.1.1. Basic fibers and their extensions

For $r \in \mathbb{C}$ (resp. $r_1, r_2 \in \mathbb{C}$) with $|r| > 0$ (resp. $|r_2| > |r_1| > 0$), we denote

$$\begin{aligned} &D(r) := \{|z| < |r|\}, \quad D_\circ(r) := \{0 < |z| < |r|\}, \quad D_\infty(r) = \{|r| < |z| < \infty\}, \\ &A(r_1, r_2) := \{|r_1| < |z| < |r_2|\}, \quad * := \{0\}. \end{aligned} \tag{3.1}$$

For any $0 < \delta < 1$, we define the δ-extensions by

$$\begin{aligned} &D^\delta(r) := D(\delta^{-1} r), \quad D^\delta_\circ(r) := D_\circ(\delta^{-1} r), \quad D^\delta_\infty(r) := D_\infty(\delta r), \\ &A^\delta(r_1, r_2) := A(\delta r_1, \delta^{-1} r_2), \quad *^\delta := *. \end{aligned} \tag{3.2}$$

For any $0 < \rho < \infty$, we define the $\{\rho\}$-extension $\mathcal{F}^{\{\rho\}}$ of $\mathcal{F}$ to be $\mathcal{F}^{\delta}$ where δ satisfies the equations

$$\begin{aligned} \rho &= \frac{2\pi\delta}{1-\delta^2} \quad \text{for } \mathcal{F} \text{ of type } D, \\ \rho &= \frac{\pi^2}{2|\log\delta|} \quad \text{for } \mathcal{F} \text{ of type } D_\circ, D_\infty, A. \end{aligned} \tag{3.3}$$

The motivation for this notation comes from the following fact, describing the hyperbolic-metric properties of a domain $\mathcal{F}$ within its $\{\rho\}$-extension.

Fact 14. *Let $\mathcal{F}$ be a domain of type $A, D, D_\circ, D_\infty$ and let S be a component of the boundary of $\mathcal{F}$ in $\mathcal{F}^{\{\rho\}}$. Then the length of S in $\mathcal{F}^{\{\rho\}}$ is at most ρ.*

3.1.2. The definition of a complex cell

Let $\mathcal{X}, \mathcal{Y}$ be sets and $\mathcal{F} : \mathcal{X} \to 2^{\mathcal{Y}}$ be a map taking points of $\mathcal{X}$ to subsets of $\mathcal{Y}$. Then we denote

$$\mathcal{X} \odot \mathcal{F} := \{(x, y) : x \in \mathcal{X}, y \in \mathcal{F}(x)\}. \tag{3.4}$$

If $r : \mathcal{X} \to \mathbb{C} \setminus \{0\}$ then for the purpose of this notation we understand $D(r)$ as the map assigning to each $x \in \mathcal{X}$ the disc $D(r(x))$, and similarly for $D_\circ, D_\infty, A$.

We now introduce the notion of a complex cell of *length* $\ell \in \mathbb{Z}_{\geq 0}$. If $U \subset \mathbb{C}^n$ is a definable domain, we denote by $\mathcal{O}_d(U)$ the space of definable holomorphic functions on U. As a shorthand we denote $\mathbf{z}_{1..\ell} = (\mathbf{z}_1, \ldots, \mathbf{z}_\ell)$.

Definition 15 (Complex cells). A complex cell $\mathcal{C}$ of *length* zero is the point $\mathbb{C}^0$. A complex cell of length $\ell + 1$ has the form $\mathcal{C}_{1..\ell} \odot \mathcal{F}$ where the *base* $\mathcal{C}_{1..\ell}$ is a cell of length ℓ, and the *fiber* $\mathcal{F}$ is one of $*, D(r), D_\circ(r), D_\infty(r), A(r_1, r_2)$ where $r \in \mathcal{O}_d(\mathcal{C}_{1..\ell})$ satisfies $|r(\mathbf{z}_{1..\ell})| > 0$ for $\mathbf{z}_{1..\ell} \in \mathcal{C}_{1..\ell}$; and $r_1, r_2 \in \mathcal{O}_d(\mathcal{C}_{1..\ell})$ satisfy $0 < |r_1(\mathbf{z}_{1..\ell})| < |r_2(\mathbf{z}_{1..\ell})|$ for $\mathbf{z}_{1..\ell} \in \mathcal{C}_{1..\ell}$.

Next, we define the notion of a δ-extension (resp. $\{\rho\}$-extension).

Definition 16. The cell of length zero is defined to be its own δ-extension. A cell $\mathcal{C}$ of length $\ell + 1$ admits a δ-extension $\mathcal{C}^\delta := \mathcal{C}^\delta_{1..\ell} \odot \mathcal{F}^\delta$ if $\mathcal{C}_{1..\ell}$ admits a δ-extension, and if the function r (resp. r_1, r_2) involved in $\mathcal{F}$ admits holomorphic continuation to $\mathcal{C}^\delta_{1..\ell}$ and satisfies $|r(\mathbf{z}_{1..\ell})| > 0$ (resp. $0 < |r_1(\mathbf{z}_{1..\ell})| < |r_2(\mathbf{z}_{1..\ell})|$) in this larger domain. The $\{\rho\}$-extension $\mathcal{C}^{\{\rho\}}$ is defined in an analogous manner.

As a shorthand, when say that $\mathcal{C}^\delta$ is a complex cell (resp. $\mathcal{C}^{\{\rho\}}$) we mean that $\mathcal{C}$ is a complex cell admitting a δ (resp $\{\rho\}$) extension.

3.1.3. The real setting

We introduce the notion of a *real* complex cell $\mathcal{C}$, which we refer to simply as *real cells* (but note that these are subsets of $\mathbb{C}^\ell$). We also define the notion of *real part* of a real cell $\mathcal{C}$ (which lies in $\mathbb{R}^\ell$), and of a *real* holomorphic function on a real cell. Below we let $\mathbb{R}_+$ denote the set of positive real numbers.

Definition 17 (Real complex cells). The cell of length zero is real and equals its real part. A cell $\mathcal{C} := \mathcal{C}_{1..\ell} \odot \mathcal{F}$ is real if $\mathcal{C}_{1..\ell}$ is real and the radii involved in $\mathcal{F}$ can be chosen to be real holomorphic functions on $\mathcal{C}_{1..\ell}$; The real part $\mathbb{R}\mathcal{C}$ (resp. positive real part $\mathbb{R}_+\mathcal{C}$) of $\mathcal{C}$ is defined to be $\mathbb{R}\mathcal{C}_{1..\ell} \odot \mathbb{R}\mathcal{F}$ (resp. $\mathbb{R}_+\mathcal{C}_{1..\ell} \odot \mathbb{R}_+\mathcal{F}$) where $\mathbb{R}\mathcal{F} := \mathcal{F} \cap \mathbb{R}$ (resp. $\mathbb{R}_+\mathcal{F} := \mathcal{F} \cap \mathbb{R}_+$) except the case $\mathcal{F} = *$, where we set $\mathbb{R}* = \mathbb{R}_+* = *$; A holomorphic function on $\mathcal{C}$ is said to be real if it is real on $\mathbb{R}\mathcal{C}$.

3.2. Cellular parametrization

We now state a result that can be viewed as a complex analog of the cell decomposition theorem. We start by introducing the notion of prepared maps.

Definition 18 (Prepared maps). Let $\mathcal{C}, \hat{\mathcal{C}}$ be two cells of length ℓ. We say that a holomorphic map $f : \mathcal{C} \to \hat{\mathcal{C}}$ is *prepared* if it takes the form $\mathbf{w}_j = \mathbf{z}_j^{q_j} + \phi_j(\mathbf{z}_{1..j-1})$ where $\phi_j \in \mathcal{O}_d(\mathcal{C}_{1..j})$ for $j = 1, \ldots, \ell$.

Since our cells are always centered at the origin, it is the images of cellular maps that should be viewed as analogous to the cells of o-minimality. The additional exponent q_j in Definition 18 is needed to handle ramification issues that are not visible in the real context.

Definition 19. For a complex cell $\mathcal{C}$ and $F \in \mathcal{O}_d(\mathcal{C})$ we say that F is *compatible* with $\mathcal{C}$ if F vanishes either identically or nowhere on $\mathcal{C}$. For a cellular map $f : \hat{\mathcal{C}} \to \mathcal{C}$, we say that f is compatible with F if f^*F is compatible with $\hat{\mathcal{C}}$.

We will be interested in covering (real) cells by prepared images of (real) cells.

Definition 20. Let $\mathcal{C}^{\{\rho\}}$ be a cell and $\{f_j : \mathcal{C}_j^{\{\sigma\}} \to \mathcal{C}^{\{\rho\}}\}$ be a finite collection of cellular maps. We say that this collection is a *cellular cover* of $\mathcal{C}$ if $\mathcal{C} \subset \bigcup_j (f_j(\mathcal{C}_j))$. Similarly, we say it is a *real* cellular cover if $\mathbb{R}_+\mathcal{C} \subset \bigcup_j (f_j(\mathbb{R}_+\mathcal{C}_j))$.

Finally, we can state our main conjecture on complex cellular parametrizations.

Conjecture 21 (Cellular Parametrization Theorem, CPT). *Let $\rho, \sigma \in (0, \infty)$. Let $\mathcal{C}^{\{\rho\}}$ be a (real) cell and $F_1, \ldots, F_M \in \mathcal{O}_d(\mathcal{C}^{\{\rho\}})$ (real) holomorphic functions, with $\mathcal{C}^{\{\rho\}}$ and each F_j having format $\mathcal{F}$ and degree D. Then there exists a (real) cellular cover $\{f_j : \mathcal{C}_j^{\{\sigma\}} \to \mathcal{C}^{\{\rho\}}\}$ such that each f_j is prepared and compatible with each F_k. The number of cells is* $\text{poly}_{\mathcal{F}}(D, M, \rho, 1/\sigma)$*, and each of them has format* $\text{const}(\mathcal{F})$ *and degree* $\text{poly}_{\mathcal{F}}(D)$.

The main result of [9] is that Conjecture 21 holds in the structure $\mathbb{R}_{\text{alg}}$ (we assume there for technical convenience that the functions are bounded rather than just definable, but this does not seem to be a serious obstacle). We remark that there are significant difficulties with extending this proof to the general sharply o-minimal case.

3.3. Analytically generated structures

We say that a sharply o-minimal structure $(\mathcal{S}, \Omega)$ is *analytically generated* if there is a collection of complex cells $\{\mathcal{C}_\alpha\}$ admitting a $1/2$-extension, and associated formats and degrees $(\mathcal{F}_\alpha, D_\alpha)$ such that $\mathcal{S}$ is generated by $\{\mathcal{C}_\alpha\}$ and Ω is generated by $\{(\mathcal{C}_\alpha, \mathcal{F}_\alpha, D_\alpha)\}$. We fix such a structure $\mathcal{S}$ below. Assuming the CPT, one can prove the following analog of Theorem 9 giving a cell decomposition by real parts of complex analytic cells.

Theorem 22. *Let $(\mathcal{S}, \Omega)$ be sharply o-minimal and assume that it satisfies the CPT. Then there exists another FD-filtration Ω' with $(\mathcal{S}, \Omega')$ sharply o-minimal such that $\Omega \leqslant \Omega'$, and in Ω' the following holds.*

Let $X_1, \dots, X_k \in \Omega'_{\mathcal{F},D}$, all subsets of $\mathbb{R}^\ell$. Then there exists a real cellular cover $\{f_j : \mathcal{C}_j^{\{\sigma\}} \to \mathbb{C}^\ell\}$ such that each f_j is prepared, and each $f_j(\mathbb{R}_+\mathcal{C}_j^{\{\sigma\}})$ is compatible with each X_i. The number of cells is $\operatorname{poly}_{\mathcal{F}}(D, k, 1/\sigma)$, *and each of them has format* $\operatorname{const}(\mathcal{F})$ *and degree* $\operatorname{poly}_{\mathcal{F}}(D)$.

In particular, the cells $f_j(\mathbb{R}_+\mathcal{C}_j) \subset \mathbb{R}^\ell$ form a cell-decomposition of $\mathbb{R}^\ell$ compatible with $X_1, \dots, X_k$. In addition, each cell admits "analytic continuation" to a complex cell $\mathcal{C}_j$ with a $\{\sigma\}$-extension.

In [9] it is shown that from a parametrization of the type provided by Theorem 22 one can produce C^r-smooth parametrizations, with the number of maps depending polynomially on both D and r. In particular, the conclusion of Theorem 22 implies Conjectures 11 and 13. It therefore seems that proving the CPT in a general analytically-generated sharply o-minimal structure provides a plausible approach to these two conjectures.

We remark that a different complex analytic approach, based on the notion of *Weierstrass polydiscs*, was employed in [8] to prove the Wilkie conjecture in the structure $\mathbb{R}^{\mathrm{RE}}$. This may also give an approach to proving Conjecture 13 in general, but it does not seem to be applicable to Conjecture 11.

3.4. Complex cells, hyperbolic geometry, and preparation theorems

The main motivation for introduction the notion of $\{\rho\}$-extensions of complex cells is that one can use the hyperbolic geometry of $\mathcal{C}$ inside $\mathcal{C}^{\{\rho\}}$ to control the geometry of holomorphic functions defined on complex cells. This is used extensively in the proof of the algebraic CPT in [9], but also gives statements of independent interest. We illustrate two of the main statements.

For any hyperbolic Riemann surface X, we denote by $\operatorname{dist}(\cdot, \cdot; X)$ the hyperbolic distance on X. We use the same notation when $X = \mathbb{C}$ and $X = \mathbb{R}$ to denote the usual Euclidean distance, and when $X = \mathbb{C}P^1$ to denote the Fubini–Study metric normalized to have diameter 1. For $x \in X$ and $r > 0$, we denote by $B(x, r; X)$ the open r-ball centered at x in X. For $A \subset X$, we denote by $B(A, r; X)$ the union of r-balls centered at all points of A.

Lemma 23 (Fundamental lemma for $\mathbb{C} \setminus \{0, 1\}$). *Let $\mathcal{C}^{\{\rho\}}$ be a complex cell and let $f : \mathcal{C}^{\{\rho\}} \to \mathbb{C} \setminus \{0, 1\}$ be holomorphic. Then one of the following holds:*

$$f(\mathcal{C}) \subset B\big(\{0, 1, \infty\}, e^{-\Omega_\ell(1/\rho)}; \mathbb{C}P^1\big) \quad \textit{or} \quad \operatorname{diam}\big(f(\mathcal{C}); \mathbb{C} \setminus \{0, 1\}\big) = O_\ell(\rho). \qquad (3.5)$$

The fundamental lemma for $\mathbb{C} \setminus \{0, 1\}$ implies the Great Picard Theorem: indeed, taking $\mathcal{C}$ to be a small punctured disc $D_\circ$ around the origin, it implies that any function

$f : D_\circ \to \mathbb{C} \setminus \{0, 1\}$ has an image of small diameter in $\mathbb{C}P^1$, hence is bounded away from some $w \in \mathbb{C}P^1$, and it follows elementarily that f is meromorphic at the origin.

If $f : \mathcal{C}^{\{\rho\}} \to \mathbb{C} \setminus \{0\}$ is a bounded holomorphic map then we may decompose it as $f = \mathbf{z}^{\boldsymbol{\alpha}(f)} \cdot U(\mathbf{z})$, where $U : \mathcal{C}^{\{\rho\}} \to \mathbb{C} \setminus \{0\}$ is a holomorphic map and the branches of $\log U : \mathcal{C}^{\{\rho\}} \to \mathbb{C}$ are univalued. The following lemma shows that U enjoys strong boundedness properties when restricted to $\mathcal{C}$.

Lemma 24 (Monomialization lemma). *Let $0 < \rho < \infty$ and let $f : \mathcal{C}^{\{\rho\}} \to \mathbb{C} \setminus \{0\}$ be a holomorphic map. If $\mathcal{C}^{\{\rho\}}, f \in \Omega_{\mathcal{F},D}$ then there exists a polynomial $P_{\mathcal{F}}(\cdot)$ such that $|\boldsymbol{\alpha}(f)| \leqslant P_{\mathcal{F}}(D)$ and*

$$\operatorname{diam}\big(\log U(\mathcal{C}); \mathbb{C}\big) < P_{\mathcal{F}}(D) \cdot \rho, \quad \operatorname{diam}\big(\operatorname{Im} \log U(\mathcal{C}); \mathbb{R}\big) < P_{\mathcal{F}}(D). \tag{3.6}$$

The monomialization lemma is proved in this form for the structure $\mathbb{R}_{\mathrm{alg}}$ in **[9]**, but the proof extends to the general sharply o-minimal case. It is also shown in **[9]** that the monomialization lemma in combination with the CPT gives an effective version of the subanalytic preparation theorem of Parusinski **[33]** and Lion–Rolin **[26]**, which is a key technical tool in the theory of the structure $\mathbb{R}_{\mathrm{an}}$.

3.5. Unrestricted exponentials

One of the milestones in the development of o-minimality is Wilkie's theorem on the model-completeness of $\mathbb{R}_{\exp}$ **[43]**, which, together with Khovanskii's theory of fewnomials, established the o-minimality of $\mathbb{R}_{\exp}$. Wilkie's methods were later used by van den Dries and Miller to establish the o-minimality of $\mathbb{R}_{\mathrm{an,exp}}$. This latter structure plays a key role in many of the applications of o-minimality to arithmetic geometry, since it contains the uniformizing maps of (mixed) Shimura varieties restricted to an appropriate fundamental domain. We conjecture a sharply o-minimal version of the theorems of Wilkie and van den Dries, Miller as follows.

Conjecture 25. *Let $(\mathcal{S}, \Omega)$ be an analytically generated sharply o-minimal structure. Let $\mathcal{S}_{\exp}$ denote the structure generated by $\mathcal{S}$ and the unrestricted exponential, and let $\Omega_{\exp}$ be the FD-filtration of $\mathcal{S}_{\exp}$ generated by Ω and by the graph of the unrestricted exponential (say with format and degree 1). Then $(\mathcal{S}_{\exp}, \Omega_{\exp})$ is sharply o-minimal.*

It is perhaps plausible to make the same conjecture even without the assumption of analytic generation. However, the analytic case appears to be sufficient for all (currently known) applications, and the availability of the tools discussed in this section make the conjecture seem somewhat more amenable in this case. In particular, Lion–Rolin **[26]** have a geometric approach to the o-minimality of $\mathbb{R}_{\mathrm{an,exp}}$ using the subanalytic preparation theorem as a basic tool. The CPT provides a sharp version of the subanalytic preparation theorem, thus suggesting a possible path to the proof of Conjecture 25.

4. SHARPLY O-MINIMAL STRUCTURES ARISING FROM GEOMETRY

The fundamental motivation for introducing the notion of sharply o-minimal structures is the expectation that structures arising naturally from geometry should indeed be tame in this stronger sense. We start by motivating the discussion with the example of Abel–Jacobi maps, and then state some general conjectures.

4.1. Abel–Jacobi maps

Recall that for C a compact Riemann surface of genus g and $\omega_1, \dots, \omega_g$ a basis of holomorphic one-forms on C, there is an associated lattice of periods $\Lambda \subset \mathbb{C}^g$, a principally polarized abelian variety $\mathrm{Jac}(C) \simeq \mathbb{C}^g/\Lambda$ and, for any choice of base point $p_0 \in C$, an *Abel–Jacobi* map

$$u_C : C \to \mathrm{Jac}(C), \quad u_C(p) = \int_{p_0}^{p} (\omega_1, \dots, \omega_g) \bmod \Lambda. \tag{4.1}$$

To discuss definability properties of u_C, we choose a semi-algebraic (or even semi-linear) fundamental domain $\Delta \subset \mathbb{C}^g$ for the Λ-action and consider u_C as a map $u_C : C \to \Delta$.

Proposition 26. *There is an analytically generated sharply o-minimal structure where every u_C is definable.*

Indeed, after covering C by finitely many charts $\phi_j : D \to C$, where ϕ_j are algebraic maps extending to some neighborhood of $\bar{D}$, it is enough to show that the structure generated by these $\phi_j^* u_C$ is sharply o-minimal. Moreover, it is enough to show instead that the lifts

$$\tilde{u}_{C,j} : D \to \mathbb{C}^g, \quad \tilde{u}_{C,j}(z) = \int_0^z \phi^*(\omega_1, \dots, \omega_g) \tag{4.2}$$

are definable. Indeed, $\tilde{u}_{C,j}(\bar{D})$ being compact meets finitely many translates of Δ, and the further projection $\mathbb{C}^g \to \Delta$ restricted to some ball containing $\tilde{u}_{C,j}(D)$ is thus definable in any sharply o-minimal structures (even in $\mathbb{R}_{\mathrm{alg}}$). The sharp o-minimality of the structure generated by all these $\tilde{u}_{C,j}$ follows from Theorem 6, since these functions, as indefinite integrals of algebraic one-forms, are restricted-Pfaffian (see, e.g., [27] for the elliptic case).

The construction above, however, is not uniform over C of a given genus. More precisely, while we do have $\tilde{u}_{j,C} \in \Omega_{\mathcal{F},D}$ for some uniform $\mathcal{F}$, D, the number of algebraic charts $\phi_j : D \to C$ may tend to infinity as C approaches the boundary of the moduli space $\mathcal{M}_g$ of compact genus g curves. However, we do have the following.

Proposition 27. *There is a sharply o-minimal structure where every $u_C \in \Omega_{\mathcal{F},D}$ for some uniform $\mathcal{F} = \mathcal{F}(g)$ and $D = D(g)$.*

To prove this, we replace the covering $\phi_j : D \to C$ by a covering $\phi_j : \mathcal{C}_j^{1/2} \to C$, where each $\mathcal{C}_j$ is a one-dimensional complex cell and $\phi_j(\mathcal{C}_j)$ covers C. By the removable singularity theorem, we may assume each $\mathcal{C}_j$ is either a disc or an annulus. Moreover, $\#\{\phi_j\}$ and their degrees are $\mathrm{poly}_{\mathcal{F}}(g)$ by the algebraic CPT. Here we use the fact that a genus g curve can always be realized as an algebraic curve of degree $d = \mathrm{poly}(g)$. The same construction as above now shows that each $\tilde{u}_{C,j} : \mathcal{C} \to \mathbb{C}^g$, if univalued, is restricted Pfaffian of format $\mathcal{F} = \mathcal{F}(g)$ and degree $D = \text{poly}_g(D)$. In general, we have

$$\tilde{u}_{C,j}(z) = u'_{C,j}(z) + (a_{C,j,1}, \dots, a_{C,j,g}) \log z \tag{4.3}$$

where $u'_{C,j}$ is univalued and $a_{C,j,k}$ is the residue of $\phi_j^* \omega_k$ around the annulus. Since $\log z$, understood for instance as having a branch cut in the negative real line, is restricted Pfaffian with uniform format and degree over every annulus, this proves the general case.

Finally, one should check that the projection $\mathbb{C}^g \to \Delta$, restricted to $\tilde{\phi}_j(\mathcal{C}_j)$ is definable (say in $\mathbb{R}_{\text{alg}}$) with format and degree depending only on g. Equivalently, one should check that $\tilde{\phi}_j(\mathcal{C}_j)$ meets finitely many translates of Δ, with the number of translates depending only on g (if $\tilde{\phi}_j(\mathcal{C}_j)$ is multivalued then one should take one of its branches). This indeed holds, provided that the fundamental domains Δ are chosen appropriately. It can be deduced, albeit ineffectively, from the definability of theta functions (in both τ and z) on an appropriate fundamental domain [34]. In the case $g = 1$, an explicit upper bound for these constants is given in [24].

The appearance of logarithmic factors in (4.3) is the reason that the structure we obtain is not analytically generated. However, the construction does prove the following.

Proposition 28. *There is an analytically generated sharply o-minimal structure* $(\mathcal{S}, \Omega)$ *where every* $u_C \in (\Omega_{\text{exp}})_{\mathcal{F},D}$ *for some uniform* $\mathcal{F} = \mathcal{F}(g)$ *and* $D = D(g)$.

According to Conjecture 25 the structure $\mathcal{S}_{\text{exp}}$ is indeed sharply o-minimal as well, but this remains open.

4.2. Uniformizing maps of abelian varieties

One can essentially repeat the construction above replacing $\text{Jac}(C)$ by an arbitrary (say principally polarized) abelian variety A of genus g. We similarly have a map $u : A \to \Delta$ where $\Delta \subset \mathbb{C}^g$ is a semilinear fundamental domain for the period lattice of A, corresponding to some fixed basis of the holomorphic ones-forms $\omega_1, \dots, \omega_g$ on A. Propositions 26, 27, and 28 extend to this more general context with essentially the same proof.

4.3. Noetherian functions

We have seen in Sections 4.1 and 4.2 that Abel–Jacobi maps and uniformizing maps of abelian varieties live in a sharply o-minimal structure (in fact, uniformly over all curves or abelian varieties of a given genus). This eventually boils down to the fact that the relevant maps are definable in $\mathbb{R}_{\text{rPfaff}}$. However, we do not believe that all functions arising from geometry are definable in this structure. For instance, we conjecture that the graph of the modular invariant $j(\tau)$ restricted to any nonempty domain is not definable in $\mathbb{R}_{\text{rPfaff}}$. We do not know how to prove this fact, but Freitag [17] has recently at least shown that $j(\tau)$ it not itself Pfaffian, on any nonempty domain, as a consequence of the strong minimality of the differential equation satisfied by $j(\tau)$ [18].

One natural extension of the notion of Pfaffian functions are the *Noetherian functions*. Let $B \subset \mathbb{R}^\ell$ be a product of finite intervals. A tuple $f_1, \dots, f_m : \bar{B} \to \mathbb{R}$ of analytic functions is called a restricted *Noetherian chain* if they satisfy a system of algebraic differential equations of the form

$$\frac{\partial f_i}{\partial x_j} = P(x_1, \dots, x_\ell, f_1, \dots, f_m), \quad \forall i, j. \tag{4.4}$$

We denote the structure generated by the restricted Noetherian functions by $\mathbb{R}_{\text{rNoether}}$. Since all restricted Noetherian functions are restricted analytic, $\mathbb{R}_{\text{rNoether}}$ is o-minimal.

Conjecture 29. *The structure $\mathbb{R}_{\text{rNoether}}$ is sharply o-minimal with respect to some FD-filtration.*

Gabrielov and Khovanskii have considered some local analogs of the theory of fewnomials for *nondegenerate* systems of Noetherian equations in [20], and made some (still local) conjectures about the general case. These conjectures are proved in [7] under a technical condition. However, these results are all local, bounding the number of zeros in some sufficiently small ball.

Despite the general Conjecture 29 being open, an effective Pila–Wilkie counting theorem was obtained in [4] for semi-Noetherian sets.

Theorem 30. *Let A be defined by finitely many restricted Noetherian equalities and inequalities. Then for every $\varepsilon > 0$, we have*

$$\#A^{\text{trans}}(g, H) \leqslant C_{g,A} H^{\varepsilon} \tag{4.5}$$

where $C_{g,A}$ can be computed explicitly from the data defining A.

Of course, provided Conjecture 29 an effective Pila–Wilkie theorem with better bounds (for instance, polynomial in the degree of A) would follow from Theorem 12. More generally, as a consequence of Conjecture 13 we would expect sharper polylogarithmic bounds as well. Some results in this direction are discussed in the following section.

4.4. Bezout-type theorems and point counting with foliations

One can think of the graphs of Noetherian functions equivalently as leafs of algebraic foliations. Partial results in the direction of Conjecture 29 have been obtained in [5] in this language. To state the result we consider an ambient quasi-projective variety $\mathbb{M}$ and a nonsingular m-dimensional foliation $\mathcal{F}$ of $\mathbb{M}$, both defined over $\bar{\mathbb{Q}}$. For $p \in \mathbb{M}$ denote by $\mathcal{L}_p$ the germ of the leaf passing through p. For a pure-dimensional variety $V \subset \mathbb{M}$, denote

$$\Sigma_V := \{p \in \mathbb{M} : \dim(V \cap \mathcal{L}_p) > m - \operatorname{codim}_{\mathbb{M}} V\}. \tag{4.6}$$

If V is defined over $\bar{\mathbb{Q}}$, we denote by δ_V the sum of the degree $\deg V$, the log-height $h(V)$, and the degree of the field of definition of V over $\mathbb{Q}$. Here the log-height is taken, for instance, to be the log-height of the point representing V in an appropriate Chow variety. In terms of this data we have the following Bezout-type theorem.

Theorem 31 ([5, THEOREM 1]). *Let $V \subset \mathbb{M}$ be defined over a number field and suppose $\operatorname{codim}_M V = m$. Let K be a compact subset of a leaf of $\mathcal{F}$. Then*

$$\#(K \cap V) \leqslant \operatorname{poly}_K\big(\delta_V, \log \operatorname{dist}^{-1}(K, \Sigma_V)\big). \tag{4.7}$$

In fact, the bound in Theorem 31 can be made more explicit giving the precise dependence on $\mathcal{F}$ and on K, and this is important in some applications, but we omit the details for brevity. The same bound without the dependence on $h(V)$ and $\log \operatorname{dist}^{-1}(K, \Sigma_V)$ would be a consequence of Conjecture 29, and establishing such a bound is probably the main step toward proving the conjecture.

As a consequence of Theorem 31 one can deduce some polylogarithmic point-counting results in the spirit of Conjecture 13. We state the simplest result of this type for illustration below.

Theorem 32 ([5, COROLLARY 6]). *Suppose $\mathcal{L}_p$ contains no germs of algebraic curves, for any $p \in \mathbb{M}$. Let K be a compact subset of a leaf of $\mathcal{F}$. Then*

$$\#K(g, h) = \operatorname{poly}_K(g, h). \tag{4.8}$$

Once again, the dependence on K can be made explicit in terms of the foliation $\mathcal{F}$ and this plays a role in some applications. In practice, Theorem 32 and its more refined forms can be used to deduce the conclusion of Conjecture 13 in most arithmetic applications, since the sets appearing in such applications are always defined in terms of leafs of some highly symmetric foliations.

4.5. Q-functions

Many important functions arising from geometry, such as period integrals, are Noetherian. Indeed, such functions arise as horizontal sections of the Gauss–Manin connection and can thus be viewed as solutions of a linear systems of differential equations. However, the structure $\mathbb{R}_{\mathrm{rNoether}}$ only contains the restrictions of such maps to compact domains. If we consider general Noetherian functions on noncompact domains, the result would not even be o-minimal (as illustrated by the sine and cosine functions, for instance). If one is to obtain an o-minimal structure, one must restrict singularities at the boundary.

One candidate class is provided by the notion of *Q-functions* considered in [10, 11]. Let $P \subset \mathbb{C}^n$ be a polydisc, $\Sigma \subset \mathbb{C}^n$ a union of coordinate hyperplanes, and ∇ the connection on $P \times \mathbb{C}^\ell$ given by

$$\nabla v = \mathrm{d}v - A \cdot v \tag{4.9}$$

where A is a matrix of one-forms holomorphic in $\bar{P} \setminus \Sigma$. Suppose that the entries of A are algebraic and defined over $\bar{Q}$, that ∇ has regular singularities along Σ, and that the monodromy of ∇ is quasiunipotent. Finally, let P° be a simply-connected domain obtained by removing from $P \setminus \Sigma$ a branch cut $\{\operatorname{Arg} x_i = \alpha_i\}$ for each of the components $\{x_i = 0\}$ of Σ and for some choice of $\alpha_i \in \mathbb{R} \bmod 2\pi$. Every solution of $\nabla v = 0$ extends as a holomorphic vector-valued function in P°. We call each component of such a function a Q-function. Denote by $\mathbb{R}_{\mathrm{QF}}$ the structure generated by all such Q-functions. This structure contains, as sections of the Gauss–Manin connection, all period integrals of algebraic families.

By the classical theory of regular–singular linear equations, every Q-function is definable in $\mathbb{R}_{\mathrm{an,exp}}$, and $\mathbb{R}_{\mathrm{QF}}$ is thus o-minimal.

Conjecture 33. *The structure $\mathbb{R}_{\mathrm{QF}}$ is sharply o-minimal with respect to some FD-filtration.*

Some initial motivation for Conjecture 33 is provided by the results of [10], which give effective bounds for the number of zeros of Q-functions restricted to any algebraic curve in P. However, treating systems of equations in several variables, and obtaining sharp bounds with respect to degrees, is still widely open.

5. APPLICATIONS IN ARITHMETIC GEOMETRY

In this section we describe some applications of sharply o-minimal structures in arithmetic geometry. For some of these, Theorem 12 suffices, while for others Conjecture 13 is necessary—in some suitable sharply o-minimal structure, such as the one conjectured to exist in Conjecture 33. However, in all cases discussed below one can actually carry out the strategy using known results, mostly Theorem 32 and its generalizations, in place of these general conjectures (though various technical difficulties must be resolved in each case). We thus hope to convince the reader that the strategy laid out below is feasible, on the one hand, and fits coherently into the general framework of sharply o-minimal structures, on the other.

5.1. Geometry governs arithmetic

Geometry governs arithmetic describes a general phenomenon in the interaction between geometry (for instance, algebraic geometry) and arithmetic: namely, that arithmetic problems often admit finitely many solutions unless there is an underlying geometric reason to expect infinitely many. Perhaps the most famous example is given by *Mordell's conjecture*, now Falting's theorem [16]: an algebraic curve $C \subset \mathbb{P}^2$ contains finitely many rational points, unless it is rational or elliptic. The two exceptions in Falting's theorem may be viewed as geometric obstructions to the finitude of rational solutions: the rational parametrization in the former, and the group law in the latter, are geometric mechanisms that can produce infinitely many rational points on the curve.

The Pila–Wilkie theorem itself may be viewed as an instance where geometry (namely the existence of an algebraic part) controls arithmetic (namely the occurrence of many rational points, as a function of height). A general strategy by Pila and Zannier [38] reduces many unlikely intersection questions to the Pila–Wilkie theorem. This has been used to prove the finiteness of solutions, under natural geometric hypotheses, to a large number of Diophantine problems. For instance, the finiteness of torsion points on a subvariety of an abelian variety (Manin–Mumford) [38]; the finiteness of maximal special points on subvariety of a Shimura variety (André–Oort) [35,41]; the finiteness of "torsion values" for sections of families of abelian surfaces (relative Manin–Mumford) [30]; the finiteness of the set of $t \in \mathbb{C}$ for which a Pell equation $P^2 - DQ^2 = 1$ with given $D \in \mathbb{Q}^{\mathrm{alg}}[X,t]$ is solvable in $P, Q \in \mathbb{C}[X]$ [2,31,32]; the finiteness of the set of values $t \in \mathbb{C}$ where an algebraic one-form $f_t = f(t,x)\mathrm{d}x$ is integrable in elementary terms [32]; and various other examples.

5.2. The Pila–Zannier strategy

Below we briefly explain the Pila–Zannier strategy in the Manin–Mumford case. Let A be an abelian variety and $V \subset A$ an algebraic subvariety containing no cosets of abelian subvarieties, both defined over a number field $\mathbb{K}$.

Let $\pi : [0,1]^{2g} \to A$ be the universal covering map of A written in period coordinates, so that rational points with common denominator N in $[0,1]^{2g}$ correspond to N-torsion points in A. One checks that under our assumptions, $X := \pi^{-1}(V)$ has no algebraic part (this can be done with the help of the Pila–Wilkie theorem as well, following a strategy of Pila in [35]). The Pila–Wilkie theorem then implies that the number of torsion points in V is at most $C(X,\varepsilon)N^{\varepsilon}$ where $C(X,\varepsilon)$ is the Pila–Wilkie constant.

On the other hand, there is $c > 0$ such that if $p \in A$ is an N-torsion point then $[\mathbb{Q}(p) : \mathbb{Q}] \gg_A N^c$ by a result of David [15]. Here the implied constant depends effectively on A. This is an example of a Galois lower bound, which in the Pila–Zannier strategy plays the yin to Pila–Wilkie's yang.

Choose $\varepsilon = c/2$ and suppose that V contains an N-torsion point p. Then it contains a fraction of $[\mathbb{K} : \mathbb{Q}]^{-1}$ of its Galois conjugates, and we obtain a contradiction as soon as $N \gg_{A,V} C(X,\varepsilon/2)^{2/c}$. We thus proved a bound for the order of any torsion point in V, and in particular the finiteness of the set of torsion points.

5.3. Point counting and Galois lower bounds

Traditionally in the Pila–Zannier strategy, the Pila–Wilkie theorem is used to obtain an upper bound on the number of special points, while the competing Galois lower bounds are obtained using other methods—usually involving a combination of height estimates and transcendence methods, such as the results of David [15] or Masser–Wüstholz [29].

In [39] Schmidt suggested an alternative approach to proving Galois lower bounds, replacing the more traditional transcendence methods by polylogarithmic counting results as in Conjecture 13. We illustrate again in the Manin–Mumford setting. Let A be an abelian variety over a number field $\mathbb{K}$ and let $p \in A$ be a torsion point. Consider now X given by the graph of the map π defined in the previous section, which is easily seen to contain no algebraic part. The points $p, 2p, \dots, Np$ correspond to N points $x_1, \dots, x_n$ on this graph. Recall that the height of a torsion point in A is $O_A(1)$ (since the Neron–Tate height is zero), and the height of the corresponding point in $[0,1]^{2g}$ is at most N. It follows that $h(x_j) \ll_A \log N$. On the other hand, the field of definition of each x_j is, by the product law of A, contained in $\mathbb{K}(p)$. We thus have

$$N \leqslant \#X\big([\mathbb{K}(p) : \mathbb{Q}], \log N\big) = \operatorname{poly}_A\big([\mathbb{K}(p) : \mathbb{Q}], \log N\big) \tag{5.1}$$

by Conjecture 13, and this readily implies $[\mathbb{Q}(p) : \mathbb{Q}] \gg_A N^c$ for some $c > 0$, giving a new proof of the Galois lower bound for torsion points—and with it a "purely point-counting" proof of Manin–Mumford. This has been carried out in [5] using Theorem 32.

The main novelty of this strategy is that it applies in contexts where we have polylog counting result, and where the more traditional transcendence techniques are not available. In [12] this idea was applied in the context of a general Shimura variety S. It is shown that if the special points $p \in S$ satisfy a discriminant-negligible height bound

$$h(p) \ll_{S,\varepsilon} \operatorname{disc}(p)^{\varepsilon}, \quad \forall \varepsilon > 0 \tag{5.2}$$

where $\operatorname{disc}(p)$ is an appropriately defined discriminant, then they also satisfy a Galois bound $[\mathbb{Q}(p) : \mathbb{Q}] \gg_S \operatorname{disc}(p)^c$ for some $c > 0$. Further, it was already known by the work of many authors based on the strategy of Pila [35] that this implies the André–Oort conjecture for S.

In the case of the Siegel modular variety $S = \mathcal{A}_g$, the height bound (5.2) was proved by Tsimerman [41] as a simple consequence of the recently proven averaged Colmez formula [1,46]. Tsimerman deduces the corresponding Galois bound from this using Masser–Wüstholz' isogeny estimates [29]. However, these estimates are proved using transcendence methods applied to abelian functions, and have no known counterpart applicable when the Shimura variety S does not parameterize abelian varieties (i.e., is not of abelian type). The result of [12] removes this obstruction.

A few months after [12] appeared on the arXiv, Pila–Shankar–Tsimerman have posted a paper [36] (with an appendix by Esnault, Groechenig) establishing the conjecture (5.2) for arbitrary Shimura varieties (by a highly sophisticated reduction to the $\mathcal{A}_g$ case where averaged Colmez applies). In combination with [12] this establishes the André–Oort conjecture for general Shimura varieties (as well as for mixed Shimura varieties by the work of Gao [22]). It is interesting to note that the proof of André–Oort now involves three distinct applications of point-pointing: for functional transcendence, for Galois lower bounds, and for the Pila–Zannier strategy.

5.4. Effectivity and polynomial time computability

In each of the problems listed at the end of Section 5.1, it is natural to ask, when the data defining the problem is given over $\bar{\mathbb{Q}}$, whether one can effectively determine the finite set of solutions; and whether one can compute the set in polynomial time (say, in the degrees and the log-heights of the algebraic data involved, for a fixed dimension). In most cases mentioned above, the use of the Pila–Wilkie theorem is the only source of ineffectivity in the proofs. In fact, for all examples above excluding the André–Oort conjecture, definability of the relevant transcendental sets in an (effective) sharply o-minimal structure is expected to imply the (effective) polynomial time computability of these finite sets. This has been carried out using Theorem 12 for Manin–Mumford [6] and using Theorem 32 for a case of relative Manin–Mumford [5], giving effective polynomial time decidability of these problems. We see no obstacles in similarly applying [5] to the other problems listed above, though this is yet to be verified in each specific case.

In the André–Oort conjecture Siegel's class number bound introduces another source of ineffectivity in the finiteness result. Nevertheless, in [5] Theorem 32 is used to prove the polynomial time decidability of André–Oort for subvarieties of $\mathbb{C}^n$ (i.e., by a polynomial-time algorithm involving a universal, undetermined Siegel constant). This is expected to extend to arbitrary Shimura varieties.

FUNDING

This research was supported by the ISRAEL SCIENCE FOUNDATION (grant No. 1167/17). This project has received funding from the European Research Council (ERC) under the European Union's Horizon 2020 research and innovation programme (grant agreement No 802107).

REFERENCES

[1] F. Andreatta, E. Z. Goren, B. Howard, and K. Madapusi Pera, Faltings heights of abelian varieties with complex multiplication. *Ann. of Math. (2)* **187** (2018), no. 2, 391–531.

[2] F. Barroero and L. Capuano, Unlikely intersections in families of abelian varieties and the polynomial Pell equation. *Proc. Lond. Math. Soc. (3)* **120** (2020), no. 2, 192–219.

[3] S. Basu, R. Pollack, and M.-F. Roy, *Algorithms in real algebraic geometry. Second edn.* Algorithms Comput. Math. 10, Springer, Berlin, 2006.

[4] G. Binyamini, Density of algebraic points on Noetherian varieties. *Geom. Funct. Anal.* **29** (2019), no. 1, 72–118.

[5] G. Binyamini, Point counting for foliations over number fields. 2020, arXiv:2009.00892.

[6] G. Binyamini, G. Jones, H. Schmidt, and M. Thomas, Effective Pila–Wilkie in the restricted sub-Pffafian structure (in preparation).

[7] G. Binyamini and D. Novikov, Multiplicities of Noetherian deformations. *Geom. Funct. Anal.* **25** (2015), no. 5, 1413–1439.

[8] G. Binyamini and D. Novikov, Wilkie's conjecture for restricted elementary functions. *Ann. of Math. (2)* **186** (2017), no. 1, 237–275.

[9] G. Binyamini and D. Novikov, Complex cellular structures. *Ann. of Math. (2)* **190** (2019), no. 1, 145–248.

[10] G. Binyamini, D. Novikov, and S. Yakovenko, On the number of zeros of Abelian integrals. *Invent. Math.* **181** (2010), no. 2, 227–289.

[11] G. Binyamini, D. Novikov, and S. Yakovenko, Quasialgebraic functions. In *Algebraic methods in dynamical systems*, pp. 61–81, Banach Center Publ. 94, Polish Acad. Sci. Inst. Math, Warsaw, 2011.

[12] G. Binyamini, H. Schmidt, and A. Yafaev, Lower bounds for Galois orbits of special points on Shimura varieties: a point-counting approach. 2021, arXiv:2104.05842.

[13] G. Binyamini and N. Vorobjov, Effective cylindrical cell decompositions for restricted sub-Pfaffian sets. *Int. Math. Res. Not. IMRN* (2020); rnaa285, DOI 10.1093/imrn/rnaa285.

[14] D. Burguet, A proof of Yomdin–Gromov's algebraic lemma. *Israel J. Math.* **168** (2008), 291–316.

[15] S. David, Fonctions thêta et points de torsion des variétés abéliennes. *Compos. Math.* **78** (1991), no. 2, 121–160.

[16] G. Faltings, Endlichkeitssätze für abelsche Varietäten über Zahlkörpern. *Invent. Math.* **73** (1983), no. 3, 349–366.

[17] J. Freitag, Not Pfaffian. 2021, arXiv:2109.09230.

[18] J. Freitag and T. Scanlon, Strong minimality and the j-function. *J. Eur. Math. Soc. (JEMS)* **20** (2018), no. 1, 119–136.

[19] A. Gabrielov, Relative closure and the complexity of Pfaffian elimination. In *Discrete and computational geometry*, pp. 441–460, Algorithms Combin. 25, Springer, Berlin, 2003.

[20] A. Gabrielov and A. Khovanskii, Multiplicity of a Noetherian intersection. In *Geometry of differential equations*, pp. 119–130, Amer. Math. Soc. Transl. Ser. 2 186, Amer. Math. Soc., Providence, RI, 1998.

[21] A. Gabrielov and N. Vorobjov, Complexity of computations with Pfaffian and Noetherian functions. In *Normal forms, bifurcations and finiteness problems in differential equations*, pp. 211–250, NATO Sci. Ser. II Math. Phys. Chem. 137, Kluwer Acad. Publ., Dordrecht, 2004.

[22] Z. Gao, About the mixed André–Oort conjecture: reduction to a lower bound for the pure case. *C. R. Math. Acad. Sci. Paris* **354** (2016), no. 7, 659–663.

[23] M. Gromov, Entropy, homology and semialgebraic geometry. *Astérisque*, **145–146** (1987), no. 5, 225–240.

[24] G. Jones and H. Schmidt, Pfaffian definitions of Weierstrass elliptic functions. *Math. Ann.* **379** (2021), no. 1–2, 825–864.

[25] A. G. Khovanskiĭ, *Fewnomials*. Transl. Math. Monogr. 88, American Mathematical Society, Providence, RI, 1991.

[26] J.-M. Lion and J.-P. Rolin, Théorème de préparation pour les fonctions logarithmico-exponentielles. *Ann. Inst. Fourier (Grenoble)* **47** (1997), no. 3, 859–884.

[27] A. Macintyre, Some observations about the real and imaginary parts of complex Pfaffian functions. In *Model theory with applications to algebra and analysis. Vol. 1*, pp. 215–223, London Math. Soc. Lecture Note Ser. 349, Cambridge Univ. Press, Cambridge, 2008.

[28] D. Masser, *Auxiliary polynomials in number theory*. Cambridge Tracts in Math. 207, Cambridge University Press, Cambridge, 2016.

[29] D. Masser and G. Wüstholz, Isogeny estimates for abelian varieties, and finiteness theorems. *Ann. of Math. (2)* **137** (1993), no. 3, 459–472.

[30] D. Masser and U. Zannier, Torsion anomalous points and families of elliptic curves. *Amer. J. Math.* **132** (2010), no. 6, 1677–1691.

[31] D. Masser and U. Zannier, Torsion points on families of simple abelian surfaces and Pell's equation over polynomial rings. *J. Eur. Math. Soc. (JEMS)* **17** (2015), no. 9, 2379–2416.

[32] D. Masser and U. Zannier, Torsion points, Pell's equation, and integration in elementary terms. *Acta Math.* **225** (2020), no. 2, 227–313.

[33] A. Parusiński, Lipschitz stratification of subanalytic sets. *Ann. Sci. Éc. Norm. Supér. (4)* **27** (1994), no. 6, 661–696.

[34] Y. Peterzil and S. Starchenko, Definability of restricted theta functions and families of abelian varieties. *Duke Math. J.* **162** (2013), no. 4, 731–765.

[35] J. Pila, O-minimality and the André–Oort conjecture for $\mathbb{C}^n$. *Ann. of Math. (2)* **173** (2011), no. 3, 1779–1840.

[36] J. Pila, A. Shankar, and J. Tsimerman, Canonical heights on shimura varieties and the André–Oort conjecture, with an appendix by H. Esnault and M. Groechenig. 2021, arXiv:2109.08788.

[37] J. Pila and A. J. Wilkie, The rational points of a definable set. *Duke Math. J.* **133** (2006), no. 3, 591–616.

[38] J. Pila and U. Zannier, Rational points in periodic analytic sets and the Manin–Mumford conjecture. *Atti Accad. Naz. Lincei Rend. Lincei Mat. Appl.* **19** (2008), no. 2, 149–162.

[39] H. Schmidt, Counting rational points and lower bounds for Galois orbits. *Atti Accad. Naz. Lincei Rend. Lincei Mat. Appl.* **30** (2019), no. 3, 497–509.

[40] A. Surroca, Sur le nombre de points algébriques où une fonction analytique transcendante prend des valeurs algébriques. *C. R. Math. Acad. Sci. Paris* **334** (2002), no. 9, 721–725.

[41] J. Tsimerman, The André–Oort conjecture for $\mathcal{A}_g$. *Ann. of Math. (2)* **187** (2018), no. 2, 379–390.

[42] L. van den Dries, *Tame topology and o-minimal structures*. London Math. Soc. Lecture Note Ser. 248, Cambridge University Press, Cambridge, 1998.

[43] A. J. Wilkie, Model completeness results for expansions of the ordered field of real numbers by restricted Pfaffian functions and the exponential function. *J. Amer. Math. Soc.* **9** (1996), no. 4, 1051–1094.

[44] Y. Yomdin, Volume growth and entropy. *Israel J. Math.* **57** (1987), no. 3, 285–300.

[45] Y. Yomdin, Local complexity growth for iterations of real analytic mappings and semicontinuity moduli of the entropy. *Ergodic Theory Dynam. Systems* **11** (1991), no. 3, 583–602.

[46] X. Yuan and S.-W. Zhang, On the averaged Colmez conjecture. *Ann. of Math. (2)* **187** (2018), no. 2, 533–638.

GAL BINYAMINI

Weizmann Institute of Science, Rehovot, Israel, gal.binyamini@weizmann.ac.il

DMITRY NOVIKOV

Weizmann Institute of Science, Rehovot, Israel, dmitry.novikov@weizmann.ac.il

RAMSEY THEORY OF HOMOGENEOUS STRUCTURES: CURRENT TRENDS AND OPEN PROBLEMS

NATASHA DOBRINEN

This paper is dedicated to Norbert Sauer for his seminal works on the partition theory of homogeneous structures, and for his mathematical and personal generosity.

ABSTRACT

This article highlights historical achievements in the partition theory of countable homogeneous relational structures, and presents recent work, current trends, and open problems. Exciting recent developments include new methods involving logic, topological Ramsey spaces, and category theory. The paper concentrates on big Ramsey degrees, presenting their essential structure where known and outlining areas for further development. Cognate areas, including infinite dimensional Ramsey theory of homogeneous structures and partition theory of uncountable structures, are also discussed.

MATHEMATICS SUBJECT CLASSIFICATION 2020

Primary 05C55; Secondary 03C15, 03E02, 03E75, 05C05, 05C15

KEYWORDS

Ramsey theory, homogeneous structure, big Ramsey degree, coding tree

ICM INTERNATIONAL CONGRESS OF MATHEMATICIANS 2022 JULY 6–14

Proc. Int. Cong. Math. 2022, Vol. 3, pp. 1462–1486
DOI 10.4171/ICM2022/108

Published by EMS Press

1. INTRODUCTION

Ramsey theory is a beautiful subject which interrelates with a multitude of mathematical fields. In particular, since its inception, developments in Ramsey theory have often been motivated by problems in logic; in turn, Ramsey theory has instigated some seminal developments in logic. The intent of this article is to provide the general mathematician with an introduction to the intriguing subject of Ramsey theory on homogeneous structures while being detailed enough to describe the state-of-the-art and the main ideas at play. We present historical highlights and discuss why solutions to problems on homogeneous structures require more than just straightforward applications of finite structural Ramsey theory. In the following sections, we map out collections of recent results and methods which were developed to overcome obstacles associated with forbidden substructures. These new methods involve applications from logic (especially forcing but also ideas from model theory), topological Ramsey spaces, and category theory.

The subject of Ramsey theory on infinite structures begins with this lovely theorem.

Theorem 1.1 (Ramsey, [58]). *Given positive integers k and r and a coloring of the k-element subsets of the natural numbers $\mathbb{N}$ into r colors, there is an infinite set of natural numbers $N \subseteq \mathbb{N}$ such that all k-element subsets of N have the same color.*

There are two natural interpretations of Ramsey's theorem in terms of infinite structures. First, letting $<$ denote the standard linear order on $\mathbb{N}$, Ramsey's theorem shows that given any finite coloring of all linearly ordered substructures of $(\mathbb{N}, <)$ of size k, there is an isomorphic substructure $(N, <)$ of $(\mathbb{N}, <)$ such that all linearly ordered substructures of $(N, <)$ of size k have the same color. Second, one may think of the k-element subsets of $\mathbb{N}$ as k-hyperedges. Then Ramsey's theorem yields that, given any finite coloring of the k-hyperedges of the complete k-regular hypergraph on infinitely many vertices, there is an isomorphic subgraph in which all k-hyperedges have the same color.

Given this, one might naturally wonder about other structures.

Question 1.2. Which infinite structures carry an analogue of Ramsey's theorem?

The rational numbers $(\mathbb{Q}, <)$ as a dense linearly ordered structure (without endpoints) was the earliest test case. It is a fun exercise to show that given any coloring of the rational numbers into finitely many colors, there is one color-class which contains a dense linear order, that is, an isomorphic subcopy of the rationals in one color. Thus, the rationals satisfy a structural pigeonhole principle known as *indivisibility*.

The direct analogy with Ramsey's theorem ends, however, when we consider pairs of rationals. It follows from the work of Sierpiński in [65] that there is a coloring of the pairs of rationals into two colors so that both colors persist in every isomorphic subcopy of the rationals. Sierpiński's coloring provides a clear understanding of one of the fundamental issues arising in partition theory of infinite structures not occurring in finite structural Ramsey theory. Let $\{q_i : i \in \mathbb{N}\}$ be a listing of the rational numbers, without repetition, and for $i < j$ define $c(\{q_i, q_j\}) =$ blue if $q_i < q_j$, and $c(\{q_i, q_j\}) =$ red if $q_j < q_i$. Then in each subset $Q \subseteq \mathbb{Q}$ forming a dense linear order, both color classes persist; that is, there are pairs of rationals in Q colored red and also pairs of rationals in Q colored blue. Since it is impossible to find an isomorphic subcopy of the rationals in which all pairsets have the same color, a direct analogue of Ramsey's theorem does not hold for the rationals.

The failure of the straightforward analogue of Ramsey's theorem is not the end, but rather just the beginning of the story. Galvin (unpublished) showed a few decades later that there is a bound on the number of unavoidable colors: Given any coloring of the pairs of rationals into finitely many colors, there is a subcopy of the rationals in which all pairs belong to the union of two color classes. Now one sees that Question 1.2 ought to be refined.

Question 1.3. For which infinite structures $\mathbf{S}$ is there a Ramsey-analogue in the following sense: Let $\mathbf{A}$ be a finite substructure of $\mathbf{S}$. Is there a positive integer T such that for any coloring of the copies of $\mathbf{A}$ into finitely many colors, there is a subcopy $\mathbf{S}'$ of $\mathbf{S}$ in which there are no more than T many colors for the copies of $\mathbf{A}$?

The least such integer T, when it exists, is denoted $T(\mathbf{A})$ and called the *big Ramsey degree* of $\mathbf{A}$ in $\mathbf{S}$, a term coined in Kechris–Pestov–Todorcevic (2005). The "big" refers to the fact that we require an isomorphic subcopy of an *infinite* structure in which the number of colors is as small as possible (in contrast to the concept of *small Ramsey degree* in finite structural Ramsey theory).

Notice how Sierpiński played the enumeration $\{q_i : i \in \mathbb{N}\}$ of the rationals against the dense linear order to construct a coloring of pairsets of rationals into two colors, each of which persists in every subcopy of the rationals. This simple, but deep idea sheds light on a fundamental difference between finite and infinite structural Ramsey theory. The interplay between the enumeration and the relations on an infinite structure has bearing on the number of colors that must persist in any subcopy of that structure. We will see examples of this at work throughout this article and explain the general principles which have been found for certain classes of structures with relations of arity at most two, even as the subject aims towards a future overarching theory of big Ramsey degrees.

2. THE QUESTIONS

Given a finite relational language $\mathcal{L} = \{R_i : i < k\}$ with each relation symbol R_i of some finite arity, say, n_i, an *$\mathcal{L}$-structure* is a tuple $\mathbf{A} = \langle A, R_0^{\mathbf{A}}, \dots, R_{k-1}^{\mathbf{A}} \rangle$, where $A \neq \emptyset$ is the *universe* of $\mathbf{A}$ and for each $i < k$, $R_i^{\mathbf{A}} \subseteq A^{n_i}$. For $\mathcal{L}$-structures $\mathbf{A}$ and $\mathbf{B}$, an *embedding* from $\mathbf{A}$ into $\mathbf{B}$ is an injection $e : A \to B$ such that for all $i < k$, $R_i^{\mathbf{A}}(a_1, \dots, a_{n_i}) \leftrightarrow R_i^{\mathbf{B}}(e(a_1), \dots, e(a_{n_i}))$. The e-image of $\mathbf{A}$ is a *copy* of $\mathbf{A}$ in $\mathbf{B}$. If e is the identity map, then $\mathbf{A}$ is a *substructure* of $\mathbf{B}$. An *isomorphism* is an embedding which is onto its image. We write $\mathbf{A} \leq \mathbf{B}$ exactly when there is an embedding of $\mathbf{A}$ into $\mathbf{B}$, and $\mathbf{A} \cong \mathbf{B}$ exactly when $\mathbf{A}$ and $\mathbf{B}$ are isomorphic.

A class $\mathcal{K}$ of finite structures for a relational language $\mathcal{L}$ is called a *Fraïssé class* if it is hereditary, satisfies the joint embedding and amalgamation properties, contains (up to isomorphism) only countably many structures, and contains structures of arbitrarily large finite cardinality. Class $\mathcal{K}$ is *hereditary* if whenever $\mathbf{B} \in \mathcal{K}$ and $\mathbf{A} \leq \mathbf{B}$, then also $\mathbf{A} \in \mathcal{K}$; $\mathcal{K}$ satisfies the *joint embedding property* if for any $\mathbf{A}, \mathbf{B} \in \mathcal{K}$, there is a $\mathbf{C} \in \mathcal{K}$ such that $\mathbf{A} \leq \mathbf{C}$ and $\mathbf{B} \leq \mathbf{C}$; $\mathcal{K}$ satisfies the *amalgamation property* if for any embeddings $f : \mathbf{A} \to \mathbf{B}$ and $g : \mathbf{A} \to \mathbf{C}$, with $\mathbf{A}, \mathbf{B}, \mathbf{C} \in \mathcal{K}$, there is a $\mathbf{D} \in \mathcal{K}$ and there are embeddings $r : \mathbf{B} \to \mathbf{D}$ and $s : \mathbf{C} \to \mathbf{D}$ such that $r \circ f = s \circ g$. A Fraïssé class $\mathcal{K}$ satisfies the *strong amalgamation property* (SAP) if given $\mathbf{A}, \mathbf{B}, \mathbf{C} \in \mathcal{K}$ and embeddings $e : \mathbf{A} \to \mathbf{B}$ and $f : \mathbf{A} \to \mathbf{C}$, there is some $\mathbf{D} \in \mathcal{K}$ and embeddings $e' : \mathbf{B} \to \mathbf{D}$ and $f' : \mathbf{C} \to \mathbf{D}$ such that $e' \circ e = f' \circ f$, and $e'[B] \cap f'[C] = e' \circ e[A] = f' \circ f[A]$. We say that $\mathcal{K}$ satisfies the *free amalgamation property* (FAP) if it satisfies the SAP and, moreover, $\mathbf{D}$ can be chosen so that $\mathbf{D}$ has no additional relations other than those inherited from $\mathbf{B}$ and $\mathbf{C}$.

Let $\mathbf{A}, \mathbf{B}, \mathbf{C}$ be $\mathcal{L}$-structures such that $\mathbf{A} \leq \mathbf{B} \leq \mathbf{C}$. We use $\binom{\mathbf{B}}{\mathbf{A}}$ to denote the set of all copies of $\mathbf{A}$ in $\mathbf{B}$. The Erdős–Rado arrow notation $\mathbf{C} \to (\mathbf{B})^{\mathbf{A}}_k$ means that for each coloring of $\binom{\mathbf{C}}{\mathbf{A}}$ into k colors, there is a $\mathbf{B}' \in \binom{\mathbf{C}}{\mathbf{B}}$ such that $\binom{\mathbf{B}'}{\mathbf{A}}$ is *monochromatic*, meaning every member of $\binom{\mathbf{B}'}{\mathbf{A}}$ has the same color.

Definition 2.1. A Fraïssé class $\mathcal{K}$ has the *Ramsey property* if for any two structures $\mathbf{A} \leq \mathbf{B}$ in $\mathcal{K}$ and any $k \geq 2$, there is a $\mathbf{C} \in \mathcal{K}$ with $\mathbf{B} \leq \mathbf{C}$ such that $\mathbf{C} \to (\mathbf{B})^{\mathbf{A}}_k$.

Many Fraïssé classes, such as the class of finite graphs, do not have the Ramsey property. However, by allowing a finite expansion of the language, often by just a linear order, the Ramsey property becomes more feasible. Letting $<$ be a binary relation symbol not in the language $\mathcal{L}$ of $\mathcal{K}$, an $\mathcal{L} \cup \{<\}$-structure is in $\mathcal{K}^<$ if and only if its universe is linearly ordered by $<$ and its $\mathcal{L}$-reduct is a member of $\mathcal{K}$. A highlight is the work of Nešetřil and Rödl in [51] and [52], proving that for any Fraïssé class $\mathcal{K}$ with FAP, its ordered version $\mathcal{K}^<$ has the Ramsey property. The recent paper [40] by Hubička and Nešetřil presents the state-of-the-art in finite structural Ramsey theory. Examples of Fraïssé classes with the Ramsey property include the class of finite linear orders, and the classes of finite ordered versions of graphs, digraphs, tournaments, triangle-free graphs, posets, metric spaces, hypergraphs, hypergraphs omitting some irreducible substructures, and many more.

A structure $\mathbf{K}$ is called *universal* for a class of structures $\mathcal{K}$ if each member of $\mathcal{K}$ embeds into $\mathbf{K}$. A structure $\mathbf{K}$ is *homogeneous* if each isomorphism between finite substructures of $\mathbf{K}$ extends to an automorphism of $\mathbf{K}$. Unless otherwise specified, we will write *homogeneous* to mean *countably infinite homogeneous*, such structures being the focus of this paper. The *age* of an infinite structure $\mathbf{K}$, denoted $\mathrm{Age}(\mathbf{K})$, is the collection of all finite structures which embed into $\mathbf{K}$. A fundamental theorem of Fraïssé from [31] shows that each Fraïssé class gives rise to a homogeneous structure via a construction called the *Fraïssé limit*. Conversely, given any countable homogeneous structure $\mathbf{K}$, $\mathrm{Age}(\mathbf{K})$ is a Fraïssé class and, moreover, the Fraïssé limit of $\mathrm{Age}(\mathbf{K})$ is isomorphic to $\mathbf{K}$. The Kechris–Pestov–Todorcevic correspondence between the Ramsey property of a Fraïssé class and extreme amenability of the automorphism group of its Fraïssé limit in [41] propelled a burst of discoveries of more Fraïssé classes with the Ramsey property.

First we state an esoteric but driving question in the area.

Question 2.2. What *is* a big Ramsey degree?

What is the essential nature of a big Ramsey degree? Why is it that given a Fraïssé class $\mathcal{K}$ satisfying the Ramsey property, its Fraïssé limit usually fails to carry the full analogue of Ramsey's Theorem 1.1 (i.e., all big Ramsey degrees being one)? A theorem of Hjorth in [37] showed that for any homogeneous structure $\mathbf{K}$ with $|\mathrm{Aut}(\mathbf{K})| > 1$, there is a structure in $\mathrm{Age}(\mathbf{K})$ with big Ramsey degree at least two. While much remains open, we now have an answer to Question 2.2 for FAP and some SAP homogeneous structures with finitely many relations of arity at most two, and these results will be discussed in the following sections.

We say that $\mathbf{S}$ has *finite big Ramsey degrees* if $T(\mathbf{A})$ exists for each finite substructure $\mathbf{A}$ of $\mathbf{S}$. We say that *exact big Ramsey degrees* are known if there is either a computation of the degrees or a characterization from which they can be computed. *Indivisibility* holds if $T(\mathbf{A}) = 1$ for each one-element substructure $\mathbf{A}$ of $\mathbf{S}$. The following questions progress in order of strength: A positive answer to (3) implies a positive answer to (2), which in turn implies a positive answer to (1).

Question 2.3. Given a homogeneous structure $\mathbf{K}$,

(1) Does $\mathbf{K}$ have finite big Ramsey degrees? That is, can one find upper bounds ensuring that big Ramsey degrees exist?

(2) If $\mathbf{K}$ has finite big Ramsey degrees, is there a characterization of the exact big Ramsey degrees via canonical partitions? If yes, calculate or find an algorithm to calculate them.

(3) Does $\mathbf{K}$ carry a big Ramsey structure?

Part (2) of this question involves finding *canonical partitions*.

Definition 2.4 (Canonical Partition, [44]). Given a Fraïssé class $\mathcal{K}$ with Fraïssé limit $\mathbf{K}$, and given $\mathbf{A} \in \mathcal{K}$, a partition $\{P_i : i < n\}$ of $\binom{\mathbf{K}}{\mathbf{A}}$ is *canonical* if the following hold: For each finite coloring of $\binom{\mathbf{K}}{\mathbf{A}}$, there is a subcopy $\mathbf{K}'$ of $\mathbf{K}$ such that for each $i < n$, all members of $P_i \cap \binom{\mathbf{K}'}{\mathbf{A}}$ have the same color; and *persistence*: For every subcopy $\mathbf{K}'$ of $\mathbf{K}$ and each $i < n$, $P_i \cap \binom{\mathbf{K}'}{\mathbf{A}}$ is nonempty.

Canonical partitions recover an exact analogue of Ramsey's theorem for each piece of the partition. In practice such partitions are characterized by adding extra structure to $\mathbf{K}$, including the enumeration of the universe of $\mathbf{K}$ and a tree-like structure capturing the relations of $\mathbf{K}$ against the enumeration.

Part (3) of Question 2.3 has to do with a connection between big Ramsey degrees and topological dynamics, in the spirit of the Kechris–Pestov–Todorcevic correspondence, proved by Zucker in [70]. A *big Ramsey structure* is essentially a finite expansion $\mathbf{K}^*$ of $\mathbf{K}$ so that each finite substructure of $\mathbf{K}^*$ has big Ramsey degree one, and, moreover, the unavoidable colorings cohere in that for $\mathbf{A}, \mathbf{B} \in \mathrm{Age}(\mathbf{K})$ with $\mathbf{A}$ embedding into $\mathbf{B}$, the canonical partition for copies of $\mathbf{B}$ when restricted to copies of $\mathbf{A}$ recovers the canonical partition for copies of **A**. Big Ramsey structures imply canonical partitions. The reverse is not known in general, but certain types of canonical partitions are known to imply big Ramsey structures (Theorem 6.10 in [8]), and it seems reasonable to the author to expect that (1)–(3) are equivalent.

Canonical partitions and big Ramsey structures are really getting at the question of whether we can find an optimal finite expansion $\mathbf{K}^*$ of a given homogeneous structure $\mathbf{K}$ so that $\mathbf{K}^*$ carries an exact analogue of Ramsey's theorem. In this sense, big Ramsey degrees are not quite so mysterious, but are rather saying that an exact analogue of Ramsey's theorem holds for an appropriately expanded structure. The question then becomes: What is the appropriate expansion?

3. CASE STUDY: THE RATIONALS

The big Ramsey degrees for the rationals were determined by 1979. Laver in 1969 (unpublished, see [10]) utilized a Ramsey theorem for trees due to Milliken [50] (Theorem 3.2) to find upper bounds. Devlin completed the picture in his PhD thesis [10], calculating the big Ramsey degrees of the rationals. These surprisingly turn out to be related to the odd coefficients in the Taylor series of the tangent function: The big Ramsey degree for n-element subsets of the rationals is $T(n) = (2n-1)!c_{2n-1}$, where c_k is the kth coefficient in the Taylor series for the tangent function, $\tan(x) = \sum_{k=0}^{\infty} c_k x^k$. As Todorcevic states, the big Ramsey degrees for the rationals "characterize the Ramsey theoretic properties of the countable dense linear ordering $(\mathbb{Q}, <)$ in a very precise sense. The numbers $T(n)$ are some sort of Ramsey degrees that measure the complexity of an arbitrary finite coloring of the n-element subsets of $\mathbb{Q}$ modulo, of course, restricting to the n-element subsets of X for some appropriately chosen dense linear subordering X of $\mathbb{Q}$." (page 143, [66], notation modified)

We present Devlin's characterization of the big Ramsey degrees of the rationals and the four main steps in his proof. (A detailed proof appears in Section 6.3 of [66].) Then we will present a method from [8] using coding trees of 1-types which bypasses nonessential constructs, providing what we see as a satisfactory answer to Question 2.2 for the rationals.

We use some standard mathematical logic notation, providing definitions as needed for the general mathematician. The set of all natural numbers $\{0, 1, 2, \dots\}$ is denoted by ω. Each natural number $k \in \omega$ is equated with the set $\{0, \dots, k-1\}$ and its natural linear ordering. For us $k \in \omega$ and $k < \omega$ are synonymous. For $k \in \omega$, $k^{<\omega}$ denotes the tree of all finite sequences with entries in $\{0, \dots, k-1\}$, and $\omega^{<\omega}$ denotes the tree of all finite sequences of natural numbers. Finite sequences with any sort of entries are thought of as functions with domain some natural number. Thus, for a finite sequence t the *length* of t, denoted $|t|$, is the domain of the function t, and for $i \in \mathrm{dom}(t)$, $t(i)$ denotes the ith entry of the sequence t. For $\ell \in \omega$, we write $t \restriction \ell$ to denote the initial segment of t of length ℓ if $\ell \le |t|$, and t otherwise. For two finite sequences s and t, we write $s \sqsubseteq t$ when s is an initial segment of t, and we write $s \sqsubset t$ when s is a proper initial segment of t, meaning that $s \sqsubseteq t$ and $s \neq t$. We write $s \wedge t$ to denote the *meet* of s and t, that is, the longest sequence which is an initial segment of both s and t. Given a subset S of a tree of finite sequences, the *meet closure* of S, denoted $\mathrm{cl}(S)$, is the set of all nodes in S along with the set of all meets $s \wedge t$, for $s, t \in S$.

A Ramsey theorem for trees, due to Milliken, played a central role in Devlin's work and has informed subsequent approaches to finding upper bounds for big Ramsey degrees. In this area, a subset $T \subseteq \omega^{<\omega}$ is called a *tree* if there is a subset $L_T \subseteq \omega$ such that $T = \{t \restriction \ell : t \in T, \ell \in L_T\}$. Thus, a tree is closed under initial segments of lengths in L_T, but not necessarily closed under all initial segments in $\omega^{<\omega}$. The *height* of a node t in T, denoted $\mathrm{ht}_T(t)$, is the order-type of the set $\{s \in T : s \sqsubset t\}$, linearly ordered by $\sqsubset$. We write $T(n)$ to denote $\{t \in T : \mathrm{ht}_T(t) = n\}$. For $t \in T$, let $\mathrm{Succ}_T(t) = \{s \restriction (|t| + 1) : s \in T \text{ and } t \sqsubset s\}$, noting that $\mathrm{Succ}_T(t) \subseteq T$ only if $|t| + 1 \in L_T$.

A subtree $S \subseteq T$ is a *strong subtree* of T if $L_S \subseteq L_T$ and each node s in S branches as widely as T will allow, meaning that for $s \in S$, for each $t \in \mathrm{Succ}_T(s)$ there is an extension $s' \in S$ such that $t \sqsubseteq s'$. For the next theorem, define $\prod_{i<d} T_i(n)$ to be the set of sequences $(t_0, \ldots, t_{d-1})$ where $t_i \in T_i(n)$, the product of the nth levels of the trees T_i. Then let

$$\bigotimes_{i<d} T_i := \bigcup_{n<\omega} \prod_{i<d} T_i(n). \tag{3.1}$$

The following is the strong tree version of the Halpern–Läuchli theorem.

Theorem 3.1 (Halpern–Läuchli, [34]). *Let d be a positive integer, $T_i \subseteq \omega^{<\omega}$ $(i < d)$ be finitely branching trees with no terminal nodes, and $r \geq 2$. Given a coloring $c : \bigotimes_{i<d} T_i \to r$, there is an increasing sequence $\langle m_n : n < \omega \rangle$ and strong subtrees $S_i \leq T_i$ such that for all $i < d$ and $n < \omega$, $S_i(n) \subseteq T_i(m_n)$, and c is constant on $\bigotimes_{i<d} S_i$.*

The Halpern–Läuchli theorem has a particularly strong connection with logic. It was isolated by Halpern and Lévy as a key juncture in their work to prove that the Boolean Prime Ideal Theorem is strictly weaker than the Axiom of Choice over the Zermelo–Fraenkel Axioms of set theory. Once proved by Halpern and Läuchli, Halpern and Lévy completed their proof in [35].

Harrington (unpublished) devised an innovative proof of the Halpern–Läuchli theorem which used Cohen forcing. The forcing helps find good nodes in the trees T_i from which to start building the subtrees S_i. From then on, the forcing is used ω many times, each time running an unbounded search for finite sets $S_i(n)$ which satisfy that level of the Halpern–Läuchli theorem. Being finite, each $S_i(n)$ is in the ground model. The proof entails neither passing to a generic extension nor any use of Shoenfield's Absoluteness Theorem.

A *k-strong subtree* is a strong subtree with k many levels. The following theorem is proved inductively using Theorem 3.1.

Theorem 3.2 (Milliken, [50]). *Let $T \subseteq \omega^{<\omega}$ be a finitely branching tree with no terminal nodes, $k \geq 1$, and $r \geq 2$. Given a coloring of all k-strong subtrees of T into r colors, there is an infinite strong subtree $S \subseteq T$ such that all k-strong subtrees of S have the same color.*

For more on the Halpern–Läuchli and Milliken theorems, see [21, 46, 66]. Now we look at Devlin's proof of the exact big Ramsey degrees of the rationals, as it has bearing on many current approaches to big Ramsey degrees.

The rationals can be represented by the tree $2^{<\omega}$ of binary sequences with the lexicographic order $\lhd$ defined as follows: Given $s, t \in 2^{<\omega}$ with $s \neq t$, and letting u denote $s \wedge t$, define $s \lhd t$ to hold if and only if ($|u| < |s|$ and $s(|u|) = 0$) or ($|u| < |t|$ and $t(|u|) = 1$). Then $(2^{<\omega}, \lhd)$ is a dense linear order. The following is Definition 6.11 in **[66]**, using the terminology of **[62]**. For $|s| < |t|$, the number $t(|s|)$ is called the *passing number* of t at s.

Definition 3.3. For $A, B \subseteq \omega^{<\omega}$, we say that A and B are *similar* if there is a bijection $f : \mathrm{cl}(A) \to \mathrm{cl}(B)$ such that for all $s, t \in \mathrm{cl}(A)$,

(a) (preserves end-extension) $s \sqsubseteq t \Leftrightarrow f(s) \sqsubseteq f(t)$,

(b) (preserves relative lengths) $|s| < |t| \Leftrightarrow |f(s)| < |f(t)|$,

(c) $s \in A \Leftrightarrow f(s) \in B$,

(d) (preserves passing numbers) $t(|s|) = f(t)(|f(s)|)$ whenever $|s| < |t|$.

Similarity is an equivalence relation; a similarity equivalence class is called a *similarity type*. We now outline the four main steps to Devlin's characterization of big Ramsey degrees in the rationals. Fix $n \geq 1$.

I. (Envelopes) Given a subset $A \subseteq 2^{<\omega}$ of size n, let k be the number of levels in $\mathrm{cl}(A)$. An *envelope* of A is a k-strong subtree $E(A)$ of $2^{<\omega}$ such that $A \subseteq E(A)$. Given any k-strong subtree S of $2^{<\omega}$, there is exactly one subset $B \subseteq S$ which is similar to A. This makes it possible to transfer a coloring of the similarity copies of A in $2^{<\omega}$ to the k-strong subtrees of $2^{<\omega}$ in a well-defined manner.

II. (Finite Big Ramsey Degrees) Apply Milliken's theorem to obtain an infinite strong subtree $T \subseteq 2^{<\omega}$ such that every similarity copy of A in T has the same color. As there are only finitely many similarity types of sets of size n, finitely many applications of Milliken's theorem results in an infinite strong subtree $S \subseteq 2^{<\omega}$ such that the coloring is monochromatic on each similarity type of size n. This achieves finite big Ramsey degrees.

III. (Diagonal Antichain for Better Upper Bounds) To obtain the exact big Ramsey degrees, Devlin constructed a particular antichain of nodes $D \subseteq 2^{<\omega}$ such that $(D, \lhd)$ is a dense linear order and no two nodes in the meet closure of D have the same length, a property called *diagonal*. He also required $(*)$: All passing numbers at the level of a terminal node or a meet node in $\mathrm{cl}(D)$ are 0, except of course the rightmost extension of the meet node. Diagonal antichains turn out to be essential to characterizing big Ramsey degrees, whereas the additional requirement $(*)$ is now seen to be nonessential when viewed through the lens of coding trees of 1-types.

IV. (Exact Big Ramsey Degrees) To characterize the big Ramsey degrees, Devlin proved that the similarity type of each subset of D of size n persists in every subset $D' \subseteq D$ such that $(D', \lhd)$ is a dense linear order. The similarity types of antichains in D thus form a canonical partition for linear orders of size n. By calculating the number of different similarity types of subsets of D of size n, Devlin found the big Ramsey degrees for the rationals.

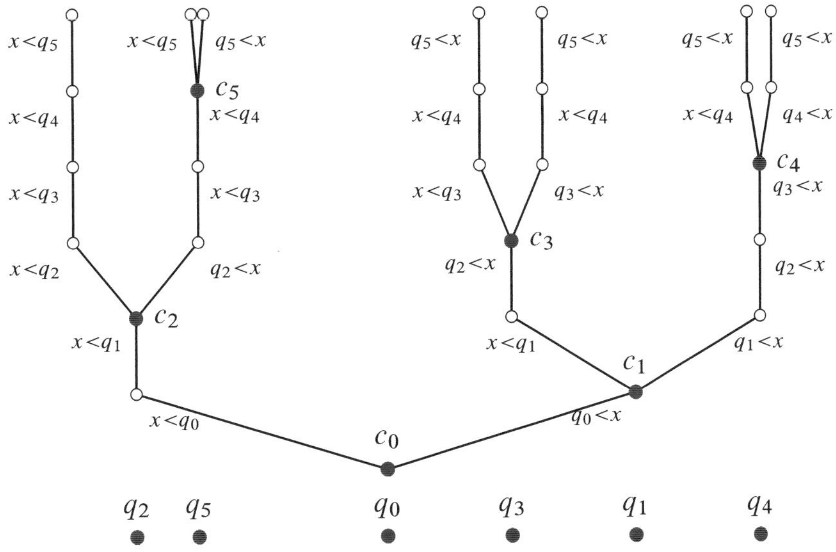

FIGURE 1
Coding tree $\mathbb{S}(\mathbb{Q})$ of 1-types for $(\mathbb{Q}, <)$ and the linear order represented by its coding nodes.

Now we present the characterization of the big Ramsey degrees for the rationals using coding trees of 1-types. Coding trees on $2^{<\omega}$ were first developed in **[13]** to solve the problem of whether or not the triangle-free homogeneous graph has finite big Ramsey degrees. The presentation given here is from **[8]**, where the notion of coding trees was honed using model-theoretic ideas. We hope that presenting this view here will set the stage for a concrete understanding of big Ramsey degree characterizations discussed in Section 5.

Fix an enumeration $\{q_0, q_1, \dots\}$ of $\mathbb{Q}$. For $n < \omega$, we let $\mathbb{Q} \restriction n$ denote the substructure $(\{q_i : i \in n\}, <)$ of $(\mathbb{Q}, <)$, which we refer to as an *initial substructure*. One can think of $\mathbb{Q} \restriction n$ as a finite approximation in a construction of the rationals. The definition of a coding tree of 1-types in **[8]** uses complete realizable quantifier-free 1-types over initial substructures. Here, we shall retain the terminology of **[8]** but (with apologies to model-theorists) will use sets of literals instead, since this will convey the important aspects of the constructions while being more accessible to a general readership. For now, we call a set of formulas $s \subseteq \{(q_i < x) : i \in n\} \cup \{(x < q_i) : i \in n\}$ a *1-type* over $\mathbb{Q} \restriction n$ if (a) for each $i < n$ exactly one of the formulas $(q_i < x)$ or $(x < q_i)$ is in s, and (b) there is some (and hence infinitely many) $j \geq n$ such that q_j *satisfies* s, meaning that replacing the variable x by the rational number q_j in each formula in s results in a true statement. In other words, s is a 1-type if s prescribes a legitimate way to extend $\mathbb{Q} \restriction n$ to a linear order of size $n + 1$.

Definition 3.4 (Coding Tree of 1-Types for $\mathbb{Q}$, **[8]**). For a fixed enumeration $\{q_0, q_1, \dots\}$ of the rationals, the *coding tree of 1-types* $\mathbb{S}(\mathbb{Q})$ is the set of all 1-types over initial substructures along with a function $c : \omega \to \mathbb{S}(\mathbb{Q})$ such that $c(n)$ is the 1-type of q_n over $\mathbb{Q} \restriction n$. The tree-ordering is simply inclusion.

Given $s \in \mathbb{S}(\mathbb{Q})$ let $|s| = j + 1$ where j is maximal such that one of $(x < q_j)$ or $(q_j < x)$ is in s. For each $i < |s|$, we let $s(i)$ denote the formula from among $(x < q_i)$ or $(q_i < x)$ which is in s. The *coding nodes* $c(n)$, in practice usually denoted by c_n, are special distinguished nodes representing the rational numbers; c_n represents the rational q_n, because c_n is the 1-type with parameters from among $\{q_i : i \in n\}$ that q_n satisfies. Notice that this tree $\mathbb{S}(\mathbb{Q})$ has at most one splitting node per level. The effect is that any antichain of coding nodes in $\mathbb{S}(\mathbb{Q})$ will automatically be diagonal. (See Figure 1, reproduced from [8].)

Fix an ordering $<_{\text{lex}}$ on the literals: For $i < j$, define $(x < q_i) <_{\text{lex}} (q_i < x) <_{\text{lex}} (x < q_j)$. Extend $<_{\text{lex}}$ to $\mathbb{S}(\mathbb{Q})$ by declaring for $s, t \in \mathbb{S}(\mathbb{Q})$, $s <_{\text{lex}} t$ if and only if s and t are incomparable and for $i = |s \wedge t|$, $s(i) <_{\text{lex}} t(i)$.

Definition 3.5. For A, B sets of coding nodes in $\mathbb{S}(\mathbb{Q})$, we say that A and B are *similar* if there is a bijection $f : \text{cl}(A) \to \text{cl}(B)$ such that for all $s, t \in \text{cl}(A)$, f satisfies (a)–(c) of Definition 3.3 and (d′) $s <_{\text{lex}} t \iff f(s) <_{\text{lex}} f(t)$,

When B is similar to A, we call B a *similarity copy* of A. Condition (d) in Definition 3.3 implies that the lexicographic order on $2^{<\omega}$ is preserved, and, moreover, that passing numbers at meet nodes and at terminal nodes are preserved. In (d′) we only need to preserve lexicographic order.

Extending Harrington's method, forcing is utilized to obtain a pigeonhole principle for coding trees of 1-types in the vein of the Halpern–Läuchli Theorem 3.1, but for colorings of finite sets of coding nodes, rather than antichains. Via an inductive argument using this pigeonhole principle, we obtain the following Ramsey theorem on coding trees.

Theorem 3.6 ([8]). *Let $\mathbb{S}(\mathbb{Q})$ be a coding tree of 1-types for the rationals. Given a finite set A of coding nodes in $\mathbb{S}(\mathbb{Q})$ and a finite coloring of all similarity copies of A in $\mathbb{S}(\mathbb{Q})$, there is a coding subtree S of $\mathbb{S}(\mathbb{Q})$ similar to $\mathbb{S}(\mathbb{Q})$ such that all similarity copies of A in S have the same color.*

Fix $n \geq 1$. By applying Theorem 3.6 once for each similarity type of coding nodes of size n, we prove finite big Ramsey degrees, accomplishing step II while bypassing step I in Devlin's proof. Upon taking *any* antichain D of coding nodes in $\mathbb{S}(\mathbb{Q})$ representing a dense linear order, we obtain better upper bounds which are then proved to be exact, accomplishing steps III and IV.

Big Ramsey degrees of the rationals. In [8], we show that given $n \geq 1$, the big Ramsey degree $T(n)$ for linear orders of size n in the rationals is the number of similarity types of antichains of coding nodes in $\mathbb{S}(\mathbb{Q})$.

What then *is* the big Ramsey degree $T(n)$ in the rationals? It is the number of different ways to order the indexes of an increasing sequence of rationals $\{q_{i_0} < q_{i_1} < \cdots < q_{i_{n-1}}\}$ with incomparable 1-types along with the number of ways to order the first differences of their 1-types over initial substructures of $\mathbb{Q}$. The first difference between the 1-types of the rationals q_i and q_j occurs at the least k such that $q_i < q_k$ and $q_k < q_j$, or vice versa. This means that q_i and q_j are in the same interval of $\mathbb{Q} \restriction k$ but in different intervals of

$\mathbb{Q} \upharpoonright (k+1)$. Concretely, $T(n)$ is the number of $<$-isomorphism classes of $(2n-1)$-tuples of integers $(i_0, \ldots, i_{n-1}, k_0, \ldots, k_{n-2})$ with the following properties: $\{q_{i_0} < q_{i_1} < \cdots < q_{i_{n-1}}\}$ is a set of rationals in increasing order, and for each $j < n-1$, $q_{i_j} < q_{k_j} < q_{i_{j+1}}$ where $k_j < \min(i_j, i_{j+1})$ and is the least integer satisfying this relation.

4. HISTORICAL HIGHLIGHTS, RECENT RESULTS, AND METHODS

We now highlight some historical achievements, and present recent results and the main ideas of their methods. For an overview of results up to the year 2000, see the appendix by Sauer in Fraïssé's book [32]; for an overview up to the year 2013, see Nguyen Van Thé's habilitation thesis [54]. Those interested in open problems intended for undergraduate research may enjoy [18].

The Rado graph is the second example of a homogeneous structure with nontrivial big Ramsey degrees which has been fully understood in terms of its partition theory. The Rado graph $\mathbf{R}$ is up to isomorphism the homogeneous graph on countably many vertices which is universal for all countable graphs. It was known to Erdős and other Hungarian mathematicians in the 1960s, though possibly earlier, that the Rado graph is indivisible. In their 1975 paper [30], Erdős, Hajnal, and Pósa constructed a coloring of the edges in $\mathbf{R}$ into two colors such that both colors persist in each subcopy of $\mathbf{R}$. Pouzet and Sauer later showed in [57] that the big Ramsey degree for edge colorings in the Rado graph is exactly two. The complete characterization of the big Ramsey degrees of the Rado graph was achieved in a pair of papers by Sauer [62] and by Laflamme, Sauer, and Vuksanovic [44], both appearing in 2006, and the degrees were calculated by Larson in [45]. The two papers [62] and [44] in fact characterized exact big Ramsey degrees for all unrestricted homogeneous structures with finitely many binary relations, including the homogeneous digraph, homogeneous tournament, and random graph with finitely many edges of different colors. Milliken's theorem was used to prove existence of upper bounds, alluding to a deep connection between big Ramsey degrees and Ramsey theorems for trees. These results are discussed in Section 5.1.

In [43], for each $n \geq 2$, Laflamme, Nguyen Van Thé, and Sauer calculated the big Ramsey degrees of $\mathbb{Q}_n$, the rationals with an equivalence relation with n many equivalence classes each of which is dense in $\mathbb{Q}$. This hinged on proving a "colored version" of Milliken's theorem, where the levels of the trees are colored, to achieve upper bounds. Applying their result for $\mathbb{Q}_2$, they calculated the big Ramsey degrees of the dense local order, denoted $\mathbf{S}(2)$. In his PhD thesis [38], Howe proved finite big Ramsey degrees for the generic bipartite graph and the Fraïssé limit of the class of finite linear orders with a convex equivalence relation.

A robust and streamlined approach applicable to a large class of homogeneous structures, and recovering the previously mentioned examples (except for $\mathbf{S}(2)$), was developed by Coulson, Patel, and the author in [8], building on ideas in [13] and [12]. In [8], it was shown that homogeneous structures with relations of arity at most two satisfying a strengthening of SAP, called SDAP^+, have big Ramsey structures which are characterized in a simple manner, and therefore their big Ramsey degrees are easy to compute. The proof proceeds via a Ramsey theorem for colorings of finite antichains of coding nodes on *diagonal coding trees of* 1*-types*. This approach bypasses any need for envelopes, the theorem producing of its own accord exact upper bounds. Moreover, the Halpern–Läuchli-style theorem, which is proved via forcing arguments to achieve a ZFC result and used as the pigeonhole principle in the Ramsey theorem, immediately yields indivisibility for all homogeneous structures satisfying SDAP^+, with relations of any arity. These results and their methods are discussed in Section 5.1.

The k-clique-free homogeneous graphs, denoted $\mathbf{G}_k$, $k \geq 3$, were constructed by Henson in his 1971 paper [36], where he proved these graphs to be weakly indivisible. In their 1986 paper [42], Komjáth and Rödl proved that $\mathbf{G}_3$ is indivisible, answering a question of Hajnal. A few years later, El-Zahar and Sauer gave a systematic approach in [24], proving that for each $k \geq 3$, the k-clique-free homogeneous graph $\mathbf{G}_k$ is indivisible. In 1998, Sauer proved in [60] that the big Ramsey degree for edges in $\mathbf{G}_3$ is two. Further progress on big Ramsey degrees of $\mathbf{G}_3$, however, needed a new approach. This was achieved by the author in [13], where the method of coding trees was first developed. In [12], the author extended this work, proving that $\mathbf{G}_k$ has finite big Ramsey degrees, for each $k \geq 3$. In [13] and [12], the author proved a Ramsey theorem for colorings of finite antichains of coding nodes in diagonal coding trees. These diagonal coding trees were designed to achieve very good upper bounds and directly recover the indivisibility results in [42] and [24], discovering much of the essential structure involved in characterizing their exact big Ramsey degrees. (Milliken-style theorems on nondiagonal coding trees which fully branch at each level do not directly prove indivisibility results, and produce looser upper bounds.) In particular, after a minor modification, the trees in [13] produced exact big Ramsey degrees for $\mathbf{G}_3$, as shown in [14]. Around the same time, exact big Ramsey degrees for $\mathbf{G}_3$ were independently proved by Balko, Chodounský, Hubička, Konečný, Vena, and Zucker, instigating the collaboration of this group with the author.

Given a finite relational language $\mathcal{L}$, an $\mathcal{L}$-structure $\mathbf{A}$ is called *irreducible* if each pair of its vertices are in some relation of $\mathbf{A}$. Given a set $\mathcal{F}$ of finite irreducible $\mathcal{L}$-structures, $\mathrm{Forb}(\mathcal{F})$ denotes the class of all finite $\mathcal{L}$-structures into which no member of $\mathcal{F}$ embeds. Fraïssé classes of the form $\mathrm{Forb}(\mathcal{F})$ are exactly those with free amalgamation. Zucker in [71] proved that for any Fraïssé class of the form $\mathrm{Forb}(\mathcal{F})$, where $\mathcal{F}$ is a finite set of irreducible substructures and all relations have arity at most two, its Fraïssé limit has finite big Ramsey degrees. His proof used coding trees which branch at each level and a forcing argument to obtain a Halpern–Läuchli-style theorem which formed the pigeonhole principle for a Milliken-esque theorem for these coding trees. An important advance in this paper is Zucker's abstract, top-down approach, providing simplified and relatively short proof of finite big Ramsey degrees for this large class of homogeneous structures. On the other hand, his Milliken-style theorem does not directly recover indivisibility (more work is needed afterwards to show this), and the upper bounds in [71] did not recover those in [13] or [12] for the homogeneous k-clique-free graphs. However, by further work done in [6], by Balko, Chodounský, Hubička, Konečný, Vena, Zucker, and the author, indivisibility results are proved and exact big Ramsey degrees are characterized. Thus, the picture for FAP classes with finitely many relations of arity at most two is now clear. These results will be discussed in Section 5.2.

Next, we look at homogeneous structures with relations of arity at most two which do not satisfy SDAP^+ and whose ages have strong (but not free) amalgamation. Nguyen Van Thé made a significant contribution in his 2008 paper [53], in which he proved that the ultrametric Urysohn space $\mathbf{Q}_S$ has finite big Ramsey degrees if and only if S is a finite distance set. In the case that S is finite, he calculated the big Ramsey degrees. Moreover, he showed that for an infinite countable distance set S, $\mathbf{Q}_S$ is indivisible if and only if S with the reverse order as a subset of the reals is well ordered. His proof used infinitely wide trees of finite height and his pigeonhole principle was actually Ramsey's theorem. All countable Urysohn metric spaces with finite distance set were proved to be indivisible by Sauer in [63], completing the work that was initiated in [55] in relation to the celebrated distortion problem from Banach space theory and its solution by Odell and Schlumprecht in [56].

Mašulović instigated the use of category theory to prove transport principles showing that finite big Ramsey degrees can be inferred from one category to another. After proving a general transport principle in [47], he applied it to prove finite big Ramsey degrees for many universal structures and also for homogenous metric spaces with finite distance sets with a certain property which he calls *compact with one nontrivial block*. Mašulović proved in [48] that in categories satisfying certain mild conditions, small Ramsey degrees are minima of big Ramsey degrees. In the paper [49] with Šobot (not using category theory), finite big Ramsey degrees for finite chains in countable ordinals were shown to exist if and only if the ordinal is smaller than ω^ω. Dasilva Barbosa in [9] proved that categorical precompact expansions grant upper bounds for big and small Ramsey degrees. As an application, he calculated the big Ramsey degrees of the circular directed graphs $\mathbf{S}(n)$ for all $n \geq 2$, extending the work in [43] for $\mathbf{S}(2)$.

Hubička recently developed a new method to handle forbidden substructures utilizing topological Ramsey spaces of parameter words due to Carlson and Simpson [7]. In [39], he applied his method to prove that the homogeneous partial order and Urysohn S-metric spaces (where S is a set of nonnegative reals with $0 \in S$ satisfying the 4-values condition) have finite big Ramsey degrees. He also showed that this method is quite broad and can be applied to yield a short proof of finite big Ramsey degrees in $\mathbf{G}_3$. Beginning with the upper bounds in [39], the exact big Ramsey degrees of the generic partial order have been characterized in [5] by Balko, Chodounský, Hubička, Konečný, Vena, Zucker, and the author. Also utilizing techniques from [39], Balko, Chodounský, Hubička, Konečný, Nešetřil, and Vena in [2] have found a condition which guarantees finite big Ramsey degrees for binary relational homogeneous structures with strong amalgamation. Examples of structures satisfying this condition include the S-Urysohn space for finite distance sets S, Λ-ultrametric spaces for a finite distributive lattice, and metric spaces associated to metrically homogeneous graphs of a finite diameter from Cherlin's list with no Henson constraints.

For homogeneous structures with free amalgamation, a recent breakthrough of Sauer proving indivisibility in [64] culminates a long line of work in [25–28, 61]. Complementary work appeared in [8], where it was proved that for finitely many relations of any arity, SDAP$^+$ implies indivisibility. On the other hand, big Ramsey degrees of structures with relations of arity greater than two has only recently seen progress, beginning with [3] and [4], where Balko, Chodounský, Hubička, Konečný, and Vena found upper bounds for the big Ramsey degrees of the generic 3-hypergraph. Work in this area is ongoing and promising.

5. EXACT BIG RAMSEY DEGREES

This section presents characterizations of exact big Ramsey degrees known at the time of writing. These hold for homogeneous structures with finitely many relations of arity at most two. Two general classes have been completely understood: Structures satisfying a certain strengthening of strong amalgamation called SDAP$^+$ (Section 5.1) and structures whose ages have free amalgamation (Section 5.2). Lying outside of these two classes, the generic partial order has been completely understood in terms of exact big Ramsey degrees and will be briefly discussed at the end of Section 5.2. These characterizations all involve the notion of a diagonal antichain, in various trees or spaces of parameter words, representing a copy of an enumerated homogeneous structure. Here, we present these notions in terms of structures, as they are independent of the representation.

Let $\mathbf{K}$ be an enumerated homogeneous structure with universe $\{v_n : n < \omega\}$. Let $\mathbf{A} \leq \mathbf{K}$ be a finite substructure of $\mathbf{K}$, and suppose that the universe of $\mathbf{A}$ is $\{v_i : i \in I\}$ for some finite set $I \subseteq \omega$. We say that $\mathbf{A}$ is an *antichain* if for each pair $i < j$ in I there is a $k(i,j) < i$ such that the set $\{k(i,j) : i, j \in I \text{ and } i < j\}$ is disjoint from I, and

$$\mathbf{K} \restriction (\{v_\ell : \ell < k(i,j)\} \cup \{v_i\}) \cong \mathbf{K} \restriction (\{v_\ell : \ell < k(i,j)\} \cup \{v_j\}), \tag{5.1}$$

$$\mathbf{K} \restriction (\{v_\ell : \ell \leq k(i,j)\} \cup \{v_i\}) \not\cong \mathbf{K} \restriction (\{v_\ell : \ell \leq k(i,j)\} \cup \{v_j\}). \tag{5.2}$$

An antichain $\mathbf{A}$ is called *diagonal* if $\{k(i,j) : i < j \leq m\}$ has cardinality m. We call $k(i,j)$ the *meet level* of the pair v_i, v_j.

The notion of diagonal antichain is central to all characterizations of big Ramsey degrees obtained so far. It seems likely that antichains will be essential to all characterizations of big Ramsey degrees. However, preliminary work shows that some homogeneous binary relational structures, such as two or more independent linear orders, will have characterizations in their trees of 1-types involving antichains which are not diagonal, but could still be characterized via products of finitely many diagonal antichains.

The indexing of the relation symbols $\{R_\ell : \ell < L\}$ in the language $\mathcal{L}$ of $\mathbf{K}$ induces a lexicographic ordering on trees representing relational structures. Here, we present this idea directly on the structures. For $m \neq n$, we declare $v_m <_{\text{lex}} v_n$ if and only if $\{v_m, v_n\}$ is an antichain and, letting k be the meet level of the pair v_m, v_n, and letting ℓ denote the least index in L such that v_m and v_n disagree on their R_ℓ-relationship with v_k, either $R_\ell(v_k, v_n)$ holds while $R_\ell(v_k, v_m)$ does not, or else $R_\ell(v_n, v_k)$ holds while $R_\ell(v_m, v_k)$ does not.

Two diagonal antichains $\mathbf{A}$ and $\mathbf{B}$ in an enumerated homogeneous structure $\mathbf{K}$ are *similar* if they have the same number of vertices, and the increasing bijection from the universe $A = \{v_{m_i} : i \leq p\}$ of $\mathbf{A}$ to the universe $B = \{v_{n_i} : i \leq p\}$ of $\mathbf{B}$ induces an isomorphism from $\mathbf{A}$ to $\mathbf{B}$ which preserves $<_{\text{lex}}$ and induces a map on the meet levels which, for each $i < j \leq p$, sends $k(m_i, m_j)$ to $k(n_i, n_j)$. This implies that the map sending the coding node c_{m_i} to c_{n_i} $(i \leq p)$ in the coding tree of 1-types $\mathbb{S}(\mathbf{K})$ (see Definition 3.4) induces a map on the meet-closures of $\{c_{m_i} : i \leq p\}$ and $\{c_{n_i} : i \leq p\}$ satisfying Definition 3.5.

Similarity is an equivalence relation, and an equivalence class is called a *similarity type*. We say that $\mathbf{K}$ has *simply characterized big Ramsey degrees* if for $\mathbf{A} \in \text{Age}(\mathbf{K})$, the big Ramsey degree of $\mathbf{A}$ is exactly the number of similarity types of diagonal antichains representing $\mathbf{A}$. In the next subsection, we will see many homogeneous structures with simply characterized big Ramsey degrees.

5.1. Exact big Ramsey degrees with a simple characterization

The decades-long investigation of the big Ramsey degrees of the Rado graph culminated in the two papers [62] and [44]. These two papers moreover characterized the big Ramsey degrees for all *unrestricted* binary relational homogeneous structures. Unrestricted binary relational structures are determined by a finite language $\mathcal{L} = \{R_0, \dots, R_{l-1}\}$ of binary relation symbols and a nonempty constraint set $\mathcal{C}$ of $\mathcal{L}$-structures with universe $\{0, 1\}$ with the following property: If $\mathbf{A}$ and $\mathbf{B}$ are two isomorphic $\mathcal{L}$-structures with universe $\{0, 1\}$, then either both are in $\mathcal{C}$ or neither is in $\mathcal{C}$. We let $\mathbf{H}_{\mathcal{C}}$ denote the homogeneous structure such that each of its substructures with universe of size two is isomorphic to one of the structures in $\mathcal{C}$. Examples of unrestricted binary relational homogeneous structures include the Rado graph, the generic directed graph, the generic tournament, and random graphs with more than one edge relation.

Given a universal constraint set $\mathcal{C}$, letting $k = |\mathcal{C}|$, Sauer showed in [62] how to form a structure, call it $\mathbf{U}_{\mathcal{C}}$, with nodes in the tree $k^{<\omega}$ as vertices, such that $\mathbf{H}_{\mathcal{C}}$ embeds into $\mathbf{U}_{\mathcal{C}}$. Fix a bijection $\lambda : \mathcal{C} \to k$. Given two nodes $s, t \in k^{<\omega}$ with $|s| < |t|$, declare that $t(|s|) = j$ if and only if the induced substructure of $\mathbf{U}_{\mathcal{C}}$ on universe $\{s, t\}$ is isomorphic to the structure $\lambda(j)$ in $\mathcal{C}$, where the isomorphism sends s to 0 and t to 1. For two nodes $s, t \in k^{<\omega}$ of the same length, declare that for s lexicographically less than t, the induced substructure of $\mathbf{U}_{\mathcal{C}}$ on universe $\{s, t\}$ is isomorphic to the structure $\lambda(0)$ in $\mathcal{C}$, where the isomorphism sends s to 0 and t to 1. As a special case, a universal graph is constructed as follows: Let each node in $2^{<\omega}$ be a vertex. Define an edge relation E between vertices by declaring that, for $s \neq t$ in $2^{<\omega}$, $s \mathrel{E} t$ if and only if $|s| \neq |t|$ and $(|s| < |t| \Longrightarrow t(|s|) = 1)$. Then $(2^{<\omega}, E)$ is universal for all countable graphs. In particular, the Rado graph embeds into the graph $(2^{<\omega}, E)$, and vice versa.

In trees of the form $k^{<\omega}$, the notion of similarity is exactly that of Definition 3.3, and steps I–IV discussed in Section 3 outline the proof of exact big Ramsey degrees contained in the pair of papers [62] and [44]. Milliken's theorem was used to prove existence of upper bounds via strong tree envelopes. For step III, Sauer constructed in [62] a diagonal antichain $D \subseteq k^{<\omega}$ such that the substructure of $\mathbf{U}_{\mathcal{C}}$ restricted to universe D is isomorphic to $\mathbf{H}_{\mathcal{C}}$, achieving upper bounds shown to be exact in [44], finishing step IV. The big Ramsey degree of a finite substructure $\mathbf{A}$ of $\mathbf{H}_{\mathcal{C}}$ is exactly the number of distinct similarity types of subsets of D whose induced substructure in $\mathbf{U}_{\mathcal{C}}$ is isomorphic to $\mathbf{A}$.

The work in [62] and [44] greatly influenced the author's development of *coding trees* and their Ramsey theorems in [13] and [12] (discussed in Section 5.2). Those papers along with a suggestion of Sauer to the author during the Banff 2018 Workshop on *Unifying Themes in Ramsey Theory*, to try moving the forcing arguments in those papers from coding trees to structures, informed the approach taken in the paper [8], which is now discussed.

Let $\mathbf{K}$ be an enumerated Fraïssé structure with vertices $\{v_n : n < \omega\}$. For $n < \omega$, we let $\mathbf{K}_n$ denote $\mathbf{K} \upharpoonright \{v_i : i < n\}$, the induced substructure of $\mathbf{K}$ on its first n vertices, and call $\mathbf{K}_n$ an *initial substructure* of $\mathbf{K}$. We write 1*-type* to mean complete realizable quantifier-free 1-type over $\mathbf{K}_n$ for some n.

Definition 5.1 (Coding Tree of 1-Types, [8]). The *coding tree of* 1*-types* $\mathbb{S}(\mathbf{K})$ for an enumerated Fraïssé structure $\mathbf{K}$ is the set of all 1-types over initial substructures of $\mathbf{K}$ along with a function $c : \omega \to \mathbb{S}(\mathbf{K})$ such that $c(n)$ is the 1-type of v_n over $\mathbf{K}_n$. The tree-ordering is simply inclusion.

A substructure $\mathbf{A}$ of $\mathbf{K}$ with universe $A = \{v_{n_0}, \dots, v_{n_m}\}$ is represented by the set of coding nodes $\{c(n_0), \dots, c(n_m)\}$ as follows: For each $i \leq m$, since $c(n_i)$ is the quantifier-free 1-type of v_{n_i} over $\mathbf{K}_{n_i}$, substituting v_{n_i} for the variable x into each formula in $c(n_i)$ which has only parameters from $\{v_{n_j} : j < i\}$ uniquely determines the relations in $\mathbf{A}$ on the vertices $\{v_{n_j} : j \leq i\}$. In [8], we formulated the following strengthening of SAP in order to extract a general property ensuring that big Ramsey degrees have simple characterizations.

Definition 5.2 (SDAP). A Fraïssé class $\mathcal{K}$ has the *Substructure Disjoint Amalgamation Property (SDAP)* if $\mathcal{K}$ has strong amalgamation, and the following holds: Given $\mathbf{A}, \mathbf{C} \in \mathcal{K}$, suppose that $\mathbf{A}$ is a substructure of $\mathbf{C}$, where $\mathbf{C}$ extends $\mathbf{A}$ by two vertices, say v and w. Then there exist $\mathbf{A}', \mathbf{C}' \in \mathcal{K}$, where $\mathbf{A}$ is a substructure of $\mathbf{A}'$ and $\mathbf{C}'$ is a disjoint amalgamation of $\mathbf{A}'$ and $\mathbf{C}$ over $\mathbf{A}$, such that letting v', w' denote the two vertices in $C' \setminus A'$ and assuming (1) and (2), the conclusion holds:

(1) Suppose $\mathbf{B} \in \mathcal{K}$ is any structure containing $\mathbf{A}'$ as a substructure, and let σ and τ be 1-types over $\mathbf{B}$ satisfying $\sigma \upharpoonright \mathbf{A}' = \mathrm{tp}(v'/\mathbf{A}')$ and $\tau \upharpoonright \mathbf{A}' = \mathrm{tp}(w'/\mathbf{A}')$,

(2) Suppose $\mathbf{D} \in \mathcal{K}$ extends $\mathbf{B}$ by one vertex, say v'', such that $\mathrm{tp}(v''/\mathbf{B}) = \sigma$.

Then there is an $\mathbf{E} \in \mathcal{K}$ extending $\mathbf{D}$ by one vertex, say, w'', such that $\mathrm{tp}(w''/\mathbf{B}) = \tau$ and $\mathbf{E} \upharpoonright (\mathrm{A} \cup \{v'', w''\}) \cong \mathbf{C}$.

This amalgamation property can, of course, be presented in terms of embeddings, but the form here is indicative of how it is utilized. A free amalgamation version called *SFAP* is obtained from SDAP by restricting to FAP classes and requiring $\mathbf{A}' = \mathbf{A}$ and $\mathbf{C}' = \mathbf{C}$. Both of these amalgamation properties are preserved under free superposition. A *diagonal subtree* of $\mathbb{S}(\mathbf{K})$ is a subtree such that at any level, at most one node branches, the branching degree is two, and branching and coding nodes never occur on the same level. Diagonal coding trees are subtrees of $\mathbb{S}(\mathbf{K})$ which are diagonal and represent a subcopy of $\mathbf{K}$. The property SDAP$^+$ holds for a homogeneous structure $\mathbf{K}$ if (a) its age satisfies SDAP, (b) there is a diagonal coding subtree of $\mathbb{S}(\mathbf{K})$, and (c) a technicality called the Extension Property which in most cases is trivially satisfied. Classes of the form $\mathrm{Forb}(\mathcal{F})$ where $\mathcal{F}$ is a finite set of 3-irreducible structures, meaning each triple of vertices is in some relation, satisfy SFAP; their ordered versions satisfy SDAP^+.

A version of the Halpern–Läuchli theorem for diagonal coding trees was proved in [8] using the method of forcing to obtain a ZFC result, with the following theorem as an immediate consequence.

Theorem 5.3 ([8]). *Let* $\mathbf{K}$ *be a homogeneous structure satisfying* SDAP^+, *with finitely many relations of any arity. Then* $\mathbf{K}$ *is indivisible.*

For relations of arity at most two, an induction proof then yields a Ramsey theorem for finite colorings of finite antichains of coding nodes in diagonal coding trees. This accomplishes steps I–III simultaneously and directly, without any need for envelopes, providing upper bounds which are then proved to be exact, finishing step IV.

Theorem 5.4 ([8]). *Let* $\mathbf{K}$ *be a homogeneous structure satisfying* SADP^+, *with finitely many relations of arity at most two. Then* $\mathbf{K}$ *admits a big Ramsey structure and, moreover, has simply characterized big Ramsey degrees.*

Theorem 5.4 provides new classes of examples of big Ramsey structures while recovering results in [10,38,43,44] and special cases of the results in [71]. Theorem 5.3 provides new classes of examples of indivisible Fraïssé structures, in particular for ordered structures such as the ordered Rado graph, while recovering results in [24,27,42] and certain cases of Sauer's results in [64] for FAP classes, while providing new SAP examples with indivisibility.

5.2. Big Ramsey degrees for free amalgamation classes

An obstacle to progress in partition theory of homogeneous structures had been the fact that Milliken's theorem is not able to handle forbidden substructures, for instance, triangle-free graphs. Most results up to 2010 had either utilized Milliken's theorem or a variation (as in [43,62]) or else used difficult direct methods (as in [60]) which did not lend naturally to generalizations. The idea of coding trees came to the author during the her stay at the Isaac Newton Institute in 2015 for the programme, *Mathematical, Foundational and Computational Aspects of the Higher Infinite*, culminating in the work [13]. The ideas behind coding trees included the following: Knowing that at the end of the process one will want a diagonal antichain representing a copy of $\mathbf{G}_3$, starting with a tree where vertices in $\mathbf{G}_3$ are represented by special nodes on different levels should not hurt the results. Further, by using special nodes to code the vertices of $\mathbf{G}_3$ into the trees, one might have a chance to prove Milliken-style theorems on a collection of trees, each of which codes a subcopy of $\mathbf{G}_3$.

The author had made a previous attempt at this problem starting early in 2012. Upon stating her interest this problem, Todorcevic (2012, at the Fields Institute Thematic Program on Forcing and Its Applications) and Sauer (2013, at the Erdős Centenary Meeting) each told the author that a new kind of Milliken theorem would need to be developed in order to handle triangle-free graphs, which intrigued her even more. Though unknown to her at the time, a key piece to this puzzle would be Harrington's forcing proof of the Halpern–Läuchli theorem, which Laver was kind enough to outline to her in 2011. (At that time, the author was unaware of the proof in [67].) While at the INI in 2015, Bartošová reminded the author of her interest in big Ramsey degrees of $\mathbf{G}_3$. Having had time by then to fill out and digest Laver's outline, it occurred to the author to try approaching the problem first with the strongest tool available, namely forcing.

Forcing is a set-theoretic method which is normally used to extend a given universe satisfying a given set of axioms (often ZFC) to a larger universe in which the same set of axioms hold while some other statement or property is different than in the original universe. The beautiful thing about Harrington's proof is that, while it does involve the method of forcing, the forcing is only used as a search engine for an object which already exists in the universe in which one lives. In the context of the Fraïssé limit $\mathbf{K}$ of a class $\mathrm{Forb}(\mathcal{F})$, where $\mathcal{F}$ is a finite set of finite irreducible structures, by carefully designing forcings on coding trees with partial orders ensuring that new levels obtained by the search engine are capable of extending a given fixed finite coding tree to a subcoding tree representing a copy of $\mathbf{K}$, one is able to prove Halpern–Läuchli-style theorems for coding trees. These form the pigeonhole principles of various Milliken-style theorems for coding trees.

As the results and main ideas of the methods in [12,13,71] have been discussed in the previous section, we now present the characterization of big Ramsey degrees in [6].

Theorem 5.5 ([6]). *Let* $\mathbf{K}$ *be a homogeneous structure with finitely many relations of arity at most two such that* $\mathrm{Age}(\mathbf{K}) = \mathrm{Forb}(\mathcal{F})$ *for some finite set* $\mathcal{F}$ *of finite irreducible structures. Then* $\mathbf{K}$ *admits a big Ramsey structure.*

Given a Fraïssé class $\mathcal{K} = \mathrm{Forb}(\mathcal{F})$ with relations of arity at most two, where $\mathcal{F}$ is a finite set of finite irreducible structures, let $\mathbf{K}$ denote an enumerated Fraïssé limit of $\mathcal{K}$. Coding trees for $\mathbf{K}$ appearing in various papers are all essentially coding trees of 1-types. The proof of Theorem 5.5 uses the upper bounds of Zucker in [71] as the starting point. It then proceeds by constructing a diagonal antichain of coding nodes which represent the structure $\mathbf{K}$, with additional requirements if there are any forbidden irreducible substructures of size three or more. While the exact characterization in its full generality is not short to state, the simpler version for the structures $\mathbf{G}_k$ include the following: All coding nodes $c_n \in \mathbf{A}$ code an edge with v_m for some $m < n$ and have the following property: If $\mathbf{B}$ is any finite graph which has the same relations over $\mathbf{G}_k \restriction |c_n|$ as c_n does, then $\mathbf{B}$ has no edges. Furthermore, changes in the sets of structures which are allowed to extend a given truncation of $\mathbf{A}$ (as a level set in the coding tree) happen as gradually as possible. From the characterization in [6], one can make an algorithm to compute the big Ramsey degrees.

As a concrete example, we present the exact characterization for triangle-free graphs. In Figure 2, on the left is the beginning of $\mathbf{G}_3$ with some fixed enumeration of the vertices as $\{v_n : n < \omega\}$. The nth *coding node* in the tree $\mathbb{S} = \mathbb{S}(\mathbf{G}_3) \subseteq 2^{<\omega}$ represents the nth vertex v_n in $\mathbf{G}_3$, where passing number 0 represents a nonedge and passing number 1 represents an edge. Equivalently, $\mathbb{S}$ is the coding tree of 1-types for $\mathbf{G}_3$, as the left branch at the level of c_n represents the literal $(x \not E v_n)$ and the right branch represents $(x E v_n)$.

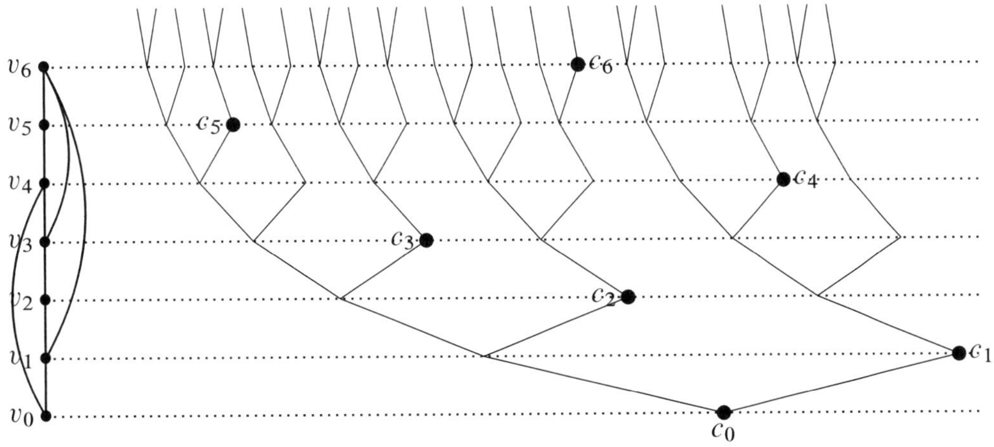

FIGURE 2
Coding tree $\mathbb{S}(\mathbf{G}_3)$ and the triangle-free graph represented by its coding nodes.

Given an antichain $\mathbf{A} \subseteq \mathbf{K}$, we say that $\mathbf{A}$ is a *diagonal substructure* if, letting I be the set of indices of vertices in $\mathbf{A}$, the following hold: (a) For each $i \in I$, v_i has an edge with v_m for some $m < i$; let m_i denote the least such m. (b) If $i < j$ are in I with $v_i \not{E} v_j$ and $m_j < i$, then there is some $n \in i$ such that $v_i E v_n$ and $v_j E v_n$, and the least such n, denoted $n(i, j)$ is not in I. (c) For each $i, j, k, \ell \in I$ (not necessarily distinct) with $i < j$, $k < \ell$, $(i, j) \neq (k, \ell)$, $n_j < i$, and $n_\ell < k$, we have $n(i, j) \neq n(k, \ell)$. Given a finite triangle-free graph $\mathbf{A}$, the big Ramsey degree $T(\mathbf{A})$ in $\mathbf{G}_3$ is the number of different diagonal substructures representing a copy of $\mathbf{A}$.

We conclude this section by mentioning the exact big Ramsey degrees in the generic partial order in [5]. This result begins with the upper bounds proved by Hubička in [39] and then proceeds by taking a diagonal antichain D representing the generic partial order with additional structure of interesting levels built into D. A level ℓ of D is *interesting* if there are exactly two nodes, say s, t, in that level so that $(*)$ for exactly one relation $\rho \in \{<, >, \perp\}$, given any $s', t' \in D$ extending s, t, respectively, $s' \rho t'$, while there is no such relation for the pair $s \restriction (\ell - 1), t \restriction (\ell - 1)$. Since an interesting level for a pair of nodes s, t predetermines the relations between any pair s', t' extending s, t, respectively, passing numbers are unnecessary to the characterization. The big Ramsey degree of a given finite partial order $\mathbf{P}$ is then the number of different diagonal antichains $A \subseteq D$ representing $\mathbf{P}$ along with the order in which the interesting levels are interspersed between the splitting levels and the nodes in A.

6. OPEN PROBLEMS AND RELATED DIRECTIONS

Section 2 laid out the guiding questions for big Ramsey degrees. Here we discuss some of the major open problems in big Ramsey degrees and ongoing research in cognate areas.

Problem 6.1. For which SAP Fraïssé classes does the Fraïssé limit have finite big Ramsey degrees?

Subquestions are the following: Given an SAP Fraïssé class with finitely many relations and a finite set of forbidden substructures, does its Fraïssé limit have finite big Ramsey degrees? Results in [40] give evidence for a positive answer to this question. For such classes with relations of arity at most two, do big Ramsey degrees always exist? We would like a general condition on SAP classes characterizing those with finite big Ramsey degrees. We point out that Problem 6.1 in its full generality is still open for small Ramsey degrees

Problem 6.2. For results whose proofs use the method of forcing, find new proofs which are purely combinatorial.

This has been done for the triangle-free graph by Hubička in [39], but new methods will be needed for k-clique-free homogeneous graphs for $k \geq 4$ and other such FAP classes.

The next problem has to do with topological dynamics of automorphism groups of homogeneous structures. The work of Zucker in [70] has established a connection but not a complete correspondence yet.

Problem 6.3. Does every homogeneous structure with finite big Ramsey degrees also carry a big Ramsey structure? Is there an exact correspondence, in the vein of the KPT-correspondence, between big Ramsey structures and topological dynamics?

The hope in Problem 6.3 is to obtain as complete a dynamical understanding of big Ramsey degrees as we have for small Ramsey degrees, where a result of [69] shows that given a Fraïssé class $\mathcal{K}$ with Fraïssé limit $\mathbf{K}$, then $\mathcal{K}$ has finite small Ramsey degrees if and only if the universal minimal flow of $\mathrm{Aut}(\mathbf{K})$ is metrizable.

Finally, we mention several areas of ongoing study related to the main focus of this paper. Computability-theoretic and reverse mathematical aspects have been investigated by Anglès d'Auriac, Cholak, Dzhafarov, Monin, and Patey. In their treatise [1], they show that the Halpern–Läuchli theorem is computably true and find reverse-mathematical strengths for various instances of the product Milliken theorem and the big Ramsey structures of the rationals and the Rado graph. As these structures both have simply characterized big Ramsey degrees, it will be interesting to see if different reverse mathematical strengths emerge for structures such as the triangle-free homogeneous graph or the generic partial order.

Extending Harrington's forcing proof to the uncountable realm, Shelah in [59] showed that it is consistent, assuming certain large cardinals, that the Halpern–Läuchli theorem holds for trees $2^{<\kappa}$, where κ is a measurable cardinal. Džamonja, Larson, and Mitchell applied a slight modification of his theorem to characterize the big Ramsey degrees for the κ-rationals and the κ-Rado graph in [22] and [23]. Their characterizations have as their basis the characterizations of Devlin and Laflamme–Sauer–Vuksanovic for the rationals and Rado graph, respectively, but also involve well-orderings of each level of the tree $2^{<\kappa}$, necessitated by κ being uncountable. The field of big Ramsey degrees for uncountable homogeneous structures is still quite open, but the fleshing out of the Ramsey theorems on trees of uncountable height has seen some recent work in [19, 20, 68].

The next problem comes from a general question in [41].

Problem 6.4. Develop infinite-dimensional Ramsey theory on spaces of copies of a homogeneous structure.

For a set $N \subseteq \omega$, let $[N]^\omega$ denote the set of all infinite subsets of N, and note that $[\omega]^\omega$ represents the Baire space. The infinite-dimensional Ramsey theorem of Galvin and Prikry [33] says that given any Borel subset $\mathcal{X}$ of the Baire space, there is an infinite set N such that $[N]^\omega$ is either contained in $\mathcal{X}$ or is disjoint from $\mathcal{X}$. Ellentuck's theorem in [29] found optimality in terms of sets with the property of Baire with respect to a finer topology. The question in [41] asks for extensions of these theorems to subspaces of $[\omega]^\omega$, where each infinite set represents a copy of some fixed homogeneous structure. A Galvin–Prikry-style theorem for spaces of copies of the Rado graph has been proved by the author in [17]. By a comment of Todorcevic in Luminy in 2019, the infinite-dimensional Ramsey theorem should ideally also recover exact big Ramsey degrees. Such a theorem is being written down by the author for structures satisfying SDAP$^+$ with relations of arity at most two. This is one instance where coding trees are necessitated to be diagonal in order for the infinite dimensional Ramsey theorem to directly recover exact big Ramsey degrees.

We close by mentioning that structural Ramsey theory has been central in investigations of ultrafilters which are relaxings of Ramsey ultrafilters in the same way that big Ramsey degrees are relaxings of Ramsey's theorem. An exposition of recent work appearing in [16] will give the reader yet another view of the power of Ramsey theory.

ACKNOWLEDGMENTS

The author is grateful to her PhD advisor Karel Prikry, and to Richard Laver, Norbert Sauer, and Stevo Todorcevic who have mentored her in areas related to this paper. She thanks Jan Hubička, Matěj Konečný, Dragan Mašulović, Lionel Nguyen Van Thé, and Andy Zucker for valuable feedback on drafts of this paper.

FUNDING

This work was partially supported by National Science Foundation Grant DMS-1901753.

REFERENCES

[1] P. E. Anglès d'Auriac, P. A. Cholak, D. D. Dzhafarov, B. Monin, and L. Patey, Milliken's tree theorem and its applications: a computability-theoretic perspective. *Mem. Amer. Math. Soc.* (to appear).

[2] M. Balko, D. Chodounský, J. Hubička, M. Konečný, J. Nešetřil, and L. Vena, Big Ramsey degrees and forbidden cycles. In *Extended abstracts EuroComb 2021*, edited by J. Nešetřil, G. Perarnau, J. Rué, and O. Serra, pp. 436–441, Springer, 2021.

[3] M. Balko, D. Chodounský, J. Hubička, M. Konečný, and L. Vena, Big Ramsey degrees of 3-uniform hypergraphs. *Acta Math. Univ. Comenian.* **LXXXVIII** (2019), no. 3, 415–422.

[4] M. Balko, D. Chodounsky, J. Hubička, M. Konečny, and L. Vena, Lluis, Big Ramsey degrees of 3-uniform hypergraphs are finite. *Combinatorica* **42** (2022), no. 5, 659–672.

[5] M. Balko, D. Chodounský, N. Dobrinen, J. Hubička, M. Konečný, L. Vena, and A. Zucker, Big Ramsey degrees of the generic partial order. In *Extended abstracts EuroComb 2021*, edited by J. Nešetřil, G. Perarnau, J. Rué, and O. Serra, pp. 637–643, Springer, 2021.

[6] M. Balko, D. Chodounský, N. Dobrinen, J. Hubička, M. Konečný, L. Vena, and A. Zucker, Exact big Ramsey degrees via coding trees. 2021, arXiv:2110.08409.

[7] T. Carlson and S. G. Simpson, A dual form of Ramsey's theorem. *Adv. Math.* **53** (1984), no. 3, 265–290.

[8] R. Coulson, N. Dobrinen, and R. Patel, Fraïssé classes with simply characterized big Ramsey structures. 2020, arXiv:2010.02034.

[9] K. Dasilva Barbosa, A categorical notion of precompact expansions. 2020, arXiv:2002.11751.

[10] D. Devlin, *Some partition theorems for ultrafilters on ω*. Ph.D. thesis, Dartmouth College, 1980.

[11] N. Dobrinen, Forcing in Ramsey theory. *RIMS Kokyuroku* **2042** (2017), 17–33.

[12] N. Dobrinen, The Ramsey theory of Henson graphs. *J. Math. Log.* (2022). DOI 10.1142/S0219061322500180.

[13] N. Dobrinen, The Ramsey theory of the universal homogeneous triangle-free graph. *J. Math. Log.* **20** (2020), no. 2, 2050012, 75 pp.

[14] N. Dobrinen, Ramsey theory of the universal homogeneous triangle-free graph, Part II: exact big Ramsey degrees. 2020, arXiv:2009.01985.

[15] N. Dobrinen, Ramsey Theory on infinite structures and the method of strong coding trees. In *Contemporary logic and computing*, edited by A. Rezus, pp. 444–467, College Publications, London, 2020.

[16] N. Dobrinen, Topological Ramsey spaces dense in forcings. In *Structure and randomness in computability and set theory*, edited by D. Cenzer and J. Zapletal, pp. 3–58, World Scientific, 2020.

[17] N. Dobrinen, Borel sets of Rado graphs and Ramsey's theorem. In *European Journal of Combinatorics, Proc. 2016 Prague DocCourse on Ramsey Th.* To appear, arXiv:1904.00266.

[18] N. Dobrinen and W. Gasarch, When Ramsey theory fails settle for more colors (Big Ramsey Degrees!). *SIGACT News Open Prob. Col.* **51** (2020), no. 4, 30–46.

[19] N. Dobrinen and D. Hathaway, The Halpern–Läuchli theorem at a measurable cardinal. *J. Symbolic Logic* **82** (2017), no. 4, 1560–1575.

[20] N. Dobrinen and D. Hathaway, Forcing and the Halpern–Läuchli theorem. *J. Symbolic Logic* **85** (2020), no. 1, 87–102.

[21] P. Dodos and V. Kanellopoulos, *Ramsey theory for product spaces*. Math. Surveys Monogr. 212, Amer. Math. Soc., Providence, 2016.

[22] M. Džamonja, J. Larson, and W. J. Mitchell, A partition theorem for a large dense linear order. *Israel J. Math.* **171** (2009), 237–284.

[23] M. Džamonja, J. Larson, and W. J. Mitchell, Partitions of large Rado graphs. *Arch. Math. Logic* **48** (2009), no. 6, 579–606.

[24] M. El-Zahar and N. Sauer, The indivisibility of the homogeneous K_n-free graphs. *J. Combin. Theory Ser. B* **47** (1989), no. 2, 162–170.

[25] M. El-Zahar and N. Sauer, Ramsey-type properties of relational structures. *Discrete Math.* **94** (1991), no. 1, 1–10.

[26] M. El-Zahar and N. Sauer, On the divisibility of homogeneous directed graphs. *Canad. J. Math.* **45** (1993), no. 2, 284–294.

[27] M. El-Zahar and N. Sauer, On the divisibility of homogeneous hypergraphs. *Combinatorica* **14** (1994), no. 2, 159–165.

[28] M. El-Zahar and N. Sauer, Indivisible homogeneous directed graphs and a game for vertex partitions. *Discrete Math.* **291** (2005), no. 1–3, 99–113.

[29] E. Ellentuck, A new proof that analytic sets are Ramsey. *J. Symbolic Logic* **39** (1974), no. 1, 163–165.

[30] P. Erdős, A. Hajnal, and L. Pósa, Strong embeddings of graphs into colored graphs. In *Infinite and finite sets (Colloq., Keszthely, 1973; dedicated to P. Erdős on his 60th birthday), vol. I*, pp. 585–595, Colloq. Math. Soc. János Bolyai 10, North-Holland, Amsterdam, 1975.

[31] R. Fraïssé, Sur l'extension aux relations de quelques propriétés des ordres. *Ann. Sci. Éc. Norm. Supér.* **71** (1954), no. 3, 363–388.

[32] R. Fraïssé, *Theory of relations. Revised edition. With an appendix by Norbert Sauer*. Stud. Logic Found. Math. 145, North-Holland Publishing Co., Amsterdam, 2000.

[33] F. Galvin and K. Prikry, Borel sets and Ramsey's theorem. *J. Symbolic Logic* **38** (1973), no. 2, 193–198.

[34] J. D. Halpern and H. Läuchli, A partition theorem. *Trans. Amer. Math. Soc.* **124** (1966), 360–367.

[35] J. D. Halpern and A. Lévy, The Boolean prime ideal theorem does not imply the axiom of choice. In *Axiomatic set theory, Proc. Sympos. Pure Math., Vol. XIII, Part I, Univ. California, Los Angeles, Calif., 1967*, pp. 83–134, Amer. Math. Soc., 1971.

[36] C. W. Henson, A family of countable homogeneous graphs. *Pacific J. Math.* **38** (1971), no. 1, 69–83.

[37] G. Hjorth, An oscillation theorem for groups of isometries. *Geom. Funct. Anal.* **18** (2008), no. 2, 489–521.

[38] J. Howe, *Big Ramsey degrees in homogeneous structures*. Ph.D. thesis, University of Leeds, Expected, 2021.

[39] J. Hubička, Big Ramsey degrees using parameter spaces. 2020, arXiv:2009.00967.

[40] J. Hubička and J. Nešetřil, All those Ramsey classes (Ramsey classes with closures and forbidden homomorphisms). *Adv. Math.* **356** (2019), 106791, 89 pp.

[41] A. Kechris, V. Pestov, and S. Todorcevic, Fraïssé limits, Ramsey theory, and topological dynamics of automorphism groups. *Geom. Funct. Anal.* **15** (2005), no. 1, 106–189.

[42] P. Komjáth and V. Rödl, Coloring of universal graphs. *Graphs Combin.* **2** (1986), no. 1, 55–60.

[43] C. Laflamme, L. Nguyen Van Thé, and N. Sauer, Partition properties of the dense local order and a colored version of Milliken's theorem. *Combinatorica* **30** (2010), no. 1, 83–104.

[44] C. Laflamme, N. Sauer, and V. Vuksanovic, Canonical partitions of universal structures. *Combinatorica* **26** (2006), no. 2, 183–205.

[45] J. Larson, Counting canonical partitions in the random graph. *Combinatorica* **28** (2008), no. 6, 659–678.

[46] J. Larson, Sets and extensions in the twentieth century. In *Handbook of the history of logic, 6, infinite combinatorics*, pp. 145–357, North-Holland, 2012.

[47] D. Mašulović, Finite big Ramsey degrees in universal structures. *J. Combin. Theory Ser. A* **170** (2020), no. 1, 105137, 30 pp.

[48] D. Mašulović, Ramsey degrees: big vs. small. *European J. Combin.* **95** (2021), 103323, 25 pp.

[49] D. Mašulović and B. Šobot, Countable ordinals and big Ramsey degrees. *Combinatorica* **41** (2021), no. 3, 425–446.

[50] K. R. Milliken, A Ramsey theorem for trees. *J. Combin. Theory Ser. A* **26** (1976), 215–237.

[51] J. Nešetřil and V. Rödl, Partitions of finite relational and set systems. *J. Combin. Theory Ser. A* **22** (1977), no. 3, 289–312.

[52] J. Nešetřil and V. Rödl, Ramsey classes of set systems. *J. Combin. Theory Ser. A* **34** (1983), no. 2, 183–201.

[53] L. Nguyen Van Thé, Big Ramsey degrees and divisibility in classes of ultrametric spaces. *Canad. Math. Bull.* **51** (2008), no. 3, 413–423.

[54] L. Nguyen Van Thé, *Structural Ramsey theory with the Kechris–Pestov–Todorcevic correspondence in mind*. Habilitation thesis, Université d'Aix-Marseille, 2013.

[55] L. Nguyen Van Thé and N. Sauer, The Urysohn sphere is oscillation stable. *Geom. Funct. Anal.* **19** (2009), no. 2, 536–557.

[56] T. Odell and T. Schlumprecht, The distortion problem. *Acta Math.* **173** (1994), no. 2, 259–281.

[57] M. Pouzet and N. Sauer, Edge partitions of the Rado graph. *Combinatorica* **16** (1996), no. 4, 505–520.

[58] F. P. Ramsey, On a problem of formal logic. *Proc. Lond. Math. Soc. (2)* **30** (1929), no. 4, 264–296.

[59] S. Saharon, Strong partition relations below the power set: consistency – was Sierpinski right? II. In *Sets, graphs and numbers (Budapest, 1991)*, pp. 637–688, Colloq. Math. Soc. János Bolyai 60, North-Holland, Amsterdam, 1992.

[60] N. Sauer, Edge partitions of the countable triangle free homogenous graph. *Discrete Math.* **185** (1998), no. 1–3, 137–181.

[61] N. Sauer, Canonical vertex partitions. *Combin. Probab. Comput.* **12** (2003), no. 6, 671–704.

[62] N. Sauer, Coloring subgraphs of the Rado graph. *Combinatorica* **26** (2006), no. 2, 231–253.

[63] N. Sauer, Vertex partitions of metric spaces with finite distance sets. *Discrete Math.* **312** (2012), no. 1, 119–128.

[64] N. Sauer, Coloring homogeneous structures. 2020, arXiv:2008.02375.

[65] W. Sierpiński, Sur une problème de lat théorie des relations. *Ann. Sc. Norm. Super. Pisa (2)* **2** (1933), 239–242.

[66] S. Todorcevic, *Introduction to Ramsey spaces*. Princeton University Press, Princeton, New Jersey, 2010.

[67] S. Todorcevic and I. Farah, *Some applications of the method of forcing*. Yenisei Ser. Pure Appl. Math., Yenisei, Moscow, 1995.

[68] J. Zhang, A tail cone version of the Halpern–Läuchli theorem at a large cardinal. *J. Symbolic Logic* **84** (2017), no. 2, 473–496.

[69] A. Zucker, Topological dynamics of automorphism groups, ultrafilter combinatorics, and the generic point problem. *Trans. Amer. Math. Soc.* **368** (2016), no. 9, 6715–6740.

[70] A. Zucker, Big Ramsey degrees and topological dynamics. *Groups Geom. Dyn.* **13** (2018), no. 1, 235–276.

[71] A. Zucker, On big Ramsey degrees for binary free amalgamation classes. *Adv. Math.* **408 A** (2022), 108585, 25 pp.

NATASHA DOBRINEN

University of Denver, Department of Mathematics, 2390 S. York St., Denver, CO 80210, USA; current address: University of Notre Dame, Department of Mathematics, 255 Hurley Building, Notre Dame, IN 46556, USA, ndobrine@nd.edu

This Page Intentionally Left Blank

MEASURABLE GRAPH COMBINATORICS

ANDREW S. MARKS

ABSTRACT

We survey some recent results in the theory of measurable graph combinatorics. We also discuss applications to the study of hyperfiniteness and measurable equidecompositions.

MATHEMATICS SUBJECT CLASSIFICATION 2020

Primary 03E15; Secondary 05C21, 22F10, 28A75, 37A20, 52B45

KEYWORDS

Descriptive set theory, measurable graph combinatorics, Borel graph, amenability, Borel equivalence relations, hyperfiniteness, asymptotic dimension, equidecomposition, tilings, Lovasz Local Lemma

INTERNATIONAL CONGRESS OF MATHEMATICIANS 2022 JULY 6–14

Proc. Int. Cong. Math. 2022, Vol. 3, pp. 1488–1502
DOI 10.4171/ICM2022/157

Published by EMS Press

1. INTRODUCTION

Measurable graph combinatorics focuses on finding measurable solutions to combinatorial problems on infinite graphs. This study involves ideas and techniques from combinatorics, ergodic theory, probability theory, descriptive set theory, and theoretical computer science. We survey some recent progress in this area, focusing on the study of *locally finite* graphs: graphs where each vertex has finitely many neighbors. We also discuss applications to the study of hyperfiniteness of Borel actions of groups, and measurable equidecompositions.

Without any constraints such as measurability conditions, combinatorial problems on locally finite graphs often simplify to studying their restriction to finite subgraphs. This is the case with the problem of graph coloring. Recall that if $G = (V, E)$ is a graph, a *(proper) Y-coloring* of G is a map $c: V \to Y$ so that for every two adjacent vertices $\{x, y\} \in E$, the colors assigned to these two vertices are distinct, $c(x) \neq c(y)$. The *chromatic number* $\chi(G)$ of G is the smallest cardinality of a set Y so there is a Y-coloring of G. A classical theorem of De Bruijn and Erdős states that for a locally finite graph G, the chromatic number of G is equal to the supremum of the chromatic number of all finite subgraphs of G. That is, $\chi(G) = \sup_{\text{finite } H \subseteq G} \chi(H)$. The proof of this theorem is a straightforward compactness argument using the Axiom of Choice.

In contrast, many phenomena can influence measurable chromatic numbers beyond just the constraints imposed by finite subgraphs. We illustrate this change in behavior with a simple example. Let S^1 be the circle, let $T: S^1 \to S^1$ be an irrational rotation, and let μ be Lebesgue measure on S^1. Consider the graph G_T with vertex set S^1 and where x, y are adjacent if $T(x) = y$ or $T(y) = x$. Every vertex in G_T has degree 2 and every connected component of G_T is infinite. Hence, by alternating between two colors, it is easy to see that the classical chromatic number of G_T is 2. However, there can be no Lebesgue measurable 2-coloring of G_T. Suppose $c: S^1 \to \{0, 1\}$ was a Lebesgue measurable coloring of G_T, and $A_0 = \{x : c(x) = 0\}$ and $A_1 = \{x : c(x) = 1\}$ were the two color sets. Then since the coloring must alternate between the two colors, we must have $T(A_0) = A_1$, and since T is measure preserving and A_0 and A_1 are disjoint and cover S^1, we therefore have $\lambda(A_0) = \lambda(A_1) = \frac{1}{2}$. However, the transformation T^2 is also an irrational rotation and hence T^2 is *ergodic*, meaning any set invariant under T^2 must be null or conull. Since $T^2(A_0) = A_0$, A_0 must be null or conull. Contradiction!

In this paper we focus on the study of combinatorial problems on *Borel graphs*: graphs where the set V of vertices is a standard Borel space and where the edge relation E is Borel as a subset of $V \times V$. In the setting where each vertex has at most countably many neighbors, this is equivalent to saying that there are countably many Borel functions $f_0, f_1, \ldots : V \to V$ that generate G in the sense that $x \mathrel{E} y$ if and only if $f_i(x) = y$ for some i. The equivalence follows from the Lusin–Novikov theorem [28, 18.15]. An important example of a Borel graph is the following type of *Schreier graph*. If a is a Borel action of a countable group Γ on a standard Borel space X and S is a symmetric set of generators for Γ, then let $G(a, S)$ be the graph on the vertex set $V = X$ where $x, y \in V$ are adjacent if there is a $\gamma \in S$ such that $\gamma \cdot x = y$. For example, the graph associated to the irrational rotation described above is a graph of this form.

For more comprehensive surveys of this area, the reader should consult the papers [30, 44]. A notable recent development we will not discuss is the connections that have been found between measurable combinatorics and the study of distributed algorithms in theoretical computer science, particularly the LOCAL model. This model of computing takes place on a large graph where each vertex represents a computer which is assigned a unique identifier, and each edge is a communication link. These processors execute the same algorithm in parallel, communicating with their neighbors in rounds to construct a global solution to some combinatorial problem. Recent work [2, 3, 6, 17] has established some precise connections between measurable combinatorics and LOCAL algorithms which have already led to new theorems in both areas (see, e.g., [2, 4]).

2. MEASURABLE COLORINGS

If G is a Borel graph, we define the Borel chromatic number $\chi_B(G)$ of G to be the smallest cardinality of a standard Borel space Y so that there is a Borel measurable Y-coloring of G. We clearly have that $\chi(G) \leq \chi_B(G)$ where $\chi(G)$ is the classical chromatic number of G. Borel chromatic numbers were first studied in a foundational paper of Kechris, Solecki, and Todorcevic [32].

Let $G = (V, E)$ be a graph. If $x \in V$ is a vertex, we let $N(x) = \{y : \{x, y\} \in E\}$ denote the set of *neighbors* of x. The *degree* of x is the cardinality of $N(x)$. We say that a graph is Δ-*regular* if every vertex has degree Δ. A basic result about graph coloring is that, given any finite graph G of finite maximum degree Δ, there is a $(\Delta + 1)$-coloring of G. This is easy to see by coloring the vertices of G one by one. If we have a partial coloring of G, then any uncolored vertex x has at most Δ neighbors so there must be a color from the set of $\Delta + 1$ colors we can use to extend this partial coloring to x. The analogous fact remains true about Borel colorings:

Theorem 2.1 (Kechris, Solecki, Todorcevic [32, PROPOSITION 4.6]). *If G is a Borel graph of finite maximum degree Δ, then G has a Borel $(\Delta + 1)$-coloring.*

One method of proving this theorem is to adapt the greedy algorithm described above. Recall that a set of vertices is *independent* if it does not contain two adjacent vertices. First, we may find a countable sequence of Borel sets A_n such that each A_n is independent, and their union is all vertices $\bigcup_n A_n = V(G)$. Then we can iteratively construct a coloring of G in countably many steps where at step n we color all the elements of A_n the least color not already used by one of its neighbors. In general, the connection between algorithms for solving combinatorial problems and measurable combinatorics is deep. Many techniques for constructing measurable colorings are based on algorithmic ideas, since algorithms for solving combinatorial problems will often yield an explicitly definable solutions to them.

The upper bound given by Theorem 2.1 is tight; a complete graph on $\Delta + 1$ vertices has maximum degree Δ and chromatic number $\Delta + 1$. Surprisingly, the upper bound of

Theorem 2.1 is also optimal even in the case of acyclic Borel graphs. Hence, for bounded degree Borel graphs, the Borel chromatic number and classical chromatic number may be very far apart since any acyclic graph has classical chromatic number at most 2.

Theorem 2.2 (Marks [38]). *For every finite Δ, there is an acyclic Borel graph of degree Δ with no Borel Δ-coloring.*

The graphs used to establish Theorem 2.2 are quite natural, and arise from Schreier graphs of actions of free products of Δ many copies of $\mathbb{Z}/2\mathbb{Z}$. Theorem 2.2 is proved using Martin's theorem of Borel determinacy [41] which states that in any infinite two-player game of perfect information with a Borel payoff set, one of the two players has a winning strategy. The direct use of Borel determinacy to prove this theorem leads to an interesting question of reverse mathematics since Borel determinacy requires a great deal of set-theoretic power to prove: the use of uncountably many iterates of the powerset of $\mathbb{R}$ [19]. We currently do not know of any simpler proof of Theorem 2.2 that avoids the use of Borel determinacy or can be proved in second-order arithmetic (which suffices for most theorems of descriptive set theory).

Problem 2.3. Is Theorem 2.2 provable in the theory Z_2 of full second-order arithmetic?

Recently, Brandt, Chang, Grebík, Grunau, Rozhoň, and Vidnyánszky [6] have shown that characterizing the set of Borel graphs of maximum degree $\Delta \geq 3$ that have no Borel $(\Delta + 1)$-coloring is as hard as possible in a precise sense: the set of such graphs is $\mathbf{\Sigma}^1_2$ complete. Here $\mathbf{\Sigma}^1_2$ completeness is a logical measurement of the complexity of this problem. The proof of their theorem combines the techniques of [39] with earlier work of Todorcevic and Vidnyánszky [48] proving $\mathbf{\Sigma}^1_2$ completeness for the set of locally finite Borel graphs generated by a single function that have finite Borel chromatic number. In contrast to the work of [6] for $\Delta \geq 3$, in the case $\Delta = 2$, a dichotomy theorem of Carroy, Miller, Schrittesser, and Vidnyánszky [8] characterizes the 2-colorable Borel graphs in a simple way as those for which there is no Borel homomorphism from a canonical non-Borel-2-colorable graph known as $\mathbb{L}_0$.

This type of theorem—proving it is hard to characterize the set of graphs with some combinatorial property—is familiar in finite graph theory via computational complexity theory. For example, it is a well-known theorem that the set of finite graphs that are k-colorable for $k \geq 3$ is NP-complete. Indeed, there are some surprising newly found connections between computational complexity theory and complexity in measurable combinatorics. Thornton [47] has used techniques adapted from the celebrated CSP (constraint satisfaction problem) dichotomy theorem [7, 51] in theoretical computer science to bootstrap the results of [6] to show many other natural combinatorial problems on locally finite Borel graphs are either $\mathbf{\Sigma}^1_2$ complete or a $\mathbf{\Pi}^1_1$. The CSP dichotomy theorem concerns a certain class of natural problems in NP: general constraint satisfaction problems like graph coloring with k colors, k-SAT, or, more generally, computing the set of finite structures X that have a homomorphism to a given fixed finite structure D. The CSP dichotomy states that all such constraint satisfaction problems are either in P (like 2-coloring or 2-SAT), or they are NP-complete (like 3-coloring or 3-SAT).

The results in [6] rule out any simple theory for understanding Borel chromatic number for locally finite Borel graphs in general. In contrast, if we weaken our measurability condition to study μ-measurable colorings with respect to some Borel probability measure μ instead of Borel colorings, the theory of μ-measurable colorings appears to have a much closer connection to finite graph theory. If μ is a Borel measure on the vertex set of a Borel graph G, let $\chi_\mu(G)$ be the least size of a set Y so there is a μ-measurable coloring of G. So $\chi(G) \leq \chi_\mu(G) \leq \chi_B(G)$, since every Borel function is μ-measurable.

For finite graphs of maximum degree Δ, a theorem of Brooks characterizes those connected graphs which have chromatic number of $\Delta + 1$. They are precisely the complete graphs on $\Delta + 1$ vertices, and odd cycles in the case $\Delta = 2$. Analogously, we have the following generalization of Brooks's theorem for μ-measurable colorings:

Theorem 2.4 (Conley, Marks, Tucker-Drob [13]). *Suppose that G is a Borel graph with degree bounded by a finite $\Delta \geq 3$. Suppose further that G contains no complete graph on $\Delta + 1$ vertices. If μ is any Borel probability measure on $V(G)$, then G admits a μ-measurable Δ-coloring.*

Several important open problems in descriptive set theory concern whether there is a difference between being able to find a Borel solution to a problem versus being able to find a μ-measurable solution with respect to every Borel probability measure μ (e.g., the hyperfiniteness vs measure hyperfiniteness problem [29, PROBLEM 8.29]). Theorems 2.2 and 2.4 are encouraging in this context because they point the way towards tools that may be able to resolve these types of questions.

The proof of Theorem 2.4 is based on a technique for finding *one-ended spanning subforests* in Borel graphs: acyclic subgraphs on the same vertex set where each connected component has exactly one end. More recently, these techniques for finding one-ended spanning subforests were applied to prove new results in the theory of cost: a real valued invariant of countable groups arising from their ergodic actions [9].

Bernshteyn has substantially strengthened Theorem 2.4 by showing for k within a factor of $\sqrt{\Delta}$ of Δ, there is a μ-measurable k-coloring of G if and only if there is any k-coloring of G.

Theorem 2.5 (Bernshteyn [2]). *There is a Δ_0 so that if G is a Borel graph with finite maximum degree $\Delta \geq \Delta_0$ and μ is a Borel probability measure on $V(G)$, then if c satisfies $c \leq \sqrt{D} - 5/2$, then G has a $(\Delta - c)$-coloring if and only if G has a μ-measurable $(\Delta - c)$-coloring.*

The above results give cases where the μ-measurable chromatic number behaves similarly to the classical chromatic number. These two quantities may still differ by a large amount, however. Let $\mathbb{F}_n$ be the free group on n generators and let $S_n \subseteq \mathbb{F}_n$ be a free symmetric generating set. Let a_n be the action of $\mathbb{F}_n$ on the space $[0,1]^{\mathbb{F}_n}$ via the *Bernoulli shift*: $(\gamma \cdot x)(\delta) = x(\gamma^{-1}\delta)$ restricted to its free part. Let $G_n = G(a_n, S_n)$ be the Schreier graph of this action, and let $\mu_n = \lambda^{\mathbb{F}_n}$ be the product of Lebesgue measure λ on $[0, 1]$. Since G_n is acyclic, the classical chromatic number is $\chi(G_n) = 2$. However, $\chi_{\mu_n}(G_n) \geq \frac{n}{\log 2n}$ which can be shown using results about the size of independent sets in random $(2n)$-regular graphs and an ultraproduct argument. This argument was first suggested by [36]; see [38] for a detailed proof. Bernshteyn has recently proven an upper bound on $\chi_{\mu_n}(G_n)$ which is within a factor of two of this lower bound [1]. However, it remains an open problem to compute the precise rate of growth of $\chi_{\mu_n}(G_n)$.

Bernshteyn's Theorem 2.5 and the above upper bound on $\chi_{\mu_n}(G_n)$ are based on an adaptation of the powerful Lovász Local Lemma (LLL) to the setting of measurable combinatorics. The LLL is a tool of probabilistic combinatorics which can show the existence of objects which are described by constraints that are local in the sense that each constraint is independent of all but a small number of other constraints, and each constraint has a high probability of being satisfied. Precisely, the *symmetric LLL* states that if $A_1, \ldots, A_n$ are events in a probability space which each occur with probability at most p, each event A_i is independent of all but at most d of the other events, and $ep(d + 1) \leq 1$, then there is a positive probability none of these events occur.

The LLL is a pure existence result, and since the desired object typically exists with exponentially small probability, it was a major open problem to find an algorithmic way to quickly find satisfying assignments where none of the events $A_1, \ldots, A_n$ happen. In particular, a naive attempt to randomly sample from the probability distribution until a solution is found would take at least exponential time. In a breakthrough result in 2009, Moser and Tardos [42] gave an efficient randomized algorithm that can quickly compute satisfying assignments for the LLL.

Adaptations of the Moser–Tardos algorithm and the LLL to the setting of measurable combinatorics began with work of Kun [33], who used a version of the Moser–Tardos algorithm to find spanning subforests to prove a strengthening of the Gaboriau–Lyons [20] theorem in ergodic theory. More recently, Csoka, Grabowski, Mathe, Pikhurko, and Tyros [14] have proved a Borel version of the symmetric LLL for Borel graphs of subexponential growth, and Bernshteyn has proved μ-measurable versions for Bernoulli shifts of groups, and probability measure preserving Borel graphs [1, 2]. These results, combined with the large literature in combinatorics using the LLL to construct colorings of graphs, are the main tool in the proof of Theorem 2.5.

It is known that there cannot be a Borel version of the symmetric LLL for bounded degree Borel graphs in general [12]. Indeed, the existence of such a theorem combined with standard coloring techniques using the LLL would contradict Theorem 2.2. However, an interesting special case remains open: a Borel version of the symmetric LLL for Borel Schreier graphs generated by Borel actions of *amenable* groups, which are defined in the next section. Such a version of the local lemma could be a useful tool for making progress on the open problems discussed in the next section.

The theorems we have described above are a small selection of what is now known about measurable chromatic numbers. We hope they give the reader some sense of the variety of results and tools of the subject.

3. CONNECTIONS WITH HYPERFINITENESS

A major research program in modern descriptive set theory has been to understand the relative complexity of equivalence relations under Borel reducibility. Precisely, if E and F are equivalence relations on standard Borel spaces X and Y, say that E is *Borel reducible* to F if there is a Borel function $f: X \to Y$ such that for all $x, y \in X$, we have $x \mathrel{E} y \iff f(x) \mathrel{F} f(y)$. Such a function induces a definable injection from X/E to Y/F. If we think of E and F as classification problems, this means E is simpler than F in the sense that any invariants that can be used to classify F can also be used to classify E. In the study of Borel reducibility of equivalence relations, there has been success both in understanding the abstract structure of all Borel equivalence relations under Borel reducibility, and also in proving particular nonclassification results of interest to working mathematicians. For example, Hjorth's theory of turbulence [26] gives a precise dichotomy for when an equivalence relation generated by a Polish group action can be classified by invariants that are countable structures, and turbulence has been applied to prove nonclassifiability results in C^* algebras [18].

A Borel equivalence relation E is said to be *countable* if every E-class is countable. The countable Borel equivalence relations are an important and well-studied subclass of Borel equivalence relations with rich connections with operator algebras and ergodic theory. One reason for this is the Feldman–Moore theorem [31, THEOREM 1.3], which states that every countable Borel equivalence relation is induced by a Borel action of a countable group. Results proved about the dynamics of measure preserving actions of countable groups have played a played an important role in our understanding of the theory of countable Borel equivalence relations.

Understanding how the descriptive-set-theoretic complexity of countable Borel equivalence relations is related to the dynamics of the group actions that generate them is a deep problem. An important simplicity notion for Borel reducibility is hyperfiniteness: a Borel equivalence relation is *hyperfinite* if it can be written as an increasing union of Borel equivalence relations whose classes are all finite. The hyperfinite equivalence relations are the simplest nontrivial class of Borel equivalence relations as made precise by the Glimm–Effros dichotomy of Harrington, Kechris, and Louveau [25]. Weiss has asked if the group-theoretic notion of amenability precisely corresponds to hyperfiniteness:

Problem 3.1 (Weiss, [50]). Suppose E is a Borel equivalence relation generated by a Borel action of a countable amenable group. Is E hyperfinite?

Amenability was defined by von Neumann in reaction to the Banach–Tarski paradox. It is a group-theoretic notion of dynamical tameness. Precisely, a group Γ is *amenable* if and only if for every $\varepsilon > 0$ and every finite $S \subseteq \Gamma$ there exists some nonempty finite $F \subseteq \Gamma$ such that $|SF \triangle F|/|F| < \varepsilon$. Such an F is called an (ε, S)-Følner set. Examples of amenable groups include finite, abelian, and solvable groups, while the free group on two generators is nonamenable. If Weiss's question has a positive answer, then amenability precisely characterizes hyperfiniteness since every nonamenable group has a nonhyperfinite Borel action.

Evidence that Weiss's question has a positive answer is given by a theorem in ergodic theory of Ornstein and Weiss [43] that every measure preserving action of an amenable group on a standard probability space is hyperfinite modulo a nullset.

Progress on Weiss's question has grown out of progress on the problem of finding Borel tilings of group actions by Følner sets. Precisely, if $a\colon \Gamma \curvearrowright X$ is an action of a finitely generated group Γ, and $F_1, \ldots, F_n \subseteq \Gamma$ are finite subsets of Γ, a *tiling* of a by the shapes $F_1, \ldots, F_n$ is a collection of subsets $A_1, \ldots, A_n \subseteq X$ so that the sets $F_1 \cdot A_1, \ldots, F_n \cdot A_n$ are pairwise disjoint and form a partition of X. Finding tilings of a group action can be thought of as a generalized coloring problem or constraint satisfaction problem of the type often studied in measurable combinatorics, and can be approached using many of the same tools. For example, Jackson, Kechris, and Louveau [27] have shown that Weiss's question has a positive answer for groups of polynomial volume growth. Their argument uses Voronoi regions around Borel maximal independent sets to make Borel tilings with desirable properties. Gao and Jackson [21] have shown that Weiss's question has a positive answer for abelian groups. Their argument centers around a more refined inductive argument to find tilings of $\mathbb{Z}^n$ by hyperrectangles. These tilings are found by iteratively adjusting the location of the boundaries of hyperrectangular tiles that cover the space until their parallel boundaries are far apart. Schneider and Seward have extended Gao and Jackson's machinery to all locally nilpotent groups [45]. All these tilings are the building blocks out of which witnesses to hyperfiniteness are constructed.

A positive answer to the following open problem would be progress towards a positive solution to Weiss's question:

Problem 3.2. Let Γ be an amenable group with finite symmetric generating set S and $a\colon \Gamma \curvearrowright X$ be a free Borel action of a on a standard Borel space X. For every $\varepsilon > 0$, do there exist (ε, S)-Følner sets $F_1, \ldots, F_n \subseteq \Gamma$ such that the action a has a Borel tiling with shapes $F_1, \ldots, F_n$?

The existence of such tilings without any measurability conditions was only recently established by Downarowicz, Huczek, and Zhang [15]. A key step in their proof is to use Hall's matching theorem to match untiled points in a Ornstein–Weiss style *quasitiling* [43] to construct an exact tiling. Recall that if $G = (V, E)$ is a graph, a *perfect matching* of G is a subset $M \subseteq E$ of edges so that each vertex $x \in V$ is incident to exactly one edge in M. Hall's matching theorem states that a bipartite graph with bipartition $A, B \subseteq V$ has a perfect matching if and only if for every finite set $F \subseteq A$ or $F \subseteq B$,

$$\left|N(F)\right| \geq |F|.$$

Recently, Problem 3.2 has been shown to have a positive answer modulo a nullset [10]. A key part of the proof is a measurable matching result proved using an idea of Lyons and Nazarov [36] that was originally used to find factor of i.i.d. perfect matchings of regular trees. Lyons and Nazarov's argument uses the *Hungarian matching algorithm* (repeatedly flipping augmenting paths) to show that if a bipartite Borel graph G satisfies a certain measure-theoretic expansion condition strengthening Hall's condition, then it has a measurable perfect matching.

Conley, Jackson, Marks, Seward, and Tucker-Drob have proven the following:

Theorem 3.3 (Conley, Jackson, Marks, Seward, Tucker-Drob [11]). *Let* Γ *be a countable group admitting a normal series where each quotient of consecutive terms is a finite group or a torsion-free abelian group with finite* $\mathbb{Q}$*-rank, except that the top quotient can be any group of uniform local polynomial volume-growth or the lamplighter group* $\mathbb{Z}_2 \wr \mathbb{Z}$*. Then every free Borel action of* Γ *is hyperfinite.*

By combining this with prior work of Seward and Schneider [45, COR. 8.2] they obtain the following corollary:

Corollary 3.4. *Weiss's question has a positive answer for polycyclic groups.*

This is the best partial result on Weiss's question that is currently known. Significantly, Corollary 3.4 applies to groups of exponential volume growth such as certain semidirect products of $\mathbb{Z}^n$. All the previous work on Weiss's question applied only to groups locally of polynomial volume growth, and this seemed an inherent limitation to previous methods.

The central idea of [11] is to adapt the machinery of Gromov's theory of asymptotic dimension of groups to the setting of descriptive set theory, making a theory of Borel asymptotic dimension. These ideas were implicitly hidden in all previous work on Weiss's question, but were first made explicit in [11]. Asymptotic dimension was introduced by Gromov as a quasiisometry invariant of metric spaces, used to study geometric group theory. The asymptotic dimension of a metric space (X, ρ) is the least d such that for every $r > 0$ there is a uniformly bounded cover U of X so that every closed r-ball intersects at most $d + 1$ sets in U. Essentially, asymptotic dimension is a "large-scale" analogue of Lebesgue covering dimension. There are actually several different ways to define asymptotic dimension whose equivalences are nontrivial to prove. Proving that these different definitions still define the same notion in the Borel context is one of the keys to the work in [11]. Alternate definitions allow the conversion between Voronoi cell-type tilings patterned on the work of Jackson, Kechris, and Louveau, and covers with far apart facial boundaries pioneered by Gao and Jackson.

Resolving Weiss's question for all amenable groups appears to be a difficult problem. In general, we have a poor understanding of the geometry and structure of Følner sets in arbitrary amenable groups. Problem 3.1 for arbitrary amenable groups seems to either require significant advances in our geometric understanding of amenable groups, or completely different descriptive-set theoretic tools for attacking the hyperfiniteness problem. One question which gets at the heart of this difficulty is the following:

Problem 3.5. Suppose G is a bounded degree Borel graph having uniformly bounded polynomial growth. Is the connectedness relation of G hyperfinite?

The obstacle in resolving Problem 3.5 is that while polynomial growth groups have tight both upper and lower bound on their growth, Problem 3.5 only posits an upper bound on the growth of G, which may consequently have much less uniformity in its growth than the Schreier graph associated to an action of a polynomial growth group. This lack of a lower bound on growth means that the techniques of Jackson, Kechris, and Louveau for proving hyperfiniteness of groups of polynomial growth cannot resolve Problem 3.5 Finding techniques for resolving Problem 3.5 where there is far less regular geometric structure would be one way of making progress towards resolving Weiss's question in general since we know little about any regular geometric structure in arbitrary amenable groups.

4. MEASURABLE EQUIDECOMPOSITIONS

If $a: \Gamma \curvearrowright X$ is an action of a group Γ on a space X, then we say sets $A, B \subseteq X$ are *a-equidecomposable* if there are a finite partition $\{A_0, \dots, A_n\}$ of A and group elements $\gamma_0, \dots, \gamma_n \in \Gamma$ so that $\gamma_0 A_0, \dots, \gamma_n A_n$ is a partition of B. For example, in this language, the Banach–Tarski paradox says that one unit ball is equidecomposable with two unit balls under the group action of isometries of $\mathbb{R}^3$. In the last few years several new results proved about these types of geometrical paradoxes with the unifying theme that the "paradoxical" sets in many classical geometrical paradoxes can surprisingly be much nicer than one would naively expect.

A classical generalization of the Banach–Tarski paradox states that any two bounded sets $A, B \subseteq \mathbb{R}^3$ with nonempty interior are equidecomposable. Of course, the pieces used in these equidecompositions must be nonmeasurable in general, since A and B may have different measure. However, a remarkable theorem of Grabowski, Máthé, and Pikhurko states that there is always an equidecomposition using measurable sets when A and B have the same Lebesgue measure.

Theorem 4.1 (Grabowski, Máthé, Pikhurko [24]). *If $A, B \subseteq \mathbb{R}^3$ are bounded sets with nonempty interior and if additionally A and B are assumed to have the same Lebesgue measure, then A and B can be equidecomposed using Lebesgue measurable pieces.*

It is an open problem whether Theorem 4.1 can be strengthened to yield a Borel equidecomposition, assuming A and B are Borel.

Key to Theorem 4.1 and other advances in measurable equidecompositions has been progress made on measurable matching problems. The connection comes from the following graph-theoretic reformulation of equidecompositions as perfect matchings. Let $a: \Gamma \curvearrowright X$ be a Borel action of a group Γ on a space X, let $A, B, \subseteq X$ be subsets of X, and let $S \subseteq \Gamma$ be finite. Let $G(A, B, S)$ be the graph whose set of vertices is the disjoint union $A \sqcup B$ and where $x \in A$ and $Y \in B$ are adjacent if there is a $\gamma \in S$ so that $\gamma \cdot x = y$. Then it is easy to see that A, B are equidecomposable using group elements from S if and only if there is a perfect matching of the graph $G(A, B, S)$.

Theorem 4.1 and other new results about measurable equidecompositions rely on combining process made on measurable matching problems with modern results about the dynamics of the group actions being studied. For example, Theorem 4.1 uses the local spectral gap of Boutonnet, Ioana, and Salehi Golsefidy [5] for certain lattices in the group $SO_3(\mathbb{R})$ of rotations in $\mathbb{R}^3$. This result is used to check that the graph $G(A, B, S)$ satisfies the expansion condition of Lyons and Nazarov [36] which ensures the existence of a measurable matching.

Some other recent theorems about measurable equidecompositions concern Tarski's famous circle squaring problem from 1925: the question of whether a disk and square of the same area in $\mathbb{R}^2$ are equidecomposable by isometries. Tarski's circle squaring problem arose from the fact that the analogue of the Banach–Tarski paradox is false in $\mathbb{R}^2$. This is because there are so-called *Banach measures* in $\mathbb{R}^2$: finitely additive isometry-invariant measures that extend Lebesgue measure. Their existence is proved using the amenability of the isometry group of $\mathbb{R}^2$. Hence, if Lebesgue measurable sets $A, B \subseteq \mathbb{R}^2$ are equidecomposable, they must have the same Lebesgue measure. The real thrust of Tarski's circle squaring problem is the converse of this: the general problem of to what extent there is an equivalence between equidecomposability and having the same measure.

In 1990, Laczkovich [34] (see also [35]) gave a positive answer to Tarski's circle squaring problem using the Axiom of Choice. His proof involved sophisticated tools from Diophantine approximation and discrepancy theory to prove strong quantitative bounds on the ergodic theorem for translation actions of the torus, as well as the graph-theoretic approach to equidecomposition described above.

Marks and Unger have shown that there is a Borel solution to Tarski's circle squaring problem, building on an earlier result of Grabowski, Máthé, and Pikhurko, [23] that the circle can be squared using Lebesgue measurable pieces.

Theorem 4.2 (Marks, Unger [40]). *Tarski's circle squaring problem has a positive solution using Borel pieces. More generally, for all $n \geq 1$, if $A, B \subseteq \mathbb{R}^n$ are bounded Borel sets with the same positive Lebesgue measure whose boundaries have upper Minkowski dimension less than n, then A and B are equidecomposable using Borel pieces.*

Hence, for Borel sets whose boundaries are not wildly fractal, having the same measure is actually equivalent to having an explicitly definable Borel equidecomposition.

Theorem 4.2 uses new techniques for constructing Borel perfect matchings in Borel graphs based on first finding a real-valued Borel *flow* as an intermediate step. Precisely, if $f: V \to \mathbb{R}$ is a function on the vertices of a graph G, then an f-flow on G is a real-valued function ϕ on the edges of G such that $\phi(x, y) = -\phi(y, x)$ for every directed edge (x, y) of G, and such that for every $x \in V$ the flow ϕ satisfies Kirchoff's law,

$$f(x) = \sum_{y \in N(x)} \phi(x, y).$$

Given a circle and square $A, B \subseteq [0, 1)^2$ of the same area, the first step in the proof of Theorem 4.2 is finding an explicit $(1_A - 1_B)$-flow of an appropriate Borel graph whose vertices are all the elements of $[0, 1)^2$ and whose edges are generated by finitely many translations.

The advantage of working with the generality of flows is twofold. First, a flow can be constructed in countably many steps, making the error in Kirchoff's law above continuously approach 0 whereas the error in a partial matching that makes it imperfect is discrete. Second, the average of f-flows is an f-flow and so it is possible to integrate families of definable flows to get another definable flow. Finally, there are well known combinatorial equivalences between flows and matchings which are used in the last step of the proof of Theorem 4.2 to "round" a real-valued flow into an integer valued flow and then use it to construct a matching.

Another key ingredient in the proof of Theorem 4.2 is the hyperfiniteness of Borel actions of abelian groups. In particular, the proof of Theorem 4.2 uses a recent refinement due to Gao, Jackson, Krohne, and Seward [22] of Gao and Jackson's [21] theorem that Borel actions of abelian groups are hyperfinite. These witnesses to hyperfiniteness are used to ensure that the Ford–Fulkerson algorithm converges when it is used to round the Borel real-valued flow described above into a Borel integer-valued flow.

This flow approach to equidecomposition problems may be useful for attacking other open questions such as the Borel–Ruziewicz problem:

Problem 4.3 (Wagon [49]). Suppose $n \geq 2$. Is Lebesgue measure the unique finitely additive rotation invariant probability measure defined on the Borel subsets of the n-sphere S^n?

This question is inspired by a theorem of Margulis [37] and Sullivan [46] ($n \geq 4$), and Drinfeld [16] ($n = 2, 3$), who proved that Lebesgue measure is the unique finitely additive rotation invariant measure on the Lebesgue measurable subsets of S^n. Wagon's proposed strengthening would be a more natural result since the Borel sets are the canonical σ-algebra to measure.

ACKNOWLEDGMENTS

The author would like to thank Alekos Kechris for helpful discussions.

FUNDING

This author is partially supported by NSF grant DMS-2054182.

REFERENCES

[1] A. Bernshteyn, Measurable versions of the Lovász Local Lemma and measurable graph colorings. *Adv. Math.* **353** (2019), 153–223.

[2] A. Bernshteyn, Distributed algorithms, the Lovász local lemma, and descriptive combinatorics. 2020, arXiv:2004.04905.

[3] A. Bernshteyn, Probabilistic constructions in continuous combinatorics and a bridge to distributed algorithms. 2021, arXiv:2102.08797.

[4] A. Bernshteyn, A fast distributed algorithm for $(\Delta + 1)$-edge-coloring. *J. Combin. Theory Ser. B* (2022).

[5] R. Boutonnet, A. Ioana, and A. Salehi Golsefidy, Local spectral gap in simple Lie groups and applications. *Invent. Math.* **208** (2017), no. 3, 715–802.

[6] S. Brandt, Y. Chang, J. Grebík, C. Grunau, V. Rozhoň, and Z. Vidnyánszky, On homomorphism graphs. 2021, arXiv:2111.03683.

[7] A. Bulatov, A dichotomy theorem for nonuniform CSPs. In *2017 IEEE 58th Annual Symposium on Foundations of Computer Science (FOCS)*, pp. 319–330, IEEE, 2017.

[8] R. Carroy, B. Miller, D. Schrittesser, and Z. Vidnyánszky, Minimal definable graphs of definable chromatic number at least three. *Forum Math. Sigma* **E7** (2021), no. 9, 1–16.

[9] C. T. Conley, D. Gaboriau, A. S. Marks, and R. D. Tucker-Drob, One-ended spanning subforests and treeability of groups. 2021, arXiv:2104.07431.

[10] C. T. Conley, S. Jackson, D. Kerr, A. S. Marks, B. Seward, and R. D. Tucker-Drob, Følner tilings for actions of amenable groups. *Math. Ann.* **371** (2018), 663–683.

[11] C. T. Conley, S. Jackson, A. S. Marks, B. Seward, and R. D. Tucker-Drob, Borel asymptotic dimension and hyperfinite equivalence relations. 2000, arXiv:2009.06721.

[12] C. T. Conley, S. Jackson, A. S. Marks, B. Seward, and R. D. Tucker-Drob, Hyperfiniteness and Borel combinatorics. *J. Eur. Math. Soc. (JEMS)* **22** (2020), no. 3, 877–892.

[13] C. T. Conley, A. S. Marks, and R. D. Tucker-Drob, Brooks's theorem for measurable colorings. *Forum Math. Sigma* **4** (2016).

[14] E. Csóka, Ł. Grabowski, A. Máthé, O. Pikhurko, and K. Tyros, Borel version of the local lemma. 2016, arXiv:1605.04877.

[15] T. Downarowicz, D. Huczek, and G. Zhang, Tilings of amenable groups. *J. Reine Angew. Math.* **747** (2009).

[16] V. G. Drinfeld, Finitely-additive measures on S^2 and S^3, invariant with respect to rotations. *Funktsional. Anal. i Prilozhen.* **18** (1984), no. 3, 77.

[17] G. Elek, Qualitative graph limit theory, Cantor dynamical systems and constant-time distributed algorithms. 2018, arXiv:1812.07511.

[18] I. Farah, Logic and operator algebras. In *Proceedings of the Seoul ICM, vol. II*, edited by Sun Young Jang et al., pp. 15–39, Kyung Moon SA, 2004.

[19] H. Friedman, Higher set theory and mathematical practice. *Ann. Math. Logic* **2** (1971), no. 3, 245–266.

[20] D. Gaboriau and R. Lyons, A measurable-group-theoretic solution to von Neumann's problem. *Invent. Math.* **177** (2009), 533–540.

[21] S. Gao and S. Jackson, Countable abelian group actions and hyperfinite equivalence relations. *Invent. Math.* **201** (2015), no. 1, 309–383.

[22] S. Gao, S. Jackson, E. Krohne, and B. Seward, Forcing constructions and countable Borel equivalence relations. 2015, arXiv:1503.07822.

[23] L. Grabowski, A. Máthé, and O. Pikhurko, Measurable circle squaring. *Ann. of Math. (2)* **185** (2017), 671–710.

[24] L. Grabowski, A. Máthé, and O. Pikhurko, Measurable equidecompositions for group actions with an expansion property. *J. Eur. Math. Soc.* (to appear).

[25] L. Harrington, A. S. Kechris, and A. Louveau, A Glimm–Effros dichotomy for Borel equivalence relations. *J. Amer. Math. Soc.* **3** (1990), no. 4, 903–928.

[26] G. Hjorth, *Classification and orbit equivalence relations*. Math. Surveys Monogr. 75, American Math Society, 2000.

[27] S. Jackson, A. S. Kechris, and A. Louveau, Countable Borel equivalence relations. *J. Math. Log.* **2** (2002), 1–80.

[28] A. S. Kechris, *Classical descriptive set theory*. Grad. Texts in Math. 156, Springer, New York, 1995.

[29] A. S. Kechris, The theory of countable Borel equivalence relations. 2021. URL http://math.caltech.edu/~kechris/papers/lectures%20on%20CBER08book.pdf, preprint.

[30] A. S. Kechris and A. S. Marks, Descriptive graph combinatorics. 2020. URL http://www.math.caltech.edu/~kechris, preprint.

[31] A. S. Kechris and B. D. Miller, *Topics in orbit equivalence*. Lecture Notes in Math. 1852, Springer, Berlin, 2004.

[32] A. S. Kechris, S. Solecki, and S. Todorcevic, Borel chromatic numbers. *Adv. Math.* **141** (1999), 1–44.

[33] G. Kun, Expanders have a spanning Lipschitz subgraph with large girth. 2013, arXiv:1303.4982.

[34] M. Laczkovich, Equidecomposability and discrepancy; a solution of Tarski's circle-squaring problem. *J. Reine Angew. Math.* **404** (1990), 77–117.

[35] M. Laczkovich, Decomposition of sets with small boundary. *J. Lond. Math. Soc.* **46** (1992), 58–64.

[36] R. Lyons and F. Nazarov, Perfect matchings as IID factors of non-amenable groups. *European J. Combin.* **32** (2011), 1115–1125.

[37] G. A. Margulis, Some remarks on invariant means. *Montash. Math.* **90** (1980), no. 3, 233–235.

[38] A. S. Marks, A determinacy approach to Borel combinatorics. *J. Amer. Math. Soc.* **29** (2016), 579–600.

[39] A. S. Marks, Uniformity, universality, and computability theory. *J. Math. Log.* **17** (2017), no. 1.

[40] A. Marks and S. Unger, Borel circle squaring. *Ann. of Math. (2)* **186** (2017), 581–605.

[41] D. A. Martin, Borel determinacy. *Ann. of Math. (2)* **102** (1975), 363–371.

[42] R. Moser and G. Tardos, A constructive proof of the general Lovász Local Lemma. *J. ACM* **57** (2010), no. 2.

[43] D. Ornstein and B. Weiss, Ergodic theory and amenable group actions, I: the Rohlin lemma. *Bull. Amer. Math. Soc.* **2** (1980), 161–164.

[44] O. Pikhurko, Borel combinatorics of locally finite graphs. In *Surveys in combinatorics 2021*, edited by K. K. Dabrowski et al., the invited volume of the 28th British Combinatorial Conference, pp. 267–319, London Math. Soc. Lecture Note Ser., Cambridge University Press, 2021.

[45] S. Schneider and B. Seward, Locally nilpotent groups and hyperfinite equivalence relations. 2013, arXiv:1308.5853.

[46] D. Sullivan, For $n > 3$ there is only one finitely additive rotationally invariant measure on the n-sphere defined on all Lebesgue measurable subsets. *Bull. Amer. Math. Soc.* **4** (1981), no. 1, 121–123.

[47] R. Thornton, The CSP dichotomy and Borel combinatorics. 2021, preprint.

[48] S. Todorcevic and Z. Vidnyanszky, A complexity problem for Borel graphs. *Invent. Math.* **226** (2021), 225–249.

[49] S. Wagon, *The Banach Tarski paradox*. University Press, Cambridge, 1986.

[50] B. Weiss, Measurable dynamics. *Contemp. Math.* **26** (1984), 395–421.

[51] D. Zhuk, A proof of the CSP dichotomy conjecture. In *2017 IEEE 58th Annual Symposium on Foundations of Computer Science (FOCS)*, pp. 331–342, IEEE, 2017.

ANDREW S. MARKS

UCLA Mathematics, BOX 951555, Los Angeles, CA 90095-1555, USA, marks@math.ucla.edu

THE PARIS–HARRINGTON PRINCIPLE AND SECOND-ORDER ARITHMETIC—BRIDGING THE FINITE AND INFINITE RAMSEY THEOREM

KEITA YOKOYAMA

ABSTRACT

The *Paris–Harrington principle* (PH) is known as one of the earliest examples of "mathematical" statements independent from the standard axiomatization of natural numbers called *Peano Arithmetic (*PA*)*. In this article, we discuss various variations of PH and examine the relations between finite and infinite Ramsey's theorem and systems of arithmetic.

MATHEMATICS SUBJECT CLASSIFICATION 2020

Primary 03F30; Secondary 03F35, 03B30, 05D10

KEYWORDS

Paris–Harrington principle, Ramsey's theorem, reverse mathematics, proof theory

ICM INTERNATIONAL CONGRESS OF MATHEMATICIANS 2022 JULY 6–14

Proc. Int. Cong. Math. 2022, Vol. 3, pp. 1504–1528
DOI 10.4171/ICM2022/106

1. INTRODUCTION

To prove a statement about natural numbers, we usually rely explicitly or implicitly on reasoning by mathematical induction. In the setting of mathematical logic, the axiomatic system for natural numbers consists of the axioms for discrete ordered semirings and the scheme of mathematical induction, which is known as *Peano Arithmetic* (PA). Within PA, one can prove many theorems in number theory or finite combinatorics, such as the existence of infinitely many prime numbers or the following finite Ramsey theorem (FRT):

(FRT) For any $n, k, m, a \in \mathbb{N}$, there exists $b \in \mathbb{N}$ such that for any $f : [[a, b)_{\mathbb{N}}]^n \to k$ there exist $H \subseteq [a, b)_{\mathbb{N}}$ and $c < k$ such that $[H]^n \subseteq f^{-1}(c)$ and $|H| = m$.

(Here, $[a, b)_{\mathbb{N}} = \{x \in \mathbb{N} : a \leq x < b\}$ and $[X]^n = \{F \subseteq X : |F| = n\}$ where $|F|$ denotes the cardinality of F. We write k for the set $[0, k)_{\mathbb{N}}$.) Thus, the question might arise: can we prove all true numerical statements within PA?

The answer is known to be negative. The famous *incompleteness theorem* by Kurt Gödel says that there is a numerical statement which is independent from PA (i.e., cannot be proved or disproved from PA). Such an independent statement is provided by diagonalization or self-reference as the liar paradox, and in particular, the numerical statement which intends to say "PA is consistent" is independent from PA. This leads to another question whether there is a "mathematical" statement which is independent from PA. The *Paris–Harrington principle* (PH) [33] is one of the earliest and most important such examples. It is a variant of the finite Ramsey theorem which states the following:

(PH) For any $n, k, a \in \mathbb{N}$, there exists $b \in \mathbb{N}$ such that for any $f : [[a, b)_{\mathbb{N}}]^n \to k$ there exist $H \subseteq [a, b)_{\mathbb{N}}$ and $c < k$ such that $[H]^n \subseteq f^{-1}(c)$ and $|H| > \min H$.

Here, a set H is said to be *relatively large* if $|H| > \min H$, so PH says "for any $a \in \mathbb{N}$, there exists a large enough finite set X above a such that any coloring on X for the Ramsey theorem has a solution which is relatively large." By some standard coding of finite sets of natural numbers as single natural numbers (e.g., by binary expansion), PH can be considered as a purely numerical statement. By easy combinatorics, one can prove PH from the infinite Ramsey theorem (RT), thus PH is a *true* statement about natural numbers.

So how can we know that PH is not provable from PA? The reason is again provided by the Gödel incompleteness, namely, PA + PH implies the consistency of PA and thus it is not provable from PA. Indeed, Paris and Harrington showed that PH is equivalent over PA to the correctness of PA with respect to ∀∃-sentences (the statement "any ∀∃-sentence provable from PA is true"), which is a strengthening of the consistency of PA.

On the other hand, many variants of the infinite Ramsey theorem are widely studied in the setting of second-order arithmetic. This is one of the central topics in the project named *reverse mathematics* whose ultimate goal is to determine the logical strength of mathematical theorems in various fields and classify them from viewpoints of several fields in logic. Typically, the strength of variants of the infinite Ramsey theorem is precisely calibrated from the viewpoints of computability and proof theory. Particularly, precise analyses for variants of the Paris–Harrington principle are important approaches to identify the consistency strength of variants of the infinite Ramsey theorem.

In this article, we will overview the relations between the Paris–Harrington principle, the infinite Ramsey theorem and correctness statements (also known as reflection principles) mainly in the setting of second-order arithmetic. For this purpose, we will work with nonstandard models of arithmetic and relate the finite and infinite Ramsey theorem in them. A brief idea here is that if a nonstandard model satisfies some variant of finite Ramsey theorem with a solution of nonstandard size, then it should include a model for infinite Ramsey theorem. This can be realized by the theory of *indicators* introduced by Kirby and Paris [23]. We reformulate their argument and connect variants of PH with the correctness of the infinite Ramsey theorem.

The structure of this article is the following. In Section 2, we set up basic definitions and review the studies on the Ramsey theorem in arithmetic. We give several formulations of the Paris–Harrington principle and their equivalents within second-order arithmetic in Sections 3 and 4. In Section 5, we see how the Paris–Harrington principle is related to the infinite Ramsey theorem by means of indicators. Some proofs in Section 5 require basic knowledge of nonstandard models of arithmetic.

2. FIRST- AND SECOND-ORDER ARITHMETIC AND THE RAMSEY THEOREM

In this section, we introduce fragments of first- and second-order arithmetic and set up basic definitions. For precise definitions, basic properties and other information, see, e.g., [16, 21] for first-order arithmetic and [17, 39] for second-order arithmetic.

We write $\mathcal{L}_1$ for the language of first-order arithmetic, which consists of constants 0, 1, function symbols $+$, $\times$, and binary relation symbols $=$, $\leq$, and write $\mathcal{L}_2$ for the language of second-order arithmetic which consists of $\mathcal{L}_1$ plus another binary relation $\in$. We use $x, y, z, \dots$ for first-order (number) variables and $X, Y, Z, \dots$ for second-order (set) variables. An $\mathcal{L}_2$-formula φ is said to be *bounded* or Σ^0_0 if it does not contain any second-order quantifiers and all first-order quantifiers are of the form $\forall x \leq t$ or $\exists x \leq t$, and it is said to be Σ^0_n (resp. Π^0_n) if it is of the form $\exists x_1 \forall x_2 \dots Q x_n \theta$ (resp. $\forall x_1 \exists x_2 \dots Q x_n \theta$) where θ is Σ^0_0. An $\mathcal{L}_2$-formula φ is said to be *arithmetical* or Σ^1_0 if it does not contain any second-order quantifiers, and it is said to be Σ^1_n (resp. Π^1_n) if it is of the form $\exists X_1 \forall X_2 \dots Q X_n \theta$ (resp. $\forall X_1 \exists X_2 \dots Q X_n \theta$) where θ is Σ^1_0. If a Σ^0_n-formula (resp. Π^0_n-formula) φ does not contain any set variables (i.e., φ is an $\mathcal{L}_1$-formula), it is said to be Σ_n (resp. Π_n). We can extend $\mathcal{L}_1$ with unary relation symbols $\vec{U} = U_1, \dots, U_k$. Here, we identify U_i's as second-order (set) constants and consider $\mathcal{L}_1 \cup \vec{U}$-formulas as Σ^1_0-formulas (with extra constants). Then, an $\mathcal{L}_1 \cup \vec{U}$-formula is said to be $\Sigma^{\vec{U}}_n$ (resp. $\Pi^{\vec{U}}_n$) if it is Σ^0_n (resp. Π^0_n).

For our discussions, we need to distinguish the actual ("standard") natural numbers from natural numbers formalized in axiomatic systems. Here, we use $\mathfrak{N}$ for the set of standard natural numbers, and $\mathbb{N}$ for natural numbers formalized in the system. When we write

"$n = 2, 3, 4, \ldots,$" it is intended that n ranges over $\mathfrak{N}$ and $n \geq 2$, while "$n \geq 2$" means that n ranges over $\mathbb{N}$ and $n \geq 2$.

2.1. The Paris–Harrington principle in first-order arithmetic

We adopt the *elementary function arithmetic* (EFA) for our base system of first-order arithmetic. It consists of the axioms of discrete ordered semirings, the totality of exponentiation[1] and the *induction axiom (*IND*)* of the form

$$\text{(IND)}\quad \varphi(0) \wedge \forall x(\varphi(x) \to \varphi(x+1)) \to \forall x \varphi(x)$$

for each Σ_0-formula $\varphi(x)$. Then, the system $\mathsf{I}\Sigma_n$ is defined as EFA plus the induction axioms for Σ_n-formulas, and the *Peano arithmetic* (PA) is defined as $\mathsf{PA} = \bigcup_{n \in \mathfrak{N}} \mathsf{I}\Sigma_n$. We may also expand EFA with unary predicates. If $\vec{U} = U_1, \ldots, U_k$ are unary predicates, $\mathsf{EFA}^{\vec{U}}$ consists of EFA plus the induction axioms for $\Sigma_0^{\vec{U}}$-formulas.

Within EFA, finite sets of natural numbers, finite sequences of natural numbers, functions on finite sets, or other finite objects on $\mathbb{N}$ are coded by numbers. We write $[\mathbb{N}]^{<\mathbb{N}}$ for the set of all (codes) of finite subsets of $\mathbb{N}$. For each $F \in [\mathbb{N}]^{<\mathbb{N}}$, we can define $|F|$ as the (unique) smallest $m \in \mathbb{N}$ such that there is a bijection between F and $m = [0, m)_{\mathbb{N}}$. In the context of the Ramsey theorem, a function of the form $c : [X]^n \to k$ is often called a *coloring*. (Recall that $[X]^n = \{F \in [\mathbb{N}]^{<\mathbb{N}} : |F| = n \wedge F \subseteq \mathbb{N}\}$.) Then, a set $H \subseteq X$ is said to be *c-homogeneous* if there exists $i < k$ such that $[H]^n \subseteq c^{-1}(i)$.

We first define the key notion introduced by Paris [32]. The following definition can be made within EFA.

Definition 2.1 (Density). Let $n \geq 1$ or $n = \infty$ and $k \geq 2$ or $k = \infty$. For given $m \in \mathbb{N}$, we define *m-density for* (n, k) as follows:

- a finite set F is said to be 0-dense(n, k) if $|F| > \min F$ (F is relatively large),
- a finite set F is said to be $(m+1)$-dense(n, k) if for any $c : [F]^{n'} \to k'$ where $n' \leq \min\{n, \min F\}$ and $k' \leq \min\{k, \min F\}$, there exists a c-homogeneous set $H \subseteq F$ such that H is m-dense(n, k). (Here, we set $\min\{\infty, a\} = a$ for $a \in \mathbb{N}$.)

Although the notion is defined inductively, the statement that F is m-dense(n, k) is Σ_0, in other words, there exists a Σ_0-formula $\psi(n, k, F, m)$ such that $\psi(n, k, F, m)$ holds if and only if F is m-dense(n, k).

Definition 2.2 (The Paris–Harrington principle). Let $n \geq 1$ or $n = \infty$, $k \geq 2$ or $k = \infty$ and $m \in \mathbb{N}$. Then, the *Paris–Harrington principle,* $m\mathrm{PH}^n_k$ and ItPH^n_k, is defined as follows:

- $m\mathrm{PH}^n_k$: $\forall a \exists b \geq a([a, b)_{\mathbb{N}}$ is m-dense$(n, k))$.
- $\mathrm{ItPH}^n_k :\equiv \forall m\, m\mathrm{PH}^n_k$.

---

1 Technically, it is not easy (but possible) to define the exponential function in this setting, see [16]. Alternatively, one may safely add an extra function symbol $\exp(x) = 2^x$ and its recursive definition.

We simply write PH^n_k for $1\mathrm{PH}^n_k$. Additionally, we usually omit ∞ and write PH^n for PH^n_∞, PH for $\mathrm{PH}^\infty_\infty$, and so on.

It is known that $\mathsf{I}\Sigma_1$ proves $\mathrm{PH}^{n+1}_2 \to \mathrm{PH}^n$. Thus there is a hierarchy of implications

$$\mathrm{PH}^1 \leq \mathrm{PH}^2_2 \leq \mathrm{PH}^2_3 \leq \cdots \leq \mathrm{PH}^2 \leq \mathrm{PH}^3_2 \leq \mathrm{PH}^3_3 \leq \cdots \leq \mathrm{PH}^3 \leq \mathrm{PH}^4_2 \leq \cdots .$$

It is known that this hierarchy is strict above PH^2 over $\mathsf{I}\Sigma_1$, whereas $\mathsf{I}\Sigma_n$ proves PH^{n+1}_k for $k = 2, 3, \ldots$ On the other hand, calibrating the strength of $m\mathrm{PH}^n_k$ for $m \geq 2$ is much harder, except for the implication $m\mathrm{PH}^n_2 \to \mathrm{PH}^n_{m+1}$ which directly follows from the definition.

We next formalize the correctness of theories of arithmetic. Within EFA, basic notions of first-order logic such as (well-formed) formulas, formal proofs (by the Hilbert-style proof system or other formal systems) are formalizable by means of Gödel numbering. Typically, we can encode the provability for first- and second-order arithmetic within EFA, namely, there exists a Σ_1-formula $\mathrm{Prov}(T, x)$ which means that a formula (encoded by) x is provable from a theory (i.e., a finite or recursive set of sentences) T.[2] On the other hand, we can also formalize the truth on $\mathbb{N}$, but only partially. By formalizing Tarski's truth definition, for each tuples of variables $\vec{Z}$ and $\vec{z}$, there exists a Π^0_1-formula $\pi(\vec{Z}, \vec{z}, x)$ such that for any unary predicates $\vec{U}$ and a $\Sigma^{\vec{U}}_0$-formula $\varphi(\vec{z})$, $\mathsf{EFA}^{\vec{U}}$ proves $\forall\vec{z}(\pi(\vec{U}, \vec{z}, \ulcorner\varphi\urcorner) \leftrightarrow \varphi(\vec{z}))$ where $\ulcorner\varphi\urcorner$ is the Gödel number encoding φ. Then, for $n = 1, 2, \ldots,$ there exists a Π^0_n-formula $\mathrm{Tr}_n(\vec{Z}, \vec{z}, x)$ such that for any unary predicates $\vec{U}$ and a $\Pi^{\vec{U}}_n$-formula $\varphi(\vec{z})$, $\mathsf{EFA}^{\vec{U}}$ proves $\forall\vec{z}(\mathrm{Tr}_n(\vec{U}, \vec{z}, \ulcorner\varphi\urcorner) \leftrightarrow \varphi(\vec{z}))$. This formula is called the Π_n*-truth predicate*. The formalized correctness statements (also known as reflection principles) are defined as follows. (Formally, π and Tr_n depend on the number of variables, but we may assume that $\vec{Z}$ and $\vec{z}$ contains all variables which will appear in the entire discussion. We may ignore variables not appearing in the formula encoded by x by substituting 0 into them.)

Definition 2.3 (Correctness). Let $n = 1, 2, \ldots,$ and let T be an $\mathcal{L}_1$- or $\mathcal{L}_2$-theory. Then the Π_n-correctness of T (Π_n-corr(T)) is the following statement:

$$\forall x(\text{“}x \text{ is (a Gödel number of) a } \Pi_n\text{-sentence”} \wedge \mathrm{Prov}(T, x) \to \mathrm{Tr}_n(x)).$$

Note that Π_n-corr(T) is a Π_n-statement, and it implies the consistency of T since it implies $\neg(0 = 1) \to \neg\mathrm{Prov}(T, \ulcorner 0 = 1\urcorner)$.

Now we are ready to state the theorem by Paris and Harrington.

Theorem 2.1 (Paris and Harrington **[32, 33]**). *The following are equivalent over* $\mathsf{I}\Sigma_1$[3]*:*

1. PH.
2. ItPH^n_k *(*$n = 3, 4, \ldots, k = 2, 3, \ldots$ *or* $k = \infty$*).*
3. Π_2-corr(PA).

---

2 We encode T, e.g., by its recursive index.

3 In **[32]**, Paris showed that ItPH^3_2 is independent of PA, while his argument implies the equivalence of statements 2 and 3. See Section 5.2.

Here, ItPH_2^3 is the original statement independent of PA introduced by Paris [32]. The equivalence of ItPH_2^3 and PH can be proved in a combinatorial way, while we see that both are equivalent to Π_2-corr(PA) in Section 5. Moreover, the Π_2-correctness of fragments of PA can be characterized by PH as well.

Theorem 2.2 (Paris, see [16]). *Let $n = 1, 2, \ldots$. Then Π_2-corr($\mathsf{I}\Sigma_n$) is equivalent to* PH^{n+1} *over* $\mathsf{I}\Sigma_1$.

There are many other combinatorial or other numerical principles known to be independent of PA such as the Kanamori–McAloon theorem (KM) [20] and the termination of the Goodstein sequence [15]. Many of them are equivalent to the Π_2-correctness of PA, while some others are strictly stronger. A typical such example is a finite variant of Kruskal's tree theorem introduced by Friedman. See [13, 38].

2.2. Second-order arithmetic and the infinite Ramsey theorem

The system of second-order induction $\mathsf{I}\Sigma_n^i$ consists of EFA plus the induction axioms for Σ_n^i-formulas. It is not difficult to see that $\mathsf{I}\Sigma_n^0$ is a conservative extension of $\mathsf{I}\Sigma_n$, in other words, they prove the same $\mathcal{L}_1$-sentences. Our base system for second-order arithmetic is RCA_0, which consists of $\mathsf{I}\Sigma_1^0$ plus the following *recursive comprehension axiom (*RCA*)*: for each pair of Σ_1^0-formulas $\varphi(x)$, $\psi(x)$,

$$\forall x(\varphi(x) \leftrightarrow \neg\psi(x)) \to \exists X \forall x(x \in X \leftrightarrow \varphi(x)).$$

The next system is WKL_0, which consists of RCA_0 plus *weak Kőnig's lemma (*WKL*)*. Here, we define WKL in a slightly stronger form (but still equivalent to the original definition over RCA_0, see [39, LEMMA IV.1.4]). A *tree* T is a family of functions of the form $p : [0, m)_{\mathbb{N}} \to \mathbb{N}$ ($m \in \mathbb{N}$) such that for any $p \in T$ and $\ell \in \mathbb{N}$ with $[[0, \ell)_{\mathbb{N}}]^n \subseteq \mathrm{dom}(p)$, $p\restriction[[0, \ell)_{\mathbb{N}}]^n$ is also a member of T. A tree T is said to be *bounded* if there exists a function $h : \mathbb{N} \to \mathbb{N}$ such that $p(i) \leq h(i)$ for any $p \in T$ and $i \in \mathrm{dom}(p)$. Then WKL asserts the following:

> for any infinite bounded tree T, there exists a function (a *path of* T) f such that $f\restriction[0, m)_{\mathbb{N}} \in T$ for any $m \in \mathbb{N}$.

Finally, the system ACA_0 consists of RCA_0 plus the *arithmetical comprehension axiom (*ACA*)*: for each Σ_0^1-formula $\varphi(x)$,

$$\exists X \forall x(x \in X \leftrightarrow \varphi(x)).$$

The strength of these three systems is precisely known and WKL_0 is strictly in-between RCA_0 and ACA_0. On the other hand, the $\mathcal{L}_1$-consequences of RCA_0 and WKL_0 are the same and they coincide with those of $\mathsf{I}\Sigma_1$, while the $\mathcal{L}_1$-consequences of ACA_0 coincide with those of PA.

Over RCA_0, the infinite Ramsey theorem is directly formalizable as follows.

Definition 2.4 (The infinite Ramsey theorem). The *infinite Ramsey theorem* RT_k^n is defined as follows:

- RT^n_k: for any $c : [\mathbb{N}]^n \to k$, there exists an infinite set $H \subseteq \mathbb{N}$ such that H is c-homogeneous ($n \geq 1$ and $k \geq 2$).
- $\mathrm{RT}^n_\infty :\equiv \forall k \mathrm{RT}^n_k$, $\mathrm{RT}^\infty_\infty :\equiv \forall n \mathrm{RT}^n_\infty$.

We usually omit ∞ and write RT^n for RT^n_∞, RT for $\mathrm{RT}^\infty_\infty$.

Within RCA_0, it is known that RT^n_k implies RT^n_{k+1} and RT^{n+1}_2 implies RT^n. Be aware that the former does not imply $\mathrm{RT}^n_2 \to \mathrm{RT}^n$ because of the lack of induction. So, we have the hierarchy

$$\mathrm{RT}^1_2 \leq \mathrm{RT}^1 \leq \mathrm{RT}^2_2 \leq \mathrm{RT}^2 \leq \mathrm{RT}^3_2 \leq \cdots.$$

However, this hierarchy collapses at the level of $n = 3$.

Theorem 2.3 (Jockusch [19], reformulated by Simpson [39]). *Let $n = 3, 4, \ldots,$ and let $k = 2, 3, \ldots$ or $k = \infty$. Then, over RCA_0, RT^n_k is equivalent to ACA_0.*

On the other hand, the full infinite Ramsey theorem RT is strictly stronger than ACA_0. This is unavoidable since RT implies PH over RCA_0, and thus it implies the consistency of PA. To prove RT, we need the system ACA_0' which consists of ACA_0 plus the assertion that for any $n \in \mathbb{N}$ and any set X, the nth Turing jump of X exists.

Theorem 2.4 (McAloon [29], see also [17]). *Over RCA_0, RT is equivalent to ACA_0'.*

The situations of RT^2_2 and RT^2 are complicated. There are many important results on the reverse mathematical and computability theoretic strength of RT^2_2 or RT^2 such as [7,8,30,37]. Typically, RT^2_2 and RT^2 are strictly in between RCA_0 and ACA_0, but still different from WKL_0 even with full induction.

Theorem 2.5 (Jockusch [19], Liu [28]). *RT^2_2 and RT^2 are incomparable with WKL_0 over $\mathsf{RCA}_0 + \mathsf{I}\Sigma^1_\infty$ (where $\mathsf{I}\Sigma^i_\infty = \{\mathsf{I}\Sigma^i_n : n \in \mathfrak{N}\}$).*

The Π^1_1-consequences (or equivalently, $\mathcal{L}_1$-consequences with second-order constants) of RT^2_2 and RT^2 are also studied precisely. A Π^1_n-formula $\forall X_1 \ldots QX_n\theta$ is said to be *restricted* Π^1_n ($\mathrm{r}\Pi^1_n$) if θ is Σ^0_2 and n is odd or θ is Π^0_2 and n is even, and $\mathrm{r}\Sigma^1_n$-formulas are defined in the dual way.

Theorem 2.6.

1. *$\mathsf{RCA}_0 + \mathrm{RT}^2_2$ proves $\mathsf{B}\Sigma^0_2$ and it is Π^1_1-conservative over $\mathsf{RCA}_0 + \mathsf{I}\Sigma^0_2$ (i.e., any Π^1_1-sentences which are provable from $\mathsf{RCA}_0 + \mathrm{RT}^2_2$ are provable from $\mathsf{RCA}_0 + \mathsf{I}\Sigma^0_2$). (Hirst [18] and Cholak/Jockusch/Slaman [7])*[4]
2. *$\mathsf{RCA}_0 + \mathrm{RT}^2_2$ is $\mathrm{r}\Pi^1_1$-conservative over RCA_0. (Patey/Yokoyama [34], see also Kołodziejczyk/Yokoyama [25])*
3. *$\mathsf{RCA}_0 + \mathrm{RT}^2$ proves $\mathsf{B}\Sigma^0_3$ and it is Π^1_1-conservative over $\mathsf{RCA}_0 + \mathsf{B}\Sigma^0_3$. (Hirst [18] and Slaman/Yokoyama [40])*

4 $\mathsf{B}\Sigma^0_n$ is called a *bounding principle*, see [16] for the definition.

The above theorem decides the consistency strength (or proof-theoretic strength) of RT^2_2 and RT^2, and more precise studies have been carried out for RT^2_2 with respect to the size of proofs [24,25]. However, the exact $\mathcal{L}_1$-consequences of $\mathsf{RCA}_0 + \mathrm{RT}^2_2$ are still not identified. Meanwhile, several hybrid approaches of computability and proof/model-theory are currently being developed such as [9,10] which may help to calibrate the $\mathcal{L}_1$-consequences of various combinatorial principles.

3. THE PARIS–HARRINGTON PRINCIPLE IN SECOND-ORDER ARITHMETIC

In this section, we consider the Paris–Harrington principle in the setting of second-order arithmetic. The main difference is that we can now consider the Paris–Harrington principle within an infinite set. Then, Theorems 2.1 and 2.2 are reformulated as Theorems 3.2–3.6.

3.1. Second-order formulations of PH

Recall that PH^n_k asserts that there exists an arbitrary large finite set which is 1-dense(n,k). Indeed, a 1-dense(n,k) set should exist within any infinite subset of $\mathbb{N}$ by the infinite Ramsey theorem (see the proof of Proposition 3.1 below). We reformulate PH^n_k based on this idea in second-order arithmetic.

Definition 3.1 (The Paris–Harrington principle, second-order form). Let $n \geq 1$ or $n = \infty$, $k \geq 2$ or $k = \infty$ and $m \in \mathbb{N}$. Then, the *Paris–Harrington principle,* $m\overline{\mathrm{PH}}^n_k$ and $\mathrm{It}\overline{\mathrm{PH}}^n_k$, is defined as follows:

- $m\overline{\mathrm{PH}}^n_k$: for any infinite set X_0, there exists a finite set $F \subseteq X_0$ such that F is m-dense(n,k).
- $\mathrm{It}\overline{\mathrm{PH}}^n_k :\equiv \forall m\, m\overline{\mathrm{PH}}^n_k$.

Just like for PH, we write $\overline{\mathrm{PH}}^n_k$ for $1\overline{\mathrm{PH}}^n_k$, $\overline{\mathrm{PH}}^n$ for $\overline{\mathrm{PH}}^n_\infty$, and so on.

We first see that any of these variants of the Paris–Harrington theorem are true since they are consequences of the infinite Ramsey theorem by the following "compactness" argument.

For given $n \geq 1$ and $k \geq 2$, an *(n,k)-coloring tree T* on a set X is a family of functions of the form $p : [m \cap X]^n \to k$ $(m \in \mathbb{N})$ such that for any $p \in T$ and $\ell \in \mathbb{N}$ with $[\ell \cap X]^n \subseteq \mathrm{dom}(p)$, $p \restriction [\ell \cap X]^n$ is also a member of T. Then, WKL_0 proves that any infinite (n,k)-coloring tree T on an infinite set X has a path $f : [X]^n \to k$ in the sense that $f \restriction [m \cap X]^n \in T$ for any $m \in \mathbb{N}$.

Proposition 3.1. *Let $n \geq 1$ or $n = \infty$, $k \geq 2$ or $k = \infty$ and $m \in \mathbb{N}$.* $\mathsf{WKL}_0 + \mathrm{RT}^n_k$ *proves* $m\overline{\mathrm{PH}}^n_k \to m + 1\overline{\mathrm{PH}}^n_k$. *In particular,* $\mathsf{WKL}_0 + \mathrm{RT}^n_k$ *proves* $\overline{\mathrm{PH}}^n_k$, *and* $\mathsf{WKL}_0 + \mathrm{RT}^n_k + \mathsf{I}\Sigma^1_1$ *proves* $\mathrm{It}\overline{\mathrm{PH}}^n_k$.

Proof. We prove for the case $n \geq 1$ and $k \geq 2$. Assume that $m+1\overline{\mathrm{PH}}^n_k$ fails on some infinite set X. Let T be an (n,k)-coloring tree on X such that $p \in T$ if and only if there is no p-homogeneous set which is m-dense(n,k). Then, T is infinite since any finite subset of X is not $m+1$-dense(n,k), and thus it has a path $f : [X]^n \to k$. By RT^n_k, there is an infinite set $H \subseteq X$ which is f-homogeneous. Then $m\mathrm{PH}^n_k$ fails on H by the definition of f. ■

Proving $\overline{\mathrm{PH}}^n_k$ just from the induction is much harder, but if $n = 1, 2, \ldots$, $\mathsf{I}\Sigma^0_n$ still proves $\overline{\mathrm{PH}}^{n+1}_k$ for $k = 2, 3, \ldots$ On the other hand, stronger induction does not help with the absence of the infinite Ramsey theorem. Indeed, $\mathsf{RCA}_0 + \mathsf{I}\Sigma^1_\infty$ does not prove $\overline{\mathrm{PH}}$ or even PH.[5]

Within RCA_0, the statement of $\mathrm{r}\Pi^1_n$-correctness of a theory T ($\mathrm{r}\Pi^1_n$-corr(T)) can be defined like in Definition 2.3, and $\mathrm{r}\Pi^1_n$-corr(T) is an $\mathrm{r}\Pi^1_n$-statement. Second-order versions of the Paris–Harrington principle are closely related to $\mathrm{r}\Pi^1_1$-correctness of the infinite Ramsey theorem and other systems, and also related to well-orderedness of ordinals, which is naturally formalizable within RCA_0. Here we summarize the relations between the Paris–Harrington principles, $\mathrm{r}\Pi^1_1$-correctness and well-foundedness of ordinals.

Theorem 3.2. *The following are equivalent over* RCA_0*:*

1. $\overline{\mathrm{PH}}^2$.
2. $\mathrm{It}\overline{\mathrm{PH}}^2_2$.
3. $\mathrm{r}\Pi^1_1$-corr$(\mathsf{I}\Sigma^0_1)$.
4. $\mathrm{r}\Pi^1_1$-corr$(\mathsf{WKL}_0 + \mathrm{RT}^2_2)$.
5. *Well-foundedness of* ω^ω.

Theorem 3.3. *The following are equivalent over* RCA_0*:*

1. $\overline{\mathrm{PH}}^3$.
2. $\mathrm{It}\overline{\mathrm{PH}}^2$.
3. $\mathrm{r}\Pi^1_1$-corr$(\mathsf{I}\Sigma^0_2)$.
4. $\mathrm{r}\Pi^1_1$-corr$(\mathsf{WKL}_0 + \mathrm{RT}^2)$.
5. *Well-foundedness of* ω^{ω^ω}.

Theorem 3.4. *The following are equivalent over* RCA_0 *(for* $n = 1, 2, \ldots$*):*

1. $\overline{\mathrm{PH}}^{n+1}$.
2. $\mathrm{r}\Pi^1_1$-corr$(\mathsf{I}\Sigma^0_n)$.
3. *Well-foundedness of* ω_{n+1}.

---

5 Indeed, $\mathsf{WKL}_0 + \mathsf{I}\Sigma^1_\infty$ is a Π^1_1-conservative extension of $\mathsf{RCA}_0 + \mathsf{I}\Sigma^0_\infty$.

Theorem 3.5. *The following are equivalent over* RCA_0*:*

1. $\overline{\mathrm{PH}}$.
2. $\mathrm{It}\overline{\mathrm{PH}}^n_k$ *(*$n = 3, 4, \dots, k = 2, 3, \dots, \infty$*).*
3. $\mathrm{r}\Pi^1_1$-corr(ACA_0).
4. *Well-foundedness of* ε_0.

Theorem 3.6. *Over* RCA_0, $\mathrm{It}\overline{\mathrm{PH}}$ *is equivalent to* $\mathrm{r}\Pi^1_1$-corr(ACA_0').

Over ACA_0, any Π^1_1-formula is equivalent to a $\mathrm{r}\Pi^1_1$-formula. Thus, $\mathsf{ACA}_0 + \overline{\mathrm{PH}}$ implies Π_n-corr(PA) for any $n \in \mathfrak{N}$, in other words, the $\mathcal{L}_1$-correctness schema of PA.

Many of the equivalences in the above theorems have been known to experts in one formulation or another for a long time, although at least some of them are hard to find in the literature. On the other hand, 3 $\leftrightarrow$ 4 of Theorems 3.2 and 3.3 are more recent, and not easy since they correspond to the study of the first-order strength of the infinite Ramsey theorem for pairs, which we have seen in Theorem 2.6. The equivalences between variants of PH and the well-orderedness of ordinals are obtained by measuring the largeness of finite sets using ordinals, as presented in the next subsection. In Section 5, we explain how to prove the equivalences between variants of PH and the correctness statements by the method of indicators.

3.2. PH and the notion of α-largeness

The Paris–Harrington principle is closely related to a notion of largeness for finite sets defined using ordinals. In [22], Ketonen and Solovay introduced the notion of α-largeness for ordinal $\alpha < \varepsilon_0$ and calibrated how large set is needed for PH.

Definition 3.2 (α-largeness, within RCA_0[6]). For $\alpha < \varepsilon_0$ and $m \in \mathbb{N}$, define $\alpha[m] = 0$ if $\alpha = 0$, $\alpha[m] = \beta$ if $\alpha = \beta + 1$, $\alpha[m] = \beta + \omega^\gamma \cdot m$ if $\alpha = \beta + \omega^{\gamma+1}$, and $\alpha[m] = \beta + \omega^{\gamma[m]}$ if $\alpha = \beta + \omega^\gamma$ and γ is a limit ordinal. Then a finite set $X = \{x_0 < \cdots < x_{\ell-1}\} \subseteq \mathbb{N}$ ($\{x_i\}_i$ is the increasing enumeration of X) is called *α-large* if $\alpha[x_0] \dots [x_{\ell-1}] = 0$.

The well-foundedness of ordinals and the notion of α-largeness is closely related. Indeed, if α is well-founded and $X = \{x_0 < x_1 < \cdots\}$ is infinite, then $\alpha[x_0][x_1]\dots$ should terminate at 0 within finitely many steps, which means that X contains an α-large set. It is not difficult to see the converse, and we have the following.

Proposition 3.7. *Let* $\alpha < \varepsilon_0$. *The following assertions are equivalent over* RCA_0*:*

1. *Any infinite set contains an* α*-large finite subset.*
2. α *is well-founded.*

---

6 Indeed, this definition still works within EFA with primitive recursive descriptions of ordinals.

The relations between PH and α-largeness are well-studied and have been the topic of ordinal analysis; see, e.g., [3–5, 22, 25, 27, 41]. Here we list several (digested) results from those papers. Let $\omega_0^\alpha = \alpha$ and $\omega_{n+1}^\alpha = \omega^{\omega_n^\alpha}$, and let $\omega_n = \omega_n^1$.

Theorem 3.8. *The following are provable within* RCA_0. *Let* $F \subseteq \mathbb{N}$ *be a finite set with* $\min F \geq 3$*, and let* $n, k \geq 1$ *and* $m \geq 0$.

1. *If* F *is* ω^{k+4}*-large, then* F *is* 1-dense$(2, k)$. *(Ketonen/Solovay* [22]*)*
2. *If* F *is* 1-dense$(2, k+1)$*, then* F *is* ω^k*-large. (folklore)*
3. *If* F *is* $\omega_n^{\omega \cdot k+1}$*-large, then* F *is* 1-dense$(n+1, k)$. *(essentially* [22]*)*
4. *If* F *is* 1-dense$(n+1, 3^n)$*, then* F *is* ω_n*-large. (Kotlarski/Piekart/Weiermann* [27]*)*
5. *If* F *is* ω^{300^m}*-large, then* F *is* m-dense$(2, 2)$. *(Kołodziejczyk/Yokoyama* [25]*)*
6. *If* F *is* ω_{3m+2}*-large, then* F *is* m-dense$(3, 2)$. *(Bigorajska/Kotlarski* [4]*)*

Many implications of Theorems 3.2–3.5 follow from the above theorem. Indeed, $1 \leftrightarrow 2 \leftrightarrow 5$ of Theorem 3.2 follows from statements 1, 2 and 5 of the above, and $5 \rightarrow 1$ of Theorem 3.3, $3 \rightarrow 1$ of Theorem 3.4, and $1 \leftrightarrow 4 \rightarrow 2$ of Theorem 3.5 follow from statements 3, 4, and 6. We see other implications in Section 5.

Well-foundedness of ordinals is also heavily related with correctness statements and their relations are widely studied. For the recent developments, see, e.g., [1, 31].

4. GENERALIZATIONS OF PH

In this section, we see several generalizations of the Paris–Harrington principle by modifying the relative largeness condition "$|H| > \min H$." They are still natural strengthenings of the finite Ramsey theorem and quickly follow from the infinite Ramsey theorem and a compactness argument of the kind presented in Proposition 3.1. Nonetheless, a strong enough form of the Paris–Harrington principle recovers the infinite Ramsey theorem (Theorem 4.5) and its iterations provide the $\mathrm{r}\Pi^1_2$-correctness of the infinite Ramsey theorem (Theorems 4.6–4.8).

4.1. Phase transition

A natural generalization of PH^n_k would be provided by changing the relative largeness condition $|H| > \min H$ to $|H| > f(\min H)$ for some function f. We write $\mathrm{PH}^n_{k,f}$ or $\overline{\mathrm{PH}}^n_{k,f}$ for the statement defined as PH^n_k or $\overline{\mathrm{PH}}^n_k$ but with $|H| > \min H$ replaced by $|H| > f(\min H)$. Unfortunately, this does not make PH stronger in most cases. Indeed, one can easily prove the following.

Proposition 4.1. 1. *Let* $n = 2, 3, \ldots$ *or* $n = \infty$*, and let* f *be a primitive recursive function. Then* $\mathsf{I}\Sigma_1 + \mathrm{PH}^n$ *proves* PH^n_f.

2. *Let f be a provably recursive function of* PA. *Then* $\mathsf{I}\Sigma_1 + \mathrm{PH}$ *proves* PH_f.

3. *Let $n = 1, 2, \ldots$ or $n = \infty$ and let $k = 2, 3, \ldots$ or $k = \infty$. Then* $\mathsf{RCA}_0 + \overline{\mathrm{PH}}^n_k$ *proves that for any function f,* $\overline{\mathrm{PH}}^n_{k,f}$ *holds.*

On the other hand, PH_f can be weaker if f is slower growing than the identity function. Indeed, if f is a constant function, then PH_f is just the finite Ramsey theorem, and thus it is provable within PA. Weiermann **[44]** revealed the border of the provability and unprovability in this context as part of his research program called *phase transition*.

Theorem 4.2 (Weiermann **[44]**). *Let* $\log_n$ *be the inverse function of the nth iterated exponential function* $\exp^n(x)$ *where* $\exp(x) = 2^x$, *and let* $\log_*$ *be the inverse function of the superexponential (tower) function* 2_x.

1. $\mathrm{PH}_{\log_n}$ *is not provable from* PA *for any* $n \geq 1$.

2. $\mathrm{PH}_{\log_*}$ *is provable from* PA.

A sharper border is revealed in **[44]**, and similar analyses have been done for KM and other principles as well **[35]**.

4.2. PH with generalized largeness

To obtain further generalization of PH, we want to consider some condition of the form $|H| > f(H)$ where f assigns some "required size" for each finite set. Inspired by Terrence Tao's blog **[43]**, Gaspar and Kohlenbach **[14]** introduced several "finitary" versions of the infinite pigeonhole principle (RT^1 in our terminology) which are formulated based on this idea. Then, Pelupessy generalizes it to the infinite Ramsey theorem as follows.

Definition 4.1 (Gaspar/Kohlenbach **[14]**, Pelupessy **[36]**). A function $f : [\mathbb{N}]^{<\mathbb{N}} \to \mathbb{N}$ is said to be *asymptotically stable* if for any increasing sequence of finite sets $F_0 \subseteq F_1 \subseteq \cdots$ $\{f(F_i)\}_{i\in\mathbb{N}}$ converges. Then, the finitary infinite Ramsey theorem FIRT^n_k states the following:

- FIRT^n_k: for any asymptotically stable function $f : [\mathbb{N}]^{<\mathbb{N}} \to \mathbb{N}$, there exists $r \in \mathbb{N}$ such that for any $c : [[0, r)_\mathbb{N}]^n \to k$, there exists a homogeneous set $H \subseteq [0, r)_\mathbb{N}$ such that $|H| > f(H)$.

- $\mathrm{FIRT}^n_\infty \equiv \forall k \mathrm{FIRT}^n_k$, $\mathrm{FIRT}^\infty_\infty \equiv \forall n \mathrm{FIRT}^n_k$.

The finitary infinite pigeonhole principle FIPP_2 in **[14]** is the same as FIRT^1_∞.

Gaspar/Kohlenbach and Pelupessy showed that FIRT^n_k is equivalent to RT^n_k over WKL_0 (we will see this in detail later). Thus, FIRT^n_k could be considered as a "finitary" rephrasing of infinite combinatorics.

Remark 4.3. In **[14]**, another form of the finitary infinite pigeonhole principle FIPP_3 is also studied, and the question is raised which is more appropriate as the finitary version of infinite pigeonhole principle. However, FIPP_3 is equivalent to ACA_0 [45], and it does not fit with the general form of the Ramsey theorem.

Then, can we consider more general statements? Remember that the original idea of the finite Ramsey theorem or the Paris–Harrington principle is that if a large enough set is given, one must find a homogeneous set which is still "large" in some sense. Here, we consider a general concept of largeness for finite sets as follows.

Definition 4.2 (Largeness notion). A family of finite sets $\mathbb{L} \subseteq [\mathbb{N}]^{<\mathbb{N}}$ is said to be a *prelargeness notion* if it is upward closed, in other words, $F_0 \in \mathbb{L}$ and $F_0 \subseteq F_1$ implies $F_1 \in \mathbb{L}$. A prelargeness notion $\mathbb{L}$ is said to be a *largeness notion* if for any infinite set $X \subseteq \mathbb{N}$, there exists a finite set $F \subseteq X$ such that $F \in \mathbb{L}$.

The idea of the above definition is that an infinite set is always large enough and thus it should contain a "large finite set" in the sense of $\mathbb{L}$. For example, $\mathbb{L}_\omega = \{F \in [\mathbb{N}]^{<\mathbb{N}} : |F| > \min F\}$ is a largeness notion. Note that "$\mathbb{L}$ is a prelargeness notion" is just a $\Pi^{\mathbb{L}}_1$-statement and thus it is available within $\mathsf{EFA}^{\mathbb{L}}$. On the other hand, "$\mathbb{L}$ is a largeness notion" is an $\mathrm{r}\Pi^1_1$-statement, so it strictly requires the second-order language. Next, we generalize the density notion. The following definition can be made within $\mathsf{EFA}^{\mathbb{L}}$.

Definition 4.3 (Density with respect to $\mathbb{L}$). Let $n \geq 1$ or $n = \infty$ and $k \geq 2$ or $k = \infty$. Let $\mathbb{L}$ be a prelargeness notion. We define the density for $(n, k, \mathbb{L})$ as follows:

- a finite set F is said to be 0-dense$(n, k, \mathbb{L})$ if $F \in \mathbb{L}$,
- a finite set F is said to be $m + 1$-dense$(n, k, \mathbb{L})$ if for any $c : [F]^{n'} \to k'$ where $n' \leq \min\{n, \min F\}$ and $k' \leq \min\{k, \min F\}$, there exists a c-homogeneous set $H \subseteq F$ such that H is m-dense$(n, k, \mathbb{L})$.

The statement that F is m-dense$(n, k, \mathbb{L})$ is $\Sigma^{\mathbb{L}}_0$.

Now we define the generalized Paris–Harrington principle. The following definition can be made within RCA_0.

Definition 4.4 (Generalized PH). Let $n \geq 1$ or $n = \infty$, $k \geq 2$ or $k = \infty$ and $m \in \mathbb{N}$. Then, the generalized Paris–Harrington principle, $m\mathrm{GPH}^n_k$ and ItGPH^n_k, is defined as follows:

- $m\mathrm{GPH}^n_k$: for any largeness notion $\mathbb{L}$ and for any infinite set X_0, there exists a finite set $F \subseteq X_0$ such that F is m-dense$(n, k, \mathbb{L})$.
- $\mathrm{ItGPH}^n_k :\equiv \forall m\, m\mathrm{GPH}^n_k$.

Just like for PH, we write GPH^n_k for $1\mathrm{GPH}^n_k$, GPH^n for GPH^n_∞ and so on.

Unlike $\overline{\mathrm{PH}}^n_k$, GPH^n_k is "iterable." Indeed, GPH^n_k states that if $\mathbb{L}$ is a largeness notion, then the family of all 1-dense$(n, k, \mathbb{L})$ sets is also a largeness notion, and thus GPH^n_k can be applied to it again. Furthermore, any infinite subset $X \subseteq \mathbb{N}$ is "isomorphic to $\mathbb{N}$" in the following sense; if $h : \mathbb{N} \to X$ is a monotone increasing bijection and $\mathbb{L}$ is a largeness notion, then $h^{-1}(\mathbb{L})$ is a largeness notion and for any $F \subseteq_{\text{fin}} \mathbb{N}$, F is 1-dense$(n, k, h^{-1}(\mathbb{L}))$ if and only if $h(F)$ is 1-dense$(n, k, \mathbb{L})$. Using these ideas, we can get the following.

Proposition 4.4. *Let $n = 1, 2, 3, \ldots$. The following are equivalent over* RCA_0*:*

1. $m\mathrm{GPH}^n_k$ *$(k = 2, 3, 4, \ldots, m = 1, 2, 3, \ldots)$.*
2. GPH^n_2 *on $\mathbb{N}$: for any largeness notion $\mathbb{L}$, there exists a finite set $F \subseteq \mathbb{N}$ such that F is* 1-dense$(n, 2, \mathbb{L})$.

To give a characterization of GPH, we consider the following variants of the infinite Ramsey theorem which was originally introduced by Flood **[11]**.

Definition 4.5 (Ramsey-type weak Kőnig's lemma). An *infinite homogeneous function* for an infinite (n, k)-coloring tree T on X_0 is a function $h : [X]^n \to k$ such that $X \subseteq X_0$ is infinite and for any $m \in \mathbb{N}$, there exists $p \in T$ such that $h \upharpoonright [X \cap m]^n = p \upharpoonright [X \cap m]^n$.

We define two forms of the *Ramsey-type weak Kőnig's lemma*, RWKL^n_k and RWKL^{n-}_k, as follows:

- RWKL^{n-}_k: for any infinite (n, k)-coloring tree T on $\mathbb{N}$, there exists an infinite homogeneous function for T ($n \geq 1$ and $k \geq 2$),
- $\mathrm{RWKL}^{n-}_\infty \equiv \forall k \mathrm{RWKL}^{n-}_k$, $\mathrm{RWKL}^{\infty-}_\infty \equiv \forall n \mathrm{RWKL}^{n-}_\infty$,
- RWKL^n_k: for any infinite (n, k)-coloring tree T on $\mathbb{N}$, there exists a constant infinite homogeneous function for T ($n \geq 1$ and $k \geq 2$),
- $\mathrm{RWKL}^n_\infty \equiv \forall k \mathrm{RWKL}^n_k$, $\mathrm{RWKL}^\infty_\infty \equiv \forall n \mathrm{RWKL}^n_\infty$.

Note that the original definition of Ramsey-type weak Kőnig's lemma by Flood is our RWKL^1_2.[7] Over RCA_0, it is strictly in-between WKL and DNR (see **[11, 12]**). Variants of Ramsey-type weak Kőnig's lemma with homogeneous functions are introduced and studied by Bienvenu, Patey, and Shafer in **[2]** and the definition of RWKL^{n-}_k is inspired by them.

Theorem 4.5. *Let $n \geq 1$ or $n = \infty$ and $k \geq 2$ or $k = \infty$. The following are equivalent over* RCA_0*:*

1. GPH^n_k.
2. FIRT^n_k.
3. RWKL^n_k.
4. $\mathrm{RT}^n_k + \mathrm{RWKL}^{n-}_k$.

Proof. It is enough to show the equivalence for the case $n \geq 1$ and $k \geq 2$. Equivalence $3 \leftrightarrow 4$ is easy from the definition. If $f : [\mathbb{N}]^{<\mathbb{N}} \to \mathbb{N}$ is asymptotically stable, then $\mathbb{L} = \{F : \exists G \subseteq$

---

7 The original name in **[11]** was "Ramsey-type Kőnig's lemma", but "Ramsey-type weak Kőnig's lemma" turned to be the standard name in the later works.

$F|G| > f(G)\}$ is a largeness notion, which implies $1 \to 2$. Conversely, if $\mathbb{L}$ is a largeness notion, then a function f defined as $f(F) = \min\{|G| - 1 : G \subseteq F \wedge G \in \mathbb{L}\} \cup \{|F|\}$ is asymptotically stable and $F \in \mathbb{L} \leftrightarrow |F| > f(F)$. This implies $2 \to 1$. Implication $3 \to 1$ is a standard compactness argument which we have seen in Proposition 3.1. To show $1 \to 3$, let T be an infinite (n, k)-coloring tree on $\mathbb{N}$ with no infinite constant homogeneous function. Define $\mathbb{L}$ as $F \in \mathbb{L}$ if there is no $p \in T$ such that p is constant on $[F]^n$. Then, one can check that $\mathbb{L}$ is a largeness notion, and hence by 1, there exists a finite set $F_0 \subseteq \mathbb{N}$ which is 1-dense$(n, k, \mathbb{L})$. Take some $p \in T$ so that $\mathrm{dom}(p) \supseteq [F_0]^n$, then there must exist $H \subseteq F_0$ such that $H \in \mathbb{L}$ and p is constant on $[H]^n$, which is a contradiction. ■

In case $n = 3, 4, 5, \ldots$, any of the statements in the above theorem is just equivalent to ACA_0, so we mostly interested in the case $n = 1$ and 2. On the other hand, unlike RT^1_2 or PH^1_2, the principle GPH^1_2 is still not trivial since RWKL^1_2 (which is equivalent to RWKL^{1-}_2) is not provable within RCA_0. This may be interpreted as saying that the generalized version of the Paris–Harrington principle cannot be proved without using some compactness argument. In general, RWKL^{n-}_k is easily implied by WKL_0, but we do not know whether it is strictly weaker than WKL over RCA_0 or not in case $n \geq 2$.

4.3. Iterations of generalized PH and correctness statements

The iterated version of GPH can be related to stronger correctness statements.

Theorem 4.6. *Let $k = 2$ or $k = \infty$. Then* ItGPH^2_k *is equivalent to* $\mathrm{r}\Pi^1_2$-corr$(\mathsf{WKL}_0 + \mathrm{RT}^2_k)$ *over* WKL_0.

Over ACA'_0, any Π^1_2-formula (of possibly nonstandard length) is equivalent to a $\mathrm{r}\Pi^1_2$-formula, and thus $\mathrm{r}\Pi^1_2$-truth predicate is actually the truth predicate for all Π^1_2-formulas. Furthermore, It is known that $\mathrm{r}\Pi^1_2$-corr(ACA_0) is equivalent to ACA'_0.[8] So we simply write Π^1_2-corr(T) for $\mathrm{r}\Pi^1_2$-corr(T) if $T \supseteq \mathsf{ACA}_0$.

Theorem 4.7. *The following are equivalent over* RCA_0*:*

1. RT.

2. GPH.

3. ItGPH^n_k *$(n = 3, 4, \ldots, k = 2, 3, \ldots, \infty)$.*

4. Π^1_2-corr(ACA_0).

Theorem 4.8. *Over* RCA_0, ItGPH *is equivalent to* Π^1_2-corr(ACA'_0).

We will see the proofs of these theorems using indicators in the next section.

The strength of ItGPH^2_2 or ItGPH^2 is rather unclear. It is not difficult to check that $\mathsf{RCA}_0 + \mathrm{ItGPH}^2_2$ implies RT^2 and $\mathsf{WKL}_0 + \mathrm{RT}^2_2 + \mathsf{I}\Sigma^1_1$ implies ItGPH^2 as in the proof of Proposition 3.1. (Note that even ItGPH does not imply $\mathsf{I}\Sigma^1_1$ since $\mathsf{I}\Sigma^1_1$ is never implied from

8 This follows from the proof of [39, THEOREM IX.4.5].

any true Π^1_2-statement.) In particular, they are true in any ω-models of $\mathsf{WKL}_0 + \mathrm{RT}^2_2$. Meanwhile, the following questions are still open.

Question 4.6.

1. Is ItGPH^2_2 equivalent to RT^2 over WKL_0?
2. Does ACA_0 imply ItGPH^2 or ItGPH^2_2?

5. INDICATORS AND CORRECTNESS STATEMENTS

The notion of indicators is introduced by Kirby and Paris [23, 32] to show several independence results from PA, and its theory is organized systematically by Kaye [21]. The argument of indicators can connect first-order objects with second-order objects by means of nonstandard models. Recently, indicators have been used to calibrate the proof-theoretic strength of the infinite Ramsey theorem in the context of reverse mathematics [6, 24, 34, 46].

5.1. Models of first- and second-order arithmetic

To introduce the argument of indicators, we first set up basic model theory of first- and second-order arithmetic. For the details, see [16, 21, 26, 39]. A structure for $\mathcal{L}_1$ is a 6-tuple $M = (M; 0^M, 1^M, +^M, \times^M, \leq^M)$. (We often omit the superscript M if it is clear from the context.) An $\mathcal{L}_1$-structure $\mathfrak{N} = (\mathfrak{N}; 0, 1, +, \times, \leq)$ where $0, 1, +, \times, \leq$ are usual is called the *standard model*, and an $\mathcal{L}_1$-structure is said to be *nonstandard* if it is not isomorphic to $\mathfrak{N}$. When we consider an expanded language $\mathcal{L}_1 \cup \vec{U}$ where $\vec{U} = U_1, \ldots, U_k$ are second-order constants, an $\mathcal{L}_1 \cup \vec{U}$-structure is a pair $(M, \vec{U}^M)$ where M is an $\mathcal{L}_1$-structure and $U_i \subseteq M$. We may consider $\mathbb{N}$ as a special second-order constant which satisfies $\forall x x \in \mathbb{N}$, in other words, $\mathbb{N}^M = M$ for any M. For second-order arithmetic, we use Henkin semantics. A structure for $\mathcal{L}_2$ is a pair (M, S) where M is an $\mathcal{L}_1$-structure and $S \subseteq \mathcal{P}(M)$. Thus, any $\mathcal{L}_1 \cup \vec{U}$-structure can be considered as an $\mathcal{L}_2$-structure.

Let M be a nonstandard model of $\mathsf{EFA}^{\vec{U}}$. We write $[M]^{<M}$ for the set of all "finite sets in M" (also called M-finite sets), in other words, $[M]^{<M} = ([\mathbb{N}]^{<\mathbb{N}})^M$. A nonempty proper subset $I \subsetneq M$ is said to be a *cut* if $a < b \wedge b \in I$ implies $a \in I$ for any $a, b \in M$ (denoted by $I \subseteq_e M$) and $a + 1 \in I$ for any $a \in I$. If I is a cut and $\varphi(x)$ is a $\Sigma^{\vec{U}}_0$-formula such that $M \models \varphi(a)$ for any $a \in I$ (resp. $a \in M \setminus I$), then there exists $a \in M \setminus I$ (resp. $a \in I$) such that $M \models \varphi(a)$. This principle is called *overspill* (resp. *underspill*). A cut $I \subseteq_e M$ is said to be *semiregular* if for any $F \in [M]^{<M}$ with $|F| \leq \min F$, $F \cap I$ is bounded in I.

In our study, models of WKL_0 play central roles. Here are two important theorems.

Theorem 5.1 (Harrington, see Section IX.2 of [39]).

1. *For any countable model* $(M, S) \models \mathsf{RCA}_0$, *there exists* $\bar{S} \supseteq S$ *such that* $(M, \bar{S}) \models \mathsf{WKL}_0$.[9]
2. WKL_0 *is* Π^1_1-*conservative over* RCA_0.

9 Model (M, S) is said to be countable if both of M and S are countable.

Theorem 5.2 (see, e.g., Theorems 7.1.5 and 7.1.7 of [26]). *Let M be a model of* EFA *and* $I \subsetneq_e M$ *be a cut. Then,* I *is semiregular if and only if* $(I, \mathrm{Cod}(M/I)) \models \mathsf{WKL}_0$, *where* $\mathrm{Cod}(M/I) = \{F \cap I : F \in [M]^{<M}\}$.

5.2. Indicators

Now we give the definition of indicators. Here, we slightly arrange the definition in [21] so as to fit better with second-order arithmetic.

Definition 5.1 (Indicators). Let $\vec{U} = U_1, \ldots, U_k$ be second-order constants, and let $T \supseteq \mathsf{EFA}$ be an $\mathcal{L}_2$-theory.

1. Let M be a countable nonstandard model of $\mathsf{EFA}^{\vec{U}}$. A $\Sigma_0^{\vec{U}}$-definable function $Y : [M]^{<M} \to M$ is said to be an *indicator for* T *on* M if for each $F, F' \in [M]^{<M}$, $Y(F) \leq \max F$, $Y(F) \leq Y(F')$ if $F \subseteq F'$, and

   (cut) $Y(F) > m$ for any $m \in \mathfrak{N}$ if and only if there exists a cut $I \subsetneq_e M$ and $S \subseteq \mathrm{Cod}(M/I)$ such that $(I, S) \models T$, $U_i^M \cap I \in S$ for each $U_i \in \vec{U}$ and $F \cap I$ is unbounded in I.

2. A $\Sigma_0^{\vec{U}}$-formula $Y(F, m)$ is said to be an *indicator for* T if for any countable nonstandard model $M \models \mathsf{EFA}^{\vec{V}}$ with $\vec{U} \subseteq \vec{V}$ ($\vec{U}$ is a subtuple of $\vec{V}$), Y defines an indicator for T on M.

For a given indicator Y, we define two statements "$Y \geq m$" and "$Y^{\mathrm{int}} \geq m$" as follows:

$$Y \geq m \equiv \forall X_0(X_0 \text{ is infinite} \to \exists F \subseteq_{\mathrm{fin}} X_0\, Y(F) \geq m),$$

$$Y^{\mathrm{int}} \geq m \equiv \forall a \exists b Y([a, b)_{\mathbb{N}}) \geq m.$$

Note that $Y \geq m$ is a $\mathrm{r}\Pi^1_1$-statement while $Y^{\mathrm{int}} \geq m$ is a Π_2-statement.

Theorem 5.3. *Define Σ_0-formulas $Y_{\mathrm{PH}^n}(F, m)$, $Y_{\mathrm{PH}}(F, m)$, and $Y_{\mathrm{ItPH}^n_k}(F, m)$ as follows:*

- $Y_{\mathrm{PH}^n}(F, m) \leftrightarrow m = \max\{k' \leq \max F : F \textit{ is } 1\text{-dense}(n, k')\} \cup \{0\}$ *($n = 2, 3, \ldots$),*
- $Y_{\mathrm{PH}}(F, m) \leftrightarrow m = \max\{n' \leq \max F : F \textit{ is } 1\text{-dense}(n', 2)\} \cup \{0\}$,
- $Y_{\mathrm{ItPH}^n_k}(F, m) \leftrightarrow m = \max\{m' \leq \max F : F \textit{ is } m'\text{-dense}(n, k)\} \cup \{0\}$ *($n = 2, 3, \ldots$ or $n = \infty$ and $k = 2, 3, 4, \ldots$ or $k = \infty$).*

Then, we have the following:

1. Y_{PH^n} *is an indicator for* $\mathsf{RCA}_0 + \mathsf{I}\Sigma^0_{n-1}$.
2. Y_{PH} *is an indicator for* ACA_0.
3. $Y_{\mathrm{ItPH}^n_k}$ *is an indicator for* $\mathsf{WKL}_0 + \mathrm{RT}^n_k$.

In addition, these facts are provable within WKL_0.

Proof. For statements 1 and 2, one can reformulate the discussions of **[21, SECTION 14.3]**. Statement 3 is essentially due to Paris **[32, EXAMPLE 2]** (see also **[6, THEOREM 1]** and **[34, LEMMA 3.2]**). We sketch the proof for statement 3 for the case $n = 2, 3, \ldots$ and $k = 2, 3, \ldots$

It is enough to check the condition (cut) for $Y_{\mathrm{ItPH}^n_k}$. The right-to-left direction follows from Proposition 3.1 and overspill. For the left-to-right direction, let M be a countable nonstandard model of $\mathsf{EFA}^{\vec{U}}$ and let $F \in [M]^{<M}$ be m-dense(n, k) for any $m \in \mathfrak{N}$. By overspill, take $d \in M \setminus \mathfrak{N}$ such that F is d-dense(n, k). We will construct a countable decreasing sequence of M-finite sets $\{F_i\}_{i \in \mathfrak{N}}$ such that F_i is $(d - i)$-dense(n, k) and

(i) if $E \in [M]^{<M}$ and $|E| \leq E$, then $E \cap [\min F_i, \max F_i)_{\mathbb{N}} = \emptyset$ for some $i \in \mathfrak{N}$,

(ii) if $p \in [M]^{<M}$ and $p : [F]^n \to k$, then for some $i \in \mathfrak{N}$, F_i is p-homogeneous.

Once such a sequence is constructed, put $I = \{a \in M : \exists i \in \mathfrak{N}(a < \min F_i)\}$. Then, $F_i \cap I$ is unbounded in I and $U_i^M \cap I \in \mathrm{Cod}(M/I)$. By Theorem 5.2, $(I, \mathrm{Cod}(M/I)) \models \mathsf{WKL}_0$ since I is a semiregular cut by (i), and (ii) implies $(I, \mathrm{Cod}(M/I)) \models \mathrm{RT}^n_k$.

Finally, we construct $\{F_i\}_{i \in \mathfrak{N}}$. Since $[M]^{<M}$ is countable, it is enough to show:

(i)′ if $E \in [M]^{<M}$, $|E| \leq \min E$ and F is $\ell + 1$-dense(n, k) then there exists $F' \subseteq F$ which is ℓ-dense(n, k) such that $E \cap [\min F_i, \max F_i)_{\mathbb{N}} = \emptyset$,

(ii)′ if $p \in [M]^{<M}$, $p : [F]^n \to k$ and F is $\ell + 1$-dense(n, k) with $\ell \geq 1$, then there exists $F' \subseteq F$ which is ℓ-dense(n, k) such that F is p-homogeneous.

Indeed, (ii)′ is trivial from the definition of density. For (i)′, define $c : [F]^2 \to 2$ as $c(\{x, y\}) = 0 \leftrightarrow [x, y)_{\mathbb{N}} \cap E = \emptyset$, and take a c-homogeneous set $F' \subseteq F$ such that F' is ℓ-dense(n, k). If $[F']^2 \subseteq c^{-1}(1)$, then put $F'' = F' \setminus \{\min F'\}$ and we have $|F''| \leq |E| \leq \min E < \min F''$, but F'' must be relatively large since it is at least 0-dense(n, k). Hence $[F']^2 \subseteq c^{-1}(0)$, which we are done. ■

For the next theorem, we want to formalize model-theoretic arguments within second-order arithmetic. Within WKL_0, one can set up basic (countable) model theory for first-order logic, and then prove Gödel's completeness theorem **[39, SECTIONS II.8 AND IV.3]**. Standard techniques for countable nonstandard models of arithmetic such as the compactness theorem, over/underspill, back and forth, recursive saturation and forcing are naturally formalizable once a countable model with a full evaluation function (truth definition) is provided. On the other hand, it is not possible in general to consider $\mathbb{N}$ itself as a model of first-order arithmetic since its truth definition is too complicated,[10] hence it is not easy to guarantee that a family of true sentences are consistent. Still, we can deal with the consistency of Π_2-sentences as follows.

---

10 Some strong enough system such as ACA_0^+ can do this, but WKL_0 is not enough.

Lemma 5.4. RCA_0 *proves the following. Let $\vec{A} = A_1, \ldots, A_k$ be sets, and let Γ be a set of true $\Pi_2^{\vec{A}}$-sentences. Then, Γ is consistent (with considering $\vec{A}$ as second-order constants).*[11]

Proof. We work within RCA_0 and show that $\mathbb{N}$ (together with $\vec{A}$) is a weak model of Γ in the sense of [39, DEFINITION II.8.9]. It is enough to construct a function $f : S^\Gamma \to 2$ which satisfies Tarski's truth definition, where S^Γ is the set of all substitution instances of subformulas of Γ. Let S_0^Γ be the set of all substitution instances of $\Sigma_0^{\vec{A}}$-subformulas of Γ. Since there is a $\Pi_1^{\vec{A}}$-formula which defines the truth of all $\Sigma_0^{\vec{A}}$-formulas, one can take a function $f : S_0^\Gamma \to 2$ which satisfies the truth definition. Then f can be expanded to S^Γ by putting the truth value 1 for all sentences in $S^\Gamma \setminus S_0^\Gamma$. (They are $\Sigma_1^{\vec{A}}$ or $\Pi_2^{\vec{A}}$ and always true.) ■

Theorem 5.5. *Let $T \supseteq \mathsf{RCA}_0$ be an $\mathcal{L}_2$-theory, and let Y be an indicator for T.*

1. *For any $\mathrm{r}\Pi_1^1$-sentence φ, $T \vdash \varphi$ if and only if $\mathsf{RCA}_0 + \{Y \geq m : m \in \mathfrak{N}\} \vdash \varphi$.*
2. *For any Π_2-sentence φ, $T \vdash \varphi$ if and only if $\mathsf{I}\Sigma_1 + \{Y^{\mathrm{int}} \geq m : m \in \mathfrak{N}\} \vdash \varphi$.*[12]

If Y is an indicator for T provably in WKL_0, we also have the following:

3. *Over RCA_0, $\mathrm{r}\Pi_1^1$-corr(T) is equivalent to $\forall m Y \geq m$.*
4. *Over $\mathsf{I}\Sigma_1$, Π_2-corr(T) is equivalent to $\forall m Y^{\mathrm{int}} \geq m$.*

Proof. We show statements 1 and 3. (Statements 2 and 4 can be shown similarly.)

The right-to-left direction of statement 1 follows from Theorem 5.1.1 and Tanaka's self-embedding theorem [42]. Indeed, if (M, S) is a countable nonstandard model of T, then there exists a model $(\bar{M}, \bar{S})$ which is isomorphic to (M, S) such that $M \subsetneq_e \bar{M}$ and $S \subseteq \mathrm{Cod}(\bar{M}/M)$. If $X \in S$ is infinite in (M, S) and $m \in \mathfrak{N}$, then there exists $\bar{F} \in [\bar{M}]^{<\bar{M}}$ such that $X = \bar{F} \cap M$. By the condition (cut), $Y(\bar{F}) \geq m$, hence there exists a set $F \in [M]^{<M}$ such that $F \subseteq X$ and $Y(F) \geq m$ by underspill.

For the left-to-right direction of statement 1, it is enough to show that if

$$\{\forall x \exists y \theta(U, x, y)\} \cup \mathsf{RCA}_0 \cup \{Y \geq m : m \in \mathfrak{N}\}$$

is consistent with a second-order constant U and a Σ_0^U-formula θ, then

$$\{\forall x \exists y \theta(U, x, y)\} \cup T$$

is consistent. Let (M, S) be a countable nonstandard model of

$$\{\forall x \exists y \theta(U, x, y)\} \cup \mathsf{RCA}_0 \cup \{Y \geq m : m \in \mathfrak{N}\}.$$

Then there exists an infinite set A in (M, S) such that for any $a, b \in A$ with $a < b$, $\forall x < a \exists y < b \theta(U, x, y)$. By overspill, there exists an M-finite set $F \subseteq A$ with $Y(F) \geq m$

11 This lemma also follows from (the relativization of) the fact that $\mathsf{I}\Sigma_1$ is equivalent to Π_3-corr(EFA). See [1].

12 For statements 1 and 2, the base theories RCA_0 and $\mathsf{I}\Sigma_1$ can be weakened to RCA_0^* and $\mathsf{EFA} + \mathsf{B}\Sigma_1$ (the proof still works using recursively saturated models).

for any $m \in \mathfrak{N}$. By (cut), take $I \subsetneq_e M$ and $S' \subseteq \mathrm{Cod}(M/I)$ such that $(I, S') \models T$ and $F \cap I$ is unbounded in I. The latter implies $(I, S') \models \forall x \exists y \theta(U, x, y)$.

For the left-to-right direction of statement 3, we first formalize the right-to-left direction of statement 1 within WKL_0. In other words, "for each $m \in \mathbb{N}$, $Y \geq m$ is provable in T" is provable within WKL_0. Thus it is provable within RCA_0 by Theorem 5.1.2 since it is a Π^0_2-statement, and hence $\mathrm{r}\Pi^1_1$-corr(T) implies $\forall m Y \geq m$.

For the right-to-left direction, again we first work within WKL_0. It is enough to show that if $\forall x \exists y \theta(U, x, y)$ holds for some set U and a Σ^U_0-formula θ, then $\{\forall x \exists y \theta(U, x, y)\} \cup T$ is consistent. Take an infinite set A such that A is Δ^U_1-definable and for any $a, b \in A$ with $a < b$, $\forall x < a \exists y < b \theta(U, x, y)$. Then, by the assumption, for any $m \in \mathbb{N}$, there exists a finite set $F \subseteq A$ such that $Y(F) \geq m$. Thus, by Lemma 5.4, a set of Π^U_2-sentences $\Gamma = \mathsf{EFA}^U \cup \{\forall a \in F \forall b \in F(a < b \to \forall x < a \exists y < b \theta(U, x, y))\} \cup \{Y(F) \geq m : m \in \mathbb{N}\}$ is consistent (consider F as a new number constant). Take a countable nonstandard model of Γ and formalize the argument for the left-to-right direction of statement 1, then we see that $\{\forall x \exists y \theta(U, x, y)\} \cup T$ is consistent.

The above argument actually showed that for any set U, "$\forall m Y \geq m$ with respect to any infinite set $A \leq_T U$" implies Σ^U_2-corr(T). This is a Π^1_1-statement provable in WKL_0, so it is also provable within RCA_0 by Theorem 5.1.2. Thus RCA_0 proves that $\forall m Y \geq m$ implies $\mathrm{r}\Pi^1_1$-corr(T). ■

Theorems 5.3 and 5.5 directly connect PH and the correctness statements, and Theorems 2.1 and 2.2 are direct consequences of them. They also imply conservation theorems. Indeed, Theorem 2.6.2 is a direct consequence of Theorems 5.3 and 5.5 plus Theorem 3.8.5 (see [25]).

Proofs of Theorems 3.2–3.6. By definitions, $\overline{\mathrm{PH}}^n$, $\overline{\mathrm{PH}}$, and $\mathrm{It}\overline{\mathrm{PH}}^n_k$ are equivalent to $\forall m Y_{\mathrm{PH}^n} \geq m$, $\forall m Y_{\mathrm{PH}} \geq m$ and $\forall m Y_{\mathrm{ItPH}^n_k} \geq m$, respectively. Then, equivalences between variants of PH and corresponding $\mathrm{r}\Pi^1_1$-correctness statements ($1 \leftrightarrow 3$ and $2 \leftrightarrow 4$ of Theorems 3.2 and 3.3, $1 \leftrightarrow 2$ of Theorem 3.4, $1 \leftrightarrow 2 \leftrightarrow 3$ of Theorem 3.5 and Theorem 3.6) follow from Theorems 5.3 and 5.5. Implications between variants of PH and well-foundedness statements follow from Theorem 3.8 (see the paragraph below Theorem 3.8). Other implications can be shown as follows: $3 \to 5$ of Theorem 3.3 and $2 \to 3$ of Theorem 3.4 are implied from the formalization of the fact that $\mathsf{RCA}_0 + \mathsf{I}\Sigma^0_n$ proves well-foundedness of ω^k_n for each $k \in \mathfrak{N}$, and $3 \leftrightarrow 4$ of Theorem 3.3 is implied from the formalization of the conservation result for $\mathsf{WKL}_0 + \mathrm{RT}^2$ in [40]. ■

5.3. Indicators corresponding to largeness notions

To obtain a characterization of $\mathrm{r}\Pi^1_2$-correctness, we modify Theorem 5.5 using indicators which can preserve largeness notions.

Given two finite sets $F_0 = \{x_0 < \cdots < x_{\ell-1}\}$ and $F_1 = \{x'_0 < \cdots < x'_{\ell'-1}\}$, define $F_0 \trianglelefteq F_1$ as $\ell \leq \ell'$ and $x_i \geq x'_i$ for any $i < \ell$. A prelargeness notion $\mathbb{L}$ is said to be *normal* if $F_0 \in \mathbb{L}$ and $F_0 \trianglelefteq F_1$ implies $F_1 \in \mathbb{L}$. It is not difficult to check that $\mathbb{L}_\omega$ is a normal largeness notion. For a given prelargeness notion $\mathbb{L}$, put $\mathbb{L}^+ = \{F \in \mathbb{L} : \forall G \subseteq [0, \max F]_{\mathbb{N}}(G \trianglerighteq F \to G \in \mathbb{L})\}$. Then $\mathbb{L}^+$ is a normal prelargeness notion.

Lemma 5.6. *The following is provable within* WKL_0. *For any largeness notion* $\mathbb{L}$, $\mathbb{L}^+$ *is a largeness notion.*

Proof. Assume that $\mathbb{L}$ is a prelargeness notion and there exists an infinite set $X = \{x_0 < x_1 < \cdots\}$ such that no finite subset of X is a member of $\mathbb{L}^+$. Define a tree $T \subseteq \mathbb{N}^{<\mathbb{N}}$ as $\sigma \in T$ if and only if σ is strictly increasing, $\{\sigma(i) : i < |\sigma|\} \trianglerighteq \{x_i : i < |\sigma|\}$ and $\{\sigma(i) : i < |\sigma|\} \notin \mathbb{L}$. Then, T is a bounded tree and T is infinite. Take a path $h \in [T]$, then $Y = \{h(i) : i \in \mathbb{N}\}$ is an infinite set and any finite subset of Y is not a member of $\mathbb{L}$. ■

Now we generalize the notion of semiregularity with a normal (pre)largeness notion and consider a variant of Theorem 5.5.

Definition 5.2 ($\mathbb{L}$-semiregularity). Let M be a nonstandard model of $\mathsf{EFA}^{\mathbb{L}}$, and let $\mathbb{L}$ be a normal prelargeness notion in M. Then, a cut $I \subseteq_e M$ is said to be $\mathbb{L}$-*semiregular* if for any finite set $F \notin \mathbb{L}$, $F \cap I$ is bounded in I, or equivalently, $\mathbb{L} \cap I$ is a normal largeness notion in $(I, \mathrm{Cod}(M/I))$.

A $\Sigma_0^{\vec{U}}$-formula $Y^{\mathbb{L}} \equiv Y(\mathbb{L}, F, m)$ (where $\mathbb{L} \in \vec{U}$) is said to be an $\mathbb{L}$-*semiregular indicator* for an $\mathcal{L}_2$-theory T if for any countable nonstandard model $M \models \mathsf{EFA}^{\vec{V}}$ with $\vec{U} \subseteq \vec{V}$ such that $\mathbb{L}$ is a normal prelargeness notion in M, $Y^{\mathbb{L}}$ defines an indicator for T on M but the condition (cut) replaced by

($\mathbb{L}$-cut) $Y(F) > m$ for any $m \in \mathfrak{N}$ if and only if there exists an $\mathbb{L}$-semiregular cut $I \subsetneq_e M$ and $S \subseteq \mathrm{Cod}(M/I)$ such that $(I, S) \models T$, $U_i^M \cap I \in S$ for each $U_i \in \vec{U}$ and $F \cap I$ is unbounded in I.

Theorem 5.7. *Let* $T \supseteq \mathsf{WKL}_0$ *be an* $\mathcal{L}_2$*-theory, and let* $Y^{\mathbb{L}}$ *be an* $\mathbb{L}$*-semiregular indicator for* T *provably in* WKL_0. *Then the following assertions are equivalent over* WKL_0*:*

1. $\mathrm{r}\Pi^1_2$-corr(T).
2. *For any* $\mathbb{L}$*, if* $\mathbb{L}$ *is a normal largeness notion, then* $\forall m Y^{\mathbb{L}} \geq m$.

Proof. Implication $1 \to 2$ follows from the same discussion as the proof for Theorem 5.5. To show $2 \to 1$, we reason within WKL_0 and show that, assuming statement 2 is true, if $\theta(U)$ holds for some set U and an $\mathrm{r}\Pi^1_1$-formula $\theta(U)$, then $\{\theta(U)\} \cup T$ is consistent. By [34, PROPOSITION 2.5], take a Σ^0_0-formula $\eta(G, F)$ such that WKL_0 proves

$$\forall V(\theta(V) \leftrightarrow \forall Z(Z \text{ is infinite} \to \exists F \subseteq_{\mathrm{fin}} Z \eta(V \cap [0, \max F]_{\mathbb{N}}, F))).$$

Define $\mathbb{L}_0 \subseteq [\mathbb{N}]^{<\mathbb{N}}$ as $G \in \mathbb{L}_0 \leftrightarrow \exists F \subseteq G \eta(U \cap [0, \max F]_{\mathbb{N}}, F)$, and let $\mathbb{L} = \mathbb{L}_0^+$. Since $\theta(U)$ holds, $\mathbb{L}$ is a normal largeness notion. By assumption, we have $Y^{\mathbb{L}} \geq m$ for any $m \in \mathbb{N}$. Thus, by Lemma 5.4, a set of $\Pi_2^{U,\mathbb{L}}$-sentences $\Gamma = \mathsf{EFA}^{U,\mathbb{L}} \cup \{\mathbb{L}$ is a normal prelargeness notion$\} \cup \{\forall G(G \in \mathbb{L} \to \exists G' \eta(U \cap [0, \max G']_{\mathbb{N}}, G'))\} \cup \{Y^{\mathbb{L}}(F) \geq m : m \in \mathbb{N}\}$ is consistent (consider F as a new number constant).

Take a countable nonstandard model $M \models \Gamma$. Then, $M \models Y^{\mathbb{L}}(F^M) \geq m$ for any $m \in \mathbb{N}$ and thus there exists an $\mathbb{L}$-semiregular cut $I \subsetneq_e M$ and $S \subseteq \mathrm{Cod}(M/I)$ such that $U^I = U^M \cap I \in S, \mathbb{L}^I = \mathbb{L}^M \cap I \in S$ and $(I, S) \models T$. Since I is $\mathbb{L}^M$-semiregular, $\mathbb{L}^I$ is a largeness notion in (I, S). Since $M \models \forall G(G \in \mathbb{L} \to \exists G'\eta(U \cap [0, \max G']_{\mathbb{N}}, G'))$, we have $(I, S) \models \theta(U^I)$. ■

Theorem 5.8. *Let* $n = 2, 3, 4 \ldots$ *or* $n = \infty$ *and* $k = 2, 3, 4, \ldots$ *or* $k = \infty$*. Define* $\Sigma_0^{\mathbb{L}}$*-formula* $Y^{\mathbb{L}}_{\mathrm{ItGPH}^n_k}$ *as follows:*

$$Y_{\mathrm{ItGPH}^n_k}(\mathbb{L}, F, m) \leftrightarrow m = \max\{m' \leq \max F : F \textit{ is } m'\text{-dense}(n, k, \mathbb{L} \cap \mathbb{L}_\omega)\} \cup \{0\}.$$

Then, $Y^{\mathbb{L}}_{\mathrm{ItGPH}^n_k}$ *is an* $\mathbb{L}$*-semiregular indicator for* $\mathsf{WKL}_0 + \mathrm{RT}^n_k$*. Moreover, this fact is provable within* WKL_0.

Proof. Essentially the same as the proof for Theorem 5.3.3. We additionally need to show the following (which is an analogous of (i)′):

> If $\mathbb{L}$ is a normal prelargeness notion, F is $\ell + 1$-dense$(n, k, \mathbb{L} \cap \mathbb{L}_\omega)$ with $\ell \geq 1$ and G is a finite set such that $G \notin \mathbb{L}$, then there exists $F' \subseteq F$ such that F' is ℓ-dense$(n, k, \mathbb{L} \cap \mathbb{L}_\omega)$ and $[\min F', \max F')_{\mathbb{N}} \cap G = \emptyset$.

Given $\ell, \mathbb{L}, F$ and G as above, define $c : [F]^2 \to 2$ as $c(\{x, y\}) = 1 \leftrightarrow [x, y)_{\mathbb{N}} \cap G \neq \emptyset$. Take a c-homogeneous set $F' \subseteq F$ such that F' is ℓ-dense$(n, k, \mathbb{L} \cap \mathbb{L}_\omega)$. If $[F']^2 \subseteq c^{-1}(0)$, we are done, so assume $[F']^2 \subseteq c^{-1}(1)$. Put $G' = G \cap [\min F', \max F')_{\mathbb{N}}$ and $F'' = F' \setminus \{\min F'\}$. Then F'' is at least 0-dense$(n, k, \mathbb{L} \cap \mathbb{L}_\omega)$ and thus $F'' \in \mathbb{L}$. On the other hand, $G' \trianglerighteq F''$ by the definition of c, and thus $G' \in \mathbb{L}$. This is a contradiction since $G' \subseteq G$ and $G \notin \mathbb{L}$. ■

Proofs of Theorems 4.6, 4.7, and 4.8. By Lemma 5.6, ItGPH^n_k is equivalent to the statement that if $\mathbb{L}$ is a normal largeness notion, then $\forall m Y^{\mathbb{L}}_{\mathrm{ItGPH}^n_k} \geq m$. Then, implications between ItGPH^n_k and $\mathrm{r}\Pi^1_2$-corr$(\mathsf{WKL}_0 + \mathrm{RT}^n_k)$ follow from Theorems 5.7 and 5.8. ■

ACKNOWLEDGMENTS

It is a pleasure to thank Leszek Kołodziejczyk for careful reading of the manuscript and helpful discussions and comments, and to Kazuyuki Tanaka for helpful suggestions and advice.

FUNDING

This work is partially supported by JSPS KAKENHI grant number 19K03601.

REFERENCES

[1] L. Beklemishev, Reflection principles and provability algebras in formal arithmetic. *Russian Math. Surveys* **60** (2005).

[2] L. Bienvenu, L. Patey, and P. Shafer, On the logical strengths of partial solutions to mathematical problems. *Trans. Lond. Math. Soc.* **4** (2017), no. 1, 30–71.

[3] T. Bigorajska and H. Kotlarski, A partition theorem for α-large sets. *Fund. Math.* **160** (1999), no. 1, 27–37.

[4] T. Bigorajska and H. Kotlarski, Some combinatorics involving ξ-large sets. *Fund. Math.* **175** (2002), no. 2, 119–125.

[5] T. Bigorajska and H. Kotlarski, Partitioning α-large sets: some lower bounds. *Trans. Amer. Math. Soc.* **358** (2006), no. 11, 4981–5001.

[6] A. Bovykin and A. Weiermann, The strength of infinitary Ramseyan principles can be accessed by their densities. *Ann. Pure Appl. Logic* **168** (2017), no. 9, 1700–1709.

[7] P. A. Cholak, C. G. Jockusch, and T. A. Slaman, On the strength of Ramsey's theorem for pairs. *J. Symbolic Logic* **66** (2001), no. 1, 1–15.

[8] C. T. Chong, T. A. Slaman, and Y. Yang, The metamathematics of Stable Ramsey's Theorem for Pairs. *J. Amer. Math. Soc.* **27** (2014), no. 3, 863–892.

[9] D. D. Dzhafarov, D. R. Hirschfeldt, and S. C. Reitzes, Reduction games, provability, and compactness. 2020, arXiv:2008.00907.

[10] M. Fiori Carones, L. A. Kołodziejczyk, T. L. Wong, and K. Yokoyama, An isomorphism theorem for models of weak Kőnig's lemma without primitive recursion. 2021, arXiv:2112.10876.

[11] S. Flood, Reverse mathematics and a Ramsey-type Kőnig's lemma. *J. Symbolic Logic* **77** (2012), no. 4, 1272–1280.

[12] S. Flood and H. Towsner, Separating principles below WKL_0. *MLQ Math. Log. Q.* **62** (2016), no. 6, 507–529.

[13] H. M. Friedman, K. McAloon, and S. G. Simpson, A finite combinatorial principle which is equivalent to the 1-consistency of predicative analysis. In *Patras Logic Symposion (Patras, 1980)*, pp. 197–230, Stud. Logic Found. Math. 109, North-Holland, Amsterdam–New York, 1982.

[14] J. Gaspar and U. Kohlenbach, On Tao's "finitary" infinite pigeonhole principle. *J. Symbolic Logic* **75** (2010), no. 1, 355–371.

[15] R. L. Goodstein, On the restricted ordinal theorem. *J. Symbolic Logic* **9** (1944), 33–41.

[16] P. Hájek and P. Pudlák, *Metamathematics of first-order arithmetic*. Perspect. Math. Log., Springer, Berlin, 1998.

[17] D. R. Hirschfeldt, *Slicing the truth*. Lect. Notes Ser. 28, Institute for Mathematical Sciences, National University of Singapore, 2014.

[18] J. L. Hirst, *Combinatorics in subsystems of second order arithmetic*. Ph.D. thesis, The Pennsylvania State University, 1987.

[19] C. G. Jockusch, Ramsey's theorem and recursion theory. *J. Symbolic Logic* **37** (1972), no. 2, 268–280.

[20] A. Kanamori and K. McAloon, On Gödel incompleteness and finite combinatorics. *Ann. Pure Appl. Logic* **33** (1987), no. 1, 23–41.

[21] R. Kaye, *Models of Peano arithmetic*. Oxford University Press, 1991.

[22] J. Ketonen and R. Solovay, Rapidly growing Ramsey functions. *Ann. of Math. (2)* **113** (1981), no. 2, 267–314.

[23] L. A. S. Kirby and J. B. Paris, Initial segments of models of Peano's axioms. In *Set theory and hierarchy theory V (Proc. Third Conf., Bierutowice, 1976)*, pp. 211–226, Lecture Notes in Math. 619, Springer, Berlin, 1977.

[24] L. A. Kołodziejczyk, T. L. Wong, and K. Yokoyama, Ramsey's theorem for pairs, collection, and proof size. 2021, arXiv:2005.06854v2.

[25] L. A. Kołodziejczyk and K. Yokoyama, Some upper bounds on ordinal-valued Ramsey numbers for colourings of pairs. *Selecta Math. (N.S.)* **26** (2020), no. 4, 56, 18 pp.

[26] R. Kossak and J. H. Schmerl, *The structure of models of Peano arithmetic*. Oxford Logic Guides 50, Oxford University Press, Oxford, 2006.

[27] H. Kotlarski, B. Piekart, and A. Weiermann, More on lower bounds for partitioning α-large sets. *Ann. Pure Appl. Logic* **147** (2007), no. 3, 113–126.

[28] J. Liu, RT^2_2 does not imply WKL_0. *J. Symbolic Logic* **77** (2012), no. 2, 609–620.

[29] K. McAloon, Paris–Harrington incompleteness and progressions of theories. In *Recursion theory (Ithaca, NY, 1982)*, pp. 447–460, Proc. Sympos. Pure Math. 42, Amer. Math. Soc., Providence, RI, 1985.

[30] B. Monin and L. Patey, SRT22 does not imply RT22 in omega-models. 2019, arXiv:1905.08427v3.

[31] F. Pakhomov and J. Walsh, Reflection ranks and ordinal analysis. 2019, arXiv:1805.02095v2.

[32] J. B. Paris, Some independence results for Peano arithmetic. *J. Symbolic Logic* **43** (1978), no. 4, 725–731.

[33] J. Paris and L. Harrington, A mathematical incompleteness in Peano arithmetic. In *Handbook of mathematical logic*, edited by J. Barwise, pp. 1133–1142, Stud. Logic Found. Math. 90, North-Holland, Amsterdam, 1977.

[34] L. Patey and K. Yokoyama, The proof-theoretic strength of Ramsey's theorem for pairs and two colors. *Adv. Math.* **330** (2018), 1034–1070.

[35] F. Pelupessy, Phase transition results for three Ramsey-like theorems. *Notre Dame J. Form. Log.* **57** (2016), no. 2, 195–207.

[36] F. Pelupessy, On the "finitary" Ramsey's theorem. 2016, arXiv:1508.02013v7.

[37] D. Seetapun and T. A. Slaman, On the strength of Ramsey's theorem. *Notre Dame J. Form. Log.* **36** (1995), no. 4, 570–582.

[38] S. G. Simpson, Unprovable theorems and fast-growing functions. In *Logic and combinatorics (Arcata, CA, 1985)*, pp. 359–394, Contemp. Math. 65, Amer. Math. Soc., Providence, RI, 1987.

[39] S. G. Simpson, *Subsystems of second order arithmetic*. Perspect. Math. Log., Springer, Berlin, 1999.

[40] T. A. Slaman and K. Yokoyama, The strength of Ramsey's theorem for pairs and arbitrarily many colors. *J. Symbolic Logic* **83** (2018), no. 4, 1610–1617.

[41] M. D. Smet and A. Weiermann, Partitioning α-large sets for $\alpha < \varepsilon_\omega$. 2010, arXiv:1001.2437.

[42] K. Tanaka, The self-embedding theorem of WKL_0 and a non-standard method. *Ann. Pure Appl. Logic* **84** (1997), no. 1, 41–49.

[43] T. Tao, Soft analysis, hard analysis, and the finite convergence principle. http://terrytao.wordpress.com, 2007, appeared in T. Tao, *Structure and randomness: pages from year one of a mathematical blog*. American Mathematical Society, 2008, 298 pp.

[44] A. Weiermann, A classification of rapidly growing Ramsey functions. *Proc. Amer. Math. Soc.* **132** (2004), no. 2, 553–561.

[45] K. Yokoyama, Unpublished note.

[46] K. Yokoyama, On the strength of Ramsey's theorem without Σ_1-induction. *MLQ Math. Log. Q.* **59** (2013), no. 1–2, 108–111.

KEITA YOKOYAMA

Mathematical Institute, Tohoku University, 6-3, Aramaki Aza-Aoba, Aoba-ku, Sendai 980-8578, Japan, keita.yokoyama.c2@tohoku.ac.jp

This Page Intentionally Left Blank

CONSTRAINT SATISFACTION PROBLEM: WHAT MAKES THE PROBLEM EASY

DMITRIY ZHUK

ABSTRACT

The Constraint Satisfaction Problem is the problem of deciding whether there is an assignment to a set of variables subject to some specified constraints. Systems of linear equations, graph coloring, and many other combinatorial problems can be expressed as Constraint Satisfaction Problems for some constraint language. In 1993 it was conjectured that for any constraint language the problem is either solvable in polynomial time, or NP-complete, and for many years this conjecture was the main open question in the area. After this conjecture was resolved in 2017, we finally can say what makes the problem hard and what makes the problem easy. In the first part of the paper, we give an elementary introduction to the area, explaining how the full classification appeared and why it is formulated in terms of polymorphisms. We discuss what makes the problem NP-hard, what makes the problem solvable by local consistency checking, and explain briefly the main idea of one of the two proofs of the conjecture. The second part of the paper is devoted to the extension of the CSP, called Quantified CSP, where we allow using both universal and existential quantifiers. Finally, we discuss briefly other variants of the CSP, as well as some open questions related to them.

MATHEMATICS SUBJECT CLASSIFICATION 2020

Primary 3B70; Secondary 08A70, 68Q17, 03D15, 68T27, 03B10

KEYWORDS

Constraint satisfaction problem, computational complexity, CSP dichotomy, quantified constraints

ICM INTERNATIONAL CONGRESS OF MATHEMATICIANS 2022 JULY 6–14

Proc. Int. Cong. Math. 2022, Vol. 3, pp. 1530–1552
DOI 10.4171/ICM2022/125

1. INTRODUCTION

Probably the main question in theoretical computer science is to understand why some computational problems are easy (solvable in polynomial time) while others are difficult (NP-hard, PSpace-hard, and so on). What is the difference between P and NP? Why a system of linear equations can be solved in polynomial time by the Gaussian elimination but we cannot check whether a graph is 3-colorable in polynomial time (if we believe that $P \neq NP$). What is the principal difference between these two problems? To work on this question, first we would like to classify the problems by whether they are solvable in polynomial time (*tractable*) or NP-complete. Even for very simple decision problems, sometimes we do not know the answer.

For example, a system of linear equations in $\mathbb{Z}_2$ can be solved by Gaussian elimination, but if we are allowed to add one linear equation with usual sum for integers then the problem becomes NP-complete [26]. Surprisingly, the complexity is not known if we can add one equation modulo 24 to a system of linear equations in $\mathbb{Z}_2$ (variables are still from $\{0, 1\}$) [17]. In the paper we give a formal definition to such problems and discuss why some of them can be solved in polynomial time, while others are NP-hard.

2. CONSTRAINT SATISFACTION PROBLEM

The above problems are known as the Constraint Satisfaction Problem (CSP), which is the problem of deciding whether there is an assignment to a set of variables subject to some specified constraints. Formally, the *Constraint Satisfaction Problem* is defined as a triple $\langle \mathbf{X}, \mathbf{D}, \mathbf{C} \rangle$, where

- $\mathbf{X} = \{x_1, \dots, x_n\}$ is a set of variables,
- $\mathbf{D} = \{D_1, \dots, D_n\}$ is a set of the respective domains,
- $\mathbf{C} = \{C_1, \dots, C_m\}$ is a set of constraints,

where each variable x_i can take on values in the nonempty domain D_i, every *constraint* $C_j \in \mathbf{C}$ is a pair (t_j, R_j) where t_j is a tuple of variables of length m_j, called the *constraint scope*, and R_j is an m_j-ary relation on the corresponding domains, called the *constraint relation*.

The question is whether there exists *a solution* to $\langle \mathbf{X}, \mathbf{D}, \mathbf{C} \rangle$, that is, a mapping that assigns a value from D_i to every variable x_i such that for each constraint C_j the image of the constraint scope is a member of the constraint relation.

To simplify the presentation, we assume that the domain of every variable is a finite set A. We also assume that all the relations are from a set Γ, which we call *the constraint language*. Then the Constraint Satisfaction Problem over a constraint language Γ, denoted $\mathrm{CSP}(\Gamma)$, is the following decision problem: given a conjunctive formula

$$R_1(v_{1,1}, \dots, v_{1,n_1}) \wedge \dots \wedge R_s(v_{s,1}, \dots, v_{s,n_s}),$$

where $R_1, \ldots, R_s \in \Gamma$, and $v_{i,j} \in \{x_1, \ldots, x_n\}$ for every i, j, decide whether this formula is satisfiable. Note that in the paper we do not distinguish between relations and predicates, and in the previous formula we write relations meaning predicates.

2.1. Examples

It is well known that many combinatorial problems can be expressed as CSP(Γ) for some constraint language Γ. Moreover, for some Γ the corresponding decision problem can be solved in polynomial time; while for others it is NP-complete. It was conjectured that CSP(Γ) is either in P or NP-complete [29]. Let us consider several examples.

System of linear equations. Let $A = \{0, 1\}$ and

$$\Gamma = \{a_1x_1 + a_2x_2 + \cdots + a_kx_k = a_0 \mid a_0, a_1, \ldots, a_k \in \mathbb{Z}_2\},$$

i.e., Γ consists of all linear equations in the field $\mathbb{Z}_2$. Then CSP(Γ) is equivalent to the problem of solving a system of linear equations, which is solvable by the Gaussian elimination in polynomial time, thus, CSP(Γ) is in P.

Graph 2-coloring. To color a graph using two colors, we just need to choose a color of every vertex so that adjacent vertices have different colors. We assign a variable to each vertex, and encode the two colors with 0 and 1. For an edge between the ith and jth vertices, we add the constraint $x_i \neq x_j$. For instance, the 5-cycle is equivalent to the CSP instance

$$(x_1 \neq x_2) \wedge (x_2 \neq x_3) \wedge (x_3 \neq x_4) \wedge (x_4 \neq x_5) \wedge (x_5 \neq x_1).$$

Hence, the problem of graph 2-coloring is equivalent to CSP(Γ) for $A = \{0, 1\}$ and $\Gamma = \{\neq\}$. This problem can be solved locally. We choose a color of some vertex, then we color their neighbors with a different color, and so on. Either we will color all the vertices, or we will find an odd cycle, which means that the graph is not colorable using two colors. Thus, this problem is solvable in polynomial time.

Graph 3-coloring. Similarly, the problem of coloring a graph using 3 colors is equivalent to CSP(Γ) for $A = \{0, 1, 2\}$ and $\Gamma = \{\neq\}$. Unlike the graph 2-coloring, this problem is known to be NP-complete [1].

NAE-satisfability and 1IN3-satisfability. Suppose $A = \{0, 1\}$. NAE is the ternary not-all-equal relation, that is, NAE $= \{0, 1\}^3 \setminus \{(0, 0, 0), (1, 1, 1)\}$. 1IN3 is the ternary 1-in-3 relation, that is, 1IN3 $= \{(0, 0, 1), (0, 1, 0), (1, 0, 0)\}$. As it is known [40], both CSP({NAE}) and CSP({1IN3}) are NP-complete.

The main goal of this paper is to explain why the first two examples are in P, while the others are NP-hard.

2.2. Reduction from one language to another

To prove the hardness result, we usually reduce a problem to a known NP-hard problem. Let us show how we can go from one constraint language to another. CSP(Γ) can be viewed as the problem of evaluating a sentence

$$\exists x_1 \dots \exists x_n \big(R_1(v_{1,1},\dots,v_{1,n_1}) \wedge \dots \wedge R_s(v_{s,1},\dots,v_{s,n_s})\big), \tag{2.1}$$

where all variables are existentially quantified. Hence, if we could express one language using conjunctions and existential quantifiers from another language, then we get a reduction from one CSP to another. Let us explain how it works on a concrete example.

Let $\mathrm{NA1} = \{0,1\}^3 \setminus \{(1,1,1)\}$, that is, a ternary relation that holds whenever not all elements are 1. Let $A = \{0,1\}$, $\Gamma_1 = \{\mathrm{NA1}, \neq\}$, and $\Gamma_2 = \{\mathrm{1IN3}\}$. Let us show that $\mathrm{CSP}(\Gamma_1)$ and $\mathrm{CSP}(\Gamma_2)$ are (polynomially) equivalent. We may check that

$$(x \neq y) = \exists u \exists v\ \mathrm{1IN3}(x,y,u) \wedge \mathrm{1IN3}(u,u,v). \tag{2.2}$$

If fact, from $\mathrm{1IN3}(u,u,v)$ we derive that $u = 0$, hence $x \neq y$. Similarly, we have

$$\begin{aligned}\mathrm{NA1}(x,y,z) = \exists x' \exists y' \exists z' \exists x'' \exists y'' \exists z''\ &\mathrm{1IN3}(x',y',z')\\ &\wedge \mathrm{1IN3}(x,x',x'') \wedge \mathrm{1IN3}(y,y',y'') \wedge \mathrm{1IN3}(z,z',z''). \end{aligned} \tag{2.3}$$

If $x = y = z = 1$, then $x' = y' = z' = 0$, which contradicts $\mathrm{1IN3}(x',y',z')$. In all other cases, we can find an appropriate assignment.

Any instance of $\mathrm{CSP}(\Gamma_1)$ can be reduced to an instance of $\mathrm{CSP}(\Gamma_2)$ in the following way. We replace each constraint $(x_i \neq x_j)$ by the right-hand side of (2.2) introducing two new variables. Also, we replace each constraint $\mathrm{NA1}(x_i,x_j,x_k)$ by the right-hand side of (2.3) introducing six new variables. This reduction is obviously polynomial (and even log-space). Similarly, we have

$$\begin{aligned}\mathrm{1IN3}(x,y,z) = \exists x' \exists y' \exists z' \big(&\mathrm{NA1}(x,y,y) \wedge \mathrm{NA1}(y,z,z) \wedge \mathrm{NA1}(z,x,x)\\ &\wedge (x \neq x') \wedge (y \neq y') \wedge (z \neq z') \wedge \mathrm{NA1}(x',y',z')\big),\end{aligned}$$

which implies a polynomial reduction from $\mathrm{CSP}(\Gamma_2)$ to $\mathrm{CSP}(\Gamma_1)$.

Let us give a formal definition for the above reduction. A formula of the form $\exists y_1 \dots \exists y_n \Phi$, where Φ is a conjunction of relations from Γ is called *a positive primitive formula (pp-formula) over* Γ. If $R(x_1,\dots,x_n) = \exists y_1 \dots \exists y_n \Phi$, then we say that R is *pp-defined* by this formula, and $\exists y_1 \dots \exists y_n \Phi$ is called its *pp-definition*.

Theorem 2.1 ([35]). *Suppose* Γ_1 *and* Γ_2 *are finite constraint languages such that each relation from* Γ_1 *is pp-definable over* Γ_2. *Then* $\mathrm{CSP}(\Gamma_1)$ *is polynomial time reducible to* $\mathrm{CSP}(\Gamma_2)$.

2.3. Polymorphisms as invariants

If we can pp-define a relation R from a constraint language Γ and $\mathrm{CSP}(\{R\})$ is NP-hard, then $\mathrm{CSP}(\Gamma)$ is also NP-hard. How to show that such a relation cannot be pp-defined? To prove that something cannot be done, we usually find some fundamental property (invariant) that is satisfied by anything we can obtain. For the relations, the operations play the role of invariants.

We say that an operation $f : A^n \to A$ *preserves* a relation R of arity m if for any tuples $(a_{1,1}, \dots, a_{1,m}), \dots, (a_{n,1}, \dots, a_{n,m}) \in R$ the tuple

$$\big(f(a_{1,1}, \dots, a_{n,1}), \dots, f(a_{1,m}, \dots, a_{n,m})\big)$$

is in R. In this case we also say that f is a *polymorphism* of R, and R is an *invariant* of f. We say that an operation *preserves a set of relations* Γ if it preserves every relation in Γ. In this case we also write f is *a polymorphism of* Γ or $f \in \mathrm{Pol}(\Gamma)$. It can be easily checked that if f preserves Γ, then f preserves any relation R pp-definable from Γ. Moreover, we can show [15, 31] that $\mathrm{Pol}(\Gamma_1) \subseteq \mathrm{Pol}(\Gamma_2)$ if and only if Γ_2 is pp-definable over Γ_1, which means that the complexity of CSP(Γ) depends only on $\mathrm{Pol}(\Gamma)$.

Example 1. Let R be the linear order relation on $\{0, 1, 2\}$, i.e.,

$$R = \begin{pmatrix} 0 & 0 & 0 & 1 & 1 & 2 \\ 0 & 1 & 2 & 1 & 2 & 2 \end{pmatrix},$$

where columns are tuples from the relation. Then "an n-ary operation f preserves R" means that for all

$$\begin{pmatrix} a_1 \\ b_1 \end{pmatrix}, \dots, \begin{pmatrix} a_n \\ b_n \end{pmatrix} \in R,$$

that is, $a_i \leq b_i$, we have

$$f\begin{pmatrix} a_1 & a_2 & \dots & a_n \\ b_1 & b_2 & \dots & b_n \end{pmatrix} := \begin{pmatrix} f(a_1, \dots, a_n) \\ f(b_1, \dots, b_n) \end{pmatrix} \in R,$$

that is, $f(a_1, \dots, a_n) \leq f(b_1, \dots, b_n)$. In other words, f is monotonic. For instance, the operations max and min are monotonic. By the above observation, we know that any relation pp-definable from R is also preserved by min and max.

Example 2. Let $A = \{0, 1\}$. Let us show that 1IN3 cannot be pp-defined from NA1 and $x \leq y$. We can check that the conjunction $x \wedge y$ (an operation on $\{0, 1\}$) preserves both NA1 and $x \leq y$. However, $x \wedge y$ does not preserve 1IN3 as we have

$$\begin{pmatrix} 1 \\ 0 \\ 0 \end{pmatrix} \wedge \begin{pmatrix} 0 \\ 1 \\ 0 \end{pmatrix} = \begin{pmatrix} 0 \\ 0 \\ 0 \end{pmatrix} \notin \mathrm{1IN3}\,.$$

For more information on polymorphisms and how they can be used to study the complexity of the CSP, see [6].

2.4. Local consistency

The first step of almost any algorithm solving a CSP instance is checking local consistency. For instance, if a constraint forces a variable to be equal to 0, then we could substitute 0 and remove this variable.

Suppose we have a CSP instance

$$R_1(v_{1,1}, \dots, v_{1,n_1}) \wedge \dots \wedge R_s(v_{s,1}, \dots, v_{s,n_s}). \tag{2.4}$$

This instance is called *1-consistent (also known as arc-consistent)*, if for any variable x any two constraints $R_i(v_{i,1}, \dots, v_{i,n_i})$ and $R_j(v_{j,1}, \dots, v_{j,n_j})$ having this variable in the scope have the same projection onto this variable. This means that for every variable x there exists $D_x \subseteq A$, called the *domain of* x, such that the projection of any constraint on x is D_x.

Sometimes we need a stronger consistency (similar to singleton-arc-consistency in [36]). We say that $z_1 - C_1 - z_2 - \cdots - C_{l-1} - z_l$ is *a path* in a CSP instance $\mathcal{I}$ if z_i, z_{i+1} are in the scope of the constraint C_i for every $i \in \{1, 2, \dots, l-1\}$. We say that *a path* $z_1 - C_1 - z_2 - \cdots - C_{l-1} - z_l$ *connects* b *and* c if there exist $a_1, a_2, \dots, a_l \in A$ such that $a_1 = b$, $a_l = c$, and the projection of each C_i onto z_i, z_{i+1} contains the tuple (a_i, a_{i+1}). A CSP instance $\mathcal{I}$ is called *cycle-consistent* if it is 1-consistent and for every variable z and $a \in D_z$ any path starting and ending with z in $\mathcal{I}$ connects a and a.

It is not hard to find a polynomial procedure making the instance 1-consistent or cycle-consistent. For 1-consistency, the idea is to find a variable where the consistency is violated, then reduce the domain D_x of this variable and reduce the corresponding relations. We repeat this while some constraints violate consistency. Finally, we either get a 1-consistent instance, or we get a contradiction (derive that $D_x = \varnothing$). For cycle-consistency, we should go deeper. For every variable x and every value $a \in D_x$, we reduce the domain of x to $\{a\}$ and check whether the remaining instance can be made 1-consistent. If not, then x cannot be equal to a, and a can be excluded from the domain D_x.

Later we will show that in some cases 1-consistency and cycle-consistency are enough to solve a CSP instance, that is, any consistent instance has a solution. See [5, 36] for more information about local consistency conditions.

2.5. CSP over a 2-element domain

The complexity of CSP(Γ) for each constraint language Γ on $\{0, 1\}$ was described in 1978 [40]. This classification can be formulated nicely using polymorphisms.

Theorem 2.2 ([34, 40]). *Suppose* $A = \{0, 1\}$, Γ *is a constraint language on* A. *Then* CSP(Γ) *is solvable in polynomial time if*

(1) 0 *preserves* Γ, *or*

(2) 1 *preserves* Γ, *or*

(3) $x \vee y$ *preserves* Γ, *or*

(4) $x \wedge y$ *preserves* Γ, *or*

(5) $xy \vee yz \vee xz$ *preserves* Γ, *or*

(6) $x + y + z$ *preserves* Γ.

CSP(Γ) *is NP-complete otherwise.*

Let us consider each case and explain how the polymorphisms make the problem easy. Note that the cases (1) and (2), (3) and (4) are dual to each other, that is why we consider only one in each pair in detail.

0 preserves Γ. This case is almost trivial. "The constant 0 preserves a relation $R \in \Gamma$" means that $R(0, 0, \dots, 0)$ holds. If 0 preserves all relations from Γ, then $(0, 0, \dots, 0)$ is always a solution of a CSP instance, which makes the problem CSP(Γ) trivial.

$x \vee y$ preserves Γ. Let us show how to solve an instance of CSP(Γ) if $x \vee y \in \mathrm{Pol}(\Gamma)$. First, we make our instance 1-consistent. Then, unless we get a contradiction, every variable x has its domain D_x which is either $\{0\}$, or $\{1\}$, or $\{0, 1\}$. We claim that if we send the variables with domain $\{0\}$ to 0, and the variables with the domain $\{1\}$ and $\{0, 1\}$ to 1, then we get a solution. In fact, if we apply $x \vee y$ to all the tuples of some constraint, we obtain a tuple consistent with the solution. Thus, 1-consistency guarantees the existence of a solution in this case.

$xy \vee yz \vee xz$ preserves Γ. The operation $xy \vee yz \vee xz$ returns the most popular value and is known as a majority operation. It is not hard to check [2] that any relation preserved by a majority operation can be represented as a conjunction of binary relations, and we may assume that Γ consists of only binary relations. As it is shown in Section 2.8, for a 2-element domain this gives a polynomial algorithm for CSP(Γ). Additionally, we can show [36, 47] that any cycle-consistent instance of CSP(Γ) has a solution. Hence to solve an instance, it is sufficient to make it cycle-consistent, and unless we obtain an empty domain (contradiction) the instance has a solution.

$x + y + z$ preserves Γ. It is known (see Lemma 2.8) that $x + y + z$ preserves a relation R if and only if the relation R can be represented as a conjunction of linear equations. Thus, CSP(Γ) is equivalent to the problem of solving of a system of linear equations in the field $\mathbb{Z}_2$, which is tractable.

2.6. CSP solvable by local consistency checking

As we see in the previous section all tractable CSPs on a 2-element domain can be solved by two algorithms. The first algorithm just checks some local consistency (1-consistency, cycle-consistency) and, if a sufficient level of consistency achieved, we know that the instance has a solution. The second algorithm is the Gaussian elimination applied to a system of linear equations. In this section we discuss when the first algorithm is sufficient and why some instances can be solved by a local consistency checking, while others require something else.

To simplify the presentation in this section, we assume that all constant relations $x = a$ are in the constraint language. In this case any polymorphism f of Γ is *idempotent*, that is, $f(x, x, \dots, x) = x$. This restriction does not affect the generality of the results because we can always consider the core of the constraint language and then add all constant relations (see [34]). Consider the following system of linear equations in $\mathbb{Z}_p$:

$$\begin{cases} x_1 + x_2 = x_3 + 0, \\ x_3 + 0 = x_4 + x_5, \\ x_4 + 0 = x_1 + x_6, \\ x_5 + x_6 = x_2 + 1. \end{cases} \tag{2.5}$$

If we calculate the sum of all equations, we will get $0 = 1$, which means that the system does not have a solution. Nevertheless, we may check that the system is cycle-consistent, which means that the cycle-consistency does not guarantee the existence of a solution for linear equations. In fact, we can show that there does not exist a local consistency condition that guarantees the existence of a solution of a system of linear equations (see [5]).

As it was shown in [5,47] if $\mathrm{CSP}(\Gamma)$ cannot be solved by cycle-consistency checking then we can express a linear equation modulo p using Γ. Since our constraint language is on a domain A, we could not expect to pp-define the relation $x_1 + x_2 = x_3 + x_4 \pmod{p}$. Instead, we claim that there exist $S \subseteq A$ and a surjective mapping $\varphi : A \to \mathbb{Z}_p^s$ such that the relation

$$\{(a_1, a_2, a_3, a_4\} \mid a_1, a_2, a_3, a_4 \in S, \varphi(a_1) + \varphi(a_2) = \varphi(a_3) + \varphi(a_4)\} \tag{2.6}$$

is pp-definable. This means that the linear equation is defined on some S modulo some equivalence relation defined by φ. To avoid such a transformation, we could introduce the notion of pp-constructability and say that $x_1 + x_2 = x_3 + x_4 \pmod{p}$ is pp-constructable from Γ. To keep everything simple, we do not define pp-constructability and use it informally hoping that the idea of this notion is clear from our example. For more details about pp-constructability, see [7].

If such a linear equation cannot be pp-defined (pp-constructed) then there should be some operation that preserves Γ but not the linear equation modulo p. An operation f is called *a Weak Near Unanimity Operation (WNU)* if it satisfies the following identity:

$$f(y, x, x, \ldots, x) = f(x, y, x, \ldots, x) = \cdots = f(x, x, \ldots, x, y).$$

It is not hard to check that an idempotent WNU of arity p does not preserve a nontrivial linear equation modulo p (see Lemma 4.9 in [47]). Thus, the existence of an idempotent p-ary WNU polymorphism of Γ guarantees that a linear equation modulo p cannot be pp-defined (pp-constructed). That is why a relation satisfying (2.6) is called *p-WNU-blocker*. Hence, if Γ has WNU polymorphisms of all arities then no linear equations can appear. The following theorem confirms that nothing but linear equations could be an obstacle for the local consistency checking.

Theorem 2.3 ([47]). *Suppose Γ is a constraint language containing all constant relations. The following conditions are equivalent:*

(1) *every cycle-consistent instance of* $\mathrm{CSP}(\Gamma)$ *has a solution;*

(2) *Γ has a WNU polymorphisms of all arities $n \geq 3$;*

(3) *there does not exist a p-WNU-blocker pp-definable from Γ.*

Thus, the fact that we cannot express (pp-define, pp-construct) a nontrivial linear equation makes the problem solvable by the cycle-consistency checking.

2.7. CSP Dichotomy Conjecture

In this subsection, we formulate a criterion for CSP(Γ) to be solvable in polynomial time. This criterion is known as the CSP Dichotomy Conjecture, it was formulated almost 30 years ago [28, 29] but was an open question until 2017 [19, 20, 42, 44].

Theorem 2.4 ([19, 20, 42, 44]). *Suppose Γ is a constraint language on a finite set A. Then*

(1) CSP(Γ) *is solvable in polynomial time if Γ is preserved by a WNU;*

(2) CSP(Γ) *is NP-complete otherwise.*

The reason why the existence of a WNU polymorphism makes the problem easy is the fact that we cannot pp-define a strong relation giving us NP-hardness. A relation $R = (B_0 \cup B_1)^3 \setminus (B_0^3 \cup B_1^3)$, where $B_0, B_1 \subseteq A$, $B_0 \neq \varnothing$, $B_1 \neq \varnothing$, and $B_0 \cap B_1 = \varnothing$, is called *a WNU-blocker*. Such relations are similar to the not-all-equal (NAE) relation on $\{0, 1\}$, where B_0 means 0 and B_1 means 1. Instead of the existence of a pp-definable WNU-blocker, we could say that the relation NAE is pp-constructable from Γ. Note that CSP($\{$NAE$\}$) and CSP($\{R\}$) for a WNU-blocker R are NP-complete problems.

We can check (see Lemma 4.8 in [47]) that a WNU operation does not preserve a WNU-blocker. Moreover, we have the following criterion.

Lemma 2.5 ([47]). *A constraint language Γ containing all constant relations is preserved by a WNU if and only if there is no WNU-blocker pp-definable from Γ.*

Thus, CSP(Γ) is solvable in polynomial time if and only if a WNU-blocker cannot be pp-defined. Hence, the fact that we cannot pp-construct the not-all-equal relation makes the problem easy, and a WNU is an operation that guarantees that this relation cannot be pp-constructed.

2.8. How to solve CSP if pp-definable relations are simple

Below we discuss how the fact that only simple relations can be pp-defined from Γ can help to solve CSP(Γ) in polynomial time. In this case we can calculate the sentence explicitly eliminating existential quantifiers one by one. I believe that a similar idea should work for any Γ preserved by a WNU, which will give us a simple algorithm for CSP(Γ).

CSP(Γ) can be viewed as the following problem. Given a sentence

$$\exists x_1 \dots \exists x_n \big(R_1(v_{1,1}, \dots, v_{1,n_1}) \wedge \dots \wedge R_s(v_{s,1}, \dots, v_{s,n_s})\big),$$

we need to check whether it holds. To do this, let us remove the quantifiers one by one. Let

$$\Delta_{n-1}(x_1, \dots, x_{n-1}) = \exists x_n \big(R_1(v_{1,1}, \dots, v_{1,n_1}) \wedge \dots \wedge R_s(v_{s,1}, \dots, v_{s,n_s})\big).$$

In general, Δ_{n-1} could be any relation of arity $n-1$, and even to write this relation we need $|A|^{n-1}$ bits. Nevertheless, we believe that if CSP(Γ) is tractable then the relation Δ_{n-1} (or the important part of it) has a compact representation and can be efficiently computed. Then we calculate $\Delta_{n-2}, \Delta_{n-3}, \dots, \Delta_0$, where $\Delta_{i-1}(x_1, \dots, x_{i-1}) = \exists x_i \Delta_i(x_1, \dots, x_i)$, and the value of Δ_0 is the answer we need.

We may check that on a 2-element domain we have

$$\begin{aligned}&\exists x_n \big(R_1(v_{1,1}, \dots, v_{1,n_1}) \wedge \cdots \wedge R_s(v_{s,1}, \dots, v_{s,n_s})\big)\\&\quad = \bigwedge_{i,j \in \{1,2,\dots,s\}} \big(\exists x_n R_i(v_{i,1}, \dots, v_{i,n_i}) \wedge R_j(v_{j,1}, \dots, v_{j,n_j})\big).\end{aligned} \tag{2.7}$$

The implication $\Rightarrow$ is obvious. To prove $\Leftarrow$ assume that the left-hand side does not hold. Then the conjunctive part does not hold on both $(x_1, \dots, x_{n-1}, 0)$ and $(x_1, \dots, x_{n-1}, 1)$. Hence, there exist i and j such that $R_i(v_{i,1}, \dots, v_{i,n_i})$ does not hold on $(x_1, x_2, \dots, x_{n-1}, 0)$ and $R_j(v_{j,1}, \dots, v_{j,n_j})$ does not hold on $(x_1, x_2, \dots, x_{n-1}, 1)$. Hence, the (i, j)-part of the right-hand side does not hold.

There are two problems if we use (2.7) to solve the CSP. First, as we mentioned above, the relation $R_{i,j}(\dots) = \exists x_n R_i(v_{i,1}, \dots, v_{i,n_i}) \wedge R_j(v_{j,1}, \dots, v_{j,n_j})$ probably does not have a compact representation. Second, if we remove the quantifiers $\exists x_n, \exists x_{n-1}, \dots, \exists x_1$ one by one, potentially we could get an exponential number of relations in the formula. Let us show how these problem are solved for concrete examples on a 2-element domain.

2.9. System of linear equations in $\mathbb{Z}_2$

Let $A = \{0, 1\}$ and let Γ consist of linear equations in $\mathbb{Z}_2$. Suppose that for every i we have

$$R_i(v_{i,1}, \dots, v_{i,n_i}) = (a_1^i x_n + a_2^i x_2 + \cdots + a_n^i x_n = a_0^i).$$

For $a_n^i = a_n^j = 1$, we have

$$\begin{aligned}&R_{i,j}(\dots)\\&\quad := \exists x_n \big(R_i(v_{i,1}, \dots, v_{i,n_i}) \wedge R_j(v_{j,1}, \dots, v_{j,n_j})\big)\\&\quad = (a_1^i x_1 + a_2^i x_2 + \cdots + a_{n-1}^i x_{n-1} + a_0^i = a_1^j x_1 + a_2^j x_2 + \cdots + a_{n-1}^j x_{n-1} + a_0^j).\end{aligned}$$

If $a_n^i = 0$ then the constraint $R_i(v_{i,1}, \dots, v_{i,n_i})$ does not depend on x_n, so we keep it as it is when remove the quantifier. Hence, in every case we have a compact representation of Δ_{n-1}. To avoid the exponential growth of the number of the constraints, we use the idea from the Gaussian elimination. Choose k such that $a_n^k = 1$, then calculate only $R_{k,1}, \dots, R_{k,s}$ and ignore all the other relations. Thus, in this case we have

$$\begin{aligned}\Delta_{n-1}(x_1, \dots, x_{n-1}) &= \exists x_n \big(R_1(v_{1,1}, \dots, v_{1,n_1}) \wedge \cdots \wedge R_s(v_{s,1}, \dots, v_{s,n_s})\big)\\&= \bigwedge_{j \in \{1,2,\dots,s\}} \big(\exists x_n R_k(v_{k,1}, \dots, v_{k,n_k}) \wedge R_j(v_{j,1}, \dots, v_{j,n_j})\big).\end{aligned} \tag{2.8}$$

Proceeding this way, we calculate $\Delta_{n-2}, \Delta_{n-3}, \dots, \Delta_0$. Note that (2.8) holds not only for linear equations but whenever a variable x_n is uniquely determined by the other variables in $R_k(v_{k,1}, \dots, v_{k,n_1})$.

2.10. 2-satisfability

Let $A = \{0, 1\}$ and let Γ consist of all binary relations. In this case $R_{i,j}$ is also binary, which means that we do not have a problem with a compact representation. Also, every time we eliminate a quantifier and caclulate Δ_i, we remove the repetitive constraints. Therefore, in each Δ_i we cannot have more than $i \cdot i \cdot 2^{2^2}$ constraints because we have i different variables and 2^{2^2} different binary relations on $\{0, 1\}$.

As we see, the main question in both examples is the existence of a compact representation. In the first example we represent any relation as a conjunction of linear equations, in the second we represent as a conjunction of binary relations. We could ask when such a compact representation exists. Let $s_\Gamma(n)$ be the number of pp-definable from Γ relations of arity n. If $\log_2 s_\Gamma(n)$ grows exponentially then we need exponential space to encode relations of arity n and we cannot expect a compact representation. We say that Γ has few subpowers if $\log_2 s_\Gamma(n) < p(n)$ for a polynomial $p(n)$. It turns out that there is a simple criterion for the constraint language to have few subpowers. An operation t is called *an edge operation* if it satisfies the following identities:

$$\begin{aligned} t(x, x, y, y, y, \dots, y, y) &= y, \\ t(x, y, x, y, y, \dots, y, y) &= y, \\ t(y, y, y, x, y, \dots, y, y) &= y, \\ t(y, y, y, y, x, \dots, y, y) &= y, \\ &\dots \\ t(y, y, y, y, y, \dots, x, y) &= y, \\ t(y, y, y, y, y, \dots, y, x) &= y. \end{aligned}$$

Theorem 2.6 ([9]). *A constraint language Γ containing all constant relations has few subpowers if and only if it has an edge polymorphism.*

We can show that if Γ has few subpowers then the pp-definable relations have a natural compact representation, which gives a polynomial algorithm for CSP(Γ) [33]. Note that two examples of an edge operation were given earlier in this paper. The first example is a majority operation satisfying $m(y, y, x) = m(y, x, y) = m(x, y, y) = y$. By adding 3 dummy variables in the beginning, we get the required properties of an edge operation. Another example is $x + y + z$ on $\{0, 1\}$. By adding dummy variables at the end, we can easily satisfy all the identities. Very roughly speaking, any few subpowers case is just a combination (probably very complicated) of the majority case and the linear case.

2.11. Strong subuniverses and a proof of the CSP Dichotomy Conjecture

In this subsection, we consider another simple idea that can solve the CSP in polynomial time. This idea is one of the two main ingredients of the proof of the CSP Dichotomy Conjecture in [42, 44].

Assume that for every variable x whose domain is D_x, $|D_x| > 1$, we can choose a subset $B_x \subsetneq D_x$ such that if the instance has a solution, then it has a solution with $x \in B_x$.

In this case we can reduce the domains iteratively until the moment when each domain has exactly one element, which usually gives us a solution.

As we saw in Section 2.5, if Γ is preserved by $x \vee y$ and the instance is 1-consistent then we can safely reduce the domain of a variable to $\{1\}$. Similarly, if Γ is preserved by the majority operation $xy \vee yz \vee xz$ and the instance is cycle-consistent, then we can safely reduce the domain $\{0,1\}$ to $\{0\}$ and $\{1\}$ [47]. It turns out that this idea can be generalized for any constraint language preserved by a WNU operation.

A unary relation $B \subseteq A$ is called *a subuniverse* if B is pp-definable over Γ. It can be easily checked that all the domains D_x that appear while checking consistency (see Section 2.4) are subuniverses. Let us define three types of strong subuniverses:

Binary absorbing subuniverse. We say that B' is *a binary absorbing subuniverse of B* if there exists a binary operation $f \in \mathrm{Pol}(\Gamma)$ such that $f(B', B) \subseteq B'$ and $f(B, B') \subseteq B'$. For example, if the operation $x \vee y$ preserves Γ then $\{1\}$ is a binary absorbing subuniverse of $\{0,1\}$ and $x \vee y$ is a binary absorbing operation.

Ternary absorbing subuniverse. We say that B' is *a ternary absorbing subuniverse of B* if there exists a ternary operation $f \in \mathrm{Pol}(\Gamma)$ such that $f(B', B', B) \subseteq B'$, $f(B', B, B') \subseteq B'$, and $f(B, B', B') \subseteq B'$. For example, if the majority operation $xy \vee yz \vee xz$ preserves Γ, then both $\{0\}$ and $\{1\}$ are ternary absorbing subuniverses of $\{0,1\}$. Since we can always add a dummy variable to a binary absorbing operation, any binary absorbing subuniverse is also a ternary absorbing subuniverse.

To define the last type of strong subalgebras we need some understanding of the Universal Algebra. We do not think a concrete definition is important here, that is why if a reader thinks the definition is too complicated, we recommend to skip it and think about the last type as something similar to the first two.

PC subuniverse. A set F of operations is called *Polynomially Complete* (PC) if any operation can be derived from F and constants using composition. We say that B' is *a PC subuniverse* of B if there exists a pp-definable equivalence relation $\sigma \subseteq B \times B$ such that $\mathrm{Pol}(\Gamma)/\sigma$ is PC.

A subset B' of B is called *a strong subuniverse* if B' is a ternary absorbing subuniverse or a PC subuniverse.

Theorem 2.7 ([47]). *Suppose Γ contains all constant relations and is preserved by a WNU operation, $B \subseteq A$ is a subuniverse. Then*

(1) *there exists a strong subuniverse $B' \subsetneq B$, or*

(2) *there exists a pp-definable nontrivial equivalence relation σ on B and $f \in \mathrm{Pol}(\Gamma)$ such that $(B; f)/\sigma \cong (\mathbb{Z}_p^k; x - y + z)$.*

As it follows from the next lemma, the second condition implies that any pp-definable relation (modulo σ) can be viewed as a system of linear equations in a field.

Lemma 2.8 ([32]). *Suppose $R \subseteq \mathbb{Z}_p^n$ preserved by $x - y + z$. Then R can be represented as a conjunction of relations of the form $a_1x_1 + \cdots + a_nx_n = a_0 \pmod{p}$.*

For CSPs solvable by the local consistency checking, strong subuniverses have the following property.

Theorem 2.9 ([47]). *Suppose*

(1) *Γ is a constraint language containing all constant relations;*

(2) *Γ is preserved by a WNU of each arity $n \geq 3$;*

(3) *$\mathcal{I}$ is a cycle-consistent instance of* CSP(Γ)*;*

(4) *D_x is the domain of a variable x;*

(5) *B is a strong subalgebra of D_x.*

Then $\mathcal{I}$ has a solution with $x \in B$.

Thus, strong subuniverses have the required property that we cannot loose all the solutions when we restrict a variable to it. As it was proved in [44], a similar theorem holds for any constraint language preserved by a WNU operation (with additional consistency conditions on the instance). We skip this result because it would require too many additional definitions.

As we see from Theorem 2.7, for every domain D_x either we have a strong subuniverse and can reduce the domain of some variable, or, modulo some equivalence relation, we have a system of linear equations in a field. If Γ has a WNU polymorphism of each arity $n \geq 3$, then we always have the first case; hence, we can iteratively reduce the domains until the moment when all the domains have just one element, which gives us a solution. That is why any cycle-consistent instance in this situation has a solution. If we always have the second case then this situation is similar to a system of linear equations, but different linear equations can be mixed which makes it impossible to apply usual Gaussian elimination. Nevertheless, the few subpowers algorithm solves the problem [33].

For many years the main obstacle was that these two situations can be mixed and at the moment we do not know an elegant way how to split them. Nevertheless, the general algorithm for tractable CSP presented in [44] is just a smart combination of these two ideas:

- if there exists a strong subalgebra, reduce
- if there exists a system of linear equations, solve it.

For more information about this approach as well as its connection with the second general algorithm see [3].

2.12. Conclusions

Even though we still do not have a simple algorithm that solves all tractable Constraint Satisfaction Problems, we understand what makes the problem hard, and what makes the problem easy. First, we know that in all the hard cases we can pp-construct (pp-define) the not-all-equal relation, which means that all the NP-hard cases have the same nature. Second, if the CSP is not solvable locally then we can pp-construct (pp-define) a linear equation in a field. Moreover, any domain of a tractable CSP either has a strong subalgebra and we can (almost) safely reduce the domain, or there exists a system of linear equations on this domain. This implies that any tractable CSP can be solved by a smart combination of the Gaussian elimination and local consistency checking, and emphasizes the exclusive role of the linear case in Universal Algebra and Computational Complexity.

Note that both CSP algorithms in [20, 44] depend exponentially on the size of the domain, and we could ask whether there exists a universal polynomial algorithm that works for any constraint language Γ admitting a WNU polymorphism.

Problem 1. Does there exist a polynomial algorithm for the following decision problem: given a conjunctive formula $R_1(v_{1,1}, \dots, v_{1,n_1}) \wedge \dots \wedge R_s(v_{s,1}, \dots, v_{s,n_s})$, where all relations $R_1, \dots, R_s$ are preserved by a WNU, decide whether this formula is satisfiable.

If the domain is fixed then the above problem can be solved by the algorithms from [19,42]. In fact, we know from [4, THEOREM 4.2] that from a WNU on a domain of size k we can always derive a WNU (and also a cyclic operation) of any prime arity greater than k. Thus, we can find finitely many WNU operations on a domain of size k such that any constraint language preserved by a WNU is preserved by one of them. It remains to apply the algorithm for each WNU and return a solution if one of them gave a solution.

3. QUANTIFIED CSP

A natural generalization of the CSP is the *Quantified Constraint Satisfaction Problem* (QCSP), where we allow to use both existential and universal quantifiers. Formally, for a constraint language Γ, QCSP(Γ) is the problem to evaluate a sentence of the form

$$\forall x_1 \exists y_1 \dots \forall x_n \exists y_n \; R_1(v_{1,1}, \dots, v_{1,n_1}) \wedge \dots \wedge R_s(v_{s,1}, \dots, v_{s,n_s}),$$

where $R_1, \dots, R_s \in \Gamma$, and $v_{i,j} \in \{x_1, \dots, x_n, y_1, \dots, y_n\}$ for every i, j (see [16,23,24,37]). Unlike the CSP, the problem QCSP(Γ) can be PSpace-hard if the constraint language Γ is powerful enough. For example, QCSP({NAE}) and QCSP({1IN3}) on the domain $A = \{0, 1\}$ are PSpace-hard [25,27], and QCSP($\{\neq\}$) for $|A| > 2$ is also PSpace-hard [16]. Nevertheless, if Γ consists of linear equations modulo p then QCSP(Γ) is tractable [16]. It was conjectured by Hubie Chen [22,24] that for any constraint language Γ the problem QCSP(Γ) is either solvable in polynomial time, or NP-complete, or PSpace-complete. Recently, this conjecture was disproved in [48], where the authors found constraint languages Γ such that QCSP(Γ) is coNP-complete (on a 3-element domain), DP-complete (on a 4-element domain), Θ_2^P-complete (on a 10-element domain). Despite the whole zoo of the complexity classes, we still hope to obtain a full classification of the complexity for each constraint language Γ.

Below we consider the main idea that makes the problem easier than PSpace.

3.1. PGP reduction for Π_2 restrictions

For simplicity let us consider the Π_2-restriction of QCSP(Γ), denoted QCSP$^2(\Gamma)$, in which the input is of the form

$$\forall x_1 \dots \forall x_n \exists y_1 \dots \exists y_m R_1(\dots) \wedge \dots \wedge R_s(\dots). \tag{3.1}$$

Such an instance holds whenever the conjunctive formula $R_1(\dots) \wedge \dots \wedge R_s(\dots)$ is solvable for any evaluation of $x_1, \dots, x_n$, which gives us a reduction of the instance to $|A|^n$ instances of CSP(Γ^*), where by Γ^* we denote $\Gamma \cup \{(x = a) \mid a \in A\}$. If we need to check $|A|^n$ tuples, which is exponentially many, this does not make the problem easier. Nevertheless, sometimes it is sufficient to check only polynomially many tuples. Let us consider a concrete example.

System of linear equations. Suppose $A = \{0, 1\}$ and Γ consists of linear equations in $\mathbb{Z}_2$. Let us check that the instance (3.1) holds for $(x_1, \dots, x_n) = (0, \dots, 0)$, and $(x_1, \dots, x_n) = (0, \dots, 0, 1, 0, \dots, 0)$ for any position of 1. To do this, we solve the CSP instance $R_1(\dots) \wedge \dots \wedge R_s(\dots) \wedge \bigwedge_{i=1}^n (x_i = 0)$, and for every $j \in \{1, 2, \dots, n\}$ we solve the instance $R_1(\dots) \wedge \dots \wedge R_s(\dots) \wedge (x_j = 1) \wedge \bigwedge_{i \neq j} (x_i = 0)$. Each instance is a system of linear equations and can be solved in polynomial time. If at least one of the instances does not have a solution, then the instance (3.1) does not hold. Assume that all of them are satisfiable, then consider the relation Δ defined by the following pp-formula over Γ:

$$\Delta(x_1, \dots, x_n) = \exists y_1 \dots \exists y_m R_1(\dots) \wedge \dots \wedge R_s(\dots).$$

Since Γ is preserved by $x + y + z$, Δ is also preserved by $x + y + z$. Applying this operation to the tuples $(0, 0, \dots, 0), (1, 0, \dots, 0), (0, 1, 0, \dots, 0), \dots, (0, 0, \dots, 0, 1) \in \Delta$ coordinatewise, we derive that $\Delta = \{0, 1\}^n$, that is, Δ contains all tuples and (3.1) holds. Thus, we showed that QCSP$^2(\Gamma)$ is solvable in polynomial time.

This idea can be generalized as follows. We say that a set of operations F (or an algebra $(A; F)$) has the *polynomially generated powers (PGP)* property if there exists a polynomial $p(n)$ such that A^n can be generated from $p(n)$ tuples using operations of F. Another behavior that might arise is that there is an exponential function f so that the smallest generating sets for A^n require size at least $f(n)$. We describe this as the *exponentially generated powers (EGP)* property. As it was proved in [43] these are the only two situations we could have on a finite domain. Moreover, it was shown that the generating set in the PGP case can be chosen to be very simple and efficiently computable. As a generating set of polynomial size, we can take the set of all tuples with at most k switches, where a switch is a position in $(a_1, \dots, a_n)$ such that $a_i \neq a_{i+1}$. This gives a polynomial reduction of QCSP$^2(\Gamma)$ to CSP(Γ^*) if Pol(Γ) has the PGP property.

3.2. A general PGP reduction

Let us show that the same idea can be applied to the general form of QCSP(Γ). First, we show how to move universal quantifiers left and convert an instance into the Π_2-form. Notice that the sentence $\exists y_1 \exists y_2 \dots \exists y_s \forall x \Phi$ is equivalent to

$$\forall x^1 \forall x^2 \dots \forall x^{|A|} \exists y_1 \exists y_2 \dots \exists y_s \Phi_1 \wedge \Phi_2 \wedge \dots \wedge \Phi_{|A|},$$

where each Φ_i is obtained from Φ by renaming x by x^i. In this way we can convert any instance $\exists y_1 \forall x_1 \dots \exists y_t \forall x_t \Phi$ of QCSP(Γ) into the Π_2-restriction by moving all universal quantifiers left:

$$\begin{aligned} &\forall x_1^1 \dots \forall x_1^{|A|} \forall x_2^1 \dots \forall x_2^{|A|^2} \dots \forall x_t^1 \dots \forall x_t^{|A|^t} \\ &\exists y_1 \exists y_2^1 \dots \exists y_2^{|A|} \dots \exists y_t^1 \dots \exists y_t^{|A|^{t-1}} \Phi_1 \wedge \Phi_2 \wedge \dots \wedge \Phi_q, \end{aligned} \tag{3.2}$$

where each Φ_i is obtained from Φ by renaming the variables. The only problem with this reduction is that the number of variables and constraints could be exponential. Nevertheless, we can apply the PGP idea to this sentence. If Pol(Γ) has the PGP property then there exists a constant k such that it is sufficient to check (3.2) only on the tuples with at most k switches. Those k switches appear in at most k original x-variables and all the remaining variables can be fixed with constants. This allows reducing QCSP(Γ) to a sentence with a constant number of universal quantifiers or even remove all of them.

Theorem 3.1 ([45]). *Suppose* Pol(Γ) *has the PGP property. Then* QCSP(Γ) *is polynomially equivalent to the modification of* QCSP^2(Γ) *where sentences have at most* $|A|$ *universally quantified variables.*

Theorem 3.2 ([45]). *Suppose* Pol(Γ) *has the PGP property. Then* QCSP(Γ) *is polynomially reduced to* CSP(Γ^*).

This idea gives us a complete classification of the complexity of QCSP(Γ) for a two-element domain.

Theorem 3.3 ([25, 27]). *Suppose* Γ *is a constraint language on* $\{0, 1\}$. *Then* QCSP(Γ) *is solvable in polynomial time if* Γ *is preserved by an idempotent WNU;* QCSP(Γ) *is PSpace-complete otherwise.*

It is known [39] that if Γ admits an idempotent WNU, then it is preserved by $x + y + z$, $x \vee y$, $x \wedge y$, or $xy \vee yz \vee xz$. Hence, to prove the above theorem, it is sufficient to check that these operations guarantee the PGP property, which by Theorem 3.2 gives a polynomial reduction to a tractable CSP. To show the PGP property, we verify that the tuples $(0, 0, \dots, 0)$, $(1, 1, \dots, 1)$, $(1, 0, \dots, 0)$, $(0, 1, 0, \dots, 0)$, $\dots$, $(0, \dots, 0, 1)$ generate $\{0, 1\}^n$ using any of the above operations.

3.3. Does EGP mean hard?

Thus, if Pol(Γ) has the PGP property then we have a nice reduction to CSP, and QCSP(Γ) belongs to NP. What can we say about the complexity of QCSP(Γ) if Pol(Γ) has the EGP property? Hubie Chen conjectured in [24] that QCSP(Γ) is PSpace-complete whenever Pol(Γ) has the EGP property.

For constraint languages Γ containing all constant relations, a characterization of Pol(Γ) that have the EGP property is given in [43], where it is shown that Γ must allow the pp-definition of relations τ_n with the following special form.

Definition 3.4. Let $\alpha \cup \beta = A$, yet neither α nor β equals D. Let $S = \alpha^3 \cup \beta^3$ and τ_n be the $3n$-ary relation given by $\bigvee_{i=1}^{n} S(x_i, y_i, z_i)$.

The complement to S represents the not-all-equal relation and the relations τ_n allow for the encoding of the complement of *Not-All-Equal 3-Satisfiability* (where $\alpha \setminus \beta$ is 0 and $\beta \setminus \alpha$ is 1). Thus, if one has polynomially computable (in n) pp-definitions of τ_n, then it is clear that QCSP(Γ) is co-NP-hard [22]. In light of this observation, it seemed that only a small step remained to proving the actual Chen Conjecture, at least with coNP-hard in place of PSpace-complete.

3.4. Surprising constraint language and the QCSP on a 3-element domain

As we saw in Section 2.7, the CSP is NP-hard if and only the we can pp-define (pp-construct) the not-all-equal relation. In the previous subsection, we mentioned that in the EGP case we can always pp-construct the complement to Not-All-Equal 3-Satisfability, which almost guarantees coNP-hardness. Surprisingly, two constraint languages Γ on $A = \{0, 1, 2\}$ were discovered in [48] for which any pp-definition of τ_n is of exponential size, which makes it impossible to use this reduction.

Theorem 3.5 ([48]). *There exists a constraint language* Γ *on* $\{0, 1, 2\}$ *such that*

(1) Pol(Γ) *has the EGP property,*

(2) τ_n *is pp-definable over* Γ

(3) *any pp-definition of* τ_n *for* $\alpha = \{0, 1\}$ *and* $\beta = \{0, 2\}$ *has at least* 2^n *variables, and*

(4) QCSP(Γ) *is solvable in polynomial time.*

The algorithm in (4) consists of the following three steps. First, it reduces an instance to a Π_2-form $\forall x_1 \dots \forall x_n \exists y_1 \dots \exists y_m \Phi$. Then, by solving polynomially many CSPs, it calculates polynomially many evaluations to $(x_1, \dots, x_n)$ we need to check. Finally, it checks that Φ has a solution for each of these evaluations. It is proved in [48] that this test guarantees that the instance holds.

This result was shocking because of several reasons. Not only it disproved the widely believed Chen Conjecture but showed that we need to worry about the existence of an efficient pp-definition. Before, if we could pp-define a strong relation (such as τ_n) then the problem was hard. Another surprising thing is that we have to calculate the evaluations of $(x_1, \dots, x_n)$ we need to check, In fact, if we do not look inside Φ then we have to check all the tuples from $\{0, 1\}^n$.

Despite the fact that we are far from having a full classification of the complexity of the QCSP, we know the complexity for any constraint language on a 3-element domain containing all constant relations. This classification is given in terms of polymorphisms.

Theorem 3.6 ([48]). *Suppose* Γ *is a finite constraint language on* $\{0, 1, 2\}$ *containing all constant relations. Then* QCSP(Γ) *is either solvable in polynomial time, NP-complete, coNP-complete, or PSpace-complete.*

3.5. Conclusions

Unlike the CSP where the complexity is known for any constraint language Γ here the complexity is wide open.

Problem 2. What is the complexity of QCSP(Γ)?

Moreover, we do not even have a conjecture describing the complexity. We know that for some constraint languages Γ the problem QCSP(Γ) is DP-complete and Θ_2^P-complete, but we do not know whether there are some other complexity classes and whether we have finitely many of them.

Problem 3. What complexity classes (up to polynomial equivalence) can be expressed as QCSP(Γ) for some constraint language Γ?

Now it is hard to believe that there will be a simple classification, that is why it is interesting to start with a 3-element domain (without constant relations) and 4-element domain. Probably, a more important problem is to describe all tractable cases assuming $P \neq NP$.

Problem 4. Describe all constraint languages Γ such that QCSP(Γ) is solvable in polynomial time.

4. OTHER VARIANTS OF CSP

The Quantified CSP is only one of many other variations and generalizations of the CSP whose complexity is still unknown. Here we list some of them.

4.1. CSP over an infinite domain

If we consider CSP(Γ) for a constraint language on an infinite domain, the situation changes significantly. As was shown in [11], every computational problem is equivalent (under polynomial-time Turing reductions) to a problem of the form CSP(Γ). In [14] the authors gave a nice example of a constraint language Γ such that CSP(Γ) is undecidable. Let Γ consist of three relations (predicates) $x + y = z$, $x \cdot y = z$ and $x = 1$ over the set of all integers $\mathbb{Z}$. Then the Hilbert's 10th problem can be expressed as CSP(Γ), which proves undecidability of CSP(Γ). Nevertheless, there are additional assumptions that send the CSP back to the class NP and make complexity classifications possible [8, 12]. For more information about the infinite domain CSP and the algebraic approach, see [10, 14].

4.2. Surjective Constraint Satisfaction Problem

A natural modification of the CSP is *the Surjective Constraint Satisfaction Problem*, where we want to find a surjective solution. Formally, for a constraint language Γ over a domain A, SCSP(Γ) is the following decision problem: given a formula

$$R_1(\dots) \wedge \dots \wedge R_s(\dots),$$

where all relations $R_1, \dots, R_s$ are from Γ, decide whether there exists a surjective solution, that is, a solution with $\{x_1, \dots, x_n\} = A$. Probably, the most natural examples of the Surjective CSP are defined as the surjective graph homomorphism problem, which is equivalent to SCSP(Γ) where Γ consists of one binary relation that is viewed as a graph. An interesting fact about the complexity of the Surjective CSP is that its complexity remained unknown for many years even for very simple graphs and constraint languages. Three most popular examples of such long-standing problems are the complexity for the reflexive 4-cycle (undirected having a loop at each vertex) [38], the complexity for the nonreflexive 6-cycle (undirected without loops) [41], and the complexity of the No-Rainbow-Problem (SCSP($\{N\}$) where $A = \{0, 1, 2\}$ and $N = \{(a, b, c) \mid \{a, b, c\} \neq A\}$) [46]. Even though these three problems turned out to be NP-complete, the complexity seems to be unknown even for graphs of size 5 and cycles.

Problem 5. What is the complexity of SCSP(Γ)?

It was shown in [46] that the complexity of SCSP(Γ) cannot be described in terms of polymorphisms, which disproved the only conjecture about the complexity of SCSP(Γ) we know. This conjecture, formulated by Hubie Chen, says that SCSP(Γ) and CSP(Γ^*) have the same complexity. Nevertheless, this conjecture still can hold for graphs.

Problem 6. Is it true that SCSP($\{R\}$) and CSP($\{R\}^*$) have the same complexity for any binary relation R?

For more results on the complexity of the SCSP, see the survey [13].

4.3. Promise CSP

A natural generalization of the CSP is *the Promise Constraint Satisfaction Problem*, where a promise about the input is given (see [18, 21]). Let $\Gamma = \{(R_1^A, R_1^B), \dots, (R_t^A, R_t^B)\}$, where R_i^A and R_i^B are relations of the same arity over the domains A and B, respectively. Then PCSP(Γ) is the following decision problem: given two conjunctive formulas

$$R_{i_1}^A(v_{1,1}, \dots, v_{1,n_1}) \wedge \dots \wedge R_{i_s}^A(v_{s,1}, \dots, v_{s,n_s}),$$
$$R_{i_1}^B(v_{1,1}, \dots, v_{1,n_1}) \wedge \dots \wedge R_{i_s}^B(v_{s,1}, \dots, v_{s,n_s}),$$

where $(R_{i_j}^A, R_{i_j}^B)$ are from Γ for every j and $v_{i,j} \in \{x_1, \dots, x_n\}$ for every i, j, distinguish between the case when both of them are satisfiable, and when both of them are not satisfiable. Thus, we are given two CSP instances and a promise that if one has a solution then another has a solution. Usually, it is also assumed that there exists a mapping (homomorphism) $h : A \to B$ such that $h(R_i^A) \subseteq R_i^B$ for every i. In this case, the satisfiability of the first formula implies the satisfiability of the second one. To make sure that the promise can actually make an NP-hard problem tractable, see Example 2.8 in [21].

The most popular example of the Promise CSP is graph (k, l)-colorability, where we need to distinguish between k-colorable graphs and not even l-colorable, where $k \leq l$. This problem can be written as follows.

Problem 7. Let $|A| = k$, $|B| = l$, $\Gamma = \{(\neq_A, \neq_B)\}$. What is the complexity of PCSP(Γ)?

Recently, it was proved [21] that (k, l)-colorability is NP-hard for $l = 2k - 1$ and $k \geq 3$ but even the complexity of $(3, 6)$-colorability is still not known.

Even for a 2-element domain the problem is wide open, but recently a dichotomy for symmetric Boolean PCSP was proved [30].

Problem 8. Let $A = B = \{0, 1\}$. What is the complexity of PCSP(Γ)?

REFERENCES

[1] S. Arora and B. Barak, *Computational complexity: a modern approach.* Cambridge University Press, 2009.

[2] K. A. Baker and F. Pixley, Polynomial interpolation and the Chinese Remainder Theorem for algebraic systems. *Math. Z.* **143** (1975), 165–174.

[3] L. Barto, Z. Brady, A. Bulatov, M. Kozik, and D. Zhuk, Minimal Taylor algebras as a common framework for the three algebraic approaches to the CSP In *2021 36th annual ACM/IEEE symposium on logic in computer science (LICS)*, pp. 1–13, IEEE, 2021.

[4] L. Barto and M. Kozik, Absorbing subalgebras, cyclic terms, and the Constraint Satisfaction Problem. *Log. Methods Comput. Sci.* **8** (2012), no. 1.

[5] L. Barto and M. Kozik, Constraint satisfaction problems solvable by local consistency methods. *J. ACM* **61** (2014), no. 1, 1–19.

[6] L. Barto, A. Krokhin, and R. Willard, *Polymorphisms, and how to use them.* Dagstuhl Follow-Ups 7, Schloss Dagstuhl-Leibniz-Zentrum fuer Informatik, 2017.

[7] L. Barto, J. Opršal, and M. Pinsker, The wonderland of reflections. *Israel J. Math.* **223** (2018), no. 1, 363–398.

[8] L. Barto and M. Pinsker, The algebraic dichotomy conjecture for infinite domain constraint satisfaction problems. In *2016 31st annual ACM/IEEE symposium on logic in computer science (LICS)*, pp. 1–8, IEEE, 2016.

[9] J. Berman, P. Idziak, P. Marković, R. McKenzie, M. Valeriote, and R. Willard, Varieties with few subalgebras of powers. *Trans. Amer. Math. Soc.* **362** (2010), no. 3, 1445–1473.

[10] M. Bodirsky, Complexity classification in infinite-domain constraint satisfaction. 2012, arXiv:1201.0856.

[11] M. Bodirsky and M. Grohe, Non-dichotomies in constraint satisfaction complexity. In *International colloquium on automata, languages, and programming*, pp. 184–196, Springer, 2008.

[12] M. Bodirsky and J. Kára, The complexity of temporal constraint satisfaction problems. *J. ACM* **57** (2010), no. 2, 9.

[13] M. Bodirsky, J. Kára, and B. Martin, The complexity of surjective homomorphism problems – a survey. *Discrete Appl. Math.* **160** (2012), no. 12, 1680–1690.

[14] M. Bodirsky and M. Mamino, *Constraint satisfaction problems over numeric domains*. Dagstuhl Follow-Ups 7, Schloss Dagstuhl-Leibniz-Zentrum fuer Informatik, 2017.

[15] V. G. Bodnarchuk, L. A. Kaluzhnin, V. N. Kotov, and B. A. Romov, Galois theory for post algebras parts I and II. *Cybernetics* **5** (1969), 243–252, 531–539. 9

[16] F. Börner, A. A. Bulatov, H. Chen, P. Jeavons, and A. A. Krokhin, The complexity of constraint satisfaction games and QCSP. *Inform. and Comput.* **207** (2009), no. 9, 923–944.

[17] J. Brakensiek, S. Gopi, and V. Guruswami, CSPs with global modular constraints: algorithms and hardness via polynomial representations. In *Proceedings of the 51st annual ACM SIGACT symposium on theory of computing*, pp. 590–601, Association for Computing Machinery, 2019.

[18] J. Brakensiek and V. Guruswami, Promise constraint satisfaction: structure theory and a symmetric boolean dichotomy. In *Proceedings of the twenty-ninth annual ACM–SIAM symposium on discrete algorithms*, pp. 1782–1801, SIAM, 2018.

[19] A. A. Bulatov, A dichotomy theorem for nonuniform CSPs. In *2017 IEEE 58th annual symposium on foundations of computer science (FOCS)*, pp. 319–330, IEEE, 2017.

[20] A. A. Bulatov, A dichotomy theorem for nonuniform CSPs. 2017, arXiv:1703.03021.

[21] J. Bulín, A. Krokhin, and J. Opršal, Algebraic approach to promise constraint satisfaction. In *Proceedings of the 51st annual ACM SIGACT symposium on theory of computing*, pp. 602–613, ACM, 2019.

[22] C. Carvalho, B. Martin, and D. Zhuk, The complexity of quantified constraints using the algebraic formulation. In *42nd international symposium on mathematical foundations of computer science, MFCS 2017, August 21–25, 2017 – Aalborg, Denmark*, pp. 27:1–27:14, Leibniz Center for Informatics, 2017.

[23] H. Chen, The complexity of quantified constraint satisfaction: Collapsibility, sink algebras, and the three-element case. *SIAM J. Comput.* **37** (2008), no. 5, 1674–1701.

[24] H. Chen, Meditations on quantified constraint satisfaction. In *Logic and program semantics – essays dedicated to Dexter Kozen on the occasion of his 60th birthday*, pp. 35–49, Springer, 2012.

[25] N. Creignou, S. Khanna, and M. Sudan, *Complexity classifications of boolean constraint satisfaction problems*. SIAM, 2001.

[26] N. Creignou, H. Schnoor, and I. Schnoor, Non-uniform boolean constraint satisfaction problems with cardinality constraint. In *International workshop on computer science logic*, pp. 109–123, Springer, 2008.

[27] V. Dalmau, Some dichotomy theorems on constant-free quantified boolean formulas *Tech. rep. LSI97-43-R. Llenguatges i Sistemes Informátics—Universitat Politécnica de Catalunya, Barcelona, Spain*, 1997.

[28] T. Feder and M. Y. Vardi, Monotone monadic SNP and constraint satisfaction. In *Proceedings of the twenty-fifth annual ACM symposium on theory of computing*, pp. 612–622, Association for Computing Machinery, 1993.

[29] T. Feder and M. Y. Vardi, The computational structure of monotone monadic SNP and constraint satisfaction: a study through datalog and group theory. *SIAM J. Comput.* **28** (1999), no. 1, 57–104.

[30] M. Ficak, M. Kozik, M. Olsak, and S. Stankiewicz, Dichotomy for symmetric boolean PCSPs. 2019, arXiv:1904.12424.

[31] D. Geiger, Closed systems of functions and predicates. *Pacific J. Math.* **27** (1968), no. 1, 95–100.

[32] W. H. Greub, *Linear algebra*. Grad. Texts in Math. 23, Springer, 2012.

[33] P. Idziak, P. Marković, R. McKenzie, M. Valeriote, and R. Willard, Tractability and learnability arising from algebras with few subpowers. *SIAM J. Comput.* **39** (2010), no. 7, 3023–3037.

[34] P. Jeavons, On the algebraic structure of combinatorial problems. *Theoret. Comput. Sci.* **200** (1998), no. 1–2, 185–204.

[35] P. Jeavons, D. Cohen, and M. C. Cooper, Constraints, consistency and closure. *Artif. Intell.* **101** (1998), no. 1–2, 251–265.

[36] M. Kozik, Weak consistency notions for all the CSPs of bounded width. In *2016 31st annual ACM/IEEE symposium on logic in computer science (LICS)*, pp. 1–9, IEEE, 2016.

[37] B. Martin, Quantified Constraints in Twenty Seventeen. In *The constraint satisfaction problem: complexity and approximability*, edited by A. Krokhin and S. Zivny, pp. 327–346, Dagstuhl Follow-Ups 7, Schloss Dagstuhl–Leibniz-Zentrum fuer Informatik, Dagstuhl, Germany, 2017.

[38] B. Martin and D. Paulusma, The computational complexity of disconnected cut and 2_{k2}-partition. *J. Combin. Theory Ser. B* **111** (2015), 17–37.

[39] E. L. Post, *The two-valued iterative systems of mathematical logic*. Ann. of Math. Stud. 5, Princeton University Press, Princeton, NJ, 1941.

[40] T. J. Schaefer, The complexity of satisfiability problems. In *Proceedings of the tenth annual ACM symposium on theory of computing, STOC'78*, pp. 216–226, ACM, New York, NY, USA, 1978.

[41] N. Vikas, Computational complexity of graph partition under vertex-compaction to an irreflexive hexagon. In *42nd international symposium on mathematical foundations of computer science, MFCS 2017*, pp. 69:1–69:14, LIPIcs 83, Leibniz Center for Informatics, 2017.

[42] D. Zhuk, A proof of CSP dichotomy conjecture. In *2017 IEEE 58th annual symposium on foundations of computer science (FOCS)*, pp. 331–342, IEEE, 2017.

[43] D. Zhuk, The size of generating sets of powers. *J. Combin. Theory Ser. A* **167** (2019), 91–103.

[44] D. Zhuk, A proof of the CSP dichotomy conjecture. *J. ACM* **67** (2020), no. 5, 1–78.

[45] D. Zhuk, The complexity of the Quantified CSP having the polynomially generated powers property. 2021, arXiv:2110.09504.

[46] D. Zhuk, No-rainbow problem and the surjective constraint satisfaction problem. In *2021 36th annual ACM/IEEE symposium on logic in computer science (LICS)*, pp. 1–7, IEEE, 2021.

[47] D. Zhuk, Strong subalgebras and the constraint satisfaction problem. *J. Mult.-Valued Logic Soft Comput.* **36** (2021).

[48] D. Zhuk and B. Martin, QCSP monsters and the demise of the Chen conjecture. In *Proceedings of the 52nd annual ACM SIGACT symposium on theory of computing*, pp. 91–104, 2020.

DMITRIY ZHUK

Department of Mechanics and Mathematics, Lomonosov Moscow State University, Vorobjovy Gory, 119899 Moscow, Russia, zhuk@intsys.msu.ru

2. ALGEBRA

A TOTALLY DISCONNECTED INVITATION TO LOCALLY COMPACT GROUPS

PIERRE-EMMANUEL CAPRACE AND
GEORGE A. WILLIS

ABSTRACT

We present a selection of results contributing to a structure theory of totally disconnected locally compact groups.

MATHEMATICS SUBJECT CLASSIFICATION 2020

Primary 22D05; Secondary 20E15, 20E32, 20E08, 22F50, 43A07

KEYWORDS

Locally compact group, profinite group, amenable group, simple group, topological dynamics, commensurated subgroup, minimising subgroup, flat group of automorphisms

INTERNATIONAL CONGRESS OF MATHEMATICIANS 2022 JULY 6–14

Proc. Int. Cong. Math. 2022, Vol. 3, pp. 1554–1576
DOI 10.4171/ICM2022/99

1. INTRODUCTION

Locally compact groups have attracted sustained attention because, on the one hand, rich classes of these groups have fruitful connections with other fields and, on the other, they have a well-developed theory that underpins those connections and delineates group structure. Salient features of this theory are the existence of a left-invariant, or Haar, measure; and the decomposition of a general group into pieces, many of which may be described concretely and in detail.

Haar measure permits representations of a general locally compact group by operators on spaces of measurable functions, and is thus the foundation for abstract harmonic analysis. Connections with partial differential equations, physics, and number theory come about through these representations. Locally compact groups are the largest class for which an invariant measure exists and for which harmonic analysis can be done in this form, as was shown by A. Weil [81].

The decomposition theory of an arbitrary locally compact group G begins with the short exact sequence

$$0 \to G^\circ \to G \to G/G^\circ \to 0,$$

in which the closed normal subgroup G° is the connected component of the identity. The Gleason–Yamabe theorem [73, TH. 6.0.11] applies to G° to show that it is a projective limit of connected Lie groups, and powerful tools from the theory of Lie groups may thus be brought to bear on G°. Groups occurring in physics and differential equations are often Lie groups. The quotient G/G° is a totally disconnected locally compact group (abbreviated *tdlc group*). Lie groups over local fields are important examples of tdlc groups having links to number theory and algebraic geometry (see, for example, [49,69]). Unlike the connected case however, many other significant tdlc groups, such as the automorphism groups of locally finite trees first studied in [76], cannot be approximated by Lie groups. While substantial progress has been made with our understanding of tdlc groups much remains to be done before it could be said that the structure theory has reached maturity. This article surveys our current state of knowledge, much of which is founded on a theorem of van Dantzig, [77], which ensures that a tdlc group G has a basis of identity neighborhoods consisting of compact open subgroups.

Decompositions of general tdlc groups are described in Section 2. This section includes a discussion of the so-called elementary groups, which are those built from discrete and compact groups by standard operations. Discrete and compact groups are large domains of study in their own right and it is seen how elementary groups can be factored out in the analysis of a general tdlc group. Simple groups are an important aspect of any decomposition theory and what is known about them is summarized in Section 3. This includes a local structure theory and the extent to which local structure determines the global structure of the group. Section 4 treats scale methods, which associate invariants and special subgroups to abelian groups of automorphisms and which in some circumstances substitute for the Lie methods available for connected groups. A unifying theme of our approach is the dynamics of the conjugation action: Section 2 is concerned with the conjugation action of G on its closed subnormal subgroups, Section 3 uses in an essential way the conjugation action of G on its closed subgroups, especially those that are locally normal, while Section 4 concerns the dynamics of the conjugation action of cyclic subgroups (and, more generally, flat subgroups) on the topological space G. Section 5 highlights a few open questions and directions for further research.

2. DECOMPOSITION THEORY

2.1. Normal subgroup structure

Finite groups, Lie groups, and algebraic groups constitute three of the most important classes of groups. Their respective structure theories are deep and far-reaching. One of the common themes consists in reducing problems concerning a given group G in one of these classes to problems about simple groups in the corresponding class, and then tackling the reduced problem by invoking classification results. Striking illustrations of this approach in the case of finite groups can be consulted in R. Guralnick's ICM address [43].

Since the category of locally compact groups contains all discrete groups, hence all groups, developing a similar theory for locally compact groups is hopeless. Nevertheless, the possibility to construct meaningful "decompositions of locally compact groups into simple pieces" has been highlighted in [23]. Wide-ranging results have subsequently been established by C. Reid and P. Wesolek in a series of papers [63, 64], some of whose contributions are summarized below. A more in-depth survey can be consulted in [62].

Given closed normal subgroups K, L of a locally compact group G, the quotient group K/L is called a *chief factor* of G if L is strictly contained in K and for every closed normal subgroup N of G with $L \leq N \leq K$, we have $N = L$ or $N = K$. Given a closed normal subgroup N of G, the quotient $Q = G/N$ is a chief factor if and only if Q is *topologically simple*, i.e., Q is nontrivial and the only closed normal subgroups of Q are $\{1\}$ and Q. More generally, every chief factor $Q = K/L$ is *topologically characteristically simple*, i.e., the only closed subgroups of Q that are invariant under all homeomorphic automorphisms of Q are $\{1\}$ and Q. A topological group is called *compactly generated* if it has a compact generating set.

Theorem 2.1 (See [64, TH. 1.3]). *Every compactly generated tdlc group G has a finite series* $\{1\} = G_0 < G_1 < G_2 < \cdots < G_n = G$ *of closed normal subgroups such that for all* $i = 1, \ldots, n$, *the quotient* G_i/G_{i-1} *is compact, or discrete infinite, or a chief factor of G which is noncompact, nondiscrete, and second countable.*

A normal series as in Theorem 2.1 is called an *essentially chief series*. The theorem obviously has no content if G is compact or discrete. Let us illustrate Theorem 2.1 with two examples.

Example 2.2. Let I be a set and for each $i \in I$, let G_i be a tdlc group and $U_i \leq G_i$ be a compact open subgroup. The *restricted product* of $(G_i, U_i)_{i \in I}$, denoted by $\bigoplus_{i \in I}(G_i, U_i)$, is the subgroup of $\prod_{i \in I} G_i$ consisting of those tuples $(g_i)_{i \in I}$ such that $g_i \in U_i$ for all but finitely many $i \in I$. It is endowed with the unique tdlc group topology such that the inclusion $\prod_{i \in I} U_i \to \bigoplus_{i \in I}(G_i, U_i)$ is continuous and open. Given a prime p, set $M(p) = \bigoplus_{n \in \mathbf{Z}}(\mathrm{PSL}_2(\mathbf{Q}_p), \mathrm{PSL}_2(\mathbf{Z}_p))$. The cyclic group $\mathbf{Z}$ naturally acts on $M(p)$ by shifting the coordinates. The semidirect product $G(p) = M(p) \rtimes \mathbf{Z}$ is a compactly generated tdlc group, with an essentially chief series given by $\{1\} < M(p) < G(p)$. The group $M(p)$ is not compactly generated. It has minimal closed normal subgroups, but does not admit any finite essentially chief series, which illustrates the necessity of the compact generation hypothesis in Theorem 2.1.

Example 2.3. A more elaborate construction in [63, §9] yields an example of a compactly generated tdlc group $G'(p)$ with an essentially chief series given by $\{1\} < H(p) < G'(p)$ such that $G'(p)/H(p) \cong \mathbb{Z}$ and $H(p)$ has a nested chain of closed normal subgroups $(H(p)_n)$ indexed by $\mathbf{Z}$, permuted transitively by the conjugation $G'(p)$-action, and such that $H(p)_n/H(p)_{n-1} \cong M(p)$ for all $n \in \mathbf{Z}$.

A tdlc group is compactly generated if and only if it is capable of acting continuously, properly, with finitely many vertex orbits, by automorphisms on a connected locally finite graph. For a given compactly generated tdlc group G, vertex-transitive actions on graphs are afforded by the following construction. Given a compact open subgroup $U < G$, guaranteed to exist by van Dantzig's theorem, and a symmetric compact generating set Σ of G, we construct a graph Γ whose vertex set is the coset space G/U by declaring that the vertices gU and hU are adjacent if $h^{-1}g$ belongs to $U\Sigma U$. The fact that Σ generates G ensures that Γ is connected. Moreover, G acts vertex-transitively by automorphisms on Γ. Since U is compact open, the set $U\Sigma U$ is a finite union of double cosets modulo U; this implies that Γ is *locally finite*, i.e., the degree of each vertex is finite. Notice that all vertices have the same degree since Γ is homogeneous. The graph Γ is called a *Cayley–Abels graph* for G, since its construction was first envisaged by H. Abels [1, BEISPIEL 5.2] and specializes to a Cayley graph when G is discrete and $U = \{1\}$. The proof of Theorem 2.1 proceeds by induction on the minimum degree of a Cayley–Abels graph.

2.2. Elementary groups

By its very nature, Theorem 2.1 highlights the special role played by compact and discrete groups. A conceptual approach to studying the role of compact and discrete groups in the structure theory of tdlc groups is provided by P. Wesolek's notion of *elementary groups*. That concept is inspired by the class of elementary amenable discrete groups introduced by M. Day [33]. It is defined as the smallest class $\mathscr{E}$ of second countable tdlc groups (abbreviated *tdlcsc*) containing all countable discrete groups and all compact tdlcsc groups, which is stable under the following two group theoretic operations:

- Given a tdlcsc group G and a closed normal subgroup N, if $N \in \mathscr{E}$ and $G/N \in \mathscr{E}$, then $G \in \mathscr{E}$. In other words $\mathscr{E}$ is *stable under group extensions*.
- Given a tdlcsc group G and a directed set $(O_i)_{i \in I}$ of open subgroups, if $O_i \in \mathscr{E}$ for all i and if $G = \bigcup_i O_i$, then $G \in \mathscr{E}$. In other words $\mathscr{E}$ is *stable under directed unions of open subgroups*.

(The class $\mathscr{E}$ has a natural extension beyond the second countable case, see **[29, §6]**. For simplicity of the exposition, we stick to the second countable case here.) Using the permanence properties of the class $\mathscr{E}$, it can be shown that every tdlcsc group G has a largest closed normal subgroup that is elementary; it is denoted by $R_{\mathscr{E}}(G)$ and called the *elementary radical* of G. It indeed behaves as a radical, in the sense that it contains all elementary closed normal subgroups, and satisfies $R_{\mathscr{E}}(G/R_{\mathscr{E}}(G)) = \{1\}$, see **[82, §7.2]**. Further properties of the quotient $G/R_{\mathscr{E}}(G)$ will be mentioned in Section 3 below.

Similarly as for elementary amenable discrete groups, the class $\mathscr{E}$ admits a canonical rank function $\xi : \mathscr{E} \to \omega_1$, taking values in the set ω_1 of countable ordinals, called the *decomposition rank*. It measures the complexity of a given group $G \in \mathscr{E}$. By convention, the function ξ is extended to all tdlcsc groups by setting $\xi(G) = \omega_1$ for each nonelementary tdlcsc group G. We refer to **[82]**, **[83]** and **[62, §5]**. Let us merely mention here that the class $\mathscr{E}$ has remarkable permanence properties (e.g., it is stable under passing to closed subgroups and quotient groups), that the rank function has natural monotonicity properties, and that a nontrivial compactly generated group $G \in \mathscr{E}$ has a nontrivial discrete quotient. It follows in particular that if G is a tdlcsc group having a closed subgroup $H \leq G$ admitting a nondiscrete compactly generated topologically simple quotient, then $G \notin \mathscr{E}$. Therefore, the only compactly generated topologically simple groups in $\mathscr{E}$ are discrete. On the other hand, the class $\mathscr{E}$ contains numerous topologically simple groups that are not compactly generated, e.g., simple groups that are *regionally elliptic*, i.e., groups that can be written as a directed union of compact open subgroups. Those groups have decomposition rank 2. Explicit examples appear in **[88, §3]** or **[19, §6]**.

2.3. More on chief factors

The existence of essentially chief series prompts us to ask whether the chief factors of G depend upon the choice of a specific normal series in Theorem 2.1. It is tempting to tackle that question by invoking arguments à la Jordan–Hölder. A technical obstruction for doing so is that the product of two closed normal subgroups need not be closed. More generally, given closed subgroups A, N in G such that N is normal, the product AN need not be closed so that the natural abstract isomorphism $A/A \cap N \to AN/N$ need not be a homeomorphism. It is a continuous injective homomorphism of the locally compact group $A/A \cap N$ to a dense subgroup of the locally compact group $\overline{AN}/N$. This illustrates the necessity of considering dense embeddings of locally compact groups. We shall come back to this theme in Section 3.1 below. In the context of chief factors, this has led Reid–Wesolek to define an equivalence relation on nonabelian chief factors of G, called *association*, defined as follows: the chief factors K_1/L_1 and K_2/L_2 are *associated* if $\overline{K_1 L_2} = \overline{K_2 L_1}$ and $K_i \cap \overline{L_1 L_2} = L_i$ for $i = 1, 2$. In that case K_1/L_1 and K_2/L_2 both embed continuously as dense normal subgroups in $\overline{K_1 K_2}/\overline{L_1 L_2}$. We also recall that the *quasicenter* of a locally compact group G, denoted by $\mathrm{QZ}(G)$, is the collection of elements whose centralizer is open. It is a topologically characteristic (not necessarily closed) subgroup of G containing all the discrete normal subgroups. It was first introduced by M. Burger and S. Mozes [14]. Every nontrivial tdlcsc group with a dense quasicenter is elementary of decomposition rank 2 (see [62, LEM. 5]).

Theorem 2.4 (See [62, COR. 5]). *Let G be a compactly generated tdlc group and let $\{1\} = A_0 < A_1 < A_2 < \cdots < A_m = G$ and $\{1\} = B_0 < B_1 < B_2 < \cdots < B_n = G$ be essentially chief series for G. Then for each $i \in \{0, 1, \ldots, m\}$, if A_i / A_{i-1} is a chief factor with a trivial quasicenter, there is a unique j such that B_j / B_{j-1} is a chief factor with a trivial quasicenter that is associated with A_i / A_{i-1}. In other words, the association relation establishes a bijection between the sets of chief factors with a trivial quasicenter appearing respectively in the two series.*

The natural next question is to ask what can be said about chief factors. By the discussion above, one should focus on properties that are invariant under the association relation. Following Reid–Wesolek, an association class of nonabelian chief factors is called a *chief block*, and a group property shared by all members of a chief block is called a *block property*. The following are shown in [63] to be block properties: compact generation, amenability, having a trivial quasicenter, having a dense quasicenter, being elementary of a given decomposition rank.

As mentioned above, every chief factor is topologically characteristically simple. In particular, a compactly generated chief factor is subjected to the following description.

Theorem 2.5 (See [23, COR. D] and [22, REM. 3.10]). *Let G be a compactly generated nondiscrete, noncompact tdlc group which is topologically characteristic simple. Then there is a compactly generated nondiscrete topologically simple tdlc group S, an integer $d \geq 1$ and an injective continuous homomorphism $S^d = S \times \cdots \times S \to G$ of the direct product of d copies of S, such that the image of each simple factor is a closed normal subgroup of G, and the image of the whole product is dense.*

In the setting of Theorem 2.5, we say that G is the *quasiproduct* d copies of the simple group S. Theorem 2.5 provides a major incentive to study the compactly generated nondiscrete topologically simple tdlc groups. We shall come back to this theme in Section 3 below.

Developing a meaningful structure theory for topologically characteristically simple tdlc groups that are not compactly generated is very challenging. Remarkably, significant results have been established by Reid–Wesolek [63] under the mild assumption of second countability (abbreviated sc). In spite of the noncompact generation, they introduce an appropriate notion of chief blocks, and show that there are only three possible configurations for the arrangement of chief blocks in a topologically characteristically simple tdlcsc group G, that they call *weak type*, *semisimple type*, and *stacking type*. Moreover, if G is of weak type, then it is automatically elementary of decomposition rank $\leq \omega + 1$. The topologically characteristically simple groups $M(p)$ and $H(p)$ appearing in Examples 2.2 and 2.3 above are respectively of semisimple type and stacking type. We refer to [63] and [62] for details.

3. SIMPLE GROUPS

Let $\mathscr{S}$ be the class of nondiscrete, compactly generated, topologically simple locally compact groups and $\mathscr{S}_{\mathrm{td}}$ be the subclass consisting of the totally disconnected members of $\mathscr{S}$. By the Gleason–Yamabe theorem [73, TH. 6.0.11], all elements of $\mathscr{S} \setminus \mathscr{S}_{\mathrm{td}}$ are connected simple Lie groups. Prominent examples of groups in $\mathscr{S}_{\mathrm{td}}$ are provided by simple algebraic groups over non-Archimedean local fields, irreducible complete Kac–Moody groups over finite fields, certain groups acting on trees and many more, see [28, APPENDIX A]. A systematic study of the class $\mathscr{S}_{\mathrm{td}}$ as a whole has been initiated by Caprace–Reid–Willis in [28], and continued with P. Wesolek in [29] and with A. Le Boudec in [22]. We now outline some of their contributions. Another survey of the properties of nondiscrete simple locally compact groups can be consulted in [17]; the present account emphasizes more recent results.

3.1. Dense embeddings and local structure

As mentioned in Section 2.3 above, the failure of the second isomorphism theorem for topological groups naturally leads one to consider *dense embeddings*, i.e., continuous injective homomorphisms with dense image. If G, H are locally compact groups and $\psi : H \to G$ is a dense embedding, and if G is a connected simple Lie group or a simple algebraic group over a local field, then H is discrete or ψ is an isomorphism (see [29, §3]). This property however generally fails for groups $G \in \mathscr{S}$; see [50] for explicit examples. Nevertheless, as soon as the group H is nondiscrete, it turns out that key structural features of G are inherited by the dense subgroup H. To state this more precisely, we recall the definition of the class $\mathscr{R}$ of robustly monolithic groups, introduced in [29]. A tdlc group G is *robustly monolithic* if the intersection M of all nontrivial closed normal subgroups of G is nontrivial, if M is topologically simple and if M has a compactly generated open subgroup without any nontrivial compact normal subgroup. The class $\mathscr{R}$ contains $\mathscr{S}_{\mathrm{td}}$ and that inclusion is strict. The following result provides the main motivation to enlarge one's viewpoint by considering $\mathscr{R}$ instead of the smaller class $\mathscr{S}_{\mathrm{td}}$.

Theorem 3.1 (See [29, TH. 1.1.2]). *Let G, H be tdlc groups and $\psi : H \to G$ be a dense embedding. If $G \in \mathscr{R}$ and H is nondiscrete, then $H \in \mathscr{R}$.*

We emphasize that in general H is not topologically simple even in the special case where $G \in \mathscr{S}_{\mathrm{td}}$.

The approach in studying the classes $\mathscr{S}_{\mathrm{td}}$ and $\mathscr{R}$ initiated in [28] is based on the concept of *locally normal subgroup*, defined as a subgroup whose normalizer is open. To motivate it, recall once more that if M, N are closed normal subgroups of a tdlc group G, then the normal subgroup MN need not be closed. On the other hand, if $U \leq G$ is a compact open subgroup, then $M \cap U$ and $N \cap U$ are closed normal subgroups of the compact group U (hence they are both locally normal), so that the product $(M \cap U)(N \cap U)$ is closed. This observation motivates the definition of the *structure lattice* $\mathcal{LN}(G)$ of a tdlc group G, first introduced in [27], defined as the set of closed locally normal subgroups of G, divided by the *local equivalence relation* $\sim$, where $H \sim K$ if $H \cap K$ is relatively open both in H

and in K. The local class of a closed locally normal subgroup K is denoted by $[K]$. We also set $0 = [\{1\}]$ and $\infty = [G]$. The structure lattice carries a natural G-invariant order relation defined by the inclusion of representatives. The poset $\mathcal{LN}(G)$ is a modular lattice (see **[27, LEM. 2.3]**). The greatest lower bound and least upper bound of two elements $\alpha, \beta \in \mathcal{LN}(G)$ are respectively denoted by $\alpha \wedge \beta$ and $\alpha \vee \beta$. When G is a p-adic Lie group, the structure lattice $\mathcal{LN}(G)$ can naturally be identified with the lattice of ideals in the $\mathbf{Q}_p$-Lie algebra of G. The theory developed in **[27]** reveals that the structure lattice is especially well-behaved when the tdlc group G is *[A]-semisimple*, i.e., $\mathrm{QZ}(G) = \{1\}$ and the only abelian locally normal subgroup of G is $\{1\}$. That term is motivated by the fact that if G is a p-adic Lie group, then it is [A]-semisimple if and only if $\mathrm{QZ}(G) = \{1\}$ and the $\mathbf{Q}_p$-Lie algebra of G is semisimple, see **[27, PROP. 6.18]**. An important result of P. Wesolek is that the quotient $G/R_{\mathscr{E}}(G)$ of every tdlcsc group G by its elementary radical is [A]-semisimple (see **[82, COR. 9.15]**), so that every nonelementary group has a nontrivial [A]-semisimple quotient. The following result shows that [A]-semisimplicity is automatically fulfilled by groups in $\mathscr{R}$.

Theorem 3.2 (See **[28, TH. A]** and **[29, TH. 1.2.5]**). *Every group $G \in \mathscr{R}$ is [A]-semisimple.*

Given an [A]-semisimple tdlc group G, two closed locally normal subgroups H, $K \leq G$ that are locally equivalent have the same centralizer; moreover, they commute if and only if their intersection is trivial (see **[27, TH. 3.19]**). This ensures that the map $\mathcal{LN}(G) \to \mathcal{LN}(G) : [K] \mapsto [K]^\perp = [C_G(K)]$ is well defined, and that $\alpha \wedge \alpha^\perp = 0$ for all $\alpha \in \mathcal{LN}(G)$. This allows one to define the *centralizer lattice* of G by setting $\mathcal{LC}(G) = \{\alpha^\perp \mid \alpha \in \mathcal{LN}(G)\}$. If G is [A]-semisimple, the centralizer lattice $\mathcal{LC}(G)$ is a Boolean algebra (see **[27, TH. II]**). We denote its Stone dual by Ω_G. Thus Ω_G is a totally disconnected compact space endowed with a canonical continuous G-action by homeomorphisms. In general, the G-action on Ω_G need not be faithful. Actually, if $\mathcal{LC}(G) = \{0, \infty\}$ then Ω_G is a singleton. This happens if and only if any two non-trivial closed locally normal subgroups of G have a nontrivial intersection. The following result shows that the dynamics of the G-action on Ω_G has remarkable features.

Theorem 3.3 (See **[28, TH. J]** and **[29, TH. 1.2.6]**). *Let $G \in \mathscr{R}$. Then the G-action on Ω_G is minimal, strongly proximal, and has a compressible open set. Moreover, the G-action on Ω_G is faithful if and only if $\mathcal{LC}(G) \neq \{0, \infty\}$.*

Recall that a compact G-space X is called *minimal* if every G-orbit is dense. It is called *strongly proximal* if the closure of each G-orbit in the space of probability measures on X contains a Dirac mass. A nonempty subset α of X is called *compressible* if for every nonempty open subset $\beta \subseteq X$ there exists $g \in G$ with $g\alpha \subseteq \beta$. Obviously, if X is a minimal strongly proximal compact G-space and if G fixes a probability measure on X, then X is a singleton. Therefore, the following consequence of Theorem 3.3 is immediate.

Corollary 3.4. *Let $G \in \mathscr{R}$. If G is amenable, then $\mathcal{LC}(G) = \{0, \infty\}$.*

A *local isomorphism* between tdlc groups G_1, G_2 is a triple (φ, U_1, U_2) where U_i is an open subgroup of G_i and $\varphi : U_1 \to U_2$ is an isomorphism of topological groups. We emphasize that the structure lattice and the centralizer lattice are *local invariants*: they only depend on the local isomorphism class of the ambient tdlc group. However, for a group $G \in \mathscr{R}$, the compact G-space Ω_G can also be characterized by global properties among all compact G-spaces. In order to be more precise, let us first recall some terminology. Given an action of a group G by homeomorphisms on a Hausdorff topological space X, we define the *rigid stabilizer* $\mathrm{Rist}_G(U)$ of a subset $U \subseteq X$ as the pointwise stabilizer of the complement of U in X. The G-action on X is called *microsupported* if for every nonempty open subset $U \subset X$ with $U \neq X$, the rigid stabilizer $\mathrm{Rist}_G(U)$ acts nontrivially on X. The term "microsupported" was first coined in [28], although the notion it designates has frequently appeared in earlier references, notably in the work of M. Rubin on reconstruction theorems (see [67] and references therein). A prototypical example of a microsupported action of a tdlc group is given by the action of the full automorphism group $\mathrm{Aut}(T)$ of a locally finite regular tree T of degree ≥ 3 on the compact space ∂T consisting of the ends of T. The following result shows that for a general group $G \in \mathscr{R}$, the G-action on Ω_G shares many dynamical properties with the $\mathrm{Aut}(T)$-action on ∂T.

Theorem 3.5 (See [28, TH. J], [29, TH. 7.3.3] and [22, TH. 7.5]). *Let $G \in \mathscr{R}$. Then the G-action on Ω_G is microsupported. Moreover, for each nonempty microsupported compact G-space X on which the G-action is faithful, there is a G-equivariant continuous surjective map $\Omega_G \to X$. In particular, the G-action on X is minimal, strongly proximal, and has a compressible open set.*

This shows that Ω_G is universal among the faithful microsupported compact G-spaces; in particular, the purely local condition that $\mathscr{LC}(G) = \{0, \infty\}$ ensures that G does not have any faithful microsupported continuous action on any compact space. Theorem 3.5 was first established for totally disconnected compact G-spaces in [28, 29], and then extended to all compact G-spaces in [22], using tools from topological dynamics. Further properties of the G-space Ω_G and on the algebraic structure of groups in $\mathscr{R}$ can be consulted in those references.

We now present another aspect of the local approach to the structure of simple tdlc groups. We define the *local prime content* of a tdlc group G, denoted by $\pi(G)$, to be the set of those primes p such that every compact open subgroup $U \leq G$ contains an infinite pro-p subgroup.

Theorem 3.6 (See [28, TH. H] and [29, COR. 1.1.4 AND TH. 1.2.1]). *The following assertions hold for any group $G \in \mathscr{R}$:*

(i) *The local prime content $\pi(G)$ is finite and nonempty.*

(ii) *For each $p \in \pi(G)$, there is a group $G_{(p)} \in \mathscr{R}$ that is locally isomorphic to a pro-p group, and a dense embedding $G_{(p)} \to G$.*

(iii) *If H is a tdlc group acting continuously and faithfully by automorphisms on G, then H is locally isomorphic to a pro-$\pi(G)$ group.*

Roughly speaking, Theorem 3.6(ii) asserts that every group in $\mathscr{R}$ can be "approximated" by a locally pro-p group in $\mathscr{R}$. The restriction on the automorphism group of a group in $\mathscr{R}$ from Theorem 3.6(iii) should be compared with the automorphism group of the restricted product $M(p)$ from Example 2.2. Indeed, the Polish group $\mathrm{Sym}(\mathbf{Z})$ embeds continuously in $\mathrm{Aut}(M(p))$ by permuting the simple factors, and every tdlcsc group continuously embeds in $\mathrm{Sym}(\mathbf{Z})$. In some sense, the construction of stacking type chief factors in Example 2.3 crucially relies on the hugeness of the group $\mathrm{Aut}(M(p))$. Theorem 3.6(iii) shows that the automorphism group of a group in $\mathscr{R}$ is considerably smaller.

Let us finish this subsection with a brief discussion of classification problems. The work of S. Smith [72] shows that $\mathscr{S}_{\mathrm{td}}$ contains uncountably many isomorphism classes; his methods of proof suggest that the isomorphism relation on $\mathscr{S}_{\mathrm{td}}$ has a similar complexity as the isomorphism relation on the class of finitely generated discrete simple groups. This provides evidence that the problem of classifying groups in $\mathscr{S}_{\mathrm{td}}$ up to isomorphism is ill-posed. The recent results on the local structure of groups in $\mathscr{S}_{\mathrm{td}}$ or in $\mathscr{R}$ may be viewed as a hint to the fact the local isomorphism relation might be better behaved (see [29, TH. 1.1.5]). At the time of this writing, we do not know whether or not the groups in $\mathscr{S}_{\mathrm{td}}$ fall into countably many local isomorphism classes. However, classifying simple groups up to isomorphism remains a pertinent problem for some significant subclasses of $\mathscr{S}_{\mathrm{td}}$. To wit, let us mention that, by [30, COR. 1.4], a group $G \in \mathscr{S}_{\mathrm{td}}$ is isomorphic to a simple algebraic group over a local field if and only if it is locally isomorphic to a *linear group*, i.e., a subgroup of $\mathrm{GL}_d(k)$ for some integer d and some locally compact field k. Lastly, a remarkable classification theorem concerning an important class of nonlinear simple groups acting on locally finite trees has been obtained by N. Radu [61]. It would be highly interesting to extend Radu's results by classifying all groups in $\mathscr{S}_{\mathrm{td}}$ acting properly and continuously by automorphisms on a given locally finite tree T in such a way that the action on the set of ends of T is doubly transitive. That class is denoted by $\mathscr{S}_T$. Results from [25] ensure that the isomorphism relation restricted to $\mathscr{S}_T$ is *smooth* (see [37, DEFINITION 5.4.1]), which means that it comes at the bottom of the hierarchy of complexity of classification problems in the formalism established by invariant descriptive set theory (see [37, CH. 15]). Let us close this discussion by mentioning that we do not know whether there is a tree T such that $\mathscr{S}_T$ contains uncountably many isomorphism classes.

3.2. Applications to lattices

The study of lattices in semisimple Lie and algebraic groups has known tremendous developments since the mid-20th century, with Margulis' seminal contributions as cornerstones. Remarkably, several key results on lattices have been established at a high level of generality, well beyond the realm of linear groups. An early illustration is provided by [13]. More recently, Y. Shalom [70] and Bader–Shalom [5] have established an extension of Margulis' Normal Subgroup Theorem valid for all irreducible cocompact lattices in products of groups in $\mathscr{S}$, while various analogues of Margulis' superrigidity for irreducible lattices in products have been established for various kinds of target spaces, see [3,4,31,36,38,55,56,70]. Those results have in common that they rely on *transcendental methods*: they use a mix of tools from ergodic theory, probability theory, and abstract harmonic analysis, but do not require any detailed consideration of the algebraic structure of the ambient group. Another breakthrough in this field was accomplished by M. Burger and S. Mozes [15], who constructed a broad family of new finitely presented infinite simple groups as irreducible lattices in products of nonlinear groups in $\mathscr{S}_{\mathrm{td}}$. Their seminal work involves a mix of transcendental methods together with a fair amount of structure theory developed in [14].

The following two recent results rely in an essential way on the properties of the class $\mathscr{S}_{\mathrm{td}}$ outlined above.

Theorem 3.7 (See [21, TH. A]). *Let $n \geq 2$ be an integer, let $G_1, \dots, G_n \in \mathscr{S}_{\mathrm{td}}$ and $\Gamma \leq G = G_1 \times \cdots \times G_n$ be a lattice such that the projection $p_i(\Gamma)$ is dense in G_i for all i. Assume that Γ is cocompact, or that G has Kazhdan's property (T). Then the set of discrete subgroups of G containing Γ is finite.*

Theorem 3.8 (See [21, TH. C]). *Let $n \geq 2$ be an integer and let $G_1, \dots, G_n \in \mathscr{S}_{\mathrm{td}}$ be compactly presented. For every compact subset $K \subset G = G_1 \times \cdots \times G_n$, the set of discrete subgroups $\Gamma \leq G$ with $G = K\Gamma$ and with $p_i(\Gamma)$ dense in G_i for all i, is contained in a union of finitely many* $\mathrm{Aut}(G)$*-orbits.*

For a detailed discussion of the notion of *compactly presented* locally compact groups, we refer to [32, CH. 8].

Theorems 3.7 and 3.8 can be viewed as respective analogues of two theorems of H. C. Wang [78,79] on lattices in semisimple Lie groups and reveal the existence of positive lower bounds on the covolume of certain families of irreducible cocompact lattices. It should be underlined that the corresponding statements fail for lattices in a single group $G \in \mathscr{S}_{\mathrm{td}}$, see [6, TH. 7.1]. Theorem 3.8 is established by combining Theorem 3.7 with recent results on local rigidity of cocompact lattices in arbitrary groups, due to Gelander–Levit [39].

3.3. Applications to commensurated subgroups

The structure theory of tdlc groups provides valuable tools in exploring the so-called commensurated subgroups of an abstract group. In this section, we recall that connection and illustrate it with several recent results. Further results on commensurated subgroups will be mentioned in Section 4 below.

Let Γ be a group. Two subgroups $\Lambda_1, \Lambda_2 \leq \Gamma$ are called *commensurate* if their intersection $\Lambda_1 \cap \Lambda_2$ has finite index both in Λ_1 and in Λ_2. The *commensurator* of a subgroup $\Lambda \leq \Gamma$, denoted by $\mathrm{Comm}_\Gamma(\Lambda)$, is the set of those $\gamma \in \Gamma$ such that Λ and $\gamma\Lambda\gamma^{-1}$ are commensurate. It is easy to see that $\mathrm{Comm}_\Gamma(\Lambda)$ is a subgroup of Γ containing the normalizer $N_\Gamma(\Lambda)$. The commensurator has naturally appeared in group theory; one of its early occurrences is in Mackey's irreducibility criterion for induced unitary representations (see [51]). It also appears in a celebrated characterization of arithmetic lattices in semisimple groups due to Margulis [53, CH. IX, TH. (B)]. A *commensurated subgroup* of Γ is a subgroup $\Lambda \leq \Gamma$ such that $\mathrm{Comm}_\Gamma(\Lambda) = \Gamma$. Clearly, every normal subgroup of Γ is commensurated; more generally, every subgroup that is commensurate to a normal subgroup is commensurated.

Those commensurated subgroups are considered as trivial. For example, finite subgroups and subgroups of finite index are always commensurated subgroups. It is however important to underline that commensurated subgroups are not all of this trivial form. Indeed, an easy but crucial observation is that compact open subgroups are always commensurated. In particular, in the simple group $\mathrm{PSL}_2(\mathbf{Q}_p)$, the subgroup $\mathrm{PSL}_2(\mathbf{Z}_p)$ (which is obviously not commensurate to any normal subgroup of $\mathrm{PSL}_2(\mathbf{Q}_p)$) is commensurated.

Let us next remark that if U is a commensurated subgroup of a group G and $\varphi : \Gamma \to G$ is a group homomorphism, then $\varphi^{-1}(U)$ is a commensurated subgroup of Γ. This is the case in particular if G is a tdlc group and $U \leq G$ is a compact open subgroup. A fundamental observation is that all commensurated subgroups of Γ arise in this way. More, precisely, a subgroup $\Lambda \leq \Gamma$ is commensurated if and only if there is a tdlc group G, a compact open subgroup $U \leq G$, and a homomorphism $\varphi : \Gamma \to G$ with dense image such that $\varphi^{-1}(U) = \Lambda$. Indeed, given a commensurated subgroup $\Lambda \leq \Gamma$, then Λ acts on the coset space Γ/Λ with finite orbits, so that the closure of the natural image of Γ in the permutation group $\mathrm{Sym}(\Gamma/\Lambda)$, endowed with the topology of pointwise convergence, is a tdlc group containing the closure of the image of Λ as a compact open subgroup. That tdlc group is called the *Schlichting completion* of the pair (Γ, Λ), denoted by $\Gamma//\Lambda$. We refer to [68], [71, SECTION 3] and [65] for more information. Let us merely mention here that a commensurated subgroup $\Lambda \leq \Gamma$ is commensurate to a normal subgroup if and only if the Schlichting completion $G = \Gamma//\Lambda$ is *compact-by-discrete*, i.e., G has a compact open normal subgroup (see [22, LEM. 5.1]).

The occurrence of nontrivial commensurated subgroups in finitely generated groups with few normal subgroups (e.g., simple groups, or *just-infinite groups*, i.e., groups all of whose proper quotients are finite) remains an intriguing phenomenon. On the empirical basis of the known examples, it seems to be rather rare. The following result provides valuable information in that context.

Theorem 3.9 (See [22, TH. 5.4]). *Let Γ be a finitely generated group. Assume that all normal subgroups of Γ are finitely generated, and that every proper quotient of Γ is virtually nilpotent. Let also X be a compact Γ-space on which the Γ-action is faithful, minimal and microsupported. Assume that at least one of the following conditions is satisfied:*

(1) *Γ is residually finite.*

(2) *Γ fixes a probability measure on X.*

Then every commensurated subgroup of Γ is commensurate to a normal subgroup.

This applies to all finitely generated branch groups, as well as to numerous finitely generated almost simple groups arising in Cantor dynamics, and whose study has known spectacular recent developments (see [34, 59] and references therein). We refer to [22] for details and a more precise description of those applications.

Let us briefly outline how the proof of Theorem 3.9 works in the case where Γ fixes a probability measure on X. Let $\Lambda \leq \Gamma$ be a commensurated subgroup and $G = \Gamma//\Lambda$ be the corresponding Schlichting completion. That Γ is finitely generated implies that G is compactly generated. The hypotheses made on the normal subgroup structure of Γ yield some restrictions on the essentially chief series of G afforded by Theorem 2.1. More precisely, assuming by contradiction that Λ is not commensurate to a normal subgroup, then the upper most chief factor K/L with trivial quasicenter in an essentially chief series for G must be compactly generated. Its structure is therefore described by Theorem 2.5. A key point in the proof, relying on various ingredients from topological dynamics and involving detailed considerations of the Chabauty space of closed subgroups of Γ and G, is to show that the given Γ-action on X gives rise to a continuous, faithful, microsupported G/L-action on a compact space Y which is closely related to the original space X. Invoking (a suitable version of) Theorem 3.5 for the chief factor K/L ensures that Y has a compressible open set, from which it follows that X has a compressible open set for the Γ-action. This finally contradicts the hypothesis of existence of a Γ-invariant probability measure.

4. SCALE METHODS

The scale of a tdlc group endomorphism, α, is a positive integer that conveys information about the dynamics of the action of α. Roughly speaking, α contracts towards the identity on one subgroup of G and expands on another, and the scale is the expansion factor. This section gives an account of properties of the scale and descriptions of the action of α on certain associated subgroups of G which, when applied to inner automorphisms, answer questions about group structure.

Let $\alpha : G \to G$ be a continuous endomorphism. The *scale* of α is

$$s(\alpha) = \min\{[\alpha(U) : \alpha(U) \cap U] \mid U \leq G \text{ compact and open}\}.$$

This value is a positive integer because $\alpha(U) \cap U$ is an open subgroup of the compact group $\alpha(U)$. Subgroups at which the minimum is attained are said to be *minimizing for* α. The following results from **[85, 86, 89]** relate minimizing subgroups to the dynamics of α.

Theorem 4.1. *Let α be a continuous endomorphism of the tdlc group G and let $U \leq G$ be compact and open. Define subgroups*

$$U_+ = \{u \in U \mid \exists \{u_n\}_{n\geq 0} \subset U \text{ with } u_0 = u \text{ and } u_n = \alpha(u_{n+1})\},$$
$$U_- = \{u \in U \mid \alpha^n(u) \in U \text{ for all } n \geq 0\}.$$

Also define the subgroup $U_{--} = \bigcup_{n\geq 0} \alpha^{-n}(U_-)$ of G.

Then U is minimizing for α if and only if

$$\text{(TA)} \quad U = U_+U_- \quad \text{and} \quad \text{(TB)} \quad U_{--} \text{ is closed.}$$

A compact open subgroup U satisfying *TA* and *TB* is said to be *tidy for* α, and $s(\alpha) = [\alpha(U_+) : U_+]$ for any such subgroup U. Tidiness has two further dynamical interpretations: (1) an α-trajectory $\{\alpha^n(g)\}_{n\geq 0}$ cannot return to a tidy subgroup once it departs; and (2) when α is an automorphism, U is tidy for α if and only if the orbit $\{\alpha^n(U)\}_{n\in\mathbb{Z}}$ is a geodesic for the metric $d(U, V) = \log[U : U \cap V] + \log[V : U \cap V]$ on the set of compact open subgroups of G.

Note that every compact open subgroup of G has a subgroup U for which *TA* holds and, if α is the inner automorphism $\alpha_g(x) := gxg^{-1}$, then property *TA* implies that $Ug^m Ug^n U = Ug^{m+n} U$ for all $m, n \geq 0$. These points were already used in [12] in the proof that a reductive group over a locally compact field of positive characteristic is type I, where they were observed to hold in such groups.

In the following compilation of results from [54, 85, 86, 89], Δ denotes the modular function on the automorphism group of G.

Theorem 4.2. *The scale* $s : \mathrm{End}(G) \to \mathbb{Z}^+$ *satisfies:*

(i) $s(\alpha) = 1$ *if and only if there is a compact open subgroup* $U \leq G$ *with* $\alpha(U) \leq U$;

(ii) $s(\alpha) = \lim_{n\to\infty}[\alpha^n(V) : \alpha^n(V) \cap V]^{\frac{1}{n}}$ *for every compact open* $V \leq G$, *and* $s(\alpha^n) = s(\alpha)^n$ *for every* $n \geq 0$; *and*

(iii) *if* α *is an automorphism, then* $\Delta(\alpha) = s(\alpha)/s(\alpha^{-1})$.

The function $s \circ \alpha_\bullet : G \to \mathbb{Z}^+$, *with* $\alpha_g(x) = gxg^{-1}$, *is continuous for the group topology on* G *and the discrete topology on* $\mathbb{Z}^+$.

Continuity of $s \circ \alpha_\bullet$ is implied by the fact that, if U is tidy for g, then U is also tidy for all $h \in UgU$ and $s(h) = s(g)$, [85, THEOREM 3].

Questions about the structure of tdlc groups may be answered with scale and tidy subgroup techniques. K. H. Hofmann and A. Mukherjea conjectured in [45] that all locally compact groups are "neat"—a property involving the conjugation action by a single element g. They used approximation by Lie groups to reduce to the totally disconnected case, and subgroups tidy for g are used in [47] to show that all groups are neat. Answering another question of K. H. Hofmann, the set $\mathrm{per}(G)$, comprising those elements of G such that the closure of $\langle g \rangle$ is compact, is shown in [84] to be closed by appealing to the properties of the scale given in Theorem 4.2.

The scale and the subgroup U_+ associated with it in Theorem 4.1 are given a concrete representation in [9]. Put $U_{++} = \bigcup_{n\geq 0} \alpha^n(U_+)$. Then U_{++} is closed if U is tidy and $U_{++} \rtimes \langle \alpha \rangle$ acts on a regular tree with valency $s(\alpha) + 1$: the image of $U_{++} \rtimes \langle \alpha \rangle$ is a closed subgroup of the isometry group of the tree; is transitive on vertices; and fixes an end of the tree. The resulting isometry groups of trees correspond to the *self-replicating groups* studied in [58]. Moreover, the semidirect product $U_{++} \rtimes \langle \alpha \rangle$ also belongs to the family of *focal hyperbolic groups* studied in [18].

4.1. Contraction and other groups

Subgroups of G defined in terms of the action of α are related to the scale and tidy subgroups. It is convenient to confine the statements to automorphisms here. Extensions to endomorphisms may be found in [16, 89].

The *contraction subgroup* for $\alpha \in \mathrm{Aut}(G)$ is

$$\mathrm{con}(\alpha) = \{x \in G \mid \alpha^n(x) \to 1 \text{ as } n \to \infty\}.$$

The next result, from **[9, 46]**, relates contraction subgroups to the scale.

Theorem 4.3. *Let $\alpha \in \mathrm{Aut}(G)$. Then $\bigcap\{U_{--} \mid U$ is tidy for $\alpha\}$ is equal to $\overline{\mathrm{con}(\alpha)}$, and $s(\alpha^{-1})$ is equal to the scale of the restriction of α^{-1} to $\overline{\mathrm{con}(\alpha)}$. Hence $s(\alpha^{-1}) > 1$ if and only if $\overline{\mathrm{con}(\alpha)}$ is not compact.*

If G is a p-adic Lie group, then $\mathrm{con}(\alpha)$ is closed for every α, **[80]**, but that is not the case if, for example, G is the isometry group of a regular tree, or a certain type of complete Kac–Moody group **[7]**, or if $\mathcal{LC}(G) \neq \{0, \infty\}$ **[28]**. The closedness of $\mathrm{con}(\alpha)$ is equivalent, by **[9, THEOREM 3.32]**, to the triviality of the *nub* subgroup,

$$\mathrm{nub}(\alpha) = \bigcap\{U \mid U \text{ tidy for } \alpha\}.$$

The nub for α is compact and is the largest α-stable subgroup of G on which α acts ergodically, which sharpens the theorem of N. Aoki in **[2]** that a totally disconnected locally compact group with an ergodic automorphism must be compact. P. Halmos had asked in **[44]** whether that was so for all locally compact groups. See **[48, 90]** for the connected case, and also **[60]**.

The structure of closed contraction subgroups $\mathrm{con}(\alpha)$ is described precisely in **[40]**. If $\mathrm{con}(\alpha)$ is closed, there is a composition series

$$\{1\} = G_0 \lhd \cdots \lhd G_n = \mathrm{con}(\alpha)$$

of α-stable closed subgroups of $\mathrm{con}(\alpha)$ such that the factors G_{i+1}/G_i have no proper, nontrivial α-stable closed subgroups. The factors appearing in any such series are unique up to permutation and isomorphism, and their isomorphism types come from a countable list: each torsion factor being a restricted product $\bigoplus_{i \in \mathbf{Z}}(G_i, U_i)$ with $G_i = F$, a finite simple group, and $U_i = F$ if $i \geq 0$ and trivial if $i < 0$, and the automorphism the shift; and each divisible factor being a p-adic vector group and the automorphism a linear transformation. Moreover, $\mathrm{con}(\alpha)$ is the direct product $T \times D$ with T a torsion and D a divisible α-stable subgroup. The divisible subgroup D is a direct product $D_{p_1} \times \cdots \times D_{p_r}$ with D_{p_i} a nilpotent p-adic Lie group for each p_i. The torsion group T may include nonabelian irreducible factors but, should it happen to be locally pro-p, then it is nilpotent too, see **[42]**. The number of nonisomorphic locally pro-p closed contraction groups is uncountable **[41]**.

Contraction groups correspond to unipotent subgroups of algebraic groups and, following **[75]**, the *Tits core*, $G^\dagger$, of the tdlc group G is defined to be the subgroup generated by all closures of contraction groups. It is shown in **[26]** that, if G is topologically simple, then $G^\dagger$ is either trivial or is abstractly simple and dense in G.

The correspondence with algebraic groups is pursued in **[9]**, where the *parabolic subgroup* for $\alpha \in \mathrm{Aut}(G)$ is defined to be

$$\mathrm{par}(\alpha) = \{x \in G \mid \{\alpha^n(x)\}_{n \geq 0} \text{ has compact closure}\},$$

and the *Levi factor* to be $\mathrm{lev}(\alpha) = \mathrm{par}(\alpha) \cap \mathrm{par}(\alpha^{-1})$. Then $\mathrm{par}(\alpha)$, and hence $\mathrm{lev}(\alpha)$, is closed in G, **[85, PROPOSITION 3]**. It may be verified that $\mathrm{con}(\alpha) \lhd \mathrm{par}(\alpha)$ and shown, see **[9]**, that $\mathrm{par}(\alpha) = \mathrm{lev}(\alpha)\,\mathrm{con}(\alpha)$.

4.2. Flat groups of automorphisms

A group, $\mathcal{H}$, of automorphisms of G is *flat* if there is a compact open subgroup, $U \leq G$, that is tidy for every $\alpha \in \mathcal{H}$. The stabilizer of U in $\mathcal{H}$ is called the *uniscalar subgroup* and denoted $\mathcal{H}_u$. The factoring of subgroups tidy for a single automorphism in Theorem 4.1 extends to flat groups as follows.

Theorem 4.4 ([87]). *Let $\mathcal{H}$ be a finitely generated flat group of automorphisms of G and suppose that U is tidy for $\mathcal{H}$. Then $\mathcal{H}_u \lhd \mathcal{H}$ and there is $r \geq 0$ such that*

$$\mathcal{H}/\mathcal{H}_u \cong \mathbb{Z}^r.$$

- *There are $q \geq 0$ and closed groups $U_j \leq U$, $j \in \{0, 1, \ldots, q\}$ such that $\alpha(U_0) = U_0$; $\alpha(U_j)$ is either a subgroup or supergroup of U_j for every $j \in \{1, \ldots, q\}$; and $U = U_0 U_1 \cdots U_q$.*
- *$\tilde{U}_j := \bigcup_{\alpha \in \mathcal{H}} \alpha(U_j)$ is a closed subgroup of G for each $j \in \{1, \ldots, q\}$.*
- *There are, for each $j \in \{1, \ldots, q\}$, an integer $s_j > 1$ and a surjective homomorphism $\rho_j : \mathcal{H} \to (\mathbb{Z}, +)$ such that $\Delta(\alpha|_{\tilde{U}_j}) = s_j^{\rho_j(\alpha)}$.*
- *The integers r and q, and integers s_j and homomorphisms ρ_j for each $j \in \{1, \ldots, q\}$, are independent of the subgroup U tidy for $\mathcal{H}$.*

The number r in Theorem 4.4 is the *flat rank* of $\mathcal{H}$. The singly-generated group $\langle \alpha \rangle$ has flat rank equal to 0 if α is uniscalar and 1 if not. Flat groups of automorphisms with rank at least 1 correspond to Cartan subgroups in Lie groups over local fields and may be interpreted geometrically in terms of apartments in isometry groups of buildings [8].

More generally, flatness of groups of automorphisms may be shown by the following converse to the fact that flat groups are abelian modulo the stablizer of tidy subgroups.

Theorem 4.5 ([87][71]). *Every finitely generated nilpotent subgroup of* Aut(G) *is flat, and every polycyclic subgroup is virtually flat.*

Flatness is used—in combination with bounded generation of arithmetic groups [57,74], the fact that almost normal subgroups are close to normal [11], and the Margulis normal subgroup theorem [53]—to prove the Margulis–Zimmer conjecture in the special case of Chevalley groups in [71] and show that there are no commensurated subgroups of arithmetic subgroups other than the natural ones.

5. FUTURE DIRECTIONS

The contributions to the structure theory of tdlc groups surveyed in this article highlight that, for a general tdlc group G, as soon as the topology is nondiscrete, its interaction with the group structure yields significant algebraic constraints. As mentioned in the introduction, we view the dynamics of the conjugation action as a unifying theme of our considerations. The results we have surveyed reveal that those dynamics tend to be richer than one might expect. This is especially the case among tdlc groups that are nonelementary. We hope that further advances will shed more light on this paradigm in the future.

Concerning decomposition theory, it is an important open problem to clarify what distinguishes elementary and nonelementary tdlc groups. A key question asks whether every nonelementary tdlcsc group G contains a closed subgroup H admitting a quotient in $\mathscr{S}_{\mathrm{td}}$. Concerning simple groups, our results yield a dichotomy, depending on whether the centralizer lattice is trivial or not. The huge majority of known examples of groups in $\mathscr{S}_{\mathrm{td}}$ (listed in **[28, APPENDIX A]**) have a nontrivial centralizer lattice, the most notable exceptions being the simple algebraic groups over local fields. Finding new groups in $\mathscr{S}_{\mathrm{td}}$ with a trivial centralizer lattice would be a decisive step forward. A fundamental source of examples of tdlc groups is provided by Galois groups of transcendental field extensions with finite transcendence degree (see **[66, TH. 2.9]**, highlighting the occurrence of topologically simple groups), but this territory remains largely unexplored from the viewpoint of structure theory of tdlc groups. Concerning scale methods, the structure of tdlc groups all of whose elements are *uniscalar* (i.e., have scale 1) is still mysterious. In particular, we do not know whether every such group is elementary. This is equivalent to asking whether a tdlc group, all of whose closed subgroups are unimodular, is necessarily elementary. A positive answer would provide a formal incarnation to the claim that the dynamics of the conjugation action is nontrivial for all nonelementary tdlc groups. We refer to **[24]** for a more extensive list of specific problems.

We believe that a good measurement of the maturity of a mathematical theory is provided by its ability to solve problems arising on the outside of the theory. For the structure theory of tdlc groups, the Margulis–Zimmer conjecture appears as a natural target. As mentioned in Section 4, partial results in the nonuniform case, relying on scale methods on tdlc groups, have already been obtained in **[71]**.

Another source of external problems is provided by abstract harmonic analysis. As mentioned in the introduction, the emergence of locally compact groups as an independent subject of study coincides with the foundation of abstract harmonic analysis. However, fundamental problems clarifying the links between the algebraic structure of a locally compact group and the properties of its unitary representations remain open. The class of amenable locally compact groups is defined by a representation theoretic property (indeed, a locally compact group is amenable if and only if every unitary representation is weakly contained in the regular), but purely algebraic characterizations of amenable groups are still missing. In particular, the following nondiscrete version of Day's problem is open and intriguing: *Is every amenable second countable tdlc group elementary (in the sense of Section 2)?* The unitary representation theory also reveals a fundamental dichotomy between locally compact groups of *type I* (roughly speaking, those for which the problem of classifying the irreducible unitary representations up to equivalence is tractable) and the others (see **[10, 35, 52]**). Algebraic characterizations of type I groups are also desirable. In particular, we underline the following question: *Does every second countable locally compact group of type I contain a cocompact amenable subgroup?* For a more detailed discussion of that problem and related results, we refer to **[20]**.

ACKNOWLEDGMENTS

It is a pleasure to take this opportunity to heartily thank our collaborators, past and present, for their ideas and companionship. We are grateful to Adrien Le Boudec and Colin Reid for their comments on a preliminary version of this article.

FUNDING

This work was written while P.-E. Caprace was a Senior Research Associate of the Belgian F.R.S.-FNRS and G. A. Willis was Australian Research Council Laureate Professor under the grant FL170100032. The support of these bodies is gratefully acknowledged.

REFERENCES

[1] H. Abels, Specker-Kompaktifizierungen von lokal kompakten topologischen Gruppen. *Math. Z.* **135** (1973/1974), 325–361.

[2] N. Aoki, Dense orbits of automorphisms and compactness of groups. *Topology Appl.* **20** (1985), no. 1, 1–15.

[3] U. Bader and A. Furman, Boundaries, Weyl groups, and superrigidity. *Electron. Res. Announc. Math. Sci.* **19** (2012), 41–48.

[4] U. Bader and A. Furman, Boundaries, rigidity of representations, and Lyapunov exponents. In *Proceedings of the International Congress of Mathematicians—Seoul 2014. Vol. III*, pp. 71–96, Kyung Moon Sa, Seoul, 2014.

[5] U. Bader and Y. Shalom, Factor and normal subgroup theorems for lattices in products of groups. *Invent. Math.* **163** (2006), no. 2, 415–454.

[6] H. Bass and R. Kulkarni, Uniform tree lattices. *J. Amer. Math. Soc.* **3** (1990), no. 4, 843–902.

[7] U. Baumgartner, J. Ramagge, and B. Rémy, Contraction groups in complete Kac–Moody groups. *Groups Geom. Dyn.* **2** (2008), no. 3, 337–352.

[8] U. Baumgartner, B. Rémy, and G. A. Willis, Flat rank of automorphism groups of buildings. *Transform. Groups* **12** (2007), no. 3, 413–436.

[9] U. Baumgartner and G. A. Willis, Contraction groups and scales of automorphisms of totally disconnected locally compact groups. *Israel J. Math.* **142** (2004), 221–248.

[10] B. Bekka and P. de la Harpe, *Unitary representations of groups, duals, and characters*. Math. Surveys Monogr. 250, American Mathematical Society, Providence, RI, 2020.

[11] G. M. Bergman and H. W. Lenstra Jr., Subgroups close to normal subgroups. *J. Algebra* **127** (1989), no. 1, 80–97.

[12] I. N. Bernšteĭn, All reductive $\mathfrak{p}$-adic groups are of type I. *Funktsional. Anal. i Prilozhen.* **8** (1974), no. 2, 3–6.

[13] I. N. Bernšteĭn and D. A. Każdan, The one-dimensional cohomology of discrete subgroups. *Funktsional. Anal. i Prilozhen.* **4** (1970), no. 1, 1–5.

[14] M. Burger and S. Mozes, Groups acting on trees: from local to global structure. *Publ. Math. Inst. Hautes Études Sci.* **92** (2000), 113–150.

[15] M. Burger and S. Mozes, Lattices in product of trees. *Publ. Math. Inst. Hautes Études Sci.* **92** (2000), 151–194.

[16] T. P. Bywaters, H. Glöckner, and S. Tornier, Contraction groups and passage to subgroups and quotients for endomorphisms of totally disconnected locally compact groups. *Israel J. Math.* **227** (2018), 691–752.

[17] P.-E. Caprace, Non-discrete simple locally compact groups. In *European Congress of Mathematics*, pp. 333–354, Eur. Math. Soc., Zürich, 2018.

[18] P.-E. Caprace, Y. Cornulier, N. Monod, and R. Tessera, Amenable hyperbolic groups. *J. Eur. Math. Soc. (JEMS)* **17** (2015), no. 11, 2903–2947.

[19] P.-E. Caprace and T. De, Medts, Simple locally compact groups acting on trees and their germs of automorphisms. *Transform. Groups* **16** (2011), no. 2, 375–411.

[20] P.-E. Caprace, M. Kalantar, and N. Monod, A type I conjecture and boundary representations of hyperbolic groups. 2021, arXiv:2110.00190.

[21] P.-E. Caprace and A. Le Boudec, Bounding the covolume of lattices in products. *Compos. Math.* **155** (2019), no. 12, 2296–2333.

[22] P.-E. Caprace, A. Le Boudec, and D. Francoeur, Commensurated subgroups and micro-supported actions. 2020, arXiv:2002.02910. To appear in *J. Eur. Math. Soc. (JEMS)*.

[23] P.-E. Caprace and N. Monod, Decomposing locally compact groups into simple pieces. *Math. Proc. Cambridge Philos. Soc.* **150** (2011), no. 1, 97–128.

[24] P.-E. Caprace and N. Monod, Future directions in locally compact groups: a tentative problem list. In *New directions in locally compact groups*, pp. 343–355, London Math. Soc. Lecture Note Ser. 447, Cambridge Univ. Press, Cambridge, 2018.

[25] P.-E. Caprace and N. Radu, Chabauty limits of simple groups acting on trees. *J. Inst. Math. Jussieu* **19** (2020), no. 4, 1093–1120.

[26] P.-E. Caprace, C. D. Reid, and G. A. Willis, Limits of contraction groups and the Tits core. *J. Lie Theory* **24** (2014), 957–967.

[27] P.-E. Caprace, C. D. Reid, and G. A. Willis, Locally normal subgroups of totally disconnected groups. Part I: general theory. *Forum Math. Sigma* **5** (2017), e11, 76.

[28] P.-E. Caprace, C. D. Reid, and G. A. Willis, Locally normal subgroups of totally disconnected groups. Part II: compactly generated simple groups. *Forum Math. Sigma* **5** (2017), e12, 89.

[29] P.-E. Caprace, C. Reid, and P. Wesolek, Approximating simple locally compact groups by their dense locally compact subgroups. *Int. Math. Res. Not. IMRN* **7** (2021), 5037–5110.

[30] P.-E. Caprace and T. Stulemeijer, Totally disconnected locally compact groups with a linear open subgroup. *Int. Math. Res. Not. IMRN* **24** (2015), 13800–13829.

[31] I. Chatterji, T. Fernós, and A. Iozzi, The median class and superrigidity of actions on CAT(0) cube complexes. *J. Topol.* **9** (2016), no. 2, 349–400.

[32] Y. Cornulier and P. de la Harpe, *Metric geometry of locally compact groups*. EMS Tracts Math. 25, European Mathematical Society (EMS), Zürich, 2016.

[33] M. M. Day, Amenable semigroups. *Illinois J. Math.* **1** (1957), 509–544.

[34] Y. de Cornulier, *Groupes pleins-topologiques (d'après Matui, Juschenko, Monod, ...)*. Astérisque **361** (2014), no. 1064, viii, 183–223.

[35] J. Dixmier, *Les C^*-algèbres et leurs représentations. Les Grands Classiques Gauthier-Villars*. [Gauthier-Villars Great Classics], Éditions Jacques Gabay, Paris, 1996.

[36] E. Fioravanti, Superrigidity of actions on finite rank median spaces. *Adv. Math.* **352** (2019), 1206–1252.

[37] S. Gao, *Invariant descriptive set theory*. Pure Appl. Math. (Boca Raton) 293, CRC Press, Boca Raton, FL, 2009.

[38] T. Gelander, A. Karlsson, and G. A. Margulis, Superrigidity, generalized harmonic maps and uniformly convex spaces. *Geom. Funct. Anal.* **17** (2008), no. 5, 1524–1550.

[39] T. Gelander and A. Levit, Local rigidity of uniform lattices. *Comment. Math. Helv.* **93** (2018), no. 4, 781–827.

[40] H. Glöckner and G. A. Willis, Classification of the simple factors appearing in composition series of totally disconnected contraction groups. *J. Reine Angew. Math.* **643** (2010), 141–169.

[41] H. Glöckner and G. A. Willis, Decompositions of locally compact contraction groups, series and extensions. *J. Algebra* **570** (2021), 164–214.

[42] H. Glöckner and G. A. Willis, Locally pro-p contraction groups are nilpotent. *J. Reine Angew. Math.* (2021). DOI 10.1515/crelle-2021-0050.

[43] R. Guralnick, Applications of the classification of finite simple groups. In *Proceedings of the International Congress of Mathematicians—Seoul 2014. Vol. II*, pp. 163–177, Kyung Moon Sa, Seoul, 2014.

[44] P. R. Halmos, *Lectures on ergodic theory*. Publ. Math. Soc Jpn. 3, The Mathematical Society of Japan, 1956.

[45] K. H. Hofmann and A. Mukherjea, Concentration functions and a class of noncompact groups. *Math. Ann.* **256** (1981), no. 4, 535–548.

[46] W. Jaworski, On contraction groups of automorphisms of totally disconnected locally compact groups. *Israel J. Math.* **172** (2009), 1–8.

[47] W. Jaworski, J. Rosenblatt, and G. A. Willis, Concentration functions in locally compact groups. *Math. Ann.* **305** (1996), no. 4, 673–691.

[48] R. Kaufman and M. Rajagopalan, On automorphisms of a locally compact group. *Michigan Math. J.* **13** (1966), 373–374.

[49] V. Lafforgue, Shtukas for reductive groups and Langlands correspondence for function fields. In *Proceedings of the International Congress of Mathematicians–Rio de Janeiro 2018. I. Plenary lectures*, pp. 635–668, World Sci. Publ., Hackensack, NJ, 2018.

[50] A. Le Boudec, Groups acting on trees with almost prescribed local action. *Comment. Math. Helv.* **91** (2016), no. 2, 253–293.

[51] G. W. Mackey, On induced representations of groups. *Amer. J. Math.* **73** (1951), 576–592.

[52] G. W. Mackey, *The theory of unitary group representations*. Chicago Lectures in Math., University of Chicago Press, Chicago, IL–London, 1976.

[53] G. A. Margulis, *Discrete subgroups of semisimple Lie groups*. Ergeb. Math. Grenzgeb. (3) [Res. Math. Related Areas (3)] 17, Springer, Berlin, 1991.

[54] R. G. Möller, Structure theory of totally disconnected locally compact groups via graphs and permutations. *Canad. J. Math.* **54** (2002), 795–827.

[55] N. Monod, Superrigidity for irreducible lattices and geometric splitting. *J. Amer. Math. Soc.* **19** (2006), no. 4, 781–814.

[56] N. Monod and Y. Shalom, Cocycle superrigidity and bounded cohomology for negatively curved spaces. *J. Differential Geom.* **67** (2004), no. 3, 395–455.

[57] D. W. Morris, Bounded generation of $SL(n, A)$. *New York J. Math.* **13** (2007), 383–421.

[58] V. Nekrashevych, *Self-similar groups*. Math. Surveys Monogr. 117, American Mathematical Society, Providence, RI, 2005.

[59] V. Nekrashevych, Simple groups of dynamical origin. *Ergodic Theory Dynam. Systems* **39** (2019), no. 3, 707–732.

[60] W. H. Previts and T. S. Wu, Dense orbits and compactness of groups. *Bull. Aust. Math. Soc.* **68** (2003), no. 1, 155–159.

[61] N. Radu, A classification theorem for boundary 2-transitive automorphism groups of trees. *Invent. Math.* **209** (2017), no. 1, 1–60.

[62] C. D. Reid, Normal subgroup structure of totally disconnected locally compact groups. In *2016 MATRIX Annals*, pp. 525–560, MATRIX Book Ser. 1, Springer, Cham, 2018.

[63] C. D. Reid and P. R. Wesolek, Dense normal subgroups and chief factors in locally compact groups. *Proc. Lond. Math. Soc. (3)* **116** (2018), no. 4, 760–812.

[64] C. D. Reid and P. R. Wesolek, The essentially chief series of a compactly generated locally compact group. *Math. Ann.* **370** (2018), no. 1–2, 841–861.

[65] C. D. Reid and P. R. Wesolek, Homomorphisms into totally disconnected, locally compact groups with dense image. *Forum Math.* **31** (2019), no. 3, 685–701.

[66] M. Rovinsky, Motives and admissible representations of automorphism groups of fields. *Math. Z.* **249** (2005), no. 1, 163–221.

[67] M. Rubin, Locally moving groups and reconstruction problems. In *Ordered groups and infinite permutation groups*, pp. 121–157, Math. Appl. 354, Kluwer Acad. Publ., Dordrecht, 1996.

[68] G. Schlichting, Operationen mit periodischen Stabilisatoren. *Arch. Math. (Basel)* **34** (1980), no. 2, 97–99.

[69] P. Scholze, p-adic geometry. In *Proceedings of the International Congress of Mathematicians–Rio de Janeiro 2018. I. Plenary lectures*, pp. 899–933, World Sci. Publ., Hackensack, NJ, 2018.

[70] Y. Shalom, Rigidity of commensurators and irreducible lattices. *Invent. Math.* **141** (2000), no. 1, 1–54.

[71] Y. Shalom and G. A. Willis, Commensurated subgroups of arithmetic groups, totally disconnected groups and adelic rigidity. *Geom. Funct. Anal.* **23** (2013), no. 5, 1631–1683.

[72] S. M. Smith, A product for permutation groups and topological groups. *Duke Math. J.* **166** (2017), no. 15, 2965–2999.

[73] T. Tao, *Hilbert's fifth problem and related topics*. Grad. Stud. Math. 153, American Mathematical Society, Providence, RI, 2014.

[74] O. I. Tavgen, Bounded generability of Chevalley groups over rings of s-integer algebraic numbers. *Izv. Akad. Nauk SSSR Ser. Mat.* **54** (1990), no. 1, 97–122, 221–222.

[75] J. Tits, Algebraic and abstract simple groups. *Ann. of Math. (2)* **80** (1964), 313–329.

[76] J. Tits, Sur le groupe des automorphismes d'un arbre. In *Essays on topology and related topics (Mémoires dédiés à Georges de Rham)*, pp. 188–211, Springer, New York, 1970.

[77] D. Van Dantzig, Zur topologischen Algebra. III. Brouwersche und Cantorsche Gruppen. *Compos. Math.* **3** (1936), 408–426.

[78] H.-C. Wang, On a maximality property of discrete subgroups with fundamental domain of finite measure. *Amer. J. Math.* **89** (1967), 124–132.

[79] H.-C. Wang, In *Topics on totally discontinuous groups*, pp. 459–487, Pure Appl. Math. 8, Dekker, New York, 1972.

[80] J. S. P. Wang, The Mautner phenomenon for p-adic Lie groups. *Math. Z.* **185** (1984), no. 3, 403–412.

[81] A. Weil, *L'intégration dans les groupes topologiques et ses applications*. Actual. Sci. Ind. [Curr. Sci. Ind. Top.] 869, Hermann et Cie, Paris, 1940.

[82] P. Wesolek, Elementary totally disconnected locally compact groups. *Proc. Lond. Math. Soc. (3)* **110** (2015), no. 6, 1387–1434.

[83] P. Wesolek, A survey of elementary totally disconnected locally compact groups. In *2016 MATRIX Annals*, pp. 593–611, MATRIX Book Ser. 1, Springer, Cham, 2018.

[84] G. Willis, Totally disconnected groups and proofs of conjectures of Hofmann and Mukherjea. *Bull. Aust. Math. Soc.* **51** (1995), no. 3, 489–494.

[85] G. A. Willis, The structure of totally disconnected, locally compact groups. *Math. Ann.* **300** (1994), no. 2, 341–363.

[86] G. A. Willis, Further properties of the scale function on a totally disconnected group. *J. Algebra* **237** (2001), no. 1, 142–164.

[87] G. A. Willis, Tidy subgroups for commuting automorphisms of totally disconnected groups: an analogue of simultaneous triangularisation of matrices. *New York J. Math.* **10** (2004), 1–35.

[88] G. A. Willis, Compact open subgroups in simple totally disconnected groups. *J. Algebra* **312** (2007), no. 1, 405–417.

[89] G. A. Willis, The scale and tidy subgroups for endomorphisms of totally disconnected locally compact groups. *Math. Ann.* **361** (2015), no. 1–2, 403–442.

[90] T. S. Wu, Continuous automorphisms on locally compact groups. *Math. Z.* **96** (1967), 256–258.

PIERRE-EMMANUEL CAPRACE

UCLouvain – IRMP, 1348 Louvain-la-Neuve, Belgium, pierre-emmanuel.caprace@uclouvain.be

GEORGE A. WILLIS

University of Newcastle, Callaghan 2308, Australia, george.willis@newcastle.edu.au

This Page Intentionally Left Blank

THE ZARISKI CANCELLATION PROBLEM AND RELATED PROBLEMS IN AFFINE ALGEBRAIC GEOMETRY

NEENA GUPTA

ABSTRACT

In this article, we shall discuss the solution to the Zarsiki Cancellation Problem in positive characteristic, various approaches taken so far towards the possible solution in characteristic zero, and several other questions related to this problem.

MATHEMATICS SUBJECT CLASSIFICATION 2020

Primary 14R10; Secondary 14R20,14R25,13B25,13F20, 13N15

KEYWORDS

Polynomial ring, cancellation Problem, embedding problem, affine fibration problem, locally nilpotent derivations

INTERNATIONAL CONGRESS OF MATHEMATICIANS 2022 JULY 6–14

Proc. Int. Cong. Math. 2022, Vol. 3, pp. 1578–1598
DOI 10.4171/ICM2022/151

Published by EMS Press

1. INTRODUCTION

"Polynomials and power series
May they forever rule the world." —*Shreeram S. Abhyankar, 1970* [4]

Right from the beginning of the 19th century, mathematicians have been involved in studying polynomial rings (over $\mathbb{C}$ and over $\mathbb{R}$). Some of the early breakthroughs on polynomial rings have led to the foundation of Commutative Algebra. One such result is the Hilbert Basis Theorem, a landmark result on the finite generation of ideals, which solved a central problem on invariant theory. This was followed by the Hilbert Nullstellensatz which connects affine varieties (zero locus of a set of polynomials) with rings of regular functions on varieties and thus enables one to make use of the algebraic machinery of commutative algebra to study geometric properties of varieties.

Affine Algebraic Geometry deals with the study of affine spaces (and certain closed subspaces), equivalently, polynomial rings (and certain quotients). There are many fundamental problems on polynomial rings which can be formulated in an elementary mathematical language but whose solutions remain elusive. Any significant progress requires the development of new and powerful methods and their ingenious applications.

One of the most challenging problems in Affine Algebraic Geometry is the Zariski Cancelation Problem (ZCP) on polynomial rings (Question $1'$ below). In this article, we shall discuss the solution to the ZCP in positive characteristic, various approaches taken so far towards the possible solution in characteristic zero, and several other questions related to this problem. For a survey on problems in Affine Algebraic Geometry, one may look at [42, 62, 69].

Throughout the article, all rings will be assumed to be commutative with unity and k will denote a field. For a ring R, R^* will denote the group of units of R. We shall use the notation $R^{[n]}$ for a polynomial ring in n variables over a commutative ring R. Thus, $E = R^{[n]}$ will mean that $E = R[t_1, \dots, t_n]$ for some elements $t_1, \dots, t_n$ in E which are algebraically independent over R. Unless otherwise stated, capital letters like $X_1, X_2, \dots, X_n, Y_1, \dots, Y_m, X, Y, Z, T$ will be used as variables of polynomial rings.

2. CANCELLATION PROBLEM

Let A be an affine (finitely generated) algebra over a field k. The k-algebra A is said to be cancellative (over k) if, for any k-algebra B, $A[X] \cong_k B[X]$ implies that $A \cong_k B$. A natural question in this regard is: which affine domains are cancellative? More precisely:

Question 1. Let A be an affine algebra over a field k. Suppose that B is a k-algebra such that the polynomial rings $A[X]$ and $B[X]$ are isomorphic as k-algebras. Does it follow that $A \cong_k B$? In other words, is the k-algebra A cancellative?

A special case of Question 1, famously known as the Zariski Cancellation Problem, asks whether affine spaces are cancellative, i.e., whether any polynomial ring in n variables over a field k is cancellative. More precisely:

Question 1′. Suppose that B is an affine k algebra satisfying $B[X] \cong_k k[X_1, \ldots, X_{n+1}]$ for some positive integer n. Does it follow that $B \cong_k k[X_1, \ldots, X_n]$? In other words, is the polynomial ring $k[X_1, \ldots, X_n]$ cancellative?

Abhyankar, Eakin, and Heinzer have shown that any domain A of transcendence degree one over any field k is cancellative [3]. In fact, they showed that, for any UFD R, the polynomial ring $R[X]$ is cancellative over R. This was further generalized by Hamann to a ring R which either contains $\mathbb{Q}$ or is a seminormal domain [52].

In 1972, Hochster demonstrated the first counterexample to Question 1 [53]. His example, a four-dimensional ring over the field of real numbers $\mathbb{R}$, is based on the fact that the projective module defined by the tangent bundle over the real sphere with coordinate ring $S = \mathbb{R}[X, Y, Z]/(X^2 + Y^2 + Z^2 - 1)$ is stably free but not a free S-module.

One of the major breakthroughs in 1970s was the establishment of an affirmative answer to Question 1′ for the case $n = 2$. This was proved over a field of characteristic zero by Fujita, Miyanishi, and Sugie [43,70] and over perfect fields of arbitrary characteristic by Russell [74]. Later, it has been shown that even the hypothesis of perfect field can be dropped [20]. A simplified proof of the cancellation property of $k[X, Y]$ for an algebraically closed field k is given by Crachiola and Makar-Limanov in [22].

Around 1989, Danielewski [26] constructed explicit two-dimensional affine domains over the field of complex numbers $\mathbb{C}$ which are not cancellative over $\mathbb{C}$. New examples of noncancellative varieties over any field k have been studied in [9,32,49]. This addresses the Cancellation Problem, as formulated in Question 1, for all dimensions.

In [45] and [47], the author settled the Zariski Cancellation Problem (Question 1′) completely for affine spaces in positive characteristic. She has first shown in [45] that a certain threefold constructed by Asanuma is a counterexample to the ZCP in positive characteristic for the affine three space. Later in [46], she studied a general threefold of the form $x^m y = F(x, z, t)$, which includes the Asanuma threefold as well as the famous Russell cubic defined below. A major theorem of [46] is stated as Theorem 5.4 of this article. In [47], using a modification of the theory developed in [46], she constructed a family of examples which are counterexamples to the ZCP in positive characteristic in all dimensions greater than 2. The ZCP is still a challenging problem in characteristic zero. A few candidate counterexamples are discussed below.

The Russell cubic. Let $A = \mathbb{C}[X, Y, Z, T]/(X^2Y + X + Z^2 + T^3)$, $V = \operatorname{Spec} A$ and let x denote the image of X in A. The ring A, known as the Russell cubic, is one of the simplest examples of the Koras–Russell threefolds, a family of threefolds which arose in the context of the problem of determining whether there exist nonlinearizable $\mathbb{C}^*$-actions on $\mathbb{C}^3$. It was an exciting open problem for some time whether $A \cong \mathbb{C}^{[3]}$. It was first observed that the ring

A (respectively the variety V) has several properties in common with $\mathbb{C}^{[3]}$ (respectively $\mathbb{C}^3$), for instance,

(i) A is a regular UFD.

(ii) There exists an injective $\mathbb{C}$-algebra homomorphism from A to $\mathbb{C}^{[3]}$. Note that $\mathbb{C}^{[3]} \hookrightarrow A$.

(iii) The variety V is homeomorphic (in fact, diffeomorphic) to $\mathbb{R}^6$.

(iv) V has logarithmic Kodaira dimension $-\infty$.

These properties appeared to provide evidence in favor of the surmise that $A \cong \mathbb{C}^{[3]}$. The establishment of an isomorphism between A and $\mathbb{C}^{[3]}$ would have led to counterexamples to the "Linearization Conjecture" on $\mathbb{C}^3$ (stated in [58]) and the Abhyankar–Sathaye Conjecture for $n = 3$ (stated in Section 5 of the present article). Indeed, if A were isomorphic to $\mathbb{C}^{[3]}$, as was then suspected, it would have shown the existence of nonlinearizable $\mathbb{C}^*$-actions on $\mathbb{C}^3$. Moreover, note that

(v) $A/(x-\lambda) = \mathbb{C}^{[2]}$ for every $\lambda \in \mathbb{C}^*$.

(vi) $A/(x) \neq \mathbb{C}^{[2]}$.

Therefore, if A were isomorphic to $\mathbb{C}^{[3]}$, then property (vi) would show that $x-\lambda$ cannot be a coordinate in A for any λ and then, by property (v), it would have yielded a counterexample to the Abhyankar–Sathaye Conjecture for $n = 3$.

However, Makar-Limanov proved [65] that $A \neq \mathbb{C}^{[3]}$; for this result, he introduced a new invariant which distinguished between A and $\mathbb{C}^{[3]}$. This invariant, which he had named AK-invariant, is now named Makar-Limanov invariant and is denoted by ML. It is defined in Section 3. Makar-Limanov proved that

(vii) $\mathrm{ML}(A) = \mathbb{C}[x]$ (Makar-Limanov [65]).

However, the Makar-Limanov invariant of $\mathbb{C}^{[n]}$ is $\mathbb{C}$ for any integer $n \geq 1$. Thus $A \not\cong \mathbb{C}^{[3]}$. Subsequently, other Koras–Russell threefolds were shown to be not isomorphic to the polynomial ring. Eventually, Kaliman–Koras–Makar-Limanov–Russell proved that every $\mathbb{C}^*$-action on $\mathbb{C}^3$ is linearizable (cf. [55]).

Now for ZCP in characteristic zero, a crucial question, still open, is whether $A^{[1]} = \mathbb{C}^{[4]}$. Because if $A^{[1]} = \mathbb{C}^{[4]}$, then A would be a counterexample to the ZCP in characteristic zero for $n = 3$. In this context, the following results have been proved:

(viii) $\mathrm{ML}(A^{[1]}) = \mathbb{C}$ (Dubouloz [30]).

(ix) V is $\mathbb{A}^1$-contractible (Dubouloz–Fasel [31], also see [33, 54]).

Note that $A^{[1]} = \mathbb{C}^{[4]}$ would imply that $\mathrm{ML}(A^{[1]}) = \mathbb{C}$ and Dubouloz's result (viii) shows that the latter indeed holds. On the other hand, Asok had suggested a program for showing that the variety V is not $\mathbb{A}^1$-contractible and hence A is not a stably polynomial ring (see [54]). However, Hoyois, Krishna, and Østvær have proved [54] that a step in his program does not hold for V. They had further shown that V is stably $\mathbb{A}^1$-contractible. In a remarkable paper [31], Dubouloz and Fasel have established that V is in fact $\mathbb{A}^1$-contractible, which seems to provide further evidence in favor of $A^{[1]} = \mathbb{C}^{[4]}$. The variety V is in fact the first example of an $\mathbb{A}^1$-contractible threefold which is not algebraically isomorphic to $\mathbb{C}^3$.

Nonrectifiable epimorphisms and Asanuma's rings. Let $m \leq n$ be two integers. A k-algebra epimorphism $\phi : k[X_1, \ldots, X_n] \twoheadrightarrow k[Y_1, \ldots, Y_m]$ is said to be *rectifiable* if there exists a k-algebra automorphism ψ of $k[X_1, \ldots, X_n]$ such that $\phi \circ \psi(X_i) = Y_i$ for $1 \leq i \leq m$ and $\phi \circ \psi(X_j) = 0$ for $m + 1 \leq j \leq n$. Equivalently, over an algebraically closed field k, a k-embedding $\Phi : \mathbb{A}_k^m \hookrightarrow \mathbb{A}_k^n$ is said to be *rectifiable* if there exists an automorphism Ψ of $\mathbb{A}_k^n$ such that $\Psi \circ \Phi$ is the canonical embedding mapping $(y_1, \ldots, y_m) \to (y_1, \ldots, y_m, 0, \ldots, 0)$.

A famous theorem of Abhyankar–Moh and Suzuki proves that any epimorphism $\phi : k[X, Y] \to k[T]$ is rectifiable in characteristic zero [5, 86]. On the other hand, in positive characteristic, there exist nonrectifiable epimorphisms from $k[X, Y]$ to $k[T]$ (see Segre [83], Nagata [71]). It is an open problem whether there exist nonrectifiable epimorphisms over the field of complex numbers (see [38]).

Asanuma has described an explicit method for constructing affine rings which are stably polynomial rings, by making use of nonrectifiable epimorphisms ([7], also see [38, PROPOSITION 3.7]). Such rings are considered to be potential candidates for counterexamples to the ZCP. For instance, when k is of positive characteristic, nonrectifiable epimorphisms from $k[X, Y]$ to $k[T]$ yield counterexamples to the ZCP.

Let $\phi : \mathbb{R}[X, Y, Z] \to \mathbb{R}[T]$ be defined by

$$\phi(X) = T^3 - 3T, \quad \phi(Y) = T^4 - 4T^2, \quad \phi(Z) = T^5 - 10T.$$

Shastri constructed the above epimorphism ϕ and proved that it defines a nonrectifiable (polynomial) embedding of the trefoil knot in $\mathbb{A}_{\mathbb{R}}^3$ [84]. Using a result of Serre [63, THEOREM 1, P. 281], one knows that $\ker(\phi) = (f, g)$ for some $f, g \in k[X, Y, Z]$. Using f and g, Asanuma constructed the ring $B = \mathbb{R}[T][X, Y, Z, U, V]/(T^d U - f, T^d V - g)$ and proved that $B^{[1]} = \mathbb{R}[T]^{[4]} = \mathbb{R}^{[5]}$ (cf. [7, COROLLARY 4.2]). He asked [7, REMARK 7.8]:

Question 2. Is $B = \mathbb{R}^{[4]}$?

The interesting aspect of the question is that once the problem gets solved, irrespective of whether the answer is "Yes" or "No," that is, either way, one would have solved a major problem in Affine Algebraic Geometry. Indeed:

If $B = \mathbb{R}^{[4]}$, then there exist nonlinearizable $\mathbb{R}^*$-actions on the affine four-space $\mathbb{A}_{\mathbb{R}}^4$.

If $B \neq \mathbb{R}^{[4]}$, then clearly B is a counterexample to the ZCP!!

3. CHARACTERIZATION PROBLEM

The Characterization Problem in affine algebraic geometry seeks a "useful characterization" of the polynomial ring or, equivalently (when the ground field is algebraically closed), an affine n-space. For instance, the following two results give respectively an algebraic and a topological characterization of $k^{[1]}$ (or $\mathbb{A}^1_{\mathbb{C}}$).

Theorem 3.1. *Let k be an algebraically closed field of characteristic zero. Then the polynomial ring $k^{[1]}$ is the only one-dimensional affine UFD with $A^* = k^*$.*

Theorem 3.2. *Let k be the field of complex numbers $\mathbb{C}$. Then the affine line $\mathbb{A}^1_{\mathbb{C}}$ is the only acyclic normal curve.*

While the Characterization Problem is one of the most important problems in affine algebraic geometry in its own right, it is also closely related to some of the challenging open problems on the affine space like the "Cancellation Problem." For instance, each of the above characterizations of $k^{[1]}$ immediately solves the Cancellation Problem in dimension one: $A^{[1]} = k^{[2]} \implies A = k^{[1]}$. The complexity of the Characterization Problem increases with the dimension of the rings.

In his attempt to solve the Cancellation Problem for the affine plane, Ramanujam obtained a remarkable topological characterization of the affine plane $\mathbb{C}^2$ in 1971 [72]. He proved that

Theorem 3.3. *$\mathbb{C}^2$ is the only contractible smooth surface which is simply connected at infinity.*

Ramanujam also constructed contractible surfaces which are not isomorphic to $\mathbb{C}^2$. Soon, in 1975, Miyanishi [67] obtained an algebraic characterization of the polynomial ring $k^{[2]}$. He proved that

Theorem 3.4. *Let k be an algebraically closed field of characteristic zero and A be a two-dimensional affine factorial domain over k. Then $A = k^{[2]}$ if and only if it satisfies the following:*

(i) $A^* = k^*$.

(ii) *There exists an element $f \in A$ and a subring B of A such that $A[f^{-1}] = B[f^{-1}]^{[1]}$.*

This algebraic characterization was used by Fujita, Miyanishi, and Sugie [43,70] to solve the Cancellation Problem for $k[X, Y]$. In 2002 [50], using methods of Mumford and Ramanujam, Gurjar gave a topological proof of the cancellation property of $\mathbb{C}[X, Y]$.

Remarkable characterizations of the affine three space were obtained by Miyanishi [68] and Kaliman [56] (also see [69] for a beautiful survey). We state below the version of Kaliman.

Theorem 3.5. *Let A be a three-dimensional smooth factorial affine domain over the field of complex numbers $\mathbb{C}$. Let $X = \operatorname{Spec} A$. Then $A = \mathbb{C}^{[3]}$ if and only if it satisfies the following:*

(i) $A^* = \mathbb{C}^*$.

(ii) *$H_3(X, \mathbb{Z}) = 0$, or X is contractible.*

(iii) *X contains a cylinder-like open set V such that* $V \cong U \times \mathbb{A}^2$ *for some curve U and each irreducible component of the complement* $X \setminus V$ *has at most isolated singularities.*

When $A^{[1]} = \mathbb{C}^{[4]}$, it is easy to see that A possesses properties (i) and (ii) of Theorem 3.5. Thus, by Theorem 3.5, the ZCP for $\mathbb{C}^{[3]}$ reduces to examining whether condition (iii) necessarily holds for a $\mathbb{C}$-algebra A satisfying $A^{[1]} = \mathbb{C}^{[4]}$.

In [29], we have obtained another characterization of the affine three-space using certain invariants of an affine domain defined by locally nilpotent derivations. We state it below.

Locally nilpotent derivations and a characterization of $\mathbb{C}^{[3]}$. Let B be an affine domain over a field k of characteristic zero. A k-linear derivation D on B is said to be a locally nilpotent derivation if, for any $a \in B$ there exists an integer n (depending on a) satisfying $D^n(a) = 0$. Let $\mathrm{LND}(B)$ denote the set of all locally nilpotent k-derivations of B and let

$$\mathrm{LND}^*(B) = \{D \in \mathrm{LND}(B) \mid Ds = 1 \text{ for some } s \in B\}.$$

Then we define

$$\mathrm{ML}(B) := \bigcap_{D \in \mathrm{LND}(B)} \ker D \quad \text{and} \quad \mathrm{ML}^*(B) := \bigcap_{D \in \mathrm{LND}^*(B)} \ker D.$$

The above $\mathrm{ML}(B)$, introduced by Makar-Limanov, is now called the Makar-Limanov invariant of B; $\mathrm{ML}^*(B)$ was introduced by Freudenburg in [41, p. 237]. We call it the Makar-Limanov–Freudenburg invariant or ML-F invariant. If $\mathrm{LND}^*(B) = \emptyset$, we define $\mathrm{ML}^*(B)$ to be B. We have obtained the following theorem [29, Theorem 4.6].

Theorem 3.6. *Let A be a three-dimensional affine factorial domain over an algebraically closed field k of characteristic zero. Then the following are equivalent:*

(I) $A = k^{[3]}$.

(II) $\mathrm{ML}^*(A) = k$.

(III) $\mathrm{ML}(A) = k$ *and* $\mathrm{ML}^*(A) \neq A$.

A similar result has also been proved in dimension two under weaker hypotheses [29, Theorem 3.8]. The above characterization of the affine three-space does not extend to higher dimensions [29, Example 5.6]. So far, no suitable characterization of the affine n-space for $n \geq 4$ is known to the author.

4. AFFINE FIBRATIONS

Let R be a commutative ring. A fundamental theorem of Bass–Connell–Wright and Suslin [10, 85] on the structure of locally polynomial algebras states that:

Theorem 4.1. *Let A be a finitely presented algebra over a ring R. Suppose that for each maximal ideal $\mathfrak{m}$ of R, $A_{\mathfrak{m}} = R_{\mathfrak{m}}^{[n]}$ for some integer $n \geq 0$. Then $A \cong \mathrm{Sym}_R(P)$ for some finitely generated projective R-module P of rank n.*

Now for a prime ideal P of R, let $k(P)$ denote the residue field R_P/PR_P. The area of affine fibrations seeks to derive information about the structure and properties of an R-algebra A from the information about the fiber rings $A \otimes_R k(P)(= A_P/PA_P)$ of A at the points P of the prime spectrum of R, i.e., at the prime ideals P of R.

An R-algebra A is said to be an $\mathbb{A}^n$-fibration over R if A is a finitely generated flat R-algebra and for each prime ideal P of R, $A \otimes_R k(P) = k(P)^{[n]}$.

The most important problem on $\mathbb{A}^n$-fibrations, due to Veĭsfeĭler and Dolgačev [87], can be formulated as follows:

Question 3. Let R be a Noetherian domain of dimension d and A be an $\mathbb{A}^n$-fibration over R.

(i) If R is regular, is $A \cong \mathrm{Sym}_R(Q)$ for some projective module Q over R? (In particular, if R is regular local, is then $A = R^{[n]}$?)

(ii) In general, what can one say about the structure of A?

Question 3 is considered a hard problem. When $n = 1$, it has an affirmative answer for all d. This has been established in the works of Kambayashi, Miyanishi, and Wright [59, 60]. Their results were further refined by Dutta who showed that it is enough to assume the fiber conditions only on generic and codimension-one fibers ([34]; also see [14, 17, 40]).

In case $n = 2$, $d = 1$, and R contains the field of rational numbers, an important theorem of Sathaye [81] gives an affirmative answer to Question 3 (i). To prove this theorem, Sathaye first generalized the Abhyankar–Moh expansion techniques originally developed over $k[[x]]$ to $k[[x_1, \dots, x_n]]$ [80]. The expansion techniques were used by Abhyankar–Moh to prove their famous epimorphism theorem. The generalized expansion techniques were further developed by Sathaye [82] to prove a conjecture of Daigle and Freudenburg. The result was a crucial step in Daigle–Freudenburg's theorem that the kernel of any triangular derivation of $k[X_1, X_2, X_3, X_4]$ is a finitely generated k-algebra [23].

When the residue field of R is of positive characteristic, Asanuma has shown in [6, THEOREM 5.1] that Question 3 (i) has a negative answer for $n = 2$, $d = 1$, and the author has generalized Asanuma's ring [47] to give a negative answer to Question 3 (i) for $n = 2$ and any $d > 1$ (also see [48]). In Theorem 5.4, the author proved that in a special situation $\mathbb{A}^2$-fibration is indeed trivial.

However, if $n = 2$, $d = 2$, and R contains the field of rational numbers, Question 3 (i) is an open problem. A candidate counterexample is discussed in Section 7.

In the context of Question 3 (ii), a deep work of Asanuma [6] provides a stable structure theorem for A. As a consequence of Asanuma's structure theorem, it follows that if R is regular local, then there exists an integer $m \geq 0$ such that $A^{[m]} = R^{[m+n]}$. Thus it is very tempting to look for possible counterexamples to the affine fibration problem in order to obtain possible counterexamples to the ZCP in characteristic zero. One can see [12, 24, 36, 37] and [38, SECTION 3.1] for more results on affine fibrations.

So far we have considered affine fibrations where the fibre rings are polynomial rings. Bhatwadekar and Dutta have obtained some nice results on rings whose fiber rings are of the form $k[X, 1/X]$ [15, 16]. Later Bhatwadekar, the author, and A. Abhyankar studied rings whose fiber rings are Laurent polynomial algebras or rings of the form $k[X, 1/f(X)]$, or of the form $k[X, Y, 1/(aX + b), 1/(cY + d)]$ for some $a, b, c, d \in k$ [1, 2, 18, 19, 44]. One of the results of Bhatwadekar and the author provides a Laurent polynomial analogue of Theorem 4.1 and the affine fibration problem Question 3. More generally, we have [19, THEOREMS A AND C]:

Theorem 4.2. *Let R be a Noetherian normal domain with field of fractions K and A be a faithfully flat R-algebra such that*

(i) $A \otimes_R K \cong K[X_1, \frac{1}{X_1}, \ldots, X_n, \frac{1}{X_n}]$,

(ii) *for each height-one prime ideal P of R,* $A \otimes_R k(P) \cong k(P)[X_1, \frac{1}{X_1}, \ldots, X_n, \frac{1}{X_n}]$.

Then A is a locally Laurent polynomial algebra in n variables over R, i.e.,

$$A_m = R_m[X_1, \frac{1}{X_1}, \ldots, X_n, \frac{1}{X_n}]$$

and is of the form $B[I^{-1}]$, where B is the symmetric algebra of a projective R-module Q of rank n, Q is a direct sum of finitely generated projective R-modules of rank one, and I is an invertible ideal of B.

5. EPIMORPHISM PROBLEM

The Epimorphism Problem for hypersurfaces asks the following fundamental question:

Question 4. Let k be a field and $f \in B = k^{[n]}$ for some integer $n \geq 2$. Suppose

$$B/(f) \cong k^{[n-1]}.$$

Does this imply that $B = k[f]^{[n-1]}$, i.e., is f a coordinate in B?

This problem is generally known as the *Epimorphism Problem*. It is an open problem and is regarded as one of the most challenging and celebrated problems in the area of affine algebraic geometry (see [38, 69, 75, 77] for useful surveys).

The first major breakthrough on Question 4 was achieved during 1974–1975, independently, by Abhyankar–Moh and Suzuki [5, 86]. They showed that Question 4 has an affirmative answer when k is a field of characteristic zero and $n = 2$. Over a field of positive characteristic, explicit examples of nonrectifiable epimorphisms from $k[X, Y]$ to $k[T]$ (referred to in Section 2) and hence explicit examples of nontrivial lines had already been demonstrated by Segre [83] in 1957 and Nagata [71] in 1971. However, over a field of characteristic zero, we have the following conjecture:

Abhyankar–Sathaye Conjecture. *Let k be a field of characteristic zero and $f \in B = k^{[n]}$ for some integer $n \geq 2$. Suppose that $B/(f) \cong k^{[n-1]}$. Then $B = k[f]^{[n-1]}$.*

In case $n = 3$, some special cases have been solved by Sathaye, Russell, and Wright [73,76,79,89]. In [79], Sathaye proved the conjecture for the linear planes, i.e., polynomials F of the form $aZ - b$, where $a, b \in k[X, Y]$. This was further extended by Russell over fields of any characteristic. They proved that

Theorem 5.1. *Let $F \in k[X, Y, Z]$ be such that $F = aZ - b$, where $a(\neq 0)$, $b \in k[X, Y]$, and $k[X, Y, Z]/(F) = k^{[2]}$. Then there exist $X_0, Y_0 \in k[X, Y]$ such that $k[X, Y] = k[X_0, Y_0]$ with $a \in k[X_0]$ and $k[X, Y, Z] = k[X_0, F]^{[1]}$.*

When k is an algebraically closed field of characteristic $p \geq 0$, Wright [89] proved the conjecture for polynomials F of the form $aZ^m - b$ with $a, b \in k[X, Y]$, $m \geq 2$ and $p \nmid m$. Das and Dutta showed [28, THEOREM 4.5] that Wright's result extends to any field k. They proved that

Theorem 5.2. *Let k be any field with $\operatorname{ch} k = p$ (≥ 0) and $F = aZ^m - b \in k[X, Y, Z]$ be such that $a(\neq 0)$, $b \in k[X, Y]$, $m \geq 2$ and $p \nmid m$. Suppose that $k[X, Y, Z]/(F) = k^{[2]}$. Then there exists $X_0 \in k[X, Y]$ such that $k[X, Y] = k[X_0, b]$ with $a \in k[X_0]$ and $k[X, Y, Z] = k[F, Z, X_0]$.*

The condition that $p \nmid m$ is necessary in Theorem 5.2 (cf. [28, REMARK 4.6]).

Most of the above cases are covered by the following generalization due to Russell and Sathaye [76, THEOREM 3.6]:

Theorem 5.3. *Let k be a field of characteristic zero and let*

$$F = a_m Z^m + a_{m-1} Z^{m-1} + \cdots + a_1 Z + a_0 \in k[X, Y, Z]$$

where $a_0, \ldots, a_m \in k[X, Y]$ are such that GCD $(a_1, \ldots, a_m) \notin k$. *Suppose that*

$$k[X, Y, Z]/(F) = k^{[2]}.$$

Then there exists $X_0 \in k[X, Y]$ such that $k[X, Y] = k[X_0, b]$ with $a_m \in k[X_0]$. Further, $k[X, Y, Z] = k[F]^{[2]}$.

Thus, for $k[X, Y, Z]$, the Abhyankar–Sathaye conjecture remains open for the case when $\mathrm{GCD}(a_1, \ldots, a_m) = 1$.

A common theme in most of the partial results proved in the Abhyankar–Sathaye conjecture for $k[X, Y, Z]$ is that, if F is considered as a polynomial in Z, then the coordinates of $k[X, Y]$ can be so chosen that the coefficient of Z becomes a polynomial in X. The Abhyankar–Sathaye conjecture for $k[X, Y, Z]$ can now be split into two parts.

Question 4A. Let k be a field of characteristic zero and let

$$F = a_m Z^m + a_{m-1} Z^{m-1} + \cdots + a_1 Z + a_0 \in k[X, Y, Z]$$

where $a_0, \ldots, a_m \in k[X, Y]$. Suppose that $k[X, Y, Z]/(F) = k^{[2]}$. Does there exist $X_0 \in k[X, Y]$ such that $k[X, Y] = k[X_0]^{[1]}$ with $a_m \in k[X_0]$?

Question 4B. Let k be a field of characteristic zero and suppose

$$F = a_m(X)Z^m + a_{m-1}Z^{m-1} + \cdots + a_1 Z + a_0 \in k[X, Y, Z]$$

where $a_0, \ldots, a_{m-1} \in k[X, Y]$ and $a_m \in k[X]$. Suppose that $k[X, Y, Z]/(F) = k^{[2]}$. Does this imply that $k[X, Y, Z] = k[F]^{[2]}$?

Sangines Garcia in his PhD thesis [78] answered Question 4A affirmatively for the case $m = 2$. In [21], Bhatwadekar and the author have given an alternative proof of this result of Garcia.

When k is any field, as a partial generalization of Theorem 5.1 and Question 4B in four variables, the author proved the Abhyankar–Sathaye conjecture for a polynomial F of the form $X^m Y - F(X, Z, T) \in k[X, Y, Z, T]$. This was one of the consequences of her general investigation on the ZCP [46]. In the process, she related it with other central problems on affine spaces like the affine fibration problem and the ZCP. The author has proved equivalence of ten statements, some of which involve an invariant introduced by Derksen, which is called the Derksen invariant.

The Derksen invariant of an integral domain B, denoted by $\mathrm{DK}(B)$, is defined as the smallest subring of B generated by the kernel of D, where D varies over the set of all locally nilpotent derivations of B.

Theorem 5.4. *Let k be a field of any characteristic and A an integral domain defined by*

$$A = k[X, Y, Z, T]/\big(X^m Y - F(X, Z, T)\big), \quad \text{where } m > 1.$$

Let x, y, z, and t denote, respectively, the images of X, Y, Z, and T in A. Set $f(Z, T) := F(0, Z, T)$ and $G := X^m Y - F(X, Z, T)$. Then the following statements are equivalent:

(i) $k[X, Y, Z, T] = k[X, G]^{[2]}$.

(ii) $k[X, Y, Z, T] = k[G]^{[3]}$.

(iii) $A = k[x]^{[2]}$.

(iv) $A = k^{[3]}$.

(v) $A^{[\ell]} \cong_k k^{[\ell+3]}$ *for some integer* $\ell \geq 0$ *and* $\mathrm{DK}(A) \neq k[x, z, t]$.

(vi) *A is an $\mathbb{A}^2$-fibration over $k[x]$ and* $\mathrm{DK}(A) \neq k[x, z, t]$.

(vii) *A is geometrically factorial over k,* $\mathrm{DK}(A) \neq k[x, z, t]$ *and the canonical map $k^* \to K_1(A)$ (induced by the inclusion $k \hookrightarrow A$) is an isomorphism.*

(viii) *A is geometrically factorial over k,* $\mathrm{DK}(A) \neq k[x, z, t]$ *and* $(A/xA)^* = k^*$.

(ix) $k[Z, T] = k[f]^{[1]}$.

(x) $k[Z, T]/(f) = k^{[1]}$ *and* $\mathrm{DK}(A) \neq k[x, z, t]$.

The equivalence of (ii) and (iv) provides an answer to Question 4 for the special case of the polynomial $X^m Y - F(X, Z, T)$. The equivalence of (i) and (iii) provides an answer to a special case of Question 4′ (stated below) for the ring $R = k[x]$. The equivalence of (iii) and (vi) answers Question 3 in a special situation. For more discussions, see [48].

In a remarkable paper Kaliman proved the following result over the field of complex numbers [56]. Later, Daigle and Kaliman extended it over any field k of characteristic zero [25].

Theorem 5.5. *Let k be a field of characteristic zero. Let $F \in k[X, Y, Z]$ be such that $k[X, Y, Z]/(F - \lambda) = k^{[2]}$ for almost every $\lambda \in k$. Then $k[X, Y, Z] = k[F]^{[2]}$.*

A general version of Question 4 can be asked as:

Question 4′. Let R be a ring and $f \in A = R^{[n]}$ for some integer $n \geq 2$. Suppose

$$A/(f) \cong R^{[n-1]}.$$

Does this imply that $A = R[f]^{[n-1]}$, i.e., is f a coordinate in A?

There have been affirmative answers to Question 4′ in special cases by Bhatwadekar, Dutta, and Das [11, 13, 28]. Bhatwadekar and Dutta had considered linear planes, i.e., polynomials F of the form $aZ - b$, where $a, b \in R[X, Y]$ over a discrete valuation ring R and proved that special cases of the linear planes are actually variables. Bhatwadekar–Dutta have also shown [12] that a negative answer to Question 4′ in the case when $n = 3$ and R is a discrete valuation ring containing $\mathbb{Q}$ will give a negative answer to the affine fibration problem (Question 3 (i)) for the case $n = 2$ and $d = 2$. An example of a case of linear planes which remains unsolved is discussed in Section 7.

6. $\mathbb{A}^n$-FORMS

Let A be an algebra over a field k. We say that A is an $\mathbb{A}^n$-form over k if $A \otimes_k L = L^{[n]}$ for some finite algebraic extension L of k. Let A be an $\mathbb{A}^n$-form over a field k.

When $n = 1$, it is well known that if $L|_k$ is a separable extension, then $A = k^{[1]}$ (i.e., trivial) and that if $L|_k$ is purely inseparable then A need not be $k^{[1]}$. An extensive study of such purely inseparable algebras was made by Asanuma in [8]. Over any field of positive characteristic, the nontrivial purely inseparable $\mathbb{A}^1$-forms can be used to give examples of nontrivial $\mathbb{A}^n$-forms for any integer $n > 1$.

When $n = 2$ and $L|_k$ is a separable extension, then Kambayashi established that $A = k^{[2]}$ [57]. However, the problem of existence of nontrivial separable $\mathbb{A}^3$-forms is open in general. A few recent partial results on the triviality of separable $\mathbb{A}^3$-forms are mentioned below.

Let A be an $\mathbb{A}^3$-form over a field k of characteristic zero and $\bar{k}$ be an algebraic closure of k. Then $A = k^{[3]}$ if it satisfies any one of the following:

(1) A admits a fixed point free locally nilpotent derivation D (Daigle and Kaliman [25, COROLLARY 3.3]).

(2) A contains an element f which is a coordinate of $A \otimes_k \bar{k}$ (Daigle and Kaliman [25, PROPOSITION 4.9]).

(3) A admits an effective action of a reductive algebraic k-group of positive dimension (Koras and Russell [61, THEOREM C]).

(4) A admits either a fixed point free locally nilpotent derivation or a nonconfluent action of a unipotent group of dimension two (Gurjar, Masuda, and Miyanishi [51]).

(5) A admits a locally nilpotent derivation D such that $\mathrm{rk}(D \otimes 1_{\bar{k}}) \leq 2$ (Dutta, Gupta, and Lahiri [39]).

Now let R be a ring containing a field k. An R-algebra A is said to be an $\mathbb{A}^n$-form over R with respect to k if $A \otimes_k \bar{k} = (R \otimes_k \bar{k})^{[n]}$, where $\bar{k}$ denotes the algebraic closure of k. A few results on triviality of separable A^n-forms over a ring R are listed below.

Let A be an $\mathbb{A}^n$-form over a ring R containing a field k of characteristic 0. Then:

(1) If $n = 1$, then A is isomorphic to the symmetric algebra of a finitely generated rank one projective module over R [35, THEOREM 7].

(2) If $n = 2$ and R is a PID containing $\mathbb{Q}$, then $A = R^{[2]}$ [35, REMARK 8].

(3) If $n = 2$, then A is an $\mathbb{A}^2$-fibration over R.

(4) If $n = 2$ and R is a one-dimensional Noetherian domain, then there exists a finitely generated rank-one projective R-module Q such that $A \cong (\mathrm{Sym}_R(Q))^{[1]}$ [39, THEOREM 3.7].

(5) If $n = 2$ and A admits has a fixed point free locally nilpotent R-derivation over any ring R, then there exists a finitely generated rank one projective R-module Q such that $A \cong (\mathrm{Sym}_R(Q))^{[1]}$ [39, THEOREM 3.8].

The result (3) above shows that an affirmative answer to the $\mathbb{A}^2$-fibration problem (Question 3 (i)) will ensure an affirmative answer to the problem of $\mathbb{A}^2$-forms over general rings. Over a field F of any characteristic, Das has shown [27] that any factorial $\mathbb{A}^1$-form A over a ring R containing F is trivial if there exists a retraction map from A to R.

We cannot say much about $\mathbb{A}^3$-forms over general rings till the time we solve it over fields.

7. AN EXAMPLE OF BHATWADEKAR AND DUTTA

The following example arose from the study of linear planes over a discrete valuation ring by Bhatwadekar and Dutta [12]. Question 5 stated below is an open problem for at least three decades. Let

$$A = \mathbb{C}[T, X, Y, Z] \quad \text{and} \quad R = \mathbb{C}[T, F] \subset A,$$

where $F = TX^2Z + X + T^2Y + TXY^2$.

Let

$$P := XZ + Y^2,$$
$$G := TY + XP,$$

and

$$H := T^2Z - 2TYP - XP^2$$

Then, we can see that

$$XH + G^2 = T^2P$$

and $F = X + TG$. Clearly, $\mathbb{C}[T, T^{-1}][F, G, H] \subseteq \mathbb{C}[T, T^{-1}][X, Y, Z]$.

Then the following statements hold:

(i) $\mathbb{C}[T, T^{-1}][X, Y, Z] = \mathbb{C}[T, T^{-1}, F, G, H] = \mathbb{C}[T, T^{-1}][F]^{[2]}$.

(ii) $\mathbb{C}[T, X, Y, Z]$ is an $\mathbb{A}^2$-fibration over $\mathbb{C}[T, F]$.

(iii) $\mathbb{C}[T, X, Y, Z]^{[1]} = \mathbb{C}[T, F]^{[3]}$.

(iv) $\mathbb{C}[T, X, Y, Z]/(F) = \mathbb{C}[T]^{[2]} = \mathbb{C}^{[3]}$.

(v) $\mathbb{C}[T, X, Y, Z]/(F - f(T)) = \mathbb{C}[T]^{[2]}$ for every polynomial $f(T) \in \mathbb{C}[T]$.

(vi) $\mathbb{C}[T, X, Y, Z][1/F] = \mathbb{C}[T, F, 1/F, G]^{[1]}$.

(vii) For any $u \in (T, F)R$, $A[1/u] = R[1/u]^{[2]}$, i.e., $\mathbb{C}[T, X, Y, Z][1/u] = \mathbb{C}[T, F, 1/u]^{[2]}$.

Question 5. (a) Is $A = \mathbb{C}[T, F]^{[2]} (= R^{[2]})$?

(b) At least is $A = \mathbb{C}[F]^{[3]}$?

If the answer is "No" to (a), then it is a counterexample to the following problems:

(1) $\mathbb{A}^2$-fibration Problem over $\mathbb{C}^{[2]}$ by (ii).

(2) Cancellation Problem over $\mathbb{C}^{[2]}$ by (iii).

(3) Epimorphism problem over the ring $\mathbb{C}[T]$ (see Question 4′) by (iv).

If the answer is "No" to (b) and hence to (a), then it is a counterexample also to the Epimorphism Problem for $\mathbb{C}^{[4]} \twoheadrightarrow \mathbb{C}^{[3]}$.

Though the above properties have been proved in several places, a proof is presented below. A variant of the Bhatwadekar–Dutta example was also constructed by Vénéreau in his thesis **[88]**; for a discussion on this and related examples, see **[24, 41, 64]**.

Proof. (i) We show that

$$\mathbb{C}[T, T^{-1}][X, Y, Z] = \mathbb{C}[T, T^{-1}][F, G, H]. \tag{1}$$

Note that

$$X = F - TG, \quad P = \frac{XH + G^2}{T^2},$$
$$Y = (G - XP)/T,$$

and

$$Z = (H + 2TYP + XP^2)/T^2,$$

and hence equation (1) follows.

(ii) Clearly, A is a finitely generated R-algebra. It can be shown by standard arguments that A is a flat R-algebra **[66, THEOREM 20.H]**. We now show that $A \otimes_R k(p) = k(p)^{[2]}$ for every prime ideal p of R. We note that $F - X \in TA$ and hence the image of F in A/TA is same as that of X. Now let p be a prime ideal of R. Then either $T \in p$ or $T \notin p$. If $T \in p$, then $A \otimes_R k(p) = k(p)[Y, Z] = k(p)^{[2]}$. If $T \notin p$, then image of T in $k(p)$ is a unit and the result follows from (i).

(iii) Let $D = A[W] = \mathbb{C}[T, X, Y, Z, W] = \mathbb{C}^{[5]}$. We shall show that

$$D = \mathbb{C}[T, F]^{[3]} = R^{[3]}.$$

Let

$$W_1 := TW + P,$$
$$G_1 := \frac{(G - FW_1)}{T} = Y - XW - (TY + XP)(TW + P) = Y - XW - GW_1,$$
$$H_1 := \frac{\{H + 2GW_1 - (F - GT)W_1^2\}}{T^2} = Z + 2YW - XW^2.$$

Now let

$$G_2 := G_1 + FW_1^2 = (Y - XW) - TW_1(Y - XW - GW_1) = Y - XW - TW_1G_1$$

and

$$W_2 := \frac{W_1 - (H_1F + G_2^2)}{T} = W + 2G_1W_1(Y - XW) - GH_1 - TG_1^2W_1^2$$

Then, it is easy to see that

$$\begin{aligned} D\left[T^{-1}\right] &= \mathbb{C}\left[T, T^{-1}\right][X, Y, Z, W] \\ &= \mathbb{C}\left[T, T^{-1}\right][F, G, H, W_1] \\ &= \mathbb{C}\left[T, T^{-1}\right][F, G_1, H_1, W_1] \\ &= \mathbb{C}\left[T, T^{-1}\right][F, G_2, H_1, W_2] \end{aligned}$$

and that $\mathbb{C}[T, F, G_2, H_1, W_2] \subseteq D$. Let $D/TD = \mathbb{C}[x, y, z, w]$, where x, y, z, w denote the images of X, Y, Z, W in D/TD. We now show that $D \subseteq \mathbb{C}[T, F, G_2, H_1, W_2]$. For this, it is enough to show that the kernel of the natural map $\phi : \mathbb{C}[T, F, G_2, H_1, W_2] \to D/TD$ is generated by T. We note that the image of ϕ is

$$\mathbb{C}\left[x, y - xw, z + 2yw - xw^2, w + 2p(y - xw - xp^2)(y - xw) - xp(z + 2yw - xw^2)\right],$$

which is of transcendence degree 4 over $\mathbb{C}$. Hence the kernel of ϕ is a prime ideal of height one and is generated by T. Therefore, $D = \mathbb{C}[T, F, G_2, H_1, W_2]$.

(iv)–(v) Let $B = \mathbb{C}[T, X, Y, Z]/(F - f(T))$ for some polynomial $f \in \mathbb{C}[T]$ and $S = \mathbb{C}[T]$. By (ii), it follows that B is an $\mathbb{A}^2$-fibration over S. Hence, by Sathaye's theorem [81], B is locally a polynomial ring over S and hence by Theorem 4.1, B is a polynomial ring over S.

(vi) Let $H_1 := \frac{FH+G^2}{T}$. Then

$$H_1 = \frac{(X + TG)(T^2Z - 2TYP - XP^2) + (TY + XP)^2}{T} = TP + GH.$$

Let $H_2 := \frac{FH_1+G^3}{T}$. Then

$$\begin{aligned} H_2 &= \frac{(X + TG)(TP + GH) + G^3}{T} \\ &= \frac{T(G^2H + TGP + XP) + G(XH + G^2)}{T} \\ &= \frac{T(G^2H + TGP + XP) + GT^2P}{T} \\ &= G^2H + XP + 2TGP. \end{aligned}$$

Let $H_3 := \frac{F(H_2-G)+G^4}{T}$. Then

$$\begin{aligned} H_3 &= \frac{F(G^2H + XP + 2TGP - XP - TY) + G^4}{T} \\ &= \frac{F(2TGP - TY) + G^2(FH + G^2)}{T} \\ &= \frac{TF(2GP - Y) + TH_1G^2}{T} \\ &= F(2GP - Y) + H_1G^2. \end{aligned}$$

Now it is easy to see that

$$\begin{aligned} \mathbb{C}[T, X, Y, Z, F^{-1}][T^{-1}] &= \mathbb{C}[T, T^{-1}][F, F^{-1}, G, H] \\ &= \mathbb{C}[T, T^{-1}][F, F^{-1}, G, H_1] \\ &= \mathbb{C}[T, T^{-1}][F, F^{-1}, G, H_2] \\ &= \mathbb{C}[T, T^{-1}][F, F^{-1}, G, H_3], \end{aligned}$$

and that the image of $\mathbb{C}[T, F, F^{-1}, G, H_2]$ in $A[F^{-1}]/TA[F^{-1}]$ is of transcendence degree 3. Hence $A[F^{-1}] = \mathbb{C}[T, F, F^{-1}, G, H_3] = \mathbb{C}[T, F, F^{-1}, G]^{[1]}$.

(vii) Let m be any maximal ideal of R other than (T, F). Then either $T \notin m$ or $F \notin m$. Thus, in either case, from (i) and (vi), we have $A_m = R_m^{[2]}$.

Let $u \in (T, F)R$. Then a maximal ideal of $R[1/u]$ is an extension of a maximal ideal of R other than $(T, F)R$. Hence $A[1/u]$ is a locally polynomial ring in two variables over $R[1/u]$. Further any projective module over $R[1/u]$ is free. Thus, by Theorem 4.1, we have $A[1/u] = R[1/u]^{[2]}$. ■

ACKNOWLEDGMENTS

The author thanks Professor Amartya Kumar Dutta for introducing and guiding her to this world of affine algebraic geometry. The author also thanks him for carefully going through this draft and improving the exposition.

REFERENCES

[1] A. M. Abhyankar and S. M. Bhatwadekar, Generically Laurent polynomial algebras over a D.V.R. which are not quasi Laurent polynomial algebras. *J. Pure Appl. Algebra* **218** (2014), no. 4, 651–660.

[2] A. M. Abhyankar and S. M. Bhatwadekar, A note on quasi Laurent polynomial algebras in n variables. *J. Commut. Algebra* **6** (2014), no. 2, 127–147.

[3] S. Abhyankar, P. Eakin, and W. Heinzer, On the uniqueness of the coefficient ring in a polynomial ring. *J. Algebra* **23** (1972), 310–342.

[4] S. S. Abhyankar, Polynomials and power series. In *Algebra, Arithmetic and Geometry with Applications*, edited by C. Christensen et al., pp. 783–784, Springer, 2004.

[5] S. S. Abhyankar and T. T. Moh, Embeddings of the line in the plane. *J. Reine Angew. Math.* **276** (1975), 148–166.

[6] T. Asanuma, Polynomial fibre rings of algebras over Noetherian rings. *Invent. Math.* **87** (1987), 101–127.

[7] T. Asanuma, Non-linearizable algebraic k^*-actions on affine spaces. *Invent. Math.* **138** (1999), no. 2, 281–306.

[8] T. Asanuma, Purely inseparable k-forms of affine algebraic curves. In *Affine algebraic geometry*, pp. 31–46, Contemp. Math. 369, Amer. Math. Soc., Providence, RI, 2005.

[9] T. Asanuma and N. Gupta, On 2-stably isomorphic four dimensional affine domains. *J. Commut. Algebra* **10** (2018), no. 2, 153–162.

[10] H. Bass, E. H. Connell, and D. L. Wright, Locally polynomial algebras are symmetric algebras. *Invent. Math.* **38** (1977), 279–299.

[11] S. M. Bhatwadekar, Generalized epimorphism theorem. *Proc. Indian Acad. Sci.* **98** (1988), no. 2–3, 109–166.

[12] S. M. Bhatwadekar and A. K. Dutta, On affine fibrations. In *Commutative algebra: Conf. Comm. Alg. ICTP (1992)*, edited by A. Simis, N. V. Trung, and G. Valla, pp. 1–17, World Sc., 1994.

[13] S. M. Bhatwadekar and A. K. Dutta, Linear planes over a discrete valuation ring. *J. Algebra* **166** (1994), no. 2, 393–405.

[14] S. M. Bhatwadekar and A. K. Dutta, On $\mathbb{A}^1$-fibrations of subalgebras of polynomial algebras. *Compos. Math.* **95** (1995), no. 3, 263–285.

[15] S. M. Bhatwadekar and A. K. Dutta, Structure of $\mathbb{A}^*$-fibrations over one-dimensional seminormal semilocal domains. *J. Algebra* **220** (1999), 561–573.

[16] S. M. Bhatwadekar and A. K. Dutta, On $\mathbb{A}^*$-fibrations. *J. Pure Appl. Algebra* **149** (2000), 1–14.

[17] S. M. Bhatwadekar, A. K. Dutta, and N. Onoda, On algebras which are locally $\mathbb{A}^1$ in codimension-one. *Trans. Amer. Math. Soc.* **365** (2013), no. 9, 4497–4537.

[18] S. M. Bhatwadekar and N. Gupta, On locally quasi $\mathbb{A}^*$ algebras in codimension-one over a Noetherian normal domain. *J. Pure Appl. Algebra* **215** (2011), 2242–2256.

[19] S. M. Bhatwadekar and N. Gupta, The structure of a Laurent polynomial fibration in n variables. *J. Algebra* **353** (2012), no. 1, 142–157.

[20] S. M. Bhatwadekar and N. Gupta, A note on the cancellation property of $k[X, Y]$. *J. Algebra Appl. (special issue in honour of Prof. Shreeram S. Abhyankar)* **14** (2015), no. 9, 15400071, 5 pp.

[21] S. M. Bhatwadekar and N. Gupta, On quadratic planes. Preprint.

[22] A. J. Crachiola and L. Makar-Limanov, An algebraic proof of a cancellation theorem for surfaces. *J. Algebra* **320** (2008), no. 8, 3113–3119.

[23] D. Daigle and G. Freudenburg, Triangular derivations of $k[X_1, X_2, X_3, X_4]$. *J. Algebra* **241** (2001), no. 1, 328–339.

[24] D. Daigle and G. Freudenburg, Families of affine fibrations. In *Symmetry and spaces*, edited by H. E. A. Campbell et al., pp. 35–43, Progr. Math. 278, Birkhäuser, 2010.

[25] D. Daigle and S. Kaliman, A note on locally nilpotent derivations and variables of $k[X, Y, Z]$. *Canad. Math. Bull.* **52** (2009), no. 4, 535–543.

[26] W. Danielewski, On a Cancellation Problem and automorphism groups of affine algebraic varieties. Preprint, Warsaw, 1989. (Appendix by K. Fieseler).

[27] P. Das, A note on factorial $\mathbb{A}^1$-forms with retractions. *Comm. Algebra* **40** (2012), no. 9, 3221–3223.

[28] P. Das and A. K. Dutta, Planes of the form $b(X, Y)Z^n - a(X, Y)$ over a DVR. *J. Commut. Algebra* **3** (2011), no. 4, 491–509.

[29] N. Dasgupta and N. Gupta, An algebraic characterisation of the affine three space. 2019, arXiv:1709.00169. To appear in *J. Commut. Algebra*. https://projecteuclid.org/journals/jca/journal-of-commutative-algebra/DownloadAcceptedPapers/20301.pdf.

[30] A. Dubouloz, The cylinder over the Koras–Russell cubic threefold has a trivial Makar-Limanov invariant. *Transform. Groups* **14** (2009), no. 3, 531–539.

[31] A. Dubouloz and J. Fasel, Families of $\mathbb{A}^1$-contractible affine threefolds. *Algebr. Geom.* **5** (2018), no. 1, 1–14.

[32] A. Dubouloz, Affine surfaces with isomorphic $\mathbb{A}^2$-cylinders. *Kyoto J. Math.* **59** (2019), no. 1, 181–193.

[33] A. Dubouloz, S. Pauli, and P. A. Østvær, $\mathbb{A}^1$-contractibility of affine modifications. *Int. J. Math.* **14** (2019), no. 30, 1950069.

[34] A. K. Dutta, On $\mathbb{A}^1$-bundles of affine morphisms. *J. Math. Kyoto Univ.* **35** (1995), no. 3, 377–385.

[35] A. K. Dutta, On separable $\mathbb{A}^1$-forms. *Nagoya Math. J.* **159** (2000), 45–51.

[36] A. K. Dutta, Some results on affine fibrations. In *Advances in algebra and geometry*, edited by C. Musili, pp. 7–24, Hindustan Book Agency, India, 2003.

[37] A. K. Dutta, Some results on subalgebras of polynomial algebras. In *Commutative algebra and algebraic geometry*, edited by S. Ghorpade et al., pp. 85–95 390, Contemp. Math., 2005.

[38] A. K. Dutta and N. Gupta, The epimorphism theorem and its generalisations. *J. Algebra Appl. (special issue in honour of Prof. Shreeram S. Abhyankar)* **14** (2015), no. 9, 15400010, 30 pp.

[39] A. K. Dutta, N. Gupta, and A. Lahiri, On Separable $\mathbb{A}^2$ and $\mathbb{A}^3$-forms. *Nagoya Math. J.* **239** (2020), 346–354.

[40] A. K. Dutta and N. Onoda, Some results on codimension-one $\mathbb{A}^1$-fibrations. *J. Algebra* **313** (2007), 905–921.

[41] G. Freudenburg, Algebraic theory of locally nilpotent derivations. In *Invariant Theory and Algebraic Transformation Groups VII*, Encyclopaedia Math. Sci. 136, Springer, Berlin, 2006.

[42] G. Freudenburg and P. Russell, Open problems in affine algebraic geometry, In *Affine algebraic geometry*, pp. 1–30, Contemp. Math. 369, 2005.

[43] T. Fujita, On Zariski problem. *Proc. Jpn. Acad.* **55** (1979), no. A, 106–110.

[44] N. Gupta, On faithfully flat fibrations by a punctured line. *J. Algebra* **415** (2014), 13–34.

[45] N. Gupta, On the Cancellation Problem for the affine space $\mathbb{A}^3$ in characteristic p. *Invent. Math.* **195** (2014), 279–288.

[46] N. Gupta, On the family of affine threefolds $x^m y = F(x, z, t)$. *Compos. Math.* **150** (2014), no. 6, 979–998.

[47] N. Gupta, On Zariski's Cancellation Problem in positive characteristic. *Adv. Math.* **264** (2014), 296–307.

[48] N. Gupta, A survey on Zariski Cancellation Problem. *Indian J. Pure Appl. Math.* **46** (2015), no. 6, 865–877.

[49] N. Gupta and S. Sen, On double Danielewski surfaces and the Cancellation Problem. *J. Algebra* **533** (2019), 25–43.

[50] R. V. Gurjar, A topological proof of cancellation theorem for $\mathbb{C}^2$. *Math. Z.* **240** (2002), no. 1, 83–94.

[51] R. V. Gurjar, K. Masuda, and M. Miyanishi, Affine space fibrations. In *Springer Proceedings in Mathematics and Statistics: Polynomial rings and Affine Algebraic Geometry, (PRAAG) 2018, Tokyo, Japan, February 12–16*, pp. 151–194, Springer, 2018.

[52] E. Hamann, On the R-invariance of $R[x]$. *J. Algebra* **35** (1975), 1–16.

[53] M. Hochster, Non-uniqueness of the ring of coefficients in a polynomial ring. *Proc. Amer. Math. Soc.* **34** (1972), no. 1, 81–82.

[54] M. Hoyois, A. Krishna, and P. A. Østvær, $\mathbb{A}^1$-contractibility of Koras–Russell threefolds. *Algebr. Geom.* **3** (2016), no. 4, 407–423.

[55] S. Kaliman, M. Koras, L. Makar-Limanov, and P. Russell, $\mathbb{C}^*$-actions on $\mathbb{C}^3$ are linearizable. *Electron. Res. Announc. Am. Math. Soc.* **3** (1997), 63–71.

[56] Sh. Kaliman, Polynomials with general $\mathbb{C}^{[2]}$-fibers are variables. *Pacific J. Math.* **203** (2002), no. 1, 161–190.

[57] T. Kambayashi, On the absence of nontrivial separable forms of the affine plane. *J. Algebra* **35** (1975), 449–456.

[58] T. Kambayashi, Automorphism group of a polynomial ring and algebraic group actions on affine space. *J. Algebra* **60** (1979), 439–451.

[59] T. Kambayashi and M. Miyanishi, On flat fibrations by affine line. *Illinois J. Math.* **22** (1978), no. 4, 662–671.

[60] T. Kambayashi and D. Wright, Flat families of affine lines are affine line bundles. *Illinois J. Math.* **29** (1985), no. 4, 672–681.

[61] M. Koras and P. Russell, Separable forms of $\mathbb{G}_m$-actions on $\mathbb{A}^3_k$. *Transform. Groups* **18** (2013), no. 4, 1155–1163.

[62] H. Kraft, Challenging problems on affine n-space. *Séminaire Bourbaki* **802** (1995), 295–317.

[63] T. Y. Lam, *Serre's problem on projective modules*. Springer, Berlin–Heidelberg, 2006.

[64] D. Lewis, Vénéreau-type polynomials as potential counterexamples. *J. Pure Appl. Algebra* **217** (2013), no. 5, 946–957.

[65] L. Makar-Limanov, On the hypersurface $x + x^2y + z^2 + t^3 = 0$ in $\mathbb{C}^4$ or a $\mathbb{C}^3$-like threefold which is not $\mathbb{C}^3$. *Israel J. Math.* **96** (1996), no. B, 419–429.

[66] H. Matsumura, *Commutative algebra. 2nd edn*. Benjamin, 1980.

[67] M. Miyanishi, An algebraic characterization of the affine plane. *J. Math. Kyoto Univ.* **15** (1975), 169–184.

[68] M. Miyanishi, An algebro-topological characterization of the affine space of dimension three. *Amer. J. Math.* **106** (1984), 1469–1486.

[69] M. Miyanishi, Recent developments in affine algebraic geometry: from the personal viewpoints of the author. In *Affine algebraic geometry*, pp. 307–378, Osaka Univ. Press, Osaka, 2007.

[70] M. Miyanishi and T. Sugie, Affine surfaces containing cylinderlike open sets. *J. Math. Kyoto Univ.* **20** (1980), 11–42.

[71] M. Nagata, *On automorphism group of* $k[X, Y]$. Kyoto Univ. Lect. Math. 5, Kinokuniya, Tokyo, 1972.

[72] C. P. Ramanujam, A topological characterization of the affine plane as an algebraic variety. *Ann. of Math.* **94** (1971), 69–88.

[73] P. Russell, Simple birational extensions of two dimensional affine rational domains. *Compos. Math.* **33** (1976), no. 2, 197–208.

[74] P. Russell, On affine-ruled rational surfaces. *Math. Ann.* **255** (1981), 287–302.

[75] P. Russell, Embedding problems in affine algebraic geometry. In *Polynomial automorphisms and related topics*, edited by H. Bass et al., pp. 113–135, Pub. House for Sc. and Tech., Hanoi, 2007.

[76] P. Russell and A. Sathaye, On finding and cancelling variables in $k[X, Y, Z]$. *J. Algebra* **57** (1979), no. 1, 151–166.

[77] P. Russell and A. Sathaye, Forty years of the epimorphism theorem. *Eur. Math. Soc. Newsl.* **90** (2013), 12–17.

[78] L. M. Sangines Garcia, *On quadratic planes*. Ph.D. Thesis, McGill Univ., 1983.

[79] A. Sathaye, On linear planes. *Proc. Amer. Math. Soc.* **56** (1976), 1–7.

[80] A. Sathaye, Generalized Newton–Puiseux expansion and Abhyankar–Moh semigroup theorem. *Invent. Math.* **74** (1983), no. 1, 149–157.

[81] A. Sathaye, Polynomial ring in two variables over a D.V.R.: a criterion. *Invent. Math.* **74** (1983), 159–168.

[82] A. Sathaye, An application of generalized Newton Puiseux expansions to a conjecture of D. Daigle and G. Freudenburg. In *Algebra, arithmetic and geometry with applications*, edited by C. Christensen et al., pp. 687–701, Springer, 2004.

[83] B. Segre, Corrispondenze di Möbius e trasformazioni cremoniane intere. *Atti Accad. Sci. Torino Cl. Sci. Fis. Mat. Natur.* **91** (1956/1957), 3–19.

[84] A. R. Shastri, Polynomial representations of knots. *Tohoku Math. J. (2)* **44** (1992), no. 1, 11–17.

[85] A. A. Suslin, Locally polynomial rings and symmetric algebras. *Izv. Akad. Nauk SSSR Ser. Mat.* **41** (1977), no. 3, 503–515 (Russian).

[86] M. Suzuki, Propriétés topologiques des polynômes de deux variables complexes, et automorphismes algébriques de l'espace $\mathbb{C}^2$. *J. Math. Soc. Japan* **26** (1974), 241–257.

[87] B. Veĭsfeĭler and I. V. Dolgačev, Unipotent group schemes over integral rings. *Izv. Akad. Nauk SSSR Ser. Mat.* **38** (1974), 757–799.

[88] S. Vénéreau, *Automorphismes et variables de ĺanneau de polynómes* $A[y_1, \dots, y_m]$. Ph.D. Thesis, Institut Fourier, Grenoble, 2001.

[89] D. Wright, Cancellation of variables of the form $bT^n - a$. *J. Algebra* **52** (1978), no. 1, 94–100.

NEENA GUPTA

Statistics and Mathematics Unit, Indian Statistical Institute, 203 B.T. Road, Kolkata 700 108, India, neenag@isical.ac.in, rnanina@gmail.com

This Page Intentionally Left Blank

THE FORMAL MODEL OF SEMI-INFINITE FLAG MANIFOLDS

SYU KATO

ABSTRACT

The formal model of semi-infinite flag manifold is a variant of an affine flag variety that was recognized from the 1980s but not studied extensively until the late 2010s. In this note, we exhibit constructions and ideas appearing in our recent study of the formal model of semi-infinite flag manifold of a simple algebraic group. Our results have some implications to the theory of rational maps from a projective line to partial flag manifolds, and also on the structures of quantum cohomologies and quantum K-groups of partial flag manifolds.

MATHEMATICS SUBJECT CLASSIFICATION 2020

Primary 14M15; Secondary 14D24, 14N35, 20G44, 22E50, 57T15

KEYWORDS

Semi-infinite flag manifold, quasi-map space, quantum K-group, Kac–Moody group, affine Lie algebra, global Weyl module

ICM INTERNATIONAL CONGRESS OF MATHEMATICIANS 2022 JULY 6–14

Proc. Int. Cong. Math. 2022, Vol. 3, pp. 1600–1622
DOI 10.4171/ICM2022/25
Published by EMS Press

1. INTRODUCTION

Compact complex-analytic spaces that admit homogeneous Lie group actions are quite rare in nature, and their classification reduces into three primitive classes: finite groups, tori, and (partial) flag manifolds. The first have discrete topology and the role of geometric consideration is rather small, in general. The second, particularly those admit polarizations, offer a major subject known as abelian varieties. The third, the (partial) flag manifolds of compact simple Lie groups, are ubiquitous in representation theory of semisimple algebraic groups and quantum groups. By the universal nature of general linear groups, flag manifolds of unitary groups are extensively studied from the geometric perspective.

In representation-theoretic considerations, we usually consider flag manifolds as projective algebraic varieties defined over an algebraically closed field (that form a family over Spec $\mathbb{Z}$). This definition naturally extends to an arbitrary Kac–Moody setting, but the resulting objects have at least two variants, thin flag varieties and thick flag manifolds (defined by Kac–Peterson [75] and Kashiwara [40], respectively). In case the Kac–Moody group is of affine type, we have a loop realization of the corresponding Kac–Moody group, essentially identifying the corresponding group with the set of $\Bbbk((z))$-valued points of a simple algebraic group over a field $\Bbbk$. This motivates us to consider yet other versions of flag manifolds of affine type that can be understood as an enhancement of arc schemes of usual flag manifolds. These are the semi-infinite flag manifolds that originate from the ideas of Lusztig [63, §11] and Drinfeld [22]. Lusztig's original idea is to construct varieties that naturally encode representation theory of simple algebraic groups over finite fields. The Lusztig program (see, e.g., [44, 63]) adds representation theory of quantum groups at roots of unity and representation theory of affine Lie algebras at negative rational levels into the picture, and Feigin–Frenkel [19] put representation theory of affine Lie algebras at the critical level into the picture. The semi-infinite flag manifolds itself have two realizations, that we refer to as the ind-model and the formal model. The geometry of the ind-model of semi-infinite flag manifolds, also known as the space of quasimaps from a projective line to a flag manifold, was studied extensively by Braverman, Finkelberg, Mirković, and their collaborators (see [8, 18, 21, 22]).

One instance of the ind-model of semi-infinite flag manifold is the space of principal bundles on an algebraic curve equipped with some reduction. This interpretation realizes some portion of the above representation-theoretic expectations [2, 31]. The formal model of semi-infinite flag manifolds is expected to add a concrete understanding of related representation-theoretic patterns [19, 22, 63]. Unfortunately, such an idea needs to be polished as its implementation faces difficulty due to its essential infinite-dimensionality. This forces us to employ affine Grassmannians instead of semi-infinite flag manifolds in some cases (see [26, 30, 78]) at the moment, that is possible by some tight connections [27, 70].

Meanwhile, it is realized that the semi-infinite flag manifold is a version of the loop space of a flag manifold, and hence it is related to its quantum cohomology [32]. In fact, the ind-model of a semi-infinite flag manifold offers a description of the quantum K-theoretic J-function of a flag manifold [9] that encodes its small quantum K-group.

In both contexts of the above two paragraphs, the Peterson isomorphism [59, 74], that connects the quantum cohomology of a flag manifold with the homology of an affine Grassmannian, should admit an interpretation using a semi-infinite flag manifold. However, such an interpretation is not known today (though we have Corollary 7.3).

The main goal of this note is to explain a realization of the formal model of semi-infinite flag manifold [46,50,52], that is reminiscent to the classical description of the original flag manifolds. Our realization is supported by recent developments in representation theory of affine Lie algebras [14,15,51], that is also reminiscent to the representation theory of simple Lie algebras. It turns out that the study of the formal model of the semi-infinite flag manifold has implications to the corresponding ind-model [50], as well as the study of quantum K-groups of partial flag manifolds and the K-groups of affine Grassmannians [45, 47, 48]. This includes an interpretation (and a proof) of the K-theoretic analogue of the Peterson isomorphism using semi-infinite flag manifolds (Theorem 8.2).

The results presented here describe the formal model of semi-infinite flag manifolds in a down-to-earth fashion, and also provide first nontrivial conclusions deduced from them. However, we have not yet reached our primary goal to understand representation theory from this perspective in a satisfactory fashion. We hope to improve this situation in the near future.

The organization of this note is as follows: We first recall the construction of flag manifolds that is parallel to our later construction in Section 2. We explain the role of quantum groups in the structure theory of Kac–Moody algebras and exhibit two versions of flag varieties of Kac–Moody groups in Section 3. In Section 4, we exhibit some representation theory of affine Lie algebras. Based on it, we explain our construction of the formal model of semi-infinite flag manifolds in Section 5. This enables us to present our idea on the Frobenius splitting of semi-infinite flag manifolds in Section 6. We explain the connection between its Richardson varieties and quasimap spaces in Section 7, and explain how they fit into the study of quantum cohomology of flag manifolds. We exhibit the K-theoretic Peterson isomorphism in Section 8. We discuss the functoriality of the quantum K-groups of partial flag manifolds in Section 9. We finish this note by discussing some perspectives in Section 10.

We assume that every field $\Bbbk$ has characteristic $\neq 2$. A variety is some algebraic-geometric object that admits singularity, and a manifold is a variety that is supposed to be smooth in some sense. An algebraic variety is a separated scheme of finite type defined over a field (i.e., our variety is not necessarily irreducible or reduced). We set $\mathbb{N} := \mathbb{Z}_{\geq 0}$.

2. FLAG MANIFOLDS VIA REPRESENTATION THEORY

Let G be a simply connected semisimple algebraic group over an algebraically closed field $\Bbbk$. Let $T \subset B$ be its maximal torus and a Borel subgroup (maximal solvable subgroup). Let W $(= N_G(T)/T)$ be the Weyl group of G. Let $\mathbb{X}$ be the set of one-dimensional rational T-characters (the set of T-weights), that admits a natural W-action. We set $\mathbb{X}_+ := \sum_{i=1}^{r} \mathbb{N}\varpi_i$, where $\varpi_1, \ldots, \varpi_r \in \mathbb{X}$ are fundamental weights with respect to B. The set of isomorphism classes of irreducible rational representations $\{L(\lambda)\}_\lambda$ of G is labeled by $\mathbb{X}_+$ in such a way that each $L(\lambda)$ contains a unique (up to scalar) B-eigenvector

$\mathbf{v}_\lambda$ with its T-weight λ. We refer λ ($\in \mathbb{X}_+$) as the highest weight of $L(\lambda)$. The flag manifold $\mathscr{B} := G/B$ of G is the maximal G-homogeneous space that is projective.

In case $\Bbbk = \mathbb{C}$, we have

$$\mathscr{B} = (\mathscr{Y} \setminus E)/T,$$

where $\mathscr{Y}$ is an affine algebraic variety with $(G \times T)$-action whose ring $\mathbb{C}[\mathscr{Y}]$ of regular functions is written as

$$\mathbb{C}[\mathscr{Y}] \cong \bigoplus_{\lambda \in \mathbb{X}_+} L(\lambda)^* \quad (\text{as } G \times T\text{-modules}), \tag{2.1}$$

and $E \subset \mathscr{Y}$ is the locus where the T-action is not free. Here, the G-action on $\mathbb{C}[\mathscr{Y}]$ is the natural actions on $L(\lambda)$, and the T-action on $\mathbb{C}[\mathscr{Y}]$ comes from the grading $\mathbb{X}_+ \subset \mathbb{X}$ in the RHS of (2.1). These data, together with the condition $E \neq \mathscr{Y}$, essentially determine $\mathbb{C}[\mathscr{Y}]$ as $\mathbb{C}$-algebras generated by $L(\varpi_i)^*$ for $1 \le i \le r$. Consider a point $x_0 \in \mathscr{Y}$ given by $\{\mathbf{v}_\lambda\}_\lambda$, seen as linear maps on $\{L(\lambda)^*\}_\lambda$. The image $[x_0]$ of this point x_0 has its G-stabilizer equal to B. This induces an inclusion

$$G/B \hookrightarrow \mathscr{B} \subset \prod_{i=1}^{r} \mathbb{P}\big(L(\varpi_i)\big)$$

induced from $B/B \mapsto [x_0]$ by the G-action. (One needs additional representation-theoretic analysis to conclude $G/B \cong \mathscr{B}$.) This consideration transfers all geometric statements relevant to $\mathscr{B}$ to algebraic statements on the space in (2.1) in principle, but most of the geometric results on $\mathscr{B}$ and its subvarieties were proved for the first time by other methods (see, e.g., **[56]**).

Note that the vector space (2.1) does not acquire the structure of a ring when $\operatorname{char} \Bbbk = p > 0$. The reason is that we do not have a map $L(\lambda)^* \otimes L(\mu)^* \to L(\lambda + \mu)^*$, or equivalently, $L(\lambda + \mu) \to L(\lambda) \otimes L(\mu)$ for general $\lambda, \mu \in \mathbb{X}_+$. One way to improve the situation is to replace $\{L(\lambda)\}_{\lambda \in \mathbb{X}_+}$ with a suitable family of modules $\{Y(\lambda)\}_{\lambda \in \mathbb{X}_+}$ with larger members such that the G-module map

$$Y(\lambda + \mu) \to Y(\lambda) \otimes Y(\mu) \tag{2.2}$$

exists uniquely (up to constant) for every $\lambda, \mu \in \mathbb{X}_+$. It yields an analogous ring of (2.1) that should be closely related to $\mathscr{B}$. A standard choice of $Y(\lambda)$ ($\lambda \in \mathbb{X}_+$) is the Weyl module $V(\lambda)$ of G, that is, the projective cover of $L(\lambda)$ in the categories of rational G-modules whose composition factors are in $\{L(\mu)\}_{\lambda \ge \mu \in \mathbb{X}_+}$, where $\ge$ is the dominance ordering on $\mathbb{X}$. This produces $\mathscr{B}$ for all characteristics.

Theorem 2.1 (Orthogonality of Weyl modules, **[36, II §4.13]**). *For each $\lambda, \mu \in \mathbb{X}_+$, we have*

$$\operatorname{Ext}^i_G\big(V(\lambda), V(\mu)^*\big) \cong \Bbbk^{\oplus \delta_{i,0}\delta_{\lambda,\mu^*}},$$

where μ^* is the highest weight of $L(\mu)^*$. By taking the Euler–Poincaré characteristic, this Ext*-orthogonality implies the orthogonality of the T-characters of $V(\lambda)$. In particular, the T-characters of $V(\lambda)$ do not depend on $\Bbbk$.*

Note that $L(\lambda) = V(\lambda)$ for char $\Bbbk = 0$ by the semisimplicity of representations, and hence Theorem 2.1 is Schur's lemma in such a case. As $V(\lambda) = \Bbbk \otimes_{\mathbb{Z}} V_{\mathbb{Z}}(\lambda)$ holds for a collection of free $\mathbb{Z}$-modules $V_{\mathbb{Z}}(\lambda)$ ($\lambda \in \mathbb{X}_+$), we find that $\mathscr{B}$ extends to a scheme flat over $\mathbb{Z}$. Another possible choice of $Y(\lambda)$ ($\lambda \in \mathbb{X}_+$), the Verma module $M(\lambda)$ of the (divided power) enveloping algebra of Lie G, produces an open dense B-orbit in $\mathscr{B}$.

3. KAC–MOODY FLAG VARIETIES

Let us keep the setting of the previous section.

3.1. Reminder on Kac–Moody algebras and their quantum groups

Let $\mathfrak{g}_C$ be the Kac–Moody algebra associated to a symmetrizable generalized Cartan matrix (= GCM) C (see [38]). In case char $\Bbbk = 0$, we have the notion of the highest weight integrable representations of $\mathfrak{g}_C$ parametrized by the set of dominant weights P_+ defined similarly to $\mathbb{X}_+$. Let $L(\Lambda)$ denote the highest weight integrable representation of $\mathfrak{g}_C$ corresponding to $\Lambda \in P_+$.

We have the quantum group (or the quantized enveloping algebra) $U_q(\mathfrak{g}_C)$ of $\mathfrak{g}_C$ originally defined by Drinfeld and Jimbo in the 1980s [17, 37]. It is an algebra defined over $\mathbb{Q}(q)$, and the specialization $q \mapsto 1$ recovers the universal enveloping algebra $U(\mathfrak{g}_C)$ of $\mathfrak{g}_C$. Kashiwara [41] and Lusztig [63] defined the canonical/global bases (of the positive/negative parts $U_q^{\pm}(\mathfrak{g}_C)$) of $U_q(\mathfrak{g}_C)$ and their integrable representations that generate their $\mathbb{Q}[q]$-lattices. The construction of Lusztig [64] clarified that quantum groups are, in fact, defined over $\mathbb{Z}[q^{\pm 1}]$ (or even over $\mathbb{N}[q^{\pm 1}]$ if one can say). In the 2010s, the categorification theorems of a quantum group and its integrable representations appeared [39, 53, 76, 77], and there every algebra that admits a categorification has a suitable $\mathbb{Z}[q]$-integral structure with distinguished bases, being the Grothendieck group of a module category of a finitely-generated graded algebras (called KLR algebras or quiver Hecke algebras). Therefore, the following is now widely recognized:

Theorem 3.1 (Lusztig [63, 64, 66] and Kashiwara [41–43]). *Assume that $\Bbbk = \mathbb{C}$. The (lower) global bases of $U_q^{\pm}(\mathfrak{g}_C)$ induce a $\mathbb{Z}$-integral form $U_{\mathbb{Z}}(\mathfrak{g}_C)$ of $U(\mathfrak{g}_C)$ via $q \mapsto 1$. For each $\Lambda \in P_+$, we have a $\mathbb{Z}$-lattice $L(\Lambda)_{\mathbb{Z}}$ of $L(\Lambda)$ obtained from the (lower) global base of the corresponding integrable highest weight module of $U_q(\mathfrak{g}_C)$. In addition, $L(\Lambda)_{\mathbb{Z}}$ is generated by the $U_{\mathbb{Z}}(\mathfrak{g}_C)$-action from a highest weight vector of $L(\Lambda)$.*

By a specialization of $L(\Lambda)_{\mathbb{Z}}$, we obtain a highest weight integrable module $L(\Lambda)$ over an arbitrary field $\Bbbk$. The module $L(\Lambda)$ is no longer irreducible when char $\Bbbk > 0$ (in general), and hence it is a $\mathfrak{g}_C$-analogue of Weyl modules rather than $L(\lambda)$ for G; it is a lack of brevity of the author to choose this notation here. We close this subsection by noting that the integral forms at the end of Section 2 coincide with the integral forms in Theorem 3.1.

3.2. Thin and thick flag varieties

Presentations of the flag varieties for general Kac–Moody groups $\mathscr{G}$ associated to a GCM C are similar to those in the previous section. A triangular decomposition of $\mathfrak{g}_C$ yields an analogous group $\mathscr{I}$ to the Borel subgroup. Let $\mathscr{T}$ be a (standard) maximal torus of $\mathscr{I}$. The highest weight vector in $L(\Lambda)$ is precisely an $\mathscr{I}$-eigenvector with its $\mathscr{T}$-weight Λ. Therefore, the construction in the previous section produces $\mathscr{G}/\mathscr{I}$ via the ring

$$\bigoplus_{\Lambda \in P_+} L(\Lambda)^\vee \subset \bigoplus_{\Lambda \in P_+} L(\Lambda)^*, \tag{3.1}$$

where $L(\Lambda)^*$ is the vector space dual of $L(\Lambda)$, and $L(\Lambda)^\vee$ is the restricted dual of $L(\Lambda)$, defined to be the direct sum of (finite-dimensional) vector space duals offered by the $\mathscr{T}$-weight decomposition of $L(\Lambda)$.

In this case, both vector spaces in (3.1) are naturally rings. This corresponds to the choice of $\mathscr{G}$. The former ring defines $\mathscr{B}_C^{\mathrm{thick}} = \mathscr{G}/\mathscr{I}$ **[40,49,71]** if we take $\mathscr{G}$ to be a version of the Kac–Moody group that is completed with respect to the opposite direction to $\mathscr{I}$. (This is the maximal Kac–Moody group, but the completion is taken in the opposite way as in the literature.) The latter ring can be seen as the projective limit of finitely-generated algebras, and the union of the spectrums of these rings yields $\mathscr{B}_C^{\mathrm{thin}} = \mathscr{G}/\mathscr{I}$ **[56,75]** if we take $\mathscr{G}$ as the uncompleted Kac–Moody group (the Kac–Peterson group or the minimal Kac–Moody group), or as the maximal Kac–Moody group completed with respect to the direction of $\mathscr{I}$. In other words, we have variants of flag manifolds of Kac–Moody groups associated to a GCM C as:

$$\bigcup_n \mathscr{B}_{C,n}^{\mathrm{thin}} = \mathscr{B}_C^{\mathrm{thin}} \subset \mathscr{B}_C^{\mathrm{thick}}. \tag{3.2}$$

The scheme $\mathscr{B}_C^{\mathrm{thick}}$ is a union of infinite-dimensional affine spaces, and hence is smooth. However, $\mathscr{B}_C^{\mathrm{thick}}$ is not compact in an essential way **[24]**. This picture is compatible with the fact that the Kac–Peterson group is defined by one-parameter generators (and relations), and hence $\mathscr{B}_C^{\mathrm{thin}}$ is a union of finite-dimensional subvarieties $\mathscr{B}_{C,n}^{\mathrm{thin}}$ consisting of points presented by a product of at most n generating elements. As such, each scheme $\mathscr{B}_{C,n}^{\mathrm{thin}}$ is singular, and hence $\mathscr{B}_C^{\mathrm{thin}}$ is understood to be singular. In fact, it does not admit an inductive limit description by finite-dimensional smooth pieces **[24]**.

4. GLOBAL WEYL MODULES AND THEIR PROJECTIVITY

Let us consider the untwisted affine Kac–Moody case hereafter, with the same conventions as in the previous sections. In particular, our Kac–Moody groups are extensions of the groups

$$G((z)) := G(\Bbbk((z))) \quad \text{and} \quad G[z^{\pm 1}] := G(\Bbbk[z^{\pm 1}])$$

by the loop rotation $\mathbb{G}_m$-actions (that we denote by $\mathbb{G}_m^{\mathrm{rot}}$) and the central extension $\mathbb{G}_m$-actions. (These correspond to the maximal/minimal realizations of the Kac–Moody groups in the previous section.) These are not (pro-)algebraic groups, and it sometimes causes difficulty. Nevertheless, each rational representation V of G induces representations

$$V(\!(z)\!) := V \otimes_{\Bbbk} \Bbbk(\!(z)\!) \quad \text{and} \quad V\big[z^{\pm 1}\big] := V \otimes_{\Bbbk} \Bbbk\big[z^{\pm 1}\big]$$

of $G(\!(z)\!)$ and $G[z^{\pm 1}]$, respectively. These representations are not of highest weight, but still integrable representations when we lift them to the central extensions of $G(\!(z)\!)$ and $G[z^{\pm 1}]$ by letting the center $\mathbb{G}_m$ act trivially (i.e., they are level-zero integrable representations viewed as representations of affine Lie algebras).

In addition to the T-action, the representation $V[z^{\pm 1}]$ carries $\mathbb{G}_m^{\mathrm{rot}}$-action. Let δ be the degree-one character of $\mathbb{G}_m^{\mathrm{rot}}$, and set $q := e^{\delta}$. By abuse of notation, we might consider q^n ($n \in \mathbb{Z}$) as the functor that twists the $\mathbb{G}_m^{\mathrm{rot}}$-action by degree n. We define a graded character of a semisimple $(T \times \mathbb{G}_m^{\mathrm{rot}})$-module U as

$$\operatorname{gch} U := \sum_{n \in \mathbb{Z}} \sum_{\lambda \in \mathbb{X}} q^n e^{\lambda} \dim \operatorname{Hom}_{T \times \mathbb{G}_m^{\mathrm{rot}}}(\mathbb{C}_{\lambda + n\delta}, U).$$

Then, $\operatorname{gch} V[z^{\pm 1}]$ makes sense as all the coefficients are in $\mathbb{Z}$. However, if we take the second symmetric power $S^2(V[z^{\pm 1}])$ of $V[z^{\pm 1}]$ over $\Bbbk$, then it contains an infinity as a coefficient. To avoid such a complication, we sometimes restrict ourselves to the subgroups

$$G[\![z]\!] := G\big(\Bbbk[\![z]\!]\big) \subset G(\!(z)\!) \quad \text{and} \quad G[z] := G\big(\Bbbk[z]\big) \subset G\big[z^{\pm 1}\big].$$

We sometimes use the subgroup $\mathbf{I} \subset G[\![z]\!]$ defined by the pullback of B under the evaluation map $\mathtt{ev}_0 : G[\![z]\!] \to G$ at $z = 0$. The group $\mathbf{I}$ is the Iwahori subgroup obtained from (the completed version of) $\mathcal{I}$ by removing $\mathbb{G}_m^{\mathrm{rot}}$ and quotient out by the central extension.

By the quotient map $\Bbbk[z] \to \Bbbk$ (and $\Bbbk[\![z]\!] \to \Bbbk$) sending $z \mapsto 0$, we can regard every rational G-module V as a $G[z]$-module or a $G[\![z]\!]$-module with (trivial) $\mathbb{G}_m^{\mathrm{rot}}$-action through $\mathtt{ev}_0$. We also have a $G[\![z]\!]$-module structure (without a $\mathbb{G}_m^{\mathrm{rot}}$-action) on $V[\![z]\!] := V \otimes \Bbbk[\![z]\!]$ that surjects onto V.

Definition 4.1 (global Weyl modules). Let $\mathcal{C}(\lambda)$ be the category of rational $G[z]$-modules M that admits a decreasing filtration

$$M = F_0 M \supset F_1 M \supset F_2 M \supset \cdots \quad \text{such that} \bigcap_{k \geq 0} F_k M = \{0\}$$

and each $F_k M / F_{k-1} M$ ($k \geq 1$) belongs to $\{q^m L(\mu)\}_{m \in \mathbb{Z}, \lambda \geq \mu \in \mathbb{X}_+}$. For each $\lambda \in \mathbb{X}_+$, we define the global Weyl module $\mathbb{W}(\lambda)$ of $G[z]$ as the projective cover of $L(\lambda)$ in $\mathcal{C}(\lambda)$.

Note that $\mathbb{W}(\lambda)$ automatically acquires a $\mathbb{G}_m^{\mathrm{rot}}$-action by its universality (as it exists).

Theorem 4.2. *For each $\lambda \in \mathbb{X}_+$ with $\lambda = \sum_{i=1}^{r} m_i \varpi_i$, we have*

$$\operatorname{End}_{G[z]} \mathbb{W}(\lambda) \cong \bigotimes_{i=1}^{r} \Bbbk[x_{i,1}, \ldots, x_{i,m_i}]^{\mathfrak{S}_{m_i}},$$

where each $x_{i,1}, \ldots, x_{i,m_i}$ is of degree one with respect to the $\mathbb{G}_m^{\mathrm{rot}}$-action. In addition, the action of $\operatorname{End}_{G[z]} \mathbb{W}(\lambda)$ *on* $\mathbb{W}(\lambda)$ *is free.*

Theorem 4.2 was proved by Fourier–Littelmann **[25]** (for $\Bbbk = \mathbb{C}$ and G of type ADE), Naoi **[72]** (for $\Bbbk = \mathbb{C}$ and G of type BCFG), and it was transferred to char $\Bbbk > 0$ in **[50]** using results from the global bases of quantum affine algebras **[4, 42]**.

By Theorem 4.2, we factor out the positive degree parts of $\operatorname{End}_{G[z]} \mathbb{W}(\lambda)$ to obtain

$$W(\lambda) := \Bbbk \otimes_{\operatorname{End}_{G[z]} \mathbb{W}(\lambda)} \mathbb{W}(\lambda), \quad \lambda \in \mathbb{X}_+.$$

We call it a local Weyl module of $G[z]$.

The following result clarifies that our global/local Weyl modules are the best possible analogues of Weyl modules for G (see Theorem 2.1):

Theorem 4.3 (Chari–Ion **[14]** for char $\Bbbk = 0$, and **[50]** $+ \varepsilon$ for char $\Bbbk > 0$). *For each $\lambda, \mu \in \mathbb{X}_+$, we have*

$$\operatorname{Ext}^i_{G[z]}\big(\mathbb{W}(\lambda), W(\mu)^*\big) \cong \Bbbk^{\oplus \delta_{i,0}\delta_{\lambda,\mu^*}}, \tag{4.1}$$

where μ^* is the highest weight of $L(\mu)^*$. By taking the graded Euler–Poincaré characteristic, (4.1) *implies the orthogonality of Macdonald polynomials with respect to the Macdonald pairing specialized to $t = 0$. In particular,* $\operatorname{gch} W(\lambda)$ *and* $\operatorname{gch} \mathbb{W}(\lambda)$ *do not depend on* $\Bbbk$.

The proof of Theorem 4.3 in **[50, §3.3]** relies on the adjoint property of the Demazure functors observed in **[20, PROPOSITION 5.7]** and systematically utilized in **[15]**. The case $\lambda = \mu^*$ and $i > 1$ in Theorem 4.3 is not recorded in **[50]**, and might appear elsewhere.

5. SEMI-INFINITE FLAG MANIFOLDS

We keep the setting of the previous section. In view of the projectivity of $\mathbb{W}(\lambda)$'s in $\mathcal{C}(\lambda)$'s, we find unique degree-zero $G[z]$-module maps

$$\mathbb{W}(\lambda + \mu) \to \mathbb{W}(\lambda) \otimes \mathbb{W}(\mu), \quad \lambda \in \mathbb{X}_+. \tag{5.1}$$

Therefore, the recipe described in Section 2 equips

$$R_G := \bigoplus_{\lambda \in \mathbb{X}_+} \mathbb{W}(\lambda)^\vee$$

with a structure of a commutative algebra compatible with the action of $G[z] \rtimes \mathbb{G}_m^{\mathrm{rot}} \times T$. Since the $\mathbb{G}_m^{\mathrm{rot}}$-degree of R_G is bounded from the above, the $G[z]$-action on R_G automatically extends to the $G[\![z]\!]$-action. We set

$$\mathbf{Q}_G := (\operatorname{Spec} R_G \setminus E)/T,$$

where E is a closed subset of $\operatorname{Spec} R_G$ on which the T-action is not free. Let us consider the $G(\!(z)\!)$-orbit of

$$\big\{[\mathbf{v}_{\varpi_i}]\big\}_{i=1}^r \in \prod_{i=1}^r \mathbb{P}\big(V(\varpi_i)(\!(z)\!)\big), \tag{5.2}$$

viewed as a set of points, that we denote by $\mathscr{Q}_G$. By examining the coefficients of the defining relations of $\mathscr{B}$ with its $\Bbbk((z))$-valued points, we find that the intersection

$$\mathscr{Q}_G \cap \prod_{i=1}^{r} \mathbb{P}\big(V(\varpi_i)[\![z]\!]z^{m_i}\big) \subset \prod_{i=1}^{r} \mathbb{P}\big(V(\varpi_i)[\![z]\!]z^{m_i}\big) \subset \prod_{i=1}^{r} \mathbb{P}\big(V(\varpi_i)((z))\big) \tag{5.3}$$

defines a closed subscheme for any choice of $m_1, \ldots, m_r \in \mathbb{Z}$. We denote this subscheme by $\mathbf{Q}_G(t_\beta)$, where $\beta = \sum_{i=1}^{r} m_i \alpha_i^\vee$ is an element of the dual lattice (coroot lattice) $\mathbb{X}^\vee$ of $\mathbb{X}$ equipped with a basis $\{\alpha_i^\vee\}_{i=1}^r$ such that $\alpha_i^\vee(\varpi_j) = \delta_{i,j}$ (i.e., $\alpha_i^\vee$ is a simple coroot). We note that $\mathbb{P}(V(\varpi_i)[\![z]\!]z^{m_i})$ is a scheme, but it is not of finite type, and $\mathbf{Q}_G(t_\beta)$ is also of infinite type.

Lemma 5.1. *We have $\mathbf{Q}_G(t_\beta) \cong \mathbf{Q}_G(t_\gamma)$ for each pair $\beta, \gamma \in \mathbb{X}^\vee$ as schemes equipped with $G[\![z]\!]$-actions. Hence, the union*

$$\mathbf{Q}_G^{\mathrm{rat}} = \bigcup_{\beta} \mathbf{Q}_G(t_\beta)$$

is a pure ind-scheme of ind-infinite type equipped with the action of $G[\![z]\!] \rtimes \mathbb{G}_m^{\mathrm{rot}}$. Moreover, the set of $G[\![z]\!]$-orbits in $\mathbf{Q}_G^{\mathrm{rat}}$ is in bijection with $\mathbb{X}^\vee$.

In effect, we have an open dense $G[\![z]\!]$-orbit $\mathbf{O}_G(t_\beta) \subset \mathbf{Q}_G(t_\beta)$ that is isomorphic to $G[\![z]\!]/(T \cdot N[\![z]\!])$. By the Bruhat decomposition, we divide $\mathbf{O}_G(t_\beta)$ into the disjoint union of $\mathbf{I}$-orbits as $\bigsqcup_{w \in W} \mathbf{O}(w t_\beta)$ such that $\mathbf{O}(t_\beta) \subset \mathbf{O}_G(t_\beta)$ is open dense. Identifying $\beta \in \mathbb{X}^\vee$ with t_β, we set $W_{\mathrm{af}} := W \ltimes \mathbb{X}^\vee$. We define

$$\mathbf{Q}_G(w) := \overline{\mathbf{O}(w)} \subset \mathbf{Q}_G^{\mathrm{rat}}, \quad w \in W_{\mathrm{af}}.$$

The inclusion relation on $\{\mathbf{Q}_G(w)\}_{w \in W_{\mathrm{af}}}$ is described by the generic Bruhat order [62]. We refer to the partial order on W_{af} induced from this closure ordering by $\leq_{\frac{\infty}{2}}$ as in [50,52] (there we sometimes called $\leq_{\frac{\infty}{2}}$ as the semi-infinite Bruhat order).

Theorem 5.2. *The scheme $\mathbf{Q}_G(w)$ is normal for each $w \in W_{\mathrm{af}}$. In addition, the ind-scheme $\mathbf{Q}_G^{\mathrm{rat}}$ is a strict ind-scheme in the sense that each inclusion is a closed immersion. The ind-scheme $\mathbf{Q}_G^{\mathrm{rat}}$ coarsely ind-represents the coset $G((z))/(T \cdot N((z)))$.*

The first two statements are proved in [52] when $\operatorname{char} \Bbbk = 0$. The proof valid for $\operatorname{char} \Bbbk \neq 2$, as well as the last assertion, are contained in [50]. This last assertion says that the (ind-)scheme $\mathbf{Q}_G^{\mathrm{rat}}$ is the universal one that maps to every (ind-)scheme whose points yield $\mathscr{Q}_G$. It follows that if we take a family $\{\mathbb{Y}(\lambda)\}_{\lambda \in \mathbb{X}_+}$ instead of $\{\mathbb{W}(\lambda)\}_{\lambda \in \mathbb{X}_+}$ to define $\mathbf{Q}_G(t_\beta)$, then the corresponding coordinate ring R'_G admits a map to R_G. Let us point out that this can be thought of as a family version of the properties of global Weyl modules discussed in Section 4, and we indeed have several reasonable choices of $\{\mathbb{Y}(\lambda)\}_{\lambda \in \mathbb{X}_+}$ other than $\{\mathbb{W}(\lambda)\}_{\lambda \in \mathbb{X}_+}$ including the coordinate ring of the arc scheme of G/N. For simplicity, we may refer to $\mathbf{Q}_G(t_0)$ as $\mathbf{Q}_G$ below.

The inclusion

$$\mathbf{Q}_G \subset \prod_{i=1}^{r} \mathbb{P}\big(V(\varpi_i)[\![z]\!]\big) \tag{5.4}$$

induces a line bundle $\mathcal{O}_{\mathbf{Q}_G}(\varpi_i)$ on $\mathbf{Q}_G$, that is, the pull-back of $\mathcal{O}(1)$ from $\mathbb{P}(V(\varpi_i)[\![z]\!])$. By taking the tensor products, we have $\mathcal{O}_{\mathbf{Q}_G}(\lambda) := \bigotimes_{i=1}^r \mathcal{O}_{\mathbf{Q}_G}(\varpi_i)^{\otimes n_i}$ for $\lambda = \sum_{i=1}^r n_i \varpi_i$ ($n_i \in \mathbb{Z}$). By Lemma 5.1, we have $\mathcal{O}_{\mathbf{Q}_G^{\mathrm{rat}}}(\lambda)$ ($\lambda \in \mathbb{X}$) on $\mathbf{Q}_G^{\mathrm{rat}}$ that yields $\mathcal{O}_{\mathbf{Q}_G}(\lambda)$ by restriction.

Theorem 5.3 ([52] for char $\Bbbk = 0$, and [50] for char $\Bbbk \neq 2$). *For each $\lambda \in \mathbb{X}$, we have*

$$H^i\big(\mathbf{Q}_G, \mathcal{O}_{\mathbf{Q}_G}(\lambda)\big)^\vee \cong \begin{cases} \mathbb{W}(\lambda), & i = 0, \lambda \in \mathbb{X}_+, \\ \{0\}, & \textit{otherwise}. \end{cases}$$

The proof of Theorem 5.3 depends on the freeness of R_G over an infinitely-many-variable polynomial ring, that yields a regular sequence of infinite length. Such a situation never occur for finite type schemes, or infinite type schemes like $\mathscr{B}_C^{\mathrm{thick}}$. In case $G = \mathrm{SL}(2)$, Theorem 5.3 reduces to an exercise in algebraic geometry by $\mathbf{Q}_G \cong \mathbb{P}(\Bbbk^2[\![z]\!])$.

Theorem 5.3 has an ind-model counterpart proved earlier [10]. The Frobenius splitting of $\mathbf{Q}_G$ (explained later) and Theorem 5.3 imply this ind-model counterpart. However, the author is uncertain whether [10] implies Theorem 5.3 (even in case char $\Bbbk = 0$) since the natural ring coming from the ind-model is a completion of R_G, and the completion operation of a ring loses information in general. We have an analogue of Theorem 5.3 for all $\mathbf{I}$-orbit closures, proved for the ind-model in [46, 50] and for the formal model in [50, 52].

6. FROBENIUS SPLITTINGS

We continue to work in the setting of the previous section. We fix a prime $p > 0$. For a scheme $\mathfrak{X}$ over $\mathbb{F}_p$, we have a Frobenius morphism $\mathsf{Fr} : \mathfrak{X} \to \mathfrak{X}$ induced from the pth power map. We have a natural map $\mathsf{Fr}^* \mathcal{O}_{\mathfrak{X}} \to \mathcal{O}_{\mathfrak{X}}$ that induces a map $\mathcal{O}_{\mathfrak{X}} \to \mathsf{Fr}_* \mathcal{O}_{\mathfrak{X}}$ by adjunction. The Frobenius splitting $\phi : \mathsf{Fr}_* \mathcal{O}_{\mathfrak{X}} \to \mathcal{O}_{\mathfrak{X}}$ is an $\mathcal{O}_{\mathfrak{X}}$-module map such that the composition

$$\mathcal{O}_{\mathfrak{X}} \to \mathsf{Fr}_* \mathcal{O}_{\mathfrak{X}} \xrightarrow{\phi} \mathcal{O}_{\mathfrak{X}}$$

is the identity. If $\mathfrak{X}$ is projective (and is of finite type) and $\mathcal{O}_{\mathfrak{X}}$ admits a Frobenius splitting, then $\mathfrak{X}$ is reduced and an ample line bundle has the higher cohomology vanishing [68].

For generality on Frobenius splittings, as well as their applications to $\mathscr{B}$ and $\mathscr{B}_C^{\mathrm{thin}}$, we refer to Brion–Kumar [12] (note that [12] has a finite type assumption, that we drop in case the proof does not require it. In the paragraph above, reducedness does not require the finite type assumption, while the higher cohomology vanishing requires the finite type assumption through the Serre vanishing). Frobenius splitting of $\mathscr{B}$ in char $\Bbbk = p$ is useful in proving that Schubert and Richardson varieties are reduced, normal, and have rational singularities. There are two major ways to construct a Frobenius splitting of $\mathscr{B}$: one is to investigate the global section of the $(1 - p)$th power of the canonical bundle, and the other is to use a Bott–Samelson–Demazure–Hansen (=BSDH) resolution of $\mathscr{B}$.

Since $\mathscr{B}_C^{\mathrm{thin}}$ is no longer smooth, we cannot use the canonical bundle to construct a Frobenius splitting. Nevertheless, a (partial) BSDH resolution does the job. The situation of $\mathscr{B}_C^{\mathrm{thick}}$ is a bit worse. The canonical bundle of $\mathscr{B}_C^{\mathrm{thick}}$ makes some sense, but the author does not know whether it has enough power to produce a Frobenius splitting. The scheme $\mathscr{B}_C^{\mathrm{thick}}$

admits a BSDH resolution, but it is a successive $\mathbb{P}^1$-fibration over an infinite-type scheme. Thus, we cannot equip $\mathscr{B}_C^{\text{thick}}$ with a Frobenius splitting by either of the above means at present. Despite this, we can transfer a Frobenius splitting of $\mathscr{B}_C^{\text{thin}}$ to $\mathscr{B}_C^{\text{thick}}$ by using the compatible splitting property of a point **[49]**, following an idea of Mathieu.

Frobenius splitting of $\mathbf{Q}_G^{\text{rat}}$ (or rather each of its ind-piece $\mathbf{Q}_G(w)$) is used below, and hence we need a recipe to produce one. However, the situation of the BSDH resolution is similar to that of $\mathscr{B}_C^{\text{thick}}$, and the canonical bundle on $\mathbf{Q}_G^{\text{rat}}$ simply does not make sense naively (e.g., its T-weight at a point must be infinity). Therefore, we need a new proof strategy. Our strategy in **[50]** is to regard R_G as a subalgebra of the corresponding coordinate ring of $\mathscr{B}_C^{\text{thick}}$, and prove that a Frobenius splitting of $\mathscr{B}_C^{\text{thick}}$ preserves R_G. For this, we first see that each $\mathbb{W}(m\lambda)$ ($m \in \mathbb{Z}_{>0}$, $\lambda \in \mathbb{X}_+$) is a quotient of $L(m\Lambda)$ for some $\Lambda \in P_+$ by twisting the $G[z^{-1}]$-action into a $G[z]$-action as $z^{-1} \mapsto z$. Let $\pi_m : L(m\Lambda) \to \mathbb{W}(m\lambda)$ be the quotient map. This embeds (a suitable $\mathbb{Z}$-graded subalgebra of) R_G into (3.1) as an algebra with $G[\![z]\!] \ltimes \mathbb{G}_m^{\text{rot}}$-action. We need to show that the map $\phi^\vee$ obtained by dualizing the Frobenius splitting of $\mathscr{B}_C^{\text{thick}}$ induces a map $\phi_{\mathbb{W}}^\vee$ in the following diagram:

$$\begin{array}{ccccc} L(m\Lambda) & \xrightarrow{\phi^\vee} & L(pm\Lambda) & \longrightarrow & L(m\Lambda) \\ \downarrow{\scriptstyle \pi_m} & & \downarrow{\scriptstyle \pi_{pm}} & & \downarrow{\scriptstyle \pi_m} \\ \mathbb{W}(m\lambda) & \overset{\phi_{\mathbb{W}}^\vee}{\dashrightarrow} & \mathbb{W}(pm\lambda) & \longrightarrow & \mathbb{W}(m\lambda). \end{array} \tag{6.1}$$

This is equivalent to seeing that $\phi^\vee(\ker \pi_m) \subset \ker \pi_{pm}$. We use the projectivity of $\mathbb{W}(m\lambda)$ in $\mathscr{C}(m\lambda)$ to assume that the $G[z]$-module generators of $\ker \pi_m$ have T-weights that do not appear in $\mathbb{W}(m\lambda)$. In view of the fact that $\ker \pi_{pm}$ contains all the T-weight spaces in $L(pm\Lambda)$ whose T-weights do not appear in $\mathbb{W}(pm\lambda)$, we have necessarily $\phi^\vee(\ker \pi_m) \subset \ker \pi_{pm}$ by the T-weight comparison of the generators.

In fact, every $L(\Lambda)$ admits a filtration by global Weyl modules when $\operatorname{char} \Bbbk = 0$ if we twist the action of $G[z]$ on global Weyl modules into $G[z^{-1}]$ **[51]**. Therefore, we indeed obtain a Frobenius splitting of $\mathbf{Q}_G$ via a novel proof based on the "universality" of the global Weyl module $\mathbb{W}(\lambda)$ explained in Section 4. In conclusion, we have:

Theorem 6.1 (**[50, THEOREM B]**). *The ind-scheme $\mathbf{Q}_G^{\text{rat}}$ admits a Frobenius splitting that is compatible with all* **I***-orbits when* $\operatorname{char} \Bbbk > 2$.

7. CONNECTION TO THE SPACE OF RATIONAL MAPS

Keep the setting as in Section 5. Let us consider the vector space embedding $\Bbbk((z)) \subset \Bbbk[\![z, z^{-1}]\!]$ into the formal power series with unbounded powers. The space $\Bbbk[\![z, z^{-1}]\!]$ no longer forms a ring. Nevertheless, we have an automorphism of $\Bbbk[\![z, z^{-1}]\!]$ by swapping z with z^{-1}. Together with the Chevalley involution of G (an automorphism of G that sends each element of T to its inverse), it induces an involution θ on the ambient space

$$\mathbf{Q}_G^{\text{rat}} \subset \prod_{i=1}^{r} \mathbb{P}\big(V(\varpi_i)[\![z, z^{-1}]\!]\big).$$

We remark that θ induces an automorphism of G such that $B \cap \theta(B) = T$. Let w_0 be the longest element in W.

Theorem 7.1 ([58, THEOREM B]). *For all $w, v \in W_{\mathrm{af}}$, the scheme-theoretic intersection $\mathbf{Q}_G(w) \cap \theta(\mathbf{Q}_G(vw_0))$ is reduced (we denote this intersection by $\mathcal{Q}_G(w, v)$ and call it a Richardson variety of $\mathbf{Q}_G^{\mathrm{rat}}$ below). It is normal when* $\mathrm{char}\, \Bbbk = 0$ *or* $\mathrm{char}\, \Bbbk \gg 0$.

The scheme $\mathcal{Q}_G(w, v)$ is always of finite type, and the case $w, v \in W$ yields a Richardson variety of $\mathscr{B}$. The normality part of the proof of Theorem 7.1 goes as follows: Our Frobenius splitting of $\mathbf{Q}_G^{\mathrm{rat}}$ induces a Frobenius splitting of $\mathcal{Q}_G(w, v)$. In particular, it is reduced and weakly normal in $\mathrm{char}\, \Bbbk > 2$. (Here a weakly normal ring is essentially a normal ring up to topology.) Then, we lift the weak normality to characteristic zero and prove the normality of the intersection by a geometric consideration. Once we deduce the normality in characteristic zero, we can reduce it to $\mathrm{char}\, \Bbbk \gg 0$ by a general result.

Let us exhibit some relevant geometric considerations here. To this end, we assume $\Bbbk = \mathbb{C}$ in the rest of this section. Recall that $H_2(\mathscr{B}, \mathbb{Z}) \cong \mathbb{X}^\vee$. Let $\mathscr{G}\mathscr{B}_{2,\beta}$ (resp. $\mathscr{B}_{2,\beta}$) be the space of genus-zero stable maps with two marked points to $(\mathbb{P}^1 \times \mathscr{B})$ (resp. $\mathscr{B}$) whose image has class $(1, \beta) \in H_2(\mathbb{P}^1 \times \mathscr{B}, \mathbb{Z})$ (resp. $\beta \in H_2(\mathscr{B}, \mathbb{Z})$), regarded as an algebraic variety with rational singularities [28]. We have a subvariety $\mathscr{G}\mathscr{B}_{2,\beta}^\flat$ such that the first marked point lands in $0 \in \mathbb{P}^1$ and the second marked point lands in $\infty \in \mathbb{P}^1$ through the composition

$$\left(C, \{x_1, x_2\}\right) \xrightarrow{f} \mathbb{P}^1 \times \mathscr{B} \xrightarrow{\mathrm{pr}_1} \mathbb{P}^1.$$

Consider the Schubert variety (a B-orbit closure) $\mathscr{B}(w) \subset \mathscr{B}$ corresponding to $w \in W$ and the opposite Schubert variety (a $\theta(B)$-orbit closure) $\mathscr{B}^{\mathrm{op}}(v) \subset \mathscr{B}$ corresponding to $v \in W$.

Let $\mathrm{ev}_i : \mathscr{G}\mathscr{B}_{2,\beta}^\flat \to \mathscr{B}$ $(i = 1, 2)$ denote the evaluation at the point x_i on C. We define

$$\mathscr{G}\mathscr{B}_\beta(w, v) := \mathrm{ev}_1^{-1}\left(\mathscr{B}(w)\right) \cap \mathrm{ev}_2^{-1}\left(\mathscr{B}^{\mathrm{op}}(v)\right).$$

Similarly, let $\mathrm{e}_i : \mathscr{B}_{2,\beta} \to \mathscr{B}$ $(i = 1, 2)$ be the evaluation maps. For all $w, v \in W$ and $\beta \in \mathbb{X}^\vee$, we set $\mathscr{B}_\beta(w, v) := (\mathrm{e}_1^{-1}(\mathscr{B}(w)) \cap \mathrm{e}_2^{-1}(\mathscr{B}^{\mathrm{op}}(v)))$. Let $\mathring{\mathcal{Q}}_G(\beta)$ denote the space of maps from $\mathbb{P}^1$ to $\mathscr{B}$ of degree β. By adding the identity map to $\mathbb{P}^1$, each point of $\mathring{\mathcal{Q}}_G(\beta)$ yields a map $\mathbb{P}^1 \to (\mathbb{P}^1 \times \mathscr{B})$ of degree $(1, \beta)$. In addition, the identification of two $\mathbb{P}^1$'s completely determines the marked points. Hence we have an inclusion $\mathring{\mathcal{Q}}_G(\beta) \subset \mathscr{G}\mathscr{B}_{2,\beta}^\flat$.

Let $\mathcal{Q}_G(\beta)$ $(\beta \in \mathbb{X}^\vee)$ denote the space of quasimaps from $\mathbb{P}^1$ to $\mathscr{B}$ of degree β [22], that is, a natural compactification of $\mathring{\mathcal{Q}}_G(\beta)$ such that

$$\mathcal{Q}_G(\beta) = \bigsqcup_{0 \le \gamma \le \beta} \mathring{\mathcal{Q}}_G(\beta - \gamma) \times (\mathbb{P}^1)^\gamma,$$

where $\gamma \le \beta$ is defined as $\beta - \gamma \in \sum_{i=1}^r \mathbb{Z}_{\ge 0} \alpha_i^\vee$, and

$$(\mathbb{P}^1)^\gamma = \prod_{i=1}^r \left((\mathbb{P}^1)^{m_i} / \mathfrak{S}_{m_i}\right) \quad \text{where } \gamma = \sum_{i=1}^r m_i \alpha_i^\vee.$$

Here $(\mathbb{P}^1)^\gamma$ records the place where the degree of the genuine map drops in which degree components (without ordering). By adding extra $\mathbb{P}^1$ components and (compatible) maps to

$\mathscr{B}$ to $\mathbb{P}^1$ in $(f : \mathbb{P}^1 \to \mathscr{B}) \in \mathring{\mathscr{Q}}_G(\beta - \gamma)$ at the places (and total degrees) recorded by $(\mathbb{P}^1)^\gamma$ (for each $0 \leq \gamma \leq \beta$), we obtain a map of topological spaces

$$\pi : \mathscr{G}\mathscr{B}^\flat_{2,\beta} \to \mathscr{Q}_G(\beta),$$

that is an identity on $\mathring{\mathscr{Q}}_G(\beta)$. Givental's main lemma asserts that this is a birational morphism of integral algebraic varieties.

Proposition 7.2 ([58, §5.2]). *For each $\beta \in \mathbb{X}^\vee$, we have*

$$\mathscr{Q}_G(\beta) \cong \mathscr{Q}_G(e, t_\beta)$$

as schemes. In addition, π restricts to a birational morphism

$$\pi_{\beta,w,v} : \mathscr{G}\mathscr{B}_\beta(w, v) \to \mathscr{Q}_G(w, vt_\beta), \quad w, v \in W.$$

In particular, we have $\mathscr{G}\mathscr{B}_\beta(w, v) \neq \emptyset$ if and only if $w \leq_{\frac{\infty}{2}} vt_\beta$, and its dimension is given by the distance between w and vt_β with respect to $\leq_{\frac{\infty}{2}}$.

In other words, the Richardson varieties of $\mathbf{Q}_G^{\mathrm{rat}}$ are precisely the spaces of quasi-maps, possibly with additional conditions imposed by the space of stable maps. According to Buch–Chaput–Mihalcea–Perrin [13], the variety $\mathscr{G}\mathscr{B}_\beta(w, v)$ is irreducible and has rational singularities if it is nonempty. Hence, we find that $\mathscr{Q}_G(w, vt_\beta)$ is irreducible in general. Proposition 7.2 and properties of the maps $\pi_{\beta,w,v}$ are used in our proof of Theorem 7.1.

Proposition 7.2 implies that $\mathscr{Q}_G(w, vt_\beta)$ is the closure (in $\mathscr{Q}_G(\beta)$) of the space of maps from $\mathbb{P}^1$ to $\mathscr{B}$ such that $0, \infty \in \mathbb{P}^1$ land in $\mathscr{B}(w)$ and $\mathscr{B}^{\mathrm{op}}(v)$, respectively. By examining the natural map $\mathscr{G}\mathscr{B}_\beta(w, v) \to \mathscr{B}_\beta(w, v)$ (obtained by forgetting the map to $\mathbb{P}^1$), we obtain:

Corollary 7.3. *For all $w, v \in W$ and $0 \neq \beta \in \mathbb{X}^\vee$, we have*

$$\dim \mathscr{B}_\beta(w, v) = \dim \mathscr{G}\mathscr{B}_\beta(w, v) - 1 \quad \textit{if } \mathscr{G}\mathscr{B}_\beta(w, v) \neq \emptyset,$$

and $\mathscr{B}_\beta(w, v) \neq \emptyset$ if and only if $\mathscr{G}\mathscr{B}_\beta(w, v) \neq \emptyset$. Moreover, we have

$$\mathscr{B}_\beta(w, v) \neq \emptyset \quad \textit{and} \quad \dim \mathscr{B}_\beta(w, v) = 0$$

if and only if $w \leq_{\frac{\infty}{2}} vt_\beta$ are adjacent with respect to $\leq_{\frac{\infty}{2}}$. In such a case, $\mathscr{B}_\beta(w, v)$ is a point.

Thanks to the dimension axiom in quantum correlators [54, (2.5)], Corollary 7.3 describes which (primary) two-point cohomological Gromov–Witten invariant of $\mathscr{B}$ with respect to the Schubert bases is nonzero (we can also tell its exact value). By the divisor axiom [54, §2.2.4] and the classical Chevalley formula [16], we find the Chevalley formula in quantum cohomology of $\mathscr{B}$ from this [29]. This clarifies the role of $\mathscr{Q}_G(w, vt_\beta)$ in the study of quantum cohomology of $\mathscr{B}$ from our perspective.

Theorem 7.4 ([47]). *Let $\beta \in \mathbb{X}^\vee$ and $w, v \in W$. The variety $\mathscr{Q}_G(w, vt_\beta)$ has rational singularities.*

Theorem 7.4 is proved by Braverman–Finkelberg [9,10] for the case $w = e$, $v = w_0$ by an analysis of Zastava spaces, which does not extend to general w, v. Theorem 7.4 is the most subtle technical point in [47] and its induction steps become possible by Theorem 7.1.

8. K-THEORETIC PETERSON ISOMORPHISM

We follow the setting of the previous section with $\Bbbk = \mathbb{C}$. We understand that the K-groups appearing here contain a suitable class of line bundles supported on subvarieties equipped with some group actions, and its scalar is extended from $\mathbb{Z}$ to $\mathbb{C}$. Let $\mathrm{Gr}_G := G((z))/G[\![z]\!]$ be the affine Grassmannian of G. The set of $\mathbf{I}$-orbits in Gr_G is in bijection with $\mathbb{X}^\vee$, while the set of $G[\![z]\!]$-orbits of Gr_G is in bijection with $\mathbb{X}^\vee_< \subset \mathbb{X}^\vee$ formed by the set of antidominant coroots. For $\beta \in \mathbb{X}^\vee$, we set $\mathring{\mathrm{Gr}}_G(\beta) \subset \mathrm{Gr}_G$ as the corresponding $\mathbf{I}$-orbit and set $\mathrm{Gr}_G(\beta) := \overline{\mathring{\mathrm{Gr}}_G(\beta)} \subset \mathrm{Gr}_G$. We normalize so that $\mathrm{Gr}_G(\beta)$ is G-stable when $\beta \in \mathbb{X}^\vee_<$, and we have $\dim \mathrm{Gr}_G(\beta) = -2|\beta|$ in such a case, where $|\beta| := \sum_{i=1}^r \beta(\varpi_i)$.

We define

$$K_T(\mathrm{Gr}_G) := \bigcup_{\beta \in \mathbb{X}^\vee} K_T\big(\mathrm{Gr}_G(\beta)\big) \quad \text{and} \quad K_G(\mathrm{Gr}_G) := \bigcup_{\beta \in \mathbb{X}^\vee_<} K_G\big(\mathrm{Gr}_G(\beta)\big).$$

These spaces are equipped with the convolution product, defined by the diagram

$$\mathrm{Gr}_G \times \mathrm{Gr}_G \stackrel{p}{\leftarrow} G((z)) \times \mathrm{Gr}_G \stackrel{q}{\rightarrow} G((z)) \times_{\mathbf{I}} \mathrm{Gr}_G \stackrel{\mathrm{mult}}{\rightarrow} \mathrm{Gr}_G$$

as follows: For all cycles $a, b \in K_T(\mathrm{Gr}_G) \cong K_{\mathbf{I}}(\mathrm{Gr}_G)$, we find a left $\mathbf{I}$-equivariant class (a, b) on $G((z)) \times_{\mathbf{I}} \mathrm{Gr}_G$ such that

$$p^*(a \boxtimes b) = q^*(a, b)$$

and set

$$a \odot' b := \sum_{i \geq 0} (-1)^i \big[\mathbb{R}^i \,\mathrm{mult}_*(a, b)\big] \in K_{\mathbf{I}}(\mathrm{Gr}_G).$$

This yields an associative product structure on $K_T(\mathrm{Gr}_G)$ that contains a zero divisor. If we restrict ourselves to $K_G(\mathrm{Gr}_G)$, then the algebra structure given by $\odot'$ becomes commutative and integrally closed. Using an isomorphism $K_T(\mathrm{pt}) \otimes_{K_G(\mathrm{pt})} K_G(\mathrm{Gr}_G) \cong K_T(\mathrm{Gr}_G)$ of $K_T(\mathrm{pt})$-modules, we find a multiplication $\odot$ of $K_T(\mathrm{Gr}_G)$ that extends $\odot'$ on $K_G(\mathrm{Gr}_G)$ as a $K_T(\mathrm{pt})$-algebra. This product $\odot$ coincides with a K-theoretic analogue of the Pontrjagin product (by the calculations in [47, §2.2]). In addition, we have

$$[\mathcal{O}_{\mathrm{Gr}_G(\beta+\gamma)}] = [\mathcal{O}_{\mathrm{Gr}_G(\beta)}] \odot [\mathcal{O}_{\mathrm{Gr}_G(\gamma)}] \quad \text{for} \quad \beta, \gamma \in \mathbb{X}^\vee_<.$$

This yields a multiplicative system in $K_T(\mathrm{Gr}_G)$, whose localization is denoted by $K_T(\mathrm{Gr}_G)_{\mathrm{loc}}$.

The (localized) small T-equivariant quantum K-group of $\mathscr{B}$ is defined as a vector space

$$qK_T(\mathscr{B})_{\mathrm{loc}} := K_T(\mathscr{B}) \otimes \mathbb{C}\mathbb{X}^\vee \quad \big(\equiv K_T(\mathscr{B}) \otimes_{\mathbb{C}} \mathbb{C}H_2(\mathscr{B}, \mathbb{Z})\big).$$

We denote the variable corresponding to $\beta \in \mathbb{X}^\vee$ as Q^β. The quantum K-theoretic product $\star$ is a binary operation on $qK_T(\mathscr{B})_{\mathrm{loc}}$, defined by Givental [33] and Lee [61], whose value (a priori) belongs to a completion of $qK_T(\mathscr{B})_{\mathrm{loc}}$. It is one of the consequence of our analysis that $\star$ preserves $qK_T(\mathscr{B})_{\mathrm{loc}}$. This is usually referred to as the *finiteness* of the quantum K-theoretic product (for $\mathscr{B}$) in the literature [1,13], and is one of the most fundamental questions in the study of $qK_T(\mathscr{B})$. Lam–Li–Mihalcea–Shimozono [58] conjectured that:

Theorem 8.1 ([47]). *We have an isomorphism of commutative algebras*

$$K_T(\mathrm{Gr}_G)_{\mathrm{loc}} \xrightarrow{\cong} qK_T(\mathscr{B})_{\mathrm{loc}}$$

such that

$$[\mathcal{O}_{\mathrm{Gr}_G(w\beta)}] \odot [\mathcal{O}_{\mathrm{Gr}_G(\gamma)}]^{-1} \mapsto [\mathcal{O}_{\mathscr{B}(w)}]Q^{\beta-\gamma}$$

for $\beta, \gamma \in \mathbb{X}^\vee_<$ such that $\beta(\varpi_i) < 0$ for every $1 \le i \le r$.

Note that a presentation of the ring $qK_T(\mathscr{B})$ for $G = \mathrm{SL}(n)$ can be read-off from Givental–Lee [34], and a presentation of the ring $K_T(\mathrm{Gr}_G)$ is obtained in Bezrukavnikov–Finkelberg–Mirković [6]. However, these are not enough to yield Theorem 8.1 (for $G = \mathrm{SL}(n)$) as the correspondence between Schubert bases is unclear.

We have an action of the nilpotent version $\mathscr{H}^{\mathrm{nil}}$ of the double affine Hecke algebra (associated to G) on $K_T(\mathrm{Gr}_G)$, coming from Kostant–Kumar [55]. In [47], we defined the T-equivariant K-group $K_T(\mathbf{Q}_G^{\mathrm{rat}})$ of $\mathbf{Q}_G^{\mathrm{rat}}$ based on the construction of the $(T \times \mathbb{G}_m^{\mathrm{rot}})$-equivariant K-group of $\mathbf{Q}_G^{\mathrm{rat}}$ in [52]. The $\mathbf{I}$-action on $\mathbf{Q}_G^{\mathrm{rat}}$ induces a $\mathscr{H}^{\mathrm{nil}}$-action on $K_T(\mathbf{Q}_G^{\mathrm{rat}})$.

The object $K_T(\mathbf{Q}_G^{\mathrm{rat}})$ needs a completion in order to admit an action of the line bundle twists by $\mathcal{O}_{\mathbf{Q}_G^{\mathrm{rat}}}(\lambda)$ $(\lambda \in \mathbb{X})$. It reflects the fact that the right-hand side of Theorem 5.3 (i.e., a global Weyl module) is infinite-dimensional in general, and hence the effect of $\otimes\mathcal{O}_{\mathbf{Q}_G^{\mathrm{rat}}}(\varpi_i)$ $(1 \le i \le r)$ requires infinitely many terms to describe.

Our main idea in the proof of Theorem 8.1 is to put $\mathbf{Q}_G^{\mathrm{rat}}$ into the picture:

Theorem 8.2 ([47, THEOREM C]). *We have a commutative diagram*

$$\begin{array}{ccc} & K_T(\mathbf{Q}_G^{\mathrm{rat}}) & \\ \overset{\Phi}{\nearrow} & & \overset{\Psi}{\nwarrow} \\ K_T(\mathrm{Gr}_G)_{\mathrm{loc}} & \longrightarrow & qK_T(\mathscr{B})_{\mathrm{loc}} \end{array}$$

that respects the Schubert bases in each object. In addition, the map Ψ is an embedding of representations of $\mathscr{H}^{\mathrm{nil}}$, and the map Ψ intertwines the tensor product with $\mathcal{O}_{\mathbf{Q}_G^{\mathrm{rat}}}(-\varpi_i)$ in $K_T(\mathbf{Q}_G^{\mathrm{rat}})$ and the quantum product of $\mathcal{O}_{\mathscr{B}}(-\varpi_i)$ on $qK_T(\mathscr{B})_{\mathrm{loc}}$ for each $1 \le i \le r$.

The completion of $K_T(\mathbf{Q}_G^{\mathrm{rat}})$ is compatible with the standard completion of $qK_T(\mathscr{B})$ via the map Ψ. Theorem 8.2 implies that the inverse of the operation $\star\mathcal{O}_{\mathscr{B}}(-\varpi_i)$ makes sense only after the completion of $qK_T(\mathscr{B})_{\mathrm{loc}}$.

Since the quantum K-theoretic correlators (see [33,61]) satisfy neither the dimension axiom nor divisor axiom as in the theory of quantum cohomology, the proof of Theorem 8.2

must be necessarily different from Corollary 7.3. Our construction of the map Ψ is based on the following two observations:

- an interpretation of the ($\mathbb{G}_m$-equivariant) quantum K-theoretic correlator

$$\chi\big(\mathcal{Q}(w, w_0 t_\beta), \mathcal{O}_{\mathcal{Q}(w,w_0 t_\beta)}(\lambda)\big) = \chi\big(\mathcal{GB}_\beta(w, w_0), \pi^*_{\beta,w,w_0} \mathcal{O}_{\mathcal{Q}(w,w_0 t_\beta)}(\lambda)\big), \tag{8.1}$$

  valued in $\mathbb{C}[T][q^{\pm 1}] = \mathbb{C}[T \times \mathbb{G}_m]$, for each $w \in W$, $\beta \in \mathbb{X}^\vee$, $\lambda \in \mathbb{X}_+$;

- an interpretation of its asymptotic behavior

$$\lim_{\beta \to \infty} \chi\big(\mathcal{Q}(w, w_0 t_\beta), \mathcal{O}_{\mathcal{Q}(w,w_0 t_\beta)}(\lambda)\big) = \chi\big(\mathbf{Q}_G(w), \mathcal{O}_{\mathbf{Q}_G(w)}(\lambda)\big) \in \mathbb{C}(\!(q^{-1})\!)[T] \tag{8.2}$$

  for each $w \in W$, $\lambda \in \mathbb{X}_+$ as an element of $K_T(\mathbf{Q}_G^{\mathrm{rat}})$.

Here we can further interpret $\chi(\mathcal{GB}_\beta(w, w_0), \pi^*_{\beta,w,w_0} \mathcal{O}_{\mathcal{Q}(w,w_0 t_\beta)}(\lambda))$ using the shift operators of line bundles in quantum K-theory **[35, PROPOSITION 2.13]**, and hence we obtain an (abstract) presentation of $qK_T(\mathcal{B})$ from (8.1) by the reconstruction theorem **[35, PROPOSITION 2.12]**. The identity (8.1) is a consequence of Theorem 7.4, and (8.2) is a consequence of compatible Frobenius splitting properties of $\mathcal{Q}_G(w, v)$s and $\mathbf{Q}_G^{\mathrm{rat}}$ in char $\Bbbk > 2$ (see the explanation about the proof of Theorem 7.1).

There is a noncommutative version of Theorem 8.2, meaning that we include $\mathbb{G}_m^{\mathrm{rot}}$ (the variable "q" above) in each item **[49]**.

9. FUNCTORIALITY OF QUANTUM K-GROUPS

We continue to work in the setting as in the previous section. In **[50]**, we have presented analogues of Theorems 5.2, 5.3, and 7.1 for partial flag manifolds of G. Let us find a standard parabolic subgroup $B \subset P \subset G$ and consider $\mathcal{B}_P := G/P$. Our parabolic version of the semi-infinite flag manifold $\mathbf{Q}_{G,P}^{\mathrm{rat}}$ has its set of $\Bbbk$-valued points $G(\!(z)\!)/(T \cdot [P, P](\!(z)\!))$. The fiber of the natural map

$$\pi_P : \mathbf{Q}_G^{\mathrm{rat}} \to \mathbf{Q}_{G,P}^{\mathrm{rat}}$$

is isomorphic to the semi-infinite flag manifold of $[L, L]$, where $L \subset P$ is the maximal semisimple subgroup of P that contains T (the standard Levi subgroup). We also have the higher cohomology vanishing of equivariant line bundles on $\mathbf{Q}_{G,P}^{\mathrm{rat}}$ (or rather $\pi_P(\mathbf{Q}_G)$) as in Theorem 5.3. These are enough to yield a morphism

$$K_{T \times \mathbb{G}_m^{\mathrm{rot}}}(\mathbf{Q}_G^{\mathrm{rat}}) \to K_{T \times \mathbb{G}_m^{\mathrm{rot}}}(\mathbf{Q}_{G,P}^{\mathrm{rat}})$$

obtained by the push-forward by π_P (up to technical reservations neglected here and below).

By transferring Theorem 7.4 to Richardson varieties of $\mathbf{Q}_{G,P}^{\mathrm{rat}}$, we find a map

$$\Psi_P : qK_T(\mathcal{B}_P)_{\mathrm{loc}} \to K_T(\mathbf{Q}_{G,P}^{\mathrm{rat}}),$$

that intertwines appropriate line bundle twists (and analogous quantum multiplications). This yields a diagram

$$\begin{array}{ccc} qK_T(\mathscr{B})_{\mathrm{loc}} & \xrightarrow{\Psi} & K_T(\mathbf{Q}_G) \\ \downarrow & & \downarrow (\pi_P)_* \\ qK_T(\mathscr{B}_P)_{\mathrm{loc}} & \xrightarrow{\Psi_P} & K_T(\mathbf{Q}_{G,P}) \end{array}$$

where we set $\mathbf{Q}_{G,P} := \pi_P(\mathbf{Q}_G)$.

The resulting map $qK_T(\mathscr{B}) \to qK_T(\mathscr{B}_P)$ is, in fact, an algebra map [48], and is easy to describe. Note that we cannot have an analogous map between ordinary K-groups because of the higher direct images. It turns out this map sends $Q^{\alpha_i^\vee}$ to 1 for a simple coroot $\alpha_i^\vee$ belonging to L, and hence is not compatible with a naive generalization of the corresponding map in the Peterson isomorphism in homology [59].

We also have a restriction map $qK_T(\mathscr{B}) \to qK_T(\mathscr{B}^L)$, where $\mathscr{B}^L := L/(L \cap B)$ is the flag manifold of a standard Levi subgroup. This map extends to algebra maps [45]

$$K_{G\times\mathbb{G}_m^{\mathrm{rot}}}(\mathrm{Gr}_G) \to K_{L\times\mathbb{G}_m^{\mathrm{rot}}}(\mathrm{Gr}_L) \to K_{T\times\mathbb{G}_m^{\mathrm{rot}}}(\mathrm{Gr}_T)$$

anticipated in Finkelberg and Tsymbaliuk [23].

10. SOME PERSPECTIVES

Compared with the theory of flag manifolds, many precise results and constructions for $\mathbf{Q}_G^{\mathrm{rat}}$ are still missing. The most accessible set of problems might be to spell out analogues of numerous explicit formulas in classical Schubert calculus purely combinatorially by admitting geometric conclusions from [3,45,47,48,52] partly explained in the previous two sections. We close this note by briefly discussing some of other problems.

10.1. Categorifications of the coordinate rings

The homogeneous coordinate rings of Schubert varieties of a usual flag manifold, that are B-stable quotient rings of (2.1), can be seen as the Grothendieck groups of suitable categories equipped with cluster structures ([60]; see also Section 3.1). Hence, it is natural to expect categorifications of the homogeneous coordinate rings of $\mathbf{Q}_G^{\mathrm{rat}}(w)$ and $\mathscr{B}_C^{\mathrm{thick}}$. See also [21] and [43] for related problems and partial answers.

10.2. Peterson isomorphism in quantum cohomology

The Peterson isomorphism in quantum cohomology [59,74] is an analogue of Theorem 8.1 for homology. We may apply Corollary 7.3 to [69] (that is an essential ingredient in [59]) to utilize $\mathbf{Q}_G^{\mathrm{rat}}$ in its proof (that looks similar to the original strategy in [74]). However, we do not know an analogue of Theorem 8.2 as we lack a proper definition of $H^\bullet(\mathbf{Q}_G)$.

10.3. Constructible sheaves on semi-infinite flags

In representation-theoretic analysis on $\mathscr{B}$, we sometimes encounter constructible sheaves that are not N-equivariant. Also, we want some notion of (co)homology of $\mathbf{Q}_G$ in

Section 10.2. Therefore, it is desirable to understand constructible sheaves on $\mathbf{Q}_G$ following [7]. The resulting objects should have connection to [30]. Note that the combinatorics that should be satisfied by the $\mathbf{I}$-equivariant sheaves (equipped with Frobenius endomorphisms) have been worked out in detail [62, 65].

10.4. Tensor product decompositions

The tensor product decomposition of rational representations of G is deeply connected with our whole story due to the presentation (2.1). In [57], the geometry of flag varieties is used to deduce subtle information on the tensor products beyond the classical Littlewood–Richardson rule. It would be interesting to pursue their analogues in $\mathbf{Q}_G^{\mathrm{rat}}$, possibly utilizing some modular interpretation [11] and connecting with the perspectives in [5].

10.5. The cotangent bundle of semi-infinite flags

A version of the cotangent bundle of $\mathbf{Q}_G^{\mathrm{rat}}$ would make it possible to compare our results with the perspectives in [21, 67, 73]. In addition, its quantization should realize some numerics in Section 10.3. The author hopes to say a bit more on this in St. Petersburg.

ACKNOWLEDGMENTS

The works presented here could not be carried out without suggestions and interest by Misha Finkelberg. The author would like to express his deepest gratitude to him. The author also would like to thank Ivan Cherednik and Thomas Lam for sharing their insights over years, and Noriyuki Abe and Toshiyuki Tanisaki for their comments.

FUNDING

This work was partially supported by JSPS KAKENHI Grant Numbers JP26287004, JP19H01782, and JPJSBP120213210.

REFERENCES

[1] D. Anderson, L. Chen, H. H. Tseng, and H. Iritani, On the finiteness of quantum K-theory of a homogeneous space. 2018, arXiv:1804.04579v3.

[2] S. Arkhipov, A. Braverman, R. Bezrukavnikov, D. Gaitsgory, and I. Mirković, Modules over the small quantum group and semi-infinite flag manifold. *Transform. Groups* **10** (2005), no. 3–4, 279–362.

[3] S. Baldwin and S. Kumar, Positivity in T-equivariant K-theory of flag varieties associated to Kac–Moody groups. II. *Represent. Theory* **21** (2017), 35–60.

[4] J. Beck and H. Nakajima, Crystal bases and two-sided cells of quantum affine algebras. *Duke Math. J.* **123** (2004), no. 2, 335–402.

[5] P. Belkale, The tangent space to an enumerative problem. In *Proceedings of the ICM 2010 (Hyderabad) II*, pp. 405–426, Hindustan Book Agency, New Delhi, 2010.

[6] R. Bezrukavnikov, M. Finkelberg, and I. Mirković, Equivariant homology and K-theory of affine Grassmannians and Toda lattices. *Compos. Math.* **141** (2005), no. 3, 746–768.

[7] A. Bouthier, Cohomologie étale des espaces d'arcs. 2015, arXiv:1509.02203v6.

[8] A. Braverman, Spaces of quasi-maps into the flag varieties and their applications. In *Proceedings of the ICM 2006 (Madrid) II*, pp. 1145–1170, Eur. Math. Soc., Zürich, 2006.

[9] A. Braverman and M. Finkelberg, Semi-infinite Schubert varieties and quantum K-theory of flag manifolds. *J. Amer. Math. Soc.* **27** (2014), no. 4, 1147–1168.

[10] A. Braverman and M. Finkelberg, Weyl modules and q-Whittaker functions. *Math. Ann.* **359** (2014), no. 1–2, 45–59.

[11] A. Braverman, M. Finkelberg, D. Gaitsgory, and I. Mirković, Intersection cohomology of Drinfeld's compactifications. *Selecta Math. (N.S.)* **8** (2002), no. 3, 381–418.

[12] M. Brion and S. Kumar, *Frobenius splitting methods in geometry and representation theory*. Progr. Math. 231, Birkhäuser Boston, Inc., Boston, MA, 2005.

[13] A. S. Buch, P.-E. Chaput, L. C. Mihalcea, and N. Perrin, Finiteness of cominuscule quantum K-theory. *Ann. Sci. Éc. Norm. Supér. (4)* **46** (2013), no. 3, 477–494.

[14] V. Chari and B. Ion, BGG reciprocity for current algebras. *Compos. Math.* **151** (2015), no. 7, 1265–1287.

[15] I. Cherednik and S. Kato, Nonsymmetric Rogers–Ramanujan sums and thick Demazure modules. *Adv. Math.* **374** (2020), 107335, 57pp.

[16] C. Chevalley, Sur les décompositions cellulaires des espaces G/B. In *Algebraic groups and their generalizations: classical methods (University Park, PA, 1991)*, pp. 1–23, Amer. Math. Soc., Providence, RI, 1994.

[17] V. G. Drinfel'd, Quantum groups. In *Proceedings of the ICM 1986 (Berkeley) Vol. 1, 2*, pp. 798–820, Amer. Math. Soc., Providence, RI, 1987.

[18] B. Feigin, M. Finkelberg, A. Kuznetsov, and I. Mirković, Semi-infinite flags. II. Local and global intersection cohomology of quasimaps' spaces. In *Differential topology, infinite-dimensional Lie algebras, and applications*, pp. 113–148, Amer. Math. Soc. Transl. Ser. 2 194, Amer. Math. Soc., Providence, RI, 1999.

[19] B. L. Feĭgin and E. V. Frenkel, Affine Kac–Moody algebras and semi-infinite flag manifolds. *Comm. Math. Phys.* **128** (1990), no. 1, 161–189.

[20] E. Feigin, S. Kato, and I. Makedonskyi, Representation theoretic realization of non-symmetric Macdonald polynomials at infinity. *J. Reine Angew. Math.* **764** (2020), 181–216.

[21] M. Finkelberg, Double affine Grassmannians and Coulomb branches of $3d$ $\mathcal{N} = 4$ quiver gauge theories. In *Proceedings of the ICM 2018 (Rio de Janeiro)*, pp. 1283–1302, World Sci. Publ., Hackensack, NJ, 2018.

[22] M. Finkelberg and I. Mirković, Semi-infinite flags. I. Case of global curve $\mathbf{P}^1$. In *Differential topology, infinite-dimensional Lie algebras, and applications*, pp. 81–112, Amer. Math. Soc. Transl. Ser. 2 194, Amer. Math. Soc., Providence, RI, 1999.

[23] M. Finkelberg and A. Tsymbaliuk, Multiplicative slices, relativistic Toda and shifted quantum affine algebras. In *Representations and nilpotent orbits of Lie algebraic systems*, edited by M. Gorelik, V. Hinich, and A. Melnikov, pp. 133–304, Progr. Math. 330, Birkhäuser, Basel, 2019.

[24] S. Fishel, I. Grojnowski, and C. Teleman, The strong Macdonald conjecture and Hodge theory on the loop Grassmannian. *Ann. of Math. (2)* **168** (2008), no. 1, 175–220.

[25] G. Fourier and P. Littelmann, Weyl modules, Demazure modules, KR-modules, crystals, fusion products and limit constructions. *Adv. Math.* **211** (2007), no. 2, 566–593.

[26] E. Frenkel and D. Gaitsgory, Localization of $\mathfrak{g}$-modules on the affine Grassmannian. *Ann. of Math. (2)* **170** (2009), no. 3, 1339–1381.

[27] E. Frenkel, D. Gaitsgory, D. Kazhdan, and K. Vilonen, Geometric realization of Whittaker functions and the Langlands conjecture. *J. Amer. Math. Soc.* **11** (1998), no. 2, 451–484.

[28] W. Fulton and R. Pandharipande, Notes on stable maps and quantum cohomology. In *Algebraic Geometry: Santa Cruz 1995*, edited by R. L. J. Kollár and D. Morrison, pp. 45–96, Proc. Sympos. Pure Math. 62, Amer. Math. Soc., Providence, RI, 1995.

[29] W. Fulton and C. Woodward, On the quantum product of Schubert classes. *J. Algebraic Geom.* **13** (2004), 641–661.

[30] D. Gaitsgory, The semi-infinite intersection cohomology sheaf. *Adv. Math.* **327** (2018), 789–868.

[31] D. Gaitsgory, The local and global versions of the Whittaker category. *Pure Appl. Math. Q.* **16** (2020), no. 3, 775–904.

[32] A. Givental, Homological geometry and mirror symmetry. In *Proceedings of the ICM 1994 (Zurich)*, pp. 472–480, Birkhäuser, Basel, 1995.

[33] A. Givental, On the WDVV equation in quantum K-theory. *Michigan Math. J.* **48** (2000), 295–304.

[34] A. Givental and Y. P. Lee, Quantum K-theory on flag manifolds, finite-difference Toda lattices and quantum groups. *Invent. Math.* **151** (2003), no. 1, 193–219.

[35] H. Iritani, T. Milanov, and V. Tonita, Reconstruction and convergence in quantum K-theory via difference equations. *Int. Math. Res. Not. IMRN* **2015** (2015), no. 11, 2887–2937.

[36] J. C. Jantzen, *Representations of algebraic groups: second edition*. Math. Surveys Monogr. 107, Amer. Math. Soc., Providence, RI, 2003.

[37] M. Jimbo, Solvable lattice models and quantum groups. In *Proceedings of the ICM 1990 (Kyoto) I, II*, pp. 1343–1352, Math. Soc. Japan, Tokyo, 1991.

[38] V. G. Kac, *Infinite-dimensional Lie algebras*. 3rd edn., Cambridge University Press, Cambridge, 1990.

[39] S. J. Kang and M. Kashiwara, Categorification of highest weight modules via Khovanov–Lauda–Rouquier algebras. *Invent. Math.* **190** (2012), no. 3, 699–742.

[40] M. Kashiwara, The flag manifold of Kac–Moody Lie algebra. In *Algebraic analysis, geometry, and number theory (Baltimore, MD, 1988)*, pp. 161–190, Johns Hopkins Univ. Press, Baltimore, MD, 1989.

[41] M. Kashiwara, Crystallizing the q-analogue of universal enveloping algebras. In *Proceedings of the ICM 1990 (Kyoto)*, pp. 791–797, Math. Soc. Japan, Tokyo, 1991.

[42] M. Kashiwara, Crystal bases of modified quantized enveloping algebra. *Duke Math. J.* **73** (1994), no. 2, 383–413.

[43] M. Kashiwara, Crystal bases and categorifications. In *Proceedings of the ICM 2018 (Rio de Janeiro)*, pp. 249–258, World Sci. Publ., Hackensack, NJ, 2018.

[44] M. Kashiwara and T. Tanisaki, On Kazhdan–Lusztig conjectures. *Sugaku Expositions* **11** (1998), no. 2, 177–195.

[45] S. Kato, Darboux coordinates on the BFM spaces. 2020, arXiv:2008.01310v2.

[46] S. Kato, Demazure character formula for semi-infinite flag varieties. *Math. Ann.* **371** (2018), no. 3, 1769–1801.

[47] S. Kato, Loop structure on equivariant K-theory of semi-infinite flag manifolds. 2018, arXiv:1805.01718v6.

[48] S. Kato, On quantum K-groups of partial flag manifolds. 2019, arXiv:1906.09343v2.

[49] S. Kato, Frobenius splitting of thick flag manifolds of Kac–Moody algebras. *Int. Math. Res. Not. IMRN* **2020** (2020), no. 17, 5401–5427.

[50] S. Kato, Frobenius splitting of Schubert varieties of semi-infinite flag manifolds. *Forum Math. Pi* **9** (2021), e5, 56 pp.

[51] S. Kato and S. Loktev, A Weyl module stratification of integrable representations. *Comm. Math. Phys.* **368** (2019), 113–141.

[52] S. Kato, S. Naito, and D. Sagaki, Equivariant K-theory of semi-infinite flag manifolds and the Pieri–Chevalley formula. *Duke Math. J.* **169** (2020), no. 13, 2421–2500.

[53] M. Khovanov and A. D. Lauda, A diagrammatic approach to categorification of quantum groups. I. *Represent. Theory* **13** (2009), 309–347.

[54] M. Kontsevich and Yu. Manin, Gromov–Witten classes, quantum cohomology, and enumerative geometry. *Comm. Math. Phys.* **164** (1994), no. 3, 525–562.

[55] B. Kostant and S. Kumar, T-equivariant K-theory of generalized flag varieties. *J. Differential Geom.* **32** (1990), no. 2, 549–603.

[56] S. Kumar, *Kac–Moody groups, their flag varieties and representation theory*. Progr. Math. 204, Birkhäuser Boston, Inc., Boston, MA, 2002.

[57] S. Kumar, Tensor product decomposition. In *Proceedings of the ICM 2010 (Hyderabad) III*, pp. 1226–1261, Hindustan Book Agency, New Delhi, 2010.

[58] T. Lam, C. Li, L. C. Mihalcea, and M. Shimozono, A conjectural Peterson isomorphism in K-theory. *J. Algebra* **513** (2018), 326–343.

[59] T. Lam and M. Shimozono, Quantum cohomology of G/P and homology of affine Grassmannian. *Acta Math.* **204** (2010), no. 1, 49–90.

[60] B. Leclerc, Cluster algebras and representation theory. In *Proceedings of the ICM 2010 (Hyderabad) IV*, pp. 2471–2488, Hindustan Book Agency, New Delhi, 2010.

[61] Y. P. Lee, Quantum K-theory. I. Foundations. *Duke Math. J.* **121** (2004), no. 3, 389–424.

[62] G. Lusztig, Hecke algebras and Jantzen's generic decomposition patterns. *Adv. Math.* **37** (1980), no. 1, 121–164.

[63] G. Lusztig, Intersection cohomology methods in representation theory. In *Proceedings of the ICM 1990 (Kyoto) I, II*, pp. 155–174, Math. Soc. Japan, Tokyo, 1991.

[64] G. Lusztig, *Introduction to quantum groups*. Progr. Math. 110, Birkhäuser, 1994.

[65] G. Lusztig, Bases in equivariant K-theory. *Represent. Theory* **2** (1998), 298–369.

[66] G. Lusztig, Study of a $\mathbf{Z}$-form of the coordinate ring of a reductive group. *J. Amer. Math. Soc.* **22** (2009), no. 3, 739–769.

[67] D. Maulik and A. Okounkov, Quantum groups and quantum cohomology. *Astérisque* **408** (2019), ix+209 pp.

[68] V. B. Mehta and A. Ramanathan, Frobenius splitting and cohomology vanishing for Schubert varieties. *Ann. of Math. (2)* **122** (1985), no. 1, 27–40.

[69] L. C. Mihalcea, On equivariant quantum cohomology of homogeneous spaces: Chevalley formulae and algorithms. *Duke Math. J.* **140** (2007), no. 2, 321–350.

[70] I. Mirković and K. Vilonen, Geometric Langlands duality and representations of algebraic groups over commutative rings. *Ann. of Math. (2)* **166** (2007), no. 1, 95–143.

[71] C. Mokler, An algebraic geometric model of an action of the face monoid associated to a Kac–Moody group on its building. *J. Pure Appl. Algebra* **219** (2017), 331–397.

[72] K. Naoi, Weyl modules, Demazure modules and finite crystals for non-simply laced type. *Adv. Math.* **229** (2012), no. 2, 875–934.

[73] A. Okounkov, On the crossroads of enumerative geometry and geometric representation theory. In *Proceedings of the ICM 2018 (Rio de Janeiro)*, pp. 839–868, World Sci. Publ., Hackensack, NJ, 2018.

[74] D. Peterson, *Quantum cohomology of* G/P. Lecture at MIT, 1997.

[75] D. H. Peterson and V. G. Kac, Infinite flag varieties and conjugacy theorems. *Proc. Natl. Acad. Sci. USA* **80** (1983), 1778–1782.

[76] R. Rouquier, Quiver Hecke algebras and 2-Lie algebras. *Algebra Colloq.* **19** (2012), no. 2, 359–410.

[77] M. Varagnolo and E. Vasserot, Canonical bases and KLR-algebras. *J. Reine Angew. Math.* **659** (2011), 67–100.

[78] G. Williamson, Parity sheaves and the Hecke category. In *Proceedings of the ICM 2018 (Rio de Janeiro)*, pp. 979–1015, World Sci. Publ., Hackensack, NJ, 2018.

SYU KATO

Department of Mathematics, Kyoto University, Oiwake Kita-Shirakawa, Sakyo, Kyoto 606-8502, Japan, syuchan@math.kyoto-u.ac.jp

This Page Intentionally Left Blank

CHARACTER ESTIMATES FOR FINITE SIMPLE GROUPS AND APPLICATIONS

MICHAEL J. LARSEN

ABSTRACT

Let G be a finite simple group, χ an irreducible complex character, and g an element of G. It is often desirable to have upper bounds for $|\chi(g)|$ in terms of $\chi(1)$ and some measure of the regularity of g. This paper reviews what is known in this direction and presents typical applications of such bounds: to proving certain products of conjugacy classes cover G, to solving word equations over G, and to counting homomorphisms from a Fuchsian group to G.

MATHEMATICS SUBJECT CLASSIFICATION 2020

Primary 20-02; Secondary 20C15, 20C30, 20C33, 20D06

INTERNATIONAL CONGRESS OF MATHEMATICIANS 2022 JULY 6–14

Proc. Int. Cong. Math. 2022, Vol. 3, pp. 1624–1635
DOI 10.4171/ICM2022/14

1. INTRODUCTION

Let G be a finite group, χ the character of an irreducible complex representation ρ of G, and g an element of G. As the eigenvalues of $\rho(g)$ are roots of unity, the bound $|\chi(g)| \leq \chi(1)$ is trivial. For central elements g, no stronger upper bound than $\chi(1)$ is possible. However, according to Schur, we know that

$$\sum_{g \in G} \chi(g)\overline{\chi(g)} = |G|,$$

and since $\chi(x) = \chi(g)$ for all x in the conjugacy class g^G, we obtain the *centralizer bound*

$$|\chi(g)| \leq \sqrt{\frac{|G|}{|g^G|}} = \sqrt{|C_G(g)|}.$$

Other known upper bounds typically hold only for special classes of groups.

This paper reviews what is known about character bounds when G is a finite simple group or is closely related to such a group. There is a substantial literature on upper bounds for character ratios $\frac{|\chi(g)|}{\chi(1)}$; see Martin Liebeck's survey **[29]** for recent results and applications in the case of groups of Lie type. These bounds are typically weakest for characters χ of low degree, which points to the desirability of *exponential bounds*, that is, bounds of the form $|\chi(g)| \leq \chi(1)^{\alpha(g)}$, where the size of $\alpha(g)$ is typically related to the size of the centralizer of g compared to $|G|$. The next two sections focus on alternating groups and groups of Lie type, respectively. The remaining sections give some applications of these results and present some open problems.

2. SYMMETRIC AND ALTERNATING GROUPS

Motivated by questions in probability theory, a number of people have considered character ratio bounds for symmetric groups. In this series of groups, unlike groups of Lie type, character ratios for nontrivial elements and nontrivial characters can be arbitrarily close to 1. The worst case for $G = \mathsf{S}_n$ is the ratio $\frac{n-3}{n-1}$, achieved when g is a transposition and χ is a character of degree $n-1$. Persi Diaconis and Mehrdad Shahshahani considered the case that g is a transposition and χ is any irreducible character, proving in **[4]** that if both the first row and the first column of the Young diagram for $\chi = \chi_\lambda$ have length $\leq n/2$, then the character ratio is less than $1/2$, while if, for instance, the first row satisfies $\lambda_1 > n/2$, then

$$0 < \frac{\chi(g)}{\chi(1)} \leq \frac{\lambda_1(\lambda_1 - 1) + (n - \lambda_1 - 1)(n - \lambda_1 - 2) - 2}{n(n-1)}.$$

A similar bound was given by Leopold Flatto, Andrew Odlyzko, and David Wales **[8, THEOREM 5.2]**.

Yuval Roichman **[39]** gave a character bound of the form

$$\frac{|\chi(g)|}{\chi(1)} \leq \max\left(\lambda_1/n, \lambda_1'/n, c\right)^{\operatorname{supp}(g)},$$

where $\operatorname{supp}(g)$ denotes the number of elements of $\{1, \ldots, n\}$ not fixed by g, and $c < 1$ is an absolute constant. This reflects the fact that elements with high support tend to have small centralizers. The bound is quite good when χ has small degree. However, for large n, most characters of S_n have degree greater than A^n for any fixed A, and for such characters, Roichman's bound is weaker than the centralizer bound for most elements $g \in G$.

Philippe Biane [3] gave character ratio bounds for elements of bounded support and "balanced" characters, namely those where $\lambda_1/\sqrt{n}$ and $\lambda'_1/\sqrt{n}$ are bounded. By the work of Logan–Shepp [34] and Veršik–Kerov [44], high degree characters are typically balanced. To be more precise, this is true for characters chosen randomly, weighted by the Plancherel measure. Amarpreet Rattan and Piotr Śniady [38] generalized Biane's character bound so it applies whenever $\operatorname{supp}(g)$ is small enough compared to n; if g cannot be expressed as the product of less than π transpositions, then

$$\frac{|\chi(g)|}{\chi(1)} \le \left(\frac{D\max(1,\pi^2/n)}{\sqrt{n}}\right)^{\pi},$$

where D depends on the sizes of $\lambda_1/\sqrt{n}$ and $\lambda'_1/\sqrt{n}$. Valentin Féray and Śniady [7] proved a bound of the form

$$\frac{|\chi(g)|}{\chi(1)} \le \left(\frac{a\max(\lambda_1,\lambda'_1,\pi)}{n}\right)^{\pi},$$

which simultaneously improves on the results of [39] and [38].

Thomas Müller and Jan-Christoph Schlage-Puchta gave a character bound of exponential type [37, THEOREM B] which is good in a wide variety of situations. They proved that $|\chi(g)| \le \chi(1)^{\alpha(g)}$, where

$$\alpha(g) = 1 - \left(\left(1-(1/\log n)\right)^{-1}\frac{12\log n}{\log(n/\operatorname{fix}(g))} + 18\right)^{-1}.$$

Being exponential, it works well whether $\chi(1)$ is large or small. The exponent is optimal, up to a multiplicative constant, for elements g consisting of many cycles, for instance, for involutions. However, it can be greatly improved upon for elements consisting of few cycles. In particular, $\alpha(g)$ is no smaller when g is an n-cycle than when it is of shape $2^{n/2}$.

Sergey Fomin and Nathan Lulov [9] gave a bound specifically for elements g of the shape $r^{n/r}$. For fixed r and varying n, it takes the form

$$|\chi(g)| = O\left(n^{\frac{r-1}{2r}}\chi(1)^{1/r}\right),$$

so it is essentially a bound of exponential type. Aner Shalev and I gave an exponential bound [22] for elements g of arbitrary shape $1^{a_1}2^{a_2}\cdots$ which is roughly comparable in strength to the Fomin–Lulov bound. Define the sequence $e_1, e_2, \ldots$ such that for all $k \ge 1$,

$$n^{e_1+\cdots+e_k} = \sum_{i=1}^{k} i a_i.$$

Then

$$|\chi(g)| \le \chi(1)^{\sum_{i=1}^{n} e_i/i + o(1)}.$$

This result is stronger than the exponential bound of Müller–Schlage-Puchta for almost all elements but inferior to it when the number of fixed points of g is very large.

None of these bounds can compete with the centralizer bound for elements consisting of very few cycles, for instance, for n-cycles, where the centralizer bound gives $|\chi(g)| \le \sqrt{n}$. For such elements, the Murnaghan–Nakayama rule asserts $|\chi(g)| \le 1$, which is obviously optimal.

From symmetric group bounds, we easily obtain alternating group bounds of comparable strength. Recall that for $\lambda \neq \lambda'$, the characters χ_λ and $\chi_{\lambda'}$ restrict to the same irreducible character of A_n. All other irreducible characters of A_n arise from partitions satisfying $\lambda = \lambda'$; for each such λ, the restriction of χ_λ to A_n decomposes as a sum of two irreducibles χ_λ^1 and χ_λ^2. The χ_λ^i take the character value $\chi_\lambda(g)/2$ for all $g \in \mathsf{S}_n \setminus C$, where C is a single S_n-conjugacy class which decomposes into two A_n-conjugacy classes. For elements of C, a theorem of Frobenius gives character values, which are of the form

$$\frac{1 \pm \sqrt{\pm n_1 \cdots n_k}}{2},$$

where $n_i = \lambda_i - i$ for $1 \le i \le k$. Character degree estimates, like those in [22], now imply that $|\chi_\lambda^i(g)| \le \chi_\lambda(1)^\varepsilon$ whenever n is sufficiently large compared to $\varepsilon > 0$.

3. GROUPS OF LIE TYPE

Character estimates for finite simple groups of Lie type go back to the work of David Gluck [13–15]. Unlike in the case of alternating and symmetric groups, there is a uniform bound [15] on character ratios for nontrivial characters and nontrivial g, namely

$$\frac{|\chi(g)|}{\chi(1)} \le \frac{19}{20}.$$

When the cardinality q of the field of definition of G is large, this upper bound can be improved; Gluck [14] gives an upper bound of the form $C/\sqrt{q}$ for large q. The q-exponent is optimal, since for odd q, $\mathrm{PSL}_2(q)$ has characters of degree $\frac{q+1}{2}$ or $\frac{q-1}{2}$, and the value of such a character at a nontrivial unipotent element g is $\frac{\pm 1 \pm \sqrt{(-1)^{\frac{q-1}{2}} q}}{2}$.

If G is a finite simple group of bounded rank, then $\chi(1) < |G| = O(q^D)$, where D denotes the dimension associated to the Lie type of G. Therefore, the Gluck bound $C/\sqrt{q}$ can be converted to an exponential bound $|\chi(g)| \le \chi(1)^\alpha$, where $\alpha < 1$ depends only on the rank. To achieve exponential bounds in general, therefore, it suffices to limit our attention to the case that G is a classical group, that is, one of $\mathrm{PSL}_{r+1}(q)$, $\mathrm{PSU}_{r+1}(q)$, $\mathrm{P\Omega}_{2r}^{\pm}(q)$, $\mathrm{PSp}_{2r}(q)$, or $\mathrm{P\Omega}_{2r+1}(q)$.

We cannot expect that character ratios go to 0 as the order of a classical group goes to infinity. For instance, let $G = \mathrm{PSL}_{r+1}(q)$. The permutation representation associated with the action of G on $\mathbb{PF}_q^r$ can be expressed as $\chi + 1$, for χ irreducible. Let g be the image of a transvection in $\mathrm{SL}_{r+1}(\mathbb{F}_q)$ in G. Then the fixed points of g form a hyperplane in $\mathbb{PF}_q^n$, and $\chi(g) = q^{n-1} + q^{n-2} + \cdots + q$. Thus,

$$\lim_{n \to \infty} \frac{\chi(g)}{\chi(1)} = \frac{1}{q}.$$

Defining the support $\operatorname{supp}(g)$ as the smallest codimension of any eigenspace of g for the natural projective representation of G, the elements g in the above example have constant support 1 even as the rank of G goes to infinity. Shalev, Pham Huu Tiep, and I proved **[24, THEOREM 4.3.6]** that as the support goes to infinity, the character ratio goes to 0:

$$\frac{|\chi(g)|}{\chi(1)} \leq q^{-\sqrt{\operatorname{supp}(g)}/481}.$$

This falls well short of a uniform exponential character bound, even for elements of maximal support. Robert Guralnick, Tiep, and I found uniform exponential bounds for elements g whose centralizer is small compared to the order of G. For instance, we proved **[16, THEOREM 1.4]** that if G is of the form $\mathrm{PSL}_n(q)$ or $\mathrm{PSU}_n(q)$ and $|C_G(g)| \leq q^{n^2/12}$, then $|\chi(g)| \leq \chi(1)^{8/9}$. More generally, but less explicitly, we proved **[17, THEOREM 1.3]** that for all $\varepsilon > 0$, there exists $\delta > 0$ such that $|C_G(g)| \leq |G|^\delta$ implies $|\chi(g)| \leq \chi(1)^\varepsilon$. However, the method of these papers applies only to elements with small centralizer, for instance, it does not give any bound at all for involutions.

This defect was remedied in the sequel **[28]**, which proved that for all positive $\delta < 1$ there exists $\varepsilon < 1$ such that $|C_G(g)| \leq |G|^\delta$ implies $|\chi(g)| \leq \chi(1)^\varepsilon$. More precisely, $|\chi(g)| \leq \chi(1)^{\alpha(g)}$ where

$$\alpha(g) = 1 - c + c\frac{\log|C_G(g)|}{\log|G|},$$

and $c > 0$ is an absolute constant, which can be made explicit (but is, unfortunately, extremely small). This theorem holds more generally for quasisimple finite groups of Lie type.

For many elements g in a classical group of rank r, much better exponents are available, thanks to the work of Roman Bezrukavnikov, Liebeck, Shalev, and Tiep **[2]**. For q odd, if the centralizer of g is a proper split Levi subgroup, then $|\chi(g)| \leq f(r)\chi(1)^{\alpha(g)}$, where $\alpha(g)$ is an explicitly computable rational number which is known to be optimal in many cases. This idea was further developed by Jay Taylor and Tiep, who proved **[43]**, among other things, that for every nontrivial element $g \in \mathrm{PSL}_n(q)$,

$$|\chi(g)| \leq h(r)\chi(1)^{\frac{n-1}{n-2}}.$$

All of these estimates are poor for elements with small centralizers, such as regular elements. A general result, due to Shelly Garion, Alexander Lubotzky, and myself, which sometimes gives reasonably good bounds for regular elements, is the following **[10, THEOREM 3]**. Let G be a finite group, not necessarily simple, and g an element of G whose centralizer A is abelian. Suppose $A_1, \ldots, A_n$ are subgroups of A not containing g such that the centralizer of every element of $A \setminus \bigcup_i A_i$ is A. Then, for every irreducible character of G,

$$|\chi(g)| \leq (4/\sqrt{3})^n [N_G(A) : A].$$

For example, this gives an upper bound of $2(n-1)^2/\sqrt{3}$ for $|\chi(g)|$ when $G = \mathrm{PSL}_n(q)$ and g is the image of an element with irreducible characteristic polynomial. It would be nice to have optimal upper bounds for $|\chi(g)|$ for general regular semisimple elements g.

4. PRODUCTS OF CONJUGACY CLASSES

If $C_1, \ldots, C_n$ are conjugacy classes of a finite group G, then the number N of n-tuples $(g_1, \ldots, g_n) \in C_1 \times \cdots \times C_n$ satisfying $g_1 g_2 \cdots g_n = 1$ is given by the Frobenius formula

$$N = \frac{|C_1| \cdots |C_n|}{|G|} \sum_{\chi} \frac{\chi(C_1) \cdots \chi(C_n)}{\chi(1)^{n-2}},$$

where χ ranges over all irreducible characters of G. In conjunction with upper bounds for the $|\chi(C_i)|$, this can sometimes be used to prove that $N \neq 0$, as the contribution from $\chi = 1$ often dominates the sum. Exponential bounds for the $\chi(C_i)$ are especially convenient, since results of Liebeck and Shalev [32] give a great deal of information about when we can expect

$$\sum_{\chi \neq 1} \chi(1)^{-s} < 1.$$

A well-known conjecture attributed to Thompson asserts that for every finite simple group G, there exists a conjugacy class C such that $C^2 = G$. Thanks to work of Erich Ellers and Nikolai Gordeev [6], we know that this is true except for a list of possible counterexamples, all finite simple groups of Lie type with $q \leq 8$. Tiep and I used our uniform exponential bounds to show that several of the infinite families on this list, in particular, the symplectic groups for all $q \geq 2$, can be eliminated in sufficiently high rank [28, THEOREM 7.7]. It would be interesting if these results could be extended to the remaining families on the list, giving an asymptotic version of Thompson's conjecture.

Andrew Gleason and Cheng-hao Xu [18,19] proved Thompson's conjecture for alternating groups, using the conjugacy class of an n-cycle if n is odd or a permutation of shape $2^1(n-2)^1$ if n is even. In [22, THEOREM 1.13], Shalev and I proved that in the limit $n \to \infty$ the probability that a randomly chosen $g \in \mathsf{A}_n$ belongs to a conjugacy class with $C^2 = \mathsf{A}_n$ rapidly approaches 1.

The analogous claim cannot be true for all finite simple groups since $C^2 = G$ implies that $C = C^{-1}$, and for, e.g., $\mathrm{PSL}_3(q)$ as $q \to \infty$, the probability that a random element is real goes to 0. However, there are several variants of this question which do not have an obvious counterexample. As the order of G tends to infinity, does the probability that a random *real* element belongs to a conjugacy class with $C^2 = G$ approach 1? Does the probability that a random element g belongs to a conjugacy class C with $C^2 \cup \{1\} = G$ approach 1? Also, as the order of G tends to infinity, does the probability that a random element belongs to a conjugacy class with $CC^{-1} = G$ approach 1?

The weaker claim that every element $g \in G$ lies in CC^{-1} for some conjugacy class (depending, perhaps, on g) is equivalent to the statement that every element of G is a commutator. This was was an old conjecture of Ore and is now a theorem of Liebeck, Eamonn O'Brien, Shalev, and Tiep [30].

One can also ask about S^2 where S is an arbitrary conjugation-invariant subset of G. On naive probabilistic grounds, it might seem plausible that given $\varepsilon > 0$ fixed, for G sufficiently large, every normal subset of G with at least $\varepsilon|G|$ elements satisfies $S^2 = G$. However, a moment's reflection shows that, unless $\varepsilon > \frac{1}{2}$, there is no reason to expect $1 \in S^2$.

Is it true, for G sufficiently large, that $S^2 \cup \{1\} = G$? For alternating groups and for groups of Lie type in bounded rank, the answer is affirmative [26], but we do not know in general.

In a different direction, given a conjugacy class C, how large must n be so that the nth power C^n is all of G? More generally, given conjugacy classes $C_1, \dots, C_n$ with sufficiently strong character bounds, the Frobenius formula can be used to show that each element of G is represented as a product $g_1 \cdots g_n$, with $g_i \in C_i$, in approximately $\frac{|C_1|\cdots|C_n|}{|G|}$ ways. For instance, it follows from the exponential character bounds given above that there exists an absolute constant k such that if G is a finite simple group of Lie type and $C_1, \dots, C_n$ are conjugacy classes in G satisfying $|C_1| \cdots |C_n| > |G|^k$, then for each $g \in G$,

$$\left|\{(g_1, \dots, g_n) \in C_1 \times \cdots \times C_n \mid g_1 \cdots g_n = g\}\right| = \bigl(1 + o(1)\bigr)\frac{|C_1| \cdots |C_n|}{|G|}.$$

Via Lang–Weil estimates, this further implies that if $\underline{C}_1, \dots, \underline{C}_n$ are conjugacy classes of a simple algebraic group $\underline{G}$, and

$$\dim \underline{C}_1 + \cdots + \dim \underline{C}_n > k \dim \underline{G},$$

then the product morphism of varieties $\underline{C}_1 \times \cdots \times \underline{C}_n \to \underline{G}$ has the property that every fiber is of dimension $\dim \underline{C}_1 + \cdots + \dim \underline{C}_n - \dim \underline{G}$.

In the special case that $C_1 = \cdots = C_n = C$, the question of the distribution of products $g_1 \cdots g_n$, $g_i \in C$, can be expressed in terms of the mixing time of the random walk on the Cayley graph of (G, C). A consequence of the exponential character bounds [28] is that for groups of Lie type, the mixing time of such a random walk is $O(\log |G| / \log |C|)$. This is the same order of growth as the diameter of the Cayley graph, thus settling conjectures of Lubotzky [35, P. 179] and Shalev [42, CONJECTURE 4.3].

The situation is different for alternating groups $G = \mathsf{A}_n$. For instance, if C is the class of 3-cycles and $n \geq 6$, then $\log |G| / \log |C| < n$, and $C^{\lfloor n/2 \rfloor} = G$ [5, THEOREM 9.1]. However, for any fixed k, the probability that the product of kn random 3-cycles g_i fixes 1 is at least the probability that each individual g_i fixes 1, which goes to e^{-3k} as $n \to \infty$. Thus the expected number of fixed points of $g_1 \cdots g_n$ grows linearly with n. It would be interesting to know, for general $C \subset \mathsf{A}_n$, what the mixing time is.

5. WARING'S PROBLEM

Waring's problem for finite simple groups originally meant the following question. Does there exist a function $f: \mathbb{N} \to \mathbb{N}$ such that for all positive integers n and all sufficiently large finite simple groups G (in terms of n), every element of G is a product of $f(n)$ nth powers? Positive solutions were given by Martinez–Zelmanov [36] and Saxl–Wilson [40].

This can be extended as follows. Let w denote a nontrivial element in any free group F_d. For every finite simple group G, w determines a function $G^d \to G$. We replace the nth powers with word values, that is, elements of G in the image of w. Liebeck and Shalev proved [31] that for G sufficiently large (in terms of w), every element of G can be written as a product of a bounded number of word values (where the bound may depend on w, just as in the classical version of Waring's problem, the minimum number of the nth powers needed to represent a given integer may depend on n).

It was therefore, perhaps, surprising when Shalev proved [41] that the Waring number for finite simple groups is uniform in w and is, in fact, at most three. This has now been improved to the optimal bound, two [23,24]. More generally, for any two nontrivial words w_1 and w_2, if G is a sufficiently large finite simple group, every element of G is a product of their word values. In fact, it is even possible [27] to choose subsets S_1 and S_2 of the sets of word values of w_1 and w_2 such that $S_1 S_2 = G$ and $|S_i| = O(|G|^{1/2} \log^{1/2} |G|)$. The set of values of any word is a union of conjugacy classes, and the basic strategy of the proof is to try to find conjugacy classes C_1 and C_2 contained in the word values of w_1 and w_2, respectively, such that $C_1 C_2 = G$ and very few elements of G have significantly fewer representations as such products than one would expect. Then a random choice of subsets $S_i \subset C_i$ of suitable size can almost always be slightly modified to work.

In general, the probability distribution on the word values of w obtained by evaluation at a uniformly distributed random element of G^d is far from uniform. For instance, for $g \in \mathsf{A}_{3n}$ uniformly distributed, the probability that $g^3 = 1$ is at least $|\mathsf{A}_{3n}|^{-1}$ times the number of elements of shape 3^n, i.e.,

$$(3n-1)(3n-2) \cdot (3n-4)(3n-5) \cdots (2)(1) > (3n-1)!^{\frac{2}{3}} > |\mathsf{A}_{3n}|^{\frac{2}{3} - \frac{1}{3n}}$$

for n sufficiently large. Thus, setting $w_1 = w_2 = x^3$, the probability that the product of cubes of two randomly chosen elements is 1 is at least $|\mathsf{A}_{3n}|^{-2/3-2/3n}$, which, for large n, makes the distribution far from uniform, at least in the L^∞ sense.

Using exponential character estimates, Shalev, Tiep, and I proved [25, **THEOREM 4**] that for any word w, there exists k such that as $|G| \to \infty$, the L^∞-deviation from uniformity in the product of k independent randomly generated values of w goes to 0. The dependence of k on w is unavoidable, as the above example suggests. On the other hand, the L^1-deviation from uniformity goes to 0 in the product of two independent randomly generated values of w, for any nontrivial word w [25, **THEOREM 1**]. I do not know what to expect for L^p-deviation for $1 < p < \infty$.

6. FUCHSIAN GROUPS

For $g, m \geq 0$, let $d_1, \ldots, d_m \geq 2$ be integers. For

$$\Gamma = \langle x_1, \ldots, x_m, y_1, \ldots, y_g, z_1, \ldots, z_g \mid x_1^{d_1}, \ldots, x_m^{d_m}, \\ x_1 \cdots x_m [y_1, z_1] \cdots [y_g, z_g] \rangle,$$

define the Euler characteristic

$$e = 2 - 2g - \sum_{i=1}^{m} (1 - d_i^{-1}).$$

Assume $e < 0$, so Γ is an oriented, cocompact Fuchsian group. Let G be a finite group, and let $C_1, \ldots, C_m$ denote conjugacy classes in G of elements whose orders divide $d_1, \ldots, d_m$, respectively. The Frobenius formula can be regarded as the $g = 0$ case of a more general formula for the number of homomorphisms $\Gamma \to G$ mapping x_i to an element of C_i for all i,

$$\left|\mathrm{Hom}_{\{C_i\}}(\Gamma, G)\right| = |G|^{2g-1}|C_1|\cdots|C_m| \sum_{\chi} \frac{\chi(C_1)\cdots\chi(C_m)}{\chi(1)^{m+2g-2}}.$$

In favorable situations, one can prove that the $\chi = 1$ term dominates all the others combined, in which case one has a good estimate for the number of such homomorphisms. Using this, Liebeck and Shalev proved [32, THEOREM 1.5] that if $g \geq 2$, and G is a simple of Lie type group of rank r, then

$$\left|\mathrm{Hom}(\Gamma, G)\right| = |G|^{1-e+O(1/r)}.$$

By the same method, employing the character bounds of [28], one obtains the same estimate whenever e is less than some absolute constant, regardless of the value of g. It would be interesting to know whether this is true in general for $e < 0$. Some evidence in favor of this idea is given in [21, 33], but for small q the problem is open.

An interesting geometric consequence of the method of Liebeck–Shalev is that if $\underline{G}$ is a simple algebraic group of rank r and $g \geq 2$, the morphism $\underline{G}^{2g} \to \underline{G}$ given by the word $[y_1, z_1]\cdots[y_g, z_g]$ has all fibers of the same dimension, $(2g - 1)\dim \underline{G}$. This has been refined by Avraham Aizenbud and Nir Avni, who proved [1] that for $g \geq 373$, the fibers of this morphism are reduced and have rational singularities. It would be interesting to extend this to the case of general Fuchsian groups. For instance, does there exist an absolute constant k such that for all simple algebraic groups $\underline{G}$ and conjugacy classes $\underline{C}_1, \ldots, \underline{C}_m$ with $\dim \underline{C}_1 + \cdots + \dim \underline{C}_m > k \dim \underline{G}$, all fibers of the multiplication morphism $\underline{C}_1 \times \cdots \times \underline{C}_m \to \underline{G}$ are reduced with rational singularities The ideas of Glazer–Hendel [11, 12] may be applicable.

For $g = 1$, we can no longer hope for equidimensional fibers, since the generic fiber dimension is $\dim \underline{G}$, while the fiber over the identity element has dimension $r + \dim \underline{G}$. However, Zhipeng Lu and I proved [20] that for $\underline{G} = \mathrm{SL}_n$, all fibers over noncentral elements have dimension $\underline{G}$. It would be interesting to know whether this is true for general simple algebraic groups $\underline{G}$.

FUNDING

This work was partially supported by the National Science Foundation.

REFERENCES

[1] A. Aizenbud and N. Avni, Representation growth and rational singularities of the moduli space of local systems. *Invent. Math.* **204** (2016), no. 1, 245–316.

[2] R. Bezrukavnikov, M. Liebeck, A. Shalev, and P. H. Tiep, Character bounds for finite groups of Lie type. *Acta Math.* **221** (2018), no. 1, 1–57.

[3] P. Biane, Representations of symmetric groups and free probability. *Adv. Math.* **138** (1998), no. 1, 126–181.

[4] P. Diaconis and M. Shahshahani, Generating a random permutation with random transpositions. *Z. Wahrsch. Verw. Gebiete* **57** (1981), no. 2, 159–179.

[5] Y. Dvir, Covering properties of permutation groups. In *Products of conjugacy classes in groups*, pp. 197–221, Lecture Notes in Math. 1112, Springer, Berlin, 1985.

[6] E. W. Ellers and N. Gordeev, On the conjectures of J. Thompson and O. Ore. *Trans. Amer. Math. Soc.* **350** (1998), no. 9, 3657–3671.

[7] V. Féray and P. Śniady, Asymptotics of characters of symmetric groups related to Stanley character formula. *Ann. of Math.* **173** (2011), no. 2, 887–906.

[8] L. Flatto, A. M. Odlyzko, and D. B. Wales, Random shuffles and group representations. *Ann. Probab.* **13** (1985), no. 1, 154–178.

[9] S. Fomin and N. Lulov, On the number of rim hook tableaux. *J. Math. Sci. (N. Y.)* **87** (1997), no. 6, 4118–4123.

[10] S. Garion, M. Larsen, and A. Lubotzky, Beauville surfaces and finite simple groups. *J. Reine Angew. Math.* **666** (2012), 225–243.

[11] I. Glazer and Y. I. Hendel (with an appendix joint with G. Kozma), On singularity properties of convolutions of algebraic morphisms—the general case. 2018, arXiv:1811.09838.

[12] I. Glazer and Y. I. Hendel, On singularity properties of word maps and applications to probabilistic Waring type problems. 2019, arXiv:1912.12556.

[13] D. Gluck, Character value estimates for groups of Lie type. *Pacific J. Math.* **150** (1991), 279–307.

[14] D. Gluck, Character value estimates for nonsemisimple elements. *J. Algebra* **155** (1993), no. 1, 221–237.

[15] D. Gluck, Sharper character value estimates for groups of Lie type. *J. Algebra* **174** (1995), no. 1, 229–266.

[16] R. M. Guralnick, M. Larsen, and P. H. Tiep, Character levels and character bounds. *Forum Math. Pi* **8** (2020), e2, 81 pp.

[17] R. M. Guralnick, M. Larsen, and P. H. Tiep, Character levels and character bounds for finite classical groups. 2019, arXiv:1904.08070.

[18] C. Hsü, The commutators of the alternating group. *Sci. Sinica* **14** (1965), 339–342.

[19] D. M. W. H. Husemoller, Ramified coverings of Riemann surfaces. *Duke Math. J.* **29** (1962), 167–174.

[20] M. Larsen and Z. Lu, Flatness of the commutator map over SL_n. *Int. Math. Res. Not. IMRN* **2021**, no. 8, 5605–5622.

[21] M. Larsen and A. Lubotzky, Representation varieties of Fuchsian groups. In *From Fourier analysis and number theory to Radon transforms and geometry*, pp. 375–397, Dev. Math. 28, Springer, New York, 2013.

[22] M. Larsen and A. Shalev, Characters of symmetric groups: sharp bounds and applications. *Invent. Math.* **174** (2008), no. 3, 645–687.

[23] M. Larsen and A. Shalev, Word maps and Waring type problems. *J. Amer. Math. Soc.* **22** (2009), no. 2, 437–466.

[24] M. Larsen, A. Shalev, and P. H. Tiep, The Waring problem for finite simple groups. *Ann. of Math.* **174** (2011), no. 3, 1885–1950.

[25] M. Larsen, A. Shalev, and P. H. Tiep, Probabilistic Waring problems for finite simple groups. *Ann. of Math.* **190** (2019), no. 2, 561–608.

[26] M. Larsen, A. Shalev, and P. H. Tiep, Products of normal subgroups and derangements. 2020, arXiv:2003.12882.

[27] M. Larsen and P. H. Tiep, A refined Waring problem for finite simple groups. *Forum Math. Sigma* **3** (2015), Paper No. e6, 22 pp.

[28] M. Larsen and P. H. Tiep, Uniform characters bounds for finite classical groups (submitted).

[29] M. W. Liebeck, Character ratios for finite groups of Lie type, and applications. In *Finite simple groups: thirty years of the Atlas and beyond*, pp. 193–208, Contemp. Math. 694, Amer. Math. Soc., Providence, RI, 2017.

[30] M. W. Liebeck, E. A. O'Brien, A. Shalev, and P. H. Tiep, The Ore conjecture. *J. Eur. Math. Soc.* **12** (2010), no. 4, 939–1008.

[31] M. W. Liebeck and A. Shalev, Diameters of simple groups: sharp bounds and applications. *Ann. of Math.* **154** (2001), 383–406.

[32] M. W. Liebeck and A. Shalev, Fuchsian groups, finite simple groups and representation varieties. *Invent. Math.* **159** (2005), no. 2, 317–367.

[33] M. W. Liebeck, A. Shalev, and P. H. Tiep, Character ratios, representation varieties and random generation of finite groups of Lie type. *Adv. Math.* **374** (2020), 107386, 39 pp.

[34] B. F. Logan and L. A. Shepp, A variational problem for random Young tableaux. *Adv. Math.* **26** (1977), no. 2, 206–222.

[35] A. Lubotzky, Cayley graphs: eigenvalues, expanders and random walks. In *Surveys in combinatorics, 1995 (Stirling)*, pp. 155–189, London Math. Soc. Lecture Note Ser. 218, Cambridge Univ. Press, Cambridge, 1995.

[36] C. Martinez and E. Zelmanov, Products of powers in finite simple groups. *Israel J. Math.* **96** (1996), part B, 469–479.

[37] T. W. Müller and J-C. Schlage-Puchta, Character theory of symmetric groups, subgroup growth of Fuchsian groups, and random walks. *Adv. Math.* **213** (2007), no. 2, 919–982.

[38] A. Rattan and P. Śniady, Upper bound on the characters of the symmetric groups for balanced Young diagrams and a generalized Frobenius formula. *Adv. Math.* **218** (2008), 673–695.

[39] Y. Roichman, Upper bound on the characters of the symmetric groups. *Invent. Math.* **125** (1996), 451–485.

[40] J. Saxl and J. S. Wilson, A note on powers in simple groups. *Math. Proc. Cambridge Philos. Soc.* **122** (1997), no. 1, 91–94.

[41] A. Shalev, Word maps, conjugacy classes, and a noncommutative Waring-type theorem. *Ann. of Math. (2)* **170** (2009), no. 3, 1383–1416.

[42] A. Shalev, Conjugacy classes, growth and complexity. In *Finite simple groups: thirty years of the atlas and beyond*, pp. 209–221, Contemp. Math. 694, Amer. Math. Soc., Providence, RI, 2017.

[43] J. Taylor and P. H. Tiep, Lusztig induction, unipotent supports, and character bounds. *Trans. Amer. Math. Soc.* **373** (2020), no. 12, 8637–8676.

[44] A. M. Veršik and S. V. Kerov, Asymptotic behavior of the Plancherel measure of the symmetric group and the limit form of Young tableaux (Russian). *Dokl. Akad. Nauk SSSR* **233** (1977), no. 6, 1024–1027.

MICHAEL J. LARSEN

Department of Mathematics, Indiana University, Bloomington, IN 47405, U.S.A., mjlarsen@indiana.edu

FINITE APPROXIMATIONS AS A TOOL FOR STUDYING TRIANGULATED CATEGORIES

AMNON NEEMAN

ABSTRACT

Small, finite entities are easier and simpler to manipulate than gigantic, infinite ones. Consequently, huge chunks of mathematics are devoted to methods reducing the study of big, cumbersome objects to an analysis of their finite building blocks. The manifestation of this general pattern, in the study of derived and triangulated categories, dates back almost to the beginnings of the subject—more precisely to articles by Illusie in SGA6, way back in the early 1970s.

What is new, at least new in the world of derived and triangulated categories, is that one gets extra mileage from analyzing more carefully and quantifying more precisely just how efficiently one can estimate infinite objects by finite ones. This leads one to the study of metrics on triangulated categories, and of how accurately an object can be approximated by finite objects of bounded size.

MATHEMATICS SUBJECT CLASSIFICATION 2020

Primary 18G80; Secondary 14F08, 55P42

KEYWORDS

Derived categories, triangulated categories, metrics, norms, t-structures

INTERNATIONAL CONGRESS OF MATHEMATICIANS 2022 JULY 6–14

Proc. Int. Cong. Math. 2022, Vol. 3, pp. 1636–1658
DOI 10.4171/ICM2022/35

1. INTRODUCTION

In every branch of mathematics, we try to solve complicated problems by reducing to simpler ones, and from antiquity people have used finite approximations to study infinite objects. Naturally, whenever a new field comes into being, one of the first developments is to try to understand what should be the right notion of finiteness in the discipline. Derived and triangulated categories were introduced by Verdier in his PhD thesis in the mid-1960s (although the published version only appeared much later in [38]). Not surprisingly, the idea of studying the finite objects in these categories followed suit soon after, see Illusie [13–15].

Right from the start there was a pervasive discomfort with derived and triangulated categories—the intuition that had been built up, in dealing with concrete categories, mostly fails for triangulated categories. In case the reader is wondering: in the previous sentence the word "concrete" has a precise, technical meaning, and it is an old theorem of Freyd [10, 11] that triangulated categories often are not concrete. Further testimony, to the strangeness of derived and triangulated categories, is that it took two decades before the intuitive notion of finiteness, which dates back to Illusie's articles [13–15], was given its correct formal definition. The following may be found in [23, DEFINITION 1.1].

Definition 1.1. Let $\mathscr{T}$ be a triangulated category with coproducts. An object $C \in \mathscr{T}$ is called *compact* if $\mathrm{Hom}(C, -)$ commutes with coproducts. The full subcategory of all compact objects will be denoted by $\mathscr{T}^c$.

Remark 1.2. I have often been asked where the name "compact" came from. In the preprint version of [23], these objects went by a different name, but the (anonymous) referee did not like it. I was given a choice: I was allowed to baptize them either "compact" or "small."

Who was I to argue with a referee?

Once one has a good working definition of what the finite objects ought to be, the next step is to give the right criterion which guarantees that the category has "enough" of them. For triangulated categories, the right definition did not come until [24, DEFINITION 1.7].

Definition 1.3. Let $\mathscr{T}$ be a triangulated category with coproducts. The category $\mathscr{T}$ is called *compactly generated* if every nonzero object $X \in \mathscr{T}$ admits a nonzero map $C \to X$, with $C \in \mathscr{T}$ a compact object.

As the reader may have guessed from the name, compactly generated triangulated categories are those in which it is often possible to reduce general problems to questions about compact objects—which tend to be easier.

All of the above nowadays counts as "classical," meaning that it is two or more decades old and there is already a substantial and diverse literature exploiting the ideas. This article explores the recent developments that arose from trying to understand how efficiently one can approximate arbitrary objects by compact ones. We first survey the results obtained to date. This review is on the skimpy side, partly because there already are other, more expansive published accounts in the literature, but mostly because we want to leave ourselves space to suggest possible directions for future research. Thus the article can be thought of as having two components: a bare-bone review of what has been achieved to date, occupying Sections 2 to 6, followed by Section 7 which comprises suggestions of avenues that might merit further development.

Our review presents just enough detail so that the open questions, making up Section 7, can be formulated clearly and comprehensibly, and so that the significance and potential applications of the open questions can be illuminated. This has the unfortunate side effect that we give short shrift to the many deep, substantial contributions, made by numerous mathematicians, which preceded and inspired the work presented here. The author apologizes in advance for this omission, which is the inescapable corollary of page limits. The reader is referred to the other surveys of the subject, where more care is taken to attribute the ideas correctly to their originators, and give credit where credit is due.

We permit ourselves to gloss over difficult technicalities, nonchalantly skating by nuances and subtleties, with only an occasional passing reference to the other surveys or to the research papers for more detail.

The reader wishing to begin with examples and applications, to keep in mind through the forthcoming abstraction, is encouraged to first look at the Introduction to [31].

2. APPROXIMABLE TRIANGULATED CATEGORIES—THE FORMAL DEFINITION AS A VARIANT ON FOURIER SERIES

It is now time to start our review, offering a glimpse of the recent progress that was made by trying to measure how "complicated" an object is, in other words, how far it is from being compact. What follows is sufficiently new for there to be much room for improvement: the future will undoubtedly see cleaner, more elegant, and more general formulations. What is presented here is the current crude state of this emerging field.

Discussion 2.1. This section is devoted to defining approximable triangulated categories, and the definition is technical and at first sight could appear artificial, maybe even forbidding. It might help therefore to motivate it with an analogy.

Let $\mathbb{S}^1$ be the circle, and let $M(\mathbb{S}^1)$ be the set of all complex-valued, Lebesgue-measurable functions on $\mathbb{S}^1$. As usual we view $\mathbb{S}^1 = \mathbb{R}/\mathbb{Z}$ as the quotient of its universal cover $\mathbb{R}$ by the fundamental group $\mathbb{Z}$; this identifies functions on $\mathbb{S}^1$ with periodic functions on $\mathbb{R}$ with period 1. In particular the function $g(x) = e^{2\pi i x}$ belongs to $M(\mathbb{S}^1)$. And, for each $\ell \in \mathbb{Z}$, we have that $g(x)^\ell = e^{2\pi i \ell x}$ also belongs to $M(\mathbb{S}^1)$. Given a norm on the space $M(\mathbb{S}^1)$, for example, the L^p-norm, we can try to approximate arbitrary $f \in M(\mathbb{S}^1)$ by Laurent polynomials in g, that is, look for complex numbers $\{\lambda_\ell \in \mathbb{C} \mid -n \leq \ell \leq n\}$ such that

$$\left\| f(x) - \sum_{\ell=-n}^{n} \lambda_\ell g(x)^\ell \right\|_p = \left\| f(x) - \sum_{\ell=-n}^{n} \lambda_\ell e^{2\pi i \ell x} \right\|_p < \varepsilon$$

with $\varepsilon > 0$ small. This leads us to the familiar territory of Fourier series.

Now imagine trying to do the same, but replacing $M(\mathbb{S}^1)$ by a triangulated category. Given a triangulated category $\mathscr{T}$, which we assume to have coproducts, we would like to pretend to do Fourier analysis on it. We would need to choose:

(1) Some analog of the function $g(x) = e^{2\pi i x}$. Our replacement for this will be to choose a compact generator $G \in \mathscr{T}$. Recall that a *compact generator* is a compact object $G \in \mathscr{T}$ such that every nonzero object $X \in \mathscr{T}$ admits a nonzero map $G[i] \to X$ for some $i \in \mathbb{Z}$.

(2) We need to choose something like a metric, the analog of the L^p-norm on $M(\mathbb{S}^1)$. For us this will be done by picking a t-structure $(\mathscr{T}^{\leq 0}, \mathscr{T}^{\geq 0})$ on $\mathscr{T}$. The heuristic is that we will view a morphism $E \to F$ in $\mathscr{T}$ as "short" if, in the triangle $E \to F \to D$, the object D belongs to $\mathscr{T}^{\leq -n}$ for large n. We will come back to this in Discussion 6.10.

(3) We need to have an analog of the construction that passes, from the function $g(x) = e^{2\pi i x}$ and the integer $n > 0$, to the vector space of trigonometric Laurent polynomials $\sum_{\ell=-n}^{n} \lambda_\ell e^{2\pi i \ell x}$.

As it happens our solution to (3) is technical. We need a recipe that begins with the object G and the integer $n > 0$, and proceeds to cook up a collection of more objects. We ask the reader to accept it as a black box, with only a sketchy explanation just before Remark 2.3.

Black Box 2.2. Let $\mathscr{T}$ be a triangulated category and let $G \in \mathscr{T}$ be an object. Let $n > 0$ be an integer. We will have occasion to refer to the following four full subcategories of $\mathscr{T}$:

(1) The subcategory $\langle G\rangle_n \subset \mathscr{T}$ is defined unconditionally, and if $\mathscr{T}$ has coproducts one can also define the larger subcategory $\overline{\langle G\rangle}_n$. Both of these subcategories are classical, the reader can find the subcategory $\langle G\rangle_n$ in Bondal and Van den Bergh **[6, THE DISCUSSION BETWEEN LEMMA 2.2.2 AND DEFINITION 2.2.3]**, and the subcategory $\overline{\langle G\rangle}_n$ in **[6, THE DISCUSSION BETWEEN DEFINITION 2.2.3 AND PROPOSITION 2.2.4]**.

(2) If the category $\mathscr{T}$ has coproducts, we will also have occasion to consider the full subcategory $\overline{\langle G\rangle}^{(-\infty,n]}$. Once again this category is classical (although the name is not). The reader can find it in Alonso, Jeremías, and Souto **[1]**, where it would go by the name "the cocomplete pre-aisle generated by $G[-n]$".

(3) Once again assume that $\mathscr{T}$ has coproducts. Then we will also look at the full subcategory $\overline{\langle G\rangle}_n^{[-n,n]}$. This construction is relatively new.

Below we give a vague description of what is going on in these constructions; but when it comes to the technicalities, we ask the reader to either accept these as black boxes, or refer to **[29, REMINDER 0.8 (VII), (XI) AND (XII)]** for detail. We mention that there is a slight clash of notation in the literature: what we call $\overline{\langle G\rangle}_n$ in (1), following Bondal and Van den Bergh, goes by a different name in [29, REMINDER 0.8 (XI)]. The name it goes by there is the case $A = -\infty$ and $B = \infty$ of the more general subcategory $\overline{\langle G\rangle}_n^{[A,B]}$.

Now for the vague explanation of what goes on in (1), (2), and (3) above: in a triangulated category $\mathscr{T}$, there are not many ways to build new objects out of old ones. One can shift objects, form direct summands, form finite direct sums (or infinite ones if coproducts exist), and one can form extensions. In the categories $\langle G\rangle_n$ and $\overline{\langle G\rangle}_n$ of (1), there is a bound on the number of allowed extensions, and the difference between the two is whether infinite coproducts are allowed. In the category $\overline{\langle G\rangle}^{(-\infty,n]}$ of (2), the bound is on the permitted shifts. And in the category $\overline{\langle G\rangle}_n^{[-n,n]}$ of (3), both the shifts allowed and the number of extensions permitted are restricted.

Remark 2.3. The reader should note that an example would not be illuminating, the categories $\langle G\rangle_n$, $\overline{\langle G\rangle}_n$, $\overline{\langle G\rangle}^{(-\infty,n]}$, and $\overline{\langle G\rangle}_n^{[-n,n]}$ are not usually overly computable. For example, let R be an associative ring, and let $\mathscr{T} = \mathbf{D}(R)$ be the unbounded derived category of complexes of left R-modules. The object $R \in \mathscr{T}$, that is, the complex which is R in degree zero and vanishes in all other degrees, is a compact generator for $\mathscr{T} = \mathbf{D}(R)$.

But if we wonder what the categories $\langle R\rangle_n$, $\overline{\langle R\rangle}_n$, $\overline{\langle R\rangle}^{(-\infty,n]}$, and $\overline{\langle R\rangle}_n^{[-n,n]}$ might turn out to be, only the category $\overline{\langle R\rangle}^{(-\infty,n]}$ is straightforward: it is the category of all cochain complexes whose cohomology vanishes in degrees $> n$. The three categories $\langle R\rangle_n$, $\overline{\langle R\rangle}_n$, and $\overline{\langle R\rangle}_n^{[-n,n]}$ are mysterious in general. In fact, the computation of $\langle G\rangle_n$ is the subject of conjectures that have attracted much interest. We will say a tiny bit about theorems in this direction in Section 4, and will mention one of the active, open conjectures in the discussion between Definition 7.7 and Problem 7.8.

Remark 2.4. In the definition of approximable triangulated categories, which is about to come, the category $\overline{\langle G\rangle}_n^{[-n,n]}$ will play the role of the replacement for the vector space of trigonometric Laurent polynomials of degree $\leq n$, which came up in the desiderata of Discussion 2.1(3). The older categories $\langle G\rangle_n$, $\overline{\langle G\rangle}_n$, and $\overline{\langle G\rangle}^{(-\infty,n]}$ will be needed later in the article.

Remark 2.5. Let us return to the heuristics of Discussion 2.1. Assume we have chosen the t-structure $(\mathscr{T}^{\leq 0}, \mathscr{T}^{\geq 0})$ as in Discussion 2.1(2), which we think of as our replacement for the L^p-norm on $M(\mathbb{S}^1)$. And we have also chosen a compact generator $G \in \mathscr{T}$ as in Discussion 2.1(1), which we think of as the analog of the exponential function $g(x) = e^{2\pi i x}$. We have declared that the subcategories $\overline{\langle G\rangle}_n^{[-n,n]}$ will be our replacement for the vector space of trigonometric Laurent polynomials of degree $\leq n$, as in Discussion 2.1(3). It is now time to start approximating functions by trigonometric Laurent polynomials.

Let us therefore assume we start with some object $F \in \mathscr{T}$, and find a good approximation of it by the object $E \in \overline{\langle G\rangle}_m^{[-m,m]}$, meaning that we find a morphism $E \to F$ such that, in the triangle $E \to F \to D$, the object D belongs to $\mathscr{T}^{\leq -M}$ for some suitably large M.

Now we can try to iterate, and find a good approximation for D. Thus we can look for a morphism $E'' \to D$, with $E'' \in \overline{\langle G\rangle}_n^{[-n,n]}$, and such that in the triangle $E'' \to D \to D'$ the object D' belongs to $\mathscr{T}^{\leq -N}$, with $N > M$ even more enormous than M. Can we combine these to improve our initial approximation of F?

To do this, let us build up the octahedron on the composable morphisms $F \to D \to D'$. We end up with a diagram where the rows and columns are triangles

$$\begin{array}{ccccc} E & \longrightarrow & E' & \longrightarrow & E'' \\ \| & & \downarrow & & \downarrow \\ E & \longrightarrow & F & \longrightarrow & D \\ & & \downarrow & & \downarrow \\ & & D' & =\!= & D' \end{array}$$

and in particular the triangle $E' \to F \to D'$ gives that E' is an even better approximation of F than E was. We are therefore interested in knowing if the triangle $E \to E' \to E''$, coupled with the fact that $E \in \overline{\langle G\rangle}_m^{[-m,m]}$ and $E'' \in \overline{\langle G\rangle}_n^{[-n,n]}$, gives any information about where E' might lie with respect to the construction of Black Box 2.2(3). Hence it is useful to know the following.

Facts 2.6. Let $\mathscr{T}$ be a triangulated category with coproducts. The construction of Black Box 2.2(3) satisfies

(1) If E is an object of $\overline{\langle G\rangle}_n^{[-n,n]}$, then the shifts $E[1]$ and $E[-1]$ both belong to $\overline{\langle G\rangle}_{n+1}^{[-n-1,n+1]}$.

(2) Given an exact triangle $E \to E' \to E''$, with $E \in \overline{\langle G\rangle}_m^{[-m,m]}$ and $E'' \in \overline{\langle G\rangle}_n^{[-n,n]}$, it follows that $E' \in \overline{\langle G\rangle}_{m+n}^{[-m-n,m+n]}$.

Combining Remark 2.5 with Facts 2.6 allows us to improve approximations through iteration. Hence part (2) of the definition below becomes natural, it iterates to provide arbitrarily good approximations.

Definition 2.7. Let $\mathscr{T}$ be a triangulated category with coproducts. It is *approximable* if there exist a t-structure $(\mathscr{T}^{\leq 0}, \mathscr{T}^{\geq 0})$, a compact generator $G \in \mathscr{T}$, and an integer $n > 0$ such that

(1) G belongs to $\mathscr{T}^{\leq n}$ and $\mathrm{Hom}(G, \mathscr{T}^{\leq -n}) = 0$;

(2) Every object $X \in \mathscr{T}^{\leq 0}$ admits an exact triangle $E \to X \to D$ with $E \in \overline{\langle G\rangle}_n^{[-n,n]}$ and with $D \in \mathscr{T}^{\leq -1}$.

Remark 2.8. While part (2) of Definition 2.7 comes motivated by the analogy with Fourier analysis, part (1) of the definition seems random. It requires the t-structure, which is our replacement for the L^p-norm, to be compatible with the compact generator, which is the analog of $g(x) = e^{2\pi i x}$. As the reader will see in Proposition 5.5, this has the effect of uniquely specifying the t-structure (up to equivalence). So maybe a better parallel would be to fix our norm to be a particularly nice one, for example, the L^2-norm on $M(\mathbb{S}^1)$.

Let me repeat myself: as with all new mathematics, Definition 2.7 should be viewed as provisional. In the remainder of this survey, we will discuss the applications as they now stand, to highlight the power of the methods. But I would not be surprised in the slightest if future applications turn out to require modifications, and/or generalizations, of the definitions and of the theorems that have worked so far.

3. EXAMPLES OF APPROXIMABLE TRIANGULATED CATEGORIES

In Section 1 we gave the definition of approximable triangulated categories. The definition combines old, classical ingredients (t-structures and compact generators) with a new construction, the category $\overline{\langle G\rangle}_n^{[-n,n]}$ of Black Box 2.2(3). The first thing to show is that the theory is nonempty: we need to produce examples, categories people care about which satisfy the definition of approximability. The current section is devoted to the known examples of approximable triangulated categories. We repeat what we have said before: the subject is in its infancy, there could well be many more examples out there.

Example 3.1. Let $\mathscr{T}$ be a triangulated category with coproducts. If $G \in \mathscr{T}$ is a compact generator such that $\mathrm{Hom}(G, G[i]) = 0$ for all $i > 0$, then the category $\mathscr{T}$ is approximable.

This example turns out to be easy, the reader is referred to [29, EXAMPLE 3.3] for the (short) proof. Special cases include

(1) $\mathscr{T} = \mathbf{D}(R\text{–Mod})$, where R is a dga with $H^i(R) = 0$ for $i > 0$;

(2) The homotopy category of spectra.

Example 3.2. If X is a quasicompact, separated scheme, then the category $\mathbf{D}_{\mathrm{qc}}(X)$ is approximable. We remind the reader of the traditional notation being used here: the category $\mathbf{D}(X)$ is the unbounded derived category of complexes of sheaves of $\mathscr{O}_X$-modules, and the full subcategory $\mathbf{D}_{\mathrm{qc}}(X) \subset \mathbf{D}(X)$ has for objects the complexes with quasicoherent cohomology.

The proof of the approximability of $\mathbf{D}_{\mathrm{qc}}(X)$ is not trivial. The category has a standard t-structure, that part is easy. The existence of a compact generator G needs proof, it may be found in Bondal and Van den Bergh [6, THEOREM 3.1.1(II)]. Their proof is not constructive, it is only an existence proof, but it does give enough information to deduce that part (1) of Definition 2.7 is satisfied by every compact generator (indeed, it is satisfied by every compact object). See [6, THEOREM 3.1.1(I)]. But it is a challenge to show that we may choose a compact generator G and an integer $n > 0$ in such a way that Definition 2.7(2) is satisfied.

If we further assume that X is of finite type over a noetherian ring R, then the (relatively intricate) proof of the approximability of $\mathbf{D}_{\mathrm{qc}}(X)$ occupies [33, SECTIONS 4 AND 5]. The little trick, that extends the result to all quasicompact and separated X, was not observed until later: it appears in [29, LEMMA 3.5].

Example 3.3. It is a theorem that, under mild hypotheses, the recollement of any two approximable triangulated categories is approximable. To state the "mild hypotheses" precisely: suppose we are given a recollement of triangulated categories

$$\mathscr{R} \rightleftarrows \mathscr{S} \rightleftarrows \mathscr{T}$$

with $\mathscr{R}$ and $\mathscr{T}$ approximable. Assume further that the category $\mathscr{S}$ is compactly generated, and any compact object $H \in \mathscr{S}$ has the property that $\text{Hom}(H, H[i]) = 0$ for $i \gg 0$. Then the category $\mathscr{S}$ is also approximable.

The reader can find the proof in [7, THEOREM 4.1], it is the main result in the paper. The bulk of the article is devoted to developing the machinery necessary to prove the theorem—hence it is worth noting that this machinery has since demonstrated usefulness in other contexts, see the subsequent articles [27,28].

There is a beautiful theory of noncommutative schemes, and a rich literature studying them. And many of the interesting examples of such schemes are obtained as recollements of ordinary schemes, or of admissible pieces of them. Thus the theorem that recollements of approximable triangulated categories are approximable gives a wealth of new examples of approximable triangulated categories.

Since this ICM is being held in St. Petersburg, it would be remiss not to mention that the theory of noncommutative algebraic geometry, in the sense of the previous paragraph, is a subject to which Russian mathematicians have contributed a vast amount. The seminal work of Bondal, Kontsevich, Kuznetsov, Lunts, and Orlov immediately springs to mind. For a beautiful introduction to the field, the reader might wish to look at the early sections of Orlov [34]. The later sections prove an amazing new theorem, but the early ones give a lovely survey of the background. In fact, the theory sketched in this survey was born when I was trying to read and understand Orlov's beautiful article.

4. APPLICATIONS: STRONG GENERATION

We begin by reminding the reader of a classical definition, going back to Bondal and Van den Bergh [6].

Definition 4.1. Let $\mathscr{T}$ be triangulated category. An object $G \in \mathscr{T}$ is called a *strong generator* if there exists an integer $\ell > 0$ with $\mathscr{T} = \langle G \rangle_\ell$, where the notation is as in Black Box 2.2(1). The category $\mathscr{T}$ is called *regular* or *strongly generated* if it contains a strong generator.

The first application of approximability is the proof of the following two theorems.

Theorem 4.2. *Let X be a quasicompact, separated scheme. The derived category of perfect complexes on X, denoted here by $\mathbf{D}^{\text{perf}}(X)$, is regular if and only if X has a cover by open subsets $\text{Spec}(R_i) \subset X$, with each R_i of finite global dimension.*

Remark 4.3. If X is noetherian and separated, then Theorem 4.2 specializes to saying that $\mathbf{D}^{\text{perf}}(X)$ is regular if and only if X is regular and finite-dimensional. Hence the terminology.

Theorem 4.4. *Let X be a noetherian, separated, finite-dimensional, quasiexcellent scheme. Then the category $\mathbf{D}^b(\text{Coh}(X))$, the bounded derived category of coherent sheaves on X, is always regular.*

Remark 4.5. The reader is referred to [33] and to Aoki [4] for the proofs of Theorems 4.2 and 4.4. More precisely, for Theorem 4.2 see [33, THEOREM 0.5]. About Theorem 4.4: if we add the assumption that every closed subvariety of X admits a regular alteration then the result may be found in [33, THEOREM 0.15], but Aoki [4] found a lovely argument that allowed him to extend the statement to all quasiexcellent X.

There is a rich literature on strong generation, with beautiful papers by many authors. In the introduction to [33], as well as in [26] and [31, SECTION 7], the reader can find an extensive discussion of (some of) this fascinating work and of the way Theorems 4.2 and 4.4 compare to the older literature. For a survey taking an entirely different tack, see Minami [22], which places in historical perspective a couple of the key steps in the proofs that [33] gives for Theorems 4.2 and 4.4.

Since all of this is now well documented in the published literature, let us focus the remainder of the current survey on the other applications of approximability. Those are all still in preprint form, see [27–29], although there are (published) surveys in [31, SECTIONS 8 AND 9] and in [30]. Those surveys are fuller and more complete than the sketchy one we are about to embark on. As we present the material, we will feel free to refer the reader to the more extensive surveys whenever we deem it appropriate.

5. THE FREEDOM IN THE CHOICE OF COMPACT GENERATOR AND T-STRUCTURE

Definition 2.7 tells us that a triangulated category $\mathscr{T}$ with coproducts is approximable if there exist, in $\mathscr{T}$, a compact generator G and a t-structure $(\mathscr{T}^{\leq 0}, \mathscr{T}^{\geq 0})$ satisfying some properties. The time has come to explore just how free we are in the choice of the compact generator and of the t-structure. To address this question we begin by formulating:

Definition 5.1. Let $\mathscr{T}$ be a triangulated category. Then two t-structures $(\mathscr{T}_1^{\leq 0}, \mathscr{T}_1^{\geq 0})$ and $(\mathscr{T}_2^{\leq 0}, \mathscr{T}_2^{\geq 0})$ are declared *equivalent* if there exists an integer $n > 0$ such that

$$\mathscr{T}_1^{\leq -n} \subset \mathscr{T}_2^{\leq 0} \subset \mathscr{T}_1^{\leq n}.$$

Discussion 5.2. Let $\mathscr{T}$ be a triangulated category with coproducts. If $G \in \mathscr{T}$ is a compact object and $\overline{\langle G \rangle}^{(-\infty,0]}$ is as in Black Box 2.2(2), then Alonso, Jeremías, and Souto [1, THEOREM A.1], building on the work of Keller and Vossieck [16], teaches us that there is a unique t-structure $(\mathscr{T}^{\leq 0}, \mathscr{T}^{\geq 0})$ with $\mathscr{T}^{\leq 0} = \overline{\langle G \rangle}^{(-\infty,n]}$. We will call this the *t-structure generated by* G, and denote it $(\mathscr{T}_G^{\leq 0}, \mathscr{T}_G^{\geq 0})$.

In Black Box 2.2(2) we asked the reader to accept, as a black box, the construction passing from an object $G \in \mathscr{T}$ to the subcategory $\overline{\langle G \rangle}^{(-\infty,0]}$. If G is compact, then [1, THEOREM A.1] allows us to express this as $\mathscr{T}_G^{\leq 0}$ for a unique t-structure. We ask the reader to accept on faith that:

Lemma 5.3. *If G and H are two compact generators for the triangulated category $\mathscr{T}$, then the two t-structures $(\mathscr{T}_G^{\leq 0}, \mathscr{T}_G^{\geq 0})$ and $(\mathscr{T}_H^{\leq 0}, \mathscr{T}_H^{\geq 0})$ are equivalent as in Definition 5.2.*

As it happens, the proof of Lemma 5.3 is easy, the interested reader can find it in **[29, REMARK 0.15]**. And Lemma 5.3 leads us to:

Definition 5.4. Let $\mathscr{T}$ be a triangulated category in which there exists a compact generator. We define the *preferred equivalence class of t-structures* as follows: a t-structure belongs to the preferred equivalence class if it is equivalent to $(\mathscr{T}_G^{\leq 0}, \mathscr{T}_G^{\geq 0})$ for some compact generator $G \in \mathscr{T}$, and by Lemma 5.3 it is equivalent to $(\mathscr{T}_H^{\leq 0}, \mathscr{T}_H^{\geq 0})$ for every compact generator H.

The following is also not too hard, and may be found in **[29, PROPOSITIONS 2.4 AND 2.6]**.

Proposition 5.5. *Let $\mathscr{T}$ be an approximable triangulated category. Then for any t-structure $(\mathscr{T}^{\leq 0}, \mathscr{T}^{\geq 0})$ in the preferred equivalence class, and for any compact generator $H \in \mathscr{T}$, there exists an integer $n > 0$ (which may depend on H and on the t-structure), satisfying*

(1) *H belongs to $\mathscr{T}^{\leq n}$ and* $\mathrm{Hom}(H, \mathscr{T}^{\leq -n}) = 0$;

(2) *Every object $X \in \mathscr{T}^{\leq 0}$ admits an exact triangle $E \to X \to D$ with $E \in \overline{\langle H \rangle}_n^{[-n,n]}$ and with $D \in \mathscr{T}^{\leq -1}$.*

Moreover, if H is a compact generator, $(\mathscr{T}^{\leq 0}, \mathscr{T}^{\geq 0})$ is a t-structure, and there exists an integer $n > 0$ satisfying (1) and (2) above, then the t-structure $(\mathscr{T}^{\leq 0}, \mathscr{T}^{\geq 0})$ must belong to the preferred equivalence class.

Remark 5.6. Strangely enough, the value of Proposition 5.5 can be that it allows us to find an explicit t-structure in the preferred equivalence class.

Consider the case where X is a quasicompact, separated scheme. By Bondal and Van den Bergh **[6, THEOREM 3.1.1(II)]**, we know that the category $\mathbf{D_{qc}}(X)$ has a compact generator, but in Example 3.2 we mentioned that the existence proof is not overly constructive, it does not give us a handle on any explicit compact generator. Let G be some compact generator. From Alonso, Jeremías, and Souto **[1, THEOREM A.1]**, we know that the subcategory $\overline{\langle G \rangle}^{(-\infty,0]}$ of Black Box 2.2(2) is equal to $\mathscr{T}_G^{\leq 0}$ for a unique t-structure $(\mathscr{T}_G^{\leq 0}, \mathscr{T}_G^{\geq 0})$ in the preferred equivalence class. But this does not leave us a whole lot wiser—the compact generator G is not explicit, hence neither is the t-structure.

However, the combination of **[33, THEOREM 5.8]** and **[29, LEMMA 3.5]** tells us that the category $\mathbf{D_{qc}}(X)$ is approximable, and it so happens that the t-structure *used in the proof*, that is, the t-structure for which a compact generator H and an integer $n > 0$ satisfying (1) and (2) of Proposition 5.5 are shown to exist, happens to be the standard t-structure. From Proposition 5.5, we now deduce that the standard t-structure is in the preferred equivalence class.

6. STRUCTURE THEOREMS IN APPROXIMABLE TRIANGULATED CATEGORIES

An approximable triangulated category $\mathscr{T}$ must have a compact generator G, and Definition 5.4 constructed for us a preferred equivalence class of t-structures—namely all those equivalent to $(\mathscr{T}_G^{\le 0}, \mathscr{T}_G^{\ge 0})$. Recall that, for any t-structure $(\mathscr{T}^{\le 0}, \mathscr{T}^{\ge 0})$, it is customary to define

$$\mathscr{T}^- = \bigcup_{n=1}^{\infty} \mathscr{T}^{\le n}, \quad \mathscr{T}^+ = \bigcup_{n=1}^{\infty} \mathscr{T}^{\ge -n}, \quad \mathscr{T}^b = \mathscr{T}^- \cap \mathscr{T}^+.$$

It is an easy exercise to show, directly from Definition 5.1, that equivalent t-structures give rise to identical $\mathscr{T}^-$, $\mathscr{T}^+$, and $\mathscr{T}^b$. Therefore triangulated categories with a single compact generator, and in particular approximable triangulated categories, have preferred subcategories $\mathscr{T}^-$, $\mathscr{T}^+$, and $\mathscr{T}^b$, which are intrinsic—they are simply those corresponding to any t-structure in the preferred equivalence class. In the remainder of this survey, we will assume that $\mathscr{T}^-$, $\mathscr{T}^+$, and $\mathscr{T}^b$ always stand for the preferred ones.

In the heuristics of Discussion 2.1(2), we told the reader that a t-structure $(\mathscr{T}^{\le 0}, \mathscr{T}^{\ge 0})$ is to be viewed as a metric on $\mathscr{T}$. In Definition 6.1 below, the heuristic is that we construct a full subcategory $\mathscr{T}_c^-$ to be the closure of $\mathscr{T}^c$ with respect to any of the (equivalent) metrics that come from t-structures in the preferred equivalence class.

Definition 6.1. Let $\mathscr{T}$ be an approximable triangulated category. The full subcategory $\mathscr{T}_c^-$ is given by

$$\operatorname{Ob}(\mathscr{T}_c^-) = \left\{ F \in \mathscr{T} \;\middle|\; \begin{array}{c} \text{For every integer } n > 0 \text{ and for every t-structure} \\ \left(\mathscr{T}^{\le 0}, \mathscr{T}^{\ge 0}\right) \text{ in the preferred equivalence class,} \\ \text{there exists an exact triangle } E \to F \to D \text{ in } \mathscr{T} \\ \text{with } E \in \mathscr{T}^c \text{ and } D \in \mathscr{T}^{\le -n} \end{array} \right\}.$$

The full subcategory $\mathscr{T}_c^b$ is defined to be $\mathscr{T}_c^b = \mathscr{T}_c^- \cap \mathscr{T}^b$.

Remark 6.2. Let $\mathscr{T}$ be an approximable triangulated category. Aside from the classical, full subcategory $\mathscr{T}^c$ of compact objects, which we encountered back in Definition 1.1, we have in this section concocted five more intrinsic, full subcategories of $\mathscr{T}$: they are $\mathscr{T}^-$, $\mathscr{T}^+$, $\mathscr{T}^b$, $\mathscr{T}_c^-$, and $\mathscr{T}_c^b$. It can be proved that all six subcategories, that is, the old $\mathscr{T}^c$ and the five new ones, are thick subcategories of $\mathscr{T}$. In particular, each of them is a triangulated category.

Example 6.3. It becomes interesting to figure out what all these categories come down to in examples.

Let X be a quasicompact, separated scheme. From Example 3.2, we know that the category $\mathscr{T} = \mathbf{D}_{\mathbf{qc}}(X)$ is approximable, and in Remark 5.6 we noted that the standard t-structure is in the preferred equivalence class. This can be used to show that, for $\mathscr{T} = \mathbf{D}_{\mathbf{qc}}(X)$, we have

$$\begin{aligned} \mathscr{T}^- &= \mathbf{D}_{\mathbf{qc}}^-(X), & \mathscr{T}^+ &= \mathbf{D}_{\mathbf{qc}}^+(X), & \mathscr{T}^b &= \mathbf{D}_{\mathbf{qc}}^b(X), \\ \mathscr{T}^c &= \mathbf{D}^{\mathbf{perf}}(X), & \mathscr{T}_c^- &= \mathbf{D}_{\mathbf{coh}}^-(X), & \mathscr{T}_c^b &= \mathbf{D}_{\mathbf{coh}}^b(X), \end{aligned}$$

where the last two equalities assume that the scheme X is noetherian, and all six categories on the right of the equalities have their traditional meanings.

The reader can find an extensive discussion of the claims above in [31], more precisely in the paragraphs between [31, PROPOSITION 8.10] and [31, THEOREM 8.16]. That discussion goes beyond the scope of the current survey, it analyzes the categories $\mathscr{T}_c^b \subset \mathscr{T}_c^-$ in the generality of non-noetherian schemes, where they still have a classical description—of course, not involving the category of coherent sheaves. After all coherent sheaves do not behave well for non-noetherian schemes.

Remark 6.4. In this survey we spent some effort introducing the notion of approximable triangulated categories. In Example 3.2 we told the reader that it is a theorem (and not a trivial one) that, as long as a scheme X is quasicompact and separated, the derived category $\mathbf{D_{qc}}(X)$ is approximable. In this section we showed that every approximable triangulated category comes with canonically defined, intrinsic subcategories $\mathscr{T}^-$, $\mathscr{T}^+$, $\mathscr{T}^b$, $\mathscr{T}^c$, $\mathscr{T}_c^-$, and $\mathscr{T}_c^b$, and in Example 6.3 we informed the reader that, in the special case where $\mathscr{T} = \mathbf{D_{qc}}(X)$, these turn out to be $\mathbf{D}^-_{\mathbf{qc}}(X)$, $\mathbf{D}^+_{\mathbf{qc}}(X)$, $\mathbf{D}^b_{\mathbf{qc}}(X)$, $\mathbf{D}^{\mathbf{perf}}(X)$, $\mathbf{D}^-_{\mathbf{coh}}(X)$, and $\mathbf{D}^b_{\mathbf{coh}}(X)$, respectively.

Big deal. This teaches us that the traditional subcategories $\mathbf{D}^-_{\mathbf{qc}}(X)$, $\mathbf{D}^+_{\mathbf{qc}}(X)$, $\mathbf{D}^b_{\mathbf{qc}}(X)$, $\mathbf{D}^{\mathbf{perf}}(X)$, $\mathbf{D}^-_{\mathbf{coh}}(X)$, and $\mathbf{D}^b_{\mathbf{coh}}(X)$ of the category $\mathbf{D_{qc}}(X)$ all have intrinsic descriptions. This might pass as a curiosity, unless we can actually use it to prove something we care about that we did not use to know.

Discussion 6.5. To motivate the next theorem, it might help to think of the parallel with functional analysis.

Let $M(\mathbb{R})$ be the vector space of Lebesgue-measurable, real-valued functions on $\mathbb{R}$. Given any two functions $f, g \in M(\mathbb{R})$, we can pair them by integrating the product, that is, we form the pairing

$$\langle f, g \rangle = \int fg \, d\mu,$$

where μ is Lebesgue measure. This gives us a map

$$M(\mathbb{R}) \times M(\mathbb{R}) \xrightarrow{\quad \langle -,- \rangle \quad} \mathbb{R} \cup \{\infty\},$$

where the integral is declared to be infinite if it does not converge.

We can restrict this pairing to subspaces of $M(\mathbb{R})$. For example, if $f \in L^p(\mathbb{R})$ and $g \in L^q(\mathbb{R})$ with $\frac{1}{p} + \frac{1}{q} = 1$ then the integral converges, that is, $\langle f, g \rangle \in \mathbb{R}$, and we deduce a map

$$L^p(\mathbb{R}) \longrightarrow \operatorname{Hom}\big(L^q(\mathbb{R}), \mathbb{R}\big)$$

which turns out to be an isometry of Banach spaces.

The category-theoretic version is that on any category $\mathscr{T}$ there is the pairing sending two objects $A, B \in \mathscr{T}$ to $\operatorname{Hom}(A, B)$. Of course, this pairing is not symmetric, we have to keep track of the position of A and of B in $\operatorname{Hom}(A, B)$. If R is a commutative ring and $\mathscr{T}$ happens to be an R-linear category, then $\operatorname{Hom}(A, B)$ is an R-module and the pairing delivers a map

$$\mathscr{T}^{\mathrm{op}} \times \mathscr{T} \xrightarrow{\quad \operatorname{Hom}(-,-) \quad} R\text{–Mod},$$

where the op keeps track of the variable in the first position. And now we are free to restrict to subcategories of $\mathscr{T}$.

If $\mathscr{T}$ happens to be approximable and R-linear, we have just learned that it comes with six intrinsic subcategories $\mathscr{T}^-$, $\mathscr{T}^+$, $\mathscr{T}^b$, $\mathscr{T}^c$, $\mathscr{T}_c^-$, and $\mathscr{T}_c^b$. We are free to restrict the Hom pairing to any couple of them. This gives us 36 possible pairings, and each of those yields two maps from a subcategory to the dual of another. There are 72 cases we could study, and the theorem below tells us something useful about four of those.

Theorem 6.6. *Let R be a noetherian ring, and let $\mathscr{T}$ be an R-linear, approximable triangulated category. Suppose there exists in $\mathscr{T}$ a compact generator G so that* $\mathrm{Hom}(G, G[n])$ *is a finite R-module for all $n \in \mathbb{Z}$. Consider the two functors*

$$\mathscr{Y} : \mathscr{T}_c^- \to \mathrm{Hom}_R\big([\mathscr{T}^c]^{\mathrm{op}}, R\text{–Mod}\big), \qquad \widetilde{\mathscr{Y}} : [\mathscr{T}_c^-]^{\mathrm{op}} \to \mathrm{Hom}_R\big(\mathscr{T}_c^b, R\text{–Mod}\big)$$

defined by the formulas $\mathscr{Y}(B) = \mathrm{Hom}(-, B)$ *and* $\widetilde{\mathscr{Y}}(A) = \mathrm{Hom}(A, -)$*, as in Discussion* 6.5. *Now consider the following composites:*

$$\mathscr{T}_c^b \xhookrightarrow{\quad i \quad} \mathscr{T}_c^- \xrightarrow{\quad \mathscr{Y} \quad} \mathrm{Hom}_R\big([\mathscr{T}^c]^{\mathrm{op}}, R\text{–Mod}\big),$$

$$[\mathscr{T}^c]^{\mathrm{op}} \xhookrightarrow{\quad \tilde{i} \quad} [\mathscr{T}_c^-]^{\mathrm{op}} \xrightarrow{\quad \widetilde{\mathscr{Y}} \quad} \mathrm{Hom}_R\big(\mathscr{T}_c^b, R\text{–Mod}\big).$$

We assert:

(1) *The functor $\mathscr{Y}$ is full, and the essential image consists of the locally finite homological functors (see Explanation* 6.7 *for the definition of locally finite functors). The composite $\mathscr{Y} \circ i$ is fully faithful, and the essential image consists of the finite homological functors (again, see Explanation* 6.7 *for the definition).*

(2) *With the notation as in Black Box* 2.2(1)*, assume*[1] *that $\mathscr{T} = \overline{\langle H \rangle}_n$ for some integer $n > 0$ and some object $H \in \mathscr{T}_c^b$. Then the functor $\widetilde{\mathscr{Y}}$ is full, and the essential image consists of the locally finite homological functors. The composite $\widetilde{\mathscr{Y}} \circ \tilde{i}$ is fully faithful, and the essential image consists of the finite homological functors.*

Explanation 6.7. In the statement of Theorem 6.6, the *locally finite* functors, either of the form $H : [\mathscr{T}^c]^{\mathrm{op}} \to R$–Mod or of the form $H : \mathscr{T}_c^b \to R$–Mod, are the functors such that

(1) $H(A[i])$ is a finite R–module for every $i \in \mathbb{Z}$ and every A in either $\mathscr{T}^c$ or $\mathscr{T}_c^b$;

(2) For fixed A, in one of $\mathscr{T}^c$ or $\mathscr{T}_c^b$, we have $H(A[i]) = 0$ if $i \ll 0$.

The *finite* functors are those for which we also have

(3) $H(A[i]) = 0$ for all $i \gg 0$.

Remark 6.8. The proof of part (1) of Theorem 6.6 may be found in **[29]**, while the proof of part (2) of Theorem 6.6 occupies **[28]**. These are not easy theorems.

Let $\mathscr{T} = \mathbf{D_{qc}}(X)$, with X a scheme proper over a noetherian ring R. Then the hypotheses of Theorem 6.6(1) are satisfied. We learn (among other things) that the natural

---

1 What's important for the current survey is that, if X is a noetherian, separated scheme, then $\mathscr{T} = \mathbf{D_{qc}}(X)$ satisfies this hypothesis provided X is finite-dimensional and quasiexcellent.

functor, taking an object $B \in \mathbf{D}^b_{\mathbf{coh}}(X)$ to the R-linear functor $\mathrm{Hom}(-, B) : \mathbf{D}^{\mathbf{perf}}(X)^{\mathrm{op}} \to$ Mod–R, is a fully faithful embedding

$$\mathbf{D}^b_{\mathbf{coh}}(X) \xrightarrow{\quad \mathscr{Y} \circ i \quad} \mathrm{Hom}_R\big(\mathbf{D}^{\mathbf{perf}}(X)^{\mathrm{op}},\ R\text{–Mod}\big)$$

whose essential image is precisely the finite homological functors.

If we further assume that the scheme X is finite-dimensional and quasiexcellent then the hypotheses of Theorem 6.6(2) are also satisfied. We learn that the functor, taking an object $A \in \mathbf{D}^{\mathbf{perf}}(X)$ to the R-linear functor $\mathrm{Hom}(A, -)$, is a fully faithful embedding

$$\mathbf{D}^{\mathbf{perf}}(X)^{\mathrm{op}} \xrightarrow{\quad \widetilde{\mathscr{Y}} \circ \tilde{i} \quad} \mathrm{Hom}_R\big(\mathbf{D}^b_{\mathbf{coh}}(X),\ R\text{–Mod}\big)$$

whose essential image is also the finite homological functors.

In [31, HISTORICAL SURVEY 8.2] the reader can find a discussion of the (algebro-geometric) precursors of Theorem 6.6. As for the applications, let us go through one of them.

Remark 6.9. Let X be a scheme proper over the field $\mathbb{C}$ of complex numbers, and let X^{an} be the underlying complex analytic space. The analytification induces a functor we will call $\mathscr{L} : \mathbf{D}^b_{\mathbf{coh}}(X) \to \mathbf{D}^b_{\mathbf{coh}}(X^{\mathrm{an}})$, it is the functor taking a bounded complex of coherent algebraic sheaves on X to the analytification, which is a bounded complex of coherent analytic sheaves on X^{an}. The pairing sending an object $A \in \mathbf{D}^{\mathbf{perf}}(X)$ and an object $B \in \mathbf{D}^b_{\mathbf{coh}}(X^{\mathrm{an}})$ to $\mathrm{Hom}(\mathscr{L}(A), B)$ delivers a map

$$\mathbf{D}^b_{\mathbf{coh}}(X^{\mathrm{an}}) \longrightarrow \mathrm{Hom}_R\big(\mathbf{D}^{\mathbf{perf}}(X)^{\mathrm{op}},\ \mathbb{C}\text{–Mod}\big).$$

Since the image lands in the finite homological functors, Theorem 6.6(1) allows us to factor this uniquely through the inclusion $\mathscr{Y} \circ i$, that is, there exists (up to canonical natural isomorphism) a unique functor $\mathscr{R}$ rendering commutative the triangle

$$\begin{array}{ccc} \mathbf{D}^b_{\mathbf{coh}}(X^{\mathrm{an}}) & \searrow & \\ {\scriptstyle \exists!\mathscr{R}}\ \big\downarrow & & \mathrm{Hom}_R\big(\mathbf{D}^{\mathbf{perf}}(X)^{\mathrm{op}},\ \mathbb{C}\text{–Mod}\big). \\ \mathbf{D}^b_{\mathbf{coh}}(X) & \underset{\mathscr{Y} \circ i}{\nearrow} & \end{array}$$

And proving Serre's GAGA theorem reduces to the easy exercise of showing that $\mathscr{L}$ and $\mathscr{R}$ are inverse equivalences, the reader can find this in the (short) [29, SECTION 8 AND APPENDIX A].

The brilliant inspiration underpinning the approach is due to Jack Hall [12], he is the person who came up with the idea of using the pairing above, coupled with representability theorems, to prove GAGA. The representability theorems available to Jack Hall at the time were not powerful enough, and Theorem 6.6 was motivated by trying to find a direct path from the ingenious, simple idea to a fullblown proof.

Discussion 6.10. In preparation for the next theorem, we give a very brief review of metrics in triangulated categories. The reader is referred to the survey article [30] for a much fuller and more thorough account.

Given a triangulated category $\mathscr{T}$, a *metric* on $\mathscr{T}$ assigns a length to every morphism. In this article the only metrics we consider are the ones arising from t-structures. If $\mathscr{T}$ is an approximable triangulated category we choose a t-structure $(\mathscr{T}^{\leq 0}, \mathscr{T}^{\geq 0})$ in the preferred equivalence class, and this induces a metric as follows. Given a morphism $f : X \to Y$, we may complete to an exact triangle $X \xrightarrow{f} Y \to D$, and the length of f is given by the formula

$$\text{Length}(f) = \inf\left\{ \frac{1}{2^n} \;\middle|\; n \in \mathbb{Z} \text{ and } D \in \mathscr{T}^{\leq -n} \right\}.$$

In this survey we allow the length of a morphism to be infinite; if the set on the right is empty then we declare $\text{Length}(f) = \infty$.

This metric depends on the choice of a t-structure, but not a lot. As all t-structures in the preferred equivalence class are equivalent, any two preferred t-structures will give rise to equivalent metrics (with an obvious definition of equivalence of metrics).

Note that if $\mathscr{T}$ is a triangulated category and $\mathscr{S}$ is a triangulated subcategory, then a metric on $\mathscr{T}$ restricts to a metric on $\mathscr{S}$. In particular, if $\mathscr{T}$ is approximable, the metric on $\mathscr{T}$ of the previous paragraph restricts to give metrics on the full subcategories $\mathscr{T}^c$ and $\mathscr{T}_c^b$. Once again these metrics are only defined up to equivalence. And, of course, a metric on $\mathscr{S}$ is also a metric on $\mathscr{S}^{\text{op}}$, thus we have specified (up to equivalence) canonical metrics on $\mathscr{T}^c$, $\mathscr{T}_c^b$, $[\mathscr{T}^c]^{\text{op}}$, and $[\mathscr{T}_c^b]^{\text{op}}$.

Suppose $\mathscr{S}$ is a triangulated category with a metric. A *Cauchy sequence* in $\mathscr{S}$ is a sequence of morphisms $E_1 \to E_2 \to E_3 \to \cdots$ which eventually become arbitrarily short. If $\mathscr{A}b$ is the category of abelian groups, then the Yoneda embedding $Y : \mathscr{S} \to \text{Mod–}\mathscr{S}$ embeds $\mathscr{S}$ into the category Mod–$\mathscr{S}$ of additive functors $\mathscr{S}^{\text{op}} \to \mathscr{A}b$. In the category Mod–$\mathscr{S}$ colimits exist, allowing us to define

(1) The category $\mathfrak{L}(\mathscr{S})$ is the full subcategory of Mod–$\mathscr{S}$, whose objects are the colimits of Yoneda images of Cauchy sequences in $\mathscr{S}$;

(2) The full subcategory $\mathfrak{S}(\mathscr{S}) \subset \mathfrak{L}(\mathscr{S})$ has for objects those functors $F \in \mathfrak{L}(\mathscr{S}) \subset$ Mod–$\mathscr{S}$ which take sufficiently short morphisms to isomorphisms. In symbols, $F \in \mathfrak{L}(\mathscr{S})$ belongs to $\mathfrak{S}(\mathscr{S})$ if there exists an $\varepsilon > 0$ such that

$$\{\text{Length}(f) < \varepsilon\} \Rightarrow \{F(f) \text{ is an isomorphism}\};$$

(3) The exact triangles in $\mathfrak{S}(\mathscr{S})$ are the colimits in Mod–$\mathscr{S}$ of Yoneda images of Cauchy sequences of exact triangles in $\mathscr{S}$, where the colimits happen to lie in $\mathfrak{S}(\mathscr{S})$.

A word of caution about (3): if we are given in $\mathscr{S}$ a Cauchy sequence of exact triangles, we can form the colimit in Mod–$\mathscr{S}$ of its Yoneda image. This colimit is guaranteed to lie in $\mathfrak{L}(\mathscr{S})$, but will not usually lie in the smaller $\mathfrak{S}(\mathscr{S})$. If it happens to lie in $\mathfrak{S}(\mathscr{S})$ then (3) declares it to be an exact triangle in $\mathfrak{S}(\mathscr{S})$.

And now we are ready for the theorem.

Theorem 6.11. *Let $\mathscr{S}$ be a triangulated category with a metric. Assume the metric is* good; *this is a technical term, see* [30, DEFINITION 10] *for the precise formulation. Then*

(1) *The category $\mathfrak{S}(\mathscr{S})$ of Discussion 6.10(2), with the exact triangles as defined in Discussion 6.10(3), is a triangulated category.*

Now let $\mathscr{T}$ be an approximable triangulated category. In Discussion 6.10 we constructed (up to equivalence) a metric on $\mathscr{T}$, and hence on its subcategories $\mathscr{T}^c$ and $[\mathscr{T}_c^b]^{\mathrm{op}}$. Those metrics are all good, and the theorem goes on to give natural, exact equivalences of triangulated categories

(2) $\mathfrak{S}(\mathscr{T}^c) \cong \mathscr{T}_c^b$. *This equivalence is unconditional.*

(3) *If the approximable triangulated category $\mathscr{T}$ happens to be* noetherian *as in* [27, DEFINITION 5.1], *then* $\mathfrak{S}([\mathscr{T}_c^b]^{\mathrm{op}}) \cong [\mathscr{T}^c]^{\mathrm{op}}$.

Remark 6.12. First of all, in Theorem 6.11(3) we assumed that the approximable triangulated $\mathscr{T}$ is noetherian as in [27, DEFINITION 5.1]. The only observation we want to make here is that if X is a noetherian, separated scheme then the approximable triangulated category $\mathscr{T} = \mathbf{D}_{\mathbf{qc}}(X)$ is noetherian. Thus, for noetherian, separated schemes X, Theorem 6.11 gives exact equivalences of triangulated categories

$$\mathfrak{S}\big(\mathbf{D}^{\mathbf{perf}}(X)\big) \cong \mathbf{D}^b_{\mathbf{coh}}(X), \quad \mathfrak{S}\big(\mathbf{D}^b_{\mathbf{coh}}(X)^{\mathrm{op}}\big) \cong \mathbf{D}^{\mathbf{perf}}(X)^{\mathrm{op}}.$$

The research paper [27] contains the proofs of the assertions in Theorem 6.11. The reader can find a skimpy survey in [31, SECTION 9] and a more extensive one in [30]. In [31, HISTORICAL SURVEY 9.1] there is a discussion of precursors of the results.

7. FUTURE DIRECTIONS

New scientific developments are tentative and unpolished; only with the passage of time do they acquire the gloss and elegance of a refined, varnished theory. And there is nothing more difficult to predict than the future. My colleague Neil Trudinger used to joke that my beard makes me look like a biblical prophet—the reader should not be deceived, appearances are notoriously misleading, the abundance of facial hair is not a reliable yardstick for measuring the gift of foresight that marks out a visionary, and I am certifiably not a clairvoyant. All I do in this section is offer a handful of obvious questions that spring to mind. The list is not meant to be exhaustive, and might well be missing major tableaux of the overall picture. It is entirely possible that the future will see this theory flourish in directions orthogonal to those sketched here.

Let us begin with what is freshest in our minds: we have just seen Theorem 6.11, part (1) of which tells us that, given a triangulated category $\mathscr{S}$ with a good metric, there is a recipe producing another triangulated category $\mathfrak{S}(\mathscr{S})$, which, as it happens, comes with an induced good metric. We can ask:

Problem 7.1. Can one formulate reasonable sufficient conditions, on the triangulated category $\mathscr{S}$ and on its good metric, to guarantee that $\mathfrak{S}(\mathfrak{S}(\mathscr{S})^{\mathrm{op}}) = \mathscr{S}^{\mathrm{op}}$? Who knows, maybe even necessary and sufficient conditions?

Motivating Example 7.2. Let $\mathscr{T}$ be an approximable triangulated category and put $\mathscr{S} = \mathscr{T}^c$, with the metric of Discussion 6.10. Theorem 6.11(2) computes for us that $\mathfrak{S}(\mathscr{T}^c) \cong \mathscr{T}_c^b$. I ask the reader to believe that the natural, induced metric on $\mathfrak{S}(\mathscr{T}^c)$ agrees with the metric on $\mathscr{T}_c^b \subset \mathscr{T}$ given in Discussion 6.10. Now Theorem 6.11(3) goes on to tell us that, as long as the approximable triangulated category $\mathscr{T}$ is noetherian, we also have that $\mathfrak{S}([\mathscr{T}_c^b]^{\mathrm{op}}) \cong [\mathscr{T}^c]^{\mathrm{op}}$; as it happens, the induced good metric on $\mathfrak{S}([\mathscr{T}_c^b]^{\mathrm{op}})$ also agrees, up to equivalence, with the metric that Discussion 6.10 created on $[\mathscr{T}^c]^{\mathrm{op}}$. Combining these we have many examples of exact equivalences of triangulated categories $\mathfrak{S}(\mathfrak{S}(\mathscr{S})^{\mathrm{op}}) \cong \mathscr{S}^{\mathrm{op}}$, which are homeomorphisms with respect to the metrics. Thus Problem 7.1 asks the reader to find the right generalization.

Next one can wonder about the functoriality of the construction. Suppose $F : \mathscr{S} \to \mathscr{T}$ is a triangulated functor, and that both $\mathscr{S}$ and $\mathscr{T}$ have good metrics. What are reasonable sufficient conditions which guarantee the existence of an induced functor $\mathfrak{S}(F)$, either from $\mathfrak{S}(\mathscr{S})$ to $\mathfrak{S}(\mathscr{T})$ or in the other direction? So far there is one known result of this genre, the reader can find the statement below in Sun and Zhang [37, THEOREM 1.1(3)].

Theorem 7.3. *Suppose we are given two triangulated categories $\mathscr{S}$ and $\mathscr{T}$, both with good metrics. Suppose we are also given a pair of functors $F : \mathscr{S} \rightleftarrows \mathscr{T} : G$ with $F \dashv G$, meaning that F is left adjoint to G. Assume further that both F and G are continuous with respect to the metrics, in the obvious sense.*

Then the functor $\hat{F} : \text{Mod–}\mathscr{T} \to \text{Mod–}\mathscr{S}$ induced by composition with F, that is, the functor taking the $\mathscr{T}$–module $H : \mathscr{T}^{\mathrm{op}} \to \mathscr{A}b$ to the $\mathscr{S}$–module $(H \circ F) : \mathscr{S}^{\mathrm{op}} \to \mathscr{A}b$, restricts to a functor which we will denote $\mathfrak{S}(F) : \mathfrak{S}(\mathscr{T}) \to \mathfrak{S}(\mathscr{S})$. That is, the functor $\mathfrak{S}(F)$ is defined to be the unique map making the square below commute

$$\begin{array}{ccc} \mathfrak{S}(\mathscr{T}) & \xrightarrow{\mathfrak{S}(F)} & \mathfrak{S}(\mathscr{S}) \\ \big\downarrow & & \big\downarrow \\ \text{Mod–}\mathscr{T} & \xrightarrow{\hat{F}} & \text{Mod–}\mathscr{S} \end{array}$$

where the vertical inclusions are given by the definition of $\mathfrak{S}(?) \subset \mathfrak{L}(?) \subset \text{Mod–}?$ of Discussion 6.10 (1) and (2).

Furthermore, the functor $\mathfrak{S}(F)$ respects the exact triangles as defined in Discussion 6.10(3).

Sun and Zhang go on to study recollements. Suppose we are given a recollement of triangulated categories

$$\mathscr{R} \;\begin{array}{c} \xleftarrow{\;I_\lambda\;} \\ \xrightarrow{\;I\;} \\ \xleftarrow[\;I_\rho\;]{} \end{array}\; \mathscr{S} \;\begin{array}{c} \xleftarrow{\;J_\lambda\;} \\ \xrightarrow{\;J\;} \\ \xleftarrow[\;J_\rho\;]{} \end{array}\; \mathscr{T}.$$

If all three triangulated categories come with good metrics, and if all six functors are continuous, then the following may be found in [37, THEOREM 1.2].

Theorem 7.4. *Under the hypotheses above, applying $\mathfrak{S}$ yields a right recollement*

$$\mathfrak{S}(\mathscr{R}) \underset{\mathfrak{S}(I)}{\overset{\mathfrak{S}(I_\lambda)}{\rightleftarrows}} \mathfrak{S}(\mathscr{S}) \underset{\mathfrak{S}(J)}{\overset{\mathfrak{S}(J_\lambda)}{\rightleftarrows}} \mathfrak{S}(\mathscr{T}).$$

In the presence of enough continuous adjoints, we deduce that a semiorthogonal decomposition of $\mathscr{S}$ gives rise to a semiorthogonal decomposition of $\mathfrak{S}(\mathscr{S})$. In view of the fact that there are metrics on $\mathbf{D}^{\mathbf{perf}}(X)$ and $\mathbf{D}^b_{\mathbf{coh}}(X)$ such that

$$\mathfrak{S}\big(\mathbf{D}^{\mathbf{perf}}(X)\big) = \mathbf{D}^b_{\mathbf{coh}}(X), \quad \mathfrak{S}\big(\mathbf{D}^b_{\mathbf{coh}}(X)^{\mathrm{op}}\big) = \mathbf{D}^{\mathbf{perf}}(X)^{\mathrm{op}},$$

it is natural to wonder how the recent theorem of Sun and Zhang [37, THEOREM 1.2] compares with the older work of Kuznetsov [19, SECTION 2.5] and [20, SECTION 4].

The above shows that, subject to suitable hypotheses, the construction taking $\mathscr{S}$ to $\mathfrak{S}(\mathscr{S})$ can preserve (some of) the internal structure on the category $\mathscr{S}$—for example, semiorthogonal decompositions. This leads naturally to

Problem 7.5. What other pieces of the internal structure of $\mathscr{S}$ are respected by the construction that passes to $\mathfrak{S}(\mathscr{S})$? Under what conditions are these preserved?

Problem 7.5 may sound vague, but it can be made precise enough. For example, there is a huge literature dealing with the group of autoequivalences of the derived categories $\mathbf{D}^b_{\mathbf{coh}}(X)$. Now, as it happens, the metrics for which Remark 6.12 gives the equivalences

$$\mathfrak{S}\big(\mathbf{D}^{\mathbf{perf}}(X)\big) \cong \mathbf{D}^b_{\mathbf{coh}}(X), \quad \mathfrak{S}\big(\mathbf{D}^b_{\mathbf{coh}}(X)^{\mathrm{op}}\big) \cong \mathbf{D}^{\mathbf{perf}}(X)^{\mathrm{op}}$$

can be given (up to equivalence) intrinsic descriptions. Note that the way we introduced these metrics, in Discussion 6.10, was to use a preferred t-structure on $\mathscr{T} = \mathbf{D}_{\mathbf{qc}}(X)$ to give on $\mathscr{T}$ a metric, unique up to equivalence, and hence induced metrics on $\mathscr{T}^c = \mathbf{D}^{\mathbf{perf}}(X)$ and on $\mathscr{T}^b_c = \mathbf{D}^b_{\mathbf{coh}}(X)$ which are also unique up to equivalence. But this description seems to depend on an embedding into the large category $\mathscr{T}$. What I am asserting now is that there are alternative descriptions of the same equivalence classes of metrics on $\mathscr{T}^c$ and on $\mathscr{T}^b_c$, which do not use the embedding into $\mathscr{T}$. The interested reader can find this in the later sections of [27]. Anyway, a consequence is that any autoequivalence, of either $\mathbf{D}^{\mathbf{perf}}(X)$ or of $\mathbf{D}^b_{\mathbf{coh}}(X)$, must be continuous with a continuous inverse. Hence the group of autoequivalences of $\mathbf{D}^b_{\mathbf{coh}}(X)$ must be isomorphic to the group of autoequivalences of $\mathbf{D}^{\mathbf{perf}}(X)$. Or more generally, assume $\mathscr{T}$ is a noetherian, approximable triangulated category, where noetherian has the meaning of [27, DEFINITION 5.1]. Then the group of exact autoequivalences of $\mathscr{T}^c$ is canonically isomorphic to the group of exact autoequivalences of $\mathscr{T}^b_c$.

Are there similar theorems about t-structures in $\mathscr{S}$ going to t-structures in $\mathfrak{S}(\mathscr{S})$? Or about stability conditions on $\mathscr{S}$ mapping to stability conditions on $\mathfrak{S}(\mathscr{S})$?

We should note that any such theorem will have to come with conditions. After all, the category $\mathbf{D}^b_{\mathbf{coh}}(X)$ always has a bounded t-structure, while Antieau, Gepner, and Heller [3,

THEOREM 1.1] show that $\mathbf{D}^{\mathrm{perf}}(X)$ does not in general. Thus it is possible for $\mathscr{S}$ to have a bounded t-structure but for $\mathfrak{S}(\mathscr{S})$ not to. And in this particular example, the equivalence class of the metric has an intrinsic description, in the sense mentioned above.

Perhaps we should remind the reader that the article **[3]**, by Antieau, Gepner, and Heller, finds a K-theoretic obstruction to the existence of bounded t-structures, more precisely if an appropriate category $\mathscr{E}$ has a bounded t-structure then $K_{-1}(\mathscr{E}) = 0$. Hence the reference to **[3]** immediately raises the question of how the construction passing from $\mathscr{S}$ to $\mathfrak{S}(\mathscr{S})$ might relate to K-theory, especially to negative K-theory. Of course, one has to be a little circumspect here. While there is a K-theory for triangulated categories (see **[25]** for a survey), this K-theory has only been proved to behave well for "nice" triangulated categories, for example, for triangulated categories with bounded t-structures. Invariants like negative K-theory have never been defined for triangulated categories, and might well give nonsense. In what follows we will assume that all the K-theoretic statements are for triangulated categories with chosen enhancements, and that K-theory means the Waldhausen K-theory of the enhancement. We recall in passing that the enhancements are unique for many interesting classes of triangulated categories, see Lunts and Orlov **[21]**, Canonaco and Stellari **[9]**, Antieau **[2]** and Canonaco, Neeman, and Stellari **[8]**.

With the disclaimers out of the way, what do the results surveyed in this article have to do with negative K-theory?

Let us begin with Schlichting's conjecture **[36, CONJECTURE 1 OF SECTION 10]**; this conjecture, now known to be false **[32]**, predicted that the negative K-theory of any abelian category should vanish. But Schlichting also proved that (1) $K_{-1}(\mathscr{A}) = 0$ for any abelian category $\mathscr{A}$, and (2) $K_{-n}(\mathscr{A}) = 0$ whenever $\mathscr{A}$ is a noetherian abelian category and $n > 0$. Now note that the $K(\mathscr{A}) = K(\mathscr{A}^{\mathrm{op}})$, hence the negative K-theory of any artinian abelian category must also vanish. And playing with extensions of abelian categories, we easily deduce the vanishing of the negative K-theory of a sizeable class of abelian categories. So while Schlichting's conjecture is false in the generality in which it was stated, there is some large class of abelian categories for which it is true. The challenge is to understand this class.

It becomes interesting to see what relation, if any, the results surveyed here have with this question.

Let us begin with Theorems 4.4 and 4.2. Theorem 4.4 tells us that, when X is a quasiexcellent, finite-dimensional, separated noetherian scheme, the category $\mathbf{D}^b_{\mathbf{coh}}(X)$ is strongly generated. This category has a unique enhancement whose K-theory agrees with the K-theory of the noetherian abelian category $\mathrm{Coh}(X)$, hence the negative K-theory vanishes. Theorem 4.2 and Remark 4.3 tell us that the category $\mathbf{D}^{\mathrm{perf}}(X)$ has a strong generator if and only if X is regular and finite-dimensional—in which case it is equivalent to $\mathbf{D}^b_{\mathbf{coh}}(X)$ and its unique enhancement has vanishing negative K-theory. This raises the question:

Problem 7.6. If $\mathscr{T}$ is a triangulated category with a strong generator, does it follow that any enhancement of $\mathscr{T}$ has vanishing negative K-theory?

Let us refine this question a little. In Definition 4.1 we learned that a *strong generator,* for a triangulated category $\mathscr{T}$, is an object $G \in \mathscr{T}$ such that there exists an integer $\ell > 0$

with $\mathscr{T} = \langle G \rangle_\ell$. Following Rouquier, we can ask for estimates on the integer ℓ. This leads us to:

Definition 7.7. Let $\mathscr{T}$ be a triangulated category. The *Rouquier dimension* of $\mathscr{T}$ is the smallest integer $\ell \geq 0$ (we allow the possibility $\ell = \infty$), for which there exists an object G with $\mathscr{T} = \langle G \rangle_{\ell+1}$. See Rouquier **[35]** for much more about this fascinating invariant.

There is a rich and beautiful literature estimating this invariant and its various cousins—see Rouquier **[35]** for the origins of the theory, and a host of other places for subsequent developments. For this survey we note only that, for $\mathbf{D}^b_{\mathbf{coh}}(X)$, the Rouquier dimension is conjectured to be equal to the Krull dimension of X. But by a conjecture of Weibel **[39]**, now a theorem of Kerz, Strunk, and Tamme **[18]**, the Krull dimension of X also has a K-theoretic description: the groups K_n of the unique enhancement of $\mathbf{D}^{\mathbf{perf}}(X)$ vanish for all $n < -\dim(X)$. Recalling that $\mathscr{S} = \mathbf{D}^b_{\mathbf{coh}}(X)$ is related to $\mathbf{D}^{\mathbf{perf}}(X)$ by the fact that the construction $\mathfrak{S}$ interchanges them (up to passing to opposite categories, which has no effect on K-theory), this leads us to ask:

Problem 7.8. Let $\mathscr{S}$ be a regular (= strongly generated) triangulated category as in Definition 4.1, and let $N < \infty$ be its Rouquier dimension. Is it true that K_n vanishes on any enhancement of $\mathfrak{S}(\mathscr{S})$, for any metric on $\mathscr{S}$ and whenever $n < -N$?

In an entirely different direction, we know that the construction $\mathfrak{S}$ interchanges $\mathbf{D}^{\mathbf{perf}}(X)$ and $\mathbf{D}^b_{\mathbf{coh}}(X)$, and that these categories coincide if and only if X is regular. This leads us to ask:

Problem 7.9. Is there a way to measure the "distance" between $\mathscr{S}$ and $\mathfrak{S}(\mathscr{S})$, in such a way that resolution of singularities can be viewed as a process reducing this distance? Who knows, maybe there is even a good metric on $\mathscr{S} = \mathbf{D}^{\mathbf{perf}}(X)$ and/or on $\mathscr{S}' = \mathbf{D}^b_{\mathbf{coh}}(X)$ such that the construction $\mathfrak{S}$ takes either $\mathscr{S}$ or $\mathscr{S}'$ to an $\mathfrak{S}(\mathscr{S})$ or $\mathfrak{S}(\mathscr{S}')$ which is $\mathbf{D}^{\mathbf{perf}}(Y) = \mathbf{D}^b_{\mathbf{coh}}(Y)$ for some resolution of singularities Y of X.

While on the subject of regularity (= strong generation):

Problem 7.10. Is there some way to understand which are the approximable triangulated categories $\mathscr{T}$ for which $\mathscr{T}^c$ and/or $\mathscr{T}^b_c$ are regular?

Theorems 4.2 and 4.4 deal with the case $\mathscr{T} = \mathbf{D}_{\mathbf{qc}}(X)$. Approximability is used in the proofs given in **[33]** and **[4]**, but only to ultimately reduce to the case of $\mathscr{T}^c = \mathbf{D}^{\mathbf{perf}}(X)$ with X an affine scheme—this case turns out to be classical, it was settled already in Kelly's 1965 article **[17]**. And the diverse precursors of Theorems 4.2 and 4.4, which we have hardly mentioned in the current survey, are also relatively narrow in scope. But presumably there are other proofs out there, yet to be discovered. And new approaches might well lead to generalizations that hold for triangulated categories having nothing to do with algebraic geometry.

Next let us revisit Theorem 6.6, the theorem identifying each of $[\mathscr{T}^c]^{\mathrm{op}}$ (respectively $\mathscr{T}^b_c$) as the finite homological functors on the other. In view of the motivating application, discussed in Remark 6.9, and of the generality of Theorem 6.6, it is natural to wonder:

Problem 7.11. Do GAGA-type theorems have interesting generalizations to other approximable triangulated categories? The reader is invited to check [29, SECTION 8 AND APPENDIX A]: except for the couple of paragraphs in [29, EXAMPLE A.2] everything is formulated in gorgeous generality and might be applicable in other contexts.

In the context of $\mathbf{D}^b_{\mathbf{coh}}(X)$, where X is a scheme proper over a noetherian ring R, there was a wealth of different-looking GAGA-statements before Jack Hall's lovely paper [12] unified them into one. In other words, the category $\mathbf{D}^b_{\mathbf{coh}}(X) = \mathscr{T}^b_c$ had many different-looking incarnations, and it was not until Hall's paper that it was understood that there was one underlying reason why they all coincided.

Hence Problem 7.11 asks whether this pattern is present for other $\mathscr{T}^b_c$, in other words for $\mathscr{T}^b_c \subset \mathscr{T}$ where $\mathscr{T}$ are some other R-linear, approximable triangulated categories.

And finally:

Problem 7.12. Is there a version of Theorem 6.6 that holds for non-noetherian rings?

There is evidence that something might be true, see Ben-Zvi, Nadler, and Preygel [5, SECTION 3]. But the author has no idea what the right statement ought to be, let alone how to go about proving it.

ACKNOWLEDGMENTS

We would like to thank Asilata Bapat, Jim Borger, Anand Deopurkar, Jack Hall, Bernhard Keller, Tony Licata, and Bregje Pauwels for corrections and improvements to earlier drafts. Needless to say, the flaws that remain are entirely the author's fault.

FUNDING

This work was partially supported by the Australian Research Council, more specifically by grants number DP150102313 and DP200102537.

REFERENCES

[1] L. Alonso Tarrío, A. Jeremías López, and M. J. Souto Salorio, Construction of t-structures and equivalences of derived categories. *Trans. Amer. Math. Soc.* **355** (2003), no. 6, 2523–2543 (electronic).

[2] B. Antieau, On the uniqueness of infinity-categorical enhancements of triangulated categories. 2018, arXiv:1812.01526.

[3] B. Antieau, D. Gepner, and J. Heller, K-theoretic obstructions to bounded t-structures. *Invent. Math.* **216** (2019), no. 1, 241–300.

[4] K. Aoki, Quasiexcellence implies strong generation. *J. Reine Angew. Math.* (published online 14 August 2021, 6 pages), see also 2020, arXiv:2009.02013.

[5] D. Ben-Zvi, D. Nadler, and A. Preygel, Integral transforms for coherent sheaves. *J. Eur. Math. Soc. (JEMS)* **19** (2017), no. 12, 3763–3812.

[6] A. I. Bondal and M. Van den Bergh, Generators and representability of functors in commutative and noncommutative geometry. *Mosc. Math. J.* **3** (2003), no. 1, 1–36, 258.

[7] J. Burke, A. Neeman, and B. Pauwels, Gluing approximable triangulated categories. 2018, arXiv:1806.05342.

[8] A. Canonaco, A. Neeman, and P. Stellari, Uniqueness of enhancements for derived and geometric categories. 2021, arXiv:2101.04404.

[9] A. Canonaco and P. Stellari, Uniqueness of dg enhancements for the derived category of a Grothendieck category. *J. Eur. Math. Soc. (JEMS)* **20** (2018), no. 11, 2607–2641.

[10] P. J. Freyd, Homotopy is not concrete. In *The Steenrod algebra and its applications*, pp. 25–34, Lecture Notes in Math. 168, Springer, 1970.

[11] P. J. Freyd, Homotopy is not concrete. *Repr. Theory Appl. Categ.* **6** (2004), 1–10 (electronic).

[12] J. Hall, GAGA theorems. 2018, arXiv:1804.01976.

[13] L. Illusie, Conditions de finitude relatives. In *Théorie des intersections et théorème de Riemann–Roch*, pp. 222–273, Lecture Notes in Math. 225, Springer, Berlin, 1971.

[14] L. Illusie, Existence de résolutions globales. In *Théorie des intersections et théorème de Riemann–Roch*, pp. 160–221, Lecture Notes in Math. 225, Springer, Berlin, 1971.

[15] L. Illusie, Généralités sur les conditions de finitude dans les catégories dérivées. In *Théorie des intersections et théorème de Riemann–Roch*, pp. 78–159, Lecture Notes in Math. 225, Springer, Berlin, 1971.

[16] B. Keller and D. Vossieck, Aisles in derived categories. *Bull. Soc. Math. Belg. Sér. A* **40** (1988), no. 2, 239–253.

[17] G. M. Kelly, Chain maps inducing zero homology maps. *Proc. Camb. Philos. Soc.* **61** (1965), 847–854.

[18] M. Kerz, F. Strunk, and G. Tamme, Algebraic K-theory and descent for blow-ups. *Invent. Math.* **211** (2018), no. 2, 523–577.

[19] A. G. Kuznetsov, Lefschetz decompositions and categorical resolutions of singularities. *Selecta Math. (N.S.)* **13** (2008), no. 4, 661–696.

[20] A. G. Kuznetsov, Base change for semiorthogonal decompositions. *Compos. Math.* **147** (2011), no. 3, 852–876.

[21] V. A. Lunts and D. O. Orlov, Uniqueness of enhancement for triangulated categories. *J. Amer. Math. Soc.* **23** (2010), no. 3, 853–908.

[22] N. Minami, From Ohkawa to strong generation via approximable triangulated categories—a variation on the theme of Amnon Neeman's Nagoya lecture series. In *Bousfield classes and Ohkawa's theorem*, pp. 17–88, Springer Proc. Math. Stat. 309, Springer, Singapore, 2020.

[23] A. Neeman, The connection between the K-theory localisation theorem of Thomason, Trobaugh and Yao, and the smashing subcategories of Bousfield and Ravenel. *Ann. Sci. Éc. Norm. Supér.* **25** (1992), 547–566.

[24] A. Neeman, The Grothendieck duality theorem via Bousfield's techniques and Brown representability. *J. Amer. Math. Soc.* **9** (1996), 205–236.

[25] A. Neeman, The K-theory of triangulated categories. In *Handbook of K-theory, vol. 1, 2*, pp. 1011–1078, Springer, Berlin, 2005.

[26] A. Neeman, Strong generation of some derived categories of schemes. In *Research perspectives CRM Barcelona*, pp. 121–127, Trends Math. 5, Springer, 2016.

[27] A. Neeman, The categories $\mathscr{T}^c$ and $\mathscr{T}_c^b$ determine each other. 2018, arXiv:1806.06471.

[28] A. Neeman, The category $[\mathscr{T}^c]^{\mathrm{op}}$ as functors on $\mathscr{T}_c^b$. 2018, arXiv:1806.05777.

[29] A. Neeman, Triangulated categories with a single compact generator and a Brown representability theorem. 2018, arXiv:1804.02240.

[30] A. Neeman, Metrics on triangulated categories. *J. Pure Appl. Algebra* **224** (2020), no. 4, 106206, 13.

[31] A. Neeman, Approximable triangulated categories. In *Representations of algebras, geometry and physics*, pp. 111–155, Contemp. Math. 769, Amer. Math. Soc., Providence, RI, 2021.

[32] A. Neeman, A counterexample to vanishing conjectures for negative K-theory. *Invent. Math.* **225** (2021), no. 2, 427–452.

[33] A. Neeman, Strong generators in $\mathbf{D}^{\mathbf{perf}}(X)$ and $\mathbf{D}^b_{\mathbf{coh}}(X)$. *Ann. of Math. (2)* **193** (2021), no. 3, 689–732.

[34] D. O. Orlov, Smooth and proper noncommutative schemes and gluing of DG categories. *Adv. Math.* **302** (2016), 59–105.

[35] R. Rouquier, Dimensions of triangulated categories. *J. K-Theory* **1** (2008), no. 2, 193–256.

[36] M. Schlichting, Negative K-theory of derived categories. *Math. Z.* **253** (2006), no. 1, 97–134.

[37] Y. Sun and Y. Zhang, Ladders and completion of triangulated categories. *Theory Appl. Categ.* **37** (2021), Paper No. 4, 95–106.

[38] J.-L. Verdier, Des catégories dérivées des catégories abéliennes. *Astérisque* (1996), no. 239, xii+253 pp. (1997). With a preface by Luc Illusie, edited and with a note by Georges Maltsiniotis.

[39] C. A. Weibel, K-theory and analytic isomorphisms. *Invent. Math.* **61** (1980), no. 2, 177–197.

AMNON NEEMAN

Mathematical Sciences Institute, Building 145, The Australian National University, Canberra, ACT 2601, Australia, Amnon.Neeman@anu.edu.au

This Page Intentionally Left Blank

SYZYGIES OVER A POLYNOMIAL RING

IRENA PEEVA

ABSTRACT

We discuss results and open problems on graded minimal free resolutions over polynomial rings.

MATHEMATICS SUBJECT CLASSIFICATION 2020

13D02

KEYWORDS

Syzygies, free resolutions, Betti numbers

INTERNATIONAL CONGRESS OF MATHEMATICIANS 2022 JULY 6–14

Proc. Int. Cong. Math. 2022, Vol. 3, pp. 1660–1680
DOI 10.4171/ICM2022/42

1. INTRODUCTION

Research on free resolutions is a core and beautiful area in Commutative Algebra. It contains a number of challenging conjectures and open problems; some of them are discussed in the book [101].

For simplicity, we will work throughout over the polynomial ring $S = \mathbb{C}[x_1, \dots, x_n]$, which is *standard graded* by $\deg(x_i) = 1$ for every i. Many of the results work in much bigger generality; for example, over any field, or over some graded quotient rings of S. We leave it to the interested reader to look for the precise generality of the results using the references. We focus on some main ideas about finite resolutions which are present over polynomial rings.

The idea to describe the structure of a module by a free resolution was introduced by Hilbert in his famous paper [76]; this approach was present in the work of Cayley [35] as well. Every finitely generated S-module T has a free resolution. If T is graded, there exists a minimal free resolution $\mathbf{F}_T$ which is unique up to an isomorphism and is contained in any free resolution of T. Hilbert's insight was that the properties of the minimal free resolution $\mathbf{F}_T$ are closely related to the invariants of the resolved module T. The key point is that the resolution can be interpreted as an exact complex of finitely generated free modules F_i so that

$$\cdots \to F_2 \xrightarrow{d_2=\begin{pmatrix}\text{minimal}\\ \text{relations}\\ \text{on the}\\ \text{relations}\\ \text{in } d_1\end{pmatrix}} F_1 \xrightarrow{d_1=\begin{pmatrix}\text{minimal}\\ \text{relations}\\ \text{on the}\\ \text{generators}\\ \text{of } T\end{pmatrix}} F_0 \xrightarrow{d_0=\begin{pmatrix}\text{minimal}\\ \text{generators}\\ \text{of } T\end{pmatrix}} T \to 0. \tag{1.1}$$

Thus, the resolution is a way of describing the structure of T.

The condition of minimality is important. The mere existence of free resolutions suffices for computing Hilbert functions and for foundational issues such as the definition of Ext and Tor. However, without minimality, resolutions are not unique, and the uniformity of constructions of nonminimal resolutions (like the Bar resolution) implies that they give little insight into the structure of the resolved modules. In contrast, the minimal free resolution $\mathbf{F}_T$ encodes a lot of properties of T; for example, the Auslander–Buchsbaum formula expresses the depth of T in terms of the length (called projective dimension) of $\mathbf{F}_T$, while nonminimal resolutions do not measure depth.

Free resolutions have applications in mathematical fields as diverse as Algebraic Geometry, Combinatorics, Computational Algebra, Invariant Theory, Mathematical Physics, Noncommutative Algebra, Number Theory, and Subspace Arrangements. For many years, they have been both central objects and fruitful tools in Commutative Algebra.

The connections of resolutions to Algebraic Geometry are especially rich, and the book [51] is focussed on that. One of the most challenging open problems in this area, which remains open to this date, is Green's conjecture; see the recent paper by Aprodu–Farkas–Papadima–Raicu–Weyman [5] for more details on this problem.

It should be noted that the world of minimal free resolutions is much wider and diverse than graded resolutions over polynomial rings. Resolutions are studied in other major situations, and there are many important and exciting results and open problems there. For example, there is an extensive research in the multigraded case, which contains resolutions of monomial ideals, resolutions of toric ideals, and resolutions of binomial edge ideals. Another fascinating and important area is the study of minimal free resolutions over quotient rings; such resolutions are usually infinite (by a theorem of Serre) and so their properties are quite different than what we see in finite resolutions over a polynomial ring. An interesting new idea is the recent introduction of virtual resolutions by Berkesch–Erman–Smith [13].

2. FREE RESOLUTIONS

A *free resolution* of a finitely generated S-module T is an exact sequence

$$\mathbf{F}: \quad \dots \to F_2 \xrightarrow{d_2} F_1 \xrightarrow{d_1} F_0 \xrightarrow{d_0} T \to 0$$

of homomorphisms of free finitely generated S-modules F_i. The maps d_i are called *differentials*.

If T is graded, there exists a minimal free resolution $\mathbf{F}_T$ which is unique up to an isomorphism and is contained in any free resolution of T (see [101, THEOREM 7.5], [101, THEOREM 3.5]). Minimality can be characterized in the following simple way: $\mathbf{F}$ is *minimal* if

$$d_{i+1}(F_{i+1}) \subseteq (x_1, \dots, x_n)F_i \quad \text{for all } i \geq 0,$$

that is, no invertible elements appear in the differential matrices.

Hilbert's intuition was that the properties of the minimal free resolution $\mathbf{F}_T$ are closely related to the invariants of the resolved module T. The key point is that the map $d_0 : F_0 \to T$ sends a basis of F_0 to a minimal system $\mathcal{G}$ of generators of T, the first differential d_1 describes the minimal relations $\mathcal{R}$ among the generators $\mathcal{G}$, the second differential d_2 describes the minimal relations on the relations $\mathcal{R}$, etc.; see (1.1). Hilbert's Syzygy Theorem 4.1 is a fundamental result on the structure of such resolutions and leads to many applications. It shows that every finitely generated graded S-module has a finite free resolution (that is, $F_j = 0$ for $j \gg 0$).

The submodule $\operatorname{Im}(d_i) = \operatorname{Ker}(d_{i-1})$ of F_{i-1} is called the ith *syzygy module* of T, and its elements are called ith *syzygies*.

3. BETTI NUMBERS

Let T be a graded finitely generated S-module. The differentials in the minimal free resolution $\mathbf{F}_T$ of T are often very intricate, and so it may be more fruitful to focus on numerical invariants. The rank of the free module F_i in $\mathbf{F}_T$ is called the ith *Betti number* and is denoted by $b_i(T)$. It may be expressed as

$$b_i(T) = \dim \operatorname{Tor}_i^S(T, \mathbb{C}) = \dim \operatorname{Ext}_S^i(T, \mathbb{C}).$$

The Betti numbers are extensively studied numerical invariants of T, and they encode a lot of information about the module.

Note that in the graded case we have graded Betti numbers $b_{i,j}(T)$: Since T is graded, it has a graded minimal free resolution, that is, the differentials preserve degree (they are homogeneous maps of degree 0). Thus, we have graded Betti numbers

$$b_{i,j}(T) = \dim \operatorname{Tor}_i^S(T, \mathbb{C})_j = \dim \operatorname{Ext}_S^i(T, \mathbb{C})_j .$$

Hilbert showed how to use them in order to compute the *Hilbert series* $\sum_{i=0} t^i \dim_{\mathbb{C}}(T_i)$ which measures the size of the module T; see [101, THEOREM 16.2].

The graded Betti numbers can be assembled in the *Betti table* $\beta(T)$, which has entry $b_{i,i+j} = b_{i,i+j}(T)$ in position i, j. Following the conventions in the computer algebra system Macaulay2 [68], the columns of $\beta(T)$ are indexed from left to right by homological degree, and the rows are indexed increasingly from top to bottom. For example, if T is generated in nonnegative degrees then the Betti table $\beta(T)$ has the form:

	0	1	2	$\dots$
0:	$b_{0,0}$	$b_{1,1}$	$b_{2,2}$	$\dots$
1:	$b_{0,1}$	$b_{1,2}$	$b_{2,3}$	$\dots$
2:	$b_{0,2}$	$b_{1,3}$	$b_{2,4}$	$\dots$
$\vdots$	$\vdots$	$\vdots$	$\vdots$	

The main general open-ended question on Betti numbers is:

Question 3.1. How do the properties of the (graded) Betti numbers relate to the structure of the minimal free resolution of T and/or the structure of T?

The BEH Conjecture is a long-standing open conjecture on Betti numbers:

BEH Conjecture 3.2 (Buchsbaum–Eisenbud, Horrocks, [20,73]). *If T is a finitely generated graded artinian S-module (artinian means that the module has finite length), then*

$$b_i(T) \geq \binom{n}{i} \quad \textit{for } i \geq 0.$$

Essentially, the conjecture states that the Koszul resolution (see [101, SECTION 14]) of the residue field $\mathbb{C}$ is the smallest minimal free resolution of an artinian module.

If the above conjecture holds, then it easily follows that we get a lower bound on the Betti numbers for any module (not necessarily artinian) in terms of its codimension; see [17]. The expository papers [17] by Boocher–Grifo and [36] by Charalambous–Evans provide nice overviews on the scarce positive results that are known so far; for example, Herzog–Kühl [74] proved the desired inequalities for linear resolutions. The best currently known result is:

Theorem 3.3 (Walker, [108]). *If T is a finitely generated graded artinian S-module, then*

$$\sum_{i=0}^{n} b_i(T) \geq 2^n,$$

and equality holds if and only if T is a complete intersection.

People have wondered how sharp the above bound is when the module is not a complete intersection (that is, T is not a quotient ring by a regular sequence):

Question 3.4 (Charalambous–Evans–Miller, [37]). If T is a finitely generated graded artinian S-module that is not a complete intersection, then do we have

$$\sum_{i=0}^{n} b_i(T) \geq 2^n + 2^{n-1}?$$

There are many questions that one may ask and study about Betti numbers when restricted to special classes of modules; most ambitiously, we would like to have a characterization of the sequences that are Betti numbers. A recent result of this kind is the Boij–Söderberg theory, which was conjectured by Boij–Söderberg [16], and proved soon after that. Eisenbud–Fløystad–Weyman proved the characteristic-zero case in [52] and then Eisenbud–Schreyer dealt with any characteristic in [56]. Later, efficient methods for such constructions were given by Berkesch, Kummini, Erman, Sam in [14] and by Fløystad in [61, SECTION 3]. The expository papers [60, 62] provide nice overviews of this topic.

4. PROJECTIVE DIMENSION

Projective dimension and regularity are the main numerical invariants that measure the complexity of a minimal free resolution. We will discuss regularity in the next section.

The *projective dimension* of a graded finitely generated S-module T is

$$\operatorname{pd}(T) = \max\{i \mid b_i(T) \neq 0\},$$

and it is the index of the last nonzero column of the Betti table $\beta(T)$, so it measures the width of the table.

Hilbert's Syzygy Theorem 4.1 (see [101, THEOREM 15.2]). *The minimal graded free resolution of a finitely generated graded S-module is finite, and its projective dimension is at most n (recall that n is the number of variables in the polynomial ring S).*

Hilbert's Syzygy Theorem 4.1 provides a nice upper bound on the projective dimension in terms of the number of variables in the polynomial ring. One may wonder if the number of minimal generators of an ideal can be used to get another nice upper bound on projective dimension. The answer turns out to be negative. A construction of Burch [21] and Kohn [79] produces ideals with 3 generators whose projective dimension is arbitrarily large. Later Bruns [18] showed that the minimal free resolutions of three-generated ideals capture all the pathology of minimal free resolutions of modules. However, the degrees of the generators in these constructions are forced to grow large. Motivated by computational complexity issues, Stillman raised the following question:

Question 4.2 (Stillman, [102, PROBLEM 3.14]). Fix an $m \geq 1$ and a sequence of natural numbers $a_1, \ldots, a_m$. Is there a number p such that $\operatorname{pd}(I) \leq p$ for every homogeneous ideal I with a minimal system of generators of degrees $a_1, \dots, a_m$ in a polynomial ring? Note that the number of variables in the polynomial ring is not fixed.

A positive answer is provided by:

Theorem 4.3 (Ananyan–Hochster, [4]). *Stillman's Question* 4.2 *has a positive answer.*

Other proofs were later given by Erman–Sam–Snowden [58] and Draisma–Lasoń–Leykin [50]. Yet, there are many open questions motivated by a desire to get better upper bounds since the known bounds are quite large. See the recent paper by Caviglia–Liang [29] for some explicit bounds.

Families of ideals with large projective dimension were constructed by McCullough in [89] and by Beder, McCullough, Núñez-Betancourt, Seceleanu, Snapp, Stone in [12]. Such constructions indicate that finding tight bounds could be difficult. Many results dealing with special cases are known in this direction. The expository papers [62, 94] provide nice overviews of this topic.

5. REGULARITY

Let L be a homogeneous ideal in S. The height of the Betti table of L is measured by the index of the last nonzero row, and is called the *(Castelnuovo–Mumford) regularity* of L, so

$$\operatorname{reg}(L) = \max\{j \mid \text{there exists an } i \text{ such that } b_{i,i+j}(L) \neq 0\}.$$

Note that $\operatorname{reg}(L) < \infty$ by Hilbert's Syzygy Theorem 4.1. An important role of regularity is that it measures the complexity of the minimal free resolution of L, in the sense that it shows up to what degree we have nonvanishing Betti numbers. It has several other important roles.

The definition of regularity implies that it provides an upper bound on the generating degree, namely

$$\operatorname{reg}(L) \geq \operatorname{maxdeg}(L),$$

where $\operatorname{maxdeg}(L)$ is the maximal degree of an element in a minimal system of homogeneous generators of L.

Another role of regularity is that it identifies how high we have to truncate an ideal in order to get a linear resolution; we say that a graded ideal has an *r-linear resolution* if the ideal is generated in degree r and the entries in the differential maps in its minimal free resolution are linear.

Theorem 5.1 (see [101, THEOREM 19.7]). *Let L be a graded ideal in S. If $r \geq \operatorname{reg}(L)$ then*

$$L_{\geq r} := L \cap \left(\bigoplus_{i \geq r} S_i\right)$$

has an r-linear minimal free resolution, equivalently,

$$\operatorname{reg}(L_{\geq r}) = r.$$

Another role of regularity is related to Gröbner basis computation. Many computer computations in Commutative Algebra and Algebraic Geometry are based on Gröbner basis theory. It is used, for example, in the computer algebra systems Cocoa [1], Macaulay2 [68], Singular [49]. It is proved by Bayer–Stillman [9] that in generic coordinates and with respect to revlex order, one has to compute up to degree $\mathrm{reg}(L)$ in order to compute a Gröbner basis of L. This means that $\mathrm{reg}(L)$ is the degree-complexity of the Gröbner basis computation.

Yet another role of regularity is that it can be defined in terms of vanishing of local cohomology modules. See the expository paper [19] for a detailed discussion.

The expository papers [38, 39] provide nice overviews of the properties of regularity.

In the rest of this section, we discuss bounds on regularity.

The projective dimension $\mathrm{pd}(L)$ of L is bounded above by the number of variables n in S by Hilbert's Syzygy Theorem 4.1. This bound is very nice in several ways: it is small, involves only one parameter, and is given by a simple formula. One may hope that similarly, a nice upper bound on regularity exists. In contrast, the upper bound on regularity involving n is doubly exponential. Bayer–Mumford (see [8, THEOREM 3.7]) and Caviglia–Sbarra [32] proved:

Theorem 5.2 (Bayer–Mumford [8], Caviglia–Sbarra [32]). *Let L be a graded ideal in S. Then*

$$\mathrm{reg}(L) \leq \big(2\,\mathrm{maxdeg}(L)\big)^{2^{n-2}},$$

where $\mathrm{maxdeg}(L)$ *is the maximal degree of an element in a minimal system of homogeneous generators of L.*

This bound is nearly sharp. The Mayr–Meyer construction [88] leads to examples of families of ideals attaining high regularity. The following three types of families of ideals attaining doubly exponential regularity were constructed by Bayer–Mumford [8], Bayer–Stillman [10], and Koh [78]:

Theorem 5.3. (1) (Bayer–Stillman, [10, THEOREM 2.6]) *For $r \geq 1$, there exists a homogeneous ideal I_r (using $d = 3$ in their notation) in a polynomial ring with $10r + 11$ variables for which*

$$\mathrm{maxdeg}(I_r) = 5,$$
$$\mathrm{reg}(I_r) \geq 3^{2^{r-1}}.$$

(2) (Bayer–Mumford, [8, PROPOSITION 3.11]) *For $r \geq 1$, there exists a homogeneous ideal I_r in $10r + 1$ variables for which*

$$\mathrm{maxdeg}(I_r) = 4,$$
$$\mathrm{reg}(I_r) \geq 2^{2^r}.$$

(3) (Koh, [78]) *For* $r \geq 1$, *there exists a homogeneous* I_r *generated by* $22r - 2$ *quadrics in a polynomial ring with* $22r$ *variables for which*

$$\operatorname{maxdeg}(I_r) = 2,$$
$$\operatorname{reg}(I_r) \geq 2^{2^{r-1}}.$$

Further examples of ideals with high regularity were produced by Beder et. al. [12], Caviglia [23], Chardin–Fall [41], and Ullery [107].

Despite these examples of high regularity, there are many important and interesting cases where regularity is bounded by (or equal to) a nice formula and is not dramatically large. As always, the following open-ended problem is of high interest:

Problem 5.4. Find important and interesting cases where regularity is bounded by (or equal to) a nice formula and is not dramatically large.

6. REGULARITY OF PRIME IDEALS

Regularity was studied in Algebraic Geometry as well. In that setting, much better bounds than the doubly-exponential bound discussed in Theorem 5.2, are expected for the regularity of the defining ideals of geometrically nice projective varieties. Lazarsfeld's book [86, SECTION 1.8] and the introduction of the paper [84] by Kwak–Park provide nice overviews of that point of view. In fact, the concept of regularity was introduced by Mumford [98] and generalizes ideas of Castelnuovo. The relation between the definitions of regularity of a coherent sheaf and regularity of a graded ideal (or module) is given in Eisenbud–Goto [53], and may be also found in [51, PROPOSITION 4.16].

Consider a *nondegenerate* projective variety $X \subset \mathbb{P}^{n-1}$, that is, X does not lie on a hyperplane in $\mathbb{P}^{n-1}$.

Some nice bounds were proved in the smooth case. The following bound follows from a more general result by Bertram–Ein–Lazarsfeld [15]:

Theorem 6.1 (Bertram–Ein–Lazarsfeld, [15]). *Let* $X \subset \mathbb{P}^{n-1}$ *be a smooth irreducible projective variety. If* X *is cut out scheme-theoretically by hypersurfaces of degree* $\leq s$, *then*

$$\operatorname{reg}(X) \leq 1 + (s-1)\operatorname{codim}(X).$$

This result was generalized in [42] and [48]. See also [38] for an overview.

Theorem 6.2 (Mumford, [8, THEOREM 3.12]). *If* $X \subset \mathbb{P}^{n-1}$ *is a nondegenerate smooth projective variety, then*

$$\operatorname{reg}(X) \leq \big(\dim(X) + 1\big)\big(\deg(X) - 2\big) + 2.$$

This bound was improved by Kwak–Park as follows:

Theorem 6.3 (Kwak–Park, [84, THEOREM C]). *If* $X \subset \mathbb{P}^{n-1}$ *is a nondegenerate smooth projective variety with* $\operatorname{codim}(X) \geq 2$, *then*

$$\operatorname{reg}(X) \leq \dim(X)\big(\deg(X) - 2\big) + 1.$$

In the influential paper [8], Bayer and Mumford wrote:

> *"...the main missing piece of information between the general case and the geometrically nice smooth case is that we do not have yet a reasonable bound on the regularity of all reduced equidimensional ideals."*

Note that the bounds in the above theorems involve two parameters; for example, $\dim(X)$ and $\deg(X)$ are used in Theorem 6.2. The following bound involving only $\deg(X)$ was first considered in the smooth case:

$$\operatorname{reg}(X) \leq \deg(X).$$

It was conjectured by Eisenbud–Goto [53] for any reduced and irreducible nondegenerate variety, and they expected that it might even hold for reduced equidimensional X which are connected in codimension 1 [8]. In fact, they conjectured the more refined bound

$$\operatorname{reg}(X) \leq \deg(X) - \operatorname{codim}(X) + 1,$$

which is sharp as equality holds for the twisted cubic curve. This is called the Regularity Conjecture. In particular, it yields the following regularity conjecture for prime ideals:

Conjecture 6.4 (Eisenbud–Goto [53], 1984). *If L is a homogeneous prime ideal in S, and $L \subset (x_1, \ldots, x_n)^2$, then*

$$\operatorname{reg}(L) \leq \deg(L).$$

In particular, L is generated in degrees $\leq \deg(L)$.

The condition $L \subset (x_1, \ldots, x_n)^2$ is equivalent to requiring that the projective variety $V(L)$ is not contained in a hyperplane in $\mathbb{P}^{n-1}$. Prime ideals that satisfy this condition are called *nondegenerate*.

The Regularity Conjecture is proved for curves by Gruson–Lazarsfeld–Peskine [69], completing fundamental work of Castelnuovo [22]; see also [67]. It is also proved for smooth surfaces by Lazarsfeld [85] and Pinkham [103]. In the smooth case, Kwak [81–83] gives bounds for regularity in dimensions 3 and 4 that are only slightly worse than the optimal ones. The conjecture also holds in the Cohen–Macaulay case by a result of Eisenbud–Goto [53]. Many other special cases and related bounds have been proved as well.

In [92] Jason McCullough and I construct counterexamples to the Regularity Conjecture. We provide a family of prime ideals P_r, depending on a parameter r, whose degree is singly exponential in r and whose regularity is doubly exponential in r. Our main theorem is much stronger:

Theorem 6.5 (McCullough–Peeva, [92]). *The regularity of nondegenerate homogeneous prime ideals is not bounded by any polynomial function of the degree (multiplicity), i.e., for any polynomial $f(x) \in \mathbb{R}[x]$ there exists a nondegenerate homogeneous prime ideal Y in a standard graded polynomial ring over $\mathbb{C}$ such that* $\operatorname{reg}(Y) > f(\deg(Y))$.

For this purpose, we introduce in [92] an approach which, starting from a homogeneous ideal I, produces a prime ideal P whose projective dimension, regularity, degree, dimension, depth, and codimension are expressed in terms of numerical invariants of I. Our approach involves two new concepts:

(1) Rees-like algebras (inspired by an example by Hochster published in [11]) which, unlike the standard Rees algebras, have well-structured defining equations and minimal free resolutions;

(2) A step-by-step homogenization technique which, unlike classical homogenization, preserves graded Betti numbers.

Further research in this direction was carried out by Caviglia–Chardin–McCullough–Peeva–Varbaro in [24]. Our expository paper [93] provides an overview of counterexamples and the techniques used to prove them.

The bound in the Regularity Conjecture is very elegant, so it is reasonable to expect that work will continue on whether it holds when we impose extra conditions on the prime ideal: for example, for smooth varieties or for toric ideals (in the sense of the definition in [101, SECTION 65]).

Instead of trying to repair the Regularity Conjecture by imposing extra conditions, one may wonder:

Question 6.6 (McCullough–Peeva, [93]). What is an optimal function $f(x)$ such that $\text{reg}(L) \leq f(\deg(L))$ for any nondegenerate homogeneous prime ideal L in a standard graded polynomial ring over $\mathbb{C}$?

Since Theorem 5.2 gives a doubly exponential bound on regularity for all homogeneous ideals, and in view of Theorem 6.5, the following question is of interest:

Question 6.7 (McCullough–Peeva, [93]). Does there exist a singly exponential bound for regularity of homogeneous nondegenerate prime ideals in a standard graded polynomial ring over $\mathbb{C}$, in terms of the multiplicity alone?

In [8, COMMENTS AFTER THEOREM 3.12] Bayer and Mumford wrote:

> *"We would conjecture that if a linear bound doesn't hold, at the least a single exponential bound, i.e.* $\text{reg}(L) \leq \text{maxdeg}(L)^{\mathcal{O}(n)}$*, ought to hold for any reduced equidimensional ideal. This is an essential ingredient in analyzing the worst-case behavior of all algorithms based on Gröbner bases."*

For prime ideals, their conjecture is:

Conjecture 6.8 (Bayer–Mumford, [8, COMMENTS AFTER THEOREM 3.12]). *If L is a homogeneous non-degenerate prime ideal in $S = \mathbb{C}[x_1, \ldots, x_n]$, then*

$$\text{reg}(L) \leq \text{maxdeg}(L)^{\mathcal{O}(n)},$$

where $\text{maxdeg}(L)$ *is the maximal degree of an element in a minimal system of homogeneous generators of L.*

7. REGULARITY OF THE RADICAL

Ravi [104] proved that in some cases the regularity of the radical of an ideal is no greater than the regularity of the ideal itself. For a long time, there was a folklore conjecture that this would hold for every homogeneous ideal. However, counterexamples were constructed by Chardin–D'Cruz [40]. They obtained examples where regularity of the radical is nearly the square (or the cube) of that of the ideal.

Theorem 7.1 (Chardin–D'Cruz, [40, EXAMPLE 2.5]). *For $m \geq 1$ and $r \geq 3$, the ideal*

$$J_{m,r} = (y^m u^2 - x^m z v,\ z^{r+1} - x u^r,\ u^{r+1} - x v^r,\ y^m v^r - x^{m-1} z u^{r-1} v)$$

in the polynomial ring $\mathbb{C}[x, y, z, u, v]$ has

$$\operatorname{reg}(J_{m,r}) = m + 2r + 1,$$
$$\operatorname{reg}(\sqrt{J_{m,r}}) = m(r^2 - 2r - 1) + 1.$$

The existence of a polynomial bound is very unclear, so perhaps it is reasonable to focus on the following folklore question which is currently open:

Question 7.2. Is there a singly exponential bound on $\operatorname{reg}(\sqrt{I})$ in terms of $\operatorname{reg}(I)$ (and possibly $\operatorname{codim}(I)$ or n) for every homogeneous ideal I in a standard graded polynomial ring over $\mathbb{C}$?

In order to form reasonable conjectures, it would be very helpful to develop methods for producing interesting examples. In [86, REMARK 1.8.33] Lazarsfeld wrote:

> *"...the absence of systematic techniques for constructing examples is one of the biggest lacunae in the current state of the theory."*

8. SHIFTS

Let T be a graded finitely generated S-module. The (upper) shifts are refinements of the numerical invariant regularity. The *(upper) shift* at step i is

$$t_i(T) = \max\{j \mid b_{i,j}(T) \neq 0\}$$

and the *adjusted shift* is

$$r_i(T) = \max\{j \mid b_{i,i+j}(T) \neq 0\},$$

so

$$r_i(T) = t_i(T) - i.$$

Note that $r_0(T)$ is the maximal degree of an element in a minimal system of generators of T, and

$$\operatorname{reg}(T) = \max_i \{r_i(T)\}.$$

Let L be a graded ideal in S. The *a, b-subadditivity condition* for L is

$$t_{a+b}(S/L) \leq t_a(S/L) + t_b(S/L). \tag{8.1}$$

Note that it is equivalent to

$$r_{a+b}(S/L) \leq r_a(S/L) + r_b(S/L).$$

We say that L satisfies the *general subadditivity condition* if (8.1) holds for every a, b. We say that L satisfies the *initial subadditivity condition* if (8.1) holds for $b = 1$ and every a. We say that L satisfies the *closing subadditivity condition* if (8.1) holds for every a, b with $a + b = \mathrm{pd}(L)$. Gorenstein ideals failing the subadditivity condition were constructed by McCullough–Seceleanu in [95].

Problem 8.1.

(1) (McCullough, [91]) It is expected that the general subadditivity condition holds for every monomial ideal L.

(2) (Avramov–Conca–Iyengar, [7]) It is conjectured that the general subadditivity condition holds if S/L is a Koszul algebra.

(3) (McCullough, [91]) It is expected that the general subadditivity condition holds for every toric ideal L.

There are supporting results in special cases; the expository paper [91] provides a nice overview of the current state of these problems. For monomial ideals, Herzog–Srinivasan [75] proved that the initial subadditivity condition holds.

Another interesting direction of using shifts is:

Problem 8.2. Find good upper bounds on regularity using the shifts in part of the minimal free resolution.

The following result shows how this may work:

Theorem 8.3 (McCullough, [90]). *Let L be a homogeneous ideal in S. Set $c = \lceil \frac{n}{2} \rceil$. Then*

$$\mathrm{reg}(S/L) \leq \sum_{i=1}^{c} t_i(S/I) + \frac{\prod_{i=1}^{h} t_i(S/I)}{(c-1)!}.$$

9. THE EGH CONJECTURE

We start with a brief introduction to Hilbert functions and lex ideals. If I is a homogeneous ideal in S, then the quotient $R := S/I$ inherits the grading by $R_i = S_i/I_i$ for all i. The size of a homogeneous ideal J in R is measured by its *Hilbert function*

$$\mathrm{Hilb}_{R/J}(i) = \dim_{\mathbb{C}}(R_i/J_i) \quad \text{for } i \in \mathbf{Z}.$$

Hilbert's insight was that $\mathrm{Hilb}_{R/J}$ is determined by finitely many of its values. He proved that there exists a polynomial (called the *Hilbert polynomial*) $g(t) \in \mathbf{Q}[t]$ such that

$$\mathrm{Hilb}_{R/J}(i) = g(i) \quad \text{for } i \gg 0.$$

If S/J (here $R = S$) is the coordinate ring of a projective algebraic variety X, then the degree of the Hilbert polynomial equals the dimension of X; the leading coefficient of the Hilbert polynomial determines another important invariant – the degree (multiplicity) of X. Hilbert functions for monomial ideals in the ring $\mathbb{C}[x_1, \ldots, x_n]/(x_1^2, \ldots, x_n^2)$ have been extensively studied in Combinatorics since each such Hilbert function counts the number of faces in a simplicial complex.

Lex ideals are fruitful tools in the study of Hilbert functions. They are monomial ideals defined in a simple way: Denote by $>_{lex}$ the lexicographic order on the monomials in S extending $x_1 > \cdots > x_n$. A monomial ideal L in S is *lex* if the following property holds: if $m \in L$ is a monomial and $q >_{lex} m$ is a monomial of the same degree, then $q \in L$ (that is, for each $i \geq 0$ the vector space L_i is either zero or is spanned by lex-consecutive monomials of degree i starting with x_1^i).

A core result in Commutative Algebra is Macaulay's Theorem 9.1, which characterizes the Hilbert functions of homogeneous ideals in the polynomial ring S:

Theorem 9.1 (Macaulay, [87]). *For every homogeneous ideal in S there exists a lex ideal with the same Hilbert function.*

The Hilbert function of a lex ideal is easy to count. This leads to an equivalent formulation of Macaulay's Theorem 9.1 which characterizes numerically (by certain inequalities) the Hilbert functions of homogeneous ideals; see [101, SECTION 49].

Lex ideals also play an important role in the study of Hilbert schemes. Grothendieck introduced the Hilbert scheme $\mathcal{H}_{r,g}$ that parametrizes subschemes of $\mathbf{P}^r$ with a fixed Hilbert polynomial g. The structure of the Hilbert scheme is known to be very complicated. In [71] Harris and Morrison state Murphy's Law for Hilbert schemes:

> *"There is no geometric possibility so horrible that it cannot be found generically on some component of some Hilbert scheme."*

The main structural result on $\mathcal{H}_{r,g}$ is Hartshorne's Theorem:

Theorem 9.2 (Hartshorne, [72]). *The Hilbert scheme $\mathcal{H}_{r,g}$ is connected.*

The situation is that every homogeneous ideal with a fixed Hilbert function h is connected by a sequence of deformations to the lex ideal with Hilbert function h. A deformation connects two ideals $J_{t=0}$ and $J_{t=1}$ in the sense that we have a family of homogeneous ideals J_t varying with the parameter $t \in [0, 1]$ so that the Hilbert function is preserved; in this case, the ideals J_t form a path on the Hilbert scheme. Hartshorne's proof [72] relies on deformations called "distractions" which use generic change of coordinates and polarization. Analyzing the paths on a Hilbert scheme may shed light on whether there exists an object with maximal Betti numbers.

Theorem 9.3 (Bigatti–Hulett–Pardue, see [100]). *A lex ideal in S has the greatest Betti numbers among all homogeneous ideals in S with the same Hilbert function.*

This result was quite surprising when it was discovered since counterexamples were known in which no ideal with a fixed Hilbert function attains minimal Betti numbers. It yields numerical upper bounds on Betti numbers as follows: the minimal free resolution of a lex ideal is the Eliahou–Kervaire resolution [57] (see [101, SECTION 28]), and it provides numerical formulas for the Betti numbers of a lex ideal.

It is natural to ask if similar results hold over other rings. For starters, we need rings over which Theorem 9.1 holds. It actually fails over most graded quotient rings of S. For example, there is no lex ideal in the ring $\mathbb{C}[x, y]/(x^2 y, xy^2)$ with the same Hilbert function as the ideal (xy).

Theorem 9.4. *Macaulay's Theorem 9.1 holds over the following rings:*

(1) (Kruskal, Katona, [77, 80]) *an exterior algebra E over $\mathbb{C}$.*

(2) (Clements–Lindström, [45]) *a Clements–Lindström ring $C := \mathbb{C}[x_1, \dots, x_n]/P$, where P is an ideal generated by powers of the variables.*

(3) (Gasharov–Murai–Peeva, [66]) *a Veronese ring $V := S/J$, where J is the defining ideal of a Veronese toric variety.*

Proving analogues of Theorem 9.3 for the above rings is difficult since minimal resolutions over exterior algebras, Clements–Lindström rings, or Veronese rings are infinite (in contrast, Theorem 9.3 is about finite resolutions) and so they are considerably more intricate. It was proved that every lex ideal has the greatest Betti numbers among all homogeneous ideals with the same Hilbert function over the following rings: over E by Aramova–Herzog–Hibi [6], over C by Murai–Peeva [99], and over V by Gasharov–Murai–Peeva [66].

Hilbert functions of ideals containing $(x_1^2, \dots, x_n^2)$ are characterized numerically (by certain inequalities) by Kruskal–Katona's Theorem [77, 80], which is a natural analogue of Macaulay's Theorem; see Theorem 9.4(1,2). Eisenbud–Green–Harris conjectured that the same numerical inequalities for the Hilbert function hold for all ideals in S containing a quadratic regular sequence:

Conjecture 9.5 (Eisenbud–Green–Harris, [54]). *Let $L \subset S$ be a homogeneous ideal containing a regular sequence of n quadratic forms. There exists an ideal N containing $x_1^2, \dots, x_n^2$ with the same Hilbert function as L.*

Kruskal–Katona's Theorem was generalized by Clements–Lindström [45] to a characterization of the Hilbert functions of ideals containing powers of the variables; see Theorem 9.4(2). In view of this, Eisenbud–Green–Harris noted in [54] that Conjecture 9.5 can be extended to cover all complete intersections as follows:

Conjecture 9.6 (Eisenbud–Green–Harris, [54]). *Let $L \subset S$ be a homogeneous ideal containing a regular sequence of forms of degrees $a_1 \leq \dots \leq a_n$. There exists an ideal N containing $x_1^{a_1}, \dots, x_n^{a_n}$ with the same Hilbert function as L.*

Conjecture 9.5 is considered to be the main case of the Eisenbud–Green–Harris Conjectures, called the EGH Conjectures.

In their original form in [54], the EGH Conjectures are stated in terms of numerical inequalities for the Hilbert function. We give an equivalent form, which follows immediately from the Clements–Lindström Theorem 9.4(2).

Eisenbud, Green, and Harris were led to the EGH Conjectures by extending a series of results and conjectures in Castelnuovo Theory in [54]. After that, they made the connection to the Cayley–Bacharach Theory in [55]. They provide in [55] a nice survey of the long history of Cayley–Bacharach theory in Algebraic Geometry.

The EGH Conjectures turned out to be very challenging. Some special cases, applications, and related results are proved in [2,3,25,27,30,33,34,43,44,46,47,54,55,59,70,96,105,106]. One of the strongest results is the recent paper [26] by Caviglia–DeStefani.

We now focus on Betti numbers related to the EGH Conjectures. Let $L \subset S$ be a homogeneous ideal containing a regular sequence of forms of degrees $a_1 \leq \cdots \leq a_n$. The concept of a lex ideal can be generalized to the concept of a *lex-plus-powers ideal* which is a monomial ideal containing $x_1^{a_1}, \ldots, x_n^{a_n}$ and otherwise is like a lex ideal. G. Evans conjectured the more general Lex-Plus-Powers Conjecture that, among all graded ideals with a fixed Hilbert function and containing a homogeneous regular sequence of degrees $a_1 \leq \cdots \leq a_n$, the lex-plus-powers ideal (which exists according to the EGH Conjectures) has the greatest Betti numbers. This conjecture was inspired by Theorem 9.3.

Theorem 9.7 (Mermin–Murai, [97]). *The Lex-Plus-Powers Conjecture holds for ideals containing pure powers.*

The general Lex-Plus-Powers Conjecture (for ideals containing a homogeneous regular sequence) is very difficult. Some special cases are proved in [31,59,63,64,105,106]. The expository papers by Caviglia–DeStefani–Sbarra [28] and by Francisco-Richert [65] provide nice overviews of this challenging topic.

FUNDING

This work was partially supported by NSF grant DMS-2001064.

REFERENCES

[1] J. Abbott, A. Bigatti, and L. Robbiano, CoCoA: a system for doing computations in Commutative Algebra, available at http://cocoa.dima.unige.it.

[2] A. Abedelfatah, On the Eisenbud–Green–Harris conjecture. *Proc. Amer. Math. Soc.* **143** (2015), 105–115.

[3] A. Abedelfatah, Hilbert functions of monomial ideals containing a regular sequence. *Israel J. Math.* **214** (2016), 857–865.

[4] T. Ananyan and M. Hochster, Small subalgebras of polynomial rings and Stillman's Conjecture. *J. Amer. Math. Soc.* **33** (2019), 291–309.

[5] M. Aprodu, G. Farkas, S. Papadima, C. Raicu, and J. Weyman, Koszul modules and Green's conjecture. *Invent. Math.* **218** (2019), 657–720.

[6] A. Aramova, J. Herzog, and T. Hibi, Gotzmann theorems for exterior algebras and combinatorics. *J. Algebra* **191** (1997), 174–211.

[7] L. Avramov, A. Conca, and S. Iyengar, Subadditivity of syzygies of Koszul algebras. *Math. Ann.* **361** (2015), 511–534.

[8] D. Bayer and D. Mumford, What can be computed in algebraic geometry? In *Computational algebraic geometry and commutative algebra. Symposia Mathematica XXXIV*, pp. 1–48, Cambridge University Press, Cambridge, 1993.

[9] D. Bayer and M. Stillman, A criterion for detecting m-regularity. *Invent. Math.* **87** (1987), 1–11.

[10] D. Bayer and M. Stillman, On the complexity of computing syzygies. Computational aspects of commutative algebra. *J. Symbolic Comput.* **6** (1988), 135–147.

[11] J. Becker, On the boundedness and the unboundedness of the number of generators of ideals and multiplicity. *J. Algebra* **48** (1977), 447–453.

[12] J. Beder, J. McCullough, L. Núnez-Betancourt. A. Seceleanu, B. Snapp, and B. Stone, Ideals with larger projective dimension and regularity. *J. Symbolic Comput.* **46** (2011), 1105–1113.

[13] C. Berkesch, D. Erman, and G. Smith, Virtual resolutions for a product of projective spaces. *Algebr. Geom.* **7** (2020), 460–481.

[14] C. Berkesch, M. Kummini, D. Erman, and S. Sam, Poset structures in Boij–Söderberg theory. *J. Eur. Math. Soc.* **15** (2013), 2257–2295.

[15] A. Bertram, L. Ein, and R. Lazarsfeld, Vanishing theorems, a theorem of Severi, and the equations defining projective varieties. *J. Amer. Math. Soc.* **4** (1991), 587–602.

[16] M. Boij and J. Söderberg, Graded Betti numbers of Cohen–Macaulay modules and the multiplicity conjecture. *J. Lond. Math. Soc.* **78** (2008), 85–106.

[17] A. Boocher and E. Grifo, Lower bounds on Betti numbers. In *Commutative algebra: Expository papers in honor of David Eisenbud's 75'th birthday*, edited by I. Peeva, Springer, 2022.

[18] W. Bruns, "Jede" endliche freie Auflösung ist freie Auflösung eines von drei Elementen erzeugten Ideals. *J. Algebra* **39** (1976), 429–439.

[19] W. Bruns, A. Conca, and M. Varbaro, Castelnuovo–Mumford regularity and powers. In *Commutative algebra: Expository papers in honor of David Eisenbud's 75'th birthday*, edited by I. Peeva, Springer, 2022.

[20] D. Buchsbaum and D. Eisenbud, Algebra structures for finite free resolutions, and some structure theorems for ideals of codimension 3. *Amer. J. Math.* **99** (1977), 447–485.

[21] L. Burch, A note on the homology of ideals generated by three elements in local rings. *Proc. Camb. Philos. Soc.* **64** (1968), 949–952.

[22] G. Castelnuovo, Sui multipli di una serie lineare di gruppi di punti appartenente ad una curva algebrica. *Rend. Circ. Mat. Palermo* **7** (1893), 89–110.

[23] G. Caviglia, *Koszul algebras, Castelnuovo–Mumford regularity and generic initial ideals*. PhD Thesis, University of Kansas, 2004.

[24] G. Caviglia, M. Chardin, J. McCullough, I. Peeva, and M. Varbaro, Regularity of prime ideals. *Math. Z.* **291** (2019), 421–435.

[25] G. Caviglia, A. Constantinescu, and M. Varbaro, On a conjecture by Kalai. *Israel J. Math.* **204** (2014), 469–475.

[26] G. Caviglia and A. De Stefani, A Cayley–Bacharach theorem for points in $\mathbf{P}^n$. 2020, arXiv:2006.14717.

[27] G. Caviglia and A. De Stefani, The Eisenbud–Green–Harris conjecture for fast-growing degree sequences. 2020, arXiv:2007.15467.

[28] G. Caviglia, A. De Stefani, and E. Sbarra, The Eisenbud–Green–Harris conjecture. In *Commutative algebra: Expository papers in honor of David Eisenbud's 75'th birthday*, edited by I. Peeva, Springer, 2022.

[29] G. Caviglia and Y. Liang, Explicit Stillman bounds for all degrees. 2020, arXiv:2009.02826.

[30] G. Caviglia and D. Maclagan, Some cases of the Eisenbud–Green–Harris conjecture. *Math. Res. Lett.* **15** (2008), 427–433.

[31] G. Caviglia and A. Sammartano, On the lex-plus-powers conjecture. *Adv. Math.* **340** (2018), 284–299.

[32] G. Caviglia and E. Sbarra, Characteristic-free bounds for the Castelnuovo–Mumford regularity. *Compos. Math.* **141** (2005), 1365–1373.

[33] G. Caviglia and E. Sbarra, Distractions of Shakin rings. *J. Algebra* **419** (2014), 318–331.

[34] G. Caviglia and E. Sbarra, The lex-plus-powers inequality for local cohomology modules. *Math. Ann.* **364** (2016), 225–241.

[35] A. Cayley, On the theory of elimination. *Cambridge and Dublin Math. J.* **3** (1848), 116–120; *Collected papers Vol. I*, pp. 116–120, Cambridge University Press, 1889.

[36] H. Charalambous and G. Evans, Problems on Betti numbers of finite length modules. In *Free resolutions in commutative algebra and algebraic geometry (Sundance, UT, 1990)*, pp. 25–33, Res. Notes Math. 2, Jones and Bartlett, Boston, 1992.

[37] H. Charalambous, G. Evans, and M. Miller, Betti numbers for modules of finite length. *Proc. Amer. Math. Soc.* **109** (1990), 63–70.

[38] M. Chardin, Bounds for Castelnuovo–Mumford regularity in terms of degrees of defining equations. In *Commutative algebra, singularities and computer algebra (Sinaia, 2002)*, pp. 67–73, NATO Sci. Ser. II Math. Phys. Chem. 115, Kluwer Acad. Publ., Dordrecht, 2003.

[39] M. Chardin, Some results and questions on Castelnuovo–Mumford regularity. In *Syzygies and Hilbert functions*. Lect. Notes Pure Appl. Math. 254, pp. 1–40, Chapman & Hall/CRC, Boca Raton, FL, 2007.

[40] M. Chardin and C. D'Cruz, Castelnuovo–Mumford regularity: Examples of curves and surfaces. *J. Algebra* **270** (2003), 347–360.

[41] M. Chardin and A. Fall, Sur la régularité de Castelnuovo–Mumford des idéaux, en dimension 2. *C. R. Math. Acad. Sci. Paris* **341** (2005), 233–238.

[42] M. Chardin and B. Ulrich, Liaison and Castelnuovo–Mumford regularity. *Amer. J. Math.* **124** (2002), 1103–1124.

[43] R.-X. Chen, Some special cases of the Eisenbud–Green–Harris conjecture. *Illinois J. Math.* **56** (2012), 661–675.

[44] K. Chong, An application of liaison theory to the Eisenbud–Green–Harris conjecture. *J. Algebra* **445** (2016), 221–231.

[45] G. Clements and B. Lindström, A generalization of a combinatorial theorem of Macaulay. *J. Combin. Theory* **7** (1969), 230–238.

[46] S. Cooper, Growth conditions for a family of ideals containing regular sequences. *J. Pure Appl. Algebra* **212** (2008), 122–131.

[47] S. Cooper, Subsets of complete intersections and the EGH conjecture. In *Progress in commutative algebra 1*, pp. 167–198, de Gruyter, Berlin, 2012.

[48] T. de Fernex and L. Ein, A vanishing theorem for log canonical pairs. *Amer. J. Math.* **132** (2010), 1205–1221.

[49] W. Decker, G.-M. Greuel, G. Pfister, and H. Schönemann, Singular 4-1-2 – A computer algebra system for polynomial computations, available at http://www.singular.uni-kl.de, 2019.

[50] J. Draisma, M. Lason, and A. Leykin, Stillman's conjecture via generic initial ideals. *Comm. Algebra* **47** (2019), 2384–2395.

[51] D. Eisenbud, *The geometry of syzygies. A second course in commutative algebra and algebraic geometry*. Grad. Texts in Math. 229, Springer, 2005.

[52] D. Eisenbud, G. Fløystad, and J. Weyman, The existence of equivariant pure free resolutions. *Ann. Inst. Fourier* **61** (2011), 905–926.

[53] D. Eisenbud and S. Goto, Linear free resolutions and minimal multiplicity. *J. Algebra* **88** (1984), 89–133.

[54] D. Eisenbud, M. Green, and J. Harris, Higher Castelnuovo theory. In *Journées de Géométrie Algébrique d'Orsay (Orsay, 1992)*, Astérisque **218** (1993), 187–202.

[55] D. Eisenbud, M. Green, and J. Harris, Cayley–Bacharach theorems and conjectures. *Bull. Amer. Math. Soc. (N.S.)* **33** (1996), 295–324.

[56] D. Eisenbud and F.-O. Schreyer, Betti numbers of graded modules and cohomology of vector bundles. *J. Amer. Math. Soc.* **22** (2009), 859–888.

[57] S. Eliahou and M. Kervaire, Minimal resolutions of some monomial ideals. *J. Algebra* **129** (1990), 1–25.

[58] D. Erman, S. Sam, and A. Snowden, Big polynomial rings and Stillman's conjecture. *Invent. Math.* **218** (2019), 413–439.

[59] G. Evans and E. Riehl, On the intersections of polynomials and the Cayley–Bacharach theorem. *J. Pure Appl. Algebra* **183** (2003), 293–298.

[60] G. Fløystad, Boij–Söderberg theory: introduction and survey. In *Progress in commutative algebra 1*, pp. 1–54, de Gruyter, 2012.

[61] G. Fløystad, Zipping Tate resolutions and exterior coalgebras. *J. Algebra* **437** (2015), 249–307.

[62] G. Fløystad, J. McCullough, and I. Peeva, Three themes of syzygies. *Bull. Amer. Math. Soc. (N.S.)* **53** (2016), 415–435.

[63] C. Francisco, Almost complete intersections and the lex-plus-powers conjecture. *J. Algebra* **276** (2004), 737–760.

[64] C. Francisco, *Hilbert functions and graded free resolutions*. PhD Thesis, Cornell University, 2004.

[65] C. Francisco and B. Richert, Lex-plus-powers ideals. In *Syzygies and Hilbert functions*, edited by I. Peeva, Lect. Notes Pure Appl. Math. 254, pp. 145–178, CRC Press, 2007.

[66] V. Gasharov, S. Murai, and I. Peeva, Hilbert schemes and maximal Betti numbers over Veronese rings. *Math. Z.* **267** (2011), 155–172.

[67] D. Giaimo, On the Castelnuovo–Mumford regularity of connected curves. *Trans. Amer. Math. Soc.* **358** (2006), 267–284.

[68] D. Grayson and M. Stillman, Macaulay2, a software system for research in algebraic geometry, available at http://www.math.uiuc.edu/Macaulay2/.

[69] L. Gruson, R. Lazarsfeld, and C. Peskine, On a theorem of Castelnuovo and the equations defining projective varieties. *Invent. Math.* **72** (1983), 491–506.

[70] T. Harima, A. Wachi, and J. Watanabe, The EGH conjecture and the Sperner property of complete intersections. *Proc. Amer. Math. Soc.* **145** (2017), 1497–1503.

[71] J. Harris and I. Morrison, *Moduli of curves*. Grad. Texts in Math. 187, Springer, New York, 1998.

[72] R. Hartshorne, Connectedness of the Hilbert scheme. *Publ. Math. Inst. Hautes Études Sci.* **29** (1966), 5–48.

[73] R. Hartshorne, Algebraic vector bundles on projective spaces: a problem list. *Topology* **18** (1979), 117–128.

[74] J. Herzog and M. Kühl, On the Betti numbers of finite pure and linear resolutions. *Comm. Algebra* **12** (1984), 1627–1646.

[75] J. Herzog and H. Srinivasan, On the subadditivity problem for maximal shifts in free resolutions. In *Commutative algebra and noncommutative algebraic geometry, Vol. II*, pp. 245–249, Math. Sci. Res. Inst. Publ. 68, Cambridge Univ. Press, New York, 2015.

[76] D. Hilbert, Ueber die Theorie der algebraischen Formen. *Math. Ann.* **36** (1890), 473–534.

[77] G. Katona, A theorem for finite sets. In *Theory of graphs*, pp. 187–207, Academic Press, New York, 1968.

[78] J. Koh, Ideals generated by quadrics exhibiting double exponential degrees. *J. Algebra* **200** (1998), 225–245.

[79] P. Kohn, Ideals generated by three elements. *Proc. Amer. Math. Soc.* **35** (1972), 55–58.

[80] J. Kruskal, The number of simplices in a complex. In *Mathematical optimization techniques*, edited by R. Bellman, pp. 251–278, University of California Press, Berkeley/Los Angeles, 1963.

[81] S. Kwak, Castelnuovo regularity for smooth subvarieties of dimensions 3 and 4. *J. Algebraic Geom.* **7** (1998), 195–206.

[82] S. Kwak, Castelnuovo–Mumford regularity bound for smooth threefolds in $\mathbf{P}^5$ and extremal examples. *J. Reine Angew. Math.* **509** (1999), 21–34.

[83] S. Kwak, Generic projections, the equations defining projective varieties and Castelnuovo regularity. *Math. Z.* **234** (2000), 413–434.

[84] S. Kwak and J. Park, A bound for Castelnuovo–Mumford regularity by double point divisors. *Adv. Math.* **364** (2020).

[85] R. Lazarsfeld, A sharp Castelnuovo bound for smooth surfaces. *Duke Math. J.* **55** (1987), 423–438.

[86] R. Lazarsfeld, *Positivity in algebraic geometry. I. II.* Ergeb. Math. Grenzgeb. (3) 48, 49, Springer, Berlin, 2004.

[87] F. Macaulay, Some properties of enumeration in the theory of modular systems. *Proc. Lond. Math. Soc.* **26** (1927), 531–555.

[88] E. Mayr and A. Meyer, The complexity of the word problem for commutative semigroups and polynomial ideals. *Adv. Math.* **46** (1982), 305–329.

[89] J. McCullough, A family of ideals with few generators in low degree and large projective dimension. *Proc. Amer. Math. Soc.* **139** (2011), 2017–2023.

[90] J. McCullough, A polynomial bound on the regularity of an ideal in terms of half of the syzygies. *Math. Res. Lett.* **19** (2012), 555–565.

[91] J. McCullough, Subadditivity of syzygies of ideals and related problems. In *Commutative algebra: Expository papers in honor of David Eisenbud's 75'th birthday*, edited by I. Peeva, Springer, 2022.

[92] J. McCullough and I. Peeva, Counterexamples to the Eisenbud–Goto regularity conjecture. *J. Amer. Math. Soc.* **31** (2018), 473–496.

[93] J. McCullough and I. Peeva, The Regularity Conjecture for prime ideals in polynomial rings. *EMS Surv. Math. Sci.* **7** (2020), 173–206.

[94] J. McCullough and A. Seceleanu, Bounding projective dimension. In *Commutative algebra. Expository papers dedicated to David Eisenbud on the occasion of his 65th birthday*, edited by I. Peeva, pp. 551–576, Springer, London, 2013.

[95] J. McCullough and A. Seceleanu, Quadratic Gorenstein algebras with many surprising properties. 2020, arXiv:2004.10237.

[96] J. Mermin, Monomial regular sequences. *Proc. Amer. Math. Soc.* **138** (2010), 1983–1988.

[97] J. Mermin and S. Murai, The lex-plus-powers conjecture holds for pure powers. *Adv. Math.* **226** (2011), 3511–3539.

[98] D. Mumford, *Lectures on curves on an algebraic surface, with a section by G. M. Bergman*. Ann. of Math. Stud. 59, Princeton University Press, Princeton, NJ, 1966.

[99] S. Murai and I. Peeva, Hilbert schemes and Betti numbers over a Clements–Lindström rings. *Compos. Math.* **148** (2012), 1337–1364.

[100] K. Pardue, Deformation classes of graded modules and maximal Betti numbers. *Illinois J. Math.* **40** (1996), 564–585.

[101] I. Peeva, *Graded syzygies*. Springer, London, 2011.

[102] I. Peeva and M. Stillman, Open problems on syzygies and Hilbert functions. *J. Commut. Algebra* **1** (2009), 159–195.

[103] H. Pinkham, A Castelnuovo bound for smooth surfaces. *Invent. Math.* **83** (1986), 321–332.

[104] M. Ravi, Regularity of ideals and their radicals. *Manuscripta Math.* **68** (1990), 77–87.

[105] B. Richert, A proof of Evans' convexity conjecture. *Comm. Algebra* **43** (2015), 3275–3281.

[106] B. Richert and S. Sabourin, The residuals of lex plus powers ideals and the Eisenbud–Green–Harris conjecture. *Illinois J. Math.* **52** (2008), 1355–1384.

[107] B. Ullery, Designer ideals with high Castelnuovo–Mumford regularity. *Math. Res. Lett.* **21** (2014), 1215–1225.

[108] M. Walker, Total Betti numbers of modules of finite projective dimension. *Ann. of Math. (2)* **186** (2017), 641–646.

IRENA PEEVA

Department of Mathematics, Cornell University, Ithaca, NY 14853, USA, ivp1@cornell.edu

3. NUMBER THEORY

SPECIAL LECTURE

SURVEY LECTURE ON ARITHMETIC DYNAMICS

JOSEPH H. SILVERMAN

ABSTRACT

Arithmetic dynamics is a relatively new field in which classical problems from number theory and algebraic geometry are reformulated in the setting of dynamical systems. Thus, for example, rational points on algebraic varieties become rational points in orbits, and torsion points on abelian varieties become points having finite orbits. Moduli problems also appear, where, for example, the complex multiplication points in the moduli space of abelian varieties correspond to the postcritically finite points in the moduli space of rational maps. In this article we give a survey of some of the major problems motivating the field of arithmetic dynamics, and some of the progress that has been made during the past 20 years.

MATHEMATICS SUBJECT CLASSIFICATION 2020

Primary 37P15; Secondary 37P05, 37P30, 37P35, 37P45

KEYWORDS

Dynamical uniform boundedness, dynamical unlikely intersection, arboreal representation, dynamical moduli space, dynamical and arithmetic degree

INTERNATIONAL CONGRESS OF MATHEMATICIANS 2022 JULY 6–14

Proc. Int. Cong. Math. 2022, Vol. 3, pp. 1682–1710
DOI 10.4171/ICM2022/67

Published by EMS Press

1. INTRODUCTION

This article is a survey of the comparatively new field of *Arithmetic Dynamics*, a field where arithmetic and dynamics join forces.[1] But the word "arithmetic" in "arithmetic dynamics" is itself short for "arithmetic geometry," a field where the venerable subjects of number theory and algebraic geometry meet. Thus arithmetic dynamics is a melting pot filled with ingredients from three classical areas of mathematics.

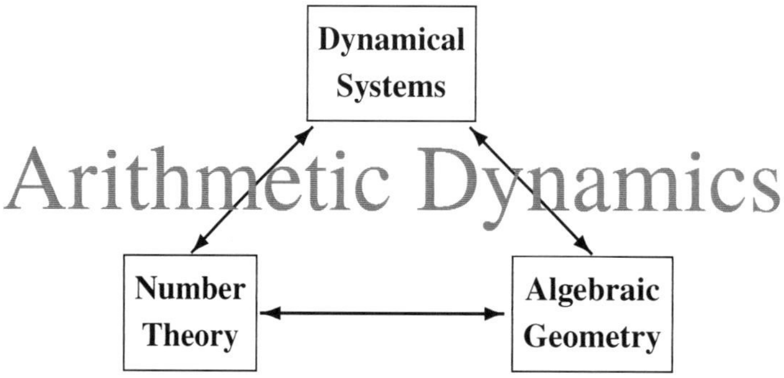

In this article we will discuss arithmetic dynamics over global fields, which for the sake of exposition we will generally take to be number fields, i.e., finite extensions of $\mathbb{Q}$. Our primary focus will be dynamical analogues and generalizations of famous theorems and conjectures in arithmetic geometry, centered around the following five major topics that have helped drive the development of arithmetic dynamics over the past few decades:

- Topic #1: Dynamical Uniform Boundedness
- Topic #2: Dynamical Moduli Spaces
- Topic #3: Dynamical Unlikely Intersections
- Topic #4: Dynatomic and Arboreal Representations
- Topic #5: Dynamical and Arithmetic Complexity

Remark 1.1. Of course, our chosen five topics do not fully cover the varied problems that fall under the rubric of arithmetic dynamics over global fields. And there are also highly active areas of arithmetic dynamics in which people study dynamical systems defined over non-archimedean fields such as $\mathbb{Q}_p$ and $\mathbb{C}_p$ and over finite fields $\mathbb{F}_q$. We refer the interested reader to the survey article [10] for a more extensive discussion.

1 As Jung might have said: "The meeting of two mathematical fields is like the contact of two chemical substances: if there is any reaction, both are transformed."

2. DEFINITIONS AND TERMINOLOGY

An abstract *dynamical system* is simply an object X and an endomorphism (self-map)[2]

$$f : X \to X.$$

The *iterates of f* are denoted by

$$f^n = \underbrace{f \circ f \circ \cdots \circ f}_{n \text{ copies of } f},$$

and the *(forward) f-orbit of an element* $x \in X$ is its image for the iterates of f,[3]

$$\mathcal{O}_f(x) = \{ f^n(x) : n \geq 0 \}.$$

We say that $x \in X$ is *f-periodic* if

$$f^n(x) = x \quad \text{for some } n \geq 1,$$

in which case the smallest such n is the *(exact) period of* x. A point $x \in X$ is *f-preperiodic* if its f-orbit $\mathcal{O}_f(x)$ is finite, or equivalently, if $f^m(x)$ is periodic for some $m \geq 0$.

Two dynamical systems $f_1, f_2 : X \to X$ are *isomorphic* if there is an automorphism $\varphi \in \operatorname{Aut}(X)$ such that

$$f_2 = f_1^{\varphi} = \varphi^{-1} \circ f_1 \circ \varphi. \tag{2.1}$$

Note that (2.1) is a good notion of isomorphism for dynamics, since it respects iteration,

$$(f^{\varphi})^n = (\varphi^{-1} \circ f \circ \varphi)^n = \varphi^{-1} \circ f^n \circ \varphi = (f^n)^{\varphi}.$$

In particular, orbits and (pre)periodic points of the isomorphic dynamical systems f and f^{φ} are more-or-less identical, since

$$\mathcal{O}_{f^{\varphi}}(x) = \varphi^{-1}\big(\mathcal{O}_f(\varphi(x))\big), \quad \operatorname{Per}(f^{\varphi}) = \varphi^{-1}\big(\operatorname{Per}(f)\big), \quad \operatorname{PrePer}(f^{\varphi}) = \varphi^{-1}\big(\operatorname{PrePer}(f)\big).$$

We conclude this section with a brief discussion of endomorphisms $f : \mathbb{P}^1 \to \mathbb{P}^1$ i.e., rational functions of one variable. For $P \in \mathbb{P}^1$, we choose a local parameter z_P at P and define P to be a *critical point of f* if

$$\frac{df}{dz_P}(P) = 0. \tag{2.2}$$

The vanishing condition (2.2) is independent of the choice of z_P, and counted with appropriate multiplicities, the map f has $2\deg(f) - 2$ critical points.[4]

---

2 To avoid complications, we always work in a subcategory of the category of sets, i.e., all of our objects are sets.

3 More generally, let $\mathcal{F} = \{f_1, \dots, f_r\}$ be a set of endomorphisms of X, and let $\langle \mathcal{F} \rangle$ be the semigroup of maps generated by arbitrary composition of elements of $\mathcal{F}$. Then the *$\mathcal{F}$-orbit of x* is the set $\mathcal{O}_{\mathcal{F}}(x) = \{ f(x) : f \in \langle \mathcal{F} \rangle \}$.

4 More precisely, this is true as long as f is separable, so in particular it is always true in characteristic 0.

Arithmetic Geometry		Dynamical Systems
rational and integral points on varieties	⟷	rational and integral points in orbits
torsion points on abelian varieties	⟷	periodic and preperiodic points of rational maps
abelian varieties with complex multiplication	⟷	postcritically finite rational maps

TABLE 1
A dictionary for Arithmetic Dynamics **[82, §6.5]**

The critical points of an endomorphism f of $\mathbb{P}^1$ are the points at which f fails to be locally bijective. Their location crucially affects the dynamics of f.

Definition 2.1. A (separable) endomorphism $f : \mathbb{P}^1 \to \mathbb{P}^1$ is *postcritically finite* (PCF) if all of its critical points are preperiodic. PCF maps play a key role in the study of dynamics on $\mathbb{P}^1$.

3. A DICTIONARY FOR ARITHMETIC DYNAMICS

Table 1 gives three fundamental analogies that are used to travel between the worlds of arithmetic geometry and dynamical systems. The associations described in the first two lines of Table 1 are fairly tight, in the sense that they may be used to reformulate many standard results and conjectures in arithmetic geometry as dynamical statements. The following two examples illustrate these connections.

Example 3.1. Let A be an abelian group, let $P \in A$, and let $f_P : A \to A$ be the translation-by-P map, i.e., $f_P(Q) = Q + P$. Then the subgroup of A generated by P is the union of two orbits

$$\mathbb{Z}P = \mathcal{O}_{f_P}(0) \cup \mathcal{O}_{f_{-P}}(0).$$

More generally, for any finite set of elements $P_1, \ldots, P_r \in A$, we let $\mathcal{P} = \{\pm P_1, \ldots, \pm P_r\}$, and then the subgroup $\langle \mathcal{P} \rangle$ generated by $P_1, \ldots, P_r$ is the generalized orbit

$$\langle \mathcal{P} \rangle = \mathcal{O}_{\mathcal{P}}(0) = \{ f_P(0) : P \in \langle \mathcal{P} \rangle \}.$$

In this way, statements about finitely generated subgroups of abelian varieties may be reformulated as statements about orbits.

Example 3.2. Let G be a group, let $d \geq 2$, and let $f_d : G \to G$ be the d-power map $f_d(g) = g^d$. Then it is an easy exercise to check that

$$\operatorname{PrePer}(f) = G_{\text{tors}},$$

i.e., the elements of G that are preperiodic for the d-power map are exactly the elements of G having finite order. In this way statements about torsion points on abelian varieties may be reformulated as statements about preperiodic points for the multiplication-by-d map.

Remark 3.3. Examples 3.1 and 3.2 help to justify the associations described in the first two lines of Table 1. The third line is a bit more nebulous. It is a rough analogy based on the following reasoning:[5]

- The CM points in the moduli space $\mathcal{A}_g$ of abelian varieties of dimension g are associated to abelian varieties that have a special algebraic property, namely their endomorphism ring is unusually large. The set of CM points is a countable, Zariski-dense set of points in $\mathcal{A}_g$ whose coordinates are algebraic numbers.
- The PCF points in the moduli space $\mathcal{M}_d^1$ of endomorphisms of $\mathbb{P}^1$ are associated to maps that have a special dynamical property, namely the orbits of their critical points are unusually small. The set of PCF points is a countable, Zariski-dense set of points in $\mathcal{M}_d^1$ whose coordinates are algebraic numbers.

Section 6 describes some progress that helps to justify the third analogy in Table 1. But we must also note that the analogy is not perfect. In particular, CM abelian varieties are abundant in all dimensions, i.e., CM points are Zariski-dense in $\mathcal{A}_g$ for all $g \geq 1$. However, evidence suggests that for $N \geq 2$, PCF maps are not Zariski dense in the moduli space $\mathcal{M}_d^N$ of endomorphisms of $\mathbb{P}^N$; cf. [34].

4. TOPIC #1: DYNAMICAL UNIFORM BOUNDEDNESS

The prototype and motivation for the dynamical uniform boundedness conjecture is the following famous theorem.

Theorem 4.1 ([54]). *Let $E/\mathbb{Q}$ be an elliptic curve defined over $\mathbb{Q}$. Then*

$$\#E(\mathbb{Q})_{\text{tors}} \leq 16.$$

Remark 4.2. Mazur's theorem was generalized by Kamienny [37] to number fields of small degree, and then by Merel [58], who proved that for all number fields $K/\mathbb{Q}$ and for all elliptic curves E/K, there is a uniform bound

$$\#E(K)_{\text{tors}} \leq C, \quad \text{where } C \text{ depends only on the degree } [K:\mathbb{Q}].$$

A long-standing conjecture says that the same should be true for abelian varieties A/K of any dimension, where the upper bound depends on $[K:\mathbb{Q}]$ and $\dim(A)$.

Using the dictionary in Table 1, the theorems of Mazur–Kamienny–Merel and the conjectural abelian variety generalization lead us to a major motivating problem in arithmetic dynamics.

5 See Section 5 for the construction of the moduli space $\mathcal{M}_d^N$.

Conjecture 4.3 (Dynamical uniform boundedness conjecture, [62]). *Fix integers $N \geq 1$, $d \geq 2$, and $D \geq 1$. There is a constant $C(N, d, D)$ such that for all degree-d morphisms $f : \mathbb{P}^N \to \mathbb{P}^N$ defined over a number field K of degree $[K : \mathbb{Q}] = D$, the number of K-rational preperiodic points is uniformly bounded,*

$$\#\mathrm{PrePer}\big(f, \mathbb{P}^N(K)\big) \leq C(N, d, D).$$

Remark 4.4. See also [79] for an earlier dynamical uniform boundedness conjecture for K3 surfaces admitting noncommuting involutions.

Remark 4.5. Although Conjecture 4.3 only deals with preperiodic points in projective space, it can be used to prove the uniform boundedness conjecture for abelian varieties alluded to in Remark 4.2 [21].

Remark 4.6. Conjecture 4.3 has been generalized to cover quite general families of dynamical systems; see [72, QUESTION 3.2].

Conjecture 4.3 seems out of reach at present. Indeed, even quite special cases present challenges that have not been overcome. We briefly summarize what is known and conjectured in the simplest nontrivial case, which is quadratic polynomials over $\mathbb{Q}$.

Theorem/Conjecture 4.7. *For $c \in \mathbb{Q}$, let $f_c(x) = x^2 + c$.*

(a) **Theorem**. *For each $n \in \{1, 2, 3\}$, there are infinitely many $c \in \mathbb{Q}$ such that $f_c(x)$ has a $\mathbb{Q}$-rational point of period n.*

(b) **Theorem**. *For all $c \in \mathbb{Q}$, the polynomial $f_c(x)$ does not have a $\mathbb{Q}$-rational point ...*

- *of order* 4 [60];
- *of order* 5 [25];
- *of order* 6, *conditional on the Birch–Swinnerton-Dyer conjecture* [86].

(c) **Conjecture**. *For all $n \geq 4$, the polynomial $f_c(x)$ does not have a $\mathbb{Q}$-rational point of period n; see* [91][6] *and* [25].

Remark 4.8. Just as there are elliptic modular curves $X_1^{\mathrm{ell}}(n)$ whose points classify pairs (E, P) consisting of an elliptic curve E and an n-torsion point P, there are so-called dynatomic modular curves $X_1^{\mathrm{dyn}}(n)$ whose points classify pairs (c, α) such that α is a point of period n for the polynomial $f_c(x) = x^2 + c$. Mazur's method for proving Theorem 4.1 is to show that $X_1^{\mathrm{ell}}(n)$ has no (noncuspidal) $\mathbb{Q}$-rational points by mapping $X_1^{\mathrm{ell}}(n)$ into a carefully chosen quotient A of its Jacobian variety and showing that the group $A(\mathbb{Q})$ is

6 Although in fairness it should be noted that [91] suggests the opposite conclusion, stating: "Are there any rational periodic orbits of a quadratic $x^2 + c$ of period greater than 3? The results for periods 1, 2, and 3 would lead one to suspect that there must be."

finite. The proof of Theorem 4.7 starts similarly using $X_1^{\mathrm{dyn}}(n)$ instead of $X_1^{\mathrm{ell}}(n)$, but in this situation, the Jacobian generally does not have a quotient whose group of rational points is finite. Current methods, such as Chabauty–Coleman, for explicitly determining the rational points on curves of high genus (barely) suffice to handle $X_1^{\mathrm{dyn}}(n)$ for $n \le 6$. The difficulty, or more concretely the difference, between the elliptic curve and dynamical settings centers around the lack of a theory of Hecke correspondences in the dynamical case. (Mea culpa: This simplified explanation is not entirely accurate, but it is meant to convey the overall strategy of the proofs.)

Remark 4.9. Contingent on an appropriate version of the $abcd$-conjecture, the uniform boundedness conjecture has been proven for the family of polynomials $x^d + c$ **[47]**, and more recently for all polynomials **[46]**. An alternative proof, also using the abc-conjecture and only valid over $\mathbb{Q}$, says that if d is sufficiently large and $c \ne -1$, then $x^d + c$ has no $\mathbb{Q}$-rational periodic points other than fixed points **[68]**.[7]

Remark 4.10. A function field analogue of the uniform boundedness conjecture for $x^d + c$ is proven in **[17,18]**. In the function field setting, the uniformity in the degree $[K : \mathbb{Q}]$ described in Conjecture 4.3 is replaced by a bound that depends on the gonality[8] of the field extension.

5. TOPIC #2: DYNAMICAL MODULI SPACES

We fix a field K and consider parameter and moduli spaces for the set of rational self-maps of $\mathbb{P}_K^N$. A rational map $f : \mathbb{P}_K^N \dashrightarrow \mathbb{P}_K^N$ of degree-d is specified by an $(N+1)$-tuple of degree-d homogeneous polynomials,

$$f = [f_0, \dots, f_N], \quad f_0, \dots, f_N \in K[X_0, \dots, X_n],$$

such that $f_0, \dots, f_N$ have no common factors. The map f is a morphism if $f_0, \dots, f_N$ have no common roots in $\mathbb{P}^N(\bar{K})$. We label the coefficients of $f_0, \dots, f_N$ in some specified order as $a_1(f), a_2(f), \dots, a_\nu(f)$, where $\nu = \nu(N,d) := \binom{N+d}{d}(N+1)$. Then each such f determines a point

$$f = \big[a_1(f), \dots, a_\nu(f)\big] \in \mathbb{P}^{\nu-1}.$$

There is a homogeneous polynomial $\mathcal{R} \in \mathbb{Z}[a_1, \dots, a_\nu]$ called the Macaulay resultant having the property that

$$f = [f_0, \dots, f_N] \text{ is a morphism} \iff \mathcal{R}\big(a_1(f), \dots, a_\nu(f)\big) \ne 0.$$

The parameter space of degree-d endomorphisms of $\mathbb{P}^N$ is

$$\mathrm{End}_d^N = \big\{f \in \mathbb{P}^{\nu-1} : \mathcal{R}(f) \ne 0\big\}.$$

---

7 We remark that it is easy to prove uniform boundedness for $x^d + c$ over $\mathbb{Q}$ when d is odd, and more generally over any field $K/\mathbb{Q}$ with a real embedding. Indeed, it is an elementary fact that if $f : \mathbb{R} \to \mathbb{R}$ is any nondecreasing function, then f has no nonfixed periodic points; cf. **[64]**.

8 The *gonality* of an algebraic curve X, or its function field, is the minimal degree of a non-constant map $X \to \mathbb{P}^1$.

The isomorphism class of dynamical systems associated to f is the set of all conjugates, i.e., the set of all

$$f^{\varphi} = \varphi^{-1} \circ f \circ \varphi, \quad \text{where } \varphi \in \operatorname{Aut}(\mathbb{P}^N) = \mathrm{PGL}_{N+1}.$$

Conjugation gives an algebraic action of PGL_{N+1} on the parameter space End_d^N via

$$\mathrm{PGL}_{N+1} \times \operatorname{End}_d^N \to \operatorname{End}_d^N, \quad (\varphi, f) \mapsto f^{\varphi}, \tag{5.1}$$

and this action extends naturally to $\mathbb{P}^{\nu}$.

Definition 5.1. The *moduli space of degree-d dynamical systems on* $\mathbb{P}^N$ is the quotient space of End_d^N for the conjugation action (5.1),

$$\mathcal{M}_d^N = \operatorname{End}_d^N / \mathrm{PGL}_{N+1}. \tag{5.2}$$

It is natural to ask whether the quotient (5.2) can be given some nice sort of structure. Geometric invariant theory (GIT) **[63]** provides a powerful tool for studying quotients of a variety (or scheme) X by an infinite algebraic group G. GIT says that there are stable and semistable loci $X^{\mathrm{s}} \subseteq X^{\mathrm{ss}} \subseteq X$ such that there exist quotient varieties (or schemes) $X^{\mathrm{s}}//G$ and $X^{\mathrm{ss}}//G$ having many agreeable properties.[9]

Theorem 5.2. *Let $N \geq 1$ and $d \geq 2$.*

(a) *The quotient space $\mathcal{M}_d^N(\mathbb{C}) = \operatorname{End}_d^N(\mathbb{C})/\mathrm{PGL}_{N+1}(\mathbb{C})$ has a natural structure as an orbifold over $\mathbb{C}$* **[59]**.

(b) *The quotient space $\mathcal{M}_d^N = \operatorname{End}_d^N / \mathrm{PGL}_{N+1}$ has a natural structure as a GIT quotient scheme over $\mathbb{Z}$; see* **[80]** *for $N = 1$ and* **[44,69]** *for $N \geq 1$.*[10]

It is clear that $\mathcal{M}_d^N$ is unirational, i.e., it is rationally finitely covered by a projective space, since End_d^N is itself an open subset of a projective space. A subtler question is whether $\mathcal{M}_d^N$ is rational.

Theorem 5.3. *Let $d \geq 2$.*

(a) *There is an isomorphism $\mathcal{M}_2^1 \cong \mathbb{A}^2$, and the semi-stable GIT compactification of $\mathcal{M}_2^1$ as the quotient of the semi-stable locus in $\mathbb{P}^5$ is isomorphic to $\mathbb{P}^2$* **[59,80]**.

(b) *The space $\mathcal{M}_d^1$ is rational, i.e., there exists a birational map $\mathbb{P}^{2d-2} \dashrightarrow \mathcal{M}_d^1$* **[44]**.

Question 5.4. Is $\mathcal{M}_d^N$ rational for all $d \geq 2$ and all $N \geq 1$?

---

9 For example, over $\mathbb{C}$ the stable GIT quotient satisfies $(X^{\mathrm{s}}//G)(\mathbb{C}) = X^{\mathrm{s}}(\mathbb{C})/G(\mathbb{C})$, i.e., the geometric points of the stable quotient $X^{\mathrm{s}}//G$ are the $G(\mathbb{C})$-orbits of the geometric points of X. And the semistable GIT quotient has the property that $(X^{\mathrm{ss}}//G)(\mathbb{C})$ is proper, i.e., it is compact, so it provides a natural compactification of the stable quotient.

10 More precisely, the parameter space End_d^N is in the GIT stable locus for the action of SL_{N+1} on $\mathbb{P}^{\nu}$ linearized relative to the line bundle $\mathcal{O}_{\mathbb{P}^{\nu}}(1)$, and thus the quotient $\mathcal{M}_d^N$ exists as a GIT quotient scheme over $\mathbb{Z}$.

Just as is done with the moduli space of abelian varieties, it is advantageous to add level structure to dynamical moduli spaces by specifying maps together with points or cycles of various shapes. We start with the case of a single periodic point, and then consider more complicated level structures.

Definition 5.5. For $N \geq 1$, $n \geq 1$, and $d \geq 2$, we write

$$\mathrm{End}_d^N[n] = \{(f, P) \in \mathrm{End}_d^N \times \mathbb{P}^N : P \text{ has exact } f\text{-period } n\}.$$

Thus the points of $\mathrm{End}_d^N[n]$ classify maps with a marked point of exact period n.

More generally, we define a *(preperiodic) portrait* $\mathcal{P}$ to be the directed graph of a self-map of a finite set of points. (See Figure 1 for an example of a portrait.) Then for a portait $\mathcal{P}$ having k vertices, we let[11]

$$\mathrm{End}_d^N[\mathcal{P}] = \left\{ (f, P_1, \ldots, P_k) \in \mathrm{End}_d^N \times (\mathbb{P}^N)^k : \begin{array}{l} P_1, \ldots, P_k \text{ are } f\text{-preperiodic and} \\ f : \{P_1, \ldots, P_k\} \to \{P_1, \ldots, P_k\} \\ \text{is a model for the portrait } \mathcal{P} \end{array} \right\}.$$

There is a natural action of $\varphi \in \mathrm{PGL}_{N+1}$ on $\mathrm{End}_d^N[\mathcal{P}]$ given by

$$(f, P_1, \ldots, P_k)^\varphi = \left(f^\varphi, \varphi^{-1}(P_1), \ldots, \varphi^{-1}(P_k)\right).$$

We denote the resulting quotient space by

$$\mathcal{M}_d^N[\mathcal{P}] = \mathrm{End}_d^N[\mathcal{P}] / \mathrm{PGL}_{N+1}. \tag{5.3}$$

If $\mathcal{C}_n$ is a portrait consisting of a single n-cycle, then $\mathrm{End}_d^N[\mathcal{C}_n] \cong \mathrm{End}_d^N[n]$, and we write $\mathcal{M}_d^N[n]$ for $\mathcal{M}_d^N[\mathcal{C}_n]$.

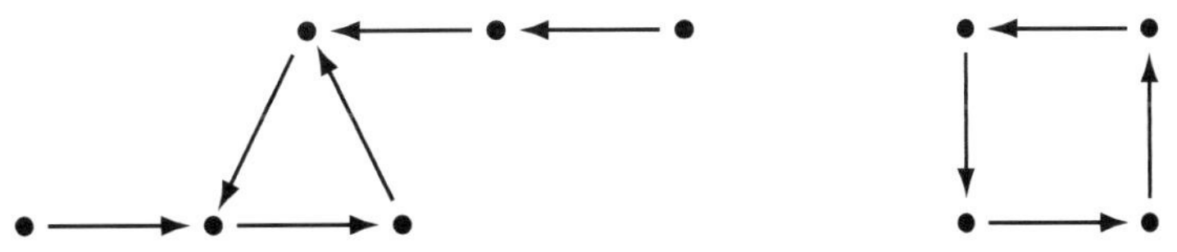

FIGURE 1
A portrait consisting of a 3-cycle, a 4-cycle, and three other preperiodic points

Theorem 5.6 (**[20]**). *Let $\mathcal{P}$ be a preperiodic portrait.*[12] *Then the quotient space $\mathcal{M}_d^N[\mathcal{P}]$ described in* (5.3) *exists*[13] *as a GIT geometric quotient scheme over $\mathbb{Z}$.*

---

11 This definition of $\mathrm{End}_d^N[\mathcal{P}]$ conveys the right idea; see **[20]** for a rigorous definition.

12 More generally, one can construct the moduli space $\mathcal{M}_d^N[\mathcal{P}]$ associated to a portrait $\mathcal{P}$ that includes nonpreperiodic points and/or whose vertices are assigned multiplicities.

13 There is a precise combinatorial-geometric characterization of the portraits $\mathcal{P}$ for which $\mathcal{M}_d^1[\mathcal{P}](\mathbb{C}) \neq \emptyset$, but analogous characterizations for $N \geq 2$ and/or in positive characteristic are not currently known.

It is known **[87]** that the moduli space $\mathcal{A}_g$ of principally polarized abelian varieties is of general type for all $g \geq 9$. Analogous results for dynamical moduli spaces are still unknown, but our dictionary yields some conjectures.[14]

Conjecture 5.7. *Let $N \geq 1$ and $d \geq 2$.*

(a) *For all $n \geq 1$, the moduli space $\mathcal{M}_d^N[n]$ is irreducible.*

(b) *For all sufficiently large n, depending on N and d, the moduli space $\mathcal{M}_d^N[n]$ is a variety of general type.*

Remark 5.8. The moduli space $\mathcal{M}_2^1[n]$ of degree-2 endomorphisms of $\mathbb{P}^1$ is a finite cover of $\mathcal{M}_2^1 \cong \mathbb{A}^2$, so it is a surface. It is known to be irreducible for all $n \geq 1$ **[48]**. For $1 \leq n \leq 5$, the surface $\mathcal{M}_2^1[n]$ is a rational surface, while $\mathcal{M}_2^1[6]$ is a surface of general type **[12]**.

Remark 5.9. Tai's proof **[87]** that $\mathcal{A}_g$ is of general type relies on the theory of theta functions, which are used to create sections of the canonical bundle. There are similarly naturally defined functions on $\mathcal{M}_d^N$, and more generally on $\mathcal{M}_d^N[n]$, that are defined using multiplier systems.[15] For $N = 1$, it is known that a multiplier system of sufficiently high degree gives a map $\mathcal{M}_d^1 \to \mathbb{A}^r$ that is (essentially) finite-to-one[16] onto its image **[55]**. So although the analogy between theta functions on $\mathcal{A}_g$ and multiplier system functions on $\mathcal{M}_d^N$ is tenuous at best, the latter currently provide one of the most natural ways to create dynamically defined functions on dynamical moduli spaces.

A map $f \in \mathrm{End}_d^N(K)$ defined over K with a K-rational n-periodic point $P \in \mathbb{P}^N(K)$ gives a K-rational point $\langle f, P \rangle \in \mathcal{M}_d^N[n](K)$. The dynamical uniform boundedness conjecture (Conjecture 4.3) is thus closely related to the question of K-rational points on dynamical moduli spaces. We formulate a uniform boundedness conjecture for such spaces.

Conjecture 5.10 (Dynamical uniform boundedness conjecture: version 2). *Fix integers $N \geq 1$, $d \geq 2$, and $D \geq 1$. There is a constant $C'(N, d, D)$ such that for all number fields K of degree $[K : \mathbb{Q}] = D$ and all preperiodic portraits $\mathcal{P}$,*

$$\big(\#\{\text{vertices of } \mathcal{P}\} \geq C'(N, d, D)\big) \Longrightarrow \mathcal{M}_d^N[\mathcal{P}](K) = \emptyset.$$

---

14 See **[10, CONJECTURE 10.13]** for a generalization of Conjecture 5.7 that deals with quite general dynamical moduli spaces that classify families of maps with marked periodic points of large period, including bounds on their number of components, Kodaira dimension, and gonality.

15 Briefly, for $N = 1$, let $k \geq 1$, let $f \in \mathrm{End}_d^1$, and let $P_1, \ldots, P_r$ be the periodic points of f with period dividing k. The derivatives $(f^k)'(P_i)$ are PGL_2-conjugate independent, and the k-level multiplier system of f is the list $\Lambda_k(f)$ of the elementary symmetric functions of $(f^k)'(P_1), \ldots, (f^k)'(P_r)$. Then $\Lambda_k(f)$ gives a well-defined morphism $\Lambda_k(f) : \mathcal{M}_d^1 \to \mathbb{A}^r$.

16 More precisely, the map is finite-to-one unless n is a square, in which case it maps the j-line of flexible Lattès maps to a single point. This is thus one of those results that's "true except in the obvious cases where it is false."

Remark 5.11. It is clear that Conjecture 5.10 implies Conjecture 4.3. The opposite implication is also true, but the proof is more difficult due to the *Field-of-Moduli versus Field-of-Definition Problem*. The key step, proven in [19] and [20, SECTIONS 16–17], is to show that every point in $\mathcal{M}_d^N[\mathcal{P}](K)$ is represented by a point in $\mathrm{End}_d^N[\mathcal{P}](L)$ defined over an extension L/K whose degree is bounded solely by d and N. When $N = 1$, one can take $[L : K] \le 2$ [32], but for $N \ge 2$ it is an open question whether $[L : K]$ needs to depend on d.

Within $\mathcal{M}_d^N$ and its GIT semistable compactification $\overline{\mathcal{M}}_d^N$ lie many interesting subvarieties. For example:

- The space of polynomial maps[17]
$$\mathrm{Poly}_d^N = \{f \in \mathcal{M}_d^N : f \text{ comes from a morphism } \mathbb{A}^N \to \mathbb{A}^N\}$$
is a subvariety of $\mathcal{M}_d^N$ satisfying $\dim(\mathrm{Poly}_d^1) = \frac{N}{N+1}\dim(\mathcal{M}_d^N)$.
- Iteration of dominant rational maps presents its own interesting challenges; see Section 8 for some examples. The set of degree d dominant rational maps $\mathbb{P}^N \dashrightarrow \mathbb{P}^N$ is a Zariski open subvariety of $\mathbb{P}^{\nu-1}(\mathbb{C})$ [81, PROPOSITION 7], but the locus of points in $(\overline{\mathcal{M}}_d^N \setminus \mathcal{M}_d^N)(\mathbb{C})$ arising from dominant rational maps is not well understood; cf. [42].

The spaces of polynomial maps and dominant rational maps have large dimension. At the other extreme are various 1-parameter families of maps that have been much studied, starting with the ubiquitous family of quadratic polynomials

$$f_c(x) = x^2 + c$$

that gives a line $\mathbb{A}^1$ in $\mathcal{M}_2^1 \cong \mathbb{A}^2$. Adding level structure leads to a dynamical analogue of the classical elliptic modular curve $X_1^{\mathrm{ell}}(n)$ that classifies pairs (E, P) consisting of an elliptic curve E and an n-torsion point P. In the dynamical setting, we replace the n-torsion point with a point of period n, but the following example shows that some care is needed.

Example 5.12. The polynomial $f(x) = x^2 - \frac{3}{4}$ has no points of exact period 2, since

$$f(x) - x = (2x+1)(2x-3) \quad \text{and} \quad f^2(x) - x = (2x+1)^3(2x-3).$$

But since $\frac{f^2(x)-x}{f(x)-x} = (2x+1)^2$, we say that $x = -\frac{1}{2}$ is a point of *formal period* 2 for $f(x)$.[18]

---

17 For example, the space $\mathrm{Poly}_d^1 \subset \mathcal{M}_d^1$ is the space of polynomials $x^d + a_2x^{d-2} + \cdots + a_d$ modulo the conjugation $x \to \zeta x$ for a primitive $(d-1)$-root of unity ζ, so Poly_d^1 is a quotient of $\mathbb{A}^{d-1}$ by a finite group.

18 In general, points of *formal period* n for the polynomial $f(x)$ are roots of the *dynatomic polynomial*

$$\Phi_f(x) := \prod_{d \mid n} \left(f^d(x) - x\right)^{\mu(n/d)},$$

where μ is the Möbius function. Dynatomic polynomials are thus dynamical analogues of classical cyclotomic polynomials, but with the caveat that $\Phi_f(x)$ may have roots of multiplicity greater than 1, even in characteristic 0. In higher dimension, the points of formal period n give a *dynatomic* 0*-cycle* whose effectivity is proven in [33].

Definition 5.13. The *level n dynatomic curve*[19] *(associated to x^2+c)* is the affine curve

$$Y_1^{\mathrm{dyn}}(n) = \{(c,\alpha) \in \mathbb{A}^2 : \alpha \text{ is a point of formal period } n \text{ for } f_c(x) = x^2 + c\}.$$

The desingularized projective completion of $Y_1^{\mathrm{dyn}}(n)$ is denoted $X_1^{\mathrm{dyn}}(n)$. The points in the complement $X_1^{\mathrm{dyn}}(n) \setminus Y_1^{\mathrm{dyn}}(n)$, which correspond to degenerate maps, are called *cusps*.[20]

Points in $Y_1^{\mathrm{dyn}}(n)(K)$ classify quadratic polynomials defined over K having a K-rational point of period n, so a version of Theorem/Conjecture 4.7 says that

$$\begin{array}{rll} \textbf{Exercise} & X_1^{\mathrm{dyn}}(n) \cong \mathbb{P}^1 & \text{for } n \in \{1,2,3\}, \\ \textbf{Theorem} & X_1^{\mathrm{dyn}}(n)(\mathbb{Q}) = \{\text{cusps}\} & \text{for } n \in \{4,5,6\}, \\ \textbf{Conjecture} & X_1^{\mathrm{dyn}}(n)(\mathbb{Q}) = \{\text{cusps}\} & \text{for all } n \geq 4. \end{array}$$

Much is known about the geometry of $X_1^{\mathrm{dyn}}(n)$, as summarized in the next result, although we note that even the proof that $X_1^{\mathrm{dyn}}(n)$ is geometrically irreducible relies on dynamical properties of x^2+c as reflected in the geometry of the Mandelbrot set.

Theorem 5.14. *Let $X_1^{\mathrm{dyn}}(n)$ be the smooth projective dynatomic curve associated to x^2+c.*

(a) *The dynatomic modular curve $X_1^{\mathrm{dyn}}(n)$ is geometrically irreducible over $\mathbb{C}$* [13, 41, 76].[21]

(b) *There is an explicit, but rather complicated, formula for the genus of $X_1^{\mathrm{dyn}}(n)$* [61]. *In any case,* $\operatorname{genus}(X_1^{\mathrm{dyn}}(n)) \to \infty$ *as $n \to \infty$.*

(c) *The gonality*[22] *of $X_1^{\mathrm{dyn}}(n)$ goes to ∞ as $n \to \infty$* [18].

6. TOPIC #3: DYNAMICAL UNLIKELY INTERSECTIONS

The guiding philosophy of unlikely intersections in arithmetic geometry is the following general, albeit somewhat vague, principle.

---

19 There are dynatomic curves associated to many other interesting 1-parameter families of maps, including, for example, the family of degree-d *unicritical polynomials* $f_{d,c}(x) = x^d + c$ and the family of degree-2 rational maps $g_b(x) = x/(x^2+b)$ that admit a nontrivial automorphism $g_b(-x) = -g_b(x)$.

20 We mention that there is a natural action of f on $Y_1^{\mathrm{dyn}}(n)$ defined by $(c,\alpha) \mapsto (c, f(\alpha))$, and that the quotient curve $Y_0^{\mathrm{dyn}}(n) = Y_1^{\mathrm{dyn}}(n)/\langle f \rangle$ and its completion $X_0^{\mathrm{dyn}}(n)$ provide analogues of the elliptic modular curve $X_0^{\mathrm{ell}}(n)$.

21 More generally, the dynatomic modular curves associated to the family of unicritical polynomials $x^d + c$ are irreducible. However, the dynatomic modular curves associated to the family $x/(x^2+b)$ turn out to be reducible for even n; see [48].

22 The *gonality* of an algebraic curve X is the minimal degree of a nonconstant map $X \to \mathbb{P}^1$.

The Tao of Unlikely Intersections
Let X be an algebraic variety, let $Y \subseteq X$ be an algebraic subvariety of X, and let $\Gamma \subset X$ be an "interesting" countable subset of X. Then

$\Gamma \cap Y$ is sparse (except when it is "obviously" not).

Slightly more precisely, if $\Gamma \cap Y$ is Zariski dense in Y, then there should be a geometric reason that explains its density.

We recall two famous unlikely intersection theorems from arithmetic geometry, which we initially state in an intuitively appealing, though somewhat whimsical, manner.

Theorem 6.1 (Mordell–Lang conjecture, [23,98]). *Let $A/\mathbb{C}$ be an abelian variety, let $Y \subseteq A$ be a subvariety of A, and let $\Gamma \subset A(\mathbb{C})$ be a finitely generated subgroup of A. Then*[23]

$\Gamma \cap Y$ is not Zariski dense in Y (except when it "obviously" is).

Theorem 6.2 (Manin–Mumford conjecture, [73,74]). *Let $A/\mathbb{C}$ be an abelian variety, and let $Y \subseteq A$ be a subvariety of A. Then*

$A_{\mathrm{tors}} \cap Y$ is not Zariski dense in Y (except when it "obviously" is).

The actual statements of Theorems 6.1 and 6.2 explain quite precisely that if Y is saturated with special points, then there is a geometric reason for that saturation.

Theorem 6.3 (Rigorous formulation of Theorems 6.1 and 6.2). *If $\Gamma \cap Y$ or $A_{\mathrm{tors}} \cap Y$ is Zariski dense in Y, then Y is necessarily a translate of an abelian subvariety of A by a torsion point of A.*

Remark 6.4. Theorems 6.1 and 6.2 may be combined and strengthened by replacing the abelian variety A with a semi-abelian variety and by replacing Γ with its divisible subgroup $\bigcup_{n\geq 0}[n]^{-1}(\Gamma)$; see [56].

Theorem 6.1 says that points in a finitely generated subgroup Γ generally do not lie on a subvariety. According to Table 1, for the dynamical analogue of Theorem 6.1 we should replace the group Γ with the points in an orbit. This leads to our first dynamical unlikely intersection conjecture.

Conjecture 6.5 (Dynamical Mordell–Lang conjecture). *Let $X/\mathbb{C}$ be a smooth quasiprojective variety, let $f : X \to X$ be a regular self-map of X, let $P \in X(\mathbb{C})$ be a point with infinite f-orbit, and let $Y \subseteq X$ be a subvariety of X. Then*

$\mathcal{O}_f(P) \cap Y$ is not Zariski dense in Y (except when it "obviously" is).

Rigorous Formulation #1. *If $\mathcal{O}_f(P) \cap Y$ is Zariski dense, then Y is f-periodic.*[24]

23 The proof of Theorem 6.1 uses methods from Diophantine approximation. An earlier proof in the case that Y is a curve of genus at least 2 used moduli-theoretic techniques [22].

24 We says that Z is *f-periodic* if there is an integer $n > 0$ such that $f^n(Z) = Z$.

Rigorous Formulation #2. *The set*

$$\{n \geq 0 : f^n(P) \in Y\}$$

is a finite union of one-sided arithmetic progressions [29].[25]

Example 6.6. Among the known cases of the dynamical Mordell–Lang conjecture, we cite the following:

> **Unramified maps.** Conjecture 6.5 is true for étale morphisms of quasiprojective varieties [7]. See the monograph [8] for additional information.
>
> **Endomorphisms of $\mathbb{A}^2$.** Conjecture 6.5 is true for all endomorphisms of $\mathbb{A}^2$ defined over $\bar{\mathbb{Q}}$ [92].
>
> **Split endomorphisms.** Conjecture 6.5 is true for split endomorphisms of $(\mathbb{P}^1)^n$, which are maps of the form $f_1(P_1) \times \cdots \times f_n(P_n)$ [11], and more generally for certain skew-split endomorphisms [31].

Remark 6.7. The dynamical Mordell–Lang conjecture has also been investigated in characteristic p, although the statement may need a tweak. For example, if f is a projective surface automorphism or a birational endomorphism of $\mathbb{A}^2$ whose dynamical degree (see Section 8) satisfies $\delta_f > 1$, then Conjecture 6.5 is true in all characteristics [94]. For other results in finite characteristic, see, for example, [8, 14, 26].

We now turn to Theorem 6.2, which asserts that torsion points generally do not lie on a subvariety. According to Table 1, for the dynamical analogue we should replace the torsion points with preperiodic points, leading to our second dynamical unlikely intersection conjecture.

Conjecture 6.8 (Dynamical Manin–Mumford conjecture). *Let $X/\mathbb{C}$ be a smooth quasi-projective variety, let $f : X \to X$ be a regular self-map of X, and let $Y \subseteq X$ be a subvariety of X. Then*

$$\mathrm{PrePer}(f) \cap Y \text{ is not Zariski dense in } Y \text{ (except when it "obviously" is).}$$

Unfortunately, the following natural rigorous formulation of Conjecture 6.8 turns out to be false.

Incorrect Rigorous Formulation of Conjecture 6.8. If $\mathrm{PrePer}(f) \cap Y$ is Zariski dense in Y, then Y is f-preperiodic.

See [38] for a counterexample, and for an alternative formulation of Conjecture 6.8 that requires more stringent hypotheses on f and Y.

Both the Mumford–Manin and Mordell–Lang conjectures concern how special points lie on subvarieties of a given variety. The André–Oort conjecture has a similar flavor,

25 A *one-sided arithmetic progression* is a set of integers of the form $\{ak + b : k \in \mathbb{N}\}$ for some fixed $a, b \in \mathbb{N}$. N.B. We allow $a = 0$.

but the ambient variety is a moduli space and the specialness of the points comes from the properties of the objects that they represent. The André–Oort conjecture is easy to state as long as we are willing to sweep some quite technical definitions under the rug![26]

Conjecture 6.9 (André–Oort conjecture). *Let $\mathcal{S}$ be a Shimura variety, let $\Gamma \subset \mathcal{S}$ be a set of special points of $\mathcal{S}$, and let $Y \subset \mathcal{S}$ be an irreducible subvariety such that $\Gamma \cap Y$ is Zariski dense in Y. Then Y is a special subvariety of $\mathcal{S}$.*

The rough idea is that $\mathcal{S}$ is a moduli space whose points classify a certain class of abelian varieties, a collection of special points $\mathcal{T} \subset \mathcal{S}$ consists of points whose associated abelian varieties have an additional special structure, and a special subvariety is one in which every associated abelian variety has the $\mathcal{T}$ property for geometric reasons. The André–Oort conjecture has been proven in many cases, including for $\mathcal{S} = \mathcal{A}_1^d$ **[70]** and for $\mathcal{S} = \mathcal{A}_g$ **[71,88]**.

We describe two sample dynamical unlikely intersection theorems that take place in the moduli space of unicritical polynomials, which are polynomials of the form $x^d + c$. We view the first as a mixed unlikely intersection, because it involves one moduli parameter and two orbit parameters.

Theorem 6.10 (**[3]**). *Let $d \geq 2$, and let $a, b \in \mathbb{C}$ be complex numbers with $a^2 \neq b^2$. Then*

$$\Big\{ \underbrace{c \in \mathbb{C}}_{\text{moduli parameter}} : \underbrace{a \text{ and } b \text{ are both preperiodic}}_{\text{special orbit parameters}} \text{ for } x^d + c \Big\} \quad \text{is a finite set.}$$

The second result has more of the flavor of the André–Oort conjecture in that it involves only moduli parameters and follows the dictionary in Table 1 by replacing complex multiplication abelian varieties with postcritically finite rational maps.

Theorem 6.11 (**[28]**). *Let $d \geq 2$, and let $Y \subset \mathbb{A}^2$ be an irreducible curve that is not a line of one of the following forms:*

$$\text{vertical line } \{(a,t) : t \in \mathbb{A}^1\}; \quad \text{horizontal line } \{(t,b) : t \in \mathbb{A}^1\};$$
$$\text{shifted diagonal line } \{(t,\zeta t) : t \in \mathbb{A}^1\}, \text{ where } \zeta^{d-1} = 1.$$

Then

$$\underbrace{\big\{(a,b) \in Y : x^2 + a \text{ and } x^2 + b \text{ are both PCF}\big\}}_{\text{special moduli parameters}} \quad \text{is a finite set.}$$

A conjectural generalization of Theorem 6.10 allows both the map $x^2 + c$ and the points a and b to vary simultaneously.

Conjecture 6.12 (**[15,27]**). *Let $d \geq 2$, let T be an irreducible curve, and let*

$$\alpha : T \to \mathbb{P}^1, \quad \beta : T \to \mathbb{P}^1, \quad \text{and} \quad f : T \to \mathrm{End}_d^1$$

be morphisms, i.e., α and β are 1-parameter families of points in $\mathbb{P}^1$ and f is a 1-parameter

---

26 See, for example, **[89]** for the precise definition of *Shimura variety*, *special point*, and *special subvariety*.

family of degree-d endomorphisms of $\mathbb{P}^1$. *Assume that the families* α *and* β *are not* f*-dynamically related.*[27] *Then*

$$\{t \in T : \alpha_t \text{ and } \beta_t \text{ are both preperiodic for } f_t\} \quad \text{is a finite set.}$$

Formulating a general dynamical André–Oort conjecture is more complicated. The first step is to construct an appropriate moduli space of rational maps with marked critical points:[28]

$$\mathcal{M}_d^{\text{crit}} := \left\{ (f, P_1, \ldots, P_{2d-2}) : \begin{array}{l} f \in \operatorname{End}_d^1 \text{ and } P_1, \ldots, P_{2d-2} \\ \text{are critical points of } f \end{array} \right\} / \operatorname{PGL}_2 .$$

Conjecture 6.13 (Dynamical André–Oort Conjecture, [4,82]). *Let* $Y \subseteq \mathcal{M}_d^{\text{crit}}$ *be an algebraic subvariety such that the PCF maps in* Y *are Zariski dense in* Y. *Then* Y *is cut out by* "critical orbit relations."

Formulas of the form $f^n(P_i) = f^m(P_j)$ define critical point relations,[29] but other relations may arise from symmetries of f, and even subtler relations may come from "hidden relations" due to subdynamical systems. See [82, REMARK 6.58] for an example due to Ingram. Thus for now we do not have a good geometric description of the phrase "critical orbit relations" in general, but there is such a description for 1-dimensional families, i.e., for Conjecture 6.13 with $\dim(Y) = 1$ [4]. In this case the conjecture has been proven for 1-dimensional families of polynomials [24], but it remains open for rational maps.

7. TOPIC #4: DYNATOMIC AND ARBOREAL REPRESENTATIONS

The focus of this section is on the arithmetic of fields generated by the coordinates of dynamically interesting points. We let $K/\mathbb{Q}$ be a number field, and we start with a motivating result from arithmetic geometry. Let E/K be an elliptic curve, and let

$$\rho_{E/K,\ell}^{\text{ell}} : \operatorname{Gal}(\bar{K}/K) \to \operatorname{Aut}\big(T_\ell(E)\big) \cong \operatorname{GL}_2(\mathbb{Z}_\ell) \tag{7.1}$$

be the representation that describes the action of the Galois group on the ℓ-power torsion points of E. A famous theorem characterizes the image.[30]

---

27 Intuitively, the families α and β are f-dynamically related if there is a relationship between the f-orbits of α and β that holds identically for all parameter values in T. However, there are some subtleties; see [10, DEFINITION 11.2] for a discussion and the precise, albeit somewhat technical, definition.

28 It is easy to construct the GIT quotient for maps f having $2d-2$ *distinct* marked critical points, but some care is needed to handle maps having higher multiplicity critical points; see [20].

29 One might view these $f^n(P_i) = f^m(P_j)$ relations as dynamical analogues of Hecke correspondences, although the analogy is somewhat tenuous.

30 A 19th century precursor to Serre's theorem is a fundamental result on cyclotomic fields. It says that the cyclotomic representation $\rho^{\text{cyclo}} : \operatorname{Gal}(\bar{K}/K) \to \operatorname{Aut}(\mu_{\ell^\infty}) \cong \mathbb{Z}_\ell^*$ describing the action of the Galois group on ℓ-power roots of unity is surjective when $K = \mathbb{Q}$, and that the image of ρ^{cyclo} has finite index in $\mathbb{Z}_\ell^*$ for all K.

Theorem 7.1 (Serre's Image-of-Galois Theorem, [77, 78]). *Assume that E does not have complex multiplication.*

(a) *For all sufficiently large primes ℓ, the Galois representation $\rho^{\mathrm{ell}}_{E/K,\ell}$ is surjective.*

(b) *For all primes ℓ, the image of the Galois representation $\rho^{\mathrm{ell}}_{E/K,\ell}$ is a subgroup of finite index in $\mathrm{GL}_2(\mathbb{Z}_\ell)$.*

There are analogous conjectures, and some theorems, for the Galois representations associated to higher-dimensional abelian varieties. We consider two analogues in arithmetic dynamics.

7.1. Topic #4(a): Dynatomic representations

Let

$$f : \mathbb{P}^N \to \mathbb{P}^N$$

be a morphism of degree $d \geq 2$ defined over K, and let

$$\mathrm{Per}^*_n(f) = \{P \in \mathbb{P}^N(\bar{K}) : P \text{ is } f\text{-periodic with exact period } n\}.$$

The action of f on $\mathrm{Per}^*_n(f)$ splits it into a disjoint union of directed n-cycles, and the action of $\mathrm{Gal}(\bar{K}/K)$ on $\mathrm{Per}^*_n(f)$ respects the cycle structure. The analogue of GL_2 in (7.1) is thus the group of automorphisms of the graph

$$\mathcal{P}_{n,\nu} = \text{a disjoint union of } \nu \text{ directed } n\text{-gons}.$$

The abstract automorphism group of the directed graph $\mathcal{P}_{n,\nu}$ is naturally described as a wreath product in which an automorphism of $\mathcal{P}_{n,\nu}$ is characterized as a permutation of the ν polygons combined with a rotation of each polygon:

$$\mathrm{Aut}(\mathcal{P}_{n,\nu}) \cong (\mathbb{Z}/n\mathbb{Z}) \wr \mathcal{S}_\nu \cong (\mathbb{Z}/n\mathbb{Z})^\nu \rtimes \mathcal{S}_\nu.$$

Definition 7.2. Let $f \in \mathrm{End}^d_N(K)$. The *$n$-level dynatomic representation of f over K* is the homomorphism

$$\rho^{\mathrm{dyn}}_{K,n,f} : \mathrm{Gal}(\bar{K}/K) \to \mathrm{Aut}(\mathcal{P}_{n,\nu(f)}), \quad \text{where } \nu(f) = \frac{1}{n}\#\mathrm{Per}^*_n(f).$$

The analogue of Serre's theorem would assert that if f has no automorphisms,[31] then $\rho^{\mathrm{dyn}}_{K,n,f}$ is surjective for sufficiently large n. It seems too much to ask that this be true for all maps, so we pose the following challenge:

Question 7.3 (Dynatomic Image-of-Galois Problem). Let $K/\mathbb{Q}$ be a number field, let $N \geq 1$, and let $d \geq 2$. Characterize the maps $f \in \mathrm{End}^N_d(K)$ for which there is a constant $C(f)$ such that for all $n \geq 1$,

$$\mathrm{Image}(\rho^{\mathrm{dyn}}_{K,f,n}) \text{ has index at most } C(f) \text{ in } \mathrm{Aut}(\mathcal{P}_{n,\nu(f)}).$$

31 The automorphism group of f is $\mathrm{Aut}(f) = \{\varphi \in \mathrm{PGL}_{N+1} : \varphi^{-1} \circ f \circ \varphi = f\}$. The elements of $\mathrm{Gal}(\bar{K}/K)$ commute with the action of $\mathrm{Aut}_K(f)$, so if $\mathrm{Aut}_K(f) \neq (1)$, then the image of $\rho^{\mathrm{dyn}}_{K,n,f}$ is restricted, just as the image of $\rho^{\mathrm{ell}}_{E/K,\ell}$ is restricted if E has CM.

7.2. Topic #4(b): Arboreal representations

The dynatomic extensions described in Section 7.1 are generated by points with finite orbits. In this section we consider arboreal extensions, which are extensions generated by backward orbits.

Example 7.4. We illustrate with the map $f(x) = x^d$.

$$\begin{bmatrix} \textit{Dynatomic extension.} \text{ Field generated by roots of } x^{d^n} = x \text{ for } n \geq 1. \\ \textit{Arboreal extension.} \text{ Field generated by roots of } x^{d^n} = a \text{ for } n \geq 1. \end{bmatrix} \tag{7.2}$$

Thus (7.2) suggests that dynatomic extensions resemble cyclotomic extensions, while the arboreal extensions resemble Kummer extensions; although we readily admit that this is far from a perfect analogy.

Definition 7.5. Let $f : \mathbb{P}^N \to \mathbb{P}^N$ be a morphism of degree $d \geq 2$ defined over K, and let $P \in \mathbb{P}^N(K)$. The *inverse image tree of f rooted at P* is the (disjoint) union of the inverse images of P by the iterates of f:

$$\mathcal{T}_{f,P} = \bigcup_{n \geq 0} f^{-n}(P) = \bigcup_{n \geq 0} \{Q \in \mathbb{P}^N(\bar{K}) : f^n(Q) = P\}.$$

We say that f is *arboreally complete at P* if $\#f^{-n}(P) = d^{nN}$ for all $n \geq 0$, in which case $\mathcal{T}_{f,P}$ is a complete rooted d^N-ary tree, where f maps the points in $f^{-n-1}(P)$ to the points in $f^{-n}(P)$. Figure 2 illustrates a complete inverse image tree for a degree-2 map $f : \mathbb{P}^1 \to \mathbb{P}^1$.

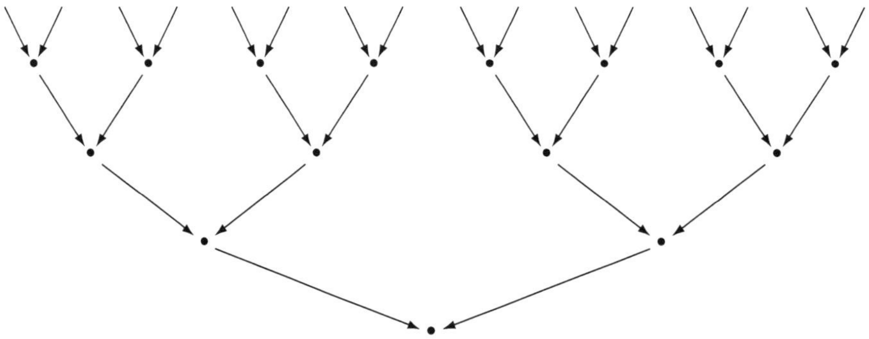

FIGURE 2
A complete binary inverse image tree

The points in the iterated inverse image of P generate a (generally infinite) algebraic extension of K, so the Galois group $\mathrm{Gal}(\bar{K}/K)$ acts on the points in $\mathcal{T}_{f,P}$. And since the action of the Galois group commutes with the map f, the action of $\mathrm{Gal}(\bar{K}/K)$ on $\mathcal{T}_{f,P}$ preserves the tree structure. Thus in this case, the analogue of GL_2 in (7.1) is the group of automorphisms of the tree $\mathcal{T}_{f,P}$, which leads us to our primary object of study.

Definition 7.6. Let $f \in \mathrm{End}_N^d(K)$, and let $P \in \mathbb{P}^N(K)$. The *arboreal representation (over K) of f rooted at P* is the homomorphism

$$\rho_{K,f,P}^{\mathrm{dyn}} : \mathrm{Gal}(\bar{K}/K) \to \mathrm{Aut}(\mathcal{T}_{f,P}).$$

The *Odoni*[32] *index over K of f at P* is the index of the image in the full tree automorphism group,

$$\iota_K(f, P) = \left[\mathrm{Aut}(\mathcal{T}_{f,P}) : \mathrm{Image}(\rho_{K,f,P}^{\mathrm{dyn}})\right].$$

As in the dynatomic case, it is again too much to hope that the image of $\rho_{K,f,P}^{\mathrm{dyn}}$ has finite index in $\mathrm{Aut}(\mathcal{T}_{f,P})$ for all f, but we might expect this to be true for most f. This leads to a number of fundamental questions.

Question 7.7 (Arboreal Image-of-Galois Problem). (a) Let $K/\mathbb{Q}$ be a number field, and let $N \geq 1$ and $d \geq 2$. Characterize the maps $f \in \mathrm{End}_d^N(K)$ and points $P \in \mathbb{P}^N(K)$ whose Odoni index $\iota_K(f, P)$ is finite, especially when f is arboreally complete at P.

(b) (Generalized Odoni conjecture) For all number fields $K/\mathbb{Q}$ and all $N \geq 1$ and $d \geq 2$, does there exist a point $P \in \mathbb{P}^N(K)$ and a map $f \in \mathrm{End}_d^N(K)$ that is arboreally complete at P such that $\iota_K(f, P) = 1$?

(c) Fix a number field $K/\mathbb{Q}$ and integers $N \geq 1$ and $d \geq 2$. Is it true that $\iota_K(f, P) = 1$ for "almost all" pairs (f, P) in $\mathrm{End}_d^N(K) \times \mathbb{P}^N(K)$ for some appropriate sense of density?

Remark 7.8. Odoni's original conjecture was both more restrictive and stronger than Question 7.7(b) in that he considered only $N = 1$ and polynomial maps. Odoni asked if for all $K/\mathbb{Q}$ and all $d \geq 2$, there exists a degree-d monic polynomial $f(x) \in K[x]$ and a point $\alpha \in K$ such that $\mathcal{T}_{f,\alpha}$ is a complete d-ary tree and such that $\iota_K(f, \alpha) = 1$. Odoni's conjecture was proven over $\mathbb{Q}$ for prime values of d in **[45]**, and then in full generality in **[85]**. We mention that Odoni originally conjectured that the statement should hold for all Hilbertian fields, but this was recently resolved in the negative **[36]**.

Remark 7.9. We close with the well-known observation that the automorphism group of an n-level complete rooted regular tree (labeling the levels $0, 1, 2, \ldots, n$) is an n-fold wreath product of the symmetric group. Hence if f is arboreally complete at P, then the automorphism group of $\mathcal{T}_{f,P}$ is the inverse limit

$$\mathrm{Aut}(\mathcal{T}_{f,P}) \cong \varprojlim \underbrace{\mathcal{S}_{d^N} \wr \mathcal{S}_{d^N} \wr \cdots \wr \mathcal{S}_{d^N}}_{n\text{-fold iterated wreath product with } n \to \infty}.$$

32 Named in honor of R. W. K. Odoni, who appears to have been the first to seriously study such problems in a series of papers **[65–67]**, in one of which he proves that $\iota_{\mathbb{Q}}(x^2 - x + 1, 0) = 1$.

The profinite group $G(\bar{K}/K)$ then acts continuously on the profinite group $\mathrm{Aut}(\mathcal{T}_{f,P})$, just as in arithmetic geometry $G(\bar{K}/K)$ acts continuously on the Tate module $T_\ell(A) = \varprojlim A[\ell^n]$ of an abelian variety A/K.

8. TOPIC #5: DYNAMICAL AND ARITHMETIC COMPLEXITY

We informally define the *complexity* of a mathematical object to be a rough estimate for how much space it takes to store the object:

$$\begin{aligned} h(\mathcal{X}) &= \text{complexity of object } \mathcal{X} \\ &\asymp \text{\# of basic storage units (e.g., bits, scalars) required to describe } \mathcal{X}. \end{aligned}$$

Example 8.1. The complexity of a nonzero integer $c \in \mathbb{Z}$ is the number of bits needed to describe c, so roughly $\log |c|$.

Example 8.2. The complexity of a nonzero polynomial $f(x) \in K[x]$ is the number of coefficients needed to describe f, so roughly $\deg(f)$.

For a sequence of objects $\boldsymbol{\mathcal{X}} = (\mathcal{X}_n)_{n\ge 1}$ whose complexity is expected to grow exponentially, we define the *sequential complexity of* $\boldsymbol{\mathcal{X}}$ to be the limit[33]

$$\sigma(\boldsymbol{\mathcal{X}}) = \lim_{n\to\infty} h(\mathcal{X}_n)^{1/n}.$$

Example 8.3. Let $f : \mathbb{P}^N \dashrightarrow \mathbb{P}^N$ be a degree-d dominant rational map, i.e., a map given by homogeneous degree-d polynomials $[f_0, \dots, f_N]$ in $\mathbb{C}[x_0, \dots, x_N]$ having no common factors. Then $h(f) = \deg(f) = d$. The sequential complexity of the sequence of iterates f^n is called the *dynamical degree of* f and is denoted

$$\delta_f = \lim_{n\to\infty} (\deg f^n)^{1/n}. \tag{8.1}$$

Example 8.4. Let $P = [c_0, \dots, c_N] \in \mathbb{P}^N(\mathbb{Q})$ be a point written with relatively prime integer coordinates. Then

$$h(P) = \log \max |c_i|. \tag{8.2}$$

More generally, if $K/\mathbb{Q}$ is a number field, then there is a well-defined Weil height function[34]

$$h : \mathbb{P}^N(K) \to [0, \infty) \tag{8.3}$$

---

33 In cases where the limit is not known to exist, we may consider the *upper and lower sequential complexities*

$$\overline{\sigma}(\boldsymbol{\mathcal{X}}) = \limsup_{n\to\infty} h(\mathcal{X}_n)^{1/n} \quad \text{and} \quad \underline{\sigma}(\boldsymbol{\mathcal{X}}) = \liminf_{n\to\infty} h(\mathcal{X}_n)^{1/n}.$$

34 The Weil height of a point $P = [a_0, \dots, a_N] \in \mathbb{P}^N(K)$ may be defined as follows: Let $d = [K : \mathbb{Q}]$, write the fractional ideal generated by $a_0, \dots, a_N$ as $\mathfrak{A}\mathfrak{B}^{-1}$ with relatively prime integral ideals $\mathfrak{A}$ and $\mathfrak{B}$, and let $\sigma_1, \dots, \sigma_d : K \hookrightarrow \mathbb{C}$ be the distinct complex embeddings of K. Then

$$h(P) = \frac{1}{d} \log |\mathrm{N}_{K/\mathbb{Q}}(\mathfrak{B})| + \frac{1}{d} \sum_{i=1}^{d} \log \max_{0\le j\le N} |\sigma_i(a_j)|.$$

that generalizes (8.2). The height of a point $P \in \mathbb{P}^N(K)$ measures the complexity of the coordinates of P.

Now let $K/\mathbb{Q}$ be a number field, let $P \in \mathbb{P}^N(K)$, and let $f : \mathbb{P}^N \dashrightarrow \mathbb{P}^N$ be a dominant rational map defined over K. Then the sequential complexity of the orbit $\mathcal{O}_f(P)$ is called the *arithmetic degree of the f-orbit of P* and is denoted

$$\alpha_f(P) = \lim_{n\to\infty} h\big(f^n(P)\big)^{1/n}. \tag{8.4}$$

The notation in Table 2 will be used throughout the remainder of this section. We will generalize the complexity measures from Examples 8.3 and 8.4 and describe a number of results and questions.

Definition 8.5. The *(first) dynamical degree* of a dominant rational map $f : X \dashrightarrow X$ is

$$\delta_f = \lim_{n\to\infty} \big(\deg_X(f^n)\big)^{1/n}. \tag{8.5}$$

The limit (8.5) converges and is independent of the choice of the ample divisor H used to define $\deg_X$ **[16]**.[35] Dynamical degrees on $\mathbb{P}^N$ were first studied in the 1990s **[2,9,75]**. A long-standing question concerning the algebraicity of the dynamical degree was recently answered in the negative.

Theorem 8.6 (**[5,6]**). *For all $N \geq 2$, there exist dominant rational maps $f : \mathbb{P}^N \dashrightarrow \mathbb{P}^N$ defined over $\mathbb{Q}$ such that δ_f is a transcendental number. For $N \geq 3$, there exist such maps that are birational automorphisms of $\mathbb{P}^N$.*

K	a number field with algebraic closure $\bar{K}$
X	a smooth projective variety of dimension d defined over K
f	a dominant rational map $f : X \dashrightarrow X$ defined over K
X_f	$= \{P \in X(\bar{K}) : f$ is well-defined at $f^n(P)$ all $n \geq 0\}$
$\deg_X(f)$	$= (f^*H) \cdot H^{d-1}$, where H is an ample divisor on X, and this formula is a d-fold intersection index
h_X	the height on X coming from a projective embedding $\iota : X \hookrightarrow \mathbb{P}^N$, i.e., $h_X = h \circ \iota$, where h is the Weil height (8.3) on $\mathbb{P}^N$
h_X^+	$= \max\{1, h_X\}$

TABLE 2
Notation for Section 8

35 The convergence of (8.5) when $X = \mathbb{P}^N$ is a fun exercise using $\deg(f \circ g) \leq (\deg f)(\deg g)$.

There is an arithmetic analogue of the dynamical degree that measures the average arithmetic complexity of the algebraic points in an orbit. But since rational maps may not be defined everywhere, the next definition must restrict attention to X_f, the points in X where the full forward orbit of f is well defined.[36]

Definition 8.7. Let $f : X \dashrightarrow X$ be a dominant rational map defined over K, and let $P \in X_f(\bar{K})$. The *arithmetic degree of the f-orbit of P* is

$$\alpha_f(P) = \lim_{n\to\infty} h_X^+\big(f^n(P)\big)^{1/n}. \tag{8.6}$$

Question 8.8. Does the limit (8.6) always exist?

In any case, we may consider the upper and lower arithmetic degrees $\underline{\alpha}_f(P)$ and $\overline{\alpha}_f(P)$ defined using, respectively, the liminf and the limsup. It is not hard to show that these quantities are independent of the choice of the complexity function h_X^+. It is also easy to show that $\overline{\alpha}_f(P)$ is finite, but more difficult to show that there is a uniform geometric bound, as in the next result.

Theorem 8.9 ([50]). *Let $f : X \dashrightarrow X$ be a dominant rational map defined over K, and let $P \in X_f(\bar{K})$. Then*

$$\overline{\alpha}_f(P) \le \delta_f.$$

Moral of Theorem 8.9. The arithmetic complexity of an orbit is no worse than the dynamical complexity of the map.

Theorem 8.9 suggests a natural question. When do the arithmetic and dynamical complexities coincide?

Conjecture 8.10 ([39, 40])). *Let $f : X \dashrightarrow X$ be a dominant rational map defined over K, and let $P \in X_f(\bar{K})$. Then*

$$\mathcal{O}_f(P) \text{ is Zariski dense in } X \Longrightarrow \overline{\alpha}_f(P) = \delta_f.$$

Moral of Conjecture 8.10. An orbit with maximal geometric complexity also has maximal arithmetic complexity.

Question 8.11. Does $X(\bar{K})$ always contain a point with Zariski dense f-orbit? The answer is clearly no. For example, if there exists a dominant rational map $\varphi : X \dashrightarrow \mathbb{P}^1$ satisfying $\varphi \circ f = \varphi$, then each f-orbit lies in a fiber of φ. Xie asks whether this is the only obstruction. An affirmative answer for certain maps in dimension 2 is given in [35, 93].

---

36 The complement $X \setminus X_f$ is a countable union of proper subvarieties, so cardinality considerations show that $X_f(\mathbb{C})$ is nonempty; but the situation is less clear for a countable field such as $\bar{\mathbb{Q}}$. It is shown in [1] that $X_f(\bar{\mathbb{Q}})$ is Zariski dense in X.

Example 8.12. It is easy to prove Conjecture 8.10 for morphisms f of $\mathbb{P}^N$, since in that case $\delta_f = \deg(f)$, and the theory of canonical heights implies that

$$\alpha_f(P) = \begin{cases} \deg(f) & \text{if } \#\mathcal{O}_f(P) = \infty, \\ 1 & \text{if } P \text{ is } f\text{-prepediodic.} \end{cases}$$

More generally, a similar argument works for endomorphisms of any smooth projective variety whose Néron–Severi group has rank 1 [38]. But the conjecture is still open for dominant rational maps of $\mathbb{P}^N$, and for morphisms of more general varieties.

Example 8.13. The past decade has been significant progress on various cases of Conjecture 8.10, especially in the case of morphisms, using an assortment of tools ranging from linear-forms-in-logarithms to canonical heights for nef divisors to the minimal model program in algebraic geometry. In particular, Conjecture 8.10 has been proven for

- group endomorphisms (homomorphisms composed with translations) of semi-abelian varieties (extensions of abelian varieties by algebraic tori) [39, 52, 83, 84],
- endomorphisms of (not necessarily smooth) projective surfaces [38, 53, 57],
- extensions to $\mathbb{P}^N$ of regular affine automorphisms of $\mathbb{A}^N$ [38],
- endomorphisms of hyperkähler varieties [43],
- endomorphisms of degree greater than 1 of smooth projective threefolds of Kodaira dimension 0 [43],
- endomorphisms of normal projective varieties such that $\mathrm{Pic}^0 \otimes \mathbb{Q} = 0$ and with nef cone generated by finitely many semi-ample integral divisors [49], and
- smooth projective threefolds having at least one int-amplified[37] endomorphism, and surjective endomorphisms of smooth rationally connected projective varieties [51].

Remark 8.14. Various generalizations of Conjecture 8.10 have been proposed. We mention in particular the *Small Arithmetic Non-Density Conjecture* [51], which says that points of small arithmetic degree are not Zariski dense when f is a morphism. However, as the authors observe, their conjecture is only for morphisms, since it may fail for dominant rational maps. The authors of [51] prove the SAND conjecture for many of the cases listed in Example 8.13.

Conjecture 8.10 is a relatively coarse estimate for the height growth of points in Zariski-dense orbits. An affirmative answer to the following question would yield a quantitative version of the conjecture.

---

37 A morphism $f : X \to X$ is *int-amplified* if there exists an ample Cartier divisor H such that $f^*H - H$ is also ample.

Question 8.15 ([10, QUESTION 14.5]). Let $f : X \dashrightarrow X$ be a dominant rational map defined over K, and let $P \in X_f(\bar{K})$ be a point whose orbit $\mathcal{O}_f(P)$ is Zariski dense in X. Do there exist (integers) $0 \le \ell_f \le N$ and $k_f \ge 0$ such that

$$h\big(f^n(P)\big) \asymp \delta_f^n \cdot n^{\ell_f} \cdot (\log n)^{k_f},$$

where the implied constants depend on f and P, but not on n? If $\delta_f > 1$, is it further true that $k_f = 0$?

ACKNOWLEDGMENTS

The author would like to thank Rob Benedetto, John Doyle, Nicole Looper, Michelle Manes, and Yohsuke Matsuzawa for their helpful comments, corrections, and suggestions.

FUNDING

This work was partially supported by Simons Collaboration Grant #712332.

REFERENCES

[1] E. Amerik, Existence of non-preperiodic algebraic points for a rational self-map of infinite order. *Math. Res. Lett.* **18** (2011), no. 2, 251–256.

[2] V. I. Arnol'd, Dynamics of complexity of intersections. *Bol. Soc. Bras. Mat. (N. S.)* **21** (1990), no. 1, 1–10.

[3] M. Baker and L. DeMarco, Preperiodic points and unlikely intersections. *Duke Math. J.* **159** (2011), no. 1, 1–29.

[4] M. Baker and L. De Marco, Special curves and postcritically finite polynomials. *Forum Math. Pi* **1** (2013), e3, 35 pp.

[5] J. Bell, J. Diller, M. Jonsson, and H. Krieger, Birational mpas with transcendental dynamical degree. 2021, arXiv:2107.04113.

[6] J. P. Bell, J. Diller, and M. Jonsson, A transcendental dynamical degree. *Acta Math.* **225** (2020), no. 2, 193–225.

[7] J. P. Bell, D. Ghioca, and T. J. Tucker, The dynamical Mordell–Lang problem for étale maps. *Amer. J. Math.* **132** (2010), no. 6, 1655–1675.

[8] J. P. Bell, D. Ghioca, and T. J. Tucker, *The dynamical Mordell–Lang conjecture*. Math. Surveys Monogr. 210, American Mathematical Society, Providence, RI, 2016.

[9] M. P. Bellon and C.-M. Viallet, Algebraic entropy. *Comm. Math. Phys.* **204** (1999), no. 2, 425–437.

[10] R. Benedetto, P. Ingram, R. Jones, M. Manes, J. H. Silverman, and T. J. Tucker, Current trends and open problems in arithmetic dynamics. *Bull. Amer. Math. Soc. (N.S.)* **56** (2019), no. 4, 611–685.

[11] R. L. Benedetto, D. Ghioca, P. Kurlberg, and T. J. Tucker, A case of the dynamical Mordell–Lang conjecture. *Math. Ann.* **352** (2012), no. 1, 1–26.

[12] J. Blanc, J. K. Canci, and N. D. Elkies, Moduli spaces of quadratic rational maps with a marked periodic point of small order. *Int. Math. Res. Not. IMRN* **23** (2015), 12459–12489.

[13] T. Bousch, *Sur quelques problèmes de dynamique holomorphe*. Ph.D. thesis, Université de Paris-Sud, Centre d'Orsay, 1992.

[14] P. Corvaja, D. Ghioca, T. Scanlon, and U. Zannier, The dynamical Mordell–Lang conjecture for endomorphisms of semiabelian varieties defined over fields of positive characteristic. *J. Inst. Math. Jussieu* **20** (2021), no. 2, 669–698.

[15] L. DeMarco, Bifurcations, intersections, and heights. *Algebra Number Theory* **10** (2016), no. 5, 1031–1056.

[16] T.-C. Dinh and N. Sibony, Une borne supérieure pour l'entropie topologique d'une application rationnelle. *Ann. of Math. (2)* **161** (2005), no. 3, 1637–1644.

[17] J. R. Doyle, H. Krieger, A. Obus, R. Pries, S. Rubinstein-Salzedo, and L. West, Reduction of dynatomic curves. *Ergodic Theory Dynam. Systems* **39** (2019), no. 10, 2717–2768.

[18] J. R. Doyle and B. Poonen, Gonality of dynatomic curves and strong uniform boundedness of preperiodic points. *Compos. Math.* **156** (2020), no. 4, 733–743.

[19] J. R. Doyle and J. H. Silverman, A uniform field-of-definition/field-of-moduli bound for dynamical systems on $\mathbb{P}^N$. *J. Number Theory* **195** (2019), 1–22.

[20] J. R. Doyle and J. H. Silverman, Moduli spaces for dynamical systems with portraits. *Illinois J. Math.* **64** (2020), no. 3, 375–465.

[21] N. Fakhruddin, Boundedness results for periodic points on algebraic varieties. *Proc. Indian Acad. Sci. Math. Sci.* **111** (2001), no. 2, 173–178.

[22] G. Faltings, Endlichkeitssätze für abelsche Varietäten über Zahlkörpern. *Invent. Math.* **73** (1983), no. 3, 349–366.

[23] G. Faltings, Diophantine approximation on abelian varieties. *Ann. of Math. (2)* **133** (1991), 349–366.

[24] C. Favre and T. Gauthier, The arithmetic of polynomial dynamical pairs. 2020, arXiv:2004.13801.

[25] E. V. Flynn, B. Poonen, and E. F. Schaefer, Cycles of quadratic polynomials and rational points on a genus-2 curve. *Duke Math. J.* **90** (1997), no. 3, 435–463.

[26] D. Ghioca, The dynamical Mordell–Lang conjecture in positive characteristic. *Trans. Amer. Math. Soc.* **371** (2019), no. 2, 1151–1167.

[27] D. Ghioca, L.-C. Hsia, and T. J. Tucker, Preperiodic points for families of polynomials. *Algebra Number Theory* **7** (2013), no. 3, 701–732.

[28] D. Ghioca, H. Krieger, K. D. Nguyen, and H. Ye, The dynamical André–Oort conjecture: unicritical polynomials. *Duke Math. J.* **166** (2017), no. 1, 1–25.

[29] D. Ghioca and T. J. Tucker, Periodic points, linearizing maps, and the dynamical Mordell–Lang problem. *J. Number Theory* **129** (2009), no. 6, 1392–1403.

[30] D. Ghioca, T. J. Tucker, and S. Zhang, Towards a dynamical Manin–Mumford conjecture. *Int. Math. Res. Not. IMRN* **22** (2011), 5109–5122.

[31] D. Ghioca and J. Xie, The dynamical Mordell–Lang conjecture for skew-linear self-maps. Appendix by Michael Wibmer. *Int. Math. Res. Not. IMRN* **21** (2020), 7433–7453.

[32] R. A. Hidalgo, A simple remark on the field of moduli of rational maps. *Q. J. Math.* **65** (2014), no. 2, 627–635.

[33] B. Hutz, Effectivity of dynatomic cycles for morphisms of projective varieties using deformation theory. *Proc. Amer. Math. Soc.* **140** (2012), no. 10, 3507–3514.

[34] P. Ingram, R. Ramadas, and J. H. Silverman, Post-critically finite maps in $\mathbb{P}^n$ for $n \geq 2$ are sparse. 2019, arXiv:1910.11290.

[35] J. Jia, J. Xie, and D.-Q. Zhang, Surjective endomorphisms of projective surfaces – the existence of infinitely many dense orbits. 2020, arXiv:2005.03628.

[36] B. Kadets, Large arboreal Galois representations. *J. Number Theory* **210** (2020), 416–430.

[37] S. Kamienny, Torsion points on elliptic curves and q-coefficients of modular forms. *Invent. Math.* **109** (1992), no. 2, 221–229.

[38] S. Kawaguchi and J. H. Silverman, Examples of dynamical degree equals arithmetic degree. *Michigan Math. J.* **63** (2014), no. 1, 41–63.

[39] S. Kawaguchi and J. H. Silverman, Dynamical canonical heights for Jordan blocks, arithmetic degrees of orbits, and nef canonical heights on abelian varieties. *Trans. Amer. Math. Soc.* **368** (2016), no. 7, 5009–5035.

[40] S. Kawaguchi and J. H. Silverman, On the dynamical and arithmetic degrees of rational self-maps of algebraic varieties. *J. Reine Angew. Math.* **713** (2016), 21–48.

[41] E. Lau and D. Schleicher, Internal addresses in the Mandelbrot set and irreducibility of polynomials. Tech. Rep. 1994/19, 1994.

[42] C. G. Lee and J. H. Silverman, GIT stability of Hénon maps. *Proc. Amer. Math. Soc.* **148** (2020), no. 10, 4263–4272.

[43] J. Lesieutre and M. Satriano, Canonical heights on hyper-Kähler varieties and the Kawaguchi–Silverman conjecture. *Int. Math. Res. Not. IMRN* **10** (2021), 7677–7714.

[44] A. Levy, The space of morphisms on projective space. *Acta Arith.* **146** (2011), no. 1, 13–31.

[45] N. Looper, Dynamical Galois groups of trinomials and Odoni's conjecture. *Bull. Lond. Math. Soc.* **51** (2019), no. 2, 278–292.

[46] N. Looper, The uniform boundedness and dynamical Lang conjectures for polynomials. 2021, arXiv:2105.05240.

[47] N. R. Looper, Dynamical uniform boundedness and the abc-conjecture. *Invent. Math.* **225** (2021), no. 1, 1–44.

[48] M. Manes, Moduli spaces for families of rational maps on $\mathbb{P}^1$. *J. Number Theory* **129** (2009), no. 7, 1623–1663.

[49] Y. Matsuzawa, Kawaguchi–Silverman conjecture for endomorphisms on several classes of varieties. *Adv. Math.* **366** (2020), 107086.

[50] Y. Matsuzawa, On upper bounds of arithmetic degrees. *Amer. J. Math.* **142** (2020), no. 6, 1797–1820.

[51] Y. Matsuzawa, S. Meng, T. Shibata, and D.-Q. Zhang, Non-density of points of small arithmetic degrees. 2020, arXiv:2002.10976.

[52] Y. Matsuzawa and K. Sano, Arithmetic and dynamical degrees of self-morphisms of semi-abelian varieties. *Ergodic Theory Dynam. Systems* **40** (2020), no. 6, 1655–1672.

[53] Y. Matsuzawa, K. Sano, and T. Shibata, Arithmetic degrees and dynamical degrees of endomorphisms on surfaces. *Algebra Number Theory* **12** (2018), no. 7, 1635–1657.

[54] B. Mazur, Modular curves and the Eisenstein ideal. *Publ. Math. Inst. Hautes Études Sci.* **47** (1977), 33–186.

[55] C. T. McMullen, Families of rational maps and iterative root-finding algorithms. *Ann. of Math. (2)* **125** (1987), no. 3, 467–493.

[56] M. McQuillan, Division points on semi-abelian varieties. *Invent. Math.* **120** (1995), no. 1, 143–159.

[57] S. Meng and D.-Q. Zhang, Kawaguchi–Silverman conjecture for surjective endomorphisms. 2019, arXiv:1908.01605.

[58] L. Merel, Bornes pour la torsion des courbes elliptiques sur les corps de nombres. *Invent. Math.* **124** (1996), no. 1–3, 437–449.

[59] J. Milnor, Geometry and dynamics of quadratic rational maps. *Exp. Math.* **2** (1993), no. 1, 37–83.

[60] P. Morton, Arithmetic properties of periodic points of quadratic maps. *Acta Arith.* **62** (1992), no. 4, 343–372.

[61] P. Morton, On certain algebraic curves related to polynomial maps. *Compos. Math.* **103** (1996), no. 3, 319–350.

[62] P. Morton and J. H. Silverman, Rational periodic points of rational functions. *Int. Math. Res. Not.* **2** (1994), 97–110.

[63] D. Mumford, J. Fogarty, and F. Kirwan, *Geometric Invariant Theory*, third edn. Ergeb. Math. Grenzgeb. (2) [Results Math. Relat. Areas (2)] 34, Springer, Berlin, 1994.

[64] W. Narkiewicz, On a class of monic binomials. *Proc. Steklov Inst. Math.* **280** (2013), no. suppl. 2, S65–S70.

[65] R. W. K. Odoni, The Galois theory of iterates and composites of polynomials. *Proc. Lond. Math. Soc. (3)* **51** (1985), no. 3, 385–414.

[66] R. W. K. Odoni, On the prime divisors of the sequence $w_{n+1} = 1 + w_1 \cdots w_n$. *J. Lond. Math. Soc. (2)* **32** (1985), no. 1, 1–11.

[67] R. W. K. Odoni, Realising wreath products of cyclic groups as Galois groups. *Mathematika* **35** (1988), no. 1, 101–113.

[68] C. Panraksa, Rational periodic points of $x^d + c$ and abc conjecture. 2021, arXiv:2105.03715.

[69] C. Petsche, L. Szpiro, and M. Tepper, Isotriviality is equivalent to potential good reduction for endomorphisms of $\mathbb{P}^N$ over function fields. *J. Algebra* **322** (2009), no. 9, 3345–3365.

[70] J. Pila, o-minimality and the André–Oort conjecture for $\mathbb{C}^n$. *Ann. of Math. (2)* **173** (2011), 1779–1840.

[71] J. Pila and J. Tsimerman, Ax-Lindemann for $\mathcal{A}_g$. *Ann. of Math. (2)* **179** (2014), no. 2, 659–681.

[72] B. Poonen, Uniform boundedness of rational points and preperiodic points. 2012, arXiv:1206.7104.

[73] M. Raynaud, Courbes sur une variété abélienne et points de torsion. *Invent. Math.* **71** (1983), no. 1, 207–233.

[74] M. Raynaud, Sous-variétés d'une variété abélienne et points de torsion. In *Arithmetic and geometry, I*, pp. 327–352, Progr. Math. 35, Birkhäuser Boston, Boston, MA, 1983.

[75] A. Russakovskii and B. Shiffman, Value distribution for sequences of rational mappings and complex dynamics. *Indiana Univ. Math. J.* **46** (1997), no. 3, 897–932.

[76] D. Schleicher, Internal addresses of the Mandelbrot set and Galois groups of polynomials. *Arnold Math. J.* **3** (2017), no. 1, 1–35.

[77] J.-P. Serre, Propriétés galoisiennes des points d'ordre fini des courbes elliptiques. *Invent. Math.* **15** (1972), no. 4, 259–331.

[78] J.-P. Serre, *Abelian l-adic representations and elliptic curves*. Res. Notes Math. 7, A K Peters Ltd., Wellesley, MA, 1998.

[79] J. H. Silverman, Rational points on K3 surfaces: a new canonical height. *Invent. Math.* **105** (1991), no. 2, 347–373.

[80] J. H. Silverman, The space of rational maps on $\mathbb{P}^1$. *Duke Math. J.* **94** (1998), no. 1, 41–77.

[81] J. H. Silverman, Height estimates for equidimensional dominant rational maps. *J. Ramanujan Math. Soc.* **26** (2011), no. 2, 145–163.

[82] J. H. Silverman, *Moduli Spaces and Arithmetic Dynamics*. CRM Monogr. Ser. 30, American Mathematical Society, Providence, RI, 2012.

[83] J. H. Silverman, Dynamical degree, arithmetic entropy, and canonical heights for dominant rational self-maps of projective space. *Ergodic Theory Dynam. Systems* **34** (2014), no. 2, 647–678.

[84] J. H. Silverman, Arithmetic and dynamical degrees on abelian varieties. *J. Théor. Nombres Bordeaux* **29** (2017), no. 1, 151–167.

[85] J. Specter, Polynomials with surjective arboreal Galois representations exist in every degree. 2018, arXiv:1803.00434.

[86] M. Stoll, Rational 6-cycles under iteration of quadratic polynomials. *LMS J. Comput. Math.* **11** (2008), 367–380.

[87] Y.-S. Tai, On the Kodaira dimension of the moduli space of abelian varieties. *Invent. Math.* **68** (1982), no. 3, 425–439.

[88] J. Tsimerman, The André–Oort conjecture for $\mathcal{A}_g$. *Ann. of Math. (2)* **187** (2018), no. 2, 379–390.

[89] E. Ullmo and A. Yafaev, Galois orbits and equidistribution of special subvarieties: towards the André–Oort conjecture. *Ann. of Math. (2)* **180** (2014), no. 3, 823–865.

[90] P. Vojta, Siegel's theorem in the compact case. *Ann. of Math. (2)* **133** (1991), no. 3, 509–548.

[91] R. Walde and P. Russo, Rational periodic points of the quadratic function $Q_c(x) = x^2 + c$. *Amer. Math. Monthly* **101** (1994), no. 4, 318–331.

[92] J. Xie, The dynamical Mordell–Lang conjecture for polynomial endomorphisms of the affine plane. *Astérisque* (2017), no. 394, vi+110 pp.

[93] J. Xie, The existence of Zariski dense orbits for polynomial endomorphisms of the affine plane. *Compos. Math.* **153** (2017), no. 8, 1658–1672.

[94] J. Xie, Remarks on algebraic dynamics in positive characteristic. 2021, arXiv:2107.03559.

JOSEPH H. SILVERMAN

Mathematics Department, Brown University, Box 1917, Providence, RI 02912, USA, joseph_silverman@brown.edu

3. NUMBER THEORY

RELATIVE TRACE FORMULAE AND THE GAN–GROSS–PRASAD CONJECTURES

RAPHAËL BEUZART-PLESSIS

ABSTRACT

This paper reports on some recent progress that have been made on the so-called Gan–Gross–Prasad conjectures through the use of relative trace formulae. In their global aspects, these conjectures, as well as certain refinements first proposed by Ichino–Ikeda, give precise relations between the central values of some higher-rank L-functions and automorphic periods. There are also local counterparts describing branching laws between representations of classical groups. In both cases, approaches through relative trace formulae have shown to be very successful and have even lead to complete proofs, at least in the case of unitary groups. However, the works leading to these definite results have also been the occasion to develop further and gain new insights on these fundamental tools of the still emerging relative Langlands program.

MATHEMATICS SUBJECT CLASSIFICATION 2020

Primary 11F70; Secondary 22E50, 22E55, 11F72

KEYWORDS

Gan-Gross-Prasad conjecture, relative trace formula, periods of automorphic forms, L-functions, fundamental lemma, endoscopy

INTERNATIONAL CONGRESS OF MATHEMATICIANS 2022 JULY 6–14

Proc. Int. Cong. Math. 2022, Vol. 3, pp. 1712–1743
DOI 10.4171/ICM2022/182

Published by EMS Press

In broad terms, the Gan–Gross–Prasad conjectures concern two interrelated questions in the fields of representation theory and automorphic forms. On the one hand, these conjectures predict highly-sophisticated descriptions of some branching laws between representations of classical groups (that is, orthogonal, symplectic/metaplectic, or unitary groups) over local fields which can be seen as direct descendants of classical results of H. Weyl on similar branching problems for compact Lie groups. The predictions are given in terms of the recently established local Langlands correspondence for these groups that provides a parameterization of the irreducible representations in terms of data of arithmetic nature. On the other hand, the Gan–Gross–Prasad conjectures also give far-reaching higher-rank generalizations of certain celebrated relations between special values of L-functions and period integrals. We start this paper by discussing two, by now well-known, examples of the former kind of relations.

First, we briefly review Hecke's integral representation for L-functions of modular forms. Let $S_2(\Gamma_1(N))$ be the space of cuspidal modular form of weight 2 for the group

$$\Gamma_1(N) := \left\{ \gamma \in \mathrm{SL}_2(\mathbb{Z}) \mid \gamma \equiv \begin{pmatrix} 1 & \star \\ 0 & 1 \end{pmatrix} \bmod N \right\}.$$

It consists in the holomorphic functions $f : \mathbb{H} \to \mathbb{C}$, where $\mathbb{H} = \{x + iy \mid x, y \in \mathbb{R}, y > 0\}$ is Poincaré upper half-plane, satisfying the functional equation

$$f\left(\frac{az+b}{cz+d}\right) = (cz+d)^2 f(z), \quad \forall \begin{pmatrix} a & b \\ c & d \end{pmatrix} \in \Gamma_1(N) \tag{0.1}$$

and that are "vanishing at the cusps," a condition imposing in some sense means that f is rapidly decreasing modulo the above symmetries. Another more geometric way to describe $S_2(\Gamma_1(N))$ is as a space of holomorphic differential forms: for $f \in S_2(\Gamma_1(N))$, the form $\omega_f = f(z)dz$ descends to the open modular curve $Y_1(N) = \Gamma_1(N)\backslash\mathbb{H}$ (a Riemann surface as soon as $N > 3$) and the vanishing at the cusps condition translates to the fact that this form extends holomorphically to the canonical compactification $X_1(N)$ of $Y_1(N)$. Moreover, the map $f \mapsto \omega_f$ yields an isomorphism $S_2(\Gamma_1(N)) \simeq \Omega^1(X_0(N))$.

It follows from the functional equation (0.1) that every $f \in S_2(\Gamma_1(N))$ is periodic of period 1 and thus admits a Fourier expansion

$$f = \sum_{n \geqslant 1} a_n q^n, \quad q = e^{2i\pi z}, \tag{0.2}$$

where the fact that the sum is restricted to positive integers is part of the assumption that f vanishes at the cusps. The Hecke L-function of f is then defined as the Dirichlet series

$$L(s, f) = \sum_{n \geqslant 1} \frac{a_n}{n^s},$$

converging absolutely in the range $\Re(s) > 2$. Hecke has shown that this can also essentially be expressed as the Mellin transform of the restriction of f to the imaginary line,

$$(2\pi)^{-s}\Gamma(s)L(s, f) = \int_0^\infty f(iy)y^{s-1}dy. \tag{0.3}$$

This formula implies at once two essential analytic properties of $L(s, f)$: its analytic continuation to the complex plane and a functional equation of the form $s \leftrightarrow 2-s$. Moreover, it also has interesting arithmetic content: when specialized to the central value $s = 1$ and combined with a theorem of Drinfeld and Manin, it allows showing that the ratio between the central value of the L-function of a (modular) elliptic curve and its (unique) real period is always rational as predicted by a refinement of the Birch–Swinnerton-Dyer conjecture.

The above formula of Hecke can be reformulated (and slightly generalized) in the language of adelic groups and automorphic forms as follows. Let π be a cuspidal automorphic representation of $\mathrm{PGL}_2(\mathbb{A})$, where $\mathbb{A} = \mathbb{R} \times \prod_p' \mathbb{Q}_p$ denotes the adele ring of the rationals. This roughly means that π is an irreducible representation realized in a space of smooth and rapidly decreasing functions on $\mathrm{PGL}_2(\mathbb{Q})\backslash \mathrm{PGL}_2(\mathbb{A})$. Then, for every $\varphi \in \pi$ we have an identity of the following shape:

$$\int_{A(\mathbb{Q})\backslash A(\mathbb{A})} \varphi(a)da \sim L\left(\frac{1}{2}, \pi\right), \tag{0.4}$$

where $A = \begin{pmatrix} \star & \\ & \star \end{pmatrix}$ is the standard split torus in PGL_2 and $L(s, \pi)$ is the L-function of π, a particular instance of the notion of automorphic L-functions defined by Langlands. For specific π's, this recovers Hecke's formula (0.3) for $s = 1$, although $L(s, \pi)$ then coincides with the L-functions of a modular form only up to a renormalization that moves its center of symmetry to $1/2$. Moreover, the $\sim$ sign means that the equality only holds up to other, arguably more elementary, multiplicative factors.

Let $E/\mathbb{Q}$ be a quadratic extension. In the 1980s, Waldspurger [46] has established another striking formula for the central value of the base-change L-function

$$L(s, \pi_E) = L(s, \pi)L(s, \pi \otimes \chi_E)$$

where $\chi_E : \mathbb{A}^\times/\mathbb{Q}^\times$ is the idele class character associated to the extension $E/\mathbb{Q}$. Waldspurger's formula roughly takes the following shape:

$$\left| \int_{T(\mathbb{Q})\backslash T(\mathbb{A})} \varphi(t)dt \right|^2 \sim L\left(\frac{1}{2}, \pi_E\right) \tag{0.5}$$

for $\varphi \in \pi$, where T is a torus in PGL_2 isomorphic to $\mathrm{Res}_{E/F}(\mathbb{G}_m)/\mathbb{G}_m$ ($\mathrm{Res}_{E/F}$ denoting Weil's restriction of scalars). This result has lead in the subsequent years to striking arithmetic applications such as to the Birch–Swinnerton-Dyer conjecture or to p-adic L-functions.

Although of a similar shape, the two formulas (0.4) and (0.5) also have important differences, e.g., the left-hand side of (0.5) is typically far more algebraic in nature, and indeed sometimes just reduces to a finite sum, whereas the formula (0.4) can be deformed to all complex number s, giving an integral representation of the L-function $L(s, \pi)$ as Hecke's original formula, and typically carries information that is more transcendental.

The left-hand sides of (0.4) and (0.5) are particular instances of *automorphic periods* that can be informally defined as the integral of an automorphic form over a subgroup. We can also consider these two period integrals in a more representation-theoretic way as giving explicit $A(\mathbb{A})$- or $T(\mathbb{A})$-invariant linear forms on π. This point of view rapidly leads to a local related problem which, given a place v of $\mathbb{Q}$, aims to describe the irreducible representations of $\mathrm{PGL}_2(\mathbb{Q}_v)$ admitting a nonzero $A(\mathbb{Q}_v)$- or $T(\mathbb{Q}_v)$-invariant linear form. It turns out that for the torus A the answer is always positive except for some degenerate one-dimensional representations. On the other hand, the answer for the torus T is far more subtle and involves local ε-factors as shown by Tunnell and Saito **[44]**.

A natural generalization of Hecke's formula (0.4) is given by the theory of so-called Rankin–Selberg convolutions as developed by Jacquet Piatetski-Shapiro and Shalika **[31]**. On the other hand, the Gan–Gross–Prasad conjectures **[23]** aim to give far-reaching higher-rank generalizations of the above result of Waldspurger as well as of the theorem of Tunnell–Saito.

There has been a lot of progress on these conjectures, as well as some refinements thereof, in recent years, in particular in the case of unitary groups. In this paper, we will survey some of these developments with a particular emphasis on the use of (various forms of) *relative trace formulae*. Actually, a point I will try to advocate here is that the long journey towards the Gan–Gross–Prasad conjectures was also the occasion to develop and discover new features of these trace formulae.

The content is roughly divided as follows. In the first section, we review the local conjectures of Gan–Gross–Prasad and discuss their proofs in some cases based on some local trace formulae. Then, in Section 2, we introduce the global conjectures for unitary groups, as well as their refinements by Ichino–Ikeda, and describe an approach to both of them through a comparison of global relative trace formulae proposed by Jacquet and Rallis. The next two sections, Sections 3 and 4, are devoted to explaining the various ingredients needed to carry out this comparison effectively. In the final Section 5, we offer few thoughts about possible future developments.

1. THE LOCAL CONJECTURES AND MULTIPLICITY FORMULAE

1.1. The branching problem

Let F be a local field (of any characteristic) and E be either a separable quadratic extension of F or F itself. In the case where $[E:F]=2$, we let c denote the nontrivial F-automorphism of E and otherwise, to obtain uniform notation, we simply set $c=1$. Let V be a Hermitian or quadratic space over E i.e. a finite dimensional E-vector space equipped with a nondegenerate c-sesquilinear form

$$h: V \times V \to E$$

satisfying $h(v,w)=h(w,v)^c$ for every $v,w \in V$. Let $W \subset V$ be a nondegenerate subspace and let $U(V)$ (resp. $U(W)$) be the group of E-linear automorphisms $g \in \mathrm{GL}_E(V)$ (resp. $g \in \mathrm{GL}_E(W)$) that preserve the form h and are of determinant one when $E=F$. In other words, $U(V)$, $U(W)$ are the unitary groups associated of the Hermitian spaces V, W when $[E:F]=2$ and the special orthogonal groups of the quadratic spaces V, W when $E=F$. Note that there is a natural embedding $U(W) \hookrightarrow U(V)$ given by extending the action of $g \in U(W)$ trivially on the orthogonal complement $Z=W^{\perp}$ of W in V. We assume that

$$Z \text{ is odd-dimensional and } U(Z) \text{ is quasisplit.} \tag{1.1}$$

Concretely, this means that there exists a basis $(z_i)_{-r\leqslant i\leqslant r}$ of Z and $\nu\in F^\times$ such that $h(z_i,z_j)=\nu\delta_{i,-j}$ for $-r\leqslant i,j\leqslant r$. Let $N\subset U(V)$ be the unipotent radical of a parabolic subgroup $P\subset U(V)$ stabilizing a maximal flag of isotropic subspaces in Z, e.g., with a basis as before, we can take the flag $Ez_r\subset Ez_r\oplus Ez_{r-1}\subset\cdots\subset Ez_r\oplus\cdots\oplus Ez_1$. Then, $U(W)$ normalizes N and Gan–Gross–Prasad construct a certain conjugacy class of $U(W)$-invariant characters $\xi:N\to\mathbb{C}^\times$. Concretely, we can take

$$\xi(u)=\psi\left(\sum_{i=0}^{r-1}h(uz_i,z_{-i-1})\right),\quad u\in N,$$

where $\psi:F\to\mathbb{C}^\times$ is a nontrivial character.

The local GGP conjectures roughly address the following branching problems: for smooth irreducible complex representations (π,V_π) and (σ,V_σ) of $U(V)$ and $U(W)$ respectively, determine the dimension (also called *multiplicity*) of the following intertwining space:

$$m(\pi,\sigma)=\dim\operatorname{Hom}_{U(W)\ltimes N}(V_\pi,V_\sigma\otimes\xi).\tag{1.2}$$

Here, when F is Archimedean by a *smooth* representation we actually mean an admissible smooth Fréchet representation of moderate growth in the sense of Casselman–Wallach **[19]**. Moreover, in this case V_π, V_σ are Fréchet spaces and by definition $\operatorname{Hom}_{U(W)\ltimes N}(V_\pi,V_\sigma\otimes\xi)$ only consists in the *continuous* $U(W)\ltimes N$-equivariant intertwining maps.

By deep theorems of Aizenbud–Gourevitch–Rallis–Schiffmann **[2]** in the p-adic case and Sun–Zhu **[42]** in the Archimedean case, the branching multiplicity $m(\pi,\sigma)$ is known to always be at most 1 (at least when F is of characteristic 0, see **[37]** for the case of positive characteristic). Thus, the question reduces to determine when $m(\pi,\sigma)$ is nonzero.

Gan, Gross, and Prasad formulated a precise answer to this question, under some restrictions on the representations π and σ, based on the so-called *Langlands correspondences* for the groups $U(V)$ and $U(W)$. More precisely, these give ways to parametrize smooth irreducible representations of those groups in terms of *L-parameters* which are certain kind of morphisms

$$\phi:\mathcal{L}_F\to {}^LU(V)\quad\text{or}\quad {}^LU(W)$$

from a group $\mathcal{L}_F$ which can be taken to be either the Weil group W_F (in the Archimedean case) or a product $W_F\times \mathrm{SL}_2(\mathbb{C})$ (in the non-Archimedean case) to a semidirect product ${}^LU(V)=\widehat{U(V)}\rtimes W_F$ or ${}^LU(W)=\widehat{U(W)}\rtimes W_F$ known as the *L-group*. In the cases at hand, the connected components $\widehat{U(V)}$ and $\widehat{U(W)}$ turn out to be either complex general linear groups (in the unitary case) or complex special orthogonal/symplectic groups (in the orthogonal case) and the relevant sets of L-parameters can be more concretely described as sets of complex representations of $\mathcal{L}_E$ of fixed dimension and satisfying certain properties of (conjugate-)self-duality. We refer the reader to **[23, §8]** for details and content ourself to briefly sketch this alternative description for unitary groups: the L-parameters for $U(V)$ can be equivalently described as isomorphism classes of $n=\dim(V)$-dimensional complex semisimple representations $\phi:\mathcal{L}_E\to\mathrm{GL}(M)$ which are conjugate-self-dual of sign $(-1)^{n-1}$. Here, ϕ is said to be *conjugate-self-dual* if there is an isomorphism $T:M\to M^{\vee\sigma}$

with its conjugate-contragredient $\phi^{\vee\sigma} : \mathcal{L}_E \to \mathrm{GL}(M^{\vee\sigma})$ obtained by twisting the contragredient by any chosen lift $\sigma \in \mathcal{L}_F \setminus \mathcal{L}_E$ of c and it is, moreover, said to be of sign $\varepsilon \in \{\pm\}$ if the isomorphism T can be chosen so that ${}^tT\phi(\sigma^2) = \varepsilon T$. Besides these L-parameters ϕ, the local Langlands correspondence is also supposed to associate to irreducible representations irreducible characters of the finite group of components

$$S_\phi = \pi_0\big(\mathrm{Cent}_{\widehat{U(V)}}(\phi)\big)$$

of the centralizer of the image of ϕ in $\widehat{U(V)}$. For the group considered here, S_ϕ is always a 2-group and moreover, once again, it also admits a more concrete description, e.g., if $U(V)$ is a unitary group and we identify ϕ with a $(-1)^{n-1}$-conjugate-self-dual representation of $\mathcal{L}_E$ as before, this can be decomposed into irreducible representations as follows:

$$\phi = \bigoplus_{i\in I} n_i\phi_i \bigoplus_{j\in J} m_j\phi_j \bigoplus_{k\in K} l_k(\phi_k \oplus \phi_k^{\vee\sigma}) \tag{1.3}$$

where the ϕ_i's (resp. ϕ_j's) are irreducible conjugate-self-dual of the same sign $(-1)^{n-1}$ (resp. of opposite sign $(-1)^n$) whereas the ϕ_k's are irreducible but not conjugate-self-dual and using this decomposition we have

$$S_\phi = \bigoplus_{i\in I} \mathbb{Z}/2\mathbb{Z}e_i. \tag{1.4}$$

We are now ready to state a version of the local Langlands correspondence, including an essential refinement by Vogan **[45]**, necessary for the local Gan–Gross–Prasad conjecture. It turns out to be more easily described if we consider more than one group at the same time: besides $U(V)$ [1] itself, we need to consider its *pure inner forms* which here consist of the groups $U(V')$ where V' runs over the isomorphism classes of Hermitian/quadratic spaces of the same dimension as V and of the same discriminant in the orthogonal case. If F is non-Archimedean, and provided V is not an hyperbolic quadratic plane, there are always two such isomorphism classes of Hermitian/quadratic spaces and thus as many pure inner forms whereas if F is Archimedean, by their classification using signatures there are $\dim(V)+1$ (resp. $\frac{\dim(V)+1}{2}$ for $\dim(V)$ odd, $\frac{\dim(V)+\mathrm{disc}(V)+1}{2}$ for $\dim(V)$ even) pure inner forms in the unitary case (resp. orthogonal case). For such a pure inner form, let us denote by $\mathrm{Irr}(U(V'))$ the set of isomorphism classes of smooth irreducible representations of $U(V')$. Then, modulo the auxilliary choice of a quasisplit pure inner form $U(V')$ and a *Whittaker datum* on it[2] that we will suppress from the notation, the local Langlands correspondence posits the existence of a natural decomposition into finite sets called *L-packets*

$$\bigsqcup_{V'} \mathrm{Irr}\big(U(V')\big) = \bigsqcup_{\phi} \Pi(\phi),$$

where the left union runs over all pure inner forms whereas the right union is over all L-parameters $\phi : \mathcal{L}_F \to {}^LU(V)$ (the pure inner forms all share the same L-group) together

1 Of course, the following discussion also applies to $U(W)$.

2 A Whittaker datum of $U(V')$ is a pair (N,θ) consisting of a maximal unipotent subgroup $N \subset U(V')$ and a generic character $\theta : N \to \mathbb{C}^\times$. This datum only matters up to conjugacy.

with bijections

$$\begin{aligned} \Pi(\phi) &\simeq \widehat{S_\phi}, \\ \pi(\phi,\chi) &\leftarrow\!\!\mid \chi, \end{aligned} \tag{1.5}$$

with the character group $\widehat{S_\phi}$ of S_ϕ. Thus, in a sense the correspondence gives a way to parameterize the admissible duals of all the pure inner forms of $U(V)$ at the same time. However, there is a precise recipe for the characters $\widehat{S_\phi}^{V'}$ corresponding to the intersection $\Pi^{V'}(\phi) = \Pi(\phi) \cap \mathrm{Irr}(U(V'))$ and therefore this also induces a parameterization of the individual admissible duals $\mathrm{Irr}(U(V'))$. Moreover, the naturality condition can be made precise through the so-called *endoscopic relations* that characterize the Langlands parameterization uniquely in terms of the known correspondence for GL_n.[3] For real groups, the correspondence was constructed long ago by Langlands and is known to satisfy the endoscopic relations thanks to the work of Shelstad. In his monumental work [7], Arthur has established, among other things, the existence of this correspondence for quasisplit special orthogonal or symplectic p-adic groups (with an important technical caveat for even special orthogonal groups $\mathrm{SO}(2n)$ where the correspondence is only proven up to conjugation by the full orthogonal group $O(2n)$). This work was subsequently extended in [39] and [34] to include unitary groups (not necessarily quasisplit).

For the purpose of stating the local Gan–Gross–Prasad conjecture, we will also need to vary the two groups $U(V)$, $U(W)$. However, we will need these to vary in a compatible way in order for the multiplicity (1.2) to still be well-defined. More precisely, the *relevant* pure inner forms of $U(V) \times U(W)$ are defined by varying the small Hermitian/quadratic space W while keeping the orthogonal complement $Z = W^\perp$ fixed: these are the groups of the form $U(V') \times U(W')$ where W' is a Hermitian/quadratic space of the same dimension as W, and same discriminant in the orthogonal case, whereas V' is given by the orthogonal sum $V' = W' \oplus^\perp Z$. Since the orthogonal complement Z is the same, for each relevant pure inner form $U(V') \times U(W')$ we can define as before a multiplicity function $(\pi,\sigma) \in \mathrm{Irr}(U(V')) \times \mathrm{Irr}(U(W')) \mapsto m(\pi,\sigma)$.

We are now ready to formulate the local Gan–Gross–Prasad conjecture except for one technical but important detail: as alluded to above, the local Langlands correspondences, and more particularly the bijections (1.5), depend on the choice of Whittaker data on some pure inner forms of $U(V)$ and $U(W)$. Actually, it turns out that there exists a unique relevant pure inner form $U(V_{qs}) \times U(W_{qs})$ which is quasisplit and on which we can fix a Whittaker datum through the choice of a nontrivial character $\psi : E \to \mathbb{C}^\times$ that is, moreover, trivial for F in the unitary case (see [23, §12] for details). With these prerequisites in place, we can now state:

Conjecture 1.1 (Gan–Gross–Prasad). *Let $\phi : \mathcal{L}_F \to {}^L U(V)$ and $\phi' : \mathcal{L}_F \to {}^L U(W)$ be L-parameters for $U(V)$ and $U(W)$, respectively. Assume that the corresponding L-packets*

3 This situation is peculiar to classical groups because those can be realized as *twisted endoscopic groups* of some GL_N.

$\Pi(\phi)$, $\Pi(\phi')$ are generic, that is, they contain one representation which is generic with respect to each Whittaker datum. Then:

(1) *There exists a unique pair*

$$(\pi, \sigma) \in \bigsqcup_{W'} \Pi^{V'}(\phi) \times \Pi^{W'}(\phi'),$$

where the union runs over relevant pure inner forms, such that $m(\pi, \sigma) = 1$.

(2) *The unique characters $\chi \in \widehat{S_\phi}$ and $\chi' \in \widehat{S_{\phi'}}$ such that $\pi = \pi(\phi, \chi)$ and $\sigma = \pi(\phi', \chi')$ are given by explicit formulas involving local root numbers, e.g., in the unitary case, identifying ϕ, ϕ' with conjugate-self-dual representations of $\mathcal{L}_E$ and using the description (1.4) of $S_{\phi'}$ in terms of the decomposition (1.3), we have*

$$\chi(e_i) = \varepsilon(\phi_i \otimes \phi', \psi_{2\delta}), \quad \textit{for all } i \in I. \tag{1.6}$$

Here δ stands for the discriminant of the odd dimensional Hermitian space among (V_{qs}, W_{qs}), $\psi_{2\delta}(z) := \psi(2\delta z)$ and $\varepsilon(\phi_i \otimes \phi', \psi_{2\delta})$ denotes the local root number of the Weil or Weil–Deligne representation $\phi_i \otimes \phi'$ associated to this additive character [43].

When $(\dim(V), \dim(W)) = (3, 2)$ (quadratic case) or $(\dim(V), \dim(W)) = (2, 1)$ (Hermitian case), the above conjecture essentially reduces to the result of Tunnell and Saito [44] on restrictions of irreducible representations of GL(2) to a maximal torus mentioned in the introduction. There has been a lot of recent progress towards Conjecture 1.1 and here is the status of what is currently known in the characteristic zero case:

Theorem 1.1. *Assume that F is of characteristic 0. Then:*

(1) *Both (1) and (2) of Conjecture 1.1 hold true in the following cases: if V, W are Hermitian spaces (i.e., in the unitary case) or if these are quadratic spaces and F is p-adic.*

(2) *Conjecture 1.1 (1) is verified when V, W are quadratic spaces and F is Archimedean.*

The first real breakthrough on Conjecture 1.1 was made by Waldspurger who established in a stunning series of papers [38, 47–49], the last one in collaboration with Mœglin, the full conjecture for p-adic special orthogonal groups under the assumption that the local Langlands correspondence is known for those groups and have expected properties. In my PhD thesis [8–10], I extended the method to deal with p-adic unitary groups therefore obtaining the conjecture under the slightly weaker assumption that the parameters ϕ, ϕ' are *tempered* which means that the corresponding L-packets consist of tempered representations. The extension to generic L-packets was carried out in the appendix to [24] using crucially a result of Heiermann. Later, I revisited Waldspurger's method which is based on a novel sort of local trace formulae, putting it on firmer grounds, and in the monograph [12] I established part (1) of the conjecture (sometimes called the *multiplicity one property for L-packets*) for unitary groups over arbitrary fields of characteristic 0, thus reproving part of my thesis in the p-adic case, and still under the assumption that L-parameters are tempered which as we will see is quite natural from the method. In the meantime, H. He [28] has developed a different approach to the conjecture based on the local θ-correspondence and very special features of the representation theory of real unitary groups (in particular, this approach cannot deal with p-adic groups) which allowed him to prove the full conjecture for those groups whenever ϕ and ϕ' are *discrete* parameters (a stronger condition than being tempered). Recently, this technique was enhanced by H. Xue [54] who was able to show the conjecture for real unitary groups without any restriction. Finally, in the recent preprint [36] Z. Luo adapted my previous work to deal with real special orthogonal groups proving the multiplicity one property for tempered L-packets.

1.2. Approach through local trace formulae

Let me give more details on the general structure of the approach taken by Waldspurger which was clarified and then further refined in [12]. It is mainly based on one completely novel ingredient that is a formula expressing the multiplicity $m(\pi, \sigma)$ in terms of the Harish-Chandra characters of π and σ. To be more specific, we recall a deep result of Harish-Chandra asserting that the distribution-character of a smooth irreducible representation π, i.e. the distribution $f \in C_c^\infty(U(V)) \mapsto \operatorname{Trace} \pi(f)$, can be represented by a locally L^1 function Θ_π known as its Harish-Chandra character. The aforementioned formula gives an identity roughly of the form:

$$m(\pi, \sigma) = \int_{\Gamma(V,W)}^{\mathrm{reg}} c_\pi(x) c_\sigma(x^{-1}) dx, \tag{1.7}$$

where $\Gamma(V, W)$ is a certain set of semisimple conjugacy classes in $U(V)$ equipped with some measure dx reminiscent of Weyl integration formula (although it is more singular than measures coming from maximal tori, e.g., singular orbits are typically not negligible for dx), $c_\pi(x)$ and $c_\sigma(x^{-1})$ are renormalized values for the characters Θ_π and Θ_σ, respectively (although these characters are smooth on open dense subsets of regular semisimple elements, they typically blow up at the singular conjugacy classes in $\Gamma(V, W)$; the renormalization is based on further results of Harish-Chandra describing the local behavior of characters near singular elements), and finally the "reg" sign indicates that the integral itself has sometimes to be regularized in a certain way (or put another way, it is *improperly* convergent). Originally, formula (1.7) was only proven to hold for *tempered* representations but through the process of reducing the general conjecture to the tempered case, it was eventually shown a posteriori to hold for every irreducible representations belonging to generic L-packets. In the degenerate case where $U(V)$ is compact, the right-hand side of the integral formula (1.7) reduces to the L^2-scalar product of $\Theta_\pi \mid_{U(W)}$ and Θ_σ and the formula itself is an easy consequence of the orthogonality relations of characters, but in general the formula looks much more mysterious.

Let us sketch very briefly how we can deduce from formula (1.7) the first part of Conjecture 1.1 for tempered parameters (multiplicity one in tempered L-packets). The idea, due to Waldspurger, is to take advantage of inner cancellations in the sum

$$\sum_{W'}\sum_{(\pi,\sigma)} m(\pi,\sigma) = \sum_{W'} \sum_{(\pi,\sigma)\in\Pi^{V'}(\phi)\times\Pi^{W'}(\phi')} \int_{\Gamma(V',W')}^{\mathrm{reg}} c_\pi(x)c_\sigma(x^{-1})dx \tag{1.8}$$

that can be deduced from certain character relations (which are basic instances of the already mentioned endoscopic relations). The first step is to rewrite the sum as

$$\sum_{W'} \int_{\Gamma(V',W')}^{\mathrm{reg}} c_\phi^{V'}(x)c_{\phi'}^{W'}(x^{-1})dx \tag{1.9}$$

where $\Theta_\phi^{V'} = \sum_{\pi\in\Pi^{V'}(\phi)} \Theta_\pi$, $\Theta_{\phi'}^{W'} = \sum_{\sigma\in\Pi^{W'}(\phi')} \Theta_\sigma$ and $c_\phi^{V'}(x)$, $c_{\phi'}^{W'}(x^{-1})$ are renormalized values for those characters as before. The first property of the Langlands correspondence that we need is that $\Theta_\phi^{V'}$, $\Theta_{\phi'}^{W'}$ are *stable*, i.e., are constant on the union of semisimple regular conjugacy classes that become the same over an algebraic closure (a so-called regular stable conjugacy class). It follows from this stability property that the renormalized functions $c_\phi^{V'}$, $c_{\phi'}^{W'}$ are also invariant under a suitable extension of stable conjugation for singular elements. Consequently, the sum of multiplicities can be further rewritten as

$$\sum_{W'}\sum_{(\pi,\sigma)} m(\pi,\sigma) = \sum_{W'} \int_{\Gamma(V',W')/\mathrm{stab}}^{\mathrm{reg}} c_\phi^{V'}(y)c_{\phi'}^{W'}(y^{-1})dy, \tag{1.10}$$

where $\Gamma(V',W')/\mathrm{stab}$ stands for the space of stable conjugacy classes in $\Gamma(V',W')$. At this point, it is convenient to make the simplifying assumption that F is p-adic and W is not a split quadratic space of dimension $\leqslant 2$. Then, there are exactly two relevant pure inner forms $U(V)\times U(W)$ and $U(V')\times U(W')$ with, say, the first one quasisplit. Moreover, the character relations in this case read

$$\Theta_\phi^V(y) = \varepsilon_V \Theta_\phi^{V'}(y') \quad \left(\text{resp. } \Theta_{\phi'}^W(y) = \varepsilon_W \Theta_{\phi'}^{W'}(y')\right)$$

for certain signs $\varepsilon_V, \varepsilon_W \in \{\pm 1\}$ satisfying $\varepsilon_V\varepsilon_W = -1$ and for every regular stable conjugacy classes y, y' in $U(V)$, $U(V')$ (resp. in $U(W)$, $U(W')$) that are related by a certain correspondence (which is just an identity of characteristic polynomials except in the even orthogonal case). This correspondence actually naturally extends to give an embedding $\Gamma(V',W')/\mathrm{stab} \hookrightarrow \Gamma(V,W)/\mathrm{stab}$, $y' \mapsto y$, for which we have

$$c_\phi^V(y)c_{\phi'}^W(y) = -c_\phi^{V'}(y')c_{\phi'}^{W'}(y').$$

This implies that in the right-hand side of (1.10), all the terms indexed by $\Gamma(V',W')/\mathrm{stab}$ can be cancelled with the corresponding terms coming from their images in $\Gamma(V,W)/\mathrm{stab}$. The only remaining contribution, it turns out, is that of the trivial conjugacy class:

$$\sum_{W'}\sum_{(\pi,\sigma)} m(\pi,\sigma) = c_\phi^V(1)c_{\phi'}^W(1) \tag{1.11}$$

which, by a result of Rodier, can be interpreted as the number of representations in the packet $\Pi^V(\phi)\otimes\Pi^W(\phi')$ that are generic with respect to a certain Whittaker datum (actually really an average of such numbers over all Whittaker data in the unitary case). By a third property of tempered L-packets (existence and unicity of a generic representation for a given Whittaker datum), this number is just 1 and this immediately implies the first part of Conjecture 1.1.

The proof of the multiplicity formula (1.7), on the other hand, is much more involved. Set $G = U(W) \times U(V)$ and $H = U(W) \ltimes N$ that we see as a subgroup of G through the natural diagonal embedding. Then, following the approach that I have developed in [12], (1.7) can be deduced from a certain simple trace formula for the "space" $X = (H, \xi)\backslash G$. More precisely, this trace formula is roughly seeking to compute the trace of the convolution operators

$$\phi \in L^2(X, \xi) \mapsto (R(f)\phi)(x) = \int_G f(g)\phi(xg)dg, \quad \text{for } f \in C_c^\infty(G),$$

where $L^2(X, \xi)$ denotes the Hilbert space of measurable functions ϕ on G satisfying $\phi(hg) = \xi(h)\phi(g)$ for $(h, g) \in H \times G$ and $\int_{H\backslash G} |\phi(x)|^2 dx < \infty$. It is classical, and easy to see, that these operators are given by kernels,

$$\big(R(f)\phi\big)(x) = \int_X K_f(x, y)\phi(y)dy, \quad \text{for } (f, \phi) \in C_c^\infty(G) \times L^2(X, \xi),$$

where $K_f(x, y) = \int_H f(x^{-1}hy)\xi(h)dh$. Thus, at a formal level (hence the quotation marks) we have

$$\text{"Trace } R(f) = \int_X K_f(x, x)dx\text{"}.$$

However, neither of the two sides above make sense in general: the convolution operator is not of trace-class and the kernel not integrable over the diagonal. The basic idea is then to restrict oneself to a subspace of test functions for which at least one of the two expressions is meaningful. A convenient such subspace is that of *strongly cuspidal* functions introduced by Waldspurger in [47]: a function $f \in C_c^\infty(G)$ is strongly cuspidal if for every proper parabolic subgroup $P = MU \subsetneq G$, we have

$$\int_U f(mu)du = 0, \quad \forall m \in M.$$

Moreover, as is shown in [12], for $f \in C_c^\infty(G)$ strongly cuspidal, the integral

$$J(f) = \int_X K_f(x, x)dx$$

is absolutely convergent (the argument of [12] is given in the context of Gan–Gross–Prasad for unitary groups but it can be adapted to a much more general context). Then, the aforementioned simple local trace formula expands the distribution $f \to J(f)$ in two different ways:

Theorem 1.2. *For every strongly cuspidal $f \in C_c^\infty(G)$, we have the identities*

$$\int_{\Gamma(V,W)}^{\text{reg}} c_f(x)dx = J(f) = \int_{\mathcal{X}(G)} m(\Pi)\widehat{\theta_f}(\Pi)d\Pi, \tag{1.12}$$

where

- *$c_f(x)$ is the renormalized value of a function $x \mapsto \theta_f(x)$ constructed from weighted orbital integrals of f in the sense of Arthur [3] and whose local behavior is similar to that of Harish-Chandra characters on the group G;*

- *$\mathcal{X}(G)$ is a certain space of virtual representations of G obtained by parabolic induction from the so-called elliptic representations (as defined in* [6]*) of Levi subgroups and $f \mapsto \widehat{\theta}_f(\Pi)$ is a weighted character in the sense of Arthur* [4]*;*
- *Finally, for an irreducible representation $\Pi = \pi \otimes \sigma$ of G, $m(\Pi)$ is defined as the multiplicity $m(\pi, \sigma^\vee)$ with $\sigma^\vee$ the smooth contragredient of σ.*

We refer the reader to [12] for precise definitions of all the terms and a proof in the case of unitary groups. This was adapted in [36] to special orthogonal groups. The deduction of the integral formula (1.7) roughly goes as follows: we first show the multiplicity formula for representations that are properly parabolically induced by expressing both sides in terms of the inducing data and applying an induction hypothesis whereas for the remaining representations, the so-called elliptic representations, the formula can be obtained by applying the trace formula (1.12) to some sort of *pseudocoefficient*.

Finally, let us say a word on how the more refined part (2) of Conjecture 1.1 can be proven using this approach (so far it has only been done for p-adic groups in [49] and [9], following the previous slightly different method of Waldspurger, but there is little doubt that the techniques developed in [12] should allow to treat the case of real groups in a similar way). For Langlands parameters ϕ, ϕ' as in Conjecture 1.1, as well as characters $\chi \in \widehat{S_\phi}$, $\chi' \in \widehat{S_{\phi'}}$, combining the multiplicity formula (1.7) with the general endoscopic character relations that characterize the Langlands correspondences for $U(V)$ and $U(W)$, we can express $m(\pi(\phi, \chi), \sigma(\phi', \chi'))$ as a sum of integrals of (renormalized) twisted characters on some products $\mathrm{GL}_n(E) \times \mathrm{GL}_m(E)$. The remaining ingredient is to relate these integrals of twisted characters to the epsilon factors of pairs defined by Jacquet–Piatetski-Shapiro–Shalika in [31]. More precisely, these expressions involve the twisted characters of tempered irreducible representations $\pi^{\mathrm{GL}}, \sigma^{\mathrm{GL}}$ of general linear groups $\mathrm{GL}_n(E), \mathrm{GL}_m(E)$, with $n \geqslant m$ of distinct parities, which are self-dual (in the orthogonal case) or conjugate-self-dual (in the unitary case). These properties of (conjugate-)self-duality imply that π^{GL} and σ^{GL} extend to representations $\pi^{\mathrm{GL}}, \sigma^{\mathrm{GL}}$ of the nonconnected groups $\mathrm{GL}_n(E) \rtimes \langle \theta_n \rangle$ and $\mathrm{GL}_m(E) \rtimes \langle \theta_m \rangle$, respectively, where θ_k ($k = n, m$) denotes the automorphism $g \mapsto {}^t(g^c)^{-1}$. The twisted characters in question are then the restrictions $\Theta_{\pi^{\mathrm{GL}}}$ and $\Theta_{\sigma^{\mathrm{GL}}}$ of the Harish-Chandra characters of π^{GL} and σ^{GL} to the nonneutral components $\widetilde{\mathrm{GL}_n}(E) = \mathrm{GL}_n(E)\theta_n$ and $\widetilde{\mathrm{GL}_m}(E) = \mathrm{GL}_m(E)\theta_m$, respectively. Replacing the functions c_π, c_σ by similar suitable renormalizations of these twisted characters at singular semisimple conjugacy classes, there is a formula very analogous to (1.7) for the ε-factor of pair $\varepsilon(\pi^{\mathrm{GL}} \times \sigma^{\mathrm{GL}}, \psi)$.

For p-adic fields, this formula was established in [48] in the self-dual case and in [8] in the conjugate-self-dual case. The proof follows closely that of (1.7) and is based on a local trace formula very similar to that of Theorem 1.2 for the natural action of $G' := \widetilde{\mathrm{GL}_n}(E) \times \widetilde{\mathrm{GL}_m}(E)$ on the homogeneous space $X' = H' \backslash G'$ where $G' = \mathrm{GL}_n(E) \times \mathrm{GL}_m(E)$ and $H' = \mathrm{GL}_m(E) \ltimes N'$ is the semidirect product with a unipotent subgroup N' whose definition is analogous to that of N. More precisely, there is also a similar unitary character ξ' of N' that is $\mathrm{GL}_m(E)$-invariant and the twisted trace formula we are mentioning is roughly trying to compute the trace of convolution operators $R(f)$ of functions $f \in C_c^\infty(G')$ on

$L^2(X', \xi')$. Rather than describing it in details, let us just explain how the ε-factors show up in the analysis. As in Theorem 1.2, one of the main ingredient on the spectral side of this trace formula is a twisted multiplicity $m(\pi^{\mathrm{GL}} \otimes \sigma^{\mathrm{GL}})$ which computes the trace of a natural operator on the space of intertwiners

$$\mathrm{Hom}_H(\pi^{\mathrm{GL}} \otimes \sigma^{\mathrm{GL}}, \xi). \tag{1.13}$$

The operator in question is given by $\ell \mapsto \ell \circ (\pi^{\mathrm{GL}} \otimes \sigma^{\mathrm{GL}})(\theta)$ where $\theta \in \widetilde{\mathrm{GL}}_n(E) \times \widetilde{\mathrm{GL}}_m(E)$ is a certain element stabilizing the pair (H, ξ) (which is anyway needed to extend the right action of G' on $L^2(X', \xi')$ to an action of G'). Actually, it turns out that the space (1.13) is always one-dimensional and a reformulation of the so-called *local functional equation* from **[31]** shows that this operator is essentially acting (for suitable normalizations of π^{GL}, σ^{GL} and up to more elementary factors) as multiplication by the ε-factor $\varepsilon(\pi \times \sigma, \psi)$.

2. THE GLOBAL GAN–GROSS–PRASAD CONJECTURES AND ICHINO–IKEDA REFINEMENTS

2.1. Statements and results

We now move to a global setting. Let E/F be a quadratic extension of number fields and $W \subset V$ be Hermitian spaces over E satisfying condition (1.1) (there are similar, and actually prior, conjectures for orthogonal groups, but here we will concentrate on unitary groups for which much more is known). By a construction similar to that from the previous section, we may obtain from these data a triple (G, H, ξ) where $G = U(V) \times U(W)$, $H = U(W) \ltimes N$ is a subgroup of G (which we will this time consider as *algebraic groups* over F) and $\xi : N(\mathbb{A}_F) \to \mathbb{C}^\times$ is a character on the adelic points of N trivial on the subgroup $N(F)$ and that extends to a character of $H(\mathbb{A}_F)$ trivial on $U(W)(\mathbb{A}_F)$.

The global analog of the previous branching problem is that of characterizing the nonvanishing of the *automorphic period* associated to the pair (H, ξ). More precisely, if $\pi = \pi_V \otimes \pi_W \hookrightarrow \mathcal{A}_{\mathrm{cusp}}(G(F)\backslash G(\mathbb{A}_F))$ is a cuspidal automorphic representation of $G(\mathbb{A}_F)$, we consider the automorphic period

$$\mathcal{P}_{H,\xi} : \pi \to \mathbb{C},$$
$$\mathcal{P}_{H,\xi}(\varphi) = \int_{[H]} \varphi(h)\xi(h)dh, \tag{2.1}$$

where here and throughout the rest of the paper, for a linear algebraic group R over F, we denote by $[R] = R(F)\backslash R(\mathbb{A}_F)$ the corresponding automorphic quotient. On the other hand, let $\pi_E = \pi_{V,E} \otimes \pi_{W,E}$ be the (weak) base-change of π to $\mathrm{GL}_n(\mathbb{A}_E) \times \mathrm{GL}_m(\mathbb{A}_E)$ where $(n, m) = (\dim(V), \dim(W))$. Here, $\pi_{V,E}$, $\pi_{W,E}$ are automorphic representations whose Satake parameters at almost every unramified places are the image by the base-change homomorphisms ${}^LU(V) \to {}^L\mathrm{Res}_{E/F}\,\mathrm{GL}_{n,E}$, ${}^LU(W) \to {}^L\mathrm{Res}_{E/F}\,\mathrm{GL}_{m,E}$ (where $\mathrm{Res}_{E/F}$ denotes Weil's restriction of scalars) of the local Satake parameters of π_V, π_W, respectively. The existence of these weak base-changes in general is one of the main results of **[34,39]**. Also, although $\pi_{V,E}$, $\pi_{W,E}$ are not always cuspidal, they are isobaric sums of cuspidal representations which implies, by a result of Jacquet and Shalika, that they are uniquely determined by their Satake parameters at almost all places hence that the weak base-change π_E is unique. We denote by

$$L(s, \pi_E) = L(s, \pi_{V,E} \times \pi_{W,E})$$

the corresponding completed Rankin–Selberg L-function associated to $\pi_{V,E}$ and $\pi_{W,E}$.

Define the *automorphic L-packet* of π as the set of cuspidal automorphic representations π' of the various pure inner form $G' = U(V') \times U(W')$ of G with the same base-change $\pi'_E = \pi_E$ as π. By the Jacquet–Shalika theorem again and injectivity of base-change homomorphisms at the level of conjugacy classes, it is equivalent to asking that π and π' are *nearly equivalent*, that is, $\pi_v \simeq \pi'_v$ for almost all places v (this makes sense since $G_v \simeq G'_v$ for almost all v). Moreover, for a relevant pure inner form G' of G, we can define a pair (H', ξ') in exactly the same way as (H, ξ). The global version of the Gan–Gross–Prasad conjecture can now be stated as follows:

Conjecture 2.1 (Gan–Gross–Prasad [23]). *Assume that π_E is generic. Then, the following assertions are equivalent:*

(1) $L(\frac{1}{2}, \pi_E) \neq 0$;

(2) *There exists a relevant pure inner form $G' = U(W') \times U(V')$ of G (see Section* 1.1 *for the definition of a relevant pure inner form), a cuspidal automorphic representation π' of $G'(\mathbb{A}_F)$ in the same automorphic L-packet as π and a form $\varphi' \in \pi'$ such that*

$$\mathcal{P}_{H',\xi'}(\varphi') \neq 0.$$

When $(\dim(V), \dim(W)) = (2, 1)$, the conjecture essentially reduces to the celebrated theorem of Waldspurger [46] on toric periods for GL_2. Actually, as explained in the introduction, Waldspurger's result is more precise and gives an explicit identity relating (the square of) $\mathcal{P}_H(\varphi)$ to the central value $L(\frac{1}{2}, \pi_E)$.

There is also a similar conjecture for special orthogonal groups which actually predates the one for unitary groups [26] (as well as other conjectures for the so-called Fourier–Jacobi periods on unitary and symplectic/metaplectic groups stated in [23]). In [30], Ichino and Ikeda have proposed a refinement of this conjecture for $SO(n) \times SO(n-1)$ in the form of a precise identity generalizing Waldspurger's formula. Subsequently, similar refinements have been proposed by R. N. Harris [27], for $U(n) \times U(n-1)$, and then by Y. Liu [35] for general Bessel periods on orthogonal or unitary groups.

In order to state this refinement, we need to introduce two extra ingredients, namely local periods and a certain finite group S_π of endoscopic nature.

We start with the local periods. We endow $H(\mathbb{A}_F)$ with its global Tamagawa measure dh (this is the measure with which we will normalize the period integral (2.1)) and we fix a factorization $dh = \prod_v dh_v$ into a product of local Haar measures. We also fix a decomposition $\pi = \bigotimes'_v \pi_v$ of π into smooth irreducible representations of the localizations

$G_v = G(F_v)$ as well as a factorization $\langle\cdot,\cdot\rangle_{\mathrm{Pet}} = \prod_v \langle\cdot,\cdot\rangle_v$ of the Petersson inner product

$$\langle\varphi,\varphi\rangle_{\mathrm{Pet}} = \int_{G(F)\backslash G(\mathbb{A}_F)} |\varphi(g)|^2 dg$$

(which we also normalize using the Tamagawa measure on $G(\mathbb{A}_F)$) into local invariant inner products. The local periods are now given by the sesquilinear forms

$$\mathcal{P}_{H,\xi,v} : \varphi_v \otimes \varphi'_v \in \pi_v \otimes \pi_v \mapsto \int_{H_v}^{\mathrm{reg}} \langle \pi_v(h_v)\varphi_v, \varphi'_v\rangle_v \xi_v(h_v)\, dh_v. \tag{2.2}$$

The above integral of matrix coefficient is actually not convergent in general and has to be regularized (hence the "reg" sign above the integral). This regularization is, moreover, only possible under the extra assumption that the local component π_v is tempered. It is expected (under the Generalized Ramanujan Conjecture) that the hypothesis of the base-change π_E being generic implies that each of the local component π_v is tempered, but this is far from being known in general. Assuming now that π_v is tempered at every place v, an unramified computation shows that for almost all places v, if $\varphi_v \in \pi_v^{G(\mathcal{O}_v)}$ is a spherical vector such that $\langle\varphi_v,\varphi_v\rangle_v = 1$, we have

$$\mathcal{P}_{H,\xi,v}(\varphi_v,\varphi_v) = \Delta_v \frac{L(\frac{1}{2},\pi_{E,v})}{L(1,\pi_v,Ad)}$$

where $L(\frac{1}{2},\pi_{E,v})$, $L(1,\pi_v,Ad)$ denote the local Rankin–Selberg and adjoint L-factors of π_E and π, respectively, whereas Δ_v stands for the product of local abelian L-factors

$$\Delta_v = \prod_{i=1}^{n} L(i,\eta^i_{E_v/F_v})$$

with η_{E_v/F_v} the quadratic character associated to the local extension E_v/F_v and $n = \dim(V)$. The *normalized* local periods are then defined by

$$\mathcal{P}^{\natural}_{H,\xi,v}(\varphi_v,\varphi_v) = \Delta_v^{-1} \frac{L(1,\pi_v,Ad)}{L(\frac{1}{2},\pi_{E,v})} \mathcal{P}_{H,\xi,v}(\varphi_v,\varphi_v).$$

Finally, writing the base-change π_{V_E} and $\pi_{W,E}$ as isobaric sums

$$\pi_{V,E} = \pi_{V,1} \boxplus \cdots \boxplus \pi_{V,k}, \quad \pi_{W,E} = \pi_{W,1} \boxplus \cdots \boxplus \pi_{W,l}$$

of cuspidal automorphic representations of some general linear groups, we set $S_\pi = (\mathbb{Z}/2\mathbb{Z})^{k+l}$. It serves as a substitute for the centralizer of the, yet nonexistent in general, global Langlands parameter of π.

Conjecture 2.2 (Ichino–Ikeda, N. Harris, Y. Liu). *Assume that for every place v of F, π_v is a tempered representation. Then, for every factorizable vector $\varphi = \bigotimes'_v \varphi_v \in \pi$, we have*

$$|\mathcal{P}_{H,\xi}(\varphi)|^2 = |S_\pi|^{-1} \Delta \frac{L(\frac{1}{2},\pi_E)}{L(1,\pi,Ad)} \prod_v \mathcal{P}^{\natural}_{H,\xi,v}(\varphi_v,\varphi_v) \tag{2.3}$$

where $\Delta = \prod_{i=1}^n L(i,\eta^i_{E/F})$ and $L(s,\pi,Ad) = \prod_v L(s,\pi_v,Ad)$ denotes the completed adjoint L-function of π.

Note that at a formal level, that is, formally expanding L-functions as Euler products outside the range of convergence, the above formula can be rewritten in the more compact way as

$$\left|\mathcal{P}_{H,\xi}(\varphi)\right|^2 = |S_\pi|^{-1} \prod_v{}' \mathcal{P}_{H,\xi,v}(\varphi_v, \varphi_v), \tag{2.4}$$

where the prime symbol on the product sign indicates that it is not convergent and has to be suitably reinterpreted "in the sense of L-functions" as identity (2.3).

Thanks to the work of many authors that we are going to summarize in the next sections, it is now relatively easy to state the current status on these two conjectures:

Theorem 2.1. *Both Conjectures 2.1 and 2.2 hold in full generality.*

The rest of this paper is devoted to reviewing the long series of works leading to the above theorem. They all stem from a strategy originally proposed by Jacquet and Rallis [32] of comparing two relative trace formulae. Let us mention here that there has actually been other fruitful approaches to the global Gan–Gross–Prasad conjecture among which one of the most notable has been the method pioneered by Ginzburg–Jiang–Rallis [25] using automorphic descent and that has recently seen much development with the work [33] of Jiang and L. Zhang proving in full generality the implication (2) $\Rightarrow$ (1) of Conjecture 2.1.

2.2. The approach of Jacquet–Rallis

In [32], Jacquet and Rallis have proposed a way to attack the Gan–Gross–Prasad conjecture for unitary groups through a comparison of relative trace formulae. They only consider the case where $\dim(W) = \dim(V) - 1$ (in which case $H = U(W)$ and the character ξ is trivial) and we assume throughout that this condition is satisfied. The global relative trace formulae considered here are powerful analytic tools introduced originally by Jacquet and that relate automorphic periods to more geometric distributions known as relative orbital integrals.

Let us be more specific in the case at hand. For $f \in C_c^\infty(G(\mathbb{A}_F))$, a global test function, we let

$$K_f(x, y) = \sum_{\gamma \in G(F)} f(x^{-1}\gamma y), \quad x, y \in G(F)\backslash G(\mathbb{A}_F),$$

be its automorphic kernel which describes the operator $R(f)$ of right convolution by f on the space of automorphic forms. The first trace formula introduced by Jacquet and Rallis is formally obtained by expanding the (usually divergent) expression

$$J(f) = \int_{[H]\times[H]} K_f(h_1, h_2) dh_1 dh_2 \tag{2.5}$$

in two different ways. More precisely, but still at a formal level, this distribution can be expanded as

$$\cdots + \sum_{\delta \in H(F)\backslash G_{\mathrm{rs}}(F)/H(F)} O(\delta, f) = J(f) = \sum_{\varphi \in \mathcal{A}_{\mathrm{cusp}}(G)} \mathcal{P}_H\big(R(f)\varphi\big)\overline{\mathcal{P}_H(\varphi)} + \cdots, \tag{2.6}$$

where the right sum runs over an orthonormal basis for the space of cuspidal automorphic forms whereas the left sum is indexed by the so-called *regular semisimple* double cosets of $H(F)$ in $G(F)$. Here, an element $\delta \in G$ is called (relatively) regular semisimple if its stabilizer under the $H \times H$-action is trivial and the corresponding orbit is (Zariski) closed. We denote by G_{rs} the nonempty Zariski open subset of regular semisimple elements and for $\delta \in G_{\mathrm{rs}}(F)$,

$$O(\delta, f) = \int_{H(\mathbb{A}_F)\times H(\mathbb{A}_F)} f(h_1 \delta h_2) dh_1 dh_2$$

denotes the corresponding *relative orbital integral* of f at δ. The left suspension points in (2.6) represent the remaining contributions from singular orbits whereas the right suspension points indicate the contribution of the continuous spectrum (both of which are somehow responsible for the divergence of the original expression (2.5)).

The second trace formula introduced by Jacquet and Rallis has to do with the following triple of groups:

$$\begin{aligned} H_1 &= \mathrm{Res}_{E/F}\,\mathrm{GL}_{n,E} \hookrightarrow G' = \mathrm{Res}_{E/F}\,\mathrm{GL}_{n+1,E} \times \mathrm{Res}_{E/F}\,\mathrm{GL}_{n,E} \hookleftarrow H_2 \\ &= \mathrm{GL}_{n+1,F} \times \mathrm{GL}_{n,F}, \end{aligned}$$

where $n = \dim(W)$, the first embedding is the diagonal one and the second embedding is the natural one. Note that G' is the group on which the base-change π_E "lives." For $f' \in C_c^\infty(G'(\mathbb{A}_F))$, we write (again formally)

$$I(f') = \int_{[H_1]\times[H_2]} K_{f'}(h_1, h_2)\eta(h_2) dh_1 dh_2 \tag{2.7}$$

where $K_{f'}$ is the automorphic kernel of f' and $\eta : [H_2] \to \{\pm 1\}$ is the automorphic character defined by $\eta(g_n, g_{n+1}) = \eta_{E/F}(\det g_n)^{n+1}\eta_{E/F}(\det g_{n+1})^n$. This formal distribution can be analogously expanded as

$$\cdots + \sum_{\gamma \in H_1(F)\backslash G'_{\mathrm{rs}}(F)/H_2(F)} O_\eta(\gamma, f') = I(f') = \sum_{\varphi \in \mathcal{A}_{\mathrm{cusp}}(G')} \mathcal{P}_{H_1}\big(R(f')\varphi\big)\overline{\mathcal{P}_{H_2,\eta}(\varphi)} + \cdots, \tag{2.8}$$

where G'_{rs} stands for the open subset of regular and semisimple elements under the $H_1 \times H_2$-action, the relative orbital integrals are given by

$$O_\eta(\gamma, f') = \int_{H_1(\mathbb{A}_F)\times H_2(\mathbb{A}_F)} f'(h_1 \gamma h_2)\eta(h_2) dh_1 dh_2$$

and $\mathcal{P}_{H_1}$, $\mathcal{P}_{H_2,\eta}$ denote the automorphic period integrals over $[H_1]$ and $[H_2]$ twisted by η, respectively.

The discussion so far is, of course, oversimplifying and ignoring serious analytical and convergence issues (we will come back to this later). However, as a motivation for considering this relative trace formula on G', we have the following results on automorphic periods:

- The period $\mathcal{P}_{H_1}$ is a Rankin–Selberg period studied by Jacquet–Piatetskii-Shapiro–Shalika that essentially represents the central value $L(\frac{1}{2}, \Pi)$ on $\Pi \hookrightarrow \mathcal{A}_{\mathrm{cusp}}(G')$;

- The period $\mathscr{P}_{H_2,\eta}$ was studied by Rallis and Flicker who have shown that it detects exactly the cuspidal automorphic Π's that come by base-change from G (i.e., it is nonzero precisely on those cuspidal representations of the form π_E, for π a cuspidal automorphic representation of G).

Thus, on a very formal and sketchy sense, the Gan–Gross–Prasad conjecture implies that the spectral sides of $I(f')$ should somehow "match" that of $J(f)$. The idea of Jacquet and Rallis was to make precise the existence of such a comparison, from which the global Gan–Gross–Prasad conjecture was eventually to be deduced, by equalling the geometric sides term by term. As a first step, they define a *correspondence* of orbits, which here takes the form of a natural embedding between regular semisimple cosets

$$H(k)\backslash G_{\mathrm{rs}}(k)/H(k) \hookrightarrow H_1(k)\backslash G'_{\mathrm{rs}}(k)/H_2(k), \quad \delta \mapsto \gamma, \tag{2.9}$$

for every field extension k/F. Using this correspondence, they then introduced a related notion of local *transfer* (or *matching*): for a place v of F, two test functions $f_v \in C_c^\infty(G_v)$ and $f'_v \in C_c^\infty(G'_v)$ are said to be transfers of each other (simply denoted by $f_v \leftrightarrow f'_v$ for short) if for every $\delta \in H(F_v)\backslash G_{\mathrm{rs}}(F_v)/H(F_v)$ we have an identity

$$O(\delta, f_v) = \Omega_v(\gamma) O_{\eta_v}(\gamma, f'_v), \tag{2.10}$$

where $\gamma \in H_1(F_v)\backslash G'_{rs}(F_v)/H_2(F_v)$ is the image of δ by the above correspondence, $O(\delta, f_v)$ and $O_{\eta_v}(\gamma, f'_v)$ are local relative orbital integrals defined in the same way as their global counterparts (replacing in the domain of integration, adelic groups by the corresponding local groups) and $\gamma \mapsto \Omega_v(\gamma)$ is a certain transfer factor which in particular has the effect of making the right-hand side above $H_1(F_v) \times H_2(F_v)$-invariant in γ.

As in the usual paradigm of endoscopy, to make this notion useful and allow for a global comparison we basically need two local ingredients: first the existence of local transfer (i.e., for every $f_v \in C_c^\infty(G_v)$ there exists $f'_v \in C_c^\infty(G'_v)$ such that $f_v \leftrightarrow f'_v$ and conversely, every f'_v admits a transfer f_v) and then a *fundamental lemma* (saying, at least, that $\mathbf{1}_{G(\mathcal{O}_v)} \leftrightarrow \mathbf{1}_{G'(\mathcal{O}_v)}$ for almost all v).

3. COMPARISON: LOCAL TRANSFER AND FUNDAMENTAL LEMMA

A first breakthrough on the Jacquet–Rallis approach to the Gan–Gross–Prasad conjecture was made in [57] by Wei Zhang who proved the existence of the local transfer at all non-Archimedean places. His strategy for doing so roughly goes as follows:

- The first step is to reduce to a statement on Lie algebras using some avatar of the exponential map (also known as Cayley map): we are then left with proving the existence of a similar transfer between the orbital integrals for the adjoint action of $U(W_v)$ on $\mathfrak{u}(V_v) = \mathrm{Lie}(U(V_v))$ and for the adjoint action of $\mathrm{GL}_n(F_v)$ on $\mathfrak{gl}_{n+1}(F_v)$.
- Then, a crucial ingredient in Zhang's proof is to show that the transfer at the Lie algebra level essentially commutes (i.e., up to some explicit multiplicative

constants) with 3 different partial Fourier transforms $\mathcal{F}_1$, $\mathcal{F}_2$, and $\mathcal{F}_3$ that can naturally be defined on the two spaces $C_c^\infty(\mathfrak{u}(V_v))$, $C_c^\infty(\mathfrak{gl}_{n+1}(F_v))$. One of them, that we will denote by $\mathcal{F}_1$, is the Fourier transform with respect to "the last row and column" on $\mathfrak{gl}_{n+1}(F_v)$ or $\mathfrak{u}(V_v)$ when realizing the latter in matrix form using a basis adapted to the decomposition $V_v = W_v \oplus W_v^\perp$. (Recall that we are assuming that $\dim(W_v^\perp) = 1$.) For this, Zhang develops some relative trace formulae for the aforementioned actions on $\mathfrak{gl}_{n+1}(F_v)$ and $\mathfrak{u}(V_v)$ and combines them with a clever induction argument.

- Finally, the proof of the existence of transfer on Lie algebras is obtained by combining the second step with a certain uncertainty principle due to Aizenbud [1], which allows reducing the construction of the transfer to functions that are supported away from the *relative nilpotent cones* (i.e., the set of elements whose orbit closure contains an element of the center of the Lie algebra), as well as a standard descent argument whose essence goes back to Harish-Chandra.

It is noteworthy to mention that this result was subsequently extended, following the same strategy, by H. Xue [53] to Archimedean places, although the final result there is slightly weaker. (More precisely, Xue was only able to show that a *dense* subspace of test functions admit a transfer but also observed that it is sufficient for all expected applications.)

The Jacquet–Rallis fundamental lemma for its part, was proven earlier by Yun [55] in the case of fields of positive characteristic following and adapting the geometric-cohomological approach based on Hitchin fibrations that was developed by Ngô in the context of the endoscopic fundamental lemma. This result was then transferred to fields of characteristic zero, but of sufficiently large residual characteristic, using model-theoretic techniques by Julia Gordon in the appendix of [55].

Later, in [14], I found a completely new and elementary proof of this fundamental lemma. The argument, despite that of Gordon–Yun, works directly in characteristic zero and is purely based on techniques from harmonic analysis. Thus, we have:

Theorem 3.1 (Yun–Gordon, Beuzart-Plessis). *Let v be a place of F of residue characteristic not 2 that is unramified in E and assume that the Hermitian spaces W_v, $W_v^\perp$ both admit self-dual lattices L_v^W and $L_v^{W^\perp}$. Set $L_v = L_v^W \oplus L_v^{W^\perp}$ (a self-dual lattice in V_v) and $K_v = \mathrm{Stab}_{G_v}(L_v \times L_v^W)$ for the stabilizer in $G_v = U(V_v) \times U(W_v)$ of the lattices L_v and L_v^W (a hyperspecial compact subgroup of G_v). Then, setting $K_v' = \mathrm{GL}_{n+1}(\mathcal{O}_{E_v}) \times \mathrm{GL}_n(\mathcal{O}_{E_v})$, we have $\mathbf{1}_{K_v} \leftrightarrow \mathbf{1}_{K_v'}$.*

More precisely, in [14] I proved a Lie algebra analog of the Jacquet–Rallis fundamental lemma (of which the original statement can easily be reduced; at least in residual characteristic not 2) stating that the relative orbital integrals of $\mathbf{1}_{\mathfrak{u}(L_v)}$ *match* those of $\mathbf{1}_{\mathfrak{gl}_{n+1}(\mathcal{O}_{F_v})}$ in a suitable sense (where $\mathfrak{u}(L_v)$ denotes the lattice in $\mathfrak{u}(V_v)$ stabilizing L_v). The argument is based on a hidden SL(2) symmetry involving a Weil representation. More specifically, we consider the Weil representations of $\mathrm{SL}(2, F_v)$ associated to the quadratic form q sending a matrix of size $n+1$,

$$X = \begin{pmatrix} A & b \\ c & \lambda \end{pmatrix},$$

either in $\mathfrak{gl}_{n+1}(F_v)$ or in $\mathfrak{u}(V_v)$, to $q(X) = cb$ (where here, A is a square-matrix, b is a column vector, and c a row vector all of size n). Using the aforementioned result of Zhang that the transfer commutes with the partial Fourier transform $\mathcal{F}_1$, it can be shown that these representations descend to spaces of relative orbital integrals on $C_c^\infty(\mathfrak{u}(V_v))$ and $C_c^\infty(\mathfrak{gl}_{n+1}(F_v))$ and coincide on their intersections (identifying the spaces of regular semisimple orbits through a correspondence similar to (2.9)). Consider then the difference

$$\Phi : X \in \mathfrak{u}(V_v)_{\mathrm{rs}}/U(W_v) \mapsto O(X, \mathbf{1}_{\mathfrak{u}(L_v^V)}) - \omega_v(Y) O_{\eta_v}(Y, \mathbf{1}_{\mathfrak{gl}_{n+1}(\mathcal{O}_{F_v})}),$$

where $\mathfrak{u}(V_v)_{\mathrm{rs}}$ denotes the Lie algebra analog of the relative regular semisimple locus, Y is the image of X by a correspondence of orbits $\mathfrak{u}(V_v)_{\mathrm{rs}}/U(W_v) \hookrightarrow \mathfrak{gl}_{n+1}(F_v)_{\mathrm{rs}}/\mathrm{GL}_n(F_v)$ similar to (2.9) and $\omega_v(Y)$ is the Lie algebra counterpart of the transfer factor. The fundamental lemma then states that Φ is identically zero. The proof proceeds roughly in three steps:

- First, we show that $\Phi(X) = 0$ for $|q(X)| \geqslant 1$. When $|q(X)| = 1$, this requires an inductive argument on n. Moreover, this vanishing can be reformulated by saying that Φ is fixed by the subgroup $\left(\begin{smallmatrix} 1 & \mathfrak{p}_{F_v}^{-1} \\ 0 & 1 \end{smallmatrix}\right)$ through the Weil representation.
- Secondly, we remark that Φ is also fixed by $w = \left(\begin{smallmatrix} 0 & -1 \\ 1 & 0 \end{smallmatrix}\right)$. This comes from the fact that the action of w descends from the partial Fourier transform $\mathcal{F}_1$ which leaves (for a suitable normalization) the functions $\mathbf{1}_{\mathfrak{u}(L_v^V)}, \mathbf{1}_{\mathfrak{gl}_{n+1}(\mathcal{O}_{F_v})}$ invariant.
- Finally, as $\mathrm{SL}_2(F_v)$ is generated by $\left(\begin{smallmatrix} 1 & \mathfrak{p}_{F_v}^{-1} \\ 0 & 1 \end{smallmatrix}\right)$ and w, we infer that Φ is fixed by $\mathrm{SL}_2(F_v)$ from which it is relatively straightforward to deduce $\Phi = 0$.

It is also worth mentioning that in a very interesting work, Jingwei Xiao [51] has shown that the Jacquet–Rallis fundamental lemma implies the (usual) endocospic fundamental lemma for unitary groups. Thus, combining his argument with the proof outlined above yields a completely elementary proof of the Langlands–Shelstad fundamental lemma for unitary groups!

The two previous results on smooth transfer and the fundamental lemma are already enough to imply the Gan–Gross–Prasad Conjecture 2.1 under some local restrictions on the cuspidal representation π (originating from the use of simple versions of the Jacquet–Rallis trace formulae, allowing to bypass all convergence issues) as was done by W. Zhang in [57]. However, to derive the refinement of Conjecture 2.2 following the same strategy, we need an extra local ingredient relating the local periods of Ichino–Ikeda to similar local distributions associated to the Rankin–Selberg and Flicker–Rallis periods. More precisely, by the work of Jacquet–Piatetskii-Shapiro–Shalika, on the one hand, and Flicker–Rallis, on the other hand, it is known that the two automorphic periods $\mathcal{P}_{H_1}$ and $\mathcal{P}_{H_2,\eta}$ admit factorizations of the form

$$\mathcal{P}_{H_1}(\varphi) = \prod_v{}' \mathcal{P}_{H_1,v}(W_{\varphi,v}), \tag{3.1}$$

$$\mathcal{P}_{H_2,\eta}(\varphi) = \frac{1}{4}\prod_v{}' \mathcal{P}_{H_2,\eta,v}(W_{\varphi,v}) \tag{3.2}$$

for φ a factorizable vector in a given cuspidal automorphic representation $\Pi = \Pi_{n+1} \otimes \Pi_n$ of $G'(\mathbb{A}_F)$, where

$$W_\varphi(g) = \int_{[N']} \varphi(ug)\psi'(u)^{-1}du = \prod_v W_{\varphi,v}(g_v)$$

denotes a factorization of the Whittaker function of φ (here N' stands for the standard maximal unipotent subgroup of G' and ψ' is a nondegenerate character of $[N']$), $\mathcal{P}_{H_1,v}$, $\mathcal{P}_{H_2,\eta,v}$ are explicit linear forms on the local Whittaker model $\mathcal{W}(\Pi_v, \psi'_v)$ of Π_v and the products in (3.1), (3.2) are to be regularized and understood "in the sense of L-functions" in a way similar to (2.4).

Based on the factorizations (3.1) and (3.2), the contribution of Π to the spectral expansion (2.8) can be shown to itself admit a factorization roughly as the product of local distributions (called *relative characters*) I_{Π_v} defined by

$$I_{\Pi_v}(f'_v) = \sum_{W_v \in \mathcal{W}(\Pi_v,\psi'_v)} \mathcal{P}_{H_1,v}\big(\Pi_v(f'_v)W_v\big)\overline{\mathcal{P}_{H_2,\eta,v}(W_v)}, \quad f'_v \in C_c^\infty(G'_v),$$

where the sum runs over a suitable orthonormal basis of the Whittaker model. On the other hand, from the Ichino–Ikeda Conjecture 2.2, we expect the contribution of $\pi \hookrightarrow \mathcal{A}_{\mathrm{cusp}}(G)$ to the spectral expansion of (2.6) to essentially factorize into the product of the local relative characters (where again the sum is taken over an orthonormal basis)

$$J_{\pi_v}(f_v) = \sum_{\varphi_v \in \pi_v} \mathcal{P}_{H,v}\big(\pi_v(f_v)\varphi_v, \varphi_v\big), \quad f_v \in C_c^\infty(G_v).$$

In **[56]**, W. Zhang has conjectured that the local Jacquet–Rallis transfer $f_v \leftrightarrow f'_v$ also satisfies certain precise spectral relations involving the above relative characters. This is exactly the extra local ingredient needed to finish the proof of the Ichino–Ikeda conjecture based on a comparison of the Jacquet–Rallis trace formulae. This conjecture was shown in **[56]** to hold for unramified and supercuspidal representations, and the method was further extended and amplified in **[13]**, allowing to prove the conjecture for all (tempered) representations at non-Archimedean places. Later, in **[15]** I gave a better proof of this conjecture which also has the advantage of working uniformly at all places (including Archimedean ones). To state the result, we introduce some terminology/notation: for a place v of F and a smooth irreducible representation π_v of G_v, we denote by $\pi_{E,v}$ the local *base-change* of π_v, that is, the smooth irreducible representation of G'_v whose L-parameter is given by composing that of π_v with the natural embedding of L-groups ${}^LG_v \to {}^LG'_v$, and, moreover, we say that π_v is *H_v-distinguished* if $\mathrm{Hom}_{H_v}(\pi_v, \mathbb{C}) \neq 0$, that is, with the notation of Section 1.1, if the multiplicity $m(\pi_v)$ equals 1.

Theorem 3.2. *There exist explicit local constants $(\kappa_v)_v$ indexed by the set of all places of F and satisfying the product formula $\prod_v \kappa_v = 1$ such that the following property is verified: for every place v, every tempered representation π_v of G_v which is H_v-distinguished and every pair $(f_v, f'_v) \in C_c^\infty(G_v) \times C_c^\infty(G'_v)$ of matching functions (that is, $f_v \leftrightarrow f'_v$), we have*

$$I_{\pi_{v,E}}(f'_v) = \kappa_v J_{\pi_v}(f_v). \tag{3.3}$$

Moreover, the above identities characterize the Jacquet–Rallis transfer, that is, if two functions $f_v \in C_c^\infty(G_v)$, $f'_v \in C_c^\infty(G'_v)$ satisfy (3.3) for every tempered irreducible representation π_v of G_v that is H_v-distinguished, then these functions are transfers of each other.

The proof given in [15] of the above theorem is mainly based on another ingredient of independent interest which is an explicit Plancherel decomposition for the space $G'_v/H_{2,v}$ or rather, decomposing this quotient as a product in a natural way, for the symmetric variety $\mathrm{GL}_n(E_v)/\mathrm{GL}_n(F_v)$. This spectral decomposition is roughly obtained by applying the Plancherel formula for the group $\mathrm{GL}_n(E_v)$ to a family of zeta integrals, depending on a complex parameter s, introduced by Flicker and Rallis [22] and that represents local Asai L-factors and taking the residue at $s = 1$ of the resulting expression. We will not describe the exact process here, but just mention that this settles in the case at hand a general conjecture of Sakellaridis–Venkatesh [41] on the spectral decomposition of spherical varieties. This Plancherel formula is then used to write the explicit spectral expansion for a local analog of the Jacquet–Rallis trace formula (2.8) which is then compared with a local counterpart of the trace formula (2.6) yielding as a consequence Theorem 3.2 above. Moreover, as another by-product of this local comparison, we also get a formula conjectured by Hiraga–Ichino–Ikeda for the formal degree of discrete series [29] in the case of unitary groups.

4. GLOBAL ANALYSIS OF JACQUET–RALLIS TRACE FORMULAE

With all the local ingredients explained in the previous section in place, the only remaining tasks to finish the program initiated by Jacquet and Rallis to prove the Gan–Gross–Prasad and Ichino–Ikeda conjectures are global. More specifically, although simple versions of the Jacquet–Rallis trace formulae have been successfully used to establish these conjectures under some local restrictions [13, 57], in order to detect all the relevant cuspidal representations of unitary groups, we need more refined versions of the geometric and spectral expansions of (2.6) and (2.8).

As a first important step in that direction, Zydor [58, 59] has completely regularized the singular contributions to the geometric sides. We can summarize his main results as follows: for all test functions $f \in C_c^\infty(G(\mathbb{A}_F))$ and $f' \in C_c^\infty(G'(\mathbb{A}_F))$, there exist "canonical" regularization of the (usually divergent) integrals (2.5) and (2.7), that we will still denote by $J(f)$ and $I(f')$, as well as decompositions

$$J(f) = \sum_{\delta \in (H\backslash G//H)(F)} O(\delta, f) \quad \text{and} \quad I(f') = \sum_{\gamma \in (H_1\backslash G'//H_2)(F)} O_\eta(\gamma, f'), \tag{4.1}$$

where $H\backslash G//H$ and $H_1\backslash G'//H_2$ stand for the corresponding categorical quotients and $O(\delta,\cdot)$, $O_\eta(\gamma,\cdot)$ are distributions supported on the union of the adelic double cosets with images δ and γ in $(H\backslash G//H)(\mathbb{A}_F)$ and $(H_1\backslash G'//H_2)(\mathbb{A}_F)$, respectively, which coincide with the previously defined relative orbital integrals when δ and γ are regular semisimple.

Zydor obtains these regularized orbital integrals by adapting a truncation procedure developed by Arthur in the context of the usual trace formula to the relative setting at hand. It should be emphasized that contrary to what happens with Arthur's trace formula, the resulting distributions are directly invariant (in a relative sense, that is, here under the natural action of $H \times H$ or $H_1 \times H_2$) and do not depend on any auxiliary choice (such as that of a maximal compact subgroup). It is in this sense that the regularizations of Zydor are really "canonical." It should be mentioned that another, different, approach to such regularization was proposed by Sakellaridis [40] in the context of general relative trace formulae. It is based on analyzing the exponents at infinity of generalized theta series together with a natural procedure to regularize integrals of multiplicative functions when the corresponding character is nontrivial.

Before we even consider the analogous, more subtle, regularization problem on the spectral side, there appears the natural question of how to compare the singular contributions to the refined geometric expansions of (4.1). This issue was completely resolved in a very long paper [20] by Chaudouard and Zydor. To state their main result, it is convenient to again consider the relevant pure inner forms of G (as defined in Section 1.1): for every Hermitian space W' of the same dimension as W, we have a relevant pure inner form $G^{W'} = U(V') \times U(W')$ with its diagonal subgroup $H^{W'} = U(W')$ where $V' = W' \oplus^\perp W^\perp$. Moreover, the correspondence of orbits (2.9) extends to an isomorphism between categorical quotients,

$$H\backslash G//H \simeq H_1\backslash G'//H_2, \tag{4.2}$$

and for every W' as before, $H^{W'}\backslash G^{W'}//H^{W'}$ can naturally be identified with $H\backslash G//H$. With these preliminaries, the main result of Chaudouard and Zydor can now be stated as follows:

Theorem 4.1 (Chaudouard–Zydor). *Assume that $f^{W'} = \prod_v f_v^{W'} \in C_c^\infty(G^{W'}(\mathbb{A}_F))$, where W' runs over all isomorphism classes of Hermitian spaces of dimension n, and $f' = \prod_v f'_v \in C_c^\infty(G'(\mathbb{A}_F))$ are factorizable test functions such that for every place v, and each W', $f_v^{W'}$ and f'_v are Jacquet–Rallis transfers of each other (that is, $f_v^{W'} \leftrightarrow f'_v$). Then, for every $\delta \in (H\backslash G//H)(F)$ with image $\gamma \in (H_1\backslash G'//H_2)(F)$ by (4.2), we have*

$$\sum_{W'} O(\delta, f^{W'}) = O_\eta(\gamma, f'). \tag{4.3}$$

It should be noted that when δ, hence also γ, is regular semisimple, the left-hand sum in (4.3) only contains one nonidentically vanishing term but that in general more than one relevant pure inner forms can contribute. Also, the above result extends to nonfactorizable test functions, provided the wording is changed suitably.

The next natural step would be to develop regularized spectral expansions similar to (4.1). As a first result in that direction, Zydor has shown decompositions of the form

$$J(f) = \sum_{\chi \in \mathcal{X}(G)} J_\chi(f) \quad \text{and} \quad I(f') = \sum_{\chi' \in \mathcal{X}(G')} I_{\chi'}(f'), \tag{4.4}$$

where $\mathcal{X}(G)$ and $\mathcal{X}(G')$ stand for the set of cuspidal data of the groups G and G' respectively, that is the sets of pairs (M, σ) where M is a Levi subgroup (of G or G') and σ is a cuspidal automorphic representation of $M(\mathbb{A}_F)$ taken up to conjugacy (by $G(F)$ or $G'(F)$). According to Langlands theory of pseudo-Eisenstein series, these sets index natural equivariant Hilbertian decompositions:

$$L^2([G]) = \widehat{\bigoplus_{\chi \in \mathcal{X}(G)}} L^2_\chi([G]), \quad L^2([G']) = \widehat{\bigoplus_{\chi' \in \mathcal{X}(G')}} L^2_{\chi'}([G']).$$

The automorphic kernels K_f, $K_{f'}$ decompose accordingly into series $K_f = \sum_\chi K_{f,\chi}$, $K_{f'} = \sum_{\chi'} K_{f',\chi'}$ where $K_{f,\chi}$ and $K_{f',\chi}$ are kernel functions representing the restrictions $R_\chi(f)$ and $R_{\chi'}(f')$ of the right convolution operators $R(f)$ and $R(f')$ to $L^2_\chi([G])$ and $L^2_{\chi'}([G'])$, respectively. The distributions $f \mapsto J_\chi(f)$ and $f' \mapsto I_{\chi'}(f')$ are then roughly defined by applying the same regularization procedure that Zydor used for the expressions $J(f)$ and $I(f')$ up to replacing the integrands by $K_{f,\chi}$ and $K_{f',\chi'}$ respectively, that is, in symbolic terms:

$$\begin{aligned} J_\chi(f) &= \int^{\mathrm{reg}}_{[H]\times[H]} K_{f,\chi}(h_1, h_2)\, dh_1\, dh_2, \\ I_{\chi'}(f') &= \int^{\mathrm{reg}}_{[H_1]\times[H_2]} K_{f',\chi'}(h_1, h_2)\eta(h_2)\, dh_1\, dh_2. \end{aligned} \tag{4.5}$$

However, the expansions (4.4) are of little use as they stand and need to be suitably refined to allow for a meaningful comparison of the trace formulae. In Arthur's terminology, (4.4) are *coarse spectral expansions* and we need *refined spectral expansions* for each of the terms $J_\chi(f)$ or $I_{\chi'}(f')$.

This problem has so far proved to be a very difficult for general cuspidal data χ and χ'. However, a recent result of mine in collaboration with Y. Liu, W. Zhang, and X. Zhu **[17]** allows isolating in the coarse spectral expansions (4.4) the only terms that are eventually of interest consequently reducing the problem to some very particular cuspidal data χ' of G'.

The result proved in **[17]** is very general so let us place ourself for one moment in the framework of an arbitrary connected reductive group G over the number field F. Let Σ be a set of non-Archimedean places of F (possibly infinite) such that for each $v \in \Sigma$, the group G_v is unramified and fix a hyperspecial compact subgroup $K_v \subset G_v$ with $K_v = G(\mathcal{O}_v)$ for almost all $v \in \Sigma$. We let $\mathcal{X}_\Sigma(G)$ be the set of Σ-unramified cuspidal data of G, that is, the cuspidal data represented by pairs (M, σ) with σ unramified at all places of $v \in \Sigma$ (with respect to K_v or, rather, the hyperspecial subgroup it induces in M_v). For $\chi \in \mathcal{X}_\Sigma(G)$, we define its Σ-*near equivalence class*, henceforth denoted by $\mathcal{N}_\Sigma(\chi)$, as the set of all cuspidal data $\chi' \in \mathcal{X}_\Sigma(G)$ such that if χ and χ' are represented by pairs (M, σ) and (M', σ') respectively, then there exist automorphic unramified characters λ and λ' of $M(\mathbb{A}_F)$

and $M'(\mathbb{A}_F)$, respectively, with the property that for every $v \in \Sigma$ the Satake parameters of the unique K_v-unramified subquotients in $I_{P_v}^{G_v}(\sigma_v \otimes \lambda_v)$ and $I_{P'_v}^{G_v}(\sigma'_v \otimes \lambda'_v)$ (where P, P' are arbitrary chosen parabolics with Levi components M, M') are isomorphic. We also fix a compact-open subgroup $K = \prod_{v \in S_f} K_v$ of $G(\mathbb{A}_f)$ (where S_f denotes the set of finite places of F and K_v coincides with the previous choice of hyperspecial subgroup when $v \in \Sigma$) and we define the *Schwartz space* of K-biinvariant functions on $G(\mathbb{A}_F)$ as the restricted tensor product

$$\mathcal{S}_K\big(G(\mathbb{A}_F)\big) = \mathcal{S}\big(G(F_\infty)\big) \otimes \bigotimes_{v \in S_f}' C_c(K_v \backslash G_v / K_v),$$

where $C_c(K_v \backslash G_v / K_v)$ denotes the space of bi-K_v-invariant compactly supported functions on G_v (that is the K_v-spherical Hecke algebra when $v \in \Sigma$), F_∞ is the product of the Archimedean completions of F and $\mathcal{S}(G(F_\infty))$ stands for the Schwartz space of the reductive Lie group $G(F_\infty)$ in the sense of [19]. More precisely, $\mathcal{S}(G(F_\infty))$ is the space of smooth functions $f : G(F_\infty) \to \mathbb{C}$ such that for every polynomial differential operator on $G(F_\infty)$, the derivatives Df is bounded or, equivalently, such that for every left- (or right-)invariant differential operator X, Xf is decreasing faster than the inverse of any polynomial on $G(F_\infty)$.

The Schwartz space $\mathcal{S}(G(F_\infty))$ is naturally a Fréchet algebra under the convolution product and we also set

$$\mathcal{M}_\infty(G) = \mathrm{End}_{\mathrm{cont}, \mathcal{S}(G(F_\infty))-\mathrm{bimod}}\big(\mathcal{S}\big(G(F_\infty)\big)\big)$$

for the space of continuous endomorphisms of $\mathcal{S}(G(F_\infty))$ seen as a bimodule over itself. This is an algebra acting on any smooth admissible Fréchet representation of moderate growth of $G(F_\infty)$ in the sense of Casselman–Wallach. Moreover, as an application of a form of Schur lemma, for every irreducible Casselman–Wallach representation π_∞ of $G(F_\infty)$ and every $\mu_\infty \in \mathcal{M}_\infty(G)$ there exists a scalar $\mu_\infty(\pi_\infty) \in \mathbb{C}$ such that $\pi_\infty(\mu_\infty) = \mu_\infty(\pi_\infty) Id$. Thus, $\mathcal{M}_\infty(G)$ can be seen as some big algebra of multipliers for $\mathcal{S}(G(F_\infty))$. We also define the algebra of Σ-*multipliers* as the restricted tensor product

$$\mathcal{M}_\Sigma(G) = \mathcal{M}_\infty(G) \bigotimes_{v \in \Sigma}' \mathcal{H}(G_v, K_v),$$

where, for $v \in \Sigma$, $\mathcal{H}(G_v, K_v) = C_c(K_v \backslash G_v / K_v)$ is the spherical Hecke algebra. Then, $\mathcal{M}_\Sigma(G)$ acts naturally on the global Schwartz space $\mathcal{S}_K(G(\mathbb{A}_F))$, and we shall denote this action as the convolution product $*$. One of the main result of [17] can now be stated as follows:

Theorem 4.2 (Beuzart-Plessis–Liu–Zhang–Zhu). *Let $\chi \in \mathfrak{X}_\Sigma(G)$. Then, there exists a multiplier $\mu_\chi \in \mathcal{M}_\Sigma(G)$ such that for every Schwartz function $f \in \mathcal{S}_K(G(\mathbb{A}_F))$ and every other cuspidal datum $\chi' \in \mathfrak{X}_\Sigma(G)$, we have*

$$R_{\chi'}(\mu_\chi * f) = \begin{cases} R_{\chi'}(f) & \text{if } \chi' \in \mathcal{N}_\Sigma(\chi), \\ 0 & \text{otherwise.} \end{cases}$$

The above theorem can be roughly paraphrased by saying that the multiplier μ_χ "isolates" the near-equivalence class $\mathcal{N}_\Sigma(\chi)$ from the other cuspidal data. A large part of the proof given in [17] consists in establishing the existence of a large subalgebra of $\mathcal{M}_\infty(G)$ which admits an explicit spectral description, that is, through its action on irreducible Casselman–Wallach representations of $G(F_\infty)$. The algebra thus constructed generalizes Arthur's multipliers [5] and, moreover, builds on previous work of Delorme [21].

Going back to the setting of the Jacquet–Rallis trace formulae, the above theorem can be applied to isolate in the expansions (4.4) the automorphic L-packet of a given cuspidal automorphic representation π of $G(\mathbb{A}_F)$, on the one hand, and the cuspidal datum χ of G' "supporting" its base-change π_E, on the other hand. Moreover, essentially using the spectral characterization of Theorem 3.2 for the transfer, this can be done by multipliers $\mu_\pi \in \mathcal{M}_\Sigma(G)$ and $\mu_\chi \in \mathcal{M}_\Sigma(G')$ that are compatible with the Jacquet–Rallis transfer in the following sense: if $f = \prod_v f_v \in \mathcal{S}_K(G(\mathbb{A}_F))$ and $f' = \prod_v f'_v \in \mathcal{S}_{K'}(G'(\mathbb{A}_F))$ are transfers of each other then so are $\mu_\pi * f$ and $\mu_\chi * f'$ (where here we take Σ to consist of almost all places that split in E and for K, K' arbitrary compact-open subgroups of $G(\mathbb{A}_f)$, $G'(\mathbb{A}_f)$ that are hyperspecial at places in Σ). All in all, applying these multipliers to global test functions f and f' that are transfers of each other, and comparing the geometric expansions (4.1), we obtain an identity of the following shape:

$$\sum_{W'} \sum_{\substack{\pi' \hookrightarrow \mathcal{A}_{\mathrm{cusp}}(G^{W'}) \\ \pi'_E = \pi_E}} J_{\pi'}(f) = I_\chi(f'),$$

where the outside left sum runs over isomorphism classes of Hermitian spaces of the same dimension as W (or, equivalently, relevant pure inner forms of G). Besides, as a consequence of the local Gan–Gross–Prasad conjecture, when π_E is generic, the left-hand side contains at most one nonzero term. Thus, as a final step to establish the Gan–Gross–Prasad and Ichino–Ikeda conjectures, it only remains to analyze the distribution I_χ. When the base-change π_E is itself cuspidal, that is, when $\chi = \{(G', \pi_E)\}$, by the works of Jacquet–Piatetski-Shapiro–Shalika and Flicker–Rallis already recalled, I_χ essentially factors as the product of the local relative characters $I_{\pi_{E,v}}$ and Theorem 3.2 then allows to conclude. However, in general a similar factorization of I_χ is far from obvious and was actually established in my joint work with Chaudouard and Zydor [16]. It is exactly of the shape predicted by the Ichino–Ikeda conjecture. More precisely:

Theorem 4.3 (Beuzart-Plessis–Chaudouard–Zydor). *Let π be a cuspidal automorphic representation of $G(\mathbb{A}_F)$ whose base-change π_E is generic. Let χ be the cuspidal datum of G' such that π_E contributes to the spectral decomposition of $L^2_\chi([G'])$. Then, for every factorizable test function $f' = \prod_v f'_v \in \mathcal{S}(G'(\mathbb{A}_F))$, we have*

$$I_\chi(f') = \frac{1}{|S_\pi|} \prod_v{}' I_{\pi_{E,v}}(f'_v), \tag{4.6}$$

where the product has to be understood, as for (2.4), "in the sense of L-functions."

In [16], two proofs are actually given of the above theorem: one using truncations operators and the other one based on the global theory of Zeta integrals. For both methods, a crucial step is to spectrally expand the restriction of the Flicker–Rallis period (that is, the integral over $[H_2]$) to functions $\varphi \in L^2_\chi([G'])$ that are sufficiently rapidly decreasing. A consequence of this computation is that this period only depends on the π_E-component of φ and it is mainly for this reason that the contribution of χ to the Jacquet–Rallis trace formula $I(f')$ is eventually discrete (although in the case at hand, $L^2_\chi([G'])$ usually has a purely continuous spectrum). For this, the truncation method is based on the work of Jacquet–Lapid–Rogawski who have defined and studied generalizations of Arthur's truncation operator to the setting of Galois periods and proved analogs of the Maass–Selberg relations in this context. On the other hand, the other method starts by expressing the Flicker–Rallis period as a residue of the integral over $[H_2]$ of φ against an Eisenstein series. Unfolding carefully this expression as in the work of Flicker–Rallis, we can rewrite it as a Zeta integral of the sort that represents Asai L-functions. The precise location of the poles of these L-functions, as well as an explicit residue computation of a family of distributions, then allows to conclude.

Finally, let me mention that in work in progress with P.-H. Chaudouard, we are able to analyze the contributions to the Jacquet–Rallis trace formula of more general cuspidal data $\chi \in \mathfrak{X}(G')$ than that appearing in Theorem [16]. The final result is similar to (4.6) except that the right-hand side has to be integrated over a certain family of automorphic representations π of $G(\mathbb{A}_F)$. More precisely, our results include some cuspidal data supporting the base-changes of automorphic representations of $G = U(V) \times U(W)$ that are Eisenstein in the first factor and cuspidal in the second. In this particular case, the contribution of the corresponding cuspidal datum to the trace formula $J(f)$ is absolutely convergent and a refined spectral expansion can readily be obtained as an integral of Gan–Gross–Prasad periods between a cusp form and an Eisenstein series. These last periods are related, by some unfolding, to Bessel periods of cusp forms on smaller unitary groups. For this reason, our extension of Theorem 4.3 with Chaudouard should have similar applications to the Gan–Gross–Prasad and Ichino–Ikeda conjectures for general Bessel periods.

5. LOOKING FORWARD

As illustrated in the previous sections, various trace formula approaches to the Gan–Gross–Prasad conjectures for unitary groups have been very successful. However, despite these favorable and definite results, these developments also raise interesting questions or have lead to fertile new research direction:

- First, there is the question of whether similar techniques can be applied to prove the global Gan–Gross–Prasad conjectures for other groups. Indeed, the original conjectures in [23] also include general Bessel periods on (product of) orthogonal groups $\mathrm{SO}(n) \times \mathrm{SO}(m)$ ($n \not\equiv m$ [2]), as well as so-called Fourier–Jacobi periods on unitary groups $U(n) \times U(m)$ ($n \equiv m$ [2]) or symplectic/metaplectic groups $\mathrm{Mp}(n) \times \mathrm{Sp}(m)$. In the case of $U(n) \times U(n)$, a relative trace formula approach has been proposed by Y. Liu and further developed by H. Xue [52]. However, the situation is not as complete as for the Jacquet–Rallis trace formulae in the case of $U(n+1) \times U(n)$. It would be interesting to see if the latest developments, in particular those from my joint work with Chaudouard and Zydor [16], can be adapted to this setting. This could possibly lead to a proof of the Gan–Gross–Prasad conjecture for general Fourier–Jacobi periods on unitary groups. The situation for orthogonal and symplectic/metaplectic groups is much less satisfactory and there is no clear approach through a comparison of relative trace formulae, yet. This is due in particular to the fact that, instead of the Flicker–Rallis periods, in these cases we are naturally lead to consider period integrals originally studied by Bump–Ginzburg that detect cuspidal automorphic representations of $\mathrm{GL}(n)$ of orthogonal type. These period integrals involve the product of two exceptional theta series on a double cover of $\mathrm{GL}(n)$ and do not have any obvious geometric realizations (except when $n = 2$). This makes the task of writing a geometric expansion for the corresponding trace formulae quite unclear. It would certainly be interesting to see if the recent Hamiltonian duality picture of Ben Zvi–Sakellaridis–Venkatesh can shed some light on this matter (in particular, by associating a Hamiltonian space to the Bump–Ginzburg periods).

- In the local setting, the new trace formulae first discovered by Waldspurger [47] and further developed in [12] seem to be of quite broad applicability to all kind of *distinction problems*. Actually, similar trace formulae have already been established in other contexts [11, 18, 50] with new applications in the spirit of "relative Langlands functorialities" each time. However, all these developments have been made on a case-by-case basis so far and it would be very interesting and instructive to elaborate a general theory. In particular, in view of the proposal by Sakellaridis–Venkatesh [41] of a general framework for the relative Langlands program, we could hope to establish general local relative trace formulae for the L^2 spaces of spherical varieties X and relate those to the dual group construction of Sakellaridis–Venkatesh.

- Finally, in a slightly different direction the general isolation Theorem 4.2 clearly has the potential to be applied in other context, e.g., it would be interesting to see if it can be used as a technical device to simplify some other known comparison of trace formulae. Another intriguing question is to look for a precise spectral description of the (abstract) multiplier algebra $\mathcal{M}_\infty(G)$ and in [17], we actually argue that $\mathcal{M}_\infty(G)$ should be seen as the natural Archimedean analog of the Bernstein center for p-adic groups.

ACKNOWLEDGMENTS

First and foremost, I thank my former PhD advisor Jean-Loup Waldspurger for introducing me to this fertile circle of ideas and for always being so generous in sharing his insights on these problems.

I am also grateful to my colleagues and coauthors Pierre-Henri Chaudouard, Patrick Delorme, Wee-Teck Gan, Yifeng Liu, Dipendra Prasad, Yiannis Sakellaridis, Chen Wan, Wei Zhang, Xinwen Zhu, and Michal Zydor for many inspiring discussions and exchanges over the years.

FUNDING

The author's work was partially supported by the Excellence Initiative of Aix-Marseille University (A*MIDEX), a French "Investissements d'Avenir" program.

REFERENCES

[1] A. Aizenbud, A partial analog of the integrability theorem for distributions on p-adic spaces and applications. *Israel J. Math.* **193** (2013), no. 1, 233–262.

[2] A. Aizenbud, D. Gourevitch, S. Rallis, and G. Schiffmann, Multiplicity one theorems. *Ann. of Math. (2)* **172** (2010), no. 2, 1407–1434.

[3] J. Arthur, A trace formula for reductive groups. I. Terms associated to classes in $G(\mathbb{Q})$. *Duke Math. J.* **45** (1978), no. 4, 911–952.

[4] J. Arthur, A trace formula for reductive groups. II. Applications of a truncation operator. *Compos. Math.* **40** (1980), no. 1, 87–121.

[5] J. Arthur, A Paley–Wiener theorem for real reductive groups. *Acta Math.* **150** (1983), no. 1–2, 1–89.

[6] J. Arthur, On elliptic tempered characters. *Acta Math.* **171** (1993), no. 1, 73–138.

[7] J. Arthur, *The endoscopic classification of representations. Orthogonal and symplectic groups.* Amer. Math. Soc. Colloq. Publ. 61, American Mathematical Society, Providence, RI, 2013, xviii+590 pp.

[8] R. Beuzart-Plessis, Expression d'un facteur epsilon de paire par une formule intégrale. *Canad. J. Math.* **66** (2014), no. 5, 993–1049.

[9] R. Beuzart-Plessis, Endoscopie et conjecture locale raffinée de Gan–Gross–Prasad pour les groupes unitaires. *Compos. Math.* **151** (2015), no. 7, 1309–1371.

[10] R. Beuzart-Plessis, La conjecture locale de Gross–Prasad pour les représentations tempérées des groupes unitaires. *Mém. Soc. Math. Fr. (N.S.)* **149** (2016), vii+191 pp.

[11] R. Beuzart-Plessis, On distinguished square-integrable representations for Galois pairs and a conjecture of Prasad. *Invent. Math.* **214** (2018), no. 1, 437–521.

[12] R. Beuzart-Plessis, A local trace formula for the Gan–Gross–Prasad conjecture for unitary groups: the Archimedean case. *Astérisque* **418** (2020), viii+299 pp.

[13] R. Beuzart-Plessis, Comparison of local relative characters and the Ichino–Ikeda conjecture for unitary groups. *J. Inst. Math. Jussieu* **20** (2021), no. 6, 1803–1854.

[14] R. Beuzart-Plessis, A new proof of the Jacquet–Rallis fundamental lemma. *Duke Math. J.* **170** (2021), no. 12, 2805–2814.

[15] R. Beuzart-Plessis, Plancherel formula for $\mathrm{GL}_n(F)\backslash \mathrm{GL}_n(E)$ and applications to the Ichino–Ikeda and formal degree conjectures for unitary groups. *Invent. Math.* **225** (2021), no. 1, 159–297.

[16] R. Beuzart-Plessis, P.-H. Chaudouard, and M. Zydor, The global Gan–Gross–Prasad conjecture for unitary groups: the endoscopic case. *Publ. Math. Inst. Hautes Études Sci.* (2022).

[17] R. Beuzart-Plessis, Y. Liu, W. Zhang, and X. Zhu, Isolation of cuspidal spectrum, with application to the Gan–Gross–Prasad conjecture. *Ann. of Math. (2)* **194** (2021), no. 2, 519–584.

[18] R. Beuzart-Plessis and C. Wan, A local trace formula for the generalized Shalika model. *Duke Math. J.* **168** (2019), no. 7, 1303–1385.

[19] W. Casselman, Canonical extensions of Harish-Chandra modules to representations of G. *Canad. J. Math.* **XLI** (1989), no. 3, 385–438.

[20] P.-H. Chaudouard and M. Zydor, Le transfert singulier pour la formule des traces de Jacquet–Rallis. *Compos. Math.* **157** (2021), no. 2, 303–434.

[21] P. Delorme, Formules limites et formules asymptotiques pour les multiplicités dans $L^2(G/\Gamma)$. *Duke Math. J.* **53** (1986), no. 3, 691–731.

[22] Y. Z. Flicker, Twisted tensors and Euler products. *Bull. Soc. Math. France* **116** (1988), no. 3, 295–313.

[23] W. T. Gan, B. H. Gross, and D. Prasad, Symplectic local root numbers, central critical L values, and restriction problems in the representation theory of classical groups. In *Sur les conjectures de Gross et Prasad. I*, pp. 1–109, Astérisque 346, Société Mathématique de France, Paris, 2012.

[24] W. T. Gan and A. Ichino, The Gross–Prasad conjecture and local theta correspondence. *Invent. Math.* **206** (2016), no. 3, 705–799.

[25] D. Ginzburg, D. Jiang, and S. Rallis, On the nonvanishing of the central value of the Rankin–Selberg L-functions. *J. Amer. Math. Soc.* **17** (2004), no. 3, 679–722.

[26] B. H. Gross and D. Prasad, On irreducible representations of $\mathrm{SO}_{2n+1} \times \mathrm{SO}_{2m}$. *Canad. J. Math.* **46** (1994), no. 5, 930–950.

[27] R. N. Harris, The refined Gross–Prasad conjecture for unitary groups. *Int. Math. Res. Not. IMRN* **2** (2014), 303–389.

[28] H. He, On the Gan–Gross–Prasad conjecture for $U(p,q)$. *Invent. Math.* **209** (2017), no. 3, 837–884.

[29] K. Hiraga, A. Ichino, and T. Ikeda, Formal degrees and adjoint γ-factors. *J. Amer. Math. Soc.* **21** (2008), no. 1, 283–304.

[30] A. Ichino and T. Ikeda, On the periods of automorphic forms on special orthogonal groups and the Gross–Prasad conjecture. *Geom. Funct. Anal.* **19** (2010), no. 5, 1378–1425.

[31] H. Jacquet, I. I. Piatetskii-Shapiro, and J. A. Shalika, Rankin–Selberg convolutions. *Amer. J. Math.* **105** (1983), no. 2, 367–464.

[32] H. Jacquet and S. Rallis, On the Gross–Prasad conjecture for unitary groups. In *On certain L-functions*, pp. 205–264, Clay Math. Proc. 13, Amer. Math. Soc., Providence, RI, 2011.

[33] D. Jiang and L. Zhang, Arthur parameters and cuspidal automorphic modules of classical groups. *Ann. of Math. (2)* **191** (2020), no. 3, 739–827.

[34] T. Kaletha, A. Minguez, S. W. Shin, and P.-J. White, Endoscopic classification of representations: inner forms of unitary groups. 2014, arXiv:1409.3731.

[35] Y. Liu, Refined global Gan–Gross–Prasad conjecture for Bessel periods. *J. Reine Angew. Math.* **717** (2016), 133–194.

[36] Z. Luo, The local Gan–Gross–Prasad conjecture for special orthogonal groups over Archimedean local fields. 2021, arXiv:2102.11404.

[37] D. Mezer, Multiplicity one theorems over positive characteristic. 2020, arXiv:2010.16112.

[38] C. Mœglin and J.-L. Waldspurger, La conjecture locale de Gross–Prasad pour les groupes spéciaux orthogonaux: le cas général. In *Sur les conjectures de Gross et Prasad. II*, pp. 167–216, Astérisque 347, Société Mathématique de France, Paris, 2012.

[39] C. P. Mok, Endoscopic classification of representations of quasi-split unitary groups. *Mem. Amer. Math. Soc.* **235** (2015), no. 1108, vi+248 pp.

[40] Y. Sakellaridis, The Schwartz space of a smooth semi-algebraic stack. *Selecta Math. (N.S.)* **22** (2016), no. 4, 2401–2490.

[41] Y. Sakellaridis and A. Venkatesh, Periods and harmonic analysis on spherical varieties. *Astérisque* **396** (2017), viii+360 pp.

[42] B. Sun and C.-B. Zhu, Multiplicity one theorems: the Archimedean case. *Ann. of Math. (2)* **175** (2012), no. 1, 23–44.

[43] J. Tate, Number theoretic background. In *Automorphic forms, representations and L-functions (Proc. Sympos. Pure Math., Oregon State Univ., Corvallis, Ore., 1977), Part 2*, pp. 3–26, Proc. Sympos. Pure Math. XXXIII, Amer. Math. Soc., Providence, R.I., 1977.

[44] J. Tunnell, Local ε-factors and characters of GL(2). *Amer. J. Math.* **105** (1983), no. 6, 1277–1307.

[45] D. A. Vogan, The local Langlands conjecture. In *Representation theory of groups and algebras*, pp. 305–379, Contemp. Math. 145, Amer. Math. Soc., Providence, RI, 1993.

[46] J.-L. Waldspurger, Sur les valeurs de certaines fonctions L automorphes en leur centre de symétrie. *Compos. Math.* **54** (1985), no. 2, 173–242.

[47] J.-L. Waldspurger, Une formule intégrale reliée à la conjecture locale de Gross–Prasad. *Compos. Math.* **146** (2010), no. 5, 1180–1290.

[48] J.-L. Waldspurger, Calcul d'une valeur d'un facteur ε par une formule intégrale. In *Sur les conjectures de Gross et Prasad. II*, pp. 1–102, Astérisque 347, Société Mathématique de France, Paris, 2012.

[49] J.-L. Waldspurger, La conjecture locale de Gross–Prasad pour les représentations tempérées des groupes spéciaux orthogonaux. In *Sur les conjectures de Gross et Prasad. II*, pp. 103–165, Astérisque 347, Société Mathématique de France, Paris, 2012.

[50] C. Wan, A local relative trace formula for the Ginzburg–Rallis model: the geometric side. *Mem. Amer. Math. Soc.* **261** (2019), no. 1263.

[51] J. Xiao, Endoscopic transfer for unitary Lie algebras. 2018, arXiv:1802.07624.

[52] H. Xue, The Gan–Gross–Prasad conjecture for $U(n) \times U(n)$. *Adv. Math.* **262** (2014), 1130–1191.

[53] H. Xue, On the global Gan–Gross–Prasad conjecture for unitary groups: approximating smooth transfer of Jacquet–Rallis. *J. Reine Angew. Math.* **756** (2019), 65–100.

[54] H. Xue, Bessel models for real unitary groups: the tempered case, 2020, https://www.math.arizona.edu/~xuehang/lggp.pdf.

[55] Z. Yun, The fundamental lemma of Jacquet and Rallis. With an appendix by Julia Gordon. *Duke Math. J.* **156** (2011), no. 2, 167–227.

[56] W. Zhang, Automorphic period and the central value of Rankin–Selberg L-function. *J. Amer. Math. Soc.* **27** (2014), no. 2, 541–612.

[57] W. Zhang, Fourier transform and the global Gan–Gross–Prasad conjecture for unitary groups. *Ann. of Math. (2)* **180** (2014), no. 3, 971–1049.

[58] M. Zydor, La variante infinitésimale de la formule des traces de Jacquet–Rallis pour les groupes unitaires. *Canad. J. Math.* **68** (2016), no. 6, 1382–1435.

[59] M. Zydor, La variante infinitésimale de la formule des traces de Jacquet–Rallis pour les groupes linéaires. *J. Inst. Math. Jussieu* **17** (2018), no. 4, 735–783.

RAPHAËL BEUZART-PLESSIS

Aix Marseille Univ, CNRS, I2M, Marseille, France, raphael.beuzart-plessis@univ-amu.fr

THE COHOMOLOGY OF SHIMURA VARIETIES WITH TORSION COEFFICIENTS

ANA CARAIANI

ABSTRACT

In this article, we survey recent work on some vanishing conjectures for the cohomology of Shimura varieties with torsion coefficients, under both local and global conditions. We discuss the p-adic geometry of Shimura varieties and of the associated Hodge–Tate period morphism, and explain how this can be used to make progress on these conjectures. Finally, we describe some applications of these results, in particular to the proof of the Sato–Tate conjecture for elliptic curves over CM fields.

MATHEMATICS SUBJECT CLASSIFICATION 2020

Primary 11G18; Secondary 11R39, 11F33, 11F70, 11F75, 11F80

KEYWORDS

Shimura varieties, locally symmetric spaces, automorphic representations, Galois representations, p-adic Langlands programme

INTERNATIONAL CONGRESS OF MATHEMATICIANS
2022 JULY 6–14

Proc. Int. Cong. Math. 2022, Vol. 3, pp. 1744–1766
DOI 10.4171/ICM2022/161
Published by EMS Press

1. INTRODUCTION

Shimura varieties are algebraic varieties defined over number fields and equipped with many symmetries, which often provide a geometric realization of the Langlands correspondence. After base change to $\mathbb{C}$, they are closely related to certain locally symmetric spaces, but the beauty of Shimura varieties lies in their rich arithmetic.

To describe a Shimura variety, one needs to start with a Shimura datum (G, X). Here, G is a connected reductive group over $\mathbb{Q}$ and X is a conjugacy class of homomorphisms $h : \mathrm{Res}_{\mathbb{C}/\mathbb{R}}\mathbb{G}_m \to G_{\mathbb{R}}$ of algebraic groups over $\mathbb{R}$. Both G and X are required to satisfy certain highly restrictive axioms, cf. **[22, §2.1]**. In particular, this allows one to give the conjugacy class X a more geometric flavor, as a variation of polarisable Hodge structures. One can show that such an X is a disjoint union of finitely many copies of Hermitian symmetric domains.

Let $K \subset G(\mathbb{A}_f)$ be a sufficiently small compact open subgroup (the precise technical condition is called "neat"). The double quotient $G(\mathbb{Q})\backslash X \times G(\mathbb{A}_f)/K$, a priori a complex manifold, comes from an algebraic variety S_K defined over a number field E, called the reflex field of the Shimura datum. The varieties S_K are smooth and quasiprojective. Their étale cohomology groups (with or without compact support) $H^*_{(c)}(S_K \times_E \overline{\mathbb{Q}}, \mathbb{Q}_\ell)$ are equipped with two kinds of symmetries. There is a Hecke symmetry coming from varying the level, i.e., the compact open subgroup K, and considering various transition morphisms between Shimura varieties at different levels. There is also a Galois symmetry, coming from the natural action of $\mathrm{Gal}(\overline{E}/E)$ on étale cohomology.

For this reason, Shimura varieties have played an important role in realizing instances of the global Langlands correspondence over number fields. Indeed, a famous conjecture of Kottwitz predicts the relationship between the Galois representations occurring in the ℓ-adic étale cohomology of the Shimura varieties for G and those Galois representations associated with (regular, C-algebraic) cuspidal automorphic representations of G. See **[64, REMARK 1.1.1]** for a modern formulation of this conjecture.

There is a complete classification of groups that admit a Shimura datum. For example, if $G = \mathrm{GSp}_{2n}$, one can take X to be the Siegel double space

$$\{Z \in \mathrm{M}_n(\mathbb{C}) \mid Z = Z^t, \mathrm{Im}(Z) \text{ positive or negative definite}\}. \tag{1.1}$$

The associated Shimura varieties are called Siegel modular varieties and they are moduli spaces of principally polarized abelian varieties. Many other Shimura varieties – those of so-called "abelian type" – can be studied using moduli-theoretic techniques, by relating them to Siegel modular varieties. See **[39]** for an excellent introduction to the subject, which is focused on examples.

In this article, we will be primarily concerned with the geometry of the Shimura varieties S_K, after base change to a p-adic field, as well as with their étale cohomology groups $H^*_{(c)}(S_K \times_E \overline{\mathbb{Q}}, \mathbb{F}_\ell)$ with torsion coefficients. These groups are much less understood than their characteristic zero counterparts. We discuss certain conjectures about when these cohomology groups are expected to vanish, under both global and local conditions. Furthermore, we explain how the geometry of the Hodge–Tate period morphism, introduced in **[53]** and refined in [17], can be used to make progress on these conjectures. Finally, we describe some applications of these results, in particular to the proof of the Sato–Tate conjecture for elliptic curves over CM fields [1].

2. A VANISHING CONJECTURE FOR LOCALLY SYMMETRIC SPACES

Let $G/\mathbb{Q}$ be a connected reductive group. We consider the symmetric space associated with the Lie group $G(\mathbb{R})$, which we define as $X = G(\mathbb{R})/K_\infty^\circ A_\infty^\circ$. Here, K_∞° is the connected component of the identity in a maximal compact subgroup $K_\infty \subset G(\mathbb{R})$, and A_∞° is the connected component of the identity inside the real points of the maximal $\mathbb{Q}$-split torus in the center of G. Given a neat compact open subgroup $K \subset G(\mathbb{A}_f)$, we can form the double quotient $X_K = G(\mathbb{Q})\backslash X \times G(\mathbb{A}_f)/K$, which we call a locally symmetric space for G. This is a smooth Riemannian manifold, which does not have a complex structure, in general.

Example 2.1. If $G = \mathrm{SL}_2/\mathbb{Q}$, we can identify $X = \mathrm{SL}_2(\mathbb{R})/\mathrm{SO}_2(\mathbb{R})$ with the upper half-plane $\mathbb{H}^2 = \{z \in \mathbb{C} \mid \mathrm{Im} z > 0\}$ equipped with the hyperbolic metric, on which $\mathrm{SL}_2(\mathbb{R})$ acts transitively by the isometries

$$z \mapsto \frac{az+b}{cz+d} \quad \text{for } z \in \mathbb{H}^2 \quad \text{and} \quad \begin{pmatrix} a & b \\ c & d \end{pmatrix} \in \mathrm{SL}_2(\mathbb{R}).$$

Under this action, $\mathrm{SO}_2(\mathbb{R})$ is the stabilizer of the point i. By strong approximation for $\mathrm{SL}_2/\mathbb{Q}$, for any compact open subgroup $K \subseteq \mathrm{SL}_2(\widehat{\mathbb{Z}})$, there is only one double coset $\mathrm{SL}_2(\mathbb{Q})\backslash \mathrm{SL}_2(\mathbb{A}_f)/K$. Write $\Gamma = \mathrm{SL}_2(\mathbb{Q}) \cap K$, which will be a congruence subgroup contained in $\mathrm{SL}_2(\mathbb{Z})$. The locally symmetric spaces X_K can be identified with quotients $\Gamma\backslash\mathbb{H}^2$. For Γ neat, these quotients inherit the complex structure on $\mathbb{H}^2$ and can be viewed as Riemann surfaces. Even more, these quotients arise from algebraic curves called *modular curves*, which are defined over finite extensions of $\mathbb{Q}$. Modular curves are examples of (connected) Shimura varieties. They represent moduli problems of elliptic curves endowed with additional structures. Even though they are some of the simplest Shimura varieties (the main complication being that they are noncompact), their geometry is already fascinating.

However, let $F/\mathbb{Q}$ be an imaginary quadratic field and take $G = \mathrm{Res}_{F/\mathbb{Q}}\mathrm{SL}_2$. Then we can identify the symmetric space $X = \mathrm{SL}_2(\mathbb{C})/\mathrm{SU}_2(\mathbb{R})$ with 3-dimensional hyperbolic space $\mathbb{H}^3$. Once again, we can identify the locally symmetric spaces X_K with quotients $\Gamma\backslash\mathbb{H}^3$, where $\Gamma = \mathrm{SL}_2(F) \cap K$ is a congruence subgroup. In this case, the locally symmetric spaces are arithmetic hyperbolic 3-manifolds and do not admit a complex structure. In particular, we cannot speak of Shimura varieties in this setting.

In general, Shimura varieties are closely related to locally symmetric spaces, as in the first example, though the latter are much more general objects. For example, the locally symmetric spaces for $G = \mathrm{Res}_{F/\mathbb{Q}}\mathrm{GL}_n$ do not arise from Shimura varieties if $n \geq 3$, and, for $n = 2$, they can only be related to Shimura varieties if F is a totally real field. In some instances, such as for $\mathrm{Res}_{F/\mathbb{Q}}\mathrm{GL}_2$ with F a totally real field, one needs to replace

$G(\mathbb{R})/K_\infty^\circ A_\infty^\circ$ by a slightly different quotient in order to obtain Shimura varieties.[1] We now define the invariants

$$l_0 = \mathrm{rk}\big(G(\mathbb{R})\big) - \mathrm{rk}(K_\infty) - \mathrm{rk}(A_\infty) \quad \text{and} \quad q_0 = \frac{1}{2}\big(\dim_{\mathbb{R}}(X) - l_0\big).$$

These were first introduced by Borel–Wallach in **[5]**. There, they show up naturally in the computation of the $(\mathfrak{g}, K_\infty)$-cohomology of tempered representations of $G(\mathbb{R})$. In the Shimura variety setting, we consider the variants $l_0 = l_0(G^{\mathrm{ad}})$ and $q_0 = q_0(G^{\mathrm{ad}})$ because of the different quotient used. In this case, $l_0(G^{\mathrm{ad}})$ can be shown to be equal to 0 by the second axiom in the definition of a Shimura datum.

As K varies, we have a tower of locally symmetric spaces $(X_K)_K$, on which a spherical Hecke algebra $\mathbb{T}$ for G acts by correspondences. The systems of Hecke eigenvalues occurring in the cohomology groups $H^*_{(c)}(X_K, \mathbb{C})$ or, equivalently, the maximal ideals of $\mathbb{T}$ in the support of these cohomology groups, can be related to automorphic representations of $G(\mathbb{A}_f)$ by work of Franke and Matsushima **[29]**. The goal of this section is to state a conjecture on the cohomology of locally symmetric spaces with torsion coefficients $\mathbb{F}_\ell$, where ℓ is a prime number. This conjecture is formulated in **[25]** (see the discussion around Conjecture 3.3) and in **[12, CONJECTURE B]**. Roughly, it says that the part of the cohomology outside the range of degrees $[q_0, q_0 + l_0]$ is somehow degenerate. Note that this range of degrees is symmetric about the middle $\frac{1}{2}\dim_{\mathbb{R}} X$ of the total range of cohomology and, in the Shimura variety case, it equals the middle degree of cohomology.

To formulate this more precisely, we use the notion of a non-Eisenstein maximal ideal in the Hecke algebra, for which we need to pass to the Galois side of the global Langlands correspondence. For simplicity, we will restrict our formulation to the case of $G = \mathrm{Res}_{F/\mathbb{Q}}\mathrm{GL}_n$ for some number field F, although the conjecture makes sense more generally. Let $\mathbb{T}$ be the abstract spherical Hecke algebra away from a finite set S of primes of F and let $\mathfrak{m} \subset \mathbb{T}$ be a maximal ideal in the support of $H^*_{(c)}(X_K, \mathbb{F}_\ell)$. Assume that there exists a continuous, semisimple Galois representation $\bar{\rho}_{\mathfrak{m}} : \mathrm{Gal}(\overline{F}/F) \to \mathrm{GL}_n(\overline{\mathbb{F}}_\ell)$ *associated with* $\mathfrak{m}$: by this, we mean that $\bar{\rho}_{\mathfrak{m}}$ is unramified at all the primes of F away from the finite set S, and that, at any prime away from S, the Satake parameters of $\mathfrak{m}$ match the Frobenius eigenvalues of $\bar{\rho}_{\mathfrak{m}}$. (The precise condition is in terms of the characteristic polynomial of $\bar{\rho}_{\mathfrak{m}}$ applied to the Frobenius at such a prime and depends on various choices of normalizations. See, for example, **[1, THEOREM 2.3.5]** for a precise formulation.) Since the Galois representation is assumed to be semisimple and we are working with $\mathrm{Res}_{F/\mathbb{Q}}\,\mathrm{GL}_n$, this property will characterize $\bar{\rho}_{\mathfrak{m}}$ by the Cebotarev density theorem and the Brauer–Nesbitt theorem. We say that $\mathfrak{m}$ is *non-Eisenstein* if such a $\bar{\rho}_{\mathfrak{m}}$ is absolutely irreducible.

The existence of $\bar{\rho}_{\mathfrak{m}}$ as above should be thought of as a mod ℓ version of the global Langlands correspondence, in the automorphic-to-Galois direction; in the case $F = \mathbb{Q}$, this

---

1 We make a small abuse of notation by using X to denote both the conjugacy class from the introduction, which is used in the definition of a Shimura datum, and the quotient $G(\mathbb{R})/K_\infty^\circ A_\infty^\circ$ considered in this section. See **[24, §2.4]** for an extended discussion of the various quotients.

existence was conjectured by Ash [4]. The striking part of this conjecture is that it should apply to torsion classes in the cohomology of locally symmetric spaces, not just to those classes that lift to characteristic zero, and which can be related to automorphic representations of G. For general number fields, the existence of such Galois representations seems out of reach at the moment, even for classes in characteristic zero!

However, let F be a CM field: using nonstandard terminology, we mean that F is either a totally real field or a totally complex quadratic extension thereof. In this case, Scholze constructed such Galois representations in the breakthrough paper [53]. This strengthened previous work [33] that applied to cohomology with $\mathbb{Q}_\ell$-coefficients. Both these results relied, in turn, on the construction of Galois representations in the self-dual case, due to many people, including Clozel, Kottwitz, Harris–Taylor [34], Shin [61], and Chenevier–Harris [21].

We can now state the promised vanishing conjecture for the cohomology of locally symmetric spaces with $\mathbb{F}_\ell$-coefficients.

Conjecture 2.2. *Assume that $\mathfrak{m} \subset \mathbb{T}$ is a non-Eisenstein maximal ideal in the support of $H^*_{(c)}(X_K, \mathbb{F}_\ell)$. Then $H^i_{(c)}(X_K, \mathbb{F}_\ell)_\mathfrak{m} \neq 0$ only if $i \in [q_0, q_0 + l_0]$.*

In the two examples discussed in Example 2.1, this conjecture can be verified "by hand," since one only needs to control cohomology in degree 0 (the top degree of cohomology can be controlled using Poincaré duality). In the case of $\mathrm{GL}_2/\mathbb{Q}$, one can show that the systems of Hecke eigenvalues $\mathfrak{m}$ in the support of $H^0(X_K, \mathbb{F}_\ell)$ satisfy

$$\bar{\rho}_\mathfrak{m} \simeq \chi \oplus \chi_{\text{cyclo}} \cdot \chi, \tag{2.1}$$

where χ is a suitable mod ℓ character of $\mathrm{Gal}(\overline{\mathbb{Q}}/\mathbb{Q})$ and $\chi_{\text{cyclo}} : \mathrm{Gal}(\overline{\mathbb{Q}}/\mathbb{Q}) \to \mathbb{F}_\ell^\times$ is the mod ℓ cyclotomic character. Later, we will introduce a local *genericity* condition at an auxiliary prime $p \neq \ell$ and we will see that the $\bar{\rho}_\mathfrak{m}$ in (2.1) also fail to satisfy genericity everywhere. In addition to these and a few more low-dimensional examples, one can also consider the analogue of Conjecture 2.2 for $H^*_{(c)}(X_K, \mathbb{Q}_\ell)$. This analogue is related to Arthur's conjectures on the cohomology of locally symmetric spaces [3] and can be verified for GL_n over CM fields using work of Franke and Borel–Wallach (see [1, THEOREM 2.4.9]).

Conjecture 2.2 is motivated by the Calegari–Geraghty enhancement [12] of the classical Taylor–Wiles method for proving automorphy lifting theorems. The classical method works well in settings where the (co)homology of locally symmetric spaces is concentrated in one degree, for example, for $\mathrm{GL}_2/\mathbb{Q}$ after localizing at a non-Eisenstein maximal ideal, or for definite unitary groups over totally real fields. In general, however, a certain numerical coincidence that is used to compare the Galois and automorphic sides breaks down. Calegari and Geraghty had a significant insight: they reinterpret the failure of the numerical coincidence in terms of the invariant l_0. More precisely, l_0 arises naturally from a computation on the Galois side, and the commutative algebra underlying the method can be adjusted if one knows that the cohomology on the automorphic side, after localizing at a non-Eisenstein maximal ideal, is concentrated in a range of degrees of length at most l_0. For an overview of the key ideas involved in the Calegari–Geraghty method, see [10, §10].

In the case of Shimura varieties, Conjecture 2.2 predicts that the non-Eisenstein part of the cohomology with $\mathbb{F}_\ell$-coefficients is concentrated in the middle degree. The initial progress on this conjecture in the Shimura variety setting had rather strong additional assumptions: for example, one needed ℓ to be an unramified prime for the Shimura datum and K_ℓ to be hyperspecial, as in the work of Dimitrov [23] and Lan–Suh [40, 41]. The theory of perfectoid Shimura varieties and their associated Hodge–Tate period morphism has been a game-changer in this area. For the rest of this article, we will discuss more recent progress on Conjecture 2.2 and related questions in the special case of Shimura varieties, as well as applications that go beyond the setting of Shimura varieties.

3. THE HODGE–TATE PERIOD MORPHISM

The Hodge–Tate period morphism was introduced by Scholze in his breakthrough paper [53] and it was subsequently refined in [17]. It gives an entirely new way to think about the geometry and cohomology of Shimura varieties. In the past decade, it had numerous striking applications to the Langlands programme: to Scholze's construction of Galois representations for torsion classes, to the vanishing theorems discussed in Sections 4 and 5, to the construction of higher Coleman theory by Boxer and Pilloni [8], and to a radically new approach to the Fontaine–Mazur conjecture due to Pan [48].

For simplicity, let us consider a Shimura datum (G, X) of *of Hodge type*. By this, we mean that (G, X) admits a closed embedding into a Siegel datum $(\tilde{G}, \tilde{X})$, where $\tilde{G} = \mathrm{GSp}_{2n}$, for some $n \in \mathbb{Z}_{\geq 1}$, and $\tilde{X}$ is as in (1.1). For example, (G, X) could be a Shimura datum of *PEL type* arising from a unitary similitude group: the corresponding Shimura varieties will represent a moduli problem of abelian varieties equipped with extra structures (polarizations, endomorphisms, and level structures). This unitary case will be the main example to keep in mind, as this will also play a central role in Section 4.

For some representative $h \in X$, we consider the *Hodge cocharacter*

$$\mu = h \times_{\mathbb{R}} \mathbb{C}|_{1\text{st } \mathbb{G}_m \text{ factor}} : \mathbb{G}_{m,\mathbb{C}} \to G_{\mathbb{C}}.$$

The axioms in the definition of the Shimura datum imply that μ is minuscule. The reflex field E is the field of definition of the conjugacy class $\{\mu\}$; it is a finite extension of $\mathbb{Q}$ and the corresponding Shimura varieties admit canonical models over E. The cocharacter μ also determines two opposite parabolic subgroups P_μ^{std} and P_μ, whose conjugacy classes are defined over E. These are given by

$$P_\mu^{\mathrm{std}} = \Big\{g \in G \mid \lim_{t\to\infty} \mathrm{ad}\big(\mu(t)\big)g \text{ exists}\Big\}, \quad P^\mu = \Big\{g \in G \mid \lim_{t\to 0} \mathrm{ad}\big(\mu(t)\big)g \text{ exists}\Big\}.$$

We let $\mathrm{Fl}^{\mathrm{std}}$ and Fl denote the associated flag varieties, which are also defined over E.

Here is a more moduli-theoretic way to think about these the two parabolics. The chosen symplectic embedding $(G, X) \hookrightarrow (\tilde{G}, \tilde{X})$ gives rise to a faithful representation V of G. The embedding also gives rise to an abelian scheme A_K over the Shimura variety S_K at some level $K = G(\mathbb{A}_f) \cap \tilde{K}$, obtained by restricting the universal abelian scheme over the Siegel modular variety at level $\tilde{K} \subset \tilde{G}(\mathbb{A}_f)$. The cocharacter μ induces a grading of $V_{\mathbb{C}}$, which in turn defines two filtrations on $V_{\mathbb{C}}$, a descending one $\mathrm{Fil}^\bullet$ and an ascending one $\mathrm{Fil}_\bullet$. The parabolic P_μ^{std} is the stabilizer of $\mathrm{Fil}^\bullet$, which is morally the Hodge–de Rham filtration on the Betti cohomology of A_K. There is a holomorphic, $G(\mathbb{R})$-equivariant embedding

$$\pi_{\mathrm{dR}} : X \hookrightarrow \mathrm{Fl}^{\mathrm{std}}(\mathbb{C}) = G(\mathbb{C})/P_\mu^{\mathrm{std}} \tag{3.1}$$

called the Borel embedding, defined by $h \mapsto \mathrm{Fil}^\bullet(\mu_h)$. The axioms of a Shimura datum imply that X is a variation of polarisable Hodge structures of abelian varieties. Moduli-theoretically, π_{dR} sends a Hodge structure, such as

$$H^1(A,\mathbb{Q}) \otimes_{\mathbb{Q}} \mathbb{C} \simeq H^0(A,\Omega_A^1) \oplus H^1(A,\mathcal{O}_A),$$

to the associated Hodge–de Rham filtration, e.g., $H^0(A,\Omega_A^1) \subset H^1(A,\mathbb{Q}) \otimes_{\mathbb{Q}} \mathbb{C}$. The embedding π_{dR} is an example of a *period morphism*. Historically, it has played an important role in the construction of canonical models of automorphic vector bundles over E (or even integrally), such as in work of Harris and Milne.

On the other hand, the parabolic subgroup P_μ is the stabilizer of the ascending filtration $\mathrm{Fil}_\bullet$. This gives rise to an antiholomorphic embedding

$$X \hookrightarrow \mathrm{Fl}(\mathbb{C}) = G(\mathbb{C})/P_\mu. \tag{3.2}$$

Morally, P_μ is the stabilizer of the Hodge–Tate filtration on the p-adic étale cohomology of A_K. The Hodge–Tate period morphism will be a p-adic analogue of the embedding (3.1) (or perhaps of the embedding (3.2), depending on one's perspective).

Let p be a rational prime, $\mathfrak{p} \mid p$ a prime of E, and let C be the completion of an algebraic closure of $E_{\mathfrak{p}}$. We consider the adic spaces $\mathcal{S}_K$ and $\mathscr{F}\ell$ over $\mathrm{Spa}(C,\mathcal{O}_C)$ corresponding to the algebraic varieties S_K and Fl over E. A striking result of Scholze shows that the tower of Shimura varieties $(\mathcal{S}_{K^pK_p})_{K_p}$ acquires the structure of a perfectoid space (in the sense of [51]) as K_p varies over compact open subgroups of $G(\mathbb{Q}_p)$. More precisely, the following result was established in [53, §3,4] and later refined in [17, §2], by correctly identifying the target of the Hodge–Tate period morphism.

Theorem 3.1. *There exists a unique perfectoid space $\mathcal{S}_{K^p}$ satisfying $\mathcal{S}_{K^p} \sim \varprojlim_{K_p} \mathcal{S}_{K^pK_p}$,*[2] *in the sense of* [55, DEFINITION 2.4.1], *and a $G(\mathbb{Q}_p)$-equivariant morphism of adic spaces*

$$\pi_{\mathrm{HT}} : \mathcal{S}_{K^p} \to \mathscr{F}\ell.$$

Moreover, π_{HT} is equivariant for the usual action of Hecke operators away from p on $\mathcal{S}_{K^p}$ and their trivial action on $\mathscr{F}\ell$.

In the Siegel case $G = \mathrm{GSp}_{2n}/\mathbb{Q}$, one can describe the Hodge–Tate period morphism π_{HT} from a moduli-theoretic perspective as follows. An abelian variety A/C, equipped with a trivialization $T_pA \simeq \mathbb{Z}_p^{2n}$ will be sent to the first piece of the Hodge–Tate filtration

$$\mathrm{Lie}\, A \subset T_pA \otimes_{\mathbb{Z}_p} C \simeq C^{2n}.$$

2 It is enough to consider the Shimura varieties as adic spaces over $E_{\mathfrak{p}}$ and the tower still acquires a perfectoid structure in a noncanonical way. We work over C for simplicity and also because this gives rise to the étale cohomology groups we want to understand.

Dually, one has the Hodge–Tate filtration on the p-adic étale cohomology of A:

$$0 \to H^1(A, \mathcal{O}_A) \to H^1_{\text{ét}}(A, \mathbb{Z}_p) \otimes_{\mathbb{Z}_p} C \to H^0(A, \Omega^1_{A/C})(-1) \to 0, \tag{3.3}$$

where (-1) denotes a Tate twist (which is important for keeping track of the Galois action). To show that the morphism defined this way on $\mathrm{Spa}(C, C^+)$-points comes from a morphism of adic spaces, it is important to know that the filtration (3.3) varies continuously. At the same time, to extend the result to Shimura varieties of Hodge type and to cut down the image to $\mathcal{F}\ell$, one needs to keep track of Hodge tensors carefully. Both problems are solved via relative p-adic Hodge theory for the morphism $\mathcal{A}_K \to \mathcal{S}_K$, where $\mathcal{A}_K$ is the restriction to $\mathcal{S}_K$ of a universal abelian scheme over an ambient Siegel modular variety. See **[13, §3]** for an overview.

Theorem 3.1 can be extended to minimal and toroidal compactifications of Siegel modular varieties, cf. **[53]** and **[49]**. Moreover, there is a natural affinoid cover of $\mathcal{F}\ell$ such that the preimage under π_{HT} of each affinoid in the cover is an affinoid perfectoid subspace of $\mathcal{S}^*_{K^p}$. The underlying reason for this is the fact that the partial minimal compactification of the ordinary locus is affine. The perfectoid structure on $\mathcal{S}^*_{K^p}$ and the affinoid nature of the Hodge–Tate period morphism play an important role in Scholze's p-adic interpolation argument, that is key for the construction of Galois representations associated with torsion classes. See also **[44]** for an exposition of the main ideas.

Theorem 3.1 can also be extended to minimal and toroidal compactifications of Shimura varieties of Hodge type and even abelian type, cf. **[32, 58]** and **[8]**, although there are some technical issues at the boundary. For example, the cleanest formulation currently available in full generality is that the relationship $\mathcal{S}^*_{K^p} = \varprojlim_{K_p} \mathcal{S}^*_{K^p K_p}$, for a perfectoid space $\mathcal{S}^*_{K^p}$, holds in Scholze's category of diamonds **[54]**.

Example 3.2. To see where the perfectoid structure on $\mathcal{S}_{K^p}$ comes from, it is instructive to consider the case of modular curves and study the geometry of their special fibers: we are particularly interested in the geometry of the so-called Deligne–Rapoport model. Set $G = \mathrm{GL}_2/\mathbb{Q}$. Let $K_p^0 = \mathrm{GL}_2(\mathbb{Z}_p)$, the hyperspecial compact open subgroup and let $\overline{S}_{K^p K_p^0}/\mathbb{F}_p$ be the special fiber of the integral model over $\mathbb{Z}_{(p)}$ of the modular curve at this level. This is a smooth curve over $\mathbb{F}_p$ that represents a moduli problem (E, α) of elliptic curves equipped with prime-to-p level structures (determined by the prime-to-p level K^p). The isogeny class of the p-divisible group $E[p^\infty]$ induces the Newton stratification

$$\overline{S}_{K^p K_p^0} = \overline{S}^{\text{ord}}_{K^p K_p^0} \sqcup \overline{S}^{\text{ss}}_{K^p K_p^0} \tag{3.4}$$

into an open dense ordinary stratum $\overline{S}^{\text{ord}}_{K^p K_p^0}$ (where $E[p^\infty]$ is isogenous to $\mu_{p^\infty} \times \mathbb{Q}_p/\mathbb{Z}_p$) and a closed supersingular stratum $\overline{S}^{\text{ss}}_{K^p K_p^0}$ consisting of finitely many points (where $E[p^\infty]$ is connected).

Now let $K_p^1 \subset \mathrm{GL}_2(\mathbb{Q}_p)$ be the Iwahori subgroup and $\overline{S}_{K^p K_p^1}/\mathbb{F}_p$ be the special fiber of the integral model of the modular curve at this level. This represents a moduli problem (E, α, D) of elliptic curves equipped with prime-to-p level structures and also with a level structure at p given by a finite flat subgroup scheme $D \subset E[p]$ of order p. Again, we have the preimage of the Newton stratification $\overline{S}_{K^pK_p^1} = \overline{S}^{\text{ord}}_{K^pK_p^1} \sqcup \overline{S}^{\text{ss}}_{K^pK_p^1}$. The modular curve at this level is not smooth, but rather a union of irreducible components that intersect transversely at the finitely many supersingular points.

The open and dense ordinary locus $\overline{S}^{\text{ord}}_{K^pK_p^1}$ is a disjoint union of two Kottwitz–Rapoport strata: the one where $D \simeq \mu_p$ and the one where $D \simeq \mathbb{Z}/p\mathbb{Z}$. Both of these Kottwitz–Rapoport strata can be shown to be abstractly isomorphic to the ordinary stratum at hyperspecial level. If we restrict the natural forgetful map $\overline{S}^{\text{ord}}_{K^pK_p^1} \to \overline{S}^{\text{ord}}_{K^pK_p^0}$ to the Kottwitz–Rapoport stratum where $D \simeq \mathbb{Z}/p\mathbb{Z}$, the map can be identified (up to an isomorphism) with the geometric Frobenius. (The restriction of the map to the Kottwitz–Rapoport stratum where $D \simeq \mu_p$ is an isomorphism.)

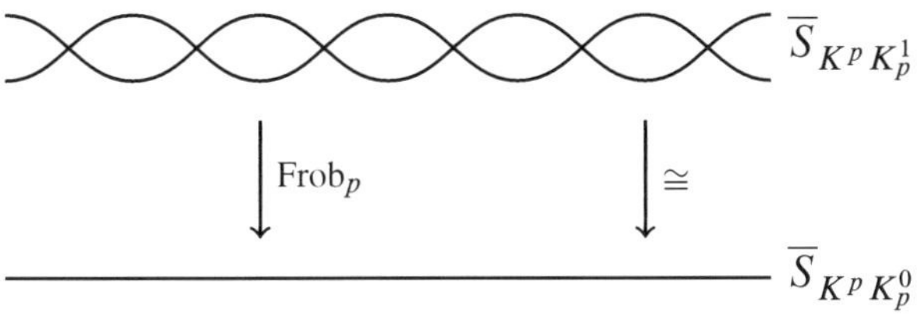

(3.5)

On the adic generic fiber, one can extend this picture to an *anticanonical* ordinary tower, where the transition morphisms reduce modulo p to (powers of) the geometric Frobenius, giving a perfectoid space in the limit. To extend beyond the ordinary locus, Scholze uses the theory of the canonical subgroup, the action of $\mathrm{GL}_2(\mathbb{Q}_p)$ at infinite level, and a rudimentary form of the Hodge–Tate period morphism that is just defined on the underlying topological spaces.

The above strategy generalizes relatively cleanly to higher-dimensional Siegel modular varieties, modulo subtleties at the boundary. To extend Theorem 3.1 to general Shimura varieties of Hodge type, Scholze considers an embedding at infinite level into a Siegel modular variety. It is surprisingly subtle to understand directly the perfectoid structure on a general Shimura variety of Hodge type (especially in the case when $G_{\mathbb{Q}_p}$ is nonsplit) and this is related to the discussion in Section 5. This is also related to the fact that the geometry of the EKOR stratification is more intricate when $G_{\mathbb{Q}_p}$ is nonsplit.

For simplicity, let us now assume that (G, X) is a Shimura datum of PEL type and that p is an unramified prime for this Shimura datum. Recall the Kottwitz set $B(G)$ classifying isocrystals with $G_{\mathbb{Q}_p}$-structure. The Hodge cocharacter μ defines a subset $B(G, \mu^{-1}) \subset B(G)$ of μ^{-1}-admissible elements. The special fiber of the Shimura variety with hyperspecial level at p admits a *Newton stratification*

$$\overline{S}_{K^pK_p^0} = \bigsqcup_{b \in B(G,\mu^{-1})} \overline{S}^{b}_{K^pK_p^0}$$

into locally closed strata indexed by this subset. This stratification is in terms of isogeny classes of p-divisible groups with $G_{\mathbb{Q}_p}$-structure and generalizes the stratification (3.4) from the modular curve case.

For each $b \in B(G, \mu^{-1})$, one can choose a (completely slope divisible) p-divisible group with $G_{\mathbb{Q}_p}$-structure $\mathbb{X}_b/\overline{\mathbb{F}}_p$ and define the corresponding *Oort central leaf*. This is a smooth closed subscheme $\mathscr{C}^{\mathbb{X}_b}$ of the Newton stratum $\overline{S}^{b}_{K^pK_p^0}$, such that the isomorphism class of the p-divisible group with $G_{\mathbb{Q}_p}$-structure over each geometric point of the leaf is constant and equal to that of $\mathbb{X}_b$:

$$\mathscr{C}^{\mathbb{X}_b} = \{x \in \overline{S}^{b}_{K^pK_p^0} \mid \overline{A}_{K^pK_p^0}[p^\infty] \times \kappa(\bar{x}) \simeq \mathbb{X}_b \times \kappa(\bar{x})\}.$$

In general, there can be infinitely many leaves inside a given Newton stratum. Over each central leaf, one has the *perfect Igusa variety* $\mathrm{Ig}^b/\overline{\mathbb{F}}_p$, a profinite cover of $\mathscr{C}^{\mathbb{X}^b}$ which parametrizes trivializations of the universal p-divisible group with $G_{\mathbb{Q}_p}$-structure.

Variants of Igusa varieties were introduced in [34] in the special case of Shimura varieties of Harris–Taylor type. They were defined more generally for Shimura varieties of PEL type by Mantovan [43] and their ℓ-adic cohomology was computed in many cases by Shin using a counting point formula [59–61]. All these authors consider Igusa varieties as profinite étale covers of central leaves, which trivialize the graded pieces of the slope filtration on the universal p-divisible group. Taking perfection gives a more elegant moduli-theoretic interpretation, while preserving ℓ-adic cohomology. However, the coherent cohomology of Igusa varieties is also important for defining and studying p-adic families of automorphic forms on G, as pioneered by Katz and Hida. Taking perfection is too crude for this purpose.

While the central leaf $\mathscr{C}^{\mathbb{X}_b}$ depends on the choice of $\mathbb{X}_b$ in its isogeny class, one can show that the perfect Igusa variety Ig^b only depends on the isogeny class: this follows from the equivalent moduli-theoretic description in [17, LEMMA 4.3.4] (see also [19, LEMMA 4.2.2], which keeps track of the extra structures more carefully). In particular, the pair (G, μ) is not determined by the Igusa variety Ig^b – it can happen that Igusa varieties that are a priori obtained from different Shimura varieties are isomorphic. See [19, THEOREM 4.2.4] for an example and [57] for a systematic analysis of this phenomenon in the function field setting.

Because $\mathrm{Ig}^b/\overline{\mathbb{F}}_p$ is perfect, the base change $\mathrm{Ig}^b \times_{\overline{\mathbb{F}}_p} \mathcal{O}_C/p$ admits a canonical lift to a flat formal scheme over $\mathrm{Spf}\,\mathcal{O}_C$. We let $\mathfrak{Ig}^b$ denote the adic generic fiber of this lift, which is a perfectoid space over $\mathrm{Spa}(C, \mathcal{O}_C)$. The spaces Ig^b and $\mathfrak{Ig}^b$ have naturally isomorphic ℓ-adic cohomology groups and they both have an action of a locally profinite group $G_b(\mathbb{Q}_p)$, where G_b is an inner form of a Levi subgroup of G.

For each $b \in B(G, \mu^{-1})$, one can also consider the associated *Rapoport–Zink space*, a moduli space of p-divisible groups with $G_{\mathbb{Q}_p}$-structure that is a local analogue of a Shimura variety. Concretely in the PEL case, one considers a moduli problem of p-divisible groups equipped with $G_{\mathbb{Q}_p}$-structure, satisfying the Kottwitz determinant condition with respect to μ, and with a modulo p quasiisogeny to the fixed p-divisible group $\mathbb{X}_b$. This moduli problem was shown by Rapoport–Zink [50] to be representable by a formal scheme over $\mathrm{Spf}\,\mathcal{O}_{\breve{E}_\mathfrak{p}}$, where $\breve{E}_\mathfrak{p}$ is the completion of the maximal unramified extension of $E_\mathfrak{p}$. We let $\mathcal{M}^b$ denote the adic generic fiber of this formal scheme,[3] base changed to $\mathrm{Spa}(C, \mathcal{O}_C)$, and let $\mathcal{M}^b_\infty$ denote the corresponding infinite-level Rapoport–Zink space. The latter object can

3 As a consequence of the comparison with moduli spaces of local shtukas in [56], one obtains a group-theoretic characterization of Rapoport–Zink spaces as local Shimura varieties determined by the tuple (G, b, μ). We suppress (G, μ) from the notation for simplicity.

be shown to be a perfectoid space using the techniques of [55], by which the infinite-level Rapoport–Zink space admits a local analogue of the Hodge–Tate period morphism

$$\pi_{\mathrm{HT}}^{b} : \mathcal{M}_{\infty}^{b} \to \mathscr{F}\ell.$$

It turns out that the geometry of π_{HT} is intricately tied up with the geometry of its local analogues π_{HT}^{b}. The following result is a conceptually cleaner, infinite-level version of the Mantovan product formula established in [43], which describes Newton strata inside Shimura varieties in terms of a product of Igusa varieties and Rapoport–Zink spaces.

Theorem 3.3. *There exists a Newton stratification*

$$\mathscr{F}\ell = \bigsqcup_{b\in B(G,\mu^{-1})} \mathscr{F}\ell^{b}$$

into locally closed strata.

For each $b \in B(G, \mu^{-1})$, one can consider the Newton stratum $\mathcal{S}_{K^p}^{\circ b}$ as a locally closed subspace of the good reduction locus $\mathcal{S}_{K^p}^{\circ}$. There exists a Cartesian diagram of diamonds over $\mathrm{Spd}(C, \mathcal{O}_C)$

$$\begin{array}{ccc} \mathcal{M}_{\infty}^{b} \times_{\mathrm{Spd}(C,\mathcal{O}_C)} \mathfrak{Ig}^{b} & \longrightarrow & \mathcal{M}_{\infty}^{b} \\ \downarrow & & \downarrow \pi_{\mathrm{HT}}^{b} \\ \mathcal{S}_{K^p}^{\circ b} & \xrightarrow{\pi_{\mathrm{HT}}} & \mathscr{F}\ell^{b}. \end{array}$$

Moreover, each vertical map is a pro-étale torsor for the group diamond $\tilde{G}_b$ (identified with $\mathrm{Aut}_G(\widetilde{\mathbb{X}}_b)$*), in the notation of* [17, §4]*).*

The decomposition into Newton strata is defined in [17, §3]. Morally, one first constructs a map of v-stacks $\mathscr{F}\ell \to \mathrm{Bun}_G$, where the latter is the v-stack of G-bundles on the Fargues–Fontaine curve. To construct this map of v-stacks, it is convenient to notice that one can identify the diamond associated to $\mathscr{F}\ell$ with the minuscule Schubert cell defined by μ inside the B_{dR}^{+}-Grassmannian for G. Once the map to Bun_G is in the picture, one uses Fargues's result that the points of Bun_G are in bijection with the Kottwitz set $B(G)$, cf. [27] (see also [2] for an alternative proof that also works in equal characteristic). Moreover, the Newton decomposition is a *stratification*, in the sense that, for $b \in B(G, \mu)$, we have

$$\overline{\mathscr{F}\ell^{b}} = \bigsqcup_{b' \geq b} \mathscr{F}\ell^{b'},$$

where $\geq$ denotes the Bruhat order. The latter fact follows from a recent result of Viehmann, see [63, THEOREM 1.1].

On rank one points, π_{HT} is compatible with the two different ways of defining the Newton stratification: via pullback from $\overline{S}_{K^p K_p^0}$ on $\mathcal{S}_{K^p}$ and via pullback from Bun_G on $\mathscr{F}\ell$. The behavior is more subtle on higher rank points. This is related to the fact that the closure relations are reversed in the two settings: the basic locus inside $\overline{S}_{K^p K_p^0}$ is the unique closed stratum, whereas each basic stratum inside Bun_G is open. On the other hand, the (μ)-ordinary locus is open and dense inside $\overline{S}_{K^p K_p^0}$, whereas it is a zero-dimensional closed stratum inside

$\mathscr{F}\ell$. The infinite-level product formula is established in [17, §4], although it is formulated in terms of functors on $\mathrm{Perf}_{\breve{E}_{\mathfrak{p}}}$.[4] This was extended to Shimura varieties of Hodge type by Hamacher [31].

Assume that the Shimura varieties S_K are compact. We have the following consequence for the fibers of π_{HT}: let $\bar{x} : \mathrm{Spa}(C, C^+) \to \mathscr{F}\ell^b$ be a geometric point. Then there is an inclusion of $\mathfrak{Ig}^b$ into $\pi_{\mathrm{HT}}^{-1}(\bar{x})$, which identifies the target with the canonical compactification of the source, in the sense of [54, PROPOSITION 18.6]. In [18, THEOREM 1.10], we extend the computation of the fibers to minimal and toroidal compactifications of (noncompact) Shimura varieties attached to quasisplit unitary groups. In this case, the fibers can be obtained from partial minimal and toroidal compactifications of Igusa varieties. It would be interesting to extend the whole infinite-level product formula to compactifications.

Example 3.4. We make the geometry of π_{HT} explicit in the case of the modular curve, i.e., for $G = \mathrm{GL}_2/\mathbb{Q}$. In this case, we identify $\mathscr{F}\ell = \mathbb{P}^{1,\mathrm{ad}}$ and we have the decomposition into Newton strata

$$\begin{array}{ccccc} \mathcal{S}^*_{K^p} & = & \mathcal{S}^{*,\mathrm{ord}}_{K^p} & \sqcup & \mathcal{S}^{\mathrm{ss}}_{K^p} \\ \downarrow{\scriptstyle \pi_{\mathrm{HT}}} & & \downarrow & & \downarrow \\ \mathbb{P}^{1,\mathrm{ad}} & = & \mathbb{P}^{1,\mathrm{ad}}(\mathbb{Q}_p) & \sqcup & \Omega. \end{array}$$

The ordinary locus inside $\mathbb{P}^{1,\mathrm{ad}}$ consists of the set of points defined over $\mathbb{Q}_p$ and the basic / supersingular locus is its complement Ω, the Drinfeld upper half-plane.

The fibers of π_{HT} over the ordinary locus are "perfectoid versions" of Igusa curves. The infinite-level version of the product formula reduces, in this case, to the statement that the ordinary locus is parabolically induced from $\mathfrak{Ig}^{\mathrm{ord}}$, as in [19, §6]. The fibers of π_{HT} over the supersingular locus are profinite sets: the corresponding Igusa varieties can be identified with double cosets $D^\times \backslash D^\times(\mathbb{A}_f^p)/K^p$, where $D/\mathbb{Q}$ is the quaternion algebra ramified precisely at ∞ and p. This precise result is established in [35], although the idea goes back to Deuring–Serre. One should be able to give an analogous description for basic Igusa varieties in much greater generality – this is closely related to Rapoport–Zink uniformization.

4. COHOMOLOGY WITH MOD ℓ COEFFICIENTS

In this section, we outline some recent strategies for computing the cohomology of Shimura varieties with modulo ℓ coefficients using the p-adic Hodge–Tate period morphism, where ℓ and p are two *distinct* primes. We emphasize the strategies developed in [17–19, 38].

We will assume throughout that (G, X) is a Shimura datum of abelian type and, in practice, we will focus on two examples: the case of Shimura varieties associated with unitary similitude groups and the case of Hilbert modular varieties. Let $\mathfrak{m} \subset \mathbb{T}$ be a max-

4 The result precedes the notion of diamonds and, in order to ensure that $\mathcal{S}^{\circ b}_{K^p}$ is a diamond, one needs to take care in defining it. At hyperspecial level, one should consider the adic generic fiber of the formal completion of the integral model of the Shimura variety along the Newton stratum indexed by b in its special fiber.

imal ideal in the support of $H^*_{(c)}(S_K(\mathbb{C}), \mathbb{F}_\ell)$. By work of Scholze (cf. [53, THEOREM 4.3.1]) and by the construction of Galois representations in the essentially self-dual case, we know in many cases how to associate a global modulo ℓ Galois representation $\bar{\rho}_{\mathfrak{m}}$ to the maximal ideal $\mathfrak{m}$. Therefore, the non-Eisenstein condition makes sense, and one can at least formulate Conjecture 2.2. In order to make progress on this conjecture, we impose a local representation-theoretic condition at the prime p, which we treat as an auxiliary prime.

Definition 4.1. Let $\mathbb{F}$ be a finite field of characteristic ℓ.

(1) Let $p \neq \ell$ be a prime, $K/\mathbb{Q}_p$ be a finite extension, and $\bar{\rho} : \mathrm{Gal}(\overline{K}/K) \to \mathrm{GL}_n(\mathbb{F})$ be a continuous representation. We say that $\bar{\rho}$ is *generic* if it is unramified and the eigenvalues (with multiplicity) $\alpha_1, \ldots, \alpha_n \in \overline{\mathbb{F}}_\ell$ of $\bar{\rho}(\mathrm{Frob}_K)$ satisfy $\alpha_i/\alpha_j \neq |\mathcal{O}_K/\mathfrak{m}_K|$ for $i \neq j$.

(2) Let F be a number field and $\bar{\rho} : \mathrm{Gal}(\overline{F}/F) \to \mathrm{GL}_n(\mathbb{F})$ be a continuous representation. We say that a prime $p \neq \ell$ is *decomposed generic* for $\bar{\rho}$ if p splits completely in F and, for every prime $\mathfrak{p} \mid p$ of F, $\bar{\rho}|_{\mathrm{Gal}(\overline{F}_{\mathfrak{p}}/F_{\mathfrak{p}})}$ is generic. We say that $\bar{\rho}$ is *decomposed generic* if there exists a prime $p \neq \ell$ which is decomposed generic for $\bar{\rho}$. (If one such prime exists, then infinitely many do.)

Remark 4.2. The condition for the local representation $\bar{\rho}$ of $\mathrm{Gal}(\overline{K}/K)$ to be generic implies that any lift to characteristic 0 of $\bar{\rho}$ corresponds under the local Langlands correspondence to a generic principal series representation of $\mathrm{GL}_n(K)$. Such a representation can never arise from a nonsplit inner form of GL_n/K via the Jacquet–Langlands correspondence. For this reason, a generic $\bar{\rho}$ cannot be the modulo ℓ reduction of the L-parameter of a smooth representation of a nonsplit inner form of GL_n/K.

A semisimple 2-dimensional representation $\bar{\rho}$ of $\mathrm{Gal}(\overline{\mathbb{Q}}/\mathbb{Q})$ is either decomposed generic or it satisfies (2.1): the case where $\bar{\rho}$ is a direct sum of two characters can be analyzed by hand, and the case where $\bar{\rho}$ is absolutely irreducible follows from the paragraph after Theorem 3.1 in [37]. More generally, the condition for a global representation $\bar{\rho}$ of $\mathrm{Gal}(\overline{F}/F)$ to be decomposed generic can be ensured when $\bar{\rho}$ has large image. For example, if $\ell > 2$, F is a totally real field, and $\bar{\rho}$ is a totally odd 2-dimensional representation with nonsolvable image, then $\bar{\rho}$ is decomposed generic (cf. [19, LEMMA 7.1.8]).

Let F be an imaginary CM field. Let $(B, *, V, \langle\cdot,\cdot\rangle)$ be a PEL datum of type A, where B is a central simple algebra with center F. We let (G, X) be the associated Shimura datum. For a neat compact open subgroup $K \subset G(\mathbb{A}_f)$, we let S_K/E be the associated Shimura variety, of dimension d. The following conjecture is a slightly different formulation of [38, CONJECTURE 1.2], with essentially the same content.

Conjecture 4.3. *Let $\mathfrak{m} \subset \mathbb{T}$ be a maximal ideal in the support of $H^i_{(c)}(S_K(\mathbb{C}), \mathbb{F}_\ell)$. Assume that $\bar{\rho}_{\mathfrak{m}}$ is decomposed generic. Then the following statements hold true:*

(1) *if $H^i_c(S_K(\mathbb{C}), \mathbb{F}_\ell)_{\mathfrak{m}} \neq 0$, then $i \leq d$;*

(2) *if $H^i(S_K(\mathbb{C}), \mathbb{F}_\ell)_{\mathfrak{m}} \neq 0$, then $i \geq d$.*

If the Shimura varieties S_K are compact, or if we additionally assume $\mathfrak{m}$ to be non-Eisenstein, Conjecture 4.3 implies a significant part of Conjecture 2.2 for Shimura varieties of PEL type A. Analogues of Conjecture 4.3 can be formulated (and are perhaps within reach) for other Shimura varieties, such as Siegel modular varieties.

Theorem 4.4 ([17] strengthened in [38]). *Assume that G is anisotropic modulo center, so that the Shimura varieties S_K are compact. Then Conjecture 4.3 holds true.*

Theorem 4.5 ([18] strengthened in [38]). *Assume that $B = F$, $V = F^{2n}$ and G is a quasisplit group of unitary similitudes. Then Conjecture 4.3 holds true.*

Remark 4.6. The more recent results of [38] have significantly fewer technical assumptions than the earlier ones of [17] and [18]. For example, Koshikawa's version of Theorem 4.5 allows F to be an imaginary quadratic field. It seems nontrivial to obtain this case with the methods of [18]. In the noncompact case, his results rely on the geometric constructions in [18], in particular on the semiperversity result for Shimura varieties attached to quasisplit unitary groups that is established there. As he notes, a generalization of this semiperversity result should lead to a full proof of Conjecture 4.3 for Shimura varieties of PEL type A. The more general semiperversity result will be obtained in the upcoming PhD thesis of Mafalda Santos.

In the case of Harris–Taylor Shimura varieties, Theorem 4.4 was first proved by Boyer [9]. Boyer's argument uses the integral models of Shimura varieties of Harris–Taylor type, but it is close in spirit to the argument carried out in [17] on the generic fiber. What is really interesting about Boyer's results is that he goes *beyond genericity*, in the following sense. Given the eigenvalues (with multiplicity) $\alpha_1, \dots, \alpha_n$ of $\bar{\rho}_{\mathfrak{m}}(\mathrm{Frob}_{\mathfrak{p}})$, with $\mathfrak{p} \mid p$ the relevant prime of F,[5] one can define a "defect" that measures how far $\bar{\rho}_{\mathfrak{m}}$ is from being generic at $\mathfrak{p}$. Concretely, set $\delta_{\mathfrak{p}}(\mathfrak{m})$ to be equal to the length of the maximal chain of eigenvalues where the successive terms have ratio equal to $|\mathcal{O}_{F_{\mathfrak{p}}}/\mathfrak{m}_{F_{\mathfrak{p}}}|$. Boyer shows that the cohomology groups $H^i_{(c)}(S_K(\mathbb{C}), \mathbb{F}_\ell)_{\mathfrak{m}}$ are nonzero at most in the range $[d - \delta_{\mathfrak{p}}(\mathfrak{m}), d + \delta_{\mathfrak{p}}(\mathfrak{m})]$. As noted by both Emerton and Koshikawa, such a result is consistent with Arthur's conjectures on the cohomology of Shimura varieties with $\mathbb{C}$-coefficients and points towards a modulo ℓ analogue of these conjectures.

Let us also discuss the analogous vanishing result in the Hilbert case. Let F be a totally real field of degree g and let $G = \mathrm{Res}_{F/\mathbb{Q}} \mathrm{GL}_2$. For a neat compact open subgroup $K \subset G(\mathbb{A}_f)$, we let $S_K/\mathbb{Q}$ be the corresponding Hilbert modular variety, of dimension g.

Theorem 4.7 ([19, THEOREM A]). *Let $\ell > 2$ and $\mathfrak{m} \subset \mathbb{T}$ be a maximal ideal in the support of $H^i_{(c)}(S_K(\mathbb{C}), \mathbb{F}_\ell)$. Assume that the image of $\bar{\rho}_{\mathfrak{m}}$ is not solvable, which implies that $\bar{\rho}_{\mathfrak{m}}$ is absolutely irreducible and decomposed generic. Then $H^i_c(S_K(\mathbb{C}), \mathbb{F}_\ell)_{\mathfrak{m}} = H^i(S_K(\mathbb{C}), \mathbb{F}_\ell)_{\mathfrak{m}}$ is nonzero only for $i = g$.*

---

5 In this special case, one does not have to impose the condition that p splits completely in F, and it suffices to have genericity at one prime $\mathfrak{p} \mid p$.

The same result holds for all quaternionic Shimura varieties, and we can even prove the analogue of Boyer's result that goes beyond genericity in all these settings. As an application, we deduce (under some technical assumptions) that the p-adic local Langlands correspondence for $\mathrm{GL}_2(\mathbb{Q}_p)$ occurs in the completed cohomology of Hilbert modular varieties, when p is a prime that splits completely in F. This uses the axiomatic approach via patching introduced in [14] and further developed in [15, 30].

We now outline the original strategy for proving Theorem 4.4, which was introduced in [17]. Let p be a prime and $K = K^p K_p \subset G(\mathbb{A}_f)$ be a neat compact open subgroup. The Hodge–Tate period morphism gives rise to a $\mathbb{T}$-equivariant diagram

$$\begin{array}{ccc} & \mathcal{S}_{K^p} & \\ \swarrow & & \searrow^{\pi_{\mathrm{HT}}} \\ \mathcal{S}_{K^p K_p} & & \mathscr{F}\ell. \end{array} \tag{4.1}$$

The standard comparison theorems between various cohomology theories allow us to identify $H^*_{(c)}(S_K(\mathbb{C}), \mathbb{F}_\ell)_{\mathfrak{m}}$ with $H^*_{(c)}(\mathcal{S}_K, \mathbb{F}_\ell)_{\mathfrak{m}}$. The arrow on the left-hand side of (4.1) is a K_p-torsor, so the Hochschild–Serre spectral sequence allows us to recover $H^*_{(c)}(\mathcal{S}_K, \mathbb{F}_\ell)_{\mathfrak{m}}$ from $H^*_{(c)}(\mathcal{S}_{K^p}, \mathbb{F}_\ell)_{\mathfrak{m}}$. The idea is now to compute $H^*_{(c)}(\mathcal{S}_{K^p}, \mathbb{F}_\ell)_{\mathfrak{m}}$ in two stages: first understand the complex of sheaves $(R\pi_{\mathrm{HT}*}\mathbb{F}_\ell)_{\mathfrak{m}}$ on $\mathscr{F}\ell$, then compute the total cohomology using the Leray–Serre spectral sequence.

Two miraculous things happen that greatly simplify the structure of $(R\pi_{\mathrm{HT}*}\mathbb{F}_\ell)_{\mathfrak{m}}$. The first is that $(R\pi_{\mathrm{HT}*}\mathbb{F}_\ell)_{\mathfrak{m}}$ behaves like a perverse sheaf on $\mathscr{F}\ell$. This is because π_{HT} is simultaneously *affinoid*, as discussed after Theorem 3.1, and *partially proper*, because the Shimura varieties were assumed to be compact. In particular, the restriction of $(R\pi_{\mathrm{HT}*}\mathbb{F}_\ell)_{\mathfrak{m}}$ to a highest-dimensional stratum in its support is concentrated in one degree. By the computation of the fibers of π_{HT}, this implies that the cohomology groups $R\Gamma(\mathfrak{Ig}^b, \mathbb{Z}_\ell)_{\mathfrak{m}}$ are concentrated in one degree and torsion-free. The second miracle is that, whenever the group $G_b(\mathbb{Q}_p)$ acting on $\mathfrak{Ig}^b$ comes from a nonquasisplit inner form, the localization $R\Gamma(\mathfrak{Ig}^b, \mathbb{Q}_\ell)_{\mathfrak{m}}$ vanishes. This uses the genericity of $\bar{\rho}_{\mathfrak{m}}$ at each $\mathfrak{p} \mid p$ and suggests that the cohomology of Igusa varieties satisfies some form of local–global compatibility. Finally, the condition that p splits completely in F guarantees that the only Newton stratum for which G_b is quasisplit is the ordinary one. Therefore, the hypotheses of Theorem 4.4 guarantee that $(R\pi_{\mathrm{HT}*}\mathbb{F}_\ell)_{\mathfrak{m}}$ is as simple as possible – it is supported in one degree on a zero-dimensional stratum!

The computation of $R\Gamma(\mathrm{Ig}^b, \mathbb{Q}_\ell)_{\mathfrak{m}}$, at least at the level of the Grothendieck group, can be done using the trace formula method pioneered by Shin [60]. This is the method used for Shimura varieties of PEL type A in [17] and [18]. For inner forms of $\mathrm{Res}_{F/\mathbb{Q}}\,\mathrm{GL}_2$, with F a totally real field, one can avoid these difficult computations, cf. [19]. In this setting, one can reinterpret results of Tian–Xiao [62] on geometric instances of the Jacquet–Langlands correspondence as giving rise to exotic isomorphisms between Igusa varieties arising from different Shimura varieties. This is what happens for the basic stratum in Example 3.4. Then one can conclude by applying the classical Jacquet–Langlands correspondence.

In [38], Koshikawa introduces a novel and complementary strategy for proving these kinds of vanishing theorems. He shows that, under the same genericity assumption in Definition 4.1, only the restriction of $(R\pi_{\mathrm{HT}*}\mathbb{F}_\ell)_\mathfrak{m}$ to the ordinary locus contributes to the total cohomology of the Shimura variety. To achieve this, he proves the analogous generic vanishing theorem for the cohomology $R\Gamma_c(\mathcal{M}^b, \mathbb{Z}_\ell)_{\mathfrak{m}_p}$ of the Rapoport–Zink space, where $\mathfrak{m}_p$ is a maximal ideal of the local spherical Hecke algebra at p. This relies on the recent work of Fargues–Scholze on the geometrization of the local Langlands conjecture [28].

Koshikawa's strategy is more flexible, allowing him to handle with ease the case where F is an imaginary quadratic field. On the other hand, the original approach also gives information about the complexes of sheaves $(R\pi_{\mathrm{HT}*}\mathbb{F}_\ell)_\mathfrak{m}$, rather than just about the cohomology groups $H^*_{(c)}(S_K(\mathbb{C}), \mathbb{F}_\ell)_\mathfrak{m}$. These complexes should play an important role for questions of local–global compatibility in Fargues's geometrization conjecture, cf. [26, §7].

5. COHOMOLOGY WITH MOD p AND p-ADIC COEFFICIENTS

The most general method for constructing p-adic families of automorphic forms from the cohomology of locally symmetric spaces is via *completed cohomology*. First introduced by Emerton in [24], this has the following definition:

$$\tilde{H}^*(K^p, \mathbb{Z}_p) = \varprojlim_n \big(\varinjlim_{K_p} H^*(X_{K^pK_p}, \mathbb{Z}/p^n)\big),$$

where $K^p \subset G(\mathbb{A}_f)$ is a sufficiently small, fixed tame level, and $K_p \subset G(\mathbb{Q}_p)$ runs over compact open subgroups. This space has an action of the spherical Hecke algebra $\mathbb{T}$, built from Hecke operators away from p, as well as an action of the group $G(\mathbb{Q}_p)$. One can make the analogous definition for completed cohomology with compact support, and a variant gives completed homology and completed Borel–Moore homology. See [25] for an excellent survey that gives motivation, examples, and sketches the basic properties of these spaces.

Motivated by heuristics from the p-adic Langlands programme, Calegari and Emerton made several conjectures about the range of degrees in which one can have nonzero completed (co)homology and about the codimension of completed homology over the completed group rings $\mathbb{Z}_p[\![K_p]\!]$. See [11, CONJECTURE 1.5] for the original formulation and [32, CONJECTURE 1.3] for a slightly different formulation, which emphasizes the natural implications between the various conjectures. In particular, Calegari–Emerton conjectured that

$$\tilde{H}^i_c(K^p, \mathbb{Z}_p) = \tilde{H}^i(K^p, \mathbb{Z}_p) = 0 \quad \text{for } i > q_0.$$

For Shimura varieties of preabelian type, the Calegari–Emerton conjectures were proved by Hansen–Johansson in [32], building on work of Scholze who proved the vanishing of completed cohomology with compact support for Shimura varieties of Hodge type [53].

We sketch Scholze's argument, which illustrates the role of p-adic geometry in this result. It is enough to show that

$$\tilde{H}^i_c(K^p, \mathbb{F}_p) = \varinjlim_{K_p} H^i_c(S_{K^pK_p}(\mathbb{C}), \mathbb{F}_p)$$

vanishes for $i > d = \dim_E S_K$. Since (G, X) is a Shimura datum of Hodge type, we are in the setting of Theorem 3.1 – in fact, we know that the minimal compactification $\mathcal{S}^*_{K^p}$ is perfectoid. The primitive comparison theorem of [52] gives an almost isomorphism between $\tilde{H}^i_c(K^p, \mathbb{F}_p) \otimes \mathcal{O}_C/p$ and $H^i_{\text{ét}}(\mathcal{S}^*_{K^p}, \mathcal{I}^+/p)$, where $\mathcal{I}^+ \subseteq \mathcal{O}^+$ is the subsheaf of sections that vanish along the boundary. On an affinoid perfectoid space, Scholze proved the almost vanishing of the étale cohomology of $\mathcal{O}^+/p$ in degree $i > 0$. With some care at the boundary, one deduces that it is enough to prove that the analytic cohomology groups $H^i_{\text{an}}(\mathcal{S}^*_{K^p}, \mathcal{I}^+/p)$ are almost 0 in degree $i > d$. This final step follows from a theorem of Scheiderer on the cohomological dimension of spectral spaces.

In [20] and [16], we study Shimura varieties with unipotent level at p. More precisely, assume that (G, X) is a Shimura datum of Hodge type and that $G_{\mathbb{Q}_p}$ is split. Choose a split model of G and a Borel subgroup B over $\mathbb{Z}_p$, and let $U \subset B$ be the unipotent radical.

Theorem 5.1 ([16, THEOREM 1.1]). *Let $H \subseteq U(\mathbb{Z}_p)$ be a closed subgroup. We have*

$$\varinjlim_{K_p \supseteq H} H^i_c(S_{K^pK_p}(\mathbb{C}), \mathbb{F}_p) = 0 \quad \text{for } i > d.$$

This result is stronger than the Calegari–Emerton conjecture for completed cohomology with compact support, since we can take $H = \{1\}$ and recover Scholze's result discussed above. In addition to the argument sketched above, the key new idea needed for Theorem 5.1 is that the *Bruhat decomposition* on the Hodge–Tate period domain $\mathcal{F}\ell$ remembers how far different subspaces of $\mathcal{S}^*_{K^pU(\mathbb{Z}_p)}$ are from being perfectoid.

Example 5.2. Assume that $G = \mathrm{GL}_2/\mathbb{Q}$, so that we are working in the modular curve case. The Bruhat decomposition is given by $\mathbb{P}^{1,\mathrm{ad}} = \mathbb{A}^{1,\mathrm{ad}} \sqcup \{\infty\}$, with the two Bruhat cells in natural bijection with the two components of the ordinary locus in (3.5). We have a morphism of sites

$$\pi_{\mathrm{HT}/U(\mathbb{Z}_p)} : (\mathcal{S}^*_{K^pU(\mathbb{Z}_p)})_{\text{ét}} \to |\mathbb{P}^{1,\mathrm{ad}}|/U(\mathbb{Z}_p),$$

where we take the quotient $|\mathbb{P}^{1,\mathrm{ad}}|/U(\mathbb{Z}_p)$ only as a topological space. The preimage of $|\mathbb{A}^{1,\mathrm{ad}}|/U(\mathbb{Z}_p)$ in $\mathcal{S}_{K^pU(\mathbb{Z}_p)}$ is a perfectoid space, as proved by Ludwig in [42]. The preimage of $|\infty|/U(\mathbb{Z}_p)$ has a $\mathbb{Z}_p$-cover that is an affinoid perfectoid space. This allows us to bound the support of each $R^i\pi_{\mathrm{HT}*/U(\mathbb{Z}_p)}(\mathcal{I}^+/p)$, and we conclude by the Leray spectral sequence.

More generally, the Bruhat decomposition $G = \bigsqcup_{w \in W^{P_\mu}} BwP_\mu$ gives a decomposition $\mathcal{F}\ell = \bigsqcup_{w \in W^{P_\mu}} \mathcal{F}\ell^w$ into locally closed Schubert cells indexed by certain Weyl group elements. For each $\mathcal{F}\ell^w/U(\mathbb{Z}_p)$, we can quantify how far its preimage in $\mathcal{S}^*_{K^pU(\mathbb{Z}_p)}$ is from being a perfectoid space, which depends on the length of the Weyl group element w. The assumption that $G_{\mathbb{Q}_p}$ is split guarantees that all the Weyl group elements lie in the ordinary locus inside $\mathcal{F}\ell$, which greatly simplifies the analysis. However, the analogue of Theorem 5.1 may hold even without the assumption that $G_{\mathbb{Q}_p}$ is split, and even when the ordinary locus is empty. There is some evidence in this direction, e.g., by using embeddings into higher-dimensional Shimura varieties attached to split groups, or by using the results of [36] to handle the Harris–Taylor case, as in the upcoming PhD thesis of Louis Jaburi.

The Bruhat decomposition on $\mathscr{F}\ell$ has more recently been used by Boxer and Pilloni to define a version of higher Coleman theory indexed by each $w \in W^{P_\mu}$ in [8]. The development of higher Coleman and higher Hida theories shows that the geometric theory of p-adic automorphic forms on Shimura varieties is much richer than previously expected. Furthermore, the Bruhat decomposition indicates the form a p-adic Eichler–Shimura isomorphism should take, relating completed cohomology to these more geometric theories. In joint work in progress with Mantovan and Newton, we use the geometry described in Example 5.2 to give a new proof of the ordinary Eichler–Shimura isomorphism due to Ohta [46, 47]. Our result decomposes the ordinary completed cohomology of the modular curve in terms of Hida theory and higher Hida theory, the latter recently developed by Boxer and Pilloni in [7].

Theorem 5.1 seems far away from Conjecture 2.2, because it concerns Shimura varieties with "infinite level" at p. However, one could ask whether a version of Theorem 5.1 holds already at level $B(\mathbb{Z}_p)$, at least after applying an ordinary idempotent, as in Hida theory. If that were the case, the control theorems in Hida theory (specifically the result known as independence of level) and a careful application of Poincaré duality would imply that an $\ell = p$ analogue of Conjecture 4.3 holds, with *generic* replaced by *ordinary*. More precisely, in this case, the "auxiliary prime" p where we impose a representation-theoretic condition is no longer auxiliary but rather equal to ℓ.

6. APPLICATIONS BEYOND SHIMURA VARIETIES

While the focus of this article has been the cohomology of Shimura varieties, Theorems 4.5 and 5.1 have surprising applications to understanding the cohomology of more general locally symmetric spaces. For example, let F be an imaginary CM field and $G = \mathrm{Res}_{F/\mathbb{Q}} \mathrm{GL}_n$. Then G can be realized as the Levi quotient of the Siegel maximal parabolic of a quasisplit unitary group $\tilde{G}$. The Borel–Serre compactification $\tilde{X}^{\mathrm{BS}}_{\tilde{K}}$ for the locally symmetric spaces associated with the unitary group $\tilde{G}$ gives rise to a Hecke-equivariant long exact sequence of the form

$$\begin{aligned} \cdots \to H^i_c(\tilde{X}_{\tilde{K}}, \mathbb{Z}/\ell^n\mathbb{Z}) \to H^i(\tilde{X}_{\tilde{K}}, \mathbb{Z}/\ell^n\mathbb{Z}) \to H^i(\partial\tilde{X}_{\tilde{K}}, \mathbb{Z}/\ell^n\mathbb{Z}) \\ \to H^{i+1}_c(\tilde{X}_{\tilde{K}}, \mathbb{Z}/\ell^n\mathbb{Z}) \to \cdots, \end{aligned} \tag{6.1}$$

where $\partial\tilde{X}_{\tilde{K}} = \tilde{X}^{\mathrm{BS}}_{\tilde{K}} \setminus \tilde{X}_{\tilde{K}}$ is the boundary of the Borel–Serre compactification. The usual and compactly supported cohomology of $\tilde{X}_{\tilde{K}}$ can be simplified to some extent by applying either of the two vanishing theorems. On the other hand, the cohomology of X_K can be shown to contribute to the cohomology of $\partial\tilde{X}_{\tilde{K}}$, in some more or less controlled fashion.

Let $\mathfrak{m} \subset \mathbb{T}$ be a non-Eisenstein maximal ideal in the support of $R\Gamma(X_K, \mathbb{Z}_\ell)$ and let $\mathbb{T}(K)_{\mathfrak{m}}$ denote the quotient of $\mathbb{T}$ that acts faithfully on $R\Gamma(X_K, \mathbb{Z}_\ell)_{\mathfrak{m}}$. In addition to the residual Galois representation $\bar{\rho}_{\mathfrak{m}}$, Scholze associates to $\mathfrak{m}$ a deformation $\rho_{\mathfrak{m}}$ valued in $\mathbb{T}(K)_{\mathfrak{m}}/I$, for some nilpotent ideal I. This was subsequently shown by Newton and Thorne in [45] to satisfy $I^4 = 0$. In [20], we used a variant of Theorem 5.1 together with the excision sequence (6.1) to eliminate this nilpotent ideal entirely, under the assumption that ℓ splits completely in the CM field F. This leads to a more natural statement on the existence of Galois representations in this setting.

The Galois representations $\rho_{\mathfrak{m}}$ are expected to satisfy a certain property known as *local–global compatibility*, which is particularly subtle to state and prove at primes above ℓ. For example, after inverting ℓ, the $\rho_{\mathfrak{m}}$ are expected to be de Rham, in the sense of Fontaine, but it is less clear what the right condition should be for torsion Galois representations. In another application, Theorem 4.5 is crucially used in [1] together with the excision sequence (6.1) to prove that $\rho_{\mathfrak{m}}$ satisfies the expected local–global compatibility at primes above ℓ in two restricted families of cases: the ordinary case and the Fontaine–Laffaille case.[6] In joint work in progress with Newton, we should be able to extend these methods to cover significantly more.

The local–global compatibility results established in [1] are already extremely useful: they help us implement the Calegari–Geraghty method unconditionally for the first time in arbitrary dimension. A striking application is the following result.

Theorem 6.1 ([1, THEOREM 1.0.1]). *Let F be a CM field and E/F be an elliptic curve that does not have complex multiplication. Then E is potentially automorphic and satisfies the Sato–Tate conjecture.*

The potential automorphy of E was established at the same time in work of Boxer–Calegari–Gee–Pilloni [6], who also showed the potential automorphy of abelian surfaces over totally real fields. Their work relies on the Calegari–Geraghty method for the *coherent cohomology* of Shimura varieties and uses a preliminary version of higher Hida theory, due to Pilloni, as a key ingredient.

ACKNOWLEDGMENTS

I am grateful to my many colleagues and collaborators who have discussed mathematics in general and Shimura varieties in particular with me – I especially want to thank Frank Calegari, Matthew Emerton, Toby Gee, Dan Gulotta, Pol van Hoften, Christian Johansson, Elena Mantovan, Sophie Morel, James Newton, Vytas Paškūnas, Vincent Pilloni, Peter Scholze, Sug Woo Shin, Matteo Tamiozzo, and Richard Taylor. I am grateful to Toby Gee, Louis Jaburi, Teruhisa Koshikawa, Peter Scholze, and Matteo Tamiozzo for their comments on an earlier draft of this article.

FUNDING

This work was partially supported by a Royal Society University Research Fellowship, by a Leverhulme Prize, and by ERC Starting Grant No. 804176.

6 Up to possibly enlarging the nilpotent ideal I. It is not clear how to remove the nilpotent ideal from the statement of local–global compatibility at $\ell = p$.

REFERENCES

[1] P. B. Allen, F. Calegari, A. Caraiani, T. Gee, D. Helm, B. V. Le Hung, J. Newton, P. Scholze, R. Taylor, and J. A. Thorne, Potential automorphy over CM fields. 2018, arXiv:1812.09999.

[2] J. Anschütz, Reductive group schemes over the Fargues–Fontaine curve. *Math. Ann.* **374** (2019), no. 3–4, 1219–1260.

[3] J. Arthur, L^2-cohomology and automorphic representations. In *In Canadian Mathematical Society. 1945–1995, Vol. 3*, pp. 1–17, Canadian Math. Soc., Ottawa, ON, 1996.

[4] A. Ash, Galois representations attached to mod p cohomology of $GL(n, Z)$. *Duke Math. J.* **65** (1992), no. 2, 235–255.

[5] A. Borel and N. Wallach, *Continuous cohomology, discrete subgroups, and representations of reductive groups*. Second edn., Math. Surveys Monogr. 67, American Mathematical Society, Providence, RI, 2000.

[6] G. Boxer, F. Calegari, T. Gee, and V. Pilloni, Abelian Surfaces over totally real fields are Potentially Modular. 2018, arXiv:1812.09269.

[7] G. Boxer and V. Pilloni, Higher Hida and Coleman theories on the modular curve. 2020, arXiv:2002.06845.

[8] G. Boxer and V. Pilloni, Higher Coleman theory. 2021, arXiv:2110.10251.

[9] P. Boyer, Sur la torsion dans la cohomologie des variétés de Shimura de Kottwitz–Harris–Taylor. *J. Inst. Math. Jussieu* **18** (2019), no. 3, 499–517.

[10] F. Calegari, Reciprocity in the Langlands program since Fermat's Last Theorem. 2021, arXiv:2109.14145.

[11] F. Calegari and M. Emerton, Completed cohomology—a survey. In *Non-abelian fundamental groups and Iwasawa theory*, pp. 239–257, London Math. Soc. Lecture Note Ser. 393, Cambridge Univ. Press, Cambridge, 2012.

[12] F. Calegari and D. Geraghty, Modularity lifting beyond the Taylor–Wiles method. *Invent. Math.* **211** (2018), no. 1, 297–433.

[13] A. Caraiani, Perfectoid Shimura varieties. In *Perfectoid spaces: Lectures from the 2017 Arizona Winter School*, pp. 245–297, American Mathematical Society, Providence, RI, 2019.

[14] A. Caraiani, M. Emerton, T. Gee, D. Geraghty, V. Paškūnas, and S. W. Shin, Patching and the p-adic local Langlands correspondence. *Camb. J. Math.* **4** (2016), no. 2, 197–287.

[15] A. Caraiani, M. Emerton, T. Gee, D. Geraghty, V. Paškūnas, and S. W. Shin, Patching and the p-adic Langlands program for $GL_2(\mathbb{Q}_p)$. *Compos. Math.* **154** (2018), no. 3, 503–548.

[16] A. Caraiani, D. R. Gulotta, and C. Johansson, Vanishing theorems for Shimura varieties at unipotent level. *Journal of the EMS* (to appear), arXiv:1910.09214.

[17] A. Caraiani and P. Scholze, On the generic part of the cohomology of compact unitary Shimura varieties. *Ann. of Math. (2)* **186** (2017), no. 3, 649–766.

[18] A. Caraiani and P. Scholze, On the generic part of the cohomology of non-compact unitary Shimura varieties. 2019, arXiv:1909.01898.

[19] A. Caraiani and M. Tamiozzo, On the étale cohomology of Hilbert modular varieties with torsion coefficients. 2021, arXiv:2107.10081.

[20] A. Caraiani, D. R. Gulotta, C.-Y. Hsu, C. Johansson, L. Mocz, E. Reinecke, and S.-C. Shih, Shimura varieties at level $\Gamma_1(p^\infty)$ and Galois representations. *Compos. Math.* **156** (2020), no. 6, 1152–1230.

[21] G. Chenevier and M. Harris, Construction of automorphic Galois representations, II. *Camb. J. Math.* **1** (2013), no. 1, 53–73.

[22] P. Deligne, Variétés de Shimura: interprétation modulaire, et techniques de construction de modèles canoniques. In *Automorphic forms, representations and L-functions (Proc. Sympos. Pure Math., Oregon State Univ., Corvallis, Ore., 1977), Part 2*, pp. 247–289, Proc. Sympos. Pure Math. XXXIII, Amer. Math. Soc., Providence, RI, 1979.

[23] M. Dimitrov, Galois representations modulo p and cohomology of Hilbert modular varieties. *Ann. Sci. Éc. Norm. Supér. (4)* **38** (2005), no. 4, 505–551.

[24] M. Emerton, On the interpolation of systems of eigenvalues attached to automorphic Hecke eigenforms. *Invent. Math.* **164** (2006), no. 1, 1–84.

[25] M. Emerton, Completed cohomology and the p-adic Langlands program. In *Proceedings of the International Congress of Mathematicians, 2014. Volume II*, pp. 319–342, Kyung Moon Sa, Seoul, 2014.

[26] L. Fargues, Geometrization of the local Langlands correspondence: an overview. 2016, arXiv:1602.00999.

[27] L. Fargues, G-torseurs en théorie de Hodge p-adique. *Compos. Math.* **156** (2020), no. 10, 2076–2110.

[28] L. Fargues and P. Scholze, Geometrization of the local Langlands correspondence. 2021, arXiv:2102.13459.

[29] J. Franke, Harmonic analysis in weighted L_2-spaces. *Ann. Sci. Éc. Norm. Supér. (4)* **31** (1998), no. 2, 181–279.

[30] T. Gee and J. Newton, Patching and the completed homology of locally symmetric spaces. *J. Inst. Math. Jussieu* (2016).

[31] P. Hamacher, The product structure of Newton strata in the good reduction of Shimura varieties of Hodge type. *J. Algebraic Geom.* **28** (2019), no. 4, 721–749.

[32] D. Hansen and C. Johansson, Perfectoid Shimura varieties and the Calegari–Emerton conjectures. 2020, arXiv:2011.03951.

[33] M. Harris, K.-W. Lan, R. Taylor, and J. Thorne, On the rigid cohomology of certain Shimura varieties. *Res. Math. Sci.* **3** (2016), no. 3, 37.

[34] M. Harris and R. Taylor, *The geometry and cohomology of some simple Shimura varieties*. Ann. of Math. Stud. 151, Princeton University Press, Princeton, NJ, 2001.

[35] S. Howe, The spectral p-adic Jacquet–Langlands correspondence and a question of Serre. *Compos. Math.* (to appear).

[36] C. Johansson, J. Ludwig, and D. Hansen, A quotient of the Lubin–Tate tower II. *Math. Ann.* **380** (2021), no. 1–2, 43–89.

[37] T. Koshikawa, Vanishing theorems for the mod p cohomology of some simple Shimura varieties. *Forum Math. Sigma* **8** (2020), Paper No. e38.

[38] T. Koshikawa, On the generic part of the cohomology of local and global Shimura varieties. 2021, arXiv:2106.10602.

[39] K.-W. Lan, An example-based introduction to Shimura varieties. 2018.

[40] K.-W. Lan and J. Suh, Vanishing theorems for torsion automorphic sheaves on compact PEL-type Shimura varieties. *Duke Math. J.* **161** (2012), no. 6, 1113–1170.

[41] K.-W. Lan and J. Suh, Vanishing theorems for torsion automorphic sheaves on general PEL-type Shimura varieties. *Adv. Math.* **242** (2013), 228–286.

[42] J. Ludwig, A quotient of the Lubin–Tate tower. *Forum Math. Sigma* **5** (2017), e17.

[43] E. Mantovan, On the cohomology of certain PEL-type Shimura varieties. *Duke Math. J.* **129** (2005), no. 3, 573–610.

[44] S. Morel, Construction de représentations galoisiennes de torsion [d'après Peter Scholze]. *Astérisque* **380** (2016). Séminaire Bourbaki. Vol. 2014/2015, Exp. No. 1102, 449–473.

[45] J. Newton and J. A. Thorne, Torsion Galois representations over CM fields and Hecke algebras in the derived category. *Forum Math. Sigma* **4** (2016), e21, 88.

[46] M. Ohta, On the p-adic Eichler–Shimura isomorphism for Λ-adic cusp forms. *J. Reine Angew. Math.* **463** (1995), 49–98.

[47] M. Ohta, Ordinary p-adic étale cohomology groups attached to towers of elliptic modular curves. II. *Math. Ann.* **318** (2000), no. 3, 557–583.

[48] L. Pan, On locally analytic vectors of the completed cohomology of modular curves. 2020, arXiv:2008.07099.

[49] V. Pilloni and B. Stroh, Cohomologie cohérente et représentations Galoisiennes. *Ann. Math. Québec* **40** (2016), no. 1, 167–202.

[50] M. Rapoport and T. Zink, *Period spaces for p-divisible groups*. Ann. of Math. Stud. 141, Princeton University Press, Princeton, NJ, 1996.

[51] P. Scholze, Perfectoid spaces. *Publ. Math. Inst. Hautes Études Sci.* **116** (2012), 245–313.

[52] P. Scholze, p-adic Hodge theory for rigid-analytic varieties. *Forum Math. Pi* **1** (2013), e1, 77.

[53] P. Scholze, On torsion in the cohomology of locally symmetric varieties. *Ann. of Math. (2)* **182** (2015), no. 3, 945–1066.

[54] P. Scholze, Étale cohomology of diamonds. 2017, arXiv:1709.07343.

[55] P. Scholze and J. Weinstein, Moduli of p-divisible groups. *Camb. J. Math.* **1** (2013), no. 2, 145–237.

[56] P. Scholze and J. Weinstein, *Berkeley Lectures on p-adic Geometry*. Ann. of Math. Stud. 207, Princeton University Press, Princeton, NJ, 2020, 264 pp.

[57] J. Sempliner, *On the almost-product structure on the moduli stacks of parahoric global G-shtuka*. PhD thesis, Princeton University, 2021.

[58] X. Shen, Perfectoid Shimura varieties of abelian type. *Int. Math. Res. Not. IMRN* **21** (2017), 6599–6653.

[59] S. W. Shin, Counting points on Igusa varieties. *Duke Math. J.* **146** (2009), no. 3, 509–568.

[60] S. W. Shin, A stable trace formula for Igusa varieties. *J. Inst. Math. Jussieu* **9** (2010), no. 4, 847–895.

[61] S. W. Shin, Galois representations arising from some compact Shimura varieties. *Ann. of Math. (2)* **173** (2011), no. 3, 1645–1741.

[62] Y. Tian and L. Xiao, On Goren–Oort stratification for quaternionic Shimura varieties. *Compos. Math.* **152** (2016), no. 10, 2134–2220.

[63] E. Viehmann, On Newton strata in the B_{dR}^{+}-Grassmannian. 2021, arXiv:2101.07510.

[64] L. Xiao and X. Zhu, Cycles on Shimura varieties via geometric Satake. 2017, arXiv:1707.05700.

ANA CARAIANI

180 Queen's Gate, London SW7 2AZ, United Kingdom, a.caraiani@imperial.ac.uk

This Page Intentionally Left Blank

ON THE BRUMER–STARK CONJECTURE AND REFINEMENTS

SAMIT DASGUPTA AND MAHESH KAKDE

ABSTRACT

We state the Brumer–Stark conjecture and motivate it from two perspectives. Stark's perspective arose in his attempts to generalize the classical Dirichlet class number formula for the leading term of the Dedekind zeta function at $s = 1$ (equivalently, $s = 0$). Brumer's perspective arose by generalizing Stickelberger's work regarding the factorization of Gauss sums and the annihilation of class groups of cyclotomic fields. These viewpoints were synthesized by Tate, who stated the Brumer–Stark conjecture in its current form.

The conjecture considers a totally real field F and a finite abelian CM extension H/F. It states the existence of p-units in H whose valuations at places above p are related to the special values of the L-functions of the extension H/F at $s = 0$. Essentially equivalently, the conjecture states that a Stickelberger element associated to H/F annihilates the (appropriately smoothed) class group of H.

We describe our recent proofs of the Brumer–Stark conjecture away from 2. The conjecture has been refined by many authors in multiple directions. We state some of these refinements and our results towards them. The key technique involved in the proofs is Ribet's method.

One of the refinements we discuss is an exact p-adic analytic formula for Brumer–Stark units stated by the first author and his collaborators. We describe this formula and highlight some salient points of its proof. Since the Brumer–Stark units along with other easily described elements generate the maximal abelian extension of a totally real field, our results can be viewed as an explicit class theory for such fields. This can be considered a p-adic version of Hilbert's 12th problem.

MATHEMATICS SUBJECT CLASSIFICATION 2020

Primary 11R37; Secondary 11R42

KEYWORDS

Stark conjectures, Brumer–Stark units, Explicit class field theory

ICM INTERNATIONAL CONGRESS OF MATHEMATICIANS 2022 JULY 6–14

Proc. Int. Cong. Math. 2022, Vol. 3, pp. 1768–1799
DOI 10.4171/ICM2022/124

1. BACKGROUND AND MOTIVATION

Dirichlet's class number formula, conjectured for quadratic fields by Jacobi in 1832 and proven by Dirichlet in 1839, is one of the earliest examples of a relationship between leading terms of L-functions and global arithmetic invariants. Let F be a number field with ring of integers O_F. The Dedekind zeta function associated with F is defined as

$$\zeta_F(s) = \sum_{0 \neq \mathfrak{a} \subset O_F} \mathrm{N}\mathfrak{a}^{-s}, \quad \mathrm{Re}(s) > 1,$$

where $\mathfrak{a}$ runs through the nonzero ideals in O_F. The function $\zeta_F(s)$ generalizes Riemann's zeta function and has a meromorphic continuation to the whole complex plane with only a simple pole at $s = 1$. Dirichlet's class number formula, which is proved using a "geometry of numbers" approach, evaluates the residue at $s = 1$:

$$\lim_{s \to 1} (s-1)\zeta_F(s) = \frac{2^{r_1}(2\pi)^{r_2} R_F h_F}{w_F \sqrt{|D_F|}}.$$

Here r_1 is the number of real embeddings of F and $2r_2$ is the number of complex embeddings of F. Further, h_F and R_F denote the class number and regulator (defined below) of F, respectively, while w_F denotes the number of roots of unity in F and D_F is the discriminant of $F/\mathbf{Q}$. The meromorphic function $\zeta_F(s)$ satisfies a functional equation relating $\zeta_F(s)$ and $\zeta_F(1-s)$. Using this functional equation, Dirichlet's class number formula can be restated as giving the leading term of the Taylor expansion of $\zeta_F(s)$ at $s = 0$:

$$\zeta_F(s) = -\frac{h_F R_F}{w_F} s^{r_1+r_2-1} + O(s^{r_1+r_2}). \tag{1.1}$$

Artin described a theory of L-functions generalizing the Dedekind zeta function. Let G_F be the absolute Galois group of F. A Dirichlet character for F (or a degree 1 Artin character of F) is a homomorphism $\chi : G_F \to \mathbf{C}^\times$ with finite image. Class field theory identifies χ with a function, again denoted by χ, from the set of nonzero ideals of O_F to $\mathbf{C}$. Define

$$L(\chi, s) = \sum_{0 \neq \mathfrak{a} \subset O_F} \chi(\mathfrak{a}) \mathrm{N}\mathfrak{a}^{-s}, \quad \mathrm{Re}(s) > 1.$$

Again $L(\chi, s)$ has a meromorphic continuation to the whole complex plane with only a simple pole at $s = 1$ if χ is trivial. If H/F is a Galois extension with finite abelian Galois group $G = \mathrm{Gal}(H/F)$, then we can view any character $\chi \in \hat{G} = \mathrm{Hom}(G, \mathbf{C}^*)$ as a Dirichlet character for F, and we have the Artin factorization formula

$$\zeta_H(s) = \prod_{\chi \in \hat{G}} L(\chi, s). \tag{1.2}$$

Dirichlet's class number formula (1.1) for the field H gives the leading term of the left-hand side of (1.2) at $s = 0$. Stark asked for an analogous formula for $L(\chi, s)$ at $s = 0$ for each character χ, thereby giving a canonical factorization of the term $h_H R_H / w_H$. This led to the formulation of the abelian Stark conjecture, which we state in Section 2. This statement involves the choice of places of F that split completely in H. After stating Stark's conjecture, we restrict in the remainder of the paper to the case that the splitting places of F are finite.

Since the associated L-values here are algebraic, one can make progress on the conjectures through p-adic techniques such as p-adic Galois cohomology. To obtain nonzero L-values (and hence have nontrivial statements), parity conditions force us to restrict to the setting that F is a totally real field and H is a CM field.

Stark's conjecture at finite places has a natural restatement in terms of annihilators of class groups as formulated in the Brumer–Stark conjecture. We recall the statement and its refinements in Sections 4–5. The rest of the paper is taken up in describing the statement and proofs of our results toward the Brumer–Stark conjecture and its refinements.

2. STARK'S CONJECTURE

Let us first reformulate Dirichlet's class number formula.

For any place w of F we normalize the absolute value $|\cdot|_w : F_w^* \to \mathbf{R}$ by

$$|u|_w = \begin{cases} |u| & \text{if } w \text{ is real,} \\ |u|^2 & \text{if } w \text{ is complex,} \\ \mathrm{N}w^{-\mathrm{ord}_w(u)} & \text{if } w \text{ is nonarchimedean.} \end{cases}$$

For a finite set of places S of F, let X_S denote the degree zero subgroup of the free abelian group on S. Let $u_1, \ldots, u_{r_1+r_2-1}$ be a set of generators of the free abelian group O_F^*/μ_F. Let S_∞ be the set of archimedean places of F.

The Dirichlet regulator map

$$O_F^*/\mu_F \to \mathbf{R}X_{S_\infty}, \quad u \mapsto \sum_{w \in S_\infty} \log |u|_w \cdot w$$

induces an isomorphism $\mathbf{R}O_F^* \to \mathbf{R}X_{S_\infty}$. Here and throughout, $\mathbf{R}X_{S_\infty}$ denotes $\mathbf{R} \otimes_{\mathbf{Z}} X_{S_\infty}$, etc. Let $w_1, \ldots, w_{r_1+r_2}$ denote the archimedean places of F. Then

$$\{w_i - w_1 : 2 \le i \le r+1\}, \quad r = r_1 + r_2 - 1 \tag{2.1}$$

is an integral basis of X_{S_∞}. Let R_F be the absolute value of the determinant of the isomorphism between $\mathbf{R}O_F^*$ and $\mathbf{R}X_{S_\infty}$ with respect to the bases $\{u_1, \ldots, u_{r_1+r_2-1}\}$ and (2.1), respectively. Up to a sign, Dirichlet's class number formula can be restated as follows:

(i) The rational structure $\mathbf{Q}O_F^*$ on the left-hand side corresponds to the structure $\zeta_F^{(r)}(0)\mathbf{Q}X_{S_\infty}$ on the right-hand side.

(ii) The integral structure O_F^*/μ_F on the left-hand side corresponds to the structure $\zeta_F^{(r)}(0)X_{S_\infty}$ on the right-hand side.

Motivated by this reformulation, we present Stark's conjecture and its integral refinement due to Rubin. For details, see [41]. Let F be a number field of degree n and let H/F be a finite Galois extension with $G = \mathrm{Gal}(H/F)$ abelian. Let S, T be two finite disjoint sets of places of F satisfying the following conditions:

(1) S contains the sets S_∞ of archimedean places and S_{ram} of places ramified in H.

(2) T contains at least two primes of different residue characteristic or at least one prime of residue characteristic larger than $n+1$, where $n = [F : \mathbf{Q}]$.

For any character $\chi \in \hat{G} = \mathrm{Hom}(G, \mathbf{C}^*)$, define the S-depleted, T-smoothed L-function

$$L_{S,T}(\chi, s) = \prod_{\mathfrak{p} \notin S} \frac{1}{1 - \chi(\mathfrak{p})\mathrm{N}\mathfrak{p}^{-s}} \prod_{\mathfrak{p} \in T} \bigl(1 - \chi(\mathfrak{p})\mathrm{N}\mathfrak{p}^{1-s}\bigr), \quad \mathrm{Re}(s) > 1.$$

The function $L_{S,T}(\chi, s)$ extends by analytic continuation to a holomorphic function on $\mathbf{C}$. The *Stickelberger element* associated to this data is the unique group-ring element $\Theta_{S,T}(H/F, s) \in \mathbf{C}[G]$ satisfying

$$\chi\bigl(\Theta_{S,T}(H/F, s)\bigr) = L_{S,T}(\chi^{-1}, s) \quad \text{for all } \chi \in \hat{G}.$$

Let S_H denote the set of places of H above those in S, and similarly for T_H. Define

$$U_{S,T} = \bigl\{x \in H^* : \mathrm{ord}_w(x) \geq 0 \text{ for all } w \notin S_H \text{ and } x \equiv 1 \ (\mathrm{mod}\ T_H)\bigr\}.$$

The condition on T ensures that $U_{S,T}$ does not have any torsion. The Galois equivariant version of Dirichlet's unit theorem gives an $\mathbf{R}[G]$-module isomorphism

$$\begin{aligned} \lambda : \mathbf{R}U_{S,T} &\to \mathbf{R}X_{S_H}, \\ \lambda(u) &= \sum_{w \in S_H} \log\bigl(|u|_w\bigr) \cdot w. \end{aligned} \tag{2.2}$$

Suppose that exactly r places $v_1, \ldots, v_r \in S$ split completely in H and $\#S \geq r+1$. The order of vanishing of $L_{S,T}(\chi, s)$ at $s = 0$ is given by

$$r(\chi) = \dim_{\mathbf{C}}(\mathbf{C}U_{S,T})^{(\chi)} = \begin{cases} \#\{v \in S : \chi(v) = 1\} & \text{if } \chi \neq 1, \\ \#S - 1 & \text{if } \chi = 1, \end{cases}$$

whence $r(\chi) \geq r$ for all $\chi \in \hat{G}$. Stark's conjecture predicts that the rth derivative $\Theta^{(r)}_{S,T}(H/F, 0)$ captures the "non-rationality" of the map λ.

Conjecture 2.1 (Stark). *We have*

$$\Theta^{(r)}_{S,T}(H/F, 0) \cdot \mathbf{Q}\bigwedge^r X_{S_H} \subset \lambda\Bigl(\mathbf{Q}\bigwedge^r U_{S,T}\Bigr).$$

Concretely, this states that for each character χ of G with $r(\chi) = r$, the nonzero number $L^{(r)}_{S,T}(\chi^{-1}, s)$ lies in the one-dimensional $\mathbf{Q}$-vector space spanned by $\lambda(\bigwedge^r(U^{(\chi)}_{S,T}))$.

Let us reformulate Conjecture 2.1 in terms of the existence of special elements. Write $X^*_{S_H} = \mathrm{Hom}(X_{S_H}, \mathbb{Z}[G])$. For $\varphi \in \bigwedge^r X^*_{S_H}$, there is a determinant map

$$\bigwedge^r X_{S_H} \times \bigwedge^r X^*_{S_H} \to \mathbb{Z}[G]$$

defined by

$$(x_1 \wedge \cdots \wedge x_r, \varphi_1 \wedge \cdots \wedge \varphi_r) \mapsto \varphi_1 \wedge \cdots \wedge \varphi_r(x_1 \wedge \cdots \wedge x_r) = \det\bigl(\varphi_i(x_j)\bigr)_{i,j}.$$

We extend the determinant map to **R**-linearizations. We fix a place w_i of H above each v_i. Let $w_i^* \in X_{S_H}^*$ be induced by

$$w_i^*(w) = \sum_{\gamma \in G : \gamma w_i = w} \gamma.$$

Conjecture 2.2 (Stark). *Put $\varphi = w_1^* \wedge \cdots \wedge w_r^*$. There exists $u \in \mathbf{Q} \bigwedge^r U_{S,T}$ such that*

$$\varphi(\lambda(u)) = \Theta_{S,T}^{(r)}(H/F, 0).$$

The equivalence of Conjectures 2.1 and 2.2 is proven in [41, PROPOSITION 2.4].

We are now ready to state the integral version of Stark's conjecture. In the rank $r = 1$ case, Stark proposed the statement that u in Conjecture 2.2 lies in $U_{S,T}$. This is the famous "rank 1 abelian Stark conjecture." In the higher rank case, the obvious generalization $u \in \bigwedge^r U_{S,T}$ is not true, as was experimentally observed by Rubin [41]. Rubin defined a lattice, nowadays called "Rubin's lattice" and conjectured that it contains the element u.

Put $U_{S,T}^* = \mathrm{Hom}_{\mathbf{Z}[G]}(U_{S,T}, \mathbf{Z}[G])$.

The rth exterior bidual of $U_{S,T}$ (see [7] for a more general study and the initiation of this terminology) is defined by

$$\bigcap^r U_{S,T} = \left(\bigwedge^r U_{S,T}^*\right)^* \cong \left\{x \in \bigwedge^r \mathbf{Q}U_{S,T} : \varphi(x) \in \mathbf{Z}[G] \text{ for all } \varphi \in \bigwedge^r U_{S,T}^*\right\}.$$

We would like to consider only the "rank r" component of this bidual. To this end, for each character $\chi \in \hat{G}$ consider the associated idempotent

$$e_\chi = \frac{1}{\#G} \sum_{g \in G} \chi(g) g^{-1} \in \mathbf{C}[G].$$

Define $e_r = \sum e_\chi \in \mathbf{Q}[G]$, where the sum extends over the set

$$\{\chi \in \hat{G} : L_{S,T}^{(r)}(\chi, 0) \neq 0\} = \{\chi \in \hat{G} : \chi(G_v) \neq 1, v \in S \setminus \{v_1, \ldots, v_r\}\}.$$

Define Rubin's lattice by

$$\mathcal{L}^{(r)} U_{S,T} = \left(\bigcap^r U_{S,T}\right) \cap e_r\left(\mathbf{Q} \bigwedge^r U_{S,T}^*\right).$$

The following is Rubin's higher rank integral Stark conjecture.

Conjecture 2.3 ([41], Conjecture B'). *Put $\varphi = w_1^* \wedge \cdots \wedge w_r^*$. There exists $u \in \mathcal{L}^{(r)} U_{S,T}$ such that*

$$\varphi(\lambda(u)) = \Theta_{S,T}^{(r)}(H/F, 0).$$

3. STARK'S CONJECTURES AT FINITE PLACES

We now assume that the totally split places $v_1, \ldots, v_r$ from the previous section are all finite. This happens only when F is a totally real field and H is totally complex. In fact, the fixed fields of characters with nonvanishing L-functions at 0 are CM fields, so we restrict to the setting where F is totally real and H is CM for the remainder of the article. We also enact a slight notational change and write the set denoted S in the previous sections as S', and let $S = S' \setminus \{v_1, \ldots, v_r\}$. The reason for this is that we now still have $S \supset S_\infty \cup S_{\mathrm{ram}}$.

As we explain, in this setting Conjecture 2.2 for S' follows from a classical rationality result of Klingen–Siegel, though the integral refinement in Conjecture 2.3 remains a nontrivial statement. For a fixed place w of H, we have

$$\log |u|_w = -\operatorname{ord}_w(u) \log \mathrm{N}w. \tag{3.1}$$

Since the Euler factors at the v_i are equal to $(1 - \mathrm{N}v_i^{-s}) = (1 - \mathrm{N}w_i^{-s})$, we also have

$$\Theta_{S',T}^{(r)}(H/F, 0) = \Theta_{S,T}(H/F, 0) \cdot \prod_{i=1}^{r} \log \mathrm{N}w_i. \tag{3.2}$$

Theorem 3.1 (Klingen–Siegel). *We have* $\Theta_{S,T} := \Theta_{S,T}(H/F, 0) \in \mathbf{Q}[G]$.

With e_r as in the previous section, we are then led to define a map over $\mathbf{Q}$

$$\lambda_{\mathbf{Q}} : e_r(\mathbf{Q}U_{S',T}) \to e_r(\mathbf{Q}X_{S'_H}), \quad \lambda_{\mathbf{Q}}(u) = \sum_{w \mid v_i \text{ some } i} \operatorname{ord}_w(u) \cdot w.$$

Note that $e_r(\mathbf{Q}X_{S'_H})$ is the $\mathbf{Q}$-vector space generated by the places of H above the v_i. The map $\lambda_{\mathbf{Q}}$ is a $\mathbf{Q}[G]$-module isomorphism, and it induces an isomorphism on the free rank one $\mathbf{Q}[G]$-modules obtained by taking rth wedge powers. In view of (3.1), the map on rth wedge powers induced by the map λ of (2.2), when restricted to $e_r(\mathbf{Q} \bigwedge^r U_{S',T})$, is equal to $(\prod_{i=1}^r \log \mathrm{N}w_i) \cdot \lambda_{\mathbf{Q}}$. Conjecture 2.2 follows from this observation together with (3.2), since Theorem 3.1 implies the existence of $u \in e_r(\mathbf{Q} \bigwedge^r U_{S',T})$ such that

$$\varphi\big(\lambda_{\mathbf{Q}}(u)\big) = \Theta_{S,T}.$$

Here $\varphi = w_1^* \wedge \cdots \wedge w_r^*$ as in the statement of the conjecture.

On the other hand, the integral statement in Conjecture 2.3 lies deeper. We first note the following celebrated theorem of Deligne–Ribet [21] and Cassou-Noguès [8] refining the Klingen–Siegel theorem. The condition on the set T stated in Section 2 is crucial in this result. We remark that Deligne–RIbet prove their result using Hilbert modular forms, as an integral refinement of the strategy of the strategy established earlier by Siegel. This theme reappears in our own work described in §6.

Theorem 3.2. *We have* $\Theta_{S,T} \in \mathbf{Z}[G]$.

Conjecture 2.3 in this setting is known as the Rubin–Brumer–Stark conjecture:

Conjecture 3.3 (Rubin–Brumer–Stark). *There exists* $u \in \mathcal{L}^{(r)}U_{S',T}$ *such that*

$$\varphi\big(\lambda_{\mathbf{Q}}(u)\big) = \Theta_{S,T}.$$

We describe in Theorem 4.3 below a strong partial result toward this conjecture.

4. THE BRUMER–STARK CONJECTURE

Having stated the higher rank Rubin–Brumer–Stark conjecture, we now wind back the clock and focus on the case $r = 1$. This case had been studied independently by Brumer and Stark and served as a motivation for Rubin's work. Writing the splitting prime v_1 as $\mathfrak{p}$, the conjecture may be stated as follows.

Conjecture 4.1 (Brumer–Stark). *Fix a prime ideal $\mathfrak{p} \subset O_F$, $\mathfrak{p} \notin S \cup T$, such that $\mathfrak{p}$ splits completely in H. Fix a prime $\mathfrak{P} \subset O_H$ above $\mathfrak{p}$. There exists a unique element $u_\mathfrak{p} \in H^*$ such that $|u_\mathfrak{p}|_w = 1$ for every place w of H not lying above $\mathfrak{p}$,*

$$\operatorname{ord}_G(u_\mathfrak{p}) := \sum_{\sigma \in G} \operatorname{ord}_\mathfrak{P}\big(\sigma(u_\mathfrak{p})\big)\sigma^{-1} = \Theta_{S,T}, \tag{4.1}$$

and $u \equiv 1 \pmod{\mathfrak{q}}$ for all $\mathfrak{q} \in T_H$.

Note that the condition $|u|_w = 1$ includes all complex places w, so $c(u_\mathfrak{p}) = u_\mathfrak{p}^{-1}$ for the unique complex conjugation $c \in G$.

As we have alread noted, Stark arrived upon this statement in the 1970s through his attempts to generalize and factorize the classical Dirichlet class number formula (though in a slightly different formulation; the statement above is due to Tate [46]). Prior to this, in the 1960s, Brumer was interested in generalizing Stickelberger's classical factorization formula for Gauss sums in cyclotomic fields. Stickelberger's result can be formulated as stating that when $H = \mathbf{Q}(\mu_N)$ is a cyclotomic field, the Stickelberger element annihilates the class group of H. Let us consider Brumer's perspective of annihilation of class groups in the case of general H/F.

4.1. Annihilation of class groups

Let $\mathrm{Cl}^T(H)$ denote the ray class group of H with conductor equal to the product of primes in T_H. This is defined as follows. Let $I_T(H)$ denote the group of fractional ideals of H relatively prime to the primes in T_H. Let $P_T(H)$ denote the subgroup of $I_T(H)$ generated by principal ideals (α) where $\alpha \in O_H$ satisfies $\alpha \equiv 1 \pmod{\mathfrak{q}}$ for all $\mathfrak{q} \in T_H$. Then

$$\mathrm{Cl}^T(H) = I_T(H)/P_T(H).$$

This T-smoothed class group is naturally a G-module.

With the notation as in Conjecture 4.1, we have

$$\mathfrak{P}^{\Theta_{S,T}} = (u_\mathfrak{p}). \tag{4.2}$$

Such an equation holds for all $\mathfrak{p} \notin S \cup T$ that split completely in H. The set of primes of H above all such $\mathfrak{p}$ generate $\mathrm{Cl}^T(H)$. Hence we deduce

$$\Theta_{S,T} \in \operatorname{Ann}_{\mathbf{Z}[G]}\big(\mathrm{Cl}^T(H)\big). \tag{4.3}$$

In fact, (4.3) is almost equivalent to Conjecture 4.1; given (4.2), the element $u_\mathfrak{p}$ satisfies all of the conditions necessary for Conjecture 4.1 except possibly $c(u_\mathfrak{p}) = u_\mathfrak{p}^{-1}$. But of course $v_\mathfrak{p} = u_\mathfrak{p}/c(u_\mathfrak{p})$ satisfies this condition and moreover satisfies $\mathfrak{P}^{2\Theta_{S,T}} = (v_\mathfrak{p})$. Therefore the only possible discrepancy between the statements is a factor of 2, which disappears when we localize away from 2 as in the rest of this paper. Let us therefore define

$$R = \mathbf{Z}[1/2][G]^- = \mathbf{Z}[1/2][G]/(c+1),$$

and for any $\mathbf{Z}[G]$-module M we write $M^- = M \otimes_{\mathbf{Z}[G]} R$. There exists an element $u_{\mathfrak{p}} \in O_H[1/\mathfrak{p}]^* \otimes \mathbf{Z}[1/2]$ satisfying Conjecture 4.1 if and only if

$$\Theta_{S,T} \in \mathrm{Ann}_R\big(\mathrm{Cl}^T(H)^-\big). \tag{4.4}$$

This is the Brumer–Stark conjecture "away from 2".

Many authors have studied (4.4) as well as refinements. The works of Burns, Greither, Kurihara, Popescu, and Sano are particularly noteworthy [4–7, 25–27, 35]. Many of these refinements involve Fitting ideals, whose definition we now recall.

Let R be a commutative ring and M an R-module with finite presentation:

$$R^m \xrightarrow{A} R^n \to X \to 0.$$

Here A is an $n \times m$ matrix over R. The ith Fitting ideal $\mathrm{Fitt}_{i,R}(M)$ is the ideal generated by the $n-i \times n-i$ minors of A. It is a standard fact [37, CHAPTER 3, THEOREM 1] that $\mathrm{Fitt}_{i,R}(M)$ does not depend on the chosen presentation of M. We write $\mathrm{Fitt}_R(M)$ for $\mathrm{Fitt}_{0,R}(M)$, and when these is no ambiguity about the choice of $i = 0$, we call this the Fitting ideal of M.

The Fitting ideal of a finitely presented module is contained in its annihilator:

$$\mathrm{Fitt}_R(M) \subset \mathrm{Ann}_R(M). \tag{4.5}$$

In view of (4.4) and (4.5), it is therefore natural to ask whether $\Theta_{S,T}$ lies in the Fitting ideal of $\mathrm{Cl}^T(H)^-$ over R. It was noticed by Popescu in the function field case [38] and by Kurihara in the number field case that while this holds sometimes, it does not always hold. Greither and Kurihara [25, 26] observed that the statement may be corrected by replacing $\mathrm{Cl}^T(H)^-$ by its Pontryagin dual

$$\mathrm{Cl}^T(H)^{-,\vee} = \mathrm{Hom}_{\mathbf{Z}}\big(\mathrm{Cl}^T(H)^-, \mathbf{Q}/\mathbf{Z}\big).$$

We endow $\mathrm{Cl}^T(H)^{-,\vee}$ with the contragradient G-action $g \cdot \varphi(x) = \varphi(g^{-1}x)$. Denote by # the involution on $\mathbf{Z}[G]$ induced by $g \mapsto g^{-1}$ for $g \in G$.

Conjecture 4.2 (Kurihara, "Strong Brumer–Stark"). *We have*

$$\Theta_{S,T}^{\#} \in \mathrm{Fitt}_R\big(\mathrm{Cl}^T(H)^{-,\vee}\big).$$

Conjecture 4.2 leads to the following natural questions:

(1) What is the Fitting ideal of $\mathrm{Cl}^T(H)^{-,\vee}$?

(2) What is the Fitting ideal of $\mathrm{Cl}^T(H)^-$?

(3) Is there a natural arithmetically defined R-module whose Fitting ideal is generated by $\Theta_{S,T}$ or $\Theta_{S,T}^{\#}$?

The precise conjectural description of the Fitting ideal of $\mathrm{Cl}^T(H/F)^{-,\vee}$ was given by Kurihara [35]; we state this in Section 5.1 below. An important fact about this statement is that when S_{ram} is nonempty, the Fitting ideal of $\mathrm{Cl}^T(H/F)^{-,\vee}$ is in general not principal (and in particular is not generated by $\Theta^{\#}_{S,T}$).

A conjectural answer to the second question above has recently been provided in a striking paper by Atsuta and Kataoka [1]. They show that their conjecture is implied by the Equivariant Tamagawa Number Conjecture.

The third question is answered by a conjecture of Burns, Kurihara, and Sano, and is the topic of Section §5.3. We note that Fitting ideals of finitely presented R-modules are rarely principal. It is therefore remarkable that Burns–Kurihara–Sano defined a natural arithmetic object whose Fitting ideal is principal.

4.2. Our results

We now describe some of our results toward these conjectures [17, THEOREM 1.4].

Theorem 4.3. *Kurihara's exact formula for* $\mathrm{Fitt}_R(\mathrm{Cl}^T(H)^{-,\vee})$ *holds (see Theorem 5.1). In particular, we have the Brumer–Stark and Strong Brumer–Stark conjectures away from 2:*

$$\Theta^{\#}_{S,T} \in \mathrm{Fitt}_R\big(\mathrm{Cl}^T(H)^{-,\vee}\big) \subset \mathrm{Ann}_R\big(\mathrm{Cl}^T(H)^{-}\big)^{\#}.$$

Finally, Rubin's higher rank Brumer–Stark conjecture holds away from 2: with notation as in Conjecture 3.3, there exists $u \in \mathcal{L}^{(r)}U_{S',T} \otimes \mathbf{Z}[1/2]$ *such that* $\varphi(\lambda_{\mathbf{Q}}(u)) = \Theta_{S,T}$.

Partial results in this direction had been obtained earlier by Burns [5] (including a $\mu = 0$ hypothesis and the assumption of the Gross–Kuz'min conjecture) and by Greither and Popescu [27] (including a $\mu = 0$ hypothesis and imprimitivity conditions on S).

Our results in [17] do not seem to directly imply the conjecture of Atsuta and Kataoka on $\mathrm{Fitt}_R(\mathrm{Cl}^T(H)^{-})$ or the conjecture of Burns. However, we prove an analogous result toward the latter, with (S,T) replaced by an alternate pair (Σ, Σ'), in Theorem 5.6. This result turns out to be strong enough to deduce Theorem 4.3.

In §6 we give a detailed summary of the proof of Theorem 5.6. Key ingredients are the $\mathbf{Z}[G]$-modules $\nabla_{\Sigma}^{\Sigma'}(H)$ defined by Ritter and Weiss, and Ribet's method of using modular forms to construct Galois cohomology classes associated to L-functions.

4.3. Explicit formula for Brumer–Stark units

We conclude this section by describing a further direction in the study of Brumer–Stark units, that of *explicit formulae* and applications to explicit class field theory. This theme was initiated by Gross in [28] and [29] and developed by the first author and collaborators over a series of papers [9, 10, 13, 15, 20].

Let $\mathfrak{p}$ be as above and write $S' = S \cup \{\mathfrak{p}\}$. Let L denote a finite abelian CM extension of F containing H that is ramified over F only at the places in S'. Write $\mathfrak{g} = \mathrm{Gal}(L/F)$ and $\Gamma = \mathrm{Gal}(L/H)$, so $\mathfrak{g}/\Gamma \cong G$. Let I denote the relative augmentation ideal associated to $\mathfrak{g}$ and G, i.e., the kernel of the canonical projection $\mathrm{Aug} : \mathbf{Z}[\mathfrak{g}] \twoheadrightarrow \mathbf{Z}[G]$. Then $\Theta_{S',T}(L/F)$

lies in I, since its image under Aug is

$$\Theta_{S',T}(H/F) = \Theta_{S,T}(H/F)(1 - \sigma_{\mathfrak{p}}) = 0. \tag{4.6}$$

Here $\sigma_{\mathfrak{p}}$ denotes the Frobenius associated to $\mathfrak{p}$ in G, and this is trivial since $\mathfrak{p}$ splits completely in H. Intuitively, if we view $\Theta_{S',T}(L/F)$ as a function on the ideals of $\mathbf{Z}[\mathfrak{g}]$, equation (4.6) states that this function "has a zero" at the ideal I; the value of the "derivative" of this function at I is simply the image of $\Theta_{S',T}(L/F)$ in I/I^2. Gross provided a conjectural algebraic interpretation of this derivative as follows. Denote by

$$\mathrm{rec}_{\mathfrak{P}} : H_{\mathfrak{P}}^* \to \Gamma$$

the composition of the inclusion $H_{\mathfrak{P}}^* \hookrightarrow \mathbf{A}_H^*$ with the global Artin reciprocity map

$$\mathbf{A}_H^* \twoheadrightarrow \Gamma.$$

Throughout this article we adopt Serre's convention **[42]** for the reciprocity map. Therefore $\mathrm{rec}(\varpi^{-1})$ is a lifting to $G_{\mathfrak{p}}^{\mathrm{ab}}$ of the Frobenius element on the maximal unramified extension of $F_{\mathfrak{p}}$ if $\varpi \in F_{\mathfrak{p}}^*$ is a uniformizer.

Conjecture 4.4 (Gross, **[29, CONJECTURE 7.6]**). *Define*

$$\mathrm{rec}_G(u_{\mathfrak{p}}) = \sum_{\sigma \in G} (\mathrm{rec}_{\mathfrak{P}}\, \sigma(u_{\mathfrak{p}}) - 1)\tilde{\sigma}^{-1} \in I/I^2, \tag{4.7}$$

where $\tilde{\sigma} \in \mathfrak{g}$ is any lift of $\sigma \in G$. Then

$$\mathrm{rec}_G(u_{\mathfrak{p}}) \equiv \Theta_{S',T}^{L/F} \quad \text{in } I/I^2.$$

The main result of **[18]** is the following.

Theorem 4.5. *Let p be an odd prime and suppose that $\mathfrak{p}$ lies above p. Gross's Conjecture 4.4 holds in $(I/I^2) \otimes \mathbf{Z}_p$.*

Our interest in this result is that by enlarging S and taking larger and larger field extensions L/F, one can use (4.7) to specify all of the $\mathfrak{p}$-adic digits of $u_{\mathfrak{p}}$. One therefore obtains an exact $\mathfrak{p}$-adic analytic formula for $u_{\mathfrak{p}}$. This formula can be described either using the *Eisenstein cocycle* or more explicitly via *Shintani's method*; the latter approach is followed in Section 7.2. In Section 7.3, we describe the argument using "horizontal Iwasawa theory" to show that Theorem 4.5 implies the conjectural exact formula. In Section 7.4 we summarize the key ingredients involved in the proof of Theorem 4.5, including an integral version of the Greenberg–Stevens $\mathscr{L}$-invariant and an associated modified Ritter–Weiss module $\nabla_{\mathscr{L}}$. In the setting of F real quadratic, Darmon, Pozzi, and Vonk have given an alternate, elegant proof of the explicit formula for the units $u_{\mathfrak{p}}$ (Section 7.5). Their approach involves p-adic deformations of modular forms, rather than the tame deformations that we consider.

One significance of the exact formula is that we show that the collection of Brumer–Stark units, together with some easily described elements, generate the maximal abelian extension of the totally real field F.

Theorem 4.6. *Let* BS *denote the set of Brumer–Stark units* $u_{\mathfrak{p}}$ *as we range over all possible CM abelian extensions* H/F *and for each extension a choice of prime* $\mathfrak{p}$ *that splits completely in* H. *Let* $\{\alpha_1, \dots, \alpha_{n-1}\}$ *denote any elements of* F^* *whose signs in* $\{\pm 1\}^n/(-1, \dots, -1)$ *under the real embeddings of* F *form a basis for this* $\mathbf{Z}/2\mathbf{Z}$*-vector space. The maximal abelian extension of* F *is generated by* BS *together with* $\sqrt{\alpha_1}, \dots, \sqrt{\alpha_{n-1}}$:

$$F^{\mathrm{ab}} = F(\mathrm{BS}, \sqrt{\alpha_1}, \dots, \sqrt{\alpha_{n-1}}).$$

It is important to stress that the exact formula for $u_{\mathfrak{p}}$ described in Sectin 7.2 can be computed without knowledge of the field H, using only the data of F, $\mathfrak{p}$, and the conductor of H/F. Furthermore, we can leave out any finite set of primes $\mathfrak{p}$ without altering the conclusion of the theorem. In this way we obtain an *explicit class field theory* for F, i.e., an analytic construction of its maximal abelian extension F^{ab} using data intrinsic only to F itself. Explicit computations of class fields of real quadratic fields generated using our formula are provided in [17, §2.3] and [23].

5. REFINEMENTS OF STARK'S CONJECTURE

In this section we recall various refinements of the strong Brumer–Stark conjecture.

5.1. The conjecture of Kurihara

In this section we describe the Fitting ideal of the minus part of the dual class group. For each v in S_{ram}, let $I_v \subset G_v \subset G$ denote the inertia and decomposition groups, respectively, associated to v. Write

$$e_v = \frac{1}{\# I_v} \mathrm{N} I_v = \frac{1}{\# I_v} \sum_{\sigma \in I_v} \sigma \in \mathbf{Q}[G]$$

for the idempotent that represents projection onto the characters of G unramified at v. Let $\sigma_v \in G_v$ denote any representative of the Frobenius coset of v. The element $1 - \sigma_v e_v \in \mathbf{Q}[G]$ is independent of choice of representative. Following [25], we define the Sinnott–Kurihara ideal, *a priori* a fractional ideal of $\mathbf{Z}[G]$, by

$$\mathrm{SKu}^T(H/F) = (\Theta^{\#}_{S_\infty, T}) \prod_{v \in S_{\mathrm{ram}}} (\mathrm{N} I_v, 1 - \sigma_v e_v).$$

Kurihara proved using the theorem of Deligne–Ribet and Cassou-Noguès that $\mathrm{SKu}^T(H/F)$ is a subset of $\mathbf{Z}[G]$ (see [17, LEMMA 3.4]). The following conjecture of Kurihara is proven in [17, THEOREM 1.4].

Theorem 5.1. *We have*

$$\mathrm{Fitt}_R\big(\mathrm{Cl}^T(H)^{-,\vee}\big) = \mathrm{SKu}^T(H/F)^-.$$

The definition of the Sinnott–Kurihara ideal is inspired by Sinnott's definition of generalized Stickelberger elements for abelian extensions of $\mathbf{Q}$ [44]. For a generalization of Sinnott's ideal to arbitrary totally real fields see [25, §2]. In general, Sinnott's ideal contains the Sinnott–Kurihara ideal but it may be strictly larger.

The plus part of the Sinnott–Kurihara ideal is not very interesting as the plus part of $\Theta^{\#}_{S_\infty,T}$ is 0. The plus part of the class group is much smaller than the minus part and seems harder to describe; for example, Greenberg's conjecture on the vanishing of lambda invariants implies that the order of the plus part is bounded up the cyclotomic tower. For abelian extensions of $\mathbf{Q}$, the plus part is described by Sinnott using cyclotomic units.

5.2. The conjecture of Atsuta–Kataoka

It is, in fact, more natural to ask about the Fitting ideal of $\mathrm{Cl}^T(H)$, as opposed to the Pontryagin dual. A conjectural answer to this question has been provided in a recent paper of Atsuta–Kataoka [1] using the theory of shifted Fitting ideals developed by Kataoka [32]. We recall this notion now.

Let M be an R-module of finite length. Take a resolution

$$0 \to N \to P_1 \to \cdots \to P_d \to M \to 0$$

with each P_i of projective dimension ≤ 1. Following [32], define the shifted Fitting ideal

$$\mathrm{Fitt}_R^{[d]}(M) = \left(\prod_{i=1}^{d} \mathrm{Fitt}_R(P_i)^{(-1)^i}\right) \mathrm{Fitt}_R(N).$$

The independence of this definition from the choice of resolution is proven in [32, THEOREM 2.6 AND PROPOSITION 2.7]. Let

$$g_v = 1 - \sigma_v + \#I_v \in \mathbf{Z}[G/I_v], \quad h_v = 1 - e_v\sigma_v + \mathrm{N}I_v \in \mathbf{Q}[G].$$

Define the $\mathbf{Z}[G]$-module

$$A_v = \mathbf{Z}[G/I_v]/(g_v).$$

Conjecture 5.2 (Atsuta–Kataoka). *We have*

$$\mathrm{Fitt}_R\big(\mathrm{Cl}^T(H)^-\big) = \Big(\prod_{w \in S_{\mathrm{ram},H}} h_v^- \,\mathrm{Fitt}_R^{[1]}(A_v^-)\Big)\Theta_{S_\infty,T}.$$

In [1], the authors give an explicit description of the ideal $h_v^- \,\mathrm{Fitt}^{[1]}_{\mathbf{Z}[G]^-}(A_v^-)$ appearing in Conjecture 5.2. Write $I_v = J_1 \times \cdots \times J_s$ for cyclic groups J_i, $1 \leq i \leq s$. For each i, put

$$\mathrm{N}_i = \mathrm{N}J_i = \sum_{\sigma \in J_i} \sigma \in \mathbf{Z}[G].$$

Furthermore, put $\mathcal{I} = \ker(\mathbf{Z}[G] \to \mathbf{Z}[G/G_v])$ for the relative augmentation ideal. For each $1 \leq i \leq s$, put Z_i for the ideal of $\mathbf{Z}[G]$ generated by $\mathrm{N}_{j_1} \cdots \mathrm{N}_{j_{s-i}}$, where $(j_1, \ldots, j_{s-i})$ runs through all tuples of integers satisfying $1 \leq j_1 \leq \cdots \leq j_{s-i} \leq s$. Define

$$\mathcal{J} = \sum_{i=1}^{s} Z_i \mathcal{I}^{i-1}.$$

Although Z_i depends on the decomposition of I_v into cyclic groups, the ideal $\mathcal{J}$ is independent (see [1, DEFINITION 1.2]).

Theorem 5.3 (Atsuta–Kataoka). *We have*

$$h_v^- \operatorname{Fitt}^{[1]}_{\mathbf{Z}[G]^-}(A_v^-) = \big(\mathrm{N}I_v, (1-e_v\sigma_v)\mathcal{J}\big)$$

as fractional ideals of $\mathbf{Z}[G]^-$.

Atsuta–Kataoka prove:

Theorem 5.4. *The Equivariant Tamagawa Number Conjecture for H/F implies Conjecture 5.2.*

5.3. The conjecture of Burns–Kurihara–Sano

The refinements mentioned above do not involve principal ideals. The method of Ribet, which attempts to show the inclusion of an arithmetically defined ideal into an analytically defined ideal, works well for principal ideals. From this point of view, it is natural to ask if there is an arithmetically defined object whose Fitting ideal is generated by the Stickelberger element $\Theta_{S,T}$. Burns, Kurihara and Sano provided a conjectural answer to this question [6]. A modification of this statement (Theorem 5.6 below) is the main technical result in [17] from which all the results of Theorem 4.3 are deduced.

We now recall the statement of the conjecture of Burns–Kurihara–Sano. Let H_T^* be the group of $x \in H^*$ such that $\operatorname{ord}_w(x-1) > 0$ for each prime $w \in T_H$. Define

$$\operatorname{Sel}_S^T(H) = \operatorname{Hom}_{\mathbf{Z}}(H_T^*, \mathbf{Z}) / \prod_{w \notin S_H \cup T_H} \mathbf{Z},$$

where the implicit map sends a tuple (x_w) to the function $\sum_w x_w \operatorname{ord}_w$. The G-action on $\operatorname{Sel}_S^T(H)$ is the contragradient G-action $(g\varphi)(x) = \varphi(g^{-1}x)$.

Conjecture 5.5 (Burns–Kurihara–Sano). *We have*

$$\operatorname{Fitt}_R\big(\operatorname{Sel}_S^T(H)^-\big) = (\Theta_{S,T}^{\#}).$$

We have proven a version of this result with altered sets S and T. Fix an odd prime p and put $R_p = \mathbf{Z}_p[G]^-$. Define

$$\Sigma = S \setminus \{v \in S : v \nmid p\}$$

and

$$\Sigma' = T \cup \{v \in S : v \nmid p\}$$

Theorem 5.6 ([17, THEOREM 3.3]). *Let* $\operatorname{Sel}_\Sigma^{\Sigma'}(H)_p^- = \operatorname{Sel}_\Sigma^{\Sigma'}(H) \otimes_{\mathbf{Z}[G]} R_p$. *We have*

$$\operatorname{Fitt}_{R_p}\big(\operatorname{Sel}_\Sigma^{\Sigma'}(H)_p^-\big) = (\Theta_{\Sigma,\Sigma'}^{\#}).$$

It turns out that Theorem 5.6 is strong enough to imply Kurihara's conjecture (Theorem 5.1). The key point is that there is a short exact sequence

$$0 \longrightarrow \operatorname{Sel}_\Sigma^T(H)^- \longrightarrow \operatorname{Sel}_\Sigma^{\Sigma'}(H)^- \longrightarrow \textstyle\prod_{w \in S'_H}\big((O_H/w)^*\big)^{\vee,-} \longrightarrow 0, \tag{5.1}$$

from which one deduces (see [17, THEOREM 3.7])

$$\mathrm{Fitt}_{R_p}\big(\mathrm{Sel}_{\Sigma}^{T}(H)_p^-\big) = (\Theta_{\Sigma,T}^{\#}) \prod_{v \in S_{\mathrm{ram}}, v \nmid p} (\mathrm{N}I_v, 1 - \sigma_v e_v).$$

Since $\mathrm{Sel}_{S_\infty}^{T}(H)^- \cong \mathrm{Cl}^T(H)^{-,\vee}$, it then remains to calculate the effect of removing the primes $v \in S_{\mathrm{ram}}$, $v \mid p$ from Σ. This is a delicate process using functorial properties of the Ritter–Weiss modules discussed in §6, and one obtains (see [17, APPENDIX B]) the desired result

$$\mathrm{Fitt}_{R_p}\big(\mathrm{Sel}_{S_\infty}^{T}(H)_p^-\big) = (\Theta_{S_\infty,T}^{\#}) \prod_{v \in S_{\mathrm{ram}}} (\mathrm{N}I_v, 1 - \sigma_v e_v).$$

6. RITTER–WEISS MODULES AND RIBET'S METHOD

In this section we summarize the proof of Theorem 5.6.

6.1. Ritter–Weiss modules

The $\mathbf{Z}[G]$-module that shows up in our constructions with modular forms is a certain *transpose* of $\mathrm{Sel}_S^T(H)$ in the sense of Jannsen [31], denoted $\nabla_S^T(H)$. This module was originally defined by Ritter and Weiss in the foundational paper [40] without the smoothing set T. We incorporated the smoothing set T and established some additional properties of $\nabla_S^T(H)$ in [17, APPENDIX A]. To describe these properties, we work over $R_p = \mathbf{Z}_p[G]^-$ and consider finite disjoint sets Σ, Σ' satisfying the following:

- $\Sigma \supset S_\infty$ and $\Sigma \cup \Sigma' \supset S_{\mathrm{ram}}$.
- Σ' satisfies the condition (2) on T in Section 2.
- The primes in $\Sigma' \cap S_{\mathrm{ram}}$ have residue characteristic $\ell \neq p$.

Note that the pair (S, T) from Section 2 and the pair (Σ, Σ') considered in Section 5.3 both satisfy these conditions. The module $\nabla_\Sigma^{\Sigma'}(H)_p^- = \nabla_\Sigma^{\Sigma'}(H) \otimes_{\mathbf{Z}[G]} R_p$ satisfies the following:

- There is a short exact sequence of R_p-modules

$$0 \longrightarrow \mathrm{Cl}_\Sigma^{\Sigma'}(H)_p^- \longrightarrow \nabla_\Sigma^{\Sigma'}(H)_p^- \longrightarrow (X_{H_\Sigma})_p^- \longrightarrow 0. \tag{6.1}$$

  Here $\mathrm{Cl}_\Sigma^{\Sigma'}(H)$ denotes the quotient of $\mathrm{Cl}^{\Sigma'}(H)$ by the image of the primes in Σ_H.

- Given a R_p-module B, a surjective R_p-module homomorphism

$$\nabla_\Sigma^{\Sigma'}(H)_p^- \twoheadrightarrow B \tag{6.2}$$

  is equivalent to the data of a cocycle $\kappa \in Z^1(G_F, B)$ and a collection of elements $x_v \in B$ for $v \in \Sigma$ satsifying the following conditions:

  - The cohomology class $[\kappa] \in H^1(G_F, B)$ is unramified outside Σ', tamely ramified at Σ', and locally trivial at Σ.
  - The x_v provide local trivializations at Σ: $\kappa(\sigma) = (\sigma - 1)x_v$ for $\sigma \in G_v$.
  - The x_v along with the image of κ generate the module B over R_p.

The tuples $(\kappa, \{x_v\})$ are taken modulo the natural notion of coboundary, i.e., $(\kappa, \{x_v\}) \sim (\kappa + dx, \{x_v + x\})$ for $x \in B$.

- The module $\nabla_\Sigma^{\Sigma'}(H)_p^-$ has a quadratic presentation, i.e., there exists an exact sequence of R_p-modules

$$M_1 \xrightarrow{A} M_2 \longrightarrow \nabla_\Sigma^{\Sigma'}(H)_p^- \longrightarrow 0 \tag{6.3}$$

  where M_1 and M_2 are free of the same finite rank.

- The module $\nabla_\Sigma^{\Sigma'}(H)_p^-$ is a transpose of $\mathrm{Sel}_\Sigma^{\Sigma'}(H)_p^-$, i.e., for a suitable quadratic presentation (6.3) of $\nabla_\Sigma^{\Sigma'}(H)_p^-$, the cokernel of the induced map

$$\mathrm{Hom}_{R_p}(M_2, R_p) \xrightarrow{A^{T,\#}} \mathrm{Hom}_{R_p}(M_1, R_p) \tag{6.4}$$

  is isomorphic to $\mathrm{Sel}_\Sigma^{\Sigma'}(H)_p^-$. Here we follow our convention of giving Hom spaces the contragradient G-action.

The quadratic presentation property (6.3) implies that $\mathrm{Fitt}_{R_p}(\nabla_\Sigma^{\Sigma'}(H)_p^-) = \det(A)$ is principal. The transpose property (6.4) implies that

$$\mathrm{Fitt}_{R_p}\big(\nabla_\Sigma^{\Sigma'}(H)_p^-\big) = \mathrm{Fitt}_{R_p}\big(\mathrm{Sel}_\Sigma^{\Sigma'}(H)_p^-\big)^\#. \tag{6.5}$$

Theorem 5.6 is therefore equivalent to

$$\mathrm{Fitt}_{R_p}\big(\nabla_\Sigma^{\Sigma'}(H)_p^-\big) = (\Theta_{\Sigma,\Sigma'}). \tag{6.6}$$

We now fix (Σ, Σ') to be the pair defined in Section 5.3. In the remainder of this section we outline how (6.6) is proved using Ribet's method. Throughout, an unadorned Θ denotes $\Theta_{\Sigma,\Sigma'}$ (and $\Theta^\#$ denotes $\Theta^\#_{\Sigma,\Sigma'}$).

6.2. Inclusion implies equality

An interesting feature of Ribet's method is that it tends to prove an inclusion in one direction, that of an algebraically defined ideal contained within an analytically defined ideal. In our setting, we use it to prove

$$\mathrm{Fitt}_{R_p}\big(\nabla_\Sigma^{\Sigma'}(H)_p^-\big) \subset (\Theta), \quad \text{equivalently,} \quad \mathrm{Fitt}_{R_p}\big(\mathrm{Sel}_\Sigma^{\Sigma'}(H)_p^-\big) \subset (\Theta^\#). \tag{6.7}$$

We then employ an analytic argument to show that this inclusion is an equality. It is important to note that the inclusion (6.7) is the *reverse direction* of that required by the Brumer–Stark and Strong Brumer–Stark conjectures. It is therefore essential for our approach that one actually has the statement of an *equality* rather than just an inclusion (and an analytic argument to deduce the equality from the reverse inclusion). For this reason, the conjecture of Burns stated in Section 5.3 (more precisely the analog of it stated in Theorem 5.6) plays an essential role in our strategy.

Let us describe the analytic argument in the special case $\Sigma = S_\infty$, i.e. there are no primes above p ramified in H/F. In this case

$$\mathrm{Sel}_\Sigma^{\Sigma'}(H)_p^- \cong \mathrm{Cl}^{\Sigma'}(H)_p^- \tag{6.8}$$

is finite and Θ is a non-zero-divisor. Using (6.7), write

$$\operatorname{Fitt}_{R_p}\big(\operatorname{Sel}_{\Sigma}^{\Sigma'}(H)_p^-\big) = (\Theta^\# \cdot z) \quad \text{for some } z \in R_p. \tag{6.9}$$

We must show that $z \in R_p^*$. An elementary argument (see **[17, §2.3]**) shows that (6.9) implies

$$\#\operatorname{Sel}_{\Sigma}^{\Sigma'}(H)_p^- = \Big(\prod_{\substack{\chi \in \hat{G} \\ \chi \text{ odd}}} \chi(\Theta^\# \cdot z)\Big)_p \tag{6.10}$$

where the subscript p on the right denotes the p-power part of an integer. Yet the analytic class number formula implies (see **[17, §2.1]**)

$$\prod_{\substack{\chi \in \hat{G} \\ \chi \text{ odd}}} \chi(\Theta^\#) = \prod_{\substack{\chi \in \hat{G} \\ \chi \text{ odd}}} L_{\Sigma,\Sigma'}(\chi, 0) \doteq \#\operatorname{Cl}^{\Sigma'}(H)^-, \tag{6.11}$$

where $\doteq$ denotes equality up to a power of 2. Combining (6.8), (6.10), and (6.11), one finds that $\chi(z)$ is a p-adic unit for each odd character χ. It follows that $z \in R_p^*$ as desired.

The generalization of this argument to arbitrary Σ requires a delicate induction and is described in **[17, §5]**.

6.3. Ribet's method

We now describe our implementation of Ribet's method to prove the inclusion (6.7). The idea is to use the Galois representations associated to Hilbert modular forms to construct an R_p-module M and a surjection $\nabla_{\Sigma}^{\Sigma'}(H)_p^- \twoheadrightarrow M$ such that $\operatorname{Fitt}_{R_p}(M) \subset (\Theta)$. The properties of Fitting ideals imply that (6.7) follows from the existence of such a surjection. As described in (6.2), a surjection from $\nabla_{\Sigma}^{\Sigma'}(H)_p^-$ can be constructed by defining a cohomology class $[\kappa] \in H^1(G_F, M)$ satisfying certain local conditions along with local trivializations at the places in Σ.

Ribet's method was described in great detail by Mazur in a beautiful article written for the celebration of Ribet's 60th birthday **[36]**. We borrow from this the following schematic diagram demonstrating the general path one follows to link L-values (in our case, the Stickelberger element Θ) to class groups (in our case, the Ritter–Weiss module $\nabla_{\Sigma}^{\Sigma'}(H)_p^-$).

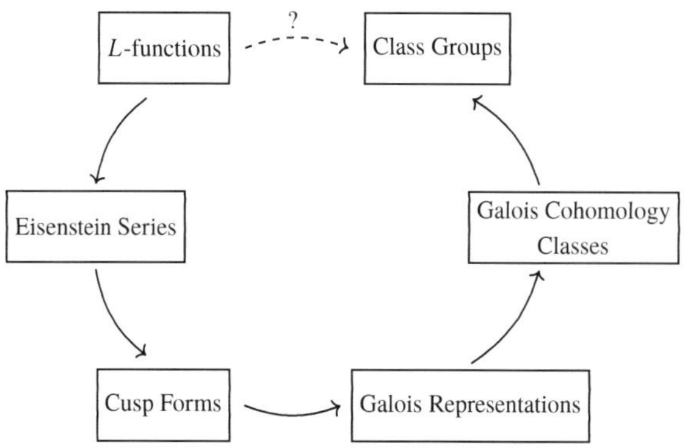

Let us now trace this path in our application.

6.3.1. L-functions to Eisenstein series

The key connection between L-functions and modular forms in Ribet's method is that L-functions appear as constant terms of Eisenstein series. We now define the space of modular forms in which our Stickelberger element Θ appears.

Let $k > 1$ be an integer such that

$$k \equiv 1 \pmod{(p-1)p^N}$$

for a large value of N. Let $\mathfrak{n} \subset O_F$ denote the conductor of H/F. Let $M_k(\mathfrak{n}; \mathbf{Z})$ denote the group of Hilbert modular forms for F of level $\mathfrak{n}$ with Fourier coefficients in $\mathbf{Z}$. For each odd character χ of G valued in $\mathbf{C}_p^*$, let

$$M_k(\mathfrak{n}, \chi) \subset M_k(\mathfrak{n}; \mathbf{Z}) \otimes \mathbf{C}_p$$

denote the subspace of forms of nebentypus χ. Let

$$\boldsymbol{\chi} : G_F \to G \to R_p^*$$

denote the canonical character, where the first arrow is projection and the second is induced by $G \hookrightarrow \mathbf{Z}[G]^*$.

Definition 6.1. The space $M_k(\mathfrak{n}, \boldsymbol{\chi}; R_p)$ of *group-ring valued Hilbert modular forms of weight* k *and level* $\mathfrak{n}$ *over* R_p consists of those $f \in M_k(\mathfrak{n}; \mathbf{Z}) \otimes R_p$ such that $\chi(f) \in M_k(\mathfrak{n}, \chi)$ for each odd character χ. Let $S_k(\mathfrak{n}, \boldsymbol{\chi}; R_p)$ denote the subspace of cusp forms. We define $M_k(\mathfrak{n}, \boldsymbol{\chi}; \mathrm{Frac}(R_p))$ and $S_k(\mathfrak{n}, \boldsymbol{\chi}; \mathrm{Frac}(R_p))$ similarly.

Hilbert modular forms f are determined by their Fourier coefficients $c(\mathfrak{m}, f)$ indexed by the nonzero ideals $\mathfrak{m} \subset O_F$ and their constant terms $c_\lambda(0, f)$ indexed by $\lambda \in \mathrm{Cl}^+(F)$, the narrow class group of F. For odd $k \geq 1$, there is an Eisenstein series $E_k(\boldsymbol{\chi}, 1) \in M_k(\mathfrak{n}, \boldsymbol{\chi})$ whose Fourier coefficients are given by

$$c\big(\mathfrak{m}, E_k(\boldsymbol{\chi}, 1)\big) = \sum_{\substack{\mathfrak{a} | \mathfrak{m} \\ (\mathfrak{m}/\mathfrak{a}, \mathfrak{n}) = 1}} \boldsymbol{\chi}(\mathfrak{m}/\mathfrak{a}) \mathrm{N}\mathfrak{a}^{k-1}.$$

To describe the constant coefficients of $E_k(\boldsymbol{\chi}, 1)$ we first set S to be minimal, i.e., the union of S_∞ and S_{ram}, where the latter is the set of primes dividing $\mathfrak{n}$. Next we assume for the remainder of the article that $\mathfrak{n} \neq 1$; the case $\mathfrak{n} = 1$ causes no difficulties but the formulas must be slightly modified. We then have

$$c_\lambda\big(0, E_1(\boldsymbol{\chi}, 1)\big) = \begin{cases} 0, & k > 1, \\ 2^{-n} \Theta_S^\#, & k = 1. \end{cases}$$

Here $\Theta_S = \Theta_{S,\phi}(H/F, 0) \in \mathbf{Q}[G]$ denotes the S-depleted but unsmoothed Stickelberger element. We have $E_k(\boldsymbol{\chi}, 1) \in M_k(\mathfrak{n}, \boldsymbol{\chi}; R_p)$ for $k > 1$ and $E_1(\boldsymbol{\chi}, 1) \in M_1(\mathfrak{n}, \boldsymbol{\chi}; \mathrm{Frac}(R_p))$ because of the possible nonintegrality of the constant term.

6.4. Eisenstein series to cusp forms

In order to define a cusp form from the Eisenstein series, one is led to consider certain linear combinations of the analogues of $E_k(\chi, 1)$ as H ranges over all its CM subfields containing F. This process also incorporates smoothing at the primes in T. We avoid stating the slightly complicated formula here (see [17, PROPOSITION 8.14]), but the end result is a group ring form $W_k(\chi, 1)$ whose constant terms are given by

$$c_\lambda\big(0, W_k(\chi, 1)\big) = \begin{cases} 0, & k > 1, \\ 2^{-n}\Theta^{\#}, & k = 1, \end{cases} \tag{6.12}$$

where we remind the reader that $\Theta^{\#} = \Theta^{\#}_{\Sigma,\Sigma'}$. Building off the computations of [19], we calculate in [17, §8] the constant terms of the $W_k(\chi, 1)$ at *all cusps*; the terms in (6.12) can be viewed as the constant terms "at infinity." Indeed, it is the attempt to cancel the constant terms at other cusps that leads naturally to the definition of the $W_k(\chi, 1)$.

In order to define a cusp form, we apply two important results of Silliman [43]. The first of these generalizes a result of Hida and Wiles and is stated below.

Theorem 6.2 ([43, THEOREM 10.7]). *Let m be a fixed positive integer. For positive integers $k \equiv 0 \pmod{(p-1)p^N}$ with N sufficiently large, there is a Hilbert modular form V_k of level 1, trivial nebentypus, and weight k defined over $\mathbf{Z}_p$ such that*

$$V_k \equiv 1 \pmod{p^m},$$

and such that the normalized constant term of V_k at every cusp is congruent to $1 \pmod{p^m}$.

The idea to construct a cusp form is to fix a very large integer m and to consider the product $W_1(\chi, 1)V_k \in M_{k+1}(\mathfrak{n}, \chi, R_p)$ with V_k as in Theorem 6.2. This series has constant terms at infinity congruent to $2^{-n}\Theta^{\#}$ modulo p^m. One then wants to subtract off $2^{-n}\Theta^{\#}H_{k+1}(\chi)$ for some group ring valued form $H_{k+1}(\chi) \in M_{k+1}(\mathfrak{n}, \chi, R_p)$ to obtain a cusp form. If there exists a prime above p dividing $\mathfrak{n}$ (i.e., $\Sigma - S_\infty$ is nonempty), then this strategy works. Silliman's second result, which generalizes a result of Chai and is stated in [43, THEOREM 10.10], implies that one can obtain a form that is cuspidal at the cusps "above infinity at p" in this fashion. Applying Hida's ordinary operator then yields a form that is cuspidal.

Theorem 6.3 ([17, THEOREM 8.18]). *Suppose* $\gcd(\mathfrak{n}, p) \neq 1$. *For positive integers $k \equiv 1 \pmod{(p-1)p^N}$ and N sufficiently large, there exists $H_k(\chi) \in M_k(\mathfrak{n}, \chi, R_p)$ such that*

$$F_k(\chi) = e_p^{\mathrm{ord}}\big(W_1(\chi, 1)V_{k-1} - \Theta^{\#}H_k(\chi)\big)$$

lies in $S_k(\mathfrak{n}p, R, \chi)$.

The significance of Theorem 6.3 is that we have now constructed a cusp form that is congruent to an Eisenstein series modulo $\Theta^{\#}$.

When $\gcd(\mathfrak{n}, p) = 1$, the construction of the cusp form is in fact more interesting, and a new feature appears. In this case, the ordinary operator at p does not annihilate the form $W_k(\chi, 1)$, and it must be incorporated into our linear combination. Moreover, this apparent cost has a great benefit—we obtain a congruence between a cusp form and Eisenstein series not only modulo $\Theta^{\#}$, but modulo a multiple $x \cdot \Theta^{\#}$ for a certain $x \in R_p$. This element x has an intuitive meaning—it represents the trivial zeroes of the p-adic L-function associated to χ, even the "mod p trivial zeroes." The precise definition is as follows.

Lemma 6.4. *Suppose* $\gcd(\mathfrak{n}, p) = 1$. *For positive* $k \equiv 1 \pmod{(p-1)p^N}$ *with* N *sufficiently large, the element*

$$x = \frac{\Theta_{S_\infty}(1-k)}{\Theta_{S_\infty}(0)} \in \operatorname{Frac}(R)$$

lies in R *and is a non-zero-divisor.*

The analogue of Theorem 6.3 for $\gcd(\mathfrak{n}, p) = 1$ is as follows.

Theorem 6.5 ([17, THEOREM 8.17]). *Suppose* $\gcd(\mathfrak{n}, p) = 1$. *For positive integers* $k \equiv 1 \pmod{(p-1)p^N}$ *and* N *sufficiently large, there exists* $H_k(\chi) \in M_k(\mathfrak{n}, \chi, R_p)$ *such that*

$$F_k(\chi) = e_p^{\mathrm{ord}}\big(xW_1(\chi, 1)V_{k-1} - W_k(\chi, 1) - x\Theta^{\#}H_k(\chi)\big)$$

lies in $S_k(\mathfrak{n}p, R, \chi)$.

The extra factor of x in our congruence between the cusp form $F_k(\chi)$ and a linear combination of Eisenstein series plays an extremely important role in showing that the Galois cohomology classes we construct are unramified at p.

We conclude this section by interpreting the congruences of Theorems 6.3 and 6.5 in terms of Hecke algebras. We consider the Hecke algebra $\tilde{\mathbf{T}}$ generated over R_p by the operators $T_{\mathfrak{q}}$ for primes $\mathfrak{q} \nmid \mathfrak{n}p$ and $U_{\mathfrak{p}}$ for primes $\mathfrak{p} \mid p$. (We ignore the operators $U_{\mathfrak{q}}$ for $\mathfrak{q} \mid \mathfrak{n}, \mathfrak{q} \nmid p$ in order to avoid issues regarding nonreducedness of Hecke algebras arising from the presence of *oldforms*.) We denote by $\mathbf{T} = e_p^{\mathrm{ord}}(\tilde{\mathbf{T}})$ Hida's ordinary Hecke algebra associated to $\tilde{\mathbf{T}}$. Let $\epsilon_{\mathrm{cyc}} : G_F \to \mathbf{Z}_p^*$ denote the p-adic cyclotomic character of F. Theorems 6.3 and 6.5 then yield:

Theorem 6.6. *Let* $x = 1$ *if* $\gcd(\mathfrak{n}, p) \neq 1$ *and let* x *be as in Lemma 6.4 if* $\gcd(\mathfrak{n}, p) = 1$. *There exists an* $R_p/x\Theta^{\#}$*-algebra* W *and a surjective* R_p*-algebra homomorphism* $\varphi : \mathbf{T} \to W$ *satisfying the following properties:*

- *The structure map* $R_p/x\Theta^{\#} \to W$ *is an injection.*
- $\varphi(T_{\mathfrak{l}}) = \epsilon_{\mathrm{cyc}}^{k-1}(\mathfrak{l}) + \chi(\mathfrak{l})$ *for* $\mathfrak{l} \nmid \mathfrak{n}p$.
- $\varphi(U_{\mathfrak{p}}) = 1$ *for* $\mathfrak{p} \mid \gcd(\mathfrak{n}, p)$.
- *Let*

$$U = \prod_{\mathfrak{p} \mid p, \mathfrak{p} \nmid \mathfrak{n}} (U_{\mathfrak{p}} - \chi(\mathfrak{p})) \in \mathbf{T}.$$

If $y \in R_p$ *and* $\varphi(U)y = 0$ *in* W, *then* $y \in (\Theta^{\#})$.

The idea of this theorem is the usual one: the homomorphism φ sends a Hecke operator to its "eigenvalue mod $x\Theta^{\#}$" acting on $F_k(\chi)$. The only subtlety is the last bullet point: the operators $U_\mathfrak{p}$ for $\mathfrak{p} \mid p$, $\mathfrak{p} \nmid \mathfrak{n}$ do not act as scalars, so a more involved argument is necessary. This explains why the ring W is not just $R_p/x\Theta^\#$. The idea behind the last statement of the theorem is that the operator $\varphi(U)$ introduces a factor of x; If xy is divisible by $x\Theta^\#$ in R_p, then y is divisible by $\Theta^\#$ since x is a non-zero-divisor. This demonstrates the essential additional ingredient provided by the "higher congruence" modulo $x\Theta^\#$ rather than just modulo $\Theta^\#$. See **[17, THEOREM 8.23]** for details.

6.5. Cusp forms to Galois representations

In this section we study the Galois representation attached to cusp forms that are congruent to Eisenstein series. Let $\mathfrak{m}$ be the intersection of the finitely many maximal ideals of $\mathbf{T}$ containing the kernel of φ. Put $\mathbf{T}_\mathfrak{m}$ for the completion of $\mathbf{T}$ with respect to $\mathfrak{m}$ and $K = \mathrm{Frac}(\mathbf{T}_\mathfrak{m})$. Then K is a finite product of fields parameterized by the $\mathbf{Q}_p$-Galois orbits of cuspidal newforms of weight k and level $\mathfrak{n}$, defined over the ring of integers in a finite extension of $\mathbf{Z}_p$, that are congruent to an Eisenstein series modulo the maximal ideal. As in **[17, §9.2]**, the work of Hida and Wiles gives a Galois representation

$$\rho : G_F \to \mathrm{GL}_2(K)$$

satisfying the following:

(1) ρ is unramified outside $\mathfrak{n}p$.

(2) For all primes $\mathfrak{l} \nmid \mathfrak{n}p$, the characteristic polynomial of $\rho(\mathrm{Frob}_\mathfrak{l})$ is given by

$$\mathrm{char}\big(\rho(\mathrm{Frob}_\mathfrak{l})\big)(x) = x^2 - T_\mathfrak{l}x + \chi(\mathfrak{l})\mathrm{N}\mathfrak{l}^{k-1}. \tag{6.13}$$

(3) For every $\mathfrak{q} \mid p$, let $G_\mathfrak{q}$ denote a decomposition group at $\mathfrak{q}$. We have

$$\rho|_{G_\mathfrak{q}} \sim \begin{pmatrix} \chi\varepsilon_{\mathrm{cyc}}^{k-1}\eta_\mathfrak{q}^{-1} & * \\ 0 & \eta_\mathfrak{q} \end{pmatrix},$$

where $\varepsilon_{\mathrm{cyc}}$ is the p-adic cyclotomic character and $\eta_\mathfrak{q}$ is an unramified character given by $\eta_\mathfrak{q}(\mathrm{rec}_\mathfrak{q}(\varpi^{-1})) = U_\mathfrak{q}$, with ϖ a uniformizer of $F_\mathfrak{q}^*$.

Let $\mathbf{I}$ denote the kernel of φ extended to $\varphi : \mathbf{T}_\mathfrak{m} \to W$. Reducing (6.13) modulo $\mathbf{I}$, and using Čebotarev to extend from $\mathrm{Frob}_\mathfrak{l}$ to all $\sigma \in G_F$, we see that the characteristic polynomial of $\rho(\sigma)$ is congruent to $(x - \chi(\sigma))(x - \epsilon_{\mathrm{cyc}}(\sigma))$ (mod $\mathbf{I}$). In particular, if $\chi(\sigma) \not\equiv \epsilon_{\mathrm{cyc}}(\sigma)$ (mod $\mathfrak{m}$), then by Hensel's lemma $\rho(\sigma)$ has distinct eigenvalues $\lambda_1, \lambda_2 \in \mathbf{T}_\mathfrak{m}$ such that $\lambda_1 \equiv \epsilon_{\mathrm{cyc}}^{k-1}(\sigma)$ (mod I) and $\lambda_2 \equiv \chi(\sigma)$ (mod $\mathbf{I}$).

To define a convenient basis for ρ, we choose $\tau \in G_F$ such that:

(1) τ restricts to the complex conjugation of G,

(2) for each $\mathfrak{q} \mid p$, the eigenspace of $\rho|_{G_\mathfrak{q}}$ projected to each factor of K is not stable under $\rho(\tau)$.

See **[17, PROPOSITION 9.3]** for the existence of such τ. Since $p \neq 2$, we have

$$\chi(\tau) = -1 \not\equiv 1 \equiv \epsilon_{\mathrm{cyc}}(\tau) \text{ (mod } \mathfrak{m}).$$

It follows from the discussion above that the eigenvalues of $\rho(\tau)$ satisfy $\lambda_1 \equiv \epsilon_{\mathrm{cyc}}(\sigma) \pmod{\mathbf{I}}$ and $\lambda_2 \equiv -1 \pmod{\mathbf{I}}$. Fix the basis consisting of eigenvectors of $\rho(\tau)$, say

$$\rho(\tau) = \begin{pmatrix} \lambda_1 & 0 \\ 0 & \lambda_2 \end{pmatrix}.$$

For a general $\sigma \in G_F$, write

$$\rho(\sigma) = \begin{pmatrix} a(\sigma) & b(\sigma) \\ c(\sigma) & d(\sigma) \end{pmatrix}.$$

For each $\mathfrak{q} \mid p$, there is a change of basis matrix

$$M_{\mathfrak{q}} = \begin{pmatrix} A_{\mathfrak{q}} & B_{\mathfrak{q}} \\ C_{\mathfrak{q}} & D_{\mathfrak{q}} \end{pmatrix} \in \mathrm{GL}_2(K)$$

satisfying

$$\begin{pmatrix} a(\sigma) & b(\sigma) \\ c(\sigma) & d(\sigma) \end{pmatrix} M_{\mathfrak{q}} = M_{\mathfrak{q}} \begin{pmatrix} \chi \varepsilon_{\mathrm{cyc}}^{k-1} \eta_{\mathfrak{q}}^{-1} & * \\ 0 & \eta_{\mathfrak{q}} \end{pmatrix}. \tag{6.14}$$

The second condition in the choice of τ ensures that $A_{\mathfrak{q}}, C_{\mathfrak{q}} \in K^*$. Furthermore, equating the upper left-hand entries in (6.14) gives

$$b(\sigma) = \frac{A_{\mathfrak{q}}}{C_{\mathfrak{q}}} \big(a(\sigma) - \chi \varepsilon_{\mathrm{cyc}}^{k-1} \eta_{\mathfrak{q}}^{-1}(\sigma)\big) \quad \text{for all } \sigma \in G_{\mathfrak{q}}. \tag{6.15}$$

6.6. Galois representations to Galois cohomology classes

We summarize [17, §9.3]. As explained above, we have

$$a(\sigma) + d(\sigma) \equiv \chi(\sigma) + \epsilon_{\mathrm{cyc}}^{k-1}(\sigma) \pmod{I} \quad \text{for all } \sigma \in G_F. \tag{6.16}$$

Applying the same rule for $\tau\sigma$ and noting that $a(\tau\sigma) = \lambda_1 a(\sigma)$, $d(\sigma\tau) = \lambda_2 d(\sigma)$, we find

$$a(\sigma)\epsilon_{\mathrm{cyc}}^{k-1}(\tau) - d(\sigma) \equiv -\chi(\sigma) + \epsilon_{\mathrm{cyc}}^{k-1}(\sigma\tau) \pmod{I}. \tag{6.17}$$

Solving the congruences (6.16) and (6.17) and once again using the fact that $\epsilon_{\mathrm{cyc}}^{k-1}(\tau) \not\equiv -1 \pmod{\mathfrak{m}}$ since $p \neq 2$, we find that $a(\sigma), d(\sigma) \in \mathbf{T}_{\mathfrak{m}}$ and

$$a(\sigma) \equiv \varepsilon_{\mathrm{cyc}}^{k-1}(\sigma) \pmod{\mathbf{I}}, \quad d(\sigma) \equiv \chi(\sigma) \pmod{\mathbf{I}} \quad \text{for all } \sigma \in G_F. \tag{6.18}$$

Let B be the $\mathbf{T}_{\mathfrak{m}}$ submodule of K generated by $\{b(\sigma) : \sigma \in G_F\} \cup \{\frac{A_{\mathfrak{q}}}{C_{\mathfrak{q}}} : \mathfrak{q} \in \Sigma \setminus S_\infty\}$. We have $\rho(\sigma\sigma') = \rho(\sigma)\rho(\sigma')$ for $\sigma, \sigma' \in G_F$. Equating the upper right entries and using equation (6.18), we obtain

$$b(\sigma\sigma') = a(\sigma)b(\sigma') + b(\sigma)d(\sigma') \equiv \varepsilon_{\mathrm{cyc}}^{k-1}(\sigma)b(\sigma') + \chi(\sigma')b(\sigma) \pmod{\mathbf{I}B}. \tag{6.19}$$

Let m be am integer such that $k \equiv 1 \pmod{(p-1)p^m}$. Let $I_{\mathfrak{q}}$ denote the inertia subgroup of G_F of a prime $\mathfrak{q}$. Put B_1 for the $\mathbf{T}_{\mathfrak{m}}$-submodule of B generated by

$$\mathbf{I}B \cup p^m B \cup \{b(\sigma) : \sigma \in I_{\mathfrak{q}} \text{ for } \mathfrak{q} \mid p, \mathfrak{q} \notin \Sigma\}.$$

Define $\overline{B} = B/B_1$. Equation (6.19) then gives that $\kappa(\sigma) = \chi^{-1}(\sigma)b(\sigma)$ is a cocyle defining a cohomology class $[\kappa]$ in $H^1(G_F, \overline{B}(\chi^{-1}))$ satisfying the following local properties:

(1) As ρ is unramified at $\mathfrak{l} \nmid \mathfrak{n}p$, so is the class $[\kappa]$.

(2) As $\overline{B}$ is pro-p, the class $[\kappa]$ is at most tamely ramified at any prime $\mathfrak{l} \mid \mathfrak{n}$ not above p.

(3) It is proven in [17, §4.1] that we may assume Σ' does not contain any primes above p. Thus $[\kappa]$ is at most tamely ramified at all primes in Σ'.

(4) By the definition of B_1, where we have included $b(I_{\mathfrak{q}})$ for primes $\mathfrak{q} \mid p$, $\mathfrak{q} \notin \Sigma$, the class $[\kappa]$ is unramified at such $\mathfrak{q}$.

(5) Equation (6.15) implies that $[\kappa]$ is locally trivial at finite primes in Σ. As p is odd, $[\kappa]$ is locally trivial at archimedian places [17, PROPOSITION 9.5].

6.7. Galois cohomology classes to class groups

The Galois cohomology class $[\kappa]$ satisfies the conditions listed after equation (6.2) and hence gives a surjection

$$\nabla_{\Sigma}^{\Sigma'}(H)_p^- \twoheadrightarrow \overline{B}(\chi^{-1}).$$

For details see [18, THEOREM 4.4]. The general properties of Fitting ideals imply

$$\mathrm{Fitt}_{R_p}\big(\nabla_{\Sigma}^{\Sigma'}(H)_p^-\big) \subset \mathrm{Fitt}_{R_p}\big(\overline{B}(\chi^{-1})\big).$$

It is therefore enough to prove that $\mathrm{Fitt}_{R_p}(\overline{B}) \subset (\Theta^\#)$. Typically in Ribet's method, one argues that the fractional ideal B is a faithful $\mathbf{T}_{\mathfrak{m}}$-module, and hence the Fitting ideal of $B/\mathbf{I}B$ is contained in $\mathbf{I}$. However, our module $\overline{B}$ is more complicated than $B/\mathbf{I}B$, so we proceed as follows. Using equation (6.15), we show that any element in $\mathrm{Fitt}_{R_p}(\overline{B})$ is annihilated by $\varphi(U)$ for the operator U from Theorem 6.6. The final assertion of this theorem then implies that $\mathrm{Fitt}_{R_p}(\overline{B})$ contained in $(\Theta^\#)$. See [17, §9.5] for details.

This concludes our summary of the proof of Theorem 5.6.

7. EXPLICIT FORMULA FOR BRUMER–STARK UNITS

In this final section of the paper, we discuss the first author's explicit formula for Brumer–Stark units as mentioned in §4.3. The conjecture in the case that F is a real quadratic field was studied in [13], and the general case was studied in [15]. Here we consider an arbitrary totally real field F, but to simplify formulas we assume that the rational prime p is inert in F. Furthermore, we let H be the narrow ray class field of some conductor $\mathfrak{n} \subset O_F$ and assume that $p \equiv 1 \pmod{\mathfrak{n}}$. This ensures that the prime $\mathfrak{p} = pO_F$ splits completely in H. Fix a prime $\mathfrak{P}$ of H above $\mathfrak{p}$. We fix $S \supset S_\infty \cup S_{\mathrm{ram}} = \{v \mid \mathfrak{n}\infty\}$. We also fix a prime ideal $\mathfrak{l} \subset O_F$ such that $\mathrm{N}\mathfrak{l} = \ell > n+1$ is a prime integer and let $T = \{\mathfrak{l}\}$.

In this setting, we will present a $\mathfrak{p}$-adic analytic formula for the image of the Brumer–Stark unit $u_{\mathfrak{p}} \in H^*$ in $H_{\mathfrak{P}}^* \cong F_{\mathfrak{p}}^*$. The most general, conceptually satisfying, and theoretically useful form of this conjecture uses the Eisenstein cocycle. This is a class in the $(n-1)$st cohomology of $\mathrm{GL}_n(\mathbf{Z})$ that has many avatars studied by several authors (see [2, 3, 9, 10, 13,

20,45]). In this paper, we avoid defining the Eisenstein cocycle and present instead the more explicit and down to earth version of the conjectural formula for $u_{\mathfrak{p}}$ stated in [15].

7.1. Shintani's method

Fixing an ordering of the n real embeddings of F yields an map $F \hookrightarrow \mathbf{R}^n$ such that the image of any fractional ideal is a cocompact lattice. We let F^* act on $\mathbf{R}^n$ by composing this embedding with componentwise multiplication and denote the action by $*$.

Let $v_1, \dots, v_r \in (\mathbf{R}^{>0})^n$, $1 \le r \le n$, be vectors in the totally positive orthant that are linearly independent over $\mathbf{R}$. The corresponding *simplicial cone* is defined by

$$C(v_1, \dots, v_r) = \left\{ \sum_{i=1}^{r} t_i v_i : 0 < t_i \right\} \subset (\mathbf{R}^{>0})^n.$$

Suppose now $r = n$. We will define a certain union of $C(v_1, \dots, v_n)$ and some of its boundary faces that we call the *Colmez closure*. Write

$$(0, 0, \dots, 1) = \sum_{i=1}^{n} q_i v_i, \quad q_i \in \mathbf{R}.$$

For each nonempty subset $J \subset \{1, \dots, n\}$, we say that J is *positive* if $q_i > 0$ for all $i \notin J$. The Colmez closure of $C(v_1, \dots, v_n)$ is defined by:

$$C^*(v_1, \dots, v_n) = \bigsqcup_{J \text{ positive}} C\big(\{v_j, j \in J\}\big).$$

Let $E(\mathfrak{n}) \subset O_F^*$ denote the subgroup of totally positive units ϵ such that $\epsilon \equiv 1 \pmod{\mathfrak{n}}$. Shintani proved that there exists a union of simplicial cones that is a fundamental domain for the action of $E(\mathfrak{n})$ on $(\mathbf{R}^{>0})^n$. For example, in the real quadratic case ($n = 2$), $E(\mathfrak{n}) = \langle \epsilon \rangle$ is cyclic and $C^*(1, \epsilon)$ is a fundamental domain. In the general case, it can be difficult to write down an explicit fundamental domain, but a nice generalization of the $n = 2$ case is obtained if we allow ourselves to consider instead a *signed fundamental domain*. For a simplicial cone C, let $\mathbf{1}_C$ denote the characteristic function of C on $(\mathbf{R}^{>0})^n$.

Definition 7.1. A *signed fundamental domain* for the action of $E(\mathfrak{n})$ on $(\mathbf{R}^{>0})^n$ is by definition a formal linear combination $D = \sum_i a_i C_i$ of simplicial cones C_i with $a_i \in \mathbf{Z}$ such that

$$\sum_{u \in E(\mathfrak{n})} \sum_i a_i \mathbf{1}_{C_i}(u * x) = 1$$

for all $x \in (\mathbf{R}_{>0})^n$.

Fix an ordered basis $\{\epsilon_1, \dots, \epsilon_{n-1}\}$ for $E(\mathfrak{n})$. Define the orientation

$$w_\epsilon = \operatorname{sign} \det\big(\log(\epsilon_{ij})\big)_{i,j=1}^{n-1}\big) = \pm 1, \tag{7.1}$$

where ϵ_{ij} denotes the jth coordinate of ϵ_i. For each permutation $\sigma \in S_{n-1}$ let

$$v_{i,\sigma} = \epsilon_{\sigma(1)} \cdots \epsilon_{\sigma(i-1)} \in E(\mathfrak{n}), \quad i = 1, \dots, n.$$

By convention, $v_{1,\sigma} = (1, 1, \ldots, 1)$ for all σ. Define

$$w_\sigma = (-1)^{n-1} w_\epsilon \operatorname{sign}(\sigma) \operatorname{sign}\bigl(\det(v_{i,\sigma})_{i=1}^n\bigr) \in \{0, \pm 1\}.$$

The following result was proven independently by Diaz y Diaz–Friedman [22] and Charollois–Dasgupta–Greenberg [10, THEOREM 1.5], generalizing the result of Colmez [11] in the case that all $w_\sigma = 1$.

Theorem 7.2. *The formal linear combination*

$$\sum_{\sigma \in S_{n-1}} w_\sigma C^*(v_{1,\sigma}, \ldots, v_{n,\sigma})$$

is a signed fundamental domain for the action of $E(\mathfrak{n})$ on $(\mathbf{R}^{>0})^n$.

7.2. The formula

Throughout this section assume that p is odd. Recall that $T = \{\mathfrak{l}\}$. Let $\mathfrak{b}$ be a fractional ideal that is relatively prime to $\mathfrak{n}\mathfrak{l}$. Let $D = \sum a_i C_i$ be the signed fundamental domain for the action of $E(\mathfrak{n})$ on $(\mathbf{R}^{>0})^n$ given in Theorem 7.2. We use all this data to define a $\mathbf{Z}$-valued measure μ on O_p, the p-adic completion of O_F. Fix an element $z \in \mathfrak{b}^{-1}$ such that $z \equiv 1 \pmod{\mathfrak{n}}$. For each compact open set $U \subset O_p$, define the *Shintani zeta function*

$$\zeta(\mathfrak{b}, U, D, s) = \sum_i a_i \sum_{\substack{\alpha \in C_i \cap \mathfrak{b}^{-1}\mathfrak{n} + z \\ \alpha \in U, (\alpha, S) = 1}} (\mathrm{N}\alpha)^{-s}.$$

Shintani proved that this sum converges for $\Re(s)$ large enough and extends to a meromorphic function on $\mathbf{C}$. Define

$$\mu_\mathfrak{b}(U) = \zeta(\mathfrak{b}, U, D, 0) - \ell \cdot \zeta(\mathfrak{b}\mathfrak{l}^{-1}, U, D, 0).$$

Using Shintani's formulas, one may show:

Theorem 7.3 ([15, PROPOSITION 3.12]). *For every compact open $U \subset O_p$, we have $\mu_\mathfrak{b}(U) \in \mathbf{Z}$.*

We may now state our conjectural exact formula for the Brumer–Stark unit u_p and all of its conjugates over F. Write

$$\Theta_{S,T} = \sum_{\sigma \in G} \zeta_{S,T}(\sigma)\sigma^{-1}, \quad \zeta_{S,T}(\sigma) \in \mathbf{Z}.$$

Define

$$u_p(\mathfrak{b})^{\mathrm{an}} = p^{\zeta_{S,T}(\sigma_\mathfrak{b})} \times\!\!\!\!\!\!\int_{O_p^*} x \, d\mu_\mathfrak{b}(x) \in F_p^*. \tag{7.2}$$

Here the crossed integral is a *multiplicative integral* in the sense of Darmon [12] and can be expressed as a limit of Riemann products:

$$\times\!\!\!\!\!\!\int_{O_p^*} x \, d\mu_\mathfrak{b}(x) := \lim_{m \to \infty} \prod_{a \in (O_p/p^m)^*} a^{\mu_\mathfrak{b}(a + p^m O_p)}.$$

Write $\sigma_\mathfrak{b} \in G$ for the Frobenius associated to $\mathfrak{b}$. In [15, THEOREM 5.15] we prove that $u_p(\mathfrak{b})^{\mathrm{an}}$ depends only on the image of $\mathfrak{b}$ in the narrow ray class group of conductor $\mathfrak{n}$, i.e., on $\sigma_\mathfrak{b} \in G$ (at least up to a root of unity in F_p^*).

Conjecture 7.4. *We have* $\sigma_{\mathfrak{b}}(u_{\mathfrak{p}}) = u_p(\mathfrak{b})^{\mathrm{an}}$ *in* F_p^*.

The expression (7.2) can be computed to high p-adic precision on a computer. See [23] for tables of narrow Hilbert class fields of real quadratic fields determined using this formula.

It is convenient to have an invariant that also satisfies $u_p(\mathfrak{b}\mathfrak{q})^{\mathrm{an}} = (u_p(\mathfrak{b})^{\mathrm{an}})^{-1}$ if $\mathfrak{q}$ is a prime such that $\sigma_{\mathfrak{q}} = c$. Conjecture 7.4 would imply such a formula, but it is unclear whether this purely analytic statement can be proved unconditionally. To this end, we fix $\mathfrak{q}$ such that $\sigma_{\mathfrak{q}} = c$ and define

$$v_p(\mathfrak{b})^{\mathrm{an}} = \left(\frac{u_p(\mathfrak{b})^{\mathrm{an}}}{u_p(\mathfrak{b}\mathfrak{q})^{\mathrm{an}}}\right)^{1/2} \in \hat{F}_p^* := F_p^* \hat{\otimes} \mathbf{Z}_p.$$

One then has

$$v_p(\mathfrak{b}\mathfrak{q})^{\mathrm{an}} = \left(v_p(\mathfrak{b})^{\mathrm{an}}\right)^{-1} \tag{7.3}$$

unconditionally, and we expect to have $v_p(\mathfrak{b})^{\mathrm{an}} = u_p(\mathfrak{b})^{\mathrm{an}}$. The following is therefore a slightly easier form of Conjecture 7.4 to study.

Conjecture 7.5. *We have* $\sigma_{\mathfrak{b}}(u_{\mathfrak{p}}) = v_p(\mathfrak{b})^{\mathrm{an}}$ *in* $\hat{F}_p^*$.

7.3. Horizontal Iwasawa theory

We now discuss the relationship between Gross's tower of fields conjecture (Conjecture 4.4) and our conjectural exact formula for Brumer–Stark units. Our goal is to prove:

Theorem 7.6. *Assume that p is odd. Gross's conjecture implies Conjecture 7.5.*

In this exposition we have assumed that the odd prime p is inert in F and that $\mathfrak{p} = pO_F$. In the case of general $\mathfrak{p}$, one must still assume that p is odd and unramified in F in the statement of Theorem 7.6.

The abelian extensions L/F to which we can apply Gross's conjecture (with $S' = S \cup \{\mathfrak{p}\}$) as in Conjecture 4.4) are those that contain H and are unramified outside $S'\infty$. Let $F_{S'}$ denote the maximal abelian extension of F unramified outside $S'\infty$. The reciprocity map of class field theory yields an explicit description of $\mathrm{Gal}(F_{S'}/H)$. For each finite $v \in S'$, let $U_{v,\mathfrak{n}} \subset O_v^*$ denote the subgroup of elements congruent to 1 modulo $\mathfrak{n}O_v$ (so $U_{f,\mathfrak{n}} = O_v^*$ for $v \nmid \mathfrak{n}$). Define $\mathbf{O}^* = \prod_{v \in S' \setminus S_\infty} U_{v,\mathfrak{n}}$. Then

$$\mathrm{Gal}(F_{S'}/H) \cong \mathbf{O}^*/\overline{E(\mathfrak{n})},$$

where $\overline{E(\mathfrak{n})}$ denotes the topological closure of $E(\mathfrak{n})$ embedded diagonally in $\mathbf{O}^*$.

For each finite extension $L \subset F_{S'}$ containing H, if we write $\Gamma = \mathrm{Gal}(L/H)$, then (4.7) yields a formula for $\mathrm{rec}_G(u_p)$ in $I/I^2 \cong \mathbf{Z}[G] \otimes \Gamma$. Under this isomorphism, the coefficient of $\sigma_{\mathfrak{b}}^{-1}$ is just the image of $\mathrm{rec}_{\mathfrak{p}}(\sigma_{\mathfrak{b}}(u_p))$ in Γ. Taking the inverse limit over all $H \subset L \subset F_{S'}$ therefore gives an equality for

$$\left(\sigma_{\mathfrak{b}}(u_p), 1, 1, \ldots, 1\right) \quad \text{in } \mathbf{O}/\overline{E(\mathfrak{n})}.$$

Here we have written O_p^* as the first component of $\mathbf{O}$.

The next key point is that the constructions of Section 7.2 can be repeated to provide a measure $\mu_{\mathfrak{b},\mathbf{O}}$ on $\mathbf{O} = \prod_{v \in S' \setminus S_\infty} O_v$ extending the measure $\mu_{\mathfrak{b}}$ on O_p. It is not hard to check that the restriction of $\mu_{\mathfrak{b},\mathbf{O}}$ to $\mathbf{O}^*$, pushed forward to $\mathbf{O}^*/\overline{E(\mathfrak{n})}$, is precisely the measure that recovers the values of the partial zeta functions of the abelian extensions L contained in $F_{S'}$. These are exactly the values appearing in Gross's conjecture. In other words, Gross's conjecture for the set S' is equivalent to

$$\left(\sigma_{\mathfrak{b}}(u_{\mathfrak{p}}), 1, 1, \ldots, 1\right) \cdot p^{-\zeta_{S,T}(\sigma_{\mathfrak{b}})} = \times\!\!\!\!\!\!\int_{\mathbf{O}^*} x \, d\mu_{\mathfrak{b},\mathbf{O}}(x) \quad \text{in } \mathbf{O}/\overline{E(\mathfrak{n})}. \tag{7.4}$$

See [15, PROPOSITION 3.4]. The next important calculation ([15, THEOREM 3.22]) is that

$$p^{\zeta_{S,T}(\sigma_{\mathfrak{b}})} \times\!\!\!\!\!\!\int_{\mathbf{O}^*} x \, d\mu_{\mathfrak{b},\mathbf{O}}(x) = \left(u_p(\mathfrak{b})^{\mathrm{an}}, 1, 1, \ldots, 1\right). \tag{7.5}$$

The first component of this is simply the compatibility of the constructions of $\mu_{\mathfrak{b}}$ and $\mu_{\mathfrak{b},\mathbf{O}}$; the interesting part of the computation is the 1's in the components away from p. Equations (7.4) and (7.5) combine to yield that the ratio $\sigma_{\mathfrak{b}}(u_p)/u_p(\mathfrak{b})^{\mathrm{an}}$ lies in the group

$$D(S) = \left\{x \in O_p^* : (x, 1, 1, \ldots, 1) \in \overline{E(\mathfrak{n})} \subset \mathbf{O}^*\right\}.$$

We can also conclude

$$\sigma_{\mathfrak{b}}(u_p)/v_p(\mathfrak{b})^{\mathrm{an}} \in D(S) \tag{7.6}$$

since $c(u_p) = u_p^{-1}$.

The final trick, inspired by the method of Taylor–Wiles, is to consider certain enlarged sets $S_Q = S \cup Q$ for a well-chosen finite set of auxiliary primes Q. Let us compare the Brumer–Stark units for S and S_Q, denoted u_p and $u_p(S_Q)$, respectively. The defining property (4.1) shows that

$$u_p(S_Q) = u_p^z, \quad \text{where } z = \prod_{\mathfrak{q} \in Q} (1 - \sigma_{\mathfrak{q}}^{-1}) \in \mathbf{Z}[G].$$

In particular, if we choose the $\mathfrak{q} \in Q$ such that $\sigma_{\mathfrak{q}} = c$, the complex conjugation of G, then $u_p(S_Q) = u_p^{2^{\#Q}}$. Using (7.3), one can similarly show that $v_p(S_Q, \mathfrak{b}) = v_p(\mathfrak{b})^{2^{\#Q}}$. Now (7.6) for S_Q implies that

$$\sigma_{\mathfrak{b}}\left(u_p(S_Q)\right)/v_p(S_Q, \mathfrak{b})^{\mathrm{an}} \in D(S_Q),$$

hence

$$\left(\sigma_{\mathfrak{b}}(u_p)/v_p(\mathfrak{b})^{\mathrm{an}}\right)^{2^{\#Q}} \in D(S_Q), \quad \text{so } \sigma_{\mathfrak{b}}(u_p)/v_p(\mathfrak{b})^{\mathrm{an}} \in D(S_Q)$$

since $p \neq 2$.

To conclude the proof of Theorem 7.6, one shows using the Čebotarev Density Theorem that one can choose the sets Q to force $D(S_Q)$ as small as desired (i.e., the intersection of $D(S_Q)$ over all possible Q is trivial). See [15, LEMMA 5.17] for details.

7.4. The Greenberg–Stevens $\mathscr{L}$-invariant

We briefly summarize our proof of the p-part of Gross's conjecture (Theorem 4.5), which as just explained implies our explicit formula for Brumer–Stark units given in Conjecture 7.5.

The work of Greenberg and Stevens [24] was a seminal breakthrough in the study of trivial zeroes of p-adic L-functions. Their perspective was highly influential in [16], where the rank one p-adic Gross–Stark conjecture was interpreted as the equality of an algebraic L-invariant $\mathscr{L}_{\mathrm{alg}}$ and an analytic L-invariant $\mathscr{L}_{\mathrm{an}}$. The analytic $\mathscr{L}$-invariant is the ratio of the leading term of the p-adic L-function at $s = 0$ to its classical counterpart,

$$\mathscr{L}_{\mathrm{an}} = -\frac{L_p'(\chi\omega, 0)}{L(\chi, 0)}. \tag{7.7}$$

The algebraic L-invariant is the ratio of the p-adic logarithm and valuation of the χ^{-1}-component of the Brumer–Stark unit,

$$\mathscr{L}_{\mathrm{alg}} = \frac{\log_p \mathrm{Norm}_{H_{\mathfrak{P}}/\mathbf{Q}_p}(u_{\mathfrak{p}}^{\chi^{-1}})}{\mathrm{ord}_{\mathfrak{P}}(u_{\mathfrak{p}}^{\chi^{-1}})}. \tag{7.8}$$

There is no difficulty in defining the ratios (7.7) and (7.8), since the quantities live in a p-adic field and the denominators are nonzero. The analogue of this situation for Gross's Conjecture 4.4 is more delicate. The role of the p-adic L-function is played by the Stickelberger element $\Theta_L := \Theta_{S',T}(L/F, 0) \in \mathbf{Z}[\mathfrak{g}]$, and the analogue of the derivative at 0 is played by the image of Θ_L in I/I^2. The role of the classical L-function is played by the element $\Theta_H := \Theta_{S,T}(H/F, 0) \in \mathbf{Z}[G]$. It is therefore not clear how to take the "ratio" of these quantities. Similarly, the role of the p-adic logarithm is played by $\mathrm{rec}_G(u_{\mathfrak{p}}) \in I/I^2$ and the role of the p-adic valuation is played by $\mathrm{ord}_G(u_{\mathfrak{p}}) \in \mathbf{Z}[G]$.

For this reason, we introduce in [18] an R-algebra $R_{\mathscr{L}}$ that is generated by an element $\mathscr{L}$ that plays the role of the analytic $\mathscr{L}$-invariant, i.e., the "ratio" between Θ_L and Θ_H. We define

$$R_{\mathscr{L}} = R[\mathscr{L}]/(\Theta_H \mathscr{L} - \Theta_L, \mathscr{L} I, \mathscr{L}^2, I^2). \tag{7.9}$$

A key nontrivial result is that this ring, in which we have adjoined a ratio $\mathscr{L}$ between Θ_L and Θ_H, is still large enough to see R/I^2.

Theorem 7.7 ([18, THEOREM 3.4]). *The kernel of the structure map $R \to R_{\mathscr{L}}$ is I^2.*

It follows from this theorem that Gross's Conjecture is equivalent to the equality

$$\mathrm{rec}_G(u_{\mathfrak{p}}) = \mathscr{L}\, \mathrm{ord}_G(u_{\mathfrak{p}}) \quad \text{in } R_{\mathscr{L}}, \tag{7.10}$$

since the right-hand side is by definition $\mathscr{L}\Theta_H = \Theta_L$.

To prove (7.10), we define a generalized Ritter–Weiss module $\nabla_{\mathscr{L}}$ over the ring $R_{\mathscr{L}}$ that can be viewed as a gluing of the modules $\nabla_S^T(H)$ and $\nabla_{S'}^T(L)$. We show in [18, THEOREM 4.6] that the Fitting ideal $\mathrm{Fitt}_{R_{\mathscr{L}}}(\nabla_{\mathscr{L}})$ is generated by the element

$$\mathrm{rec}_G(u_{\mathfrak{p}}) - \mathscr{L}\, \mathrm{ord}_G(u_{\mathfrak{p}}) \in I/I^2,$$

and hence that (7.10) is equivalent to

$$\mathrm{Fitt}_{R_{\mathscr{L}}}(\nabla_{\mathscr{L}}) = 0. \tag{7.11}$$

(For the sake of accuracy, we remark that in reality we do all of this with (S, T) replaced by the pair (Σ, Σ') defined in Section 5.3, as in Section 6.)

The vanishing of $\mathrm{Fitt}_{R_{\mathscr{L}}}(\nabla_{\mathscr{L}})$ is proven following the methods of Section 6. We interpret surjective homomorphisms from $\nabla_{\mathscr{L}}$ to $R_{\mathscr{L}}$-modules M in terms of Galois cohomology classes satisfying certain local conditions. We construct a suitable Galois cohomology class valued in a module M using an explicit construction with group-ring valued Hilbert modular forms and their associated Galois representations. The module M is shown to be large enough that its Fitting ideal over $R_{\mathscr{L}}$ vanishes, whence the same is true for $\nabla_{\mathscr{L}}$ since it has M as a quotient.

7.5. The method of Darmon–Pozzi–Vonk

We conclude by describing a proof of Conjecture 7.4 in the case that F is a real quadratic field in the beautiful work of Darmon, Pozzi, and Vonk [14]. Their method is purely p-adic (i.e., "vertical"), rather than involving the introduction of auxiliary primes (i.e., "horizontal"). The strategy follows a rich history of arithmetic formulas proven by exhibiting both sides of an equation as certain Fourier coefficients in an equality of modular forms. For instance, Katz gave an elegant proof of Leopoldt's evaluation of the Kubota–Leopoldt p-adic L-function at $s = 1$ by exhibiting an equality of p-adic modular forms, one of whose constant terms is the p-adic L-value and the other is the p-adic logarithm of a unit (see [33, §10.2]). The proof of Darmon–Pozzi–Vonk follows a similar strategy.

Let F be a real quadratic field, p an odd prime, and H a narrow ring class field extension of F (so, in particular, $p\mathcal{O}_F$ splits completely in H). Darmon–Pozzi–Vonk demonstrate an equality of certain classical modular forms of weight 2 on $\Gamma_0(p) \subset \mathrm{SL}_2(\mathbf{Z})$ that we denote f_1 and f_2.

This first of these forms f_1 is obtained by considering a Hida family of Hilbert modular cusp forms for F specializing in weight 1 to a p-stabilized Eisenstein series. The constant term of this weight 1 Eisenstein series vanishes because of the trivial zero of the corresponding p-adic L-function. Pozzi has described explicitly the Fourier coefficients of the derivative of this family with respect to the weight variables [39]. The key idea of Darmon–Pozzi–Vonk is to restrict the derivative in the antiparallel direction along the diagonal and take the ordinary projection to obtain a classical modular form of weight 2 for $\Gamma_0(p)$. The idea of taking the derivative of a family of modular forms at a point of vanishing and applying a "holomorphic projection" operator has its roots in the seminal work of Gross–Zagier [30], and appears more recently in Kudla's program for incoherent Eisenstein series [34].

Pozzi's work relates the pth Fourier coefficient of this diagonal restriction to the p-adic logarithm of the Brumer–Stark unit $\sigma_{\mathfrak{b}}(u_p)$ for the extension H. To obtain the desired weight 2 form f_1 on $\Gamma_0(p)$, one must take a certain linear combination with the diagonal restrictions of the two ordinary families of Eisenstein series passing through this weight 1 point.

The second form f_2 is defined as a generating series attached to a certain *rigid analytic theta cocycle*. These are classes in $H^1(\mathrm{SL}_2(\mathbf{Z}[1/p]), \mathcal{A}^*/\mathbf{C}_p^*)$, where $\mathcal{A}^*$ denotes the group of rigid analytic nonvanishing functions on the p-adic upper half plane. Darmon–Pozzi–Vonk construct classes in this space explicitly, and study their image under the *logarithmic annular residue map*

$$H^1\big(\mathrm{SL}_2(\mathbf{Z}[1/p]), \mathcal{A}^*/\mathbf{C}_p^*\big) \to H^1\big(\Gamma_0(p), \mathbf{Z}_p\big).$$

They compute the spectral expansion of the form f_2 and thereby show that its nonconstant Fourier coefficients are equal to those of f_1. Meanwhile, the constant coefficient is equal to the p-adic logarithm of $u_p(\mathfrak{b})^{\mathrm{an}}$. The equality of the non-constant coefficients implies that $f_1 = f_2$, and hence that the constant coefficients are equal as well, i.e.,

$$\log_p\big(\sigma_{\mathfrak{b}}(u_p)\big) = \log_p\big(u_p(\mathfrak{b})^{\mathrm{an}}\big)$$

as desired. It is a tantalizing problem to generalize this strategy to arbitrary totally real fields.

ACKNOWLEDGMENTS

We would like to thank the many mathematicians whose work has been highly influential in the development of the perspective that we have described here. Our work is the continuation of a long line of research connecting L-functions, modular forms, Galois representations, and Fitting ideals of class groups. In particular, we would like to thank Armand Brumer, David Burns, John Coates, Pierre Charollois, Pierre Colmez, Henri Darmon, Cornelius Greither, Benedict Gross, Masato Kurihara, Cristian Popescu, Alice Pozzi, Kenneth Ribet, Jurgen Ritter, Karl Rubin, Takamichi Sano, Michael Spiess, John Tate, Jan Vonk, Alfred Weiss, and Andrew Wiles.

FUNDING

The first author is supported by a grant from the National Science Foundation, DMS-1901939. The second author is supported by DST-SERB grant SB/SJF/2020-21/11, SERB SUPRA grant SPR/2019/000422 and SERB MATRICS grant MTR/2020/000215.

REFERENCES

[1] M. Atsuta and T. Kataoka, Fitting ideals of class groups for CM abelian extensions. 2021, arXiv:2104.14765.

[2] A. Beilinson, G. Kings, and A. Levin, Topological polylogarithms and p-adic interpolation of L-values of totally real fields. *Math. Ann.* **371** (2018), no. 3–4, 1449–1495.

[3] N. Bergeron, P. Charollois, and L. Garcia, Transgressions of the Euler class and Eisenstein cohomology of $\mathrm{GL}_N(\mathbf{Z})$. *Jpn. J. Math.* **15** (2020), no. 2, 311–379.

[4] D. Burns, On derivatives of Artin L-series. *Invent. Math.* **186** (2011), no. 2, 291–371.

[5] D. Burns, On derivatives of p-adic L-series at $s = 0$. *J. Reine Angew. Math.* **762** (2020), 53–104.

[6] D. Burns, M. Kurihara, and T. Sano, On zeta elements for $\mathbb{G}_m$. *Doc. Math.* **21**, (2016), 555–626.

[7] D. Burns and T. Sano, On the theory of higher rank Euler, Kolyvagin and Stark systems. *Int. Math. Res. Not.* **2021** (2021), no. 13, 10118–10206.

[8] P. Cassou-Noguès, Valeurs aux entiers négatifs des fonctions zêta et fonctions zêta p-adiques. *Invent. Math.* **51** (1979), no. 1, 29–59.

[9] P. Charollois and S. Dasgupta, Integral Eisenstein cocycles on $\mathbf{GL}_n$, I: Sczech's cocycle and p-adic L-functions of totally real fields. *Cambridge J. Math.* **2** (2014), no. 1, 49–90.

[10] P. Charollois, S. Dasgupta, and M. Greenberg, Integral Eisenstein cocycles on GL_n, II: Shintani's method. *Comment. Math. Helv.* **90** (2015), no. 2, 435–477.

[11] P. Colmez, Résidu en $s = 1$ des fonctions zêta p-adiques. *Invent. Math.* **91** (1988), no. 2, 371–389.

[12] H. Darmon, Integration on $\mathcal{H}_p \times \mathcal{H}$ and arithmetic applications. *Ann. of Math.* **154** (2001), no. 3, 589–639.

[13] H. Darmon and S. Dasgupta, Elliptic units for real quadratic fields. *Ann. of Math.* **163** (2006), no. 1, 301–346.

[14] H. Darmon, A. Pozzi, and J. Vonk, The values of the Dedekind–Rademacher cocycle at real multiplication points. 2021, arXiv:2103.02490.

[15] S. Dasgupta, Shintani zeta functions and Gross-Stark units for totally real fields. *Duke Math. J.* **143** (2008), no. 2, 225–279.

[16] S. Dasgupta, H. Darmon, and R. Pollack, Hilbert modular forms and the Gross–Stark conjecture. *Ann. of Math.* **174** (2011), no. 1, 439–484.

[17] S. Dasgupta and M. Kakde, On the Brumer–Stark conjecture. 2020, arXiv:2010.00657.

[18] S. Dasgupta and M. Kakde, Brumer–Stark units and Hilbert's 12th Problem. 2021, arXiv:2103.02516.

[19] S. Dasgupta and M. Kakde, On constant terms of Eisenstein series. *Acta Arith.* **200** (2021), no. 2, 119–147.

[20] S. Dasgupta and M. Spiess, Partial zeta values, Gross's tower of fields conjecture, and Gross–Stark units. *J. Eur. Math. Soc. (JEMS)* **20** (2018), no. 11, 2643–2683.

[21] P. Deligne and K. Ribet, Values of abelian L-functions at negative integers over totally real fields. *Invent. Math.* **59** (1980), no. 3, 227–286.

[22] F. Diaz, y. Diaz, and E. Friedman, Signed fundamental domains for totally real number fields. *Proc. Lond. Math. Soc.* **108** (2014), no. 4, 965–988.

[23] M. Fleischer and Y. Liu, Computations of elliptic units. 2021, https://github.com/~liuyj8526/Computation-of-Elliptic-Units.

[24] R. Greenberg and G. Stevens, p-adic L-functions and p-adic periods of modular forms. *Invent. Math.* **111** (1993), no. 2, 407–447.

[25] C. Greither, Determining Fitting ideals of minus class groups via the equivariant Tamagawa number conjecture. *Compos. Math.* **143** (2007), no. 6, 1399–1426.

[26] C. Greither and M. Kurihara, Stickelberger elements, Fitting ideals of class groups of CM-fields, and dualisation. *Math. Z.* **260** (2008), no. 4, 905–930.

[27] C. Greither and C. Popescu, An equivariant main conjecture in Iwasawa theory and applications. *J. Algebraic Geom.* **24** (2015), no. 4, 629–692.

[28] B. Gross, p-adic L-series at $s = 0$. *J. Fac. Sci., Univ. Tokyo, Sect. 1A, Math.* **28** (1981), no. 3, 979–994.

[29] B. Gross, On the values of abelian L-functions at $s = 0$. *J. Fac. Sci., Univ. Tokyo, Sect. 1A, Math.* **35** (1988), no. 1, 177–197.

[30] B. Gross and D. Zagier, Heegner points and derivatives of L-series. *Invent. Math.* **84** (1986), no. 2, 225–320.

[31] U. Jannsen, Iwasawa modules up to isomorphism. In *Algebraic number theory*, edited by J. Coates, R. Greenberg, B. Mazur, and I. Satake, pp. 171–207, Adv. Stud. Pure Math. 17, Academic Press, Boston, MA, 1989.

[32] T. Kataoka, Fitting invariants in equivariant Iwasawa theory. In *Development of Iwasawa theory – the centennial of K. Iwasawa's birth*, edited by M. Kurihara, K. Bannai, T. Ochiai, and T. Tsuji, pp. 413–465, Adv. Stud. Pure Math. 86, Mathematical Society of Japan, Tokyo, 2020.

[33] N. Katz, p-adic interpolation of real analytic Eisenstein series. *Ann. of Math.* **104** (1976), no. 3, 459–571.

[34] S. Kudla, Central derivatives of Eisenstein series and height pairings. *Ann. of Math.* **146** (1997), no. 3, 545–646.

[35] M. Kurihara, Notes on the dual of the ideal class groups of CM-fields. *J. Théor. Nombres Bordeaux* (to appear).

[36] B. Mazur, How can we construct abelian Galois extensions of basic number fields? *Bull. Amer. Math. Soc. (N.S.)* **48** (2011), no. 2, 155–209.

[37] D. Northcott, *Finite free resolutions*. Cambridge Univ. Press, Cambridge–New York, 1976.

[38] C. Popescu, Stark's question and a refinement of Brumer's conjecture extrapolated to the function field case. *Compos. Math.* **140** (2004), no. 3, 631–646.

[39] A. Pozzi, *The eigencurve at weight one Eisenstein points*. Ph.D. thesis, McGill University, Montreal, 2018.

[40] J. Ritter and A. Weiss, A Tate sequence for global units. *Compos. Math.* **102** (1996), no. 2, 147–178.

[41] K. Rubin, A Stark conjecture "over **Z**" for abelian L-functions with multiple zeros. *Ann. Inst. Fourier (Grenoble)* **46** (1996), no. 1, 33–62.

[42] J-P. Serre, *Corps locaux, Publications de l'Institut de Mathématique de l'Université de Nancago, VIII*. Actual. Sci. Ind. 1296, Hermann, Paris, 1962.

[43] J. Silliman, Group ring valued Hilbert modular forms. 2020, arXiv:2009.14353.

[44] W. Sinnot, On the Stickelberger ideal and the circular units of an abelian field. *Invent. Math.* **62** (1980), no. 2, 181–234.

[45] M. Spiess, Shintani cocycles and the order of vanishing of p-adic Hecke L-series at $s = 0$. *Math. Ann.* **359** (2014), no. 1–2, 239–265.

[46] J. Tate, On Stark's conjectures on the behavior of $L(s, \chi)$ at $s = 0$. *J. Fac. Sci., Univ. Tokyo, Sect. 1A, Math.* **28** (1981), no. 3, 963–978.

SAMIT DASGUPTA

Duke University, Department of Mathematics, Campus Box 90320, Durham, NC 27708-0320, USA, dasgupta@math.duke.edu

MAHESH KAKDE

Department of Mathematics, Indian Institute of Science, Bangalore 560012, India, maheshkakde@iisc.ac.in

ARITHMETIC AND DYNAMICS ON VARIETIES OF MARKOFF TYPE

ALEXANDER GAMBURD

ABSTRACT

The Markoff equation $x^2 + y^2 + z^2 = 3xyz$, which arose in his spectacular thesis in 1879, is ubiquitous in a tremendous variety of contexts. After reviewing some of these, we discuss Hasse principle, asymptotics of integer points, and, in particular, recent progress towards establishing forms of strong approximation on varieties of Markoff type, as well as ensuing implications, diophantine and dynamical.

MATHEMATICS SUBJECT CLASSIFICATION 2020

Primary 14G12; Secondary 11N36, 11D45, 37P55

KEYWORDS

Markoff triples, strong approximation, nonlinear dynamics

INTERNATIONAL CONGRESS OF MATHEMATICIANS 2022 JULY 6–14

Proc. Int. Cong. Math. 2022, Vol. 3, pp. 1800–1836
DOI 10.4171/ICM2022/191

Published by EMS Press

Important though the general concepts and propositions may be with which the modern industrious passion for axiomatizing and generalizing has presented us, in algebra perhaps more than anywhere else, nevertheless I am convinced that the special problems in all their complexity constitute the stock and core of mathematics; and to master their difficulties requires on the whole the harder labor.

Hermann Weyl, The Classical Groups, 1939

1. INTRODUCTION

1.1. Andrei Andreevich Markov is one of the towering peaks of the illustrious Saint Petersburg school of number theory, alongside with Chebyshev and Linnik. A singular characteristic of this school is a deep, often subterranean, interaction between arithmetic/combinatorics and probability/dynamics. While Markov is perhaps most widely known today for the chains named after him, it is in the context of his arguably deepest work on the minima of binary quadratic forms and badly approximable numbers[1] that the following equation, now bearing his name, was born:

$$x_1^2 + x_2^2 + x_3^2 = 3x_1x_2x_3, \tag{1.1}$$

describing a Markoff surface $X \subset \mathbb{A}^3$. *Markoff triples* $\mathcal{M}$ are the solutions of (1.1) with positive integral coordinates. *Markoff numbers* $\mathbb{M} \subset \mathbb{N}$ are obtained as coordinates of elements of $\mathcal{M}$. The *Markoff sequence* $\mathbb{M}^s$ is the set of largest coordinates of an $m \in \mathcal{M}$ counted with multiplicity; the uniqueness conjecture of Frobenius [62] asserts that $\mathbb{M} = \mathbb{M}^s$.

All elements of $\mathcal{M}$ are gotten from the root solution $r = (1, 1, 1)$ by repeated applications of an element in a set S, consisting of $\sigma \in \Sigma_3$, the permutations of the coordinates of (x_1, x_2, x_3), and of the Vieta involutions R_1, R_2, R_3 of $\mathbb{A}^3$, with $R_1(x_1, x_2, x_3) = (3x_2x_3 - x_1, x_2, x_3)$ and R_2, R_3 defined similarly. Denoting by Γ the **nonlinear** group of affine morphisms of $\mathbb{A}^3$ generated by S, the set of Markoff triples $\mathcal{M}$ can be identified with the **orbit** of the root r under the action of Γ, that is to say, $\mathcal{M} = \Gamma \cdot r$, giving rise to the *Markoff tree* [8]:

(34,1,89) < ⋮
(13,1,34) <
(13,34,1325) < ⋮
(5, 1, 13) (
(194,13,7561) < ⋮
(5,13,194) <
(5,194,2897) < ⋮
(1, 1, 1) − (1, 1, 2) − (2, 1, 5) (
(433,5,6466) < ⋮
(29,5,433) <
(29,433,37666) < ⋮
(2, 5, 29) (
(169,29,14701) < ⋮
(2,29,169) <
(2,169,985) < ⋮

1 This work of Markoff and some of the subsequent appearances of his equation in a tremendous variety of different contexts are briefly discussed in Section 2.

The first few members of $\mathbb{M}$ are

$$1, 2, 5, 13, 29, 34, 89, 169, 194, 233, 433, 610, 985, \ldots$$

The sequence $\mathbb{M}^s$ is sparse, as shown by Zagier [147]:

$$\sum_{\substack{m \in \mathbb{M}^s \\ m \leq T}} 1 \sim c(\log T)^2 \quad \text{as } T \to \infty \ (c > 0). \tag{1.2}$$

1.2. The origins of investigations which underlie "the stock and core" of this report date back to August of 2005 and involve a "special problem" pertaining to Markoff numbers; here is Peter Sarnak's recollection [126]: "For me the starting point of this investigation was in 2005 when Michel and Venkatesh asked me about the existence of poorly distributed closed geodesics on the modular surface. It was clear that Markov's constructions of his geodesics using his Markov equation provided what they wanted but they preferred quadratic forms with square free discriminant. This raised the question of sieving in this context of an orbit of a group of (nonlinear) morphisms of affine space. The kind of issues that one quickly faces in attempting to execute such a sieve are questions of the image of the orbit when reduced mod q and interestingly whether certain graphs associated with these orbits are expander families.[2] Gamburd in his thesis had established the expander property in some simpler but similar settings and he and I began a lengthy investigation into this sieving problem in the simpler setting when the group of affine morphisms acts linearly (or what we call now the affine linear sieve)."

The question posed by Michel and Venkatesh arose in the course of their joint work with Einsiedler and Lindenstrauss [58, 59] on generalizations of Duke's theorem [57]; formulated in terms of Markoff numbers, it leads to the following:

Conjecture 1. *There are infinitely many square-free Markoff numbers.*

As detailed in [21], an application of sieve methods in the setting of affine orbits leads to and demands an affirmative answer to the question as to whether *Markoff graphs*, obtained as a modular reduction of the Markoff tree,[3] form a family of expanders. Numerical experiments by de Courcy-Ireland and Lee [55], as well as results detailed in Section 2.5, are compelling in favor of the following *superstrong approximation conjecture for Markoff graphs*:

Conjecture 2. *The family of Markoff graphs $X^*(p)$ forms a family of expanders.*

Before attacking this conjecture, asserting high connectivity of Markoff graphs, one has to confront the question of their connectivity, that is to say, the issue of the *strong approximation for Markoff graphs*:

2 See [125] and [81] for definition and properties of expanders.

3 Let p be a large prime and denote by $X^*(p) = X(p) \backslash (0, 0, 0)$ the solutions of (1.1) modulo p with the removal of $(0, 0, 0)$. The Markoff graphs are obtained by joining each x in $X^*(p)$ to $R_j(x)$, $j = 1, 2, 3$. They were considered first by Arthur Baragar in his thesis [3].

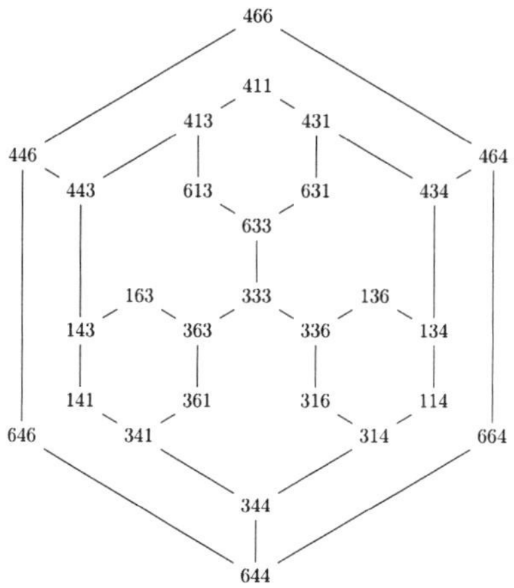

FIGURE 1
Markoff graph mod 7. In **[54]** it is proved that the Markoff graphs are not planar for primes greater than 7.

Conjecture 3. *The map $\pi_p : \mathcal{M} \to X^*(p)$ is onto, that is to say, Markoff graphs $X^*(p)$ are connected.*

While Conjectures 1 and 2 have withstood our protracted attack over the past 17 years, much progress has been made on parallel questions in the case of affine linear maps. We will return to the recent resolution of Conjecture 3, and resulting progress on diophantine properties of Markoff numbers in Section 1.5.

1.3. Before describing the general setting of Affine Linear Sieve, it is instructive to briefly examine an example which is in many ways parallel to the Markoff situation, namely integral Apollonian packings **[63,127]**. A theorem of Descartes asserts that $x_1, x_2, x_3, x_4 \in \mathbb{R}^4$ are the curvatures of four mutually tangent circles in the plane if

$$2(x_1^2 + x_2^2 + x_3^2 + x_4^2) = (x_1 + x_2 + x_3 + x_4)^2. \tag{1.3}$$

Given an initial configuration of 4 such circles, we fill in repeatedly the lune regions with the unique circle which is tangent to 3 sides (which is possible by a theorem of Apollonius). In this way we get a packing of the outside circle by circles giving an Apollonian packing. The interesting diophantine feature is that if the initial curvatures are integral then so are the curvatures of the entire packing.

The numbers in the circles in Figure 2 indicate their curvatures; note that by convention the outer circle has negative curvature. Viewing equation (1.3) as a quadratic equation in x_1, we see that the two solutions are related as $x_1 + x_1' = 2x_2 + 2x_3 + 2x_4$, the crucial point being that the Vieta involutions in this case are given by linear maps A_1, A_2, A_3, A_4 where $A_j(e_k) = -3e_k + 2(e_1 + e_2 + e_3 + e_4)$ if $k = j$ and $A_j(e_k) = e_k$ if $k \neq j$ (e_1, e_2, e_3, e_4 are the standard basis vectors). The configurations of 4 mutually tangent circles in the packing with initial configuration $a = (a_1, a_2, a_3, a_4)$ consist of points x in the orbit $\mathcal{O} = \Lambda \cdot a$ where $\Lambda = \langle A_1, A_2, A_3, A_4 \rangle$ is the Apollonian group. The elements A_j

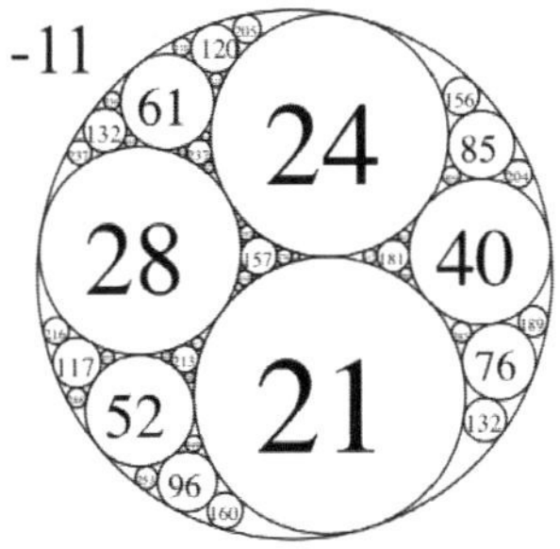

FIGURE 2
Integral Apollonian packing $(-11, 21, 24, 28)$.

preserve F given by

$$F(x_1, x_2, x_3, x_4) = 2(x_1^2 + x_2^2 + x_3^2 + x_4^2) - (x_1 + x_2 + x_3, +x_4)^2,$$

and hence $\Lambda \le O_F(\mathbb{Z})$. The group Λ is Zariski dense in O_F, but it is thin in $O_F(\mathbb{Z})$. For example, $|\{\gamma \in O_F(\mathbb{Z}) : ||\gamma|| \le T\}| \sim c_1 T^2$ as $T \to \infty$, while $|\{\gamma \in \Lambda : ||\gamma|| \le T\}| \sim c_1 T^\delta$, where[4] $\delta = 1.3\ldots$ is the Hausdorff dimension of the limit set of Λ.

The general setting of Affine Linear Sieve, introduced in **[20, 21]**, is as follows. For $j = 1, 2, \ldots, k$, let A_j be invertible integer coefficient polynomial maps from $\mathbb{Z}^n$ to $\mathbb{Z}^n$ (here $n \ge 1$ and the inverses of A_j's are assumed to be of the same type). Let Λ be the group generated by $A_1, \ldots, A_k$ and let $\mathcal{O} = \Lambda b$ be the orbit of some $b \in \mathbb{Z}^n$ under Λ. Given a polynomial $f \in Q[x_1, \ldots, x_n]$ which is integral on $\mathcal{O}$, the aim is to show that there are many points $x \in \mathcal{O}$ at which $f(x)$ has few or even the least possible number of prime factors, in particular that such points are Zariski dense in the Zariski closure, $\mathrm{Zcl}(\mathcal{O})$ of $\mathcal{O}$. Let $\mathcal{O}(f, r)$ denote the set of $x \in \mathcal{O}$ for which $f(x)$ has at most r prime factors. As $r \to \infty$, the sets $\mathcal{O}(f, r)$ increase and potentially at some point become Zariski dense. Define the *saturation number* $r_0(\mathcal{O}, f)$ to be the least integer r such that $\mathrm{Zcl}(\mathcal{O}(f, r)) = \mathrm{Zcl}(\mathcal{O})$. It is by no means obvious that it is finite or even if one should expect it to be so, in general. If it is finite, we say that the pair $(\mathcal{O}, f)$ saturates. In the case of linear maps, the theory by now is quite advanced and the basic result pertaining to the finiteness of the saturation number in all cases where it is expected to hold, namely in the case of the Levi factor of $G = \mathrm{Zcl}(\Lambda)$ being semisimple,[5] has been established **[123]**. Both strong and superstrong approximation, particularly for thin

4 This result can be deduced from the work of Lax and Phillips **[93]**. A beautiful overview of striking developments pertaining to dynamics on geometrically finite hyperbolic manifolds with applications to Apollonian circle packings (and beyond) is contained in Hee Oh's ICM report **[114]**.

5 On the other hand, as detailed in **[21, 85, 123]**, when torus intervenes, the saturation most likely fails. Tori pose particularly difficult problems, in terms of sparsity of elements in an orbit, strong approximation and diophantine properties: see **[104]** for a discussion of Artin's Conjecture in the context of strong approximation.

groups such as the Apollonian group, are crucial ingredients in executing Brun combinatorial sieve in this setting.

1.4. The strong approximation for $\mathrm{SL}_n(\mathbb{Z})$, asserting that the reduction π_q modulo q is onto, is a consequence of the Chinese remainder theorem; its extension to arithmetic groups is far less elementary but well understood [118]. If S is a finite symmetric generating set of $\mathrm{SL}_n(\mathbb{Z})$, strong approximation is equivalent to the assertion that the Cayley graphs $\mathcal{G}(\mathrm{SL}_n(\mathbb{Z}/q\mathbb{Z}), \pi_q(S))$ are connected. The quantification of this statement, asserting that they are in fact highly-connected, that is to say, form a family of expanders, is what we mean by superstrong approximation. The proof of the expansion property for $\mathrm{SL}_2(\mathbb{Z})$ has its roots in Selberg's celebrated lower bound [131] of $\frac{3}{16}$ for the first eigenvalue of the Laplacian on the hyperbolic surfaces associated with congruence subgroups of $\mathrm{SL}_2(\mathbb{Z})$. The generalization of the expansion property to $G(\mathbb{Z})$ where G is a semisimple matrix group defined over $\mathbb{Q}$ is also known thanks to developments towards the general Ramanujan conjectures that have been established; this expansion property is also referred to as *property* τ for congruence subgroups [133].

Let Γ be a finitely generated subgroup of $\mathrm{GL}_n(\mathbb{Z})$ and let $G = \mathrm{Zcl}(\Gamma)$. The discussion of the previous paragraph applies if Γ is of finite index in $G(\mathbb{Z})$. However, if Γ is *thin*, that is to say, of infinite index in $G(\mathbb{Z})$, then $\mathrm{vol}(G(\mathbb{R})\backslash\Gamma) = \infty$ and the techniques used to prove both of these properties do not apply. It is remarkable that, under suitable natural hypothesis, strong approximation continues to hold in this thin context, as proved by Matthews, Vaserstein, and Weisfeiler in 1984 [105,143]. That the expansion property might continue to hold for thin groups was first suggested by Lubotzky and Weiss in 1993 [101]; for $\mathrm{SL}_2(\mathbb{Z})$, the issue is neatly encapsulated in the following 1–2–3 question of Lubotzky [99]. For a prime $p \geq 5$ and $i = 1, 2, 3$, let us define $S_p^i = \{\left(\begin{smallmatrix} 1 & i \\ 0 & 1 \end{smallmatrix}\right), \left(\begin{smallmatrix} 1 & 0 \\ i & 1 \end{smallmatrix}\right)\}$. Let $\mathcal{G}_p^i = \mathcal{G}(\mathrm{SL}_2(\mathbb{Z}/p\mathbb{Z}), S_p^i)$, the Cayley graph of $\mathrm{SL}_2(\mathbb{Z}/p\mathbb{Z})$ with respect to S_p^i. By Selberg's theorem, $\mathcal{G}_p^1$ and $\mathcal{G}_p^2$ are families of expander graphs. However, the group $\langle\left(\begin{smallmatrix} 1 & 3 \\ 0 & 1 \end{smallmatrix}\right), \left(\begin{smallmatrix} 1 & 0 \\ 3 & 1 \end{smallmatrix}\right)\rangle$ has infinite index in $\mathrm{SL}_2(\mathbb{Z})$ and thus does not come under the purview of Selberg's theorem.

In my thesis [66], extending the work of Sarnak and Xue [129], [128] for cocompact arithmetic lattices, a generalization of Selberg's theorem for infinite index "congruence" subgroups of $\mathrm{SL}_2(\mathbb{Z})$ was proved; for such subgroups with a high enough Hausdorff dimension of the limit set, a spectral gap property was established. Following the groundbreaking work of Helfgott [77] (which builds crucially on sum–product estimate in $\mathbb{F}_p$ due to Bourgain, Katz, and Tao [27]), Bourgain and Gamburd [13] gave a complete answer to Lubotzky's question. The method introduced in [12,13] and developed in a series of papers [14–19] became known as "Bourgain–Gamburd expansion machine"; thanks to a number of major developments by many people [22,28,35,82,91,115,120,122,124], the general superstrong approximation for thin groups is now known. The state-of-the-art is summarized in *Thin groups and superstrong approximation* [36] which contains an expanded version of most of the invited lectures from the eponymous MSRI "Hot Topics" workshop, in the surveys by Breuillard [33] and Helfgott [78], and in the book by Tao *"Expansion in finite simple groups of Lie type"* [140].

1.5. We return to the progress on Conjecture 3 [23–26]. Our first result [25] asserts that there is a very large orbit.

Theorem 1. *Fix $\varepsilon > 0$. Then for p large prime, there is a Γ orbit $\mathcal{C}(p)$ in $X^*(p)$ for which*

$$|X^*(p) \backslash \mathcal{C}(p)| \leq p^{\varepsilon} \tag{1.4}$$

(note that $|X^*(p)| \sim p^2$), and any Γ orbit $\mathcal{D}(p)$ satisfies[6]

$$|\mathcal{D}(p)| \gg (\log p)^{\frac{1}{3}}. \tag{1.5}$$

The proof, discussed in section 3, establishes the strong approximation conjecture, unless $p^2 - 1$ is a very smooth number. In particular, the set of primes for which the strong approximation conjecture fails is very small.

Theorem 2. *Let E be the set of primes for which the strong approximation conjecture fails. For $\varepsilon > 0$, the number of primes $p \leq T$ with $p \in E$ is at most T^{ε}, for T large.*

Very recently, in a remarkable breakthrough, using geometric techniques involving Hurwitz stacks, degeneration, and some Galois theory, William Chen [45] proved the following result:

Theorem 3. *Every Γ orbit $\mathcal{D}(p)$ has size divisible by p.*

Combining Theorems 1 and 3 establishes Conjecture 3 for all sufficiently large primes; in combination with the following result established in [26], namely

Theorem 4. *Assume that $X^*(\mathbb{Z}/p\mathbb{Z})$ is connected. Then $X^*(\mathbb{Z}/p^k\mathbb{Z})$ is connected for all k.*

it yields

Theorem 5. *For all sufficiently large primes p, the group Γ acts minimally on $X^*(\mathbb{Z}_p)$.*

We remark that Theorem 5 is not true for $X^*(\mathbb{R})$; cf. section 4.1. While Conjecture 1 remains out of reach, the progress on strong approximation allows us to establish the following result on the diophantine[7] properties of Markoff numbers [25]:

Theorem 6. *Almost all Markoff numbers are composite, that is,*

$$\sum_{\substack{p \in M^s \\ p \text{ prime},\, p \leq T}} 1 = o\Big(\sum_{\substack{m \in M^s \\ m \leq T}} 1 \Big).$$

It is worth contrasting this result with the state of knowledge regarding the sequence $H_n = 2^n + b$, which is just a little more sparse than the sequence of Markoff numbers, for which, by Zagier's result (1.2), we have $M_n \sim A^{\sqrt{n}}$. Even assuming the generalized Riemann Hypothesis, which allowed Hooley [79] to give a conditional proof of Artin's conjecture (cf. footnote 5), was not sufficient to establish that almost all members of the sequence H_n are composite: the conditional proof in [80] necessitated postulating additional "Hypothesis A."

6 The exponent $\frac{1}{3}$ in (1.5) has been improved to $\frac{7}{9}$ in [87].

7 We remark that in [52] Corvaja and Zannier showed that the greatest prime factor of xy for a Markoff triple (x, y, z) tends to infinity.

1.6. The methods of proof of Theorems 1, 2, 4 discussed in Section 3 are robust enough to enable handling their extension to more general Markoff-type cubic surfaces, namely

$$X_k : \Phi(x_1, x_2, x_3) = x_1^2 + x_2^2 + x_3^2 - x_1x_2x_3 = k, \tag{1.6}$$

where the real dynamics was studied by Goldman [73], as discussed in Section 4.1; the family of surfaces $S_{A,B,C,D} \subset \mathbb{C}^3$ given by

$$x_1^2 + x_2^2 + x_3^2 + x_1x_2x_3 = Ax_1 + Bx_2 + Cx_3 + D, \tag{1.7}$$

where the real dynamics was studied by Cantat [38], as discussed in Section 4.2; those in [60] and even the general such nondegenerate cubic surface

$$Y = Y(\alpha, \beta, \gamma, \delta) : \sum_{i,j=1}^{3} \alpha_{ij} x_i x_j + \sum_{j=1}^{3} \beta_j x_j + \gamma = \delta x_1 x_2 x_3, \tag{1.8}$$

with $\alpha_{ij}, \beta_j, \gamma, \delta$ being integers.

The group Γ_Y is again generated by the corresponding Vieta involutions R_1, R_2, R_3. For such a Y and action Γ_Y, one must first show that there are only finitely many finite orbits in $Y(\bar{\mathbb{Q}})$, and that these may be determined effectively. The analogue of Conjecture 1 for Y is that for p large, Γ_Y has one big orbit on $Y(\mathbb{Z}/p\mathbb{Z})$ and that the remaining orbits, if there are any, correspond to one of the finite $\bar{\mathbb{Q}}$ orbits determined above.

The determination of the finite orbits of Γ on $X_k(\bar{\mathbb{Q}})$ and on $S_{A,B,C,D}(\bar{\mathbb{Q}})$ has been carried out in [56] and [96], respectively. Remarkably for these, the Γ action on affine 3-space corresponds to the (nonlinear) monodromy group for Painlevé VI equations on their parameter spaces. In this way the finite orbits in question turn out to correspond bijectively to those Painlevé VI's which are algebraic functions of their independent variable.

In this setting of the more general surfaces Y in (1.8), strong approximation for $Y(\mathbb{Z}_S)$, where S is the set of primes dividing $\alpha_{11}, \alpha_{22}, \alpha_{33}$ (so that Γ_Y preserves the S-integers $\mathbb{Z}_S$), will follow from Conjecture 1 for Y (and the results we can prove towards it, as in Theorem 2) once we have a point of infinite order in $Y(\mathbb{Z}_S)$. If there is no such point, we can increase S or replace $\mathbb{Z}$ by $\mathcal{O}_K$, the ring of integers in a number field $K/\mathbb{Q}$, to produce such a point and with it strong approximation for $Y\big((\mathcal{O}_K)_S\big)$.

Vojta's conjectures and the results proven towards them [51,141] assert that cubic and higher-degree affine surfaces typically have few S-integral points. In the rare cases where these points are Zariski dense, such as tori (e.g., $N(x_1, x_2, x_3) = k$ where N is the norm form of a cubic extension of $\mathbb{Q}$), strong approximation fails. So these Markoff surfaces appear to be rather special affine cubic surfaces not only having a Zariski dense set of integral points, but also a robust strong approximation.

1.7. Zagier's result (1.2) can be viewed as a statement about asymptotic growth of integral points on the Markoff variety, $|X(\mathbb{Z}) \cap B(T)| \sim (\log T)^2$. In Section 5 we discuss the work in [68], establishing an asymptotic formula for the number of integer solutions to the Markoff–Hurwitz equation

$$x_1^2 + x_2^2 + \cdots + x_n^2 = ax_1x_2 \cdots x_n + k, \tag{1.9}$$

giving an interpretation of the exponent of growth, which for $n > 3$ is not integral, in terms of the unique parameter for which there exists a certain conformal measure on a projective space.

1.8. The issue of the existence of a *single* integral solution to (1.9) for general a and k, even for $n = 3$, is quite subtle; see [112,130]. In the work of Ghosh and Sarnak [71], the Hasse principle is established to hold for Markoff-type cubic surfaces $X(k)$ given by (1.6) for almost all k, but it also fails to hold for infinitely many k; this work is discussed in Section 6.

1.9. Regrettably, the space/time constraints prevented us from covering cognate results pertaining to arithmetic and dynamics on K3 surfaces; see [37,65,106,108,109,135] and references therein. The Markoff equation over quadratic imaginary fields is studied in [134]. Potential cryptographic applications of Markoff graphs are discussed in [64].

1.10. To conclude this introduction, let us note that X_k is the relative character variety of representations of the fundamental group of a surface of genus 1 with one puncture to SL_2. The action of the mapping class group is that of Γ. More generally, the (affine) relative character variety V_k of representation of $\pi_1(\Sigma_{g,n})$, a surface of genus g with n punctures, into SL_2 is defined over $\mathbb{Z}$, and one can study the diophantine properties of $V_k(\mathbb{Z})$. In the work of Whang [144–146], it was shown that V_k has a projective compactification relative to which V_k is "log-Calabi–Yau." According to the conjectures of Vojta, this places V_k as being in the same threshold setting as affine cubic surfaces. Moreover, $V_k(\mathbb{Z})$ has a full descent in that the mapping class group acts via nonlinear morphisms on $V_k(\mathbb{Z})$ with finitely many orbits. These and more general character varieties connected with higher Teichmüller theory offer a rich family of threshold affine varieties for which one can approach the study of integral points.

2. THE UNREASONABLE(?) UBIQUITY OF MARKOFF EQUATION

Markoff equation and numbers appear in a surprising variety of contexts: see, for example, [1] (subtitled *Mathematical Journey from Irrational Numbers to Perfect Matchings*) and the references therein.

2.1. The Markoff chain. Equation (1.1) was discovered by Markoff in 1879 in his work on badly approximable numbers. As the sentiment[8] expressed by Frobenius [62] in 1913 seems to remain true today, we briefly review the context and statement of Markoff's theorem.

Let α be an irrational number. A celebrated theorem of Hurwitz asserts that α admits infinitely many rational approximations p/q such that $|\alpha - \frac{p}{q}| < \frac{1}{\sqrt{5}q^2}$, and, moreover, that if α is $\mathrm{GL}_2(\mathbb{Z})$-equivalent to the Golden Ratio $\theta_1 = (1+\sqrt{5})/2$, in the sense that $\alpha = \frac{a\theta_1+b}{c\theta_1+d}$

8 "*Trotz der außerordentlich merkwürdigen und wichtigen Resultate scheinen diese schwiergen Untersuchungen wenig bekannt zu sein*" [In spite of the extraordinarily noteworthy and important results these difficult investigations seem to be little known]

for some integers a, b, c, d with $ad - bc = \pm 1$, the above result is sharp and the constant $\frac{1}{\sqrt{5}}$ cannot be replaced by any smaller.

Suppose next that α is not $\mathrm{GL}_2(\mathbb{Z})$-equivalent to θ_1. Then the result of Markoff's doctoral advisors, Korkine and Zolotareff, [88] asserts that α admits infinitely many rational approximations p/q such that $|\alpha - \frac{p}{q}| < \frac{1}{\sqrt{8}q^2}$, and, moreover, that the constant $\frac{1}{\sqrt{8}}$ is sharp if and only if α is $\mathrm{GL}_2(\mathbb{Z})$-equivalent $\theta_2 = 1 + \sqrt{2}$.

The general result found by Markoff in his *Habilitation* and published in 1879 and 1880 in *Mathematische Annalen* is as follows.

Markoff's Theorem. *Let* $\mathbb{M} = \{1, 2, 5, 13, 29, 34, 89, 169, 194, \dots\}$ *be the sequence of Markoff numbers. There is a sequence of associated quadratic irrationals* $\theta_i \in \mathbb{Q}(\sqrt{\Delta_i})$, *where* $\Delta_i = 9m_i^2 - 4$ *and* m_i *is the* i*th element of the sequence, with the following property. Let* α *be a real irrational, not* $GL_2(\mathbb{Z})$*-equivalent to any of the numbers* θ_i *whenever* $m_i < m_j$. *Then* α *admits infinitely many rational approximations* p/q *with* $\left|\alpha - \frac{p}{q}\right| < \frac{m_j}{\sqrt{\Delta_j}\, q^2}$; *the constant* $m_j/\sqrt{\Delta_j}$ *is sharp if and only if* α *is* $GL_2(\mathbb{Z})$*-equivalent to* θ_h, *for some* h *such that* $m_h = m_j$.

2.2. Continued fractions and binary quadratic forms. The first paper by Markoff [102] used the theory of continued fractions, while the second memoir [103] was based on the theory of binary indefinite quadratic forms, with the final result stated as a theorem on minima of binary indefinite quadratic forms.

The alternative approach based on indefinite binary quadratic forms was the subject of an important memoir by Frobenius [62] and complete details were finally provided by Remak [121] and much simplified by Cassels [39, 40].

2.3. The geometry of Markoff numbers. A third way of looking at the problem, via hyperbolic geometry, was introduced by Gorshkov [74] in his thesis of 1953, but published only in 1977. The connection with hyperbolic geometry was rediscovered, in a somewhat different way, by Cohn [46]. The paper by Caroline Series [132] contains a beautiful exposition of the problem in this context.

2.4. Cohn tree and Nielsen transformations. Cohn is also credited for the interpretation of the problem [47] in the context of primitive words in F_2, the free group on two generators. Its automorphism group $\Phi_2 = \mathrm{Aut}(F_2)$ is generated by the following *Nielsen transformations:* $(a, b)^P = (b, a)$, $(a, b)^\sigma = (a^{-1}, b)$, $(a, b)^U = (a^{-1}, ab)$. Let $V = \sigma U$. Then $(a, b)^V = (a, ab)$.

The *Cohn tree* is a binary tree with root ab, branching to the top with U and to the bottom with V,

- ab
  - abb
    - $abbb <$
      - $abbbb < \vdots$
      - $aababab < \vdots$
    - $aabab <$
      - $ababbabb < \vdots$
      - $aaabaab < \vdots$
  - aab
    - $ababb <$
      - $abbabbb < \vdots$
      - $aabaabab < \vdots$
    - $aaab <$
      - $abababb < \vdots$
      - $aaaab < \vdots$

Markoff numbers are obtained from the Cohn tree by taking a third of the trace of the matrix obtained by substituting the matrices $A = \begin{pmatrix} 5 & 2 \\ 2 & 1 \end{pmatrix}$ and $B = \begin{pmatrix} 2 & 1 \\ 1 & 1 \end{pmatrix}$ in place of a, b and performing the matrix multiplication.

2.5. Nielsen systems and product replacement graphs. Conjecture 3 is a special case of *Conjecture Q* made by McCullough and Wanderley **[107]** in the context of Nielsen systems and product replacement graphs.

Given a group G, the *product replacement graph* $\Gamma_k(G)$ introduced in **[42]** in connection with computing in finite groups is defined as follows. The vertices of $\Gamma_k(G)$ consist of all k-tuples of generators $(g_1, \dots, g_k)$ of the group G. For every (i, j), $1 \le i$, $j \le k$, $i \ne j$, there is an edge corresponding to transformations $L^{\pm}_{i,j}$ and $R^{\pm}_{i,j}$, where $R^{\pm}_{i,j} : (g_1, \dots, g_i, \dots, g_k) \to (g_1, \dots, g_i \cdot g_j^{\pm 1}, \dots, g_k)$ and $L^{\pm}_{i,j}$ defined similarly. The graphs $\Gamma_k(G)$ are regular, of degree $4k\,(k-1)$, possibly with loops and multiple edges. The connectivity of $\Gamma_k(G)$ has been the subject of intensive recent investigations; for $G = \mathrm{SL}_2(p)$ and $k \ge 3$, it was established by Gilman in **[72]**.

In the case of the free group F_k, the moves $L^{\pm}_{i,j}$ and $R^{\pm}_{i,j}$ defined above correspond to Nielsen moves on $\Gamma_k(F_k)$. For every group G, the set $\Gamma_k(G)$ can be identified with $E = \mathrm{Epi}(F_k, G)$, the set of epimorphisms from F_k onto G, and the group $A = \mathrm{Aut}(F_k)$ acts on E in the following way: if $\alpha \in A$ and $\varphi \in E$, $\alpha(\varphi) = \varphi \cdot \alpha^{-1}$. A long-standing problem is whether $\mathrm{Aut}(F_k)$ has property (T) for $k \ge 4$; in **[100]** Lubotzky and Pak observed that a positive answer to this problem implies the expansion of $\Gamma_k(G)$ for all G and proved that $\Gamma_k(G)$ are expanders when G is nilpotent of class l and both k and l are fixed. Property (T) for $\mathrm{Aut}(F_k)$ for $k \ge 5$ was recently established in **[84]**.[9] Note that $\mathrm{Aut}(F_2)$ and $\mathrm{Aut}(F_3)$ do not satisfy property (T), while the problem is still open for $k = 4$.

In a joint work with Pak **[69]**, we established a connection between the expansion coefficient of the product replacement graph $\Gamma_k(G)$ and the minimal expansion coefficient of a Cayley graph of G with k generators, and, in particular, proved that for $k > 3$ the product

---

9 The proof stems from the groundbreaking observation by Ozawa **[116]** that G satisfies Kazhdan's property (T) if there exist $\lambda > 0$ and finitely many elements ξ_i of $\mathbb{R}[G]$ such that $\Delta^2 - \lambda\Delta = \sum_i \xi_i^* \xi_i$ where Δ is the Laplacian of the finite symmetric generating set of G.

replacement graphs $\Gamma_k(\mathrm{SL}(2,p))$ form an expander family under assumption of strong uniform expansion of $\mathrm{SL}(2,p)$ on k generators. In a joint work with Breuillard **[34]**, combining the "expansion machine" **[13]** with the uniform Tits Alternative[10] established by Breuillard **[32]**, we proved that Cayley graphs of $\mathrm{SL}(2,p)$ are strongly uniformly expanding for infinitely many primes of density one. Consequently, the following form of *nonlinear superstrong approximation* is obtained:

Theorem 7. *Let $k > 3$. The family of product replacement graphs $\{\Gamma_k(\mathrm{SL}(2,p_n))\}_n$ forms a family of expanders for infinitely many primes p_n of density one.*

As detailed in **[107]**, the situation is different for the product replacement graph of $\mathrm{SL}(2,\mathbb{F}_p)$ on 2 generators, due to Fricke identity for 2×2 matrices M and N:

$$\mathrm{tr}(M)^2 + \mathrm{tr}(N)^2 + \mathrm{tr}(MN)^2 = \mathrm{tr}(M)\mathrm{tr}(N)\mathrm{tr}(MN) + \mathrm{tr}([M,N]) + 2. \tag{2.1}$$

Letting $x_1 = \mathrm{tr}(M)$, $x_2 = \mathrm{tr}(N)$, $x_3 = \mathrm{tr}(MN)$, the Q conjecture[11] in **[107]** amounts to the assertion of the strong approximation for the surfaces

$$X_k : \Phi(x_1,x_2,x_3) = k, \tag{2.2}$$

$$\Phi(x_1,x_2,x_3) = x_1^2 + x_2^2 + x_3^2 - x_1x_2x_3, \tag{2.3}$$

and $k = \mathrm{tr}([M,N]) + 2$, with Markoff surface[12] being the special case corresponding to $\mathrm{tr}([M,N]) = -2$.

3. STRONG APPROXIMATION

We give a brief overview of the methods and tools used in the proof of Theorems 1 and 2 and some comments about their extensions to the setting of more general surfaces of Markoff type. Theorem 1, in the weaker form that $|\mathcal{C}(p)| \sim |X^*(p)|$ as $p \to \infty$, can be viewed as the finite field analogue of **[73]** where it is shown that the action of Γ on the compact real components of the relative character variety of the mapping class group of the once punctured torus is ergodic. As in **[73]** our proof makes use of the rotations $\tau_{ij} \circ R_j$, $i \neq j$, where τ_{ij} permutes x_i and x_j. These preserve the conic sections gotten by intersecting $X^*(p)$ with the planes $y_k = x_k$ (k different from i and j). If $\tau_{ij} \circ R_j$ has order t_1 (here $t_1 | p(p-1)(p+1)$), then x and these t_1 points of the conic section are connected (i.e., are in the same Γ orbit). If t_1 is maximal (i.e., is p, $p-1$, or $p+1$), then this entire conic section is connected and such conic sections in different planes which intersect are also connected. This leads to a large component which we denote by $\mathcal{C}(p)$.

---

10 This states that if the subgroup of $\mathrm{GL}_d(K)$ (where K is an algebraic number field) generated by F is not virtually solvable, then there is some $N \in \mathbb{N}$, depending only on d, such that $(F \cup F^{-1} \cup \{1\})^N$ contains two elements that generate a nonabelian free group.

11 See the paper of Will Chen **[45]** for the discussion of the relation between this conjecture and the connectivity properties of the moduli spaces of elliptic curves with $G = \mathrm{SL}(2,p)$ structures.

12 Note that the congruence $x^2 + y^2 + z^2 \equiv xyz \pmod 3$ has no nontrivial solutions.

If our starting rotation has order t_1 which is not maximal, then the idea is to ensure that among the t_1 points to which it is connected, at least one has a corresponding rotation of order $t_2 > t_1$, and then to repeat. To ensure that one can progress in this way, a critical equation over $\mathbb{F}_p$ intervenes:

$$\begin{cases} x + \frac{b}{x} = y + \frac{1}{y}, \quad b \neq 1, \\ x \in H_1, y \in H_2 \text{ with } H_1, H_2 \text{ subgroups of } \mathbb{F}_p^* \text{ (or } \mathbb{F}_{p^2}^*). \end{cases} \tag{3.1}$$

If $t_1 = |H_1| \geq p^{1/2+\delta}$ (with δ small and fixed), one can apply the proven Riemann Hypothesis for curves over finite fields [142] to count the number of solutions to (3.1). Together with a simple inclusion/exclusion argument, this shows that one of the t_1 points connected to our starting x has a corresponding maximal rotation and hence x is connected to $\mathcal{C}(p)$.

If $|H_1| \leq p^{1/2+\delta}$ then RH for these curves is of little use (their genus is too large), and we have to proceed using other methods. We assume that $|H_1| \geq |H_2|$ so that the trivial upper bound for the number of solutions to (3.1) is $2|H_2|$. What we need is a power saving in this upper bound in the case that $|H_2|$ is close to $|H_1|$, that is, a bound of the form $C_\tau |H_1|^\tau$, with $\tau < 1$, $C_\tau < \infty$ (both fixed).

In the prime modulus case, there are several ways to proceed. The first and second methods are related to "elementary" proofs of the Riemann Hypothesis for curves. One can use auxiliary polynomials as in Stepanov's proof [137] of the Riemann Hypothesis for curves to give the desired power saving with an explicit τ (cf. [76] which deals with $x + y = 1$ and $|H_1| = |H_2|$ in (3.1)). The second method, giving the best upper bound, namely $20 \max\{(|H_1|.|H_2|)^{1/3}, \frac{|H_1|.|H_2|}{p}\}$, is due to Corvaja and Zannier [53]. It uses their method for estimating the greatest common divisor of $u - 1$ and $v - 1$ in terms of the degrees of u and v and their supports, as well as (hyper) Wronskians.

The third method is based on Szemerédi–Trotter theorem for modular hyperbolas [11], whose proof uses crucially expansion and L^2-flattening lemma in $\mathrm{SL}_2(\mathbb{Z}/p\mathbb{Z})$ [16].

Theorem 8. *Let* $\Phi : \mathbb{F}_p \to \mathrm{Mat}_2(\mathbb{F}_p)$ *be such that* $\det \Phi$ *does not vanish identically and* $\mathrm{Im}\Phi \cap PGL_2(\mathbb{F}_p)$ *is not contained in a set of the form* $\mathbb{F}_p^* \cdot gH$ *for some* $g \in \mathrm{SL}_2(\mathbb{F}_p)$ *and* H *a proper subgroup of* $\mathrm{SL}_2(\mathbb{F}_p)$. *Then the following holds:*

Given $\varepsilon > 0, r > 1$, *there is* $\delta > 0$ *such that if* $A \subset P^1(\mathbb{F}_p)$ *and* $L \subset \mathbb{F}_p$ *satisfy*

$$1 \ll |A| < p^{1-\varepsilon}, \tag{3.2}$$

$$\log|A| < r \log|L|, \tag{3.3}$$

then

$$|\{(x, y, t) \in A \times A \times L; y = \tau_{\Phi(t)}(x)\}| < |A|^{1-\delta}|L|, \tag{3.4}$$

where for $g = \begin{pmatrix} a & b \\ c & d \end{pmatrix}$, $\tau_g(x) = \frac{ax+b}{cx+d}$.

While producing poor exponents τ, this method is robust and works in the generality that the superstrong approximation for $\mathrm{SL}_2(\mathbb{Z}/q\mathbb{Z})$ has been established; in particular, the analogue of Theorem 8 for $\mathbb{Z}/p^n\mathbb{Z}$, which follows from expansion in $\mathrm{SL}_2(\mathbb{Z}/p^n\mathbb{Z})$, established[13] in [16], plays crucial role in the proof of Theorem 4 in [26].

The above leads to a proof of part 1 of Theorem 1. To continue, one needs to deal with t_1 which is very small (here $|H_1| = t_1$ which divides $p^2 - 1$).

To handle these, we lift to characteristic zero and examine the finite orbits of Γ in $X(\bar{\mathbb{Q}})$. In fact, by the Chebotarev Density Theorem, a necessary condition for Conjecture 3 to hold is that there are no such orbits other than $\{0\}$. Again using the rotations in the conic sections by planes, one finds that any such finite orbit must be among the solutions with t_j's roots of unity to

$$(t_1 + t_1^{-1})^2 + (t_2 + t_2^{-1})^2 + (t_3 + t_3^{-1})^2 = (t_1 + t_1^{-1})(t_2 + t_2^{-1})(t_3 + t_3^{-1}). \tag{3.5}$$

For this particular surface X, one can show using the inequality between the geometric and arithmetic means, that (3.5) has no nontrivial solutions for complex numbers with $|t_j| = 1$. For the more general surfaces X_k, $S_{A,B,C,D}$, and those in (1.8), there is a variety of solutions with $|t_j| = 1$. However, Lang's $\mathbb{G}_m$ Conjecture which is established effectively (see [2, 92]) yields that there are only finitely many solutions to these equations in roots of unity. This allows for an explicit determination of the finite orbits of Γ_Y in $Y(\bar{\mathbb{Q}})$ (as noted earlier for the cubic surfaces $S_{A,B,C,D}$, the long list of these orbits [96] correspond to the algebraic Painlevé VI's). This $\bar{\mathbb{Q}}$ analysis leads to part 2 of Theorem 1 and, combined with the discussion above, it yields a proof of Conjecture 3, at least if $p^2 - 1$ is not very smooth. To prove Theorem 2, we need to show that there are very few primes for which the above arguments fail. This is done by extending the arguments and results in [43] and [44] concerning points (x, y) on irreducible curves over $\mathbb{F}_p$ for which $\mathrm{ord}(x) + \mathrm{ord}(y)$ is small (here $\mathrm{ord}(x)$ is the order of x in $\mathbb{F}_p^*$).

The proof of Theorem 6 in the stronger form that all Markoff numbers are highly composite, that is, for every $\nu \geq 1$, as $T \to \infty$,

$$\sum_{\substack{m \in M^s, m \leq T \\ m \text{ has at most} \\ \nu \text{ distinct prime factors}}} 1 = o\Big(\sum_{\substack{m \in M^s \\ m \leq T}} 1 \Big),$$

makes use of counting points on $X^*(\mathbb{Z})$ of height at most T and, in particular, Mirzakhani's orbit equidistribution [111], as well as the transitivity properties of Γ on $X^*(q)$ for q a product of suitable primes p. The latter are provided by the results of Meiri and Puder [110]. For $p \equiv 1(4)$ for which the induced permutation action of Γ on $X^*(p)$ is transitive, they show that the resulting permutation group is essentially the full symmetric or alternating group on $X^*(p)$. Applying Goursat's (disjointness) Lemma leads to the Γ-action on $X^*(p_1 p_2 \cdots p_k)$ being transitive for any such primes $p_1 < p_2 < \cdots < p_k$.

13 The proof of this expansion result, in turn, builds crucially on Bourgain's sum-product theorem in $\mathbb{Z}/p^n\mathbb{Z}$ in [10], which is intimately realted to his discretized sum-product theorem [9]; the origins, nature and impact of the latter are discussed in [67].

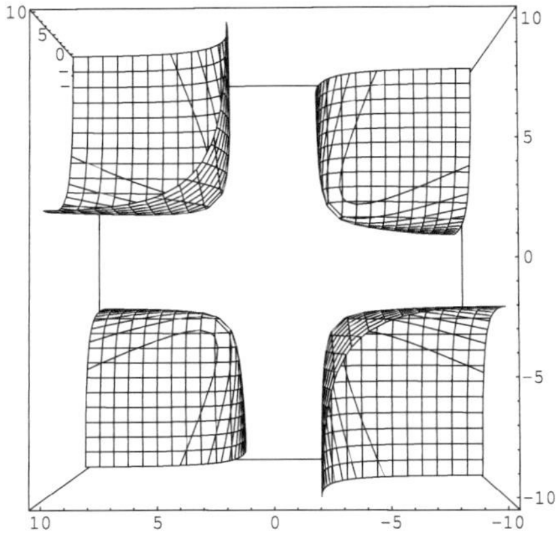

FIGURE 3
Level set $\kappa = -2.1$.

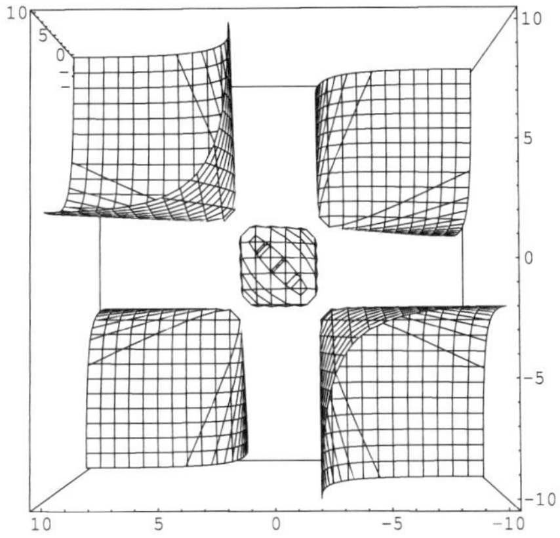

FIGURE 4
Level set $\kappa = -1.9$.

4. REAL DYNAMICS ON SURFACES OF MARKOFF TYPE

4.1. In this section we discuss the work of Goldman [73] pertaining to modular group action on real SL(2)-characters of a one-holed torus. The fundamental group π of the one-holed torus is the free group of rank two. The mapping class group of the 1-holed torus is isomorphic to the outer automorphism group $\mathrm{Out}(\pi) \cong \mathrm{GL}(2, \mathbb{Z})$ of π and acts on the moduli space of equivalence classes of $\mathrm{SL}(2, \mathbb{C})$-representations of π; this moduli space identifies naturally with affine 3-space $\mathbb{C}^3$, using the traces of two generators of π and of their product as coordinates. In these coordinates, the trace of the commutator of the two generators (representing the boundary curve of the torus) is given by $\kappa(x, y, z) = x^2 + y^2 + z^2 - xyz - 2$, which is preserved under the action of $\mathrm{Out}(\pi)$, and the action of $\mathrm{Out}(\pi)$ on $\mathbb{C}^3$ is commensurable with the action of the group Γ of polynomial automorphisms of $\mathbb{C}^3$ which preserve κ. Figures 3–8 show level sets of κ.

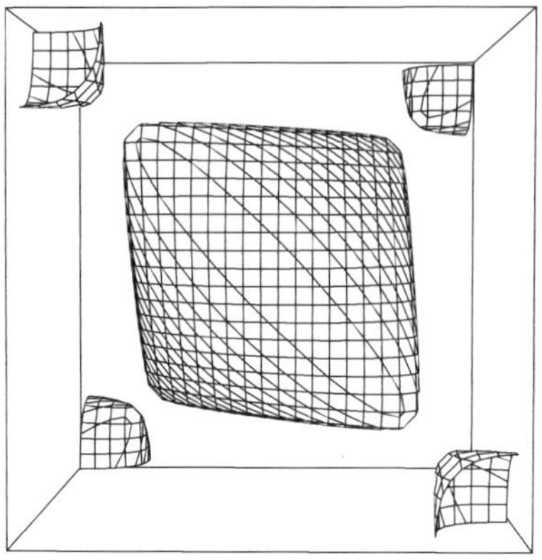

FIGURE 5
Level set $\kappa = 1.9$.

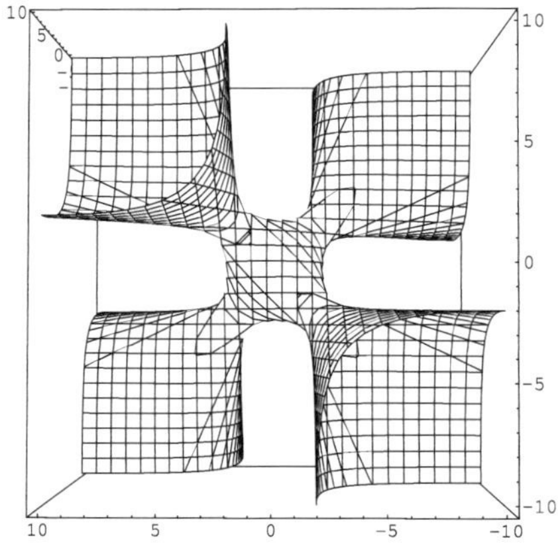

FIGURE 6
Level set $\kappa = 2.1$.

In [73] Goldman studied the dynamics of the Γ-action on the set of real points of this moduli space, and more precisely on the level sets $\kappa^{-1}(t) \cap \mathbb{R}^3$, for $t \in \mathbb{R}$. The action of Γ preserves a Poisson structure defining a Γ-invariant area form on each level set. It is shown that for $t < 2$ the Γ-action is properly discontinuous on the four contractible components of each level set and ergodic on the compact component (which is empty if $t < -2$); the contractible components correspond to Teichmüller spaces of complete hyperbolic structures on a one-holed torus if $t \leq -2$, and of a torus with a single cone point singularity if $-2 < t < 2$. For $t = 2$, the level set consists of characters of reducible representations and comprises two ergodic components, for $2 < t \leq 18$ the action of Γ on a level set is ergodic, and for $t > 18$ the moduli space contains characters of discrete representations uniformizing a three-holed sphere and the action is ergodic on the complement.

4.2. The main objective of [38] is the dynamical description of elements of the mapping class group of the four-punctured sphere acting on two-dimensional slices of its

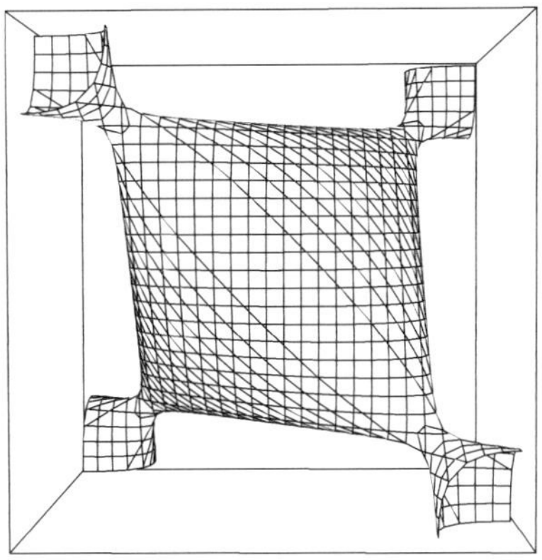

FIGURE 7
Level set $\kappa = 2.1$.

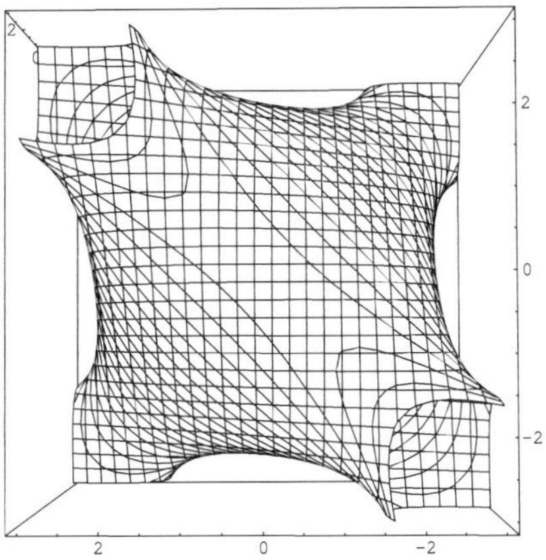

FIGURE 8
Level set $\kappa = 4$.

character variety. It also contains three striking applications of this analysis to the dynamics of the mapping class group on the character variety, to the spectrum of certain discrete Schrödinger equations, and to Painlevé sixth equation. Cantat considers the space of representations of the free group given by the presentation $F_3 = \langle \alpha, \beta, \gamma, \delta | \alpha\beta\gamma\delta = 1 \rangle$ into $\mathrm{SL}(2, \mathbb{C})$ modulo conjugacy. By fixing the trace of the images of the four generators, one obtains a space that is naturally parameterized by a cubic surface $S_{A,B,C,D} \subset \mathbb{C}^3$ given by $x^2 + y^2 + z^2 + xyz = Ax + By + Cz + D$ for some parameters $A, B, C, D \in \mathbb{C}$. This surface admits three natural involutions s_x, s_y, s_z which fix two out of the three coordinates and transform the last to the other root of the quadratic. These involutions generate a group Γ of affine automorphisms. Automorphisms of F_3 act by composition on the space of representations by preserving the trace, and the group of outer automorphisms of F_3 acts on $S_{A,B,C,D}$ in such a way that its image contains Γ as a finite index subgroup.

An element $f \in \Gamma$ is called hyperbolic if it corresponds to a pseudo-Anosov automorphism in the mapping class group, or, equivalently, if it is not conjugated to the product of

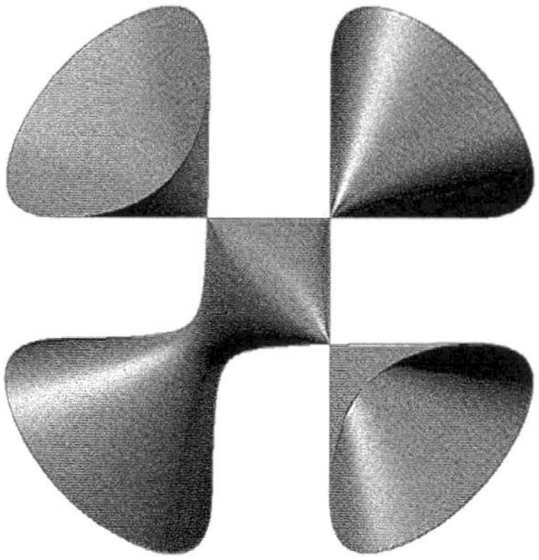

FIGURE 9
$S_{(-0.2,-0.2,-0.2,4.39)}$.

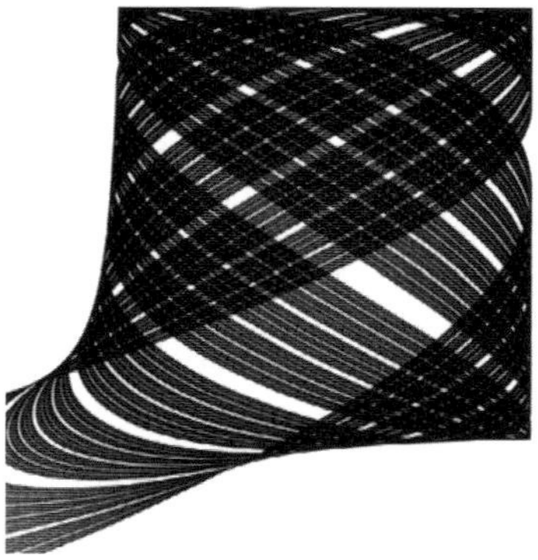

FIGURE 10
Projection of the stable manifold.

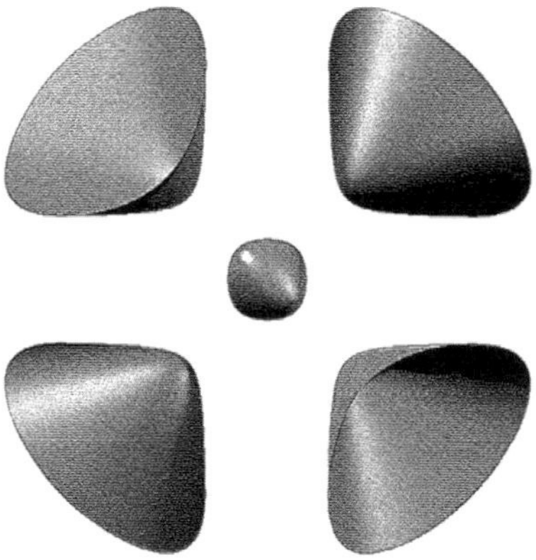

FIGURE 11
$S_{(0,0,0,3)}$.

FIGURE 12
Projection of the intersection of the stable manifold with the upper part of the surface.

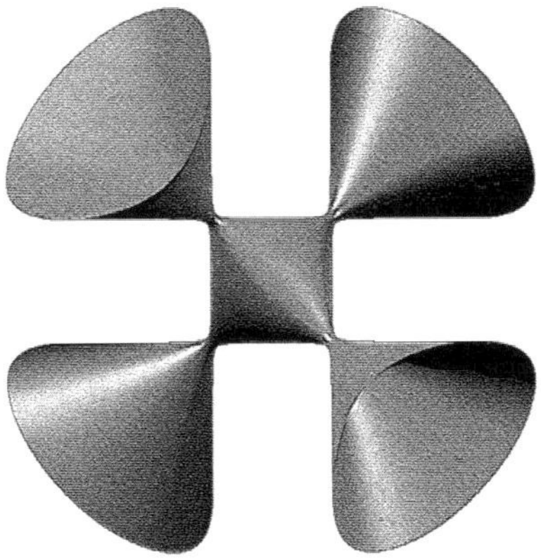

FIGURE 13
$S_{(0,0,0,4.1)}$.

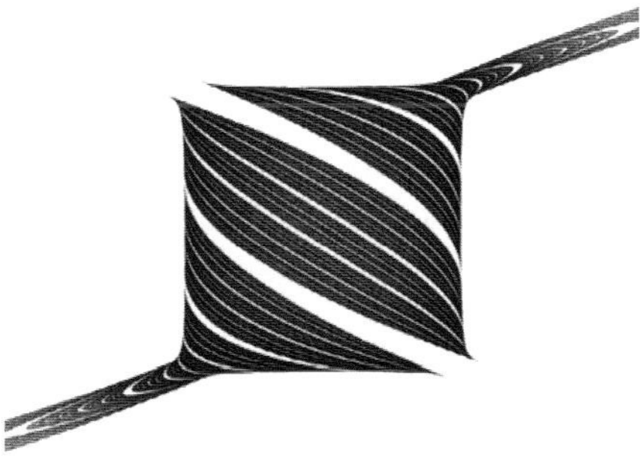

FIGURE 14
Projection of the intersection of the stable manifold with the upper part of the surface.

one or two involutions as above. Take $f \in \Gamma$ of hyperbolic type, and compactify $S_{A,B,C,D}$ by taking its closure in $\mathbb{P}^3$. The divisor at infinity is a cycle of three rational curves. By conjugating f in a suitable way, one can make it algebraically stable in the sense of Fornaess and Sibony [61], so that it contracts all curves at infinity to a single superattracting fixed point. One can then prove that at that point the map is locally conjugated to a monomial map whose spectral radius $\lambda(f)$ is greater than 1. This enables one to define the Green functions $G_{\pm} f = \lim_n \lambda(f)^{-n} \log^+ |f^{\pm n}|$, and show that they are plurisubharmonic, continuous, and possess natural invariance properties. It follows that the positive measure $\mu_f = dd^c G^+ f \wedge dd^c G^- f$ is well defined and f-invariant. Moreover, μ_f turns out to be mixing and the unique measure of maximal entropy equal to $\log \lambda(f)$. All these properties are reminiscent of the dynamics of Hénon mappings in the complex plane, and are proved analogously. Next, assume all coefficients A, B, C, D are real, and suppose the real part $S_{A,B,C,D}(\mathbb{R})$ is connected (in which case it is homeomorphic to the sphere minus four points). The main theorem of the paper states that the support of the measure μ_f is then included in $S_{A,B,C,D}(\mathbb{R})$ and that the induced map on $S_{A,B,C,D}(\mathbb{R})$ is uniformly hyperbolic on its nonwandering set. The proof of this striking theorem uses deep results by Bedford and Smillie [6] on the characterization of nonhyperbolic real Hénon maps having the same entropy as their complexification and relies on a delicate geometrical analysis of the possibilities for the intersection of stable and unstable manifolds in $S_{A,B,C,D}(\mathbb{R})$.

5. AN ASYMPTOTIC FORMULA FOR INTEGER POINTS ON MARKOFF–HURWITZ VARIETIES

For integer parameters $n \geq 3$, $a \geq 1$, and $k \in \mathbb{Z}$, consider the Diophantine equation

$$x_1^2 + x_2^2 + \cdots + x_n^2 = a x_1 x_2 \cdots x_n + k. \tag{5.1}$$

We call this the generalized[14] *Markoff–Hurwitz equation.* In this section we count solutions to (5.1) in integers, which we call *Markoff–Hurwitz tuples*. More precisely, let V be the affine subvariety of $\mathbb{C}^n$ cut out by (5.1). In a joint work with Magee and Ronan [68], we investigated the asymptotic size of the set $V(\mathbb{Z}) \cap B(R)$ where $B(R)$ is the ball of radius R in the ℓ^∞-norm on $\mathbb{R}^n \subset \mathbb{C}^n$. Perhaps somewhat surprisingly, the asymptotic growth for $n \geq 4$ is not of the order $(\log R)^{n-1}$, as was first noticed by Baragar [4], who subsequently in [5] proved that there is a number $\beta = \beta(n)$ such that when $k = 0$, if $V(\mathbb{Z}) - \{(0, 0, \ldots, 0)\}$ is nonempty then

$$|V(\mathbb{Z}) \cap B(R)| = (\log R)^{\beta + o(1)} \tag{5.2}$$

as $R \to \infty$.

14 Hurwitz [83] considered the case $k = 0$.

In [5] the following bounds for the exponents $\beta(n)$ were also obtained:

$$\begin{aligned}
&\beta(3) = 2,\\
&\beta(4) \in (2.430, 2.477),\\
&\beta(5) \in (2.730, 2.798),\\
&\beta(6) \in (2.963, 3.048),
\end{aligned} \tag{5.3}$$

and, in general,

$$\frac{\log(n-1)}{\log 2} < \beta(n) < \frac{\log(n-1)}{\log 2} + o(n^{-0.58}).$$

The following problems were posed by Silverman in 1995 [136] (in the setting of $k = 0$):

Problem 1. Is there is a true asymptotic formula for $|V(\mathbb{Z}) \cap B(R)|$ with main term proportional to $\log(R)^{\beta}$?

Problem 2. Is $\beta(n)$ irrational?

In [68] a complete answer to Problem 1 was obtained by extending Baragar's exponential rate of growth estimate to a true asymptotic formula.[15]

When $k > 0$, there are certain exceptional families of solutions to (5.1) that have a different quality of growth and, for fixed k, a, n, we write $\mathcal{E}$ for the set of exceptional tuples. We obtain the following theorem for the asymptotic number of Markoff–Hurwitz tuples:

Theorem 9. *For each (n, a, k) with $V(\mathbb{Z}) - \mathcal{E}$ infinite, there is a positive constant $c = c(n, a, k)$ such that*

$$|(V(\mathbb{Z}) - \mathcal{E}) \cap B(R)| = c(\log R)^{\beta} + o((\log R)^{\beta}).$$

Here β is the same constant as in (5.2).

After renormalizing (5.1), which allows us to set $a = 1$, and rearranging entries, Markoff–Hurwitz transformations induce the moves

$$\lambda_j(z_1, \dots, z_n) = \left(z_1, \dots, \widehat{z_j}, \dots, z_n, \left(\prod_{i \neq j} z_i\right) - z_j\right), \quad 1 \le j \le n-1, \tag{5.4}$$

on ordered tuples of real numbers. Above, $\widehat{\bullet}$ denotes omission. If sufficiently many of the z_i are large, the move λ_j can be approximated by

$$z \mapsto (z_1, \dots, \widehat{z_j}, \dots, z_n, \prod_{i \neq j} z_i)$$

15 The techniques in [5] "were inspired in part by Boyd's work on the Apollonian packing problem [29–31]". Boyd's result was extended to a true asymptotic formula in the work of Kontorovich and Oh [86].

to high accuracy relative to the largest entries of z. When the z_i are positive, at the level of logarithms this corresponds to

$$(\log z_1, \log z_2, \ldots, \log z_n) \mapsto (\log z_1, \ldots, \widehat{\log z_j}, \ldots, \log z_n, \sum_{i \neq j} \log z_i).$$

Thus one is naturally led to study the linear semigroup generated by linear maps

$$\gamma_j(y_1, y_2, \ldots, y_n) = (y_1, \ldots, \widehat{y_j}, \ldots, y_n, \sum_{i \neq j} y_i) \tag{5.5}$$

on ordered n-tuples $(y_1, \ldots, y_n)$.

Let

$$\Gamma = \langle \gamma_1, \ldots, \gamma_{n-1} \rangle_+,$$

where we have written a "+" to indicate we are generating a semigroup, not a group.

An important idea in [68] that explains why we are able to make progress on the counting problem is that we replace the generators of Γ with the countably infinite generating set $T_\Gamma = \{\gamma_{n-1}^A \gamma_j : A \in \mathbb{Z}_{\geq 0}, 1 \leq j \leq n-2\}$ and then consider the semigroup $\Gamma' = \langle T_\Gamma \rangle_+$.

Both Γ and Γ' preserve the nonnegative ordered hyperplane

$$\mathcal{H} \equiv \{(y_1, \ldots, y_n) \in \mathbf{R}^n_{\geq 0} : y_1 \leq y_2 \leq \cdots \leq y_n, \sum_{j=1}^{n-1} y_j = y_n\} \subset \mathbf{R}^n_{\geq 0}; \tag{5.6}$$

any element of Γ maps ordered tuples in $\mathbf{R}^n_{\geq 0}$ into $\mathcal{H}$. Therefore the study of orbits of Γ and Γ' on ordered tuples boils down to the study of orbits in $\mathcal{H}$. We can use the basis

$$e_j = (0, \ldots, 0, \underbrace{1}_{j}, 0, \ldots, 0, 1)$$

for the subspace spanned by $\mathcal{H}$. This basis clarifies the action of Γ'.

When $n = 3$, the linear map $\sigma : \mathcal{H} \to \mathcal{H}$ defined by

$$\sigma(a, b, a+b) = \mathsf{order}(b-a, a, b), \tag{5.7}$$

where order puts a tuple in ascending order from left to right, is such that for $j = 1, 2$ we have $\sigma\gamma_j.y = y$ for all $y \in \mathcal{H}$. Repeatedly applying the map σ to a triple $(a, b, a+b)$ with $a \leq b \in \mathbb{Z}$ performs the Euclidean algorithm on a, b. However, one application of σ corresponds in general to less than one step of the algorithm. Replacing Γ with Γ' corresponds to speeding this up so one whole step of the Euclidean algorithm corresponds to one semigroup generator. As for counting, the orbit of $(0, 1, 1)$ under Γ is precisely those $(a, b, a+b)$ with $(a, b) = 1$ and thus can be counted by elementary methods.

When $n = 3$, the semigroup Γ' is generated by

$$g_A := \gamma_2^A \gamma_1 = \begin{pmatrix} 0 & 1 \\ 1 & A+1 \end{pmatrix}$$

with respect to the basis $\{e_1, e_2\}$. These generators are classically connected with continued fractions by the formulae

$$\begin{pmatrix} 0 & 1 \\ 1 & A_1 \end{pmatrix}\begin{pmatrix} 0 & 1 \\ 1 & A_2 \end{pmatrix} \cdots \begin{pmatrix} 0 & 1 \\ 1 & A_k \end{pmatrix} = \begin{pmatrix} \star & b \\ \star & d \end{pmatrix}, \quad \frac{b}{d} = \cfrac{1}{A_1 + \cfrac{1}{A_2 + \ddots \frac{1}{A_k}}}.$$

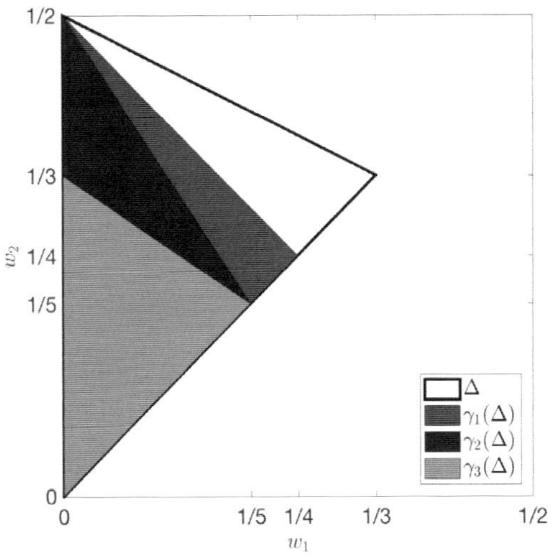

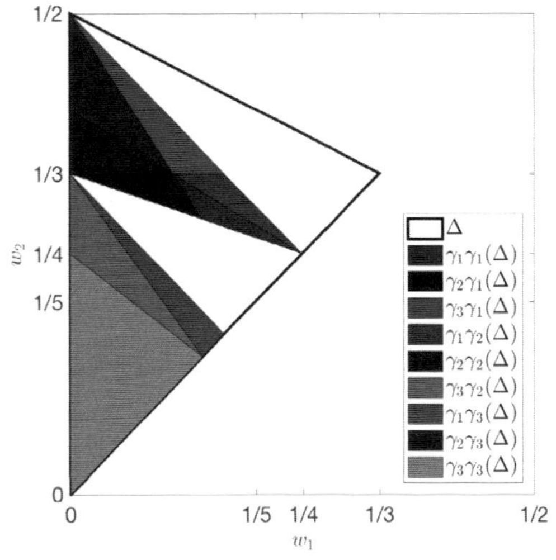

FIGURE 15

When $n = 4$, the semigroup elements map $\Delta = \mathbb{H}/\mathbb{R}_+$ into a strictly smaller subset. After iteration, this leads to more and more empty space (see also Figure 16). This does not occur when $n = 3$, as one can also see from the picture: the action of the group elements γ_2 and γ_3 on the vertical coordinate axis is a copy of the $n = 3$ dynamics.

When $n = 4$, the semigroup Γ acts in the basis given by the e_i as

$$\gamma_1 = \begin{pmatrix} 0 & 1 & 0 \\ 0 & 0 & 1 \\ 1 & 1 & 1 \end{pmatrix}, \quad \gamma_2 = \begin{pmatrix} 1 & 0 & 0 \\ 0 & 0 & 1 \\ 1 & 1 & 1 \end{pmatrix}, \quad \gamma_3 = \begin{pmatrix} 1 & 0 & 0 \\ 0 & 1 & 0 \\ 1 & 1 & 1 \end{pmatrix}.$$

This semigroup appears naturally in different areas of mathematics. In most situations that this semigroup appears, as is the case in **[68]**, the dynamics of the projective linear action of Γ on $\mathbb{R}^3_+/\mathbb{R}_+$ becomes relevant. Up to the minor modification of possibly multiplying the generators on the left or right by permutation matrices, the iterated function system given by the projective linear action of Γ on $\mathbb{R}^3_+/\mathbf{R}_+$ has a fractal attracting set that is known as the *Rauzy gasket* **[95]**.

So the semigroups Γ and Γ' are natural extensions of the Euclidean algorithm and continued fractions semigroup to higher dimensions. Writing $\Delta = \mathcal{H}/\mathbb{R}_+$, we can view Δ as a subset of $\mathbb{R}^{n-2}$. The key distinction that appears when $n \geq 4$ is that

$$\Delta \neq \bigcup_{j=1}^{n-1} \gamma_j(\Delta)$$

and so the induced dynamics on $\mathcal{H}/\mathbb{R}_+$ has "holes" as we illustrate in Figure 15.

Structure of the proof and the difficulties that arise. Here we highlight some of the main difficulties that must be overcome during the proof of Theorem 9. It is illuminating to recall the methods used by Lalley[16] in **[90]** where the action of a Schottky subgroup G

16 See Mark Policott's ICM report **[119]** for an overview of recent developments pertaining to dynamical zeta functions and thermodynamic formalism.

of $\mathrm{SL}_2(\mathbf{R})$ on the hyperbolic upper half-plane $\mathbb{H}$ is considered. Lalley obtains that, for any $x \in \mathbb{H}$, the number $\mathcal{N}(x, r)$ of elements γ of G such that

$$d_{\mathbb{H}}(i, \gamma x) - d_{\mathbb{H}}(i, x) \leq r,$$

where $d_{\mathbb{H}}$ is hyperbolic distance, satisfies $\mathcal{N}(x, r) \approx Ce^{\delta r}$, where $\delta = \delta(G)$ is the Hausdorff dimension of the limit set of G, and $C = C(G, x) > 0$. Lalley's proof incorporates at various stages the following arguments:

Shell argument. By repeated application of a "renewal equation," the quantity $\mathcal{N}(x, r)$ is related to a sum of $\mathcal{N}(y, r')$, where the sum is over y on a shell of radius $\approx cr$ in a Cayley tree of G, and r' is a translate of r that corrects for the passage between x and y. The purpose of this shell argument is that now, the points y lie close to $\partial\mathbb{H}$.

Passage to the boundary. Each of the resulting $\mathcal{N}(y, r')$ is compared to an analogous quantity $\mathcal{N}^*(y^*, r')$ where y^* is a point in $\partial\mathbb{H}$ close to y. Because each y is close to $\partial\mathbb{H}$, the errors incurred are acceptable.

Transfer operator techniques. Asymptotic formulas for $\mathcal{N}^*(y^*, r')$ are obtained using the renewal method and spectral estimates for transfer operators. This gives asymptotic formulas for $\mathcal{N}(y, r')$. The main terms of the asymptotic formulas satisfy recursive relationships between different y.

Recombination. One finally has to recombine all the asymptotic formulas obtained for $\mathcal{N}(y, r')$ to obtain an asymptotic formula for $\mathcal{N}(x, r)$. This is done using the recursive formulas obtained in the previous step.

Trying to follow the method outlined above for this orbital counting problem, we first need a suitable replacement for $\partial\mathbb{H}$. Our idea is to use the projectivization of the hyperplane $\mathcal{H}$; we call this set Δ. We compare points in the orbit of Λ (generated by λ_j in (5.4) to points in Δ by taking logarithms of all coordinates and then projectivizing. This process does not necessarily lead to a point in Δ; there is an important parameter $\alpha(z) = \prod_{j=1}^{n-2} z_j$ that appears throughout the paper and measures how good the fit is. If $\alpha(z)$ is large, then one can, in analogy with Lalley's setting, think of z as being "close to the boundary."

For Lalley, the word length of γ is roughly proportional to the quantity $d_{\mathbb{H}}(i, \gamma x) - d_{\mathbb{H}}(i, x)$ with respect to which he counts. This implies, during the shell argument, that all the elements of the shell are roughly the same distance from $\partial\mathbb{H}$. However, for us, there are arbitrarily long words in the generators of Λ for which $\alpha(z)$ is small. We solve this problem using "acceleration," by replacing Λ by Λ', and instead aim to follow Lalley's argument for orbits of Λ'. This has the immediate benefit that we can guarantee that elements z of shells of radius L, with respect to Λ', have large $\alpha(z)$, if we make L appropriately large.

However, the acceleration also has some costs to be paid. The first issue arising is that now Λ' has countably many generators, so shells for word length on Λ' are not finite. Instead of using shells, we use intersections of shells with the elements of the Λ'-orbit whose coordinates are not too large. The second issue is that the original Λ-orbit breaks up

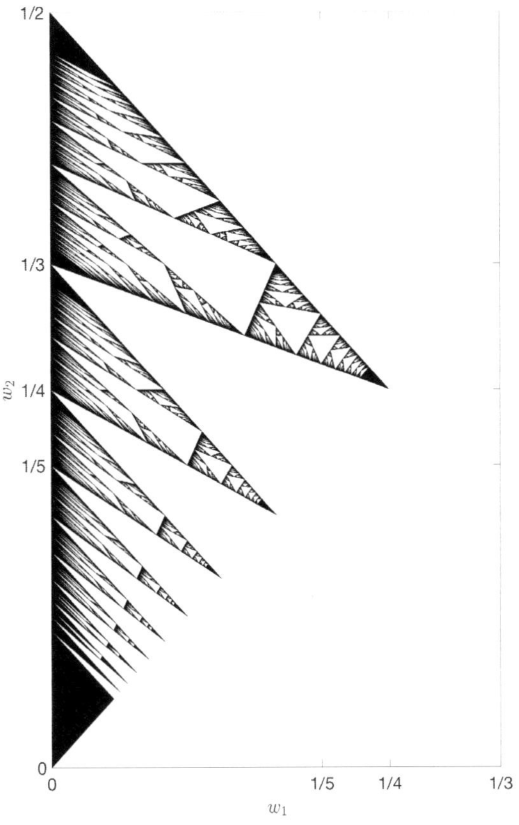

FIGURE 16
In the same setting ($n = 4$) of Figure 15, we show in black the images of Δ under the action of all words of length 10 in the generators $\{\gamma_1, \gamma_2, \gamma_3\}$.

into countably many Λ'-orbits. So we not only have to perform the recombination argument for Λ', but then have to perform an extra summation over the countably many Λ'-orbits.

After setting up our shell argument appropriately, we must perform the passage to the boundary (i.e., Δ). To this end, we compare orbits of Λ' to orbits of Γ', where Γ' is the linear semigroup. To get this to work, we must exploit the following "shadowing" feature of the map log that takes logarithms of all entries of a vector. It says (roughly) that if $\log(z)$ is within ϵ of $y \in \mathscr{H}$, with ϵ on the scale of $\alpha(z)^{-2}$, then for all $\lambda \in \Lambda'$, $\log(\lambda(z))$ is within ϵ of $\gamma(\log(z))$, where $\gamma \in \Gamma'$ is matched with λ in a natural way.

The completion of the proof relies on spectral estimates for transfer operators associated to the projective linear action of Γ' on Δ. There are three key issues arising here. First, to obtain the spectral estimates we need, we must establish that the action of Γ' on Δ is uniformly contracting; it is important to note that this argument would not work if the acceleration had not been performed previously. Secondly, we need to establish that the relevant "log-Jacobian" cocycle over the dynamical system is not cohomologous to a lattice cocycle. Finally, but importantly, we must obtain spectral estimates for transfer operators acting on $C^1(\Delta)$ which is accomplished by adapting Liverani's approach to spectral estimates from [97]. See section 4 of [68], and references therein, for the discussion of Gauss map and Gauss measure [70, 89] in this context.

The question of whether β is irrational (Problem 2) remains a tantalizing open question, and one may wonder whether it is even algebraic. Our methods do give some partial insight into the nature of this mysterious number in terms of the action of Γ' on $\mathcal{H}/\mathbf{R}_+$.

Theorem 10. *The number β is the unique parameter in $(1, \infty)$ such that there exists a probability measure ν_β on $\Delta = \mathcal{H}/\mathbf{R}_+$ with the property*

$$\int_{w\in\Delta} f(w)d\nu_\beta(w) = \sum_{\gamma\in T_\Gamma}\int_{w\in\Delta} f(\gamma.w)|\mathrm{Jac}_w(\gamma)|^{\frac{\beta}{n-1}} d\nu_\beta(w)$$

for all $f \in C^0(\Delta)$. We call ν_β a conformal measure.

Theorem 10 can be viewed as a partial analog of the connection between the exponent of growth of a finitely generated Fuchsian group and the Hausdorff dimension of its limit set as a result of Patterson–Sullivan theory [117, 138, 139]. In our setting, the lack of any symmetric space means the parameter β is not in any obvious way connected to the Hausdorff dimension of the compact Γ'-invariant subset of Δ.

The issue of the existence of a *single* integral solution for general a and k is very subtle, even for $n = 3$, as discussed in the next section.

6. HASSE PRINCIPLE ON SURFACES OF MARKOFF TYPE

Little is known about the values at integers assumed by *affine cubic forms*[17] F in three variables. For $k \neq 0$, set

$$V_{k,F} = \{\mathbf{x} = (x_1, x_2, x_3) : F(\mathbf{x}) = k\}. \tag{6.1}$$

The basic question is for which k is $V_{k,F}(\mathbb{Z}) \neq \emptyset$, or, more generally, infinite or Zariski dense in $V_{k,F}$?

A prime example is $F = S$, the sum of three cubes,

$$S(x_1, x_2, x_3) = x_1^3 + x_2^3 + x_3^3. \tag{6.2}$$

There are obvious local congruence obstructions, namely that $V_{k,S}(\mathbb{Z}) = \emptyset$ if $k \equiv 4, 5 \pmod 9$, but beyond that, it is possible that the answers to all three questions are yes for all the other k's, which we call the *admissible* values (see [50, 113]). It is known that strong approximation in its strongest form fails for $V_{k,S}(\mathbb{Z})$; the global obstruction coming from an application of cubic reciprocity [41, 49, 75]). Moreover, the authors of [94] and [7] show that $V_{1,S}(\mathbb{Z})$ is Zariski dense in $V_{1,S}$.

In [71] Ghosh and Sarnak investigate the Markoff form $F = M$,

$$M(\mathbf{x}) = x_1^2 + x_2^2 + x_3^2 - x_1x_2x_3. \tag{6.3}$$

---

17 By an *affine form* f in n variables we mean $f \in \mathbb{Z}[x_1, \ldots, x_n]$ whose leading homogeneous term f_0 is nondegenerate and such that $f - k$ is (absolutely) irreducible for all constants k.

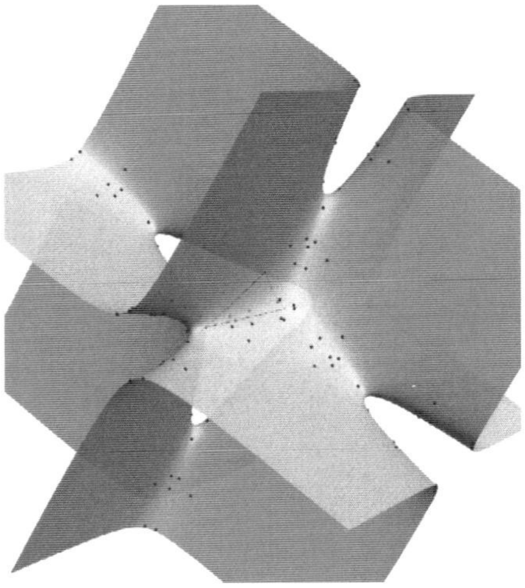

FIGURE 17
Lattice points and fundamental set for $k = 3685$.

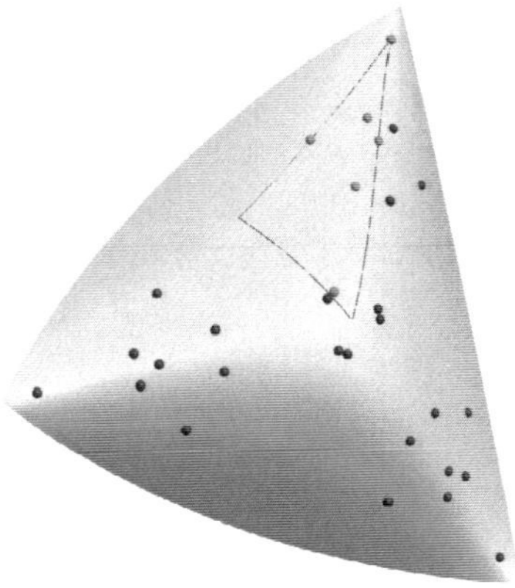

FIGURE 18
Closeup of fundamental set for $k = 3685$.

Except for the case of the Cayley cubic with $k = 4$, $V_{k;M}(\mathbb{Z})$ decomposes into a finite number $\mathfrak{h}_M(k)$ of Γ-orbits. For example, if $k = 0$, then $\mathfrak{h}_M(k) = 2$ corresponds to the orbits of $(0, 0, 0)$ and $(3, 3, 3)$. In order to study $\mathfrak{h}_M(k)$ both theoretically and numerically, they give an explicit reduction (descent) for the action of Γ on $V_{k,M}(\mathbb{Z})$. For this purpose, it is convenient to remove an explicit set of special admissible k's, namely those for which there is a point in $V_{k,M}(\mathbb{Z})$ with $|x_j| = 0$, 1 or 2. These k's take the form (i) $k = u^2 + v^2$, (ii) $4(k - 1) = u^2 + 3v^2$, or (iii) $k = 4 + u^2$. The number of these special k's (referred to as *exceptional*) with $0 \leq k \leq K$ is asymptotic to $C' \frac{K}{\sqrt{\log K}}$. The remaining admissible k's are called *generic* (all negative admissible k's are generic). For them Ghosh and Sarnak give the following elegant reduced forms:

FIGURE 19
Lattice points and fundamental set for $k = -3691$.

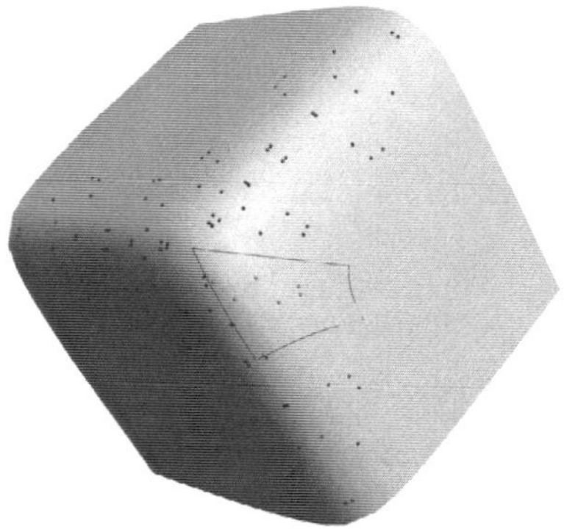

FIGURE 20
Closeup of fundamental set for $k = -3691$.

Proposition 11. (1) *Let $k \geq 5$ be generic and consider the compact set*

$$\mathfrak{F}_k^+ = \{\mathbf{u} \in \mathbb{R}^3 : 3 \leq u_1 \leq u_2 \leq u_3\,,\ u_1^2 + u_2^2 + u_3^2 + u_1 u_2 u_3 = k\}.$$

The points in $\mathfrak{F}_k^+(\mathbb{Z}) = \mathfrak{F}_k^+ \cap \mathbb{Z}^3$ are Γ-inequivalent, and any $\mathbf{x} \in V_{k,M}(\mathbb{Z})$ is Γ-equivalent to a unique point $\mathbf{u}' = (-u_1, u_2, u_3)$ with $\mathbf{u} = (u_1, u_2, u_3) \in \mathfrak{F}_k^+(\mathbb{Z})$.

(2) *Let $k < 0$ be admissible and consider the compact set*

$$\mathfrak{F}_k^- = \{\mathbf{u} \in \mathbb{R}^3 : 3 \leq u_1 \leq u_2 \leq u_3 \leq \frac{1}{2} u_1 u_2\,,\ u_1^2 + u_2^2 + u_3^2 - u_1 u_2 u_3 = k\}.$$

The points in $\mathfrak{F}_k^-(\mathbb{Z}) = \mathfrak{F}_k^- \cap \mathbb{Z}^3$ are Γ-inequivalent, and any $\mathbf{x} \in V_{k,M}(\mathbb{Z})$ is Γ-equivalent to a unique point $\mathbf{u} = (u_1, u_2, u_3) \in \mathfrak{F}_k^-(\mathbb{Z})$.

Some consequences of this are as follows: As $k \to \pm\infty$, we have

$$\mathfrak{h}_M(k) \ll_\varepsilon |k|^{\frac{1}{3}+\varepsilon}.$$

This follows from the fact that when considering the values taken by the corresponding indefinite quadratic form in the y and z variables, for each fixed x, the units are bounded in number due to the restrictions imposed by the fundamental sets.

Let $\mathfrak{h}_M^{\pm}(k) = |\mathfrak{F}_k^{\pm}(\mathbb{Z})|$ where $\pm = \operatorname{sgn}(k)$, this being defined for any k. Then for generic k, $\mathfrak{h}_M^{\pm}(k) = \mathfrak{h}_M(k)$ while otherwise $\mathfrak{h}_M(k) \le \mathfrak{h}_M^{\pm}(k)$. We have

$$\sum_{\substack{k \neq 4 \\ |k| \le K}} \mathfrak{h}_M^{\pm}(k) \sim C^{\pm} K (\log K)^2, \tag{6.4}$$

where $C^{\pm} > 0$ and $K \to \infty$.

So on average the numbers $\mathfrak{h}_M(k)$ are small. The explicit fundamental domains allow for the numerical computations; these indicate that

$$\sum_{\substack{0 < k \le K \\ k \text{ admissible} \\ \mathfrak{h}_M(k) = 0}} 1 \sim C_0 K^{\theta}, \tag{6.5}$$

with $C_0 > 0$ and $\theta \approx 0.8875\ldots$.

The main result in [71] concerns the values assumed by M and the Hasse failures in (6.5):

Theorem 12. (i) *There are infinitely many Hasse failures. More precisely, the number of* $0 < k \le K$ *and* $-K \le k < 0$ *for which the Hasse principle fails is at least* $\sqrt{K}(\log K)^{-\frac{1}{4}}$ *for* K *large.*

(ii) *Fix* $t \ge 0$. *Then as* $K \to \infty$,

$$\#\{|k| \le K : k \text{ admissible}, \ \mathfrak{h}_M(k) = 0\} = o(K).$$

Hasse failures are produced by an obstruction via quadratic reciprocity. They come in two types: one via direct use of reciprocity and the second also incorporating the descent group. Recently Colliot-Thélène, Wei, and Xu [48] and, independently, Loughran and Mitankin [98] have shown that the obstruction of the first (but not the second type) can be explained in terms of integral Brauer–Manin obstruction. For example, if $k = 4 + 2\nu^2$, with ν having all of its prime factors congruent to ± 1 (mod 8) and ν congruent to $0, \pm 3, \pm 4$ (mod 9), then k is admissible but $V_{k,M}(\mathbb{Z}) = \emptyset$.

Part (ii) of the theorem is proved by comparing the number of points on $V_{k,M}(\mathbb{Z})$ in certain tentacled regions gotten by special plane sections, with the expected number of solutions according to a product of local densities; the crucial point being that the variance of this comparison goes to zero on averaging $|k| \le K$. As detailed in [71], this moving plane quadric method applies to more general cubic surfaces including those that do not carry morphisms.

ACKNOWLEDGMENTS

I would like to thank Jean Bourgain, Emmanuel Breuillard, Michael Magee, Igor Pak, Ryan Ronan, and Peter Sarnak. My various collaborations with them form the core of this report.

Figure 1 is courtesy of Matthew de Courcy-Ireland.
Figure 2 is courtesy of Elena Fuchs.
Figures 3–8 are courtesy of William Goldman.
Figures 9–14 are courtesy of Serge Cantat.
Figures 17–20 are courtesy of Amit Ghosh.

FUNDING

The author was supported, in part, by NSF award DMS-1603715.

REFERENCES

[1] M. Aigner, *Markov's Theorem and 100 years of the Uniqueness Conjecture.* Springer, 2013.

[2] I. Aliev and C. Smyth, Solving Algebraic equations in roots of unity. *Forum Math.* **24** (2012), 641–645.

[3] A. Baragar, *The Markoff equation and equations of Hurwitz.* PhD Thesis, Brown University, RI, 1991.

[4] A. Baragar, Asymptotic growth of Markoff–Hurwitz numbers. *Compos. Math.* **94** (1994), no. 1, 1–18.

[5] A. Baragar, The exponent for the Markoff–Hurwitz equations. *Pacific J. Math.* **182** (1998), no. 1, 1–21.

[6] E. Bedford and J. Smillie, Real polynomial diffeomorphisms with maximal entropy: Tangencies. *Ann. of Math. (2)* **160** (2004), no. 1, 1–26.

[7] F. Beukers, Ternary form equations. *J. Number Theory* **54** (1995), 113–133.

[8] E. Bombieri, Continued fractions and the Markoff tree. *Expo. Math.* **25** (2007), no. 3, 187–213.

[9] J. Bourgain, On the Erdös-Volkmann and Katz-Tao ring conjecture, *Geom. Funct. Anal.* **13** (2003), 334–365.

[10] J. Bourgain, The sum-product theorem in $\mathbb{Z}_q$ with q arbitrary. *J. Anal. Math.* **106**, 1–93 (2008)

[11] J. Bourgain, A modular Szemeredi–Trotter theorem for hyperbolas. *C. R. Acad. Sci. Paris Sér. 1* **350** (2012), 793–796.

[12] J. Bourgain and A. Gamburd, New results on expanders. *C. R. Math. Acad. Sci. Paris* **342** (2006).

[13] J. Bourgain and A. Gamburd, Uniform expansion bounds for Cayley graphs of $SL_2(\mathbb{F}_p)$. *Ann. of Math.* **167** (2008), 625–642.

[14] J. Bourgain and A. Gamburd, Random walks and expansion in $\mathrm{SL}_d(\mathbb{Z}/p^n\mathbb{Z})$. *C. R. Math. Acad. Sci. Paris* **346** (2008), no. 11–12, 619–623.

[15] J. Bourgain and A. Gamburd, On the spectral gap for finitely-generated subgroups of SU(2). *Invent. Math.* **171** (2008), no. 1, 83–121.

[16] J. Bourgain and A. Gamburd, Expansion and random walks in $\mathrm{SL}_d(\mathbb{Z}/p^n\mathbb{Z})$. I. *J. Eur. Math. Soc. (JEMS)* **10** (2008), no. 4, 987–1011.

[17] J. Bourgain and A. Gamburd, Expansion and random walks in $\mathrm{SL}_d(\mathbb{Z}/p^n\mathbb{Z})$. II. With an appendix by Bourgain. *J. Eur. Math. Soc. (JEMS)* **11** (2009), no. 5, 1057–1103.

[18] J. Bourgain and A. Gamburd, Spectral gaps in $\mathrm{SU}(d)$. *C. R. Math. Acad. Sci. Paris* **348** (2010), no. 11–12, 609–611.

[19] J. Bourgain and A. Gamburd, A spectral gap theorem in $\mathrm{SU}(d)$. *J. Eur. Math. Soc. (JEMS)* **14** (2012), no. 5, 1455–1511.

[20] J. Bourgain, A. Gamburd, and P. Sarnak, Sieving and expanders. *C. R. Acad. Sci. Paris, Sér. I* **343** (2006).

[21] J. Bourgain, A. Gamburd, and P. Sarnak, Affine linear sieve, expanders and sum product. *Invent. Math.* **179** (2010), 559–644.

[22] J. Bourgain, A. Gamburd, and P. Sarnak, Generalization of Selberg's $\frac{3}{16}$ theorem and affine sieve. *Acta Math.* **207** (2011), no. 2, 255–290.

[23] J. Bourgain, A. Gamburd, and P. Sarnak, Markoff surfaces and strong approximation: 1. 2016, arXiv:1607.01530.

[24] J. Bourgain, A. Gamburd, and P. Sarnak, Markoff triples and strong approximation. *C. R. Math. Acad. Sci. Paris* **354** (2016), no. 2, 131–135.

[25] J. Bourgain, A. Gamburd, and P. Sarnak, Strong approximation and diophantine properties of Markoff numbers, preprint.

[26] J. Bourgain, A. Gamburd, and P. Sarnak, Strong approximation for varieties of Markoff type, preprint.

[27] J. Bourgain, N. Katz, and T. Tao, A sum-product estimate in finite fields, and applications. *Geom. Funct. Anal.* **14** (2004), no. 1, 27–57.

[28] J. Bourgain and P. P. Varjú, Expansion in $SL_d(Z/qZ)$, q arbitrary. *Invent. Math.* **188** (2012).

[29] D. W. Boyd, The disk-packing constant. *Aequationes Math.* **7** (1971), 182–193.

[30] D. W. Boyd, Improved bounds for the disk-packing constant. *Aequationes Math.* **9** (1973), 99–106.

[31] D. W. Boyd, The sequence of radii of the Apollonian packing. *Math. Comp.* **39** (1982), no. 159, 249–254.

[32] E. Breuillard, A strong Tits alternative. 2008, arXiv:0804.1395.

[33] E. Breuillard, Approximate subgroups and superstrong approximation. In *Groups St Andrews 2013*, pp. 1–50, London Math. Soc. Lecture Note Ser. 422, Cambridge University Press, Cambridge, 2015.

[34] E. Breuillard and A. Gamburd, Strong uniform expansion in $SL_2(\mathbb{F}_p)$. *Geom. Funct. Anal.* **20** (2010), no. 5, 1201–1209.

[35] E. Breuillard, B. Green, and T. Tao, Approximate subgroups of linear groups. *Geom. Funct. Anal.* **21** (2011).

[36] E. Breuillard and H. Oh, *Thin groups and superstrong approximation*. Math. Sci. Res. Inst. Publ. 61, Cambridge University Press, Cambridge, 2014.

[37] S. Cantat, Dynamique des automorphismes des surfaces K3. *Acta Math.* **187** (2001), no. 1, 1–57

[38] S. Cantat, Bers, Hénon, Painlevé and Schrödinger. *Duke Math. J.* **149** (2009), no. 3, 411–460.

[39] J. W. S. Cassels, The Markoff chain. *Ann. of Math. (2)* **50** (1949), 676–685.

[40] J. W. S. Cassels, *An Introduction to Diophantine Approximation*. Cambridge Tracts in Mathematics and Mathematical Physics 45, Cambridge University Press, New York, 1957, x+166 pp.

[41] J. W. S. Cassels, A note on the Diophantine equation $x^3 + y^3 + z^3 = 3$. *Math. Comp.* **44** (1985), 265–266.

[42] F. Celler, C. R. Leedham-Green, S. Murray, A. Niemeyer, and E. A. O'Brien, Generating random elements of a finite group. *Comm. Algebra* **23** (1995), 4931–4948.

[43] M.-C. Chang, Elements of large order in prime finite fields. *Bull. Aust. Math. Soc.* **88** (2013).

[44] M.-C. Chang, B. Kerr, I. Shparlinski, and U. Zannier, Elements of large orders on varieties over prime finite fields. *J. Théor. Nombres Bordeaux* **26** (2014).

[45] W. Chen, Nonabelian level structures, Nielsen equivalence, and Markoff triples. 2021, arXiv:2011.12940.

[46] H. Cohn, Approach to Markoff's minimal forms through modular functions. *Ann. of Math.* **61** (1955), 1–12.

[47] H. Cohn, Markoff forms and primitive words. *Math. Ann.* **196** (1972), 8–22.

[48] J.-L. Colliot-Thélène, D. Wei, and F. Xu, Brauer–Manin obstruction for Markoff surfaces. Annali della Scuola Normale Superiore di Pisa – Classe di Scienze Volume XXI/2021, no. 3 série V (2021).

[49] J. Colliot-Thélène and O. Wittenberg, Groupe de Brauer et points entiers de deux familles de surfaces cubiques affines. *Amer. J. Math.* **134** (2012), no. 5, 1303–1327.

[50] W. Conn and L. N. Vaserstein, *On sums of three integral cubes*. pp. 285–294, Contemp. Math. 166, Amer. Math. Soc., 1994.

[51] P. Corvaja and U. Zannier, On integral points on surfaces. *Ann. of Math. (2)* **160** (2004).

[52] P. Corvaja and U. Zannier, On the greatest prime factor of Markov pairs. *Rend. Semin. Mat. Univ. Padova* **116** (2006), 253–260.

[53] P. Corvaja and U. Zannier, Greatest common divisors of $u - 1$, $v - 1$ in positive characteristic and rational points on curves over finite fields. *J. Eur. Math. Soc. (JEMS)* **15** (2013).

[54] M. de Courcy-Ireland, Non-planarity of Markoff graphs mod p. 2022, arXiv:2105.12411v3.

[55] M. de Courcy-Ireland and S. Lee, Experiments with the Markoff surface. *Exp. Math.* (2020). DOI 10.1080/10586458.2019.1702123.

[56] B. Dubrovin and M. Mazzocco, Monodromy of certain Painlevé-VI transcendents and reflection groups. *Invent. Math.* **141** (2000), 55–147.

[57] W. Duke, Hyperbolic distribution problems and half-integral weight Maass forms. *Invent. Math.* **92** (1988).
[58] M. Einsiedler, E. Lindenstrauss, P. Michel, and A. Venkatesh, Distribution of periodic torus orbits on homogeneous spaces. *Duke Math. J.* **148** (2009).
[59] M. Einsiedler, E. Lindenstrauss, P. Michel, and A. Venkatesh, The distribution of closed geodesics on the modular surface, and Duke's theorem. *Enseign. Math. (2)* **58** (2012).
[60] M. H. El-Huti, Cubic surfaces of Markov type. *Math. USSR, Sb.* **22** (1974), no. 3, 333–348.
[61] J. Fornaess and N. Sibony, Complex dynamics in higher dimension. II. In: *Modern methods in complex analysis (Princeton, NJ, 1992)*. Ann. of Math. Stud. 137, Princeton University Press, Princeton, NJ, 1995.
[62] G. Frobenius, Über die Markoffschen Zahlen. *Akad. Wiss. Berlin* (1913), 458–487.
[63] E. Fuchs, Counting problems in Apollonian packings. *Bull. Amer. Math. Soc. (N.S.)* **50** (2013), no. 2, 229–266.
[64] E. Fuchs, K. Lauter, M. Litman, A. Tran, A Cryptographic Hash Function from Markoff Triples. 2021, arXiv:2107.10906.
[65] E. Fuchs, M. Litman, J. Silverman, A. Tran, Orbits on K3 Surfaces of Markoff Type. 2022, arXiv:2201.12588.
[66] A. Gamburd, Spectral gap for infinite index "congruence" subgroups of $\mathrm{SL}_2(\mathbb{Z})$. *Israel J. Math.* **127** (2002).
[67] A. Gamburd, Singular adventures of Baron Bourgain in the labyrinth of the continuum. *Notices Amer. Math. Soc.* **67** (2020), no. 11, 1716–1733.
[68] A. Gamburd, M. Magee, and R. Ronan, An asymptotic formula for integer points on Markoff–Hurwitz varieties. *Ann. of Math. (2)* **190** (2019), no. 3, 751–809.
[69] A. Gamburd and I. Pak, Expansion of product replacement graphs. *Combinatorica* **26** (2006), no. 4, 411–429.
[70] C. F. Gauss, Brief an Laplace vom 30 Jan. 1812, Werke X1, 1812, pp. 371–374.
[71] A. Ghosh and P. Sarnak, Integral points on Markoff type cubic surfaces. 2022, arXiv:1706.06712v1.
[72] R. Gilman, Finite quotients of the automorphism group of a free group. *Canad. J. Math.* **29** (1977), 541–551.
[73] W. Goldman, The modular group action on real $SL(2)$-characters of a one-holed torus. *Geom. Topol.* **7** (2003), 443–486.
[74] D. S. Gorshkov, Geometry of Lobachevskii in connection with certain questions of arithmetic. *Zap. Nauch. Sem. Lenin. Otd. Math. Inst. V.A. Steklova AN SSSR* **67** (1977), 39–85. English translation in *J. Sov. Math.* **16** (1981) 788–820.
[75] D. R. Heath-Brown, The density of zeros of forms for which weak approximation fails. *Math. Comp.* **59** (1992), no. 200, 613–623.
[76] R. Heath-Brown and S. Konyagin, New bounds for Gauss sums derived from k-th powers and for Heilbronn's exponential sum. *Quart. J. Math.* (2000), 221–235.

[77] H. A. Helfgott, Growth and generation in $\mathrm{SL}_2(\mathbb{Z}/p\mathbb{Z})$. *Ann. of Math. (2)* **167** (2008), no. 2, 601–623.

[78] H. A. Helfgott, Growth in groups: ideas and perspectives. *Bull. Amer. Math. Soc. (N.S.)* **52** (2015), no. 3, 357–413.

[79] C. Hooley, On Artin's conjecture. *J. Reine Angew. Math.* **225** (1967).

[80] C. Hooley, *Applications of sieve methods to the theory of numbers*. Cambridge Tracts in Math. 70, Cambridge, 1976.

[81] S. Hoory, N. Linial, and A. Wigderson, Expander graphs and their applications. *Bull. Am. Math. Soc.* (2006).

[82] E. Hrushovski, Stable group theory and approximate subgroups. *J. Amer. Math. Soc.* **25** (2012).

[83] A. Hurwitz, Über eine Aufgabe der unbestimmten Analysis. *Arch. Math. Phys.* **11** (1907), no. 3, 185–196.

[84] M. Kaluba, D. Kielak, and P. W. Nowak, On property (T) for $\mathrm{Aut}(F_n)$ and $\mathrm{SL}_n(\mathbb{Z})$. *Ann. of Math. (2)* **193** (2021), no. 2, 539–562.

[85] A. Kontorovich and J. Lagarias, On the expected number of prime factors in the affine sieve with toral Zariski closure, *Exp. Math.* **30** (2021), no. 4, 575–586.

[86] A. Kontorovich and H. Oh, Apollonian circle packings and closed horospheres on hyperbolic 3-manifolds. *J. Amer. Math. Soc.* **24** (2011), no. 3, 603–648. With an appendix by Oh and Nimish Shah.

[87] S. V. Konyagin, S. V. Makarychev, I. E. Shparlinski, and I. V. Vyugin, On the structure of graphs of Markoff triples. *Q. J. Math.* **71** (2020), no. 2, 637–648.

[88] M. A. Korkine and G. Zolotareff, Sur les formes quadratiques. *Math. Ann.* **6** (1873), 279–284.

[89] R. O. Kuzmin, On a problem of Gauss. *Atti Congr. Int. Mat., Bologna* **6** (1932), 83–89.

[90] S. P. Lalley, Renewal theorems in symbolic dynamics, with applications to geodesic flows, non-Euclidean tessellations and their fractal limits. *Acta Math.* **163** (1989), no. 1–2, 1–55.

[91] M. J. Larsen and R. Pink, Finite subgroups of algebraic groups. *J. Amer. Math. Soc.* **24** (2011).

[92] M. Laurent, Exponential diophantine equations. *C. R. Acad. Sci.* **296** (1983), 945–947.

[93] P. D. Lax and R. S. Phillips, The asymptotic distribution of lattice points in Euclidean and non-Euclidean spaces. *J. Funct. Anal.* **46** (1982), no. 3, 280–350.

[94] D. H. Lehmer, On the Diophantine equation $x^3 + y^3 + z^3 = 1$. *J. Lond. Math. Soc.* **s1-31** (1956), no. 3, 275–280.

[95] G. Levitt, La dynamique des pseudogroupes de rotations. *Invent. Math.* **113** (1993), no. 3, 633–670.

[96] O. Lisovyy and Y. Tykhyy, Algebraic solutions of the sixth Painlevé equation. *J. Geom. Phys.* **85** (2014), 124–163.

[97] C. Liverani, Decay of correlations. *Ann. of Math. (2)* **142** (1995), no. 2, 239–301.

[98] D. Loughran and V. Mitankin, Integral Hasse principle and strong approximation for Markoff surfaces. *Int. Math. Res. Not. IMRN* **18** (2021), 14086–14122.

[99] A. Lubotzky, Cayley graphs: Eigenvalues, expanders and random walks. In *Surveys in Combinatorics*, edited by P. Rowbinson, London Math. Soc. Lecture Note Ser. 218, Cambridge University Press, 1995.

[100] A. Lubotzky and I. Pak, The product replacement algorithm and Kazhdan's property (T). *J. Amer. Math. Soc.* **14** (2001).

[101] A. Lubotzky and B. Weiss, Groups and Expanders. In *DIMACS Series in Disc. Math. and Theor. Comp. Sci.* 10, edited by J. Friedman, pp. 95–109, 1993.

[102] A. Markoff, Sur les formes quadratiques binaires indéfinies. *Math. Ann.* **15** (1879).

[103] A. Markoff, Sur les formes quadratiques binaires indéfinies. *Math. Ann.* **17** (1880).

[104] C. R. Matthews, Counting points modulo p for some finitely generated subgroups of algebraic groups. *Bull. Lond. Math. Soc.* **14** (1982), 149–154.

[105] C. Matthews, L. Vaserstein, and B. Weisfeiler, Congruence properties of Zariski dense groups. *Proc. Lond. Math. Soc.* **48** (1984), 514–532.

[106] Mazur, Barry The topology of rational points. *Experiment. Math.* **1** (1992), no. 1, 35–45.

[107] D. McCullough and M. Wanderley, Nielsen equivalence of generating pairs in $SL(2, q)$. *Glasg. Math. J.* **55** (2013), 481–509.

[108] C. T. McMullen, Dynamics on K3 surfaces: Salem numbers and Siegel disks. *J. Reine Angew. Math.* **545** (2002), 201–233

[109] C. T. McMullen, Automorphisms of projective K3 surfaces with minimum entropy. *Invent. Math.* **203** (2016), no. 1, 179–215

[110] C. Meiri and D. Puder, The Markoff group of transformations in prime and composite moduli. *Duke Math. J.* **167** (2018), no. 14, 2679–2720.

[111] M. Mirzakhani, Counting mapping class group orbits on hyperbolic surfaces. 2016, arXiv:1601.03342v1.

[112] L. J. Mordell, On the integer solutions of the equation $x^2 + y^2 + z^2 + 2xyz = n$. *J. Lond. Math. Soc.* **28** (1953), 500–510.

[113] L. J. Mordell, *Diophantine equations*. Academic Press, London, New York, 1969.

[114] H. Oh, Dynamics on geometrically finite hyperbolic manifolds with applications to Apollonian circle packings and beyond. In *Proceedings of the International Congress of Mathematicians* III, pp. 1308–1331, Hindustan Book Agency, New Delhi, 2010.

[115] H. Oh and D. Winter, Uniform exponential mixing and resonance free regions for convex cocompact congruence subgroups of $\mathrm{SL}_2(\mathbb{Z})$. *J. Amer. Math. Soc.*, **29** (2016), 1069–1115

[116] N. Ozawa, $\mathrm{Aut}(F_5)$ has property (T). *J. Inst. Math. Jussieu* **15** (2016), no. 1, 85–90.

[117] S. J. Patterson, The limit set of a Fuchsian group. *Acta Math.* **136** (1976), no. 3–4, 241–273.

[118] V. Platonov and A. Rapinchuk, *Algebraic groups and number theory*. Pure Appl. Math. 139, Academic Press Inc., Boston, MA, 1994.

[119] M. Pollicott, Zeta functions for Anosov flows. In *Proceedings of the International Congress of Mathematicians—Seoul 2014. Vol. III*, pp. 661–681, Kyung Moon Sa, Seoul, 2014.

[120] L. Pyber and E. Szabo, Growth in finite simple groups of Lie type. *J. Amer. Math. Soc.* **29** (2016), no. 1, 95–146.

[121] R. Remak, Über indefinite binare quadratische Minimalformen. *Math. Ann.* **92** (1924), 155–182.

[122] A. Salehi Golsefidy, Super-approximation, I: $\mathfrak{p}$-adic semisimple case. *Int. Math. Res. Not. IMRN* **23** (2017), 7190–7263.

[123] A. Salehi Golsefidy and P. Sarnak, The affine sieve. *J. Acad. Mark. Sci.* **4** (2013), 1085–1105.

[124] A. Salehi Golsefidy and P. P. Varju, Expansion in perfect groups. *Geom. Funct. Anal.* **22** (2012).

[125] P. Sarnak, What is an expander? *Not. Amer. Math. Soc.* **51** (2004), 762–763.

[126] P. Sarnak, Affine sieve lecture slides, 2010, http://publications.ias.edu/sarnak/paper/508.

[127] P. Sarnak, Integral Apollonian packings. *Amer. Math. Monthly* **118** (2011), no. 4, 291–306.

[128] P. C. Sarnak, Diophantine problems and linear groups. In *Proceedings of the International Congress of Mathematicians, Vol. I, II (Kyoto, 1990)*, pp. 459–471, Math. Soc. Japan, Tokyo, 1991.

[129] P. Sarnak and X. X. Xue, Bounds for multiplicities of automorphic representations. *Duke Math. J.* **64** (1991), no. 1, 207–227.

[130] H. Schwartz and H. T. Muhly, On a class of cubic Diophantine equations. *J. Lond. Math. Soc.* **32** (1957), 379–382.

[131] A. Selberg, On the estimation of Fourier coefficients of modular forms. *Proc. Sympos. Pure Math.* **VII** (1965), 1–15.

[132] C. Series, The geometry of Markoff numbers. *Math. Intelligencer* **7** (1985), no. 3, 20–29.

[133] Y. Shalom, The algebraization of Kazhdan's property (T). In *International Congress of Mathematicians. Vol. II*, pp. 1283–1310, Eur. Math. Soc., Zürich, 2006.

[134] J. H. Silverman, The Markoff equation $X^2 + Y^2 + Z^2 = aXYZ$ over quadratic imaginary fields. *J. Number Theory* **35** (1990), no. 1, 72–104.

[135] J. H. Silverman, Rational points on K3 surfaces: a new canonical height. *Invent. Math.* **105** (1991), no. 2, 347–373.

[136] J. H. Silverman, Counting integer and rational points on varieties. In *Columbia University Number Theory Seminar*, pp. 4, 223–236, Astérisque 228, Math. Soc. France, Paris, 1992.

[137] S. A. Stepanov, The number of points of a hyperelliptic curve over a prime field. *Math. USSR, Izv.* **3** (1969), no. 5, 1103–1114.

[138] D. Sullivan, The density at infinity of a discrete group of hyperbolic motions. *Publ. Math. Inst. Hautes Études Sci.* **50** (1979), 171–202.

[139] D. Sullivan, Entropy, Hausdorff measures old and new, and limit sets of geometrically finite Kleinian groups. *Acta Math.* **153** (1984), no. 3–4, 259–277.

[140] T. Tao, *Expansion in finite simple groups of Lie type*. Grad. Stud. Math. 164, American Mathematical Society, Providence, RI, 2015.

[141] P. A. Vojta, generalization of theorems of Faltings and Thue–Siegel–Roth–Wirsing. *J. Amer. Math. Soc.* **25** (1992).

[142] A. Weil, On the Riemann Hypothesis in function fields. *Proc. Natl. Acad. Sci. USA* **27** (1941), 345–347.

[143] B. Weisfeiler, Strong approximation for Zariski-dense subgroups of semisimple algebraic groups. *Ann. of Math. (2)* **120** (1984), no. 2, 271–315.

[144] J. P. Whang, Arithmetic of curves on moduli of local systems. *Algebra Number Theory* **14** (2020), no. 10, 2575–2605.

[145] J. P. Whang, Global geometry on moduli of local systems for surfaces with boundary. *Compos. Math.* **156** (2020), no. 8, 1517–1559.

[146] J. P. Whang, Nonlinear descent on moduli of local systems. *Israel J. Math.* **240** (2020), no. 2, 935–1004.

[147] D. Zagier, On the number of Markoff numbers below a given bound. *Math. Comp.* **39** (1982), no. 160, 709–723.

ALEXANDER GAMBURD

The Graduate Center, CUNY, New York, NY, USA, agamburd@gmail.com

This Page Intentionally Left Blank

THE NUMBER OF RATIONAL POINTS ON A CURVE OF GENUS AT LEAST TWO

PHILIPP HABEGGER

ABSTRACT

The Mordell Conjecture states that a smooth projective curve of genus at least 2 defined over number field F admits only finitely many F-rational points. It was proved by Faltings in the 1980s and again using a different strategy by Vojta. Despite there being two different proofs of the Mordell Conjecture, many important questions regarding the set of F-rational points remain open. This survey concerns recent developments towards upper bounds on the number of rational points in connection with a question of Mazur.

MATHEMATICS SUBJECT CLASSIFICATION 2020

Primary 14G05; Secondary 11G30, 11G50, 14G25, 14K15

KEYWORDS

Mordell conjecture, smooth projective curves, rational points, Néron–Tate height, Weil height, family of abelian varieties

ICM INTERNATIONAL CONGRESS OF MATHEMATICIANS 2022 JULY 6–14

Proc. Int. Cong. Math. 2022, Vol. 3, pp. 1838–1869
DOI 10.4171/ICM2022/79

Published by EMS Press

1. INTRODUCTION

Mordell's Conjecture asserts the finiteness of the set of rational solutions

$$\{(x, y) \in \mathbb{Q}^2 : P(x, y) = 0\}$$

for certain bivariate polynomials $P \in \mathbb{Q}[X, Y]$.

To make the statement and results precise, we will adopt the language of projective algebraic curves. Indeed, for the study of the zero set, we may assume that P is irreducible, even as a polynomial in $\mathbb{C}[X, Y]$. Moreover, its homogenization defines a projective curve in the projective plane. The classical procedure of normalization allows us to resolve any singularities. The result is an irreducible smooth projective curve defined over $\mathbb{Q}$. Its complex points define a compact Riemann surface of genus $g \in \{0, 1, 2, \ldots\}$.

Conversely, let us assume we are presented with a smooth projective curve C defined over $\mathbb{Q}$ that is irreducible as a curve taken over $\mathbb{C}$. The genus g of $C(\mathbb{C})$ taken as a Riemann surface has important consequences for arithmetic questions on $C(\mathbb{Q})$. Indeed, Mordell's Conjecture, proved by Faltings [25], states that $\#C(\mathbb{Q})$ is finite if $g \geq 2$.

We begin by formulating the Mordell Conjecture in slightly higher generality. We then discuss the history of results towards this conjecture. Finally, we give an overview of the proof of a joint work by Ziyang Gao, Vesselin Dimitrov, and the author towards a question of Mazur regarding upper bounds for the cardinality $\#C(\mathbb{Q})$. The upper bound will depend on the genus g and the Mordell–Weil rank of the Jacobian of C. For a special case of this result that does not make reference to Jacobians, we refer to Section 6.

1.1. The Mordell Conjecture

We begin by recalling Faltings's Theorem [25], a finiteness statement originally conjectured by Mordell [48]. By a curve we mean a geometrically irreducible projective variety of dimension 1. Throughout, we let F denote a number field and $\overline{F}$ a fixed algebraic closure of F.

Theorem 1.1 (Faltings [25]). *Let C be a smooth curve of genus at least 2 defined over a number field F. Then $C(F)$ is finite.*

If the genus of C is small, then one cannot expect finiteness. Indeed, the set $C(F)$ is nonempty after replacing F by a suitable finite extension. If C has genus 0, then C is isomorphic to the projective line and thus $C(F)$ is infinite. If C has genus 1, then C together with a point in $C(F)$ is an elliptic curve. In particular, we obtain an algebraic group. After possibly extending F again, we may assume that $C(F)$ contains a point of infinite order. So $C(F)$ is infinite.

To prove the Mordell Conjecture, Faltings first proved the Shafarevich Conjecture for abelian varieties. At the time, the latter was known to imply the Mordell Conjecture thanks to a construction of Kodaira–Parshin.

Later, Vojta [62] gave a different proof of the Mordell Conjecture that is rooted in diophantine approximation. Bombieri [8] then simplified Vojta's proof. We will recall Vojta's approach for curves in Section 3. The technical heart is the Vojta inequality which we formulate below as Theorem 3.1.

Faltings generalized Vojta's proof of the Mordell Conjecture to cover subvarieties of any dimension of an abelian variety. Indeed, Faltings [26,27] and Hindry [36] proved the Mordell–Lang Conjecture for subvarieties of abelian varieties. Let A be an abelian variety defined over F and suppose Γ is a subgroup of $A(\overline{F})$. The division closure of Γ is the subgroup

$$\{P \in A(\overline{F}) : \text{there exists an integer } n \geq 1 \text{ with } nP \in \Gamma\}$$

of $A(\overline{F})$. For example, the division closure of the trivial subgroup $\Gamma = \{0\}$ is the subgroup A_{tors} of all points of finite order of $A(\overline{F})$. The following theorem holds for all base fields of characteristic 0.

Theorem 1.2 (Mordell–Lang conjecture, Faltings, Hindry). *Let A be an abelian variety defined over F and let Γ be the division closure of a finitely generated subgroup of $A(\overline{F})$. If V is an irreducible closed subvariety of A, then the Zariski closure of $V(\overline{F}) \cap \Gamma$ in V is a finite union of translates of algebraic subgroups of A.*

The special case when $\Gamma = A_{\text{tors}}$ is called the Manin–Mumford Conjecture and was proved by Raynaud [53].

More recently, Lawrence and Venkatesh [41] gave yet another proof of the Mordell Conjecture. It was inspired by Faltings's original approach and the method of Chabauty–Kim. We refer to the survey [6] on these developments.

In this survey we concentrate mainly on the case of curves and comment on possible extensions to the higher dimensional case.

1.2. Some remarks on effectivity

Despite the variety of approaches to the Mordell Conjecture, no *effective* proof is known. For example, if the curve C is presented explicitly as the vanishing locus of homogeneous polynomial equations with rational coefficients, say, then in full generality we know no algorithm that produces the finite list of rational points of C. The question of effectivity is already open in genus 2, for example, for the family $Y^2 = X^5 + t$ parametrized by t. Proving an effective version of the Mordell Conjecture is among the most important outstanding problems in diophantine geometry.

Although no general algorithm that determines the set of rational points is currently known, it is sometimes possible to determine the set of rational points. For example, we refer to the Chabauty–Coleman method [13,15] which provides a clean upper bound for the number of rational points subject to a hypothesis on the Mordell–Weil rank of the Jacobian of C. In several applications, this bound equals a lower bound for the number of rational points coming from a list of known rational points. Moreover, aspects of Kim's generalization of the Chabauty method were used by Balakrishnan, Dogra, Müller, Tuitman, and Vonk [5] to compute all rational points of the split Cartan modular curve of level 13 which appears in relation to Serre's uniformity question. A different approach motivated by work of Demjanenko and the theory of unlikely intersections was developed in a program by Checcoli, Veneziano, Viada [14]. Here too a condition on the rank of the curve's Jacobian is required for the method to apply. An remarkable aspect to this approach is that the authors obtain an explicit upper bound for the height of a rational point.

1.3. The number of rational points: conjectures and results

Given C and F as in Theorem 1.1, which invariants of C need to appear in an upper bound for $\#C(F)$?

Example 1.3. (i) Consider the hyperelliptic curve C presented by

$$y^2 = (x-1)\cdots(x-2022).$$

Its genus equals $(2022-2)/2 = 1010$. Then C contains the rational points $(1,0),\ldots,(2022,0)$. Together with the two points at infinity, we obtain at least 2024 rational points. This example easily generalizes to higher genus. For any $g \geq 2$ and square-free $f \in \mathbb{Q}[X]$ of degree $2g+2$, the equation $y^2 = f(x)$ determines a hyperelliptic curve C of genus g. If f splits into (pairwise distinct) linear factors over $\mathbb{Q}$, then $\#C(\mathbb{Q}) \geq 2g+4$. So any upper bound for $\#C(\mathbb{Q})$ must depend on the genus.

This lower bound is far from the truth. Stoll discovered a genus 2 curve defined over $\mathbb{Q}$ with 642 rational points in a family of such curves constructed by Elkies. Mestre showed that for all $g \geq 2$ there is a smooth curve of genus g defined over $\mathbb{Q}$ with at least $8g+16$ rational points.

(ii) Let us now fix the curve C and let the number field F vary. We take C as the genus 2 hyperelliptic curve presented by $y^2 = x^5 + 1$. Consider an integer $n \geq 0$ and the points $\{(m, \pm(m^5+1)^{1/2}) : m \in \{0,\ldots,n\}\}$. So $C(F)$ has at least $2n+2+1$ elements where $F = \mathbb{Q}((m^5+1)^{1/2})_{m\in\{1,\ldots,n\}}$. Any upper bound $C(F)$, even for C fixed, must depend on F.

Gabriel Dill pointed out that the number of F-points grows at least logarithmically in the degree $[F:\mathbb{Q}]$ in this case. Indeed, $[F:\mathbb{Q}] \leq 2^n$, so $\#C(F) \geq 2n \geq 2(\log[F:\mathbb{Q}])/\log 2$.

Let us consider the modular curve $X_0(37)$ which has genus 2 and is defined over $\mathbb{Q}$. Let p be one of the infinitely many prime numbers for which the Legendre symbol satisfies $(-p/37) = 1$; so 37 splits in the quadratic field $K = \mathbb{Q}(\sqrt{-p})$. Let F denote the Hilbert Class Field of K. There is an elliptic curve E defined over F with complex multiplication by the ring of integers of K. Moreover, E admits an isogeny of degree 37 to an elliptic curve defined over F. Thus $X_0(37)$ has an F-rational point. The Galois group $\mathrm{Gal}(F/K)$ acts on the F-rational points of $X_0(37)$. It is also known to act transitively on the moduli of elliptic curves with the same endomorphism ring as E. Thus $\#X_0(37)(F)$ is no less than $[F:K]$ which equals the class number of K by Class Field Theory. So $\#X_0(37)(F) \geq [F:K] = [F:\mathbb{Q}]/2$. By the Landau–

Siegel Theorem, $[F : \mathbb{Q}] \to \infty$ as $p \to \infty$. In particular, any upper bound for $\#X_0(37)(F)$ must grow at least linearly in $[F : \mathbb{Q}]$.

The Uniformity Conjecture by Caporaso–Harris–Mazur [11] predicts that the genus and base field of a curve are the only invariants required for a general upper bound.

Conjecture 1.4 (Caporaso–Harris–Mazur). *Let $g \geq 2$ be an integer and F a number field. There exists $c(g, F) \geq 1$ such that if C is a smooth curve of genus g defined over F, then $\#C(F) \leq c(g, F)$.*

Caporaso, Harris, and Mazur showed that the Uniformity Conjecture follows from the Weak Lang Conjecture which is an extension of the Mordell Conjecture to higher dimension. It states that if V is a smooth projective variety defined over F of positive dimension and general type, then $V(F)$ is not Zariski dense in V. Pacelli [50] showed that $\#C(F)$ is bounded from above in function of g and $[F : \mathbb{Q}]$ under the Weak Lang Conjecture after Abramovich [1] treated the case of quadratic and cubic extensions earlier. A refined version of the Weak Lang Conjecture implies, again by work of Caporaso–Harris–Mazur, that $\#C(F)$ can be bounded from above in function of the genus, if we omit finitely many F-isomorphism classes of C, see also [12] for a correction. Rémond [56] has evidence towards this stronger version of the Uniformity Conjecture. Alpoge [2] showed that, on average, a smooth curve of genus 2 defined over $\mathbb{Q}$ has a uniformly bounded number of rational points.

Mazur [46] posed the following question, which is a weaker version of the Uniformity Conjecture. We let $\mathrm{Jac}(C)$ denote the Jacobian of a smooth curve C defined over a field. Then $\mathrm{Jac}(C)$ is a principally polarized abelian variety whose dimension equals the genus of C. If the base field is a number field F, then $\mathrm{Jac}(C)(F)$ is a finitely generated abelian group by the Mordell–Weil Theorem.

Question 1.5 (Mazur [46, P. 223]). Let $g \geq 2$ and r be integers and let F be a number field. There exists $c(g, r, F) \geq 1$ such that if C is a smooth curve of genus g defined over F such that the rank of the Mordell–Weil group satisfies $\mathrm{rk}\,\mathrm{Jac}(C)(F) \leq r$, then $\#C(F) \leq c(g, r, F)$.

Let us review some work on upper bounds for $\#C(F)$. Parshin [59] showed how to extract an upper bound for the number of rational points from Faltings's theorem. In his original paper, Vojta [62] gave a blueprint on how to bound from above the number of rational points for a general C. This bound was refined by Bombieri [8] and de Diego [19]. However, these works did not provide an answer to Mazur's question.

To formulate de Diego's results we need some additional notation. We also require the Weil height on projective space and the Néron–Tate (or canonical) height, they are both defined in Section 2. Let S be an irreducible, smooth, quasiprojective variety defined over a number field F and presented with an immersion $S \subseteq \mathbb{P}^n$ defined over F. De Diego's Theorem holds for a smooth family of curves parametrized by the base S. Indeed, let $\mathscr{C} \to S$ be a smooth morphism such that each fiber is a smooth curve of genus $g \geq 2$. We write $\mathscr{C}_s$ for the fiber of $\mathscr{C} \to S$ above $s \in S(\overline{F})$. This is a smooth curve defined over $\overline{F}$. Let $\mathcal{K}_s$ denote the canonical class on $\mathscr{C}_s$, we identify it with a divisor class modulo linear equivalence of degree $2g-2$. If $P \in \mathscr{C}_s(\overline{F})$, then $(2g-2)[P] - \mathcal{K}_s$ is well defined as a divisor class of degree 0. So it represents a point in $\mathrm{Jac}(\mathscr{C}_s)$. In this way we obtain a morphism

$$j_s : \mathscr{C}_s \to \mathrm{Jac}(\mathscr{C}_s) \quad \text{given by } P \mapsto \big((2g-2)[P] - \mathcal{K}_s\big)/\sim .$$

Let θ_s denote the theta divisor on $\mathrm{Jac}(\mathscr{C}_s)$ and $\hat{h}_s = \hat{h}_{\mathscr{C}_s,\theta_s}$ the canonical height on $\mathrm{Jac}(\mathscr{C}_s)$ attached to this divisor.

Theorem 1.6 (de Diego [19]). *There exists $c(\mathscr{C}) > 1$ such that if F'/F is a finite extension and $s \in S(F')$, then*

$$\#\big\{P \in \mathscr{C}_s(F') : \hat{h}_s\big(j_s(P)\big) \geq c(\mathscr{C})\big(1 + h(s)\big)\big\} \leq \frac{55}{2} \cdot 7^{\mathrm{rk}\,\mathrm{Jac}(C)(F')}.$$

Roughly speaking, this theorem tells us that the number of points of $\mathscr{C}_s$ of sufficiently large canonical height is bounded as in Mazur's question. We will often call these points *large points*. It is striking that the constant 7 is admissible for all genera; a fact that already appeared in Bombieri's work [8]. For smooth curves of genus 2 defined over $\mathbb{Q}$ with a marked Weierstrass point, Alpoge [2] improved 7 to 1.872.

Observe that

$$\big\{P \in \mathscr{C}_s(F') : \hat{h}_s\big(j_s(P)\big) < c(\mathscr{C})\big(1 + h(s)\big)\big\} \tag{1.1}$$

is finite by the Northcott property for height functions which we will review in Section 2. To obtain a positive answer to Mazur's question we need, roughly speaking, to get a similar bound as in Theorem 1.6 for the cardinality of (1.1). There are quantitative versions of Northcott's Theorem. Estimating the cardinality (1.1) with these does, however, introduce a dependence on $h(s)$.

Work of David–Philippon [17] and Rémond [54] further clarified the other value $c(\mathscr{C})$. Indeed, David and Philippon proved a lower bound for the canonical height that, when combined with Rémond's explicit version of the Vojta inequality, yields the next theorem. To state their result, we momentarily shift our focus from families of smooth curves and their Jacobians to a curve immersed in an abelian variety.

Theorem 1.7 (Rémond [17, P. 643]). *Let A be a g-dimensional principally polarized abelian variety defined over F. Let Γ be the division closure of a finitely generated subgroup of $A(\overline{F})$ of rank r and let $C \subseteq A$ be a curve that is not smooth of genus 1. Then $C(\overline{F}) \cap \Gamma$ is finite of cardinality at most*

$$\big(2^{34} h_0(A) \deg C\big)^{g^{20}(r+1)}.$$

Here $\deg C$ is the degree of C with respect to the principal polarization. Moreover, $h_0(A)$ is a height of the abelian variety A whose definition involves classical theta functions and the degree $[F : \mathbb{Q}]$. Roughly speaking, $h_0(A)$ encodes a bound for the coefficients needed to reconstruct the abelian variety A. Mazur's question does not allow for a dependence on $h_0(A)$. The hypothesis that C is not smooth of genus 1 is natural and cannot be dropped in general. It is equivalent to stating that C is not a translate of an algebraic subgroup of A.

David and Philippon's contribution to Theorem 1.7 was their lower bound for the canonical height, see [17, THÉORÈME 1.4]. Rémond [54] made Vojta's approach (and an inequality of Mumford) completely explicit. David–Philippon and Rémond have a result for subvarieties of A of any dimension. In other words, they provide an explicit version of the Mordell–Lang Conjecture.

David and Philippon's approach to Mazur's question and its higher-dimensional counterparts is via a strong quantitative version of the Bogomolov Conjecture on points of small height. A suitable version is Conjecture 1.5 [18] where the lower bound for the canonical height grows linearly in the Faltings height. We refer to [18, THÉORÈME 1.11] regarding the connection to rational points and more generally the Mordell–Lang Conjecture.

David and Philippon were able to strengthen their height lower bound when A is a power of an elliptic curve. This provided more evidence towards a positive answer for Mazur's question. Here is a version of their result for curves; their general result holds for subvarieties of a power of an elliptic curve.

Theorem 1.8 (David and Philippon [18, THÉORÈME 1.13]). *Let E be an elliptic curve defined over F and let $g \geq 2$ be an integer. Suppose Γ is the division closure of a finitely generated subgroup of $E^g(\overline{F})$ of rank r. If $C \subseteq E^g$ is a curve that is not smooth of genus 1, then $\#C(\overline{F}) \cap \Gamma \leq \deg(C)^{7g^{18}(1+r)}$.*

Thanks to a specialization argument, David and Philippon extended the above result to include the case where F is an arbitrary field of characteristic 0. David, Nakamaye, and Philippon [16] then proved the existence of a $(g-2)$-dimensional family of curves of genus g for which Mazur's question has a positive answer.

We now very briefly turn to some cardinality estimates using the Chabauty–Coleman method, which is based on p-adic analysis. It can produce finiteness of $C(F)$ with a clean cardinality estimate subject to a restriction on the rank of the Mordell–Weil group.

Theorem 1.9 (Coleman [15]). *Suppose C is a smooth curve of genus $g \geq 2$ defined over $\mathbb{Q}$ with $\operatorname{rk}\operatorname{Jac}(C)(\mathbb{Q}) \leq g-1$. If $p > 2g$ is a prime number where C has good reduction $\tilde{C}$, then $\#C(\mathbb{Q}) \leq 2g - 2 + \#\tilde{C}(\mathbb{F}_p)$.*

In combination with the Hasse–Weil bound $\#\tilde{C}(\mathbb{F}_p) \leq p + 1 + 2g\sqrt{p}$, the estimate above yields a bound for $\#C(\mathbb{Q})$ in terms of g and p alone. Observe that a dependence in the arithmetic of C appears through the prime p. Stoll was able to remove this dependence for hyperelliptic curves at the cost of a stronger restriction on the rank of the Mordell–Weil group.

Theorem 1.10 (Stoll [58]). *Let $g \geq 2$ and $d \geq 1$ be integers. There exists $c(g,d) > 0$ with the following property. Suppose C is a smooth hyperelliptic curve of genus g defined over F with $[F:\mathbb{Q}] \leq d$. If $\operatorname{rk}\operatorname{Jac}(C)(\mathbb{Q}) \leq g-3$, then $\#C(F) \leq c(g,d)$.*

Later, Katz, Rabinoff, and Zureick-Brown dropped the hyperellipticity condition.

Theorem 1.11 (Katz, Rabinoff, and Zureick-Brown [38]). *Let $g \geq 2$ and $d \geq 1$ be integers. There exists $c(g,d) > 0$ with the following property. Suppose C is a smooth curve of genus g defined over F with $[F:\mathbb{Q}] \leq d$. If $\operatorname{rk}\operatorname{Jac}(C)(\mathbb{Q}) \leq g-3$, then $\#C(F) \leq c(g,d)$.*

After this detour to the Chabauty–Coleman method, we return to Vojta's method. Vesselin Dimitrov, Ziyang Gao, and the author have recently proved a lower bound for the canonical height that can be used as a replacement for the lower bounds by David and Philippon [17, 18] in the context of Mordell's Conjecture. We recall this height inequality in Section 4.2 below. Indeed, it led to a positive answer to a strengthening of Mazur's question. The following result is new already in genus 2.

Theorem 1.12 (Dimitrov, Gao, and Habegger [24, THEOREM 1.1]). *Let $g \geq 2$ and $d \geq 1$ be integers, there exist $c'(g,d) > 1$ and $c(g,d) > 1$ with the following property. Suppose C is a smooth curve of genus g defined over a number field F such that $[F:\mathbb{Q}] \leq d$. Then*

$$\#C(F) \leq c'(g,d) \cdot c(g,d)^{\operatorname{rk}\operatorname{Jac}(C)(F)}.$$

Regarding the Caporaso–Harris–Mazur Uniformity Conjecture, we ask

Question 1.13. Can the cardinality $\#C(F)$ be bounded from above by a function that is polynomial in $[F:\mathbb{Q}]$ and g?

No one currently knows an algorithm that computes the rank of the Mordell–Weil group $\operatorname{Jac}(C)(F)$. However, upper bounds for this rank follow, for example, from the Ooe–Top Theorem [49]. We discuss this in more depth in Section 6.

Our results also cover points on C that lie in the division closure of a finitely generated subgroup. Let $\overline{\mathbb{Q}}$ denote the algebraic closure of $\mathbb{Q}$ in $\mathbb{C}$. The Jacobian $\operatorname{Jac}(C)$ of a smooth curve C of genus g defined over $\overline{\mathbb{Q}}$ corresponds to a $\overline{\mathbb{Q}}$-point of the coarse moduli space $\mathbb{A}_g$ of g-dimensional principally polarized abelian varieties. We let $[\operatorname{Jac}(C)]$ denote the point of $\mathbb{A}_g(\overline{\mathbb{Q}})$ corresponding to $\operatorname{Jac}(C)$ with its canonical principal polarization.

For example, if $g=1$ then $\mathbb{A}_g = \mathbb{A}^1$ is the affine line. If E is an elliptic curve defined over $\overline{\mathbb{Q}}$, then $[E]$ is the j-invariant of E.

In general, $\mathbb{A}_g$ is a quasiprojective variety of dimension $g(g+1)/2$ defined over $\mathbb{Q}$. We may fix an immersion $\iota : \mathbb{A}_g \hookrightarrow \mathbb{P}^n$ into projective space. Then the absolute logarithmic Weil height h, see Section 2 for a definition, pulls back to a function $h \circ \iota : \mathbb{A}_g(\overline{\mathbb{Q}}) \to [0,\infty)$.

If C is a smooth curve of genus $g \geq 1$ defined over $\overline{\mathbb{Q}}$ and if $P_0 \in C(\overline{\mathbb{Q}})$, then a point $P \in C(\overline{\mathbb{Q}})$ defines a divisor $[P]-[P_0]$ of degree 0. One obtains a closed immersion

$$C \hookrightarrow \operatorname{Jac}(C) \quad \text{from } P \mapsto ([P]-[P_0])/\sim$$

where $\sim$ again denotes linear equivalence, induces a closed immersion. We will write $C - P_0$ for the image of C in $\operatorname{Jac}(C)$.

Theorem 1.14 (Dimitrov, Gao, and Habegger [24, THEOREM 1.2]). *Let $g \geq 2$ be an integer. There exist $c(g,\iota) > 1$, $c'(g,\iota) > 0$, and $c''(g,\iota) > 0$ that depend on g and the immersion ι with the following property. Suppose C is a smooth curve of genus g defined over $\overline{\mathbb{Q}}$ and let*

$P_0 \in C(\overline{\mathbb{Q}})$. *Let* Γ *be the division closure of a finitely generated subgroup of* $\mathrm{Jac}(C)(\overline{\mathbb{Q}})$ *of rank* r. *If*

$$h\big(\iota([\mathrm{Jac}(C)])\big) \geq c''(g,\iota) \quad \textit{then}\ \#(C - P_0)(\overline{\mathbb{Q}}) \cap \Gamma \leq c'(g,\iota)c(g,\iota)^r.$$

In particular, we may take $\Gamma = \mathrm{Jac}(C)_{\mathrm{tors}}$ and $r = 0$. Thus the theorem yields a uniform bound for the number of torsion points that lie on $C - P_0$ if the height of $\iota([\mathrm{Jac}(C)])$ is sufficiently large.

Suppose that C is defined over a number field F. Then $[\mathrm{Jac}(C)]$ is an F-rational point of the moduli space $\mathbb{A}_g$. If we impose also $h(\iota([\mathrm{Jac}(C)])) < c''(g,\iota)$, then $[\mathrm{Jac}(C)]$ lies in a finite set by the Northcott property. Thus $\mathrm{Jac}(C)$ is in one of at most finitely many $\overline{\mathbb{Q}}$-isomorphism classes and so is C by the Torelli Theorem.

Raynaud proved the following result which is the Manin–Mumford Conjecture for curves.

Theorem 1.15 (Raynaud [52]). *Let* C *be smooth curve defined over* $\mathbb{C}$ *of genus at least* 2. *Then* $(C - P_0) \cap \mathrm{Jac}(C)_{\mathrm{tors}}$ *is finite.*

Theorem 1.14 gives evidence towards the Uniform Manin–Mumford Conjecture which states that $(C - P_0) \cap \mathrm{Jac}(C)_{\mathrm{tors}}$ is bounded from above in terms of the genus g only for any smooth curve C of genus $g \geq 2$ defined over any field in characteristic 0.

Using a different approach involving equidistribution and motivated by dynamical systems, DeMarco, Krieger, and Ye [20] had made substantial progress towards the Uniform Manin–Mumford Conjecture. They proved it for smooth curves of genus 2 defined over $\mathbb{C}$ that are double covers of an elliptic curve when the base point P_0 is a Weierstrass point.

In a preprint, Kühne [39] complemented the method in [24] using ideas from equidistribution to prove the Uniform Manin–Mumford Conjecture.

Theorem 1.16 (Kühne [39]). *Let* $g \geq 2$ *be an integer. There exist* $c(g) > 1$ *and* $c'(g) > 1$ *that depend on* g *with the following property. Suppose* C *is a smooth curve of genus* g *defined over* $\mathbb{C}$ *and let* $P_0 \in C(\mathbb{C})$. *Let* Γ *be the division closure of a finitely generated subgroup of* $\mathrm{Jac}(C)(\mathbb{C})$ *of rank* r. *Then* $\#(C - P_0)(\mathbb{C}) \cap \Gamma \leq c'(g)c(g)^r$.

In contrast to Theorem 1.14, Kühne is able to handle curves C defined over $\overline{\mathbb{Q}}$ for which $[\mathrm{Jac}(C)]$ has height less than $c''(g,\iota)$. Once uniformity is established for all curves over $\overline{\mathbb{Q}}$, Kühne is able to pass to the base field $\mathbb{C}$ using a specialization argument laid out by Dimitrov, Gao, and the author [22] which relies on a result of Masser [43]. Kühne thus answers an older question of Mazur, see the top of page 234 [45], and obtains the full Mordell–Lang variant for curves.

DeMarco and Mavraki's [21] work on a relative version of the Bogomolov conjecture, see [72] and [22], motivated Kühne [39,40] to extend the reach of Arakelovian equidistribution methods of Szpiro–Ullmo–Zhang [60] and Yuan [65] to families of abelian varieties over a quasiprojective base. For algebraic curves, this settles the uniform Bogomolov and the uniform Manin–Mumford conjectures.

Yuan [66] recently gave another proof of Theorem 1.16. His method also runs via a uniform Bogomolov theorem and thus contains aspects related to height lower bounds. However, Yuan's approach relies on arithmetic bigness, rather than on equidistribution. It is independent of the approaches mentioned above and uses a new theory of adelic line bundles over quasiprojective varieties developed by Yuan and Zhang [67] which generalizes Zhang's theory [70] in the projective case. They derive a height inequality for a polarized dynamical system, see Theorem 1.3.2 and Section 6 [67], that extends our own bound. One aspect of Yuan's method is that it works for global fields in any characteristic.

We come to some questions regarding the base constant $c(g)$ in the estimates above. In the context of Mordell's Conjecture, Bombieri observed that the number of large points is bounded by a multiple of $7^{\operatorname{rk} \operatorname{Jac}(C)(F)}$.

Question 1.17. Can the base 7 in the estimate for the number of large points as in Theorem 1.6 be replaced by a function in g that tends to 1 for $g \to \infty$?

Alpoge [2] used the Kabatiansky–Levenshtein estimates on spherical codes to improve on the constant 7 in genus 2. It is quite possible that Alpoge's approach will shed light on this last question.

Concerning the constant $c(g)$ in Theorem 1.16, we pose the following two questions which also cover the moderate, i.e., nonlarge, points. They were inspired by questions of Helfgott.

Question 1.18. Can we choose the $c(g)$ in Theorem 1.16 such that there exists $B \geq 1$ with $c(g) \leq B$ for all integers $g \geq 2$?

Question 1.19. Can we choose the $c(g)$ in Theorem 1.16 with $\lim_{g \to \infty} c(g) = 1$?

Recently, Gao, Ge, and Kühne [32] completed the proof of the Uniform Mordell–Lang Conjecture for a subvariety V of a polarized abelian variety A of any dimension. Uniformity here amounts to bounding the number of irreducible components of the Zariski closure in Theorem 1.2 from above by $c'(\dim A, \deg V) c(\dim A, \deg V)^r$. Their result holds over all base fields in characteristic 0.

We refer to the comprehensive survey by Gao [31] that gives an overview of these recent developments and how they are interlinked.

Here is a brief overview of this survey. In Section 2 we recall some fundamental properties of two height functions: the Weil and Néron–Tate heights. They play a central role in the proof of Theorem 1.12. Then in Section 3 we describe Vojta's approach to the Mordell Conjecture. Later we return to the two height functions and describe their interactions on a family of abelian varieties. This is done in Section 4. Here we also describe the Betti map, an important analytic tool. In Section 5 we sketch how all this fits together in the proof of Theorem 1.12. In the final section we discuss an estimate for the number of rational points on a hyperelliptic curve that does not make reference to Jacobians.

2. HEIGHTS

Height functions are at the heart of Vojta's proof of the Mordell Conjecture and subsequent results such as Theorem 1.12. We will review two flavors of heights. The first one is the absolute logarithmic Weil height which is defined on algebraic points of the projective space. One can also use it to define a class of height functions on a projective variety equipped with an invertible sheaf. The second height function is the canonical or Néron–Tate height on an abelian variety, also equipped with an invertible sheaf. The latter is compatible with the group structure on the abelian variety.

2.1. The absolute logarithmic Weil height

We review here briefly the main properties of the Weil height. For a thorough treatment, we refer to [9, CHAPTERS 1 AND 2] or [37, PART B].

We begin by defining the height of a rational point on projective space $\mathbb{P}^n$.

Definition 2.1. Let $P \in \mathbb{P}^n(\mathbb{Q})$. There exist projective coordinates $(x_0, \dots, x_n) \in \mathbb{Z}^{n+1} \setminus \{0\}$ of $P = [x_0 : \cdots : x_n]$ with $\gcd(x_0, \dots, x_n) = 1$. Then we set

$$h(P) = \log \max\{|x_0|, \dots, |x_n|\}.$$

The vector $(x_0, \dots, x_n)$ is uniquely determined up to a sign, and so $h(P)$ is well defined. For example, $h([2 : 4 : 6]) = h([1 : 2 : 3]) = h([1/3 : 2/3 : 1]) = \log 3$.

The following theorem is a straightforward consequence of the definition of the Weil height.

Theorem 2.2 (Northcott property). *The set* $\{P \in \mathbb{P}^n(\mathbb{Q}) : h(P) \le B\}$ *is finite for all* B.

Defining the height of an algebraic point in $\mathbb{P}^n(\overline{\mathbb{Q}})$ requires some basic algebraic number theory. Indeed, let K be a number field. We let M_K denote the set of absolute values $|\cdot| : K \to [0, \infty)$ that extend either the standard absolute value on $\mathbb{Q}$ or a p-adic absolute value for some prime p. Then M_K is called the set of places of K. For each $v \in M_K$, one sets $d_v = [K_v : \mathbb{Q}_w]$ where K_v is a completion of K with respect to v and $\mathbb{Q}_w$ is the completion of $\mathbb{Q}$ in K_v with respect to $w = v|_{\mathbb{Q}}$.

Definition 2.3. Let $P \in \mathbb{P}^n(\overline{\mathbb{Q}})$ and let K be a number field such that $P = [x_0 : \cdots : x_n]$ where $(x_0, \dots, x_n) \in K^{n+1} \setminus \{0\}$. The absolute logarithmic Weil height, or just Weil height, is

$$h(P) = \frac{1}{[K : \mathbb{Q}]} \sum_{v \in M_K} d_v \log \max\{|x_0|_v, \dots, |x_n|_v\}. \tag{2.1}$$

The normalization constants d_v are chosen such that the product formula

$$\prod_{v \in M_K} |x|_v^{d_v} = 1$$

holds for all $x \in K \setminus \{0\}$. This guarantees that the right-hand side of (2.1) is independent of the choice of projective coordinates of P. In particular, we may assume that some projective coordinate of P equals 1. Thus $h(P) \ge 0$ for all $P \in \mathbb{P}^n(\overline{\mathbb{Q}})$. Moreover, $h(P)$ is independent of the field K containing the projective coordinates.

For applications to diophantine geometry, it is useful to have a height function defined on algebraic points of an irreducible projective variety V defined over $\overline{\mathbb{Q}}$. But without additional data there is no reasonable way to define a height on $V(\overline{\mathbb{Q}})$.

However, if V is a subvariety of the projective space $\mathbb{P}^n$, then we may restrict the Weil height $h : \mathbb{P}^n(\overline{\mathbb{Q}}) \to \mathbb{R}$ to a function $V(\overline{\mathbb{Q}}) \to \mathbb{R}$. Slightly more generally, if $V \to \mathbb{P}^n$ is an immersion, then we may pull back the Weil height to $V(\overline{\mathbb{Q}})$.

Recall that an immersion $V \to \mathbb{P}^n$ is induced by a tuple of $(n+1)$ global sections of a very ample invertible sheaf on V. Conversely, given a very ample invertible sheaf $\mathcal{L}$ on V, we can fix a basis of the vector space of global sections of $\mathcal{L}$ and obtain an immersion $\iota_{\mathcal{L}} : V \to \mathbb{P}^n$. So we obtain a function $h \circ \iota_{\mathcal{L}} : V(\overline{\mathbb{Q}}) \to [0, \infty)$. There is a wrinkle here, this function depends not only on $(V, \mathcal{L})$ but also on the basis of the vector space of global sections. A different basis will lead to a function $V(\overline{\mathbb{Q}}) \to [0, \infty)$ that differs from $h \circ \iota_{\mathcal{L}}$ by a bounded function on $V(\overline{\mathbb{Q}})$. We define $h_{V,\mathcal{L}}$ to be the equivalence class of functions $V(\overline{\mathbb{Q}}) \to \mathbb{R}$ modulo bounded functions that contains $h \circ \iota_{\mathcal{L}}$.

If $\mathcal{L}$ is an ample invertible sheaf on V, then there exists an integer $n \geq 1$ such that $\mathcal{L}^{\otimes n}$ is very ample. We then define $h_{V,\mathcal{L}} = \frac{1}{n} h_{V,\mathcal{L}^{\otimes n}}$; this is again only defined up to a bounded function on $V(\overline{\mathbb{Q}})$. The equivalence class does not depend on the choice of n.

Finally, an arbitrary invertible sheaf $\mathcal{L}$ in the Picard group $\mathrm{Pic}(V)$ of V is of the form $\mathcal{F} \otimes \mathcal{M}^{\otimes(-1)}$ with $\mathcal{F}$ and $\mathcal{M}$ ample on V. The difference $h_{V,\mathcal{F}} - h_{V,\mathcal{M}}$ is well defined up to a bounded function on $V(\overline{\mathbb{Q}})$. It does not depend on the pair $\mathcal{F}, \mathcal{M}$ with difference $\mathcal{L}$, and we denote it by $h_{V,\mathcal{L}}$. It is called the Weil height attached to $(V, \mathcal{L})$.

Theorem 2.4. *Let us keep the notation above. In particular, V is an irreducible projective variety defined over $\overline{\mathbb{Q}}$.*

(i) *The association $\mathcal{L} \mapsto h_{V,\mathcal{L}}$ is a group homomorphism with target the group of real-valued maps $V(\overline{\mathbb{Q}}) \to \mathbb{R}$ modulo bounded functions.*

(ii) *For V equal to projective space and $\mathcal{L}$ the hyperplane bundle $\mathcal{O}(1)$, the Weil height from Definition 2.3 represents $h_{\mathbb{P}^n,\mathcal{O}(1)}$.*

(iii) *Suppose W is a further irreducible projective variety defined over $\overline{\mathbb{Q}}$ and $f : W \to V$ is a morphism. For all $\mathcal{L} \in \mathrm{Pic}(V)$ we have $h_{V,\mathcal{L}} \circ f = h_{W,f^*\mathcal{L}}$. As usual, this equality is understood as an equality of equivalence classes of functions.*

(iv) *Suppose $\mathcal{L} \in \mathrm{Pic}(V)$ admits a nonzero global section s. Then $h_{V,\mathcal{L}}$ is bounded from below on the complement of the vanishing locus of s. In particular, $h_{V,\mathcal{L}}$ is bounded from below on a Zariski open and dense subset of V.*

Suppose that V is defined over a number field $F \subseteq \overline{\mathbb{Q}}$ and $\mathcal{L} \in \mathrm{Pic}(V)$ is ample. Then the Northcott property holds for points of bounded degree, i.e.,

$$\left\{P \in V(\overline{F}) : h'_{V,\mathcal{L}}(P) \leq B \text{ and } \left[F(P) : F\right] \leq D\right\}$$

is finite where $h'_{V,\mathcal{L}}$ denotes any representative of $h_{V,\mathcal{L}}$.

Remark 2.5.

Let V be an irreducible projective variety defined over $\overline{\mathbb{Q}}$. We conclude this section by discussing a powerful tool to translate geometric information, here on intersection numbers, into an inequality of heights. The basic question is the following. Given invertible sheaves $\mathcal{F}$ and $\mathcal{M}$ on V, under what conditions can one bound $h_{V,\mathcal{M}}$ from above in terms of $h_{V,\mathcal{F}}$?

(i) We first consider the special case $V = \mathbb{P}^n$. As $\mathrm{Pic}(\mathbb{P}^n)$ is isomorphic to $\mathbb{Z}$, any Weil height is some integral multiple of $h_{\mathbb{P}^n,\mathcal{O}(1)}$. So $h_{V,\mathcal{F}}$ and $h_{V,\mathcal{M}}$ are $\mathbb{Z}$-linearly dependent.

(ii) Let us again suppose that V is general and that $\mathcal{F}$ is ample. Then there exists an integer $k \geq 1$ such that $\mathcal{F}^{\otimes k} \otimes \mathcal{M}^{\otimes(-1)}$ is ample. So for some positive integer $l \geq 1$ the power $\mathcal{F}^{\otimes kl} \otimes \mathcal{M}^{\otimes(-l)}$ is very ample. In particular, it admits a global section that does not vanish at a prescribed point of $V(\overline{\mathbb{Q}})$. Theorem 2.4, parts (i) and (iv), imply

$$klh_{V,\mathcal{F}} - lh_{V,\mathcal{M}} = h_{V,\mathcal{F}^{\otimes kl}\otimes\mathcal{M}^{\otimes(-l)}} \geq 0;$$

this must be parsed as an inequality between functions on $V(\overline{\mathbb{Q}})$ defined up to addition of a bounded function. We conclude

$$h_{V,\mathcal{M}} \leq kh_{V,\mathcal{F}}. \tag{2.2}$$

(iii) For some applications, such as Theorem 1.12, the ampleness hypothesis on $\mathcal{F}$ in (i) is not flexible enough. Moreover, we would like some way to estimate the factor k in (2.2) from above. We now describe a criterion of Siu that provides a solution to these two issues.

An invertible sheaf $\mathcal{L} \in \mathrm{Pic}(V)$ is called big if

$$\liminf_{k\to\infty} \frac{\dim H^0(V,\mathcal{L}^{\otimes k})}{k^{\dim V}} > 0;$$

here $H^0(V,\mathcal{L})$ denotes the vector space of global sections of $\mathcal{L}$.

If $\mathcal{L}$ is a big invertible sheaf, then $\mathcal{L}^{\otimes k}$ has a nonzero global section for some $k \geq 1$. Then using (i) and (iv) of Theorem 2.4 we see that $h_{V,\mathcal{L}} = \frac{1}{k}h_{V,\mathcal{L}^{\otimes k}}$ is bounded from below on a Zariski open and dense subset of V.

For example, if $\mathcal{L} = \mathcal{F} \otimes \mathcal{M}^{\otimes(-1)}$ is big, then, again by Theorem 2.4(i), we find $h_{V,\mathcal{F}} \geq h_{V,\mathcal{M}}$ on a Zariski open and dense subset of V.

We now come Siu's Criterion; it ensures that $\mathcal{F} \otimes \mathcal{M}^{\otimes(-1)}$ is big. An invertible sheaf $\mathcal{L} \in \mathrm{Pic}(V)$ is called nef, or numerically effective, if $(\mathcal{L} \cdot [C]) \geq 0$ for all irreducible curves $C \subseteq V$.

Siu's Criterion requires that $\mathcal{F}$ and $\mathcal{M}$ are both nef and that the intersection numbers on V satisfy $(\mathcal{F}^{\cdot \dim V}) > (\dim V)(\mathcal{F}^{\cdot(\dim V-1)} \cdot \mathcal{M})$. With these hypotheses $\mathcal{F} \otimes \mathcal{M}^{\otimes(-1)}$ is big; see **[42, THEOREM 2.2.15]**.

Say $\mathcal{F}$ and $\mathcal{M}$ are nef and $(\mathcal{F}^{\cdot \dim V}) > 0$. Let k and l be positive integers with

$$(\dim V)\frac{(\mathcal{F}^{\cdot(\dim V-1)} \cdot \mathcal{M})}{(\mathcal{F}^{\cdot \dim V})} < \frac{k}{l}$$

then $\mathcal{F}^{\otimes k} \otimes \mathcal{M}^{\otimes(-l)}$ is big. So

$$h_{V,\mathcal{M}}|_U \leq \frac{k}{l} h_{V,\mathcal{F}}|_U$$

holds on some Zariski open and dense $U \subseteq V$.

This allows us to compare the heights $h_{V,\mathcal{M}}$ and $h_{V,\mathcal{F}}$ if we have information on the intersection numbers, at least on a rather large subset of $V(\overline{\mathbb{Q}})$.

Yuan [65] proved an arithmetic version of this criterion in his work on equidistribution. The author [34] used Siu's Criterion to study unlikely intersections in abelian varieties.

2.2. The canonical height on an abelian variety

Let $F \subseteq \overline{\mathbb{Q}}$ be a number field and A an abelian variety defined over F. If $\mathcal{L}$ is an invertible sheaf on A, then we have the Weil height $h_{A,\mathcal{L}}$ from Section 2.1. Recall that $h_{A,\mathcal{L}}$ is only defined up to addition of a bounded function on $A(\overline{\mathbb{Q}})$. For abelian varieties, there is a canonical choice of function in the equivalence class $h_{A,\mathcal{L}}$ called the canonical or Néron–Tate height. A general reference for this section is [9, CHAPTER 9].

For an integer $n \in \mathbb{Z}$, let $[n]$ denote the multiplication-by-n endomorphism of A. Then $\mathcal{L}$ is called even or symmetric if there is an isomorphism $[-1]^*\mathcal{L} \cong \mathcal{L}$. It is called odd or antisymmetric if $[-1]^*\mathcal{L} \cong \mathcal{L}^{\otimes(-1)}$. If $\mathcal{L}$ is any ample invertible sheaf on A, then $\mathcal{L} \otimes [-1]^*\mathcal{L}$ is ample and even. So any abelian variety admits an even, ample invertible sheaf.

Suppose that $\mathcal{L}$ is even. Then $[2]^*\mathcal{L} \cong \mathcal{L}^{\otimes 4}$ is a consequence of the Theorem of the Cube. So Theorem 2.4 implies $h_{A,\mathcal{L}} \circ [2] = 4h_{A,\mathcal{L}}$ as classes and by iteration $h_{A,\mathcal{L}} \circ [2^k] = 4^k h_{A,\mathcal{L}}$ for all $k \geq 1$. We fix a representative $h'_{A,\mathcal{L}}$ of $h_{A,\mathcal{L}}$ and find $h'_{A,\mathcal{L}} \circ [2^k] = 4^k h'_{A,\mathcal{L}} + O_k(1)$ on $A(\overline{\mathbb{Q}})$. Tate's Limit Argument is used to show convergence in the following definition.

Definition 2.6. Let $\mathcal{L}$ be an even invertible sheaf on A and let $P \in A(\overline{\mathbb{Q}})$. Then the limit

$$\hat{h}_{A,\mathcal{L}}(P) = \lim_{k\to\infty} \frac{h'_{A,\mathcal{L}}(P)}{4^k} \tag{2.3}$$

exists and is independent of the choice of representative $h'_{A,\mathcal{L}}$ of $h_{A,\mathcal{L}}$. The real-valued function $P \mapsto \hat{h}_{A,\mathcal{L}}(P)$ is called the canonical or Néron–Tate height (on A attached to $\mathcal{L}$).

If $\mathcal{L}$ is even, then (2.3) immediately implies $\hat{h}_{A,\mathcal{L}}([2](P)) = 4\hat{h}_{A,\mathcal{L}}(P)$ for all $P \in A(\overline{\mathbb{Q}})$. If P has finite order, then $[2^m](P) = [2^n](P)$ for distinct integers $0 \leq m < n$ by the Pigeonhole Principle. Thus $\hat{h}_{A,\mathcal{L}}(P) = 0$.

There is nothing special about [2]. Indeed, one can replace [2] by $[m]$ in (2.3) for any integer $m \geq 2$; one then needs to replace 4^k in the denominator by m^{2k}.

What happens if $\mathcal{L}$ is an odd invertible sheaf? In this case, $[2]^*\mathcal{L} \cong \mathcal{L}^{\otimes 2}$. Then a similar limit (2.3) exists, but now we need to divide by 2^k.

The set of odd invertible sheaves is a divisible subgroup of $\mathrm{Pic}(A)$. From this, one can show that, after possibly extending the base field F, any invertible sheaf $\mathcal{L}$ on A decomposes as $\mathcal{L}_+ \otimes \mathcal{L}_-$ with $\mathcal{L}_+$ even and $\mathcal{L}_-$ odd. One then defines $\hat{h}_{A,\mathcal{L}} = \hat{h}_{A,\mathcal{L}_+} + \hat{h}_{A,\mathcal{L}_-}$; the decomposition of $\mathcal{L}$ is not quite unique, but this ambiguity does not affect $\hat{h}_{A,\mathcal{L}}$.

For our purposes, we often restrict to even invertible sheaves.

Let us collect the some important facts about the Néron–Tate height.

Theorem 2.7. *Let us keep the notation above. In particular, A is an abelian variety defined over a number field $F \subseteq \overline{\mathbb{Q}}$.*

(i) *Then association $\mathcal{L} \mapsto \hat{h}_{A,\mathcal{L}}$ is a group homomorphism from* $\mathrm{Pic}(V)$ *to the additive group of real-valued maps $A(\overline{\mathbb{Q}}) \to \mathbb{R}$.*

Suppose $\mathcal{L}$ is an invertible sheaf on A.

(ii) *The Néron–Tate height $\hat{h}_{A,\mathcal{L}}$ represents the Weil height $h_{A,\mathcal{L}}$.*

(iii) *If $\mathcal{L}$ is even, then the parallelogram equality*

$$\hat{h}_{A,\mathcal{L}}(P+Q) + \hat{h}_{A,\mathcal{L}}(P-Q) = 2\hat{h}_{A,\mathcal{L}}(P) + 2\hat{h}_{A,\mathcal{L}}(Q)$$

holds for all $P, Q \in A(\overline{\mathbb{Q}})$.

(iv) *If $\mathcal{L}$ is even and ample, then $\hat{h}_{A,\mathcal{L}}$ takes nonnegative values and vanishes precisely on A_{tors}.*

(v) *If $\mathcal{L}$ is even and ample, then $\hat{h}_{A,\mathcal{L}}$ induces a well-defined map $A(\overline{\mathbb{Q}}) \otimes \mathbb{R} \to [0, \infty)$. It is the square of a norm $\|\cdot\|$ on the $\mathbb{R}$-vector space $A(\overline{\mathbb{Q}}) \otimes \mathbb{R}$ and satisfies the parallelogram equality.*

The norm $\|\cdot\|$ allows us to do geometry in the $\mathbb{R}$-vector space $A(\overline{\mathbb{Q}}) \otimes \mathbb{R}$ (which is infinite dimensional if $\dim A \geq 1$). Indeed, for $z, w \in A(\overline{\mathbb{Q}}) \otimes \mathbb{R}$, we define

$$\langle P, Q \rangle = \frac{1}{2}\left(\|P+Q\|^2 - \|P\|^2 - \|Q\|^2\right).$$

Then $\langle \cdot, \cdot \rangle$ is a positive definite, symmetric, bilinear form.

By abuse of notation, we also write $\|P\|$ and $\langle P, Q \rangle$ for $P, Q \in A(\overline{\mathbb{Q}})$. In this notation we have $\langle P, P \rangle = \hat{h}_{A,\mathcal{L}}(P)$.

The Mordell–Weil Theorem implies that $A(F) \otimes \mathbb{R}$ is finite dimensional. We will see that $\|\cdot\|$ is a suitable norm to do Euclidean geometry in $A(F) \otimes \mathbb{R}$.

3. VOJTA'S APPROACH TO THE MORDELL CONJECTURE

Recall that the Mordell Conjecture was proved first by Faltings. In this section we briefly describe Vojta's approach to the Mordell Conjecture [62]. At the core is the deep Vojta inequality which we state here for a curve in an abelian variety.

Let A be an abelian variety defined over a number field $F \subseteq \overline{\mathbb{Q}}$. Let $\mathcal{L}$ be an ample and even invertible sheaf on A. We write $\|\cdot\| = \hat{h}_{A,\mathcal{L}}^{1/2}$ for the norm on $A(\overline{\mathbb{Q}}) \otimes \mathbb{R}$ defined in Theorem 2.7.

Theorem 3.1 (Vojta's inequality). *Let $C \subseteq A$ be a curve that is defined over F and that is not a translate of an algebraic subgroup of A. There are $c_1 > 1$, $c_2 > 1$, and $c_3 > 0$ with the following property. If $P, Q \in C(\overline{\mathbb{Q}})$ satisfy*

$$\langle P, Q \rangle \geq \left(1 - \frac{1}{c_1}\right) \|P\| \|Q\|$$

and

$$\|Q\| \geq c_2 \|P\|,$$

then $\|P\| \leq c_3$.

We refer also to Rémond's work [55] for a completely explicit version of Vojta's inequality.

The values c_1, c_2, c_3 depend on the curve C. One remarkable aspect is that Vojta's inequality is a statement about pairs of $\overline{\mathbb{Q}}$-points of the curve C. So all three values c_1, c_2, c_3 are "absolute," i.e., we can take them as independent of the base field F of A and C. Both c_1 and c_2 are of "geometric nature." They depend only on the degree of C with respect to $\mathcal{L}$ and other discrete data attached to A and C. In contrast, c_3 is of "arithmetic nature." Roughly speaking, it depends on suitable heights of coefficients that define the curve C in some projective embedding.

Let us now sketch a proof of Mordell's Conjecture using the Vojta inequality and the classical Mordell–Weil Theorem.

Suppose C has genus $g \geq 2$. Without loss of generality, $C(F) \neq \emptyset$. So we fix a base point $P_0 \in C(F)$, then $P \mapsto P - P_0$ induces an immersion $C \to \mathrm{Jac}(C)$. So we may assume that C is a curve inside the g-dimensional $A = \mathrm{Jac}(C)$. Note that C is not a translate of an algebraic subgroup of its Jacobian since $g \geq 2$.

We observe that $C(F) = C(\overline{\mathbb{Q}}) \cap \mathrm{Jac}(C)(F)$.

By the Northcott property, stated below Theorem 2.4, combined with Theorem 2.7(ii) we find that the "ball"

$$\{P \in C(F) : \|P\|^2 \leq B\} \tag{3.1}$$

is finite for all B.

We split the set of points $C(F)$ into two subsets:

$$\{P \in C(F) : \|P\|^2 > c_3\} \quad \textbf{(large points),}$$
$$\{P \in C(F) : \|P\|^2 \leq c_3\} \quad \textbf{(moderate points).}$$

By the finiteness statement around (3.1), it suffices to show that there are at most finitely many large points.

For any $z \in \mathrm{Jac}(C)(F) \otimes \mathbb{R}$, we define the truncated cone

$$T(z) = \{w \in \mathrm{Jac}(C)(F) \otimes \mathbb{R} : \langle z, w \rangle \geq (1 - 1/c_1)\|z\|\|w\| \text{ and } \|w\|^2 > c_3\} \subseteq \mathrm{Jac}(C)(F) \otimes \mathbb{R}.$$

By the Mordell–Weil Theorem, $\mathrm{Jac}(C)(F) \otimes \mathbb{R}$ is a finite-dimensional $\mathbb{R}$-vector space. So the unit sphere with respect to the norm $\|\cdot\|$ coming from the Néron–Tate height is compact. Therefore, $\{w \in \mathrm{Jac}(C)(F) \otimes \mathbb{R} : \|w\|^2 > c_3\}$ is covered by a finite union $T(z_1) \cup \cdots \cup T(z_N)$. Using a sphere packing argument, one can arrange that N is bounded from above by $c' \cdot c^{\mathrm{rk}\,\mathrm{Jac}(C)(F)}$ where $c' > 0$ and $c > 1$ depend only on c_1. This observation will be important for deriving uniform bounds for $\#C(F)$.

Any large point in $C(F)$ has image in some $T(z_j)$ from above. After possibly adjusting N, one can arrange that each z_j is the image of a point $P_j \in C(F)$ with $\|P_j\|^2 > c_3$ for all $j \in \{1, \ldots, N\}$. If $Q \in C(F)$ has image in $T(z_j)$, then Vojta's inequality implies $\hat{h}_{\mathcal{L}}(Q)^{1/2} = \|Q\| \le c_2\|P_j\|$. But then $Q \in C(F)$ lies in a finite ball as in (3.1). So the number of possible Q that come to lie in a single $T(z_j)$ is finite. Thus $C(F)$ is finite.

The constants c_1, c_2, and c_3 in Vojta's inequality can be made effective in terms of A and C. Yet, the proof as a whole is ineffective. Indeed, the height bound for Q depends on the hypothetical point P_j. However, there is no guarantee that P_j exists and if it does not, there is no known way to know for sure.

Using Mumford's Gap Principle, one can show that the number of large points $C(F)$ that come to lie in a single $T(z_j)$ is bounded from above by $c' \cdot c^{\mathrm{rk}\,\mathrm{Jac}(C)(F)}$, after possibly increasing the constants. Now we need to introduce dependency on c_2. But the base c will remain geometric in nature, it depends on the genus of g. But it does not depend on c_3 or other arithmetic properties of C that encode the heights of coefficients defining the said curve. Finally, as observed by Bombieri, 7 is admissible for c for any genus. Indeed, he showed that 4 is admissible for c_1.

Recall that Vojta's inequality with the same values of c_1, c_2, c_3 applies to points in $C(F')$ for all finite extensions F'/F. The upshot is that the number of large points of $C(F')$ is bounded by

$$c' \cdot c^{\mathrm{rk}\,\mathrm{Jac}(C)(F')}$$

where c, c' depend on C, but not on F'.

The dichotomy between large and moderate points was already visible in Vojta's work. But its origin is older and already appears in modified form in work of Thue, Siegel, Mahler, and Roth on diophantine approximation.

Rémond's explicit Théorème 2.1 **[54]** gives a recipe how to bound the total number of rational points using a bound for the number of moderate points.

With our eyes set on Mazur's question, we aim to obtain good bounds for the number of moderate points. In the coming two sections we explain our general approach to the proof of Theorem 1.12.

4. COMPARING WEIL AND NÉRON–TATE HEIGHTS

The interplay between the Weil and Néron–Tate heights on a family of abelian varieties leads to powerful results including Silverman's Specialization Theorem **[57]** and more recent work by Masser and Zannier towards the relative Manin–Mumford Conjecture **[44]**. This interaction also plays a central role in the proof of Theorem 1.12 that resolved Mazur's question.

Having worked with a fixed abelian variety in Sections 2.2 and 3, we now shift gears and work in a family of abelian varieties.

Example 4.1. Let $Y(2) = \mathbb{P}^1 \setminus \{0, 1, \infty\}$. For $\lambda \in Y(2)(\mathbb{C})$, we have an elliptic curve $\mathcal{E}_\lambda \subseteq \mathbb{P}^2$ determined by

$$y^2 z = x(x-z)(x-\lambda z)$$

where the origin is $[0:1:0]$. The total space $\mathcal{E}$ is a surface presented with a closed immersion $\mathcal{E} \hookrightarrow \mathbb{P}^2 \times Y(2)$. It is called the Legendre family of elliptic curves and is an abelian scheme over $Y(2)$. So we can add two complex points of $\mathcal{E}$ if they are in the same fiber above $Y(2)$. More precisely, there is an addition morphism $\mathcal{E} \times_S \mathcal{E} \to \mathcal{E}$ over S, as well as an inversion morphism $\mathcal{E} \to \mathcal{E}$ over S. Finally, the zero section of $\mathcal{E}$ is given by $\lambda \mapsto ([0:1:0], \lambda)$.

Consider a geometrically irreducible smooth quasiprojective variety S defined over a number field $F \subseteq \overline{\mathbb{Q}}$. Let $\pi : \mathcal{A} \to S$ be an abelian scheme over S. So each fiber $\mathcal{A}_s = \pi^{-1}(s)$, where $s \in S(\overline{\mathbb{Q}})$, is an abelian variety. We have an addition morphism on the fibered square $\mathcal{A} \times_S \mathcal{A} \to \mathcal{A}$ and an inversion morphism $\mathcal{A} \to \mathcal{A}$; both are relative over S. Addition induces a multiplication-by-n morphism $[n] : \mathcal{A} \to \mathcal{A}$ over S for all $n \in \mathbb{Z}$.

For simplicity, we assume that $\mathcal{A}$ is presented with an immersion $\mathcal{A} \hookrightarrow \mathbb{P}^n \times S$ over S, much as in Example 4.1 above. Let $\mathcal{L}$ be the restriction of the hyperplane bundle $\mathcal{O}(1)$ on $\mathbb{P}^n \times S = \mathbb{P}^n_S$ to $\mathcal{A}$. We also assume that $\mathcal{L}$ is even, that is $[-1]^*\mathcal{L} \cong \mathcal{L}$. This allows us to define a fiberwise Néron–Tate height on $\mathcal{A}(\overline{\mathbb{Q}})$ which we abbreviate by $\hat{h}_{\mathcal{A}}$.

Let $s \in S(\overline{\mathbb{Q}})$. Then $\mathcal{A}_s$ is an abelian variety in $\mathbb{P}^n$. We have two functions, $\hat{h}_{\mathcal{A}}|_{\mathcal{A}_s(\overline{\mathbb{Q}})}$ and $h|_{\mathcal{A}_s(\overline{\mathbb{Q}})}$; the latter is the restriction of the Weil height on $\mathbb{P}^n$. By Theorem 2.7(ii), their difference is bounded in absolute value in function of s.

In the example of the Legendre family, the point $[\lambda : 0 : 1] \in \mathcal{E}_\lambda$ is of order 2 for all λ. So its Néron–Tate height vanishes, but its Weil height equals $h([\lambda : 1])$ and is thus unbounded as λ varies.

We would like to understand the difference between Néron–Tate and Weil heights on $\mathcal{A}$ as the base point $s \in S(\overline{\mathbb{Q}})$ varies. As suggested by the Legendre case, the key is the Weil height on the base S. To keep things concrete, we will assume that S comes with an immersion $S \hookrightarrow \mathbb{P}^m$. We identify S with a Zariski locally closed subset of $\mathbb{P}^m$. So S need not be projective, but its Zariski closure $\overline{S}$ in $\mathbb{P}^m$ is. We write h_S for $h|_{\overline{S}(\overline{\mathbb{Q}})} : \overline{S}(\overline{\mathbb{Q}}) \to [0, \infty)$ where h is the Weil height on $\mathbb{P}^m(\overline{\mathbb{Q}})$. In the language of Section 2.1, h_S represents the Weil height attached to $(\overline{S}, \mathcal{O}(1)|_{\overline{S}})$.

The difference between Weil and Néron–Tate heights on the total space $\mathcal{A}(\overline{\mathbb{Q}})$ was clarified in work of Zimmer [73] in the elliptic setting and Manin–Zarhin [69] and Silverman–Tate [57] in the more general setting. In our case the latter result amounts to

$$\hat{h}_{\mathcal{A}}|_{\mathcal{A}_s(\overline{\mathbb{Q}})} = h|_{\mathcal{A}_s(\overline{\mathbb{Q}})} + O\big(\max\{1, h_S(s)\}\big) \tag{4.1}$$

for all $s \in S(\overline{\mathbb{Q}})$.

We introduce a final player, a geometrically-irreducible subvariety V of $\mathcal{A}$ defined over F.

Theorem 4.2 (Silverman **[57]**). *Suppose S and $V \subseteq \mathcal{A}$ are curves such that V dominates S. Then*

$$\lim_{\substack{P \in V(\overline{\mathbb{Q}}) \\ h(\pi(P)) \to \infty}} \frac{\hat{h}_{\mathcal{A}}(P)}{h(\pi(P))} \tag{4.2}$$

exists. Suppose, in addition, that the geometric generic fiber of $\mathcal{A} \to S$ has trivial trace over $\overline{\mathbb{Q}}$. Then the limit vanishes if and only if V is an irreducible component of $\ker[N]$ *for some $N \geq 1$. Otherwise the limit is positive.*

Silverman computed the limit in terms of the Néron–Tate height of V restricted to the generic fiber $\mathcal{A} \to S$.

The "if direction" is straightforward: in this case all P in question are of finite order and their Néron–Tate height vanishes; see Theorem 2.7(iv). The "only if" direction is deeper and has many applications: Silverman's Specialization Theorem, Theorem C in **[57]**, as well as applications to unlikely intersections by Masser and Zannier, see **[44]** and **[68]** for an overview and more results.

What happens if V has dimension > 1 and S remains a curve? In this case the limit (4.2) does not make sense. Indeed, for fixed $s \in S(\overline{\mathbb{Q}})$, the set of $P \in V(\overline{\mathbb{Q}})$ that map to s has positive dimension and thus unbounded Néron–Tate height.

Motivated by Theorem 4.2, the author showed the next theorem. It may serve as a higher-dimensional substitute for Silverman's Theorem 4.2. For an irreducible subvariety of V of $\mathcal{A}$ that dominates S, we write $V_{\overline{\eta}}$ for the geometric generic fiber of $\pi|_V : V \to S$. This is a possibly reducible subvariety of the geometric generic fiber $\mathcal{A}_{\overline{\eta}}$ of $\mathcal{A} \to S$.

Theorem 4.3 (**[35]**). *Suppose $S = Y(2)$ and let $\mathcal{A} = \mathcal{E}^{[g]}$ be the g-fold fibered power of the Legendre family of elliptic curves. Suppose $V \subseteq \mathcal{E}^{[g]}$ dominates $Y(2)$ and*

$$V_{\overline{\eta}} \text{ is not a finite union of irreducible components of algebraic subgroups of } \mathcal{A}_{\overline{\eta}}. \tag{4.3}$$

Then there exist $c(V) > 0$ and a Zariski open and dense subset $U \subseteq V$ with

$$h_{Y(2)}\big(\pi(P)\big) \leq c(V) \max\big\{1, \hat{h}_{\mathcal{A}}(P)\big\} \quad \text{for all } P \in U(\overline{\mathbb{Q}}). \tag{4.4}$$

Say (4.3) holds. If $P \in U(\overline{\mathbb{Q}})$ has finite order as a point in its respective fiber, we find $h_{Y(2)}(\pi(p)) \leq c(V)$ and the total Weil height of P is bounded from above by (4.1). This simple observation led to the resolution of several "special points" problems **[35]** in the spirit of the André–Oort Conjecture. For example, torsion points $P \in V(\overline{\mathbb{Q}})$ that lie in a fiber with complex multiplication are not Zariski dense in V. The proof of Theorem 4.3 makes use of Siu's Criterion, see Remark 2.5(iii), and an investigation of monodromy in $\mathcal{E}^{[g]}$.

The Zariski open U cannot in general be taken to equal V. But there is a natural description of this set in geometric terms through unlikely intersections.

The hypothesis (4.3) is necessary and essentially rules out that V itself is a family of abelian subvarieties.

Gao and the author **[33]** then generalized Theorem 4.3 to an abelian scheme when the base is again a smooth curve S defined over $\overline{\mathbb{Q}}$. Here more care is needed in connection with the hypothesis (4.3). Indeed, if $\mathcal{A} = A \times S$ is a constant abelian scheme, where A is an abelian variety, then (4.4) cannot hold generically for $V = Y \times S$. Roughly speaking, the condition in [33] that replaces (4.3) also needs to take into account a possible constant part of $\mathcal{A}_{\overline{\eta}}$. If $\mathcal{A}_{\overline{\eta}}$ has no constant part, i.e., if its $\overline{\mathbb{Q}}(\overline{\eta})/\overline{\mathbb{Q}}$-trace is 0, then (4.3) suffices for S a curve. The case of a higher-dimensional base requires even more care, as we will see.

There were two applications of the height bound in [33].

First, and in the same paper, we proved new cases of the geometric Bogomolov Conjecture for an abelian variety defined over the function field of the curve S. This approach relied on Silverman's Theorem 4.2. It was used earlier in [35] to give a new proof of the Geometric Bogomolov Conjecture in a power of an elliptic curve. The number field case of the Bogomolov Conjecture was proved by Ullmo [61] and Zhang [71] in the 1990s. Progress in the function field case was later made by Cinkir, Faber, Moriwaki, Gubler, and Yamaki. For the state of the Geometric Bogomolov Conjecture as of 2017, we refer to a survey of Yamaki [64]. Gubler's strategy works in arbitrary characteristic and was expanded on by Yamaki. In joint work [10] with Cantat, Gao, and Xie, the author later established the Geometric Bogomolov Conjecture in characteristic 0 by bypassing the height inequality (4.4). Very recently, Xie and Yuan [63] announced a proof of the Geometric Bogomolov Conjecture in arbitrary characteristic. Their approach builds on the work of Gubler and Yamaki.

Second, and in later joint work with Dimitrov and Gao [23], we established uniformity for the number of rational points in the spirit of Mazur's question for curves parametrized by the 1-dimensional base S.

As we shall see, the proof of Theorem 1.12 requires a height comparison result like (4.4) for abelian schemes over a base S of any dimension. But now the correct condition to impose on V is more sophisticated and cannot be easily read off of the geometric generic fiber as in (4.3). The condition relies on the Betti map, which we introduce in the next section.

4.1. Degenerate subvarieties and the Betti map

In this section, S is a smooth irreducible quasiprojective variety over $\mathbb{C}$. Let $\pi : \mathcal{A} \to S$ again be an abelian scheme over S of relative dimension $g \geq 1$.

For each $s \in S(\mathbb{C})$, the fiber $\mathcal{A}_s(\mathbb{C})$ is a complex torus of dimension g. Forgetting the complex structure, each g-dimensional complex torus is diffeomorphic to $(\mathbb{R}/\mathbb{Z})^{2g}$ as a real Lie group. By Ehresmann's Theorem, this diffeomorphism extends locally in the analytic topology on the base. That is, there is a contractible open neighborhood U of s in $S(\mathbb{C})$ and a diffeomorphism $\mathcal{A}_U = \pi^{-1}(U) \to (\mathbb{R}/\mathbb{Z})^{2g} \times U$ over U. Fiberwise this diffeomorphism can be arranged to be a group isomorphism above each point of U. Thus we can locally trivialize the abelian scheme at the cost of sacrificing the complex-analytic structure.

The trivialization is not entirely unique as we can let a matrix in $\mathrm{GL}_{2g}(\mathbb{Z})$ act in the natural way on the real torus $(\mathbb{R}/\mathbb{Z})^{2g}$. But since U is connected, this is the only ambiguity. It is harmless for what follows.

The Betti map β_U attached to U is the composition of the trivialization followed by the projection

$$\overbrace{\mathcal{A}_U \longrightarrow (\mathbb{R}/\mathbb{Z})^{2g} \times U \longrightarrow (\mathbb{R}/\mathbb{Z})^{2g}}^{\beta_U}.$$

This map has appeared implicitly in diophantine geometry in work of Masser and Zannier **[44]**. We also refer to more recent work of André, Corvaja, and Zannier **[3]** for a systematic study of the Betti map.

We list some of the most important properties:

(i) For all $s \in U$, the restriction $\beta_U|_{\mathcal{A}_s(\mathbb{C})} : \mathcal{A}_s(\mathbb{C}) \to (\mathbb{R}/\mathbb{Z})^{2g}$ is a diffeomorphism of real Lie groups. In particular, $P \in \mathcal{A}_U$ has finite order in its respective fiber if and only if $\beta_U(P) \in (\mathbb{Q}/\mathbb{Z})^{2g}$.

(ii) For all $P \in U$ the fiber $\beta_U^{-1}(\beta_U(P))$ is a complex-analytic subset of $\mathcal{A}_U$.

Definition 4.4. An irreducible closed subvariety $V \subseteq \mathcal{A}$ that dominates S is called degenerate if for all U and β_U as above and all smooth points P of $V_U = \pi|_V^{-1}(U)$ the differential of $\mathrm{d}_P(\beta_U|_{V_U})$ satisfies

$$\operatorname{rk} \mathrm{d}_P(\beta_U|_{V_U}) < 2 \dim V. \tag{4.5}$$

It has become customary to call V degenerate if it is not nondegenerate.

For all smooth points P of V_U, the left-hand side of (4.5) is at most the right-hand side, which equals the real dimension of V_U. It is also at most $2g$, the real dimension of a fiber of $\mathcal{A} \to S$. Moreover, if the maximal rank of $\mathrm{d}\beta_U$ on V_U is attained at P then the maximal rank is attained also in a neighborhood of P in V_U. Being nondegenerate is a local property.

Let us consider some examples.

Example 4.5.

(i) If S is a point, then $\mathcal{A}$ is an abelian variety and an arbitrary subvariety $V \subseteq \mathcal{A}$ is nondegenerate because β_S is a diffeomorphism.

(ii) Suppose $\dim V > g$. Then $\operatorname{rk} \mathrm{d}_P(\beta_U|_{V_U}) \le 2g < 2 \dim V$ for all smooth P and so V is degenerate. In particular, $\mathcal{A}$ is a degenerate subvariety of $\mathcal{A}$ if $\dim S \ge 1$.

(iii) Suppose $\mathcal{A} = A \times S$ is a constant abelian scheme with A an abelian variety. If $Y \subseteq A$ is a closed irreducible subvariety and if $\dim S \ge 1$, then $Y \times S$ is degenerate. Indeed, the rank is at most $2 \dim Y < 2 \dim Y \times S$.

(iv) Suppose V is an irreducible component of $\ker[N]$ for some integer $N \ge 1$. Any point in $V(\mathbb{C})$ has order dividing N (and, in fact, equal to N). So the image of $\beta_U|_{V_U}$ is finite and hence V is degenerate if $\dim S \ge 1$.

(v) Suppose V is the image of a section $S \to \mathcal{A}$. If the geometric generic fiber of $\mathcal{A} \to S$ has trivial trace, then $(\beta_U)|_{V_U}$ is constant if and only if V is an irreducible component of $\ker[N]$ for some $N \ge 1$. This is Manin's Theorem of the Kernel, we refer to Bertrand's article **[7]** for the history of this theorem.

(vi) Suppose $\mathcal{A} = \mathcal{E}^{[g]}$ and V are as in Theorem 4.3. One step in the proof of this theorem consisted in verifying that V, subject to hypothesis (4.3), is nondegenerate. Crucial input came from the monodromy action of the fundamental group of the base $Y(2) = \mathbb{P}^1 \setminus \{0, 1, \infty\}$ on the first homology of a fiber $\mathcal{A}_s$ with s in general position. In this case the monodromy action is unipotent at the cusps 0 and 1 of $Y(2)$. This enabled the author to use a result of Kronecker from diophantine approximation. Already Masser and Zannier **[44]** used the monodromy action in their earlier work for V a curve.

(vii) If S is a curve, then the monodromy action of the fundamental group of $S(\mathbb{C})$ on the homology of fibers of $\mathcal{A} \to S$ is locally quasiunipotent. But if S is projective, then there are no cusps. So exploiting monodromy in this setting required a different approach. In **[33]** Gao and the author used o-minimal geometry and the Pila–Wilkie Counting Theorem **[51]**. A related case was solved by Cantat, Gao, and Xie in collaboration with the author **[10]**; we used dynamical methods.

(viii) Finally, we consider the case of an abelian scheme $\mathcal{A}$ over a base S of arbitrary dimension. This setting was studied recently in work of André–Corvaja–Zannier **[3]**. Moreover, the work of Gao on the Ax–Schanuel Theorem **[30]** for the universal family of abelian varieties led him to formulate a geometric condition **[29]** that guarantees nondegeneracy. It proves crucial in the application to Mazur's question and we will return to this point. Gao's result also relied on o-minimal geometry and the Pila–Wilkie Theorem.

4.2. Comparing the Weil and Néron–Tate heights on a subvariety

We now come to the generalization of Theorem 4.3 to nondegenerate subvarieties. We retain the notation introduced in Section 4.1. So S is a smooth irreducible quasiprojective variety defined over $\overline{\mathbb{Q}}$ equipped with an immersion in $\mathbb{P}^m$. We have a height h_S on $S(\overline{\mathbb{Q}})$. Moreover, $\pi : \mathcal{A} \to S$ is an abelian scheme over S presented with an immersion $\mathcal{A} \to \mathbb{P}^n \times S$ over S. Finally, $\mathcal{L}$ is as in Section 4 and $\hat{h}_{\mathcal{A}}$ is the fiberwise Néron–Tate height on $\mathcal{A}(\overline{\mathbb{Q}})$.

We assume that $\mathcal{A}$ carries symplectic level-ℓ structure for some fixed $\ell \geq 3$ and that $\mathcal{L}$ induces a principal polarization. For the proof of Theorem 1.12, it suffices to have the following height bound under these conditions. We also refer to **[24, THEOREM B.1]** for a version that relaxes some of the conditions.

Theorem 4.6 (**[24, THEOREM 1.6]**). *Let V be a nondegenerate irreducible subvariety of $\mathcal{A}$ that dominates S. There exist $c(V) > 0$, $c'(V) \geq 0$, and a Zariski open and dense subset $U \subseteq V$ with*

$$h_S(\pi(P)) \leq c(V)\hat{h}_{\mathcal{A}}(P) + c'(V) \quad \text{for all } P \in U(\overline{\mathbb{Q}}).$$

We refer to Yuan and Zhang's Theorem 6.2.2 **[67]** for a height inequality in the dynamical setting.

Here are just a few words on the proof of Theorem 4.6. Siu's Criterion, see Remark 2.5, is used to compare the Weil height of $\pi(P)$ with a Weil height of P. The nondegeneracy hypothesis is used to extract a volume estimate. The upshot is a lower bound for the top self-intersection number in Siu's Criterion. The predecessor of Theorem 4.6 in the earlier works [33, 35] was proved by counting torsion points using the Geometry of Number; volumes played an important role here as well. Passing from the Weil to the Néron–Tate height introduces an additional dependency on the height of $\pi(P)$, see (4.1). However, this contribution can be eliminated by using Masser's "ruthless strategy of killing Zimmer constants" [68, APPENDIX C]. This task is done by repeated iteration of the duplication morphism [2] which has the effect of truncating Tate's Limit Process after finitely many steps. Our ambient group scheme $\mathcal{A}$ is quasiprojective but in general not projective. So a suitable compactification is required that admits some compatibility with the duplication morphism.

The positive constant $c(V)$ in Theorem 4.6 ultimately comes from the application of Siu's Criterion. As such it can expressed in geometric terms.

5. APPLICATION TO MODERATE POINTS ON CURVES

In this section we sketch the main lines of the proof of Theorem 1.12. It will be enough to bound the number of moderate points, see Section 3.

5.1. The Faltings–Zhang morphism

Smooth curves of genus $g \geq 2$ defined over $\overline{\mathbb{Q}}$ are classified by the $\overline{\mathbb{Q}}$-points of a quasiprojective variety, the coarse moduli space. For us it is convenient to work with symplectic level-ℓ structure on the Jacobian for some fixed integer $\ell \geq 3$. With this extra data, we obtain a fine moduli space $\mathbb{M}_g$, together with a universal family $\mathfrak{C}_g \to \mathbb{M}_g$. Fibers of this family are smooth curves of genus g with the said level structure on the Jacobian. Then $\mathbb{M}_g$ carries the structure of a smooth quasiprojective variety of dimension $3g-3$ defined over a cyclotomic field. For convenience, we replace $\mathbb{M}_g$ by an irreducible component by choosing a complex root of unity of order ℓ and consider it as defined over $\overline{\mathbb{Q}}$.

The Torelli morphism $\tau : \mathbb{M}_g \to \mathbb{A}_g$ takes a smooth curve to its Jacobian with the level structure; here $\mathbb{A}_g$ denotes the fine moduli space of g-dimensional abelian varieties with a principal polarization and symplectic level-ℓ structure.

Let $M \geq 0$ be an integer and consider $M+1$ points $P_0, \ldots, P_M \in \mathfrak{C}_g(\mathbb{C})$ in the same fiber C of $\mathfrak{C}_g \to \mathbb{M}_g$. The differences $[P_1]-[P_0], \ldots, [P_M]-[P_0]$ are divisors of degree 0 on C. We obtain M complex points in the Jacobian of C and so M complex points of $\mathfrak{A}_g$. We obtain a commutative diagram

$$\begin{array}{ccc} \mathfrak{C}_g^{[M+1]} & \xrightarrow{\mathscr{D}} & \mathfrak{A}_g^{[M]} \times_{\mathbb{A}_g} \mathbb{M}_g \\ & \searrow & \downarrow \\ & & \mathbb{M}_g \end{array}$$

of morphisms of schemes; here the exponent $[M]$ denotes taking the Mth fibered power over the base. The morphism $\mathscr{D}$ is called the Faltings–Zhang morphism; see **[26, LEMMA 4.1]** and **[71, LEMMA 3.1]** for important applications to diophantine geometry of variants of this morphism. The morphism $\mathscr{D}$ is proper.

A modified version of this construction is also useful. Say $S \to \mathbb{M}_g$ is a quasifinite morphism with S an irreducible quasiprojective variety defined over $\overline{\mathbb{Q}}$. We obtain a proper morphism $\mathscr{D} : \mathfrak{C}_g^{[M+1]} \times_{\mathbb{M}_g} S \to \mathfrak{A}_g^{[M]} \times_{\mathbb{A}_g} S$, again called Faltings–Zhang morphism.

Gao, using his Ax–Schanuel Theorem for the universal family $\mathfrak{A}_g$ **[30]** and a characterization **[28]** of bialgebraic subvarieties of $\mathfrak{A}_g$, obtained

Theorem 5.1 (Gao **[29]**). *Let $S \to \mathbb{M}_g$ be as above, i.e., a quasifinite morphism from an irreducible quasiprojective variety S defined over $\overline{\mathbb{Q}}$ and $g \geq 2$. If $M \geq \dim \mathbb{M}_g + 1 = 3g - 2$, then $\mathscr{D}(\mathfrak{C}_g^{[M+1]} \times_{\mathbb{M}_g} S)$ is a nondegenerate subvariety of $\mathfrak{A}_g^{[M]} \times_{\mathbb{A}_g} S$.*

Mok, Pila, and Tsimerman **[47]** earlier proved an Ax–Schanuel Theorem for Shimura varieties. Gao's result **[30]** is a "mixed" version in the abelian setting. We refer to the survey **[4]** on recent developments in functional transcendence.

The hypothesis $g \geq 2$ is crucial. The definition of the Faltings–Zhang morphism makes sense for $g = 1$. But it will be surjective and the image is degenerate expect in the (for our purposes uninteresting) case $\dim S = 0$.

We consider here for simplicity only the case $S = \mathbb{M}_g$.

Using basic dimension theory, we see $\dim \mathscr{D}(\mathfrak{C}_g^{[M+1]}) \leq M + 1 + \dim \mathbb{M}_g$. The image lies in the fibered power $\mathfrak{A}_g^{[M]}$ where the relative dimension is Mg. A necessary condition for $\mathscr{D}(\mathfrak{C}_g^{[M+1]})$ to be nondegenerate is $\dim \mathscr{D}(\mathfrak{C}_g^{[M+1]}) \leq Mg$, see Example 4.5(ii). This inequality follows from

$$M + 3g - 2 = M + 1 + \dim \mathbb{M}_g \leq Mg. \tag{5.1}$$

If $M \leq 3$, the numerical condition (5.1) is not satisfied for any $g \geq 2$. For this reason, we cannot hope to work with the image of $\mathfrak{C}_g \times_{\mathbb{M}_g} \mathfrak{C}_g$ in $\mathfrak{A}_g$ by taking differences. Moreover, there seems to be no reasonable way to work with a single copy of $\mathfrak{C}_g$, where the relations between dimensions would be even worse. The numerical condition (5.1) is satisfied for all $M \geq 4$ and all $g \geq 2$. Gao's Theorem implies that $M \geq 3g - 2$ is sufficient to guarantee nondegeneracy.

We can thus apply Theorem 4.6 to the image $\mathscr{D}(\mathfrak{C}_g^{[M+1]})$ of the Faltings–Zhang morphism in $\mathfrak{A}_g^{[M]} \times_{\mathbb{A}_g} \mathbb{M}_g$. Let $M = 3g - 2$, then

$$h_{\mathbb{M}_g}(s) \leq c(g)\big(\hat{h}_{\mathfrak{A}_g}(P_1 - P_0) + \cdots + \hat{h}_{\mathfrak{A}_g}(P_M - P_0)\big) + c'(g) \tag{5.2}$$

for all $(P_0, \ldots, P_M) \in U(\overline{\mathbb{Q}})$ above $s \in \mathbb{M}_g(\overline{\mathbb{Q}})$ where U is a Zariski open and dense subset of $\mathscr{D}(\mathfrak{C}_g^{[M+1]})$. The constants $c(g) > 0$ and $c'(g) \geq 0$ depend on the various choices made regarding projective immersions of $\mathbb{M}_g$ and $\mathfrak{A}_g$. Ultimately, they depend only on g once these choices have been made.

The Zariski open U cannot be replaced by $\mathscr{D}(\mathfrak{C}_g^{[M+1]})$. Indeed, the right-hand side of (5.2) vanishes on the diagonal $P_0 = P_1 = \cdots = P_M$ whereas the left-hand side is unbounded as s varies.

Let us shift back to using $\|\cdot\|$ to denote the square root of the Néron–Tate height, see Section 2.2. Let us assume that

$$h_{\mathbb{M}_g}(s) \geq 2c'(g). \tag{5.3}$$

As $2M = 6g - 4$ we find

$$h_{\mathbb{M}_g}(s) \leq c(g)(6g-4) \max_{1\leq j\leq M} \|P_j - P_0\|^2 \quad \text{for all } (P_0, \ldots, P_M) \in U(\overline{\mathbb{Q}}). \tag{5.4}$$

Morally, (5.4) states that among a $(3g-1)$-tuple of points on a curve of genus g in general position, there must be a pair that repels one another with respect to the norm $\|\cdot\|$. The squared distance of such a pair is larger than a positive multiple, depending only on g, of the modular height $h_{\mathbb{M}_g}(s)$; this is the key to bounding the number of moderate points from Section 3.

As stated at the end of Section 4.2, the value $c(g)$ can be expressed in terms of geometry properties of the image of $\mathfrak{C}_g^{[M+1]}$ under the Faltings–Zhang morphism.

Question 5.2. What is an admissible value for $c(g)$?

5.2. Bounding the number of moderate points—a sketch

Recall that, by the discussion at the end of Section 3, we need to bound the number of moderate points.

We retain the notation of Sections 3 and 5. The curve C from Section 3 can be equipped with suitable level structure over a field F'/F with $[F' : F]$ bounded in terms of g. The rank of $\mathrm{Jac}(C)(F')$ may be dangerously larger than the rank of $\mathrm{Jac}(C)(F)$. But recall that we are interested in bounding $\#C(F)$ from above, so only the group $\mathrm{Jac}(C)(F)$ will be relevant. Moreover, c_1, c_2, and c_3 from a suitable version of Vojta's inequality are unaffected by extending F. The effect is that we may identify C with a fiber of $\mathfrak{C}_g$ above some point $s \in \mathbb{M}_g(F')$. For simplicity, we assume $F = F'$ for this proof sketch.

We require some additional information on $c_3(C)$. It turns out that we can take $c_3 = c_4(g)\max\{1, h_{\mathbb{M}_g}(s)\}$ where $c_4(g) > 0$ depends on g. This follows Rémond's work **[55]** on the Vojta inequality. A similar dependency is apparent in de Diego's result, Theorem 1.6.

Suppose now that (5.3) holds, so $h_{\mathbb{M}_g}(s)$ is sufficiently large in terms of g. We fix an auxiliary base point $P' \in C(F)$. We must bound from above the number of points in

$$B(R) = \{P \in C(F) : \|P - P'\|^2 \leq R^2\} \quad \text{with } R = \left(c_4(g)h_{\mathbb{M}_g}(s)\right)^{1/2}$$

where $\|\cdot\|$ denotes the square root of the Néron–Tate height on the fiber of $\mathfrak{A}_g \to \mathbb{A}_g$ associated to the Jacobian of C.

Recall $M = 3g - 2$ and suppose $P_0, \ldots, P_M \in C(F)$. If the tuple $(P_0, \ldots, P_M)$ is in general position, i.e., $(P_1 - P_0, \ldots, P_M - P_0)$ lies in $U(\overline{\mathbb{Q}})$ from (5.4), then there is i with

$$P_i \notin B(P_0, r) = \{P \in C(F) : \|P - P_0\|^2 \leq r^2\} \quad \text{with } r = \left(c_5(g)h_{\mathbb{M}_g}(s)\right)^{1/2}.$$

If we had a guarantee that such $(M+1)$-tuples of pairwise distinct points are always in general position, then $\#B(P_0, r) < M = 3g-2$. By sphere packing, we can cover the image of $B(R)$ in $\mathrm{Jac}(C)(F) \otimes \mathbb{R}$ by at most $(1+2R/r)^{\mathrm{rk}\,\mathrm{Jac}(C)(F)}$ closed balls in $\mathrm{Jac}(C)(F)$ of radius r. One can even arrange for the ball centers to arise as points of $C(F)$. The modular height $h_{\mathbb{M}_g}(s)$ cancels out in the quotient

$$\frac{R}{r} = \left(\frac{c_4(g)}{c_5(g)}\right)^{1/2}.$$

This would complete the proof of Theorem 1.12 except that there is no reason to believe that $(P_1 - P_0, \ldots, P_M - P_0) \in U(\overline{\mathbb{Q}})$ (even if the P_j are pairwise distinct). Treating points with image in the complement of U requires induction on the dimension. Here we rely on the freedom to replace $\mathbb{M}_g$ by a subvariety in that Gao's Theorem 5.1.

Let us briefly explain the resulting induction step. Observe that the dimension of this exceptional set is at most $\dim \mathscr{D}(\mathfrak{C}_g^{[M+1]}) - 1 \leq M + \dim \mathbb{M}_g$. There are two cases for $(P_0, \ldots, P_M)$ with image in the exceptional set $(\mathscr{D}(\mathfrak{C}_g^{[M+1]}) \setminus U)(\overline{\mathbb{Q}})$ on which we do not have the height inequality. For the case study, recall that $s \in \mathbb{M}_g(\overline{\mathbb{Q}})$ denotes the point below all the P_j and $\tau(s) \in \mathbb{A}_g(\overline{\mathbb{Q}})$ is its image under the Torelli morphism τ.

First, assume that the fiber of $\mathscr{D}(\mathfrak{C}_g^{[M+1]}) \setminus U \to \mathbb{A}_g$ above $\tau(s)$ has dimension at most M. This fiber contains $(P_1 - P_0, \ldots, P_M - P_0)$. This case is solved using a zero estimate motivated by the following simple lemma.

Lemma 5.3. *Suppose C is an irreducible curve defined over $\mathbb{C}$ and W a proper Zariski closed subset of C^M. If $\Sigma \subseteq C(\mathbb{C})$ with $\Sigma^M \subseteq W(\mathbb{C})$, then Σ is finite.*

This statement can be quantified if C is presented as a curve in some projective space. Using Bézout's Theorem, one can show that $\#\Sigma$ is bounded from above in terms of the degrees of C and W. In our application, both degrees will be uniformly bounded as all varieties arise in algebraic families. This ultimately leads to the desired uniformity estimates.

The second case is if the fiber of $\mathscr{D}(\mathfrak{C}_g^{[M+1]}) \setminus U \to \mathbb{A}_g$ above $\tau(s)$ has dimension at least $M+1$. For dimension reasons, s lies in a proper subvariety S of $\mathbb{M}_g$. Here we apply induction on the dimension and replace $\mathbb{M}_g$ by its subvariety S.

This completes the proof sketch.

Kühne [39] combined ideas from equidistribution with the approach laid out in [24] to get a suitable uniform estimate for $\#B(P_0, r)$ without the restriction (5.3) on $h_{\mathbb{M}_g}(s)$. Yuan's Theorem 1.1 [66] does so as well, but he follows a different approach. He obtains a more general estimate that works also in the function field setting and allows for a larger R.

6. HYPERELLIPTIC CURVES

A hyperelliptic curve is a smooth curve of genus at least 2 that admits a degree 2 morphism to the projective line. Hyperelliptic curves have particularly simple planar models. Indeed, if the base field is a number field F, then a hyperelliptic curve of genus g can be represented by a hyperelliptic equation

$$Y^2 = f(X) \quad \text{with } f \in F[X] \text{ monic and square-free of degree } 2g+1 \text{ or } 2g+2.$$

In this section we determine consequences of Theorem 1.12 for hyperelliptic curves. Our aim is to leave the world of curves and Jacobians and to present a bound for the number of rational solutions of $Y^2 = f(X)$ that can be expressed in terms of f. We refer to Section 6 of **[23]** for a similar example in a 1-parameter family of hyperelliptic curves.

To keep technicalities to a minimum, we assume that our base field is $F = \mathbb{Q}$ and that $f \in \mathbb{Z}[X]$ is monic of degree $d = 2g+1$ and factors into linear factors in $\mathbb{Q}[X]$. The curve represented by the hyperelliptic equation has a marked Weierstrass point "at infinity." These assumptions can be loosened with some extra effort. For example, if f does not factor in $\mathbb{Q}[X]$, then the class number of the splitting field will play a part.

Say, $f = X^d + f_{d-1}X^{d-1} + \cdots + f_0$. By the assumption above, $f = (X-\alpha_1)\cdots(X-\alpha_d)$ with $\alpha_1,\ldots,\alpha_d \in \mathbb{Q}$ which are necessarily integers. The discriminant of f is

$$\Delta_f = \prod_{1\le i<j\le d} (\alpha_j - \alpha_i)^2 \in \mathbb{Z}\setminus\{0\}.$$

The Mordell Conjecture applied to the hyperelliptic curve represented by $Y^2 = f(X)$ states

$$\#\{(x,y)\in\mathbb{Q}^2 : y^2 = f(x)\} < \infty.$$

We have the following estimate for the cardinality. Below $\omega(n)$ denotes the number of distinct prime divisors of $n \in \mathbb{Z}\setminus\{0\}$.

Theorem 6.1. *Let $g \ge 2$. There exist $c(g) > 1$ and $c'(g) > 0$ with the following property. Suppose $f \in \mathbb{Z}[X]$ is monic of degree $2g+1$, square-free, and factors into linear factors in $\mathbb{Q}[X]$. Then*

$$\#\{(x,y)\in\mathbb{Q}^2 : y^2 = f(x)\} \le c'(g)c(g)^{\omega(\Delta_f)}. \tag{6.1}$$

Proof. The hyperelliptic equation represents a curve C defined over $\mathbb{Q}$ of genus g.

If p is a prime number with $p \nmid \Delta_f$, then the α_i are pairwise distinct modulo p. If p is also odd, then the equation $Y^2 = f(X)$ reduced modulo p defines a hyperelliptic curve over $\mathbb{F}_p$. So C has good reduction at all primes that do not divide $2\Delta_f$. Thus the Jacobian $\mathrm{Jac}(C)$ has good reduction at the same primes.

We may embed C into its Jacobian $\mathrm{Jac}(C)$ by sending the marked Weierstrass point to 0. Each root α_i of f corresponds to a rational point in $C(\mathbb{Q})$ and it is sent to a point of order 2 in $\mathrm{Jac}(C)$. Moreover, these points generate the 2-torsion in $\mathrm{Jac}(C)_{\mathrm{tors}}$. In particular, all points of order 2 in $\mathrm{Jac}(C)_{\mathrm{tors}}$ are rational.

Next we bound the rank of $\mathrm{Jac}(C)(\mathbb{Q})$ from above. Indeed, we could use the work of Ooe–Top **[49]** or **[37, THEOREM C.1.9]**. The latter applied to $\mathrm{Jac}(C)$, $k = \mathbb{Q}$, $m = 2$, and S the prime divisors of $2\Delta_f$ yields $\mathrm{rk}\,\mathrm{Jac}(C)(\mathbb{Q}) \le 2g\#S \le 2g\omega(2\Delta_f) \le 2g + 2g\omega(\Delta_f)$. Here we use that $\mathbb{Q}$ has a trivial class group; in a more general setup, the class group of the splitting field of f will enter at this point. The estimate (6.1) follows from Theorem 1.12 in the case $F = \mathbb{Q}$ with adjusted constants. ■

It is tempting to average (6.1) over the f bounded in a suitable way, e.g., by bounding the maximal modulus of the roots by a parameter X. As pointed out to the author by Christian Elsholtz and Martin Widmer, this average will be unbounded as $X \to \infty$.

ACKNOWLEDGMENTS

Much of the work presented here was obtained together with my coauthors. In particular, I thank Serge Cantat, Vesselin Dimitrov, Ziyang Gao, and Junyi Xie for the fruitful collaboration. I also thank Gabriel Dill, Ziyang Gao, Lars Kühne, and David Masser for comments on this survey.

FUNDING

The author has received funding from the Swiss National Science Foundation project n° 200020_184623.

REFERENCES

[1] D. Abramovich, Uniformité des points rationnels des courbes algébriques sur les extensions quadratiques et cubiques. *C. R. Acad. Sci. Paris Sér. I Math.* **321** (1995), no. 6, 755–758.

[2] L. Alpoge, The average number of rational points on genus two curves is bounded. 2018, arXiv:1804.05859.

[3] Y. André, P. Corvaja, and U. Zannier, The Betti map associated to a section of an abelian scheme (with an appendix by Z. Gao). *Invent. Math.* **222** (2020), 161–202.

[4] B. Bakker and J. Tsimerman, Lectures on the Ax–Schanuel conjecture. In *Arithmetic geometry of logarithmic pairs and hyperbolicity of moduli spaces*, edited by M. H. Nicole, pp. 1–68, CRM Short Courses, Springer, 2020.

[5] J. Balakrishnan, N. Dogra, J. S. Müller, J. Tuitman, and J. Vonk, Explicit Chabauty–Kim for the split Cartan modular curve of level 13. *Ann. of Math. (2)* **189** (2019), no. 3, 885–944.

[6] J. S. Balakrishnan, A. J. Best, F. Bianchi, B. Lawrence, J. S. Müller, N. Triantafillou, and J. Vonk, Two recent p-adic approaches towards the (effective) Mordell Conjecture. In *Arithmetic L-functions and differential geometric methods*, edited by P. Charollois, G. Freixas i Montplet, V. Maillot, pp. 31–74, Progr. Math. 338, Birkhäuser, 2021.

[7] D. Bertrand, Manin's theorem of the kernel: a remark on a paper of C-L. Chai, preprint, 2008, https://webusers.imj-prg.fr/~daniel.bertrand/Recherche/rpdf/Manin_Chai.pdf.

[8] E. Bombieri, The Mordell Conjecture revisited. *Ann. Sc. Norm. Super. Pisa Cl. Sci. (4)* **17** (1990), no. 4, 615–640.

[9] E. Bombieri and W. Gubler, *Heights in diophantine geometry*. Cambridge University Press, 2006.

[10] S. Cantat, Z. Gao, P. Habegger, and J. Xie, The geometric Bogomolov conjecture. *Duke Math. J.* **170** (2021), no. 2, 247–277.

[11] L. Caporaso, J. Harris, and B. Mazur, Uniformity of rational points. *J. Amer. Math. Soc.* **10** (1997), no. 1, 1–35.

[12] L. Caporaso, J. Harris, and B. Mazur, Corrections to uniformity of rational points. 2021, arXiv:2012.14461.

[13] C. Chabauty, Sur les points rationnels des courbes algébriques de genre supérieur à l'unité. *C. R. Acad. Sci. Paris* **212** (1941), 882–885.

[14] S. Checcoli, F. Veneziano, and E. Viada, The explicit Mordell Conjecture for families of curves. *Forum Math. Sigma* **7** (2019), e31, 62 pp.

[15] R. F. Coleman, Effective Chabauty. *Duke Math. J.* **52** (1985), no. 3, 765–770.

[16] S. David, M. Nakamaye, and P. Philippon, Bornes uniformes pour le nombre de points rationnels de certaines courbes. In *Diophantine geometry*, pp. 143–164, CRM Series 4, Ed. Norm, Pisa, 2007.

[17] S. David and P. Philippon, Minorations des hauteurs normalisées des sous-variétés de variétés abeliennes. II. *Comment. Math. Helv.* **77** (2002), no. 4, 639–700.

[18] S. David and P. Philippon, Minorations des hauteurs normalisées des sous-variétés des puissances des courbes elliptiques. *Int. Math. Res. Pap.* **3** (2007), rpm006, 113 pp.

[19] T. de Diego, Points rationnels sur les familles de courbes de genre au moins 2. *J. Number Theory* **67** (1997), no. 1, 85–114.

[20] L. DeMarco, H. Krieger, and H. Ye, Uniform Manin–Mumford for a family of genus 2 curves. *Ann. of Math.* **191** (2020), 949–1001.

[21] L. DeMarco and N. M. Mavraki, Variation of canonical height and equidistribution. *Amer. J. Math.* **142** (2020), no. 2, 443–473.

[22] V. Dimitrov, Z. Gao, and P. Habegger, A consequence of the relative Bogomolov conjecture. *J. Number Theory*, **230** (2022), 146–160.

[23] V. Dimitrov, Z. Gao, and P. Habegger, Uniform bound for the number of rational points on a pencil of curves. *Int. Math. Res. Not. IMRN* **2** (2021), 1138–1159.

[24] V. Dimitrov, Z. Gao, and P. Habegger, Uniformity in Mordell–Lang for curves. *Ann. of Math. (2)* **194** (2021), no. 1, 237–298.

[25] G. Faltings, Endlichkeitssätze für abelsche Varietäten über Zahlkörpern. *Invent. Math.* **73** (1983), 349–366.

[26] G. Faltings, Diophantine approximation on abelian varieties. *Ann. of Math. (2)* **133** (1991), no. 3, 549–576.

[27] G. Faltings, The general case of S. Lang's conjecture. In *Barsotti symposium in algebraic geometry (Abano Terme, 1991)*, pp. 175–182, Perspect. Math. 15, Academic Press, San Diego, CA, 1994.

[28] Z. Gao, A special point problem of André–Pink–Zannier in the universal family of Abelian varieties. *Ann. Sc. Norm. Super. Pisa Cl. Sci. (5)* **17** (2017), no. 1, 231–266.

[29] Z. Gao, Generic rank of Betti map and unlikely intersections. *Compos. Math.* **156** (2020), no. 12, 2469–2509.

[30] Z. Gao, Mixed Ax–Schanuel for the universal abelian varieties and some applications. *Compos. Math.* **156** (2020), no. 11, 2263–2297.

[31] Z. Gao, Recent developments of the uniform Mordell–Lang conjecture. 2021, arXiv:2104.03431.

[32] Z. Gao, T. Ge, and L. Kühne, The uniform Mordell–Lang conjecture. 2021, arXiv:2105.15085.

[33] Z. Gao and P. Habegger, Heights in families of Abelian varieties and the geometric Bogomolov conjecture. *Ann. of Math.* **189** (2019), no. 2, 527–604.

[34] P. Habegger, Intersecting subvarieties of abelian varieties with algebraic subgroups of complementary dimension. *Invent. Math.* **176** (2009), no. 2, 405–447.

[35] P. Habegger, Special points on fibered powers of elliptic surfaces. *J. Reine Angew. Math.* **685** (2013), 143–179.

[36] M. Hindry, Autour d'une conjecture de Serge Lang. *Invent. Math.* **94** (1988), no. 3, 575–603.

[37] M. Hindry and J. H. Silverman, *Diophantine geometry: an introduction*. Springer, 2000.

[38] E. Katz, J. Rabinoff, and D. Zureick-Brown, Uniform bounds for the number of rational points on curves of small Mordell–Weil rank. *Duke Math. J.* **165** (2016), no. 16, 3189–3240.

[39] L. Kühne, Equidistribution in families of Abelian varieties and uniformity. 2021, arXiv:2101.10272.

[40] L. Kühne, The relative Bogomolov conjecture for fibered products of elliptic surfaces. 2021, arXiv:2103.06203.

[41] B. Lawrence and A. Venkatesh, Diophantine problems and p-adic period mappings. *Invent. Math.* **221** (2020), no. 3, 893–999.

[42] R. Lazarsfeld, *Positivity in algebraic geometry. I.* Ergeb. Math. Grenzgeb. (3). Ser. Mod. Surveys Math. [Res. Math. Relat. Areas (3)] 48, Springer, Berlin, 2004.

[43] D. Masser, Specializations of finitely generated subgroups of abelian varieties. *Trans. Amer. Math. Soc.* **311** (1989), no. 1, 413–424.

[44] D. Masser and U. Zannier, Torsion points on families of squares of elliptic curves. *Math. Ann.* **352** (2012), no. 2, 453–484.

[45] B. Mazur, Arithmetic on curves. *Bull. Amer. Math. Soc.* **14** (1986), no. 2, 207–259.

[46] B. Mazur, Abelian varieties and the Mordell–Lang conjecture. In *Model theory, algebra, and geometry*, pp. 199–227, Math. Sci. Res. Inst. Publ. 39, Cambridge University Press, Cambridge, 2000.

[47] N. Mok, J. Pila, and J. Tsimerman, Ax–Schanuel for Shimura varieties. *Ann. of Math. (2)* **189** (2019), no. 3, 945–978.

[48] L. J. Mordell, On the rational solutions of the indeterminate equations of the third and fourth degrees. *Proc. Camb. Philos. Soc.* **21** (1922), 179–192.

[49] T. Ooe and J. Top, On the Mordell–Weil rank of an abelian variety over a number field. *J. Pure Appl. Algebra* **58** (1989), no. 3, 261–265.

[50] P. L. Pacelli, Uniform boundedness for rational points. *Duke Math. J.* **88** (1997), no. 1, 77–102.

[51] J. Pila and A. J. Wilkie, The rational points of a definable set. *Duke Math. J.* **133** (2006), no. 3, 591–616.

[52] M. Raynaud, Courbes sur une variété abélienne et points de torsion. *Invent. Math.* **71** (1983), no. 1, 207–233.

[53] M. Raynaud, Sous-variétés d'une variété abélienne et points de torsion. In *Arithmetic and geometry, Vol. I*, pp. 327–352, Progr. Math. 35, Birkhäuser Boston, Boston, MA, 1983.

[54] G. Rémond, Décompte dans une conjecture de Lang. *Invent. Math.* **142** (2000), no. 3, 513–545.

[55] G. Rémond, Inégalité de Vojta en dimension supérieure. *Ann. Sc. Norm. Super. Pisa Cl. Sci. (4)* **29** (2000), no. 1, 101–151.

[56] G. Rémond, Borne générique pour le problème de Mordell–Lang. *Manuscripta Math.* **118** (2005), no. 1, 85–97.

[57] J. H. Silverman, Heights and the specialization map for families of abelian varieties. *J. Reine Angew. Math.* **342** (1983), 197–211.

[58] M. Stoll, Uniform bounds for the number of rational points on hyperelliptic curves of small Mordell–Weil rank. *J. Eur. Math. Soc. (JEMS)* **21** (2019), no. 3, 923–956.

[59] L. Szpiro, Un peu d'effectivité. *Astérisque* **127** (1985), 275–287.

[60] L. Szpiro, E. Ullmo, and S. Zhang, Équirépartition des petits points. *Invent. Math.* **127** (1997), no. 2, 337–347.

[61] E. Ullmo, Positivité et discrètion des points algébriques des courbes. *Ann. of Math. (2)* **147** (1998), no. 1, 167–179.

[62] P. Vojta, Siegel's theorem in the compact case. *Ann. of Math. (2)* **133** (1991), no. 3, 509–548.

[63] Y. Xie and X. Yuan, Geometric Bogomolov conjecture in arbitrary characteristics. 2021, arXiv:2108.09722.

[64] K. Yamaki, Survey on the geometric Bogomolov conjecture. In *Actes de la Conférence "Non-Archimedean analytic geometry: theory and practice"*, pp. 137–193, Publ. Math. Besançon Algèbre Théorie Nr. 2017/1, Presses Univ. Franche-Comté, Besançon, 2017.

[65] X. Yuan, Big line bundles over arithmetic varieties. *Invent. Math.* **173** (2008), no. 3, 603–649.

[66] X. Yuan, Arithmetic bigness and a uniform Bogomolov-type result. 2021, arXiv:2108.05625.

[67] X. Yuan and S. Zhang, Adelic line bundles over quasi-projective varieties. 2021, arXiv:2105.13587.

[68] U. Zannier, *Some problems of unlikely intersections in arithmetic and geometry*. Ann. of Math. Stud. 181, Princeton University Press, Princeton, NJ, 2012.

[69] J. G. Zarhin and J. I. Manin, Height on families of abelian varieties. *Mat. Sb. (N.S.)* **89** (1972), no. 131, 171–181, 349.

[70] S. Zhang, Small points and adelic metrics. *J. Algebraic Geom.* **4** (1995), no. 2, 281–300.

[71] S. Zhang, Equidistribution of small points on abelian varieties. *Ann. of Math. (2)* **147** (1998), no. 1, 159–165.

[72] S. Zhang, Small points and Arakelov theory. In *Proceedings of the International Congress of Mathematicians, Vol. II (Berlin, 1998)*, Doc. Math., pp. 217–225, 1998. Extra Vol. II.

[73] H. Zimmer, On the difference of the Weil height and the Néron–Tate height. *Math. Z.* **147** (1976), no. 1, 35–51.

PHILIPP HABEGGER

Department of Mathematics and Computer Science, University of Basel, Spiegelgasse 1, 4051 Basel, Switzerland, philipp.habegger@unibas.ch

THETA LIFTING AND LANGLANDS FUNCTORIALITY

ATSUSHI ICHINO

ABSTRACT

We review various aspects of theta lifting and its role in studying Langlands functoriality. In particular, we discuss realizations of the Jacquet–Langlands correspondence and the Shimura–Waldspurger correspondence in terms of theta lifting and their arithmetic applications.

MATHEMATICS SUBJECT CLASSIFICATION 2020

Primary 11F70; Secondary 11F27, 11F67, 22E50

KEYWORDS

Automorphic representations, Langlands functoriality, theta lifting

INTERNATIONAL CONGRESS OF MATHEMATICIANS 2022 JULY 6–14

Proc. Int. Cong. Math. 2022, Vol. 3, pp. 1870–1892
DOI 10.4171/ICM2022/82

1. INTRODUCTION

Langlands functoriality is a principle relating two different kinds of automorphic forms and plays a pivotal role in number theory. Before Langlands formulated this principle in [42], this phenomenon was already observed in the following classical example discovered by Eichler [15] and developed by Shimizu [48]. Consider the space

$$S_k(\Gamma_0(N))$$

of elliptic cusp forms of weight k and level N, where k and N are positive integers and $\Gamma_0(N)$ is the congruence subgroup given by

$$\Gamma_0(N) = \left\{ \begin{pmatrix} a & b \\ c & d \end{pmatrix} \in \mathrm{SL}_2(\mathbb{Z}) \mid c \equiv 0 \bmod N \right\}.$$

This space consists of holomorphic functions f on the upper half-plane $\mathfrak{H}$ which satisfy

$$f(\gamma z) = j(\gamma, z)^k f(z)$$

for all $\gamma \in \Gamma_0(N)$ and $z \in \mathfrak{H}$ and which vanish at all cusps. Here $\mathrm{SL}_2(\mathbb{R})$ acts on $\mathfrak{H}$ by linear fractional transformations and $j\left(\left(\begin{smallmatrix} a & b \\ c & d \end{smallmatrix}\right), z\right) = cz + d$ is the factor of automorphy. It is also equipped with the action of Hecke operators T_n for all positive integers n, which is a central tool in the arithmetic study of automorphic forms. On the other hand, to every indefinite quaternion division algebra B over $\mathbb{Q}$, we may associate a space

$$S_k(\Gamma_B)$$

of modular forms, where Γ_B is the group of norm-one elements in B. Namely, this space is defined similarly by replacing $\Gamma_0(N)$ by Γ_B (which can be regarded as a subgroup of $(B \otimes_{\mathbb{Q}} \mathbb{R})^\times \cong \mathrm{GL}_2(\mathbb{R})$) and is equipped with the action of Hecke operators T_n^B. Now assume that N is the product of an even number of distinct primes and B is ramified precisely at the primes dividing N. Then by the works of Eichler and Shimizu, the trace of T_n^B on $S_k(\Gamma_B)$ coincides with the trace of T_n on the new part of $S_k(\Gamma_0(N))$ for all n prime to N.

This remarkable relation was thoroughly studied by Jacquet–Langlands [37] in the framework of automorphic representations. Let F be a number field with adèle ring $\mathbb{A}$. Let B be a quaternion division algebra over F. Then Jacquet–Langlands proved that for any irreducible automorphic representation $\pi^B \cong \otimes_v \pi_v^B$ of $B^\times(\mathbb{A})$, there exists a unique irreducible automorphic representation $\pi \cong \otimes_v \pi_v$ of $\mathrm{GL}_2(\mathbb{A})$ such that

$$\pi_v \cong \pi_v^B$$

for almost all places v of F. Moreover, they described the image of this map $\pi^B \mapsto \pi$ precisely.

The Jacquet–Langlands correspondence gives a basic example of Langlands functoriality. To explain this, let G be a connected reductive group over F. Let ${}^LG$ be the L-group of G, which was introduced by Langlands and which should govern automorphic representations of $G(\mathbb{A})$. Explicitly, ${}^LG$ is defined as a semiproduct $\hat{G} \rtimes \Gamma_F$, where $\hat{G}$ is the complex dual group of G, $\Gamma_F = \mathrm{Gal}(\bar{F}/F)$ is the absolute Galois group of F, and the action of

Γ_F on $\hat{G}$ is inherited from the action of Γ_F on the root datum of G. To motivate it, let us admit for the moment the existence of the hypothetical Langlands group $\mathcal{L}_F$ over F, which is equipped with a surjection $\mathcal{L}_F \to \Gamma_F$. Then it is conjectured that irreducible automorphic representations of $G(\mathbb{A})$ are classified in terms of certain L-homomorphisms $\mathcal{L}_F \to {}^L G$, i.e., homomorphisms commuting with the projections to Γ_F. (Strictly speaking, we consider here packets of tempered automorphic representations.) Now suppose that we have another connected reductive quasisplit group G' over F and an L-homomorphism

$$r : {}^L G \to {}^L G'.$$

Let π be an irreducible automorphic representation of $G(\mathbb{A})$ which should correspond to an L-homomorphism

$$\phi : \mathcal{L}_F \to {}^L G.$$

Then Langlands functoriality predicts the existence of an irreducible automorphic representation π' of $G'(\mathbb{A})$ which should correspond to the L-homomorphism

$$r \circ \phi : \mathcal{L}_F \to {}^L G'.$$

This conjectural relation between π and π' can be formulated without assuming the existence of $\mathcal{L}_F$ as follows. Recall that for almost all places v of F, the local component π_v of π at v is unramified, so that it determines and is determined by a $\hat{G}$-conjugacy class $c(\pi_v)$ in ${}^L G$ via the Satake isomorphism. Then π' should satisfy

$$c(\pi'_v) = r\big(c(\pi_v)\big)$$

for almost all v. Note that the Jacquet–Langlands correspondence mentioned above is the special case when $G = B^\times$, $G' = \mathrm{GL}_2$ (so that ${}^L G = {}^L G' = \mathrm{GL}_2(\mathbb{C}) \times \Gamma_F$), and r is the identity map.

Although Langlands functoriality is out of reach in general, it led to substantial developments in the theory of automorphic forms. For example, the trace formula was developed by Arthur to study automorphic representations, culminating in his book [1] which establishes the case when G is a symplectic group or a quasisplit special orthogonal group, G' is a general linear group, and r is the standard embedding. There are also other methods to attack Langlands functoriality, such as the converse theorem [12,13], the automorphic descent [24], and the theta lifting. In this report, we will discuss various aspects of the theta lifting, which can be viewed as an explicit realization in the case when (G, G') is a certain pair of classical groups.

2. THETA LIFTING

In this section, we recall the notion of the theta lifting, with emphasis on the realization of the Jacquet–Langlands correspondence. We also review some of its applications to explicit formulas for automorphic periods in terms of special values of L-functions.

2.1. Basic definitions and properties

Let F be a number field with adèle ring $\mathbb{A} = \mathbb{A}_F$. Let W be a symplectic space over F equipped with a nondegenerate bilinear alternating form $(\cdot,\cdot)_W$ and let $\mathrm{Sp}(W)$ denote the symplectic group of W. Similarly, let V be a quadratic space over F equipped with a nondegenerate bilinear symmetric form $(\cdot,\cdot)_V$ and let $\mathrm{O}(V)$ denote the orthogonal group of V. Then the pair

$$\big(\mathrm{Sp}(W), \mathrm{O}(V)\big)$$

is an example of a reductive dual pair introduced by Howe [30]. Namely, if we consider the symplectic space $\mathbb{W} = W \otimes_F V$ equipped with the form $(\cdot,\cdot)_W \otimes (\cdot,\cdot)_V$ and the natural homomorphism

$$\mathrm{Sp}(W) \times \mathrm{O}(V) \to \mathrm{Sp}(\mathbb{W}),$$

then $\mathrm{Sp}(W)$ and $\mathrm{O}(V)$ are mutual commutants in $\mathrm{Sp}(\mathbb{W})$.

Roughly speaking, the theta lifting is an integral transform with kernel given by a particular automorphic form on $\mathrm{Sp}(\mathbb{W})(\mathbb{A})$ restricted to $\mathrm{Sp}(W)(\mathbb{A}) \times \mathrm{O}(V)(\mathbb{A})$. To be precise, we need to consider the metaplectic group $\mathrm{Mp}(\mathbb{W})(\mathbb{A})$, which is a nontrivial topological central extension

$$1 \to \{\pm 1\} \to \mathrm{Mp}(\mathbb{W})(\mathbb{A}) \to \mathrm{Sp}(\mathbb{W})(\mathbb{A}) \to 1.$$

(Here we have abused notation since $\mathrm{Mp}(\mathbb{W})(\mathbb{A})$ is not the group of $\mathbb{A}$-valued points of an algebraic group over F.) This extension splits over $\mathrm{Sp}(\mathbb{W})(F)$ canonically, so that we may speak of automorphic forms on $\mathrm{Mp}(\mathbb{W})(\mathbb{A})$. We are interested in a particular representation ω of $\mathrm{Mp}(\mathbb{W})(\mathbb{A})$ (depending on a choice of a nontrivial additive character of $\mathbb{A}/F$), called the Weil representation [61], which is a representation theoretic incarnation of theta functions. This representation has an automorphic realization, i.e., there is an $\mathrm{Mp}(\mathbb{W})(\mathbb{A})$-equivariant map $\varphi \mapsto \theta_\varphi$ from ω to the space of automorphic forms on $\mathrm{Mp}(\mathbb{W})(\mathbb{A})$. On the other hand, there exists a dotted arrow making the following diagram commute:

$$\begin{array}{ccc} \mathrm{Mp}(W)(\mathbb{A}) \times \mathrm{O}(V)(\mathbb{A}) & \dashrightarrow & \mathrm{Mp}(\mathbb{W})(\mathbb{A}) \\ \downarrow & & \downarrow \\ \mathrm{Sp}(W)(\mathbb{A}) \times \mathrm{O}(V)(\mathbb{A}) & \longrightarrow & \mathrm{Sp}(\mathbb{W})(\mathbb{A}) \end{array}$$

(Note that it descends to a homomorphism from the bottom left corner if and only if $\dim V$ is even.) Thus we may regard θ_φ as an automorphic form on $\mathrm{Mp}(W)(\mathbb{A}) \times \mathrm{O}(V)(\mathbb{A})$ by restriction and associate to an automorphic form f on $\mathrm{Mp}(W)(\mathbb{A})$ an automorphic form $\theta_\varphi(f)$ on $\mathrm{O}(V)(\mathbb{A})$ by setting

$$\theta_\varphi(f)(h) = \int_{\mathrm{Sp}(W)(F)\backslash \mathrm{Mp}(W)(\mathbb{A})} \theta_\varphi(g,h)\overline{f(g)}\,dg,$$

provided the integral converges, e.g., if f is cuspidal.

For any irreducible cuspidal automorphic representation π of $\mathrm{Mp}(W)(\mathbb{A})$, we define the theta lift $\theta(\pi)$ of π as the automorphic representation of $\mathrm{O}(V)(\mathbb{A})$ spanned by $\theta_\varphi(f)$ for all $\varphi \in \omega$ and $f \in \pi$. We only consider the case when π descends (resp. does not descend) to a representation of $\mathrm{Sp}(W)(\mathbb{A})$ if $\dim V$ is even (resp. odd); otherwise $\theta(\pi)$ is always zero. To describe $\theta(\pi)$, we need to introduce the local analog of the theta lifting. First, note that the map $(\varphi, f) \mapsto \theta_\varphi(f)$ defines an element in

$$\mathrm{Hom}_{\mathrm{Mp}(W)(\mathbb{A})\times \mathrm{O}(V)(\mathbb{A})}\big(\omega \otimes \bar{\pi}, \theta(\pi)\big) \cong \mathrm{Hom}_{\mathrm{Mp}(W)(\mathbb{A})\times \mathrm{O}(V)(\mathbb{A})}\big(\omega, \pi \otimes \theta(\pi)\big)$$

since π is unitary. Recall that ω can be regarded as the restricted tensor product of the local Weil representations ω_v of $\mathrm{Mp}(\mathbb{W})(F_v)$ via the surjection $\prod'_v \mathrm{Mp}(\mathbb{W})(F_v) \to \mathrm{Mp}(\mathbb{W})(\mathbb{A})$, where $\mathrm{Mp}(\mathbb{W})(F_v)$ is the metaplectic cover of $\mathrm{Sp}(\mathbb{W})(F_v)$. Similarly, π can be decomposed as $\pi \cong \otimes_v \pi_v$, where π_v is an irreducible representation of $\mathrm{Mp}(W)(F_v)$. We define the local theta lift $\theta(\pi_v)$ of π_v as an irreducible representation of $\mathrm{O}(V)(F_v)$ such that

$$\mathrm{Hom}_{\mathrm{Mp}(W)(F_v)\times \mathrm{O}(V)(F_v)}\big(\omega_v, \pi_v \otimes \theta(\pi_v)\big) \neq 0,$$

which is unique (if it exists) by the Howe duality [23, 31, 55]. (When such a representation does not exist, we interpret $\theta(\pi_v)$ as zero.) Now assume that $\theta(\pi)$ is nonzero and cuspidal. Then it follows from the Howe duality that $\theta(\pi)$ is irreducible and can be decomposed as

$$\theta(\pi) \cong \otimes_v \theta(\pi_v).$$

Remark 2.1. We may extend the Weil representation and define the theta lifting for the pair $(\mathrm{GSp}(W), \mathrm{GO}(V))$, where $\mathrm{GSp}(W)$ and $\mathrm{GO}(V)$ are the similitude groups of W and V, respectively.

2.2. Explicit realization of the Jacquet–Langlands correspondence

From now on, we mainly consider the case when

$$\dim W = 2, \quad \dim V = 4,$$

and the discriminant of V is trivial. Then we may identify W with the space F^2, equipped with the form $((x_1, x_2), (y_1, y_2))_W = x_1 y_2 - x_2 y_1$, so that

$$\mathrm{GSp}(W) = \mathrm{GL}_2.$$

We may also identify V with a quaternion algebra B over F equipped with the form $(x, y)_V = \mathrm{Tr}_{B/F}(xy^*)$, where $\mathrm{Tr}_{B/F}$ is the reduced trace and $*$ is the main involution, so that

$$\mathrm{GO}(V)^0 = (B^\times \times B^\times)/F^\times.$$

Here $\mathrm{GO}(V)^0$ is the identity component of $\mathrm{GO}(V)$ and $B^\times \times B^\times$ acts on V by left and right multiplication.

Let π be an irreducible cuspidal automorphic representation of $\mathrm{GL}_2(\mathbb{A})$. We regard the theta lift $\theta(\pi)$ of π (restricted to $\mathrm{GO}(V)^0(\mathbb{A})$) as an automorphic representation of $B^\times(\mathbb{A}) \times B^\times(\mathbb{A})$. Then Shimizu [49] proved that

$$\theta(\pi) = \pi^B \otimes \pi^B,$$

where π^B is the Jacquet–Langlands transfer of π to $B^\times(\mathbb{A})$. (When π does not transfer to $B^\times(\mathbb{A})$, we interpret π^B as zero.)

Remark 2.2. In [36], we gave the following variant of the above realization. Let B, B_1, B_2 be three quaternion division algebras over F such that $B = B_1 \cdot B_2$ in the Brauer group. We consider a 1-dimensional Hermitian space W over B and a 2-dimensional skew-Hermitian space V over B such that

$$\mathrm{GU}(W) = B^\times, \quad \mathrm{GU}(V)^0 = (B_1^\times \times B_2^\times)/F^\times,$$

where $\mathrm{GU}(W)$ and $\mathrm{GU}(V)$ are the unitary similitude groups of W and V, respectively. Let π^B be an irreducible automorphic representation of $B^\times(\mathbb{A})$ such that its Jacquet–Langlands transfer to $\mathrm{GL}_2(\mathbb{A})$ is cuspidal. Then we have

$$\theta(\pi^B) = \pi^{B_1} \otimes \pi^{B_2},$$

where π^{B_1} and π^{B_2} are the Jacquet–Langlands transfers of π^B to $B_1^\times(\mathbb{A})$ and $B_2^\times(\mathbb{A})$, respectively. We believe that this realization is useful to study integral period relations.

2.3. Seesaw identities

One of the advantages of the theta lifting is that it produces various period relations in a simple way, which was observed by Kudla [39]. Suppose that we have two reductive dual pairs (G, H) and (G', H') in the same symplectic group such that $G \subset G'$ and $H \supset H'$. This can be illustrated by the following picture, called a seesaw diagram:

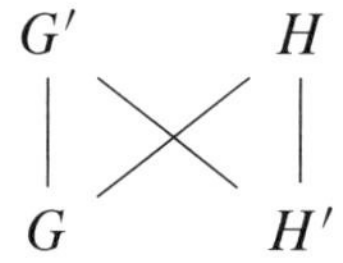

Let f and f' be automorphic forms on $G(\mathbb{A})$ and $H'(\mathbb{A})$, respectively. Then the theta lifting produces automorphic forms $\theta_\varphi(f)$ and $\theta_\varphi(f')$ on $H(\mathbb{A})$ and $G'(\mathbb{A})$, respectively, and the so-called seesaw identity

$$\begin{aligned}
\langle \theta_\varphi(f)|_{H'(\mathbb{A})}, f' \rangle &= \int_{H'(F)\backslash H'(\mathbb{A})} \theta_\varphi(f)(h') \overline{f'(h')}\, dh' \\
&= \int_{G(F)\backslash G(\mathbb{A})} \int_{H'(F)\backslash H'(\mathbb{A})} \theta_\varphi(g, h') \overline{f(g)} f'(h')\, dg\, dh' \\
&= \int_{G(F)\backslash G(\mathbb{A})} \overline{f(g)} \theta_\varphi(f')(g)\, dg = \overline{\langle f, \theta_\varphi(f')|_{G(\mathbb{A})} \rangle},
\end{aligned}$$

provided the double integral converges absolutely. Here $\langle \cdot, \cdot \rangle$ denotes the Petersson inner product.

As an example of this identity, we recall Waldspurger's formula for torus periods. We keep the setup of the previous subsection. Fix a quadratic extension E of F which embeds into B and write $B = E \oplus Ej$ with a trace zero element j in B. Let $V = V_1 \oplus V_2$ be the corresponding decomposition of quadratic spaces, so that

$$\mathrm{GO}(V_1)^0 = \mathrm{GO}(V_2)^0 = E^\times.$$

Then the identification $W \otimes_F V = (W \otimes_F V_1) \oplus (W \otimes_F V_2)$ gives rise to the following seesaw diagram:

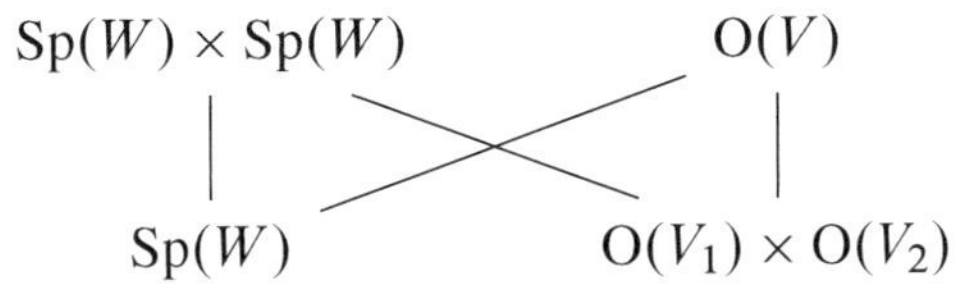

Let π^B be an irreducible automorphic representation of $B^\times(\mathbb{A})$ such that its Jacquet–Langlands transfer π to $\mathrm{GL}_2(\mathbb{A})$ is cuspidal. Let χ be an automorphic character of $\mathbb{A}_E^\times$. Assume that the product of the central character of π^B and the restriction of χ to $\mathbb{A}_F^\times$ is trivial and consider the torus period

$$P(f, \chi) = \int_{E^\times \mathbb{A}_F^\times \backslash \mathbb{A}_E^\times} f(h)\chi(h)\, dh$$

for a decomposable vector $f \in \pi^B$. Then using the above seesaw diagram, Waldspurger [54] proved that

$$\left|P(f, \chi)\right|^2 = \frac{1}{4} \frac{\zeta(2) L(1/2, \pi_E \times \chi)}{L(1, \pi, \mathrm{Ad}) L(1, \mu_{E/F})} \prod_v \alpha_v(f_v, \chi_v),$$

where

- $\zeta(s)$ is the completed Dedekind zeta function of F,
- $L(s, \pi_E \times \chi)$ is the standard L-function of the base change π_E of π to $\mathrm{GL}_2(\mathbb{A}_E)$ twisted by χ,
- $L(s, \pi, \mathrm{Ad})$ is the adjoint L-function of π,
- $L(s, \mu_{E/F})$ is the Hecke L-function of the quadratic automorphic character $\mu_{E/F}$ of $\mathbb{A}_F^\times$ associated to E/F by class field theory,
- $\alpha_v(f_v, \chi_v)$ is a certain normalized local integral of matrix coefficients.

As another example, we consider the 6-dimensional symplectic space $W' = W^3$ over F. Then the identification $W' \otimes_F V = (W \otimes_F V)^3$ gives rise to the following seesaw diagram:

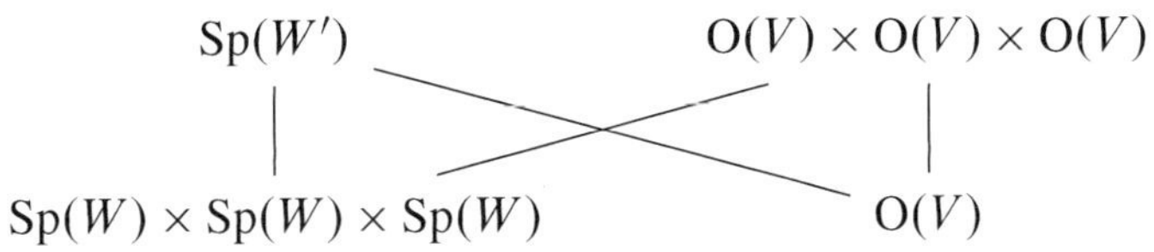

Let $\pi_1^B, \pi_2^B, \pi_3^B$ be irreducible automorphic representations of $B^\times(\mathbb{A})$ such that their Jacquet–Langlands transfers π_1, π_2, π_3 to $\mathrm{GL}_2(\mathbb{A})$ are cuspidal. Assume that the product of the central characters of $\pi_1^B, \pi_2^B, \pi_3^B$ is trivial and consider the trilinear period

$$P(f_1, f_2, f_3) = \int_{B^\times \mathbb{A}^\times \backslash B^\times(\mathbb{A})} f_1(h) f_2(h) f_3(h)\, dh$$

for decomposable vectors $f_1 \in \pi_1^B$, $f_2 \in \pi_2^B$, $f_3 \in \pi_3^B$. Then following the work of Harris–Kudla [27] and using the above seesaw diagram, we proved in [32] that

$$\left|P(f_1, f_2, f_3)\right|^2 = \frac{1}{8} \frac{\zeta(2)^2 L(1/2, \pi_1 \times \pi_2 \times \pi_3)}{L(1, \pi_1, \mathrm{Ad}) L(1, \pi_2, \mathrm{Ad}) L(1, \pi_3, \mathrm{Ad})} \prod_v \alpha_v(f_{1,v}, f_{2,v}, f_{3,v}),$$

where

- $L(s, \pi_1 \times \pi_2 \times \pi_3)$ is the triple product L-function of π_1, π_2, π_3,
- $\alpha_v(f_{1,v}, f_{2,v}, f_{3,v})$ is a certain normalized local integral of matrix coefficients.

Remark 2.3. The above two formulas are special cases of the Gross–Prasad conjecture [25,26] and its refinement [33]. This conjecture (for special orthogonal groups) was extended to all classical groups by Gan–Gross–Prasad [17], and after the breakthrough of Zhang [63,64], the global conjecture for unitary groups has been proved in a series of works [8–11,62] using the relative trace formula. We should also mention the stunning work of Waldspurger [57–60], which led to the proof of the local Gan–Gross–Prasad conjecture for Bessel models [5–7,45] and Fourier–Jacobi models [2,19] in the p-adic case, where the theta lifting is used to deduce the latter from the former.

3. THE SHIMURA–WALDSPURGER CORRESPONDENCE

In this section, we review some applications of the theta lifting to automorphic forms on metaplectic groups.

3.1. Modular forms of half-integral weight

The theta function

$$\theta(z) = \sum_{n=-\infty}^{\infty} e^{2\pi i n^2 z}$$

is a modular form of weight $1/2$ and its significance is well known. Thus it is natural to study modular forms of half-integral weight, but Hecke [28, P. 152] realized the difficulty in developing the arithmetic theory; the Hecke operator T_n is zero unless n is a square. In 1973, Shimura [50] revolutionized the theory of modular forms of half-integral weight by relating them to modular forms of integral weight, i.e., he constructed a modular form of weight $2k$ from a cusp form of weight $k + 1/2$ by using the converse theorem, where k is a positive integer. Soon after the discovery of this correspondence, Niwa [46] and Shintani [51] gave an alternative construction using the theta lifting. This was further investigated by Waldspurger [53, 56] in the framework of automorphic representations. Namely, he established a correspondence between automorphic representations of $\mathrm{Mp}_2(\mathbb{A})$ (where $\mathrm{Mp}_2(\mathbb{A})$ is the metaplectic cover of $\mathrm{SL}_2(\mathbb{A})$) and those of $\mathrm{PGL}_2(\mathbb{A})$, which can be viewed as an example of Langlands functoriality.

3.2. Global correspondence

Now we discuss a generalization of the Shimura–Waldspurger correspondence to metaplectic groups of higher rank. Let F be a number field with adèle ring $\mathbb{A}$. We denote by Sp_{2n} the symplectic group of rank n over F and by $\mathrm{Mp}_{2n}(\mathbb{A})$ the metaplectic cover of $\mathrm{Sp}_{2n}(\mathbb{A})$. Recall that this cover splits over $\mathrm{Sp}_{2n}(F)$ canonically, so that we may speak of the unitary representation of $\mathrm{Mp}_{2n}(\mathbb{A})$ on the Hilbert space

$$L^2\big(\mathrm{Sp}_{2n}(F)\backslash\mathrm{Mp}_{2n}(\mathbb{A})\big)$$

given by right translation. Since we are interested in genuine automorphic representations of $\mathrm{Mp}_{2n}(\mathbb{A})$, i.e., those which do not descend to representations of $\mathrm{Sp}_{2n}(\mathbb{A})$, we only consider its subspace

$$L^2(\mathrm{Mp}_{2n})$$

on which the central subgroup $\{\pm 1\}$ acts by the nontrivial character. Write

$$L^2(\mathrm{Mp}_{2n}) = L^2_{\mathrm{disc}}(\mathrm{Mp}_{2n}) \oplus L^2_{\mathrm{cont}}(\mathrm{Mp}_{2n})$$

for the decomposition into the discrete part and the continuous part. Then the theory of Eisenstein series gives an explicit description of $L^2_{\mathrm{cont}}(\mathrm{Mp}_{2n})$ in terms of automorphic discrete spectra of proper Levi subgroups of Mp_{2n}, i.e., $\mathrm{GL}_{n_1} \times \cdots \times \mathrm{GL}_{n_k} \times \mathrm{Mp}_{2n_0}$ with $n_1 + \cdots + n_k + n_0 = n$ and $n_0 < n$. Thus the problem is to describe the irreducible decomposition of $L^2_{\mathrm{disc}}(\mathrm{Mp}_{2n})$.

To attack this problem, it is better to divide it into two parts as follows:

(1) Describe the decomposition of $L^2_{\mathrm{disc}}(\mathrm{Mp}_{2n})$ into near equivalence classes. Here we say that two irreducible genuine representations $\pi \cong \otimes_v \pi_v$ and $\pi' \cong \otimes_v \pi'_v$ of $\mathrm{Mp}_{2n}(\mathbb{A})$ are nearly equivalent if π_v and π'_v are equivalent for almost all places v of F. (In particular, if π and π' are equivalent, then they are nearly equivalent.) Note that π_v is unramified for almost all v, so that it determines and is determined by a semisimple conjugacy class $c_{\psi_v}(\pi_v)$ in $\mathrm{Sp}_{2n}(\mathbb{C})$ (depending on a choice of a nontrivial additive character ψ_v of F_v) via the Satake isomorphism. In other words, the near equivalence classes of irreducible genuine representations of $\mathrm{Mp}_{2n}(\mathbb{A})$ can be parametrized by families of semisimple conjugacy classes

$$\{c_v\}_v$$

in $\mathrm{Sp}_{2n}(\mathbb{C})$, where we identify two families if they are equal for almost all v. Thus we want to describe the families $\{c_v\}_v$ which correspond to the near equivalence classes in $L^2_{\mathrm{disc}}(\mathrm{Mp}_{2n})$.

(2) Describe the irreducible decomposition of each near equivalence class. Namely, for any near equivalence class C in $L^2_{\mathrm{disc}}(\mathrm{Mp}_{2n})$ and any irreducible genuine representation π of $\mathrm{Mp}_{2n}(\mathbb{A})$, we want to give an explicit formula for the multiplicity of π in C in terms of the classification of representations.

In [20], we solved (1) completely and (2) partially; we described the families $\{c_v\}_v$ as above in terms of automorphic representations of general linear groups, and admitting that Arthur's endoscopic classification [1] can be extended to nonsplit odd special orthogonal groups, we established the multiplicity formula for the tempered part of $L^2_{\mathrm{disc}}(\mathrm{Mp}_{2n})$.

Now we state the first result precisely.

Theorem 3.1 ([20]). *Fix a nontrivial additive character $\psi = \otimes_v \psi_v$ of $\mathbb{A}/F$. Then we have a decomposition*

$$L^2_{\mathrm{disc}}(\mathrm{Mp}_{2n}) = \bigoplus_{\phi} L^2_{\phi}(\mathrm{Mp}_{2n}),$$

where ϕ runs over elliptic A-parameters for Mp_{2n}. Here an elliptic A-parameter for Mp_{2n} is defined to be a formal finite direct sum

$$\bigoplus_i \phi_i \otimes S_{d_i}$$

(which is a substitute for a hypothetical L-homomorphism $\mathcal{L}_F \times \mathrm{SL}_2(\mathbb{C}) \to \mathrm{Sp}_{2n}(\mathbb{C})$), where

- *ϕ_i is an irreducible self-dual cuspidal automorphic representation of $\mathrm{GL}_{n_i}(\mathbb{A})$ (which is hypothetically identified with an n_i-dimensional irreducible representation of $\mathcal{L}_F$),*
- *S_{d_i} is the d_i-dimensional irreducible representation of $\mathrm{SL}_2(\mathbb{C})$,*
- *if d_i is odd, then ϕ_i is symplectic, i.e., the exterior square L-function $L(s, \phi_i, \wedge^2)$ has a pole at $s = 1$ (and hence n_i is even),*
- *if d_i is even, then ϕ_i is orthogonal, i.e., the symmetric square L-function $L(s, \phi_i, \mathrm{Sym}^2)$ has a pole at $s = 1$,*
- *if $i \neq j$, then $(\phi_i, d_i) \neq (\phi_j, d_j)$,*
- *$\sum_i n_i d_i = 2n$.*

Also, $L^2_{\phi}(\mathrm{Mp}_{2n})$ is defined as the near equivalence class in $L^2_{\mathrm{disc}}(\mathrm{Mp}_{2n})$ which corresponds to the family of semisimple conjugacy classes

$$\{c_v(\phi_v)\}_v$$

in $\mathrm{Sp}_{2n}(\mathbb{C})$ given as follows (so that any irreducible summand π of $L^2_{\phi}(\mathrm{Mp}_{2n})$ satisfies $c_{\psi_v}(\pi_v) = c_v(\phi_v)$ for almost all v). Suppose that v is finite and $\phi_{i,v}$ is unramified for all i. Let $c_v(\phi_{i,v})$ be the semisimple conjugacy class in $\mathrm{GL}_{n_i}(\mathbb{C})$ which corresponds to $\phi_{i,v}$ and put

$$Q_v(d) = \begin{pmatrix} q_v^{(d-1)/2} & & & \\ & q_v^{(d-3)/2} & & \\ & & \ddots & \\ & & & q_v^{-(d-1)/2} \end{pmatrix}$$

for any positive integer d, where q_v is the cardinality of the residue field of F_v. We regard $c_v(\phi_{i,v}) \otimes Q_v(d_i)$ as a semisimple conjugacy class in $\mathrm{Sp}_{n_i d_i}(\mathbb{C})$. Then we set

$$c_v(\phi_v) = \bigoplus_i c_v(\phi_{i,v}) \otimes Q_v(d_i).$$

To state the second result precisely, we need to introduce more notation. For each place v of F, let $\mathcal{W}_{F_v}$ be the Weil group of F_v and put

$$\mathcal{L}_{F_v} = \begin{cases} \mathcal{W}_{F_v} & \text{if } v \text{ is infinite,} \\ \mathcal{W}_{F_v} \times \mathrm{SL}_2(\mathbb{C}) & \text{if } v \text{ is finite.} \end{cases}$$

Let $\phi = \bigoplus_i \phi_i \otimes S_{d_i}$ be an elliptic A-parameter for Mp_{2n}. We regard its local component $\phi_v = \bigoplus_i \phi_{i,v} \otimes S_{d_i}$ at v as a local A-parameter

$$\phi_v : \mathcal{L}_{F_v} \times \mathrm{SL}_2(\mathbb{C}) \to \mathrm{Sp}_{2n}(\mathbb{C})$$

via the local Langlands correspondence. Note that

$$c_v(\phi_v) = \phi_v \left(\mathrm{Fr}_v, \begin{pmatrix} q_v^{1/2} & \\ & q_v^{-1/2} \end{pmatrix} \right)$$

for almost all v, where Fr_v is a Frobenius element at v. We denote by $\mathcal{S}_{\phi_v}$ the component group of the centralizer of ϕ_v in $\mathrm{Sp}_{2n}(\mathbb{C})$, which is an elementary abelian 2-group, and by

$$\mathcal{S}_\phi = \bigoplus_i (\mathbb{Z}/2\mathbb{Z}) a_i$$

the global component group of ϕ, which is formally defined as an elementary abelian 2-group with a basis $\{a_i\}_i$ indexed by $\{\phi_i \otimes S_{d_i}\}_i$. Then we have a natural homomorphism $\mathcal{S}_\phi \to \mathcal{S}_{\phi_v}$ for all v. We also consider the compact abelian group $\mathcal{S}_{\phi,\mathbb{A}} = \prod_v \mathcal{S}_{\phi_v}$ and the diagonal map

$$\Delta : \mathcal{S}_\phi \to \mathcal{S}_{\phi,\mathbb{A}}.$$

Theorem 3.2 ([20]). *Assume that ϕ is tempered, i.e., $d_i = 1$ for all i. Then we have a decomposition*

$$L^2_\phi(\mathrm{Mp}_{2n}) \cong \bigoplus_\eta m_\eta \pi_\eta,$$

where $\eta = \otimes_v \eta_v$ runs over continuous characters of $\mathcal{S}_{\phi,\mathbb{A}}$. Here π_η is defined as the restricted tensor product of representations π_{η_v} in the local L-packets

$$\Pi_{\phi_v}(\mathrm{Mp}_{2n}(F_v))$$

associated to ϕ_v (depending on ψ_v), which consist of irreducible genuine representations of $\mathrm{Mp}_{2n}(F_v)$ indexed by characters of $\mathcal{S}_{\phi_v}$. Also, if we define a character ε_ϕ of $\mathcal{S}_\phi$ by

$$\varepsilon_\phi(a_i) = \varepsilon(1/2, \phi_i),$$

where $\varepsilon(s, \phi_i)$ is the standard ε-function of ϕ_i, then m_η is given by

$$m_\eta = \begin{cases} 1 & \textit{if } \eta \circ \Delta = \varepsilon_\phi, \\ 0 & \textit{otherwise.} \end{cases}$$

Remark 3.3. In fact, we gave another proof of the result of Waldspurger for Mp_2 [53, 56], noting that an irreducible cuspidal automorphic representation of $\mathrm{GL}_2(\mathbb{A})$ is symplectic if and only if its central character is trivial.

Remark 3.4. If we denote by SO_{2n+1} the split odd special orthogonal group of rank n over F and by $L^2_{\mathrm{disc}}(\mathrm{SO}_{2n+1})$ the discrete part of $L^2(\mathrm{SO}_{2n+1}(F)\backslash \mathrm{SO}_{2n+1}(\mathbb{A}))$, then the decomposition of $L^2_{\mathrm{disc}}(\mathrm{Mp}_{2n})$ is similar to that of $L^2_{\mathrm{disc}}(\mathrm{SO}_{2n+1})$ given by Arthur [1], except that the condition $\eta \circ \Delta = \varepsilon_\phi$ in the former has to be replaced by $\eta \circ \Delta = 1$ in the latter.

In the proof of his result for $n = 1$, Waldspurger used the theta lifting between Mp_2 and (inner forms of) $\mathrm{PGL}_2 \cong \mathrm{SO}_3$. Thus, in general, it would be natural to use the theta lifting between Mp_{2n} and (inner forms of) SO_{2n+1}, and then transfer Arthur's endoscopic classification from SO_{2n+1} to Mp_{2n}. However, there is a serious obstacle in this approach. Indeed, if π is an irreducible genuine cuspidal automorphic representation of $\mathrm{Mp}_{2n}(\mathbb{A})$ and its standard L-function $L(s,\pi)$ vanishes at $s = 1/2$, then the theta lift of π to $\mathrm{SO}_{2n+1}(\mathbb{A})$ is zero. When $n = 1$, Waldspurger proved that the twisted standard L-function $L(s,\pi,\chi)$ does not vanish at $s = 1/2$ for some quadratic automorphic character χ of $\mathbb{A}^\times$ and could use the twisted theta lifting to establish the desired correspondence. But for general n, the existence of such a character χ is considered extremely difficult to prove.

To circumvent this difficulty, we used the theta lifting in the so-called stable range studied by Li [44]. More precisely, for any irreducible genuine representation π of $\mathrm{Mp}_{2n}(\mathbb{A})$, we consider its (abstract) theta lift

$$\theta^{\mathrm{abs}}(\pi) = \otimes_v \theta(\pi_v)$$

to $\mathrm{SO}_{2r+1}(\mathbb{A})$ with $r \gg 2n$. Then it follows from the result of Li that if π occurs in $L^2_{\mathrm{disc}}(\mathrm{Mp}_{2n})$, then $\theta^{\mathrm{abs}}(\pi)$ occurs in $L^2_{\mathrm{disc}}(\mathrm{SO}_{2r+1})$. Combining this with the analytic theory of standard L-functions, we may deduce Theorem 3.1 from Arthur's endoscopic classification for SO_{2r+1}. Moreover, if π is an irreducible summand of the tempered part of $L^2_{\mathrm{disc}}(\mathrm{Mp}_{2n})$, then we proved that

$$m\big(\theta^{\mathrm{abs}}(\pi)\big) = m(\pi),$$

where $m(\cdot)$ denotes the multiplicity in the automorphic discrete spectrum. (We expect that this equality holds for any irreducible summand π of $L^2_{\mathrm{disc}}(\mathrm{Mp}_{2n})$.) Using this and describing the local theta lifting between Mp_{2n} and SO_{2r+1} explicitly, we may deduce Theorem 3.2.

Remark 3.5. When $n = 2$ and ϕ is nontempered, we proved a similar decomposition of $L^2_\phi(\mathrm{Mp}_4)$ in [21]. Note that π_η is not necessarily irreducible and ε_ϕ has to be modified in this case.

3.3. Local correspondence

There is a local analog of the above correspondence, called the local Shimura correspondence. For simplicity, we only consider the p-adic case and write F for a finite extension of $\mathbb{Q}_p$. We are interested in the set

$$\mathrm{Irr}\,\mathrm{Mp}_{2n}(F)$$

of equivalence classes of irreducible genuine representations of the metaplectic group $\mathrm{Mp}_{2n}(F)$. Recall that there are precisely two $(2n+1)$-dimensional quadratic spaces V^+ and V^- over F with trivial discriminant (up to isometry). Let $\mathrm{SO}(V^+)$ and $\mathrm{SO}(V^-)$ denote the special orthogonal groups of V^+ and V^-, respectively. Then the local Shimura correspondence, which was established by Gan–Savin [22] in the p-adic case, says that there is a bijection (depending on a choice of a nontrivial additive character ψ of F)

$$\theta : \operatorname{Irr} \mathrm{Mp}_{2n}(F) \to \operatorname{Irr} \mathrm{SO}(V^+) \sqcup \operatorname{Irr} \mathrm{SO}(V^-)$$

given by the local theta lifting. Namely, for any irreducible genuine representation π of $\mathrm{Mp}_{2n}(F)$, $\theta(\pi)$ is defined as the unique irreducible representation of $\mathrm{SO}(V^\varepsilon)$ with the unique sign $\varepsilon = \pm$ such that

$$\operatorname{Hom}_{\mathrm{Mp}_{2n}(F)\times \mathrm{SO}(V^\varepsilon)}\big(\omega^\varepsilon, \pi \otimes \theta(\pi)\big) \neq 0,$$

where ω^ε is the Weil representation of $\mathrm{Mp}_{2n}(F) \times \mathrm{SO}(V^\varepsilon)$ (depending on ψ). Moreover, they proved various natural properties:

- θ preserves the square-integrability,
- θ preserves the temperedness,
- θ is compatible with the theory of R-groups,
- θ is compatible with the Langlands classification,

and used θ to transfer the local Langlands correspondence from $\mathrm{SO}(V^\varepsilon)$ to $\mathrm{Mp}_{2n}(F)$. (In particular, this defines the local L-packets in the statement of Theorem 3.2.)

Remark 3.6. The local theta lifting has also been described for other reductive dual pairs in terms of the local Langlands correspondence. See [3, 4, 19] for recent progress.

As in Section 2.3, the local theta lifting can produce various relations between local analogs of periods. For example, we consider an irreducible genuine square-integrable representation π of $\mathrm{Mp}_{2n}(F)$ and its formal degree $d(\pi)$. Recall that $d(\pi)$ is defined as the positive real number for which the Schur orthogonality relation

$$\int_{\mathrm{Mp}_{2n}(F)} \big\langle \pi(g)v_1, v_2\big\rangle \overline{\big\langle \pi(g)v_3, v_4\big\rangle}\, dg = \frac{1}{d(\pi)} \langle v_1, v_3\rangle \overline{\langle v_2, v_4\rangle}$$

holds for all $v_1, \dots, v_4 \in \pi$, where $\langle \cdot, \cdot \rangle$ is an invariant Hermitian inner product on π. Note that $d(\pi)$ depends on the choice of a Haar measure dg on $\mathrm{Mp}_{2n}(F)$, but we take the measure determined by a Chevalley basis of the Lie algebra of $\mathrm{Mp}_{2n}(F)$ and a fixed nontrivial additive character ψ of F. The above relation suggests that $d(\pi)$ is a generalization of the dimension of an irreducible representation of a compact group, so that it is a fundamental invariant of a representation. Since a matrix coefficient $g \mapsto \langle \pi(g)v_1, v_2\rangle$ is a local analog of an automorphic form, we may also interpret $d(\pi)^{-1}$ as a local period. Then we proved in [18] that

$$d\big(\theta(\pi)\big) = d(\pi),$$

which is a local analog of the Rallis inner product formula [47], by using the doubling seesaw diagram:

$$\begin{array}{ccc} \mathrm{Mp}_{4n} & & \mathrm{SO}_{2n+1} \times \mathrm{SO}_{2n+1} \\ | & \times & | \\ \mathrm{Mp}_{2n} \times \mathrm{Mp}_{2n} & & \mathrm{SO}_{2n+1} \end{array}$$

Remark 3.7. Recall from Section 2.3 that some automorphic periods can be expressed in terms of special values of L-functions. Similarly, formal degrees should be expressed in terms of arithmetic invariants as follows. Let G be a connected reductive group over F. For simplicity, we assume that G is a pure inner form of a quasisplit group and the center of G is anisotropic. Let π be an irreducible square-integrable representation of $G(F)$. Let $d(\pi)$ denote the formal degree of π with respect to the Haar measure on $G(F)$ determined by a Chevalley basis of the Lie algebra of the split form of G and a fixed nontrivial additive character ψ of F. Then the formal degree conjecture [29] says that

$$d(\pi) = \frac{\dim \eta}{|\mathcal{S}_\phi|} \big|\gamma(0, \mathrm{Ad} \circ \phi, \psi)\big|,$$

where

- $\phi : \mathcal{L}_F \to {}^L G$ is the L-parameter (conjecturally) associated to π,
- $\mathcal{S}_\phi$ is the component group of the centralizer of ϕ in $\hat{G}$,
- η is the irreducible representation of $\mathcal{S}_\phi$ (conjecturally) associated to π,
- Ad is the adjoint representation of ${}^L G$ on its Lie algebra,
- $\gamma(s, \mathrm{Ad} \circ \phi, \psi)$ is the local γ-factor given by

$$\gamma(s, \mathrm{Ad} \circ \phi, \psi) = \varepsilon(s, \mathrm{Ad} \circ \phi, \psi) \frac{L(1-s, \mathrm{Ad} \circ \phi)}{L(s, \mathrm{Ad} \circ \phi)}.$$

In [34], we proved this conjecture for (inner forms of) SO_{2n+1} and its analog for Mp_{2n} by using the main identity of Lapid–Mao [43] and the above relation between formal degrees.

4. GEOMETRIC REALIZATION OF THE JACQUET–LANGLANDS CORRESPONDENCE

Let π be an irreducible automorphic representation of $G(\mathbb{A})$ and suppose that π is cohomological, so that π occurs in the cohomology $H^*(X, \mathbb{C})$, where G is a connected reductive group and X is a locally symmetric space for G. Then it is natural to ask whether functorial transfers of π can be realized geometrically. In this section, we discuss the simplest example, i.e., the Jacquet–Langlands correspondence for GL_2 and its inner forms.

Let F be a totally real number field. Let $\mathbb{A}$ and $\mathbb{A}_f$ denote the rings of adèles and finite adèles of F, respectively. Let π be an irreducible cuspidal automorphic representation of $\mathrm{GL}_2(\mathbb{A})$ such that π_v is the discrete series of weight 2 for all infinite places v of F. For simplicity, we assume that the central character of π is trivial, the level of π is square-free, and the Hecke eigenvalues of π lie in $\mathbb{Q}$. Let B be a quaternion division algebra over F. For each place v of F, put $B_v = B \otimes_F F_v$. Let $\mathcal{V}_B$ be the set of infinite places v of F such that B_v is split. Assume that $\mathcal{V}_B \neq \varnothing$ and put $d = |\mathcal{V}_B|$. We denote by X_B the Shimura variety for $B^\times$ (with respect to some neat open compact subgroup K_f of $B^\times(\mathbb{A}_f)$), which is a d-dimensional smooth projective variety over the reflex field F', so that

- F' is the number field contained in $\mathbb{C}$ such that

  $$\mathrm{Gal}(\bar{\mathbb{Q}}/F') = \{\sigma \in \mathrm{Gal}(\bar{\mathbb{Q}}/\mathbb{Q}) \mid \sigma\mathcal{V}_B = \mathcal{V}_B\},$$

  where $\bar{\mathbb{Q}}$ denotes the algebraic closure of $\mathbb{Q}$ in $\mathbb{C}$,

- the $\mathbb{C}$-valued points of X_B are given by

  $$X_B(\mathbb{C}) = B^\times \backslash (\mathfrak{H}^\pm)^d \times B^\times(\mathbb{A}_f)/K_f,$$

  where $\mathfrak{H}^\pm$ is the union of the upper and lower half-planes.

Now assume that the Jacquet–Langlands transfer π^B of π to $B^\times(\mathbb{A})$ exists, which is the case if and only if π_v is a discrete series for all v at which B is ramified, and that K_f is chosen appropriately so that $\dim(\pi_f^B)^{K_f} = 1$. Here π_f^B is the finite component of π^B and $(\pi_f^B)^{K_f}$ is the space of K_f-fixed vectors in π_f^B. Then it follows from Matsushima's formula that π_B occurs in the cohomology $H^*(X_B, \mathbb{C})$. More precisely, we consider the rational cohomology $H^*(X_B, \mathbb{Q})$ and its π-isotypic component

$$H^*(X_B, \mathbb{Q})_\pi = \{\alpha \in H^*(X_B, \mathbb{Q}) \mid T_v\alpha = \chi_{\pi_v}(T_v)\alpha \text{ for all } T_v \in \mathcal{H}_v \text{ and almost all } v\},$$

where $\mathcal{H}_v = \mathbb{Q}[K_v \backslash B_v^\times / K_v]$ is the Hecke algebra with respect to the standard maximal compact subgroup K_v of $B_v^\times \cong \mathrm{GL}_2(F_v)$ and $\chi_{\pi_v} : \mathcal{H}_v \to \mathbb{Q}$ is the character by which $\mathcal{H}_v$ acts on $\pi_v^{K_v}$. Then $H^*(X_B, \mathbb{Q})_\pi$ is concentrated in the middle degree d and $H^d(X_B, \mathbb{Q})_\pi$ is a 2^d-dimensional vector space over $\mathbb{Q}$. Moreover, for any prime ℓ, the ℓ-adic representation of $\mathrm{Gal}(\bar{\mathbb{Q}}/F')$ on

$$H^d(X_B, \mathbb{Q})_\pi \otimes_{\mathbb{Q}} \mathbb{Q}_\ell \cong H^d_{\text{ét}}(X_B \times_{F'} \bar{\mathbb{Q}}, \mathbb{Q}_\ell)_\pi$$

is given by the so-called tensor induction of $\rho_{\pi,\ell}$, where $\rho_{\pi,\ell}$ is the 2-dimensional ℓ-adic representation associated to π. We remark that it depends only on $\rho_{\pi,\ell}$ and $\mathcal{V}_B$.

Suppose that we have two quaternion algebras B_1 and B_2 as above such that

$$\mathcal{V}_{B_1} = \mathcal{V}_{B_2}.$$

Write $X_1 = X_{B_1}$ and $X_2 = X_{B_2}$ for the corresponding Shimura varieties, which are of the same dimension $d = |\mathcal{V}_{B_1}| = |\mathcal{V}_{B_2}|$ over the same reflex field F'. Then we have

(1) an (abstract) isomorphism

$$H^d(X_1, \mathbb{C})_\pi \cong H^d(X_2, \mathbb{C})_\pi$$

which preserves the Hodge decomposition,

(2) an (abstract) isomorphism

$$H^d(X_1, \mathbb{Q}_\ell)_\pi \cong H^d(X_2, \mathbb{Q}_\ell)_\pi$$

of ℓ-adic representations of $\mathrm{Gal}(\bar{\mathbb{Q}}/F')$ for all ℓ.

Conjecturally, these isomorphisms are obtained from a single isomorphism

$$H^d(X_1, \mathbb{Q})_\pi \cong H^d(X_2, \mathbb{Q})_\pi$$

given as follows. By (2) and the Künneth formula, the space of $\mathrm{Gal}(\bar{\mathbb{Q}}/F')$-fixed vectors in

$$H^{2d}(X_1 \times X_2, \mathbb{Q}_\ell(d))_{\pi \otimes \pi}$$

is nonzero. Hence the Tate conjecture predicts the existence of an algebraic cycle $Z \in \mathrm{CH}^d(X_1 \times X_2)$ which realizes (1) and (2). Namely, let p_1 and p_2 be the two projections

$$\begin{array}{ccccc} & & X_1 \times X_2 & & \\ & \swarrow^{p_1} & & \searrow^{p_2} & \\ X_1 & & & & X_2 \end{array}$$

and consider the following map:

$$\begin{array}{ccc} H^d(X_1, \mathbb{Q}) \xrightarrow{p_1^*} & H^d(X_1 \times X_2, \mathbb{Q}) & \\ & \downarrow \cup[Z] & \\ & H^{3d}\big(X_1 \times X_2, \mathbb{Q}(d)\big) & \xrightarrow{p_{2*}} H^d(X_2, \mathbb{Q}). \end{array}$$

Then it induces an isomorphism

$$\iota_Z : H^d(X_1, \mathbb{Q})_\pi \to H^d(X_2, \mathbb{Q})_\pi$$

such that

- $\iota_Z \otimes \mathrm{id}_{\mathbb{C}}$ preserves the Hodge decomposition,
- $\iota_Z \otimes \mathrm{id}_{\mathbb{Q}_\ell}$ is $\mathrm{Gal}(\bar{\mathbb{Q}}/F')$-equivariant for all ℓ.

When $d = 1$, the existence of Z in fact follows from the result of Faltings [16]. But for general d, this remains an open problem. On the other hand, noting that the Hodge conjecture reduces it to finding a Hodge cycle on $X_1 \times X_2$, i.e., an element in

$$H^{2d}(X_1 \times X_2, \mathbb{Q}) \cap H^{d,d}(X_1 \times X_2),$$

we gave the following evidence.

Theorem 4.1 ([35]). *Assume that B_1 and B_2 are ramified at some infinite place v of F. Then there exists a Hodge cycle ξ on $X_1 \times X_2$ which induces an isomorphism*

$$\iota_\xi : H^d(X_1, \mathbb{Q})_\pi \to H^d(X_2, \mathbb{Q})_\pi$$

such that

- $\iota_\xi \otimes \mathrm{id}_{\mathbb{C}}$ *preserves the Hodge decomposition,*
- $\iota_\xi \otimes \mathrm{id}_{\mathbb{Q}_\ell}$ *is* $\mathrm{Gal}(\bar{\mathbb{Q}}/F')$*-equivariant for all* ℓ.

Our proof proceeds as follows. First, we choose an ambient variety X equipped with an embedding $j : X_1 \times X_2 \hookrightarrow X$. Then we construct a class $\Xi \in H^{d,d}(X)$ such that the $(\pi \otimes \pi)$-isotypic component $(j^*\Xi)_{\pi\otimes\pi}$ of the pullback $j^*\Xi \in H^{d,d}(X_1 \times X_2)$ is nonzero. Finally, we modify Ξ in such a way that Ξ lies in $H^{2d}(X, \mathbb{Q})$ and $\xi = (j^*\Xi)_{\pi\otimes\pi}$ is the desired Hodge cycle.

More precisely, fix a totally imaginary quadratic extension E of F which embeds into B_1 and B_2. For $i = 1, 2$, let $\mathbf{V}_i = B_i$ be the 2-dimensional Hermitian space over E such that

$$\mathrm{GU}(\mathbf{V}_i) = (B_i^\times \times E^\times)/F^\times.$$

Then we may replace X_i by the Shimura variety for $\mathrm{GU}(\mathbf{V}_i)$. Consider the 4-dimensional Hermitian space $\mathbf{V} = \mathbf{V}_1 \oplus \mathbf{V}_2$ over E and put $\mathbf{G} = \mathrm{GU}(\mathbf{V})$. Note that if we write $v_1, \dots, v_d$ (resp. $v_{d+1}, \dots, v_{[F:\mathbb{Q}]}$) for the infinite places of F at which B_1 and B_2 are split (resp. ramified), then we have

$$\mathbf{G}(F_{v_i}) = \begin{cases} \mathrm{GU}(2,2) & \text{if } i \le d, \\ \mathrm{GU}(4) & \text{if } i > d. \end{cases}$$

Put $\mathbf{G}_\infty = \mathbf{G}(F \otimes_{\mathbb{Q}} \mathbb{R})$ and let $\mathfrak{g}_\infty$ denote the complexified Lie algebra of $\mathbf{G}_\infty$. Let $\mathbf{K}_\infty$ be the standard maximal connected compact modulo center subgroup of $\mathbf{G}_\infty$. Let X be the Shimura variety for $\mathbf{G}$ (with respect to some neat open compact subgroup $\mathbf{K}_f$ of $\mathbf{G}(\mathbb{A}_f)$), which is equipped with the embedding $j : X_1 \times X_2 \hookrightarrow X$ induced by the natural embedding

$$\mathrm{G}\big(\mathrm{U}(\mathbf{V}_1) \times \mathrm{U}(\mathbf{V}_2)\big) \hookrightarrow \mathrm{GU}(\mathbf{V}),$$

where the left-hand side is the subgroup of $\mathrm{GU}(\mathbf{V}_1) \times \mathrm{GU}(\mathbf{V}_2)$ which consists of elements with the same similitude factor. Then Matsushima's formula says that

$$H^*(X, \mathbb{C}) \cong \bigoplus_\sigma m(\sigma) H^*(\mathfrak{g}_\infty, \mathbf{K}_\infty; \sigma_\infty) \otimes \sigma_f^{\mathbf{K}_f},$$

where

- σ runs over irreducible unitary representations of $\mathbf{G}(\mathbb{A})$,
- σ_∞ and σ_f are the infinite and finite components of σ, respectively,
- $m(\sigma)$ is the multiplicity of σ in the automorphic discrete spectrum of $\mathbf{G}$,
- $H^*(\mathfrak{g}_\infty, \mathbf{K}_\infty; \sigma_\infty)$ is the relative Lie algebra cohomology,
- $\sigma_f^{\mathbf{K}_f}$ is the space of $\mathbf{K}_f$-fixed vectors in σ_f.

Hence, to construct a class Ξ as above, we need to find an irreducible automorphic representation σ of $\mathbf{G}(\mathbb{A})$ which satisfies the following properties:

(a) To achieve the condition $\Xi \in H^{d,d}(X)$, we require that

$$H^{d,d}(\mathfrak{g}_\infty, \mathbf{K}_\infty; \sigma_\infty) \neq 0.$$

If this is the case, then it follows from the result of Vogan–Zuckerman [52] that σ_{v_i} (restricted to $\mathrm{U}(\mathbf{V})(F_{v_i})$) is equal to

$$\begin{cases} \mathbf{1} \text{ or } A_{\mathfrak{q}} & \text{if } i \leq d, \\ \mathbf{1} & \text{if } i > d. \end{cases}$$

Here $\mathbf{1}$ denotes the trivial representation and $A_{\mathfrak{q}}$ is the cohomological representation of $\mathrm{U}(2,2)$ associated to the θ-stable parabolic subalgebra $\mathfrak{q}$ with Levi component $\mathfrak{u}(1,1) \oplus \mathfrak{u}(1,1)$. We further require that $\sigma_{v_i} = A_{\mathfrak{q}}$ if $i \leq d$ in order not to make σ 1-dimensional.

(b) To achieve the condition $(j^*\Xi)_{\pi\otimes\pi} \neq 0$, we require the nonvanishing of the automorphic period

$$\sigma \otimes \overline{(\pi^{B_1} \otimes \pi^{B_2})} \to \mathbb{C}.$$

For (a), we use the following variant of the theta lifting from SL_2 to $\mathrm{SO}(4,2) \sim \mathrm{U}(2,2)$ or $\mathrm{SO}(6) \sim \mathrm{U}(4)$, where $\sim$ denotes an isogeny. Let B be the quaternion algebra over F such that $B = B_1 \cdot B_2$ in the Brauer group, so that B is split at all infinite places of F. We may regard $V = \wedge^2 \mathbf{V}$ as a 3-dimensional skew-Hermitian space over B such that

$$\mathrm{GU}(V)^0/F^\times \cong \mathrm{GU}(\mathbf{V})/E^\times.$$

Let $W = B$ be the 1-dimensional Hermitian space over B such that

$$\mathrm{GU}(W) = B^\times.$$

Then any σ as in (a) with trivial central character is a theta lift of an irreducible cuspidal automorphic representation τ of $\mathrm{GU}(W)(\mathbb{A})$ such that τ_v is the discrete series of weight 3 for all infinite places v of F. For (b), we can easily find the corresponding τ by using the following seesaw diagram:

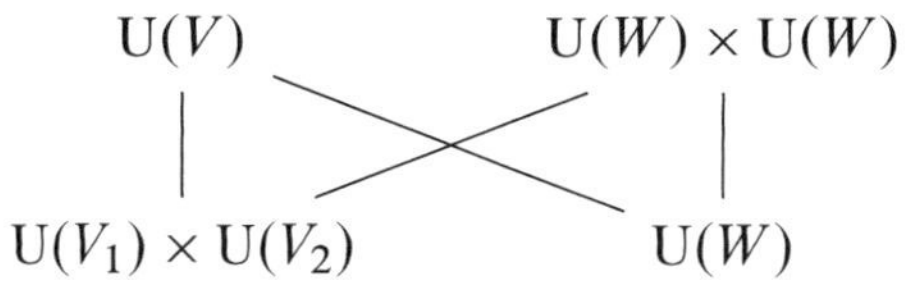

Here $V = V_1 \oplus V_2$ is a decomposition into 1- and 2-dimensional skew-Hermitian spaces over B such that

$$\mathrm{GU}(V_1)^0 = E^\times, \quad \mathrm{GU}(V_2)^0 = (B_1^\times \times B_2^\times)/F^\times,$$

and

$$\mathrm{G}\big(\mathrm{U}(V_1) \times \mathrm{U}(V_2)\big)^0/F^\times \cong \mathrm{G}\big(\mathrm{U}(\mathbf{V}_1) \times \mathrm{U}(\mathbf{V}_2)\big)/E^\times.$$

For simplicity, we further assume that the Hecke eigenvalues of σ lie in $\mathbb{Q}$. Thus we obtain a class $\Xi \in H^{d,d}(X)$ such that $\xi = (j^*\Xi)_{\pi\otimes\pi}$ induces an isomorphism

$$H^d(X_1, \mathbb{C})_\pi \cong H^d(X_2, \mathbb{C})_\pi.$$

To be precise, we need to use the theta lifting valued in cohomology developed by Kudla–Millson [40, 41]. On the other hand, we can determine the near equivalence class of σ and prove that

$$H^{2d}(X, \mathbb{C})_\sigma \subset H^{d,d}(X).$$

Hence we can modify Ξ in such a way that Ξ lies in $H^{2d}(X, \mathbb{Q})_\sigma$, so that it is a Hodge cycle. Finally, it follows from the result of Kisin–Shin–Zhu [38] that

$$H^{2d}(X, \mathbb{Q}_\ell)_\sigma \cong \mathbb{Q}_\ell(-d)^m$$

for some positive integer m, from which Theorem 4.1 follows immediately.

Remark 4.2. It is desirable to upgrade ξ to an absolute Hodge cycle in the sense of Deligne [14], but this remains an open problem.

ACKNOWLEDGMENTS

The author would like to thank Tamotsu Ikeda for his constant support and encouragement. The author would also like to thank Wee Teck Gan, Erez Lapid, and Kartik Prasanna for their generosity in sharing their ideas and numerous valuable discussions over the years.

FUNDING

This work was partially supported by JSPS KAKENHI Grant Number 19H01781.

REFERENCES

[1] J. Arthur, *The endoscopic classification of representations. Orthogonal and symplectic groups*. Amer. Math. Soc. Colloq. Publ. 61, American Mathematical Society, Providence, RI, 2013.

[2] H. Atobe, The local theta correspondence and the local Gan–Gross–Prasad conjecture for the symplectic-metaplectic case. *Math. Ann.* **371** (2018), no. 1–2, 225–295.

[3] H. Atobe and W. T. Gan, Local theta correspondence of tempered representations and Langlands parameters. *Invent. Math.* **210** (2017), no. 2, 341–415.

[4] P. Bakić and M. Hanzer, Theta correspondence for p-adic dual pairs of type I. *J. Reine Angew. Math.* **776** (2021), 63–117.

[5] R. Beuzart-Plessis, Expression d'un facteur epsilon de paire par une formule intégrale. *Canad. J. Math.* **66** (2014), no. 5, 993–1049.

[6] R. Beuzart-Plessis, Endoscopie et conjecture locale raffinée de Gan–Gross–Prasad pour les groupes unitaires. *Compos. Math.* **151** (2015), no. 7, 1309–1371.

[7] R. Beuzart-Plessis, La conjecture locale de Gross–Prasad pour les représentations tempérées des groupes unitaires. *Mém. Soc. Math. Fr. (N.S.)* **149** (2016).

[8] R. Beuzart-Plessis, Plancherel formula for $\mathrm{GL}_n(F)\backslash\mathrm{GL}_n(E)$ and applications to the Ichino–Ikeda and formal degree conjectures for unitary groups. *Invent. Math.* **225** (2021), no. 1, 159–297.

[9] R. Beuzart-Plessis, Comparison of local relative characters and the Ichino–Ikeda conjecture for unitary groups. *J. Inst. Math. Jussieu* **20** (2021), no. 6, 1803–1854.

[10] R. Beuzart-Plessis, P.-H. Chaudouard, and M. Zydor, The global Gan–Gross–Prasad conjecture for unitary groups: the endoscopic case. 2020, arXiv:2007.05601.

[11] R. Beuzart-Plessis, Y. Liu, W. Zhang, and X. Zhu, Isolation of cuspidal spectrum, with application to the Gan–Gross–Prasad conjecture. *Ann. of Math. (2)* **194** (2021), no. 2, 519–584.

[12] J. W. Cogdell, H. H. Kim, I. I. Piatetski-Shapiro, and F. Shahidi, On lifting from classical groups to GL_N. *Publ. Math. Inst. Hautes Études Sci.* **93** (2001), 5–30.

[13] J. W. Cogdell, H. H. Kim, I. I. Piatetski-Shapiro, and F. Shahidi, Functoriality for the classical groups. *Publ. Math. Inst. Hautes Études Sci.* **99** (2004), 163–233.

[14] P. Deligne, Hodge cycles on abelian varieties (Notes by J. S. Milne). In *Hodge cycles, motives, and Shimura varieties*, pp. 9–100, Lecture Notes in Math. 900, Springer, Berlin–New York, 1982.

[15] M. Eichler, Quadratische Formen und Modulfunktionen. *Acta Arith.* **4** (1958), 217–239.

[16] G. Faltings, Endlichkeitssätze für abelsche Varietäten über Zahlkörpern. *Invent. Math.* **73** (1983), no. 3, 349–366.

[17] W. T. Gan, B. H. Gross, and D. Prasad, Symplectic local root numbers, central critical L values, and restriction problems in the representation theory of classical groups. *Astérisque* **346** (2012), 1–109.

[18] W. T. Gan and A. Ichino, Formal degrees and local theta correspondence. *Invent. Math.* **195** (2014), no. 3, 509–672.

[19] W. T. Gan and A. Ichino, The Gross–Prasad conjecture and local theta correspondence. *Invent. Math.* **206** (2016), no. 3, 705–799.

[20] W. T. Gan and A. Ichino, The Shimura–Waldspurger correspondence for Mp_{2n}. *Ann. of Math. (2)* **188** (2018), no. 3, 965–1016.

[21] W. T. Gan and A. Ichino, The automorphic discrete spectrum of Mp_4. *Int. Math. Res. Not. IMRN* **3** (2021), 1603–1677.

[22] W. T. Gan and G. Savin, Representations of metaplectic groups I: epsilon dichotomy and local Langlands correspondence. *Compos. Math.* **148** (2012), no. 6, 1655–1694.

[23] W. T. Gan and S. Takeda, A proof of the Howe duality conjecture. *J. Amer. Math. Soc.* **29** (2016), no. 2, 473–493.

[24] D. Ginzburg, S. Rallis, and D. Soudry, *The descent map from automorphic representations of* $\mathrm{GL}(n)$ *to classical groups*. World Scientific Publishing Co. Pte. Ltd., Hackensack, NJ, 2011.

[25] B. H. Gross and D. Prasad, On the decomposition of a representation of SO_n when restricted to SO_{n-1}. *Canad. J. Math.* **44** (1992), no. 5, 974–1002.

[26] B. H. Gross and D. Prasad, On irreducible representations of $\mathrm{SO}_{2n+1} \times \mathrm{SO}_{2m}$. *Canad. J. Math.* **46** (1994), no. 5, 930–950.

[27] M. Harris and S. S. Kudla, The central critical value of a triple product L-function. *Ann. of Math. (2)* **133** (1991), no. 3, 605–672.

[28] E. Hecke, Neuere Fortschritte in der Theorie der elliptischen Modulfunktionen. In *Comptes Rendus du Congrès International des Mathématiciens, Tome I (Oslo, 1936)*, pp. 140–156, A. W. Brøggers Boktrykkeri A/S, Oslo, 1937.

[29] K. Hiraga, A. Ichino, and T. Ikeda, Formal degrees and adjoint γ-factors. *J. Amer. Math. Soc.* **21** (2008), no. 1, 283–304.

[30] R. Howe, θ-series and invariant theory. In *Automorphic forms, representations and L-functions, Part 1 (Proc. Sympos. Pure Math., Oregon State Univ., Corvallis, Ore., 1977)*, pp. 275–285, Proc. Sympos. Pure Math. XXXIII, Amer. Math. Soc., Providence, RI, 1979.

[31] R. Howe, Transcending classical invariant theory. *J. Amer. Math. Soc.* **2** (1989), no. 3, 535–552.

[32] A. Ichino, Trilinear forms and the central values of triple product L-functions. *Duke Math. J.* **145** (2008), no. 2, 281–307.

[33] A. Ichino and T. Ikeda, On the periods of automorphic forms on special orthogonal groups and the Gross–Prasad conjecture. *Geom. Funct. Anal.* **19** (2010), no. 5, 1378–1425.

[34] A. Ichino, E. Lapid, and Z. Mao, On the formal degrees of square-integrable representations of odd special orthogonal and metaplectic groups. *Duke Math. J.* **166** (2017), no. 7, 1301–1348.

[35] A. Ichino and K. Prasanna, Hodge classes and the Jacquet–Langlands correspondence. 2018, arXiv:1806.10563.

[36] A. Ichino and K. Prasanna, *Periods of quaternionic Shimura varieties. I.* Contemp. Math. 762, American Mathematical Society, Providence, RI, 2021.

[37] H. Jacquet and R. P. Langlands, *Automorphic forms on* GL(2). Lecture Notes in Math. 114, Springer, Berlin–New York, 1970.

[38] M. Kisin, S. W. Shin, and Y. Zhu, The stable trace formula for certain Shimura varieties of abelian type. 2021, arXiv:2110.05381.

[39] S. S. Kudla, Seesaw dual reductive pairs. In *Automorphic forms of several variables (Katata, 1983)*, pp. 244–268, Progr. Math. 46, Birkhäuser, Boston, MA, 1984.

[40] S. S. Kudla and J. J. Millson, The theta correspondence and harmonic forms. I. *Math. Ann.* **274** (1986), no. 3, 353–378.

[41] S. S. Kudla and J. J. Millson, Intersection numbers of cycles on locally symmetric spaces and Fourier coefficients of holomorphic modular forms in several complex variables. *Publ. Math. Inst. Hautes Études Sci.* **71** (1990), 121–172.

[42] R. P. Langlands, Problems in the theory of automorphic forms. In *Lectures in modern analysis and applications, III*, pp. 18–61, Lecture Notes in Math. 170, Springer, Berlin, 1970.

[43] E. Lapid and Z. Mao, On an analogue of the Ichino–Ikeda conjecture for Whittaker coefficients on the metaplectic group. *Algebra Number Theory* **11** (2017), no. 3, 713–765.

[44] J.-S. Li, Automorphic forms with degenerate Fourier coefficients. *Amer. J. Math.* **119** (1997), no. 3, 523–578.

[45] C. Mœglin and J.-L. Waldspurger, La conjecture locale de Gross–Prasad pour les groupes spéciaux orthogonaux: le cas général. *Astérisque* **347** (2012), 167–216.

[46] S. Niwa, Modular forms of half integral weight and the integral of certain theta-functions. *Nagoya Math. J.* **56** (1975), 147–161.

[47] S. Rallis, Injectivity properties of liftings associated to Weil representations. *Compos. Math.* **52** (1984), no. 2, 139–169.

[48] H. Shimizu, On zeta functions of quaternion algebras. *Ann. of Math. (2)* **81** (1965), 166–193.

[49] H. Shimizu, Theta series and automorphic forms on GL_2. *J. Math. Soc. Japan* **24** (1972), 638–683.

[50] G. Shimura, On modular forms of half integral weight. *Ann. of Math. (2)* **97** (1973), 440–481.

[51] T. Shintani, On construction of holomorphic cusp forms of half integral weight. *Nagoya Math. J.* **58** (1975), 83–126.

[52] D. A. Jr. Vogan and G. J. Zuckerman, Unitary representations with nonzero cohomology. *Compos. Math.* **53** (1984), no. 1, 51–90.

[53] J.-L. Waldspurger, Correspondance de Shimura. *J. Math. Pures Appl. (9)* **59** (1980), no. 1, 1–132.

[54] J.-L. Waldspurger, Sur les valeurs de certaines fonctions L automorphes en leur centre de symétrie. *Compos. Math.* **54** (1985), no. 2, 173–242.

[55] J.-L. Waldspurger, Démonstration d'une conjecture de dualité de Howe dans le cas p-adique, $p \neq 2$. In *Festschrift in honor of I. I. Piatetski-Shapiro on the occasion of his sixtieth birthday, Part I (Ramat Aviv, 1989)*, pp. 267–324, Israel Math. Conf. Proc. 2, Weizmann, Jerusalem, 1990.

[56] J.-L. Waldspurger, Correspondances de Shimura et quaternions. *Forum Math.* **3** (1991), no. 3, 219–307.

[57] J.-L. Waldspurger, Une formule intégrale reliée à la conjecture locale de Gross–Prasad. *Compos. Math.* **146** (2010), no. 5, 1180–1290.

[58] J.-L. Waldspurger, Calcul d'une valeur d'un facteur ϵ par une formule intégrale. *Astérisque* **347** (2012), 1–102.

[59] J.-L. Waldspurger, La conjecture locale de Gross–Prasad pour les représentations tempérées des groupes spéciaux orthogonaux. *Astérisque* **347** (2012), 103–165.

[60] J.-L. Waldspurger, Une formule intégrale reliée à la conjecture locale de Gross–Prasad, 2e partie: extension aux représentations tempérées. *Astérisque* **346** (2012), 171–312.

[61] A. Weil, Sur certains groupes d'opérateurs unitaires. *Acta Math.* **111** (1964), 143–211.

[62] H. Xue, On the global Gan–Gross–Prasad conjecture for unitary groups: approximating smooth transfer of Jacquet–Rallis. *J. Reine Angew. Math.* **756** (2019), 65–100.
[63] W. Zhang, Automorphic period and the central value of Rankin–Selberg L-function. *J. Amer. Math. Soc.* **27** (2014), no. 2, 541–612.
[64] W. Zhang, Fourier transform and the global Gan–Gross–Prasad conjecture for unitary groups. *Ann. of Math. (2)* **180** (2014), no. 3, 971–1049.

ATSUSHI ICHINO
Department of Mathematics, Kyoto University, Kitashirakawa Oiwake-cho, Sakyo-ku, Kyoto 606-8502, Japan, ichino@math.kyoto-u.ac.jp

This Page Intentionally Left Blank

RATIONAL APPROXIMATIONS OF IRRATIONAL NUMBERS

DIMITRIS KOUKOULOPOULOS

ABSTRACT

Given quantities $\Delta_1, \Delta_2, \dots \geqslant 0$, a fundamental problem in Diophantine approximation is to understand which irrational numbers x have infinitely many reduced rational approximations a/q such that $|x - a/q| < \Delta_q$. Depending on the choice of Δ_q and of x, this question may be very hard. However, Duffin and Schaeffer conjectured in 1941 that if we assume a "metric" point of view, the question is governed by a simple zero–one law: writing φ for Euler's totient function, we either have $\sum_{q=1}^{\infty} \varphi(q)\Delta_q = \infty$ and then almost all irrational numbers (in the Lebesgue sense) are approximable, or $\sum_{q=1}^{\infty} \varphi(q)\Delta_q < \infty$ and almost no irrationals are approximable. We will present the history of the Duffin–Schaeffer conjecture and the main ideas behind the recent work of Koukoulopoulos–Maynard that settled it.

MATHEMATICS SUBJECT CLASSIFICATION 2020

Primary 11J83; Secondary 05C40

KEYWORDS

Diophantine approximation, Duffin–Schaeffer conjecture, Metric Number Theory, compression arguments, density increment, graph theory

ICM INTERNATIONAL CONGRESS OF MATHEMATICIANS 2022 JULY 6–14

Proc. Int. Cong. Math. 2022, Vol. 3, pp. 1894–1916
DOI 10.4171/ICM2022/84

1. DIOPHANTINE APPROXIMATION

Let x be an irrational number. In many settings, practical and theoretical, it is important to find fractions a/q of small numerator and denominator that approximate it well. This fundamental question lies at the core of the field of *Diophantine approximation*.

1.1. First principles

The "high-school way" of approximating x is to use its decimal expansion. This approach produces fractions $a/10^n$ such that $|x - a/10^n| \approx 10^{-n}$ typically. However, the error can be made much smaller if we allow more general denominators [14, THEOREM 2.1].

Theorem 1.1. *If $x \in \mathbb{R} \setminus \mathbb{Q}$, then $|x - a/q| < q^{-2}$ for infinitely many pairs $(a, q) \in \mathbb{Z} \times \mathbb{N}$.*

Dirichlet (c. 1840) gave a short and clever proof of this theorem. However, his argument is nonconstructive because it uses the pigeonhole principle. This gap is filled by the theory of continued fractions (which actually precedes Dirichlet's proof).

Given any $x \in \mathbb{R} \setminus \mathbb{Q}$, we may write $x = n_0 + r_0 \approx n_0$, where $n_0 = \lfloor x \rfloor$ is the integer part of x and $r_0 = \{x\}$ is its fractional part. We then let $n_1 = \lfloor 1/r_0 \rfloor$ and $r_1 = \{1/r_0\}$, so that $x = n_0 + 1/(n_1 + r_1) \approx n_0 + 1/n_1$. If we repeat this process $j - 1$ more times, we find that

$$x \approx n_0 + \cfrac{1}{n_1 + \cfrac{1}{\dots + \frac{1}{n_j}}} \quad \text{with } n_i = \left\lfloor \frac{1}{r_{i-1}} \right\rfloor, \; r_i = \left\{ \frac{1}{r_{i-1}} \right\} \text{ for } i = 1, \dots, j. \tag{1.1}$$

If we write this fraction as a_j/q_j in reduced form, then a calculation reveals that

$$\begin{aligned} a_j &= n_j a_{j-1} + a_{j-2} \quad (j \geqslant 2), \quad a_1 = n_0 n_1 + 1, \quad a_0 = n_0, \\ q_j &= n_j q_{j-1} + q_{j-2} \quad (j \geqslant 2), \quad q_1 = n_1, \quad q_0 = 1. \end{aligned} \tag{1.2}$$

When $j \to \infty$, the right-hand side of (1.1), often denoted by $[n_0; n_1, \dots, n_j]$, converges to x. The resulting representation of x is called its *continued fraction expansion*. The quotients a_j/q_j are called the *convergents* of this expansion and they have remarkable properties [17]. We list some of them below, with the first giving a constructive proof of Theorem 1.1.

Theorem 1.2. *Assume the above set-up and notations.*

(a) *For each $j \geqslant 0$, we have $1/(2q_j q_{j+1}) \leqslant |x - a_j/q_j| \leqslant 1/(q_j q_{j+1})$.*

(b) *For each $j \geqslant 0$, we have $|x - a_j/q_j| = \min\{|x - a/q| : 1 \leqslant q \leqslant q_j\}$.*

(c) *If $|x - a/q| < 1/(2q^2)$ with a and q coprime, then $a/q = a_j/q_j$ for some $j \geqslant 0$.*

1.2. Improving Dirichlet's approximation theorem

It is natural to ask when a qualitative improvement of Theorem 1.1 exists. Inverting this question leads us to the following definition: we say that a real number x is *badly approximable* if there is $c = c(x) > 0$ such that $|x - a/q| \geqslant cq^{-2}$ for all $(a, q) \in \mathbb{Z} \times \mathbb{N}$.

We can characterize approximable numbers in terms of their continued fraction expansion. Indeed, Theorem 1.2(a) and relation (1.2) imply that $1/4 \leqslant n_{j+1} q_j^2 |x - a_j/q_j| \leqslant 1$.

Hence, together with Theorem 1.2(c), this implies that x is badly approximable if and only if the sequence $(n_j)_{j=0}^{\infty}$ is bounded. Famously, Lagrange proved that the quadratic irrational numbers are in one-to-one correspondence with the continued fractions that are eventually periodic [17, §10]. In particular, all such numbers are badly approximable.

A related concept to badly approximable numbers is the *irrationality measure*. For each $x \in \mathbb{R}$, we define it to be

$$\mu(x) := \sup\{\nu \geqslant 0 : 0 < |x - a/q| < q^{-\nu} \text{ for infinitely many pairs } (a,q) \in \mathbb{Z} \times \mathbb{N}\}.$$

Note that $\mu(x) = 1$ if $x \in \mathbb{Q}$, whereas $\mu(x) \geqslant 2$ if $x \in \mathbb{R} \setminus \mathbb{Q}$ by Theorem 1.1. Moreover, $\mu(x) = 2$ if x is badly approximable. In particular, $\mu(x) = 2$ for all quadratic irrationals x. Remarkably, Roth [25] proved that $\mu(x) = 2$ for all algebraic irrational numbers x.

Determining the irrationality measure of various famous transcendental constants is often very hard. We do know that $\mu(e) = 2$, where e denotes Euler's constant. However, determining $\mu(\pi)$ is a famous open problem. Towards it, Zeilberger and Zudilin [29] proved that $\mu(\pi) \leqslant 7.10320533\ldots$ It is widely believed that $\mu(\pi) = 2$.

Instead of trying to reduce the error term in Dirichlet's approximation theorem, we often require a different type of improvement: restricting the denominators q to lie in some special set $\mathcal{S}$. The theory of continued fractions is of limited use for such problems because the denominators it produces satisfy rigid recursive relations (cf. (1.2)).

For rational approximation with prime or square denominators, the best results at the moment are due to Matomäki [21] and Zaharescu [28], respectively.

Theorem 1.3 (Matomäki (2009)). *Let x be an irrational number and let $\varepsilon > 0$. There are infinitely many integers a and prime numbers p such that $|x - a/p| < p^{-4/3+\varepsilon}$.*

Theorem 1.4 (Zaharescu (1995)). *Let x be an irrational number and let $\varepsilon > 0$. There are infinitely many pairs $(a,q) \in \mathbb{Z} \times \mathbb{N}$ such that $|x - a/q^2| < q^{-8/3+\varepsilon}$.*

Two important open problems are to show that Theorems 1.3 and 1.4 remain true even if we replace the constants $4/3$ and $8/3$ by 2 and 3, respectively.

2. METRIC DIOPHANTINE APPROXIMATION

Unable to answer simple questions about the rational approximations of specific numbers, a lot of research adopted a more statistical point of view. For example, given $M > 2$, what *proportion* of real numbers have irrationality measure $\geqslant M$? This new perspective gives rise to the theory of *metric Diophantine approximation*, which has a much more analytic and probabilistic flavor than the classical theory of Diophantine approximation. As we will see, the ability to ignore small pathological sets of numbers leads to a much more robust theory that provides simple and satisfactory answers to very general questions.

In order to give precise meaning to the word "proportion," we shall endow $\mathbb{R}$ with a measure. Here, we will mainly use the Lebesgue measure (denoted by "meas").

2.1. The theorems of Khinchin and Jarník–Besicovitch

The foundational result in the field of metric Diophantine approximation was proven by Khinchin in his seminal 1924 paper [18]. It is a rather general result: given a sequence $\Delta_1, \Delta_2, \dots, \geqslant 0$ of "permissible margins of error," we wish to determine for which real numbers x there are infinitely many pairs $(a, q) \in \mathbb{Z} \times \mathbb{N}$ such that $|x - a/q| < \Delta_q$. Clearly, if x has this property, so does $x + 1$. Hence, we may focus on studying

$$\mathcal{A} := \{x \in [0,1] : |x - a/q| < \Delta_q \text{ for infinitely many pairs } (a,q) \in \mathbb{Z} \times \mathbb{N}\}. \tag{2.1}$$

Theorem 2.1 (Khinchin (1924)). *Let $\Delta_1, \Delta_2, \dots \geqslant 0$ and let $\mathcal{A}$ be defined as in (2.1).*

(a) *If $\sum_{q=1}^{\infty} q\Delta_q < \infty$, then* $\operatorname{meas}(\mathcal{A}) = 0$.

(b) *If $\sum_{q=1}^{\infty} q\Delta_q = \infty$ and the sequence $(q^2\Delta_q)_{q=1}^{\infty}$ is decreasing, then* $\operatorname{meas}(\mathcal{A}) = 1$.

Corollary 2.2. *For almost all $x \in \mathbb{R}$, we have $|x - a/q| \leqslant 1/(q^2 \log q)$ for infinitely many $(a,q) \in \mathbb{Z} \times \mathbb{N}$. On the other hand, if $c > 1$ is fixed, then for almost every $x \in \mathbb{R}$, the inequality $|x - a/q| \leqslant 1/(q^2 \log^c q)$ admits only finitely many solutions $(a,q) \in \mathbb{Z} \times \mathbb{N}$.*

In particular, Corollary 2.2 implies that the set of badly approximable numbers has null Lebesgue measure. On the other hand, it also says that almost all real numbers have irrationality measure equal to 2. This last result is the main motivation behind the conjecture that $\mu(\pi) = 2$: we expect π to behave like a "typical" real number.

Naturally, the fact that $\mathcal{W}_M := \{x \in \mathbb{R} : \mu(x) \geqslant M\}$ has null Lebesgue measure for $M > 2$ raises the question of determining its Hausdorff dimension (denoted by $\dim(\mathcal{W}_M)$). Jarník [16] and Besicovitch [5] answered this question independently of each other.

Theorem 2.3 (Jarník (1928), Besicovitch (1934)). *We have $\dim(\mathcal{W}_M) = 2/M$ for all $M \geqslant 2$.*

2.2. Generalizing Khinchin's theorem

Following the publication of Khinchin's theorem, research focused on weakening the assumption that $q^2\Delta_q \searrow$ in part (b). Importantly, doing so would open the door to understanding rational approximations using only a restricted set of denominators. Indeed, if $q^2\Delta_q \searrow$, then either $\Delta_q > 0$ for all q, or there is q_0 such that $\Delta_q = 0$ for all $q \geqslant q_0$. The second case is trivial, since it implies $\mathcal{A} = \emptyset$. So, if we wish to understand Diophantine approximation with a restricted set of denominators $\mathcal{S}$ (which would require $\Delta_q = 0$ for $q \notin \mathcal{S}$), then we must prove a version of Theorem 2.1(b) without the assumption that $q^2\Delta_q \searrow$.

In order to understand better the forces at play here, it is useful to recast Khinchin's theorem in probabilistic terms. For each q, let us define the set

$$\begin{aligned}\mathcal{A}_q &:= \left\{x \in [0,1] : \text{there is } a \in \mathbb{Z} \text{ such that } |x - \tfrac{a}{q}| < \Delta_q\right\} \\ &= [0,1] \cap \bigcup_{0 \leqslant a \leqslant q} \left(\frac{a}{q} - \Delta_q, \frac{a}{q} + \Delta_q\right).\end{aligned} \tag{2.2}$$

Then $\mathcal{A} = \{x \in [0,1] : x \in \mathcal{A}_q \text{ infinitely often}\}$, which we often write as $\mathcal{A} = \limsup_{q\to\infty} \mathcal{A}_q$. We may thus view $\mathcal{A}$ as the event that for a number chosen uniformly at random from $[0,1]$, an infinite number of the events $\mathcal{A}_1, \mathcal{A}_2, \dots$ occur. A classical result from probability theory due to Borel and Cantelli [14, LEMMAS 1.2 & 1.3] studies precisely this kind of questions.

Theorem 2.4. *Let $(\Omega, \mathcal{F}, \mathbb{P})$ be a probability space, let $E_1, E_2, \dots$ be events in that space, and let $E = \limsup_{j\to\infty} E_j$ be the event that infinitely many of the E_j's occur.*

(a) *(The first Borel–Cantelli lemma) If $\sum_{j=1}^{\infty} \mathbb{P}(E_j) < \infty$, then $\mathbb{P}(E) = 0$.*

(b) *(The second Borel–Cantelli lemma) If $\sum_{j=1}^{\infty} \mathbb{P}(E_j) = \infty$ and the events $E_1, E_2, \dots$ are mutually independent, then $\mathbb{P}(E) = 1$.*

Remark. Let N be the random variable that counts how many of the events $E_1, E_2, \dots$ occur. We have $\mathbb{E}[N] = \sum_{j=1}^{\infty} \mathbb{P}(E_j)$. Hence, Theorem 2.4 says that, under certain assumptions, $N = \infty$ almost surely if, and only if, $\mathbb{E}[N] = \infty$.

To use the above result in the setup of Khinchin's theorem, we let $\Omega = [0,1]$ and equip it with the Lebesgue measure as its probability measure. The relevant events E_j are the sets $\mathcal{A}_q$. Notice that if $\Delta_q > 1/(2q)$, then $\mathcal{A}_q = [0,1]$, in which case $\mathcal{A}_q$ occurs immediately for all $x \in [0,1]$. In order to avoid these trivial events, we will assume from now on that

$$\Delta_q \leqslant 1/(2q) \quad \text{for all } q \geqslant 1, \quad \text{whence } \operatorname{meas}(\mathcal{A}_q) = 2q\Delta_q. \tag{2.3}$$

In particular, we see that part (a) of Khinchin's theorem is a direct consequence of the first Borel–Cantelli lemma. On the other hand, the second Borel–Cantelli lemma relies crucially on the assumption that the events E_j are independent of each other, something that fails generically for the events $\mathcal{A}_q$. However, there are variations of the second Borel–Cantelli lemma, where the assumption of independence can be replaced by weaker quasi-independence conditions on the relevant events (cf. Section 3.1). From this perspective, part (b) of Khinchin's theorem can be seen as saying that the condition that the sequence $(q^2\Delta_q)_{q=1}^{\infty}$ is decreasing guarantees enough approximate independence between the events $\mathcal{A}_q$ so that the conclusion of the second Borel–Cantelli lemma remains valid.

In 1941, Duffin and Schaeffer published a seminal paper [8] that studied precisely what is the right way to generalize Khinchin's theorem so that the simple zero–one law of Borel–Cantelli holds. Their starting point was the simple observation that certain choices of the quantities Δ_q create many dependencies between the sets $\mathcal{A}_q$, thus rendering many of the denominators q redundant. Indeed, note, for example, that if $\Delta_3 = \Delta_{15}$, then $\mathcal{A}_3 \subseteq \mathcal{A}_{15}$ because each fraction with denominator 3 can also be written as a fraction with denominator 15. By exploiting this simple idea, Duffin and Schaeffer proved the following result:

Proposition 2.5. *There are $\Delta_1, \Delta_2, \dots \geqslant 0$ such that $\sum_{q=1}^{\infty} q\Delta_q = \infty$ and yet $\operatorname{meas}(\mathcal{A}) = 0$.*

Proof. Let $p_1 < p_2 < \cdots$ be the primes in increasing order, let $q_j = p_1 \cdots p_j$, and let $\mathcal{S}_j = \{dp_j : d \mid q_{j-1}\}$. We then set $\Delta_q = (q_j j \log^2 j)^{-1}$ if $q \in \mathcal{S}_j$ for some $j \geqslant 2$; otherwise, we set $\Delta_q = 0$. We claim that this choice satisfies the needed conditions.

Since $\mathcal{A}_q \subseteq \mathcal{A}_{q_j}$ for all $q \in \mathcal{S}_j$, we have $\mathcal{A} = \limsup_{j\to\infty} \mathcal{A}_{q_j}$. In addition, since $\sum_{j=1}^{\infty} q_j \Delta_{q_j} < \infty$, we have $\operatorname{meas}(\limsup_{j\to\infty} \mathcal{A}_{q_j}) = 0$ by Theorem 2.1(a). Hence, $\operatorname{meas}(\mathcal{A}) = 0$, as needed. On the other hand, we have that

$$\sum_{q\geqslant 1} q\Delta_q = \sum_{j\geqslant 2} \sum_{d\,|q_{j-1}} dp_j \cdot \frac{1}{q_j\, j \log^2 j} = \sum_{j\geqslant 2} \frac{1}{j \log^2 j} \prod_{i\leqslant j-1} \left(1 + \frac{1}{p_i}\right).$$

By the Prime Number Theorem [19, THEOREM 8.1], the last product is $\geqslant c \log j$ for some absolute constant $c > 0$. Consequently, $\sum_{q=1}^{\infty} q\Delta_q = \infty$, as claimed. ■

In order to avoid the above kind of counterexamples to the generalized Khinchin theorem, Duffin and Schaeffer were naturally led to consider a modified setup, where only reduced fractions are used as approximations. They thus defined

$$\mathcal{A}^* := \{x \in [0,1] : |x - a/q| < \Delta_q \text{ for infinitely many reduced fractions } a/q\}. \tag{2.4}$$

We may write $\mathcal{A}^*$ as the lim sup of the sets

$$\mathcal{A}_q^* := [0,1] \cap \bigcup_{\substack{0\leqslant a\leqslant q \\ \gcd(a,q)=1}} \left(\frac{a}{q} - \Delta_q, \frac{a}{q} + \Delta_q\right). \tag{2.5}$$

Assuming that (2.3) holds, we readily find that

$$\operatorname{meas}(\mathcal{A}_q^*) = 2\varphi(q)\Delta_q$$

where

$$\varphi(q) := \#\{1 \leqslant a \leqslant q : \gcd(a,q) = 1\}$$

is Euler's totient function. They then conjectured that the sets $\mathcal{A}_q^*$ have enough mutual quasi-independence so that a simple zero–one law holds, as per the Borel–Cantelli lemmas.

The Duffin–Schaeffer conjecture. *Let $\Delta_1, \Delta_2, \dots \geqslant 0$ and let $\mathcal{A}^*$ be defined as in* (2.4).

(a) *If $\sum_{q=1}^{\infty} \varphi(q)\Delta_q < \infty$, then* $\operatorname{meas}(\mathcal{A}^*) = 0$.

(b) *If $\sum_{q=1}^{\infty} \varphi(q)\Delta_q = \infty$, then* $\operatorname{meas}(\mathcal{A}^*) = 1$.

Of course, part (a) follows from Theorem 2.4(a); the main difficulty is to prove (b).

The Duffin–Schaeffer conjecture is strikingly simple and general. Nonetheless, it does not answer our original question: what is the correct generalization of Khinchin's theorem, where we may use nonreduced fractions? This gap was filled by Catlin [7].

Catlin's conjecture. *Let $\Delta_1, \Delta_2, \dots \geqslant 0$, let $\Delta_q' = \sup_{m\geqslant 1} \Delta_{qm}$, and let $\mathcal{A}$ be as in* (2.1).

(a) *If $\sum_{q=1}^{\infty} \varphi(q)\Delta_q' < \infty$, then* $\operatorname{meas}(\mathcal{A}) = 0$.

(b) *If $\sum_{q=1}^{\infty} \varphi(q)\Delta_q' = \infty$, then* $\operatorname{meas}(\mathcal{A}) = 1$.

As Catlin noticed, his conjecture is a direct corollary of that by Duffin and Schaeffer. Indeed, let us consider the set

$$\mathcal{A}' = \{x \in [0,1] : |x - a/q| < \Delta_q' \text{ for infinitely many reduced fractions } a/q\}.$$

This is the set $\mathcal{A}^*$ with the quantities Δ_q replaced by Δ'_q, so we may apply the Duffin–Schaeffer conjecture to it. In addition, it is straightforward to check that

$$\mathcal{A} \setminus \mathbb{Q} = \mathcal{A}' \setminus \mathbb{Q} \tag{2.6}$$

when $\Delta_q \to 0$. This settles Catlin's conjecture in this case. On the other hand, if $\Delta_q \not\to 0$, then $\mathcal{A} = [0, 1]$ and $\sum_{q=1}^{\infty} \varphi(q)\Delta'_q = \infty$, so that Catlin's conjecture is trivially true.

Just like in Theorem 2.3 of Jarník and Besicovitch, it would be important to also have information about the Hausdorff dimension of the sets $\mathcal{A}$ and $\mathcal{A}^*$ in the case when they have null Lebesgue measure. In light of relation (2.6), it suffices to answer this question for the latter set. Beresnevich and Velani [4] proved the remarkable result that the Duffin–Schaeffer conjecture implies a Hausdorff measure version of itself. This is a consequence of a much more general *Mass Transfer Principle* that they established, and which allows transfering information concerning the Lebesgue measure of certain lim sup sets to the Hausdorff measure of rescaled versions of them. As a corollary, they proved:

Theorem 2.6 (Beresnevich–Velani (2006)). *Assume that the Duffin–Schaeffer conjecture is true. Let $\Delta_1, \Delta_2, \dots \geqslant 0$ be such that $\sum_{q=1}^{\infty} \varphi(q)\Delta_q < \infty$. Then the Hausdorff dimension of the set $\mathcal{A}^*$ defined by (2.4) equals the infimum of the set of $s > 0$ such that $\sum_{q=1}^{\infty} \varphi(q)\Delta_q^s < \infty$.*

2.3. Progress towards the Duffin–Schaeffer conjecture

Since its introduction in 1941, the Duffin–Schaeffer conjectured has been the subject of intensive research activity, with various special cases proven over the years. This process came to a conclusion recently with the proof of the full conjecture [28].

Theorem 2.7 (Koukoulopoulos–Maynard (2020)). *The Duffin–Schaeffer conjecture is true.*

We will outline the main ideas of the proof of Theorem 2.7 in Section 3. But first we give an account of the work that preceeded it.

Notation. Given two functions $f, g : X \to \mathbb{R}$, we write $f(x) \ll g(x)$ (or $f(x) = O(g(x))$) for all $x \in X$ to mean that there is a constant C such that $|f(x)| \leqslant Cg(x)$ for all $x \in Y$.

In the same paper where they introduced their conjecture, Duffin and Schaeffer proved the first general case of it:

Theorem 2.8 (Duffin–Schaeffer (1941)). *The Duffin–Schaeffer conjecture is true for all sequences $(\Delta_q)_{q=1}^{\infty}$ such that*

$$\limsup_{Q\to\infty} \frac{\sum_{q\leqslant Q} \varphi(q)\Delta_q}{\sum_{q\leqslant Q} q\Delta_q} > 0. \tag{2.7}$$

To appreciate this result, we must make a few comments about condition (2.7). Note that its left-hand side is the average value of $\varphi(q)/q$ over $q \in [1, Q]$, where q is weighted by $w_q := q\Delta_q$. In particular, we may restrict our attention to q with $\Delta_q > 0$. Now, we know

$$\frac{\varphi(q)}{q} = \prod_{p|q}\Big(1 - \frac{1}{p}\Big).$$

In particular, $\varphi(q)/q \leqslant 1$, and the only way this ratio can become much smaller than 1 is if q is divisible by lots of small primes. To see this, let us begin by observing that q can have at most $\log q/\log 2$ prime factors in total. Therefore,

$$\prod_{p|q,\ p>\log q}\left(1-\frac{1}{p}\right) \geqslant \left(1-\frac{1}{\log q}\right)^{\log q/\log 2} \geqslant \frac{1}{5} \tag{2.8}$$

for q large enough. In addition, we have

$$\prod_{\substack{(\log q)^{0.01}<p\leqslant\log q\\ p|q}}\left(1-\frac{1}{p}\right) \geqslant \prod_{(\log q)^{0.01}<p\leqslant\log q}\left(1-\frac{1}{p}\right) \geqslant \frac{1}{200} \tag{2.9}$$

for q large enough by Mertens' estimate **[19, THEOREM 3.4]**. Already the above inequalities show that only the primes $\leqslant (\log q)^{0.01}$ can affect the size of $\varphi(q)/q$. But more is true: $\varphi(q)/q$ is small only if q is divided by *many* primes $\leqslant (\log q)^{0.01}$. Imagine, for example, that

$$\#\{p \mid q : e^{j-1} < p \leqslant e^j\} \leqslant e^j/j^2 + 1000 \tag{2.10}$$

for $j = 1, 2, \ldots, 1 + \lfloor 0.01 \log\log q \rfloor$. We would then have

$$\prod_{\substack{e^{j-1}<p\leqslant e^j\\ p|q}}\left(1-\frac{1}{p}\right) \geqslant \left(1-\frac{1}{e^j}\right)^{e^j/j^2+1000} = \exp\left(-1/j^2 + O\left(e^{-j}\right)\right).$$

Multiplying this over all j, we deduce that $\varphi(q)/q \geqslant c$ for some $c > 0$ independent of q.

We have thus proven that for (2.7) to fail, the main contribution to the weighted sum $\sum_{q\leqslant Q} w_q$ with $w_q = q\Delta_q$ must come from integers for which (2.10) fails. As a matter of fact, (2.10) must fail for lots of j's. This is an extremely rare event if we choose q *uniformly at random* from $[1, Q]$ (or even if we choose it uniformly at random from various "nice" subsets of $[1, Q]$, such as the primes, or the values of a monic polynomial with integer coefficients). A simple way to see this is to calculate the average value of the function $\#\{p \mid q : e^{j-1} < p \leqslant e^j\}$ with respect to the uniform counting measure on $[1, Q]$. We have

$$\frac{1}{Q}\sum_{q\leqslant Q}\#\{p \mid q : e^{j-1} < p \leqslant e^j\} = \sum_{e^{j-1}<p\leqslant e^j}\frac{\#\{q \leqslant Q : p \mid q\}}{Q} \leqslant \sum_{e^{j-1}<p\leqslant e^j}\frac{1}{p} \ll \frac{1}{j}$$

by Mertens' theorem **[19, THEOREM 3.4]**. This is *much* smaller than e^j/j^2, so (2.10) should fail rarely as $j \to \infty$. (For instance, we may use Markov's inequality to see this claim.)

In conclusion, Theorem 2.8 settles the Duffin–Schaeffer conjecture when Δ_q is mainly supported on "normal" integers, without too many small prime factors. In particular, it implies a significant improvement of Theorems 1.3 and 1.4 for almost all $x \in \mathbb{R}$.

Corollary 2.9. *For almost all $x \in \mathbb{R}$, there are infinitely many reduced fractions a/p and b/q^2 such that p is prime, $|x - a/p| < p^{-2}$ and $|x - b/q^2| < q^{-3}$.*

The next important step towards the Duffin–Schaeffer conjecture is a remarkable *zero–one law* due to Gallagher **[12]**.

Theorem 2.10 (Gallagher (1961)). *If* $\mathcal{A}^*$ *is as in* (2.4), *then* $\operatorname{meas}(\mathcal{A}^*) \in \{0, 1\}$.

Gallagher's theorem says *grosso modo* that either we chose the quantities Δ_q to be "too small" and thus missed almost all real numbers, or we chose them "sufficiently large" so that almost all numbers have the desired rational approximations. The Duffin–Schaeffer conjecture is then the simplest possible criterion to decide in which case we are.

The proof of Theorem 2.10 is a clever adaptation of an ergodic-theoretic argument due to Cassels [6] in the simpler setting of nonreduced rational approximations. We give Cassel's proof and refer the interested readers to [12, 14] for the proof of Theorem 2.10.

Theorem 2.11 (Cassels (1950)). *If* $\mathcal{A}$ *is as in* (2.1), *then* $\operatorname{meas}(\mathcal{A}) \in \{0, 1\}$.

Proof. We need the following fact [14, LEMMA 2.1] that uses Lebesgue's Density Theorem: Let $I_1, I_2, \ldots, J_1, J_2, \ldots$ be intervals of lengths tending to 0, and let $c > 0$. For all k, suppose $J_k \subseteq I_k$ and $\operatorname{meas}(J_k) \geqslant c \operatorname{meas}(I_k)$. Then $\operatorname{meas}(\limsup_{k\to\infty} I_k \setminus \limsup_{k\to\infty} J_k) = 0$.

Now, for each $r \geqslant 1$, let $\mathcal{A}^{(r)}$ be defined as in (2.1) but with Δ_q/r in place of Δ_q. Hence, $\operatorname{meas}(\mathcal{A} \setminus \mathcal{A}^{(r)}) = 0$ by the above fact. Therefore, if $\mathcal{A}^{(\infty)} := \bigcap_{n=1}^{\infty} \mathcal{A}^{(n)}$, then $\operatorname{meas}(\mathcal{A} \setminus \mathcal{A}^{(\infty)}) = 0$. Now, consider the map $\psi : [0, 1] \to [0, 1]$ defined by $\psi(x) := \{2x\}$, and note that $\psi(\mathcal{A}^{(\infty)}) \subseteq \mathcal{A}^{(\infty)}$. In particular, $\frac{1}{N}\sum_{n=0}^{N-1} 1_{\mathcal{A}^{(\infty)}}(\psi^n(x)) = 1$ for all $x \in \mathcal{A}^{(\infty)}$ and all $N \in \mathbb{N}$. Since ψ is ergodic with respect to the Lebesgue measure [26, P. 293 & 305–6], Birkhoff's Ergodic Theorem [26, CH. 6, COR. 5.6] implies that $\operatorname{meas}(\mathcal{A}^{(\infty)}) \in \{0, 1\}$. ■

The first significant step towards establishing the Duffin–Schaeffer conjecture for irregular sequences Δ_q, potentially supported on integers with lots of small prime factors, was carried out by Erdős [10] and Vaaler [27].

Theorem 2.12 (Erdős (1970), Vaaler (1978)). *The Duffin–Schaeffer conjecture is true for all sequences* $(\Delta_q)_{q=1}^{\infty}$ *such that* $\Delta_q = O(1/q^2)$ *for all* q.

This theorem is, of course, most interesting when $\sum_{q=1}^{\infty} \varphi(q)\Delta_q = \infty$. Since $\Delta_q = O(1/q^2)$ and $\varphi(q) \leqslant q$, we find $\sum_{q\in\mathcal{S}} 1/q = \infty$ with $\mathcal{S} = \{q : \Delta_q > 0\}$. In particular, $\mathcal{S}$ must be somewhat dense in $\mathbb{N}$. Therefore, Theorem 2.12 has the advantage over Theorem 2.8 that $\mathcal{S}$ can contain many irregular integers, and the disadvantage that it has to be quite dense.

The Duffin–Schaeffer conjecture has a natural analogue in $\mathbb{R}^k$ with $k \geqslant 2$: given $\Delta_1, \Delta_2, \cdots \geqslant 0$, let $\mathcal{A}^*(k)$ be the set of $\vec{x} = (x_1, \ldots, x_k) \in \mathbb{R}^k$ for which there are infinitely many k-tuples $(a_1/q, \ldots, a_k/q)$ of reduced fractions with $|x_j - a_j/q| < \Delta_q$ for all j. Then $\mathcal{A}^*(k)$ should contain almost no or almost all $\vec{x} \in \mathbb{R}^k$, according to whether the series $\sum_{q=1}^{\infty} (\varphi(q)\Delta_q)^k$ converges or diverges. This was proven by Pollington and Vaughan [22].

Theorem 2.13 (Pollington–Vaughan (1990)). *The* k*-dimensional Duffin–Schaeffer conjecture is true for all* $k \geqslant 2$.

Following this result, a lot of research focused on proving the Duffin–Schaeffer conjecture when the series $\sum_{q=1}^{\infty} \varphi(q)\Delta_q$ diverges fast enough (see, e.g., [14, THEOREM 3.7(III)], [3, 15]). Aistleitner, Lachmann, Munsch, Technau, and Zafeiropoulos [2] proved the Duffin–

Schaeffer conjecture when $\sum_{q=1}^{\infty} \varphi(q)\Delta_q/(\log q)^{\varepsilon} = \infty$ for some $\varepsilon > 0$. A report by Aistleitner [1], announced at the same time as [20], explains how to replace $(\log q)^{\varepsilon}$ by $(\log\log q)^{\varepsilon}$.

3. THE MAIN INGREDIENTS OF THE PROOF OF THE DUFFIN–SCHAEFFER CONJECTURE

3.1. Borel–Cantelli without independence

Recall the definition of the sets $\mathcal{A}_q^*$ in (2.5). Let us assume that $\Delta_q \leqslant 1/(2q)$ for all q (cf. (2.3)) so that $\operatorname{meas}(\mathcal{A}_q^*) = 2\varphi(q)\Delta_q \in [0,1]$, and let us also suppose that $\sum_{q=1}^{\infty} \varphi(q)\Delta_q = \infty$. The first technical difficulty we must deal with is how to prove an analogue of the second Borel–Cantelli lemma (cf. Theorem 2.4(b)) without assuming that the events $\mathcal{A}_q^*$ are independent. We follow an idea due to Turan, which is already present in [8].

By Gallagher's zero–one law, it is enough to show that $\operatorname{meas}(\mathcal{A}^*) > 0$. Since $\bigcup_{q \geqslant Q} \mathcal{A}_q^* \searrow \mathcal{A}^*$, we may equivalently prove that there is some constant $c > 0$ such that $\operatorname{meas}(\bigcup_{q\geqslant Q} \mathcal{A}_q^*) \geqslant c$ for all large Q. In order to limit the potential overlap among the sets $\mathcal{A}_q^*$, we only consider an appropriate subset of them. Since $\operatorname{meas}(\mathcal{A}_q^*) = 2\varphi(q)\Delta_q \in [0,1]$ for all q, and since $\sum_{q\geqslant Q} \varphi(q)\Delta_q = \infty$, there must exist some $R \geqslant Q$ such that

$$1 \leqslant \sum_{q\in[Q,R]} \operatorname{meas}(\mathcal{A}_q^*) \leqslant 2. \tag{3.1}$$

We will only use the events $\mathcal{A}_q^*$ with $q \in [Q,R]$. We trivially have the union bound

$$\operatorname{meas}\Big(\bigcup_{q\in[Q,R]} \mathcal{A}_q^*\Big) \leqslant \sum_{q\in[Q,R]} \operatorname{meas}(\mathcal{A}_q^*) \leqslant 2.$$

If we can show that the sets $\mathcal{A}_q^*$ with $q \in [Q,R]$ do not overlap too much, so that

$$\operatorname{meas}\Big(\bigcup_{q\in[Q,R]} \mathcal{A}_q^*\Big) \geqslant c \sum_{q\in[Q,R]} \operatorname{meas}(\mathcal{A}_q^*) \geqslant c, \tag{3.2}$$

we will be able to deduce that $\operatorname{meas}(\bigcup_{q\in[Q,R]} \mathcal{A}_q^*) \geqslant c$ and *a fortiori* that $\operatorname{meas}(\bigcup_{q\geqslant Q} \mathcal{A}_q^*) \geqslant c$. As the following lemma shows, (3.2) is true under (3.1) as long as we can control the correlations of the events $\mathcal{A}_q^*$ *on average*.

Lemma 3.1. *Let $E_1, \ldots, E_k$ be events in the probability space $(\Omega, \mathcal{F}, \mathbb{P})$. We then have that*

$$\mathbb{P}\Big(\bigcup_{j=1}^{k} E_j\Big) \geqslant \frac{(\sum_{j=1}^{k} \mathbb{P}(E_j))^2}{\sum_{i,j=1}^{k} \mathbb{P}(E_i \cap E_j)}.$$

Proof. Let $N = \sum_j 1_{E_j}$. We then have $\mathbb{E}[N] = \sum_j \mathbb{P}(E_j)$. On the other hand, the Cauchy–Schwarz inequality implies that

$$\mathbb{E}[N]^2 = \mathbb{E}[1_{N>0} \cdot N]^2 \leqslant \mathbb{P}(\operatorname{supp}(N)) \cdot \mathbb{E}[N^2].$$

Since $\operatorname{supp}(N) = \bigcup_j E_j$ and $N^2 = \sum_{i,j} 1_{E_i \cap E_j}$, the lemma follows. ■

The following proposition summarizes the discussion of this section.

Proposition 3.2. *Let $\Delta_1, \Delta_2, \dots \geqslant 0$, and let $\mathcal{A}_q^*$ be as in (2.5).*

(a) *If $C > 0$ and $R \geqslant Q \geqslant 1$ are such that*

$$1 \leqslant \sum_{q\in[Q,R]} \operatorname{meas}(\mathcal{A}_q^*) \leqslant 2 \quad \text{and} \quad \sum_{Q\leqslant q<r\leqslant R} \operatorname{meas}(\mathcal{A}_q^* \cap \mathcal{A}_r^*) \leqslant C, \tag{3.3}$$

then $\operatorname{meas}(\bigcup_{q\in[Q,R]} \mathcal{A}_q^*) \geqslant 1/(2+2C)$.

(b) *If there are infinitely many disjoint intervals $[Q, R]$ satisfying (3.3) with the same constant $C > 0$, then* $\operatorname{meas}(\limsup_{q\to\infty} \mathcal{A}_q^*) = 1$.

3.2. A bound on the pairwise correlations

As per Proposition 3.2, we need to control the correlations of the events $\mathcal{A}_q^*$. To this end, we have a lemma of Pollington–Vaughan [22] (see also [10,27]).

Lemma 3.3. *Let q, r be two distinct integers $\geqslant 2$, let $\Delta_q, \Delta_r \geqslant 0$, let $\mathcal{A}_q^*, \mathcal{A}_r^*$ be as in (2.5), and let $M(q,r) = 2\max\{\Delta_q, \Delta_r\}\operatorname{lcm}[q,r]$. If $M(q,r) \leqslant 1$, then $\mathcal{A}_q^* \cap \mathcal{A}_r^* = \emptyset$. Otherwise,*

$$\operatorname{meas}(\mathcal{A}_q^* \cap \mathcal{A}_r^*) \ll \varphi(q)\Delta_q \cdot \varphi(r)\Delta_r \cdot \exp\Bigg(\sum_{\substack{p\mid qr/\gcd(q,r)\\ p>M(q,r)}} \frac{1}{p}\Bigg).$$

Proof. Let $\Delta = \max\{\Delta_q, \Delta_r\}$, $\delta = \min\{\Delta_q, \Delta_r\}$ and $M = M(q,r)$. The intervals $I_a = (\frac{a}{q} - \Delta_q, \frac{a}{q} + \Delta_q)$ and $J_b = (\frac{b}{r} - \Delta_r, \frac{b}{r} + \Delta_r)$ intersect only if $2\Delta > |\frac{a}{q} - \frac{b}{r}|$. Since the right-hand side is $\geqslant 1/\operatorname{lcm}[q,r]$ when $\gcd(a,q) = \gcd(b,r) = 1$, we infer that $\mathcal{A}_q^* \cap \mathcal{A}_r^* = \emptyset$ if $M \leqslant 1$.

Now, assume that $M > 1$. Since $\operatorname{meas}(I_a \cap J_b) \leqslant 2\delta$ for all a, b, we have

$$\operatorname{meas}(\mathcal{A}_q^* \cap \mathcal{A}_r^*) \leqslant 2\delta \cdot \#\left\{\begin{matrix} 1 \leqslant a \leqslant q, & \gcd(a,q) = 1 \\ 1 \leqslant b \leqslant r, & \gcd(b,r) = 1 \end{matrix} : \left|\frac{a}{q} - \frac{b}{r}\right| < 2\Delta\right\}.$$

Let $a/q - b/r = m/\operatorname{lcm}[q,r]$. Then $1 \leqslant |m| \leqslant M$ and $\gcd(m, q_1 r_1) = 1$, where $q_1 = q/d$ and $r_1 = r/d$ with $d = \gcd(q,r)$. For each such m, a straightforward application of the Chinese Remainder Theorem gives that the number of admissible pairs (a,b) is

$$\leqslant d \prod_{p\mid\gcd(d,q_1r_1m)} \left(1 - \frac{1}{p}\right) \prod_{p\mid d,\ p\nmid q_1r_1m} \left(1 - \frac{2}{p}\right) \leqslant d \frac{\prod_{p\mid d}(1-\frac{1}{p})^2}{\prod_{p\mid\gcd(d,q_1r_1)}(1-\frac{1}{p})} \cdot \frac{|m|}{\varphi(|m|)},$$

where we used that $1 - 2/p \leqslant (1-1/p)^2$. We then sum this inequality over m and use Lemma 3.4 below to complete the proof. (For full details, see [22] or [14, LEMMA 2.8].) ∎

Lemma 3.4. *Fix $C \geqslant 1$, and let $(a_p)_{p \text{ prime}}$ be a sequence taking values in $[0, C]$. Then*

$$\sum_{n\leqslant x} \prod_{p\mid n} a_p \ll_C x \exp\Bigg(\sum_{p\leqslant x} \frac{a_p - 1}{p}\Bigg) \quad \textit{for all } x \geqslant 1.$$

Proof. See Theorem 14.2 in [19]. ∎

3.3. Generalizing the Erdős–Vaaler argument

The next step is to study averages of $\exp(\sum_{p|qr/\gcd(q,r)^2,\ p>M(q,r)} 1/p)$. This gets a bit too technical in general, so we focus on the following special case:

Theorem 3.5. *Let $Q \geqslant N \geqslant 2$, and let $\mathcal{S} \subseteq \{Q \leqslant q \leqslant 2Q : q \text{ square-free}\}$ be such that*

$$N/2 \leqslant \sum_{q\in\mathcal{S}} \frac{\varphi(q)}{q} \leqslant N. \tag{3.4}$$

We then have

$$\sum_{q,r\in\mathcal{S}} \frac{\varphi(q)\varphi(r)}{qr} \exp\Bigg(\sum_{\substack{p|qr/\gcd(q,r)^2 \\ p>Q/[N\gcd(q,r)]}} \frac{1}{p}\Bigg) \ll N^2. \tag{3.5}$$

In particular, if $\mathcal{A}_q^*$ is as in (2.5) with $\Delta_q = 1/(qN)$, then $\text{meas}(\bigcup_{q\in\mathcal{S}} \mathcal{A}_q^*) \gg 1$.

Remark. To see the last assertion, recall the notation $M(q,r) = 2\max\{\Delta_q, \Delta_r\}\operatorname{lcm}[q,r]$ from Lemma 3.3. By the assumptions of the theorem, we have $M(q,r) \geqslant 2Q/[N\gcd(q,r)]$ for $q, r \in \mathcal{S}$. Hence, if (3.5) holds, then $\sum_{q,r\in\mathcal{S}} \text{meas}(\mathcal{A}_q^* \cap \mathcal{A}_r^*) \ll 1$ by Lemma 3.3. We may then apply Proposition 3.2 to deduce that $\text{meas}(\bigcup_{q\in\mathcal{S}} \mathcal{A}_q^*) \gg 1$.

When $N \gg Q$, Theorem 3.5 follows from the work of Erdős and Vaaler (Theorem 2.12), but when $N = o(Q)$ it was not known prior to [20] in this generality. The proof begins by adapting the Erdős–Vaaler argument to this more general setup.

First, we must control the sum over primes in (3.5). Using (2.10) turns out to be too crude, so we modify our approach. Let $t_j = \exp(2^j)$ and j_0 be such that $\sum_{t<p\leqslant t^2} 1/p \leqslant 1$ for $t \geqslant t_{j_0}$ (j_0 exists by Mertens' theorems [19, THEOREM 3.4].) Moreover, let

$$\mathcal{L}(q,r) = \sum_{\substack{p|qr/\gcd(q,r)^2 \\ p>Q/[N\gcd(q,r)]}} \frac{1}{p}, \quad \lambda_t(q) = \sum_{\substack{p|q \\ p>t}} \frac{1}{p}, \quad L_t(q,r) = \sum_{\substack{p|qr/\gcd(q,r)^2 \\ p>t}} \frac{1}{p}.$$

If $L_{t_{j_0}}(q,r) \leqslant 101$, then obviously $\mathcal{L}(q,r) \ll 1$. Otherwise, there is an integer $j \geqslant j_0$ such that $L_{t_j}(q,r) > 101 \geqslant L_{t_{j+1}}(q,r)$. Since $j \geqslant j_0$, we then also have $L_{t_{j+1}}(q,r) > 100$. Now, note that if $Q/[N\gcd(q,r)] \geqslant t_{j+1}$, then $\mathcal{L}(q,r) \leqslant L_{t_{j+1}}(q,r) \leqslant 101$.

To sum up, $\mathcal{L}(q,r) \ll 1$, unless $(q,r) \in \mathcal{B}_{t_{j+1}}$ for some $j \geqslant j_0$, where

$$\mathcal{B}_t := \big\{(q,r) \in \mathcal{S}\times\mathcal{S} : \gcd(q,r) > Q/(Nt),\ L_t(q,r) > 100\big\}.$$

We study the contribution of such pairs to the left-hand side of (3.5): if $(q,r) \in \mathcal{B}_{t_{j+1}}$, then

$$\mathcal{L}(q,r) \leqslant 101 + \sum_{p\leqslant t_{j+1}} \frac{1}{p} \leqslant \log\log t_{j+1} + O(1) = j\log 2 + O(1)$$

by Mertens' estimate. In conclusion, Theorem 3.5 will follow if we can show that

$$\sum_{(q,r)\in\mathcal{B}_t} \frac{\varphi(q)\varphi(r)}{qr} \ll \frac{N^2}{t} \quad \text{for all } t \geqslant t_{j_0+1}. \tag{3.6}$$

Now, let us consider the special case when $N \gg Q$, which corresponds to the Erdős–Vaaler theorem. The inequality $\gcd(q,r) > Q/(Nt)$ is then basically trivially, so we must prove (3.6) by exploiting the condition $L_t(q,r) > 100$. Indeed, writing $d = \gcd(q,r)$, $q = dq_1$ and $r = dr_1$, we find that $\lambda_t(q_1) > 50$ or $\lambda_t(r_1) > 50$. By symmetry, we have

$$\#\mathcal{B}_t \leqslant 2 \sum_{d \leqslant 2Q} \#\{r_1 \leqslant 2Q/d\} \cdot \#\{q_1 \leqslant 2Q/d : \lambda_t(q_1) > 50\}.$$

The number of r_1's is of course $\leqslant 2Q/d$. Moreover, using Chernoff's inequality and Lemma 3.4 with $a_p = \exp(1_{p>t} \cdot t/p)$, we find that

$$\#\{q_1 \leqslant 2Q/d : \lambda_t(q_1) > 50\} \leqslant \sum_{q_1 \leqslant 2Q/d} \exp(-50t + t\lambda_t(p)) \ll e^{-50t} Q/d.$$

Putting everything together, we conclude that

$$\#\mathcal{B}_t \ll e^{-t} Q^2 \quad \text{for all } t \geqslant 1. \tag{3.7}$$

In particular, (3.6) holds, thus proving Theorem 3.5 when $N \gg Q$.

On the other hand, if $N = o(Q)$, the condition that $\gcd(q,r) > Q/(Nt)$ for all $(q,r) \in \mathcal{B}_t$ is nontrivial and we must understand it and exploit it to prove Theorem 3.5. Indeed, if we treat the weights $\varphi(q)/q$ as roughly constant in (3.4), we see that $\mathcal{S}$ contains about N integers from $[Q, 2Q]$, i.e., it is a rather sparse set. On the other hand, if t is not too large, then (3.7) gives no savings compared to the trivial upper bound $\#\mathcal{B}_t \leqslant \#\mathcal{S}^2 \approx N^2$.

Since the condition that $L_t(q,r) > 100$ is insufficient, let us ignore it temporarily and focus on the condition that $\gcd(q,r) > Q/(Nt)$ for all $(q,r) \in \mathcal{B}_t$. There is an obvious way in which this condition can be satisfied for many pairs $(q,r) \in \mathcal{S} \times \mathcal{S}$, namely if there is some *fixed* integer $d > Q/(Nt)$ that divides a large proportion of integers in $\mathcal{S}$. Notice that the number of total multiples of d in $[Q, 2Q]$ is about $Q/d < t \cdot N$, which is compatible with (3.4). We thus arrive at the following key question:

Model Problem. Let $D \geqslant 1$ and $\delta \in (0, 1]$, and let $\mathcal{S} \subseteq [Q, 2Q] \cap \mathbb{Z}$ be a set of $\gg \delta Q/D$ elements such that there are $\geqslant \delta \#\mathcal{S}^2$ pairs $(q,r) \in \mathcal{S} \times \mathcal{S}$ with $\gcd(q,r) > D$. Must there be an integer $d > D$ that divides $\gg \delta^{100} Q/D$ elements of $\mathcal{S}$?

It turns out that the answer to the Model Problem as stated is negative. However, a technical variant of it is true, that takes into account the weights $\varphi(q)/q$ in (3.4) and (3.6), and that is asymmetric in q and r. We shall explain this in the next section.

For now, let us assume that the Model Problem as stated has an affirmative answer, and let us see how this yields Theorem 3.5. Suppose (3.6) fails for some t. By the Model Problem, there must exist an integer $d > N/(Qt)$ dividing $\gg t^{-100} \#\mathcal{S}$ members of $\mathcal{S}$. We might then also expect that $\#\mathcal{B}_t \gg t^{-200} \#\{(dm, dn) \in \mathcal{B}_t : m, n \geqslant 1\}$. But note that if $(q,r) = (dm, dn) \in \mathcal{B}_t$, then $m, n \leqslant 2Q/d < 2tN$ and $qr/\gcd(q,r)^2 = mn/\gcd(m,n)^2$. In particular, $L_t(m,n) > 100$, so the argument leading to (3.7) implies that the number of $(dm, dn) \in \mathcal{B}_t$ is $\ll e^{-t} t^2 N^2$. Hence, $\mathcal{B}_t \ll e^{-t} t^{202} N^2 \ll N^2/t$, as needed.

3.4. An iterative compression algorithm

To attack the Model Problem, we view it as a question in graph theory: consider the graph G, with vertex set $\mathcal{S}$ and edge set $\mathcal{B}_t$. If the edge density of G is $\geqslant 1/t$, must there exist a dense subgraph G' all of whose vertices are divisible by an integer $> Q/(Nt)$?

To locate this "structured" subgraph G', we use an iterative "compression" argument, roughly inspired by the papers of Erdős–Ko–Rado [11] and Dyson [9]. With each iteration, we pass to a smaller set of vertices, where we have additional information about which primes divide them. Of course, we must ensure that we end up with a sizeable graph. We do this by judiciously choosing the new graph at each step so that it has at least as many edges as what the qualitative parameters of the old graph might naively suggest. This way the new graph will have improved "structure" *and* "quality." When the algorithm terminates, we will end up with a fully structured subset of $\mathcal{S}$, where we know that all large GCDs are due to a large fixed common factor. This will then allow us to exploit the condition that $L_t(q,r) > 100$ for all edges (q,r). Importantly, our algorithm will also control the set $\mathcal{B}_t$ in terms of the terminal edge set. Hence the savings from the condition $L_t(q,r) > 100$ in the terminal graph will be transferred to $\mathcal{B}_t$, establishing (3.6).

One technical complication is that the iterative algorithm necessitates to view G as a bipartite graph. In addition, it is important to use the weights $\varphi(q)/q$. We thus set

$$\mu(\mathcal{V}) = \sum_{v\in\mathcal{V}} \frac{\varphi(v)}{v} \quad \text{for } \mathcal{V} \subset \mathbb{N}; \quad \mu(\mathcal{E}) = \sum_{(v,w)\in\mathcal{E}} \frac{\varphi(v)\varphi(w)}{vw} \quad \text{for } \mathcal{E} \subset \mathbb{N}^2.$$

Let us now explain the algorithm in more detail. We set $\mathcal{V}_0 = \mathcal{W}_0 = \mathcal{S}$ and construct two decreasing sequences of sets $\mathcal{V}_0 \supseteq \mathcal{V}_1 \supseteq \mathcal{V}_2 \supseteq \cdots$ and $\mathcal{W}_0 \supseteq \mathcal{W}_1 \supseteq \mathcal{W}_2 \supseteq \cdots$, as well as a sequence of distinct primes $p_1, p_2, \ldots$ such that either p_j divides all elements of $\mathcal{V}_j$, or it is coprime to all elements of $\mathcal{V}_j$ (and similarly with $\mathcal{W}_j$). Since $\mathcal{S}$ consists solely of square-free integers, there are integers a_j, b_j dividing $p_1 \cdots p_j$, and such that $\gcd(v, p_1 \cdots p_j) = a_j$ and $\gcd(w, p_1 \cdots p_j) = b_j$ for all $v \in \mathcal{V}_j$ and all $w \in \mathcal{W}_j$.

Assume we have constructed $\mathcal{V}_i, \mathcal{W}_i, p_i$ as above for $i = 1, \ldots, j$. Let $\mathcal{E}_i = \mathcal{B}_t \cap (\mathcal{V}_i \times \mathcal{W}_i)$ be the edge sets. We then pick a new prime p_{j+1} that occurs as common factor of $\gcd(v,w)$ for at least one edge $(v,w) \in \mathcal{E}_j$. (If there is no such prime, the algorithm terminates.) Then, we pick $\mathcal{V}_{j+1}$ to be either $\mathcal{V}_j^{(1)} := \{v \in \mathcal{V}_j : p_{j+1} \mid v\}$ or $\mathcal{V}_j^{(0)} := \{v \in \mathcal{V}_j : p_{j+1} \nmid v\}$ (and similarly with $\mathcal{W}_{j+1}$). Deciding how to make this choice is the most crucial part of the proof and we will analyze it in more detail below. At any rate, it is clear that after a finite number of steps, this process will terminate. We will thus arrive at sets of vertices $\mathcal{V}_J$ and $\mathcal{W}_J$ where $a = a_J$ divides all elements of $\mathcal{V}_J$, $b = b_J$ divides all elements of $\mathcal{W}_J$, and $\gcd(v,w) = \gcd(a,b)$ for all edges $(v,w) \in \mathcal{E}_J \subseteq \mathcal{B}_t$. In particular, $\gcd(a,b) > Q/(Nt)$, as long as $\mathcal{E}_J \neq \emptyset$. We have thus found our fixed large common divisor, so that we can use the Erdős–Vaaler argument as in Section 3.3 to control the size of $\mathcal{E}_J$. If we can ensure that $\mathcal{E}_J$ is a large enough portion of $\mathcal{E}_0 = \mathcal{B}_t$, we will have completed the proof.

Let us now explain how to make the choice of which subgraph to focus on each time. Let $G_j = (\mathcal{V}_j, \mathcal{W}_j, \mathcal{E}_j)$ be the bipartite graph at the jth iteration. Because we will use an unbounded number of iterations, it is important to ensure that G_{j+1} has more edges than "what the qualitative parameters of G_j would typically predict." One way to assign meaning to this vague phrase is to use the edge density $\#\mathcal{E}_j/(\#\mathcal{V}_j\#\mathcal{W}_j)$. Actually, in our case, we should use the weighted density

$$\delta_j = \frac{\mu(\mathcal{E}_j)}{\mu(\mathcal{V}_j)\mu(\mathcal{W}_j)}.$$

Naively, we might guess that $\mu(\mathcal{E}_{j+1}) \approx \delta_j \mu(\mathcal{V}_{j+1})\mu(\mathcal{W}_{j+1})$, meaning that $\delta_{j+1} \approx \delta_j$. So we might try to choose G_{j+1} so that $\delta_{j+1} \geqslant \delta_j$. This would be analogous to Roth's "density increment" strategy **[23, 24]**. Unfortunately, such an argument loses all control over the size of $\mathcal{E}_j$, so we cannot use information on $\mathcal{E}_J$ to control $\mathcal{E}_0 = \mathcal{B}_t$ (which is our end goal).

In a completely different direction, we can use the special GCD structure of our graphs to come up with another "measure of quality" of our new graph compared to the old one. Recall the integers a_{j+1} and b_{j+1}. We then have

$$\#\mathcal{E}_j \leqslant \#\left\{m \leqslant \frac{2Q}{a_j},\ n \leqslant \frac{2Q}{b_j} : \gcd(m,n) > \frac{Q/(Nt)}{\gcd(a_j,b_j)}\right\}. \tag{3.8}$$

If all pairs (m,n) on the right-hand side of (3.8) were due to a fixed divisor of size $> [Q/(Nt)]/\gcd(a_j,b_j)$, then we would conclude that

$$\#\mathcal{E}_j \ll t^2 N^2 \cdot \frac{\gcd(a_j,b_j)^2}{a_j b_j}.$$

Actually, Green and Walker **[13]** proved recently that this bound is true, even without the presence of a universal divisor. So it makes sense to consider the quantity $\#\mathcal{E}_j \cdot a_j b_j / \gcd(a_j,b_j)^2$ as a qualitative measure of G_j. As a matter of fact, since we are weighing v with $\varphi(v)/v$, and we have $\varphi(v)/v \leqslant \varphi(a_j)/a_j$ whenever $a_j|v$, we may even consider

$$\lambda_j := \frac{a_j b_j}{\gcd(a_j,b_j)^2} \cdot \frac{a_j b_j}{\varphi(a_j)\varphi(b_j)} \cdot \mu(\mathcal{E}_j).$$

Let us see a different argument for why this quantity might be a good choice, by studying the effect of each prime $p \in \{p_1,\dots,p_j\}$ to the parameters Q/a_j, Q/b_j and $[Q/(Nt)]/\gcd(a_j,b_j)$ that control the size of m, n, and $\gcd(m,n)$, respectively, in (3.8):

- *Case 1: $p|a_j$ and $p|b_j$.* Then p reduces the upper bounds on the size of both m and n by a factor $1/p$. On the other hand, it also reduces the lower bound on their GCD (that affects both m and n) by $1/p$. Hence, we are in a balanced situation.

- *Case 2: $p \nmid a_j$ and $p \nmid b_j$.* In this case, p affects no parameters.

- *Case 3: $p|a_j$ and $p \nmid b_j$.* Then p reduces the upper bound on m by a factor $1/p$, but it does not affect the bound on n nor on $\gcd(m,n)$. This is an advantageous situation, gaining us a factor of p compared to what we had. Accordingly, λ_j is multiplied by p in this case. This gain allows us to afford a big loss of vertices when falling in this "asymmetric" case (a proportion of $1 - O(1/p)$).

- *Case 4: $p \nmid a_j$ and $p|b_j$.* Then we gain a factor of p as in the previous case.

Iteratively increasing λ_j would be adequate for showing (3.6), by mimicking the Erdős–Vaaler argument from Section 3.3. Unfortunately, it is not possible to guarantee that

λ_j increases at each stage because it is not sensitive enough to the edge density, and so this proposal also fails. However, we will show that (a small variation of) the hybrid quantity

$$q_j := \delta_j^9 \lambda_j \tag{3.9}$$

can be made to increase at each step, while keeping track of the sizes of the vertex sets. We call q_j the *quality* of G_j.

Let us now discuss how we might carry out the "quality increment" strategy. Given $\mathcal{V}_j$ and p_{j+1}, we wish to set $\mathcal{V}_{j+1} = \mathcal{V}_j^{(k)}$ and $\mathcal{W}_{j+1} = \mathcal{W}_j^{(\ell)}$ for some $k, \ell \in \{0, 1\}$. Let us call $G_j^{(k,\ell)}$ each of the four potential choices for G_{j+1}. For their quality $q_j^{(k,\ell)}$, we have:

$$\frac{q_j^{(1,1)}}{q_j} = \left(\frac{\delta^{(1,1)}}{\delta_j}\right)^{10} \alpha\beta\left(1-\frac{1}{p}\right)^{-2}, \qquad \frac{q_j^{(1,0)}}{q_j} = \left(\frac{\delta_j^{(1,0)}}{\delta_j}\right)^{10} \alpha(1-\beta)p\left(1-\frac{1}{p}\right)^{-1},$$
$$\frac{q_j^{(0,1)}}{q_j} = \left(\frac{\delta_j^{(0,1)}}{\delta_j}\right)^{10} (1-\alpha)\beta p\left(1-\frac{1}{p}\right)^{-1}, \quad \frac{q_j^{(0,0)}}{q_j} = \left(\frac{\delta_j^{(0,0)}}{\delta_j}\right)^{10} (1-\alpha)(1-\beta),$$

where $\delta_j^{(k,\ell)}$ is the edge density of $G_j^{(k,\ell)}$, $\alpha = \mu(\mathcal{V}_j^{(1)})/\mu(\mathcal{V}_j)$ is the proportion of vertices in $\mathcal{V}_j$ that are divisible by p_{j+1}, and similarly $\beta = \mu(\mathcal{W}_j^{(1)})/\mu(\mathcal{W}_j)$. In addition, we have

$$\delta_j^{(1,1)}\alpha\beta + \delta_j^{(1,0)}\alpha(1-\beta) + \delta_j^{(0,1)}(1-\alpha)\beta + \delta_j^{(0,0)}(1-\alpha)(1-\beta) = \delta_j,$$

so that if one of the $\delta_j^{(k,\ell)}$'s is smaller than δ, some other must be larger. Such an unbalanced situation is advantageous, so let us assume that $\delta_j^{(k,\ell)} \sim \delta_j$ for all k, ℓ.

Notice that we have an extra factor p in the asymmetric cases $(0, 1)$ and $(1, 0)$. We can then easily obtain a quality increment in one of these two cases, unless $\alpha, \beta \ll 1/p$, or if $\alpha, \beta \geqslant 1 - O(1/p)$. It turns out that the former case can be treated with a trick. So, the real difficulty is to obtain a quality increment when α, β are both close to 1. As a matter of fact, the critical case is when $\alpha, \beta \sim 1 - 1/p$. Indeed, we then have $q_j^{(k,\ell)} \sim 1$ in all four cases, so we maintain a constant quality no matter which subgraph we choose to focus on.

It is important to remark here that the factor $(1 - 1/p)^{-2}$ in the case $(k, \ell) = (1, 1)$ is essential (the factors $(1 - 1/p)^{-1}$ in the asymmetric cases are less important as it turns out). Without this extra factor, we would not have been able to guarantee that the quality stays at least as big as q_j. Crucially, this factor originates from the weights $\varphi(v)/v$ of the vertices that are naturally built in the Duffin–Schaeffer conjecture and that dampen down contributions from integers with too many prime divisors.

We conclude this discussion by going back to the Model Problem. We mentioned in Section 3.3 that this problem is false. The reason is a counterexample due to Sam Chow, a square-free version of which is given by $\mathcal{S} = \{P/j : j \mid P,\ x/2 \leqslant j \leqslant x\}$ with $P = \prod_{p \leqslant x} p$. Indeed, all pairwise GCDs here are $\geqslant P/x^2$, but there is no fixed integer of size $\gg P/x^2$ dividing a positive proportion of elements of this set. In addition, note that if $p \leqslant x/\log x$, then the proportion of $\mathcal{S}$ divisible by p is $\sim 1 - 1/p$, just like in the critical case discussed above.

3.5. The quality increment argument

We now discuss the formal details of our iterative algorithm. We must first set up some notation. We say that $G = (\mathcal{V}, \mathcal{W}, \mathcal{E}, \mathcal{P}, a, b)$ is a *square-free GCD graph* if:

- $\mathcal{V}$ and $\mathcal{W}$ are nonempty, finite sets of square-free integers;
- $(\mathcal{V}, \mathcal{W}, \mathcal{E})$ is a bipartite graph, meaning that $\mathcal{E} \subseteq \mathcal{V} \times \mathcal{W}$;
- $\mathcal{P}$ is a finite set of primes, and a and b divide $\prod_{p \in \mathcal{P}} p$;
- $a|v$ and $b|w$ for all $(v, w) \in \mathcal{V} \times \mathcal{W}$;
- if $(v, w) \in \mathcal{E}$ and $p \in \mathcal{P}$, then $p|\gcd(v, w)$ precisely when $p|\gcd(a, b)$.

We shall refer to $(\mathcal{P}, a, b)$ as the *multiplicative data* of G. Furthermore, we defined the *edge density* of G by $\delta(G) := \frac{\mu(\mathcal{E})}{\mu(\mathcal{V})\mu(\mathcal{W})}$, and its *quality* by

$$q(G) := \delta(G)^9 \cdot \mu(\mathcal{E}) \cdot \frac{ab}{\gcd(a,b)^2} \cdot \frac{ab}{\varphi(ab)} \cdot \prod_{p \in \mathcal{P}} \left(1 - \frac{1}{p^{3/2}}\right)^{-10}.$$

In addition, we define the set of "remaining large primes" of G by

$$\mathcal{R}(G) := \left\{p \notin \mathcal{P} : p > 5^{100},\ p \mid \gcd(v, w) \text{ for some } (v, w) \in \mathcal{E}\right\}.$$

Finally, if $G' = (\mathcal{V}', \mathcal{W}', \mathcal{E}', \mathcal{P}', a', b')$ is another square-free GCD graph, we call it a *subgraph* of G if $\mathcal{V}' \subseteq \mathcal{V}$, $\mathcal{W}' \subseteq \mathcal{W}$, $\mathcal{E}' \subseteq \mathcal{E}$, $\mathcal{P}' \supseteq \mathcal{P}$, $\prod_{p|a',\, p \in \mathcal{P}} p = a$, $\prod_{p|b',\, p \in \mathcal{P}} p = b$.

Lemma 3.6 (The quality increment argument). *Let $G = (\mathcal{V}, \mathcal{W}, \mathcal{E}, \mathcal{P}, a, b)$ be a square-free GCD graph, let $p \in \mathcal{R}(G)$, and let $\alpha = \frac{\mu(\{v \in \mathcal{V}: p|v\})}{\mu(\mathcal{V})}$ and $\beta = \frac{\mu(\{w \in \mathcal{W}: p|w\})}{\mu(\mathcal{W})}$.*

(a) *If $\min\{\alpha, \beta\} \leqslant 1 - 5^{12}/p$, then there is a subgraph G' of G with multiplicative data $(\mathcal{P} \cup \{p\}, ap^k, bp^\ell)$ for some $k, \ell \in \{0, 1\}$ satisfying $\delta(G')^m q(G') \geqslant 2^{k \neq \ell} \delta(G)^m q(G)$ for $m \in \{0, 1\}$.*

(b) *If $\min\{\alpha, \beta\} > 1 - 5^{12}/p$, then there is a subgraph G' of G with set of primes $\mathcal{P} \cup \{p\}$ and with quality $q(G') \geqslant q(G)$.*

Proof. Each pair of $k, \ell \in \{0, 1\}$ defines a subgraph of G with multiplicative data $(\mathcal{P} \cup \{p\}, ap^k, bp^\ell)$. Indeed, we merely need to focus on the vertex subsets $\mathcal{V}_k = \{v \in \mathcal{V} : p^k \| v\}$ and $\mathcal{W}_\ell = \{w \in \mathcal{W} : p^\ell \| w\}$. This new GCD graph is formally given by the sextuple $G_{k,\ell} = (\mathcal{V}_k, \mathcal{W}_\ell, \mathcal{E}_{k,\ell}, \mathcal{P} \cup \{p\}, ap^k, bp^\ell)$, where $\mathcal{E}_{k,\ell} = \mathcal{E} \cap (\mathcal{V}_k \times \mathcal{W}_\ell)$. Let $\delta_{k,\ell} = \frac{\mu(\mathcal{E}_{k,\ell})}{\mu(\mathcal{E})}$, $\alpha_k = \frac{\mu(\mathcal{V}_k)}{\mu(\mathcal{V})}$ and $\beta_\ell = \frac{\mu(\mathcal{W}_\ell)}{\mu(\mathcal{W})}$, so that $\alpha_0 = 1 - \alpha$, $\alpha_1 = \alpha$, $\beta_0 = 1 - \beta$ and $\beta_1 = \beta$. We then have

$$\frac{\delta(G_{k,\ell})^m q(G_{k,\ell})}{\delta(G)^m q(G)} = \frac{\delta_{k,\ell}^{10+m} (\alpha_k \beta_\ell)^{-9-m} p^{1_{k \neq \ell}}}{(1 - 1/p)^{k+\ell} (1 - p^{-3/2})^{10}}. \tag{3.10}$$

(a) We claim that there exist choices of $k, \ell \in \{0, 1\}$ such that

$$\delta_{k,\ell} \geqslant \begin{cases} (\alpha_k \beta_k)^{9/10} & \text{if } k = \ell, \\ \dfrac{\alpha(1-\beta) + (1-\alpha)\beta}{5} & \text{if } k \neq \ell. \end{cases} \tag{3.11}$$

To prove this claim, it suffices to show that

$$(\alpha\beta)^{9/10} + \big((1-\alpha)(1-\beta)\big)^{9/10} + \frac{2}{5}\cdot\big[\alpha(1-\beta)+(1-\alpha)\beta\big] \leqslant 1. \tag{3.12}$$

Let $u = \max\{\alpha\beta, (1-\alpha)(1-\beta)\}$. Then

$$(\alpha\beta)^{9/10} + \big((1-\alpha)(1-\beta)\big)^{9/10} \leqslant u^{2/5}\big[(\alpha\beta)^{1/2} + \big((1-\alpha)(1-\beta)\big)^{1/2}\big] \leqslant u^{2/5}$$

by the Cauchy–Schwarz inequality. On the other hand, we have

$$\alpha(1-\beta) + (1-\alpha)\beta = 1 - \alpha\beta - (1-\alpha)(1-\beta) \leqslant 1 - u.$$

In conclusion, the left-hand side of (3.12) is $\leqslant u^{2/5} + 2(1-u)/5 \leqslant 1$, as needed.

Now, if (3.11) is true with $k = \ell$, part (a) of the lemma follows immediately by (3.10) upon taking $G' = G_{k,k}$. Assume then that (3.11) fails when $k = \ell$. We separate two cases.

Case 1: $\max\{\alpha,\beta\} > 5^{12}/p$. We know that (3.11) holds for some choice of $k \neq \ell$. Suppose that $k = 1$ and $\ell = 0$ for the sake of concreteness; the other case is similar. Then, (3.10) implies

$$\begin{aligned}\frac{\delta(G_{1,0})^m q(G_{1,0})}{\delta(G)^m q(G)} &\geqslant \left(\frac{\alpha(1-\beta)+\beta(1-\alpha)}{5}\right)^{10+m} \frac{p}{(\alpha(1-\beta))^{9+m}} \\ &\geqslant \frac{\alpha(1-\beta)+\beta(1-\alpha)}{5^{11}}p\end{aligned}$$

for $m \leqslant 1$. The proof is complete by taking $G' = G_{1,0}$, unless $\alpha(1-\beta)+\beta(1-\alpha) < 2\cdot 5^{11}/p$. In this case, we claim that either $\max\{\alpha,\beta\} < 5^{12}/p$ or $\min\{\alpha,\beta\} > 1 - 5^{12}/p$ (both of which we have assumed are false). By symmetry, we may assume $\alpha \leqslant 1/2$. Then $\beta/2 \leqslant \beta(1-\alpha) < 2\cdot 5^{11}/p$, as needed. In particular, $\beta \leqslant 1/2$ (because $p > 5^{100}$), and thus $\alpha/2 \leqslant \alpha(1-\beta) < 2\cdot 5^{11}/p$. We have thus reached a contradiction. This proves the lemma in this case.

Case 2: $\max\{\alpha,\beta\} \leqslant 5^{12}/p$. We must then have $\delta_{1,1} \leqslant (\alpha\beta)^{9/10} \leqslant 5^{22}p^{-9/5}$. Let us now define the GCD subgraph $G' = (\mathcal{V}, \mathcal{W}, \mathcal{E}', \mathcal{P}\cup\{p\}, a, b)$, where $\mathcal{E}' = \mathcal{E}\setminus(\mathcal{V}_1\times\mathcal{W}_1)$. Notice that we trivially have $a|v$ and $b|w$. In addition, since we have removed all edges (v,w) where p divides both v and w, we must have that $p \nmid \gcd(v,w)$ whenever $(v,w)\in\mathcal{E}'$. So, indeed, we see that G' is a GCD subgraph of G. Moreover,

$$\frac{\delta(G')^m q(G')}{\delta(G)^m q(G)} = \left(\frac{\mu(\mathcal{E}')}{\mu(\mathcal{E})}\right)^{10+m}\big(1-p^{-3/2}\big)^{-10} = (1-\delta_{1,1})^{10+m}\big(1-p^{-3/2}\big)^{-10} \geqslant 1$$

for $m \leqslant 1$, because $\delta_{1,1} \leqslant 5^{22}p^{-9/5}$ and $p > 5^{100}$. This proves the lemma in this case, too.

(b) Let $c = (1-p^{-3/2})^{-1}$. Using (3.10), we get a quality increment by letting $G' = G_{k,\ell}$ if one of the following inequalities holds:

$$\begin{aligned} c\delta_{1,1} &\geqslant (\alpha\beta)^{9/10}(1-1/p)^{2/10}, & c\delta_{0,0} &\geqslant \big((1-\alpha)(1-\beta)\big)^{9/10}, \\ c\delta_{1,0} &\geqslant \big(\alpha(1-\beta)\big)^{9/10}p^{-1/10}, & c\delta_{0,1} &\geqslant \big((1-\alpha)\beta\big)^{9/10}p^{-1/10}. \end{aligned} \tag{3.13}$$

Let $\alpha = 1 - A/p$ and $\beta = 1 - B/p$ with $A, B \in [0, 5^{12}]$. It suffices to show that

$$c \geqslant \left(1-\frac{A}{p}\right)^{\frac{9}{10}}\left(1-\frac{B}{p}\right)^{\frac{9}{10}}\left(1-\frac{1}{p}\right)^{\frac{2}{10}} + \frac{(AB)^{\frac{9}{10}}}{p^{9/5}} + \frac{(1-\frac{A}{p})^{\frac{9}{10}}B^{\frac{9}{10}} + A^{\frac{9}{10}}(1-\frac{B}{p})^{\frac{9}{10}}}{p}. \tag{3.14}$$

Indeed, the right-hand side of (3.14) is

$$\begin{aligned}&\leqslant \exp\left(-\frac{0.9A+0.9B+0.2}{p}\right) + \frac{5^{22}}{p^{9/5}} + \frac{A^{9/10}+B^{9/10}}{p}\\ &\leqslant 1 - \frac{0.A+0.9B+0.2}{p} + \frac{(0.9A+0.9B+0.2)^2}{2p^2} + \frac{5^{22}}{p^{9/5}} + \frac{A^{9/10}+B^{9/10}}{p},\end{aligned}$$

where we used the inequalities $0 \leqslant 1-x \leqslant e^{-x} \leqslant 1-x+x^2/2$, valid for all $x \in [0,1]$. By the inequality of arithmetic and geometric means, we have $0.9A + 0.1 \geqslant A^{9/10}$ and $0.9B + 0.1 \geqslant B^{9/10}$. Hence, the right-hand side of (3.14) is

$$\leqslant 1 + \frac{(0.9A+0.9B+0.2)^2}{2p^2} + \frac{5^{11}}{p^{9/5}} \leqslant 1 + \frac{5^{25}}{p^2} + \frac{5^{22}}{p^{9/5}} \leqslant 1 + \frac{1}{p^{3/2}} \leqslant c$$

for $p > 5^{100}$. This completes the proof of the part (b) of the lemma. ■

3.6. Proof of Theorem 3.5

Let Q, N, $\mathcal{S}$ and $\mathcal{B}_t$ with $t \geqslant t_{j_0+1}$ be as in Section 3.3. We want to prove (3.6). We may assume that $\mu(\mathcal{B}_t) \geqslant \mu(\mathcal{S})^2/t$; otherwise, (3.6) is trivially true.

Consider the GCD graph $G_0 := (\mathcal{S}, \mathcal{S}, \mathcal{B}_t, \emptyset, \emptyset, \emptyset)$, and note that $\delta(G_0) \geqslant 1/t$. We repeatedly apply part (a) of Lemma 3.6 to create a sequence of distinct primes $p_1, p_2, \dots$ and of square-free GCD graphs $G_j = (\mathcal{V}_j, \mathcal{W}_j, \mathcal{E}_j, \{p_1, \dots, p_j\}, a_j, b_j)$, $j = 1, 2, \dots$, with G_j a subgraph of G_{j-1}. Assuming we have applied Lemma 3.6(a) j times, we may apply it once more if there is $p \in \mathcal{R}(G_j)$ dividing a proportion $\leqslant 1 - 5^{12}/p$ of $\mathcal{V}_j$ and $\mathcal{W}_j$.

Naturally, the above process will terminate after a finite time, say after J_1 steps and we will arrive at a GCD graph G_{J_1} such that if $p \in \mathcal{R}(G_{J_1})$, then p divides a proportion $> 1 - 5^{12}/p$ of the vertex sets $\mathcal{V}_{J_1}$ and $\mathcal{W}_{J_1}$. In addition, the sequence of GCD graphs produced is such that $\delta(G_j)^m q(G_j) \geqslant 2^{1_{j\in\mathcal{D}}}\delta(G_{j-1})^m q(G^{(j-1)})$ for $m \in \{0, 1\}$, where

$$\mathcal{D} = \{j \leqslant J_1 : p_j \text{ divides } a_{J_1}b_{J_1}/\gcd(a_{J_1}, b_{J_1})^2\}.$$

In particular,

$$\delta(G_{J_1})^m q(G_{J_1}) \geqslant 2^{\#\mathcal{D}}\delta(G_0)^m q(G_0) \quad \text{for } m \in \{0, 1\}. \tag{3.15}$$

To proceed, we must separate two cases.

Case 1: $q(G_{J_1}) \geqslant t^{30}q(G_0)$. We apply repeatedly Lemma 3.6 (either part (a) or (b), according to whether the condition $\min\{\alpha, \beta\} \leqslant 1 - 5^{12}/p$ holds or fails) to create a sequence of primes $p_{J_1+1}, p_{J_1+2}, \dots$ that are distinct from each other and from $p_1, \dots, p_{J_1}$, and of square-free GCD graphs $G_j = (\mathcal{V}_j, \mathcal{W}_j, \mathcal{E}_j, \{p_1, \dots, p_j\}, a_j, b_j)$, $j = J_1 + j, J_1 + 2, \dots,$

with G_j a subgraph of G_{j-1}. As before, this process will terminate, say after $J_2 - J_1$ steps, and we will arrive at a GCD graph G_{J_2} with $\mathcal{R}(G_{J_2}) = \emptyset$. By construction, we have

$$q(G_{J_2}) \geqslant q(G_{J_2-1}) \geqslant \cdots \geqslant q(G_{J_1}) \geqslant t^{30} q(G_0).$$

In addition, we have

$$q(G_0) = \frac{\mu(\mathcal{B}_t)^{10}}{\mu(\mathcal{S})^{18}} \geqslant \frac{\mu(\mathcal{B}_t)^{10}}{N^{18}} \tag{3.16}$$

by (3.4). (In particular, note that $q(G_0) > 0$, so $q(G_{J_2}) > 0$ and thus $\mathcal{E}_{J_2} \neq \emptyset$.) On the other hand, if we let $a = a_{J_2}$ and $b = b_{J_2}$, then $\gcd(v,w) \mid \gcd(a,b)P$ with $P = \prod_{p \leqslant 5^{100}} p$ for all $(v,w) \in \mathcal{E}_{J_2}$. In particular, $\gcd(a,b) > Q/(PNt)$. Moreover,

$$\mu(\mathcal{E}_{J_2}) \leqslant \sum_{m \leqslant 2Q/a,\ n \leqslant 2Q/b}\sum \frac{\varphi(am)}{am} \cdot \frac{\varphi(bn)}{bn} \leqslant \frac{\varphi(a)\varphi(b)}{ab} \cdot \frac{4Q^2}{ab}.$$

Since $\delta(G_{J_2}) \leqslant 1$ and $\prod_p (1 - 1/p^{3/2})^{-10} < \infty$, we then have

$$q(G_{J_2}) \ll \mu(\mathcal{E}_{J_2}) \frac{ab}{\gcd(a,b)^2} \frac{ab}{\varphi(a)\varphi(b)} \ll t^2 N^2. \tag{3.17}$$

Recalling that $q(G_{J_2}) \geqslant t^{30} q(G_0)$, relations (3.16) and (3.17) complete the proof of (3.6), and thus of Theorem 3.5 in this case.

Case 2: $q(G_{J_1}) < t^{30} q(G_0)$. In this case, we do not have such a big quality gain, so we need to use that $L_t(v,w) > 100$ for all $(v,w) \in \mathcal{B}_t$. But we must be very careful because this condition might be dominated by the prime divisors of the fixed integers a and b we are constructing. Before we proceed, note that (3.15) implies that

$$\delta(G_{J_1}) \geqslant \delta(G_0) \cdot \frac{q(G_0)}{q(G_{J_1})} \geqslant \frac{1}{t} \cdot \frac{1}{t^{30}} = \frac{1}{t^{31}}. \tag{3.18}$$

Let $\mathcal{R} = \mathcal{R}(J_1)$ and let $p \in \mathcal{R}$. By the construction of G_{J_1}, p divides a proportion $> 1 - 5^{12}/p$ of the vertex sets $\mathcal{V}_{J_1}$ and $\mathcal{W}_{J_1}$. Therefore,

$$\mu\big(\{(v,w) \in \mathcal{E}_{J_1} : p \mid vw/\gcd(v,w)^2\}\big) \leqslant \frac{2 \cdot 5^{12}}{p} \mu(\mathcal{V}_{J_1}) \mu(\mathcal{W}_{J_1}) \leqslant \frac{5^{13} t^{31}}{p} \mu(\mathcal{E}_{J_1}),$$

where we used (3.18). As a consequence, we find

$$\sum_{(v,w) \in \mathcal{E}_{J_1}} \frac{\varphi(v)\varphi(w)}{vw} \sum_{\substack{p > t^{32},\ p \in \mathcal{R} \\ p \mid vw/\gcd(v,w)^2}} \frac{1}{p} \leqslant \sum_{p > t^{32}} \frac{5^{13} t^{31} \mu(\mathcal{E}_{J_1})}{p^2} \leqslant \frac{\mu(\mathcal{E}_{J_1})}{100}.$$

Hence, if we let

$$\mathcal{E}_{J_1}^{\text{good}} = \Bigg\{(v,w) \in \mathcal{E}_{J_1} : \sum_{\substack{p > t^{32},\ p \in \mathcal{R} \\ p \mid vw/\gcd(v,w)^2}} \frac{1}{p} \leqslant 1\Bigg\},$$

Markov's inequality implies that $\mu(\mathcal{E}_{J_1}^{\text{good}}) \geqslant 0.99 \mu(\mathcal{E}_{J_1})$. We then define the GCD graph $G_0' = (\mathcal{V}_{J_1}, \mathcal{W}_{J_1}, \mathcal{E}_{J_1}^{\text{good}}, \mathcal{P}, a_{J_1}, b_{J_1})$. Note that

$$g(G_0') = \left(\frac{\mu(\mathcal{E}_{J_1}^{\text{good}})}{\mu(\mathcal{E}_{J_1})}\right)^{10} q(G_{J_1}) \geqslant \frac{q(G_{J_1})}{2} \geqslant \frac{q(G_0)}{2}.$$

Next, we apply repeatedly Lemma 3.6 to create a sequence of distinct primes $p_{J_1+1}, p_{J_1+2}, \ldots \in \mathcal{R}$ and of GCD graphs $G'_j = (\mathcal{V}'_j, \mathcal{W}'_j, \mathcal{E}'_j, \{p_1, \ldots, p_{J_1+j}\}, a'_j, b'_j)$, $j = 1, \ldots$, with G'_j a subgraph of G'_{j-1}. This process will terminate, say after K steps, and we will arrive at a GCD graph G'_K with $\mathcal{R}(G'_K) = \emptyset$. By construction, we have

$$q(G'_K) \geqslant q(G'_{K-1}) \geqslant \cdots \geqslant q(G'_0) \geqslant q(G_0)/2 > 0. \tag{3.19}$$

In particular, $\mathcal{E}'_K \neq \emptyset$. It remains to give an upper bound on $q(G'_K)$.

Let $a' = a'_K$ and $b = b'_K$, and recall that $P = \prod_{p \leqslant 5^{100}} p$. Then $\gcd(v, w) | \gcd(a', b')P$ for all $(v, w) \in \mathcal{E}'_K$. In particular, $\gcd(a', b') > Q/(PNt)$. Moreover, if $(v, w) \in \mathcal{E}'_K$ and we let $v = a'm$ and $w = b'n$, then

$$100 < L_t(v, w) \leqslant 5 + L_{t^{32}}(v, w) \leqslant 6 + \sum_{\substack{p > t^{32},\ p \notin \mathcal{R} \\ p | vw/\gcd(v,w)^2}} \frac{1}{p} \leqslant 6 + \frac{\#\mathcal{D}}{t^{32}} + L_{t^{32}}(m, n),$$

where the first inequality is true because $(v, w) \in \mathcal{B}_t$, the second because $\sum_{y < p \leqslant y^2} 1/p \leqslant 1$ for $y \geqslant t_{j_0}$, the third because $(v, w) \in \mathcal{E}^{\mathrm{good}}_{J_1}$, and the fourth because if p divides $a'b'/\gcd(a', b')^2$ and $p \notin \mathcal{R}$, then $p \in \mathcal{D}$. Now, since $2^{\#\mathcal{D}} \leqslant q(G_{J_1})/q(G_0) \leqslant t^{30}$, we have that $L_{t^{32}}(m, n) > 93$. Therefore,

$$\mu(\mathcal{E}'_K) \leqslant \sum\sum_{\substack{m \leqslant 2Q/a',\ n \leqslant 2Q/b' \\ L_{t^{32}}(m,n) > 93}} \frac{\varphi(a')\varphi(b')}{a'b'} \ll \frac{\varphi(a')\varphi(b')}{a'b'} \cdot \frac{4Q^2}{a'b'} e^{-t^{32}},$$

by arguing as in the proof of (3.7). We may then insert this inequality into the definition of $q(G'_K)$ and conclude that $q(G'_K) \ll e^{-t^{32}} t^2 N^2$. Together with (3.19) and (3.16), this completes the proof of (3.6), and thus of Theorem 3.5 in this last case as well.

ACKNOWLEDGMENTS

The author is grateful to James Maynard for his comments on a preliminary version of the paper.

FUNDING

The author is supported by the *Courtois Chair II in fundamental research* of the Université de Montréal, the *Natural Sciences and Engineering Research Council of Canada* (Discovery Grant 2018-05699), and the *Fonds de recherche du Québec – Nature et technologies* (projets de recherche en équipe 256442 and 300951).

REFERENCES

[1] C. Aistleitner, Decoupling theorems for the Duffin–Schaeffer problem. Progress report, 2019, 24 pp. arXiv:1907.04590.

[2] C. Aistleitner, T. Lachmann, M. Munsch, N. Technau, and A. Zafeiropoulos, The Duffin–Schaeffer conjecture with extra divergence. *Adv. Math.* **356** (2019), 106808, 11 pp.

[3] V. Beresnevich, G. Harman, A. K. Haynes, and S. Velani, The Duffin–Schaeffer conjecture with extra divergence II. *Math. Z.* **275** (2013), no. 1–2, 127–133.

[4] V. Beresnevich and S. Velani, A mass transference principle and the Duffin–Schaeffer conjecture for Hausdorff measures. *Ann. of Math. (2)* **164** (2006), no. 3, 971–992.

[5] A. S. Besicovitch, Sets of fractional dimensions (IV): on rational approximation to real numbers. *J. Lond. Math. Soc.* **9** (1934), no. 2, 126–131.

[6] J. W. S. Cassels, Some metrical theorems in Diophantine approximation. I. *Proc. Camb. Philos. Soc.* **46** (1950), 209–218.

[7] P. A. Catlin, Two problems in metric Diophantine approximation. I. *J. Number Theory* **8** (1976), no. 3, 282–288.

[8] R. J. Duffin and A. C. Schaeffer, Khinchin's problem in metric Diophantine approximation. *Duke Math. J.* **8** (1941), 243–255.

[9] F. J. Dyson, A theorem on the densities of sets of integers. *J. Lond. Math. Soc.* **20** (1945), 8–14.

[10] P. Erdős, On the distribution of the convergents of almost all real numbers. *J. Number Theory* **2** (1970), 425–441.

[11] P. Erdős, C. Ko, and R. Rado, Intersection theorems for systems of finite sets. *Quart. J. Math. Oxf. Ser. (2)* **12** (1961), 313–320.

[12] P. Gallagher, Approximation by reduced fractions. *J. Math. Soc. Japan* **13** (1961), 342–345.

[13] B. Green and A. Walker, Extremal problems for GCDs. 2020, arXiv:2012.02078.

[14] G. Harman, *Metric number theory*. London Math. Soc. Monogr. New Ser. 18, The Clarendon Press, Oxford University Press, New York, 1998.

[15] A. K. Haynes, A. D. Pollington, and S. L. Velani, The Duffin–Schaeffer conjecture with extra divergence. *Math. Ann.* **353** (2012), no. 2, 259–273.

[16] V. Jarník, Zur metrischen Theorie der diophantischen Approximationen. *Proc. Mat. Fiz.* **36** (1928), 91–106.

[17] A. Khinchin, *Continued fractions*. Dover Publications, Inc., Mineola, NY, 1997.

[18] A. Khintchine, Einige Sätze über Kettenbrüche, mit Anwendungen auf die Theorie der Diophantischen Approximationen. *Math. Ann.* **92** (1924), no. 1–2, 115–125.

[19] D. Koukoulopoulos, *The distribution of prime numbers*. Grad. Stud. Math. 203, American Mathematical Society, Providence, RI, 2019.

[20] D. Koukoulopoulos and J. Maynard. *Ann. of Math. (2)* **192** (2020), no. 1, 251–307.

[21] K. Matomäki, The distribution of αp modulo one. *Math. Proc. Cambridge Philos. Soc.* **147** (2009), no. 2, 267–283.

[22] A. D. Pollington and R. C. Vaughan, The k-dimensional Duffin and Schaeffer conjecture. *Mathematika* **37** (1990), no. 2, 190–200.

[23] K. F. Roth, Sur quelques ensembles d'entiers. *C. R. Acad. Sci. Paris* **234** (1952), 388–390.

[24] K. F. Roth, On certain sets of integers. *J. Lond. Math. Soc.* **28** (1953), 104–109.

[25] K. F. Roth, Rational approximations to algebraic numbers. *Mathematika* **2** (1955), 1–20; corrigendum, 168.

[26] E. M. Stein and R. Shakarchi, *Real analysis. Measure theory, integration, and Hilbert spaces*. Princeton Lect. Anal. 3, Princeton University Press, Princeton, NJ, 2005.

[27] J. D. Vaaler, On the metric theory of Diophantine approximation. *Pacific J. Math.* **76** (1978), no. 2, 527–539.

[28] A. Zaharescu, Small values of $n^2\alpha$ mod 1. *Invent. Math.* **121** (1995), no. 2, 379–388.

[29] D. Zeilberger and W. Zudilin, The irrationality measure of π is at most 7.103205334137… Mosc. *J. Comb. Number Theory* **9** (2020), no. 4, 407–419.

DIMITRIS KOUKOULOPOULOS

Département de mathématiques et de statistique, Université de Montréal, CP 6128 succ. Centre-Ville, Montréal, QC H3C 3J7, Canada, dimitris.koukoulopoulos@umontreal.ca

This Page Intentionally Left Blank

EULER SYSTEMS AND THE BLOCH–KATO CONJECTURE FOR AUTOMORPHIC GALOIS REPRESENTATIONS

DAVID LOEFFLER AND SARAH LIVIA ZERBES

ABSTRACT

We survey recent progress on the Bloch–Kato conjecture, relating special values of L-functions to cohomology of Galois representations, via the machinery of Euler systems. This includes new techniques for the construction of Euler systems, via the étale cohomology of Shimura varieties, and new methods for proving explicit reciprocity laws, relating Euler systems to critical values of L-functions. These techniques have recently been used to prove the Bloch–Kato conjecture for critical values of the degree 4 L-function of GSp_4, and we survey ongoing work aiming to apply this result to the Birch–Swinnerton-Dyer conjecture for modular abelian surfaces, and to generalise it to a range of other automorphic L-functions.

MATHEMATICS SUBJECT CLASSIFICATION 2020

Primary 11G40; Secondary 11F67, 11F80, 11G18

KEYWORDS

Bloch–Kato conjecture, Euler system, Selmer group

INTERNATIONAL CONGRESS OF MATHEMATICIANS 2022 JULY 6–14

Proc. Int. Cong. Math. 2022, Vol. 3, pp. 1918–1939
DOI 10.4171/ICM2022/50

Published by EMS Press

1. WHAT IS THE BLOCH–KATO CONJECTURE?

The Bloch–Kato conjecture, formulated in [11], relates the cohomology of global Galois representations to the special values of L-functions. We briefly recall a weak form of the conjecture, which will suffice for this survey. Let $L/\mathbf{Q}_p$ be a finite extension, let K be a number field, and let V be a representation of $\Gamma_K = \mathrm{Gal}(\overline{K}/K)$ on a finite-dimensional L-vector space. We suppose V is unramified outside finitely many primes and de Rham at the primes above p. Then we may attach to V the following two objects:

- Its *L-function*, which is the formal Euler product
$$L(V,s) = \prod_v P_v\big(V, \mathbf{N}(v)^{-s}\big)^{-1},$$
where v varies over (finite) primes of K, and $P_v(V, X) \in L[X]$ is a local Euler factor depending on the restriction of V to a decomposition group at v. It is conjectured that, for any choice of isomorphism $\overline{L} \cong \mathbf{C}$, this product converges for $\Re(s) \gg 0$ and and has meromorphic continuation to all of $\mathbf{C}$.
- Its *Selmer group* $H^1_{\mathrm{f}}(K, V)$, a certain (finite-dimensional) subspace of the Galois cohomology group $H^1(K, V)$ determined by local conditions at each prime, defined in [11].

The *weak Bloch–Kato conjecture* asserts that
$$\mathrm{ord}_{s=1} L\big(V^*, s\big) = \dim H^1_{\mathrm{f}}(K, V) - \dim H^0(K, V).$$
The full conjecture as formulated in [11] also determines the leading term of $L(V^*, s)$ at $s = 1$ up to a p-adic unit, in terms of the cohomology of an integral lattice $T \subset V$.

This conjecture includes as special cases a wide variety of well-known results and conjectures. For example, when V is the 1-dimensional trivial representation, the weak conjecture states that $\zeta_K(s)$ has a simple pole at $s = 1$; and the strong conjecture (for all p at once) is equivalent to the analytic class number formula, relating the residue at this pole to the class group and unit group of K. If $V = T_p(E) \otimes \mathbf{Q}_p$, where E is an elliptic curve over K and $T_p(E)$ is its Tate module, then $L(V^*, s)$ is the Hasse–Weil L-function $L(E/K, s)$, and we recover the Birch and Swinnerton-Dyer conjecture for E over K.

Critical values

The L-function $L(V^*, s)$ is expected to satisfy a functional equation relating $L(V, s)$ and $L(V^*, 1-s)$, after multiplying by a suitable product of Γ-functions $L_\infty(V^*, s)$ (determined by the Hodge–Tate weights of V at p and the action of complex conjugation). These Γ-factors may have poles at $s = 1$, forcing $L(V^*, 1)$ to vanish.

Following [49], we say V is *r-critical*, for some $r \geqslant 0$, if $L_\infty(V^*, 1-s)$ has a pole of order r at $s = 0$, and $L_\infty(V, s)$ is holomorphic there. In particular, V is 0-critical if $L(V^*, 1)$ is a critical value in the sense of Deligne [18]. The most interesting cases of the Bloch–Kato conjecture are when V is 0-critical, and it is these which our main theorems below will address; but 1-critical Galois representations will also play a crucial auxiliary role in our strategy.

Iwasawa theory

The Bloch–Kato conjecture is closely related to the *Iwasawa main conjecture*, in which the finite-dimensional Selmer group $H^1_{\mathrm{f}}(K, V)$ is replaced by a finitely-generated module over an Iwasawa algebra. This connection with Iwasawa theory, together with the proof of the Iwasawa main conjecture in this context by Mazur and Wiles, plays an important role in Huber and Kings' proof **[33]** of the Bloch–Kato conjecture for 1-dimensional representations of $\Gamma_{\mathbf{Q}}$.

2. WHAT IS AN EULER SYSTEM?

For K a number field and V a Γ_K-representation as in Section 1, we have the notion of an *Euler system* for V, defined as follows. Let S be a finite set of places of K containing all infinite places, all primes above p and all primes at which V is ramified.

We define $\mathcal{R}$ to be the collection of integral ideals of K of the form $\mathfrak{m} = \mathfrak{a} \cdot \mathfrak{b}$, where $\mathfrak{a}$ is a square-free product of primes of K not in S, and $\mathfrak{b}$ divides p^∞. For each $\mathfrak{m} \in \mathcal{R}$, let $c[\mathfrak{m}]$ be the ray class field modulo $\mathfrak{m}$. Then an Euler system for (T, S) is a collection of classes

$$\mathbf{c} = \{c[\mathfrak{m}] \in H^1(K[\mathfrak{m}], T) : \mathfrak{m} \in \mathcal{R}\},$$

satisfying the *norm-compatibility relation*

$$\mathrm{cores}^{K[\mathfrak{m}\mathfrak{q}]}_{K[\mathfrak{m}]}\big(c[\mathfrak{m}\mathfrak{q}]\big) = \begin{cases} P_{\mathfrak{q}}(V^*(1), \sigma_{\mathfrak{q}}^{-1}) \cdot c[\mathfrak{m}] & \text{if } \mathfrak{q} \notin S, \\ c[\mathfrak{m}] & \text{if } \mathfrak{q} \mid p, \end{cases}$$

where cores denotes the Galois corestriction (or norm) map, and $\sigma_{\mathfrak{q}}$ is the image of $\mathrm{Frob}_{\mathfrak{q}}$ in $\mathrm{Gal}(K[\mathfrak{m}]/K)$. By an *Euler system for* V, we mean an Euler system for some (T, S). (These general definitions are due to Kato, Perrin-Riou, and Rubin, building on earlier work of Kolyvagin; the standard reference is **[56]**.)

The crucial application of Euler systems is the following: if an Euler system exists for V whose image in $H^1(K, V)$ is non-zero (and V satisfies some auxiliary technical hypotheses), then we obtain a bound for the so-called *relaxed* Selmer group[1]

$$H^1_{\mathrm{rel}}(K, V) := \ker\Big(H^1(K, V) \to \prod_{v \nmid p} H^1_{\mathrm{f}}(K_v, V)\Big).$$

The relaxed Selmer group differs from the Bloch–Kato Selmer group in that we impose no local conditions at p. More generally, under additional assumptions on V and $\mathbf{c}$, we can obtain finer statements taking into account local conditions at p, and hence control the dimension of the Bloch–Kato Selmer group itself.

Euler systems are hence extremely powerful tools for bounding Selmer groups, as long as we can understand whether the image of $\mathbf{c}$ in $H^1(K, V)$ is non-vanishing. In order to

---

1 See **[49]** for this formulation. Theorem 2.2.3 of **[56]** is an equivalent result, but expressed in terms of a Selmer group for $V^*(1)$, which is related to that of V by Poitou–Tate duality.

use an Euler system to prove new cases of the Bloch–Kato conjecture, one needs to establish a so-called *explicit reciprocity law*, which is a criterion for the non-vanishing of the Euler system in terms of the value $L(V^*, 1)$.

Challenges. In order to use Euler system theory to approach the Bloch–Kato conjecture, and other related problems such as the Iwasawa main conjecture, there are two major challenges to be overcome:

(1) Can we construct "natural" examples of Euler systems (satisfying appropriate local conditions), for interesting global Galois representations V?

(2) Can we prove reciprocity laws relating the images of these Euler systems in $H^1(K, V)$ to the values of L-functions?

This was carried out by Kato [35] for the Galois representations arising from modular forms; but Kato's approach to proving explicit reciprocity laws has turned out to be difficult to generalise. More recently, in a series of works with various co-authors beginning with [40] (building on earlier work of Bertolini–Darmon–Rotger [6]), we developed a general strategy for overcoming these challenges, for Galois representations arising from automorphic forms for a range of reductive groups. We will describe this strategy in the remainder of this article.

Variants. A related concept is that of an *anticyclotomic Euler system*, in which K is a CM field, and we replace the ray-class fields $c[\mathfrak{m}]$ with *ring* class fields associated to ideals of the real subfield K^+. These arise naturally when V is *conjugate self-dual*, i.e. $V^c = V^*(1)$ where c denotes complex conjugation. The most familiar example is Kolyvagin's Euler system of Heegner points [39]; for more recent examples, see, e.g. [12, 15, 25]. Many of the techniques explained here for constructing and studying (full) Euler systems are also applicable to anticyclotomic Euler systems, and we shall discuss examples of both below.

A rather more distant cousin is the concept of a *bipartite Euler system*, which arises naturally in the context of level-raising congruences; cf. [31] for a general account, and [43] for a dramatic recent application to the Bloch–Kato conjecture. These require a rather different set of techniques, and we shall not discuss them further here.

The 1-critical condition. We conjectured in [49] that, in order to construct Euler systems for V by geometric means (i.e. as the images of motivic cohomology classes), we need to impose a condition on V: it needs to be *1-critical*.

However, our intended applications involve the Bloch–Kato conjecture for critical values of L-functions; so we need to construct Euler systems for representations that are 0-critical, rather than 1-critical. So we shall construct Euler systems for these representations in two stages: firstly, we shall construct Euler systems for auxiliary 1-critical representations V, using motivic cohomology; secondly, we shall "p-adically deform" our Euler systems, in order to pass from these 1-critical V to others which are 0-critical. This will be discussed in Section 4 below.

3. EULER SYSTEMS FOR SHIMURA VARIETIES

Shimura varieties. Let $(G, \mathcal{X})$ be a Shimura datum, with reflex field E. For a level $K \subset G(\mathbf{A}_f)$, we write $Y_G(K)$ for the Shimura variety $\mathrm{Sh}_K(G, \mathcal{X})$. Our first goal will be to define Euler systems, either full or anticyclotomic, for Galois representations appearing in the étale cohomology of $Y_G(K)$. We shall attempt to give a systematic general treatment, but the reader should bear the following examples in mind:

(1) $G = \mathrm{GL}_2 \times_{\mathrm{GL}_1} \mathrm{GL}_2$, as in [40];

(2) $G = \mathrm{Res}_{F/\mathbf{Q}} \mathrm{GL}_2$ for F real quadratic, as in [26, 41];

(3) $G = \mathrm{GSp}_4$, as in [47];

(4) $G = \mathrm{GSp}_4 \times_{\mathrm{GL}_1} \mathrm{GL}_2$, as in [32];

(5) $G = \mathrm{GU}(2, 1)$, as in [48];

(6) $G = \mathrm{U}(2n-1, 1)$ for $n \geqslant 1$, as in [25].

Each of these groups is naturally equipped with a Shimura datum $(G, \mathcal{X})$. In examples (1)–(4), the reflex field E is $\mathbf{Q}$; in (5) and (6), it is the imaginary quadratic field used to define the unitary group. (One can also retrospectively interpret Kato's construction [35] in these terms, taking $G = \mathrm{GL}_2$; and similarly Kolyvagin's anticyclotomic Euler system [39], which is in effect the $n = 1$ case of example (6).)

Étale cohomology. If π is a cuspidal automorphic representation which contributes to $H^*_{\text{ét}}(Y_G(K)_{\bar{E}}, V_\lambda)$ for some level K, where V_λ is the étale local system associated to the representation of G of highest weight λ, then we say π is *cohomological in weight* λ. It is conjectured that if this holds, then there exists a p-adic representation ρ_π of Γ_E, for each prime p and embedding $\overline{\mathbf{Q}} \hookrightarrow \overline{\mathbf{Q}}_p$, whose local Euler factors are determined by the Satake parameters of π at finite places, and whose Hodge–Tate weights are determined by λ.

For all of the above groups, the existence of such a ρ_π is known, and, moreover, if π is of "general type" (i.e. not a functorial lift from a smaller group), then the π_f-eigenspace in étale cohomology is concentrated in degree $d = \dim(\mathcal{X})$ and isomorphic to $\pi_f \otimes \rho_\pi$. So we can find projection maps $H^d(Y_G(K)_{\bar{E}}, V_\lambda) \twoheadrightarrow \rho_\pi$, for a suitable choice of K. Via the Hochschild–Serre spectral sequence

$$H^i\big(E, H^j_{\text{ét}}\big(Y_G(K)_{\bar{E}}, V_\lambda\big)\big) \Rightarrow H^{i+j}_{\text{ét}}\big(Y_G(K)_E, V_\lambda\big),$$

we can thus obtain classes in the Galois cohomology of ρ_π as the images of classes in the π_f-eigenspace of $H^{d+1}_{\text{ét}}(Y_G(K)_E, V_\lambda)$. (For simplicity, we shall sketch the construction below assuming $\lambda = 0$, and refer to the original papers for the case of general coefficients.)

Motivic cohomology. In order to construct classes in $H^{d+1}_{\text{ét}}(Y_G(K)_E, V_\lambda)$, we shall use two other, related cohomology theories:

- *Motivic cohomology* (see [3]), which takes values in $\mathbf{Q}$-vector spaces (or $\mathbf{Z}$-lattices in them), and is closely related to algebraic K-theory and Chow groups;

- *Deligne–Beilinson* (or *absolute Hodge*) cohomology (see [34]), which takes values in $\mathbf{R}$-vector spaces, and has a relatively straightforward presentation in terms of pairs (ω, σ), where ω is an algebraic differential form, and σ a real-analytic antiderivative of $\operatorname{Re}(\omega)$.

There is no direct relation between Deligne–Beilinson cohomology and p-adic étale cohomology – one would not expect to compare vector spaces over $\mathbf{R}$ and over $\mathbf{Q}_p$ – but both of these cohomology theories have natural maps ("realisation maps") from motivic cohomology. So we shall use the following strategy, whose roots go back to [3]: we will write down elements of motivic cohomology whose images in Deligne–Belinson cohomology are related to values of L-functions; and we will consider the images of the same motivic cohomology classes in étale cohomology.

Pushforward maps. If $(H, \mathcal{Y}) \hookrightarrow (G, \mathcal{X})$ is the inclusion of a sub-Shimura datum (with the same reflex field E), then we obtain finite morphisms of algebraic varieties over E,

$$Y_H(K \cap H)_E \to Y_G(K)_E,$$

where E is the reflex field of $(H, \mathcal{Y})$. More generally, for each $g \in G(\mathbf{A}_\mathrm{f})$ we have a map

$$\iota_g : Y_H(gKg^{-1} \cap H)_E \to Y_G(gKg^{-1})_E \xrightarrow{g} Y_G(K),$$

where the latter arrow is translation by g. So we have associated *pushforward maps* in all of our cohomology theories, namely

$$\iota_{g,\star} : H^j_{\mathrm{mot}}\big(Y_H(K \cap H)_E, \mathbf{Z}(t)\big) \to H^{j+2c}_{\mathrm{mot}}\big(Y_G(K)_E, \mathbf{Z}(t+c)\big)$$

for $j, r \in \mathbf{Z}$, where $c = \dim \mathcal{X} - \dim \mathcal{Y}$ (and similarly for étale cohomology with $\mathbf{Z}_p$ coefficients, or Deligne–Beilinson cohomology with $\mathbf{R}$ coefficients, compatibly with the realisation maps relating the theories).

We shall define motivic cohomology classes for $Y_G(K)$ using the maps $\iota_{g,\star}$. The compatibility of these classes with realisation functors allows us to compute the images of such classes in Deligne–Beilinson cohomology: the projection of such a class to the π_f-eigenspace will be computed using integrals over $Y_H(K \cap H)(\mathbf{C})$, involving the pullbacks of differential forms associated to cusp forms in the dual automorphic representation $\pi^\vee$.

Cycle classes and Siegel units. As an input to the above construction, we need a supply of "interesting" classes in $H^j_{\mathrm{mot}}(Y_H(K \cap H), \mathbf{Z}(r))$ for some k, r which are in the image of motivic cohomology.

One possibility is to start with the *identity class* $1 \in H^0_{\mathrm{mot}}(Y_H(K \cap H), \mathbf{Z}(0))$. The image of this class under $\iota_{g,\star}$ is the *cycle class* associated to the image of ι_g, a so-called "special cycle". This case is by no means trivial: indeed, these special cycles are the input used to define anticyclotomic Euler systems, such as Heegner points.

More subtly, one can obtain motivic cohomology classes from *units* in the coordinate ring of Y_H, using the relation

$$H^1_{\mathrm{mot}}\big(Y, \mathbf{Z}(1)\big) = \mathcal{O}(Y)^\times$$

for any variety Y. If Y is a modular curve (i.e. a Shimura variety for GL_2), then we have a canonical family of units: if $Y_1(N)$ is the Shimura variety of level $\{(\begin{smallmatrix} * & * \\ 0 & 1 \end{smallmatrix}) \bmod N\}$, then we have the *Siegel unit*

$$z_N \in \mathcal{O}\big(Y_1(N)\big)^{\times},$$

denoted ${}_c g_{0,1/N}$ in the notation of [35] (where c is an auxiliary integer coprime to the level). Crucially, we have an explicit formula for the image of this class in Deligne–Beilinson cohomology; it is given by

$$\big(\mathrm{dlog}\, z_N, \log|z_N|\big) = \big(E_2, E_0^{\mathrm{an}}(0)\big) \tag{3.1}$$

where E_2 is a weight 2 Eisenstein series, and $E_0^{\mathrm{an}}(s)$ is a family of real-analytic Eisenstein series depending on a parameter $s \in \mathbf{C}$. (See also [38] for analogues with coefficients, related to Eisenstein series of higher weights.)

Rankin–Eisenstein classes and Rankin–Selberg integrals. We shall consider the following general setting: we consider a Shimura datum $(H, \mathcal{Y})$ equipped with an embedding $\iota : (H, \mathcal{Y}) \to (G, \mathcal{X})$, and also with a family of maps

$$\psi = (\psi_1, \dots, \psi_t) : (H, \mathcal{Y}) \to (\mathrm{GL}_2, \mathbb{H})^t,$$

where $\mathbb{H}$ is the standard GL_2 Shimura datum and $t \geqslant 1$. We then have a collection of classes

$$z_N^H = \psi_1^*(z_N) \cup \dots \cup \psi_t^*(z_N) \in H^t_{\mathrm{mot}}\big(Y_{\mathcal{H}}\big(K_{H,1}(N)\big), \mathbf{Z}(t)\big),$$

for some level $K_{H,1}(N)$, which we call *Eisenstein classes* for H.

Remark 3.1. One might hope for a broader range of "Eisenstein classes" in motivic cohomology, associated to Eisenstein series on other groups which are not just copies of GL_2's. However, this question seems to be very difficult; see [21] for some results in this direction for symplectic groups. If we could construct motivic classes associated to Eisenstein series for the Siegel parabolic of GSp_{2n} (rather than the Klingen parabolic as in [21]), or for the analogous parabolic subgroup in the unitary group $\mathrm{U}(n,n)$, then it would open the way towards a far wider range of Euler system constructions. ■

By a *motivic Rankin–Eisenstein class* for $(G, \mathcal{X})$ (with trivial coefficients), we shall mean a class of the form

$$\iota_{g,\star}\big({}_c z_N^H\big) \in H^{2c+t}_{\mathrm{mot}}\big(Y_G(K)_E, \mathbf{Z}(c+t)\big),$$

for some N and some g and level K. If we choose our data $(H, \mathcal{Y})$ such that $2c + t = 1 + d$, then these classes land in the cohomological degree we want. The twist $c + t$ is then equal to $\frac{d+1+t}{2}$; hence, using the Hochschild–Serre spectral sequence, we can project the étale realisations of these classes into the groups $H^1(E, V_\pi)$, where

$$V_\pi = \rho_\pi\left(\frac{d+1+t}{2}\right).$$

Choosing the data

To define a Rankin–Eisenstein class, we need to choose the group H, and the maps $\iota : H \to G$ and $\psi : H \to (\mathrm{GL}_2)^t$. To guide us in choosing these, we shall use "Rankin–Selberg-type" integral formulas for L-functions of automorphic representations. There are a wide range of such formulas, relating automorphic L-functions to integrals of the form

$$\int_{H(\mathbf{Q})Z_G(\mathbf{A})\backslash H(\mathbf{A})} \iota^*(\phi)\psi^*\big(E^{\mathrm{an}}(s_1) \times \cdots \times E^{\mathrm{an}}(s_n)\big)\mathrm{d}h, \tag{3.2}$$

where $E^{\mathrm{an}}(s_i)$ are real-analytic GL_2 Eisenstein series, and ϕ is a cuspform in the space of π. We call these *period integrals*. Typically, one expects such an integral to evaluate to a product of one or more copies of the L-function of π, evaluated at some linear combination of the parameters s_i. For instance, the Rankin–Selberg integral for $\mathrm{GL}_2 \times \mathrm{GL}_2$ is of this form, as is Novodvorsky's formula for the L-functions of GSp_4 and $\mathrm{GSp}_4 \times \mathrm{GL}_2$.

Using the explicit formula (3.1) relating Siegel units to Eisenstein series, one can often show that the Deligne–Beilinson realisations of Rankin–Eisenstein classes also lead to integrals of the form (3.2), for suitably chosen ϕ and s_i. When this applies, we can use it to relate our motivic Rankin–Eisenstein classes to special values of L-functions[2] (as was carried out in Beilinson's original paper **[3]** for the L-functions of GL_2 and $\mathrm{GL}_2 \times \mathrm{GL}_2$; see, e.g. **[36, 42]** for more recent examples).

This gives one a guide to constructing "interesting" Rankin–Eisenstein classes for a given $(G, \mathcal{X})$: one first searches for a Rankin–Selberg integral describing the relevant L-function, and then attempts to breathe motivic life into this real-analytic formula, interpreting it as the Deligne–Beilinson realisation of a motivic Rankin–Eisenstein class. One should hence interpret Rankin–Eisenstein classes as "motivic avatars" of Rankin–Selberg integral formulae.

In the anti-cyclotomic ($t = 0$) case, the period integral (3.2) is more mysterious; but there are still a number of results and conjectures predicting that these period integrals should be related to values of L-functions. For instance, the Gan–Gross–Prasad conjecture **[22]** gives such a relation in the important cases $\mathrm{SO}(n) \hookrightarrow \mathrm{SO}(n) \times \mathrm{SO}(n+1)$ and $U(n) \hookrightarrow U(n) \times U(n+1)$. This conjecture has recently been proved in the unitary case **[9]**, although the orthogonal case is still open. We refer to Sakellaridis–Venkatesh **[58]** for conjectural generalisations to other pairs (G, H).

---

2 More precisely, we obtain a relation to the first derivative $L'(V_\pi^*, 1)$, with V_π being 1-critical. Unfortunately, this computation does not give us any information about the étale class in $H^1(E, V_\pi)$, since the motivic class might be in the kernel of the étale realisation map. (This is the fundamental obstruction to proving the Bloch–Kato conjecture for 1-critical Galois representations.)

Example 3.2. In our examples (1)–(6) above, we choose H and t as follows:

	G	H	t
(1)	$\mathrm{GL}_2 \times \mathrm{GL}_2$	GL_2	1
(2)	$\mathrm{Res}_{F/\mathbf{Q}}\mathrm{GL}_2$	GL_2	1
(3)	GSp_4	$\mathrm{GL}_2 \times_{\mathrm{GL}_1} \mathrm{GL}_2$	2
(4)	$\mathrm{GSp}_4 \times \mathrm{GL}_2$	$\mathrm{GL}_2 \times_{\mathrm{GL}_1} \mathrm{GL}_2$	1
(5)	$GU(2,1)$	$\mathrm{GL}_2 \times_{\mathrm{GL}_1} \mathrm{Res}_{E/\mathbf{Q}}\mathrm{GL}_1$	1
(6)	$U(2n-1,1)$	$U(n-1,1) \times U(n,0)$	0

The integral formulae for L-functions underlying examples (1) and (2) are, respectively, the classical Rankin–Selberg integral and Asai's integral formula for quadratic Hilbert modular forms. Cases (3) and (4) are related to Novodvorsky's integral formula for $\mathrm{GSp}_4 \times \mathrm{GL}_2$ L-functions (with an additional Eisenstein series on the GL_2 factor in the former case); and case (5) to an integral studied by Gelbart and Piatetski-Shapiro in [23]. Example (6) is related to conjectures of Getz–Wambach [24] on Friedberg–Jacquet periods for automorphic representations of unitary groups. ■

Rankin–Eisenstein classes and norm relations. In order to build Euler systems (either full or anticyclotomic) from Rankin–Eisenstein classes, we need the following conditions to hold:

- ("Open orbit" condition) The group H has an open orbit on the product

$$(G/B_G) \times (\mathbf{P}^1)^t,$$

  where B_G is a Borel subgroup of G, and H acts on G/B_G via ι, and on $(\mathbf{P}^1)^t$ via ψ.

- ("Small stabiliser" condition) For a point u in the open orbit, let S_u be the subgroup of H which fixes u and acts trivially on the fibre at u of the tautological $(\mathbf{G}_m)^t$-bundle over $(\mathbf{P}^1)^t$. Then we require that the image of S_u has small image in the maximal torus quotient of H.

The role of the "small stabiliser" condition is to allow us to construct classes over field extensions. Since the connected components of the Shimura variety Y_G are defined over abelian extensions of E, and the Galois action on the component group is described by class field theory, we can modify the Rankin–Eisenstein classes to define elements in $H^{d+1}_{\text{ét}}(Y_G(K)_F, \mathbf{Z}_p(\frac{d+1+t}{2}))$ for a fixed level K and varying abelian extensions F/E. The class of abelian extensions that arise will depend on the image of S_u in the maximal torus quotient; in the examples (1)–(5) above, since $S_u = \{1\}$ and the splitting field of the Galois action is the full maximal abelian extension of E, so we obtain classes over all ray class fields of E. On the other hand, in example (6) we obtain only the anticyclotomic extension (as one would expect, since $t = 0$ in this case).

The "open orbit" condition allows us to prove a so-called *vertical norm relation*, showing that after applying Hida's ordinary projector, the Rankin–Eisenstein classes form norm-compatible families over the tower $E[p^\infty]/E$, with uniformly bounded denominators relative to the étale cohomology with $\mathbf{Z}_p$-coefficients. This machinery is worked out in considerable generality in [44,46]; the arguments also simultaneously show that the Rankin–Eisenstein classes interpolate in Hida-type p-adic families (in which the weight λ of π varies).

A much more subtle problem is that of *horizontal norm relations*, comparing classes over $E[\mathfrak{m}]$ and $E[\mathfrak{m}\mathfrak{q}]$ for auxiliary primes $\mathfrak{q} \nmid \mathfrak{m}$, with the Euler factors $P_{\mathfrak{q}}$ appearing in the comparison. The strategy developed in [47] and refined in [48] is to use multiplicity-one results in smooth representation theory to reduce the norm relation to a purely local calculation with zeta-integrals, which can then be computed explicitly to give the Euler factor. These multiplicity-one results are themselves closely bound up with the open-orbit condition; see [57].

Remark 3.3. The open-orbit condition, together with the assumption that $2c + t = 1 + d$, amount to stating that the diagonal map $(\iota, \psi) : (H, \mathcal{Y}) \hookrightarrow (G, \mathcal{X}) \times (\mathrm{GL}_2, \mathbb{H})^t$ is a *special pair* of Shimura data in the sense of [59, DEFINITION 3.1]. We can thus interpret the "small stabiliser" condition, at least for $t = 0$, as a criterion for when the special cycles studied in [59] extend to norm-compatible families over field extensions. ∎

4. DEFORMATION TO CRITICAL VALUES

Critical values. The above methods allow us to define Euler systems for the automorphic Galois representations $V_\pi = \rho_\pi(\frac{d+1+t}{2})$, where π is cohomological in weight 0; and there are generalisations to representations which are cohomological for a certain range of nonzero weights λ, determined by branching laws for the restriction of algebraic representations from $\tilde{G} = G \times (\mathrm{GL}_2)^t$ to H. Let us write Σ_1 for the set of weights λ which are accessible by these methods, for some specific choice of H and ψ; this is a convex polyhedron in the weight lattice of G, cut out by finitely many linear inequalities. In the examples (1)–(5), one checks that for any π whose weight lies in Σ_1, the representation V_π is 1-critical, consistently with the conjectures of [49].

However, our goal is to prove the Bloch–Kato conjecture for critical L-values; so we are interested in those λ such that, for π of weight λ, the representation V is 0-critical, so $L(\pi^\vee, \frac{1-t}{2})$ is a critical value. The set of such λ is a finite disjoint union of polyhedral regions; and we let Σ_0 be one of these regions, chosen to be adjacent to Σ_1. In order to approach the Bloch–Kato conjecture, we need to find a way of "deforming" our Euler systems from Σ_1 to Σ_0.

Example 4.1. For $G = \mathrm{GL}_2 \times \mathrm{GL}_2$, the Galois representations associated to cohomological representations of G have the form $(\rho_f \otimes \rho_g)(n)$, where f, g are modular forms (of some weights $k + 2, \ell + 2$ with $k, \ell \geqslant 0$) and n is an arbitrary integer. If we set $j = k + \ell + 1 - n$, then the set Σ_1 is given by the inequalities

$$\{0 \leqslant j \leqslant \min(k, \ell)\},$$

and there are two candidates for the set Σ_0, namely

$$\{\ell + 1 \leqslant j \leqslant k\} \quad \text{and} \quad \{k + 1 \leqslant j \leqslant \ell\}. \qquad \blacksquare$$

Remark 4.2. A slightly different, but related, numerology applies for anticyclotomic Euler systems. In these cases, the relevant L-value is always critical, but it lies at the centre of the functional equation, so it may be forced to vanish for sign reasons. Since the local root numbers at the infinite places depend on λ, we have some ranges of weights where the root number is $+1$ (where we expect interesting central L-values) and others where it is -1 (where we expect anticyclotomic Euler systems). These play the roles of the 0-critical and 1-critical regions in the case of full Euler systems. ■

The Bertolini–Darmon–Prasanna strategy. Although the "0-critical" and "1-critical" weight ranges are disjoint, we can relate them together p-adically, using a strategy introduced by Bertolini, Darmon and Prasanna in [5].

The weights λ of cohomological representations can naturally be seen as points of a p-adic analytic space $\mathcal{W}$ (parametrising characters $T(\mathbf{Z}_p) \to \mathbf{C}_p^\times$, where T is a maximal torus in G). This space is isomorphic to a finite union of n-dimensional open discs, where n is the rank of G. Crucially, both Σ_0 and Σ_1 are Zariski-dense in $\mathcal{W}$.

Hida theory shows that there is a finite flat covering $\mathcal{E} \to \mathcal{W}$, the *ordinary eigenvariety* of G, whose points above a dominant integral weight λ ("classical points") biject with automorphic representations π of G which are cohomological of weight λ and p-ordinary.

We thus have two separate families of objects, indexed by different sets of classical points on $\mathcal{E}$:

- at points whose weights lie in Σ_0, we have the critical values of the complex L-function;
- at points whose weights lie in Σ_1, we have Euler systems arising from motivic cohomology.

Our first goal will be to "analytically continue" the Euler system classes from Σ_1 into Σ_0.

This is not all that we require, however, since we also need a relation between the resulting Euler system for each 0-critical V and the L-value $L(V^*, 1)$. Relations of this kind are known as *explicit reciprocity laws*, and they are the crown jewels of Euler system theory. Following a strategy initiated in [5] and further developed in [37], in order to prove explicit reciprocity laws, we shall use a second kind of p-adic deformation: besides deforming Euler system classes from Σ_1 to Σ_0, we shall also deform L-values from Σ_0 into Σ_1. The strategy consists of the following steps:

(i) We shall construct a function on the eigenvariety – an "analytic p-adic L-function" – whose values in Σ_0 are critical L-values (modified by appropriate periods and Euler factors).

(ii) Using the Perrin-Riou regulator map of p-adic Hodge theory, we construct a second analytic function on the eigenvariety – a "motivic p-adic L-function" – whose value at some cohomological π measures the non-triviality of Euler system classes for π locally at p.
Note that the motivic p-adic L-function has no *a priori* reason to be related to complex L-values; however, its values in Σ_1 are by definition related to the Euler system classes (which arise from motivic cohomology, hence the terminology).

(iii) We shall prove a "p-adic regulator formula", showing that the values of the analytic p-adic L-function in at points in Σ_1 are related to the localisations of the Euler system classes at p.

(iv) Using the regulator formula of step (iii), we can deduce that the motivic and analytic p-adic L-function coincide at points in Σ_1. Since weights lying in Σ_1 are Zariski-dense in $\mathcal{E}$, this implies the two p-adic L-functions coincide in Σ_0 as well. Since the values of the analytic p-adic L-function in Σ_0 are complex L-values, we obtain the sought-for explicit reciprocity law.

At the time of writing, this strategy has only been fully carried out for the examples (1) and (3) in our list, and partially for (4). However, the remaining cases are being treated in ongoing work of members of our research groups; and we expect the strategy to extend to many other Euler systems (both full and anticyclotomic) besides these.

5. CONSTRUCTING p-ADIC L-FUNCTIONS

Coherent cohomology

To construct the analytic p-adic L-function, we shall use the integral formula (3.2). Previously, for weights in Σ_1, we interpreted this integral as a cup-product in Deligne–Beilinson cohomology. We shall now give a different cohomological interpretation of the same formula, for weights in the range Σ_0. Following a strategy introduced by Harris [28,29], we can choose the cusp-form ϕ, and the Eisenstein series, to be harmonic differential forms (with controlled growth at the boundary) representing Dolbeault cohomology classes valued in automorphic vector bundles. These can then be interpreted algebraically, via the comparison between Dolbeault cohomology and Zariski sheaf cohomology. The upshot is that $L(\pi^\vee, \frac{1-t}{2})$ can be related to a cup-product in the cohomology of coherent sheaves on a smooth toroidal compactification $\mathrm{Sh}_K(H, \mathcal{Y})^{\mathrm{tor}}_{\Sigma}$ of $\mathrm{Sh}_K(H, \mathcal{Y})$.

Interpolation

In order to construct a p-adic L-function, we need to show that the cohomology classes appearing in our formula for the L-function interpolate in Hida-type p-adic families, and that the cup-product of these families makes sense.

Until recently, there was a fundamental limitation in the available techniques: we could only interpolate cohomology classes corresponding to holomorphic automorphic forms (i.e. degree 0 coherent cohomology), or (via Serre duality) those in the top-degree cohomology, which correspond to anti-holomorphic forms. This is an obstacle for our intended applications, since the integral formulas relevant for Euler systems always involve coherent cohomology in degrees close to the middle of the possible range. (More precisely, the relevant degree is $\frac{d+t-1}{2}$, where t is the number of Eisenstein series present, which is typically $0, 1, 2$.) So unless d is rather small, using holomorphic or anti-holomorphic classes will not work.

A slightly wider range of "product type" examples arises when $(G, \mathcal{X})$ is a product of two Shimura data $(G_1, \mathcal{X}_1) \times (G_2, \mathcal{X}_2)$ of approximately equal dimension, with $\dim(\mathcal{X}_1) - \dim(\mathcal{X}_2) = t - 1$; then we can build a class in the correct degree as the product of an anti-holomorphic form on $\mathcal{X}_1$ and a holomorphic one on $\mathcal{X}_2$, and the resulting cup-products can often be understood as Petersson-type scalar products in Hida theory. For instance, the Rankin–Selberg integral formula can be analysed in this way [38]. However, for $G = \mathrm{GSp}_4$ (with $t = 2$ and $d = 3$), we need to work with a class in coherent H^2, and these are not seen by orthodox Hida theory.

Higher Hida theory. A beautiful solution to this problem is provided by the "higher Hida theory" developed in [55]. Pilloni's work shows that degree 1 coherent cohomology for the GSp_4 Shimura variety interpolates in a "partial" Hida family, with one weight fixed and the other varying p-adically.

A key ingredient in this work is to consider a certain stratification of the mod p fibre of the GSp_4 Shimura variety Y_G (for some level structure unramified at p). This space parametrises abelian surfaces A with a principal polarisation and some prime-to-p level structure. There is an open subspace Y_G^{ord}, whose complement has codimension 1, where A is ordinary; and a slightly larger open set, with complement of codimension 2, where the p-rank of A (the dimension of the multiplicative part of $A[p]$) is $\geqslant 1$. This stratification can be extended to a toroidal compactification X_G of Y_G; and Pilloni's approach to studying H^1 in p-adic families is based on restricting to the tube of this p-rank $\geqslant 1$ locus in the p-adic completion $\mathfrak{X}_G$ of X_G. (In contrast, orthodox Hida theory for GSp_4 involves working over the ordinary locus; this is very effective for studying H^0 but disastrous for studying H^1, since the ordinary locus is affine in the minimal compactification, so its cuspidal cohomology vanishes in positive degrees.)

In [45] we carried out a (slightly delicate) comparison of stratifications, showing that we can find an embedding ι_g of an H-Shimura variety, for a carefully chosen g, so that the preimage of the p-rank $\geqslant 1$ locus in $\mathfrak{X}_G$ is the ordinary locus in $\mathfrak{X}_H$, that is, the image of $\mathfrak{X}_H$ "avoids" the locus where the p-rank is exactly 1. Using this, we constructed pushforward maps from the orthodox (H^0) Hida theory for H to Pilloni's H^1 theory for G, interpolating the usual coherent-cohomology pushforward maps for varying weights. This is the tool we need to construct analytic p-adic L-functions for GSp_4 and for $\mathrm{GSp}_4 \times \mathrm{GL}_2$.

At present higher Hida theory, in the above sense, is only available for a few specific groups, although these include many of the ones relevant for Euler systems: besides GSp_4, the group $GU(2,1)$ is treated in [53], and Hilbert modular groups in [27] (in both cases assuming G is locally split at p). In the GSp_4 and $GU(2,1)$ cases the results are also slightly weaker than one might ideally hope, since we only obtain families in which one component of the weight is fixed and the others vary (so the resulting p-adic L-functions have one variable fewer than one would expect). However, we expect that these restrictions will be lifted in future work.

Remark 5.1. A related theory, *higher Coleman theory*, has been developed by Boxer and Pilloni in [13]. This theory also serves to interpolate higher-degree cohomology in families, with all components of the weight varying; and the theory applies to any Shimura variety of abelian type. However, unlike the higher Hida theory of [55], this theory only applies to cohomology classes satisfying an "overconvergence" condition. This rules out the 2-parameter GL_2 Eisenstein family which plays a prominent role in the constructions of [45], as this Eisenstein series is not overconvergent. It may be possible to work around this problem by combining the higher Coleman theory of [13] with the theory of families of nearly-overconvergent modular forms for GL_2 introduced by Andreatta–Iovita [1]; but the technical obstacles in carrying this out would be formidable. ■

6. P-ADIC REGULATORS

We now turn to step (iii) of the BDP strategy: relating values of the analytic p-adic L-function in the range Σ_1 to the localisations at p of the Euler system classes.

Syntomic cohomology. For all but finitely many primes, the Shimura variety has a smooth integral model over $\mathbf{Z}_p$, and the motivic Rankin–Eisenstein classes can be lifted to the cohomology of this integral model. This allows us to study them via another cohomology theory, Besser's *rigid syntomic cohomology* [7]. This is a cohomology theory for smooth $\mathbf{Z}_p$-schemes $\mathcal{Y}$, which has two vital properties:

- Via works of Fontaine–Messing and Niziol, one can define a comparison map relating syntomic cohomology of $\mathcal{Y}$ to étale cohomology of its generic fibre Y; and this is compatible with motivic cohomology, in the sense that we have a commutative diagram (see [8]):

$$\begin{array}{ccc} H^*_{\mathrm{mot}}(\mathcal{Y}, n) & \longrightarrow & H^*_{\mathrm{mot}}(Y, n) \\ \downarrow & & \downarrow r_{\text{ét}} \\ H^*_{\mathrm{syn}}(\mathcal{Y}, n) & \xrightarrow{\mathrm{FM}} & H^*_{\text{ét}}(Y, n) \end{array}$$

  where the map $r_{\text{ét}}$ is the étale realisation map.

  The Fontaine–Messing–Niziol map induces the Bloch–Kato exponential map on Galois cohomology; so, for a class in $H^*_{\text{ét}}(Y, n)$ in the image of motivic cohomology of $\mathcal{Y}$, one can express its Bloch–Kato logarithm via cup-products in syntomic cohomology ("syntomic regulators").

- Rigid syntomic cohomology and its variant, fp-cohomology, were defined by Besser as a generalisation of Coleman's theory of p-adic integration. It is computed by an explicit complex of sheaves which is a p-adic analogue of the real-analytic Deligne–Beilinson complex: sections of this complex are pairs (ω, σ), where ω is an algebraic differential form, and σ is an overconvergent rigid–analytic differential form such that $d\sigma = (1 - \varphi)\omega$, where φ is a local lift of the Frobenius of the special fibre.

In a series of works, beginning with the breakthrough [16] by Darmon–Rotger (see also [6,10,38]), rigid syntomic cohomology has been systematically exploited to compute the Bloch–Kato logarithms of Rankin–Eisenstein classes when G is a product of copies of GL_2, in terms of Petersson products of (non-classical) p-adic modular forms. These can then be interpreted as values of p-adic L-functions in a "1-critical" region Σ_1. All of these p-adic L-functions are "product type" settings in the sense explained above, involving coherent cohomology in either top or bottom degree.

Remark 6.1. A key role in these constructions is played by an explicit formula for the image of the Siegel unit in the syntomic cohomology of the ordinary locus of the modular curve, which is the p-adic counterpart of equation (3.1): it is represented by the pair

$$\left(\operatorname{dlog} z_N, (1 - \varphi) \log_p z_N\right) = \left(E_2, E_0^{(p)}\right)$$

where $E_0^{(p)}$ is a p-adic Eisenstein series of weight 0.

We can thus understand these syntomic regulator formulae as p-adic counterparts of the integral formula (3.2), with the integral understood via Coleman's p-adic integration theory, and the real-analytic Eisenstein class replaced by a p-adic one. ■

The GSp_4 regulator formula. The approach to computing regulators of étale classes via syntomic cohomology generalises to Euler systems for other Shimura varieties, such as GSp_4: one can always express the image of the Euler system class under the Bloch–Kato logarithm, paired against a suitable de Rham cohomology class (lying in the $\pi^\vee$-eigenspace), as a cup product in syntomic cohomology.

However, syntomic cohomology of the whole Shimura variety is not well-suited to explicit computations, since there is generally no global lift of the Frobenius of the special fibre. The first major problem is hence to express the pairing in terms of the syntomic cohomology of certain open subschemes of the Shimura variety which do possess an explicit Frobenius lift. This requires some results on the Hecke eigenspaces appearing in the rigid cohomology of Newton strata of the special fibre, which are the $\ell = p$ counterparts of the vanishing theorems proved by Caraiani–Scholze [14] for ℓ-adic cohomology for $\ell \neq p$.

The second major problem is to establish a link between rigid syntomic cohomology and coherent cohomology, for varieties admitting a Frobenius lifting. We succeeded in proving such a relation via a new a spectral sequence (the so-called *Poznań spectral sequence*) which is a syntomic analogue of the Hodge–de Rham spectral sequence: its E_1 page is given by the mapping fibre of $1 - \varphi$ on coherent cohomology, and its abutment is rigid syntomic cohomology. In the case of the ordinary locus of the modular curve, where all coherent cohomology in positive degrees vanishes, this reduces to the description of a syntomic class as a pair of global sections (ω, σ) as described above.

Thanks to this new spectral sequence, we were able to express the syntomic regulator of our Rankin–Eisenstein class for GSp_4 as a pairing in coherent cohomology, which we could identify as a specialisation of the pairing in higher Hida theory defining the p-adic L-function. We can hence identify the logarithm of the GSp_4 Euler system class with a non-critical value of a p-adic L-function. This is the first example of a p-adic regulator formula where the p-adic L-function is *not* of product type. We expect this strategy to be applicable to all the other Euler systems mentioned in Section 3 above. Cases (2) and (6) are currently work in progress by Giada Grossi, and by Andrew Graham and Waqar Shah, respectively; and case (5) is being explored by some members of our research groups.

7. DEFORMATION TO CRITICAL VALUES

We can now proceed to the final step of the Bertolini–Darmon–Prasanna strategy: deforming from Σ_1 to Σ_0.

First, we must show that Euler systems interpolate over the eigenvariety. The étale cohomology eigenspaces attached to cohomological, p-ordinary automorphic representations are known to interpolate in families, giving rise to sheaves of Galois representations over $\mathcal{E}$. With this in hand, the machinery of [44, 46] then shows that the Euler system classes themselves interpolate, giving families of Euler systems taking values in these sheaves.

A generalisation of Coleman and Perrin-Riou's theory of "big logarithm" maps (cf. [37]) also allows us to define a *motivic p-adic L-function* $\mathcal{L}^{\mathrm{mot}}$ associated to the bottom class in our family of Euler systems. Perrin-Riou's local reciprocity formula implies that $\mathcal{L}^{\mathrm{mot}}$ has an interpolation property both in Σ_0 and in Σ_1. For classical points π whose weights lie in Σ_1, the value of $\mathcal{L}^{\mathrm{mot}}$ interpolates the Bloch–Kato logarithm of the geometrically-defined Euler system class for V_π. Much more subtly, if we evaluate $\mathcal{L}^{\mathrm{mot}}$ at points π whose weights lie in Σ_0, it computes the image under the *dual-exponential* map of the bottom class in the Euler system for V_π which we have just defined using analytic continuation.

We would like to make the following argument: "the regulator formula shows that $\mathcal{L}^{\mathrm{mot}}$ and the analytic p-adic L-function $\mathcal{L}$ agree at points in Σ_1, and these are Zariski-dense; so $\mathcal{L} = \mathcal{L}^{\mathrm{mot}}$ everywhere". This is essentially how we proved an explicit reciprocity law for $\mathrm{GL}_2 \times \mathrm{GL}_2$ in [37]. Unfortunately, there are two subtle technical hitches which occur in making this argument precise for GSp_4.

The first is that $\mathcal{L}$ and $\mathcal{L}^{\mathrm{mot}}$ take values in different line bundles over the eigenvariety $\mathcal{E}$ (one interpolating coherent cohomology, and the other $\mathbf{D}_{\mathrm{cris}}$ of a certain subquotient of étale cohomology). At each classical point of $\mathcal{E}$, we have a canonical isomorphism between the fibres of these two line bundles; but it is far from obvious *a priori* that these "pointwise" isomorphisms at classical points interpolate into an isomorphism of line bundles. For the

GL_2 ordinary eigenvariety, we do have such an isomorphism, the p-adic Eichler–Shimura isomorphism of Ohta [54] (extended to non-ordinary families in [2]). However, the case of higher-dimensional Shimura varieties such as GSp_4 is more difficult: one expects several Eichler–Shimura isomorphisms, each capturing coherent cohomology in a different degree, and at present only the case of H^0 is available in the literature [19]. For the problem at hand, it is the coherent H^1 (and dually H^2) which is relevant.

The second is that, while Σ_1 is indeed Zariski-dense in the eigenvariety, the function $\mathcal{L}$ is only defined on a lower-dimensional "slice" of the eigenvariety (on which the GSp_4 form has weight (r_1, r_2), with r_1 varying and r_2 fixed), and the intersection of each individual slice with Σ_1 is not Zariski-dense in the slice.

In [50], we circumvented these problems in a somewhat indirect way, by appealing to a second, independent construction of an analytic p-adic L-function, defined using Shalika models for GL_4 [20]. As written this construction shares with [45] the shortcoming of requiring r_2 to be fixed, but the methods of [44] can be applied in order to extend this construction by varying r_2 as well. Using this we were able to

The lack of an Eichler–Shimura isomorphism in families – or, more precisely, of an isomorphism between the sheaves in which $\mathcal{L}^{\mathrm{mot}}$ and the GL_4 p-adic L-function take values – can be dealt with via the so-called "leading term argument". This proceeds as follows. There is clearly a *meromorphic* isomorphism between these sheaves which maps one p-adic L-function to the other (since both are clearly non-zero).[3] If this meromorphic isomorphism degenerates to zero at some "bad" 0-critical π, then the bottom class in our Euler system for π lies in the kernel of the Perrin-Riou regulator. However, this would also apply to all the classes $c[m]$ in this Euler system, for all values of m. So we obtain an Euler system satisfying a very strong local condition at p; and a result of Mazur–Rubin [52] shows that this condition is so strong that it forces the entire Euler system to be zero. Hence we can replace all of these classes by their derivatives in the weight direction, which amounts to renormalising the Eichler–Shimura map to reduce its order of vanishing by 1.

Iterating this process, we eventually obtain a non-trivial Euler system for π; and if $L(\pi^\vee, \frac{1-t}{2}) \neq 0$, the bottom class of this Euler system is non-zero. We can now deduce the vanishing of $H^1_{\mathrm{f}}(\mathbf{Q}, V_\pi)$, where $V_\pi = \rho_\pi(\frac{d+1+t}{2})$, as predicted by the Bloch–Kato conjecture.

Non-regular weights

The above strategy can also be used to study automorphic forms which are not cohomological (so π does not contribute to étale cohomology), as long as π contributes to *coherent* cohomology in the correct degree. For instance, this applies to weight 1 modular forms, which is crucial in several works such as [17] which use Euler systems to study the Birch–Swinnerton-Dyer conjecture for Artin twists of elliptic curves. It also applies to

3 This argument can be used to *construct* an Eichler–Shimura isomorphism in families for GSp_4, which interpolates the classical H^1 comparison isomorphism at almost all classical points – see [51].

paramodular Siegel modular forms for GSp_4 of parallel weight 2, such as those corresponding to paramodular abelian surfaces.

In this situation, if π is ordinary at p, it follows from the results of [13] that it defines a point on the eigenvariety $\mathcal{E}$. However, in contrast to the case of cohomological weights, it is not clear if the eigenvariety is smooth, or étale over weight space, at π; results of Bellaiche–Dimitrov show that this can fail even for GL_2 [4].

If A is a paramodular abelian surface over $\mathbf{Q}$ which is ordinary at p, and has analytic rank 0, then we can use the above approach to prove the finiteness of $A(\mathbf{Q})$ (as predicted by the Birch–Swinnerton-Dyer conjecture), and of the p-part of the Tate–Shafarevich group, under the assumption that the GSp_4 eigenvariety be smooth at the point corresponding to A. This is work in progress.

REFERENCES

[1] F. Andreatta and A. Iovita, Triple product p-adic L-functions associated to finite slope p-adic families of modular forms (with an appendix by E. Urban). *Duke Math. J.* **170** (2021), no. 9, 1989–2083. DOI 10.1215/00127094-2020-0076

[2] F. Andreatta, A. Iovita, and G. Stevens, Overconvergent Eichler–Shimura isomorphisms. *J. Inst. Math. Jussieu* **14** (2015), no. 2, 221–274. DOI 10.1017/S1474748013000364

[3] A. Beĭlinson, Higher regulators and values of L-functions. In *Current problems in mathematics*, pp. 181–238, Itogi Nauki i Tekhniki, Akad. Nauk SSSR Vsesoyuz. Inst. Nauchn. i Tekhn. 24, Inform, Moscow, 1984. DOI 10.1007/BF02105861

[4] J. Bellaïche and M. Dimitrov, On the eigencurve at classical weight one points. *Duke Math. J.* **165** (2016), no. 2, 245–266. DOI 10.1215/00127094-3165755

[5] M. Bertolini, H. Darmon, and K. Prasanna, Generalized Heegner cycles and p-adic Rankin L-series (with an appendix by Brian Conrad). *Duke Math. J.* **162** (2013), no. 6, 1033–1148. DOI 10.1215/00127094-2142056

[6] M. Bertolini, H. Darmon, and V. Rotger, Beilinson–Flach elements and Euler systems I: syntomic regulators and p-adic Rankin L-series. *J. Algebraic Geom.* **24** (2015), no. 2, 355–378. DOI 10.1090/S1056-3911-2014-00670-6

[7] A. Besser, Syntomic regulators and p-adic integration. I. Rigid syntomic regulators. *Israel J. Math.* **120** (2000), no. 1, 291–334. DOI 10.1007/BF02834843

[8] A. Besser, On the syntomic regulator for K_1 of a surface. *Israel J. Math.* **190** (2012), no. 1, 29–66. DOI 10.1007/s11856-011-0188-0

[9] R. Beuzart-Plessis, Y. Liu, W. Zhang, and X. Zhu, Isolation of cuspidal spectrum, with application to the Gan–Gross–Prasad conjecture. *Ann. of Math. (2)* **19** (2021), no. 2, 519–584. DOI 10.4007/annals.2021.194.2.5

[10] I. Blanco-Chacón and M. Fornea, Twisted triple product p-adic L-functions and Hirzebruch–Zagier cycles. *J. Inst. Math. Jussieu* **19** (2020), no. 6, 1947–1992. DOI 10.1017/s1474748019000021

[11] S. Bloch and K. Kato, L-functions and Tamagawa numbers of motives. In *The Grothendieck Festschrift, Vol. I*, pp. 333–400, Progr. Math. 86, Birkhäuser, Boston, MA, 1990. DOI 10.1007/978-0-8176-4574-8

[12] R. Boumasmoud, E. H. Brooks, and D. Jetchev, Vertical distribution relations for special cycles on unitary Shimura varieties. *Int. Math. Res. Not.* **2018** (2018), no. 06. DOI 10.1093/imrn/rny119

[13] G. Boxer and V. Pilloni, Higher Coleman theory. 2020, https://perso.ens-lyon.fr/vincent.pilloni/HigherColeman.pdf, preprint.

[14] A. Caraiani and P. Scholze, On the generic part of the cohomology of compact unitary Shimura varieties. *Ann. of Math. (2)* **186** (2017), no. 3, 649–766. DOI 10.4007/annals.2017.186.3.1

[15] C. Cornut, An Euler system of Heegner type. 2018, https://webusers.imj-prg.fr/~christophe.cornut/papers/ESHT.pdf, preprint.

[16] H. Darmon and V. Rotger, Diagonal cycles and Euler systems I: a p-adic Gross–Zagier formula. *Ann. Sci. Éc. Norm. Supér. (4)* **47** (2014), no. 4, 779–832. DOI 10.24033/asens.2227

[17] H. Darmon and V. Rotger, Diagonal cycles and Euler systems II: the Birch and Swinnerton-Dyer conjecture for Hasse–Weil–Artin L-series. *J. Amer. Math. Soc.* **30** (2016), 601–672. DOI 10.1090/jams/861

[18] P. Deligne, Valeurs de fonctions L et périodes d'intégrales (with an appendix by N. Koblitz and A. Ogus). In *Automorphic forms, representations and L-functions (Corvallis, 1977)*, pp. 313–346, Proc. Sympos. Pure Math. 33.2, Amer. Math. Soc., Providence, RI, 1979. https://www.ams.org/books/pspum/033.2/546622

[19] H. Diao, G. Rosso, and J.-F. Wu, Perfectoid overconvergent Siegel modular forms and the overconvergent Eichler–Shimura morphism. 2021, arXiv:2106.00094.

[20] M. Dimitrov, F. Januszewski, and A. Raghuram, L-functions of $\mathrm{GL}(2n)$: p-adic properties and non-vanishing of twists. *Compos. Math.* **156** (2020), no. 12, 2437–2468. DOI 10.1112/S0010437X20007551

[21] G. Faltings, Arithmetic Eisenstein classes on the Siegel space: some computations. In *Number fields and function fields—two parallel worlds*, pp. 133–166, Progr. Math. 239, Birkhäuser, Boston, MA, 2005. DOI 10.1007/0-8176-4447-4_8

[22] W. T. Gan, B. H. Gross, and D. Prasad, Symplectic local root numbers, central critical L values, and restriction problems in the representation theory of classical groups. In *Sur les conjectures de Gross et Prasad. I*, pp. 1–109, Astérisque 346, 2012.

[23] S. Gelbart and I. I. Piatetski-Shapiro, Automorphic forms and L-functions for the unitary group. In *Lie group representations, II (College Park, MD, 1982/1983)*, pp. 141–184, Lecture Notes in Math. 1041, Springer, Berlin, 1984. DOI 10.1007/BFb0073147

[24] J. R. Getz and E. Wambach, Twisted relative trace formulae with a view towards unitary groups. *Amer. J. Math.* **136** (2014), no. 1, 1–58. DOI 10.1353/ajm.2014.0002

[25] A. Graham and S. W. A. Shah, Anticyclotomic Euler systems for unitary groups. 2020, arXiv:2001.07825.

[26] G. Grossi, On norm relations for Asai–Flach classes. *Int. J. Number Theory* **16** (2020), no. 10, 2311–2377. DOI 10.1142/S1793042120501183

[27] G. Grossi, Higher Hida theory for Hilbert modular varieties in the totally split case. 2021, arXiv:2106.05666.

[28] M. Harris, Automorphic forms of $\overline{\partial}$-cohomology type as coherent cohomology classes. *J. Differential Geom.* **32** (1990), no. 1, 1–63. DOI 10.4310/jdg/1214445036

[29] M. Harris, Occult period invariants and critical values of the degree four L-function of GSp(4). In *Contributions to automorphic forms, geometry, and number theory (in honour of J. Shalika)*, pp. 331–354, Johns Hopkins Univ. Press, 2004.

[30] H. Hida, A p-adic measure attached to the zeta functions associated with two elliptic modular forms. I. *Invent. Math.* **79** (1985), no. 1, 159–195. DOI 10.1007/BF01388661

[31] B. Howard, Bipartite Euler systems. *J. Reine Angew. Math.* **597** (2006), 1–25. DOI 10.1515/CRELLE.2006.062

[32] C.-Y. Hsu, Z. Jin, and R. Sakamoto, An Euler system for $\mathrm{GSp}(4) \times \mathrm{GL}(2)$. 2020, arXiv:2011.12894.

[33] A. Huber and G. Kings, Bloch–Kato conjecture and main conjecture of Iwasawa theory for Dirichlet characters. *Duke Math. J.* **119** (2003), no. 3, 393–464. DOI 10.1215/S0012-7094-03-11931-6

[34] U. Jannsen, Deligne homology, Hodge-$\mathfrak{D}$-conjecture, and motives. In *Beĭlinson's conjectures on special values of L-functions*, pp. 305–372, Perspect. Math. 4, Academic Press, Boston, MA, 1988. http://wwwmath.uni-muenster.de/u/pschnei/publ/beilinson-volume/Jannsen.pdf

[35] K. Kato, P-adic Hodge theory and values of zeta functions of modular forms. In *Cohomologies p-adiques et applications arithmétiques (III)*, pp. 117–290, Astérisque 295, 2004. http://smf4.emath.fr/en/Publications/Asterisque/2004/295/html/smf_ast_295_117-290.html

[36] G. Kings, Higher regulators, Hilbert modular surfaces, and special values of L-functions. *Duke Math. J.* **92** (1998), no. 1, 61–127. DOI 10.1215/S0012-7094-98-09202-X

[37] G. Kings, D. Loeffler, and S. L. Zerbes, Rankin–Eisenstein classes and explicit reciprocity laws. *Cambridge J. Math.* **5** (2017), no. 1, 1–122. DOI 10.4310/CJM.2017.v5.n1.a1

[38] G. Kings, D. Loeffler, and S. L. Zerbes, Rankin–Eisenstein classes for modular forms. *Amer. J. Math.* **142** (2020), no. 1, 79–138. DOI 10.1353/ajm.2020.0002

[39] V. A. Kolyvagin, Euler systems. In *The Grothendieck Festschrift, Vol. II*, pp. 435–483, Progr. Math. 87, Birkhäuser, Boston, MA, 1990. DOI 10.1007/978-0-8176-4575-5_11

[40] A. Lei, D. Loeffler, and S. L. Zerbes, Euler systems for Rankin–Selberg convolutions of modular forms. *Ann. of Math. (2)* **180** (2014), no. 2, 653–771. DOI 10.4007/annals.2014.180.2.6

[41] A. Lei, D. Loeffler, and S. L. Zerbes, Euler systems for Hilbert modular surfaces. *Forum Math. Sigma* **6** (2018), no. e23. DOI 10.1017/fms.2018.23

[42] F. Lemma, On higher regulators of Siegel threefolds II: The connection to the special value. *Compos. Math.* **153** (2017), no. 5, 889–946. DOI 10.1112/S0010437X16008320

[43] Y. Liu, Y. Tian, L. Xiao, W. Zhang, and X. Zhu, On the Beilinson–Bloch–Kato conjecture for Rankin–Selberg motives. 2019, arXiv:1912.11942.

[44] D. Loeffler, Spherical varieties and norm relations in Iwasawa theory. *J. Théor. Nombres Bordeaux* (2019), Iwasawa 2019 special issue, to appear. arXiv:1909.09997.

[45] D. Loeffler, V. Pilloni, C. Skinner, and S. L. Zerbes, Higher Hida theory and p-adic L-functions for GSp(4). *Duke Math. J.* **170** (2021), no. 18, 4033–4121. DOI 10.1215/00127094-2021-0049

[46] D. Loeffler, R. Rockwood, and S. L. Zerbes, Spherical varieties and p-adic families of cohomology classes. 2021, arXiv:2106.16082.

[47] D. Loeffler, C. Skinner, and S. L. Zerbes, Euler systems for GSp(4). *J. Eur. Math. Soc. (JEMS)*, published online 2021, print version to appear. DOI 10.4171/JEMS/1124

[48] D. Loeffler, C. Skinner, and S. L. Zerbes, Euler systems for GU(2, 1). *Math. Ann.*, published online 2021, print version to appear. DOI 10.1007/s00208-021-02224-4

[49] D. Loeffler and S. L. Zerbes, Euler systems with local conditions. In *Development of Iwasawa theory – the centennial of K. Iwasawa's birth*, pp. 1–26, Adv. Stud. Pure Math. 86, Math. Soc. Japan, 2020. DOI 10.2969/aspm/08610001

[50] D. Loeffler and S. L. Zerbes, On the Bloch–Kato conjecture for GSp(4). 2020, arXiv:2003.05960.

[51] D. Loeffler and S. L. Zerbes, On the Bloch–Kato conjecture for GSp(4) $\times$ GL(2). 2021, arXiv:2106.14511.

[52] B. Mazur and K. Rubin, Kolyvagin systems. *Mem. Amer. Math. Soc.* **168** (2004), no. 799. DOI 10.1090/memo/0799

[53] M. T. Nguyen, *Higher Hida theory on unitary group* GU(2, 1). PhD thesis, Université de Lyon, 2020, URL https://tel.archives-ouvertes.fr/tel-02940906.

[54] M. Ohta, Ordinary p-adic étale cohomology groups attached to towers of elliptic modular curves. II. *Math. Ann.* **318** (2000), no. 3, 557–583. DOI 10.1007/s002080000119

[55] V. Pilloni, Higher coherent cohomology and p-adic modular forms of singular weights. *Duke Math. J.* **169** (2020), no. 9, 1647–1807. DOI 10.1215/00127094-2019-0075

[56] K. Rubin, *Euler systems*. Ann. of Math. Stud. 147, Princeton Univ. Press, 2000.

[57] Y. Sakellaridis, Spherical functions on spherical varieties. *Amer. J. Math.* **135** (2013), no. 5, 1291–1381. DOI 10.1353/ajm.2013.0046

[58] Y. Sakellaridis and A. Venkatesh, *Periods and harmonic analysis on spherical varieties*. Astérisque 396, Soc. Math. France, 2017.

[59] W. Zhang, Periods, cycles, and L-functions: a relative trace formula approach. In *Proceedings of the International Congress of Mathematicians—Rio de Janeiro 2018. Vol. II. Invited lectures*, pp. 487–521, World Sci. Publ., Hackensack, NJ, 2018.

DAVID LOEFFLER

Mathematics Institute, University of Warwick, Coventry CV4 7AL, UK, d.a.loeffler@warwick.ac.uk

SARAH LIVIA ZERBES

Department of Mathematics, University College London, London WC1E 6BT, UK, and ETH Zürich, Switzerland, s.zerbes@ucl.ac.uk

COUNTING PROBLEMS: CLASS GROUPS, PRIMES, AND NUMBER FIELDS

LILLIAN B. PIERCE

ABSTRACT

Each number field has an associated finite abelian group, the class group, that records certain properties of arithmetic within the ring of integers of the field. The class group is well studied, yet also still mysterious. A central conjecture of Brumer and Silverman states that for each prime ℓ, every number field has the property that its class group has very few elements of order ℓ, where "very few" is measured relative to the absolute discriminant of the field. This paper surveys recent progress toward this conjecture, and outlines its close connections to counting prime numbers, counting number fields of fixed discriminant, and counting number fields of bounded discriminant.

MATHEMATICS SUBJECT CLASSIFICATION 2020

Primary 11R29; Secondary 11R45, 11N05

KEYWORDS

Class groups, counting number fields, distribution of primes

INTERNATIONAL CONGRESS OF MATHEMATICIANS 2022 JULY 6–14

Proc. Int. Cong. Math. 2022, Vol. 3, pp. 1940–1965
DOI 10.4171/ICM2022/104

1. HISTORICAL PRELUDE

In a 1640 letter to Mersenne, Fermat stated that an odd prime p satisfies $p = x^2 + y^2$ if and only if $p \equiv 1 \pmod 4$. Roughly 90 years later, Euler learned of Fermat's statement through correspondence with Goldbach, and by 1749, he worked out a proof. This fits into a bigger question, which Euler studied as well: for each $n \geq 1$, which primes can be written as $p = x^2 + ny^2$? Even more generally: which binary quadratic forms $ax^2 + bxy + cy^2$ represent a given integer m? This question also motivated work of Lagrange and Legendre, and then appeared in Gauss's celebrated 1801 work *Disquisitiones Arithmeticae*; see [26].

Gauss partitioned binary quadratic forms of discriminant $D = b^2 - 4ac$ into equivalence classes under $\mathrm{SL}_2(\mathbb{Z})$ changes of variable. (Here we will speak only of *fundamental* discriminants D; for notes on the original setting, see [84].) Gauss showed that for each D there are finitely many such classes (the cardinality is the *class number*, denoted $h(D)$), and verified that the classes obey a group law (composition). Based on extensive computation, Gauss noticed that as $D \to -\infty$, small class numbers stopped appearing, writing: "*Nullum dubium esse videtur, quin series adscriptae revera abruptae sint…Demonstrationes autem* rigorosae *harum observationum perdifficiles esse videntur.*" ("It seems beyond doubt that the sequences written down do indeed break off… However, *rigorous* proofs of these observations appear to be most difficult" [43, P. 13].) As $D \to +\infty$, a quite different behavior seemed to appear, leading to a conjecture that $h(D) = 1$ for infinitely many $D > 0$.

It is hard to exaggerate the interest these two conjectures have generated. In the 1830s, Dirichlet proved a class number formula, relating the class number $h(D)$ of a (fundamental) discriminant D to the value $L(1, \chi)$ of an L-function associated to a real primitive character χ modulo D. Consequently, throughout the 1900s, Gauss's questions were studied via the theory of the complex-variable functions $L(s, \chi)$. A remarkable series of works by Hecke, Deuring, Mordell, and Heilbronn confirmed that for $D < 0$ the class number $h(D)$ attains any value only finitely many times. How many times? Famously, the work of Heegner, Baker, and Stark proved that there are 9 (fundamental) discriminants $D < 0$ with class number 1. In full generality, Goldfeld showed an effective lower bound for $h(D)$ when $D < 0$ would follow from a specific case of the Birch–Swinnerton-Dyer conjecture, which was then verified by Gross and Zagier; see [42]. Now, for each $1 \leq N \leq 100$, one may find the number of discriminants $D < 0$ with $h(D) = N$ in [93]. As for the other conjecture, that infinitely many (fundamental) discriminants $D > 0$ have class number 1, this remains open, and very mysterious. These historical antecedents hint at the intertwined currents of "counting" and the analytic study of L-functions, which will also be present in the work we will survey.

We briefly mention another historical motivation for the study of class numbers, namely the failure of unique factorization. For example, in the ring $\mathbb{Z}[\sqrt{-5}]$, $21 = 3 \cdot 7$ but it also factors into irreducible, nonassociated factors as $(1 + 2\sqrt{-5})(1 - 2\sqrt{-5})$. Here is a problem where the failure of unique factorization has an impact. Suppose one is searching for solutions $x, y, z \in \mathbb{N}$ to the equation $x^p + y^p = z^p$ for a prime $p \geq 3$. If a nontrivial solution (x, y, z) exists, then for ζ_p a pth root of unity, we could write

$$y \cdot y \cdots y = (z - x)(z - \zeta_p x) \cdots (z - \zeta_p^{p-1} x).$$

If $\mathbb{Z}[\zeta_p]$ possesses unique factorization, two such factorizations cannot exist, so (x, y, z) cannot exist—verifying Fermat's Last Theorem for this exponent p. But to the disappointment of many, unique factorization fails in $\mathbb{Z}[\zeta_p]$ for infinitely many p. As Neukirch writes, "Realizing the failure of unique factorization in general has led to one of the grand events in the history of number theory, the discovery of ideal theory by Eduard Kummer" **[69, CH. I §3]**.

1.1. The class group

Let $K/\mathbb{Q}$ be a number field of degree n, with associated ring of integers $\mathcal{O}_K$. Every proper integral ideal $\mathfrak{a} \subset \mathcal{O}_K$ factors into a product of prime ideals $\mathfrak{p}_1 \cdots \mathfrak{p}_k$ in a unique way (salvaging the notion of unique factorization). Moreover, the fractional ideals of K form an abelian group J_K, the free abelian group on the set of nonzero prime ideals of $\mathcal{O}_K$. In the case that every ideal in J_K belongs to the subgroup P_K of principal ideals, $\mathcal{O}_K$ is a principal ideal domain, and unique factorization holds in $\mathcal{O}_K$. But more typically, some "expansion" occurs when passing to ideals; the class group of K is defined to measure this.

The class group of K is the quotient group

$$\mathrm{Cl}_K = J_K / P_K.$$

The elements in Cl_K are ideal classes, and the cardinality $|\mathrm{Cl}_K|$ is the class number. The quotient J_K/P_K is trivial (so that every ideal is a principal ideal, and $|\mathrm{Cl}_K| = 1$) precisely when unique factorization holds in $\mathcal{O}_K$. (Thus the above strategy for Fermat's Last Theorem works for p if $|\mathrm{Cl}_{\mathbb{Q}(\zeta_p)}| = 1$. In fact, Kummer showed that as long as the class number of $\mathbb{Q}(\zeta_p)$ is indivisible by p, the argument can be salvaged; see **[31]**. Such a prime is called a regular prime. Here is an open question: are there infinitely many regular prime numbers?)

By a result of Minkowski in the geometry of numbers, every ideal class in Cl_K contains an integral ideal $\mathfrak{b}$ with norm $\mathfrak{N}(\mathfrak{b}) = (\mathcal{O}_K : \mathfrak{b})$ satisfying

$$\mathfrak{N}(\mathfrak{b}) \le (2/\pi)^s \sqrt{D_K}, \tag{1.1}$$

where $D_K = |\mathrm{Disc}(K/\mathbb{Q})|$ and s counts the pairs of complex embeddings of K. As there are finitely many integral ideals of any given norm, Landau deduced (see **[68, THM. 4.4]**):

$$|\mathrm{Cl}_K| \ll_n D_K^{1/2} \log^{n-1} D_K. \tag{1.2}$$

In particular, the class group of a number field K is always a finite abelian group. (Throughout, $A \ll_\kappa B$ indicates that there exists a constant C_κ such that $|A| \le C_\kappa B$.)

When $K = \mathbb{Q}(\sqrt{D})$ is a quadratic field, this relates in a precise way to Gauss's construction of the class number for binary quadratic forms of discriminant D (see **[8]**). In modern terms, Gauss asked whether for each $h \in \mathbb{N}$, there are finitely many imaginary quadratic fields K with $|\mathrm{Cl}_K| = h$? (Yes.) Are there infinitely many real quadratic fields K with $|\mathrm{Cl}_K| = 1$? (We do not know.) In fact, here is an open question: are there infinitely many number fields, of arbitrary degrees, with class number 1? Here is another open question: are there infinitely many number fields, of arbitrary degrees, with bounded class number? These difficult questions must consider the regulator R_K of the field K, due to the (ineffective) inequalities by Siegel (for quadratic fields) and Brauer (in general) [68, CH. 8]:

$$D_K^{1/2-\varepsilon} \ll_{n,\varepsilon} |\mathrm{Cl}_K| R_K \ll_{n,\varepsilon} D_K^{1/2+\varepsilon}, \quad \text{for all } \varepsilon > 0.$$

2. THE ℓ-TORSION CONJECTURE

In addition to studying the size of the class group, it is also natural to study its structure. We will focus on the ℓ-torsion subgroup, defined for each integer $\ell \geq 2$ by

$$\mathrm{Cl}_K[\ell] = \{[\mathfrak{a}] \in \mathrm{Cl}_K : [\mathfrak{a}]^\ell = \mathrm{Id}\}.$$

For example, the class number is divisible by a prime ℓ precisely when $|\mathrm{Cl}_K[\ell]| > 1$. Related problems include studying the exponent of the class group, or counting how many number fields of a certain degree have class number divisible, or indivisible, by a given prime ℓ. Such problems are addressed for imaginary quadratic fields in [4, 44, 45, 82].

In this survey, we will focus on upper bounds for the ℓ-torsion subgroup. The Minkowski bound (1.2) provides an upper bound for any field of degree n, and all ℓ:

$$1 \leq |\mathrm{Cl}_K[\ell]| \leq |\mathrm{Cl}_K| \ll_{n,\varepsilon} D_K^{1/2+\varepsilon}, \quad \text{for all } \varepsilon > 0. \tag{2.1}$$

Our subject is a conjecture on the size of the ℓ-torsion subgroup, which suggests that (2.1) is far from the truth. We will focus primarily on cases when ℓ is prime, since $|\mathrm{Cl}_K[m]|$ is multiplicative as a function of m, and for a prime ℓ, $|\mathrm{Cl}_K[\ell^t]| \leq |\mathrm{Cl}_K[\ell]|^t$ (see [73]).

Conjecture 2.1 (ℓ-torsion conjecture). *Fix a degree $n \geq 2$ and a prime ℓ. Every number field $K/\mathbb{Q}$ of degree n satisfies $|\mathrm{Cl}_K[\ell]| \ll_{n,\ell,\varepsilon} D_K^\varepsilon$ for all $\varepsilon > 0$.*

This conjecture is due to Brumer and Silverman, in the more precise form: is it always true that $\log_\ell |\mathrm{Cl}_K[\ell]| \ll_{n,\ell} \log D_K / \log\log D_K$ [17, QUESTION $\mathrm{CI}(\ell, d)$]? Brumer and Silverman were motivated by counting elliptic curves of fixed conductor. Subsequently, this conjecture has appeared in many further contexts, including bounding the ranks of elliptic curves [34, §1.2]; bounding Selmer groups and ranks of hyperelliptic curves [10]; counting number fields [29, P. 166]; studying equidistribution of CM points on Shimura varieties [98, CONJECTURE 3.5]; and counting nonuniform lattices in semisimple Lie groups [6, THM. 7.5].

Conjecture 2.1 is known to be true for the degree $n = 2$ and the prime $\ell = 2$, when it follows from the genus theory of Gauss (see [68, CH. 8.3]). This is the only case in which it is known. Nevertheless, starting in the early 2000s, significant progress has been made. The purpose of this survey is to give some insight into the wide variety of methods developed in recent work toward the conjecture. As an initial measure of progress, we define:

Property $\mathbf{C}_{n,\ell}(\Delta)$. *Fix a degree $n \geq 2$ and a prime ℓ. Property $\mathbf{C}_{n,\ell}(\Delta)$ holds if for all number fields $K/\mathbb{Q}$ of degree n, $|\mathrm{Cl}_K[\ell]| \ll_{n,\ell,\Delta,\varepsilon} D_K^{\Delta+\varepsilon}$ for all $\varepsilon > 0$.*

Gauss proved that $\mathbf{C}_{2,2}(0)$ holds. Until recently, no other case with $\Delta < 1/2$ was known.

The first progress was for imaginary quadratic fields. Suppose $K = \mathbb{Q}(\sqrt{-d})$ for a square-free integer $d > 1$, and suppose that $[\mathfrak{a}]$ is a nontrivial element in $\mathrm{Cl}_K[\ell]$ for a prime $\ell \geq 3$; thus $[\mathfrak{a}]^\ell$ is the principal ideal class. Then by the Minkowski bound (1.1), there exists an integral ideal $\mathfrak{b}$ in $[\mathfrak{a}]$ such that $\mathfrak{N}(\mathfrak{b}) \ll d^{1/2}$. Moreover, $\mathfrak{b}^\ell$ is principal, say, generated by $(y + z\sqrt{-d})/2$ for some integers y, z, and so $(\mathfrak{N}(\mathfrak{b}))^\ell = \mathfrak{N}(\mathfrak{b}^\ell) = (y^2 + dz^2)/4$. Consequently, $|\mathrm{Cl}_K[\ell]|$ can be dominated (up to a factor d^ε) by the number of integral solutions to

$$4x^\ell = y^2 + dz^2, \quad \text{with } x \ll d^{1/2}, y \ll d^{\ell/4}, z \ll d^{\ell/4-1/2}. \tag{2.2}$$

When $\ell = 3$, this can be interpreted in several ways: counting solutions to a congruence $y^2 = 4x^3 \pmod{d}$; counting perfect square values of the polynomial $f(x,z) = 4x^3 - dz^2$; or counting integral points on a family of Mordell elliptic curves $y^2 = 4x^3 - D$, with $D = dz^2$. Pierce used the first two perspectives, and Helfgott and Venkatesh used the third perspective, to prove for the first time that property $\mathbf{C}_{2,3}(\Delta)$ holds for some $\Delta < 1/2$ **[48,70, 71]**. (The Scholz reflection principle shows that $\log_3 |\mathrm{Cl}_{\mathbb{Q}(\sqrt{-d})}[3]|$ and $\log_3 |\mathrm{Cl}_{\mathbb{Q}(\sqrt{3d})}[3]|$ differ by at most 1, so results for 3-torsion apply comparably to both real and imaginary quadratic fields **[76]**.) When $\ell \geq 5$, the region in which x, y, z lie in (2.2) becomes inconveniently large relative to the trivial bound (2.1). Here is an open question: for a prime $\ell \geq 5$, are there at most $\ll d^\Delta$ integral solutions to (2.2), for some $\Delta < 1/2$?

Recently, Bhargava, Taniguchi, Thorne, Tsimerman, and Zhao made a breakthrough on property $\mathbf{C}_{n,2}(\Delta)$ for all $n \geq 3$. Fix a prime ℓ and a number field K of degree n. Given any nontrivial ideal class $[\mathfrak{a}] \in \mathrm{Cl}_K[\ell]$, they show it contains an integral ideal $\mathfrak{b}$ with $\mathfrak{b}^\ell$ a principal ideal generated by an element β lying in a well-proportioned "box." By an ingenious geometry of numbers argument, they show the number of such generators β in the box is $\ll D_K^{\ell/2-1/2}$. If $\ell \geq 3$, this far exceeds the trivial bound (2.1), but if $\ell = 2$, it slightly improves it. The striking refinement comes by recalling that any β of interest must also have $|N_{K/\mathbb{Q}}(\beta)| = \mathfrak{N}(\mathfrak{b}^\ell) = (\mathfrak{N}(\mathfrak{b}))^\ell$ be a perfect ℓth power of an integer, say, y^ℓ. For $\ell = 2$, they apply a celebrated result of Bombieri and Pila to count integral solutions (x, y) to the degree n equation $N_{K/\mathbb{Q}}(\beta + x) = y^2$ **[15]**. This strategy proves that property $\mathbf{C}_{n,2}(1/2 - 1/2n)$ holds for all degrees $n \geq 3$. Further refinements for degrees 3, 4 show $\mathbf{C}_{3,2}(0.2785\ldots)$ and $\mathbf{C}_{4,2}(0.2785\ldots)$ hold; see **[10]**.

Only two further nontrivial cases of property $\mathbf{C}_{n,\ell}(\Delta)$ are known, and for these we introduce the Ellenberg–Venkatesh criterion.

2.1. The Ellenberg–Venkatesh criterion

An important criterion for bounding ℓ-torsion in the class group of a number field K relies on counting small primes that are noninert in K. The germ of the idea, which has been credited independently to Soundararajan and Michel, goes as follows. Suppose, for example, that $K = \mathbb{Q}(\sqrt{-d})$ is an imaginary quadratic field with d square-free, and ℓ is an odd prime. Let H denote $\mathrm{Cl}_K[\ell]$. Then $|H| = |\mathrm{Cl}_K|/[\mathrm{Cl}_K : H]$, and to show that $|H|$ is small, it suffices to show that the index $[\mathrm{Cl}_K : H]$ is large. Now suppose that $p_1 \neq p_2$ are rational primes not dividing $2d$ that both split in K, say, $p_1 = \mathfrak{p}_1\mathfrak{p}_1^\sigma$ and $p_2 = \mathfrak{p}_2\mathfrak{p}_2^\sigma$, where σ is the nontrivial automorphism of K. We claim that as long as p_1, p_2 are sufficiently small, $\mathfrak{p}_1$ and $\mathfrak{p}_2$ must represent different cosets of H. Indeed, supposing to the contrary that $\mathfrak{p}_1 H = \mathfrak{p}_2 H$, one deduces that $\mathfrak{p}_1 \mathfrak{p}_2^\sigma \in H$ so that $(\mathfrak{p}_1 \mathfrak{p}_2^\sigma)^\ell$ is a principal ideal, say, generated by $(y + z\sqrt{-d})/2$, for some $y, z \in \mathbb{Z}$. Taking norms shows

$$4(p_1 p_2)^\ell = y^2 + dz^2. \tag{2.3}$$

If $p_1, p_2 < (1/4)d^{1/(2\ell)}$, this forces $z = 0$, which yields a contradiction, since $4(p_1 p_2)^\ell$ cannot be a perfect square. This proves the claim. In particular, if there are M such distinct primes $p_1, \dots, p_M < (1/4)d^{1/2\ell}$ with $p_j \nmid 2d$ and p_j split in K, then $|\mathrm{Cl}_K[\ell]| \leq |\mathrm{Cl}_K| M^{-1}$.

Ellenberg and Venkatesh significantly generalized this strategy to prove an influential criterion, which we state in the case of extensions of $\mathbb{Q}$ [34]. (Throughout this survey, we will focus for simplicity on extensions of $\mathbb{Q}$, but many of the theorems and questions we mention have analogues in the literature over any fixed number field.)

Ellenberg–Venkatesh criterion. *Suppose $K/\mathbb{Q}$ is a number field of degree $n \geq 2$, fix an integer $\ell \geq 2$, and fix $\eta < \frac{1}{2\ell(n-1)}$. Suppose that there are M prime ideals $\mathfrak{p}_1, \dots, \mathfrak{p}_M \subset \mathcal{O}_K$ such that each $\mathfrak{p}_j$ has norm $\mathfrak{N}(\mathfrak{p}_j) < D_K^\eta$, $\mathfrak{p}_j$ is unramified in K and $\mathfrak{p}_j$ is not an extension of a prime ideal from any proper subfield of K. Then*

$$\left|\mathrm{Cl}_K[\ell]\right| \ll_{n,\ell,\varepsilon} D_K^{\frac{1}{2}+\varepsilon} M^{-1}, \quad \textit{for all } \varepsilon > 0. \tag{2.4}$$

(A prime ideal $\mathfrak{p} \subset \mathcal{O}_K$ lying above a prime $p \in \mathbb{Q}$ is unramified in $K/\mathbb{Q}$ if $\mathfrak{p}^2 \nmid p\mathcal{O}_K$; a prime ideal $\mathfrak{p} \subset \mathcal{O}_K$ is an extension of a prime ideal in a proper subfield $K_0 \subset K$ if there exists a prime ideal $\mathfrak{p}_0 \subset \mathcal{O}_{K_0}$ such that $\mathfrak{p} = \mathfrak{p}_0 \mathcal{O}_K$.) For example, if $p < D_K^\eta$ is a rational prime that splits completely in K, so that $p\mathcal{O}_K = \mathfrak{p}_1 \cdots \mathfrak{p}_n$ for distinct prime ideals $\mathfrak{p}_j$, then each $\mathfrak{p}_j$ satisfies the hypotheses of the criterion. In particular, if M rational (unramified) primes $p_1, \dots, p_M < D_K^\eta$ split completely in K, then (2.4) holds. Alternatively, it suffices to exhibit prime ideals $\mathfrak{p}_j \subset \mathcal{O}_K$ of degree 1, since such a prime ideal cannot be an extension of a prime ideal from a proper subfield.

Here is one of Ellenberg and Venkatesh's striking applications, which shows that $\mathbf{C}_{2,3}(1/3)$ holds—the current record for $n = 2$, $\ell = 3$. Fix a large square-free integer $d > 1$. Any prime $p \nmid 6d$ that is inert in $\mathbb{Q}(\sqrt{-3})$ must split either in $\mathbb{Q}(\sqrt{d})$ or in $\mathbb{Q}(\sqrt{-3d})$. Thus for any $\eta < 1/6$, at least one field $K \in \{\mathbb{Q}(\sqrt{d}), \mathbb{Q}(\sqrt{-3d})\}$ has a positive proportion of the primes $(1/2)d^\eta \leq p \leq d^\eta$ split in K. By the Ellenberg–Venkatesh criterion (2.4), this field K then has the property that $|\mathrm{Cl}_K[3]| \ll D_K^{1/3+\varepsilon}$ for all $\varepsilon > 0$. By the Scholz reflection principle, this bound also applies to the other field in the pair, and $\mathbf{C}_{2,3}(1/3)$ holds.

The Scholz reflection principle has also been generalized by Ellenberg and Venkatesh to bound ℓ-torsion (for odd primes ℓ) in class groups of even-degree extensions of certain number fields. In particular, by pairing their criterion with a reflection principle, they show that $\mathbf{C}_{3,3}(1/3)$ holds and $\mathbf{C}_{4,3}(\Delta)$ holds for some $\Delta < 1/2$ [34, COR. 3.7]. This concludes the list of degrees n and primes ℓ for which property $\mathbf{C}_{n,\ell}(\Delta)$ is known for some $\Delta < 1/2$.

Here are open problems: reduce the value $\Delta < 1/2$ for which $\mathbf{C}_{n,\ell}(\Delta)$ holds, when $n \geq 3$ and $\ell = 2$, or when $n = 2, 3$ or 4 and $\ell = 3$. For $n = 2, 3$ or 4 and a prime $\ell \geq 5$, prove for the first time that $\mathbf{C}_{n,\ell}(\Delta)$ holds for some $\Delta < 1/2$. For $n \geq 5$ and a prime $\ell \geq 3$, prove for the first time that $\mathbf{C}_{n,\ell}(\Delta)$ holds for some $\Delta < 1/2$.

The Ellenberg–Venkatesh criterion underlies most of the significant recent progress on bounding ℓ-torsion in class groups. What is the best result it can imply? Assuming the Generalized Riemann Hypothesis, given any number field $K/\mathbb{Q}$ of degree n with D_K sufficiently large, a positive proportion of primes $p < D_K^\eta$ split completely in K, implying

$$|\mathrm{Cl}_K[\ell]| \ll_{n,\ell,\varepsilon} D_K^{\frac{1}{2}-\frac{1}{2\ell(n-1)}+\varepsilon}, \quad \text{for all } \varepsilon > 0. \tag{2.5}$$

As this is a useful benchmark, we will call this the GRH-bound, and for convenience set $\Delta_{\mathrm{GRH}} = \frac{1}{2} - \frac{1}{2\ell(n-1)}$ once n, ℓ have been fixed. Thus if GRH is true, for each n, ℓ, property $\mathbf{C}_{n,\ell}(\Delta_{\mathrm{GRH}})$ holds. There has been intense interest in proving this without assuming GRH, and this will be our next topic.

3. FAMILIES OF FIELDS

So far we have considered, for each degree n, the "family" of number fields $K/\mathbb{Q}$ of degree n. Let us formalize this, letting $\mathscr{F}_n(X)$ be the set of all degree n extensions K of $\mathbb{Q}$, with $D_K = |\mathrm{Disc}(K/\mathbb{Q})| \leq X$; let $\mathscr{F}_n = \mathscr{F}_n(\infty)$. It is helpful at this point to consider more specific families of fields of a fixed degree. For example, we could define $\mathscr{F}_2^-(X)$ to be the set of imaginary quadratic fields K with $D_K \leq X$, and similarly $\mathscr{F}_2^+(X)$ for real quadratic fields. In general, given a transitive subgroup $G \subset S_n$, define the family

$$\mathscr{F}_n(G; X) = \{K/\mathbb{Q} : \deg K/\mathbb{Q} = n, \mathrm{Gal}(\tilde{K}/\mathbb{Q}) \simeq G, D_K \leq X\}, \tag{3.1}$$

where all K are in a fixed algebraic closure $\overline{\mathbb{Q}}$, $\tilde{K}$ is the Galois closure of $K/\mathbb{Q}$, the Galois group is considered as a permutation group on the n embeddings of K in $\overline{\mathbb{Q}}$, and the isomorphism with G is one of permutation groups. When $\mathscr{F}$ is such a family, we define:

Property $\mathbf{C}_{\mathscr{F},\ell}(\Delta)$ *holds if for all fields* $K \in \mathscr{F}$, $|\mathrm{Cl}_K[\ell]| \ll_{n,\ell,\Delta,\varepsilon} D_K^{\Delta+\varepsilon}$ *for all* $\varepsilon > 0$.

Since Property $\mathbf{C}_{\mathscr{F},\ell}(\Delta)$ remains out of reach for almost all families, we also consider:

Property $\mathbf{C}^*_{\mathscr{F},\ell}(\Delta)$ *holds if for almost all fields* $K \in \mathscr{F}$, $|\mathrm{Cl}_K[\ell]| \ll_{n,\ell,\Delta,\varepsilon} D_K^{\Delta+\varepsilon}$ *for all* $\varepsilon > 0$. We say that a result holds for "almost all" fields in a family $\mathscr{F}$ if the subset $E(X)$ of possible exceptions is density zero in $\mathscr{F}(X)$, in the sense that

$$\frac{|E(X)|}{|\mathscr{F}(X)|} \to 0 \quad \text{as } X \to \infty.$$

Here too, the first progress came for imaginary quadratic fields. Soundararajan observed that among imaginary quadratic fields with discriminant in a dyadic range $[-X, -2X]$, at most one can fail to satisfy $|\mathrm{Cl}_K[\ell]| \ll D_K^{1/2-1/2\ell+\varepsilon}$ [82]. This verified $\mathbf{C}^*_{\mathscr{F}_2^-,\ell}(\Delta_{\mathrm{GRH}})$ for all primes ℓ. For $\ell = 3$ and quadratic fields, Wong observed that $\mathbf{C}^*_{\mathscr{F}_2^\pm,3}(1/4)$ holds [96]. For any odd prime ℓ, Heath-Brown and Pierce went below the GRH-bound, proving $\mathbf{C}^*_{\mathscr{F}_2^-,\ell}(1/2 - 3/(2\ell+2))$ [46]. They used the large sieve to show that aside from at most $O(X^\varepsilon)$ exceptions, all discriminants $-d \in [-X, -2X]$ have $|\mathrm{Cl}_{\mathbb{Q}(\sqrt{-d})}[\ell]|$

controlled by counting the number of distinct primes p_1, p_2 of a certain size such that (2.3) has a nontrivial integral solution (y, z). Then they showed there can be few such solutions, while averaging nontrivially over d. These methods relied heavily on the explicit nature of methods for imaginary quadratic fields. Fields of higher degree need a different approach.

3.1. Dual problems: counting primes, counting fields

To apply the Ellenberg–Venkatesh criterion, we face a question such as: "Given a field, how many small primes split completely in it?" This question is very difficult in general (and is related to the Generalized Riemann Hypothesis). There is a dual question: "Given a prime, in how many fields does it split completely?" Ellenberg, Pierce, and Wood devised a method to apply the Ellenberg–Venkatesh criterion by tackling the dual question instead [33]. The idea goes like this: suppose that each prime splits completely in a positive proportion of fields in a family $\mathscr{F}$. Then the mean number of primes $p \leq x$ that split completely in each field should be comparable to $\pi(x)$, and unless the primes conspire, almost all fields in $\mathscr{F}$ should have close to the mean number of primes split completely in them. To prove that the primes cannot conspire, Ellenberg, Pierce, and Wood developed a sieve method, modeled on the Chebyshev inequality from probability.

As input the sieve requires precise counts for the cardinality

$$N_{\mathscr{F}}(X; p) = \big|\{K \in \mathscr{F}(X) : p \text{ splits completely in } K\}\big|.$$

It also requires analogous counts $N_{\mathscr{F}}(X; p, q)$ for when two primes $p \neq q$ split completely in K. Suppose one can prove that for some $\sigma > 0$ and $\tau < 1$, for all distinct primes p, q,

$$N_{\mathscr{F}}(X; p, q) = \delta(pq)\big|\mathscr{F}(X)\big| + O\big((pq)^{\sigma}\big|\mathscr{F}(X)\big|^{\tau}\big), \tag{3.2}$$

for a multiplicative density function $\delta(pq)$ taking values in $(0, 1)$. Then Ellenberg, Pierce, and Wood prove that there exists $\Delta_0 > 0$ (depending on τ, σ) such that the mean number of primes $p \leq X^{\Delta_0}$ that split completely in fields in $\mathscr{F}(X)$ is comparable to $\pi(X^{\Delta_0})$. Moreover, there can be at most $O(|\mathscr{F}(X)|^{1-\Delta_0})$ exceptional fields K in $\mathscr{F}(X)$ such that fewer than half the mean number of primes split completely in K. Consequently, for any family $\mathscr{F}$ for which the crucial count (3.2) can be proved, combining this sieve with the Ellenberg–Venkatesh criterion proves that $\mathbf{C}^*_{\mathscr{F},\ell}(\Delta)$ holds for every integer $\ell \geq 2$, where $\Delta = \max\{\frac{1}{2} - \Delta_0, \Delta_{\mathrm{GRH}}\}$.

For which families of fields can (3.2) be proved? Counting number fields is itself a difficult question. For each integer $D \geq 1$, there are a finite number of extensions $K/\mathbb{Q}$ of degree n and discriminant exactly D, by Hermite's finiteness theorem [78, §4.1]. Let $N_n(X)$ denote the number of degree n extensions $K/\mathbb{Q}$ with $D_K \leq X$ (counted up to isomorphism). A folk conjecture, sometimes associated to Linnik, states that

$$N_n(X) \sim c_n X \quad \text{as } X \to \infty. \tag{3.3}$$

When $n = 2$, this is essentially equivalent to counting square-free integers (see [33, APPENDIX]). For degree $n = 3$, this is a deep result of Davenport and Heilbronn [28]. For degree $n = 4$, it is known by celebrated results of Cohen, Diaz y Diaz, and Olivier (counting quartic fields

K with $\mathrm{Gal}(\tilde{K}/\mathbb{Q}) \simeq D_4$), and Bhargava (counting non-D_4 quartic fields) [7,20]. For degree $n = 5$, it is known by landmark work of Bhargava [9].

The sieve method of Ellenberg, Pierce, and Wood requires an even more refined count (3.2), with prescribed local conditions and a power-saving error term with explicit dependence on p, q. Power saving error terms for $N_n(X)$ were found for $n = 3$ by Belabas, Bhargava, and Pomerance [5], Bhargava, Shankar, and Tsimerman [11], Taniguchi and Thorne [85]; for $n = 4$ (non-D_4) by Belabas, Bhargava, and Pomerance [5]; and for $n = 5$ by Shankar and Tsimerman [79]. These results can be refined to prove (3.2). Ellenberg, Pierce, and Wood used this strategy to prove that when $\mathscr{F}$ is the family of fields of degree $n = 2, 3, 4$ (non-D_4), or 5, $\mathbf{C}^*_{\mathscr{F},\ell}(\Delta_{\mathrm{GRH}})$ holds for all sufficiently large primes ℓ. (For the few remaining small ℓ, $\mathbf{C}^*_{\mathscr{F},\ell}(\Delta)$ holds with a slightly larger $\Delta < 1/2$.) Counting quartic D_4-fields with local conditions, ordered by discriminant, remains an interesting open problem.

The probabilistic method of Ellenberg–Pierce–Wood uses the property that the density function $\delta(pq)$ in (3.2) is multiplicative (i.e., local conditions at p and q are asymptotically independent). Frei and Widmer have adapted this approach to prove $\mathbf{C}^*_{\mathscr{F},\ell}(\Delta_{\mathrm{GRH}})$ for all sufficiently large ℓ, for $\mathscr{F}$ a family of totally ramified cyclic extensions of k [40]. (That is, $\mathscr{F}$ comprises cyclic extensions K/k of degree n in which every prime ideal of $\mathcal{O}_k$ not dividing n is either unramified or totally ramified in K). This family is chosen since the density function $\delta(pq)$ is multiplicative. It would be interesting to investigate whether a probabilistic method can rely less strictly upon multiplicativity of the density function.

There is a great obstacle to expanding the above approach to the family of all fields of degree n when $n \geq 6$. Then, even the asymptotic (3.3) is not known. For each $n \geq 6$,

$$N_n(X) \leq a_n X^{c_0 (\log n)^2} \tag{3.4}$$

is the best-known bound, with $c_0 = 1.564$, by Lemke Oliver and Thorne [61]; this improves on Couveignes [25], Ellenberg and Venkatesh [36], and Schmidt [75]. For lower bounds, in general the record is $N_n(X) \gg X^{1/2+1/n}$, for all $n \geq 7$ [12]. For any n divisible by $p = 2, 3$ or 5, Klüners (personal communication) has observed that $N_n(X) \gg X$, since there exists a field $F/\mathbb{Q}$ of degree n/p such that degree p S_p-extensions of F exhibit linear asymptotics.

Tackling the problem of counting primes with certain splitting conditions in a specific field via the dual problem of counting fields with certain local conditions at specific primes seems out of reach for higher degree fields. How about tackling the problem of counting primes directly?

4. COUNTING PRIMES WITH L-FUNCTIONS

The prime number theorem states that the number $\pi(x)$ of primes $p \leq x$ satisfies $\pi(x) \sim \mathrm{Li}(x)$ as $x \to \infty$. To count *small* primes, or primes in short intervals, requires understanding the error term, as well as the main term. For each $1/2 \leq \Delta < 1$, the statement

$$\pi(x) = \mathrm{Li}(x) + O(x^{\Delta+\varepsilon}) \quad \text{for all } \varepsilon > 0 \tag{4.1}$$

is essentially equivalent to the statement that the Riemann zeta function $\zeta(s)$ is zero-free for $\Re(s) > \Delta$ [27, CH. 18]. The Riemann Hypothesis conjectures this is true for $\Delta = 1/2$; it is not known for any $\Delta < 1$. The best known Vinogradov–Korobov zero-free region is:

$$\sigma \geq 1 - \frac{C}{(\log t)^{2/3}(\log\log t)^{1/3}}, \quad t \geq 3, \tag{4.2}$$

with an absolute constant $C > 0$ computed by Ford [37].

To count primes with a specified splitting type in a Galois extension $L/\mathbb{Q}$ of degree $n_L \geq 2$, consider the counting function

$$\pi_{\mathscr{C}}(x, L/\mathbb{Q}) = \left|\left\{p \leq x : p \text{ unramified in } L, \left[\frac{L/\mathbb{Q}}{p}\right] = \mathscr{C}\right\}\right|, \tag{4.3}$$

in which $[\frac{L/\mathbb{Q}}{p}]$ is the Artin symbol and $\mathscr{C}$ is any fixed conjugacy class in $G = \mathrm{Gal}(L/\mathbb{Q})$. For example, when $L = \mathbb{Q}(e^{2\pi i/q})$, this can be used to count primes in a fixed residue class modulo q. Or, for example, for any Galois extension $L/\mathbb{Q}$, when $\mathscr{C} = \{\mathrm{Id}\}$, this counts primes that split completely in L. By the celebrated Chebotarev density theorem [88],

$$\pi_{\mathscr{C}}(x, L/\mathbb{Q}) \sim \frac{|\mathscr{C}|}{|G|}\mathrm{Li}(x), \quad \text{as } x \to \infty. \tag{4.4}$$

But just as for $\pi(x)$, to count *small* primes accurately requires more quantitative information. A central goal is to prove an asymptotic for $\pi_{\mathscr{C}}(x, L/\mathbb{Q})$ that is valid for x very small relative to $D_L = |\mathrm{Disc} L/\mathbb{Q}|$, and with an effective error term. This requires exhibiting a zero-free region for the Dedekind zeta function $\zeta_L(s)$. This is more complicated than (4.2), due to the possibility of an exceptional Landau–Siegel zero: within the region

$$\sigma \geq 1 - (4\log D_L)^{-1}, \quad |t| \leq (4\log D_L)^{-1}, \tag{4.5}$$

$\zeta_L(\sigma + it)$ can contain at most one (real, simple) zero, denoted β_0 if it exists. (As observed by Heilbronn and generalized by Stark, if β_0 exists then it must "come from" a quadratic field, in the sense that L contains a quadratic subfield F with $\zeta_F(\beta_0) = 0$ [47,83].)

Lagarias and Odlyzko used the zero-free region (4.5) to prove there exist absolute, computable constants C_1, C_2 such that for all $x \geq \exp(10 n_L (\log D_L)^2)$,

$$\left|\pi_{\mathscr{C}}(x, L/\mathbb{Q}) - \frac{|\mathscr{C}|}{|G|}\mathrm{Li}(x)\right| \leq \frac{|\mathscr{C}|}{|G|}\mathrm{Li}(x^{\beta_0}) + C_1 x \exp\bigl(-C_2 n_L^{-1/2}(\log x)^{1/2}\bigr), \tag{4.6}$$

in which the β_0 term is present only if β_0 exists (see [60], and Serre [77]). This was the first effective Chebotarev density theorem. It can be difficult to apply to questions of interest because of the mysterious β_0 term, and because x must be a large power of D_L (certainly at least $x \geq D_L^{10 n_L}$). In contrast, to apply the Ellenberg–Venkatesh criterion to a field K of degree n, we aim to exhibit primes $p < D_K^\eta$ that split completely in the Galois closure $\tilde{K}$ (and hence in K), with $\eta \approx 1/(2\ell(n-1)) \to 0$ as $n, \ell \to \infty$. (These primes are even smaller relative to $D_{\tilde{K}}$, since $D_K^{|G|/n} \ll_G D_{\tilde{K}} \ll_G D_K^{|G|/2}$, where $G = \mathrm{Gal}(\tilde{K}/\mathbb{Q})$ [72].)

If GRH holds for $\zeta_L(s)$, then $\zeta_L(s)$ is zero-free for $\Re(s) > 1/2$, and Lagarias and Odlyzko improve (4.6) in three ways: (i) it is valid for $x \geq 2$; (ii) the β_0 term is not present; (iii) the remaining error term is $O(x^{1/2}\log(D_L x^{n_L}))$. Properties (i) and (ii) show that for every $\eta > 0$, for every degree n extension $K/\mathbb{Q}$ with D_K sufficiently large, at least $\gg \pi(D_K^\eta)$ primes $p \leq D_K^\eta$ split completely in the Galois closure $\tilde{K}$ (and hence in K). When input into the Ellenberg–Venkatesh criterion, this is the source of the GRH-bound (2.5) for all integers $\ell \geq 2$.

Here is a central goal: improve the Chebotarev density theorem (4.6) without assuming GRH, so that (i′) for any $\eta > 0$ it is valid for x as small as $x \geq D_L^{\eta}$ (for all D_L sufficiently large) and (ii) the β_0 term is not present. (For many applications, the final error term in (4.6) suffices as is.) If this held for $L = \tilde{K}$ the Galois closure of a field K, the Ellenberg–Venkatesh criterion would imply the GRH-bound (2.5) for ℓ-torsion in Cl_K for all integers $\ell \geq 2$, without assuming GRH. Recently, Pierce, Turnage-Butterbaugh, and Wood showed that the key improvements (i′) and (ii) hold if for some $0 < \delta \leq 1/4$, $\zeta_L(s)/\zeta(s)$ is zero-free for $s = \sigma + it$ in the box

$$1 - \delta \leq \sigma \leq 1, \quad |t| \leq \log D_L^{2/\delta}. \tag{4.7}$$

Proving this for any particular L-function $\zeta_L(s)/\zeta(s)$ of interest is out of reach. Instead, it can be productive to study a family of L-functions. In particular, if $\mathscr{F} = \mathscr{F}_n(G;X)$ is a family of degree n fields with fixed Galois group of the Galois closure, property $\mathbf{C}^*_{\mathscr{F},\ell}(\Delta_{\mathrm{GRH}})$ will follow (for all integers $\ell \geq 2$) if it is true for *almost all* fields $K \in \mathscr{F}_n(G;X)$, that $\zeta_{\tilde{K}}(s)/\zeta(s)$ is zero-free in the box (4.7). This was the strategy Pierce, Turnage-Butterbaugh, and Wood developed in [72], which we will now briefly sketch.

4.1. Families of L-functions

There is a long history of estimating the density of zeroes within a certain region, for a family of L-functions. If we can show there are fewer possible zeroes in the region than there are L-functions in the family, then some of the L-functions must be zero-free in that region. We single out a result of Kowalski and Michel, who used the large sieve to prove a zero density result for families of cuspidal automorphic L-functions [56]. In particular, for suitable families, their result implies that almost all L-functions in the family must be zero-free in a box analogous to (4.7).

There are two fundamental barriers to applying this to our problem of interest: the representation underlying $\zeta_{\tilde{K}}(s)/\zeta(s)$ is not always cuspidal, and it is not always known to be automorphic. Suppose G has irreducible complex representations $\rho_0, \rho_1, \ldots, \rho_r$, with ρ_0 the trivial representation. Then for $K \in \mathscr{F}_n(G;X)$, $\zeta_{\tilde{K}}$ is a product of Artin L-functions,

$$\zeta_{\tilde{K}}(s)/\zeta(s) = \prod_{j=1}^{r} L(s, \rho_j, \tilde{K}/\mathbb{Q})^{\dim \rho_j}. \tag{4.8}$$

The Artin (holomorphy) conjecture posits that for each nontrivial irreducible representation ρ_j, $L(s, \rho_j, \tilde{K}/\mathbb{Q})$ is entire. The (strong) Artin conjecture posits that for each nontrivial irreducible representation ρ_j, there is an associated cuspidal automorphic representation $\pi_{\tilde{K},j}$ of $\mathrm{GL}(m_j)/\mathbb{Q}$, and $L(s, \pi_{\tilde{K},j}) = L(s, \rho_j, \tilde{K}/\mathbb{Q})$. This is known for certain types of representations of certain groups, but otherwise is a deep open problem (see recent work in [19]). For the moment, we will proceed by assuming the strong conjecture. Then the factorization (4.8) naturally slices the family $\zeta_{\tilde{K}}(s)/\zeta(s)$, as K varies over $\mathscr{F}_n(G;X)$, into r families $\mathscr{L}_1(X), \mathscr{L}_2(X), \ldots, \mathscr{L}_r(X)$, where each $\mathscr{L}_j(X)$ is the set of cuspidal automorphic representations $\pi_{\tilde{K},j}$ associated to the representation ρ_j. Kowalski and Michel's result applies to each family $\mathscr{L}_j(X)$ individually. This proves that every representation $\pi \in \mathscr{L}_j(X)$ has associated L-function $L(s,\pi)$ being zero-free in the box (4.7)—except for a possible subset of "bad" representations π, of density zero in $\mathscr{L}_j(X)$, for which $L(s,\pi)$ could have a zero in the box. (Of course, no such zero exists if GRH is true, but we are not assuming GRH.)

Now a crucial difficulty arises: if there were a "bad" representation $\pi \in \mathscr{L}_j(X)$, in how many products (4.8) could it appear, as K varies over $\mathscr{F}_n(G;X)$? Each field K for which the "bad" factor $L(s,\pi)$ appears could have a zero of $\zeta_{\tilde{K}}(s)/\zeta(s)$ in (4.7). Thus the crucial question is: for a fixed nontrivial irreducible representation ρ of G, how many fields $K_1, K_2 \in \mathscr{F}_n(G;X)$ have $L(s,\rho,\tilde{K}_1/\mathbb{Q}) = L(s,\rho,\tilde{K}_2/\mathbb{Q})$? This can be stated a different way. Given a subgroup H of G, let $\tilde{K}^H$ denote the subfield of $\tilde{K}$ fixed by H. It turns out that the question can be transformed into: how many fields $K_1, K_2 \in \mathscr{F}_n(G;X)$ have $\tilde{K}_1^{\mathrm{Ker}(\rho)} = \tilde{K}_2^{\mathrm{Ker}(\rho)}$? Let us call this a collision. If a positive proportion of fields in $\mathscr{F}_n(G;X)$ can collide for ρ_j, then via the factorization (4.8), the possible existence of even one "bad" element in $\mathscr{L}_j(X)$ could allow a positive proportion of the functions $\zeta_{\tilde{K}}(s)/\zeta(s)$ to have a zero in (4.7). In particular, then this approach would fail to prove $\mathbf{C}^*_{\mathscr{F},\ell}(\Delta_{\mathrm{GRH}})$ for the family $\mathscr{F} = \mathscr{F}_n(G;X)$. To rule this out, we aim to show that for each nontrivial irreducible representation ρ_j of G, collisions are rare.

We define the "collision problem" for the family $\mathscr{F}_n(G;X)$: how big is

$$\max_{\rho} \max_{K_1 \in \mathscr{F}_n(G;X)} \left|\left\{K_2 \in \mathscr{F}_n(G;X) : \tilde{K}_1^{\mathrm{Ker}(\rho)} = \tilde{K}_2^{\mathrm{Ker}(\rho)}\right\}\right|? \tag{4.9}$$

Here the maximum is over the nontrivial irreducible representations ρ of G with $\mathrm{Ker}(\rho)$ a proper normal subgroup of G. Suppose for a particular family $\mathscr{F}_n(G;X)$, the collisions (4.9) number at most $\ll X^{\alpha}$. Then the strategy sketched here ultimately shows that aside from at most $\ll X^{\alpha+\varepsilon}$ exceptional fields (for any $\varepsilon > 0$), every field in $K \in \mathscr{F}_n(G;X)$ has the property that an improved Chebotarev density theorem with properties (i′) and (ii) holds for its Galois closure $\tilde{K}$. If we can prove simultaneously that $|\mathscr{F}_n(G;X)| \gg X^{\beta}$ for some $\beta > \alpha$, then the improved Chebotarev density theorem holds for almost all fields in the family. Consequently, we would obtain property $\mathbf{C}^*_{\mathscr{F},\ell}(\Delta_{\mathrm{GRH}})$, for all integers $\ell \geq 2$.

Thus the goal of bounding ℓ-torsion in class groups of fields in the family $\mathscr{F}_n(G;X)$ has been transformed into a question of counting how often certain fields share a subfield. For which families can the collision problem (4.9) be controlled? For some groups, the number of collisions can be $\gg |\mathscr{F}_n(G;X)|$ (for example, $G = \mathbb{Z}/4\mathbb{Z}$). On the other hand, if G is a simple group, or if all nontrivial irreducible representations of G are faithful, the number of collisions is $\ll 1$ (but a lower bound $|\mathscr{F}_n(G;X)| \gg X^{\beta}$ for some $\beta > 0$ may not be known, yet). In general, controlling the collision problem is difficult.

One idea is to restrict attention to an advantageously chosen subfamily of fields, call it $\mathscr{F}_n^*(G;X) \subset \mathscr{F}_n(G;X)$. To bound (4.9) within a subfamily it suffices to count

$$\max_{H} \max_{\substack{F \\ \deg(F/\mathbb{Q})=[G:H]}} \left|\left\{K \in \mathscr{F}_n^*(G;X) : \tilde{K}^H = F\right\}\right|. \tag{4.10}$$

Here H ranges over the proper normal subgroups of G that appear as the kernel of some nontrivial irreducible representation. For some groups G, if $\mathscr{F}_n^*(G;X)$ is defined appropriately, this can be further transformed into counting number fields with *fixed* discriminant.

Let us see how this goes in the example $G = S_n$ with $n = 3$ or $n \geq 5$, so that A_n is the only nontrivial proper normal subgroup (the kernel of the sign representation). Consider the subfamily $\mathscr{F}_n^*(S_n; X)$ of fields with square-free discriminant. (These are a positive proportion of all degree n S_n-fields for $n \leq 5$ and conjecturally so for $n \geq 6$.) Then for $H = A_n$ and F a fixed quadratic field, it can be shown that any field K counted in (4.10) must have the property that $D_K = D_F$ (up to some easily controlled behavior of wildly ramified primes). Under this very strong identity of discriminants, (4.10) is dominated by

$$\max_{D \geq 1} \left| \left\{ K \in \mathscr{F}_n^*(S_n; X) : D_K = D \right\} \right|. \tag{4.11}$$

This strategy transforms the collision problem into counting fields of *fixed* discriminant.

For certain other groups G, (4.10) can also be dominated by a quantity analogous to (4.11) if the subfamily $\mathscr{F}_n^*(G; X)$ is defined by specifying that each prime that is tamely ramified in K has its inertia group generated by an element in a carefully chosen conjugacy class $\mathscr{I}$ of G. For such a group G, the final step in this strategy for proving improved Chebotarev density theorems for almost all fields in the family $\mathscr{F}_n^*(G; X)$ is to bound (4.11). If $|\mathscr{F}_n^*(G; X)| \gg X^\beta$ is known, it suffices to prove (4.11) is $\ll X^\alpha$ for some $\alpha < \beta$. In general, counting number fields with fixed discriminant is very difficult—we will return to this problem later. But for some families $\mathscr{F}_n^*(G; X)$, (4.11) can be controlled sufficiently well, relative to a known lower bound for $|\mathscr{F}_n^*(G; X)|$.

This is the strategy developed by Pierce, Turnage-Butterbaugh, and Wood in [72]. The result is an improved Chebotarev density theorem, with properties (i′) and (ii), that holds unconditionally for almost all fields in the following families: (a) $\mathscr{F}_p(C_p; X)$ cyclic extensions of any prime degree; (b) $\mathscr{F}_n^*(C_n; X)$ totally ramified cyclic extensions of any degree $n \geq 2$; (c) $\mathscr{F}_p^*(D_p; X)$ prime degree dihedral extensions, $\mathscr{I}$ being the class of order 2 elements; (d) $\mathscr{F}_n^*(S_n; X)$ fields of square-free discriminant, $n = 3, 4$; and (e) $\mathscr{F}_4^*(A_4; X)$, $\mathscr{I}$ being either class of order 3 elements. Conditional on the strong Artin conjecture, they proved the improved Chebotarev density theorem also holds for almost all fields in the following families: (f) $\mathscr{F}_5^*(S_5; X)$ quintic fields of square-free discriminant; and (g) $\mathscr{F}_n(A_n; X)$, for all $n \geq 5$. (There are other families, such as $\mathscr{F}_n^*(S_n; X)$ for $n \geq 6$, to which the strategy applies, but the current upper bound known for (4.11) is larger than the known lower bound for $|\mathscr{F}_n^*(S_n; X)|$.) As a consequence, Pierce, Turnage-Butterbaugh, and Wood proved for each family (a)–(e) that $\mathbf{C}^*_{\mathscr{F},n}(\Delta_{\mathrm{GRH}})$ holds unconditionally for all integers $\ell \geq 2$, and it holds for each family (f)–(g) under the strong Artin conjecture. This was the first time such a result was proved for families of fields of arbitrarily large degree.

4.2. Further developments

Since the work outlined above, many interesting new developments have followed, relating to zero density results for families of L-functions, Chebotarev density theorems for families of fields, and ℓ-torsion in class groups of fields in specific families.

First, there has been renewed interest in zero density results for families of L-functions, concerning potential zeroes in regions close to the line $\Re(s) = 1$, and extending the perspective of Kowalski and Michel [56]; see, for example, [18, 49, 87].

Second, several new strategies have focused on the problem of proving effective Chebotarev density theorems for almost all fields in a family. The work in [72] raised several desiderata. Some groups G have the property that no ramification restriction exists that allows the "collision problem" in the form (4.10) to be transformed into a "discriminant multiplicity problem" in the form (4.11). For example, this occurs for any noncyclic abelian group, or D_4. These cases remain open; instead, An recently proved a Chebotarev density theorem for almost all fields in a family of quartic D_4-fields associated to a fixed biquadratic field [2]. Another significant desideratum was to remove the dependence on the strong Artin conjecture. Thorner and Zaman recently achieved this, by proving a zero density estimate directly for Dedekind zeta functions, without passing through the factorization (4.8) [86]. But that work is still explicitly conditional on the ability to control a collision problem similar to (4.9), for which the best known strategy is still the approach of [72].

Most recently, the collision problem has been bypassed for certain groups G by interesting new work of Lemke Oliver, Thorner, and Zaman [62]. Their key idea when studying fields in a family $\mathscr{F}_n(G;X)$ is to prove a zero-free region not for $\zeta_{\tilde{K}}/\zeta$ but for $\zeta_{\tilde{K}}/\zeta_{\tilde{K}^N}$ where N is a nontrivial normal subgroup of G. This allows them to replace a collision problem like (4.9) by an "intersection multiplicity problem," bounding

$$\max_{K_1 \in \mathscr{F}_n(G;X)} \left|\left\{K_2 \in \mathscr{F}_n(G;X) : \tilde{K}_1 \cap \tilde{K}_2 \neq \tilde{K}_1^N \cap \tilde{K}_2^N\right\}\right|. \tag{4.12}$$

The number of exceptional fields, for which a desired Chebotarev-type theorem cannot be verified, is then dominated by (4.12) (up to X^{ε}). This is advantageous if G has a unique minimal nontrivial normal subgroup N, so that (4.12) is $\ll 1$. But as a trade-off, one no longer obtains an effective Chebotarev density theorem for each conjugacy class $\mathscr{C}$ in G.

Let $\pi_K(x)$ count prime ideals $\mathfrak{p} \subset \mathcal{O}_K$ with $\mathfrak{N}_{K/\mathbb{Q}}\mathfrak{p} \leq x$. Let $\mathscr{F}$ represent either of the two following families: degree p fields $K/\mathbb{Q}$ for p prime, or degree n S_n-fields $K/\mathbb{Q}$, for any $n \geq 2$. Lemke Oliver, Thorner, and Zaman prove that except for at most $\ll X^{\varepsilon}$ exceptional fields, every $K \in \mathscr{F}(X)$ has $|\pi_K(x) - \pi(x)| \leq C_1 x \exp(-C_2\sqrt{\log x})$ for every $x \geq (\log D_K)^{C_3(n,\varepsilon)}$. In either family $\mathscr{F}$, they obtain results on ℓ-torsion by applying the Ellenberg–Venkatesh criterion using prime ideals of degree 1. If $\pi_K^*(x)$ counts only prime ideals of degree 1, then $\pi_K^*(x) = \pi_K(x) + O_n(\sqrt{x})$, so the above result exhibits many small prime ideals of degree 1. Thus for either family, $\mathbf{C}^*_{\mathscr{F},n}(\Delta_{\mathrm{GRH}})$ holds unconditionally for all ℓ (and the exceptional set is very small). (They also exhibit infinitely many degree n S_n-fields K with Cl_K as large as possible, but $|\mathrm{Cl}_K[\ell]|$ bounded by (2.5) for all ℓ; and infinitely many totally real degree n S_n-fields K with Cl_K containing an element of exact order ℓ and $|\mathrm{Cl}_K[\ell]|$ bounded by (2.5).) What happens when G does not have a unique minimal nontrivial normal subgroup? Here is an open question: in general, when N is a nontrivial normal subgroup of G (not necessarily unique or minimal), what is the true order of growth of (4.12) as $X \to \infty$? Questions about this "intersection multiplicity" are gathered in [62].

Third, increased attention has turned to bounding ℓ-torsion in class groups for *all* fields in special families specified by the Galois group: that is, proving property $\mathbf{C}_{\mathscr{F},\ell}(\Delta)$ for some $\Delta < 1/2$. First, Klüners and Wang have proved $\mathbf{C}_{\mathscr{F},p}(0)$ for the family $\mathscr{F}_{p^r}(G;X)$ for any p-group G; this generalizes the application of genus theory to prove $\mathbf{C}_{2,2}(0)$ [54].

Second, let $G = (\mathbb{Z}/p\mathbb{Z})^r$ be an elementary abelian group of rank $r \geq 2$, with p prime. Wang has shown that for every ℓ, within the family of Galois G-fields $K/\mathbb{Q}$, property $\mathbf{C}_{\mathscr{F},\ell}(1/2 - \delta(\ell, p))$ holds for some $\delta(\ell, p) > 0$ **[91]**. Since the savings $\delta(\ell, p)$ is independent of the rank, for r sufficiently large this is better than $\mathbf{C}_{\mathscr{F},\ell}(\Delta_{\mathrm{GRH}})$. The method of proof plays off the interaction of three facts arising from the precise structure of G: first, $|\mathrm{Cl}_K[\ell]|$ factors as a product of $|\mathrm{Cl}_F[\ell]|$ where F varies over the $\approx p^{r-1}$ many degree p subfields of K, so it suffices to bound one of these factors nontrivially. Second, any rational prime splits completely in $\approx p^{r-2}$ of these subfields, so at least one subfield has a positive proportion of primes splitting completely in it. Third, the sizes of the discriminants of the subfields can be played against each other, so that known prime-counting results (which may *a priori* seem to count primes that are "too large") suffice for the application of the Ellenberg–Venkatesh criterion. This is an interesting counterpoint to the methods described earlier. In another direction, Wang has developed the notion of a forcing extension; certain nilpotent groups can be built from elementary p-groups via forcing extensions. If G' is constructed from G by a forcing extension, then $\mathbf{C}_{\mathscr{F}',\ell}(\Delta')$ can be deduced from $\mathbf{C}_{\mathscr{F},\ell}(\Delta)$, for some $\Delta, \Delta' < 1/2$, where $\mathscr{F}$ is the family of G-extensions and $\mathscr{F}'$ is the family of G'-extensions **[89]**.

All of the results mentioned in this section (except where genus theory suffices) directly apply or build on the Ellenberg–Venkatesh criterion. Can this criterion be strengthened? Ellenberg has suggested some possible improvements in **[32]**. In particular, let $\eta(K) := \inf\{H_K(\alpha) : K = \mathbb{Q}(\alpha)\}$ denote the minimum (relative) multiplicative Weil height of a generating element of K. Roughly speaking, Ellenberg notes the criterion (2.4) can actually allow prime ideals with norms as large as $\eta(K)^{1/\ell}$. The restriction to norms $< D_K^{\frac{1}{2\ell(n-1)}}$ in (2.4) was made since the lower bound $\eta(K) \geq D_K^{\frac{1}{2(n-1)}}$ holds for all fields **[88]**. Widmer, also with Frei, has shown that $\eta(K)$ can be enlarged for almost all fields in certain families, leading to improved upper bounds for ℓ-torsion in those fields **[41, 95]**. That is, they improve the very notion of the "GRH-bound" (2.5), and show that the parameter we have called Δ_{GRH} can actually be taken smaller for some fields. Their work raises interesting open questions: what upper and lower bounds hold for $\eta(K)$, for all (or almost all) fields in a family? Ruppert **[74]** has conjectured uniform upper bounds $\eta(K) \leq D_K^{1/2}$ (now proved for almost all fields in some families by **[72]**). If this is true, the Ellenberg–Venkatesh criterion would hit a barrier, for most fields, with a result like $|\mathrm{Cl}_K[\ell]| \ll D_K^{1/2-1/2\ell+\varepsilon}$ for any degree n, still far from the ℓ-torsion Conjecture. It would be very interesting to find a new, different criterion.

5. WHY DO WE EXPECT THE ℓ-TORSION CONJECTURE TO BE TRUE?

Recall that the ℓ-torsion Conjecture 2.1 is still known only in the case stemming from Gauss's work, namely for $n = 2$, $\ell = 2$. It is a good idea to affirm why we believe the ℓ-torsion Conjecture should be true. We will consider this from three perspectives.

5.1. From the perspective of the Cohen–Lenstra–Martinet heuristics

So far, when we have mentioned a result for almost all fields in a family, we have not focused on the size of a potential exceptional set, other than showing it is smaller than the size of the full family. But to understand the ℓ-torsion Conjecture, we must quantify a potential exceptional set, and show that for all sufficiently large discriminants, it is empty.

Let us abstract this, for a family $\mathscr{F}_0(X)$ of fields K with D_K in a dyadic range $(X/2, X]$, from which more general results can easily be deduced by summing over $\ll \log X$ dyadic ranges. Suppose $f : \mathscr{F}_0(X) \to \mathbb{N}$ is a function with $f(K) \leq D_K^a$ for all K. Suppose that for some $\Delta < a$ we can improve this to $f(K) \leq D_K^\Delta$ for all K outside of some exceptional set $E_0^\Delta(X) \subset \mathscr{F}_0(X)$. Then

$$\sum_{K\in\mathscr{F}_0(X)} f(K) = \sum_{K\in\mathscr{F}_0(X)\setminus E_0^\Delta(X)} f(K) + \sum_{K\in E_0^\Delta(X)} f(K) \leq \left|\mathscr{F}_0(X)\right|X^\Delta + \left|E_0^\Delta(X)\right|X^a. \tag{5.1}$$

As long as $|E_0^\Delta(X)| \ll |\mathscr{F}_0(X)|X^{-(a-\Delta)}$, this shows that $f(K) \ll X^\Delta$ on average. On the other hand, suppose we know $\sum_{K\in\mathscr{F}_0(X)} f(K) \leq X^b$. Then a potential set of exceptions $E_0^\Delta(X) = \{K \in \mathscr{F}_0(X) : f(K) > D_K^\Delta\}$ can be controlled by

$$X^\Delta\left|E_0^\Delta(X)\right| \ll \sum_{K\in E_0^\Delta(X)} f(K) \leq \sum_{K\in\mathscr{F}_0(X)} f(K) \leq X^b. \tag{5.2}$$

Thus $|E_0^\Delta(X)| \ll X^{b-\Delta}$, and exceptional fields are density zero in $\mathscr{F}_0(X)$, provided $X^{b-\Delta} = o(|\mathscr{F}_0(X)|)$. That is, a nontrivial upper bound on ℓ-torsion for "almost all" fields in a family $\mathscr{F}$ is essentially equivalent to the same upper bound "on average."

To verify the ℓ-torsion Conjecture, we wish to show a "pointwise" bound: for every $\varepsilon > 0$, there exists D_ε such that when $D_K \geq D_\varepsilon$, there are *no* exceptions to the bound $|\mathrm{Cl}_K[\ell]| \leq D_K^\varepsilon$. The key is to consider not averages but arbitrarily high kth moments. In the general setting above, suppose that we know $\sum_{K\in\mathscr{F}_0(X)} f(K)^k \leq X^b$, for a real number $k \geq 1$. Then for any fixed $\Delta > 0$, adapting the argument (5.2) shows that $|E_0^\Delta(X)| \ll X^{b-k\Delta}$. If the kth moment is uniformly bounded by X^b for a sequence of $k \to \infty$, then for each $\Delta > 0$, we can take k sufficiently large to conclude that the set of exceptions is empty.

This perspective has been applied by Pierce, Turnage-Butterbaugh, and Wood in [73] to prove that the ℓ-torsion Conjecture holds for *all* fields in a family $\mathscr{F}(X)$ if there is a real number $\alpha \geq 1$ such that for a sequence of arbitrarily large k,

$$\sum_{K\in\mathscr{F}(X)} \left|\mathrm{Cl}_K[\ell]\right|^k \ll_{n,\ell,k,\alpha} \left|\mathscr{F}(X)\right|^\alpha, \quad \text{for all } X \geq 1. \tag{5.3}$$

The Cohen–Lenstra–Martinet heuristics predict that (5.3) holds, in the form of an even stronger asymptotic with $\alpha = 1$, for all integers $k \geq 1$, for families of Galois G-extensions, at least for all primes $\ell \nmid |G|$. The appropriate moment formulation can be found in [21] for degree 2 fields and in [92] for higher degrees, building on [22]. This confirms that the ℓ-torsion Conjecture follows from another well-known set of conjectures.

The Cohen–Lenstra–Martinet heuristics are a subject of intense interest and much recent activity. Here are some spectacular successes most closely related to our topic. Davenport and Heilbronn [28] have proved

$$\sum_{\substack{\deg(K)=2\\0<D_K\le X}} |\mathrm{Cl}_K[3]| \sim \left(\frac{2}{3\zeta(2)}+\frac{1}{\zeta(2)}\right)X; \tag{5.4}$$

second-order terms have been found in [5, 11, 85]. Bhargava [7] has proved

$$\sum_{\substack{\deg(K)=3\\0<D_K\le X}} |\mathrm{Cl}_K[2]| \sim \left(\frac{5}{48\zeta(3)}+\frac{3}{8\zeta(3)}\right)X, \tag{5.5}$$

in which each isomorphism class of fields is counted once. Very recently, [63] obtained analogues of (5.4) for averages over $\mathscr{F}_{2^m}(G;X)$ for any permutation group $G\subset S_{2^m}$ that is a transitive permutation 2-group containing a transposition. See also the work of Smith on the distribution of 2^k-class groups in imaginary quadratic fields [81]; Koymans and Pagano on ℓ^k-class groups of degree ℓ cyclic fields [59]; Klys on moments of p-torsion in cyclic degree p fields (conditional on GRH for $p\ge 5$) [55]; Milovic and Koymans on 16-rank in quadratic fields [57, 58]; Bhargava and Varma [13, 14] elaborating on (5.4) and (5.5).

The perspective of moments (5.3) provides a strong motivation to prove the kth moment bounds for ℓ-torsion. Fouvry and Klüners have proved an asymptotic for the kth moments related to 4-torsion when K is quadratic, for all integers $k\ge 1$ [38]. Heath-Brown and Pierce have proved nontrivial bounds for the kth moments of ℓ-torsion for imaginary quadratic fields, for all odd primes ℓ [46]. For example, they establish second moment bounds

$$\sum_{\substack{K=\mathbb{Q}(\sqrt{\pm D})\\D\le X}} |\mathrm{Cl}_K[3]|^2 \ll X^{23/18}, \qquad \sum_{\substack{K=\mathbb{Q}(\sqrt{-D})\\D\le X}} |\mathrm{Cl}_K[\ell]|^2 \ll X^{2-\frac{3}{\ell+1}}, \quad \ell\ge 5 \text{ prime}, \tag{5.6}$$

as well as results for the kth moments for all $k\ge 1$. In general, proving tighter control on the size of an exceptional family $E_0^{\Delta}(X)$ can be used to deduce a better moment bound for $|\mathrm{Cl}_K[\ell]|$, similar to (5.1). This has recently been exploited by Frei and Widmer, in combination with refinements of the Ellenberg–Venkatesh criterion, to improve moment bounds on ℓ-torsion for the families of fields studied in [72] (if ℓ is sufficiently large); see [41].

Let us mention a connection to elliptic curves; this was after all the setting in which Brumer and Silverman initially posed the ℓ-torsion Conjecture. Let $E(q)$ denote the number of isomorphism classes of elliptic curves over $\mathbb{Q}$ with conductor q. Brumer and Silverman have conjectured that $E(q)\ll_\varepsilon q^\varepsilon$ for every $q\ge 1$, $\varepsilon>0$ [17]. Conditionally, this follows from GRH combined with a weak form of the Birch–Swinnerton-Dyer conjecture. They also showed this follows from the 3-torsion Conjecture for quadratic fields, by proving

$$E(q)\ll_\varepsilon q^\varepsilon \max_{1\le D\le 1728q} |\mathrm{Cl}_{\mathbb{Q}(\sqrt{\pm D})}[3]|, \quad \text{for all } \varepsilon>0. \tag{5.7}$$

Duke and Kowalski have combined this with the celebrated asymptotic (5.4) to bound $\sum_{1\le q\le Q} E(q)\ll Q^{1+\varepsilon}$ for every $\varepsilon>0$ [30]. (See also [39] for ordering by discriminant.) Pierce, Turnage-Butterbaugh, and Wood have recently proved that for all $k\ge 1$, the kth moment of 3-torsion in quadratic fields dominates the γkth moment of $E(q)$, for a numerical constant $\gamma\approx 1.9745\ldots$ coming from [48], which sharpened the relation (5.7). Thus new moment bounds for $E(q)$ can be obtained from (5.6), for example. Here is an open problem: prove that $\sum_{1\le q\le Q} E(q) = o(Q)$. This would show for the first time that integers that are the conductor of an elliptic curve have density zero in $\mathbb{Z}$. In fact, it is conjectured by Watkins that this average is asymptotic to $cQ^{5/6}$ for a certain constant c **[94]** (building on an analogous conjecture by Brumer–McGuinness for ordering by discriminant **[16]**).

To conclude, in this section we saw that the truth of the ℓ-torsion Conjecture is implied by the truth of the well-known Cohen–Lenstra–Martinet heuristics on the distribution of class groups.

5.2. From the perspective of counting number fields of fixed discriminant

Let $K/\mathbb{Q}$ be a degree n extension. The Hilbert class field H_K is the maximal abelian unramified extension of K, and Cl_K is isomorphic to $\mathrm{Gal}(H_K/K)$. A second way to motivate the ℓ-torsion Conjecture is to count intermediate fields between K and H_K.

Here is an argument recorded by Pierce, Turnage-Butterbaugh, and Wood in **[73]**. Fix a prime ℓ and write Cl_K additively, so that $\mathrm{Cl}_K[\ell] \simeq \mathrm{Cl}_K / \ell\, \mathrm{Cl}_K$. Now define the fixed field $L = H_K^{\ell\,\mathrm{Cl}_K}$ lying between K and H_K, so $\mathrm{Gal}(L/K) \simeq \mathrm{Cl}_K[\ell]$. Each surjection $\mathrm{Cl}_K[\ell] \to \mathbb{Z}/\ell\mathbb{Z}$ generates an intermediate field M, with $K \subset M \subset L$ and $\deg(M/\mathbb{Q}) = n\ell$. If $|\mathrm{Cl}_K[\ell]| = \ell^r$, say, this produces $\approx \ell^{r-1}$ such fields M. The crucial point is that since H_K is an unramified extension, all these fields satisfy a rigid discriminant identity $D_M = D_K^{\ell}$. Consequently, if we can count how many number fields of degree $n\ell$ can share the same *fixed* discriminant, then we can bound ℓ-torsion in Cl_K. (We have seen this problem before.) We formalize the problem of counting number fields of fixed discriminant as follows:

Property $\mathbf{D}_n(\Delta)$. *Fix a degree $n \ge 2$. Property $\mathbf{D}_n(\Delta)$ holds if for every $\varepsilon > 0$ and for every fixed integer $D > 1$, at most $\ll_{n,\varepsilon} D^{\Delta+\varepsilon}$ fields $K/\mathbb{Q}$ of degree n have $D_K = D$.*

The strategy sketched above ultimately proves that property $\mathbf{D}_{n\ell}(\Delta)$ implies $\mathbf{C}_{n,\ell}(\ell\Delta)$. This leads inevitably to the question: is property $\mathbf{D}_{n\ell}(0)$ true? Here is a conjecture:

Conjecture 5.1 (Discriminant multiplicity conjecture). *For each $n \ge 2$, for every $\varepsilon > 0$, and for every integer $D > 1$, at most $\ll_{n,\varepsilon} D^{\varepsilon}$ fields $K/\mathbb{Q}$ of degree n have $D_K = D$.*

This conjecture has been recorded by Duke **[29]**. It implies the ℓ-torsion Conjecture, a link noted in **[29, 35]** and quantified in **[73]**. Recall the conjecture (3.3) for counting all fields of degree n and discriminant $D_K \le X$. The Discriminant Multiplicity Conjecture for degree n would immediately imply $N_n(X) \ll X^{1+\varepsilon}$, which indicates its level of difficulty. Of course, in general, property $\mathbf{D}_n(\Delta)$ implies $N_n(X) \ll X^{1+\Delta+\varepsilon}$ for all $\varepsilon > 0$. (In terms of lower bounds, Ellenberg and Venkatesh have noted there can be $\gg D^{c/\log\log D}$ extensions $K/\mathbb{Q}$ with a fixed Galois group and fixed discriminant D **[35]**.)

The Discriminant Multiplicity Conjecture posits that $\mathbf{D}_n(0)$ holds for each $n \ge 2$. This is true for $n = 2$, but it is not known for any other degree. For degrees $n = 3, 4, 5$, the best-known results currently are $\mathbf{D}_3(1/3)$ by **[34]**; $\mathbf{D}_4(1/2)$ as found in **[52, 72, 73, 97]**; $\mathbf{D}_5(199/200)$ as found in **[33]**, building on **[9, 79]**. Currently for $n \ge 6$, the only result for

$\mathbf{D}_n(\Delta)$ is a trivial consequence of counting fields of bounded discriminant, as in (3.4), so in particular $\Delta = c_0(\log n)^2 > 1$ in those cases. It would be very interesting to improve the exponent known for $\mathbf{D}_n(\Delta)$, for any fixed degree $n \geq 3$.

As is the case for many of the problems surveyed in this paper, it can also be profitable to study the problem within a family $\mathscr{F}$ of degree n extensions:

Property $\mathbf{D}_{\mathscr{F},n}(\Delta)$. *Fix a degree $n \geq 2$. Property $\mathbf{D}_{\mathscr{F},n}(\Delta)$ holds if for every $\varepsilon > 0$ and for every fixed integer $D > 1$, at most $\ll_{n,\varepsilon} D^{\Delta+\varepsilon}$ fields $K/\mathbb{Q}$ in the family $\mathscr{F}$ have $D_K = D$.*

This is the type of property Pierce, Turnage-Butterbaugh, and Wood used to control the collision problem, in the form (4.11) [72]. Property $\mathbf{D}_{\mathscr{F},n}(0)$ has recently been proved by Klüners and Wang, for the family $\mathscr{F} = \mathscr{F}_n(G; X)$ of degree n G-extensions for any nilpotent group G. This was built from the truth of property $\mathbf{C}_{\mathscr{F},p}(0)$ for $\mathscr{F}$ being the family of Galois H-extensions for H a p-group, in [54]. There are many other cases where it is an interesting open problem to improve the known bound for Property $\mathbf{D}_{\mathscr{F},n}(\Delta)$.

To conclude, in this section we saw that the ℓ-torsion Conjecture follows from the Discriminant Multiplicity Conjecture. Now, recall that we saw in the context of bounding ℓ-torsion that uniform bounds for arbitrarily high moments can imply strong "pointwise" results for *every* field. Can the method of moments be used to approach the Discriminant Multiplicity Conjecture too? We turn to this idea next.

5.3. From the perspective of counting number fields of bounded discriminant

We come to a third motivation to believe the ℓ-torsion Conjecture. Recall the definition (3.1) of a family $\mathscr{F}_n(G; X)$ of degree n fields $K/\mathbb{Q}$ with $\mathrm{Gal}(\tilde{K}/\mathbb{Q})$ isomorphic (as a permutation group) to a nontrivial transitive subgroup $G \subseteq S_n$. Each element $g \in G$ has an index defined by $\mathrm{ind}(g) = n - o_g$, where o_g is the number of orbits of g when it acts on a set of n elements. Define $a(G)$ according to $a(G)^{-1} = \min\{\mathrm{ind}(g) : 1 \neq g \in G\}$; we see that $\frac{1}{n-1} \leq a(G) \leq 1$. Malle has made a well-known conjecture [65]:

Conjecture 5.2 (Malle). *For each $n \geq 2$, for each transitive subgroup $G \subseteq S_n$,*

$$\left|\mathscr{F}_n(G; X)\right| \ll_{G,\varepsilon} X^{a(G)+\varepsilon}, \quad \text{for all } \varepsilon > 0. \tag{5.8}$$

Also, $|\mathscr{F}_n(G; X)| \gg_G X^{a(G)}$.

The full statement of this conjecture is an open problem. Its difficulty is indicated by the fact that it implies a positive solution to the inverse Galois problem for number fields. (A refinement in [66] specified a power of $\log X$ in place of X^ε; counterexamples to this refinement have been found in [50], but the upper bound in (5.8) is expected to be true.)

Malle's Conjecture has been proved for abelian groups, with a strategy by Cohn [24], and asymptotic counts by Mäki [64], Wright [97]. For $n = 3, 4, 5$, it is known for S_n by the asymptotic (3.3), and for D_4 by Baily [3] (refined to an asymptotic in [20]). It is known for $C_2 \wr H$ under mild conditions on H (in particular, for at least one group of order n for every even n) by [53], and for $S_n \times A$ with A an abelian group by [67, 90]. For prime degree p D_p-fields, upper and lower bounds are closely related to p-torsion in class groups of quadratic fields, and have been studied in [23, 41, 51].

For many groups, it is a difficult open problem to prove upper or lower bounds approaching Malle's prediction. In many results surveyed here, proving a lower bound for $|\mathscr{F}_n(G;X)|$ has been an important step, to verify a result applies to "almost all" fields in a family. For many groups G, it is not even known that $|\mathscr{F}_n(G;X)| \gg X^\beta$ for some $\beta > 0$ as $X \to \infty$. Here is a tool to prove such a result: suppose $f(X, T_1, \ldots, T_s) \in \mathbb{Q}[X, T_1, \ldots, X_s]$ is a regular polynomial of total degree d in the T_i and of degree m in X with transitive Galois group $G \subset S_n$ over $\mathbb{Q}(T_1, \ldots, T_s)$. Then $|\mathscr{F}_n(G;X)| \gg_{f,\varepsilon} X^{\beta-\varepsilon}$ for every $\varepsilon > 0$, with $\beta = \frac{1-|G|^{-1}}{d(2m-2)}$; this is proved in [72]. For $G = A_n$, a polynomial f exhibited by Hilbert can be input to this criterion, implying that $|\mathscr{F}_n(A_n;X)| \gg X^{\beta_n+\varepsilon}$ for some $\beta_n > 0$, providing the first lower bound that grows like a power of X. Here is an open problem: for many groups G, no such polynomial f has yet been exhibited.

Now we focus on the conjectured upper bound (5.8) for counting fields with bounded discriminant. For any family $\mathscr{F} = \mathscr{F}_n(G;X)$ of fields, the strong "pointwise" property $\mathbf{D}_{\mathscr{F},n}(0)$ implies Malle's "average" upper bound (5.8) for the group G; see [54]. What is more surprising is that there is a converse to this. This relates to our question: can the method of moments be used to deduce the Discriminant Multiplicity Conjecture? Formally, it can. Given a family $\mathscr{F}$ of fields, for each integer $D \geq 1$ let $m(D)$ denote the number of fields $K \in \mathscr{F}$ with $D_K = D$. If arbitrarily high kth moment bounds are known for the function $m(D)$, the Discriminant Multiplicity Conjecture follows; see [73]. But the first moment of $m(D)$ is the subject of the Malle Conjecture (5.8), so the method of moments certainly seems a difficult avenue to pursue. Yet interestingly, Ellenberg and Venkatesh have shown that in this context the kth moments can be repackaged as *averages*.

Informally, the idea is to replace bounding the kth moment of the function $m(D)$ for G-Galois fields in a family $\mathscr{F}$ by counting fields in a family $\mathscr{F}^{(k)}$ of G^k-Galois fields. Ellenberg and Venkatesh order the fields in $\mathscr{F}^{(k)}$ not by discriminant D_K, but (roughly speaking) by the square-free kernel $D_K^\#$ of the discriminant. They generalize the Malle Conjecture to posit that in this ordering, $\ll X^{1+\varepsilon}$ fields in $\mathscr{F}^{(k)}$ have $D_K^\# \leq X$, *uniformly* for all integers $k \geq 1$. Assuming this conjecture, suppose there are $m(D)$ many G-Galois fields $K_1, \ldots, K_{m(D)}$ with $D_{K_i} = D$. Taking composita of k of these generates at least $\gg_k m(D)^k$ many G^k-Galois fields in the family $\mathscr{F}^{(k)}$, with $D_K^\# \leq D$. If we suppose $m(D) \geq D^\alpha$ for some $\alpha > 0$ and a sequence of $D \to \infty$, under the generalized Malle Conjecture it must be that $\alpha k \leq 1$ for all $k \geq 1$. Hence α must be arbitrarily small, as desired.

In full generality, Ellenberg and Venkatesh propose a generalized Malle Conjecture in terms of an f-discriminant, for any rational class function f, and an appropriate generalization $a_G(f)$ of the exponent in (5.8). They verify that for a particular choice of f, this implies the Discriminant Multiplicity Conjecture. More recently, Klüners and Wang have shown directly that Malle's Conjecture (5.8) for all groups G implies the Discriminant Multiplicity Conjecture (also over any number field) [54].

Let us sum up: the upper bound (5.8) in Malle's Conjecture for all groups G implies the Discriminant Multiplicity Conjecture. The Discriminant Multiplicity Conjecture implies the ℓ-torsion Conjecture. Also, the Discriminant Multiplicity Conjecture for $\mathscr{F}_n(G;X)$ (that is, property $\mathbf{D}_{\mathscr{F},n}(0)$) implies Malle's Conjecture for $\mathscr{F}_n(G;X)$. Moreover, there is one more converse: Alberts has shown that if the ℓ-torsion Conjecture is true for all solvable extensions and all primes ℓ (even just in an average sense), then Malle's upper bound (5.8) holds for all solvable groups [1]. Thus Malle's Conjecture, the Discriminant Multiplicity Conjecture, and the ℓ-torsion Conjecture are truly equivalent, when restricted to solvable groups. These relationships provide clear motivation for why so many methods described in this survey have involved counting number fields.

In conclusion, we have seen from three different perspectives that the ℓ-torsion Conjecture should be true. But as Gauss wrote, "*Demonstrationes autem* rigorosae *harum observationum perdifficiles esse videntur*."

FUNDING

This work was partially supported by NSF CAREER DMS-1652173.

REFERENCES

[1] B. Alberts, The weak form of Malle's conjecture and solvable groups. *Res. Number Theory* **6** (2020), no. 1, 23, Paper No. 10.

[2] C. An, ℓ-torsion in class groups of certain families of D_4-quartic fields. *J. Théor. Nombres Bordeaux* **32** (2020), no. 1, 1–23.

[3] A. M. Baily, On the density of discriminants of quartic fields. *J. Reine Angew. Math.* **315** (1980), 190–210.

[4] O. Beckwith, Indivisibility of class numbers of imaginary quadratic fields. *Res. Math. Sci.* **4** (2017), 11, Paper No. 20.

[5] K. Belabas, M. Bhargava, and C. Pomerance, Error estimates for the Davenport–Heilbronn theorems. *Duke Math. J.* **153** (2010), no. 1, 173–210.

[6] M. Belolipetsky and A. Lubotzky, Counting non-uniform lattices. *Israel J. Math.* **232** (2019), no. 1, 201–229.

[7] M. Bhargava, The density of discriminants of quartic rings and fields. *Ann. of Math.* **162** (2005), no. 2, 1031–1063.

[8] M. Bhargava, Higher composition laws and applications. In *ICM Proceedings. Vol. II*, pp. 271–294, Eur. Math. Soc., Zürich, 2006.

[9] M. Bhargava, The density of discriminants of quintic rings and fields. *Ann. of Math.* **172** (2010), no. 3, 1559–1591.

[10] M. Bhargava, A. Shankar, T. Taniguchi, F. Thorne, J. Tsimerman, and Y. Zhao, Bounds on 2-torsion in class groups of number fields and integral points on elliptic curves. *J. Amer. Math. Soc.* **33** (2020), no. 4, 1087–1099.

[11] M. Bhargava, A. Shankar, and J. Tsimerman, On the Davenport–Heilbronn theorems and second order terms. *Invent. Math.* **193** (2013), no. 2, 439–499.

[12] M. Bhargava, A. Shankar, and X. Wang, Squarefree values of polynomial discriminants. 2016, arXiv:1611.09806v2.

[13] M. Bhargava and I. Varma, On the mean number of 2-torsion elements in the class groups, narrow class groups, and ideal groups of cubic orders and fields. *Duke Math. J.* **164** (2015), 1911–1933.

[14] M. Bhargava and I. Varma, The mean number of 3-torsion elements in the class groups and ideal groups of quadratic orders. *Proc. Lond. Math. Soc.* **112** (2016), 235–266.

[15] E. Bombieri and J. Pila, The number of integral points on arcs and ovals. *Duke Math. J.* **59** (1989), no. 2, 337–357.

[16] A. Brumer and O. McGuinness, The behavior of the Mordell–Weil group of elliptic curves. *Bull. Amer. Math. Soc. (N.S.)* **23** (1990), no. 2, 375–382.

[17] A. Brumer and J. H. Silverman, The number of elliptic curves over $\mathbb{Q}$ with conductor N. *Manuscripta Math.* **91** (1996), no. 1, 95–102.

[18] F. Brumley, J. Thorner, and A. Zaman, Zeros of Rankin–Selberg L-functions at the edge of the critical strip. *J. Eur. Math. Soc. (JEMS)* (2021). DOI 10.4171/JEMS/1134.

[19] F. Calegari, The Artin conjecture for some S_5-extensions. *Math. Ann.* **356** (2013), no. 1, 191–207.

[20] H. Cohen, F. Diaz y Diaz, and M. Olivier, Enumerating quartic dihedral extensions of $\mathbb{Q}$. *Compos. Math.* **133** (2002), no. 1, 65–93.

[21] H. Cohen and H. W. Lenstra Jr., Heuristics on class groups of number fields. In *Number theory*, pp. 33–62, Lecture Notes in Math., Springer, Berlin, 1984.

[22] H. Cohen and J. Martinet, Étude heuristique des groupes de classes des corps de nombres. *J. Reine Angew. Math.* **404** (1990), 39–76.

[23] H. Cohen and F. Thorne, On D_ℓ-extensions of odd prime degree ℓ. *Proc. Lond. Math. Soc. (3)* **121** (2020), no. 5, 1171–1206.

[24] H. Cohn, The density of abelian cubic fields. *Proc. Amer. Math. Soc.* **5** (1954), 476–477.

[25] J.-M. Couveignes, Enumerating number fields. *Ann. of Math.* **192** (2020), no. 2, 487–497.

[26] D. A. Cox, *Primes of the form $x^2 + ny^2$. 2nd edn.* Pure Appl. Math., John Wiley & Sons, Inc., Hoboken, NJ, 2013.

[27] H. Davenport, *Multiplicative number theory. 3rd edn.* Grad. Texts in Math. 74, Springer, 2000.

[28] H. Davenport and H. Heilbronn, On the density of discriminants of cubic fields II. *Proc. R. Soc. Lond. A* **322** (1971), 405–420.

[29] W. Duke, Bounds for arithmetic multiplicities. In *Proceedings of the International Congress of Mathematicians, Doc. Math., Jahresber. Dtsch. Math.-Ver., Vol. II (Berlin, 1998)*, pp. 163–172, 1998.

[30] W. Duke and E. Kowalski, A problem of Linnik for elliptic curves and mean-value estimates for automorphic representations. *Invent. Math.* **139** (2000), 1–39.

[31] H. M. Edwards, *Fermat's last theorem.* Grad. Texts in Math. 50, Springer, New York, 1996.

[32] J. Ellenberg, Points of low height on $\mathbb{P}^1$ over number fields and bounds for torsion in class groups. In *Computational arithmetic geometry*, pp. 45–48, Contemp. Math. 463, AMS, Providence, 2008.

[33] J. Ellenberg, L. B. Pierce, and M. M. Wood, On ℓ-torsion in class groups of number fields. *Algebra Number Theory* **11** (2017), 1739–1778.

[34] J. Ellenberg and A. Venkatesh, Reflection principles and bounds for class group torsion. *Int. Math. Res. Not. IMRN* **1** (2007), rnm002, 18 pp.

[35] J. Ellenberg and A. Venkatesh, Counting extensions of function fields with bounded discriminant and specified Galois group. In *Prog. Math*, pp. 151–168, 235, Birkhäuser Boston, Boston, MA, 2005.

[36] J. Ellenberg and A. Venkatesh, The number of extensions of a number field with fixed degree and bounded discriminant. *Ann. of Math.* **163** (2006), no. 2, 723–741.

[37] K. B. Ford, Vinogradov's integral and bounds for the Riemann zeta function. *Proc. Lond. Math. Soc. (3)* **85** (2002), no. 3, 565–633.

[38] É. Fouvry and J. Klüners, On the 4-rank of class groups of quadratic number fields. *Invent. Math.* **167** (2006), no. 3, 455–513.

[39] É. Fouvry, M. Nair, and G. Tenenbaum, L'ensemble exceptionnel dans la conjecture de Szpiro. *Bull. Soc. Math. France* **120** (1992), no. 4, 485–506.

[40] C. Frei and M. Widmer, Average bounds for the ℓ-torsion in class groups of cyclic extensions. *Res. Number Theory* **4** (2018), no. 3, 25, Paper No. 34.

[41] C. Frei and M. Widmer, Averages and higher moments for the ℓ-torsion in class groups. *Math. Ann.* **379** (2021), no. 3–4, 1205–1229.

[42] D. Goldfeld, Gauss's class number problem for imaginary quadratic fields. *Bull. Amer. Math. Soc. (N.S.)* **13** (1985), no. 1, 23–37.

[43] C. Goldstein, N. Schappacher, and J. Schwermer (eds.), The shaping of arithmetic after C. F. Gauss's "Disquisitiones arithmeticae", Springer, Berlin, 2007.

[44] D. R. Heath-Brown, Quadratic class numbers divisible by 3. *Funct. Approx. Comment. Math.* **37(part 1)** (2007), 203–211.

[45] D. R. Heath-Brown, Imaginary quadratic fields with class group exponent 5. *Forum Math.* **20** (2008), 275–283.

[46] D. R. Heath-Brown and L. B. Pierce, Averages and moments associated to class numbers of imaginary quadratic fields. *Compos. Math.* **153** (2017), 2287–2309.

[47] H. Heilbronn, On real simple zeros of Dedekind ζ-functions. In *Proceedings of the number theory conference*, pp. 108–110, Univ. Colorado, Boulder, 1972.

[48] H. A. Helfgott and A. Venkatesh, Integral points on elliptic curves and 3-torsion in class groups. *J. Amer. Math. Soc.* **19** (2006), no. 3, 527–550.

[49] P. Humphries and J. Thorner, Zeros of Rankin–Selberg L-functions in families. 2021, arXiv:2103.05634.

[50] J. Klüners, A counterexample to Malle's conjecture on the asymptotics of discriminants. *C. R. Math. Acad. Sci. Paris* **340** (2005), 411–414.

[51] J. Klüners, Asymptotics of number fields and the Cohen–Lenstra heuristics. *J. Théor. Nombres Bordeaux* **18** (2006), 607–615.

[52] J. Klüners, The number of S_4-fields with given discriminant. *Acta Arith.* **122** (2006), no. 2, 185–194.

[53] J. Klüners, The distribution of number fields with wreath products as Galois groups. *Int. J. Number Theory* **8** (2012), no. 3, 845–858.

[54] J. Klüners and J. Wang, ℓ-torsion bounds for the class group of number fields with an ℓ-group as galois group. 2020, arXiv:2003.12161.

[55] J. Klys, The distribution of p-torsion in degree p cyclic fields. *Algebra Number Theory* **14** (2020), no. 4, 815–854.

[56] E. Kowalski and P. Michel, Zeros of families of automorphic L-functions close to 1. *Pacific J. Math.* **207** (2002), no. 2, 411–431.

[57] P. Koymans and D. Milovic, On the 16-rank of class groups of $\mathbb{Q}(\sqrt{-2p})$ for primes $p \equiv 1 \mod 4$. *Int. Math. Res. Not. IMRN* **23** (2019), 7406–7427.

[58] P. Koymans and D. Milovic, Joint distribution of spins. *Duke Math. J.* **170** (2021), no. 8, 1723–1755.

[59] P. Koymans and C. Pagano, On the distribution of $\mathrm{Cl}(K)[\ell^\infty]$ for degree ℓ cyclic fields. 2018, arXiv:1812.06884.

[60] J. Lagarias and A. Odlyzko, Effective versions of the Chebotarev density theorem. In *Proc. Sympos.*, pp. 409–464, Univ. Durham, Durham, 1975.

[61] R. Lemke Oliver and F. Thorne, Upper bounds on number fields of given degree and bounded discriminant. 2020, arXiv:2005.14110.

[62] R. Lemke Oliver, J. Thorner, and A. Zaman, An approximate form of Artin's holomorphy conjecture and non-vanishing of Artin L-functions. 2021, arXiv:2012.14422v3.

[63] R. Lemke Oliver, J. Wang, and M. M. Wood, The average size of 3-torsion in class groups of 2-extensions. 2021, arXiv:2110.07712v2.

[64] S. Mäki, On the density of abelian number fields. *Ann. Acad. Sci. Fenn. Ser. A I Math. Diss.* **54** (1985), 104.

[65] G. Malle, On the distribution of Galois groups. *J. Number Theory* **92** (2002), 315–219.

[66] G. Malle, On the distribution of Galois groups II. *Exp. Math.* **13** (2004), 129–135.

[67] R. Masri, F. Thorne, W.-L. Tsai, and J. Wang, Malle's Conjecture for $G \times A$, with $G = S_3, S_4, S_5$. 2020, arXiv:2004.04651.

[68] W. Narkiewicz, *Elementary and analytic theory of algebraic numbers. 2nd edn.* Springer, Berlin, 1990.

[69] J. Neukirch, *Algebraic number theory*. Springer, 1999.

[70] L. B. Pierce, The 3-part of class numbers of quadratic fields. *J. Lond. Math. Soc.* **71** (2005), 579–598.

[71] L. B. Pierce, A bound for the 3-part of class numbers of quadratic fields by means of the square sieve. *Forum Math.* **18** (2006), 677–698.

[72] L. B. Pierce, C. Turnage-Butterbaugh, and M. M. Wood, An effective Chebotarev density theorem for families of number fields, with an application to ℓ-torsion in class groups. *Invent. Math.* **219** (2020), 701–778.

[73] L. B. Pierce, C. Turnage-Butterbaugh, and M. M. Wood, On a conjecture for ℓ-torsion in class groups of number fields: from the perspective of moments. *Math. Res. Lett.* **28** (2021), 575–621.

[74] W. M. Ruppert, Small generators of number fields. *Manuscripta Math.* **96** (1998), no. 1, 17–22.

[75] W. M. Schmidt, Number fields of given degree and bounded discriminant. *Astérisque* **(228):4** (1995), 189–195.

[76] A. Scholz, Über die Beziehung der Klassenzahlen quadratischer Körper. *J. Reine Angew. Math.* **166** (1932), 201–203.

[77] J.-P. Serre, Quelques applications du théorème de densité de Chebotarev. *Publ. Math. Inst. Hautes Études Sci.* **54** (1982), 123–201.

[78] J.-P. Serre, *Lectures on the Mordell–Weil theorem. 3rd edn*. Friedr. Vieweg and Sohn, Braunschweig, 1997.

[79] A. Shankar and J. Tsimerman, Counting S_5-fields with a power saving error term. *Forum Math. Sigma* **2** (2014).

[80] J. H. Silverman, Lower bounds for height functions. *Duke Math. J.* **51** (1984), no. 2, 395–403.

[81] A. Smith, 2^∞-Selmer groups, 2^∞-class groups, and Goldfeld's conjecture. 2017, arXiv:1702.02325v2.

[82] K. Soundararajan, Divisibility of class numbers of imaginary quadratic fields. *J. Lond. Math. Soc. (2)* **61** (2000), no. 3, 681–690.

[83] H. M. Stark, Some effective cases of the Brauer–Siegel theorem. *Invent. Math.* **23** (1974), 135–152.

[84] H. M. Stark, The Gauss class-number problems. In *Analytic number theory*, pp. 247–256, Clay Math. Proc. 7, Amer. Math. Soc., 2007.

[85] T. Taniguchi and F. Thorne, The secondary term in the counting function for cubic fields. *Duke Math. J.* **162** (2013), 2451–2508.

[86] J. Thorner and A. Zaman, A zero density estimate for Dedekind zeta functions. 2019, arXiv:1909.01338.

[87] J. Thorner and A. Zaman, An unconditional GL_n large sieve. *Adv. Math.* **378** (2021), 107529.

[88] N. Tschebotareff, Die Bestimmung der Dichtigkeit einer Menge von Primzahlen, welche zu einer gegebenen Substitutionsklasse gehören. *Math. Ann.* **95** (1926), no. 1, 191–228.

[89] J. Wang, Pointwise bound for ℓ-torsion in class groups II: nilpotent extensions. 2020, arXiv:2006.10295.

[90] J. Wang, Malle's conjecture for $S_n \times A$ for $n = 3, 4, 5$. *Compos. Math.* **157** (2021), no. 1, 83–121.

[91] J. Wang, Pointwise bound for ℓ-torsion in class groups: elementary abelian extensions. *J. Reine Angew. Math.* **773** (2021), 129–151.

[92] W. Wang and M. M. Wood, Moments and interpretations of the Cohen–Lenstra–Martinet heuristics. *Comment. Math. Helv.* **96** (2021), no. 2, 339–387.

[93] M. Watkins, Class numbers of imaginary quadratic fields. *Math. Comp.* **73** (2004), no. 246, 907–938.

[94] M. Watkins, Some heuristics about elliptic curves. *Exp. Math.* **17** (2008), no. 1, 105–125.

[95] M. Widmer, Bounds for the ℓ-torsion in class groups. *Bull. Lond. Math. Soc.* **50** (2018), no. 1, 124–131.

[96] S. Wong, Exponents of class groups and elliptic curves, corrigendum. *J. Number Theory* **90** (2001), no. 2, 376–377.

[97] D. Wright, Distribution of discriminants of abelian extensions. *Proc. Lond. Math. Soc.* **58** (1989), 17–50.

[98] S.-W. Zhang, Equidistribution of CM-points on quaternion Shimura varieties. *Int. Math. Res. Not.* **59** (2005), 3657–3689.

LILLIAN B. PIERCE

Mathematics Department, Duke University, Durham, NC 27708, USA, pierce@math.duke.edu

POINTS ON SHIMURA VARIETIES MODULO PRIMES

SUG WOO SHIN

ABSTRACT

We survey recent developments on the Langlands–Rapoport conjecture for Shimura varieties modulo primes and an analogous conjecture for Igusa varieties. We discuss resulting implications on the automorphic decomposition of the Hasse–Weil zeta functions and ℓ-adic cohomology of Shimura varieties, along with further applications to the Langlands correspondence and related problems.

MATHEMATICS SUBJECT CLASSIFICATION 2020

Primary 11G18; Secondary 11F72, 11R39

KEYWORDS

Shimura varieties, Igusa varieties, trace formula, cohomology

INTERNATIONAL CONGRESS OF MATHEMATICIANS 2022 JULY 6–14

Proc. Int. Cong. Math. 2022, Vol. 3, pp. 1966–1988
DOI 10.4171/ICM2022/62

1. INTRODUCTION

Shimura varieties have been vital to number theory for their intrinsic beauty and wide-ranging applications. They are simultaneously locally symmetric spaces and quasi-projective varieties over number fields, serving as a geometric bridge between automorphic forms and arithmetic. This feature has been particularly fruitful in the Langlands program.

This paper concentrates on the problem of understanding the Hasse–Weil zeta functions and ℓ-adic cohomology of Shimura varieties following the approach due to Langlands, Kottwitz, Rapoport, and others. As such, we are naturally led to study integral models and special fibers of Shimura varieties at each prime (Section 2), as epitomized by the Langlands–Rapoport (LR) conjecture (Section 3). Below is a partial summary of this article:

$$\begin{array}{ccccccc}
\mathrm{LR}_1(\mathrm{Sh}) & \Longleftarrow & \boxed{\mathrm{LR}_0(\mathrm{Sh})} & \Longrightarrow & \mathrm{TF}(\mathrm{Sh}) & \Longrightarrow & \mathrm{app}(\mathrm{Sh}) \\
 & & \Big\Downarrow & & & & \Big\updownarrow \\
 & & \mathrm{LR}_0(\mathrm{Ig}) & \Longrightarrow & \mathrm{TF}(\mathrm{Ig}) & \Longrightarrow & \mathrm{app}(\mathrm{Ig}).
\end{array}$$

Central to this paper is Theorem 3.2, asserting that $\mathrm{LR}_0(\mathrm{Sh})$, a version of the LR conjecture, is true for Shimura varieties of abelian type with good reduction. This is a strengthening of another version $\mathrm{LR}_1(\mathrm{Sh})$ which was previously verified by Kisin. Even though $\mathrm{LR}_0(\mathrm{Sh})$ is weaker than the original LR conjecture (still wide open), it opens doors for most applications. Indeed, the diagram shows how $\mathrm{LR}_0(\mathrm{Sh})$ implies a (stabilized) trace formula for cohomology of Shimura varieties, designated as $\mathrm{TF}(\mathrm{Sh})$, which in turn leads to interesting applications (Section 4). In Sections 5–6, we survey related problems and directions in the bad reduction case. Finally in Section 7, we review a parallel story for Igusa varieties, where $\mathrm{LR}_0(\mathrm{Sh})$ provides a key ingredient for proving the analogous assertion $\mathrm{LR}_0(\mathrm{Ig})$ for Igusa varieties. The dotted vertical arrow suggests that interactions occur between certain applications to Shimura and Igusa varieties, e.g., through Mantovan's formula.

Conventions

Unless otherwise stated, cohomology means the ℓ-adic étale cohomology with $\overline{\mathbb{Q}}_l$-coefficients. For an inverse limit of varieties $X = (X_i)$ over a field k, we write $H(X, \overline{\mathbb{Q}}_l)$ for $\varinjlim_i H(X_i \times_k \bar{k}, \overline{\mathbb{Q}}_l)$, with $\bar{k}$ a separable closure of k; likewise for cohomology with compact support. We adhere to cohomology with constant coefficients for simplicity, though the discussed results are valid more generally. In the LR conjecture, we omit $Z_G(\mathbb{Q}_p)$-equivariance to keep the statements simple. If Γ is a topological group, $\mathcal{H}(\Gamma)$ is the Hecke algebra of locally constant compactly supported functions on Γ. We write $\mathbb{A}^\infty = \hat{\mathbb{Z}} \otimes_{\mathbb{Z}} \mathbb{Q}$ for the ring of finite adèles, and $\mathbb{A}^{\infty,p}$ for the analogous ring without the p-component. Put $\breve{\mathbb{Z}}_p := W(\overline{\mathbb{F}}_p)$ for the Witt ring of $\overline{\mathbb{F}}_p$, and $\breve{\mathbb{Q}}_p := \breve{\mathbb{Z}}_p[1/p]$. Denote by σ the Frobenius operator on $\breve{\mathbb{Q}}_p$ or a finite unramified extension of $\mathbb{Q}_p$. When we have cohomology spaces $H^i(X)$ (supported on finitely many i's) with a group action, denote by $[H(X)] = \sum_{i \geq 0} (-1)^i H^i(X)$ the alternating sum viewed in a suitable Grothendieck group of representations. For an algebraic group G over $\mathbb{Q}$ and a field k over $\mathbb{Q}$, write $G_k := G \times_{\operatorname{Spec} \mathbb{Q}} \operatorname{Spec} k$. We quietly fix field embeddings $\overline{\mathbb{Q}} \hookrightarrow \mathbb{C}$, $\overline{\mathbb{Q}}_v \hookrightarrow \mathbb{C}$ at each place v of $\mathbb{Q}$, and identify the residue field of $\overline{\mathbb{Q}}_p$ with $\overline{\mathbb{F}}_p$.

2. SHIMURA VARIETIES WITH GOOD REDUCTION

Let G be a connected reductive group over $\mathbb{Q}$, and X a $G(\mathbb{R})$-conjugacy class of $\mathbb{R}$-group morphisms $\mathrm{Res}_{\mathbb{C}/\mathbb{R}}\mathbb{G}_m \to G_{\mathbb{R}}$. We say that (G, X) is a Shimura datum if it satisfies axioms (2.1.1.1)–(2.1.1.3) of **[10]**. Each (G, X) determines a conjugacy class of cocharacters $\mu : \mathbb{G}_m \to G_{\mathbb{C}}$ over $\mathbb{C}$, whose field of definition is a number field $E = E(G, X) \subset \mathbb{C}$. There is an obvious notion of morphisms between Shimura data.

Thanks to Shimura, Deligne, Borovoi, and Milne, we have a $G(\mathbb{A}^\infty)$-scheme Sh over E (in the sense of **[10, 2.7.1]**, cf. **[31, 1.5.1]**), which is a projective limit of quasiprojective varieties over E with a $G(\mathbb{A}^\infty)$-action. If (G, X) is a Siegel datum, i.e., $G = \mathrm{GSp}_{2n}$ and X is realized by the Siegel half-spaces of genus n for some $n \in \mathbb{Z}_{\geq 1}$, then we obtain (a projective limit of) Siegel modular varieties as output. There is a hierarchy of Shimura data:

$$\text{(PEL type)} \subset \text{(Hodge type)} \subset \text{(abelian type)} \subset \text{(all)}.$$

Roughly speaking, Shimura varieties coming from PEL-type data are realized as moduli spaces of abelian varieties with polarizations (P), endomorphisms (E), and level (L) structures.[1] This case includes modular curves and, more generally, Siegel modular varieties. A Shimura datum of Hodge type embeds in a Siegel datum by definition, and the corresponding Shimura varieties embed in Siegel modular varieties. Abelian-type data are generalized from those of Hodge type to cover the case when the Dynkin diagram of $G_{\overline{\mathbb{Q}}}$ consists of only types A, B, C, and D, with a small exception in the type D case, cf. **[10, §2.3]**.

Now we turn to integral models of Shimura varieties in the good reduction case. A starting point is an *unramified Shimura datum* $(G, X, p, \mathcal{G})$, where (G, X) is a Shimura datum, p is a prime, and $\mathcal{G}$ is a reductive model of G over $\mathbb{Z}_p$. The existence of $\mathcal{G}$ is equivalent to the condition that $G_{\mathbb{Q}_p}$ is an unramified group (i.e., quasisplit over $\mathbb{Q}_p$ and split over an unramified extension of $\mathbb{Q}_p$). Now we put $K_p := \mathcal{G}(\mathbb{Z}_p)$ and consider the $G(\mathbb{A}^{\infty,p})$-scheme Sh_{K_p} over E, which is similar to Sh as above but has a fixed level K_p at p (while the level subgroup away from p varies). Kisin **[28]** ($p > 2$) and Kim–Madapusi Pera **[27]** ($p = 2$) proved the following fundamental result.

Theorem 2.1. *If (G, X) is of abelian type, then there exists a canonical integral model $\mathscr{S}_{K_p}$, which is an $\mathcal{O}_{E,(p)}$-scheme with a $G(\mathbb{A}^{\infty,p})$-action, such that the generic fiber of $\mathscr{S}_{K_p}$ is $G(\mathbb{A}^{\infty,p})$-equivariantly isomorphic to Sh_{K_p}.*

Here "canonical" means that $\mathscr{S}_{K_p}$ is formally smooth over $\mathcal{O}_{E,(p)}$ and satisfies the extension property of **[28, (2.3.7)]**, which characterizes $\mathscr{S}_{K_p}$ uniquely up to a unique isomorphism. The proof of the theorem reduces to the Hodge-type case and utilizes the

1 A caveat is that such a moduli space is in general a finite disjoint union of Shimura varieties due to a possible failure of the Hasse principle for G. See **[34, §8]** for details.

known canonical integral models in the Siegel case. Kisin constructs $\mathscr{S}_{K_p}$ by normalizing the closure of Sh_{K_p} in an ambient Siegel modular variety. The key point is to show formal smoothness of $\mathscr{S}_{K_p}$ over $\mathcal{O}_{E,(p)}$ by deformation theory and integral p-adic Hodge theory. The existence of canonical integral models is completely open beyond the abelian-type case.

3. THE LANGLANDS–RAPOPORT CONJECTURE

Given an unramified Shimura datum, the Langlands–Rapoport (LR) conjecture consists of two parts: (i) the existence of canonical integral models and (ii) a group-theoretic description of $\overline{\mathbb{F}}_p$-points of such integral models. We already addressed (i) in Section 2, which is a prerequisite for discussing (ii) in this section. There is an instructive analogy between (ii) and a description of $\mathbb{C}$-points **[47, §16]**. See Section 6.1 below for the case of bad reduction. We recommend the introduction of **[31]** for a more detailed survey of the content in this section.

3.1. Galois gerbs

Let k be a perfect field with an algebraic closure $\bar{k}$. A *Galois gerb* over k consists of a pair $(G, \mathfrak{G})$, where G is a connected linear algebraic group over $\bar{k}$, and $\mathfrak{G}$ is a topological group extension (with discrete topology on $G(\bar{k})$ and profinite topology on $\mathrm{Gal}(\bar{k}/k)$),

$$1 \to G(\bar{k}) \stackrel{i}{\to} \mathfrak{G} \stackrel{\pi}{\to} \mathrm{Gal}(\bar{k}/k) \to 1, \tag{3.1}$$

such that (i) for every $g \in \mathfrak{G}$, the conjugation by g on $G(\bar{k})$ is induced by a $\bar{k}$-group isomorphism $\pi(g)^*G \xrightarrow{\sim} G$, and (ii) there exists a finite extension K/k in $\bar{k}$ such that π admits a continuous section over $\mathrm{Gal}(\bar{k}/K)$. If G is a torus, then (ii) determines a model of G over k. We often refer to $(G, \mathfrak{G})$ as $\mathfrak{G}$ and write $\mathfrak{G}^{\Delta}$ for G.

There is a natural notion of morphisms between Galois gerbs over k. Passing to projective limits, we define pro-Galois gerbs $(G, \mathfrak{G})$ over k, which still fit in (3.1) but with G a pro-algebraic group over $\bar{k}$. When $\mathfrak{G}$ is a (pro-)Galois gerb over $\mathbb{Q}$, we can localize it at each place v of $\mathbb{Q}$ to obtain a (pro-)Galois gerb over $\mathbb{Q}_v$, to be denoted by $\mathfrak{G}(v)$.

The most basic example is the neutral Galois gerb $\mathfrak{G}_G$ which arises when G is already defined over k. By definition, $\mathfrak{G}_G := G(\bar{k}) \rtimes \mathrm{Gal}(\bar{k}/k)$ as a semidirect product with the natural action of $\mathrm{Gal}(\bar{k}/k)$ on $G(k)$.

We introduce a (pro-)Galois gerb $\mathfrak{G}_v$ over $\mathbb{Q}_v$ at each v. Take $\mathfrak{G}_\infty$ to be the real Weil group (in particular, $\mathfrak{G}_\infty^{\Delta} = \mathbb{G}_{m,\mathbb{C}}$); the definition of $\mathfrak{G}_p$ is involved but intended to encode isocrystals. For $v \neq p, \infty$, put $\mathfrak{G}_v := \mathrm{Gal}(\overline{\mathbb{Q}}_v/\mathbb{Q}_v)$, namely the trivial neutral Galois gerb.

Central to the LR conjecture is a quasimotivic pro-Galois gerb $\mathfrak{Q}$ over $\mathbb{Q}$ whose algebraic part $\mathfrak{Q}^{\Delta}$ is a pro-torus. The gerb $\mathfrak{Q}$ comes equipped with morphisms $\zeta_v : \mathfrak{G}_v \to \mathfrak{Q}(v)$, and the datum $(\mathfrak{Q}, \{\zeta_v\})$ is uniquely characterized up to a suitable equivalence. A quasimotivic gerb (more precisely, its quotient called a pseudomotivic gerb) is devised as a substitute for the Galois gerb which should arise via Tannaka duality from the category of motives over $\overline{\mathbb{F}}_p$. The morphisms ζ_v should come from the fiber functors on the latter category coming from cohomology and polarization structures. See Langlands–Rapoport **[41, §§3–4]** (complemented by **[56, §8]**) and **[58, B2.7, B2.8]** for further information.

For each torus T over $\mathbb{Q}$ and each cocharacter $\mu : \mathbb{G}_m \to T$ (defined over a finite extension of $\mathbb{Q}$), there is a recipe **[29, (3.1.10)]** to define a morphism

$$\Psi_{T,\mu} : \mathfrak{Q} \to \mathfrak{G}_T. \tag{3.2}$$

As a special case, if (T, h) is a toral Shimura datum, then we obtain Ψ_{T,μ_h} with $\mu_h : \mathbb{G}_m \to T$ coming from h. In terms of the heuristics for $\mathfrak{Q}$, the construction of Ψ_{T,μ_h} mirrors the operation of taking the mod p fiber of a CM abelian variety in characteristic 0.

3.2. Versions of the LR conjecture

Let $(G, X, p, \mathcal{G})$ be an unramified Shimura datum. Write $\mathfrak{p}$ for the prime of E over p, determined by the field embeddings in Section 1, with residue field $k(\mathfrak{p})$. A canonical integral model $\mathscr{S}_{K_p}$ over $\mathcal{O}_{E_\mathfrak{p}}$ is available in the abelian-type case (Theorem 2.1) and conjectured to exist in general. For the moment, we assume (G, X) to be of Hodge type. Then we can take the partition

$$\mathscr{S}_{K_p}(\overline{\mathbb{F}}_p) = \coprod_{\mathscr{I} \in \mathbb{I}} S(\mathscr{I}) \tag{3.3}$$

according to the set $\mathbb{I}$ of isogeny classes, and then parametrize the set $S(\mathscr{I})$ consisting of points in each isogeny class $\mathscr{I}$ relative to a "base point" of choice in $\mathscr{I}$. This was obtained by Kisin **[29, §1.4]**, where a subtlety in the notion of p-power isogenies was handled by a result on the connected components of affine Deligne–Lusztig varieties **[9]**. Each $S(\mathscr{I})$ is $\Phi^{\mathbb{Z}} \times G(\mathbb{A}^{\infty,p})$-stable, where Φ acts as the geometric Frobenius over $k(\mathfrak{p})$, and

$$S(\mathscr{I}) \cong \varprojlim_{K^p \subset G(\mathbb{A}^{\infty,p})} I_{\mathscr{I}}(\mathbb{Q}) \backslash \big(X_p(\mathscr{I}) \times X^p(\mathscr{I})/K^p\big), \tag{3.4}$$

as a right $\Phi^{\mathbb{Z}} \times G(\mathbb{A}^{\infty,p})$-set, where $X_p(\mathscr{I})$ and $X^p(\mathscr{I})$ account for p-power and prime-to-p isogenies (from a base point). The quotient by $I_{\mathscr{I}}(\mathbb{Q})$ takes care of redundant counting up to self-isogenies. Since (G, X) is of Hodge type, (3.4) simplifies as $I_{\mathscr{I}}(\mathbb{Q}) \backslash (X_p(\mathscr{I}) \times X^p(\mathscr{I}))$.

We return to general unramified Shimura data. Following **[29, (3.3.6)]**, one defines *admissible morphisms* as morphisms $\phi : \mathfrak{Q} \to \mathfrak{G}_G$ satisfying certain conditions to ensure that ϕ contributes to $\mathscr{S}_{K_p}(\overline{\mathbb{F}}_p)$. In analogy with isogeny classes above, ϕ gives rise to a right $G(\mathbb{A}^{\infty,p})$-torsor $X^p(\phi)$ and a nonempty affine Deligne–Lusztig variety $X_p(\phi)$ with a $\Phi^{\mathbb{Z}}$-action, where Φ is the Frobenius operator. Write I_ϕ for the $\mathbb{Q}$-group of automorphisms of ϕ. Then $I_\phi(\mathbb{A}^\infty)$ naturally acts on $X_p(\phi) \times X^p(\phi)$. (This is analogous to the self-isogeny of an abelian variety over $\overline{\mathbb{F}}_p$ acting on its étale cohomology away from p and crystalline cohomology at p.) Now let I_ϕ^{ad} denote the $\mathbb{Q}$-group of inner automorphisms of I_ϕ. Each $\tau \in I_\phi^{\mathrm{ad}}(\mathbb{A}^\infty)$ can be used to twist the natural action of $I(\mathbb{Q})$ on $X_p(\phi) \times X^p(\phi)$:

$$I(\mathbb{Q}) \subset I\big(\mathbb{A}^\infty\big) \xrightarrow{\tau} I\big(\mathbb{A}^\infty\big) \circlearrowright X_p(\phi) \times X^p(\phi).$$

Taking the left quotient by this action (denoted $\backslash_\tau$ below), we define a $\Phi^{\mathbb{Z}} \times G(\mathbb{A}^{\infty,p})$-set

$$S_\tau(\phi) := \varprojlim_{K^p \subset G(\mathbb{A}^{\infty,p})} I_\phi(\mathbb{Q}) \backslash_\tau \big(X_p(\phi) \times X^p(\phi)/K^p\big). \tag{3.5}$$

We just write $S(\phi)$ if τ is trivial. The isomorphism class of $S_\tau(\phi)$ depends only on $[\tau] \in \mathscr{H}(\phi) := I_\phi^{\mathrm{ad}}(\mathbb{Q})\backslash I_\phi^{\mathrm{ad}}(\mathbb{A}^\infty)/I_\phi(\mathbb{A}^\infty)$ represented by τ. (The right quotient is taken with respect to the multiplication through the natural map $I_\phi \to I_\phi^{\mathrm{ad}}$.) If ϕ, ϕ' are $G(\overline{\mathbb{Q}})$-conjugate, then $S_\tau(\phi) \cong S_\tau(\phi')$ and canonically $\mathscr{H}(\phi) \cong \mathscr{H}(\phi')$. Denoting by $\mathbb{J}$ the set of $G(\overline{\mathbb{Q}})$-conjugacy classes of admissible morphisms, we write $\mathscr{H}(\mathcal{J})$ and $S_\tau(\mathcal{J})$ respectively for $\mathscr{H}(\phi)$ and $S_\tau(\phi)$, when $\mathcal{J}$ is the $G(\overline{\mathbb{Q}})$-conjugacy class of ϕ.

It is convenient to name a "rationality" condition on the adelic element $\tau \in I_\phi^{\mathrm{ad}}(\mathbb{A}^\infty)$ that is technical but useful. For each maximal torus T of I_ϕ over $\mathbb{Q}$, we have the maps

$$I_\phi^{\mathrm{ad}}(\mathbb{A}^\infty) \xrightarrow{\partial} H^1(\mathbb{A}^\infty, Z_{I_\phi}) \to H^1(\mathbb{A}^\infty, T),$$

where ∂ is the connecting homomorphism, and the second map is induced by $Z_{I_\phi} \subset T$. We say that τ is *tori-rational* if the image of τ in $H^1(\mathbb{A}^\infty, T)$ lies in the subset of the image of $H^1(\mathbb{Q}, T) \to H^1(\mathbb{A}^\infty, T)$ which maps trivially into the abelianized cohomology of G, for every T. This condition depends only on $[\tau] \in \mathscr{H}(\phi)$.

We are ready to state versions of the LR conjecture in increasing order of strength. (To be precise, the conjecture requires extra compatibility conditions on $\tau(\mathcal{J})$ under cohomological twistings of $\mathcal{J}$ in (LR_1) and (LR_0), but we avoid mentioning them explicitly in this exposition. See **[31, §§2.6–2.7]**, where these conditions correspond to $\underline{\tau} \in \Gamma(\mathscr{H})_1$ and $\underline{\tau} \in \Gamma(\mathscr{H})_0$, respectively. With this correction, the Langlands–Rapoport-τ conjecture therein is exactly (LR_0) below.)

Conjecture 3.1. *The following assertions hold true:*

(LR_1) *There exists a $\Phi^{\mathbb{Z}} \times G(\mathbb{A}^{\infty,p})$-equivariant bijection*

$$\mathscr{S}_{K_p}(\overline{\mathbb{F}}_p) \cong \coprod_{\mathcal{J} \in \mathbb{J}} S_{\tau(\mathcal{J})}(\mathcal{J}),$$

for some family of elements $\{\tau(\mathcal{J}) \in \mathscr{H}(\mathcal{J})\}_{\mathcal{J}\in\mathbb{J}}$.

(LR_0) *The conclusion of* (LR_1) *holds with $\tau(\mathcal{J})$ tori-rational for every $\mathcal{J} \in \mathbb{J}$.*

(LR) *The conclusion of* (LR_1) *holds with $\tau(\mathcal{J})$ trivial for every $\mathcal{J} \in \mathbb{J}$.*

Statement (LR) is nothing but the original LR conjecture. In the Hodge-type case (to which the abelian-type case can be reduced in practice), a natural approach in view of (3.3) is to establish a bijection $\mathcal{I} \in \mathbb{I} \leftrightarrow \mathcal{J} \in \mathbb{J}$ such that there exists a $\Phi^{\mathbb{Z}} \times G(\mathbb{A}^{\infty,p})$-equivariant bijection $S(\mathcal{I}) \cong S_{\tau(\mathcal{J})}(\mathcal{J})$ with constraints on $\tau(\mathcal{J})$ as in the conjecture.

It is known (**[31, §3]**, cf. **[42, 46]**) that (LR) implies (3.7) below, which is the gateway to applications, but (LR) remains to be completely open even for Siegel modular varieties (of genus ≥ 2). Milne proved that the original LR conjecture follows from the Hodge conjecture for CM abelian varieties (see **[48, P. 4]** and the references therein), but the latter conjecture is also wide open.

On a positive note, Kisin **[29]** made a major breakthrough to prove (LR_1) for all unramified Shimura data $(G, X, p, \mathcal{G})$ of abelian type. Unfortunately, (LR_1) by itself is not strong enough for the next steps. This motivated us to formulate and prove the strengthening

(LR_0) in [31], which suffices for the trace formula (Section 3.3) and applications (Section 4) below.

Theorem 3.2. *For every unramified Shimura datum $(G, X, p, \mathcal{G})$ of abelian type, Conjecture* (LR_0) *holds true.*

Let us sketch some ideas of proof when (G, X) is of Hodge type. The reduction to this case is nontrivial and convoluted, cf. [31, §6]. Already in [29], Kisin proved a refinement of (LR_1) in order to propagate (LR_1) through Deligne's formalism of connected Shimura varieties. With that said, we focus on the Hodge-type setting for simplicity.

The proof consists of two parts: (i) constructing a bijection $\mathscr{I} \in \mathbb{I} \leftrightarrow \mathscr{J} \in \mathbb{J}$ and (ii) showing that $S(\mathscr{I}) \cong S_{\tau(\mathscr{J})}(\mathscr{J})$ with some control over $\tau(\mathscr{J})$. A crucial idea is to use special point data, namely toral Shimura data (T, h_T) with embeddings into (G, X), to probe both sides of the bijection. Such data can be mapped into $\mathbb{I}$ by taking mod p of the corresponding special points on Sh_{K_p}, and to $\mathbb{J}$ by composing (3.2) with the induced embedding $\mathfrak{G}_T \hookrightarrow \mathfrak{G}_G$. The map to $\mathbb{I}$ is onto by Kisin [29], generalizing Honda's result on CM lifting of an abelian variety over $\overline{\mathbb{F}}_p$ up to isogeny. The surjectivity onto $\mathbb{J}$ is due to Langlands–Rapoport [41]:

$$\begin{array}{ccccc} \{(T, h_T) \hookrightarrow (G, X)\} & \xrightarrow{\text{Kisin}} \!\!\!\!\rightarrow & \mathbb{I} = \{\text{isog. classes}\} & \longrightarrow & \{(\gamma_0, \gamma, \delta)\}/\sim \\ \text{special points data} & & \Big\updownarrow & \nearrow & \text{Kottwitz triples} \\ & \underset{\text{Langlands–Rapoport}}{\searrow\!\!\!\searrow} & \mathbb{J} = \{\text{conj. classes}\} & & \end{array} \tag{3.6}$$

From each of $\mathbb{I}$ and $\mathbb{J}$, Kisin [29] constructed Kottwitz triples consisting of certain conjugacy classes on G up to an equivalence, and showed that the outer diagram above commutes. This determines a bijection $\mathbb{I} \cong \mathbb{J}$ up to a finite ambiguity since the maps to Kottwitz triples have finite fibers. However, $S(\mathscr{I})$ and $S(\mathscr{J})$ need not be isomorphic through this bijection, since Kottwitz triples forget part of their structures. The possible deviation is recorded by $\tau(\mathscr{J})$, which is a priori under little control. This is still enough for deducing (LR_1).

To prove (LR_0), various refinements and improvements are made on both (i) and (ii) of the argument. Since Shimura varieties from toral Shimura data have canonical base points, a special point datum not only determines $\mathscr{I} \in \mathbb{I}$ but also a distinguished point on $S(\mathscr{I})$. Similarly, we have a base point on $S(\mathscr{J})$ as well if $\mathscr{J}$ comes from the same special point datum. The two points on $S(\mathscr{I})$ and $S(\mathscr{J})$ are difficult to relate, but they are shown to be compatible on the level of the maximal abelian quotient G^{ab}, based on integral p-adic Hodge theory of crystalline lattices in G-valued Galois representations, among other things. (A relevant technical issue is that the $\mathbb{Q}_p$-embedding $T \hookrightarrow G$ does not extend to a $\mathbb{Z}_p$-map from the Néron model of T to $\mathcal{G}$ over $\mathbb{Z}_p$, but this is fine if $G, \mathcal{G}$ are replaced with $G^{\mathrm{ab}}, \mathcal{G}^{\mathrm{ab}}$.) Further arguments (sketched in [31, §0.5]) amplify this compatibility to (LR_0).

3.3. From the LR conjecture to a stabilized trace formula

Here we return to a general unramified Shimura datum $(G, X, p, \mathcal{G})$ (possibly not of abelian type). Set $r \in \mathbb{Z}$ to be the inertia degree of $\mathfrak{p}$ over p. As indicated above, (LR_0)

is designed as a substitute for (LR) to imply the following formula predicted by [32,41]. The implication is shown in [31, §3] (refer to the latter for undefined notation):

$$\mathrm{tr}\big(f^{\infty,p} \times \Phi_{\mathfrak{p}}^{j} \mid \big[H_c(\mathrm{Sh}_{K_p}, \overline{\mathbb{Q}}_l)\big]\big) = \sum_{\mathfrak{c}} c(\mathfrak{c}) O_{\gamma(\mathfrak{c})}(f^{\infty,p}) TO_{\delta(\mathfrak{c})}(\phi^{(j)}), \quad j \gg 1. \tag{3.7}$$

Here $\phi^{(j)}$ is an explicit function in the unramified Hecke algebra of $G(\mathbb{Q}_{p^{jr}})$ (with respect to $\mathcal{G}(\mathbb{Z}_{p^{jr}})$), and the sum runs over certain group-theoretic data $\mathfrak{c}$ (called Kottwitz parameters) fibered over the set of stable conjugacy classes in $G(\mathbb{Q})$ which are elliptic in $G(\mathbb{R})$. Here $\mathfrak{c}$ determines an explicit constant $c(\mathfrak{c}) \in \mathbb{Q}$, $\gamma(\mathfrak{c}) \in G(\mathbb{A}^{\infty,p})$ up to conjugacy, and $\delta(\mathfrak{c}) \in G(\mathbb{Q}_{p^{jr}})$ up to σ-conjugacy. In particular, the orbital integral $O_{\gamma(\mathfrak{c})}(f^{\infty,p})$ on $G(\mathbb{A}^{\infty,p})$ and the σ-twisted orbital integral $TO_{\delta(\mathfrak{c})}(\phi^{(j)})$ on $G(\mathbb{Q}_{p^{jr}})$ are well defined. Stabilizing the right-hand side, we arrive at the following, which is a rough version of [31, THM. 3 AND 4].

Theorem 3.3. *Assume that* (LR_0) *is true. For every* $f^{\infty,p} \in \mathcal{H}(G(\mathbb{A}^{\infty,p}))$, *there exists a constant* j_0 *such that for every* $j \geq j_0$, *a formula of the following form holds:*

$$\mathrm{tr}\big(f^{\infty,p} \times \Phi^{j} \mid \big[H_c(\mathrm{Sh}_{K_p}, \overline{\mathbb{Q}}_l)\big]\big) = \sum_{\mathfrak{e} \in \mathcal{E}_{\mathrm{ell}}(G)} \mathrm{ST}_{\mathrm{ell}}^{\mathfrak{e}}\big(f^{\mathfrak{e},\infty,p} f_p^{\mathfrak{e},(j)} f_\infty^{\mathfrak{e}}\big), \tag{3.8}$$

where $\mathcal{E}_{\mathrm{ell}}(G)$ *is the set of elliptic endoscopic data for G up to isomorphism, and* $\mathrm{ST}_{\mathrm{ell}}^{\mathfrak{e}}$ *is the stable elliptic distribution associated with the endoscopic datum* $\mathfrak{e}$.

In light of Theorem 3.2, the conclusion of the theorem is unconditionally true for (G, X) of abelian type. We can easily allow a nonconstant coefficient as done in [31].

The proof of (3.7) from (LR_0) is mostly close to the deduction from (LR) (cf. [46]), and starts from the fixed-point formula for (improper) varieties over finite fields due to Fujiwara and Varshavsky [13,70]; this explains the condition on j. To compute the cohomology of the generic fiber via that of the special fiber, we apply Lan–Stroh's result [39]. Tori-rationality in (LR_0) is the main point to ensure that the fixed-point counting is not affected by the presence of $\tau(\mathcal{J})$ even if $\tau(\mathcal{J})$ is nontrivial. The stabilization from (3.7) to Theorem 3.3 follows the argument in [32] with small improvements to work without technical hypotheses. We note that $f^{\mathfrak{e},\infty,p}$ is the Langlands–Shelstad transfer of $f^{\infty,p}$ whereas $f_p^{\mathfrak{e},(j)}$ and $f_\infty^{\mathfrak{e}}$ are constructed differently. (See [31, §8.2].) As usual in endoscopy, auxiliary z-extensions are chosen if the derived subgroup of G is not simply connected, and the right-hand side of (3.8) should be interpreted appropriately.

Remark 3.4. Sometimes it is possible to obtain (3.8) bypassing any version of the LR conjecture. When (G, X) is of PEL type A or C, this is done by Kottwitz [34]; for Hodge-type data, this is worked out by Lee [42]. It is unclear how their methods interact with connected components of Shimura varieties, so their results do not easily extend to the abelian-type setup. In contrast, the formalism of the LR conjecture is well suited to such extensions.

Remark 3.5. If the adjoint quotient G/Z_G is isotropic over $\mathbb{Q}$ or, equivalently, if Sh is not proper over E (at each fixed level), it is desirable to prove the analogue of (3.8) for the intersection cohomology of the Satake–Baily–Borel compactification; see [51, §§4–5] for what new problems need to be solved. This has been carried out for certain unitary and orthogonal Shimura varieties, as well as Siegel modular varieties, in [50,52,73].

4. APPLICATIONS

4.1. The Hasse–Weil zeta functions and ℓ-adic cohomology

As pioneered by Eichler, Shimura, Deligne, Kuga, Sato, and Ihara, a central problem on Shimura varieties is to compute their ζ-functions and ℓ-adic cohomology. The goals are (i) to express the ζ-function as a quotient of products of automorphic L-functions (thereby deduce a meromorphic continuation and a functional equation when the L-functions are sufficiently understood), cf. [6, CONJ. 5.2], and (ii) to decompose the ℓ-adic cohomology according to automorphic representations and identify the Galois action on each piece. To this end, Langlands and Kottwitz developed a robust method in a series of papers in the 1970–1980s (from [40] to [32]). At the heart is a comparison between the Arthur–Selberg trace formula and a conjectural trace formula for the Hecke–Frobenius action on the cohomology at good primes p, where the latter should come from a fixed-point formula for the special fiber of Shimura varieties modulo p.

When G/Z_G is anisotropic over $\mathbb{Q}$ (equivalently, when Sh is an inverse limit of projective varieties), Theorem 3.3 should be sufficient for the goals (i) and (ii) (up to semisimplifying the Galois action), by following the outline in [32, §§8–10]. We say "should" for two reasons. Firstly, we do not have enough knowledge about automorphic representations in general (e.g., endoscopic classification, cf. [32, §8]). Thus complete details have not been worked out apart from low-rank examples, some special cases such as [33], or under simplifying hypotheses. Secondly, we typically need a positive answer to the following problem to proceed.[2] The reason is that ST^{e} should admit a relatively clean spectral expansion in terms of the discrete automorphic spectrum of endoscopic groups for G, but the spectral interpretation of $\mathrm{ST}^{\mathrm{e}}_{\mathrm{ell}}$ is expected to be quite complicated in general.

Problem 4.1. Assume that G/Z_G is $\mathbb{Q}$-anisotropic. In (3.8), prove that

$$\mathrm{ST}^{\mathrm{e}}_{\mathrm{ell}}\big(f^{e,\infty,p} f_p^{e,(j)} f_\infty^e\big) = \mathrm{ST}^{\mathrm{e}}\big(f^{e,\infty,p} f_p^{e,(j)} f_\infty^e\big), \quad \forall e \in \mathcal{E}_{\mathrm{ell}}(G),$$

where ST^{e} stands for the stable distribution as defined in [52, §5.4].

Although this problem is open, there are quite a few examples where it is known either by the nature of G or under a simplifying hypothesis on the test function. This provides a starting point for the Langlands correspondence (Section 4.2 below).

Now we remove the assumption on G/Z_G. In fact, the argument outlined in [32, §§8–10] is given in this generality, conditional on an affirmative answer to the following.

2 A shortcut getting around Problem 4.1 is possible when G has "no endoscopy," e.g., if G is a form of GL_2 or a certain unitary similitude group as in [33].

Problem 4.2. Prove a formula of the form

$$\mathrm{tr}\big(f^{\infty,p} \times \Phi^j \mid [\mathrm{IH}(\overline{\mathrm{Sh}}, \overline{\mathbb{Q}}_l)]\big) = \sum_{e \in \mathcal{E}_{\mathrm{ell}}(G)} \mathrm{ST}^e\big(f^{e,\infty,p} f_p^{e,(j)} f_\infty^e\big), \tag{4.1}$$

where $\mathrm{IH}(\overline{\mathrm{Sh}}, \overline{\mathbb{Q}}_l)$ is the intersection cohomology of the Satake–Baily–Borel compactification of Sh (see [51, 3.4–3.5], for instance).

To obtain (4.1) from Theorem 3.3, one has to match the nonelliptic terms in ST^e with the contribution to $[\mathrm{IH}(\overline{\mathrm{Sh}}, \overline{\mathbb{Q}}_l)]$ from the boundaries. As a special case, if G/Z_G is $\mathbb{Q}$-anisotropic, then $\mathrm{IH}^i(\overline{\mathrm{Sh}}, \overline{\mathbb{Q}}_l) = H_c^i(\mathrm{Sh}, \overline{\mathbb{Q}}_l)$ for each $i \geq 0$, and the identity of Problem 4.1 should hold since there are no boundaries. In this sense, Problem 4.2 generalizes Problem 4.1. Problem 4.2 has been solved for Siegel modular varieties and certain unitary/orthogonal Shimura varieties by Morel and Zhu [50, 52, 73].

4.2. The global Langlands correspondence

The computation of ℓ-adic cohomology in Section 4.1 often leads to new instances of the global Langlands correspondence satisfying a local–global compatibility in the direction from automorphic representations to Galois representations, roughly stated as follows. Refer to Buzzard–Gee [7] for the definition of L-algebraicity and a full discussion of the conjecture, including a variant conjecture for C-algebraic representations.

Conjecture 4.3. *Let F be a number field, and $\pi = \otimes'_v \pi_v$ an L-algebraic cuspidal automorphic representation of $G(\mathbb{A}_F)$. Then for each prime ℓ and each isomorphism $\iota : \overline{\mathbb{Q}}_l \cong \mathbb{C}$, there exists a continuous representation $\rho_{\ell,\iota} : \mathrm{Gal}(\overline{F}/F) \to {}^L G(\overline{\mathbb{Q}}_l)$ such that the restriction of $\rho_{\ell,\iota}$ to $\mathrm{Gal}(\overline{F}_v/F_v)$ is isomorphic to the unramified Langlands parameter of π_v at almost all finite places v of F (where π_v is unramified).*

The relevance of Shimura varieties to the conjecture is as follows. A Shimura datum (G, X) determines a representation $r_X : {}^L G \to \mathrm{GL}(V)$ (up to isomorphism). Then one expects that the Galois representation $r_X \circ \rho_{\ell,\iota}$ is realized in the ℓ-adic cohomology of the associated Shimura varieties (more precisely, the π^∞-isotypic part thereof), with several caveats including normalization issues (e.g., C-algebraic vs L-algebraic), Arthur packets, and endoscopic problems. These caveats often present much difficulty, and even if they are ignored, it is generally a subtle group-theoretic problem to recover $\rho_{\ell,\iota}$ from $r \circ \rho_{\ell,\iota}$ for a set of representations r of ${}^L G$. (Over global function fields, V. Lafforgue [38] solved the analogous problem in a revolutionary way via generalized pseudocharacters.)

The most fundamental case of Conjecture 4.3 is when $G = \mathrm{GL}_n$. When F is a totally real or CM field and π satisfies a suitable self-duality condition, then the conjecture is proven in a series of papers making use of PEL-type Shimura varieties arising from a unitary similitude group by Clozel, Kottwitz, and others. (See [67] for a discussion and references.) The duality condition allows π to "descend" to an automorphic representation on the unitary similitude group as first observed by Clozel. The self-duality condition was later removed independently by Harris–Lan–Taylor–Thorne and Scholze [20, 61], by exquisite p-adic congruences which are beyond the scope of this article.

The above results for GL_n imply new cases of Conjecture 4.3 (or a weaker form) for quasisplit unitary, symplectic, or special orthogonal groups G over a totally real or CM field via twisted endoscopy by Arthur and Mok [1,49]. However, the conjecture for symplectic or orthogonal *similitude* groups does not follow easily. To get a feel for the difference, note that the dual groups of Sp_{2n}, SO_{2n} are SO_{2n+1}, SO_{2n}, which are embeddable in GL_{2n+1}, GL_{2n}. In contrast, the dual groups of GSp_{2n} and GSO_{2n} are $GSpin_{2n+1}$ and $GSpin_{2n}$, whose faithful representations have dimensions at least 2^n (achieved by the spin representation). In [35,37], Conjecture 4.3 is verified for GSp_{2n} and a (possibly outer) form of GSO_{2n} over a totally real field F under a simplifying hypothesis on π. The basic input comes from Shimura varieties of abelian type associated with a form of $\mathrm{Res}_{F/\mathbb{Q}}GSp_{2n}$, resp. $\mathrm{Res}_{F/\mathbb{Q}}GSO_{2n}$, where r_X is essentially the spin representation, resp. a half-spin representation. Both problems in Section 4.1 have positive answers in that setup.

5. SHIMURA VARIETIES WITH BAD REDUCTION, PART I

Let (G, X) be a Shimura datum. At each prime p such that (G, X) can be promoted to an unramified Shimura datum $(G, X, p, \mathcal{G})$, we discuss three methods to study the cohomology $H_c(\mathrm{Sh}, \overline{\mathbb{Q}}_l)$ as a $G(\mathbb{A}^\infty) \times \mathrm{Gal}(\overline{E}_\mathfrak{p}/E_\mathfrak{p})$-module at each prime $\mathfrak{p}$ of E over p.[3] The "bad reduction" in the section title means that the level subgroup $K'_p \subset K_p = \mathcal{G}(\mathbb{Z}_p)$ at p is allowed to be arbitrarily small, in which case integral models typically have bad reduction mod p. The complicated geometry may be understood better through stratifications.

There are several stratifications of interest on Shimura varieties (cf. [22]) but the most relevant to us is the Newton stratification. In the Hodge-type case, this yields a partition of the mod p Shimura variety into finitely many locally closed subsets, which can be equipped with the reduced subscheme structure, according to the isogeny class of p-divisible groups with additional structure. The unique closed stratum is called the *basic* stratum and corresponds to the p-divisible group that is "most supersingular" under the given constraint.

The first method is a p-adic uniformization of Shimura varieties as pioneered by Čerednik and Drinfeld, and further developed by Rapoport–Zink, Fargues, Kim, and Howard–Pappas [11,23,26,57]. Let $\mathrm{Sh}^{\mathrm{basic}}_{K^pK'_p}$ denote the basic locus in the rigid analytification of $\mathrm{Sh}_{K^pK'_p}$ over $E_\mathfrak{p}$, defined to be the preimage of the basic stratum under the specialization map towards the special fiber. The fundamental result asserts that $\mathrm{Sh}^{\mathrm{basic}}_{K^pK'_p}$ is uniformized by the Rapoport–Zink space with level K'_p arising from the corresponding basic isogeny class. A prominent application is to prove new cases of the Kottwitz conjecture on the cohomology of basic Rapoport–Zink spaces and their generalizations [11,19,25]. Hansen's work [19] points to a synergy between the global method here and Fargues–Scholze's purely local geometric construction of the local Langlands correspondence [12].

---

3 Sometimes these methods extend beyond the stated hypotheses. For example, [18] also works for Kisin–Pappas models (Section 6), and [26] proves a uniformization result also in the nonbasic case. However, we will not try to present the methods in their maximally general settings.

Next we discuss the Harris–Taylor method [21, CHAPS. IV–V] based on a product structure, namely coverings of Newton strata by the products of Igusa varieties and Rapoport–Zink spaces. The outcome is known as Mantovan's formula [44] (generalizing [21, CHAP. IV]), which expresses the cohomology of Newton strata in terms of that of Igusa varieties and Rapoport–Zink spaces. In the basic case, this is closely related to the p-adic uniformization. Hamacher–Kim [18] extended Mantovan's formula and the product structure to Hodge-type Shimura varieties.

To go further, it is desirable to understand the cohomology of Igusa varieties – we address this problem in Section 7 below. Granting this, and putting different Newton strata together, we have a formula relating $[H_c(\mathrm{Sh}, \overline{\mathbb{Q}}_l)]$ to the cohomology of Rapoport–Zink spaces. Then our knowledge about $[H_c(\mathrm{Sh}, \overline{\mathbb{Q}}_l)]$ tells us something nontrivial about the cohomology of Rapoport–Zink spaces, and vice versa. This observation turned out to be useful for proving local–global compatibility, i.e., identifying the local Galois action for the Galois representations in Conjecture 4.3 at ramified primes (see [21, CHAP. VII], [65]) and also for understanding the cohomology of basic/nonbasic Rapoport–Zink spaces [2–4, 66].

Last but not least, there is Scholze's extension of the Langlands–Kottwitz approach from the hyperspecial level at p to arbitrarily small level subgroup at p. (See Section 6.2 below for another generalization.) One seeks for the following analogue of (3.7), where $\tau \in W_{E_\mathfrak{p}}$ is a Weil group element with positive valuation, $h \in \mathcal{H}(G(\mathbb{Q}_p))$ has support contained in $\mathcal{G}(\mathbb{Z}_p)$, and the sum is over the same set of $\mathfrak{c}$:

$$\mathrm{tr}\big(f^{\infty,p} \times h \times \tau \mid [H_c(\mathrm{Sh}, \overline{\mathbb{Q}}_l)]\big) = \sum_{\mathfrak{c}} c(\mathfrak{c}) O_{\gamma(\mathfrak{c})}(f^{\infty,p}) TO_{\delta(\mathfrak{c})}(\phi_{\tau,h}). \tag{5.1}$$

This has been verified by Scholze [59] for PEL-type data and by Youcis [71] for abelian-type data. As an application, Scholze gave a new proof and characterization of the local Langlands correspondence for GL_n over p-adic fields [60] via a base-change transfer of $\phi_{\tau,h}$. A generalization of the latter to other groups was conjectured in [62] and partially proved for unitary groups by Bertoloni Meli and Youcis [5].

In the proof of (5.1), one can push-forward from arbitrarily small level K'_p down to hyperspecial level K_p at the expense of complicating the coefficient sheaf. Applying the fixed-point formula to this, one can exploit knowledge of the fixed-points (coming from results on the LR conjecture). The main problem is to identify the local terms, which are shown to be encoded by a locally constant compactly supported function $\phi_{\tau,h}$ at p constructed from deformation spaces of p-divisible groups with additional structures.

6. SHIMURA VARIETIES WITH BAD REDUCTION, PART II

Let (G, X) be a Shimura datum, and p a prime. In this section, we survey generalizations of Sections 2–5 in the setup where $G_{\mathbb{Q}_p}$ is allowed to be ramified (thus there may be no unramified Shimura datum of the form $(G, X, p, \mathcal{G})$). We recommend the articles [14, 53, 56] for introductions to the contents of this section.

6.1. The LR conjecture in the parahoric case

From now until the end of Section 6, assume that (G, X) is of abelian type. We fix $K_p \subset G(\mathbb{Q}_p)$ a parahoric subgroup, and $\mathfrak{p}$ a place of E over p. In this setting, Kisin–Pappas [30] constructed an integral model $\mathscr{S}_{K_p}$ over $\mathcal{O}_{E_\mathfrak{p}}$ under a mild hypothesis, which are canonical in the sense of [54].

With the integral model as above, we can state versions of the Langlands–Rapoport conjecture analogous to Conjecture 3.1, cf. [56, §9].[4] One can extend the notion of isogeny classes on $\mathscr{S}_{K_p}(\overline{\mathbb{F}}_p)$ and admissible morphisms $\phi : \mathfrak{Q} \to \mathfrak{G}_G$ to the parahoric setup, following [30] and [56], respectively. Thus we can consider the set $\mathbb{I}$ of isogeny classes and the set $\mathbb{J}$ of conjugacy classes of admissible morphisms. The set $S(\mathcal{I})$ of $\overline{\mathbb{F}}_p$-points in each isogeny class $\mathcal{I} \in \mathbb{I}$ is still described by (3.4), with $X_p(\mathcal{I})$ a suitable affine Deligne–Lusztig variety at the parahoric level K_p. Analogously we define $S_\tau(\phi)$ and $S_\tau(\mathcal{J})$ for each admissible ϕ and $\mathcal{J} \in \mathbb{J}$, with $X_p(\phi)$ in (3.5) also adapted to the parahoric level K_p.

Conjecture 6.1. *With the above definitions, there exists a $G(\mathbb{A}^{\infty,p}) \times \Phi^{\mathbb{Z}}$-equivariant bijection $\mathscr{S}_{K_p}(\overline{\mathbb{F}}_p) \cong \coprod_{\mathcal{J}\in\mathbb{J}} S_{\tau(\mathcal{J})}(\mathcal{J})$ such that the exact analogue of* (LR$_1$), *resp.* (LR$_0$) *and* (LR), *holds true.*

Van Hoften and Zhou [68,72] proved the following theorem.

Theorem 6.2. *Statement* (LR$_1$) *of Conjecture* 6.1 *is true if $G_{\mathbb{Q}_p}$ is quasisplit, under a mild technical hypothesis.*

The stronger statement (LR$_0$) is expected to be within reach under the same hypothesis, by extending the argument from [31] to the parahoric setting.

To prove Theorem 6.2, the essential case is when (G, X) is of Hodge type. Zhou proves the very special parahoric case of the conjecture. Van Hoften deduces the case of general parahoric subgroup K'_p, contained in a very special parahoric K_p, by studying the localization maps from Shimura varieties of level K'_p and K_p, to their respective moduli spaces of local Shtukas. The maps are roughly given by assigning to each abelian variety the associated p-divisible group in terms of the moduli problems. Via the forgetful maps from level K'_p down to level K_p, one can form a commutative square diagram. The central claim is that the diagram is Cartesian, from which the LR conjecture at level K'_p can be deduced from the known case at level K_p. The proof of the claim eventually rests on understanding the irreducible components of Kottwitz–Rapoport strata in the situation of the diagram.

6.2. Semisimple zeta functions and Haines–Kottwitz's test function conjecture

At primes of bad reduction, it is useful to compute the semisimple local zeta factor at p instead of the (true) local factor of the Hasse–Weil zeta function (of Shimura varieties) as the former is more amenable to computation. The latter can be recovered from the former in the cases where the weight-monodromy conjecture is known [55].

---

4 One can remove the assumption in [56] that G^{der} is simply connected, by adopting Kisin's formulation in [29] via strictly monoidal categories.

Just like the local zeta factor at p can be computed in terms of the trace of powers of Frobenius on the Frobenius-invariant subspace of the cohomology with compact support, the semisimple local factor can be described in terms of the trace of powers of Frobenius on the derived Frobenius-invariants; such a trace is called the semisimple trace and will be denoted by $\mathrm{tr}^{\mathrm{ss}}$ (see **[55, §2]**, **[15, §3.1]**). Thus a key is to establish the following generalization of (3.7), which recovers (3.7) if K_p is hyperspecial, due to Haines and Kottwitz **[14, §6.1]**. The summation is over the same set as in (3.7).

Conjecture 6.3. *Let $f^{\infty,p} \in \mathcal{H}(G(\mathbb{A}^{\infty,p}))$. For all sufficiently large integers $j \gg 1$, there exist test functions $\phi_{\mathrm{HK}}^{(j)} \in \mathcal{H}(G(\mathbb{Q}_{p^{jr}}))$ such that*

$$\mathrm{tr}^{\mathrm{ss}}\big(f^{\infty,p} \times \Phi_{\mathfrak{p}}^{j} \mid [H_c(\mathrm{Sh}_{K_p}, \overline{\mathbb{Q}}_l)]\big) = \sum_{\mathfrak{c}} c(\mathfrak{c}) O_{\gamma(\mathfrak{c})}(f^{\infty,p}) TO_{\delta(\mathfrak{c})}\big(\phi_{\mathrm{HK}}^{(j)}\big). \tag{6.1}$$

Moreover, $\phi_{\mathrm{HK}}^{(j)}$ may be given by an explicit recipe only in terms of local data at p.

When K_p is a parahoric subgroup of $G(\mathbb{Q}_p)$, one can be more concrete about $\phi_{\mathrm{HK}}^{(j)}$ following **[14, §7]**: $\phi_{\mathrm{HK}}^{(j)}$ should admit a geometric construction via nearby cycles on local models, as well as a representation-theoretic description in terms of the Langlands correspondence. That the two descriptions for $\phi_{\mathrm{HK}}^{(j)}$ coincide is the test function conjecture verified by Haines–Richarz **[16, 17]** under a very mild hypothesis. (See **[14, §8]** for prior and related results.) The proof is based on geometry of mixed-characteristic affine Grassmanians and the geometric Satake equivalence. It remains to combine their theorem with the results in Section 6.1 to obtain new cases of Conjecture 6.3 and its stabilized form, so as to determine the semisimple zeta factor at p. This requires an endoscopic understanding of $\phi_{\mathrm{HK}}^{(j)}$, cf. **[14, §6.2]**; a simple exemplary case is demonstrated in **[14, §6.3]**, where endoscopic problems disappear.

In a related but somewhat different direction (cf. the last two paragraphs in **[14, §8.4]**), the Langlands–Kottwitz–Scholze approach discussed in Section 5 should extend to the current setup despite the absence of hyperspecial subgroups at p, at least when the results of Section 6.1 are available for some parahoric subgroups.

Problem 6.4. Prove the analogue of (5.1) for general Shimura data (G, X) and primes p.

7. IGUSA VARIETIES

Igusa curves were introduced to understand the geometry of modular curves modulo p when the level is divisible by a prime p **[24]**. The construction has been generalized by Harris–Taylor **[21]** and Mantovan **[44]** in the PEL-type case, and most recently to the setup of Kisin–Pappas models for Hodge-type Shimura varieties by Hamacher–Kim **[18]**. Igusa varieties have a variety of applications to p-adic and mod p modular forms, cohomology of Shimura varieties, the Langlands correspondence, and some more. We refer to the introduction of **[36]** for further details and references. In this section, we concentrate on computing the ℓ-adic cohomology of Igusa varieties via an analogue of the LR conjecture.

7.1. The LR conjecture for Igusa varieties

Let $(G, X, p, \mathscr{G})$ be an unramified Shimura datum of Hodge type, with a fixed embedding of (G, X) into a Siegel Shimura datum. Put $K_p := \mathscr{G}(\mathbb{Z}_p)$, a hyperspecial subgroup of $G(\mathbb{Q}_p)$. Let $\mathcal{A}$ denote the abelian scheme over $\mathscr{S}_{K_p}$ pulled back from the universal abelian scheme over the ambient Siegel moduli scheme. Thus we have a p-divisible group $\mathcal{A}[p^\infty]$ over $\mathscr{S}_{K_p}$ with $G_{\mathbb{Q}_p}$-structure in some precise sense.

Now fix a p-divisible group Σ over $\overline{\mathbb{F}}_p$ with $G_{\mathbb{Q}_p}$-structure. By Dieudonné theory, this determines $b \in G(\breve{\mathbb{Q}}_p)$ (up to replacing b with $g^{-1}b_x\sigma(g)$ for $g \in \mathscr{G}(\breve{\mathbb{Z}}_p)$). Write $J_b(\mathbb{Q}_p)$ for the group of self-quasi-isogenies of Σ respecting the $G_{\mathbb{Q}_p}$-structure. As a $\mathbb{Q}_p$-algebraic group, J_b is known to be an inner form of a Levi subgroup of $G_{\mathbb{Q}_p}$. We define Ig_b to be the parameter space (in the category of perfect $\overline{\mathbb{F}}_p$-schemes) of $G_{\mathbb{Q}_p}$-structure-preserving isomorphisms between $\mathcal{A}[p^\infty]$ and (the constant family of) Σ over $\mathscr{S}_{K_p,\overline{\mathbb{F}}_p}$. The scheme Ig_b is nonempty if and only if the image of b in Kottwitz's set $B(G)$ lies in the finite subset $B(G, \mu_X^{-1})$, where μ_X is the Hodge cocharacter of G determined by (G, X) up to conjugacy. (The same set $B(G, \mu_X^{-1})$ labels the Newton strata, cf. Section 5.) There is a natural action of $G(\mathbb{A}^{\infty,p}) \times J_b(\mathbb{Q}_p)$ on Ig_b, thus also on its ℓ-adic cohomology.

We may and will replace Σ with an isogenous p-divisible group which is completely slope divisible, since the isomorphism class of Ig_b with the group action is invariant under such a replacement. The advantage of doing so is that Ig_b can be written as the projective limit of finite-type varieties (up to taking perfection, which does not affect cohomology) by trivializing only a finite p-power torsion subgroup and by fixing a level subgroup $K^p \subset G(\mathbb{A}^{\infty,p})$ away from p at a time. With some care, the projective system of varieties can be defined over a common finite field. This enables us to apply the Fujiwara–Varshavsky fixed-point formula to compute the cohomology (with compact support) at each finite level, provided that we understand the structure of $\mathrm{Ig}_b(\overline{\mathbb{F}}_p)$. Thus we are prompted to think about the analogue of the LR conjecture for Igusa varieties.

In analogy with (3.3) and (3.4), keeping the same definition of $\mathbb{I}$ and $\mathbb{J}$, we have the partition $S^{\mathrm{Ig}_b}(\overline{\mathbb{F}}_p) = \coprod_{\mathcal{I}\in\mathbb{I}} S^{\mathrm{Ig}_b}(\mathcal{I})$ according to isogeny classes of abelian varieties, with

$$S^{\mathrm{Ig}_b}(\mathcal{I}) \cong I_{\mathcal{I}}(\mathbb{Q})\backslash\big(X_p^{\mathrm{Ig}_b}(\mathcal{I}) \times X^p(\mathcal{I})\big). \tag{7.1}$$

The $G(\mathbb{A}^{\infty,p})$-set $X^p(\mathcal{I})$ is the same as before, but the difference from Section 3 is that $X_p^{\mathrm{Ig}_b}(\mathcal{I})$ is no longer an affine Deligne–Lusztig variety but a right $J_b(\mathbb{Q}_p)$-torsor. Turning to the other side of the LR conjecture, let $\mathcal{J} \in \mathbb{J}$. It is natural to impose the so-called b-admissibility condition on $\mathcal{J}$ at p, which is the group-theoretic analogue of the condition that a p-divisible group is isogenous to Σ (with $G_{\mathbb{Q}_p}$-structure). For b-admissible $\mathcal{J}$, we set

$$S_\tau^{\mathrm{Ig}_b}(\mathcal{J}) := I_\phi(\mathbb{Q})\backslash_\tau\big(X_p^{\mathrm{Ig}_b}(\mathcal{J}) \times X^p(\mathcal{J})\big),$$

with the same $X^p(\mathcal{J})$ as in Section 3, and a suitably defined right $J_b(\mathbb{Q}_p)$-torsor $X_p^{\mathrm{Ig}_b}(\mathcal{J})$, where the τ-twisted quotient can be defined again as in Section 3. We are ready to state versions of the LR conjecture for Igusa varieties in parallel with Conjecture 3.1, for unramified Shimura data $(G, X, p, \mathscr{G})$ of Hodge type.

Conjecture 7.1. *There is a bijection of right $J_b(\mathbb{Q}_p) \times G(\mathbb{A}^{\infty,p})$-sets*

$$\mathrm{Ig}_b(\overline{\mathbb{F}}_p) \simeq \coprod_{\substack{\mathcal{J}\in\mathbb{J}\\ b\text{-adm.}}} S^{\mathrm{Ig}_b}_{\tau(\mathcal{J})}(\mathcal{J}),$$

where $\{\tau(\mathcal{J})\}$ over the set of b-admissible $\mathcal{J}$ satisfies the conditions in (LR_1)*, resp.* (LR_0) *and* (LR).

Mack-Crane proved the following theorem in his thesis [43], where c, $c(\mathrm{c})$, and $\gamma(\mathrm{c})$ are the same as in Section 3, but we impose a b-admissibility condition on the Kottwitz parameter c inherited from the similar condition on $\mathcal{J}$, and each c gives rise to a conjugacy class of $\delta'(\mathrm{c})$ in $J_b(\mathbb{Q}_p)$, along which we compute the (untwisted) orbital integral $O_{\delta'(\mathrm{c})}(\phi'_p)$.

Theorem 7.2. *The* (LR_0)*-version of Conjecture* 7.1 *is true. Moreover, the following analogue of* (3.7) *holds true for $f^{\infty,p} \in \mathcal{H}(G(\mathbb{A}^{\infty,p}))$ and sufficiently many functions $\phi'_p \in \mathcal{H}(J_b(\mathbb{Q}_p))$:*

$$\mathrm{tr}\big(f^{\infty,p} \times \phi'_p \mid \big[H_c(\mathrm{Ig}_b, \overline{\mathbb{Q}}_l)\big]\big) = \sum_{\mathrm{c}:\, b\text{-adm.}} c(\mathrm{c}) O_{\gamma(\mathrm{c})}(f^{\infty,p}) O_{\delta'(\mathrm{c})}(\phi'_p). \tag{7.2}$$

By "sufficiently many," we mean that the traces for such a set of functions are enough to determine $[H_c(\mathrm{Ig}_b, \overline{\mathbb{Q}}_l)]$ in the Grothendieck group of $G(\mathbb{A}^{\infty,p}) \times J_b(\mathbb{Q}_p)$-representations. (The precise condition has to do with twisting by a high power of Frobenius in the Fujiwara–Varshavsky formula.) The proof of the theorem proceeds by carefully adapting the methods of [29, 31] but with significant changes occurring at p, thus often requiring different techniques and arguments.

Formula (7.2) was obtained for some simple PEL-type Shimura varieties in [21, CHAP. 5] and [63] without formulating and proving the LR conjecture. In contrast, the above theorem represents the first LR-style approach to Igusa varieties, giving it two advantages. Firstly, the new approach makes the similarities between Shimura and Igusa varieties transparent. An important consequence is that the hard-won statement (LR_0) for Shimura varieties can be transferred to the Igusa side. (If we had the full (LR) for Shimura varieties, then that would carry over to Igusa varieties, too.) Going from (LR_0) for Igusa varieties to (7.2) is mostly the same as for Shimura varieties. Secondly, just like for Shimura varieties, the LR-style approach makes it feasible to extend the theorem to the abelian-type case, cf. Remark 3.4. (This extension has not been worked out, yet.) It should also be possible to go beyond good reduction and work in the setup of Kisin–Pappas models (Section 6). For the sake of proposing a problem, we can be even more general but still stick to (LR_0) rather than (LR) as this should suffice for most applications:

Problem 7.3. Construct Igusa varieties modulo p for all Shimura data (G, X) and all primes p. Prove the (LR_0)-version of Conjecture 7.1, thereby deduce formula (7.2).

Assuming a positive answer (known in the setting of Theorem 7.2), the next step is to unconditionally stabilize (7.2) into the following form:

$$\mathrm{tr}\big(f^{\infty,p} \times \phi'_p \mid \big[H_c(\mathrm{Ig}_b, \overline{\mathbb{Q}}_l)\big]\big) = \sum_{e\in\mathcal{E}_{\mathrm{ell}}(G)} \mathrm{ST}^e_{\mathrm{ell}}\big(f^{e,\infty,p} f^{e,\prime}_p f^e_\infty\big). \tag{7.3}$$

The formula is an exact analogue of (3.8). Indeed, $f^{e,\infty,p}$ and f^{e}_{∞} are constructed in the same way. However, $f_p^{e,\prime}$ is constructed from ϕ'_p via a "nonstandard" transfer of functions – this is the main novelty in the stabilization. Even when $G_{\mathbb{Q}_p}$ is a product of general linear groups so that local endoscopy at p disappears, the transfer goes to $G_{\mathbb{Q}_p}$ from an inner form of a Levi subgroup of $G_{\mathbb{Q}_p}$. The transfer was constructed and studied in **[64]** in a somewhat ad hoc manner, and later streamlined in **[2]**. Unfortunately, both papers make a set of technical hypotheses, to be removed in the work in progress with Bertoloni Meli.

7.2. Applications

The stabilization (7.3) is a significant step towards the following:

Problem 7.4. Obtain a decomposition of $[H_c(\mathrm{Ig}_b, \overline{\mathbb{Q}}_l)]$ according to automorphic representations of G and its endoscopic groups, and describe each piece in the decomposition.

Just from the definition of Igusa varieties, it is not even clear whether the entirety of $[H_c(\mathrm{Ig}_b, \overline{\mathbb{Q}}_l)]$ can be understood through automorphic representations. For Shimura varieties over $\mathbb{C}$, the connection is made through Matsushima's formula and its generalizations, but there is no analogue for Igusa varieties.

We have a concrete answer for some simple PEL-type Shimura varieties arising from (G, X) such that (i) endoscopy for G disappears over $\mathbb{Q}$ and $\mathbb{Q}_p$, (ii) G is anisotropic modulo center over $\mathbb{Q}$, and (iii) $G_{\mathbb{Q}_p}$ is a product of general linear groups. Recall that J_b is an inner form of a $\mathbb{Q}_p$-rational Levi subgroup, say M_b, of $G_{\mathbb{Q}_p}$. In fact, b determines a particular parabolic subgroup P_b^{op} containing M_b as a Levi component. Write Red^b for the composite morphism on the Grothendieck group of representations

$$\mathrm{Red}^b : \mathrm{Groth}\big(G(\mathbb{Q}_p)\big) \to \mathrm{Groth}\big(M_b(\mathbb{Q}_p)\big) \to \mathrm{Groth}\big(J_b(\mathbb{Q}_p)\big), \tag{7.4}$$

where the first map is the Jacquet module relative to P_b^{op} (up to a character twist), and the second is Badulescu's Langlands–Jacquet map. The answer to Problem 7.4 in this setting is given by **[21, THM. V.5.4]** and **[66, THM. 6.7]**:

$$\big[H_c(\mathrm{Ig}_b, \overline{\mathbb{Q}}_l)\big] = \big[\mathrm{Red}^b H_c(\mathrm{Sh}, \overline{\mathbb{Q}}_l)\big] \quad \text{in } \mathrm{Groth}\big(G(\mathbb{A}^{\infty,p}) \times J_b(\mathbb{Q}_p)\big). \tag{7.5}$$

Since $H_c(\mathrm{Sh}, \overline{\mathbb{Q}}_l)$ is well understood by Matsushima's formula via relative Lie algebra cohomology, (7.5) is indeed a satisfactory answer for $[H_c(\mathrm{Ig}_b, \overline{\mathbb{Q}}_l)]$. Through Mantovan's formula, (7.5) sheds light on the cohomology of Rapoport–Zink spaces **[3, 66]**, cf. Section 5.

When $G_{\mathbb{Q}_p}$ is still a product of general linear groups but G exhibits endoscopy over $\mathbb{Q}$, the formula for $[H_c(\mathrm{Ig}_b, \overline{\mathbb{Q}}_l)]$ is no longer as simple as (7.5). Computing the formula in certain endoscopic cases was crucial in the proof of local–global compatibility in **[65]**, cf. Section 5. (See **[67, §6]** for an expository account.)

In general, Problem 7.4 seems out of reach. Firstly, just like for Shimura varieties, the lack of endoscopic classification is a major obstacle. Secondly, a new difficulty in the Igusa setup is that the analogue of Problem 4.1 has no conceptual reason to have a positive answer (cf. last paragraph of Section 4.1), and the analogue of Problem 4.2 is even less clear. (The second point is related to the question at the end of this section.) Assuming that both issues go away, a conjectural formula for $[H_c(\mathrm{Ig}_b, \overline{\mathbb{Q}}_l)]$ has been given in **[2]** under some hypotheses on G, resembling Kottwitz's conjectural formula for $[\mathrm{IH}(\overline{\mathrm{Sh}}, \overline{\mathbb{Q}}_l)]$ in **[32, §10]**. The formula for $[H_c(\mathrm{Ig}_b, \overline{\mathbb{Q}}_l)]$ is far more complicated than (7.5) and involves endoscopic versions of Red^b. (In the stable case, which is the simplest, the correct analogue of (7.4) is the Jacquet module followed by a stable transfer between inner forms.) A main observation in **[2]** is that the endoscopic versions of Red^b should interact with the cohomology of Rapoport–Zink spaces (and their generalizations) in an interesting way by a global reason.

For an unconditional result towards Problem 7.4, we managed to compute the $G(\mathbb{A}^{\infty,p}) \times J_b(\mathbb{Q}_p)$-module $H^0(\mathrm{Ig}_b, \overline{\mathbb{Q}}_l)$ for unramified Shimura data $(G, X, p, \mathcal{G})$ of Hodge type in joint work with Kret **[36]**, when b is nonbasic[5] (in every $\mathbb{Q}$-simple factor of the adjoint group of G). In analogy with (7.5), the result may be expressed as

$$H^0(\mathrm{Ig}_b, \overline{\mathbb{Q}}_l) = \mathrm{Red}^b H^0(\mathrm{Sh}, \overline{\mathbb{Q}}_l) \quad \text{as} \quad G(\mathbb{A}^{\infty,p}) \times J_b(\mathbb{Q}_p)\text{-modules.} \tag{7.6}$$

Here $H^0(\mathrm{Sh}, \overline{\mathbb{Q}}_l)$ has a well-known description in terms of 1-dimensional automorphic representations of $G(\mathbb{A})$, and Red^b is unequivocally defined for 1-dimensional representations of $G(\mathbb{Q}_p)$ (which are always stable). The proof starts from the results of Section 7.1. The main point is to get around the two essential difficulties mentioned in the last paragraph, by incorporating asymptotic and inductive arguments to extract the H^0-part from a very complicated identity coming from (stabilized) trace formulas. The study of H^0 was motivated by geometric applications to the irreducibility of Igusa varieties and to the discrete Hecke orbit conjecture. The reader is referred to **[36]** for details and further references. Similar geometric results were independently obtained by van Hoften and Xiao **[68, 69]** via a more geometric approach without using automorphic forms or trace formulas.

We conclude this section with an unrefined question. Igusa varieties (at finite level) are almost never proper varieties, so the answer to Problem 7.4 does not determine $H_c^i(\mathrm{Ig}_b, \overline{\mathbb{Q}}_l)$ for $i \geq 0$ due to possible cancelations in the Grothendieck group. (For improper Shimura varieties, the intersection cohomology is free from such a cancelation thanks to purity.) Thus we can ask whether there are useful compactifications of Igusa varieties to help us understand the cohomology more precisely. This was undertaken by Mantovan **[45]** in a special case (with a different goal). In general, it is unclear how to proceed even when Shimura varieties are proper. If a strategy is found in that case, it may be possible to deal with improperness of Shimura varieties by virture of Caraiani–Scholze's partial compactification **[8]**.

ACKNOWLEDGMENTS

Special thanks are due to Michael Harris, Robert Kottwitz, and Richard Taylor for introducing the author to Shimura varieties and the Langlands program, and patiently answering his questions over many years. The author is grateful to Mark Kisin, Arno Kret, and Yihang Zhu for sharing their insight through years of recent and ongoing collaboration

5 If b is basic then Igusa varieties are 0-dimensional and the picture is quite different from (7.6), cf. **[36, §1.6]**.

that this paper is largely based on. He is also thankful for exciting mathematical journeys to the other coworkers: Sara Arias-de Reyna, Ana Caraiani, Luis Dieulefait, Matthew Emerton, Jessica Fintzen, Toby Gee, David Geraghty, Wushi Goldring, Julia Gordon, Junehyuk Jung, Tasho Kaletha, Julee Kim, Keerthi Madapusi Pera, Simon Marshall, Alberto Minguez, Vytautas Paškūnas, Peter Sarnak, Peter Scholze, Nicolas Templier, Paul-James White, and Gabor Wiese. Finally, he thanks Alex Youcis for comments to improve the exposition of the paper.

FUNDING

This work was partially supported by NSF grant DMS-1802039/2101688, NSF RTG grant DMS-1646385, and a Miller Professorship.

REFERENCES

[1] J. Arthur, *The endoscopic classification of representations*. Amer. Math. Soc. Colloq. Publ. 61, American Mathematical Society, Providence, RI, 2013.

[2] A. Bertoloni Meli, An averaging formula for the cohomology of PEL-type Rapoport–Zink spaces. 2021, arXiv:2102.10690.

[3] A. Bertoloni Meli, The cohomology of Rapoport–Zink spaces of EL-type and Harris's conjecture. *J. Inst. Math. Jussieu* (online firstview) (2021), 1–56.

[4] A. Bertoloni Meli and K. H. Nguyen, The Kottwitz conjecture for unitary PEL-type Rapoport–Zink spaces. 2021, arXiv:2104.05912.

[5] A. Bertoloni Meli and A. Youcis, The Scholze–Shin conjecture for unramified unitary groups, 2019, http://www-personal.umich.edu/~abertolo/Scholze_Shin_Conjecture_for_Unitary_Groups__No_Endoscopy.pdf, preprint.

[6] D. Blasius and J. D. Rogawski, Zeta functions of Shimura varieties. In *Motives (Seattle, WA, 1991)*, pp. 525–571, Proc. Sympos. Pure Math. 55, Amer. Math. Soc., Providence, RI, 1994.

[7] K. Buzzard and T. Gee, The conjectural connections between automorphic representations and Galois representations. In *Automorphic forms and Galois representations. Vol. 1*, pp. 135–187, London Math. Soc. Lecture Note Ser. 414, Cambridge Univ. Press, Cambridge, 2014.

[8] A. Caraiani and P. Scholze, On the generic part of the cohomology of non-compact unitary Shimura varieties. 2019, arXiv:1909.01898.

[9] M. Chen, M. Kisin, and E. Viehmann, Connected components of affine Deligne–Lusztig varieties in mixed characteristic. *Compos. Math.* **151** (2015), no. 9, 1697–1762.

[10] P. Deligne, Variétés de Shimura: interprétation modulaire, et techniques de construction de modèles canoniques. In *Automorphic forms, representations and L-functions (Corvallis, OR, 1977), Part 2*, pp. 247–289, Proc. Sympos. Pure Math. XXXIII, Amer. Math. Soc., Providence, RI, 1979.

[11] L. Fargues, Cohomologie des espaces de modules de groupes p-divisibles et correspondances de Langlands locales. *Astérisque* (2004), no. 291, 1–199.

[12] L. Fargues and P. Scholze, Geometrization of the local Langlands correspondence. 2021, arXiv:2102.13459.

[13] K. Fujiwara, Rigid geometry, Lefschetz–Verdier trace formula and Deligne's conjecture. *Invent. Math.* **127** (1997), no. 3, 489–533.

[14] T. J. Haines, The stable Bernstein center and test functions for Shimura varieties. In *Automorphic forms and Galois representations. Vol. 2*, pp. 118–186, London Math. Soc. Lecture Note Ser. 415, Cambridge Univ. Press, Cambridge, 2014.

[15] T. Haines and B. C. Ngô, Nearby cycles for local models of some Shimura varieties. *Compos. Math.* **133** (2002), no. 2, 117–150.

[16] T. J. Haines and T. Richarz, The test function conjecture for local models of Weil-restricted groups. *Compos. Math.* **156** (2020), no. 7, 1348–1404.

[17] T. J. Haines and T. Richarz, The test function conjecture for parahoric local models. *J. Amer. Math. Soc.* **34** (2021), no. 1, 135–218.

[18] P. Hamacher and W. Kim, l-adic étale cohomology of Shimura varieties of Hodge type with non-trivial coefficients. *Math. Ann.* **375** (2019), no. 3–4, 973–1044.

[19] D. Hansen, On the supercuspidal cohomology of basic local Shimura varieties. 2021, http://www.davidrenshawhansen.com/middle.pdf, preprint.

[20] M. Harris, K.-W. Lan, R. Taylor, and J. Thorne, On the rigid cohomology of certain Shimura varieties. *Res. Math. Sci.* **3** (2016), Paper No. 37, 308.

[21] M. Harris and R. Taylor, *The geometry and cohomology of some simple Shimura varieties*. Ann. of Math. Stud. 151, Princeton University Press, Princeton, NJ, 2001.

[22] X. He and M. Rapoport, Stratifications in the reduction of Shimura varieties. *Manuscripta Math.* **152** (2017), no. 3–4, 317–343.

[23] B. Howard and G. Pappas, Rapoport–Zink spaces for spinor groups. *Compos. Math.* **153** (2017), no. 5, 1050–1118.

[24] J-i. Igusa, On the algebraic theory of elliptic modular functions. *J. Math. Soc. Japan* **20** (1968), 96–106.

[25] T. Ito and Y. Mieda, Cuspidal representations in the ℓ-adic cohomology of the Rapoport–Zink space for GSp(4). 2010, arXiv:1005.5619.

[26] W. Kim, Rapoport–Zink uniformization of Hodge-type Shimura varieties. *Forum Math. Sigma* **6** (2018), Paper No. e16, 36.

[27] W. Kim and K. Madapusi Pera, 2-adic integral canonical models. *Forum Math. Sigma* **4** (2016), e28, 34.

[28] M. Kisin, Integral models for Shimura varieties of abelian type. *J. Amer. Math. Soc.* **23** (2010), no. 4, 967–1012.

[29] M. Kisin, mod p points on Shimura varieties of abelian type. *J. Amer. Math. Soc.* **30** (2017), no. 3, 819–914.

[30] M. Kisin and G. Pappas, Integral models of Shimura varieties with parahoric level structure. *Publ. Math. Inst. Hautes Études Sci.* **128** (2018), 121–218.

[31] M. Kisin, S. W. Shin, and Y. Zhu, The stable trace formula for Shimura varieties of abelian type. 2021, preprint.

[32] R. E. Kottwitz, Shimura varieties and λ-adic representations. In *Automorphic forms, Shimura varieties, and L-functions, Vol. I (Ann Arbor, MI, 1988)*, pp. 161–209, Perspect. Math. 10, Academic Press, Boston, MA, 1990.

[33] R. E. Kottwitz, On the λ-adic representations associated to some simple Shimura varieties. *Invent. Math.* **108** (1992), no. 3, 653–665.

[34] R. E. Kottwitz, Points on some Shimura varieties over finite fields. *J. Amer. Math. Soc.* **5** (1992), no. 2, 373–444.

[35] A. Kret and S. W. Shin, Galois representations for even general special orthogonal groups. 2020, arXiv:2010.08408.

[36] A. Kret and S. W. Shin, H^0 of Igusa varieties via automorphic forms. 2021, arXiv:2102.10690.

[37] A. Kret and S. W. Shin, Galois representations for general symplectic groups. *J. Eur. Math. Soc. (JEMS)* (to appear), arXiv:1609.04223.

[38] V. Lafforgue, Chtoucas pour les groupes réductifs et paramétrisation de Langlands globale. *J. Amer. Math. Soc.* **31** (2018), no. 3, 719–891.

[39] K.-W. Lan and B. Stroh, Nearby cycles of automorphic étale sheaves. *Compos. Math.* **154** (2018), no. 1, 80–119.

[40] R. P. Langlands, Some contemporary problems with origins in the Jugendtraum. In *Mathematical developments arising from Hilbert problems (Proc. Sympos. Pure Math., Vol. XXVIII, Northern Illinois Univ., De Kalb, IL, 1974)*, pp. 401–418, Amer. Math. Soc., Providence, R.I., 1976.

[41] R. P. Langlands and M. Rapoport, Shimuravarietäten und Gerben. *J. Reine Angew. Math.* **378** (1987), 113–220.

[42] D. U. Lee, Galois gerbs and Lefschetz number formula for Shimura varieties of Hodge type. 2018, arXiv:1801.03057.

[43] S. Mack-Crane, *Counting points on Igusa varieties of Hodge type*. PhD thesis, UC Berkeley, 2021.

[44] E. Mantovan, On the cohomology of certain PEL-type Shimura varieties. *Duke Math. J.* **129** (2005), no. 3, 573–610.

[45] E. Mantovan, A compactification of Igusa varieties. *Math. Ann.* **340** (2008), no. 2, 265–292.

[46] J. S. Milne, The points on a Shimura variety modulo a prime of good reduction. In *The zeta functions of Picard modular surfaces*, pp. 151–253, Univ. Montréal, Montreal, QC, 1992.

[47] J. S. Milne, Introduction to Shimura varieties. In *Harmonic analysis, the trace formula, and Shimura varieties*, pp. 265–378, Clay Math. Proc. 4, Amer. Math. Soc., Providence, RI, 2005.

[48] J. S. Milne, Points on Shimura varieties over finite fields: the conjecture of Langlands and Rapoport. 2007, arXiv:0707.3173.

[49] C. P. Mok, Endoscopic classification of representations of quasi-split unitary groups. *Mem. Amer. Math. Soc.* **235** (2015), no. 1108, vi+248.

[50] S. Morel, Complexes pondérés sur les compactifications de Baily–Borel: le cas des variétés de Siegel. *J. Amer. Math. Soc.* **21** (2008), no. 1, 23–61.

[51] S. Morel, The intersection complex as a weight truncation and an application to Shimura varieties. In *Proceedings of the International Congress of Mathematicians. Volume II*, pp. 312–334, Hindustan Book Agency, New Delhi, 2010.

[52] S. Morel, *On the cohomology of certain noncompact Shimura varieties*. Ann. of Math. Stud. 173, Princeton University Press, Princeton, NJ, 2010.

[53] G. Pappas, Arithmetic models for Shimura varieties. In *Proceedings of the International Congress of Mathematicians—Rio de Janeiro 2018. Vol. II. Invited lectures*, pp. 377–398, World Sci. Publ., Hackensack, NJ, 2018.

[54] G. Pappas, On integral models of Shimura varieties. 2020, arXiv:2003.13040.

[55] M. Rapoport, On the bad reduction of Shimura varieties. In *Automorphic forms, Shimura varieties, and L-functions, Vol. II (Ann Arbor, MI, 1988)*, pp. 253–321, Perspect. Math. 11, Academic Press, Boston, MA, 1990.

[56] M. Rapoport, A guide to the reduction modulo p of Shimura varieties. *Astérisque* **298** (2005), 271–318.

[57] M. Rapoport and T. Zink, *Period spaces for p-divisible groups*. Ann. of Math. Stud. 141, Princeton University Press, Princeton, NJ, 1996.

[58] H. Reimann, *The semi-simple zeta function of quaternionic Shimura varieties*. Lecture Notes in Math. 1657, Springer, Berlin, 1997.

[59] P. Scholze, The Langlands–Kottwitz method and deformation spaces of p-divisible groups. *J. Amer. Math. Soc.* **26** (2013), no. 1, 227–259.

[60] P. Scholze, The local Langlands correspondence for GL_n over p-adic fields. *Invent. Math.* **192** (2013), no. 3, 663–715.

[61] P. Scholze, On torsion in the cohomology of locally symmetric varieties. *Ann. of Math. (2)* **182** (2015), no. 3, 945–1066.

[62] P. Scholze and S. W. Shin, On the cohomology of compact unitary group Shimura varieties at ramified split places. *J. Amer. Math. Soc.* **26** (2013), no. 1, 261–294.

[63] S. W. Shin, Counting points on Igusa varieties. *Duke Math. J.* **146** (2009), no. 3, 509–568.

[64] S. W. Shin, A stable trace formula for Igusa varieties. *J. Inst. Math. Jussieu* **9** (2010), no. 4, 847–895.

[65] S. W. Shin, Galois representations arising from some compact Shimura varieties. *Ann. of Math. (2)* **173** (2011), no. 3, 1645–1741.

[66] S. W. Shin, On the cohomology of Rapoport–Zink spaces of EL-type. *Amer. J. Math.* **134** (2012), no. 2, 407–452.

[67] S. W. Shin, Construction of automorphic Galois representations: the self-dual case. In *Shimura varieties*, pp. 209–250, London Math. Soc. Lecture Note Ser. 457, Cambridge University Press, 2020.

[68] P. van Hoften, Mod p points on Shimura varieties of parahoric level. 2020, arXiv:2010.10496.

[69] P. van Hoften and L. X. Xiao, Monodromy and irreducibility of Igusa varieties. 2021, arXiv:2102.09870.

[70] Y. Varshavsky, Lefschetz–Verdier trace formula and a generalization of a theorem of Fujiwara. *Geom. Funct. Anal.* **17** (2007), no. 1, 271–319.

[71] A. Youcis, The Langlands–Kottwitz–Scholze method for deformation spaces of abelian type, in progress.

[72] R. Zhou, Mod p isogeny classes on Shimura varieties with parahoric level structure. *Duke Math. J.* **169** (2020), no. 15, 2937–3031.

[73] Y. Zhu, The stabilization of the Frobenius–Hecke traces on the intersection cohomology of orthogonal Shimura varieties. 2018, arXiv:1801.09404.

SUG WOO SHIN

Department of Mathematics, UC Berkeley, Berkeley, CA 94720, USA, and Korea Institute for Advanced Study, Seoul 02455, Republic of Korea, sug.woo.shin@berkeley.edu

This Page Intentionally Left Blank

THE CONGRUENT NUMBER PROBLEM AND ELLIPTIC CURVES

YE TIAN

ABSTRACT

The Birch and Swinnerton-Dyer (BSD) conjecture and Goldfeld conjecture are fundamental problems in the arithmetic of elliptic curves. The congruent number problem (CNP) is one of the oldest problems in number theory which is, for each integer n, to find all the rational right triangles of area n. It is equivalent to finding all rational points on the elliptic curve $E^{(n)} : ny^2 = x^3 - x$. The BSD conjecture for $E^{(n)}$ solves CNP, and Goldfeld conjecture for this elliptic curve family solves CNP for integers with probability one. In this article, we introduce some recent progress on these conjectures and problems.

MATHEMATICS SUBJECT CLASSIFICATION 2020

Primary 11-06; Secondary 11G05, 11G40

KEYWORDS

Congruent numbers, BSD conjecture, Goldfeld conjecture, quadratic twists, Selmer group, L-function

INTERNATIONAL CONGRESS OF MATHEMATICIANS 2022 JULY 6–14

Proc. Int. Cong. Math. 2022, Vol. 3, pp. 1990–2010
DOI 10.4171/ICM2022/85

1. CONGRUENT NUMBER PROBLEM

A positive rational number n is called a *congruent number* if the following equivalent conditions hold:

(i) There exists a rational number x such that $x^2 \pm n$ are squares of rational numbers.

(ii) There exists a right triangle with rational side lengths (called a rational right triangle) whose area is n.

In his book *Liber Quadratorum* published in 1225, Fibonacci (1175–1250) named an integer satisfying (i) a "congruum" from the Latin, which means to meet together, since the three squares $x^2 - n, x^2$, and $x^2 + n$ are congruent modulo n.

The congruent number problem (CNP, for short) is to determine, in finitely many steps, whether or not a given rational number is a congruent number, and, if it is, find all the corresponding x in (i) or rational right triangles in (ii). No such algorithm has ever been found. The Persian mathematician Al-Karaji (953–1029), perhaps the first mathematician, stated this problem in terms of (i). A similar question appeared in his Arabic translation of the work of Diophantus in Greek. In an Arab manuscript of the tenth century, Mohammed Ben Alhocain realized the equivalence between (i) and (ii) and stated that this problem is "the principal object of the theory of rational right triangles" (see Dickson's book **[20, CHAP. XVI, P. 459]**).

Recall that any rational Pythagorean triple has the following form:

$$2abt, \quad (a^2 - b^2)t, \quad (a^2 + b^2)t$$

for a unique (a, b, t), where t is a positive rational number and $a > b$ are two coprime positive integers with $2 \nmid (a + b)$. We call a rational Pythagorean triple primitive if $t = 1$, i.e., its triangle has coprime integral side lengths. It follows that n is a congruent number if and only if n has the same square-free part as $ab(a + b)(a - b)$, for some integers a and b. For example, by taking $(a, b) = (5, 4)$, $(2, 1)$, and $(16, 9)$, note that $5, 6, 7$ are congruent numbers with corresponding triangles $(20/3, 3/2, 41/6)$, $(3, 4, 5)$, and $(24/5, 35/12, 337/60)$. To consider CNP, it is enough to consider square-free integers. In *Liber Quadratorum*, Fibonacci constructed these right triangles and also claimed that 1 is not a congruent number, but did not give a proof.

In 1640, Fermat discovered his infinite descent method to show that $1, 2, 3$ are noncongruent numbers. The same method could be employed to find more noncongruent numbers, for example, any prime $p \equiv 3 \pmod 8$. In fact, suppose such a prime p is a congruent number, then there exists a primitive Pythagorean triple $(a^2 - b^2, 2ab, a^2 + b^2)$ whose area $ab(a + b)(a - b)$ has the square-free part p. Assume the area is minimal. Since $a, b, a + b, a - b$ are coprime to each other, by modulo 8 consideration, we have

$$a = r^2, \quad b = ps^2, \quad a + b = u^2, \quad a - b = v^2$$

for some positive integers r, s, u, v. Note that the Pythagorean triple $(u - v, u + v, 2r)$ is with smaller area, a contradiction.

More examples of congruent and noncongruent numbers (gray for non-congruent numbers) were found:

n mod 8	1	2	3	5	6	7
n	1	2	3	5	6	7
	9	10	11	13	14	15
	17	18	19	21	22	23
	25	26	27	29	30	31
	33	34	35	37	38	39
	41	42	43	45	46	47
	$\vdots$	$\vdots$	$\vdots$	$\vdots$	$\vdots$	$\vdots$
	217	218	219	221	222	223
	$\vdots$	$\vdots$	$\vdots$	$\vdots$	$\vdots$	$\vdots$

From the table, one may conjecture that all the positive integers congruent to $5, 6, 7$ modulo 8 are congruent numbers (conjectured by Alter, Curtz, and Kubota in [2]) and the density of positive integers congruent to $1, 2, 3$ modulo 8 being non-congruent is one.

The arithmetic of elliptic curves, in particular the BSD conjecture and the Goldfeld conjecture, provides a systematical and deeper point of view to study CNP. We now recall these conjectures to introduce notation.

For an elliptic curve A over a number field F, the set $A(F)$ of rational points has a finitely generated abelian group structure by Mordell–Weil theorem. Its rank is denoted by $\mathrm{rank}_{\mathbb{Z}} A(F)$. The Hasse–Weil L-function $L(s, A/F)$ of A is defined as an Euler product and conjectured to be entire and to satisfy a functional equation. The vanishing order of $L(s, A/F)$ at $s = 1$ denoted by $\mathrm{ord}_{s=1} L(s, A/F)$ is called the analytic rank of A/F. When $F = \mathbb{Q}$, the conjecture is known by the work of Wiles [54], et al., and the functional equation is given by

$$\Lambda(s, A/\mathbb{Q}) := N_A^{s/2} \cdot 2(2\pi)^{-s}\Gamma(s)L(s, A/\mathbb{Q}) = \epsilon(A)\Lambda(2-s, A/\mathbb{Q}),$$

where $N_A \in \mathbb{Z}_{\geq 1}$ is the conductor of $A/\mathbb{Q}$, and $\epsilon(A) \in \{\pm 1\}$ is the root number.

Conjecture 1 (BSD). *Let A be an elliptic curve over a number field F. Then the following holds:*

(1) $\mathrm{rank}_{\mathbb{Z}} A(F) = \mathrm{ord}_{s=1} L(s, A/F)$.

(2) *The Tate–Shafarevich group $Ш(A/F)$ is finite. For $r = \mathrm{ord}_{s=1} L(s, A/F)$,*

$$\frac{L^{(r)}(1, A/F)}{r!\Omega_A R_A/\sqrt{|D_F|}} = \frac{\prod_v c_v \cdot \#Ш(A/F)}{\#A(F)_{\mathrm{tors}}^2}.$$

Here D_F is the discriminant of F, while Ω_A, R_A, and c_v are the Néron period, the regulator, and the Tamagawa number of A at place v, respectively.

For a prime p, we call the equality of p-valuation on both sides the p-part BSD formula.

One significant fact related to the BSD conjecture for an elliptic curve A over $\mathbb{Q}$ is that if it holds, then there will be an effective algorithm to compute generators of $A(\mathbb{Q})$ [39]. It is easy to see that a positive integer n is a congruent number if and only if the elliptic curve (called a congruent elliptic curve)

$$E^{(n)} : ny^2 = x^3 - x$$

has Mordell–Weil group $E^{(n)}(\mathbb{Q})$ of positive rank. There exists a one-to-one correspondence between rational right triangles with area n and nontorsion rational points of $E^{(n)}$. In particular, the BSD conjecture for $E^{(n)}$ would solve the CNP.

A fundamental result on the BSD conjecture was obtained by Coates–Wiles [18], Rubin [43], Gross–Zagier [24], and Kolyvagin [36]: If $\mathrm{ord}_{s=1} L(s, A/\mathbb{Q}) \leq 1$, then

$$\mathrm{rank}_{\mathbb{Z}} A(\mathbb{Q}) = \mathrm{ord}_{s=1} L(s, A/\mathbb{Q})$$

and $\#Ш(A/\mathbb{Q}) < \infty$. There are several results on the p-part BSD formula, including Rubin [43], Kato [33], Kolyvagin [36], Skinner–Urban [46], Zhang [56], Jetchev–Skinner–Wan [31]. The full BSD conjecture was verified for a subfamily of congruent elliptic curves, which have both algebraic and analytic rank one.

Theorem 2 ([37]). *Let $n \equiv 5 \pmod 8$ be a square-free positive integer, all of whose prime factors are congruent to 1 modulo 4. Assume that $\mathbb{Q}(\sqrt{-n})$ has no ideal class of order 4, then $E^{(n)} : y^2 = x^3 - n^2 x$ has both algebraic and analytic rank 1 and the full BSD conjecture holds.*

For the above congruent number elliptic curves, the 2-part of the BSD formula is proved in [51], [58]. The p-part of the BSD formula, when $p \geq 3$, $p \nmid n$, is the consequence of works by Perrin-Riou [42], Kobayashi [35], etal. The p-part of the BSD formula, when $p \equiv 1 \pmod 4$, $p \mid n$, is proved in Li-Liu-Tian [37]. The generalization of Kobayashi's work to potential supersingular primes together with the argument of Perrin-Riou [42], also implies the p-part BSD formula for primes p of potential supersingular reduction (see [41]).

There is a conjecture on statistical behaviors of analytic ranks for a quadratic twist family of elliptic curves. For an elliptic curve $y^2 = f(x)$ over F, its quadratic twist family consists of elliptic curves $ny^2 = f(x)$ with $n \in F^\times$. Based on minimalist principle, Goldfeld proposed the following:

Conjecture 3 (Goldfeld [14,23]). *Let $\varepsilon \in \{\pm 1\}$ and $\mathcal{A}$ be a quadratic twist family of elliptic curves over F. Then, ordered by norms of conductors, among the quadratic twists $A \in \mathcal{A}$ with $\epsilon(A) = \varepsilon$,*

$$\mathrm{Prob}\big(\mathrm{ord}_{s=1} L(s, A/F) = 0\big) \quad \big(\textit{resp.}\ \mathrm{Prob}\big(\mathrm{ord}_{s=1} L(s, A/F) = 1\big)\big)$$

is one if $\varepsilon = +1$ (resp. -1). In particular, if $F = \mathbb{Q}$, as A runs over a quadratic twist family of elliptic curves,

$$\mathrm{Prob}\big(\mathrm{ord}_{s=1} L(s, A/\mathbb{Q}) = r\big)$$

is equal to $1/2$ for $r = 0, 1$, and 0 for $r \geq 2$.

We refer to $\epsilon = 1$ (resp. -1) case of the conjecture as the even (resp. odd) parity Goldfeld conjecture. The significance of Goldfeld conjecture is that, together with the Gross–Zagier formula (see Section 2), it solves the problem of finding generators of $A(\mathbb{Q})$ for density-one elliptic curves A in a quadratic twist family.

Conjecture 4 (Goldfeld [23], Katz–Sarnak [34], etc.). *Let A run over all elliptic curves over a fixed number field F as ordered by height, then*

$$\text{Prob}\big(\text{ord}_{s=1} L(s, A/F) = r\big)$$

is equal to $1/2$ for $r = 0, 1$, and 0 for $r \geq 2$.

For $n \in \mathbb{Z}_{\geq 0}$, we have

$$\epsilon\big(E^{(n)}\big) = \begin{cases} +1, & n \equiv 1, 2, 3 \pmod 8, \\ -1, & n \equiv 5, 6, 7 \pmod 8). \end{cases}$$

The central L-value of $E^{(n)}$ is related to the following ternary quadratic equation by Tunnell [52]: For a positive square-free integer n, let $a = 1$ if n is odd and $a = 2$ if n is even. Consider the equation

$$2ax^2 + y^2 + 8z^2 = n/a, \quad x, y, z \in \mathbb{Z}.$$

Let $\Sigma(n)$ be the set of its solutions and let

$$\mathcal{L}(n) = \#\big\{(x, y, z) \in \Sigma(n) \mid 2 \mid z\big\} - \#\big\{(x, y, z) \in \Sigma(n) \mid 2 \nmid z\big\}.$$

It is easy to see that $\mathcal{L}(n) = 0$ for positive $n \equiv 5, 6, 7 \pmod 8$. Tunnell proved that for n positive square-free, $\mathcal{L}(n) \neq 0$ if and only if $L(1, E^{(n)}) \neq 0$. The BSD conjecture predicts the following:

Conjecture A. *A positive square-free integer n is a congruent number if and only if $\mathcal{L}(n) = 0$. In particular, any positive integer $n \equiv 5, 6, 7 \pmod 8$ is a congruent number.*

One can determine whether $\mathcal{L}(n) = 0$ in finitely many steps, yet there is no algorithm to find all the rational points of $E^{(n)}$. Tunnell's work was recently generalized to any given quadratic twist family of elliptic curves over $\mathbb{Q}$ in [26].

The even Goldfeld conjecture for the family $E^{(n)}$ can be stated as follows:

Conjecture B1. *Among all square-free positive integers $n \equiv 1, 2, 3 \pmod 8$, the subset of n with $\mathcal{L}(n) \neq 0$ has density one.*

For an elliptic curve $A/\mathbb{Q}$ with root number -1, the BSD conjecture predicts that $A(\mathbb{Q})$ has an infinite-order point. Heegner point construction provides a systematic method to construct rational points. We now give a concrete construction for congruent elliptic curves $E^{(n)}$ with $n \equiv 5, 6, 7 \pmod 8$. Denote by E the elliptic curve $y^2 = x^3 - x$ that has conductor 32. The Abel–Jacobi map induces the complex uniformization

$$E(\mathbb{C}) \simeq \mathbb{C}/\Lambda_E, \quad z \mapsto \int_O^z dx/2y,$$

where $\Lambda_E = \{\int_\gamma dx/2y \mid \gamma \in H_1(E(\mathbb{C}), \mathbb{Z})\} \subset \mathbb{C}$ is the period lattice. Denote by ϕ the newform of weight 2 and level $\Gamma_0(32)$ associated to E. Let f be the analytic map

$$f : X_0(32)(\mathbb{C}) \to E(\mathbb{C})$$

induced by the above complex uniformization $E(\mathbb{C}) \simeq \mathbb{C}/\Lambda_E$ and

$$\mathcal{H} \to \mathbb{C} : \tau \mapsto \int_{i\infty}^{\tau} 2\pi i \phi(z) dz.$$

We now give a construction of Heegner points. For n a positive square-free integer $\equiv 5, 6, 7 \pmod 8$, let $K = \mathbb{Q}(\sqrt{-n})$, let $\mathcal{O}$ be its ring of integers, and H its Hilbert class field. Let c be the complex conjugation and let $E(K)^{c=-1} \subset E(K)$ be the subgroup on which c acts by -1, then we naturally have $E(K)^{c=-1} \simeq E^{(n)}(\mathbb{Q})$.

Define the Heegner point, which lies in $E^{(n)}(\mathbb{Q}) \otimes \mathbb{Q}$, as follows:

$$y_n := \begin{cases} \mathrm{tr}_{H/K}(1 - [i]) \cdot f(\tau_n), \\ \mathrm{tr}_{H/K}[i] \cdot f(\tau_n), \\ 2\,\mathrm{tr}_{H/K}\, f(\tau_n), \end{cases} \quad \text{where } \mathcal{H} \ni \tau_n = \begin{cases} \frac{1}{4(1-\sqrt{-n})}, \\ \frac{-1}{4\sqrt{-n}}, \\ \frac{-2}{\epsilon+\sqrt{-n}}, \end{cases} \quad \text{if } n \equiv \begin{cases} 5 \pmod 8, \\ 6 \pmod 8, \\ 7 \pmod 8. \end{cases}$$

Here ϵ is an integer such that $\epsilon^2 \equiv -n \pmod{128}$. The construction is natural from an automorphic representation point of view, which will be described in Section 2.

Conjecture B2. *Among all positive integers* $n \equiv 5, 6, 7 \pmod 8$, *the subset of* n *with* y_n *being nontorsion has density one.*

The Gross–Zagier formula (see Section 2) implies that y_n is nontorsion if and only if $L'(1, E^{(n)}) \neq 0$. Furthermore, Kolyvagin's work shows that if y_n is nontorsion, then the rank of $E^{(n)}(\mathbb{Q})$ is one [36]. The Gross–Zagier formula also helps compute y_n and therefore a generator of $E^{(n)}(\mathbb{Q})$ [19].

Remark 5. The combination of *Conjectures B1* and *B2* is equivalent to Goldfeld conjecture for congruent elliptic curves, which would solve the CNP for integers with probability one.

Example 1. For $n = 101, 102,$ and 103, the Heegner point y_n is given by

$$\left(\frac{-3975302500}{442723681}, \frac{2808122994457950}{9315348971921}\right), \ (5100, 364140),$$

and

$$\left(\frac{-777848715219380607}{8780605285453456}, \frac{406939902409963977921570495}{822785599723202981879104}\right).$$

And right triangles with area n corresponding to y_n have side lengths

$$\left(\frac{267980280100}{44538033219}, \frac{44538033219}{1326635050}, \frac{20152424629497600 01961}{59085715926389725950}\right), \ \left(\frac{20}{7}, \frac{357}{5}, \frac{2501}{35}\right),$$

and

$$\left(\frac{16286253110943816}{441394452081515}, \frac{45463628564396045}{8143126555471908}, \frac{134130664938047228374702001079697}{3594330884182957394223708580620}\right).$$

Heegner [28] in 1952 showed that any prime or double prime $n \equiv 5, 6, 7 \pmod 8$ is a congruent number. Later on, based on Heegner's method, Monsky [40] in 1990 proved that for $(p_1, p_2) \equiv (1, 5) \pmod 8$ (resp. $(p_1, p_2) \equiv (1, 7) \pmod 8$), two primes such that $(\frac{p_1}{p_2}) = -1$, the product $p_1 p_2$ (resp. $2p_1 p_2$) is a congruent number. A natural question is to seek congruent numbers with many prime factors. The following was first conjectured by Monsky in [40].

Theorem 6 (Tian [50]). *Let n be the product of an odd number of primes $\equiv 5 \pmod 8$ that are not quadratic residues to each other, then n is a congruent number.*

Theorem 7 (Burungale–Tian [11]). Conjecture B1 *is true, namely among all square-free positive integers $n \equiv 1, 2, 3 \pmod 8$, the subset of n with $\mathcal{L}(n) \neq 0$ has density one. In particular, the density of noncongruent numbers among square-free positive integers $\equiv 1, 2, 3 \pmod 8$ is one.*

Let S be the subset of positive square-free integers $n \equiv 5, 6, 7 \pmod 8$ so that $\dim_{\mathbb{F}_2} \mathrm{Sel}_2(E^{(n)}/\mathbb{Q})/E(\mathbb{Q})[2] = 1$. By the results on distribution of 2-Selmer groups [27, 32, 49], the density of S for all the positive square-free integers $n \equiv 5, 6, 7 \pmod 8$ is

$$2\prod_{j\geq 1}(1+2^{-j})^{-1}.$$

Theorem 8 ([47, 50, 51]). *There is a density-$\frac{2}{3}$ subset of S so that the analytic rank of $E^{(n)}$ is one and the 2-part BSD formula holds. In particular, among all the square-free positive integers $n \equiv 5, 6, 7 \pmod 8$, the density of congruent numbers is greater than*

$$\frac{4}{3}\prod_{j\geq 1}(1+2^{-j})^{-1} \quad \left(> \frac{1}{2}\right).$$

The strategy of the proof of Theorems 6 and 8 (resp. Theorem 7) will be given in Section 2 (resp. Section 3).

2. HEEGNER POINT AND EXPLICIT GROSS–ZAGIER FORMULA

Heegner points and Gross–Zagier formula play an important role in the study of elliptic curves. The work of Yuan, Zhang, and Zhang [55] gives the general construction of Heegner points on Shimura curves over totally real fields and establishes the general Gross–Zagier formula. Some arithmetic applications require an explicit form of the formula such as that in [24]. In this section, we introduce the explicit Gross–Zagier formula from [15] and its application to CNP.

We assume as given:

- A—an elliptic curve over $\mathbb{Q}$ with conductor N,
- K—an imaginary quadratic field with discriminant D,
- χ—a ring class character of K with conductor c, which can be viewed as a character on $\mathrm{Gal}(H_c/K)$, where H_c is the ring class field of K with conductor c,

characterized by the reciprocity law

$$t : \mathrm{Gal}(H_c/K) \xrightarrow{\sim} K^\times \backslash \hat{K}^\times / \hat{\mathcal{O}}_c^\times.$$

Here, $\mathcal{O}_c = \mathbb{Z} + c\mathcal{O}_K$ is an order of $\mathcal{O}_K$. For any abelian group M, denote $\hat{M} = M \otimes_{\mathbb{Z}} \hat{\mathbb{Z}}$ with $\hat{\mathbb{Z}} = \prod_p \mathbb{Z}_p$.

Assume that the Rankin–Selberg L-series $L(s, A, \chi)$ associated to (A, χ) has sign -1 in its function equation.

In the following, we shall introduce the construction of the Heegner points and the explicit Gross–Zagier formula for (A, χ) under the assumption $(c, N) = 1$. Let B be the unique indefinite quaternion algebra over $\mathbb{Q}$ whose ramified places are given by all p such that

$$\epsilon_p(A, \chi) = -\chi_p \eta_p(-1),$$

where $\epsilon_p(A, \chi)$ is the local root number of $L(s, A, \chi)$ at p and η_p is the quadratic character of $\mathbb{Q}_p^\times$ associated to $K_p/\mathbb{Q}_p$. In particular, there exists an embedding of K into B. Fix such an embedding once and for all. An order R of B is called *admissible* with respect to (A, χ) if the discriminant of R is N and $R \cap K = \mathcal{O}_c$. Such an order exists and is unique up to conjugaction by $\hat{K}^\times$. Fix such an admissible order R.

Denote by $X_{\hat{R}^\times}$ the Shimura curve over $\mathbb{Q}$ associated to B of level $\hat{R}^\times$. Under an isomorphism $B(\mathbb{R}) \simeq M_2(\mathbb{R})$, it has the following complex uniformization:

$$X_{\hat{R}^\times}(\mathbb{C}) = B^\times \backslash \mathcal{H}^\pm \times \hat{B}^\times / \hat{R}^\times \sqcup \{\text{cusps}\}.$$

Denote by $[z, g]_{\hat{R}^\times}$ the image of $(z, g) \in \mathcal{H}^\pm \times \hat{B}^\times$ in $X_{\hat{R}^\times}(\mathbb{C})$. Let $\xi_{\hat{R}^\times} \in \mathrm{Pic}(X_{\hat{R}^\times}) \otimes \mathbb{Q}$ be the normalized Hodge class with degree 1 on each connected component of $X_{\hat{R}^\times, \bar{\mathbb{Q}}}$ (see [55, SECTION 3.1.3]).

The following proposition follows from the modularity theorem and the Jacquet–Langlands correspondence.

Proposition 9 ([15, PROPOSITION 3.8]). *Up to scalars, there is a unique nonconstant morphism $f : X_{\hat{R}^\times} \to A$ over $\mathbb{Q}$ satisfying the following properties:*

- *f sends $\xi_{\hat{R}^\times}$ to the identity O of A in the sense that if $\xi_{\hat{R}^\times}$ is represented by a divisor $\sum n_i x_i$ on $X_{\hat{R}^\times, \overline{\mathbb{Q}}}$, then $\sum n_i f(x_i) = O$.*

- *For each place $p|(N, D)$,*

$$T_{\varpi_p} f = \chi_p^{-1}(\varpi_p) f.$$

Here, T_{ϖ_p} is the automorphism of $X_{\hat{R}^\times}$, which on $X_{\hat{R}^\times}(\mathbb{C})$ is given by $[z, g]_{\hat{R}^\times} \mapsto [z, g\varpi_p]_{\hat{R}^\times}$, with $\varpi_p \in K_p^\times$ being any uniformizer of K_p.

Let $z \in \mathcal{H}$ be the unique point fixed by $K^\times$ and let $P = [z, 1]_{\hat{R}^\times}$. By the theory of complex multiplication, P is defined over the ring class field H_c of K with conductor c and the Galois action is given by

$$[z, 1]_{\hat{R}^\times}^\sigma = [z, t_\sigma]_{\hat{R}^\times}, \quad \sigma \in \mathrm{Gal}(H_c/K),$$

where $\sigma \mapsto t_\sigma$ is the reciprocity map.

Define the Heegner point

$$P_\chi(f) = \sum_{\sigma \in \mathrm{Gal}(H_c/K)} f(P^\sigma)\chi(\sigma) \in A(H_c) \otimes \mathbb{Q}(\chi).$$

Here $\mathbb{Q}(\chi)$ is the field over $\mathbb{Q}$ generated by image of χ.

Theorem 10 ([15, 55]). *Assume $(N, c) = 1$. Then*

$$L'(1, A, \chi) = 2^{-\mu(N,D)} \frac{8\pi^2(\phi, \phi)_{\Gamma_0(N)}}{u^2\sqrt{|Dc^2|}} \cdot \frac{\hat{h}_K(P_\chi(f))}{\deg f}.$$

Here, ϕ is the newform associated to A with

$$(\phi, \phi)_{\Gamma_0(N)} = \iint_{\Gamma_0(N)\backslash\mathcal{H}} |\phi(x + iy)|^2 dxdy,$$

$u = [\mathcal{O}_c^\times : \mathbb{Z}^\times]$, *$\mu(N, D)$ is the number of common prime factors of N and D, $\hat{h}_K$ is the Néron–Tate height on A over K, and* $\deg(f)$ *is the degree of the morphism f.*

Remark 11.

(1) To compute Heegner points (if non-torsion) via CM theory and modular parameterization, one only gets an approximation. The precise computation can be carried out since one knows the height of Heegner point via the above formula (see [53]).

(2) One may use different Heegner points to construct rational points on A by choosing different K. The case $(D, N) \neq 1$ sometimes provides points with smaller height. The above formula with $(D, N) \neq 1$ was conjectured by Gross and Hayashi in [25] and employed in [53] for the computation of rational points.

Some arithmetic problems lead to the situation $(c, N) \neq 1$. Consider the following: a nonzero rational number is called a *cube sum* if it is of form $a^3 + b^3$ with $a, b \in \mathbb{Q}^\times$. For any $n \in \mathbb{Q}^\times$, consider the elliptic curve $C_n : x^3 + y^3 = 2n$. If n is not a cube, then $2n$ is a cube sum if and only if the rank of $C_n(\mathbb{Q})$ is positive.

Theorem 12 ([16]). *For any odd integer $k \geq 1$, there exist infinitely many cube-free odd integers n with exactly k distinct prime factors such that*

$$\mathrm{rank}_\mathbb{Z}\, C_n(\mathbb{Q}) = 1 = \mathrm{ord}_{s=1} L(s, C_n).$$

Here, a certain Heegner point is considered for the pair (A, χ) where $A = X_0(36) : x^2 = y^3 + 1$ is an isogeny to C_1 and χ is a certain cubic ring class character over the imaginary quadratic field $\mathbb{Q}(\sqrt{-3})$ with conductor $3n^*$ where n^* is the product of prime factors in n. In particular, the pair (A, χ) has joint ramification at the prime 3.

In fact, the explicit Gross–Zagier formula is proved for any pair (π, χ) where

- π is a cuspidal automorphic representation on GL_2 over a totally real field F with central character ω_π, discrete series of weight 2 at all archimedean places,

- $\chi : K^\times \backslash \hat{K}^\times \to \mathbb{C}^\times$ is a character of finite order for a totally imaginary quadratic extension K over F such that

  (i) $\omega_\pi \cdot \chi|_{\mathbb{A}_F^\times} = 1$,

  (ii) the root number of the Rankin–Selberg L-series $L(s, \pi, \chi)$ is -1.

Based on the work of Yuan–Zhang–Zhang **[55]**, the above explicit formula is established via generalizing Gross–Prasad local test vector theory.

The relevant problem in local harmonic analysis is the following. Let $\mathcal{B}$ be a quaternion algebra over a local field $\mathcal{F}$ with a quadratic sub-$\mathcal{F}$-algebra $\mathcal{K}$. Let π be an irreducible smooth admissible representation on $\mathcal{B}^\times$ which is of infinite dimension if $\mathcal{B}$ is split. Let χ be a character of $\mathcal{K}^\times$ such that $\chi|_{\mathcal{F}^\times}$ is the inverse of the central character of π. Consider the functional space

$$\mathcal{P}(\pi, \chi) := \mathrm{Hom}_{\mathcal{K}^\times}(\pi, \chi^{-1}).$$

In general, $\dim_{\mathbb{C}} \mathcal{P}(\pi, \chi) \leq 1$. In the case $\mathcal{P}(\pi, \chi) \neq 0$, a vector φ is called a *test vector* for (π, χ) if $\ell(\varphi) \neq 0$ for any nonzero $\ell \in \mathcal{P}(\pi, \chi)$.

Moreover, for an unitary (π, χ), $\mathcal{P}(\pi, \chi)$ is nonzero if and only if the bilinear form $\alpha \in \mathcal{P}(\pi, \chi) \otimes \mathcal{P}(\bar{\pi}, \bar{\chi})$ defined as the toric integral of matrix coefficients

$$\alpha(\varphi_1 \otimes \varphi_2) = \int_{\mathcal{F}^\times \backslash \mathcal{K}^\times} \langle \pi(t)\varphi_1, \varphi_2 \rangle \chi(t) dt, \quad \varphi_1 \in \pi, \varphi_2 \in \bar{\pi}$$

is nonzero. Here, $\bar{\pi}$ (resp. $\bar{\chi}$) is the complex conjugate of π (resp. χ) and $\langle \cdot, \cdot \rangle$ is a nondegenerate invariant pairing on $\pi \otimes \bar{\pi}$. In particular, if $\mathcal{P}(\pi, \chi) \neq 0$, φ is a test vector for (π, χ) if and only if $\alpha(\varphi, \bar{\varphi}) \neq 0$.

For any pair (π, χ) as above, in **[15]** we find an *admissible* order $\mathcal{R}$ for (π, χ) which is unique up to $\mathcal{K}^\times$-conjugacy. The invariant subspace $\pi^{\mathcal{R}^\times}$ of π by $\mathcal{R}^\times$ is at most of dimension 2. By studying the toric integral α, there is a line in $\pi^{\mathcal{R}^\times}$ containing test vectors for (π, χ).

Our explicit Gross–Zagier formula satisfies the following properties: First, the test vector only depends on the local type π_v, χ_v, for v dividing the conductor of π. It is useful when considering horizontal variation (quadratic twist, for example), see **[7, 13]**, or vertical variation (in Iwasawa theory) of the character χ. We also have a so-called S-version formula which says that for a different choice of a pure tensor test vector, for example, at a finite set of places S, the new explicit formula can be obtained by modifying the original one by local computations at S, for example, see **[16]**.

In the rest of this section, we sketch a proofs of Theorems 6 and 8. In Heegner's work, the point y_n is not 2-divisible. In **[50, 51]**, Heegner's results were generalized to many prime factors by induction on 2-divisibility of Heegner points via the Waldspurger and Gross–Zagier formulas.

For $E : y^2 = x^3 - x$, $K = \mathbb{Q}(\sqrt{-n})$, $n \equiv 5, 6, 7$ positive square-free, the explicit Gross–Zagier formula for (E, K) gives

$$\hat{h}_{\mathbb{Q}}(y_n) = 2^a \frac{L'(1, E^{(n)})}{\Omega_{E^{(n)}}}$$

where $a = 1, 0, 1$ in the case $n \equiv 5, 6, 7 \pmod 8$, respectively. Now if y_n is nontorsion, then the BSD conjecture becomes

$$\left[E^{(n)}(\mathbb{Q})/E^{(n)}(\mathbb{Q})_{\mathrm{tor}} : \mathbb{Z} \cdot y_n\right] = 2^{\mu(n)-1}\sqrt{\#\text{Ш}(E^{(n)}/\mathbb{Q})},$$

where $\mu(n)$ is the number of odd prime factors of n. If n is a prime then the 2-part of the BSD conjecture is equivalent to 2-indivisibility of y_n, this is exactly Heegner's case. As $\mu(n)$ becomes large, the 2 divisibility of y_n becomes high and the original Heegner's argument does not work directly. Whenever $\dim_{\mathbb{F}_2} \mathrm{Sel}_2(E^{(n)}/\mathbb{Q})/E^{(n)}(\mathbb{Q})[2] = 1$, the 2-divisibility of y_n fully comes from Tamagawa numbers. The 2-divisibility can be proved via induction; to do this, one employs various relations between different Heegner points.

We employ the induction method (see **[50]**) in the case $n \equiv 5 \pmod 8$ with all prime factors $\equiv 1 \pmod 4$. Let $z_n := f(\tau_n)$ and let y_n be the Heegner points as in the Section 1. Denote by H the Hilbert class field of $K = \mathbb{Q}(\sqrt{-n})$ and let $H_0 \subset H$ be the genus subfield determined by $\mathrm{Gal}(H_0/K) \simeq 2\,\mathrm{Cl}(K)$. For each $d|n$ with the same above property as n, let $y_0 = \mathrm{tr}_{H/H_0} z_n$ and $y_{d,0} = \mathrm{tr}_{H/K(\sqrt{-d})} z_n$. Then these points satisfy the following relation:

$$y_n + \sum_{\substack{1 \le d|n, d \ne n,\\ d \equiv 5 \pmod 8}} y_{d,0} = 2^{\mu(n)} y_0 \pmod{E[2]}.$$

Furthermore, the Gross–Zagier formula implies that whenever $y_{d,0}$ is nontorsion, both $y_{d,0}$ and y_d lie in the one-dimensional space $E(\mathbb{Q}(\sqrt{-d}))^{c=-1} \otimes \mathbb{Q}$ and

$$[y_{d,0} : y_d]^2 = \frac{L^{\mathrm{alg}}(1, E^{(n/d)})}{L^{\mathrm{alg}}(1, E)}, \quad \text{where } L^{\mathrm{alg}}(1, E^{(n/d)}) = \frac{L(1, E^{(n/d)})}{\Omega_{E^{(n/d)}}}.$$

By induction on the 2 divisibility of y_d and 2 divisibility of $\frac{L^{\mathrm{alg}}(1,E^{(n/d)})}{L^{\mathrm{alg}}(1,E)}$, one gets the following 2 divisibility of y_n whenever $\mathrm{Cl}(K)$ has no element of order 4:

$$y_n \in \left(2^{\mu(n)-1} E(K)^{c=-1} + E(K)_{\mathrm{tor}}\right) \setminus \left(2^{\mu(n)} E(K)^{c=-1} + E(K)_{\mathrm{tor}}\right).$$

Thus Theorem 6 follows.

The above induction argument was improved in **[51]** to handle the general case. For a positive integer d, let $g(d) = \#2\,\mathrm{Cl}(\mathbb{Q}(\sqrt{-d}))$ be the genus class number of $\mathbb{Q}(\sqrt{-d})$. For $n \equiv 5, 6, 7 \pmod 8$, let

$$\mathcal{L}(n) = \begin{cases} [L'(1, E^{(n)})/(2^{2\mu(n)-2-a(n)} \cdot \Omega_{E^{(n)}} R_{E^{(n)}})]^{1/2}, & \text{if } \mathrm{ord}_{s=1} L(1, E^{(n)}) = 1, \\ 0, & \text{otherwise}, \end{cases}$$

where $a(n) = 0$ if n is even and $a(n) = 1$ if n is odd.

Then the BSD conjecture for $E^{(n)}$ is equivalent to $\mathcal{L}(n)^2 = \#\text{Ш}(E^{(n)}/\mathbb{Q})$ whenever $L'(1, E^{(n)}) \neq 0$. We have the following criterion for the 2-indivisibility of $\mathcal{L}(n)$:

Theorem 13 ([51]). *For $n \equiv 5, 6, 7 \pmod 8$ positive square-free, $\mathcal{L}(n)$ is an integer and $2^{-\rho(n)}\mathcal{L}(n)$ is odd if*

$$\begin{cases} \sum_{\substack{n=d_0\cdots d_\ell \\ d_i\equiv 1 \pmod 8, i>0}} \prod_i g(d_i) \equiv 1 \pmod 2 \text{ or} \\ \sum_{\substack{n=d_0\cdots d_\ell, \\ d_0\equiv 5,7 \pmod 8 \\ d_1\equiv 1,3 \pmod 8 \\ d_i\equiv 1 \pmod 8, i>1}} \prod_i g(d_i) \equiv 1 \pmod 2, & \text{if } n \equiv 5, 7 \pmod 8, \\ \sum_{\substack{n=d_0\cdots d_\ell, \\ d_0\equiv 5,6,7 \pmod 8 \\ d_1\equiv 1,2,3 \pmod 8 \\ d_i\equiv 1 \pmod 8, i>1}} \prod_i g(d_i) \equiv 1 \pmod 2, & \text{if } n \equiv 6 \pmod 8, \end{cases}$$

where $\rho(n)$ is a positive integer (defined in [51]) arising from an isogeny between $E^{(n)}$ and $2ny^2 = x^3 + x$.

Let

$$\begin{aligned} s(n) &= \dim_{\mathbb{F}_2} \mathrm{Sel}_2\big(E^{(n)}/\mathbb{Q}\big)/E^{(n)}(\mathbb{Q})[2] \\ &= \mathrm{rank}_{\mathbb{Z}}\, E^{(n)}(\mathbb{Q}) + \dim_{\mathbb{F}_2} Ш\big(E^{(n)}/\mathbb{Q}\big)[2]. \end{aligned}$$

Consider the following sets for $i = 5, 6, 7$:

- Σ_i—the set of all square-free positive integers $n \equiv i \pmod 8$,
- $\Sigma'_i \subset \Sigma_i$—the subset of n with $s(n) = 1$,
- $\Sigma''_i \subset \Sigma_i$—the subset of n satisfying the conditions in the Theorem 13.

Theorem 14 (Heath-Brown [27], Swinnerton-Dyer [49], Kane [32]). *The density of Σ'_i in Σ_i is*

$$2\prod_{k=1}^{\infty}\big(1+2^{-k}\big)^{-1} = 0.8388\ldots$$

Theorem 15 (Smith [47]). *The set Σ''_i is contained in Σ'_i with density $\frac{3}{4}$, $\frac{1}{2}$, $\frac{3}{4}$ for $i = 5, 6, 7$, respectively.*

Observe that Theorem 8 is a consequence of Theorems 13, 14, and 15.

3. SELMER GROUPS: P-CONVERSE AND DISTRIBUTION

The n-Selmer group for an elliptic curve A over a number F is defined by

$$\mathrm{Sel}_n(A/F) = \ker\Big(H^1\big(F, A[n]\big) \to \prod_v H^1(F_v, A)\Big)$$

and fits into the short exact sequence

$$0 \to A(F)/nA(F) \to \mathrm{Sel}_n(A/F) \to Ш(A/F)[n] \to 0.$$

The group $\mathrm{Hom}(\mathrm{Sel}_{p^\infty}(A/F), \mathbb{Q}_p/\mathbb{Z}_p)$ is known to be a finitely generated $\mathbb{Z}_p$-module, its rank of free part is called p^∞-Selmer corank of A, denoted by $\mathrm{corank}_{\mathbb{Z}_p} \mathrm{Sel}_{p^\infty}(A/F)$.

Conjecture 16 (BSD, reformulation). *Let A/F be an elliptic curve over a number field, $r \in \mathbb{Z}_{>0}$, and p be a prime. Then the following are equivalent:*

(1) $\operatorname{ord}_{s=1} L(s, A/F) = r$,

(2) $\operatorname{rank}_{\mathbb{Z}} A(F) = r$ *and* $Ш(A/F)$ *is finite,*

(3) $\operatorname{corank}_{\mathbb{Z}_p} \operatorname{Sel}_{p^\infty}(A/F) = r$.

Notation (*p*-**converse**). *The implication*

$$\operatorname{corank}_{\mathbb{Z}_p} \operatorname{Sel}_{p^\infty}(A/F) = r \Longrightarrow \operatorname{ord}_{s=1} L(s, A/F) = r$$

is referred to as rank r p-converse.

We can also consider a Selmer variant of Goldfeld's Conjecture 3. The following was conjectured by Bhargava–Kane–Lenstra–Poonen–Rains.

Conjecture 17 ([4]). *Let F be a global field, p be a prime, and G a finite symplectic p-group. If all elliptic curves A over F are ordered by height, then for $r = 0, 1$ we have*

$$\operatorname{Prob}\big(\operatorname{Sel}_{p^\infty}(A/F) \simeq (\mathbb{Q}_p/\mathbb{Z}_p)^r \oplus G\big) = \frac{1}{2} \cdot \frac{(\#G)^{1-r}}{\#\operatorname{Sp}(G)} \cdot \prod_{i \geq r} \big(1 - p^{1-2i}\big).$$

In particular, the density of rank 0 (or 1) elliptic curves over F is $\frac{1}{2}$.

3.1. Distribution of Selmer groups and Goldfeld conjecture

The following Smith's result shows that even for a quadratic twist family, the distribution follows the same pattern as the above conjecture.

Theorem 18 (Smith [48]). *Let $A/\mathbb{Q}$ be an elliptic curve satisfying A has full rational 2-torsion, but no rational cyclic subgroup of order 4. Then, among the quadratic twists $A^{(d)}$ of A, a distribution law as in Conjecture 17 holds for $p = 2$.*

In particular, among all quadratic twists $A^{(d)}$ with sign $+1$ (resp. -1), there is a subset of density one with $\operatorname{corank}_{\mathbb{Z}_2} \operatorname{Sel}_{2^\infty}(A^{(d)}/\mathbb{Q}) = 0$ *(resp. 1).*

Remark 19. Smith's work is based on the following results of Heath-Brown, Swinnerton-Dyer, and Kane on distribution of 2-Selmer groups, which is the first step to understand the distribution of 2^∞-Selmer groups.

Theorem 20 ([27, 32, 49]). *Let $A/\mathbb{Q}$ be an elliptic curve satisfying*

- *A has full rational 2-torsion, but no rational cyclic subgroup of order 4.*

Then for $r \in \mathbb{Z}_{\geq 0}$, among the quadratic twists $A^{(d)}$ of A,

$$\operatorname{Prob}\big(\dim_{\mathbb{F}_2} \operatorname{Sel}_2(A^{(d)}/\mathbb{Q})/A^{(d)}(\mathbb{Q})[2] = r\big) = \prod_{j=0}^{\infty} \big(1 + 2^{-j}\big)^{-1} \prod_{i=1}^{r} \frac{2}{2^i - 1}.$$

In general, for a quadratic twist family of elliptic curves over $\mathbb{Q}$, its distribution of 2-Selmer groups may exhibit new behavior. For example, the quadratic twist family of Tiling number elliptic curves has

$$A^{(d)} : dy^2 = x(x-1)(x+3) \quad \text{with } A^{(1)}(\mathbb{Q})_{\text{tor}} \cong \mathbb{Z}/2\mathbb{Z} \times \mathbb{Z}/4\mathbb{Z}.$$

Perhaps surprisingly, in light of the presence of such rational 4-torsion, the distribution of 2-Selmer groups no longer seems to be as in Theorem 20. For example, if $d \neq 1$, $d \equiv 1 \pmod{12}$ is positive square-free, then

$$\dim_{\mathbb{F}_2} \mathrm{Sel}_2\big(A^{(-d)}/\mathbb{Q}\big)/A^{(-d)}(\mathbb{Q})[2] \geq 2.$$

A preliminary study suggests that for such elliptic curves, the distribution of 2-Selmer groups may look more like that of the 4-ranks of ideal class groups of the underlying imaginary quadratic fields.

Theorem 21 ([22]). *Let A be the elliptic curve $y^2 = x(x-1)(x+3)$. Among the set of positive square-free integers $d \equiv 7 \pmod{24}$ (resp. $d \equiv 3 \pmod{24}$), the subset of d such that both of $A^{(\pm d)}$ have $\mathrm{Sel}_2(A^{(\pm d)}/\mathbb{Q})/A^{(\pm d)}(\mathbb{Q})[2]$ trivial has density $\frac{1}{2}\prod_{i=1}^{\infty}(1-2^{-i}) >$ 14.4% (resp. of density $\prod_{i=1}^{\infty}(1-2^{-i}) > 28.8\%$).*

In general, for $r \in \mathbb{Z}_{\geq 0}$, for the set of positive square-free integers $d \equiv 3 \pmod{24}$,

$$\mathrm{Prob}\big(\dim_{\mathbb{F}_2} \mathrm{Sel}_2\big(A^{(d)}/\mathbb{Q}\big)/A^{(d)}(\mathbb{Q})[2] = 2r\big)$$
$$= \left(\sum_{k=0}^{r} 2^{-(r+k)(3r+3k-1)/2} \prod_{i=1}^{r+k}\big(1-2^{-i}\big)^{-2} \prod_{i=0}^{2k-1}\big(2^{r+k-i}-1\big) \prod_{i=1}^{k} \frac{4^{i-1}}{4^i-1}\right) \cdot \prod_{i=1}^{\infty}\big(1-2^{-i}\big).$$

An approach to Goldfeld conjecture. The Goldfeld conjecture for a quadratic twist family of elliptic curves over $\mathbb{Q}$ is a consequence of the following steps:

(1) Distribution of p^{∞}-Selmer groups in the quadratic twist family, which should be a certain variant of general distribution law for all elliptic curves in [4].

(2) The rank zero and rank one p-converse.

Proof of Theorem 7. It is a direct consequence of Tunnell's work on quadratic twist L-values of congruent elliptic curves [52], Theorem 18 of Smith on distribution of 2^{∞}-Selmer groups, and Theorem 22 below on the rank zero p-converse for CM elliptic curves for $p = 2$. ■

3.2. Recent progress: p-converse

In the remaining part of this section, we discuss the p-converse theorem in the CM case. For a few other p-converse theorems, see [6, 8–10]. Fix a prime p.

Theorem 22 (Rubin [43, 44], Burungale–Tian [11]). *Let $A/\mathbb{Q}$ be a CM elliptic curve. Then,*

$$\mathrm{corank}_{\mathbb{Z}_p} \mathrm{Sel}_{p^{\infty}}(A/\mathbb{Q}) = 0 \implies \mathrm{ord}_{s=1} L(s, A/\mathbb{Q}) = 0.$$

Remark 23. Assume that $A/\mathbb{Q}$ has CM by K and $p \nmid \#\mathcal{O}_K^{\times}$. Then the above theorem is due to Rubin [43, 44].

Remark 24. Skinner–Urban [46] established the rank zero p-converse for certain elliptic curves over $\mathbb{Q}$ without CM.

Theorem 25 (W. Zhang [56], Skinner [45], Castella–Wan [17]). *Let $A/\mathbb{Q}$ be a non-CM elliptic curve and $p \geq 3$. Then,*

$$\operatorname{corank}_{\mathbb{Z}_p} \operatorname{Sel}_{p^\infty}(A/\mathbb{Q}) = 1 \implies \operatorname{ord}_{s=1} L(s, A/\mathbb{Q}) = 1,$$

under certain assumptions.

Their methods essentially excludes the CM case. For CM elliptic curves:

Theorem 26 (Burungale–Tian [12], Burungale–Skinner–Tian [8]). *Let A be a CM elliptic curve over $\mathbb{Q}$ and $p \nmid 6N_A$ a prime. Then*

$$\operatorname{corank}_{\mathbb{Z}_p} \operatorname{Sel}_{p^\infty}(A/\mathbb{Q}) = 1 \implies \operatorname{ord}_{s=1} L(s, A/\mathbb{Q}) = 1.$$

3.2.1. Rank zero CM p-converse

We outline the proof of Theorem 22. Unconventionally for the CM elliptic curves, this approach is based on Kato's main conjecture [33], which we recall now.

Let $f \in S_k(\Gamma_0(N))$ be an elliptic newform of even weight $k \geq 2$, level $\Gamma_0(N)$, and Hecke field F. Fix an embedding $\iota_p : \overline{\mathbb{Q}} \to \overline{\mathbb{Q}}_p$. Let λ be the place of F induced by ι_p, F_λ be the completion of F at λ, and O_λ the integer ring. Let $V_{F_\lambda}(f)$ be the two-dimensional representation of $G_{\mathbb{Q}}$ over F_λ associated to f. We first introduce the related Iwasawa cohomology. For $n \in \mathbb{Z}_{\geq 0}$, let

$$G_n = \operatorname{Gal}\big(\mathbb{Q}(\zeta_{p^n})/\mathbb{Q}\big), \quad G_\infty = \varprojlim_n G_n = \operatorname{Gal}\big(\mathbb{Q}(\zeta_{p^\infty})/\mathbb{Q}\big).$$

Let $\Lambda = O_\lambda[[G_\infty]]$ be a two-dimensional complete semilocal ring. For $q \in \mathbb{Z}_{\geq 0}$, consider the $\Lambda_{\mathbb{Q}_p} = \Lambda \otimes \mathbb{Q}$-module

$$\mathbb{H}^q\big(V_{F_\lambda}(f)\big) = \varprojlim_n H^q\big(\mathbb{Z}[\zeta_{p^n}, 1/p], T\big) \otimes \mathbb{Q},$$

where $T \subset V_{F_\lambda}(f)$ is any $G_{\mathbb{Q}}$-stable O_λ-lattice and H^q denotes the étale cohomology. The following holds [33]:

(1) $\mathbb{H}^2(V_{F_\lambda}(f))$ is a torsion $\Lambda_{\mathbb{Q}_p}$-module, and

(2) $\mathbb{H}^1(V_{F_\lambda}(f))$ is a free $\Lambda_{\mathbb{Q}_p}$-module of rank one.

Now we introduce the submodule of $\mathbb{H}^1(V_{F_\lambda}(f))$ generated by Beilinson–Kato elements. We have the following existence of zeta elements for the p-adic Galois representation corresponding to an elliptic newform [33, THM. 12.5]:

(1) There exists a nonzero F_λ-linear morphism

$$V_{F_\lambda}(f) \to \mathbb{H}^1\big(V_{F_\lambda}(f)\big); \ \gamma \mapsto z_\gamma(f).$$

(2) Let $Z(f)$ be the $\Lambda_{\mathbb{Q}_p}$-submodule of $\mathbb{H}^1(V_{F_\lambda}(f))$ generated by $z_\gamma(f)$ for all $\gamma \in V_{F_\lambda}(f)$. Then $\mathbb{H}^1(V_{F_\lambda}(f))/Z(f)$ is a torsion $\Lambda_{\mathbb{Q}_p}$-module.

Remark 27. For a characterizing property of the morphism $\gamma \mapsto z_\gamma(f)$ in terms of the underlying critical L-values, we refer to [33, THM. 12.5 (1)].

Conjecture 28 (Kato's Main Conjecture [33]). *The following equality of ideals holds in* $\Lambda_{\mathbb{Q}_p}$*:*

$$\mathrm{Char}\big(\mathbb{H}^2\big(V_{F_\lambda}(f)\big)\big) = \mathrm{Char}\big(\mathbb{H}^1\big(V_{F_\lambda}(f)\big)/Z(f)\big).$$

Theorem 29 (Burungale–Tian [11]). *Kato's main conjecture holds for any CM modular form* f *and any prime* p.

As observed by Kato [33], the CM case of Kato's main conjecture is closely related to an equivariant main conjecture for the underlying imaginary quadratic field. This is based on an intrinsic relation between the Beilinson–Kato elements and elliptic units.

As a consequence of Theorem 29, we have the following result, which implies Theorem 22.

Theorem 30 ([11]). *Assume that* f *is CM. Let* $H^1_{\mathrm{f}}(\mathbb{Q}, V_{F_\lambda}(f)(k/2))$ *be the corresponding Bloch–Kato Selmer group (see Kato [33]). Then,*

$$H^1_{\mathrm{f}}\big(\mathbb{Q}, V_{F_\lambda}(f)(k/2)\big) = 0 \implies \mathrm{ord}_{s=k/2} L(s, f) = 0.$$

3.3. Rank one CM p-converse

In the following, we focus on the proof of the rank one CM p-converse theorem for ordinary primes p. The key is an auxiliary Heegner point main conjecture (HPMC, for short).

Classically, HPMC is only formulated for pairs (A, K') where $A/\mathbb{Q}$ is an elliptic curve and K' is an imaginary quadratic field satisfying the Heegner hypothesis. To show the rank 1 p-converse for a CM elliptic curve, we utilize a certain anticyclotomic Iwasawa theory over the CM field. The key is to construct relevant Heegner points for auxiliary Rankin–Selberg pairs, and consider the underlying HPMC.

Let $A/\mathbb{Q}$ be a CM elliptic curve with CM by K and with p^∞-Selmer corank one. Let λ be the associated Hecke character over K and θ_λ the corresponding theta series.

Lemma 31. *There exists a finite order Hecke character* χ *over* K *such that* $L(1, \lambda^*/\chi^* \cdot \chi) \neq 0$, *where* $*$ *is the involution given by nontrivial automorphism of* K, *so that the* L*-function for the Rankin pair* $(f := \theta_{\lambda/\chi}, \chi)$,

$$L(s, f \times \chi) = L(s, \lambda) L\big(s, \lambda^*/\chi^* \cdot \chi\big),$$

has sign -1 *and the same vanishing order at the center as* $L(s, \lambda) = L(s, A/\mathbb{Q})$.

We have the relevant Heegner point $P_0 \in B(K) \otimes \mathbb{Q}$ on the abelian variety $B := A_{f\times\chi}$ associated to the pair $(f = \theta_{\lambda/\chi}, \chi)$. The Gross–Zagier formula of Yuan–Zhang–Zhang [55] implies that $0 \neq P_0 \in B(K) \otimes \mathbb{Q}$ if and only if $\mathrm{ord}_{s=1} L(s, f \times \chi) = 1$.

Note that p is split in K. Let K_∞/K be the anticyclotomic extension with Galois group $\Gamma \cong \mathbb{Z}_p$. For each $n \geq 1$, let $K_n \subset K_\infty$ be the degree p^n subextension over K.

One can construct a family of norm compatible Heegner points $P_n \in B(K_n)$. Denote by $\Lambda = \mathcal{O}_{\mathfrak{p}}[[\Gamma]]$ and $\Lambda_{\mathbb{Q}_p} = \Lambda \otimes \mathbb{Q}$. Here $\mathcal{O}$ is the endomorphism ring of B (viewed as a subring of $\overline{\mathbb{Q}}$), $\mathfrak{p}|p$ the prime ideal of $\mathcal{O}$ induced by ι_p, and $\mathcal{O}_{\mathfrak{p}}$ the completion of $\mathcal{O}$ at $\mathfrak{p}$.

Proposition 32. *The $\Lambda_{\mathbb{Q}_p}$-modules*

$$S(B/K_\infty) := \big(\varprojlim_n \varprojlim_m \mathrm{Sel}_{\mathfrak{p}^m}(B/K_n)\big)_{\mathbb{Q}_p}, \quad X(B/K_\infty) := \big(\varinjlim_n \varinjlim_m \mathrm{Sel}_{\mathfrak{p}^m}(B/K_n)\big)^\vee_{\mathbb{Q}_p}$$

are finitely generated $\Lambda_{\mathbb{Q}_p}$-modules of rank one. Moreover, the element

$$\kappa = (P_n) \in S(B/K_\infty)$$

is not $\Lambda_{\mathbb{Q}_p}$-torsion so that $S(B/K_\infty)/\Lambda_{\mathbb{Q}_p} \cdot \kappa$ is a finitely generated torsion $\Lambda_{\mathbb{Q}_p}$-module.

Conjecture 33 (HPMC). *With the above notations,*

$$\big(\mathrm{Char}\big(S(B/K_\infty)/(\kappa)\big)\big)^2 = \mathrm{Char}\big(X(B/K_\infty)_{\mathrm{tor}}\big).$$

The control theorem gives

$$\#\big(S(B/K_\infty)/(\kappa)\big)_\Gamma < \infty \Rightarrow 0 \neq P_0 \in B(K),$$
$$\mathrm{corank}_{\mathcal{O}_\mathfrak{p}} \mathrm{Sel}_{\mathfrak{p}^\infty}(B/K) = 1 \Rightarrow \#\big(X(B/K_\infty)_{\mathrm{tor}}\big)_\Gamma < \infty.$$

The rank one p-converse in the CM case is a consequence of HPMC. In fact, by descent,

$$\#\big(X(B/K_\infty)_{\mathrm{tor}}\big)_\Gamma < \infty \Leftrightarrow \#\big(S(B/K_\infty)/(\kappa)\big)_\Gamma < \infty. \quad (*)$$

Now $\mathrm{corank}_{\mathbb{Z}_p} \mathrm{Sel}_{p^\infty}(A/\mathbb{Q}) = 1$ implies that the left-hand side of $(*)$ holds. On the other hand, under the Gross–Zagier formula, $\mathrm{ord}_{s=1} L(s, A/\mathbb{Q}) = 1$ is a consequence of the right-hand side of $(*)$.

First proof of HPMC. The two variable Rankin–Selberg p-adic L-function $\mathcal{L}_p(f \times \chi)$ (see [21]) associated to (f, χ) has a decomposition in terms of $\mathcal{L}_p(\lambda)$ and $\mathcal{L}_p(\lambda^*/\chi^* \cdot \chi)$, where $\mathcal{L}_p(\lambda)$, $\mathcal{L}_p(\lambda^*/\chi^* \cdot \chi)$ are the Katz p-adic L-functions (see [29]) associated to λ and $\lambda^*/\chi^* \cdot \chi$, respectively. Note that $\mathcal{L}_p(f \times \chi)$ and $\mathcal{L}_p(\lambda)$ vanish along the anticyclotomic line, thus we may consider their derivatives with respect to the cyclotomic variable, i.e.,

$$\big(\mathcal{L}'_p(f \times \chi)\big) = \big(\mathcal{L}'_p(\lambda)\big)\big(\mathcal{L}_p(\lambda^*/\chi^* \cdot \chi)\big).$$

The HPMC is based on Λ-adic Gross–Zagier formula and Rubin's main conjecture.

On the one hand, the Λ-adic Gross–Zagier formula [21] connects Heegner point with $\mathcal{L}'_p(f \times \chi)$ as

$$\big(\mathcal{L}'_p(f \times \chi)\big) = \big(\langle\kappa, \kappa\rangle\big) = \mathrm{Char}\big(S(B/K_\infty)/(\kappa)\big) R(f \times \chi),$$

where $\langle \cdot, \cdot \rangle$ is the Λ-adic height pairing and $R(f \times \chi)$ is the Λ-adic regulator of $f \times \chi$ which is nonzero by the rigidity principle [5]. On the other hand, Rubin's main conjecture [1, 43] implies

$$\big(\mathcal{L}'_p(\lambda)\big) = \mathrm{Char}\big(X(\lambda)_{\mathrm{tor}}\big) R(\lambda), \quad \big(\mathcal{L}_p(\lambda^*/\chi^* \cdot \chi)\big) = \mathrm{Char}\big(X\big(\lambda^*/\chi^* \cdot \chi\big)\big),$$

where $R(\lambda)$ is the Λ-adic regulator which is nonzero [5], $X(\lambda)$ and $X(\lambda^*/\chi^* \cdot \chi)$ are certain anticyclotomic Selmer groups. Then, the HPMC follows from the decomposition

$$\mathrm{Char}\big(X(B/K_\infty)_{\mathrm{tor}}\big) = \mathrm{Char}\big(X(\lambda)_{\mathrm{tor}}\big)\, \mathrm{Char}\big(X\big(\lambda^*/\chi^* \cdot \chi\big)\big)$$

and the comparison of Λ-adic regulators, $R(\lambda) = R(f \times \chi)$. ■

Second proof of HPMC. Via Λ-adic Waldspurger formula and nontriviality of κ, the HPMC is equivalent to the BDP main conjecture [45]. Let $p = v\overline{v}$ where v is determined by ι_p. ■

Proposition 34. *The* $\Lambda_{\mathbb{Q}_p}$*-modules*

$$S_v(B/K_\infty) := \big(\varprojlim_n \varprojlim_m \mathrm{Sel}_{\mathfrak{p}^m,v}(B/K_n)\big)_{\mathbb{Q}_p},$$

$$X_v(B/K_\infty) := \big(\varinjlim_n \varinjlim_m \mathrm{Sel}_{\mathfrak{p}^m,v}(B/K_n)\big)^\vee_{\mathbb{Q}_p}$$

are finitely generated torsion $\Lambda_{\mathbb{Q}_p}$*-modules. Here* $\mathrm{Sel}_{\mathfrak{p}^m,v}$ *is the* $\mathfrak{p}^m$*-Selmer group with* v*-relaxed and* $\overline{v}$*-strict local Selmer condition* [45].

Let $\mathcal{L}_v(B/K_\infty)$ be the anticyclotomic BDP p-adic L-function in [3, 38].

Conjecture 35 (BDP Main Conjecture). $\mathrm{Char}(X_v(B/K_\infty)) = (\mathcal{L}_v(B/K_\infty))$.

The $\Lambda_{\mathbb{Q}_p}$-modules $S_v(B/K_\infty)$, $X_v(B/K_\infty)$, $(\mathcal{L}_v(B/K_\infty))$ can be decomposed in terms of Selmer groups and p-adic L-functions of λ, $\lambda^*/\chi^* \cdot \chi$. Then, we approach the BDP main conjecture based on Iwasawa main conjecture for imaginary quadratic fields proved by Rubin [43].

The second approach generalizes to CM elliptic curves over totally real field [6, 30].

ACKNOWLEDGMENTS

The author thanks Ashay Burungale, John Coates, Henri Darmon, Yifeng Liu, Christopher Skinner, Xinyi Yuan, Shou-Wu Zhang, and Wei Zhang for their helpful comments, discussions, and constant support.

FUNDING

This work was partially supported by NSFC grant #11688101.

REFERENCES

[1] A. Agboola and B. Howard, Anticyclotomic Iwasawa theory of CM elliptic curves. *Ann. Inst. Fourier (Grenoble)* **56** (2006), no. 4, 1001–1048.

[2] R. Alter, T. B. Curtz, and K. K. Kubota, Remarks and results on congruent numbers. In *Proceedings of the Third Southeastern Conference on Combinatorics, Graph Theory and Computing (Florida Atlantic Univ., Boca Raton, FL, 1972)*, pp. 27–35, Florida Atlantic Univ., Boca Raton, FL, 1972.

[3] M. Bertolini, H. Darmon, and K. Prasanna, Generalized Heegner cycles and p-adic Rankin L-series. With an appendix by Brian Conrad. *Duke Math. J.* **162** (2013), no. 6, 1033–1148.

[4] M. Bhargava, D. Kane, H. Lenstra, B. Poonen, and E. Rains, Modeling the distribution of ranks, Selmer groups, and Shafarevich–Tate groups of elliptic curves. *Camb. J. Math.* **3** (2015), 275–321.

[5] A. Burungale, On the μ-invariant of the cyclotomic derivative of a Katz p-adic L-function. *J. Inst. Math. Jussieu* **14** (2015), no. 1, 131–148.

[6] A. Burungale, F. Castella, C. Skinner, and Y. Tian, p-converse to a theorem of Gross–Zagier and Kolyvagin: CM elliptic curves over totally real fields. Preprint, 2019.

[7] A. Burungale, H. Hida, and Y. Tian, Horizontal variation of Tate–Shafarevich groups. 2017, arXiv:1712.02148.

[8] A. Burungale, C. Skinner, and Y. Tian, p-converse to a theorem of Gross–Zagier, Kolyvagin and Rubin, II (in progress).

[9] A. Burungale, C. Skinner, and Y. Tian, Elliptic curves and Beilinson–Kato elements: rank one aspects. Preprint, 2020.

[10] A. Burungale, C. Skinner, and Y. Tian, The Birch and Swinnerton-Dyer conjecture: a brief survey. (English summary) Nine mathematical challenges, *Proc. Sympos. Pure Math.* **104** (2021), 11–29.

[11] A. Burungale and Y. Tian, A rank zero p-converse to a theorem of Gross–Zagier, Kolyvagin and Rubin. Preprint, 2019.

[12] A. Burungale and Y. Tian, p-converse to a theorem of Gross–Zagier, Kolyvagin and Rubin. *Invent. Math.* **220** (2020), no. 1, 211–253.

[13] A. Burungale and Y. Tian, Horizontal non-vanishing of Heegner points and toric periods. *Adv. Math.* **362** (2020), 106938, 35 pp.

[14] A. Burungale and Y. Tian, The even parity Goldfeld conjecture: congruent number elliptic curves. *J. Number Theory* **230** (2022), 161–195.

[15] L. Cai, J. Shu, and Y. Tian, Explicit Gross–Zagier formula and Waldspurger formula. *Algebra Number Theory* **8** (2014), no. 10, 2523–2572.

[16] L. Cai, J. Shu, and Y. Tian, Cube sum problem and an explicit Gross–Zagier formula. *Amer. J. Math.* **139** (2017), no. 3, 785–816.

[17] F. Castella and W. Xin, Perrin-Riou's main conjecture for elliptic curves at supersingular primes. arXiv:1607.02019.

[18] J. Coates and A. Wiles, On the conjecture of Birch and Swinnerton-Dyer. *Invent. Math.* **39** (1977), no. 3, 223–251.

[19] J. E. Cremona, *Algorithms for modular elliptic curves*. 2nd edn. Cambridge University Press, Cambridge, 1997, vi+376 pp. ISBN: 0-521-59820-6.

[20] L. E. Dickson, *History of the theory of numbers, vol. II: diophantine analysis*. Chelsea Publishing Co., New York, 1966, xxv+803 pp.

[21] D. Disegni, The p-adic Gross–Zagier formula on Shimura curves. *Compos. Math.* **153** (2017), no. 10, 1987–2074.

[22] K. Feng, Q. Liu, J. Pan, and Y. Tian, Toric periods and non-tiling numbers. Preprint, 2021.

[23] D. Goldfeld, Conjectures on elliptic curves over quadratic fields. In *Number theory, Carbondale 1979 (Proc. Southern Illinois Conf., Southern Illinois Univ., Carbondale, Ill., 1979)*, pp. 108–118, Lecture Notes in Math. 751, Springer, Berlin, 1979.

[24] B. Gross and D. Zagier, Heegner points and derivatives of L-series. *Invent. Math.* **84** (1986), no. 2, 225–320.

[25] Y. Hayashi, The Rankin's L-function and Heegner points for general discriminants. *Proc. Japan Acad. Ser. A Math. Sci.* **71** (1995), no. 2, 30–32.

[26] W. He, Y. Tian, and W. Xiong, Explicit theta lifting and quadratic twist L-values. Preprint, 2021.

[27] D. R. Heath-Brown, The size of Selmer groups for congruent number problem, II. *Invent. Math.* **118** (1994), no. 2, 331–370.

[28] K. Heegner, Diophantische Analysis und Modulfunktionen. *Math. Z.* **56** (1952), 227–253.

[29] H. Hida and J. Tilouine, Anti-cyclotomic Katz p-adic L-functions and congruence modules. *Ann. Sci. Éc. Norm. Supér. (4)* **26** (1993), no. 2, 189–259.

[30] M. L. Hsieh, Eisenstein congruence on unitary groups and Iwasawa main conjectures for CM fields. *J. Amer. Math. Soc.* **27** (2014), no. 3, 753–862.

[31] D. Jetchev, C. Skinner, and X. Wan, The Birch and Swinnerton-Dyer formula for elliptic curves of analytic rank one. *Camb. J. Math.* **5** (2017), no. 3, 369–434.

[32] D. Kane, On the ranks of the 2-Selmer groups of twists of a given elliptic curve. *Algebra Number Theory* **7** (2013), no. 5, 1253–1279.

[33] K. Kato, p-adic Hodge theory and values of zeta functions of modular forms, Cohomologies p-adiques et applications arithmétiques. III. *Astérisque* **295** (2004), 117–290.

[34] N. Katz and P. Sarnak, *Random matrices, Frobenius eigenvalues, and monodromy*. Amer. Math. Soc. Colloq. Publ. 45, American Mathematical Society, Providence, RI, 1999, xii+419 pp.

[35] S. Kobayashi, The p-adic Gross–Zagier formula for elliptic curves at supersingular primes. *Invent. Math.* **191** (2013), no. 3, 527–629.

[36] V. Kolyvagin, Euler systems. In *The Grothendieck Festschrift, Vol. II*, pp. 435–483, Progr. Math. 87, Birkhäuser Boston, Boston, MA, 1990.

[37] Y. X. Li, Y. Liu, and Y. Tian, On The Birch and Swinnerton-Dyer Conjecture for CM Elliptic Curves over $\mathbb{Q}$. arXiv:1605.01481.

[38] Y. F. Liu, S. W. Zhang, and W. Zhang, A p-adic Waldspurger formula. *Duke Math. J.* **167** (2018), no. 4, 743–833.

[39] Y. Manin, Cyclotomic fields and modular curves. *Uspekhi Mat. Nauk* **26** (1971), no. 6(162), 7–71 (Russian).

[40] P. Monsky, Mock Heegner points and congruent numbers. *Math. Z.* **204** (1990), no. 1, 45–67.

[41] J. Pan, *The p-adic BSD conjecture for elliptic curves over $\mathbb{Q}$ at potentially supersingular primes*. Ph.D. Thesis, 2017 (in Chinese).

[42] B. Perrin-Riou, Points de Heegner et dérivées de fonctions L p-adiques. *Invent. Math.* **89** (1987), no. 3, 455–510.

[43] K. Rubin, The "main conjectures" of Iwasawa theory for imaginary quadratic fields. *Invent. Math.* **103** (1991), no. 1, 25–68.

[44] K. Rubin, p-adic variants of the Birch and Swinnerton-Dyer conjecture for elliptic curves with complex multiplication. In *p-adic monodromy and the Birch and Swinnerton-Dyer conjecture (Boston, MA, 1991)*, pp. 71–80, Contemp. Math. 165, Amer. Math. Soc., Providence, RI, 1994.

[45] C. Skinner, A converse to a theorem of Gross, Zagier and Kolyvagin. *Ann. of Math. (2)* **191** (2020), no. 2, 329–354.

[46] C. Skinner and E. Urban, The Iwasawa main conjectures for GL_2. *Invent. Math.* **195** (2014), no. 1, 1–277.

[47] A. Smith, The congruent numbers have positive natural density. 2016, arXiv:1603.08479.

[48] A. Smith, 2^∞-Selmer groups, 2^∞-class groups, and Goldfeld's conjecture. 2017, arXiv:1702.02325.

[49] P. Swinnerton-Dyer, The effect of twisting on 2-Selmer groups. *Math. Proc. Cambridge Philos. Soc.* **145** (2008), 513–526.

[50] Y. Tian, Congruent numbers and Heegner points. *Camb. J. Math.* **2** (2014), 117–161.

[51] Y. Tian, X. Yuan, and S. Zhang, Genus periods, genus points and congruent number problem. *Asian J. Math.* **21** (2017), no. 4, 721–773.

[52] J. B. Tunnell, A classical Diophantine problem and modular forms of weight 3/2. *Invent. Math.* **72** (1983), no. 2, 323–334.

[53] M. Watkins, Some remarks on Heegner point computations. In *Explicit methods in number theory*, pp. 81–97, Panor. Synthèses 36, Soc. Math. France, Paris, 2012.

[54] A. Wiles, Modular elliptic curves and Fermat's last theorem. *Ann. of Math. (2)* **141** (1995), no. 3, 443–551.

[55] X. Yuan, S.-W. Zhang, and W. Zhang, The Gross-Zagier formula on Shimura curves. *Ann. of Math. Stud.* **184** (2013), viii+272 pages.

[56] W. Zhang, Selmer groups and the indivisibility of Heegner points. *Camb. J. Math.* **2** (2014), no. 2, 191–253.

YE TIAN

Academy of Mathematics and Systems Science, Chinese Academy of Sciences, Beijing 100190, China, ytian@math.ac.cn

This Page Intentionally Left Blank

ARITHMETIC AND GEOMETRIC LANGLANDS PROGRAM

XINWEN ZHU

ABSTRACT

We explain how the geometric Langlands program inspires some recent new prospectives of classical arithmetic Langlands program and leads to the solutions of some problems in arithmetic geometry.

MATHEMATICS SUBJECT CLASSIFICATION 2020

Primary 11R39; Secondary 11S37, 14D24, 11G18, 11G40

KEYWORDS

Classical Langlands correspondence, geometric Langlands program, Shimura varieties

INTERNATIONAL CONGRESS OF MATHEMATICIANS 2022 JULY 6–14

Proc. Int. Cong. Math. 2022, Vol. 3, pp. 2012–2045
DOI 10.4171/ICM2022/176

The classical Langlands program, originated by Langlands in 1960s [41], systematically studies reciprocity laws in the framework of representation theory. Very roughly speaking, it predicts the following deep relations between number theory and representation theory:

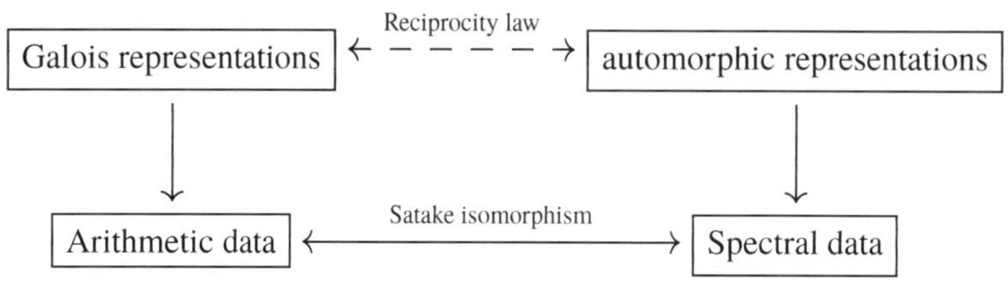

A special case of this correspondence, known as the Shimura–Tanniyama–Weil conjecture, implies Fermat's last theorem (see [62]).

The geometric Langlands program [42], initiated by Drinfeld and Laumon, arose as a generalization of Drinfeld's approach [20] to the global Langlands correspondence for GL_2 over function fields. In the geometric theory, the fundamental object to study shifts from the space of automorphic forms of a reductive group G to the category of sheaves on the moduli space of G-bundles on an algebraic curve.

For a long time, developments of the geometric Langlands were inspired by problems and techniques from the classical Langlands, with another important source of inspiration from quantum physics. The basic philosophy is known as categorification/geometrization. In recent years, however, the geometric theory has found fruitful applications to the classical Langlands program and some related arithmetic problems. Traditionally, one applies Grothendieck's sheaf-to-function dictionary to "decategorify" sheaves studied in geometric theory to obtain functions studied in arithmetic theory. This was used in Drinfeld's approach to the Langlands correspondence for GL_2, as mentioned above. Another celebrated example is Ngô's proof of the fundamental lemma [55]. In recent years, there appears another passage from the geometric theory to the arithmetic theory, again via a trace construction, but is of different nature and is closely related to ideas from physics. V. Lafforgue's work on the global Langlands correspondence over function fields [39] essentially (but implicitly) used this idea.

In this survey article, we review (a small fraction of) the developments of the geometric Langlands program, and discuss some recent new prospectives of the classical Langlands inspired by the geometric theory, which in turn lead solutions of some concrete arithmetic problems. The following diagram can be regarded as a road map:

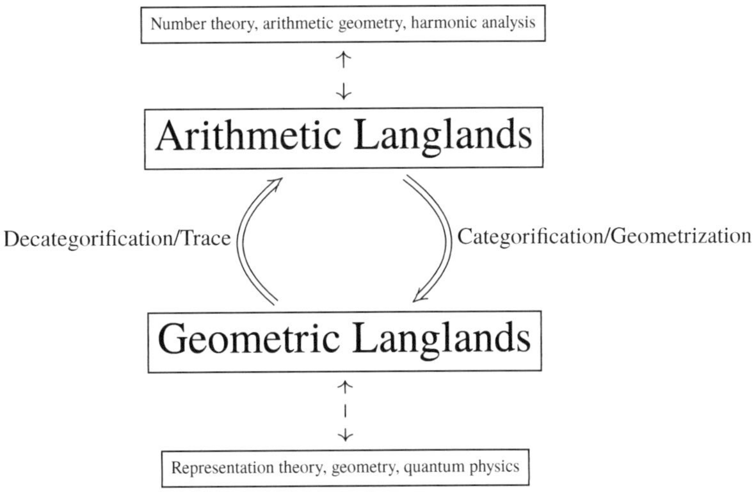

Notations. We use the following notations throughout this article. For a field F, let $\Gamma_{\tilde{F}/F}$ be the Galois group of a Galois extension $\tilde{F}/F$. Write $\Gamma_F = \Gamma_{\overline{F}/F}$, where $\overline{F}$ is a separable closure of F. Often in the article F will be either a local or a global field. In this case, let W_F denote the Weil group of F. Let cycl denote the cyclotomic character.

For a group A of multiplicative type over a field F, let $\mathbb{X}^\bullet(A) = \mathrm{Hom}(A_{\overline{F}}, \mathbb{G}_m)$ denote the group of its characters, and $\mathbb{X}_\bullet(A) = \mathrm{Hom}(\mathbb{G}_m, A_{\overline{F}})$ the group of its cocharacters.

For a prime ℓ, let Λ be $\mathbb{F}_\ell$, $\mathbb{Z}_\ell$, $\mathbb{Q}_\ell$ or a finite (flat) extension of such rings. It will serve as the coefficient ring of our sheaf theory.

1. FROM CLASSICAL TO GEOMETRIC LANGLANDS CORRESPONDENCE

In this section, we review some developments of the geometric Langlands theory inspired from the classical theory, with another important source of inspiration from quantum physics. The basic idea is categorification/geometrization, which is a process of replacing set-theoretic statements with categorical analogues

$$\text{Numbers} \dashrightarrow \text{Vector spaces} \dashrightarrow \text{Categories} \dashrightarrow \text{2-Categories} \dashrightarrow \cdots. \tag{1.1}$$

We illustrate this process by some important examples.

1.1. The geometric Satake

The starting point of the Langlands program is (Langlands' interpretation of) the Satake isomorphism, in which the Langlands dual group appears mysteriously. Similarly, the starting point of the geometric Langlands theory is the geometric Satake equivalence, which is a tensor equivalence between the category of perverse sheaves on the (spherical) local Hecke stack of a connected reductive group and the category of finite-dimensional algebraic representations of its dual group. This is a vast generalization of the classical Satake isomorphism (via the sheaf-to-function dictionary), and arguably gives a conceptual explanation why the Langlands dual group (in fact, the C-group) should appear in the Langlands correspondence.

We follow [83, SECT. 1.1] for notations and conventions regarding dual groups. Let G denote a connected reductive group over a field F. Let $(\hat{G}, \hat{B}, \hat{T}, \hat{e})$ be a pinned Langlands dual group of G over $\mathbb{Z}$. There is a finite Galois extension $\tilde{F}/F$, and a natural injective map $\xi : \Gamma_{\tilde{F}/F} \subset \mathrm{Aut}(\hat{G}, \hat{B}, \hat{T}, \hat{e})$, induced by the action of Γ_F on the root datum of G. Let ${}^L G = \hat{G} \rtimes \Gamma_{\tilde{F}/F}$ denote the usual L-group of G, and ${}^c G = \hat{G} \rtimes (\mathbb{G}_m \times \Gamma_{\tilde{F}/F})$ the group defined in [83], which is isomorphic to the C-group of G introduced by Buzzard–Gee. We write $d : {}^c G \to \mathbb{G}_m \times \Gamma_{\tilde{F}/F}$ for the projection with the kernel $\hat{G}$.

Now let F be a nonarchimedean local field with $\mathcal{O}$ being its ring of integers and $k = \mathbb{F}_q$ its residue field. That is, F is a finite extension of $\mathbb{Q}_p$ or is isomorphic to $\mathbb{F}_q((\varpi))$. Let σ be the *geometric* q-Frobenius of k. Assume that G can be extended to a connected reductive group over $\mathcal{O}$ (such G is called unramified), and we fix such an extension to have $G(\mathcal{O}) \subset G(F)$, usually called a hyperspecial subgroup of $G(F)$. With a basis of open neighborhoods of the unit given by finite-index subgroups of $G(\mathcal{O})$, the group $G(F)$ is a locally compact topological group. The classical spherical Hecke algebra is the space of compactly supported $G(\mathcal{O})$-biinvariant $\mathbb{C}$-valued functions on $G(F)$, equipped with the convolution product

$$(f * g)(x) = \int_{G(F)} f(y) g\big(y^{-1}x\big) dy, \tag{1.2}$$

where dy is the Haar measure on $G(F)$ such that $G(\mathcal{O})$ has volume 1. Note that if both f and g are $\mathbb{Z}$-valued, so is $f * g$. Therefore, the subset $H^{\mathrm{cl}}_{G(\mathcal{O})}$ of $\mathbb{Z}$-valued functions forms a $\mathbb{Z}$-algebra.[1]

On the dual side, under the unramifiedness assumption, $\Gamma_{\tilde{F}/F}$ is a finite cyclic group generated by σ. Note that $\hat{G}$ acts on ${}^c G|_{d=(q,\sigma)}$, the fiber of d at $(q, \sigma) \in \mathbb{G}_m \times \Gamma_{\tilde{F}/F}$, by conjugation. Then the classical Satake isomorphism establishes a canonical isomorphism of $\mathbb{Z}[q^{-1}]$-algebras

$$\mathrm{Sat}^{\mathrm{cl}} : \mathbb{Z}\big[q^{-1}\big]\big[{}^c G|_{d=(q,\sigma)}\big]^{\hat{G}} \cong H^{\mathrm{cl}}_{G(\mathcal{O})} \otimes \mathbb{Z}\big[q^{-1}\big]. \tag{1.3}$$

Remark 1.1.1. In fact, as explained in [83], there is a Satake isomorphism over $\mathbb{Z}$ (without inverting q), in which the C-group ${}^c G$ is replaced by certain affine monoid containing it as the group of invertible elements. On the other hand, if we extend the base ring from $\mathbb{Z}[q^{-1}]$ to $\mathbb{Z}[q^{\pm\frac{1}{2}}]$, one can rewrite (1.3) as an isomorphism

$$\mathbb{Z}\big[q^{\pm\frac{1}{2}}\big][\hat{G}\sigma]^{\hat{G}} \cong H^{\mathrm{cl}}_{G(\mathcal{O})} \otimes \mathbb{Z}\big[q^{\pm\frac{1}{2}}\big], \tag{1.4}$$

where $\hat{G}$ acts on $\hat{G}\sigma \subset {}^L G$ by the usual conjugation (e.g., see [83] for the discussion). This is the more traditional formulation of the Satake isomorphism, which is slightly noncanonical, but suffices for many applications.

In the geometric theory, where instead of thinking $G(F)$ as a topological group and considering the space of $G(\mathcal{O})$-biinvariant compactly supported functions on it, one regards

1 Here $(-)^{\mathrm{cl}}$ stands for the classical Hecke algebra, as opposed to the derived Hecke algebra mentioned in (2.2).

$G(F)$ as a certain algebro-geometric object and studies the category of $G(\mathcal{O})$-biequivariant sheaves on it. In the rest of the section, we allow F to be slightly more general. Namely, we assume that F is a local field complete with respect to a discrete valuation, with ring of integers $\mathcal{O}$ and a *perfect* residue field k of characteristic $p > 0$.[2] Let $\varpi \in \mathcal{O}$ be a uniformizer.

We work in the realm of perfect algebraic geometry. Recall that a k-algebra R is called perfect if the Frobenius endomorphism $R \to R$, $r \mapsto r^p$, is a bijection. Let $\mathbf{Aff}_k^{\mathrm{pf}}$ denote the category of perfect k-algebras. By a perfect presheaf (or, more generally, a perfect prestack), we mean a functor from $\mathbf{Aff}_k^{\mathrm{pf}}$ to the category **Sets** of sets (or, more generally, a functor from $\mathbf{Aff}_k^{\mathrm{pf}}$ to the ∞-category **Spc** of spaces). Many constructions in usual algebraic geometry work in this setting. For example, one can endow $\mathbf{Aff}_k^{\mathrm{pf}}$ with Zariski, étale, or fpqc topology as usual and talk about corresponding sheaves and stacks. One can then define perfect schemes, perfect algebraic spaces, perfect algebraic stacks, etc., as sheaves (stacks) with certain properties. It turns out that the category of perfect schemes/algebraic spaces defined this way is equivalent to the category of perfect schemes/algebraic spaces in the usual sense. Some foundations of perfect algebraic geometry can be found in [78, APPENDIX A], [13] and [64, SECT. A.1].

For a perfect k-algebra R, let $W_{\mathcal{O}}(R)$ denote the ring of Witt vectors in R with coefficient in $\mathcal{O}$. If $\mathrm{char}\, F = \mathrm{char}\, k$, then $W_{\mathcal{O}}(R) \simeq R[[\varpi]]$. If $\mathrm{char}\, F \neq \mathrm{char}\, k$, see [78, SECT. 0.5]. If $R = \bar{k}$, we denote $W_{\mathcal{O}}(\bar{k})$ by $\mathcal{O}_{\breve{F}}$ and $W_{\mathcal{O}}(\bar{k})[1/\varpi]$ by $\breve{F}$. We write $D_R = \mathrm{Spec} W_{\mathcal{O}}(R)$ and $D_R^* = \mathrm{Spec} W_{\mathcal{O}}(R)[1/\varpi]$ which are thought as a family of (punctured) discs parameterized by $\mathrm{Spec} R$.

We denote by L^+G (resp. LG) the *jet group* (resp. *loop group*) of G. As presheaves on $\mathbf{Aff}_k^{\mathrm{pf}}$,

$$L^+G(R) = G\big(W_{\mathcal{O}}(R)\big), \quad LG(R) = G\big(W_{\mathcal{O}}(R)[1/\varpi]\big).$$

Note that $L^+G(k) = G(\mathcal{O})$ and $LG(k) = G(F)$. Let

$$\mathrm{Hk}_G := L^+G \backslash LG / L^+G$$

be the étale stack quotient of LG by the left and right L^+G-action, sometimes called the (spherical) local Hecke stack of G. As a perfect prestack, it sends R to triples $(\mathcal{E}_1, \mathcal{E}_2, \beta)$, where $\mathcal{E}_1, \mathcal{E}_2$ are two G-torsors on D_R, and $\beta : \mathcal{E}_1|_{D_R^*} \simeq \mathcal{E}_2|_{D_R^*}$ is an isomorphism.

For $\ell \neq p$, the modern developments of higher category theory allow one to define the ∞-category of étale $\mathbb{F}_\ell$-sheaves on any prestack (e.g., see [35]). In particular, for $\Lambda = \mathbb{F}_\ell, \mathbb{Z}_\ell, \mathbb{Q}_\ell$ (or finite extension of these rings), it is possible to define the ∞-category $\mathbf{Shv}(\mathrm{Hk}_G, \Lambda)$ of Λ-sheaves on Hk_G, which is the categorical analogue of the space of $G(\mathcal{O})$-biinvariant functions on $G(F)$. But without knowing some geometric properties of Hk_G, very little can be said about $\mathbf{Shv}(\mathrm{Hk}_G, \Lambda)$. The crucial geometric input is the following theorem.

2 If $\mathrm{char}\, F = \mathrm{char}\, k$ (the equal characteristic case), this assumption on k is not necessary. We impose it here to have a uniform treatment of both equal and mixed characteristic (i.e., $\mathrm{char} F \neq \mathrm{char} k$) cases. For the same reason, we work with perfect algebraic geometry below even in equal characteristic.

Theorem 1.1.2. *Let* $\mathrm{Gr}_G := LG/L^+G$ *be the étale quotient of* LG *by the (right)* L^+G*-action, which admits the left* L^+G*-action. Then* Gr_G *can be written as an inductive limit of* L^+G*-stable subfunctors* $\varinjlim X_i$*, with each* X_i *being a perfect projective variety and* $X_i \to X_{i+1}$ *being a closed embedding.*

The space Gr_G is usually called the affine Grassmannian of G. See [4,23] for the equal characteristic case and [13,78] for the mixed characteristic case, and see [77,80] for examples of closed subvarieties in Gr_G. The theorem allows one to define the category of constructible and perverse sheaves on Hk_G, and to formulate the geometric Satake, as we discuss now.

First, the (left) quotient by L^+G-action induces a map $\mathrm{Gr}_G \to \mathrm{Hk}_G$. Roughly speaking, a sheaf on Hk_G is perverse (resp. constructible) if its pullback to Gr_G comes from a perverse (resp. constructible) sheaf on some X_i. Then inside $\mathbf{Shv}(\mathrm{Hk}_G, \Lambda)$ we have the categories $\mathbf{Perv}(\mathrm{Hk}_G, \Lambda) \subset \mathbf{Shv}_c(\mathrm{Hk}_G, \Lambda)$ of perverse and constructible sheaves on Hk_G. They can be regarded as categorical analogues of the space of $G(\mathcal{O})$-biinvariant compactly supported functions on $G(F)$. In addition, $\mathbf{Perv}(\mathrm{Hk}_G, \Lambda)$ is an abelian category, semisimple if Λ is a field of characteristic zero, called the Satake category of G. For simplicity, we assume that Λ is a field in the sequel.[3]

There is also a categorical analogue of the convolution product (1.2). Namely, there is the convolution diagram

$$\mathrm{Hk}_G \times \mathrm{Hk}_G \xleftarrow{\mathrm{pr}} L^+G\backslash LG \times^{L^+G} LG/L^+G \xrightarrow{m} \mathrm{Hk}_G,$$

and the convolution of two sheaves $\mathcal{A}, \mathcal{B} \in \mathbf{Shv}(\mathrm{Hk}_G, \Lambda)$ is defined as

$$\mathcal{A} \star \mathcal{B} := m_! \mathrm{pr}^*(\mathcal{A} \boxtimes \mathcal{B}). \tag{1.5}$$

This convolution product makes $\mathbf{Shv}(\mathrm{Hk}_G, \Lambda)$ into a monoidal ∞-category containing $\mathbf{Perv}(\mathrm{Hk}_G, \Lambda) \subset \mathbf{Shv}_c(\mathrm{Hk}_G, \Lambda)$ as monoidal subcategories.

Remark 1.1.3. The above construction of the Satake category as a monoidal category is essentially equivalent to the more traditional approach, in which the Satake category is defined as the category of L^+G-equivariant perverse sheaves on Gr_G (e.g., see [80] for an exposition).

Let $\mathbf{Coh}(\mathbb{B}\hat{G}_\Lambda)^\heartsuit$ denote the abelian monoidal category of coherent sheaves on the classifying stack $\mathbb{B}\hat{G}_\Lambda$ over Λ[4], which is equivalent to the category of algebraic representations of $\hat{G}$ on finite dimensional Λ-vector spaces. This following theorem is usually known as the geometric Satake equivalence.

Theorem 1.1.4. *There is a canonical equivalence of monoidal abelian categories*

$$\mathrm{Sat}_G : \mathbf{Coh}(\mathbb{B}\hat{G}_\Lambda)^\heartsuit \cong \mathbf{Perv}(\mathrm{Hk}_G \otimes \bar{k}, \Lambda).$$

3 The formulation for $\Lambda = \mathbb{Z}_\ell$ is slightly more complicated, as the right-hand side of (1.5) may not be perverse when $\mathcal{A}$ and $\mathcal{B}$ are perverse.

4 In the dual group side, we always work in the realm of usual algebraic geometry, so $\mathbb{B}\hat{G}$ is an Artin stack in the usual sense.

Geometric Satake is really one of the cornerstones of the geometric Langlands program, and has found numerous applications to representation theory, mathematical physics, and (arithmetic) algebraic geometry. When $F = k((\varpi))$, this theorem grew out of works of Lusztig, Ginzburg, Beilinson–Drinfeld and Mirković–Vilonen (cf. [5, 51, 53]). In mixed characteristic, it was proved in [69, 78], with the equal characteristic case as an input, and in [25] by mimicking the strategy in equal characteristic. We conclude this subsection with a few remarks.

Remark 1.1.5. (1) As mentioned before, the geometric Satake can be regarded as the conceptual definition of the Langlands dual group $\hat{G}$ of G, namely as the Tannakian group of the Tannakian category $\mathbf{Perv}(\mathrm{Hk}_G \otimes \bar{k}, \Lambda)$. In addition, as explained in [72, 76], the group $\hat{G}$ is canonically equipped with a pinning $(\hat{B}, \hat{T}, \hat{e})$. In the rest of the article, by the pinned Langlands dual group $(\hat{G}, \hat{B}, \hat{T}, \hat{e})$ of G, we mean the quadruple defined by the geometric Satake.

(2) For arithmetic applications, one needs to understand the Γ_k-action on $\mathbf{Perv}(\mathrm{Hk}_G \otimes \bar{k}, \Lambda)$ in terms of the dual group side. It turns out that such an action is induced by an action of Γ_k on $\hat{G}$, preserving $(\hat{B}, \hat{T})$ but not $\hat{e}$. See [76, 80], or [77] from the motivic point of view. This leads to the appearance of the group ${}^cG$. See [76, 80, 83] for detailed discussions.

(3) There is also the derived Satake equivalence [11], describing $\mathbf{Shv}_c(\mathrm{Hk}_G \otimes \bar{k}, \Lambda)$ in terms of the dual group, at least when Λ is a field of characteristic zero. We mention that the category in the dual side is not the derived category of coherent sheaves on $\mathbb{B}\hat{G}_\Lambda$.

(4) In fact, for many applications, it is important to have a family version of the geometric Satake. For a (nonempty) finite set S, there is a local Hecke stack Hk_{G,D^S} over D^S, the self-product of the disc $D = \mathrm{Spec}\mathcal{O}$ over S, which, roughly speaking, classifies quadruples $(\{x_s\}_{s\in S}, \mathcal{E}, \mathcal{E}', \beta)$, where $\{x_s\}_{s\in S}$ is an S-tuple of points of D, $\mathcal{E}$ and $\mathcal{E}'$ are two G-torsors on D, and β is an isomorphism between $\mathcal{E}$ and $\mathcal{E}'$ on $D - \bigcup_s \{x_s\}$. In equal characteristic, one can regard D as the formal disc at a k-point of an algebraic curve X over k and Hk_{G,D^S} is the restriction along $D^S \to X^S$ of the stack

$$\mathrm{Hk}_{G,X^S} = (L^+G)_{X^S} \backslash (LG)_{X^S} / (L^+G)_{X^S},$$

where $(LG)_{X^S}$ and $(L^+G)_{X^S}$ are family versions of LG and L^+G over X^S (e.g., see [80, SECT. 3.1] for precise definitions). In mixed characteristic, the stack Hk_{G,D^S} (and in fact D^S itself) does not live in the world of (perfect) algebraic geometry, but rather in the world of perfectoid analytic geometry as developed by Scholze (see [25, 59]). In both cases, one can consider certain category $\mathbf{Perv}^{\mathrm{ULA}}(\mathrm{Hk}_{G,D^S} \otimes \bar{k}, \Lambda)$ of (ULA) perverse sheaves on $\mathrm{Hk}_{G,D^S} \otimes \bar{k}$. In addition, for a map $S \to S'$ of finite sets, restriction along $\mathrm{Hk}_{G,D^{S'}} \to \mathrm{Hk}_{G,D^S}$ gives a functor $\mathbf{Perv}^{\mathrm{ULA}}(\mathrm{Hk}_{G,D^S} \otimes \bar{k}, \Lambda) \to \mathbf{Perv}^{\mathrm{ULA}}(\mathrm{Hk}_{G,D^{S'}} \otimes \bar{k}, \Lambda)$.[5] We refer to the above mentioned references for details.

---

5 Such restriction functor defines the so-called fusion product, a key concept in the geometric Satake equivalence. The terminology "fusion" originally comes from conformal field theory.

On the other hand, let $\hat{G}^S$ be the S-power self-product of $\hat{G}$ over Λ. Then for $S \to S'$, the restriction along $\mathbb{B}\hat{G}^{S'} \to \mathbb{B}\hat{G}^S$ gives a functor $\mathbf{Coh}(\mathbb{B}\hat{G}^S_\Lambda)^\heartsuit \to \mathbf{Coh}(\mathbb{B}\hat{G}^{S'}_\Lambda)^\heartsuit$. Now a family version of the geometric Satake gives a system of functors

$$\mathrm{Sat}_S : \mathbf{Coh}\big(\mathbb{B}\hat{G}^S_\Lambda\big)^\heartsuit \to \mathbf{Perv}^{\mathrm{ULA}}(\mathrm{Hk}_{G,D^S} \otimes \bar{k}, \Lambda), \tag{1.6}$$

compatible with restriction functors on both sides induced by maps between finite sets (see [28, 80]).

(5) For applications, it is important to have the geometric Satake in different sheaf-theoretic contents over different versions of local Hecke stacks. Besides the above mentioned ones, we also mention a D-module version [5], and an arithmetic D-module version [66].

1.2. Tamely ramified local geometric Langlands correspondence

We first recall the classical theory. Assume that F is a finite extension of $\mathbb{Q}_p$ or is isomorphic to $\mathbb{F}_q((\varpi))$, and for simplicity assume that G extends to a connected reductive group over $\mathcal{O}$. (In fact, results in the subsection hold in appropriate forms for quasi-split groups that are split over a tamely ramified extension of F.) In addition, we fix a pinning (B, T, e) of G over $\mathcal{O}$.

The classical local Langlands program aims to classify (smooth) irreducible representations of $G(F)$ (over $\mathbb{C}$) in terms of Galois representations. From this point of view, the Satake isomorphism (1.3) gives a classification of irreducible unramified representations of $G(F)$, i.e., those that have nonzero vectors fixed by $G(\mathcal{O})$, as such representations are in one-to-one correspondence with simple modules of $H^{\mathrm{cl}}_{G(\mathcal{O})} \otimes \mathbb{C}$, which via the Satake isomorphism (1.3) are parameterized by semisimple $\hat{G}$-conjugacy classes in ${}^cG$. (For an irreducible unramified representation π, the corresponding $\hat{G}$-conjugacy class in ${}^cG$ is usually called the Satake parameter of π.)

The next important class of irreducible representations are those that have nonzero vectors fixed by an Iwahori subgroup $G(F)$. For example, under the reduction mod ϖ map $G(\mathcal{O}) \to G(k)$, the preimage I of $B(k) \subset G(k)$ is an Iwahori subgroup of $G(F)$. As in the unramified case, the $\mathbb{Z}$-valued I-biinvariant functions form a $\mathbb{Z}$-algebra H^{cl}_I with multiplication given by convolution (1.2) (with the Haar measure dg chosen so that the volume of I is one), and the set of irreducible representations of $G(F)$ that have nonzero I-fixed vectors are in bijection with the set of simple $(H^{\mathrm{cl}}_I \otimes \mathbb{C})$-modules. Just as the Satake isomorphism, Kazhdan–Lusztig gave a description of $H^{\mathrm{cl}}_I \otimes \mathbb{C}$ in terms of geometric objects associated to $\hat{G}$.

Let $\hat{U} \subset \hat{B}$ denote the unipotent radical of $\hat{B}$. The natural morphism $\hat{U}/\hat{B} \to \hat{G}/\hat{G}$ is usually called the Springer resolution. Let

$$S^{\mathrm{unip}}_{\hat{G}} = (\hat{U}/\hat{B}) \times_{\hat{G}/\hat{G}} (\hat{U}/\hat{B}),$$

which we call the (unipotent) Steinberg stack of $\hat{G}$[6]. Over $\mathbb{C}$, there is a $\mathbb{G}_{m,\mathbb{C}}$-action on $\hat{U}_{\mathbb{C}}$ and therefore on $S^{\mathrm{unip}}_{\hat{G},\mathbb{C}}$, by identifying $\hat{U}_{\mathbb{C}}$ with its Lie algebra via the exponential map. Then

6 As $\hat{U}/\hat{B} \to \hat{G}/\hat{G}$ is not flat, the fiber product needs to considered in derived sense so $S^{\mathrm{unip}}_{\hat{G}}$ should be understood as a derived algebraic stack.

one can form the quotient stack $S^{\mathrm{unip}}_{\hat{G},\mathbb{C}}/\mathbb{G}_{m,\mathbb{C}}$. In the sequel, for an Artin stack X (of finite presentation) over $\mathbb{C}$, we let $K(X)$ denote the K-group of the (∞-)category of coherent sheaves on X.

Kazhdan–Lusztig [36] constructed (under the assumption that G is split with connected center) a canonical isomorphism (after choosing a square root of $\sqrt{q}$ of q)

$$K\big(S^{\mathrm{unip}}_{\hat{G},\mathbb{C}}/\mathbb{G}_{m,\mathbb{C}}\big) \otimes_{K(\mathbb{B}\mathbb{G}_{m,\mathbb{C}})} \mathbb{C} \cong H_I^{\mathrm{cl}} \otimes \mathbb{C}, \tag{1.7}$$

where the map $K(\mathbb{B}\mathbb{G}_{m,\mathbb{C}}) \to \mathbb{C}$ sends the class corresponding to the tautological representation of $\mathbb{G}_{m,\mathbb{C}}$ to $\sqrt{q}$. In addition, the isomorphism induces the Bernstein isomorphism

$$K(\mathbb{B}\hat{G}_{\mathbb{C}}) \otimes \mathbb{C} \cong Z\big(H_I^{\mathrm{cl}} \otimes \mathbb{C}\big), \tag{1.8}$$

where $Z(H_I^{\mathrm{cl}} \otimes \mathbb{C})$ is the center of $H_I^{\mathrm{cl}} \otimes \mathbb{C}$, and the map $K(\mathbb{B}\hat{G}_{\mathbb{C}}) \to K(S^{\mathrm{unip}}_{\hat{G},\mathbb{C}}/\mathbb{G}_{m,\mathbb{C}})$ is induced by the natural projection $S^{\mathrm{unip}}_{\hat{G}}/\mathbb{G}_m \to \mathbb{B}\hat{G}$.

Remark 1.2.1. It would be interesting to give a description of the $\mathbb{Z}$-algebra H_I^{cl} in terms of the geometry involving $\hat{G}$, which would generalize the integral Satake isomorphism from [83].

It turns out that the Kazhdan–Lusztig isomorphism (1.7) also admits a categorification, usually known as the Bezrukavnikov equivalence, which gives two realizations of the affine Hecke category. Again, when switching to the geometric theory, we allow F to be a little bit more general as in Section 1.1. We also assume that G extends to a connected reductive group over $\mathcal{O}$ and fix a pinning of G over $\mathcal{O}$. Let $L^+G \to G_k$ be the natural "reduction mod ϖ" map, and let $\mathrm{Iw} \subset L^+G$ be the preimage of $B_k \subset G_k$. This is the geometric analogue of I. Then as in the unramified case discussed in Section 1.1, one can define the Iwahori local Hecke stack $\mathrm{Hk}_{\mathrm{Iw}} = \mathrm{Iw}\backslash LG/\mathrm{Iw}$ and the monoidal categories $\mathbf{Shv}_c(\mathrm{Hk}_{\mathrm{Iw}} \otimes \bar{k}, \Lambda) \subset \mathbf{Shv}(\mathrm{Hk}_{\mathrm{Iw}} \otimes \bar{k}, \Lambda)$. The category $\mathbf{Shv}_c(\mathrm{Hk}_{\mathrm{Iw}} \otimes \bar{k}, \Lambda)$ can be thought as a categorical analogue of H_I^{cl}, usually called the affine Hecke category.

Recall that we let $\breve{F} = W_{\mathcal{O}}(\bar{k})[1/\varpi]$. The inertia $I_F := \Gamma_{\breve{F}}$ of F has a tame quotient I_F^t isomorphic to $\prod_{\ell \neq p} \mathbb{Z}_\ell(1)$.

Theorem 1.2.2. *For every choice of a topological generator τ of the tame inertia I_F^t, there is a canonical equivalence of monoidal ∞-categories*

$$\mathrm{Bez}^{\mathrm{unip}}_G : \mathbf{Coh}\big(S^{\mathrm{unip}}_{\hat{G},\mathbb{Q}_\ell}\big) \cong \mathbf{Shv}_c(\mathrm{Hk}_{\mathrm{Iw}} \otimes \bar{k}, \mathbb{Q}_\ell).$$

In fact, Bezrukavnikov proved such equivalence when $F = k(\!(\varpi)\!)$ in [9]. Yun and the author deduced the mixed characteristic case from the equal characteristic case. It would be interesting to know whether the new techniques introduced in [25, 59] can lead a direct proof of this equivalence in mixed characteristic. (See [1] for some progress in this direction.)

Remark 1.2.3. Again, for arithmetic applications, one needs to describe the action of Γ_k on $\mathbf{Shv}_c(\mathrm{Hk}_{\mathrm{Iw}} \otimes \bar{k}, \Lambda)$ in terms of the dual group side. See [9, 35] for a discussion.

We explain an important ingredient in the proof of Theorem 1.2.2 (when $F = k(\!(\varpi)\!)$). There is a smooth affine group scheme $\mathcal{G}$ (called the Iwahori group scheme) over $\mathcal{O}$ such that $\mathcal{G} \otimes F = G$ and $L^+\mathcal{G} = \mathrm{Iw}$. Then there is a local Hecke stack $\mathrm{Hk}_{\mathcal{G},D}$ over D, analogous to Hk_{G,D^S} as discussed at the end of Section 1.1 (here $S = \{1\}$). In addition, $\mathrm{Hk}_{\mathcal{G},D}|_{D^*} \cong \mathrm{Hk}_{G,D}|_{D^*}$ and $\mathrm{Hk}_{\mathcal{G},D}|_0 = \mathrm{Hk}_{\mathrm{Iw}}$, where $0 \in D$ is the closed point. Then taking nearby cycles gives

$$\mathcal{Z} : \mathbf{Coh}(\mathbb{B}\hat{G}_\Lambda)^{\heartsuit} \xrightarrow{\mathrm{Sat}_{\{1\}}} \mathbf{Perv}(\mathrm{Hk}_{\mathcal{G},D}|_{D^*_{\bar{k}}}, \Lambda) \xrightarrow{\Psi} \mathbf{Perv}(\mathrm{Hk}_{\mathrm{Iw}} \otimes \bar{k}, \Lambda). \tag{1.9}$$

This is known as Gaitsgory's central functor [27,75], which can be regarded as a categorification of (1.8). We remark this construction is motivated by the Kottwitz conjecture originated from the study of mod p geometry of Shimura varieties. See Section 3.1 for some discussions.

Theorem 1.2.2 admits a generalization to the tame level. We consider the following diagram:

$$\hat{G}/\hat{G} \leftarrow \hat{B}/\hat{B} \xrightarrow{q_{\hat{B}}} \hat{T}/\hat{T},$$

where the left morphism is the usual Grothendieck–Springer resolution. Let χ be a Λ-point of $\hat{T}/\hat{T}$, where Λ is a finite extension of $\mathbb{Q}_\ell$. Let $(\hat{B}/\hat{B})_\chi = q_{\hat{B}}^{-1}(\chi)$, and let

$$S^{\chi}_{\hat{G},\Lambda} := (\hat{B}/\hat{B})_\chi \times_{\hat{G}/\hat{G}} (\hat{B}/\hat{B})_\chi.$$

Note that if $\chi = 1$, this reduces to $S^{\mathrm{unip}}_{\hat{G},\Lambda}$. On the other hand, a (torsion) Λ-point $\chi \in \hat{T}/\hat{T}$ defines a one-dimensional character sheaf $\mathcal{L}_\chi$ on $\mathrm{Iw} \otimes \bar{k}$. Then one can define the monoidal category of bi-$(\mathrm{Iw}, \mathcal{L}_\chi)$-equivariant constructible sheaves on $LG_{\bar{k}}$, denoted as $\mathbf{Shv}_{\mathrm{cons}}({}_\chi(\mathrm{Hk}_{\mathrm{Iw}})_\chi, \Lambda)$. If $\chi = 1$, so $\mathcal{L}_\chi$ is the trivial character sheaf on Iw, this reduces to the affine Hecke category $\mathbf{Shv}_c(\mathrm{Hk}_{\mathrm{Iw}} \otimes \bar{k}, \Lambda)$. The following generalization of Theorem 1.2.2 is conjectured in [9] and will be proved in a forthcoming joint work with Dhillon–Li–Yun [18].

Theorem 1.2.4. *Assume that* $\mathrm{char}\, F = \mathrm{char}\, k$. *There is a canonical monoidal equivalence*

$$\mathrm{Bez}^{\chi}_{\hat{G}} : \mathbf{Coh}(\hat{S}^{\chi}_{\hat{G},\Lambda}) \cong \mathbf{Shv}_c\big({}_\chi(\mathrm{Hk}_{\mathrm{Iw}})_\chi, \Lambda\big).$$

Remark 1.2.5. It is important to establish a version of equivalences in Theorems 1.2.2 and 1.2.4 for $\mathbb{Z}_\ell$-sheaves.

Remark 1.2.6. The local geometric Langlands correspondence beyond the tame ramification has not been fully understood, although certain wild ramifications have appeared in concrete problems (e.g., [31,79]). It is widely believed that the general local geometric Langlands should be formulated as 2-categorical statement, predicting the 2-category of module categories under the action of (appropriately defined) category of sheaves on LG is equivalent to the 2-category of categories over the stack of local geometric Langlands parameters. The precise formulation is beyond the scope of this survey, but, roughly speaking, it implies (and is more or less equivalent to saying) that the Hecke category for appropriately chosen "level" of LG is (Morita) equivalent to the category of coherent sheaves on some stack of the form $X \times_Y X$, where Y is closely related to the moduli of local geometric Langlands parameters.

1.3. Global geometric Langlands correspondence

As mentioned at the beginning of the article, the (global) geometric Langlands program originated from Drinfeld's proof of Langlands conjecture for GL_2 over function fields. Early developments of this subject mostly focused on the construction of Hecke eigensheaves associated to Galois representations of a global function field F (or, equivalently, local systems on a smooth algebraic curve X), e.g., see [20, 26, 42].

The scope of the whole program then shifted after the work [5], in which Beilinson–Drinfeld formulated a rough categorical form of the global geometric Langlands correspondence. The formulation then was made precise by Arinkin–Gaitsgory in [2], which we now recall. Let X be a smooth projective curve over $F = \mathbb{C}$. On the automorphic side, let $\mathbf{D}_c(\mathrm{Bun}_G)$ be the ∞-category of coherent D-modules on the moduli stack Bun_G of principal G-bundles on X. On the Galois side, let $\mathbf{Coh}(\mathrm{Loc}_{\hat{G}})$ be the ∞-category of coherent sheaves on the moduli stack $\mathrm{Loc}_{\hat{G}}$ of de Rham $\hat{G}$-local systems (also known as principal $\hat{G}$-bundles with flat connection) on X.

Conjecture 1.3.1. *There is a canonical equivalence of ∞-categories*

$$\mathbb{L}_G : \mathbf{Coh}(\mathrm{Loc}_{\hat{G}}) \cong \mathbf{D}_c(\mathrm{Bun}_G),$$

satisfying a list of natural compatibility conditions.

We briefly mention the most important compatibility condition, and refer to [2] for the rest. Note that both sides admit actions by a family of commuting operators labeled by $x \in X$ and $V \in \mathbf{Coh}(\mathbb{B}\hat{G}_{\mathbb{C}})^{\heartsuit}$. Namely, for every point $x \in X$, there is the evaluation map $\mathrm{Loc}_{\hat{G}} \to \mathbb{B}\hat{G}_{\mathbb{C}}$ so every $V \in \mathbf{Coh}(\mathbb{B}\hat{G}_{\mathbb{C}})^{\heartsuit}$ gives a vector bundle $\tilde{V}_x$ on $\mathrm{Loc}_{\hat{G}}$ by pullback, which then acts on $\mathbf{Coh}(\mathrm{Loc}_{\hat{G}})$ by tensoring. On the other hand, there is the Hecke operator $H_{V,x}$ that acts on $\mathbf{D}_c(\mathrm{Bun}_G)$ by convolving the sheaf $\mathrm{Sat}_{\{1\}}(V)|_x$ from the (D-module version of) the geometric Satake (1.6). Then the equivalence $\mathbb{L}_G$ should intertwine the actions of these operators.

Although the conjecture remains widely open, it is known that the category of perfect complexes $\mathbf{Perf}(\mathrm{Loc}_{\hat{G}})$ on $\mathrm{Loc}_{\hat{G}}$ acts on $\mathbf{D}_c(\mathrm{Bun}_G)$, usually called the spectral action, in a way such that the action of $\tilde{V}_x \in \mathbf{Perf}(\mathrm{Loc}_{\hat{G}})$ on $\mathbf{D}_c(\mathrm{Bun}_G)$ is given by the Hecke operator $H_{V,x}$.

Nowadays, Conjecture 1.3.1 sometimes is referred as the de Rham version of the global geometric Langlands conjecture, as there are some other versions of such conjectural equivalences, which we briefly mention.

First, in spirit of the nonabelian Hodge theory, there should exist a classical limit of Conjecture 1.3.1, sometimes known as the Dolbeault version of the global geometric Langlands. While the precise formulation is unknown (to the author), generically, it amounts to the duality of Hitchin fibrations for G and $\hat{G}$ (in the sense of mirror symmetry), and was established "generically" in [15, 19]. By twisting/deforming such duality in positive characteristic, one can prove a characteristic p analogue of Conjecture 1.3.1 (of course, only "generically," see [10, 14, 15]). Interestingly, this "generic" characteristic p equivalence can be used to give a new proof of the main result of [5] (at least for $G = \mathrm{GL}_n$, see [12]).

The work [5] (and therefore the de Rham version of the global geometric Langlands) was strongly influenced by conformal field theory. On the other hand, motivated by topological field theory, Ben-Zvi and Nadler [7] proposed a Betti version of Conjecture 1.3.1, where on the automorphic side the category of D-modules on Bun_G is replaced with the category of sheaves of $\mathbb{C}$-vector spaces on (the analytification of) Bun_G and on the Galois side $\mathrm{Loc}_{\hat{G}}$ is replaced by the moduli of Betti $\hat{G}$-local systems (also known as $\hat{G}$-valued representations of fundamental group of X).

The Riemann–Hilbert correspondence allows passing between the de Rham $\hat{G}$-local systems and Betti $\hat{G}$-local systems, but in a transcendental way. So Conjecture 1.3.1 and its Betti analogue are not directly related. Recently, Arinkin et al. [3] proposed yet another version of Conjecture 1.3.1, which directly relates both de Rham and Betti versions, and at the same time includes a version in terms of ℓ-adic sheaves. So it is more closely related to the classical Langlands correspondence over function fields, as will be discussed in Section 2.2.

2. FROM GEOMETRIC TO CLASSICAL LANGLANDS PROGRAM

In the previous section, we discussed how the ideas of categorification and geometrization led to the developments of the geometric Langlands program. On the other hand, the ideas of quantum physics allow one to reverse arrows in (1.1) by evaluating a (topological) quantum field theory at manifolds of different dimensions. Such ideas are certainly not new in geometry and topology. But, surprisingly, they also lead to a new understanding of the classical Langlands program. Indeed, it has been widely known that there is an analogy between global fields and 3-manifolds, and under such analogy Frobenius corresponds to the fundamental group of a circle. Then "compactification of field theories on a circle" leads to the categorical trace method (e.g., see [3, 6, 77]), which plays a more and more important role in the geometric representation theory.

2.1. Categorical arithmetic local Langlands

In this subsection, let F be either a finite extension of $\mathbb{Q}_p$ or isomorphic to $\mathbb{F}_q((\varpi))$. The classical local Langlands correspondence seeks a classification of smooth irreducible representations of $G(F)$ in terms of Galois data. The precise formulation, beyond the $G = \mathrm{GL}_n$ case (which is a theorem by [38, 43]), is complicated. However, the yoga that the local geometric Langlands is 2-categorical (see Remark 1.2.6) suggests that the even the classical local Langlands correspondence should and probably needs to be categorified.

The first ingredient needed to formulate the categorical arithmetic local Langlands is the following result, due independently to [17, 25, 82]. We take the formulation from [82] and refer for the notion of (strongly) continuous homomorphisms to the same reference.

Theorem 2.1.1. *The prestack sending a $\mathbb{Z}_\ell$-algebra A to the space of (strongly) continuous homomorphisms $\rho : W_F \to {}^cG(A)$ such that $d \circ \rho = (\mathrm{cycl}^{-1}, \mathrm{pr})$ is represented by a (classical) scheme $\mathrm{Loc}^{\square}_{{}^cG}$, which is a disjoint union of affine schemes that are flat, of finite type, and of locally complete intersection over $\mathbb{Z}_\ell$.*

The conjugation action of $\hat{G}$ on ${}^cG$ induces an action of $\hat{G}$ on $\mathrm{Loc}^{\square}_{{}^cG}$, and we call the quotient stack $\mathrm{Loc}_{{}^cG} = \mathrm{Loc}^{\square}_{{}^cG}/\hat{G}$ the stack of local Langlands parameters, which, roughly speaking, classifies the groupoid of the above ρ's up to $\hat{G}$-conjugacy.

In the categorical version of the local Langlands correspondence, on the Galois side it is natural to consider the (∞-)category $\mathbf{Coh}(\mathrm{Loc}_{{}^cG})$ of coherent sheaves on $\mathrm{Loc}_{{}^cG}$. On the representation side, one might naively consider the (∞-)category $\mathbf{Rep}(G(F), \Lambda)$ of smooth representations of $G(F)$. But in fact, this category needs to be enlarged. This can be seen from different point of view. Indeed, it is a general wisdom shared by many people that in the classical local Langlands correspondence, it is better to study representations of G together with a collection of groups that are (refined version of) its inner forms. There are various proposals of such collections. Arithmetic geometry (i.e., the study of p-adic and mod p geometry of Shimura varieties and moduli of Shtukas) and geometric representation theory (i.e., the categorical trace construction) suggest studying a category glued from the categories of representations of a collection of groups $\{J_b(F)\}_{b\in B(G)}$ arising from the Kottwitz set

$$B(G) = G(\breve{F})/\sim, \quad \text{where } g \sim g' \text{ if } g' = h^{-1}g\sigma(h) \text{ for some } h \in G(\breve{F}).$$

Here for $b \in B(G)$ (lifted to an element in $G(\breve{F})$), the group J_b is an F-group defined by assigning and F-algebra the group $J_b(R) = \{h \in G(\breve{F} \otimes_F R) \mid h^{-1}b\sigma(h) = b\}$. In particular, if $b = 1$ then $J_b = G$. In general, there is a well-defined embedding $(J_b)_{\overline{F}} \to G_{\overline{F}}$ up to conjugacy, making J_b a refinement of an inner form of a Levi subgroup of G (say, G is quasisplit).

There are two ways to make this idea precise. One is due to Fargues–Scholze **[25]**, who regard $B(G)$ as the set of points of the v-stack Bun_G of G-bundles on the Fargues–Fontaine curve and consider the category $D_{\mathrm{lis}}(\mathrm{Bun}_G, \Lambda)$ of appropriately defined étale sheaves on Bun_G, which indeed glues all $\mathbf{Rep}(J_b(F), \Lambda)$'s together. We mention that this approach relies on Scholze's work on ℓ-adic formalism of diamond and condensed mathematics.

In another approach **[35,64,77,82]**, closely related to the idea of categorical trace, the set $B(G)$ is regarded as the set of points of the (étale) quotient stack

$$\mathfrak{B}(G) := LG/\mathrm{Ad}_\sigma LG,$$

where Ad_σ denotes the Frobenius twisted conjugation given by

$$\mathrm{Ad}_\sigma : LG \times LG \to LG, \quad (h, g) \mapsto hg\sigma(h)^{-1}.$$

Then we have the category of Λ-sheaves $\mathbf{Shv}(\mathfrak{B}(G) \otimes \bar{k}, \Lambda)$ as mentioned before. Although $\mathfrak{B}(G)$ is a wild object in the traditional algebraic geometry, there are still a few things one can say about its geometry, and the category $\mathbf{Shv}(\mathfrak{B}(G) \otimes \bar{k}, \Lambda)$ is quite reasonable. In addition, it is possible to define the category $\mathbf{Shv}_c(\mathfrak{B}(G) \otimes \bar{k}, \Lambda)$ of constructible sheaves on $\mathfrak{B}(G) \otimes \bar{k}$, as we now briefly explain and refer to **[35]** for careful discussions.

For every algebraically closed field Ω over k, the groupoid of Ω-points of $\mathfrak{B}(G)$ is the groupoid of F-isocrystals with G-structure over Ω and the set of its isomorphism classes can be identified with the Kottwitz set $B(G)$. However, $\mathfrak{B}(G)$ is not merely a disjoint union of its points. Rather, it admits a stratification, known as the Newton stratification, labeled by $B(G)$. Namely, the set $B(G)$ has a natural partial order and, roughly speaking, for each $b \in B(G)$ those Ω-points corresponding to $b' \leq b$ form a closed substack $i_{\leq b} : \mathfrak{B}(G)_{\leq b} \subset \mathfrak{B}(G) \otimes \bar{k}$ and those Ω-points corresponding to b form an open substack $j_b : \mathfrak{B}(G)_b \subset \mathfrak{B}(G)_{\leq b}$. In particular, basic elements in $B(G)$ (i.e., minimal elements with respect to the partial order $\leq$) give closed strata. We also mention that if b is basic, J_b is a refinement of an inner form of G, usually called an extended pure inner form of G.

In the rest of this subsection, we simply denote $\mathfrak{B}(G) \otimes \bar{k}$ by $\mathfrak{B}(G)$. We write $i_b = i_{\leq b} j_b : \mathfrak{B}(G)_b \hookrightarrow \mathfrak{B}(G)$ for the locally closed embedding. For b, let $\mathbf{Rep}_{\mathrm{f.g.}}(J_b(F), \Lambda)$ be the full subcategory of $\mathbf{Rep}(J_b(F), \Lambda)$ generated (under finite colimits and retracts) by compactly induced representations

$$\delta_{K,\Lambda} := c\text{-}\mathrm{ind}_K^{J_b(F)}(\Lambda)$$

from the trivial representation of open compact subgroups $K \subset J_b(F)$. The following theorem from [35] summarizes some properties of $\mathbf{Shv}_c(\mathfrak{B}(G), \Lambda)$.

Theorem 2.1.2. (1) *An object in* $\mathbf{Shv}(\mathfrak{B}(G), \Lambda)$ *is constructible if and only if its !-restriction to each* $\mathfrak{B}(G)_b$ *is constructible and is zero for almost all* b*'s. If* Λ *is a field of characteristic zero,* $\mathbf{Shv}_c(\mathfrak{B}(G), \Lambda)$ *consist of compact objects in* $\mathbf{Shv}(\mathfrak{B}(G), \Lambda)$.

(2) *For every* $b \in B(G)$*, there is a canonical equivalence* $\mathbf{Shv}_c(\mathfrak{B}(G)_b, \Lambda) \cong \mathbf{Rep}_{\mathrm{f.g.}}(J_b(F), \Lambda)$*. There are fully faithful embeddings* $i_{b,*}, i_{b,!} : \mathbf{Shv}_c(\mathfrak{B}(G)_b, \Lambda) \to \mathbf{Shv}_c(\mathfrak{B}(G), \Lambda)$ *(which coincide when b is basic), inducing a semiorthogonal decomposition of* $\mathbf{Shv}_c(\mathfrak{B}(G), \Lambda)$ *in terms of* $\{\mathbf{Shv}_c(\mathfrak{B}(G)_b, \Lambda)\}_b$.

(3) *There is a self-duality functor* $\mathbb{D}^{\mathrm{coh}}: \mathbf{Shv}_c(\mathfrak{B}(G), \Lambda) \simeq \mathbf{Shv}_c(\mathfrak{B}(G), \Lambda)^\vee$ *obtained by gluing cohomological dualities (in the sense of Bernstein–Zelevinsky) on various* $\mathbf{Rep}_{\mathrm{f.g.}}(J_b(F), \Lambda)$*'s.*

(4) *There is a natural perverse t-structure obtained by gluing (shifted) t-structures on various* $\mathbf{Rep}_{\mathrm{f.g.}}(J_b(F), \Lambda)$*'s, preserved by* $\mathbb{D}^{\mathrm{coh}}$ *if* Λ *is a field.*

The following categorical form of the arithmetic local Langlands conjecture [82, SECT. 4.6] is inspired by the global geometric Langlands conjecture as discussed in Section 1.3.

Conjecture 2.1.3. *Assume that G is quasisplit over F equipped with a pinning* (B, T, e) *and fix a nontrivial additive character* $\psi : F \to \mathbb{Z}_\ell[\mu_{p^\infty}]^\times$*. There is a canonical equivalence of categories*

$$\mathbb{L}_G : \mathbf{Coh}(\mathrm{Loc}_{{}^cG} \otimes \Lambda) \cong \mathbf{Shv}_c\big(\mathfrak{B}(G), \Lambda\big).$$

Remark 2.1.4. (1) There is a closely related version of the conjecture, with $\mathbf{Shv}_c(\mathfrak{B}(G), \Lambda)$ replaced by $\mathbf{Shv}(\mathfrak{B}(G), \Lambda)$ and with $\mathbf{Coh}(\mathrm{Loc}_{{}^cG} \otimes \Lambda)$ replaced by its ind-completion (with certain support condition imposed) (see [82, SECT. 4.6]). Fargues–Scholze [25] make a conjecture parallel to this version, with the category $\mathbf{Shv}(\mathfrak{B}(G), \Lambda)$ replaced by $D_{\mathrm{lis}}(\mathrm{Bun}_G, \Lambda)$ as mentioned above.

(2) It is also explained in [82] a motivic hope to have a version of such equivalence over $\mathbb{Q}$.

One consequence of the conjecture is that for every b there should exist a fully faithful embedding

$$\mathfrak{A}_{J_b} : \mathbf{Rep}_{\mathrm{f.g.}}\big(J_b(F), \Lambda\big) \to \mathbf{Coh}(\mathrm{Loc}_{^cG} \otimes \Lambda),$$

obtained as the restriction of a quasiinverse of $\mathbb{L}_G$ to $i_{b,!}(\mathbf{Rep}_{\mathrm{f.g.}}(J_b(F), \Lambda))$. The existence of such functor is closely related to the idea of local Langlands in families and has also been considered (in the case $J_b = G$ is split and Λ is a field of characteristic zero) in [6, 32].

In particular, for every open compact subgroup $K \subset J_b(F)$ there should exist a coherent sheaf

$$\mathfrak{A}_{K,\Lambda} := \mathfrak{A}_{J_b}(\delta_{K,\Lambda}) \tag{2.1}$$

on $\mathrm{Loc}_{^cG} \otimes \Lambda$ such that

$$(R\mathrm{End}_{\mathbf{Coh}(\mathrm{Loc}_{^cG} \otimes \Lambda)} \mathfrak{A}_{K,\Lambda})^{\mathrm{op}} \cong (R\mathrm{End}_{\mathbf{Rep}(G(F),\Lambda)} \delta_{K,\Lambda})^{\mathrm{op}} =: H_{K,\Lambda}. \tag{2.2}$$

The algebra $H_{K,\Lambda}$ is sometimes called the derived Hecke algebra as it might not concentrate on cohomological degree zero (when $\Lambda = \mathbb{Z}_\ell$ or $\mathbb{F}_\ell$). See [82, SECTS. 4.3–4.5] for conjectural descriptions of $\mathfrak{A}_{K,\Lambda}$ in various cases.

As in the global geometric Langlands conjecture, the equivalence from Conjecture 2.1.3 should satisfy a set of compatibility conditions. For example, it should be compatible with parabolic inductions on both sides, and should be compatible with the duality $\mathbb{D}^{\mathrm{coh}}$ on $\mathbf{Shv}_c(\mathfrak{B}(G), \Lambda)$ and the (modified) Grothendieck–Serre duality of $\mathbf{Coh}(\mathrm{Loc}_{^cG} \otimes \Lambda)$. We refer to [35, 82] for more details.

On the other hand, Conjecture 2.1.3 predicts an action of the category $\mathbf{Perf}(\mathrm{Loc}_{^cG} \otimes \Lambda)$ of perfect complexes on $\mathrm{Loc}_{^cG} \otimes \Lambda$ on $\mathbf{Shv}_c(\mathfrak{B}(G), \Lambda)$, analogous to the spectral action as mentioned in Section 1.3. One of the main results of [25] is the construction of such action in their setting. Currently the existence of such a spectral action on $\mathbf{Shv}_c(\mathfrak{B}(G), \Lambda)$ is not known. But there are convincing evidences that Conjecture 2.1.3 should still be true.

We assume that G extends to a reductive group over $\mathcal{O}$ as before. Then there are closed substacks

$$\mathrm{Loc}^{\mathrm{ur}}_{^cG} \subset \mathrm{Loc}^{\mathrm{unip}}_{^cG} \subset \mathrm{Loc}_{^cG},$$

usually called the stack of unramified parameters (resp. unipotent parameters), classifying those ρ such that $\rho(I_F)$ is trivial (resp. $\rho(I_F)$ is unipotent). For $\Lambda = \mathbb{Q}_\ell$, $\mathrm{Loc}^{\mathrm{unip}}_{^cG} \otimes \mathbb{Q}_\ell$ is a connected component of $\mathrm{Loc}_{^cG} \otimes \mathbb{Q}_\ell$.

On the other hand, there is the unipotent subcategory $\mathbf{Shv}^{\mathrm{unip}}_c(\mathfrak{B}(G), \mathbb{Q}_\ell) \subset \mathbf{Shv}_c(\mathfrak{B}(G), \mathbb{Q}_\ell)$, which roughly speaking is the glue of categories $\mathbf{Rep}^{\mathrm{unip}}_{\mathrm{f.g.}}(J_b(F), \mathbb{Q}_\ell)$

of unipotent representations of $J_b(F)$ (introduced in [52]) for all $b \in B(G)$. We have the following theorem from [35], deduced from Theorem 1.2.2 by taking the Frobenius-twisted categorical trace.

Theorem 2.1.5. *For a reductive group G over $\mathcal{O}$ with a fixed pinning (B, T, e), there is a canonical equivalence*

$$\mathbb{L}_G^{\mathrm{unip}} : \mathbf{Coh}(\mathrm{Loc}^{\mathrm{unip}}_{^cG} \otimes \mathbb{Q}_\ell) \cong \mathbf{Shv}_c^{\mathrm{unip}}(\mathfrak{B}(G), \mathbb{Q}_\ell).$$

For arithmetic applications, it is important to match specific objects under the equivalence. We give a few examples and refer to [35] for many more of such matchings (see also [82, SECTS. 4.3–4.5]).

Example 2.1.6. The equivalence $\mathbb{L}_G^{\mathrm{unip}}$ gives the the conjectural coherent sheaf in (2.1) for all parahoric subgroups $K \subset G(F)$ (in the sense of Bruhat–Tits) such that (2.2) holds. For example, we have $\mathfrak{A}_{G(\mathcal{O}),\mathbb{Q}_\ell} \cong \mathcal{O}_{\mathrm{Loc}^{\mathrm{ur}}_{^cG}} \otimes \mathbb{Q}_\ell$, which gives

$$(R\mathrm{End}_{\mathbf{Coh}(\mathrm{Loc}_{^cG})}\mathcal{O}_{\mathrm{Loc}^{\mathrm{ur}}_{^cG}})^{\mathrm{op}} \otimes \mathbb{Q}_\ell \cong (R\mathrm{End}\, \delta_{G(\mathcal{O}),\mathbb{Q}_\ell})^{\mathrm{op}} = H_{G(\mathcal{O}),\mathbb{Q}_\ell} \cong H^{\mathrm{cl}}_{G(\mathcal{O})} \otimes \mathbb{Q}_\ell. \quad (2.3)$$

As $\mathrm{Loc}^{\mathrm{ur}}_{^cG} \cong ({}^cG|_{d=(q,\sigma)})/\hat{G}$, taking the 0th cohomology recovers the Satake isomorphism (1.3). In addition, it implies that the left-hand side has no higher cohomology, which is not obvious. We mention that it is conjectured in [82, SECT. 4.3] that $\mathfrak{A}_{G(\mathcal{O}),\mathbb{Z}_\ell} \cong \mathcal{O}_{\mathrm{Loc}^{\mathrm{ur}}_{^cG}}$ so the first isomorphism in (2.3) should hold over $\mathbb{Z}_\ell$, known as the (conjectural) derived Satake isomorphism. (But $H_{G(\mathcal{O}),\mathbb{Z}_\ell} \neq H^{\mathrm{cl}}_{G(\mathcal{O})} \otimes \mathbb{Z}_\ell$ in general.)

There is also a pure Galois side description of $\mathfrak{A}_{I,\mathbb{Q}_\ell}$, known as the unipotent coherent Springer sheaf as defined in [6, 82] (see also [32]).

Example 2.1.7. By construction, there is a natural morphism of stacks $\mathrm{Loc}_{^cG} \to \mathbb{B}\hat{G}$ over $\mathbb{Z}_\ell$. For a representation of $\hat{G}$ on a finite projective Λ-module, regarded as a vector bundle on $\mathbb{B}\hat{G}_\Lambda$, let $\tilde{V}$ be its pullback to $\mathrm{Loc}_{^cG} \otimes \Lambda$, and let $\tilde{V}^? \in \mathbf{Perf}(\mathrm{Loc}^?_{^cG} \otimes \Lambda)$ be its restriction of $\mathrm{Loc}^?_{^cG} \otimes \Lambda$ for $? = \mathrm{ur}$ or unip. Note that for $\Lambda = \mathbb{Q}_\ell$, $\tilde{V}^{\mathrm{ur}} \cong \tilde{V} \otimes \mathfrak{A}_{G(\mathcal{O}),\mathbb{Q}_\ell}$. We have

$$\mathbb{L}_G^{\mathrm{unip}}(\tilde{V}^{\mathrm{ur}}) \cong \mathrm{Nt}_! r^! \mathrm{Sat}(V) =: \mathcal{S}_V,$$

where r and Nt are maps in the following correspondence:

$$\mathrm{Hk}_G = L^+G\backslash LG/L^+G \xleftarrow{r} LG/\mathrm{Ad}_\sigma L^+G \xrightarrow{\mathrm{Nt}} LG/\mathrm{Ad}_\sigma LG = \mathfrak{B}(G).$$

In particular, for two representations V and W of $\hat{G}$, there is a morphism

$$R\mathrm{Hom}_{\mathrm{Loc}^{\mathrm{ur}}_{^cG}\otimes\mathbb{Q}_\ell}(\tilde{V}^{\mathrm{ur}}, \tilde{W}^{\mathrm{ur}}) \to R\mathrm{Hom}_{\mathbf{Shv}_c(\mathfrak{B}(G),\mathbb{Q}_\ell)}(\mathcal{S}_V, \mathcal{S}_W) \quad (2.4)$$

compatible with compositions. Such map was first constructed in [64, 77] and (the version for underived Hom spaces) was then extended to $\mathbb{Z}_\ell$-coefficient in [70]. It has significant arithmetic applications, as will be explained in Section 3.

Remark 2.1.8. It is likely that Theorem 2.1.5 can be extended to the tame level by taking the Frobenius-twisted categorical trace of the equivalence from Theorem 1.2.4. On the other hand, as mentioned in Remark 1.2.5, it is important to extend these equivalences to $\mathbb{Z}_\ell$-coefficient.

2.2. Global arithmetic Langlands for function fields

Next we turn to global aspects of the arithmetic Langlands correspondence. As mentioned at the beginning, its classical formulation, very roughly speaking, predicts a natural correspondence between the set of (irreducible) Galois representations and the set of (cuspidal) automorphic representations. As in the local case, beyond the GL_n case (which is a theorem by [38]), such a formulation is not easy to be made precise. On the other hand, the global geometric Langlands conjecture from Section 1.3 and philosophy of decategorification/trace suggest that the global arithmetic Langlands can and probably should be formulated as an isomorphism between two vector spaces, arising from the Galois and the automorphic side, respectively. In this subsection, we formulate such a conjecture in the global function field case.

Let $F = \mathbb{F}_q(X)$ be the function field of a geometrically connected smooth projective curve X over $\mathbb{F}_q$. We write $\eta = \operatorname{Spec} F$ for the generic point of X and $\overline{\eta}$ for a geometric point over η. Let $|X|$ denote the set of closed points of X. For $v \in |X|$, let $\mathcal{O}_v$ denote the complete local ring of X at v and F_v its fractional field. Let $\mathbb{O}_F = \prod_{v\in|X|} \mathcal{O}_v$ be the integral adèles, and $\mathbb{A}_F = \prod'_{v\in|X|} F_v$ the ring of adèles. For a finite nonempty set of places Q, let $W_{F,Q}$ denote the Weil group of F, unramified outside Q.

Let G be a connected reductive group over F. Similarly to the local situation, the first step to formulate our global conjecture is the following theorem from [82].

Theorem 2.2.1. *Assume that $\ell \nmid 2p$. The prestack sending a $\mathbb{Z}_\ell$-algebra A to the space of (strongly) continuous homomorphisms $\rho : W_{F,Q} \to {}^cG(A)$ such that $d \circ \rho = (\mathrm{cycl}^{-1}, \mathrm{pr})$ is represented by a derived scheme $\mathrm{Loc}^{\square}_{{}^cG,Q}$, which is a disjoint union of derived affine schemes that are flat and of finite type over $\mathbb{Z}_\ell$. If $Q \neq \emptyset$, $\mathrm{Loc}^{\square}_{{}^cG,Q}$ is quasismooth.*

We then define the stack of global Langlands parameters as $\mathrm{Loc}_{{}^cG,Q} = \mathrm{Loc}^{\square}_{{}^cG,Q}/\hat{G}$. Similar to the local case (see Example 2.1.7), for a representation of $\hat{G}_\Lambda$ on a finite projective Λ-module, regarded as a vector bundle on $\mathbb{B}\hat{G}_\Lambda$, let $\tilde{V}$ be its pullback to $\mathrm{Loc}_{{}^cG,Q} \otimes \Lambda$. If V is the restriction of a representation of $({}^cG)^S$ along the diagonal embedding $\hat{G} \to ({}^cG)^S$, then there is a natural (strongly) continuous $W^S_{F,Q}$-action on $\tilde{V}$ (see [82, SECT. 2.4]). For a place v of F, let $\mathrm{Loc}_{{}^cG,v}$ denote the stack of local Langlands parameters for G_{F_v}. Let

$$\mathrm{res} : \mathrm{Loc}_{{}^cG,Q} \to \prod_{v\in Q} \mathrm{Loc}_{{}^cG,v}$$

denote the map by restricting global parameters to local parameters (induced by the map $W_{F_v} \to W_{F,Q}$). Later on, we will consider the !-pullback of coherent sheaves on $\prod_{v\in Q} \mathrm{Loc}_{{}^cG,v}$ along this map.

Remark 2.2.2. (1) In fact, when $Q = \emptyset$, the definition of $\mathrm{Loc}_{{}^cG,Q}$ needs to be slightly modified.

(2) Unlike the local situation, $\mathrm{Loc}_{{}^cG,Q}$ has nontrivial derived structure in general (see [82, REMARK 3.4.5]). Let ${}^{cl}\mathrm{Loc}_{{}^cG,Q}$ denote the underlying classical stack.

(3) A different definition of $\mathrm{Loc}_{{}^cG,Q} \otimes \mathbb{Q}_\ell$ is given by [3].

Next we move to the automorphic side. For simplicity, we assume that G is split over $\mathbb{F}_q$ in this subsection. Fix a level, i.e., an open compact subgroup $K \subset G(\mathbb{O}_F)$. Let Q be the set of places consisting of those v such that $K_v \neq G(\mathcal{O}_v)$. For a finite set S, let $\mathrm{Sht}_K(G)_{(X-Q)^S}$ denote the ind-Deligne–Mumford stack over $(X-Q)^S$ of the moduli of G-shtukas on X with S-legs in $X-Q$ and K-level structure. (For example, see **[39]** for basic constructions and properties of this moduli space.) Its base change along the diagonal map $\overline{\eta} \to (X-Q) \xrightarrow{\Delta} (X-Q)^S$ is denoted by $\mathrm{Sht}_K(G)_{\Delta(\overline{\eta})}$. For every representation V of $({}^cG)^S$ on a finite projective Λ-module, the geometric Satake (1.6) (with D replaced by $X-Q$ and with $\Lambda = \mathbb{Z}_\ell$ allowed) provides a perverse sheaf $\mathrm{Sat}_S(V)$ on $\mathrm{Sht}_K(G)_{(X-Q)^S}$. Let $C_c(\mathrm{Sht}_K(G)_{\Delta(\overline{\eta})}, \mathrm{Sat}_S(V))$ denote the (cochain complex of the) total compactly supported cohomology of $\mathrm{Sht}_K(G)_{\Delta(\overline{\eta})}$ with coefficient in $\mathrm{Sat}_S(V)$. It admits a (strongly) continuous action of $W_{F,Q}^S$ (see **[34]** for the construction of such action at the derived level, based on **[67, 68]**), as well as an action of the corresponding global (derived) Hecke algebra (with coefficients in Λ)

$$H_{K,\Lambda} = \big(R\mathrm{End}\big(c\text{-ind}_K^{G(\mathbb{A}_F)}(\Lambda)\big)\big)^{\mathrm{op}}. \tag{2.5}$$

For example, if $V = \mathbf{1}$ is the trivial representation, then (under our assumption that G is split)

$$C_c\big(\mathrm{Sht}_K(G)_{\Delta(\overline{\eta})}, \mathrm{Sat}_{\{1\}}(\mathbf{1})\big) = C_c\big(G(F)\backslash G(\mathbb{A})/K, \Lambda\big).$$

Here $G(F)\backslash G(\mathbb{A})/K$ is regarded as a discrete DM stack over $\overline{\eta}$, and $C_c(G(F)\backslash G(\mathbb{A})/K, \Lambda)$ denotes its compactly supported cohomology. When $\Lambda = \mathbb{Q}_\ell$, this is the space of compactly supported functions on $G(F)\backslash G(\mathbb{A})/K$.

We will fix a pinning (B, T, e) of G and a nondegenerate character $\psi : F\backslash\mathbb{A} \to \mathbb{Z}_\ell[\mu_p]^\times$, which gives the conjectural equivalence $\mathbb{L}_v$ as in Conjecture 2.1.3 for every $v \in Q$. In particular, corresponding to $K_v \subset G(F_v)$ there is a conjectural coherent sheaf $\mathfrak{A}_{K_v}$ (see (2.1)) on $\mathrm{Loc}_{{}^cG,v}$.

Conjecture 2.2.3. *There is a natural* $(W_{F,Q}^S \times H_{K,\Lambda})$*-equivariant isomorphism*

$$R\Gamma\big(\mathrm{Loc}_{{}^cG,Q} \otimes \Lambda, \tilde{V} \otimes \mathrm{res}^!(\boxtimes_{v\in Q}\mathfrak{A}_{K_v})\big) \cong C_c\big(\mathrm{Sht}_K(G)_{\Delta(\overline{\eta})}, \mathrm{Sat}_S(V)\big).$$

We refer to **[82, SECT. 4.7]** for more general form of the conjecture (where "generalized level structures" are allowed) and examples of such conjecture in various special cases. This conjecture could be regarded a precise form of the global Langlands correspondence for function fields. Namely, it gives a precise recipe to match Galois representations and automorphic representations. (For example, V. Lafforgue's excursion operators are encoded in such isomorphism, see below.) Moreover, such an isomorphism fits in the Arthur–Kottwitz multiplicity formula and at the same time extends such a formula to the integral level and therefore relates to automorphic lifting theories.

The most appealing evidence of this conjecture is the following theorem **[40, 82]**, as suggested (at the heuristic level) by Drinfeld as an interpretation of Lafforgue's construction.

Theorem 2.2.4. *For each i, there is a quasicoherent sheaf $\mathfrak{A}^i_K$ on ${}^{cl}\mathrm{Loc}_{{}^cG,Q} \otimes \mathbb{Q}_\ell$, equipped with an action of $H_{K,\mathbb{Q}_\ell}$, such that for every finite dimensional $\mathbb{Q}_\ell$-representation V of $({}^cG)^S$, there is a natural $(W^S_{F,Q} \times H_{K,\mathbb{Q}_\ell})$-equivariant isomorphism*

$$\Gamma({}^{cl}\mathrm{Loc}_{{}^cG,Q} \otimes \mathbb{Q}_\ell, \tilde{V} \otimes \mathfrak{A}^i_K) \cong H^i_c(\mathrm{Sht}_K(G)_{\Delta(\overline{\eta})}, \mathrm{Sat}_S(V)).$$

We mention that this theorem actually was proved for any G in **[40, 82]**. In addition, when K is everywhere hyperspecial, (2.2.4) holds at the derived level by **[3]**.

The isomorphism (2.2.4) induces an action of $\Gamma({}^{cl}\mathrm{Loc}_{{}^cG,Q} \otimes \mathbb{Q}_\ell, \mathcal{O})$ on the right-hand side. This is exactly the action by V. Lafforgue's excursion operators, which induces the decomposition of the right-hand side (in particular, $C_c(G(F)\backslash G(\mathbb{A}_F)/K, \mathbb{Q}_\ell)$) in terms of semisimple Langlands parameters. As explained **[40]**, over an elliptic Langlands parameter, such an isomorphism is closely related to the Arthur–Kottwitz multiplicity formula. In the case of $G = \mathrm{GL}_n$, it gives the following corollary, generalizing **[38]**.

Corollary 2.2.5. *Let π be a cuspidal automorphic representation of GL_n, with the associated irreducible Galois representation $\rho_\pi : W_{F,Q} \to \mathrm{GL}_n(\Lambda)$ for some finite extension $\Lambda/\mathbb{Q}_\ell$ and with $\mathfrak{m}_\pi$ the corresponding maximal ideal of $\Gamma({}^{cl}\mathrm{Loc}_{{}^cG,Q} \otimes \Lambda, \mathcal{O})$. Then there is an $(W^S_{F,Q} \times H_K)$-equivariant isomorphism*

$$H^*_c(\mathrm{Sht}_K(G)_{\Delta(\overline{\eta})}, \mathrm{Sat}_S(V))/\mathfrak{m}_\pi \cong V_{\rho_\pi} \otimes \pi^K.$$

In particular, the left-hand side only concentrates in cohomological degree zero.

2.3. Geometric realization of Jacquet–Langlands transfer

The global Langlands correspondence for number fields is far more complicated. In fact, there are analytic part of the theory which currently seems not to fit the categorification/decategorification framework. Even if we just restrict to the algebraic/arithmetic part of the theory, there are complications coming from the place at ℓ and at ∞. In particular, the categorical forms of the local Langlands correspondence at ℓ and ∞ are not yet fully understood.

Nevertheless, in a forthcoming joint work with Emerton and Emerton–Gee **[21, 22]**, we will formulate conjectural Galois theoretical descriptions for the cohomology of Shimura varieties and even cohomology for general locally symmetric space, parallel to Conjecture 2.2.3. In this subsection, we just review a conjecture from **[82]** on the geometric realization of Jacquet–Langlands transfer via cohomology of Shimura varieties and discuss results from **[35, 64]** towards this conjecture.

We fix a few notations and assumptions. We fix a prime p in this subsection. Let $\mathbb{A}_f = \prod'_q \mathbb{Q}_q$ denote the ring of finite adèles of $\mathbb{Q}$, and $\mathbb{A}^p_f = \prod'_{q\neq p} \mathbb{Q}_p$. We write $\overline{\eta} = \mathrm{Spec}\,\overline{\mathbb{Q}}$, where $\overline{\mathbb{Q}}$ is the algebraic closure of $\mathbb{Q}$ in $\mathbb{C}$. For a Shimura datum (G, X), let μ be the (minuscule) dominant weight of $\hat{G}$ (with respect to $(\hat{B}, \hat{T})$) determined by (G, X) in the usual way and let V_μ denote the minuscule representation of $\hat{G}$ of highest weight μ. Let $E \subset \overline{\mathbb{Q}} \subset \mathbb{C}$ be the reflex field of (G, X) and write $d_\mu = \dim X$. For a level (i.e., an open compact subgroup) $K = K_pK^p \subset G(\mathbb{Q}_p)G(\mathbb{A}^p_f)$, let $\mathrm{Sh}_K(G)$ be the corresponding

Shimura variety of level K (defined over the reflex field E), and let $\mathrm{Sh}_K(G)_{\overline{\eta}}$ denote its base change along $E \to \overline{\mathbb{Q}}$. Let v be a place of E above p. By a specialization $\mathrm{sp} : \overline{\eta} \to \overline{v}$, we mean a morphism from $\overline{\eta}$ to the strict henselianization of $\mathcal{O}_E$ at v.

To avoid many complications from Galois cohomology (e.g., the difference between extended pure inner forms and inner forms) and also some complications from geometry (e.g., the relation between Shimura varieties and moduli of Shtukas), we assume that G is of adjoint type in the rest of this subsection, and refer to [64] for general G. See also [82] with less restrictions on G.

Definition 2.3.1. Let G be a connected reductive group over $\mathbb{Q}$. A prime-to-p (resp. finitely) trivialized inner form of G is a G-torsor β over $\mathbb{Q}$ equipped with a trivialization β over $\mathbb{A}_f^p$ (resp. over $\mathbb{A}_f$). Then $G' := \mathrm{Aut}(\xi)$ is an inner form of G (so the dual group of G and G' are canonically identified), equipped with an isomorphism $\theta : G(\mathbb{A}_f^p) \cong G'(\mathbb{A}_f^p)$ (resp. $\theta : G(\mathbb{A}_f) \cong G'(\mathbb{A}_f)$).

Now let (G, X) and (G', X') be two Shimura data, with G' a prime-to-p trivialized inner form of G. Via θ, one can transport $K^p \subset G(\mathbb{A}_f^p)$ to an open compact subgroup $K'^p \subset G'(\mathbb{A}_f^p)$. We identify the prime-to-p (derived) Hecke algebra $H_{K^p,\Lambda}$ (defined in the same way as in (2.5)) with $H_{K'^p,\Lambda}$ and simply write them as $H_{K^p,\Lambda}$. Let $K'_p \subset G'(\mathbb{Q}_p)$ be an open compact subgroup and write $K' = K'_p K'^p$ for the corresponding level.

We fix a quasisplit inner form $G^*_{\mathbb{Q}_p}$ of $G_{\mathbb{Q}_p}$ and $G'_{\mathbb{Q}_p}$ equipped with a pinning $(B^*_{\mathbb{Q}_p}, T^*_{\mathbb{Q}_p}, e^*)$, and realize $G_{\mathbb{Q}_p}$ as J_b and $G'_{\mathbb{Q}_p}$ as $J_{b'}$ for $b, b \in B(G^*_{\mathbb{Q}_p})$. Under our assumption that G and G' are adjoint, such b, b' exist and are unique. Then we have the conjectural coherent sheaf $\mathfrak{A}_{K_p,\Lambda}$ and $\mathfrak{A}_{K'_p,\Lambda}$ as in (2.1) on the stack $\mathrm{Loc}_{{}^cG,p} \otimes \Lambda$ of local Langlands parameters for $G^*_{\mathbb{Q}_p}$ over Λ.

Conjecture 2.3.2. *For every choice of specialization map* $\mathrm{sp} : \overline{\eta} \to \overline{v}$, *there is a natural map*

$$\begin{aligned} &R\mathrm{Hom}_{\mathbf{Coh}(\mathrm{Loc}_{{}^cG,p}\otimes\Lambda)}(\widetilde{V_\mu} \otimes \mathfrak{A}_{K_p,\Lambda}, \widetilde{V_{\mu'}} \otimes \mathfrak{A}_{K'_p,\Lambda}) \\ &\quad \to R\mathrm{Hom}_{H_{K^p,\Lambda}}\big(C_c\big(\mathrm{Sh}_K(G)_{\overline{\eta}}, \Lambda[d_\mu]\big), C_c\big(\mathrm{Sh}_{K'}(G')_{\overline{\eta}}, \Lambda[d_{\mu'}]\big)\big), \end{aligned} \tag{2.6}$$

compatible with compositions. In particular, there is an (E_1-)*algebra homomorphism*

$$S : R\mathrm{End}_{\mathbf{Coh}(\mathrm{Loc}_{{}^cG,p}\otimes\Lambda)}(\widetilde{V_\mu} \otimes \mathfrak{A}_{K_p,\Lambda}) \to R\mathrm{End}_{H_{K^p,\Lambda}}\big(C_c\big(\mathrm{Sh}_K(G)_{\overline{\eta}}, \Lambda\big)\big), \tag{2.7}$$

compatible with (2.6). *In addition, the induced action*

$$H_{K_p,\Lambda} \overset{(2.2)}{\cong} R\mathrm{End}(\mathfrak{A}_{K_p,\Lambda}) \to R\mathrm{End}(\widetilde{V_\mu} \otimes \mathfrak{A}_{K_p,\Lambda}) \xrightarrow{S} R\mathrm{End}_{H_{K^p,\Lambda}}\big(C_c\big(\mathrm{Sh}_K(G)_{\overline{\eta}}, \Lambda\big)\big) \tag{2.8}$$

coincides with the natural Hecke action of $H_{K_p,\Lambda}$ *on* $C_c(\mathrm{Sh}_K(G)_{\overline{\eta}}, \Lambda)$ *(and therefore is independent of the specialization map* sp*).*

This conjecture would be a consequence of a Galois theoretic description of $C_c(\mathrm{Sh}_K(G)_{\overline{\eta}}, \Lambda)$ similar to Conjecture 2.2.3, but its formulation does not require the existence of the stack of global Langlands parameters for $\mathbb{Q}$. In any case, a step towards a Galois-theoretical description of $C_c(\mathrm{Sh}_K(G)_{\overline{\eta}}, \Lambda)$ might require Conjecture 2.3.2 as an input. We also remark that as in the function field case, there is a more general version of such conjecture in [82, SECT. 4.7], allowing "generalized level structures," so that the cohomology of Igusa varieties could appear.

The following theorem verifies the conjecture in special cases.

Theorem 2.3.3. *Suppose that the Shimura data* (G, X) *and* (G', X') *are of abelian type, with* G' *a finitely trivialized inner form of* G. *Suppose that* $G_{\mathbb{Q}_p}$ *is unramified (and therefore so is* $G'_{\mathbb{Q}_p}$*).*

(1) *The map* (2.6) *(and therefore* (2.7)*) exists when* $\Lambda = \mathbb{Q}_\ell$ *and* $K_p \subset G(\mathbb{Q}_p)$ *and* $K'_p \subset G'(\mathbb{Q}_p)$ *are parahoric subgroups (in the sense of Bruhat–Tits).*

(2) *If* K_p *is hyperspecial, then the map* (2.6) *(and therefore* (2.7)*) exists when* $\Lambda = \mathbb{Z}_\ell$, *at least for underived* Hom *spaces. In addition, the action of* $H^{\mathrm{cl}}_{K_p}$ *on* $H^*_c(\mathrm{Sh}_K(G)_{\overline{\eta}}, \Lambda)$ *via* (2.8) *coincides with the natural action of* $H^{\mathrm{cl}}_{K_p}$.

Part (1) is proved in [35,64]. The proof contains two ingredients. One is the construction of physical correspondences between mod p fibers of $\mathrm{Sh}_K(G)$ and $\mathrm{Sh}_{K'}(G')$ by [64] (this is where we currently need to assume that G and G' are unramified at p). The other ingredient is Theorem 2.1.5 (and therefore requires $\Lambda = \mathbb{Q}_\ell$). When K_p is hypersepcial, one can work with $\mathbb{Z}_\ell$-coefficient, as (the underived version of) (2.4) exists for $\mathbb{Z}_\ell$-coefficient thanks to [70]. In fact, in this case one can allow nontrivial local systems on the Shimura varieties (see [70]). The last statement is known as the $S = T$ for Shimura varieties. The case when $d_\mu = \dim \mathrm{Sh}_K(G) = 0$ is contained in [64]. The general case is proved in [63,74] using foundational works from [25,59].

3. APPLICATIONS TO ARITHMETIC GEOMETRY

Besides the previously mentioned directly applications of (ideas from) geometric Langlands to the classical Langlands program, we discuss some further arithmetic applications, mostly related to Shimura varieties and based on the author's works. We shall mention that there are many other remarkable applications of (ideas of) geometric Langlands to arithmetic problems, such as [28,31,44,66,71], to name a few.

3.1. Local models of Shimura varieties

The theory of integral models of Shimura varieties (with parahoric level) started (implicitly in the work of Kronecker) with understanding of the mod p reduction of elliptic modular curves with $\Gamma_0(p)$-level. We discuss a small fraction of this theory concerning étale local structures of these integral models via the theory of local models. The recent developments of the theory of local models are greatly influenced by the geometric Langlands program.

We use notations from Section 2.3 for Shimura varieties (but we do not assume that G is of adjoint type in this subsection). Let (G, X) be a Shimura datum and K a chosen level with $K_p = \mathcal{G}(\mathbb{Z}_p)$ for some parahoric group scheme $\mathcal{G}$ (in the sense of Bruhat–Tits) of

$G_{\mathbb{Q}_p}$ over $\mathbb{Z}_p$. Then for a place v of E over p, a local model diagram is a correspondence of quasiprojective schemes over $\mathcal{O}_{E_v}$,

$$\mathscr{S}_K \leftarrow \tilde{\mathscr{S}}_K \xrightarrow{\tilde{\varphi}} M^{\mathrm{loc}}_{\mathcal{G}}, \tag{3.1}$$

where $\mathscr{S}_K$ is an integral model of $\mathrm{Sh}_K(G)$ over $\mathcal{O}_{E_v}$, $\tilde{\mathscr{S}}_K$ is a $\mathcal{G}_{\mathcal{O}_{E_v}}$-torsor over $\mathscr{S}_K$, $M^{\mathrm{loc}}_{\mathcal{G}}$ is the so-called local model, which is a flat projective scheme over $\mathcal{O}_{E_v}$ equipped with a $\mathcal{G}_{\mathcal{O}_{E_v}}$-action, and $\tilde{\varphi}$ is a $\mathcal{G}_{\mathcal{O}_{E_v}}$-equivariant smooth morphism of relative dimension $\dim G$. Therefore, $M^{\mathrm{loc}}_{\mathcal{G}}$ models étale local structure of $\mathscr{S}_K$. On the other hand, the existence of $\mathcal{G}_{\mathcal{O}_{E_v}}$-action on $M^{\mathrm{loc}}_{\mathcal{G}}$ makes it easier than $\mathscr{S}_K$ to study.

The original construction of local models is based on realization of a parahoric group scheme as (the neutral connected component of) the stabilizer group of a self-dual lattice chain in a vector space (over a division algebra over F) with a bilinear form, e.g., see [57] for a survey and references. This approach is somehow ad hoc and is limited the so-called (P)EL (local) Shimura data. A new approach, based on the construction of an $\mathbb{Z}_p$-analogue of the stack $\mathrm{Hk}_{\mathcal{G},D}$ from Section 1.2, was systematically introduced in [58] (under the tameness assumption of G which was later lifted in [46, 50]). In [58] the construction of such a $\mathbb{Z}_p$-analogue (or rather the corresponding Beilinson–Drinfeld-type affine Grassmannian $\mathrm{Gr}_{\mathcal{G},\mathbb{Z}_p}$ over $\mathbb{Z}_p$) is based on the construction of certain "two dimensional parahoric" group scheme $\tilde{\mathcal{G}}$ over $\mathbb{Z}_p[\varpi]$ whose restriction along $\mathbb{Z}_p[\varpi] \xrightarrow{\varpi \mapsto p} \mathbb{Z}_p$ recovers $\mathcal{G}$. (See [81] for a survey.) A more direct construction of a different p-adic version of such affine Grassmannian $\mathrm{Gr}_{\mathcal{G},\mathrm{Spd}\mathbb{Z}_p}$ was given in [59] in the analytic perfectoid world. In either case, the local model is defined as the flat closure of the Schubert variety in the generic fiber corresponding to μ. In addition, the recent work [1] shows that the two constructions agree. The following theorem from [1] is the most up-to-date result on the existence of local models and about their properties.

Theorem 3.1.1. *Let G be a connected reductive group over a p-adic field F. Except the odd unitary case when $p = 2$ and triality case when $p = 3$, for every parahoric group scheme $\mathcal{G}$ of G over $\mathcal{O}$, and a conjugacy class of minuscule cocharacters μ of G defined over a finite extension E/F of F, there is a normal flat projective scheme $M^{\mathrm{loc}}_{\mathcal{G},\mu}$ over $\mathcal{O}_E$, equipped with a $\mathcal{G}_{\mathcal{O}_E}$-action such that $M^{\mathrm{loc}}_{\mathcal{G},\mu} \otimes E$ is G_E-equivariantly isomorphic to the partial flag variety $\mathcal{F}\ell_\mu$ of G_E corresponding to μ, and that $M^{\mathrm{loc}}_{\mathcal{G}} \otimes k_E$ is $(\mathcal{G} \otimes k_E)$-equivariantly isomorphic to the (canonical deperfection of the) union over the μ-admissible set of Schubert varieties in $LG/L^+\mathcal{G} \otimes k_E$. In addition, $M^{\mathrm{loc}}_{\mathcal{G}}$ is normal, Cohen–Macaulay and each of its geometric irreducible components in its special fiber is normal and Cohen–Macaulay.*

We end this subsection with a few remarks.

Remark 3.1.2. (1) Once the local model diagram (3.1) is established, this theorem also gives the corresponding properties of the integral models of Shimura varieties.

(2) A key ingredient in the study of special fibers of local models is the coherence conjecture by Pappas–Rapoport [56], proved in [75] (and the proof uses the idea of fusion).

(3) One important motivation/application of the theory of local models is the Haines–Kottwitz conjecture [29], which predicts certain central element in the parahoric Hecke algebra $H^{\mathrm{cl}}_{K_p}$ should be used as the test function in the trace formula computing the Hasse–Weil zeta function of $\mathrm{Sh}_K(G)$. As mentioned in Section 1.2, this conjecture motivated Gaitsgory's central sheaf construction (1.9). With the local Hecke stack $\mathrm{Hk}_{\mathcal{G},\mathbb{Z}_p}$ over $\mathbb{Z}_p$ constructed (either the version from [58] or from [59]), one can mimic the construction (1.9) in mixed characteristic to solve the Kottwitz conjecture. Again, see [1] for the up-to-date result.

3.2. The congruence relation

We use notations and (for simplicity) keep assumptions from Section 2.3 regarding Shimura varieties. Let (G, X) be a Shimura datum abelian type, and let K be a level such that K_p is hyperspecial. Let $v \mid p$ be the place of E. Then $\mathrm{Sh}_K(G)$ has a canonical integral model $\mathscr{S}_K$ defined over $\mathcal{O}_{E,(v)}$ [37]. Let $\overline{\mathscr{S}}_K$ be its mod p fiber, which is a smooth variety defined over the residue field k_v of v. Let σ_v denote the geometric Frobenius in Γ_{k_v}. Theorem 2.3.3 gives an action of $\mathrm{End}_{\mathrm{Loc}^{\mathrm{ur}}_{{}^cG,p}}(\widetilde{V_\mu})$ on $H^*_c(\overline{\mathscr{S}}_{K,\overline{k}_v}, \mathbb{Z}_\ell)$, which as we shall see has significant consequences.

The congruence relation conjecture (also known as the Blasius–Rogawski conjecture), generalizing the classical Eichler–Shimura congruence relation $\mathrm{Frob}_p = T_p + V_p$ for modular curves, predicts that in the Chow group of $\overline{\mathscr{S}}_K \times \overline{\mathscr{S}}_K$, the Frobenius endomorphism of $\overline{\mathscr{S}}_K$ satisfies a polynomial whose coefficients are mod p reduction of certain Hecke correspondences. Theorem 2.3.3, together with [65, SECT. 6.3], implies this conjecture at the level of cohomology.

For every representation V of ${}^c(G_{\mathbb{Q}_p})$, its character χ_V (regarded as a $\hat{G}$-invariant function on ${}^cG|_{d=(p,\sigma_p)}$) gives an element $h_V \in H^{\mathrm{cl}}_{G(\mathbb{Z}_p)}$ via the Satake isomorphism (1.3).

Theorem 3.2.1. *The following identity,*

$$\sum_{i=0}^{n} (-1)^j h_{\chi_{\wedge^j V}} \sigma_v^{\dim V - j} = 0, \tag{3.2}$$

holds in $\mathrm{End}(H^*_c(\overline{\mathscr{S}}_{K,\overline{k}_v}, \mathbb{Z}_\ell))$, *where* $V = \mathrm{Ind}_{{}^c(G_{E_v})}^{{}^c(G_{\mathbb{Q}_p})} V_\mu$ *is the tensor induction of* V_μ.

Indeed, by [65, SECT. 6.3], such an equality holds with $h_{\chi_{\wedge^i V}}$ replaced by $S(\chi_{\wedge^i V})$, where S is from Theorem 2.3.3 (1). Then part (2) of that theorem allows one to replace $S(\chi_{\wedge^i V})$ by $h_{\chi_{\wedge^i V}}$. This approach to (3.2) is the Shimura variety analogue of V. Lafforgue's approach to the Eichler–Shimura relation for $\mathrm{Sht}_K(G)$ [39]. Traditionally, there is another approach to the congruence relation conjecture for Shimura varieties by directly studying reduction mod p of Hecke operators, starting from [24] for the Siegel modular variety case.

See [45] for the latest progress and related references. This approach would give (3.2) at the level of algebraic correspondences.

Now suppose $(G, X) = (\mathrm{Res}_{F^+/\mathbb{Q}}(G_0)_{F^+}, \prod_{\varphi: F^+\to\mathbb{R}} X_0)$, where (G_0, X_0) is a Shimura datum and F^+ is a totally real field. As before, let p be a prime such that K_p is hyperspecial. In particular, p is unramified in F^+. In addition, for simplicity we assume that $G_{0,\mathbb{Q}_p}$ is split (so for a place v of E above p, $E_v = \mathbb{Q}_p$). We let $\mathbb{F}$ denote an algebraic closure of $\mathbb{F}_p$. Let $\{w_i\}_i$ be the set of primes of F^+ above p, and let k_i denote the residue field of w_i. For each i, we also fix an embedding $\rho_i : k_i \to \mathbb{F}$. Then there is a natural map

$$\prod_i (\mathbb{Z}^{f_i} \rtimes \mathfrak{S}_{f_i}) \to \mathrm{End}_{\mathrm{Loc}^{\mathrm{ur}}_{^cG,p}}(\widetilde{V_\mu}),$$

where $\mathfrak{S}_{f_i}$ is the permutation group on f_i letters. Together with Theorem 2.3.3, one obtains the following result [64].

Theorem 3.2.2. *There is an action of* $\prod_i (\mathbb{Z}^{f_i} \rtimes \mathfrak{S}_{f_i})$ *on* $H^*_c(\overline{\mathscr{S}}_{K,\overline{\mathbb{F}}}, \mathbb{Z}_\ell)$ *such that action of* σ_p *factors as* $\sigma_p = \prod_i \sigma_{p,i}$, *where* $\sigma_{p,i} = ((1,0,\dots,0),(12\cdots f_i)) \in \mathbb{Z}^{f_i} \rtimes \mathfrak{S}_{f_i}$. *Each* $\sigma_{p,i}^{f_i}$ *satisfies a polynomial equation similar to* (3.2).

This theorem gives some shadow of the plectic cohomology conjecture of Nekovář–Scholl [54].

3.3. Generic Tate cycles on mod p fibers of Shimura varieties

In [64], we applied Theorem 2.3.3 to verify "generic" cases of Tate conjecture for the mod p fibers of many Shimura varieties. We use notations and (for simplicity) keep assumptions from Section 3.2. Let $(\overline{\mathscr{S}}_{K,\overline{k_v}})^{\mathrm{pf}}$ denote the perfection of $\overline{\mathscr{S}}_{K,\overline{k_v}}$ (i.e., regard it as a perfect presheaf over $\mathbf{Aff}^{\mathrm{pf}}_{\overline{k_v}}$), then by attaching to every point of $\overline{\mathscr{S}}_{K,\bar{k}}$ an F-isocrystal with G-structure (see [37, 64]), one can define the so-called Newton map

$$\mathrm{Nt} : (\overline{\mathscr{S}}_{K,\overline{k_v}})^{\mathrm{pf}} \to \mathfrak{B}(G_{\mathbb{Q}_p})_{\overline{k_v}}.$$

Then the Newton stratification of $\mathfrak{B}(G_{\mathbb{Q}_p})_{\overline{k_v}}$ (see Section 2.1) induces a stratification of $\overline{\mathscr{S}}_{K,\overline{k_v}}$ by locally closed subvarieties. It is known that the image of Nt contains a unique basic element b and the corresponding subvarieties in $\overline{\mathscr{S}}_{K,\overline{k_v}}$ is closed, called the basic Newton stratum, and denoted by $\overline{\mathscr{S}}_b$.

Let m be the order of the action of the geometric Frobenius σ_p on $\mathbb{X}^\bullet(\hat{T})$. Let

$$\Lambda_p^{\mathrm{Tate}} = \left\{\lambda \in \mathbb{X}^\bullet(\hat{T}) \mid \sum_{i=0}^{m-1} \sigma_p^i(\lambda) = 0\right\} \subset \mathbb{X}^\bullet(\hat{T}).$$

For a representation V of $\hat{G}_{\mathbb{Q}_\ell}$ and $\lambda \in \mathbb{X}^\bullet(\hat{T})$, let $V(\lambda)$ denote the λ-weight subspace of V (with respect to $\hat{T}$), and let

$$V^{\mathrm{Tate}} = \bigoplus_{\lambda \in \Lambda_p^{\mathrm{Tate}}} V(\lambda).$$

We are in particular interested in the condition $V_\mu^{\mathrm{Tate}} \neq 0$. As explained in the introduction of [64], under the conjectural Galois theoretic description of the cohomology of the

Shimura varieties (analogous to Conjecture 2.2.3), for a Hecke module π_f whose Satake parameter at p is general enough, certain multiple $a(\pi_f)$ of the dimension of this vector space should be equal to the dimension of the space of Tate classes in the π_f-component of the middle dimensional compactly-supported cohomology of $\overline{\mathscr{S}}_{K,\overline{k}_v}$. In addition, this space is usually large. For example, in the case G is an odd (projective) unitary group of signature $(i, n-i)$ over a quadratic imaginary field, the dimension of this space at an inert prime is $\binom{\frac{n+1}{2}}{i}$.

For a (not necessarily irreducible) algebraic variety Z of dimension d over an algebraically closed field, let $H_{2d}^{\mathrm{BM}}(Z)(-d)$ denote the $(-d)$-Tate twist of the top degree Borel–Moore homology, which is the vector space spanned by the irreducible components of Z. Now let X be a smooth variety of dimension $d+r$ defined over a finite field k of q elements, and let $Z \subseteq X_{\overline{k}}$ be a (not necessarily irreducible) projective subvariety of dimension d. There is the cycle class map

$$\mathrm{cl}: H_{2d}^{\mathrm{BM}}(Z)(-d) \to \bigcup_{j \geq 1} H_c^{2d}\big(X_{\overline{k}}, \mathbb{Q}_\ell(d)\big)^{\sigma_q^j} =: T_\ell^d(X).$$

Theorem 3.3.1. *We write $d_\mu = \dim X = 2d$ and $r = \dim V_\mu^{\mathrm{Tate}}$.*

(1) *The basic Newton stratum $\overline{\mathscr{S}}_b$ of $\overline{\mathscr{S}}_{K,\overline{k}_v}$ is pure of dimension d. In particular, d is always an integer. In addition, there is an $H_{K,\mathbb{Q}_\ell}$-equivariant isomorphism*

$$H_{2d}^{\mathrm{BM}}(\overline{\mathscr{S}}_b)(-d) \cong C\big(G'(\mathbb{Q})\backslash G'(\mathbb{A}_f)/K, \mathbb{Q}_\ell\big)^{\oplus r},$$

where G' is the finitely trivialized inner form of G with $G'_{\mathbb{R}}$ is compact.

(2) *Let π_f be an irreducible module of $H_{K,\overline{\mathbb{Q}}_\ell}$, and let*

$$H_{2d}^{\mathrm{BM}}(\overline{\mathscr{S}}_b)[\pi_f] = \mathrm{Hom}_{H_{K,\overline{\mathbb{Q}}_\ell}}\big(\pi_f, H_{2d}^{\mathrm{BM}}(\overline{\mathscr{S}}_b)(-d)_{\overline{\mathbb{Q}}_\ell}\big) \otimes \pi_f$$

be the π_f-isotypical component. Then the cycle class map

$$\mathrm{cl}: H_{2d}^{\mathrm{BM}}(\overline{\mathscr{S}}_b)(-d) \to T_\ell^d(\overline{\mathscr{S}}_K)$$

restricted to $H_{2d}^{\mathrm{BM}}(\overline{\mathscr{S}}_b)[\pi_f]$ is injective if the Satake parameter of $\pi_{f,p}$ (the component of π_f at p) is V_μ-general.

(3) *Assume that $\mathrm{Sh}_K(G)$ is (essentially) a quaternionic Shimura variety or a Kottwitz arithmetic variety. Then the π_f-isotypical component of the cycle class map is surjective to $T_\ell^d(\overline{\mathscr{S}}_K)[\pi_f]$ if the Satake parameter of $\pi_{f,p}$ is strongly V_μ-general. In particular, the Tate conjecture holds for these π_f.*

Remark 3.3.2. (1) For a representation V of $\hat{G}$, the definitions of "V-general" and "strongly V-general" Satake parameters can be found in [64, DEFINITION 1.4.2]. Regular semisimple elements in ${}^cG|_{d=(p,\sigma_p)}$ are always V-general, but not the converse. See [64, REMARK 1.4.3].

(2) Some special cases of the theorem were originally proved in [33, 60].

The proof of this theorem relies on several different ingredients. Via the Rapoport–Zink uniformization of the basic locus of a Shimura variety, part (2) can be reduced a question about irreducible components of certain affine Deligne–Lusztig varieties, which was studied in **[64, §3]**. The most difficult is part (2), which we proved by calculating the intersection numbers among all d-dimensional cycles in $\overline{\mathscr{S}}_b$. These numbers can be encoded in an $r \times r$-matrix with entries in $H^{\mathrm{cl}}_{K_p}$. In general, it seems hopeless to calculate this matrix directly and explicitly. However, this matrix can be understood as the composition of certain morphisms in $\mathbf{Coh}(\mathrm{Loc}^{\mathrm{ur}}_{{}^cG,p})$. Namely, first we realize $G'(\mathbb{Q})\backslash G'(\mathbb{A})/K$ as a Shimura set with $\mu' = 0$ its Shimura cocharacter. Then using Theorem 2.3.3 (and the Satake isomorphism (2.3)), this matrix can be calculated as

$$\mathrm{Hom}_{\mathbf{Coh}(\mathrm{Loc}^{\mathrm{ur}}_{{}^cG,p})}(\mathcal{O}, \widetilde{V_\mu}) \otimes \mathrm{Hom}_{\mathbf{Coh}(\mathrm{Loc}^{\mathrm{ur}}_{{}^cG,p})}(\widetilde{V_\mu}, \mathcal{O}) \to \mathrm{Hom}_{\mathbf{Coh}(\mathrm{Loc}^{\mathrm{ur}}_{{}^cG,p})}(\mathcal{O}, \mathcal{O}) \cong H^{\mathrm{cl}}_{K_p} \otimes \mathbb{Q}_\ell.$$

Then one needs to determine when this pairing is nondegenerate, which itself is an interesting question in representation theory, whose solution relies on the study of the Chevellay's restriction map for vector-valued functions. The determinant of this matrix was calculated in **[65]**. Finally, part (3) was proved by comparing two trace formulas, the Lefschetz trace formula for G and the Arthur–Selberg trace formula for G'.

Example 3.3.3. Let $G = \mathrm{U}(1, 2r)$ be the unitary group[7] of $(2r+1)$-variables associated to an imaginary quadratic extension $E/\mathbb{Q}$, whose signature is $(1, 2r)$ at infinity. It is equipped with a standard Shimura datum, giving a Shimura variety (after fixing a level $K \subset G(\mathbb{A}_f)$). In particular, if $r = 1$, this is (essentially) the Picard modular surface. Let p be a prime inert in E such that K_p is hyperspecial. In this case $\overline{\mathscr{S}}_b$ is a union of certain Deligne–Lusztig varieties, parametrized by $G'(\mathbb{Q})\backslash G'(\mathbb{A}_f)/K$, where $G' = \mathrm{U}(0, 2r+1)$ that is isomorphic to G at all finite places. The intersection patterns of these cycles inside $\overline{\mathscr{S}}_b$ were (essentially) given in **[61]** but the intersection numbers between these cycles are much harder to compute. In fact, we do not know how to compute them directly for general r, except applying Theorem 2.3.3 to this case. (The case $r = 1$ can be handled directly.)

We have $\hat{G} = \mathrm{GL}_{2r+1}$ on which σ_p acts as $A \mapsto J(A^T)^{-1}J$, where J is the antidiagonal matrix with all entries along the antidiagonal being 1. The representation V_μ is the standard representation of GL_{2r+1}. One checks that $\dim V_\mu^{\mathrm{Tate}} = 1$ (which is consistent with the above mentioned parameterization of irreducible components of $\overline{\mathscr{S}}_b$ by $G'(\mathbb{Q})\backslash G'(\mathbb{A}_f)/K$). We identify the weight lattice of $\hat{G}$ as $\mathbb{Z}^{2r+1}$ as usual. Then $\mathrm{Hom}_{\mathbf{Coh}(\mathrm{Loc}^{\mathrm{ur}}_{{}^cG,p})}(\mathcal{O}, \widetilde{V_\mu})$ is a free rank one module over $\mathrm{Hom}_{\mathbf{Coh}(\mathrm{Loc}^{\mathrm{ur}}_{{}^cG,p})}(\mathcal{O}, \mathcal{O}) = H^{\mathrm{cl}}_{K_p} \otimes \mathbb{Q}_\ell$. Then a generator $\mathbf{a}_{\mathrm{in}}$ induces an $H_{K,\mathbb{Q}_\ell}$-equivariant homomorphism

$$S(\mathbf{a}_{\mathrm{in}}) : C\big(G'(\mathbb{Q})\backslash G'(\mathbb{A}_f)/K\big) \to H^{2r}_c\big(\overline{\mathscr{S}}_{K,\overline{k}_v}, \mathbb{Q}_\ell(r)\big),$$

7 This is not an adjoint group so the example is not consistent with our assumption. But it is more convenient for the discussion here. The computations are essentially the same.

realizing the cycle class map of $\overline{\mathscr{S}}_b$ (up to a multiple). The module $\mathrm{Hom}_{\mathbf{Coh}(\mathrm{Loc}^{\mathrm{ur}}_{^cG,p})}(\widetilde{V_\mu}, \mathcal{O})$ is also free of rank one over $H_{K_p,\mathbb{Q}_\ell}$. For a chosen generator $\mathbf{a}_{\mathrm{out}}$, the composition

$$S(\mathbf{a}_{\mathrm{out}}) \circ S(\mathbf{a}_{\mathrm{in}}) = S(\mathbf{a}_{\mathrm{out}} \circ \mathbf{a}_{\mathrm{in}})$$

calculates the intersection matrix of those cycles from the irreducible components of $\overline{\mathscr{S}}_b$.

The element $h := \mathbf{a}_{\mathrm{out}} \circ \mathbf{a}_{\mathrm{in}} \in H_{K_p,\mathbb{Q}_\ell}$ was explicitly computed in [65, EXAMPLE 6.4.2] (up to obvious modification and also via the Satake isomorphism (1.4)). Namely,

$$h = p^{r(r+1)} \sum_{i=0}^{r} (-1)^i (2i+1) p^{(i-r)(r+i+1)} \sum_{j=0}^{r-i} \begin{bmatrix} 2r+1-2j \\ r-i-j \end{bmatrix}_{t=-p} T_{p,j}. \tag{3.3}$$

Here, $T_{p,j} = 1_{K_p \lambda_j(p) K_p}$, with $\lambda_i = (1^i, 0^{2r-2i+1}, (-1)^i)$, and $\begin{bmatrix} n \\ m \end{bmatrix}_t$ is the t-analogue of the binomial coefficient given by

$$[0]_t = 1, \quad [n]_t = \frac{t^n - 1}{t - 1}, \quad [n]_t! = [n]_t [n-1]_t \cdots [1]_t, \quad \begin{bmatrix} n \\ m \end{bmatrix}_t = \frac{[n]_t!}{[n-m]_t! [m]_t!}.$$

In other words, the intersection matrix of cycles in $\overline{\mathscr{S}}_b$ in this case is calculated by the Hecke operator (3.3).

On interesting consequence is this computation is the following consequence on the intersection theory of the finite Deligne–Lusztig varieties, for which we do not know a direct proof. Let W be a $(2r+1)$-dimensional nondegenerate hermitian space over $\mathbb{F}_{p^2}$. Consider the following r-dimensional Deligne–Lusztig variety

$$\mathrm{DL}_r := \{ H \subset W \text{ of dimension } r \mid H \subseteq (H^{(p)})^{\perp} \},$$

where $H^{(p)}$ the pullback of H along the Frobenius. Let $\mathcal{H}$ denote the corresponding universal subbundle of rank r. Let $\mathcal{E} = \mathcal{H}^{(p)} \otimes ((\mathcal{H}^{(p)})^{\perp} / \mathcal{H})$. Then we have

$$\int_{\mathrm{DL}_r} c_r(\mathcal{E}) = \sum_{i=0}^{r} (-1)^i (2i+1) p^{i^2+i} \begin{bmatrix} 2r+1 \\ r-i \end{bmatrix}_{t=-p}. \tag{3.4}$$

3.4. The Beilinson–Bloch–Kato conjecture for Rankin–Selberg motives

Let M be a rational pure Chow motive of weight -1 over a number field F. The Beilinson–Bloch–Kato conjecture, which is a far reaching generalization of the Birch and Swinnerton-Dyer conjecture, predicts deep relations between certain algebraic, analytic, and cohomological invariants attached to M:

- the rational Chow group $\mathrm{CH}(M)^0$ of homologically trivial cycles of M;
- the L-function $L(s, M)$ of M;
- the Bloch–Kato Selmer group $H^1_f(F, H_\ell(M))$ of the ℓ-adic realization $H_\ell(M)$ of M.

The Beilinson–Bloch conjecture predicts an equality

$$\dim_{\mathbb{Q}} \mathrm{CH}(M)^0 = \mathrm{ord}_{s=0} L(s, M)$$

between the dimension of $\mathrm{CH}(M)^0$ and the vanishing order of the L-function at the central point, while the Bloch–Kato conjecture predicts

$$\mathrm{ord}_{s=0} L(s, M) = \dim_{\mathbb{Q}_\ell} H^1_f\big(F, H_\ell(M)\big).$$

In addition, the so-called ℓ-adic Abel–Jacobi map

$$\mathrm{AJ}_\ell : \mathrm{CH}(M)^0 \otimes \mathbb{Q}_\ell \to H^1_f\big(F, H_\ell(M)\big)$$

should be an isomorphism.

This conjecture seems to be completely out of reach at the moment. For example, for a general motive it is still widely open whether the L-function has a meromorphic continuation to the whole complex plane so that the vanishing order of $L(s, M)$ at $s = 0$ makes sense. (This would follow from the Galois-to-automorphic direction of the Langlands correspondence for number fields.) Despite this, there have been many works testing this conjecture in various special cases, mostly for motives M of small rank. In the work [49], we verify certain cases of the above conjecture for Rankin–Selberg motives, which consist of a sequence of motives of arbitrarily large rank.

We assume that F/F^+ is a (nontrivial) CM extension with F^+ totally real in the sequel.

Theorem 3.4.1. *Let A_1, A_2 be two elliptic curves over F^+. Assume that*

(1) $\mathrm{End}_{\overline{F}} A_i = \mathbb{Z}$ *and* $\mathrm{Hom}_{\overline{F}}(A_1, A_2) = 0$*;*

(2) $\mathrm{Sym}^{n-1} A_1$ *and* $\mathrm{Sym}^n A_2$ *are modular;*

(3) $F^+ \neq \mathbb{Q}$ *if* $n \geq 3$.

Under these assumption, if $L(n, \mathrm{Sym}^{n-1} A_1 \times \mathrm{Sym}^n A_2) \neq 0$, *then for almost* ℓ,

$$\dim_{\mathbb{Q}_\ell} H^1_f\big(F, \mathrm{Sym}^{n-1} V_\ell(A_1) \otimes \mathrm{Sym}^n V_\ell(A_2)(1-n)\big) = 0.$$

Here $V_\ell(A_i)$ denotes the rational Tate module of A_i as usual.

This theorem is in fact a consequence of a more general result concerning Bloch–Kato Selmer groups of Galois representations associated to certain Rankin–Selberg automorphic representations, which we now discuss.

Recall that for an irreducible regular algebraic conjugate self-dual cuspidal (RACSDC) automorphic representation Π of $\mathrm{GL}_n(\mathbb{A}_F)$, one can attach a compatible system of Galois representations $\rho_{\Pi,\lambda} : \Gamma_F \to \mathrm{GL}_n(E_\lambda)$, where $E \subset \mathbb{C}$ is a large enough number field and λ is a prime of E (see [16]). Such E is called a strong coefficient field of Π, which in the situation considered below can be taken as the number field generated by Hecke eigenvalues of Π.

Theorem 3.4.2. *Suppose that $F^+ \neq \mathbb{Q}$ if $n \geq 3$. Let Π_n (resp. Π_{n+1}) be an RACSDC automorphic representation of $\mathrm{GL}_n(\mathbb{A}_F)$ (resp. $\mathrm{GL}_{n+1}(\mathbb{A}_F)$) with trivial infinitesimal character. Let $E \subseteq \mathbb{C}$ be a strong coefficient field for both Π_n and Π_{n+1}. Let λ be an admissible prime of E with respect to $\Pi := \Pi_0 \times \Pi_1$. Let $\rho_{\Pi,\lambda} := \rho_{\Pi_n,\lambda} \otimes_{E_\lambda} \rho_{\Pi_{n+1},\lambda}$.*

(1) *If the Rankin–Selberg L-value* $L(\frac{1}{2},\Pi)\neq 0$[8], *then* $H^1_f(F,\rho_{\Pi,\lambda}(n))=0$.

(2) *If certain element* $\Delta_\lambda\in H^1_f(F,\rho_{\Pi,\lambda}(n))$ *(to be explained below) is non-zero, then* $H^1_f(F,\rho_{\Pi,\lambda}(n))$ *is generated by* Δ_λ *as an* E_λ*-vector space.*

The notion of admissible primes appearing in the theorem consists of a long list of assumptions, some of which are rather technical. Essentially, it guarantees that the Galois representation $\rho_{\Pi,\lambda}$ has a well defined $\mathcal{O}_{E,\lambda}$-lattice (still denoted by $\rho_{\Pi,\lambda}$ in the sequel) and the reduction mod λ representation is suitably large and contains certain particular elements. (This is also related to the notion of V-general from Theorem 3.3.1.) Fortunately, in some favorable situations, one can show that all but finitely many primes are admissible. For example, this is the case considered in Theorem 3.4.1. For another case in pure automorphic setting, see **[49, THM. 1.1.7]**.

The proof of the theorem uses several different ingredients. The initial step for case (1) is to translate the analytic condition $L(\frac{1}{2},\Pi)\neq 0$ into a more algebraic condition via the global Gan–Gross–Prasad (GGP) conjecture. Namely, the GGP conjecture predicts that in this case, there exist a pair of hermitian spaces (V_n,V_{n+1}) over F that are totally positive definite at ∞, where $V_{n+1}=V_n\oplus Fv$ with $(v,v)=1$, and a tempered cuspidal automorphic representation $\pi=\pi_n\times\pi_{n+1}$ of the product of unitary groups $G=U(V_n)\times U(V_{n+1})$, such that the period integral

$$[\Delta_H]: C_c^*\big(\mathrm{Sh}(G),E\big)[\pi]\to E$$

does not vanish, where $H:=U(V_n)$ embeds into G diagonally, which induces an embedding $\Delta_H:\mathrm{Sh}(H)\hookrightarrow \mathrm{Sh}(G)$ of corresponding Shimura varieties (in fact, Shimura sets) with appropriately chosen level structures (here and later we omit level structures from the notations). We denote by $[\Delta_H]$ the homology class of $\mathrm{Sh}(G)$ given by $\mathrm{Sh}(H)$ and write $C_c^*(\mathrm{Sh}(G),E)[\pi]$ for the π-isotypical component of the cohomology (i.e., functions) of $\mathrm{Sh}(G)$. This conjecture was first proved in **[73]** under some local restrictions which are too restrictive for arithmetic applications. Those restrictions are all lifted in our recent work through some new techniques in the study of trace formulae **[8]**.

The strategy then is to construct, for every $m\geq 1$, (infinitely many) cohomology classes $\{\Theta_m^p\}_p\subset H^1(F,(\rho_{\Pi,\lambda}/\lambda^m)^*(1))$, where p are appropriately chosen primes and $(-)^*(1)$ denotes the usual Pontryagin duality twisted by the cyclotomic character, such that the local Tate pairing at p between Θ_m^p and Selmer classes of the Galois representation $\rho_{\Pi,\lambda}/\lambda^m$ is related to the above period integral. Then one can use Kolyvagin type argument (amplified in **[47, 49]**), with $\{\Theta_m^p\}$ served as annihilators of the Selmer groups, to conclude.

The construction of Θ_m^p uses the diagonal embedding of Shimura varieties

$$\Delta_{H'}:\mathrm{Sh}(H')\hookrightarrow \mathrm{Sh}(G')$$

where $H'\hookrightarrow G'$ are prime-to-p trivialized (extended pure) inner forms of $H\subset G$ (see Definition 2.3.1). These Shimura varieties have parahoric level structures at p, and using

8 Here we use the automorphic normalization of the L-function.

the theory of local models (Section 3.1) one can show that their integral models are poly-semistable at p and compute the sheaf of nearby cycles on their mod p fibers. Using many ingredients, including the understanding of (integral) cohomology of $\mathrm{Sh}(G')$ over $\overline{F}$, the computations from Example 3.3.3 (in particular, (3.3) and (3.4)), and the Taylor–Wiles patching method [48], one shows that $(\rho_{\Pi,\lambda}/\lambda^m)^*(1)$ does appear in the cohomology of $\mathrm{Sh}(G')$ (the so-called arithmetic level raising for Π), and that the diagonal cycle $\Delta_{H'}$, when localized at $(\rho_{\Pi,\lambda}/\lambda^m)^*(1)$, does give the desired class Θ_m^p. We shall mention that this is consistent with conjectures in Sections 2.1 and 2.3, as coherent sheaves on $\mathrm{Loc}_{{}^cG,p} \otimes \mathcal{O}_E/\lambda^m$ corresponding to $c\text{-ind}_{K_p}^{G(\mathbb{Q}_p)}(\mathcal{O}_E/\lambda^m)$ and $c\text{-ind}_{K'_p}^{G'(\mathbb{Q}_p)}(\mathcal{O}_E/\lambda^m)$ are expected to be related exactly in this way.

We could also explain the class Δ_λ appearing in case (2). Namely, in this case we start with an embedding of Shimura varieties $\Delta_H : \mathrm{Sh}(H) \hookrightarrow \mathrm{Sh}(G)$, where G is a product of unitary groups such that Π descends to a tempered cuspidal automorphic representation π appearing in the middle dimensional cohomology of Sh_G. Then the π-isotypical component of the cycle Δ_H is homologous to zero, and we let $\Delta_\lambda = \mathrm{AJ}_\lambda(\Delta_H[\pi])$. The strategy to prove case (2) then is to reduce it to case (1) via some similar arguments as before.

ACKNOWLEDGMENTS

The author would like to thank all of his collaborators, without whom many works reported in this survey article would not be possible. Besides, he would also like to thank Edward Frenkel, Dennis Gaitsgory, Xuhua He, Michael Rapoport, Peter Scholze for teaching him and for discussions on various parts of the Langlands program over years.

FUNDING

The work is partially supported by NSF under agreement No. DMS-1902239 and a Simons fellowship.

REFERENCES

[1] J. Anschütz, I. Gleason, J. Lourenço, and T. Richarz, On the p-adic theory of local models. 2022, arXiv:2201.01234.

[2] D. Arinkin and D. Gaitsgory, Singular support of coherent sheaves, and the geometric Langlands conjecture. *Selecta Math. (N.S.)* **21** (2015), 1–199.

[3] D. Arinkin, D. Gaitsgory, D. Kazhdan, S. Raskin, N. Rozenblyum, and Y. Varshavsky, The stack of local systems with restricted variation and geometric Langlands theory with nilpotent singular support. 2020, arXiv:2010.01906.

[4] A. Beauville and Y. Laszlo, Conformal blocks and generalized theta functions. *Comm. Math. Phys.* **164** (1994), no. 2, 385–419.

[5] A. Beilinson and V. Drinfeld, Quantization of Hitchin's integrable system and Hecke eigensheaves, http://math.uchicago.edu/~drinfeld/langlands.html

[6] D. Ben-Zvi, H. Chen, D. Helm, and D. Nadler, Coherent Springer theory and the categorical Deligne–Langlands correspondence. 2021, arXiv:2010.02321.

[7] D. Ben-Zvi and D. Nadler, Betti geometric Langlands. In *Algebraic Geometry: Salt Lake City 2015*, edited by T. de Fernex et al., pp. 3–41, Proc. Sympos. Pure Math. 97.2, Amer. Math. Soc., Providence, RI, 2018.

[8] R. Beuzart-Plessis, Y. Liu, W. Zhang, and X. Zhu, Isolation of cuspidal spectrum, with application to the Gan–Gross–Prasad conjecture. *Ann. of Math.* **194** (2021), no. 2, 519–584.

[9] R. Bezrukavnikov, On two geometric realizations of an affine Hecke algebra. *Publ. Math. IHÉS* **123** (2016), 1–67.

[10] R. Bezrukavnikov and A. Braverman, Geometric Langlands conjecture in characteristic p: The GL_n case. *Pure Appl. Math. Q.* **3** (2007), no. 1. Special Issue: In honor of Robert D. MacPherson, Part 3, 153–179.

[11] R. Bezrukavnikov and M. Finkelberg, Equivariant Satake category and Kostant–Whittaker reduction. *Mosc. Math. J.* **8** (2008), no. 1, 39–72.

[12] R. Bezrukavnikov, R. Travkin, T.-H. Chen, and X. Zhu, Quantization of Hitchin integrable system via positive characteristic. 2020, arXiv:1603.01327.

[13] B. Bhatt and P. Scholze, Projectivity of Witt vector affine Grassmannians. *Invent. Math.* **209** (2017), no. 2, 329–423.

[14] T.-H. Chen and X. Zhu, Non-abelian Hodge theory for curves in characteristic p. *Geom. Funct. Anal.* **25** (2015), 1706–1733.

[15] T.-H. Chen and X. Zhu, Geometric Langlands in prime characteristic. *Compos. Math.* **153** (2017), 395–452.

[16] G. Chenevier and M. Harris, Construction of automorphic Galois representations, II. *Cambridge J. Math.* **1** (2013), no. 1, 53–73.

[17] J. F. Dat, D. Helm, R. Kurinczuk, and G. Moss, Moduli of Langlands parameters. 2020, arXiv:2009.06708.

[18] G. Dhillon, Y. Li, Z. Yun, and X. Zhu, in preparation.

[19] R. Donagi and T. Pantev, Langlands duality for Hitchin systems. *Invent. Math.* **189** (2012), 653–735.

[20] V. G. Drinfeld, Two-dimensional ℓ-adic representations of the fundamental group of a curve over a finite field and automorphic forms on GL(2). *Amer. J. Math.* **105** (1983), 85–114.

[21] M. Emerton, T. Gee, and X. Zhu, in preparation.

[22] M. Emerton and X. Zhu, in preparation.

[23] G. Faltings, Algebraic loop groups and moduli spaces of bundles. *J. Eur. Math. Soc. (JEMS)* **5** (2003), no. 1, 41–68.

[24] G. Faltings and C. L. Chai, *Degeneration of abelian varieties*. Ergeb. Math. Grenzgeb. (3) 22, Springer, Berlin, 1990.

[25] L. Fargues and P. Scholze, Geometrization of the local Langlands correspondence. 2021, arXiv:2102.13459.

[26] E. Frenkel, D. Gaitsgory, and K. Vilonen, On the geometric Langlands conjecture. *J. Amer. Math. Soc.* **15** (2001), 367–417.

[27] D. Gaitsgory, Construction of central elements in the affine Hecke algebra via nearby cycles. *Invent. Math.* **144** (2001), 253–280.

[28] D. Gaitsgory, On De Jong's conjecture. *Israel J. Math.* **157** (2007), 155–191.

[29] T. Haines, The stable Bernstein center and test functions for Shimura varieties. In *Automorphic forms and Galois representations. Vol. 2*, edited by F. Diamond, P. L. Kassaei, and M. Kim, pp. 118–186, London Math. Soc. Lecture Note Ser. 415, Cambridge University Press, Cambridge, 2014.

[30] M. Harris and R. Taylor, *The Geometry and Cohomology of Some Simple Shimura Varieties*. Ann. of Math. Stud. 151, Princeton University Press, 2001.

[31] J. Heinloth, B. C. Ngô, and Z. Yun, Kloosterman sheaves for reductive groups. *Ann. of Math.* **177** (2013), 241–310.

[32] E. Hellmann, On the derived category of the Iwahori–Hecke algebra. 2021, arXiv:2006.03013.

[33] D. Helm, Y. Tian, and L. Xiao, Tate cycles on some unitary Shimura varieties mod p. *Algebr. Number Theory* **11** (2017), 2213–2288.

[34] T. Hemo, T. Richarz, and J. Scholbach, Constructible sheaves on schemes and a categorical Künneth formula. 2021, arXiv:2012.02853.

[35] T. Hemo and X. Zhu, Unipotent categorical local Langlands correspondence, Preprint.

[36] D. Kazhdan and G. Lusztig, Proof of the Deligne–Langlands conjecture for Hecke algebras. *Invent. Math.* **87** (1987), 153–215.

[37] M. Kisin, Integral models for Shimura varieties of abelian type. *J. Amer. Math. Soc.* **23** (2010), 967–1012.

[38] L. Lafforgue, Chtoucas de Drinfeld et correspondance de Langlands. *Invent. Math.* **147** (2002), 1–241.

[39] V. Lafforgue, Chtoucas pour les groupes réductifs et paramètrisation de Langlands globale. *J. Amer. Math. Soc.* **31** (2018), 719–891.

[40] V. Lafforgue and X. Zhu, Décomposition au-dessus des paramètres de Langlands elliptiques. 2019, arXiv:1811.07976.

[41] R. Langlands, Letter to André Weil, https://publications.ias.edu/rpl/paper/43.

[42] G. Laumon, Correspondance de Langlands géométrique pour les corps de fonctions. *Duke Math. J.* **54** (1987), 309–359.

[43] G. Laumon, M. Rapoport, and U. Stuhler, D-elliptic sheaves and the Langlands correspondence. *Invent. Math.* **113** (1993), 217–338.

[44] D. Le, B. V. Le Hung, B. Levin, and S. Morra, Local models for Galois deformation rings and applications. 2020, arXiv:2007.05398.

[45] S. Y. Lee, Eichler–Shimura relations for Shimura varieties of Hodge type. 2021, arXiv:2006.11745.

[46] B. Levin, Local models for Weil-restricted groups. *Compos. Math.* **152** (2016), 2563–2601.

[47] Y. Liu, Hirzebruch–Zagier cycles and twisted triple product Selmer groups. *Invent. Math.* **205** (2016), no. 3, 693–780.

[48] Y. Liu, Y. Tian, L. Xiao, W. Zhang, and X. Zhu, Deformation of rigid conjugate self-dual Galois representations. 2021, arXiv:2108.06998.

[49] Y. Liu, Y. Tian, L. Xiao, W. Zhang, and X. Zhu, On the Beilinson–Bloch–Kato conjecture for Rankin–Selberg motives. *Invent. Math.* **228** (2022), 107–375.

[50] J. Lourenço, Grassmanniennes affines tordues sur les entiers. 2021, arXiv:1912.11918.

[51] G. Lusztig, Singularities, character formulas, and a q-analog of weight multiplicities. In *Analysis and topology on singular spaces, II, III (Luminy, 1981)*, pp. 208–229, Astérisque 101–102, Soc. Math. France, Paris, 1983.

[52] G. Lusztig, Classification of unipotent representations of simple p-adic groups. *Int. Math. Res. Not.* (1995), 517–589.

[53] I. Mirković and K. Vilonen, Geometric Langlands duality and representations of algebraic groups over commutative rings. *Ann. of Math.* **166** (2007), 95–143.

[54] J. Nekovář and A. J. Scholl, Introduction to plectic cohomology. In *Advances in the theory of automorphic forms and their L-functions*, pp. 321–337, Contemp. Math. 664, Amer. Math. Soc., Providence, RI, 2016.

[55] B. C. Ngô, Le lemme fondamental pour les algèbres de Lie. *Publ. Math. Inst. Hautes Études Sci.* **111** (2010), 1–169.

[56] G. Pappas and M. Rapoport, Local models in the ramified case. III. Unitary groups. *J. Inst. Math. Jussieu* **8** (2009), 507–564.

[57] G. Pappas, M. Rapoport, and B. D. Smithling, Local models of Shimura varieties, I. Geometry and combinatorics. In *Handbook of Moduli, Vol. III*, pp. 135–217, International Press, Somerville, MA, 2013.

[58] G. Pappas and X. Zhu, Local models of of Shimura varieties and a conjecture of Kottwitz. *Invent. Math.* **194** (2013), 147–254.

[59] P. Scholze and J. Weinstein, *Berkeley lectures on p-adic geometry*. Ann. of Math. Stud. 207, Princeton University Press, 2020.

[60] Y. Tian and L. Xiao, Tate cycles on quaternionic Shimura varieties over finite fields. *Duke Math. J.* **168** (2019), 1551–1639.

[61] I. Vollaard and T. Wedhorn, The supersingular locus of the Shimura variety of $GU(1, n-1)$, II. *Invent. Math.* **184** (2011), 591–627.

[62] A. Wiles, Modular elliptic curves and Fermat's Last Theorem. *Ann. of Math.* **141** (1995), 443–551.

[63] Z. Wu, $S = T$ for Shimura varieties and p-adic Shtukas. 2021, arXiv:2110.10350.

[64] L. Xiao and X. Zhu, Cycles on Shimura varieties via geometric Satake. 2017, arXiv:1707.05700.

[65] L. Xiao and X. Zhu, On vector-valued twisted conjugate invariant functions on a group. In *Representations of reductive groups*, pp. 361–426, Proc. Sympos. Pure Math. 101, Amer. Math. Soc., Providence, RI, 2019.

[66] D. Xu and X. Zhu, Bessel F-isocrystals for reductive groups. *Invent. Math.* **227** (2022), 997–1092.

[67] C. Xue, Finiteness of cohomology groups of stacks of shtukas as modules over Hecke algebras, and applications. *Épijournal Géom. Algébr.* **4** (2020).

[68] C. Xue, Smoothness of cohomology sheaves of stacks of Shtukas. 2020, arXiv:2012.12833.

[69] J. Yu, The integral geometric Satake equivalence in mixed characteristic. 2019, arXiv:1903.11132.

[70] J. Yu, A geometric Jacquet–Langlands transfer for automorphic forms of higher weights. *Trans. Amer. Math. Soc.* 2022.

[71] Z. Yun, Motives with exceptional Galois groups and the inverse Galois problem. *Invent. Math.* **196** (2014), 267–337.

[72] Z. Yun and X. Zhu, Integral homology of loop groups via Langlands dual groups. *Represent. Theory* **15** (2011), 347–369.

[73] W. Zhang, Fourier transform and the global Gan–Gross–Prasad conjecture for unitary groups. *Ann. of Math.* **180** (2014), 971–1049.

[74] X. Zhu, $S = T$ for Shimura varieties, Preprint.

[75] X. Zhu, On the coherence conjecture of Pappas and Rapoport. *Ann. of Math.* **180** (2014), 1–85.

[76] X. Zhu, The geometric Satake correspondence for ramified groups. *Ann. Sci. Éc. Norm. Supér.* **48** (2015), 409–451.

[77] X. Zhu, Geometric Satake, categorical traces, and arithmetic of Shimura varieties. In *Current developments in mathematics*, pp. 145–206, International Press, Boston, MA, 2016.

[78] X. Zhu, Affine Grassmannians and the geometric Satake in mixed characteristic. *Ann. of Math.* **185** (2017), 403–492.

[79] X. Zhu, Frenkel–Gross' irregular connection and Heinloth–Ngô–Yun's are the same. *Selecta Math. (N.S.)* **23** (2017), no. 1, 245–274.

[80] X. Zhu, An introduction to affine Grassmannians and the geometric Satake equivalence. *IAS/Park City Math. Ser.* **24** (2017), 59–154.

[81] X. Zhu, A survey of local models of Shimura varieties. In *Proceedings of the Sixth International Congress of Chinese Mathematicians*, pp. 247–265, Adv. Lectures Math. 36, International Press, Boston, MA, 2017.

[82] X. Zhu, Coherent sheaves on the stack of Langlands parameters. 2021, arXiv:2008.02998.

[83] X. Zhu, A note on integral Satake isomorphisms. 2021, arXiv:2005.13056.

XINWEN ZHU

Department of Mathematics, California Institute of Technology, Pasadena, CA 91125, USA, xzhu@caltech.edu

4. ALGEBRAIC AND COMPLEX GEOMETRY

SPECIAL LECTURE

MOTIVIC COHOMOLOGY

MARC LEVINE

ABSTRACT

We give a survey of the development of motivic cohomology, motivic categories, and some of their recent descendants.

MATHEMATICS SUBJECT CLASSIFICATION 2020

Primary 14F42; Secondary 14C15, 19D55, 19D45, 55P42

KEYWORDS

Motivic cohomology, motives, K-theory, algebraic cycles, motivic homotopy theory

ICM INTERNATIONAL CONGRESS OF MATHEMATICIANS 2022 JULY 6–14

Proc. Int. Cong. Math. 2022, Vol. 3, pp. 2048–2106
DOI 10.4171/ICM2022/112

1. INTRODUCTION

Motivic cohomology arose out of a marriage of Grothendieck's ideas about motives with a circle of conjectures about special values of zeta functions and L-functions. It has since taken on a very active life of its own, spawning a multitude of developments and applications. My intention in this survey is to present some of the history of motivic cohomology and the framework that supports it, its current state, and some thoughts about its future directions. I will say very little about the initial impetus given by the conjectures about zeta functions and L-functions, as there are many others who are much better qualified to tell that story. I will also say next to nothing about the many applications motivic cohomology has seen: I think this would be like writing about the applications of cohomology up to, say, 1950, and would certainly make this already lengthy survey completely unmanageable.

My basic premise is that motivic cohomology is supposed to be universal cohomology for algebro-geometric objects. As "universal" depends on the universe one happens to find oneself in, motivic cohomology is an ever-evolving construct. My plan is to give a path through some of the various universes that have given rise to motivic cohomologies, to describe the resulting motivic cohomologies and put them in a larger, usually categorical, framework. Our path will branch into several directions, reflecting the different aspects of algebraic and arithmetic geometry that have been touched by this theory. We begin with the conjectures of Beilinson and Lichtenbaum about motivic complexes that give rise to the universal Bloch–Ogus cohomology theory on smooth varieties over a field, and the candidate complexes constructed by Bloch and Suslin. We then take up Voevodsky's triangulated category of motives over a field and the embedding of the motivic complexes and motivic cohomology in this framework. The next developments moving further in this direction give us motivic homotopy categories that tell us about "generalized motivic cohomology" for a much wider class of schemes, analogous to the development of the stable homotopy category and generalized cohomology for spaces; this includes a number of candidate theories for motivic cohomology over a general base-scheme. We conclude with three variations on our theme:

- Milnor–Witt motives and Milnor–Witt motivic cohomology, incorporating information about quadratic forms,
- Motives with modulus, relaxing the usual condition of homotopy invariance with respect to the affine line, and
- p-adic, étale motivic cohomology in mixed characteristic $(0, p)$, with its connection to p-adic Hodge theory.

This last example does not yet, as far as I know, have a categorical framework, while one for a motivic cohomology with modulus is still in development.

There is already an extensive literature on the early development of motives and motivic cohomology. It was not my intention here to cover this part in detail, but I include a section on this topic to give a quick overview for the sake of background, and to isolate a few main ideas so the reader could see how they have influenced later developments.

I would like to thank all those who helped me prepare this survey, especially Tom Bachmann, Federico Binda, Dustin Clausen, Thomas Geisser, Wataru Kai, Akhil Mathew, Hiroyasu Miyazaki, Matthew Morrow, and Shuji Saito. In spite of their efforts, I feel certain that a number of errors have crept in, which are, of course, all my responsibility. I hope that the reader will be able to repair them and continue on.

2. BACKGROUND AND HISTORY

2.1. The conjectures of Beilinson and Lichtenbaum

Beilinson pointed out in his 1983 paper "Higher regulators and values of L-functions" [13] that the existence of Gillet's Chern character [53] from algebraic K-theory to an arbitrary Bloch–Ogus cohomology theory [30] with coefficients in a $\mathbb{Q}$-algebra implies that one can form the universal Bloch–Ogus cohomology $H^a_\mu(-, \mathbb{Q}(b))$ with $\mathbb{Q}$-coefficients by decomposing algebraic K-theory into its weight spaces for the Adams operations ψ_k. In terms of the indexing, one has

$$H^a\big(X, \mathbb{Q}(b)\big) := K_{2b-a}(X)^{(b)}$$

where $K_{2b-a}(X)^{(b)} \subset K_{2b-a}(X)_\mathbb{Q}$ is the weight b eigenspace for the Adams operations

$$K_{2b-a}(X)^{(b)} := \big\{x \in K_{2b-a}(X)_\mathbb{Q} \mid \psi_k(x) = k^b \cdot x\big\}.$$

This raised the question of finding the universal integral Bloch–Ogus cohomology theory. Let Sch_k denote the category of separated finite-type k-schemes with full subcategory Sm_k of smooth k-schemes. Beilinson [13] and Lichtenbaum [87] independently conjectured that this universal theory $H^a_\mu(-, \mathbb{Z}(b))$ should arise as the hypercohomology of a complex of sheaves $X \mapsto \Gamma_X(b)$ on Sm_k (for the Zariski or étale topology)

$$H^a_\mu\big(X, \mathbb{Z}(b)\big) := \mathbb{H}^a\big(X, \Gamma_X(b)\big),$$

with the $\Gamma_X(b)$ satisfying a number of axioms. We give Beilinson's list of axioms for motivic complexes in the Zariski topology (axiom iv(p) was added by Milne [90, §2]):

(i) In the derived category of sheaves on X, $\Gamma(0)$ is the constant sheaf $\mathbb{Z}$ on Sm_k, $\Gamma(1) = \mathbb{G}_m[-1]$ and $\Gamma(n) = 0$ for $n < 0$.

(ii) The graded object $\Gamma(*) := [X \mapsto \bigoplus_{n\geq 0} \Gamma_X(n)]$ is a commutative graded ring in the derived category of sheaves on Sm_k.

(iii) The cohomology sheaves $\mathcal{H}^m(\Gamma(n))$ are zero for $m > n$ and for $m \leq 0$ if $n > 0$; $\mathcal{H}^n(\Gamma(n))$ is the sheaf of Milnor K-groups $X \mapsto \mathcal{K}^M_{n,X}$.

(iv)(a) Letting $\alpha : \mathrm{Sm}_{k,\text{ét}} \to \mathrm{Sm}_{k,\mathrm{Zar}}$ be the change of topology morphism, the étale sheafification $\Gamma(n)_{\text{ét}} := \alpha^*\Gamma(n)$ of $\Gamma(n)$ satisfies $\Gamma(n)_{\text{ét}}/m \cong \mu_m^{\otimes n}$ for m prime to the characteristic, where μ_m is the étale sheaf of mth roots of unity.

(iv)(b) For m prime to the characteristic, the natural map $\Gamma(n)/m \to R\alpha_*\Gamma(n)_{\text{ét}}/m$ induces an isomorphism $\Gamma(n)/m \to \tau_{\leq n} R\alpha_*\Gamma(n)_{\text{ét}}/m$. Integrally,

$\mathbb{Z}(n) \to R\alpha_*\Gamma(n)_{\text{ét}}$ induces an isomorphism $\Gamma(n) \to \tau_{\leq n} R\alpha_*\Gamma(n)_{\text{ét}}$ and

$$R^{n+1}\alpha_*\Gamma(n)_{\text{ét}} = 0.$$

(iv)(p) For k of characteristic $p > 0$, let $W_\nu\Omega^n_{\log}$ denote the ν-truncated logarithmic de Rham–Witt sheaf. The $d\log$ map $d\log : \mathcal{K}^M_n/p^n \to W_\nu\Omega^n_{\log}$ induces via (ii) a map $\Gamma(n)/p^\nu \to W_\nu\Omega^n_{\log}[-n]$, which is an isomorphism.

One then defines motivic cohomology by

$$H^p(X, \mathbb{Z}(q)) := \mathbb{H}^p(X_{\text{Zar}}, \Gamma_X(q)).$$

(v) There should also be a spectral sequence starting with integral motivic cohomology and converging to algebraic K-theory, analogous to the Atiyah–Hirzebruch spectral sequence from singular cohomology to topological K-theory. Explicitly, this should be

$$E_2^{p,q} := H^{p-q}(X, \mathbb{Z}(-q)) \Rightarrow K_{-p-q}(X).$$

This spectral sequence should degenerate rationally, and give an isomorphism

$$H^p(X, \mathbb{Q}(q)) := H^p(X, \mathbb{Z}(q)) \otimes_{\mathbb{Z}} \mathbb{Q} \cong K_{2q-p}(X)^{(q)}.$$

The vanishing $\mathcal{H}^m(\Gamma(n)) = 0$ for $n > 0$ and $m \leq 0$ is the *integral Beilinson–Soulé vanishing conjecture*. The mod m-part of axiom (iv)(b) is known as the *Beilinson–Lichtenbaum conjecture*; this implies the integral part of (iv)(b) with the exception of the vanishing of $R^{n+1}\alpha_*\Gamma(n)_{\text{ét}}$, which is known as *Hilbert's theorem 90* for the motivic complexes. In weight $n = 1$, with the identity $\Gamma(1) = \mathbb{G}_m[-1]$, this translates into the classical Hilbert theorem 90

$$H^1_{\text{ét}}(\mathcal{O}, \mathbb{G}_m) = 0$$

for $\mathcal{O}$ a local ring, while the mod m part of (iv)(b) follows from the Kummer sequence of étale sheaves

$$1 \to \mu_m \to \mathbb{G}_m \xrightarrow{\times m} \mathbb{G}_m \to 1.$$

In light of axiom (iii), the Merkurjev–Suslin theorem **[89, THEOREM 14.1]** settled the degree ≥ 2 part of (iv)(b) for $n = 2$ even before the complex $\Gamma(2)$ was defined.

Beilinson **[14, §5.10]** rephrased and refined these conjectures to a categorical statement, invoking a conjectural category of mixed motivic sheaves, and an embedding of the hypercohomology of the Beilinson–Lichtenbaum complexes into a categorical framework.

In this framework, motivic cohomology should arise via an abelian tensor category of motivic sheaves on Sch_S, $X \mapsto \text{Sh}^{\text{mot}}(X)$, admitting the six functor formalism of Grothendieck, $f^*, f_*, f_!, f^!, \mathcal{H}om, \otimes$, on the derived categories. There should be Tate objects $\mathbb{Z}_X(n) \in \text{Sh}^{\text{mot}}(X)$, and objects $M(X) := p_{X!}p_X^!\mathbb{Z}_S(0)$ in the derived category of $\text{Sh}^{\text{mot}}(S)$, $p_X : X \to S$ the structure morphism, and motivic cohomology should arise as the Hom-groups

$$H^a_\mu(X, \mathbb{Z}(b)) = \text{Hom}_{D(\text{Sh}^{\text{mot}}(S))}(M(X), \mathbb{Z}_S(b)[a]).$$

For X smooth over S, this gives the identity

$$H^a_\mu(X, \mathbb{Z}(b)) = \mathrm{Ext}^a_{\mathrm{Sh}^{\mathrm{mot}}(X)}(\mathbb{Z}_X(0), \mathbb{Z}_X(b)).$$

This is a very strong statement, with implications that have not been verified to this day. For instance, the vanishing of $\mathrm{Ext}^a_{\mathcal{A}}(-,-)$ for an abelian category $\mathcal{A}$ and for $a < 0$ gives a vanishing $\mathcal{H}^m(\mathbb{Z}_S(n)) = 0$ for $m < 0$. The stronger vanishing posited by axiom (iii) above (with $\mathbb{Q}$-coefficients) is the *Beilinson–Soulé vanishing conjecture,* and even the weak version is only known for weight $n = 1$ (for which the strong version holds).

Beilinson's conjecture on categories of motivic sheaves is still an open problem. However, other than the integral Beilinson–Soulé vanishing conjecture, the axioms do not rely on the existence of an abelian category of motivic sheaves, and can be framed in the setting of a functorial assignment $X \mapsto \mathrm{DM}(X)$ from S-schemes to tensor-triangulated categories. Such a functor has been constructed and the axioms (except for the vanishing conjectures) have been verified. We will discuss this construction in Section 2.4.

2.2. Bloch's higher Chow groups and Suslin homology

The first good definition of motivic cohomology complexes was given by Spencer Bloch, in his landmark 1985 paper "Algebraic cycles and higher Chow groups" [24]. As suggested by the title, the starting point was the classical Chow group $\mathrm{CH}_*(X)$ of algebraic cycles modulo rational equivalence.

For X a finite type k-scheme, recall that the group of dimension d algebraic cycles on X, $\mathrm{Z}_d(X)$, is the free abelian group on the integral closed subschemes Z of X of dimension d over k. The group of cycles modulo rational equivalence, $\mathrm{CH}_d(X)$, has the following presentation. Let $n \mapsto \Delta^n$ be the cosimplicial scheme of *algebraic n-simplices*

$$\Delta^n := \mathrm{Spec}\, \mathbb{Z}[t_0, \dots, t_n] / \sum_{i=0}^{n} t_i - 1 \cong \mathbb{A}^n_{\mathbb{Z}}.$$

The coface and codegeneracy maps are defined just as for the usual real simplices $\Delta^n_{\mathrm{top}} \subset \mathbb{R}^n$. A *face* of Δ^n is a closed subscheme defined by the vanishing of some of the t_i. Let $z_d(X, n)$ be the subgroup of the $(n + d)$-dimensional algebraic cycles $Z_{n+d}(X \times \Delta^n)$ generated by the integral closed $W \subset X \times \Delta^n$ such that $\dim W \cap X \times F = m + d$ for each m-dimensional face F (or the intersection is empty). For cycles $w \in z_d(X, n)$, the face condition gives a well-defined pullback $(\mathrm{Id}_X \times g)^* : z_d(X, n) \to z_d(X, m)$ for each map $g : \Delta^m \to \Delta^n$ in the cosimplicial structure, forming the simplicial abelian group $n \mapsto z_d(X, n)$ and giving the associated chain complex $z_d(X, *)$, Bloch's cycle complex. The degree 0 and 1 terms of $z_d(X, *)$ give our promised presentation of $\mathrm{CH}_d(X)$,

$$H_0(z_d(X, *)) = \mathrm{CH}_d(X),$$

and Bloch defines his higher Chow group $\mathrm{CH}_d(X, n)$ as

$$\mathrm{CH}_d(X, n) := H_n(z_d(X, *)).$$

If X has pure dimension N over k, we index by codimension

$$z^q(X, *) := z_{N-q}(X, *); \quad \mathrm{CH}^q(X, n) := \mathrm{CH}_{N-q}(X, n).$$

With some technical difficulties due to the necessity of invoking moving lemmas to allow for pullback morphisms, the assignment

$$X \mapsto z^q(X, 2q - *)$$

can be modified via isomorphisms in the derived category to a presheaf of cohomological complexes $\mathbb{Z}_{Bl}(q)$ on Sm_k.

Following a long series of works [25, 29, 43, 52, 94, 96, 112, 113, 115–117, 120, 121, 123–125, 127] (see also [56, 102] for detailed discussions of the Bloch–Kato conjecture, the essential point in axiom (iv)(b) and the most difficult of the Beilinson axioms to prove), it has been shown that the complexes $\mathbb{Z}_{\mathrm{Bl}}(q)$ satisfy all the Beilinson–Lichtenbaum–Milne axioms, except for the Beilinson–Soulé vanishing conjecture in axiom (iii).

After Bloch introduced his cycle complexes, Suslin [111] constructed an algebraic version of singular homology. For a k-scheme X, instead of a naive generalization of the singular chain complex of a topological space by taking the free abelian group on the morphisms $\Delta^n_k \to X$, Suslin's insight was to enlarge this to the abelian group of *finite correspondences*.

A subvariety W of a product $Y \times X$ of varieties (with Y smooth) defines an irreducible finite correspondence from Y to X if $p_1 : W \to Y$ is finite and surjective to some irreducible component of Y. The association $y \mapsto p_2(p_1^{-1}(y))$ can be thought of as a multivalued map from Y to X.

The group of finite correspondences $\mathrm{Cor}_k(Y, X)$ is defined as the free abelian group on the irreducible finite correspondences. Given a morphism $f : Y' \to Y$, there is a pullback map $f^* : \mathrm{Cor}_k(Y, X) \to \mathrm{Cor}_k(Y', X)$, compatible with the interpretation as multivalued functions, and making $\mathrm{Cor}_k(-, X)$ into a contravariant functor from smooth varieties over k to abelian groups.

Suslin defines $C^{\mathrm{Sus}}_n(X) := \mathrm{Cor}_k(\Delta^n_k, X)$; the structure of Δ^*_k as smooth cosimplicial scheme makes $n \mapsto C^{\mathrm{Sus}}_n(X)$ a simplicial abelian group. As above, we have the associated complex $C^{\mathrm{Sus}}_*(X)$, the *Suslin complex* of X, whose homology is the *Suslin homology* of X:

$$H^{\mathrm{Sus}}_n(X, \mathbb{Z}) := \pi_n\big(\big|m \mapsto C^{\mathrm{Sus}}_m(X)\big|\big) = H_n\big(C^{\mathrm{Sus}}_*(X)\big).$$

In fact, the monoid of the $\mathbb{N}$-linear combinations of irreducible correspondences $W \subset X \times Y$ is the same as the monoid of morphisms

$$\phi : X \to \bigsqcup_{n \geq 0} \mathrm{Sym}^n Y$$

where $\mathrm{Sym}^n Y$ is the quotient Y^n / Σ_n of Y^n by the symmetric group permuting the factors, with the monoid structure induced by the sum map

$$\mathrm{Sym}^n Y \times \mathrm{Sym}^m Y \to \mathrm{Sym}^{m+n} Y.$$

Suslin's complex and his definition of algebraic homology can thus be thought of as an algebraic incarnation of the theorem of Dold–Thom [34, SATZ 6.4], that identifies the homotopy groups of the infinite symmetric product of a pointed CW complex T with the reduced homology of T. The main result of [112] gives an isomorphism of the mod n Suslin homology, $H_*^{\mathrm{Sus}}(X, \mathbb{Z}/n)$, for X of finite type over $\mathbb{C}$, with the mod n singular homology of $X(\mathbb{C})$, a first major success of the theory.

Let Δ^n_{top} denote the usual n-simplex

$$\Delta^n_{\mathrm{top}} := \left\{(t_0, \dots, t_n) \in \mathbb{R}^{n+1} \mid \sum_i t_i = 1, t_i \geq 0\right\},$$

with the inclusion $\Delta^n_{\mathrm{top}} \subset \Delta^n(\mathbb{C})$.

Theorem 2.1 ([112, THEOREM 8.3]). *Let X be separated finite type scheme over $\mathbb{C}$ and let $n \geq 2$ be an integer. Then the map*

$$\mathrm{Hom}\Big(\Delta^*_{\mathbb{C}}, \bigsqcup_{d\geq 0} \mathrm{Sym}^d X\Big) \to \mathrm{Hom}_{\mathrm{top}}\Big(\Delta^*_{\mathrm{top}}, \bigsqcup_{d\geq 0} \mathrm{Sym}^d X(\mathbb{C})\Big)$$

induced by the inclusions $\Delta^m_{\mathrm{top}} \subset \Delta^m(\mathbb{C})$ gives rise to an isomorphism $H_*^{\mathrm{Sus}}(X, \mathbb{Z}/n) \to H_*^{\mathrm{sing}}(X(\mathbb{C}), \mathbb{Z}/n)$.

There is also a corresponding statement for X over an arbitrary algebraically closed field k of characteristic zero in terms of étale cohomology [112, THEOREM 7.8]; this extends to characteristic $p > 0$ and n prime to p by using alterations.

2.3. Quillen–Lichtenbaum conjectures

Quillen's computation of the higher algebraic K-theory of finite fields and of number rings led to a search for a relation of higher algebraic K-theory with special values of zeta-functions and L-functions. We will not go into this in detail here, but to large part, this was responsible for the Beilinson–Lichtenbaum conjectures on the existence of motivic complexes computing the conjectural motivic cohomology. Going back to K-theory, this suggested that algebraic K-theory with mod-ℓ coefficients should be the same as mod-ℓ étale K-theory (a purely algebraic version of mod-ℓ topological K-theory, see [35]), at least in large enough degrees. This is more precisely stated as the Quillen–Lichtenbaum conjecture

Conjecture 2.2 ([101], [42, CONJECTURE 3.9]). *Let ℓ be a prime and let X be a regular, noetherian scheme with ℓ invertible on X. Suppose X has finite ℓ-étale cohomological dimension $\mathrm{cd}_\ell(X)$. Then the canonical map*

$$K_n(X; \mathbb{Z}/\ell^r) \to K_n^{\acute{e}t}(X; \mathbb{Z}/\ell^r)$$

is an isomorphism for $n \geq \mathrm{cd}_\ell(X) - 1$ and is injective for $n = \mathrm{cd}_\ell(X) - 2$.

Here $K_n^{\acute{e}t}(X; \mathbb{Z}/\ell)$ is the étale K-theory developed by Dwyer and Friedlander [35, 41, 42].

Conjecture 2.2 for a smooth k-scheme is essentially a consequence of the Beilinson–Lichtenbaum axioms (without Beilinson–Soulé vanishing). The Beilinson–Lichtenbaum conjecture (iv)(a,b) says that the comparison map $\Gamma(q)/\ell^r \to R\alpha_*\mu_{\ell^r}^{\otimes q}$ induces an isomorphism on cohomology sheaves up to degree q. Combining the local–global spectral sequence for some $X \in \mathrm{Sm}_k$ with the Atiyah–Hirzebruch spectral sequences from motivic cohomology to K-theory (axiom (v)) and from étale cohomology to étale K-theory, and keeping track of the cohomological bound in the Beilinson–Lichtenbaum conjecture gives the result.

2.4. Voevodsky's category DM and modern motivic cohomology

One can almost realize Beilinson's ideas of a categorical framework for motivic cohomology by working in the setting of triangulated categories, viewed as a replacement for the derived category of Beilinson's conjectured abelian category of motivic sheaves. Once this is accomplished, one could hope that an abelian category of mixed motives could be constructed out of the triangulated category as the heart of a suitable t-structure.

Constructions of a triangulated category of mixed motives over a perfect base-field were given by Hanamura [57–59], Voevodsky [127], and myself [83]. All three categories yield Bloch's higher Chow groups as the categorical motivic cohomology, however, Voevodsky's sheaf-theoretic approach has had the most far-reaching consequences and has been widely adopted as the correct solution. The construction of a motivic t-structure is still an open problem.[1] There are also constructions of triangulated categories of mixed motives by the method of compatible realizations, such as by Huber [64], or Nori's construction of an abelian category of mixed motives, described in [65, PART II]; we will not pursue these directions here. We also refer the reader to Jannsen's survey on mixed motives [68].

Voevodsky's triangulated category of motives over k, $\mathrm{DM}(k)$, is based on the category of *finite correspondences* on Sm_k, a refinement of Grothendieck's composition law for correspondences on smooth projective varieties. Grothendieck had constructed categories of motives for smooth projective varieties, with the morphisms from X to Y given by the group of cycles modulo rational equivalence $\mathrm{CH}_{\dim X}(X \times Y)$. The composition law is given by

$$W' \circ W = p_{XZ*}\big(p_{XY}^*(W) \cdot p_{YZ}^*(W')\big), \tag{2.1}$$

with $W \in \mathrm{CH}_{\dim X}(X \times Y)$ and $W' \in \mathrm{CH}_{\dim Y}(Y \times Z)$; one needs to pass to cycle classes to define $p_{XY}^*(W) \cdot p_{YZ}^*(W')$ and the projection p_{XZ} needs to be proper (that is, Y needs to be proper over k) to define p_{XZ*}.

Voevodsky's key insight was to restrict to finite correspondences, so that all the operations used in the composition law of correspondence classes would be defined on the level of the cycles themselves, without needing to pass to rational equivalence classes, and without needing the varieties involved to be proper. Voevodsky's idea of having a well-defined composition law on a restricted class of correspondences has been modified and applied in a wide range of different contexts, somewhat similar to the use of various flavors of bordism theories in topology.

Let X and Y be in Sm_k. Recall from Section 2.2 the subgroup $\mathrm{Cor}_k(X, Y) \subset Z_{\dim X}(X \times Y)$ generated by the integral $W \subset X \times Y$ that are finite over X and map surjectively to a component of X.

---

1 Voevodsky showed this is not possible integrally, so the best one can hope for is a t-structure with $\mathbb{Q}$-coefficients.

Lemma 2.3. *Let X, Y, Z be smooth k-varieties and take $\alpha \in \mathrm{Cor}_k(X, Y)$, $\beta \in \mathrm{Cor}_k(Y, Z)$. Then*

(i) *The cycles $p^*_{YZ}(\beta)$ and $p^*_{XY}(\alpha)$ intersect properly on $X \times Y \times Z$, so the intersection product $p^*_{YZ}(\beta) \cdot p^*_{XY}(\alpha)$ exists as a well-defined cycle on $X \times Y \times Z$.*

(ii) *Letting $|\alpha| \subset X \times Y$, and $|\beta| \subset Y \times Z$ denote the support of α and β, respectively, each irreducible component of the intersection $X \times |\beta| \cap |\alpha| \times Z$ is finite over $X \times Z$, and maps surjectively onto some component of X.*

In other words, the formula

$$\beta \circ \alpha = p_{XZ*}(p^*_{YZ}\beta \cdot p^*_{XY}\alpha)$$

makes sense for $\alpha \in \mathrm{Cor}_k(X, Y)$ and $\beta \in \mathrm{Cor}_k(Y, Z)$, and the resulting cycle on $X \times Z$ is in $\mathrm{Cor}_k(X, Z)$. This defines the composition law in Voevodsky's *category of finite correspondences*, Cor_k, with objects as for Sm_k, and morphisms $\mathrm{Hom}_{\mathrm{Cor}_k}(X, Y) = \mathrm{Cor}_k(X, Y)$. Sending a usual morphism $f : X \to Y$ of smooth varieties to its graph defines a faithful functor $[-] : \mathrm{Sm}_k \to \mathrm{Cor}_k$.

Once one has the category Cor_k, the path to $\mathrm{DM}(k)$ is easy to describe. One takes the category of additive presheaves of abelian groups on Cor_k, the category of *presheaves with transfer* $\mathrm{PST}(k)$. Inside $\mathrm{PST}(k)$ is the category $\mathrm{NST}(k)$ of Nisnevich sheaves with transfer, that is, a presheaf that is a Nisnevich sheaf when restricted to $\mathrm{Sm}_k \subset \mathrm{Cor}_k$. Each $X \in \mathrm{Sm}_k$ defines an object $\mathbb{Z}_{\mathrm{tr}}(X) \in \mathrm{NST}(k)$, as the representable (pre)sheaf $Y \mapsto \mathrm{Cor}_k(Y, X)$. Inside the derived category $D(\mathrm{NST}(k))$ is the full subcategory of complexes K whose homology presheaves $\underline{h}_i(K)$ are $\mathbb{A}^1$-*homotopy invariant*: $\underline{h}_i(K)(X) \cong \underline{h}_i(K)(X \times \mathbb{A}^1)$ for all $X \in \mathrm{Sm}_k$. This is the category of effective motives $\mathrm{DM}^{\mathrm{eff}}(k)$. The Suslin complex construction, $\mathcal{P} \mapsto C^{\mathrm{Sus}}_*(\mathcal{P})$, with

$$C^{\mathrm{Sus}}_*(\mathcal{P})(X) := \mathcal{P}(X \times \Delta^*)$$

extends to a functor $RC^{\mathrm{Sus}}_* : D(\mathrm{NST}(k)) \to \mathrm{DM}^{\mathrm{eff}}(k)$, and realizes $\mathrm{DM}^{\mathrm{eff}}(k)$ as the localization of $D(\mathrm{NST}(k))$ with respect to the complexes $\mathbb{Z}_{\mathrm{tr}}(X \times \mathbb{A}^1) \xrightarrow{p_*} \mathbb{Z}_{\mathrm{tr}}(X)$. Via RC^{Sus}_*, $\mathrm{DM}^{\mathrm{eff}}(k)$ inherits a tensor structure $\otimes$ from $D(\mathrm{NST}(k))$. The functor $\mathbb{Z}_{\mathrm{tr}} : \mathrm{Sm}_k \to \mathrm{NST}(k)$ defines the functor $M^{\mathrm{eff}} := RC^{\mathrm{Sus}}_* \circ \mathbb{Z}_{\mathrm{tr}}$,

$$M^{\mathrm{eff}} : \mathrm{Sm}_k \to \mathrm{DM}^{\mathrm{eff}}(k).$$

The *Tate object* $\mathbb{Z}(1) \in \mathrm{DM}^{\mathrm{eff}}(k)$ is the image of the complex $\mathbb{Z}_{\mathrm{tr}}(\mathrm{Spec}\, k) \xrightarrow{i_{\infty *}} \mathbb{Z}_{\mathrm{tr}}(\mathbb{P}^1)$ (with $\mathbb{Z}_{\mathrm{tr}}(\mathbb{P}^1)$ in degree 2) via RC^{Sus}_*. One forms the triangulated tensor category $\mathrm{DM}(k)$ as the category of $- \otimes \mathbb{Z}(1)$-spectrum objects in $\mathrm{DM}^{\mathrm{eff}}(k)$, inverting the endofunctor $- \otimes \mathbb{Z}(1)$; for $M \in \mathrm{DM}(k)$, one has the Tate twists $M(n) := M \otimes Z(1)^{\otimes n}$ for $n \in \mathbb{Z}$; in particular, we have the Tate objects $\mathbb{Z}(n)$. The functor M^{eff} induces the functor $M : \mathrm{Sm}_k \to \mathrm{DM}(k)$.

Bloch's higher Chow groups, Suslin homology, and the motivic complexes $\mathbb{Z}_{\mathrm{Bl}}(q)$ are represented in $\mathrm{DM}(k)$ via canonical isomorphisms

$$\mathrm{CH}^q(X, 2q-p) = \mathbb{H}^p\big(X_{\mathrm{Zar}}, \mathbb{Z}_{\mathrm{Bl}}(q)\big) \cong \mathrm{Hom}_{\mathrm{DM}(k)}\big(M(X), \mathbb{Z}(q)[p]\big),$$
$$H_n^{\mathrm{Sus}}(X, \mathbb{Z}) = H_n\big(C_*^{\mathrm{Sus}}(X)\big) \cong \mathrm{Hom}_{\mathrm{DM}(k)}\big(\mathbb{Z}[n], M(X)\big).$$

In addition, one has the presheaf of complexes $\mathbb{Z}_V(q)$ on Sm_k

$$\mathbb{Z}_V(q)(X) := C_{-*}^{\mathrm{Sus}}\big(\mathbb{Z}_{\mathrm{tr}}(\mathbb{G}_m)^{\otimes q}[-q]\big)(X),$$

where $\mathbb{Z}_{\mathrm{tr}}(\mathbb{G}_m)$ is the quotient presheaf $\mathbb{Z}_{\mathrm{tr}}(\mathbb{A}^1 \setminus \{0\})/\mathbb{Z}_{\mathrm{tr}}(\{1\})$. The complexes $\mathbb{Z}_V(q)$ and $\mathbb{Z}_{\mathrm{Bl}}(q)$ define isomorphic objects in $\mathrm{DM}^{\mathrm{eff}}(k)$, in particular, are isomorphic in the derived category of Nisnevich sheaves on Sm_k. The details of these constructions and results are carried out in [127] (with a bit of help from [117]).

2.5. Motivic homotopy theory

Although Voevodsky's triangulated category of motives does give motivic cohomology a categorical foundation, this is really a halfway station on the way to a really suitable categorical framework. As analogy, embedding the Beilinson–Lichtenbaum/Bloch–Suslin theory of motivic complexes in $\mathrm{DM}(k)$ is like considering the singular chain or cochain complex of a topological space as an object in the derived category of abelian groups. A much more fruitful framework for singular (co)homology is to be found in the stable homotopy category SH.

A parallel representability for motivic cohomology for schemes over a base-scheme B in a wider category of good cohomology theories is to be found in the *motivic stable homotopy category over* B, $\mathrm{SH}(B)$. This, together with the motivic unstable homotopy category, $\mathcal{H}(B)$, gives the proper setting for the deeper study of motivic cohomology, besides placing this theory on a equal footing with all cohomology theories on algebraic varieties that satisfy a few natural axioms.

Just as the category $\mathrm{DM}(k)$ starts out as a category of presheaves, the category $\mathrm{SH}(B)$ starts out with the category of presheaves of simplicial sets on Sm_B. The construction of the unstable motivic homotopy category $\mathcal{H}(B)$ over a general base-scheme B as a suitable localization of this presheaf category was achieved by Morel–Voevodsky [94] and the stable version $\mathrm{SH}(B)$ was described by Voevodsky in his ICM address [116]. The important six-functor formalism of Grothendieck was sketched out by Voevodsky and realized in detail by Ayoub [5,6]. A general theory of motivic categories with a six-functor formalism, including $\mathrm{SH}(-)$, was established by Cisinski–Déglise [33], and Hoyois [62] gave a construction on the level of infinity categories for an equivariant version. A new point of view, the approach of *framed correspondences*, also first sketched by Voevodsky [126], is a breakthrough in our understanding of the infinite loop objects in the motivic setting, and concerning our main interest, motivic cohomology, has led to a natural construction of motivic cohomology over a general base-scheme.

In topology, the representation of singular (co)homology via the singular (co)chain complexes is placed in the setting of stable homotopy theory through the construction of the

Eilenberg–MacLane spectra, giving a natural isomorphism for each abelian group A,

$$H^n(X, A) \cong \mathrm{Hom}_{\mathrm{SH}}\big(\Sigma^\infty X_+, \Sigma^n \mathrm{EM}(A)\big),$$

with the Eilenberg–MacLane spectrum $\mathrm{EM}(A)$ being characterized by its stable homotopy groups

$$\pi_n^s\big(\mathrm{EM}(A)\big) = \begin{cases} A & \text{for } n = 0, \\ 0 & \text{else.} \end{cases}$$

The assignment $A \mapsto \mathrm{EM}(A)$ extends to a fully faithful embedding $\mathrm{EM} : D(\mathbf{Ab}) \to \mathrm{SH}$. This realizes the ordinary (co)homology as being represented by the derived category $D(\mathbf{Ab})$ via its Eilenberg–MacLane embedding in SH, which in turn is to be viewed as the category of all cohomology theories on reasonable topological spaces.

The stable homotopy category SH is the stabilization of the unstable pointed homotopy category $\mathcal{H}_\bullet$ with respect to the suspension operator $\Sigma X := S^1 \wedge X$, which becomes an invertible endofunctor on SH. The resulting functor of $\mathcal{H}_\bullet$ to its stabilization is the infinite suspension functor Σ^∞ and gives us the "effective" subcategory $\mathrm{SH}^{\mathrm{eff}} \subset \mathrm{SH}$, as the smallest subcategory containing $\Sigma^\infty(\mathcal{H}_\bullet)$ and closed under homotopy cofibers and small coproducts. This in turn gives a decreasing filtration on SH by the subcategories $\Sigma^n \mathrm{SH}^{\mathrm{eff}}$, $n \in \mathbb{Z}$. This rather abstract looking filtration is simply the filtration by connectivity: E is in $\Sigma^n \mathrm{SH}^{\mathrm{eff}}$ if and only if $\pi_m^s E = 0$ for $m < n$. The layers in this filtration are isomorphic to the category $\mathbf{Ab}$, by the functor $E \mapsto \pi_n E$, and in fact, this filtration is the one given by a natural t-structure on SH with heart $\mathbf{Ab}$; concretely, the 0th truncation $\tau_0 E$ is given by the Eilenberg–MacLane spectrum $\mathrm{EM}(\pi_0(E))$.

A central example is the sphere spectrum $\mathbb{S} := \Sigma^\infty S^0$. Since

$$\pi_0^s \mathbb{S} = \mathrm{colim}_m \pi_m(S^m) = \mathbb{Z},$$

we have $\tau_0 \mathbb{S} = \mathrm{EM}(\mathbb{Z})$, establishing the natural relation between homology and homotopy.

In the motivic world, we have a somewhat parallel picture. The pointed unstable category $\mathcal{H}_\bullet(B)$ has a natural 2-parameter family of "spheres." Let S^n denote the constant presheaf with value the pointed n-sphere, and let $\mathbb{G}_m$ denote the representable presheaf $\mathbb{A}^1 \setminus \{0\}$ pointed by 1. Define

$$S^{a,b} := S^{a-b} \wedge \mathbb{G}_m^{\wedge b}$$

for $a \geq b \geq 0$. We consider $\mathbb{P}^1$ as the representable presheaf, pointed by 1; there is a canonical isomorphism $\mathbb{P}^1 \cong S^{2,1}$ in $\mathcal{H}_\bullet(B)$.

In order to achieve the analog of Spanier–Whitehead duality in the motivic setting, one needs to use spectra with respect to $\mathbb{P}^1$-suspension rather than with respect to S^1-suspension. The category $\mathrm{SH}(B)$ is constructed as a homotopy category of $\mathbb{P}^1$-spectra in $\mathcal{H}_\bullet(B)$, so $\mathbb{P}^1$-suspension becomes invertible and our family of spheres extends to a family of invertible suspension endofunctors

$$\Sigma^{a,b} : \mathrm{SH}(B) \to \mathrm{SH}(B), \quad a, b \in \mathbb{Z}.$$

Each $E \in \mathrm{SH}(B)$ gives the bigraded cohomology theory on Sm_B by

$$E^{a,b}(X) := \mathrm{Hom}_{\mathrm{SH}(B)}(\Sigma^\infty_{\mathbb{P}^1} X_+, \Sigma^{a,b} \wedge E).$$

Note that the translation in $\mathrm{SH}(B)$ is given by S^1-suspension, not $\mathbb{P}^1$-suspension.

The effective subcategory $\mathrm{SH}^{\mathrm{eff}}(B)$ is defined as the localizing subcategory (i.e., a triangulated subcategory closed under small coproducts) generated by the $\mathbb{P}^1$-infinite suspension spectra $\Sigma^\infty_{\mathbb{P}^1}\mathcal{X}$ for $\mathcal{X} \in \mathcal{H}_\bullet(B)$. We replace the filtration of SH with respect to S^1-connectivity with the filtration on $\mathrm{SH}(B)$ with respect to $\mathbb{P}^1$-connectivity, via the subcategories $\Sigma^n_{\mathbb{P}^1}\mathrm{SH}^{\mathrm{eff}}(B)$. This is Voevodsky's slice filtration, with associated nth truncation denoted f_n, giving for each $E \in \mathrm{SH}(B)$ the tower

$$\cdots \to f_{n+1}E \to f_n E \to \cdots \to E.$$

One has the layers $s_n E$ of this tower, fitting into a distinguished triangle

$$f_{n+1}E \to f_n E \to s_n E \to f_{n+1}E[1] = \Sigma^{1,0} f_{n+1}E.$$

An important difference from the topological case is that this is a filtration by triangulated subcategories; the $\mathbb{P}^1$-suspension is not the shift in the triangulated structure on $\mathrm{SH}(B)$, and so the slice filtration does not arise from a t-structure.

We concentrate for a while on the case $B = \mathrm{Spec}\, k$, k a perfect field. There is an Eilenberg–MacLane functor

$$\mathrm{EM} : \mathrm{DM}(k) \to \mathrm{SH}(k),$$

giving the motivic cohomology spectrum $\mathrm{EM}(\mathbb{Z}(0)) \in \mathrm{SH}(k)$ representing motivic cohomology as

$$H^p\big(X, \mathbb{Z}(q)\big) = \mathrm{EM}\big(\mathbb{Z}(0)\big)^{p,q}(X).$$

One has the beautiful internal description of motivic cohomology via Voevodsky's isomorphism [122]

$$s_0\mathbb{S}_k \cong \mathrm{EM}\big(\mathbb{Z}(0)\big); \tag{2.2}$$

see also [85, THEOREM 10.5.1] and the recent paper of Bachmann–Elmanto [9]. In other words, the 0th slice truncation of the motivic sphere spectrum represents motivic cohomology. Röndigs–Østvær [103] show that the homotopy category of $\mathrm{EM}(\mathbb{Z}(0))$-modules in $\mathrm{SH}(k)$ is equivalent to $\mathrm{DM}(k)$ and represents the Eilenberg–MacLane functor as the forgetful functor, right-adjoint to the free $\mathrm{EM}(\mathbb{Z}(0))$ functor

$$\begin{array}{ccc} \mathrm{EM}\big(\mathbb{Z}(0)\big) \wedge - : \mathrm{SH}(k) & \rightleftarrows & \mathrm{EM}(\mathbb{Z}(0))\text{-}\mathbf{Mod} \;:\mathrm{EM} \\ & & \| \\ & & \mathrm{DM}(k) \end{array}$$

This is the triangulated motivic analog of the classical result, that the heart of the t-structure on SH is $\mathbf{Ab}$.

2.6. Motivic cohomology and the rational motivic stable homotopy category

In classical homotopy theory, the Eilenberg–MacLane functor $\mathrm{EM} : D(\mathbf{Ab}) \to \mathrm{SH}$ has a nice structural property: after $\mathbb{Q}$-localization, the functor $\mathrm{EM}_{\mathbb{Q}} : D(\mathbf{Ab})_{\mathbb{Q}} \to \mathrm{SH}_{\mathbb{Q}}$ is an equivalence. Does the same happen for the motivic Eilenberg–MacLane functor $\mathrm{EM} : \mathrm{DM}(k) \to \mathrm{SH}(k)$? In general, the answer is no, and the reason goes back to Morel's $\mathbb{C}$–$\mathbb{R}$ dichotomy for $\mathrm{SH}(k)$.

We discuss the case of a characteristic zero field k as base. Suppose that k admits a real embedding $\sigma : k \to \mathbb{R}$. The embedding σ induces a *realization functor*

$$\mathfrak{R}^{\sigma}_{\mathbb{R}} : \mathrm{SH}(k) \to \mathrm{SH},$$

which sends the $\mathbb{P}^1$-suspension spectrum $\Sigma^{\infty}_{\mathbb{P}^1} X_+$ of a smooth k-scheme X to the infinite suspension spectrum of the real manifold of real points $X(\mathbb{R})$. For an embedding $\sigma : k \to \mathbb{C}$, one has the realization functor $\mathfrak{R}^{\sigma}_{\mathbb{C}} : \mathrm{SH}(k) \to \mathrm{SH}$, sending $\Sigma^{\infty}_{\mathbb{P}^1} X_+$ to $\Sigma^{\infty} X(\mathbb{C})_+$. If we take $X = \mathbb{P}^1$, the real embedding gives you $\Sigma\mathbb{S}$ and the complex embedding yields $\Sigma^2\mathbb{S}$, since $\mathbb{P}^1(\mathbb{R}) = S^1$, $\mathbb{P}^1(\mathbb{C}) = S^2$. This has the effect that the switch map $\tau : \mathbb{P}^1 \wedge \mathbb{P}^1 \to \mathbb{P}^1 \wedge \mathbb{P}^1$ induces an automorphism of $\mathbb{S}_k$ that maps to -1 under the real embedding and to $+1$ under the complex embedding. Thus, if we invert 2 and decompose the motivic sphere spectrum into ± 1 eigenfactors with respect to τ, we decompose $\mathrm{SH}(k)[1/2]$ into corresponding summands $\mathrm{SH}(k)_{\pm}$, with all of $\mathrm{SH}(k)_+$ going to zero under the real embedding and all of $\mathrm{SH}(k)_-$ going to zero under the complex one (after inverting 2 in SH).

Alternatively, the minus part is $\mathrm{SH}(k)[1/2, \eta^{-1}]$, where η is the $\mathbb{P}^1$-stabilization of the algebraic Hopf map

$$\eta : \mathbb{A}^2 \setminus \{0\} \to \mathbb{P}^1, \quad \eta(x, y) = [x : y].$$

A motivic spectrum $E \in \mathrm{SH}(k)$ is *orientable* if E has a good theory of Thom classes. For $V \to X$ a vector bundle with 0-section $s_0 : X \to V$, we have the Thom space $\mathrm{Th}(V) := V/(V \setminus s_0(X)) \in \mathcal{H}_{\bullet}(k)$ (defined as the quotient of representable presheaves). An orientation for E consists of giving a class

$$\mathrm{th}(V) \in E^{2r,r}\big(\mathrm{Th}(V)\big)$$

for each rank r vector bundle $V \to X$ over $X \in \mathrm{Sm}_k$, satisfying axioms parallel to the notion of a $\mathbb{C}$-orientation in topology; a choice of Thom classes defines E as an oriented cohomology theory. After inverting 2, all the orientable E live in the plus part; this includes motivic cohomology, as well as algebraic K-theory and algebraic cobordism. These theories E all have the property that η induces zero on E-cohomology.

Theories that live in the minus part will contrariwise invert η (after inverting 2); these include things like Witt theory or cohomology of the sheaf of Witt groups. The real and complex avatars of this are seen by noting that the complex realization of the algebraic Hopf map is the usual Hopf map, which is the 2-torsion element of stable π_1 of the sphere spectrum, while the real realization is the multiplication map $\times 2 : S^1 \to S^1$.

The analog of the fact that $\mathrm{EM}_{\mathbb{Q}} : D(\mathbb{Q}) \to \mathrm{SH}_{\mathbb{Q}}$ is an equivalence is the following result of Cisinski–Déglise

Theorem 2.4 ([33, THEOREM 16.2.13]). *The unit map* $\mathbb{S}_k \to \mathrm{EM}(\mathbb{Z}(0))$ *induces an isomorphism*

$$\mathrm{SH}(k)_{+\mathbb{Q}} \to \mathrm{DM}(k)_{\mathbb{Q}}$$

with inverse the Eilenberg–MacLane functor followed by the plus-projection

$$\mathrm{DM}(k)_{\mathbb{Q}} \to \mathrm{SH}(k)_{\mathbb{Q}} \to \mathrm{SH}(k)_{+\mathbb{Q}}.$$

The rational minus part is also a homotopy category of modules over a suitable cohomology theory, namely Witt sheaf cohomology. For a field F, we have the Witt ring $W(F)$ of virtual non-degenerate quadratic forms, modulo the hyperbolic form. This extends to a sheaf $\mathcal{W}$ on Sm_k, and the functor $X \mapsto H^p_{\mathrm{Nis}}(X, \mathcal{W})$ is represented in $\mathrm{SH}(k)$ by a suitable spectrum $\mathrm{EM}(\mathcal{W})$. We have

Theorem 2.5 ([3, THEOREM 4.2, COROLLARY 4.4]). *The functor* $E \mapsto \mathrm{EM}(\mathcal{W})_{\mathbb{Q}} \wedge E$ *induces a natural isomorphism of* $\mathrm{SH}(k)_{-\mathbb{Q}}$ *with the homotopy category* $\mathrm{EM}(\mathcal{W})_{\mathbb{Q}}$*-modules.*

From this point of view, one can see the $\mathbb{Z}$-graded cohomology theory

$$X \mapsto \bigoplus_{n \geq 0} H^n_{\mathrm{Nis}}(X, \mathcal{W})$$

as the motivic cohomology for the minus part; this theory picks up information about the real points of schemes. To get the complete theory, one also needs to include twists of $\mathcal{W}$ by line bundles, an analog of orientation local systems in the topological setting. We will say more about this in Section 4.

2.7. Slice tower and motivic Atiyah–Hirzebruch spectral sequences

The classical Atiyah–Hirzebruch spectral sequence for a spectrum $E \in \mathrm{SH}$ is the spectral sequence of the Postnikov tower of E, and looks like

$$E_2^{p,q} := H^p(X, \pi_{-q}E) \Rightarrow E^{p+q}(X).$$

This comes by identifying the qth layer in the Postnikov tower with the shifted Eilenberg–MacLane spectrum $\Sigma^q \mathrm{EM}(\pi_q(E))$.

Together with results of Pelaez [99] and Gutierrez–Röndigs–Spitzweck, Voevodsky's isomorphism (2.2) has a structural expression, namely, for any $E \in \mathrm{SH}(k)$, each slice $s_q(E)$ has a canonical structure of an $\mathrm{EM}(\mathbb{Z}(0))$-module. We write corresponding object of $\mathrm{DM}(k)$ as $\pi_q^{\mathrm{mot}}(E)$, satisfying

$$s_q(E) = \Sigma^q_{\mathbb{P}^1}\mathrm{EM}\big(\pi_q^{\mathrm{mot}}(E)\big) = S^{2q,q} \wedge \mathrm{EM}\big(\pi_q^{\mathrm{mot}}(E)\big),$$

This gives the motivic Atiyah–Hirzebruch spectral sequence

$$E_2^{p,q}(n) := H^{p-q}\big(X, \pi_{-q}^{\mathrm{mot}}(E)(n-q)\big) \Rightarrow E^{p+q,n}(X).$$

These slices have been explicitly identified in a number of important cases. The first case was algebraic K-theory, $\mathrm{KGL} \in \mathrm{SH}(k)$. Voevodsky [118, 119] and Levine [85] show

$$s_q(\mathrm{KGL}) = \mathrm{EM}\big(\mathbb{Z}(q)[2q]\big) = \Sigma^q_{\mathbb{P}^1}\mathrm{EM}\big(\mathbb{Z}(0)\big)$$

so

$$\pi_q^{\mathrm{mot}}(\mathrm{KGL}) = \mathbb{Z}(0),$$

corresponding to classical computation for topological K-theory,

$$\pi_q^s KU = \begin{cases} \mathbb{Z} & \text{for } q \text{ even}, \\ 0 & \text{for } q \text{ odd}. \end{cases}$$

Using "algebraic Bott periodicity" for KGL: $\mathrm{KGL}^{a+2n,b+n}(X) = \mathrm{KGL}^{a,b}(X) = K_{2b-a}(X)$, this yields the Atiyah-Hirzebruch spectral sequence of the Beilinson-Lichtenbaum axiom (v),

$$E_2^{p,q} := H^{p-q}\big(X, \mathbb{Z}(-q)\big) \Rightarrow K_{-p-q}(X).$$

There is also a corresponding spectral sequence with $\mathbb{Z}/m$-coefficients.

This Atiyah–Hirzebruch spectral sequence for algebraic K-theory was first constructed for X the spectrum of a field by Bloch and Lichtenbaum **[29]**, by a completely different approach and without recourse to motivic homotopy theory or Voevodsky's slice tower. Their construction was generalized to general $X \in \mathrm{Sm}_k$ by Friedlander–Suslin **[43]**, also without using the categorical machinery. The rough idea is to give a filtration by codimension of support on $X \times \Delta^*$ (with additional conditions), and then identify the layers with a suitable complex of cycles. Another approach, by Grayson **[54]**, relies on the K-theory of exact categories with commuting isomorphisms. For smooth finite-type schemes over a perfect field, all these approaches yield the same spectral sequence (see **[85, THEOREM 7.1.1, THEOREM 9.0.3]**, **[44]**)).

3. MOTIVIC COHOMOLOGY OVER A GENERAL BASE

It is natural to ask if this picture of a good motivic cohomology theory for schemes over a perfect field can be extended to more general base-schemes, not just as an interesting technical question but for a wide range of applications, especially in arithmetic. Over a perfect field, we have a number of different constructions that all lead to the same groups, each of which have their advantages and disadvantages: Bloch's higher Chow groups, the cohomology of a suitable Suslin complex, the morphisms in $\mathrm{DM}(k)$, the cohomology theory represented in $\mathrm{SH}(k)$ by $\mathrm{EM}(\mathbb{Z}(0))$, or by $s_0\mathbb{S}_k$, or by $s_0\mathrm{KGL}$.

One would expect motivic cohomology to be an *absolute* theory, like algebraic K-theory, that is, its value on a given scheme should not depend on the choice of base-scheme. In terms of a spectrum $H\mathbb{Z}_S \in \mathrm{SH}(S)$ that would represent our putative theory, this is the *cartesian condition*: there should be canonical isomorphisms $H\mathbb{Z}_T \cong f^*H\mathbb{Z}_S$ for each morphism of schemes $f : T \to S$.

The identity (2.2) raises the possibility of defining motivic cohomology over a general base-scheme B by this formula. One problem here is that the slice filtration has only a limited functoriality: for $f : C \to B$ a map of schemes, one does not in general have a natural isomorphism $f^* \circ s_0 \cong s_0 \circ f^*$. For the cartesian property to hold for a motivic cohomology defined via the slice filtration, one would want the compatibility of the slices with pullback; this latter is in fact the case for $f : C \to B$ is a morphism of separated, finite type schemes over a field k of characteristic zero (or assuming resolution of singularities for separated, finite type k-schemes), by results of Pelaez [100, COROLLARY 4.3]. This compatibility also holds for arbitrary smooth f, but is not known in general.

Another concrete candidate for the motivic Borel–Moore homology is given by the hypercohomology of a version of Bloch's cycle complex, suitably extended to the setting of finite type schemes over a Dedekind domain. This theory is nearly absolute, as it depends only on a good notion of dimension or codimension, which one would have for say equi-Krull-dimensional schemes. In general, however, this theory lacks a full functoriality under pullback and also lacks a multiplicative structure.

There is a $\mathbb{P}^1$-spectrum $\mathrm{KGL}_S \in \mathrm{SH}(S)$ that represents the so-called homotopy invariant K-theory over an arbitrary base and is cartesian, so one could try $s_0\mathrm{KGL}$ as a representing spectrum. Again, the problem is the functoriality of the slice filtration, but perhaps KGL would be easier to handle than the sphere spectrum in this regard.

3.1. Cisinski–Déglise motivic cohomology

Over an base-scheme S that is noetherian and of finite Krull dimension, Cisinski–Déglise [33, §11] have followed the program of Voevodsky to define a triangulated category of motives $\mathrm{DM}_{\mathrm{CD}}(S)$, with Tate objects $\mathbb{Z}_S(n)$, and with a "motives functor"

$$M : \mathrm{Sm}_S \to \mathrm{DM}_{\mathrm{CD}}(S); X \mapsto M(X) \in \mathrm{DM}_{\mathrm{CD}}(S).$$

This extends Voevodsky's construction of $\mathrm{DM}(k)$ for a perfect field k. The main point is that the notion of a finite correspondence for smooth finite type schemes over a field extends to a corresponding notion over a general base-scheme (see [33, §8]). This gives rise to a theory of motivic cohomology generalizing Voevodsky's definition as

$$H^{p,q}(X, \mathbb{Z}) := \mathrm{Hom}_{\mathrm{DM}_{\mathrm{CD}}(S)}\big(M(X), \mathbb{Z}_S(q)[p]\big)$$

for X smooth over S. They show that the assignment $S \mapsto \mathrm{DM}_{\mathrm{CD}}(S)$ defines a functor to the category of triangulated tensor categories, $\mathrm{DM}_{\mathrm{CD}}(-) : \mathrm{Sch}_B^{\mathrm{op}} \to \mathbf{Tr}^{\otimes}$, admitting a six-functor formalism. There are also Tate twists $M \mapsto M(n)$. This gives a definition of motivic cohomology of an general scheme Y by

$$H^{p,q}(Y, \mathbb{Z}) := \mathrm{Hom}_{\mathrm{DM}_{\mathrm{CD}}(Y)}\big(\mathbb{Z}_Y(0), \mathbb{Z}_Y(q)[p]\big),$$

which for $Y \in \mathrm{Sm}_S$ agrees with the definition given above.

They construct an adjunction

$$\phi^* : \mathrm{SH}(Y) \rightleftarrows \mathrm{DM}_{\mathrm{CD}}(Y) : \phi_*,$$

with ϕ_* playing the role of the Eilenberg–MacLane functor, giving rise to the spectrum $\mathcal{M}\mathbb{Z}_Y \in \mathrm{SH}(Y)$ representing $H^{*,*}(Y, \mathbb{Z})$ [33, DEFINITION 11.2.17]. They discuss the question of whether $Y \mapsto \mathcal{M}\mathbb{Z}_Y$ is cartesian (see [33, CONJECTURE 11.2.22, PROPOSITION 11.4.7]), without reaching a general resolution.

Cisinski–Déglise have a different approach for representing motivic cohomology with $\mathbb{Q}$-coefficients, much in the same spirit as Beilinson's construction of universal cohomology using algebraic K-theory. Using the spectrum $\mathrm{KGL}_S \in \mathrm{SH}(S)$, which represents homotopy invariant algebraic K-theory, they use the Adams operations to decompose $\mathrm{KGL}_{S\mathbb{Q}}$ into summands, $\mathrm{KGL}_{S\mathbb{Q}} = \bigoplus_i \mathrm{KGL}_S^{(i)}$, with $\mathrm{KGL}_S^{(i)}$ representing the ith graded piece of K-theory for the γ-filtration. This gives them a nice commutative monoid object (i.e., commutative ring spectrum) $H_S^{\mathbb{B}} := \mathrm{KGL}_S^{(0)} \in \mathrm{SH}(S)_{\mathbb{Q}}$, whose module category they call the category of *Beilinson motives over S*. This construction is cartesian, gives a good theory of motivic cohomology with $\mathbb{Q}$-coefficients over a general base-scheme and agrees with $\mathrm{DM}_{\mathrm{CD}}(S)_{\mathbb{Q}}$ for S a uni-branch scheme. See **[33, §14]** for details.

3.2. Spitzweck's motivic cohomology

In **[110]**, Spitzweck constructs a motivic cohomology theory over an arbitrary base-scheme. The Bloch cycle complex gives rise to a general version of Bloch's higher Chow groups for finite type schemes over a Dedekind domain, which has nice localization properties (by **[25]** and **[84]**), but has poor functoriality and lacks a multiplicative structure. On the other hand, using the Bloch–Kato conjectures, established by Voevodsky et al., the ℓ-completed higher Chow groups are recognized as a truncated ℓ-adic étale cohomology, for ℓ prime to all residue characteristics. The theorem of Geisser–Levine **[52]** describes the p-completed higher Chow groups in characteristic $p > 0$ in terms of logarithmic de Rham–Witt sheaves. Finally, there is the good theory with $\mathbb{Q}$-coefficients given by Beilinson motivic cohomology of Cisinski–Déglise, as described above.

Each of these three theories, namely the ℓ-adic étale cohomology, the cohomology of the logarithmic de Rham–Witt sheaves, and the rational Beilinson motivic cohomology, has good functoriality and multiplicative properties. Gluing the ℓ-adic, p-adic, and rational theories together via their respective comparisons with the Bloch cycle complex, Spitzweck constructs a theory with good functoriality and multiplicative properties, and which is described by a presheaf of complexes on smooth schemes over a given Dedekind domain as base-scheme. The corresponding theory agrees with Voevodsky's motivic cohomology for smooth schemes over a perfect field, and is given additively by the hypercohomology of the Bloch complex for smooth schemes over a Dedekind domain (even in mixed characteristic).

Taking the base-scheme to be Spec $\mathbb{Z}$, Spitzweck's construction yields a representing object $M\mathbb{Z}_{\mathbb{Z}}$ in $\mathrm{SH}(\mathbb{Z})$ and one can thus define absolute motivic cohomology for smooth schemes over a given base-scheme S by pulling back $M\mathbb{Z}_{\mathbb{Z}}$ to $M\mathbb{Z}_S \in \mathrm{SH}(S)$. The resulting motivic cohomology agrees with Voevodsky's for smooth schemes of finite type over a perfect base-field, and with the hypercohomology of the Bloch cycle complex for smooth finite type schemes over a Dedekind domain. This gives rise to a triangulated category of motives $\mathrm{DM}_{\mathrm{Sp}}(S)$ over a base-scheme S, defined as the homotopy category of $M\mathbb{Z}_S$-modules, and the functor $S \mapsto \mathrm{DM}_{\mathrm{Sp}}(S)$ inherits a Grothendieck six-functor formalism from that of $S \mapsto \mathrm{SH}(S)$.

3.3. Hoyois' motivic cohomology

Spitzweck's construction gives a solution to the problem of constructing a triangulated category of motives over an arbitrary base, admitting a six-functor formalism and thus yielding a good theory of motivic cohomology. His construction is a bit indirect and it would be nice to have a direct construction of a representing motivic ring spectrum $H\mathbb{Z}_S \in \mathrm{SH}(S)$ for each base-scheme S, still satisfying the cartesian condition.

Hoyois has constructed such a theory of motivic cohomology over an arbitrary base-scheme by using a recent breakthrough in our understanding of the motivic stable homotopy categories $\mathrm{SH}(S)$. This is a new construction of $\mathrm{SH}(S)$ more in line with Voevodsky construction of $\mathrm{DM}(k)$. The basic idea is sketched in notes of Voevodsky [126], which were realized in a series of works by Ananyevskiy, Garkusha, Panin, Neshitov [2,4,45–48](authorship in various combinations). Building on these works, Elmanto, Hoyois, Khan, Sosnilo, and Yakerson [36–38] construct an infinity category of framed correspondences, and use the basic program of Voevodsky's construction of $\mathrm{DM}(k)$ to realize $\mathrm{SH}(S)$ as arising from presheaves of spectra *with framed transfers*, just as objects of $\mathrm{DM}(k)$ arise from presheaves of complexes of sheaves with transfers for finite correspondences. It is not our purpose here to give a detailed discussion of this beautiful topic; we content ourselves with sketching some of the basic principles.

An integral closed subscheme $Z \subset X \times Y$ that defines a finite correspondence from X to Y can be thought of a special type of a *span* via the two projections

$$\begin{array}{ccccc} & & Z & & \\ & \swarrow^{p_1} & & \searrow^{p_2} & \\ X & & & & Y \end{array}$$

For X and Y smooth and finite type over a given base-scheme S, a framed correspondence from X to Y is also a span,

$$\begin{array}{ccccc} & & Z & & \\ & \swarrow^{p} & & \searrow^{q} & \\ X & & & & Y \end{array}$$

satisfying certain conditions, together with some additional data (the framing). For simplicity, assume that X is connected. The morphism p is required to be a finite, flat, local complete intersection (lci) morphism, called a finite *syntomic* morphism (the terminology was introduced by Mazur). The lci condition means that p factors as closed immersion $i : Z \to P$ followed by a smooth morphism $f : P \to X$, and the closed subscheme $i(Z)$ of P is locally defined by exactly $\dim_X P - \dim_X Z$ equations forming a regular sequence. The morphism p factored in this way has a relative cotangent complex $\mathbb{L}_p$ admitting a simple description, namely

$$\mathbb{L}_p = \left[\mathcal{I}_Z/\mathcal{I}_Z^2 \xrightarrow{d} i^*\Omega_{P/X}\right];$$

the conditions on i and p say that both $\mathcal{I}_Z/\mathcal{I}_Z^2$ and $i^*\Omega_{P/X}$ are locally free coherent sheaves on Z of rank $\dim_X P - \dim_X Z$ and $\dim_X P$, respectively. For p an lci morphism, the perfect complex $\mathbb{L}_p$ defines a point $\{\mathbb{L}_p\}$ in the space $\mathcal{K}(Z)$ defining the K-theory of Z; in the case of a finite syntomic morphism, the virtual rank of $\{\mathbb{L}_p\}$ is zero.

A *framing* for a syntomic map $p : Z \to X$ is a choice of a path $\gamma : [0,1] \to \mathcal{K}(Z)$ connecting $\{\mathbb{L}_p\}$ with the base-point $0 \in \mathcal{K}(Z)$. For a framing to exist, the class $[\mathbb{L}_p] \in K_0(Z)$ must be zero, but the choice of γ is additional data. The morphism $q : Z \to Y$ is arbitrary.

One has the usual notion of a composition of spans:

$$\begin{array}{c} & Z' & \\ {}^{p'}\swarrow & & \searrow^{q'} \\ Y & & W \end{array} \circ \begin{array}{c} & Z & \\ {}^{p}\swarrow & & \searrow^{q} \\ X & & Y \end{array} := \begin{array}{c} & Z\times_Y Z' & \\ {}^{p\circ p_1}\swarrow & & \searrow^{q'\circ p_2} \\ X & & W \end{array}$$

which preserves the finite syntomic condition. However, one needs a higher categorical structure to take care of associativity constraints. The composition of paths is even trickier, since we are dealing here with actual paths, not paths up to homotopy. In the end, this produces an infinity category $\mathbf{Corr}^{\mathrm{fr}}(\mathrm{Sm}_S)$ of framed correspondences on smooth S-schemes, rather than a category; roughly speaking, the composition is only defined "up to homotopy and coherent higher homotopies."

Via the infinity category $\mathbf{Corr}^{\mathrm{fr}}(\mathrm{Sm}_S)$, we have the infinity category of framed motivic spaces, $\mathbf{H}^{\mathrm{fr}}(S)$, this being the infinity category of $\mathbb{A}^1$-invariant, Nisnevich sheaves of spaces on $\mathbf{Corr}^{\mathrm{fr}}(\mathrm{Sm}_S)$. There is a stable version, $\mathbf{SH}^{\mathrm{fr}}(S)$, an infinite suspension functor $\Sigma^\infty_{\mathrm{fr}} : \mathbf{H}^{\mathrm{fr}}(S) \to \mathbf{SH}^{\mathrm{fr}}(S)$, and an equivalence of infinity categories $\gamma_* : \mathbf{SH}^{\mathrm{fr}}(S) \to \mathbf{SH}(S)$, where $\mathbf{SH}(S)$ is the infinity category version of the triangulated category $\mathrm{SH}(S)$, that is, the homotopy category of $\mathbf{SH}(S)$ is $\mathrm{SH}(S)$. The equivalence γ_* can be thought of as a version of the construction of infinite loop spaces from Segal's Γ-spaces, with a framed correspondence $X \leftarrow Z \to Y$ of degree n over X being viewed as a generalization of the map $[n]_+ \to [0]_+$ in Γ^{op}.

With this background, we can give a rough idea of Hoyois' construction of the spectrum representing motivic cohomology over S in **[63]**. He considers spans $X \overset{p}{\leftarrow} Z \overset{q}{\to} Y$, $X, Y \in \mathrm{Sm}_S$, with $p : Z \to X$ a finite morphism such that $p_*\mathcal{O}_Z$ is a locally free $\mathcal{O}_X$-module; note that this condition is satisfied if p is a syntomic morphism, but not conversely. These spans form a category $\mathbf{Corr}^{\mathrm{flf}}(\mathrm{Sm}_S)$ under span composition ("flf" stands for "finite, locally free") and forgetting the paths γ defines a morphism of (infinity) categories $\pi_{\mathrm{ad}} : \mathbf{Corr}^{\mathrm{fr}}(\mathrm{Sm}_S) \to \mathbf{Corr}^{\mathrm{flf}}(\mathrm{Sm}_S)$.

Given a commutative monoid A, the constant Nisnevich sheaf on Sm_S with value A extends to a functor

$$A_S : (\mathbf{Corr}^{\mathrm{flf}})^{\mathrm{op}} \to \mathbf{Ab},$$

where the pullback from Y to X by $X \overset{p}{\leftarrow} Z \overset{q}{\to} Y$ is given by multiplication by $\mathrm{rnk}_{\mathcal{O}_X}\mathcal{O}_Z$ if X and Y are connected; one extends to general smooth X and Y by additivity. This gives us the presheaf (of abelian monoids) with framed transfers $A_S^{\mathrm{fr}} := A_S \circ \pi_{\mathrm{ad}}^{\mathrm{op}}$, and the machinery of **[36–38]** converts this into the motivic spectrum $\gamma_*\Sigma^\infty_{\mathrm{fr}} A_S^{\mathrm{fr}} \in \mathrm{SH}(S)$. Hoyois shows **[63,**

LEMMA 20] that this construction produces a cartesian family, and that taking $A = \mathbb{Z}$ recovers Spitzweck's family $S \mapsto M\mathbb{Z}_S$ **[63, THEOREM 21]**.

This gives us a conceptually simple construction of a motivic Eilenberg–MacLane spectrum, and the corresponding motivic category $\mathrm{DM}_H(S)$, much in the spirit of Voevodsky original construction of $\mathrm{DM}(k)$ and the Röndigs–Østvær theorem identifying $\mathrm{DM}(k)$ with the homotopy category of $\mathrm{EM}(\mathbb{Z}(0))$-modules.

4. MILNOR–WITT MOTIVIC COHOMOLOGY

The classical Chow group $\mathrm{CH}^n(X)$ of codimension n algebraic cycles modulo rational equivalence on a smooth variety X is part of the motivic cohomology of X via the isomorphism $\mathrm{CH}^n(X) = H^{2n}(X, \mathbb{Z}(n))$. Barge and Morel **[12]** have introduced a refinement of the Chow groups, the *Chow–Witt groups*, that incorporates information about quadratic forms. Their construction has been embedded in a larger theory of *Milnor–Witt motives* and *Milnor–Witt motivic cohomology*, which we describe in this section. The quadratic information given by the Chow–Witt groups, Milnor–Witt motivic cohomology and related theories has proven useful in recent efforts to give quadratic refinements for intersection theory and enumerative geometry; see **[10, 11, 21, 61, 76, 77, 86]** for some examples. We refer the reader to **[8, 31, 39, 92]** for details on the theory described in this section.

4.1. Milnor–Witt K-theory and the Chow–Witt groups

A codimension n algebraic cycle $Z := \sum_i n_i Z_i$ can be thought of as the set of its generic points z_i together with the $\mathbb{Z}$-valued function n_i on z_i, from which we can write the group $Z^n(X)$ of codimension n algebraic cycles as

$$Z^n(X) = \bigoplus_{z \in X^{(n)}} \mathbb{Z},$$

where $X^{(n)}$ is the set of points $z \in X$ with closure $Z := \bar{z} \subset X$ of codimension n.

Let $\mathrm{GW}(F)$ denote the Grothendieck–Witt ring of virtual non-degenerate quadratic forms over F and let $W(F) = \mathrm{GW}(F)/(H)$ where H is the hyperbolic form $H(x, y) = x^2 - y^2$ (we assume throughout that the characteristic is $\neq 2$ to avoid technical difficulties); $W(F)$ is the *Witt ring* of anisotropic quadratic forms over F (see **[107]**).

One can consider a finite set of codimension n points $z_i \in X^{(n)}$, together with a collection of classes $\{q_i \in \mathrm{GW}(k(z_i))\}$; one recovers a $\mathbb{Z}$-valued function on z_i by taking the rank of q_i. This gives the group

$$\tilde{Z}^n(X) := \bigoplus_{z \in X^{(n)}} \mathrm{GW}\big(k(z)\big)$$

with rank homomorphism $\mathrm{rnk} : \tilde{Z}^n(X) \to Z^n(X)$. In contrast with integer-valued functions, an element $q \in \mathrm{GW}(k(z))$ does not always extend to all of $\bar{z}$; there is an obstruction given by a certain boundary map

$$\partial : \mathrm{GW}\big(k(z)\big) \to \bigoplus_{w \in \bar{z} \cap X^{(n+1)}} W\big(k(w)\big).$$

This starts to look more like classical homology, in that one should consider $\tilde{Z}^n(X)$ as a group of chains rather than a group of cycles.

This is not enough, as one needs a quadratic refinement for the classical relation given by rational equivalence. The original construction of Barge–Morel defined this relation, but later developments put their construction in a rather more natural form, which we now describe.

We recall that the Milnor K-theory ring $K_*^M(F) := \bigoplus_{n\geq 0} K_n^M(F)$ of a field F is defined as the quotient of the tensor algebra on the abelian group of units $F^\times$, modulo the Steinberg relation

$$K_*^M(F) := (F^\times)^{\otimes_{\mathbb{Z}} *}/\big(\{a \otimes (1-a) \mid a \in F \setminus \{0,1\}\}\big).$$

The quadratic refinement of $K_*^M(F)$ is the Hopkins–Morel *Milnor–Witt K-theory* of F.

Definition 4.1 (Hopkins–Morel [92, DEFINITION 6.3.1]). Let F be a field. The *Milnor–Witt K-theory of F*, $K_*^{\mathrm{MW}}(F) := \bigoplus_{n\in\mathbb{Z}} K_n^{\mathrm{MW}}(F)$, is the $\mathbb{Z}$-graded associative algebra defined by the following generators and relations.

Generators

(G1) For each $u \in F^\times$, we have the generator $[u]$ of degree 1;

(G2) There is an additional generator η of degree -1.

Relations

(R0) $\eta \cdot [u] = [u] \cdot \eta$;

(R1) $[uv] = [u] + [v] + \eta \cdot [u] \cdot [v]$;

(R2) $[u] \cdot [1-u] = 0$ for $u \in F \setminus \{0,1\}$;

(R3) Let $h = (2 + \eta \cdot [-1])$. Then $\eta \cdot h = 0$.

It follows directly that sending $[u]$ to $\{u\} \in K_1^M(F)$ and sending η to zero defines a surjective graded algebra homomorphism $K_*^{MW}(F) \to K_*^M(F)$ with kernel (η). We write $[u_1, \ldots, u_n]$ for the product $[u_1]\cdots[u_n]$.

Theorem 4.2 (Hopkins–Morel [92, THEOREM 6.4.5]). *Let $I(F) \subset \mathrm{GW}(F)$ be the kernel of the rank homomorphism $\mathrm{GW}(F) \to \mathbb{Z}$, with the nth power ideal $I^n(F) \subset \mathrm{GW}(F)$ for $n > 0$. Define $I^n(F) = W(F)$ for $n \leq 0$. Then for each $n \in \mathbb{Z}$, the surjection $K_n^{\mathrm{MW}}(F) \to K_n^M(F)$ extends to an exact sequence*

$$0 \to I^{n+1}(F) \to K_n^{\mathrm{MW}}(F) \to K_n^M(F) \to 0.$$

For $n = 0$, $K_0^M(F) = \mathbb{Z}$, $K_0^{\mathrm{MW}}(F)$ is isomorphic to $\mathrm{GW}(F)$ and the above sequence is isomorphic to the defining sequence for $I(F)$. For $n < 0$, $K_n^M(F) = 0$ and $K_n^{\mathrm{MW}}(F) \cong$

$W(F)$. Finally, we have, for each $n < 0$, a commutative diagram

$$\begin{array}{ccc} K_n^{\mathrm{MW}}(F) & \xrightarrow{\sim} & W(F) \\ \downarrow{\scriptstyle \times\eta} & & \| \\ K_{n-1}^{\mathrm{MW}}(F) & \xrightarrow{\sim} & W(F) \end{array}$$

and, for $n = 0$, the commutative diagram

$$\begin{array}{ccc} K_0^{\mathrm{MW}}(F) & \xrightarrow{\sim} & \mathrm{GW}(F) \\ \downarrow{\scriptstyle \times\eta} & & \downarrow{\scriptstyle \pi} \\ K_{-1}^{\mathrm{MW}}(F) & \xrightarrow{\sim} & W(F) \end{array}$$

where π is the canonical surjection.

The isomorphism $\mathrm{GW}(F) \xrightarrow{\sim} K_0^{\mathrm{MW}}(F)$ sends $\langle u\rangle$ to $1 + \eta[u]$, where $\langle u\rangle$ is the rank one form $\langle u\rangle(x) := ux^2$; since a quadratic form over F is diagonalizable ($\mathrm{char} F \neq 2$), the isomorphism is completely determined by its value on the forms $\langle u\rangle$. Given a 1-dimensional F-vector space L, we have the $\mathrm{GW}(F)$-module $\mathrm{GW}(F; L)$ of non-degenerate, L-valued quadratic forms $q : V \to L$; each vector space isomorphism $\phi : L \to F$ gives an isomorphism of $\mathrm{GW}(F)$-modules $\mathrm{GW}(F; L) \cong \mathrm{GW}(F)$. Since $K_*^{\mathrm{MW}}(F)$ is a $\mathbb{Z}$-graded $K_0^{\mathrm{MW}}(F) = \mathrm{GW}(F)$-module, we can form the $\mathbb{Z}$-graded $K_*^{\mathrm{MW}}(F)$-module $K_*^{\mathrm{MW}}(F; L) := \mathrm{GW}(F; L) \otimes_{\mathrm{GW}(F)} K_*^{\mathrm{MW}}(F)$.

Given a dvr $\mathcal{O}$ with residue field k, quotient field F, and generator t for the maximal ideal, one has the map

$$\partial_t : K_n^{\mathrm{MW}}(F) \to K_{n-1}^{\mathrm{MW}}(k)$$

determined by the formulas

$$\partial_t([t, u_2, \ldots, u_n]) = [\bar{u}_2, \ldots, \bar{u}_n], \quad \partial_t([u_1, u_2, \ldots, u_n]) = 0, \quad \partial_t(\eta \cdot x) = \eta \cdot \partial_t(x)$$

for $u_1, \ldots, u_n \in \mathcal{O}^\times$, and $x \in K_{n+1}^{\mathrm{MW}}(F)$, where $\bar{u}_i$ is the image of u_i in $k^\times$. This is similar to the well-known boundary map $\partial : K_n^M(F) \to K_{n-1}^M(k)$, with the difference, that ∂ does not depend on the choice of t while ∂_t does. To get a boundary map that is independent of the choice of parameter t, one needs to include the twisting. This yields the well-defined boundary map

$$\partial : K_n^{\mathrm{MW}}(F; L \otimes_{\mathcal{O}} F) \to K_{n-1}^{\mathrm{MW}}\big(k; L \otimes_{\mathcal{O}} (\mathfrak{m}/\mathfrak{m}^2)^\vee\big)$$

for L a free rank-one $\mathcal{O}$-module, independent of the choice of generator for the maximal ideal $\mathfrak{m}$, where ∂ is defined by choosing a generator t and an $\mathcal{O}$-basis λ for L, and setting

$$\partial(x \otimes \lambda) := \partial_t(x) \otimes \lambda \otimes \partial/\partial t.$$

Definition 4.3. Let X be a smooth finite type k-scheme, and let $\mathcal{L}$ be an invertible sheaf on X. The nth *$\mathcal{L}$-twisted Rost–Schmid complex* for Milnor–Witt K-theory is the complex $\mathrm{RS}^*(X, \mathcal{L}, n)$ with

$$\mathrm{RS}^m(X, \mathcal{L}, n) := \bigoplus_{x \in X^{(m)}} K_{n-m}^{\mathrm{MW}}\Big(k(x); \mathcal{L}_x \otimes_{\mathcal{O}_{X,x}} \bigwedge^m (\mathfrak{m}_x/\mathfrak{m}_x^2)^\vee\Big)$$

and boundary map $\partial^m : \mathrm{RS}^m(X, \mathcal{L}, n) \to \mathrm{RS}^{m+1}(X, \mathcal{L}, n)$ the sum of the maps

$$\partial_{w,x} : K^{\mathrm{MW}}_{n-m}\Big(k(x); \mathcal{L}_x \otimes_{\mathcal{O}_{X,x}} \bigwedge^m (\mathfrak{m}_x/\mathfrak{m}_x^2)^\vee\Big) \to K^{\mathrm{MW}}_{n-m-1}\Big(k(w); \mathcal{L}_x \otimes_{\mathcal{O}_{X,x}} \bigwedge^{m+1} (\mathfrak{m}_w/\mathfrak{m}_w^2)^\vee\Big)$$

associated to the normalization of the local ring $\mathcal{O}_{\bar{x},w}$ for $w \in \bar{x} \cap X^{(m+1)}$. Here we have cheated a bit in the definition of $\partial_{w,x}$. This is correct if $\mathcal{O}_{\bar{x},w}$ is a dvr, which is the case outside of finitely many points $w \in \bar{x} \cap X^{(m+1)}$; in general, one needs to use a push-forward map in Milnor–Witt K-theory for finite field extensions to define $\partial_{w,x}$.

The twisted Milnor–Witt sheaf $\mathcal{K}^{\mathrm{MW}}_n(\mathcal{L})_X$ is the Nisnevich sheaf on X associated to the presheaf

$$U \mapsto H^0\big(\mathrm{RS}^*(U, \mathcal{L} \otimes \mathcal{O}_U, n)\big).$$

The codimension n twisted Chow–Witt group of X, $\widetilde{\mathrm{CH}}^n(X; \mathcal{L})$, is defined as

$$\widetilde{\mathrm{CH}}^n(X; \mathcal{L}) := H^n\big(\mathrm{RS}^*(X, \mathcal{L}, n)\big)$$

For details, see **[93, CHAP. 5]** or **[31, CHAP. 2]**.

For Milnor K-theory, one has the Gersten complex $G^*(X, n)$,

$$G^*(X, n) := \bigoplus_{x \in X^{(0)}} K^{\mathrm{MW}}_n\big(k(x)\big) \xrightarrow{\partial^0} \cdots \xrightarrow{\partial^{n-m+1}} \bigoplus_{x \in X^{(m)}} K^M_{n-m}\big(k(x)\big) \xrightarrow{\partial^{n-m}} \cdots \xrightarrow{\partial^{n-1}} \bigoplus_{x \in X^{(n)}} K^M_0\big(k(x)\big),$$

with essentially the same definition as the Rost–Schmid complex, without the twisting. This gives us the Milnor K-theory sheaf $\mathcal{K}^M_{n,X} := \ker \partial^0$, and it follows easily from the definitions that $\mathrm{CH}^n(X) = H^n(G^*(X, n))$. The same ideas that lead to the Bloch–Kato formula **[78]**

$$\mathrm{CH}^n(X) \cong H^n(X_{\mathrm{Nis}}, \mathcal{K}^M_{n,X})$$

give the isomorphism

$$\widetilde{\mathrm{CH}}^n(X; \mathcal{L}) \cong H^n\big(X_{\mathrm{Nis}}, \mathcal{K}^{\mathrm{MW}}_n(\mathcal{L})_X\big)$$

(see the discussion following **[31, DEFINITION 3.1]** for details). The maps $\mathcal{K}^{\mathrm{MW}}_n \to \mathcal{K}^{\mathrm{M}}_n$ give the map of complexes $\mathrm{RS}^*(X, \mathcal{L}, n) \to G^*(X, n)$ and the corresponding map $\mathrm{rnk}_{X,n} : \widetilde{\mathrm{CH}}^n(X; \mathcal{L}) \to \mathrm{CH}^n(X)$.

The twists by an invertible sheaf are not just a device for defining the Rost–Schmid complexes and the Chow–Witt groups, they play an integral part in the structure of the overall theory. The Chow groups of smooth varieties admit the functorialities of a Borel–Moore homology theory: they have functorial pullback maps $f^* : \mathrm{CH}^n(Y) \to \mathrm{CH}^n(X)$ for each morphism $f : X \to Y$ in Sm_k, and for $f : X \to Y$ a proper morphism of relative dimension d, one has the functorial proper push-forward map $f_* : \mathrm{CH}^n(X) \to \mathrm{CH}^{n-d}(Y)$. The Chow–Witt groups also have a contravariant functoriality; for $f : X \to Y$, and $\mathcal{L}$ an invertible sheaf on Y, one has the functorial pullback

$$f^* : \widetilde{\mathrm{CH}}^n(Y, \mathcal{L}) \to \widetilde{\mathrm{CH}}^n(X, f^*\mathcal{L}).$$

But for the proper push-forward, one needs to include the orientation sheaf, this being the usual relative dualizing sheaf $\omega_f := \omega_{X/k} \otimes f^*\omega_{Y/k}^{-1}$, where $\omega_{X/k} := \det \Omega^1_{X/k}$ is the sheaf of top-dimensional forms. The push-forward takes the form

$$f_* : \widetilde{\mathrm{CH}}^n(X, \omega_f \otimes f^*\mathcal{L}) \to \widetilde{\mathrm{CH}}^{n-d}(Y, \mathcal{L}).$$

This limits the possible twists $\widetilde{\mathrm{CH}}^n(X, \mathcal{M})$ for which a push-forward f_* is even defined; this type of restricted push-forward is typical of so-called SL-oriented theories, such as hermitian K-theory. See [1] for a detailed discussion of SL-oriented theories and [31, CHAP. 3] for the details concerning the push-forward in $\widetilde{\mathrm{CH}}^*$.

4.2. The homotopy t-structure and Morel's theorem

Building on the Bloch–Kato formula, $\mathrm{CH}^n(X) \cong H^n(X_{\mathrm{Nis}}, \mathcal{K}^M_{n,X})$, one can construct a good bigraded cohomology theory $\mathrm{EM}(\mathcal{K}^M_*)^{**}$ by using all the cohomology groups. To get the correct bigrading, one should set

$$\mathrm{EM}(\mathcal{K}^M_*)^{a,b}(X) := H^{a-b}(X_{\mathrm{Nis}}, \mathcal{K}^M_b),$$

giving in particular $\mathrm{EM}(\mathcal{K}^M_*)^{2n,n}(X) = \mathrm{CH}^n(X)$. It was recognized early on that this theory is not the sought-after motivic cohomology, for instance, for $X = \mathrm{Spec}\, F$, F a field, one gets exactly the Milnor K-theory of F, and none of the other parts of the K-theory of F. In spite of this, this theory and the similarly defined theory for Milnor–Witt K-theory have a natural place in the universe of motivic cohomology theories, which we now explain.

The classical stable homotopy category SH is a triangulated category with a natural t-structure measuring connectedness, mentioned in Section 2.5. For SH, the truncations give the terms in the Moore–Postnikov tower

$$\cdots \to \tau_{\geq n+1}E \to \tau_{\geq n}E \to \cdots \to E$$

with $\tau_{\geq n}E \to E$ characterized by inducing an isomorphism on π_m for $m \geq n$ and with $\pi_m\tau_{\geq n}E = 0$ for $m < n$. The heart of SH is the category of spectra E with $\pi_m E = 0$ for $m \neq 0$, which are just the Eilenberg–MacLane spectra $\mathrm{EM}(A)$, A an abelian group. Thus, the heart of SH is **Ab** and the cohomology theory represented by $\tau_0 E$ is

$$\mathrm{EM}(\pi_0 E)^n(X) := H^n(X, \pi_0 E),$$

singular cohomology with coefficients in the abelian group $\pi_0 E$.

We have a parallel t-structure on SH(k), introduced by Morel [92, §5.2], called the *homotopy t-structure* (and *not* coming from Voevodsky's slice tower discussed in Section 2.5). This is similar to the t-structure on SH, where one takes into account the fact that one has bigraded homotopy sheaves $\pi_{a,b}E$ for $E \in \mathrm{SH}(k)$, rather than a $\mathbb{Z}$-graded family of homotopy groups $\pi_n E$ for $E \in \mathrm{SH}$. The truncation $\tau_{\geq n}E$ is characterized by

$$\pi_{a,b}(\tau_{\geq n}E) = \begin{cases} \pi_{a,b}(E) & \text{if } a-b \geq n, \\ 0 & \text{if } a-b < n \end{cases}$$

Recalling that the sphere $S^{a,b}$ is $S^{a-b} \wedge \mathbb{G}_m^b$, the homotopy t-structure on $\mathrm{SH}(k)$ is measuring S^1-connectedness, instead of the $\mathbb{P}^1$-connectedness measured by Voevodsky's slice tower.

We denote the 0th truncation $\tau_0 E$ for $E \in \mathrm{SH}(k)$ by $\mathrm{EM}(\pi_{-*,-*}E)$; the notation comes from Morel's identification of the heart with his category of *homotopy modules*; for details, see [92, §5.2]. The corresponding cohomology theory satisfies, for $X \in \mathrm{Sm}_k$,

$$\mathrm{EM}(\pi_{-*,-*}E)^{a,b}(X) = H^{a-b}\big(X_{\mathrm{Nis}}, \pi_{-b,-b}(E)\big).$$

Here we have Morel's fundamental theorem [92, THEOREM 6.4.1] computing τ_0 of the sphere spectrum $1_k \in \mathrm{SH}(k)$.

Theorem 4.4 (Morel). *Let k be a perfect field. Then there are canonical isomorphisms of sheaves on* Sm_k

$$\pi_{-n,-n}(1_k) = \mathcal{K}_n^{\mathrm{MW}}.$$

Consequently,

$$\tau_0 1_k = \mathrm{EM}(\mathcal{K}_*^{\mathrm{MW}})$$

and

$$\mathrm{EM}(\mathcal{K}_*^{\mathrm{MW}})^{a,b}(X) = H^{a-b}(X_{\mathrm{Nis}}, \mathcal{K}_{b,X}^{\mathrm{MW}}).$$

Going back in time a bit, we have the theorem of Totaro [115] and Nesterenko–Suslin [96]

$$H^n\big(F, \mathbb{Z}(n)\big) \cong K_n^M(F)$$

for F a field. Combined with the isomorphism

$$s_0 1_k \cong H\mathbb{Z}$$

of [9,85,122], we have

Theorem 4.5. *Let k be a perfect field. Then*

$$\tau_0 s_0 1_k = \tau_0 H\mathbb{Z} = \mathrm{EM}(\mathcal{K}_*^M)$$

and

$$\mathrm{EM}(\mathcal{K}_*^M)^{a,b}(X) = H^{a-b}(X_{\mathrm{Nis}}, \mathcal{K}_{b,X}^M)$$

for $X \in \mathrm{Sm}_k$.

Bachmann proves an extension of this result. Recall Voevodsky's slice tower

$$\cdots \to f_{n+1}E \to f_n E \to \cdots \to f_0 E \to \cdots \to E$$

with $s_n E$ the layer given by the distinguished triangle

$$f_{n+1}E \to f_n E \to s_n E \to f_{n+1}E[1].$$

Recall that this is *not* the truncation tower of a t-structure, as the subcategories defined by the layers $s_n := f_n/f_{n+1}$ are triangulated categories, not abelian categories.

Proposition 4.6 ([7, LEMMA 12]). *Let $1_k \to \mathrm{EM}(\mathcal{K}^M_*)$ be the composition $1_k \to \tau_0 1_k = \mathrm{EM}(\mathcal{K}^{\mathrm{MW}}_*) \to \mathrm{EM}(\mathcal{K}^M_*)$, the latter map induced by the surjection $\mathcal{K}^{\mathrm{MW}}_* \to \mathcal{K}^M_*$. Then the induced maps*

$$s_0(1_k) \to s_0\mathrm{EM}(\mathcal{K}^M_*) \leftarrow f_0\mathrm{EM}(\mathcal{K}^M_*) = f_0\tau_0 H\mathbb{Z}$$

are all isomorphisms, so all of these objects are isomorphic to the motivic cohomology spectrum $H\mathbb{Z}$.

The truncation functors for the homotopy t-structure and for the Voevodsky slice tower do not commute. Since 1_k is effective, we have $f_0 1_k = 1_k$ and so $\tau_0 f_0 1_k = \tau_0 1_k = \mathrm{EM}(\mathcal{K}^{\mathrm{MW}}_*)$. The truncations in the other order give us something new.

4.3. Milnor–Witt motivic cohomology

Definition 4.7 ([7, NOTATION, P. 1134, JUST BEFORE LEMMA 12]). Let k be a perfect field. Define the *Milnor–Witt motivic cohomology spectrum* $\tilde{H}\mathbb{Z} \in \mathrm{SH}(k)^{\mathrm{eff}}$ by

$$\tilde{H}\mathbb{Z} := f_0(\tau_0 1_k) = f_0\mathrm{EM}(\mathcal{K}^{\mathrm{MW}}_*).$$

The canonical map $\tau_0 1_k \to \tau_0 s_0 1_k = \tau_0 H\mathbb{Z}$ induces the map

$$\tilde{H}\mathbb{Z} = f_0(\tau_0 1_k) \xrightarrow{\Xi} f_0\tau_0 H\mathbb{Z} = H\mathbb{Z}.$$

For $X \in \mathrm{Sm}_k$, the *Milnor–Witt motivic cohomology in bidegree* (a, b) is defined as $\tilde{H}\mathbb{Z}^{a,b}(X)$.

Remarkably, one can compute $\tilde{H}\mathbb{Z}^{a,b}(X)$ in terms of the Milnor–Witt sheaves, at least for some of the indices (a, b); one also recovers the Chow–Witt groups. For $X = \mathrm{Spec}\, F$, the spectrum of a field F, one has a complete computation in terms of the Milnor–Witt K-groups and the usual motivic cohomology $H\mathbb{Z}^{a,b}(X) := H^a(X, \mathbb{Z}(b))$.

Theorem 4.8 (Bachmann). *For $X \in \mathrm{Sm}_k$ and $b \leq 0$, there are natural isomorphisms*

$$\tilde{H}\mathbb{Z}^{a,b}(X) \cong H^{a-b}(X_{\mathrm{Nis}}, \mathcal{K}^{\mathrm{MW}}_{b,X}) = \begin{cases} H^{a-b}(X_{\mathrm{Nis}}, \mathcal{W}_X) & \text{for } b < 0, \\ H^{a-b}(X_{\mathrm{Nis}}, \mathcal{GW}_X) & \text{for } b = 0. \end{cases}$$

Here $\mathcal{W}_X$ is the sheaf of Witt groups and $\mathcal{GW}_X$ is the sheaf of Grothendieck–Witt rings.

For $X \in \mathrm{Sm}_k$ and $n \in \mathbb{Z}$, we have

$$\tilde{H}\mathbb{Z}^{2n,n}(X) \cong \widetilde{\mathrm{CH}}^n(X).$$

For F a field, we have isomorphisms

$$\tilde{H}\mathbb{Z}^{a,b}(\mathrm{Spec}\, F) \cong \begin{cases} K^{\mathrm{MW}}_n(F) & \text{for } a = b = n, \\ H\mathbb{Z}^{a,b}(\mathrm{Spec}\, F) & \text{for } a \neq b. \end{cases}$$

This follows from

Theorem 4.9 ([7, THEOREM 17]). *Let $\tilde{\mathcal{H}}\mathbb{Z}^{a,b}$, $\mathcal{H}\mathbb{Z}^{a,b}$ denote the respective homotopy sheaves $\pi_{-a,-b}(\tilde{H}\mathbb{Z})$, $\pi_{-a,-b}(H\mathbb{Z})$. Then for $a \neq b$, the map*

$$\Xi^{a,b} : \tilde{\mathcal{H}}\mathbb{Z}^{a,b} \to \mathcal{H}\mathbb{Z}^{a,b}$$

is an isomorphism. Moreover, we have canonical isomorphisms $\tilde{\mathcal{H}}\mathbb{Z}^{b,b} = \mathcal{K}_b^{\mathrm{MW}}$, $\mathcal{H}\mathbb{Z}^{b,b} = \mathcal{K}_b^M$, *and* $\Xi^{a,b} : \tilde{\mathcal{H}}\mathbb{Z}^{b,b} \to \mathcal{H}\mathbb{Z}^{b,b}$ *is canonical surjection* $\mathcal{K}_b^{\mathrm{MW}} \to \mathcal{K}_b^M$.

To prove Theorem 4.8, one applies this to the local–global spectral sequence

$$E_2^{p,q}(n) := H^p(X_{\mathrm{Nis}}, \tilde{\mathcal{H}}\mathbb{Z}^{q,n}) \Rightarrow \tilde{H}\mathbb{Z}^{p+q,n}(X),$$

noting that $\mathcal{H}\mathbb{Z}^{q,n} = 0$ for $n < 0$. This implies that the Gersten resolution of $\mathcal{H}\mathbb{Z}^{q,n}$ has length $\le n$ and thus $H^p(X_{\mathrm{Nis}}, \mathcal{H}\mathbb{Z}^{q,n}) = 0$ for $p > n$.

In general, one can approximate $\tilde{H}\mathbb{Z}^{a,b}(X)$ using the local–global sequence. Combined with Theorem 4.9 and the exact sheaf sequence

$$0 \to \mathcal{I}^{n+1} \to \mathcal{K}_n^{\mathrm{MW}} \to \mathcal{K}_n^M \to 0,$$

this tells us that the Milnor–Witt cohomology of X is built out of the usual motivic cohomology combined with information arising from quadratic forms.

4.4. Milnor–Witt motives

Rather than pulling the Milnor–Witt cohomology out of the motivic stable homotopy hat, there is another construction that is embedded in a Voevodsky-type triangulated category built out of a modified category of correspondences. We refer to [8] and [31] for details.

The Chow–Witt groups on a smooth X have been defined using the Rost–Schmid complex; one can also define Chow–Witt cycles with a fixed support using a modified version of the Rost–Schmid complex.

Definition 4.10. Let X be a smooth k-scheme, $\mathcal{L}$ an invertible sheaf on X, and $T \subset X$ a closed subset. The *nth $\mathcal{L}$-twisted Rost–Schmid complex* with supports in T, $\mathrm{RS}_T^*(X, n; \mathcal{L})$, is the subcomplex of $\mathrm{RS}^*(X, \mathcal{L}, n)$ with

$$\mathrm{RS}_T^m(X, \mathcal{L}, n) := \bigoplus_{x \in T \cap X^{(p)}} K_{n-m}^{\mathrm{MW}}\Big(k(x); \mathcal{L}_x \otimes_{\mathcal{O}_{X,x}} \bigwedge^m (\mathfrak{m}_x/\mathfrak{m}_x^2)^\vee\Big) \subset \mathrm{RS}^m(X, \mathcal{L}, n).$$

The usual arguments used to prove Gersten's conjecture yield the following result.

Lemma 4.11. *Let X be a smooth k-scheme, $\mathcal{L}$ an invertible sheaf on X, and $T \subset X$ a closed subset. The cohomology with support $H_T^p(X, \mathcal{K}_n^{\mathrm{MW}}(\mathcal{L})_X)$ is computed as*

$$H_T^p(X, \mathcal{K}_n^{\mathrm{MW}}(\mathcal{L})_X) = H^p(\mathrm{RS}_T^*(X, \mathcal{L}, n)).$$

Suppose T has pure codimension n on X. Then $X^{(m)} \cap T = \emptyset$ for $m < n$, $X^{(n)} \cap T$ is the finite set of generic points $T^{(0)}$ of T and $X^{(n+1)} \cap T = T^{(1)}$ is the set of codimension one points of T. This gives us the exact sequence

$$\begin{aligned} 0 \to H_T^n(X, \mathcal{K}_n^{\mathrm{MW}}(\mathcal{L})_X) &\to \bigoplus_{z \in T^{(0)}} \mathrm{GW}\big(k(z), \det{}^{-1}(\mathfrak{m}_z/\mathfrak{m}_z^2) \otimes \mathcal{L}\big) \\ &\to \bigoplus_{z \in T^{(1)}} W\big(k(z), \det{}^{-1}(\mathfrak{m}_z/\mathfrak{m}_z^2) \otimes \mathcal{L}\big) \end{aligned}$$

which allows us to think of $H^n_T(X, \mathcal{K}^{\mathrm{MW}}_n(\mathcal{L})_X)$ as the group of "Grothendieck–Witt cycles" supported on T, whose definition we hinted at in the beginning of this section. We write this as $\tilde{Z}^n_T(X, \mathcal{L}, n)$, with the warning that this is only defined for T a closed subset of a smooth X of pure codimension n.

Note that the fact that T has pure codimension n implies that there are no relations in $H^n_T(X, \mathcal{K}^{\mathrm{MW}}_n(\mathcal{L})_X)$ coming from $K^{\mathrm{MW}}_1(k(w))$ for w a codimension $n-1$ point of X. For similar reasons, the corresponding group for the Chow groups, $H^n_T(X, \mathcal{K}^M_{n,X})$, is just the subgroup $Z^n_T(X)$ of $Z^n(X)$ freely generated by the irreducible components of T, that is, the group of codimension n cycles on X with support contained in T.

For $T \subset T' \subset X$, two codimension-n closed subsets, we have the evident map $\tilde{Z}^n_T(X, \mathcal{L}, n) \to \tilde{Z}^n_{T'}(X, \mathcal{L}, n)$. The rank map $\mathrm{GW}(-) \to \mathbb{Z}$ gives the homomorphism $\tilde{Z}^n_T(X, \mathcal{L}, n) \to Z^n_T(X)$.

Definition 4.12. For X, Y in Sm_k, let $\mathcal{A}(X, Y)$ be the set of closed subsets $T \subset X \times Y$ such that each component of T is finite over X and maps surjectively onto an irreducible component of X. We make $\mathcal{A}(X, Y)$ a poset by the inclusion of closed subsets.

Note that if Y is irreducible of dimension n, then a closed subset $T \subset X \times Y$ is in $\mathcal{A}(X, Y)$ if and only if T is finite over X and has pure codimension n on $X \times Y$.

Definition 4.13 (Calmès–Fasel [31, §4.1]). Let X, Y be in Sm_k and suppose Y is irreducible of dimension n. Define

$$\widetilde{\mathrm{Corr}}_k(X, Y) = \mathrm{colim}_{T \in \mathcal{A}(X,Y)} \tilde{Z}^n_T(X \times Y, p_2^* \omega_{Y/k}).$$

Extend the definition to general Y by additivity.

Using the functorial properties of pullback, intersection product and proper push-forward for the Chow–Witt groups with support, we have a well-defined composition law

$$\widetilde{\mathrm{Corr}}_k(Y, Z) \times \widetilde{\mathrm{Corr}}_k(X, Y) \to \widetilde{\mathrm{Corr}}_k(X, Z)$$

via the same formula used to define the composition in Cor_k,

$$Z_2 \circ Z_1 := p_{XZ*}\big(p^*_{YZ}(Z_2) \cap p^*_{XY}(Z_1)\big).$$

The twisting by the relative dualizing sheaf in the definition of $\widetilde{\mathrm{Corr}}_k(-,-)$ is exactly what is needed for the push-forward map p_{XZ*} to be defined.

This defines the additive category $\widetilde{\mathrm{Corr}}_k$ with objects Sm_k and morphisms $\widetilde{\mathrm{Corr}}_k(X, Y)$. The rank map gives an additive functor

$$\mathrm{rnk} : \widetilde{\mathrm{Corr}}_k \to \mathrm{Cor}_k.$$

One then follows the program used by Voevodsky to define the abelian category of Nisnevich sheaves with Milnor–Witt transfers, $\mathrm{Sh}^{\mathrm{MWtr}}_{\mathrm{Nis}}(k)$, and then $\widetilde{\mathrm{DM}}^{\mathrm{eff}}(k) \subset D(\mathrm{Sh}^{\mathrm{MWtr}}_{\mathrm{Nis}}(k))$ as the full subcategory of complexes with strictly $\mathbb{A}^1$-homotopy invariant cohomology sheaves. One has the localization functor

$$\tilde{L}_{\mathbb{A}^1} : D\big(\mathrm{Sh}^{\mathrm{MWtr}}_{\mathrm{Nis}}(k)\big) \to \widetilde{\mathrm{DM}}^{\mathrm{eff}}(k)$$

constructed using the Suslin complex, the representable sheaves $\tilde{\mathbb{Z}}^{\mathrm{tr}}(X)$ for $X \in \mathrm{Sm}_k$, their corresponding motives $\tilde{M}^{\mathrm{eff}}(X) := \tilde{L}_{\mathbb{A}^1}(\tilde{\mathbb{Z}}^{\mathrm{tr}}(X)) \in \widetilde{\mathrm{DM}}^{\mathrm{eff}}(k)$ and the Tate motives $\tilde{\mathbb{Z}}(n)$ arising from the reduced motive of $\mathbb{P}^1$. Finally, one constructs $\widetilde{\mathrm{DM}}(k)$ as a category of $\tilde{\mathbb{Z}}(1)$-spectra in $\widetilde{\mathrm{DM}}^{\mathrm{eff}}(k)$ and we have the motive $\tilde{M}(X)$ defined as the suspension spectrum of $\tilde{M}^{\mathrm{eff}}(X)$.

Definition 4.14. For $X \in \mathrm{Sm}_k$, categorical Milnor–Witt cohomology is

$$H^p\big(X, \tilde{\mathbb{Z}}(q)\big) := \mathrm{Hom}_{\widetilde{\mathrm{DM}}(k)}\big(\tilde{M}(X), \tilde{\mathbb{Z}}(q)[p]\big).$$

Theorem 4.15. *There is a natural isomorphism*

$$H^p\big(X, \tilde{\mathbb{Z}}(q)\big) \cong H\tilde{\mathbb{Z}}^{p,q}(X).$$

The proof is very much the same as for motivic cohomology. One shows there is an equivalence of $\widetilde{\mathrm{DM}}(k)$ with the homotopy category of $H\tilde{\mathbb{Z}}$-modules (this is **[8, THEOREM 5.2]**). This gives an adjunction

$$H\tilde{\mathbb{Z}} \wedge - : \mathrm{SH}(k) \rightleftarrows \widetilde{\mathrm{DM}}(k) : \widetilde{\mathrm{EM}}$$

with $H\tilde{\mathbb{Z}} \wedge -$ the free $H\tilde{\mathbb{Z}}$ module functor and the Eilenberg–MacLane functor $\widetilde{\mathrm{EM}}$ the forgetful functor. This gives $\widetilde{\mathrm{EM}}(\tilde{\mathbb{Z}}(0)) = H\tilde{\mathbb{Z}}$, $\tilde{M}(X) = H\tilde{\mathbb{Z}} \wedge \Sigma^\infty_{\mathbb{P}^1} X_+$, and induces the isomorphism

$$\begin{aligned} H^p\big(X, \tilde{\mathbb{Z}}(q)\big) &= \mathrm{Hom}_{\widetilde{\mathrm{DM}}(k)}\big(\tilde{M}(X), \tilde{\mathbb{Z}}(q)[p]\big) \\ &\cong \mathrm{Hom}_{\mathrm{SH}(k)}(\Sigma^\infty_{\mathbb{P}^1} X_+, \Sigma^{p,q} H\tilde{\mathbb{Z}}) = H\tilde{\mathbb{Z}}^{p,q}(X). \end{aligned}$$

5. CHOW GROUPS AND MOTIVIC COHOMOLOGY WITH MODULUS

Up to now, all the version of motivic cohomology we have considered share the $\mathbb{A}^1$-homotopy invariance property, namely, that $H^*(X, \mathbb{Z}(*)) \cong H^*(X \times \mathbb{A}^1, \mathbb{Z}(*))$; essentially by construction, this property is enjoyed by all theories that are represented in the motivic stable homotopy category. Although this is a fundamental property controlling a large collection of cohomology theories, this places a serious restriction in at least two naturally occurring areas.

One is the use of deformation theory. This relies on having useful invariants defined on non-reduced schemes, but a cohomology theory that satisfies $\mathbb{A}^1$-invariance will not distinguish between a scheme and its reduced closed subscheme. The second occurs in ramification theory. An $\mathbb{A}^1$-homotopy invariant theory will not detect Artin–Schreyer covers, and would not give invariants that detect wild ramification.

Fortunately, we have an interesting cohomology theory that is not $\mathbb{A}^1$-homotopy invariant, namely, algebraic K-theory, that we can use as a model for a general theory. Algebraic K-theory does satisfy the $\mathbb{A}^1$-invariance property when restricted to regular schemes, but in general this fails. Besides allowing K-theory to have a role in deformation theory and ramification theory, this lack of $\mathbb{A}^1$-invariance gives rise to interesting invariants of singularities.

5.1. Higher Chow groups with modulus

The theory of Chow groups with modulus attempts to refine the classical theory of the Chow groups to be useful in both of these areas. This is still a theory in the process of development; just as in the early days of motivic cohomology, many approaches are inspired by properties of algebraic K-theory.

The tangent space to the functor $X \mapsto \mathcal{O}_X^\times$ is given by the structure sheaf, $X \mapsto \mathcal{O}_X$, via the isomorphism

$$\mathcal{O}^\times_{X[\varepsilon]/(\varepsilon^2)} \cong \mathcal{O}_X^\times \oplus \varepsilon \cdot \mathcal{O}_X.$$

Via the isomorphism $\operatorname{Pic}(X) \cong H^1(X, \mathcal{O}_X^\times)$, this shows that the tangent space at X to the functor $\operatorname{Pic}(-)$ is $H^1(X, \mathcal{O}_X)$.

In [23], Bloch computes the tangent space to K_2 (on local $\mathbb{Q}$-algebras), giving the isomorphism of sheaves on X_{Zar} (for X a $\mathbb{Q}$-scheme)

$$\mathcal{K}_{2,X[\varepsilon]/(\varepsilon^2)} \cong \mathcal{K}_{2,X} \oplus \Omega_X$$

where Ω_X is the sheaf of absolute Kähler differentials. Bloch then uses his formula from [22],

$$H^2(X_{\mathrm{Zar}}, \mathcal{K}_2) \cong \mathrm{CH}^2(X),$$

to justify defining $\mathrm{CH}^2(X[\varepsilon]/(\varepsilon^2))$ as $H^2(X[\varepsilon]/(\varepsilon^2)_{\mathrm{Zar}}, \mathcal{K}_2)$, giving

$$\mathrm{CH}^2\big(X[\varepsilon]/(\varepsilon^2)\big) = \mathrm{CH}^2(X) \oplus H^2(X, \Omega_X).$$

For X a smooth projective surface over $\mathbb{C}$ with $H^2(X, \mathcal{O}_X) \neq 0$, the exact sheaf sequence

$$0 \to \Omega_{\mathbb{C}/\mathbb{Q}} \otimes_{\mathbb{C}} \mathcal{O}_X \to \Omega_X \to \Omega_{X/\mathbb{C}} \to 0$$

along the fact that $\Omega_{\mathbb{C}/\mathbb{Q}}$ is a $\mathbb{C}$-vector space of uncountable dimension show that $\Omega_{\mathbb{C}/\mathbb{Q}} \otimes_{\mathbb{C}} H^2(X, \mathcal{O}_X)$ makes a huge contribution to the tangent space $H^2(X, \Omega_X)$ of $\mathrm{CH}^2(-)$ on X. This is reflected in Mumford's result [95], that if $H^2(X, \mathcal{O}_X) \cong H^0(X, \Omega^2_{X/\mathbb{C}})$ is nonzero, then $\mathrm{CH}^2(X)$ is "infinite-dimensional," and gives some evidence for Bloch's conjecture [23, CONJECTURE (0.4)] on 0-cycles on surfaces X with $H^0(X, \Omega^2_{X/\mathbb{C}}) = 0$.

The algebraic cycles have disappeared in this approach to Chow groups of non-reduced schemes. Bloch and Esnault [26] gave the first construction of a cycle-theoretic theory that could say something interesting about higher cycles on the non-reduced scheme $\operatorname{Spec} k[\varepsilon]/(\varepsilon^2)$. In a second paper [27], they modified and extended this construction to give a theory of *additive higher Chow groups with modulus* m, for the field k. This was motivated by Bloch's earlier use of K-theory on the affine line, relative to $\{0, 1\}$, to study K_3. Letting 1 tend to 0, they were led to consider the relative K-theory space $K(k[\varepsilon], (\varepsilon^2))$, this being the homotopy fiber of the restriction map $K(k[\varepsilon]) \to K(k[\varepsilon]/\varepsilon^2)$, whose homotopy groups are the relative K-theory groups $K_n(k[\varepsilon], (\varepsilon^2))$. Replacing 2 with $m \geq 2$ gives the relative K-theory groups $K_n(k[\varepsilon], (\varepsilon^m))$. This led to the consideration of a complex of cycles on $\operatorname{Spec} k[\varepsilon]$, with an additional condition imposed on the mth order limiting behavior of the cycles; an explicit construction of such a cycle complex with modulus, $z^q(k, *, m)$ was given in [27]. The homology $\mathrm{ACH}^q(k, p, m) := H_p(z^q(k, *, m))$ defines the *additive codimension* *q higher Chow groups with modulus m for* Spec k. Bloch–Esnault recover the computation $\mathrm{ACH}^n(k, n-1, 2) \cong \Omega_k^{n-1}$ from [26], and relate the additive analogue of weight two K_3, $\mathrm{ACH}^2(k, 2, 2)$, with the additive dilogarithm of Cathelineau [32].

Rülling [104] studied the projective system

$$\cdots \to \mathrm{ACH}^n(k, n-1, m+1) \to \mathrm{ACH}^n(k, n-1, m) \to \cdots .$$

He showed this is endowed with additional endomorphisms F_n and V_n, and the graded group $\bigoplus_n \mathrm{ACH}^n(k, n-1, *+1)_{*\geq 2}$ has the structure of a pro-differential graded algebra. In fact, we have

Theorem 5.1 (Rülling). *Let k be a field of characteristic $\neq 2$. The pro-dga $\bigoplus_n \mathrm{ACH}^n(k, n-1, *+1)$, with F_n as Frobenius and V_n as Verschiebung, is isomorphic to the de Rham–Witt complex of Madsen–Hesselholt,*

$$\bigoplus_n \mathrm{ACH}^n(k, n-1, *+1) \cong \bigoplus_n W_* \Omega_k^{n-1}.$$

With essentially the same definition as given by Bloch–Esnault, the additive cycle complex and additive Chow groups were extended to arbitrary k-schemes Y by Park [98], replacing $\mathbb{A}^1$ and divisor $m \cdot 0$ with the scheme $Y \times \mathbb{A}^1$ and divisor $m \cdot Y \times 0$. Binda and Saito [20] went one step further, defining complexes $z^q(X, D, *)$ for a pair (X, D) of a finite type separated k-scheme X and a Cartier divisor D, using essentially the same definition as before. The homology is the *higher Chow group with modulus*

$$\mathrm{CH}^q(X, D, p) := H_p\big(z^q(X, D, *)\big).$$

The constructions of Bloch–Esnault, Park, and Binda–Saito all use a cubical model of Bloch's cycle complex. Here one replaces the algebraic n-simplex, $\Delta_k^n = \mathrm{Spec}\, k[t_0, \ldots, t_n]/\sum_i t_i - 1$, with the algebraic n-cube

$$\square^n := \big(\mathbb{P}^1 \setminus \{1\}, 0, \infty\big)^n.$$

The notation means that one considers $(\mathbb{P}^1 \setminus \{1\})^n \cong \mathbb{A}^n$ with its "faces" defined by setting some of the coordinates equal to 0 or ∞. The corresponding cycle complex $z^q(X, *)_c$ has degree n component $z^q(X, n)_c$ the codimension q cycles on $X \times \square^n$ that intersect $X \times F$ properly for all faces F of $\square^n$; one also needs to quotient out by the degenerate cycles, these being the ones that come by pullback via projection to a $\square^m$ with $m < n$. The differential is again an alternating sum of restrictions to the maximal faces $t_i = 0$ and $t_i = \infty$.

This complex also computes the motivic cohomology of X, just as Bloch's simplicial cycle complex does. In the Binda–Saito construction, the modulus condition arises by considering the closed box $\overline{\square}^n := (\mathbb{P}^1)^n$. Let $F_n^i \subset (\mathbb{P}^1)^n$ be the divisor defined by $t_i = 1$ and let $F_n = \sum_{i=1}^n F_n^i$. In $(\mathbb{P}^1)^n \times X$ we have two distinguished Cartier divisors, $(\mathbb{P}^1)^n \times D$ and $F_n \times X$. A subvariety $Z \subset (\mathbb{P}^1 \setminus \{1\})^n \times X$ that is in $z^q(X, n)_c$ satisfies the modulus condition if

$$p^*(F_n \times X) \geq p^*\big((\mathbb{P}^1)^n \times D\big)$$

where $p : \bar{Z}^N \to (\mathbb{P}^1)^n \times X$ is the normalization of the closure of Z in $\overline{\square}^n \times X$. Restricting to the subgroup of $Z^q(\square^n \times X)$ generated by codimension q subvarieties $Z \subset \square^n \times X$ that intersect faces properly and satisfy the modulus condition yields the cycle complex with modulus $z^q(X; D, *) \subset z^q(X, *)_c$; the higher Chow groups with modulus is then defined as

$$\mathrm{CH}^q(X; D, p) := H_p\big(z^q(X; D, *)\big).$$

The second construction of Bloch–Esnault, and Park's generalization, are recovered as the special cases $X = \mathbb{A}^1_k$ and $D = m \cdot 0$ in the Bloch–Esnault version and $X = Y \times \mathbb{A}^1$, $D = m \cdot Y \times 0$ in Park's version.

For X a finite type k-scheme, recall the Bloch motivic complex $\mathbb{Z}_{\mathrm{Bl}}(q)^*_X$ defined as the Zariski sheafification of the presheaf $U \mapsto z^q(X, 2q - *)$ (this is already a Nisnevich sheaf). Bloch's cycle complexes satisfy an important localization property: the natural maps to Zariski and Nisnevich hypercohomology

$$H^p\big(z^q(X, 2q - *)\big) \to \mathbb{H}^p\big(X_{\mathrm{Zar}}, \mathbb{Z}_{\mathrm{Bl}}(q)^*_X\big) \to \mathbb{H}^p\big(X_{\mathrm{Nis}}, \mathbb{Z}_{\mathrm{Bl}}(q)^*_X\big)$$

are isomorphisms. This fails for the cycle complex with modulus, although the comparison between the Zariski and Nisnevich hypercohomology seems to be still an open question.

Iwasa and Kai consider the Nisnevich sheafification $\mathcal{Z}(q)^*_{(X;D)}$ of the presheaf

$$U \mapsto z^q(U; D \times_X U, 2q - *).$$

We call $\mathbb{H}^p(X_{\mathrm{Nis}}, \mathcal{Z}(q)^*_{(X;D)})$ the *motivic cohomology with modulus* for (X, D). Kai **[74]** shows that this sheafified version has contravariant functoriality. Iwasa and Kai **[67]** construct Chern class maps from relative K-theory

$$c_{p,q} : K_{2q-p}(X; D) \to \mathbb{H}^p\big(X_{\mathrm{Nis}}, \mathcal{Z}(q)^*_{X,\mathrm{Nis}}\big).$$

5.2. 0-cycles with modulus and class field theory

There is a classical theory of 0-cycles on a smooth complete curve C with a modulus condition at a finite set of points S, due to Rosenlicht and Serre **[109, 111]**. The idea is quite simple, instead of relations coming from divisors (zeros minus poles) of an arbitrary rational function f, f is required to have a power series expansion at each point $p \in S$, with leading term 1 and the next nonzero term of the form $ut_p^{n_p}$, with $u(p) \neq 0$, t_p a local coordinate at p and the integer $n_p > 0$ being the "modulus." This is applied to the class field theory of a smooth open curve $U \subset C$ over a finite field **[109, THEOREM 4]**, that identifies the inverse limit of the groups of degree 0 cycle classes on U, with modulus supported in $C \setminus U$, with the kernel of the map $\pi_1^{\text{ét}}(U)^{ab} \to \mathrm{Gal}(\bar{k}/k)$.

In their class field theory for higher-dimensional varieties, Kato and Saito **[80]** introduce a group of 0-cycles on a k-scheme X with modulus D, defined by

$$\mathrm{CH}_0(X, D) := H^n(X, \mathcal{K}^M_{n,(X,D)})$$

with $\mathcal{K}^M_{n,(X,D)}$ a relative version of the Milnor K-theory sheaf, recalling Kato's isomorphism $H^n(X, \mathcal{K}^M_n) \cong \mathrm{CH}^n(X)$ for X a smooth k-scheme **[78]**. Kerz and Saito give a different definition of a group of relative 0-cycles $C(X, D)$ on a normal k-scheme X with effective Cartier divisor D such that $X \setminus D$ is smooth. It follows from their comments in [81, DEFINITION 1.6] that $C(X, D) = \mathrm{CH}^n(X; D, 0)$ for X of dimension n, and it is easy to see that the Kato–Saito and Kerz–Saito relative 0-cycles agree with the Rosenlicht–Serre groups in the case of curves.

Kerz and Saito consider a smooth finite-type k-scheme U, choose a normal compactification X and define the topological group $C(U) := \lim_D C(X, D)$, as D runs over effective Cartier divisors on X, supported in $X \setminus U$, and with each $C(X, D)$ given the discrete topology. They show that $C(U)$ is independent of the choice of X, and their main result generalizes class field theory for smooth curves over a finite field as described above.

Theorem 5.2 ([81, THEOREM 3.3]). *Let k be a finite field of characteristic $\neq 2$ and let U be a smooth variety over k, Then $C(U)$ is isomorphic as topological group to a dense subgroup of the abelianized étale fundamental group $\pi_1^{ét}(U)^{ab}$ and this isomorphism induces an isomorphism of the degree 0 part $C(U)^0$ of $C(U)$ with the kernel $\pi_1^{ét}(U)_0^{ab}$ of $\pi_1^{ét}(U)^{ab} \to \pi_1^{ét}(k)$.*

5.3. Categories of motives with modulus

There has been a great deal of interest in constructing a categorical framework for motivic cohomology with modulus. A central issue is the lack of $\mathbb{A}^1$-homotopy invariance for this theory, which raises the question of what type of homotopy invariance should replace this.

One direction has been the construction of a reasonable replacement for the category of homotopy invariant Nisnevich sheaves with transfers. A non-homotopy invariant version has been developed via the theory of *reciprocity sheaves*, the name coming from the reciprocity laws in class field theory of curves and its relation to the group of 0-cycles with modulus of Rosenlicht–Serre. We will say a bit about reciprocity sheaves later on, in the context of motives for log schemes Section 5.4.

For now, we will look at categories of motives with modulus constructed on the Voevodsky model by introducing a new notion of correspondence and a suitable replacement for $\mathbb{A}^1$-homotopy invariance.

Looking at algebraic K-theory, the closest replacement for $\mathbb{A}^1$-homotopy invariance seems to be the $\mathbb{P}^1$-bundle formula

$$K_n(X \times \mathbb{P}^1) \cong K(X) \cdot [\mathcal{O}_{X \times \mathbb{P}^1}] \oplus K(X) \cdot \big[\mathcal{O}_{X \times \mathbb{P}^1}(-1)\big],$$

valid for a general scheme X. This has led to attempts to create a category of motives with modulus based on a notion of "$\overline{\square}$-invariance."

Here one has the problem that $\mathbb{P}^1$ does not have the structure of an interval, a structure enjoyed by $\mathbb{A}^1$. One considers $\mathbb{A}^1$ together with "endpoints" 0, 1. Following the general theory of a site with interval, as developed by Morel–Voevodsky [94, CHAP. 2], one needs the multiplication map $m : \mathbb{A}^1 \times \mathbb{A}^1 \to \mathbb{A}^1$ to allow one to consider $(\mathbb{A}^1, 0, 1)$ as an abstract interval. In the construction of the cycle complex with modulus, one identifies $(\mathbb{A}^1, 0, 1)$ with $(\mathbb{P}^1 \setminus \{1\}, 0, \infty)$, and the corresponding multiplication map $m' : (\mathbb{P}^1 \setminus \{1\}) \times (\mathbb{P}^1 \setminus \{1\}) \to \mathbb{P}^1 \setminus \{1\}$ only extends as a rational map $\mathbb{P}^1 \times \mathbb{P}^1 \dashrightarrow \mathbb{P}^1$. However, m' becomes a morphism after blowing up the point $(1, 1)$, which suggests that one should consider the closure of the graph of m' in $\mathbb{P}^1 \times \mathbb{P}^1 \times \mathbb{P}^1$ as an allowable correspondence from $\mathbb{P}^1 \times \mathbb{P}^1$ to $\mathbb{P}^1$.

With this as starting point, Kahn, Miyazaki, Saito, and Yamazaki **[69–71]** follow Voevodsky's program, defining a category of modulus correspondences $\underline{\mathrm{M}}\mathrm{Cor}_k$. Objects are pairs $(\bar{M}, M^\infty)$ with $\bar{M}$ a separated finite-type k-scheme and M^∞ an effective Cartier divisor on $\bar{M}$ such that the open complement $M^\circ := \bar{M} \setminus M^\infty$ is smooth. The morphism group $\underline{\mathrm{M}}\mathrm{Cor}_k((\bar{M}, M^\infty), (\bar{N}, N^\infty))$ is the subgroup of $\mathrm{Cor}_k(M^\circ, N^\circ)$ generated by subvarieties Z (finite and surjective over a component of M°) such that

(i) The closure $\bar{Z}$ of Z in $\bar{M} \times \bar{N}$ is proper over $\bar{M}$ (not necessarily finite).

(ii) Let $f : \bar{Z}^N \to \bar{M} \times \bar{N}$ be the normalization of $\bar{Z}$. Then

$$f^* p_1^* M^\infty \geq f^* p_2^* N^\infty.$$

The composition law in Cor_k preserves conditions (i) and (ii), giving the category $\underline{\mathrm{M}}\mathrm{Cor}_k$ with functor $\underline{\mathrm{M}}\mathrm{Cor}_k \to \mathrm{Cor}_k$ sending $(\bar{M}, M^\infty)$ to M° and with $\underline{\mathrm{M}}\mathrm{Cor}_k((\bar{M}, M^\infty), (\bar{N}, N^\infty)) \to \mathrm{Cor}_k(M^\circ, N^\circ)$ the inclusion. The product of pairs makes $\underline{\mathrm{M}}\mathrm{Cor}_k$ a symmetric monoidal category and the functor to Cor_k is symmetric monoidal.

Let $\overline{\square}$ be the object $(\mathbb{P}^1, \{1\})$. As hinted above, the closure of the graph of $m' : (\mathbb{P}^1 \setminus \{1\}) \times (\mathbb{P}^1 \setminus \{1\}) \to \mathbb{P}^1 \setminus \{1\}$ defines a morphism $m : \overline{\square} \times \overline{\square} \to \overline{\square}$ in $\underline{\mathrm{M}}\mathrm{Cor}_k$.

They then consider the abelian category of additive presheaves of abelian groups on $\underline{\mathrm{M}}\mathrm{Cor}_k$, $\underline{\mathrm{M}}\mathrm{PST}_k := \mathrm{PreSh}^{\mathbf{Ab}}(\underline{\mathrm{M}}\mathrm{Cor}_k)$. There is also a version MCor_k of *proper* modulus pairs (X, D), with X a proper k-scheme, as a full subcategory of $\underline{\mathrm{M}}\mathrm{Cor}_k$, with its presheaf category MPST_k.

They define a category of *effective proper motives with modulus*, $\mathrm{MDM}^{\mathrm{eff}}(k)$, by localizing the derived category $D(\mathrm{MPST}_k)$. Roughly speaking, they follow the Voevodsky program, replacing the $\mathbb{A}^1$-localization with $\overline{\square}$ localization. To get the proper Nisnevich localization is a bit technical; we refer the reader to **[71, DEFINITION 1.3.9]** for details.

There is still quite a bit that is not known. One central problem is how to realize the various constructions of the higher Chow groups with modulus as morphisms in a suitable triangulated category. There is a connection, at least for the modulus version of Suslin homology and the Suslin complex, which we now describe.

One can show that the cubical version of the Suslin complex

$$C_*^{\mathrm{Sus}}(X)_c(Y) := \mathrm{Hom}_{\mathrm{Cor}_k}(Y \times \square^*, X)/\,\mathrm{degn}$$

is naturally quasi-isomorphic to the simplicial version $C_*^{\mathrm{Sus}}(X)(Y)$, where $/\,\mathrm{degn}$ means taking the quotient by the image of the pullback maps via the projections $Y \times \square^n \to Y \times \square^{n-1}$. For a modulus pair (X, D), one can similarly form the *naive Suslin complex*

$$C_*^{\mathrm{Sus}}(X, D)(Y, E) := \mathrm{Hom}_{\underline{\mathrm{M}}\mathrm{Cor}_k}\big((Y, E) \otimes \overline{\square}^*, X\big)/\,\mathrm{degn}.$$

Taking $(Y, E) = (\mathrm{Spec}\, k, \emptyset)$, we have the complex

$$C_*^{\mathrm{Sus}}(X, D) := C_*^{\mathrm{Sus}}(X, D)(\mathrm{Spec}\, k, \emptyset).$$

Next, there is a *derived* Suslin complex $RC_*^{\mathrm{Sus}}(X,D)_c(-)$ with a natural map of presheaves

$$C_*^{\mathrm{Sus}}(X,D)_c(-) \to RC_*^{\mathrm{Sus}}(X,D)_c(-).$$

By [71, THEOREM 2], for (X,D) a proper modulus pair, $RC_*^{\mathrm{Sus}}(X,D)_c(-)$ computes the maps in $\mathrm{MDM}^{\mathrm{eff}}(k)$ as

$$H_n\big(RC_*^{\mathrm{Sus}}(X,D)_c(\operatorname{Spec} k,\emptyset)\big) = \operatorname{Hom}_{\mathrm{MDM}^{\mathrm{eff}}(k)}\big((\operatorname{Spec} k,\emptyset), M^{\mathrm{eff}}(X,D)\big).$$

However, one should not expect that the Suslin complex or its derived version should yield a version of the higher Chow groups. If one looks back at the setting of $\mathrm{DM}(k)$, the object that most naturally yields the higher Chow groups for an arbitrary finite type k-scheme X is the motive with compact supports $M^c(X)$. This is defined as $C_*^{\mathrm{Sus}}(\mathbb{Z}_{\mathrm{tr}}^c(X))$, where $\mathbb{Z}_{\mathrm{tr}}^c(X)$ is the presheaf with transfers with $\mathbb{Z}_{\mathrm{tr}}^c(X)(Y)$ the free abelian group on integral $W \subset Y \times X$, with $W \to Y$ *quasi-finite* and dominant over a component of $Y \in \mathrm{Sm}_k$. See [127, CHAP. 5, PROPOSITION 4.2.9] for the relation of $M^c(X)$ with Bloch's higher Chow groups.

One can define a similar version with modulus as the object $M^c(X,D)$ associated to the presheaf $\mathbb{Z}_{\mathrm{tr}}^c(X,D)$, with $\mathbb{Z}_{\mathrm{tr}}^c(X,D)(Y,E) \subset Z_{\dim Y}(Y \times X)$ the subgroup generated by closed subvarieties $W \subset (Y \setminus E) \times (X \setminus D)$ that are quasi-finite and dominant over Y, and with the usual modulus condition, that the normalization $\nu : \bar{W}^N \to Y \times X$ of the closure of W in $Y \times X$ satisfies

$$\nu^*(E \times X) \geq \nu^*(Y \times D).$$

There is an analog of Suslin's comparison theorem in the affine case, due to Kai–Miyazaki [75]: They define an equi-dimensional cycle complex with modulus

$$z_d^{\mathrm{equi}}(X,D,*) \subset z_d(X,D,*)$$

which for $d=0$ is the Suslin complex with modulus $C_*^{\mathrm{Sus}}(\mathbb{Z}_{\mathrm{tr}}^c(X,D))(\operatorname{Spec} k,\emptyset)$

Theorem 5.3 (Kai–Miyazaki). *Let (X,D) be a modulus pair, with X affine. Then there is a pro-isomorphism*

$$\big\{H_*\big(z_d^{\mathrm{equi}}(X,mD,*)\big)\big\}_m \cong \big\{\mathrm{CH}_d(X,mD,*)\big\}_m.$$

Miyazaki [91] has defined objects $z^{\mathrm{equi}}(X,D,d) \in \underline{\mathrm{M}}\mathrm{NST}_k$, with $\mathbb{Z}_{\mathrm{tr}}^c(X,D) = z^{\mathrm{equi}}(X,D,0)$. The sheaf $z^{\mathrm{equi}}(X,D,r)$ is defined similarly to $\mathbb{Z}_{\mathrm{tr}}^c(X,D)$, with $z^{\mathrm{equi}}(X,D,d)(Y,E)$ the group of cycles on $(Y \setminus E) \times (X \setminus D)$ generated by closed, integral $W \subset (Y \setminus E) \times (X \setminus D)$ that are equi-dimensional of dimension d over $Y \setminus E$, dominate a component of $Y \setminus E$, and with $\nu : \bar{W}^N \to Y \times X$ satisfying the modulus condition

$$\nu^*(E \times X) \geq \nu^*(Y \times D).$$

Moreover, for an arbitrary modulus pair (X,D), one has

$$z_d^{\mathrm{equi}}(X,D,*) = C_*^{\mathrm{Sus}}\big(z^{\mathrm{equi}}(X,D,d)\big)(\operatorname{Spec} k,\emptyset),$$

and there is the canonical map

$$C_*^{\mathrm{Sus}}\big(z^{\mathrm{equi}}(X,D,d)\big) \to RC_*^{\mathrm{Sus}}\big(z^{\mathrm{equi}}(X,D,d)\big).$$

Letting $\mathrm{CH}_q^{\mathrm{equi}}(X,D,p) = H_p(z_q^{\mathrm{equi}}(X,D,*))$, we have the natural map

$$\mathrm{CH}_q^{\mathrm{equi}}(X,D,p) \to \mathrm{CH}_q(X,D,p)$$

which is an isomorphism for X affine, and we have the natural maps for (X,D) a proper modulus pair

$$\begin{aligned}\mathrm{CH}_q^{\mathrm{equi}}(X,D,p) &\to H_p\big(RC_*^{\mathrm{Sus}}\big(z^{\mathrm{equi}}(X,D,q)\big)(\mathrm{Spec}\,k,\emptyset)\big)\\ &\to \mathrm{Hom}_{\mathrm{MDM}^{\mathrm{eff}}(k)}\big(M(\mathrm{Spec}\,k,\emptyset)[p], RC_*^{\mathrm{Sus}}\big(z^{\mathrm{equi}}(X,D,q)\big)\big).\end{aligned}$$

For a proper modulus pair, let $M^c(X,D)$ denote the image of $\mathbb{Z}_{\mathrm{tr}}^c(X,D)$ in $\mathrm{MDM}^{\mathrm{eff}}(k)$. One can ask if there are analogs of the theorem of Kahn–Miyazaki–Saito–Yamazaki.

Question 5.4. *For (X,D) a proper modulus pair, are the maps*

$$H_p\big(RC_*^{\mathrm{Sus}}(\mathbb{Z}_{tr}^c(X,D))(\mathrm{Spec}\,k,\emptyset)\big) \to \mathrm{Hom}_{\mathrm{MDM}^{\mathrm{eff}}(k)}\big(M(\mathrm{Spec}\,k,\emptyset)[p], M^c(X,D)\big)$$

isomorphisms? More generally, are the maps

$$\begin{aligned}&H_p\big(RC_*^{\mathrm{Sus}}\big(z^{\mathrm{equi}}(X,D,q)\big)(\mathrm{Spec}\,k,\emptyset)\big)\\ &\quad\to \mathrm{Hom}_{\mathrm{MDM}^{\mathrm{eff}}(k)}\big(M(\mathrm{Spec}\,k,\emptyset)[p], RC_*^{\mathrm{Sus}}\big(z^{\mathrm{equi}}(X,D,q)\big)\big)\end{aligned}$$

isomorphisms?

It is also not clear if the map

$$\mathrm{CH}_q^{\mathrm{equi}}(X,D,p) \to H_p\big(RC_*^{\mathrm{Sus}}\big(z^{\mathrm{equi}}(X,D,q)\big)(\mathrm{Spec}\,k,\emptyset)\big)$$

should be an isomorphism. Possibly one should also consider the Nisnevich hypercohomology $\mathbb{H}^{-p}(X_{\mathrm{Nis}}, \mathcal{Z}_q^{\mathrm{equi}}(X,D,*))$, with $\mathcal{Z}_q^{\mathrm{equi}}(X,D,*)$ defined by sheafifying $U \mapsto z_q^{\mathrm{equi}}(U, U\cap D,*)$.

For Voevodsky motives, and for X a finite type k-scheme, the motivic Borel-Moore homology is defined by

$$\begin{aligned}H_p^{\mathrm{B.M.}}\big(X,\mathbb{Z}(q)\big) &:= \mathrm{Hom}_{\mathrm{DM}^{\mathrm{eff}}(k)}\big(\mathbb{Z}(q)[p], M(X)^c\big)\\ &\cong H_{p-2q}\big(z_q^{\mathrm{equi}}(X,*)\big) \cong H_{p-2q}\big(z_q(X,*)\big) = \mathrm{CH}_q(X,p-2q).\end{aligned}$$

This uses the duality $M(X)^c \cong M(X)^\vee(d)[2d]$ for X of dimension d (valid in characteristic zero, or after inverting p in characteristic $p>0$), and the extension of Suslin's quasi-isomorphism $z_q^{\mathrm{equi}}(X,*) \hookrightarrow z_q(X,*)$ to arbitrary X. Moreover, we have $M(X)^c = M(X)$ for X smooth and proper.

However, a corresponding motivic *cohomology* of modulus pairs seems to need a larger category. This is hinted at by the use of the duality (in $\mathrm{DM}(k)$) $M(X)^c \cong M(X)^\vee(d)[2d]$ in the computations described above. This says in particular that each motive $M(X)$ admits a "twisted" dual in $\mathrm{DM}^{\mathrm{eff}}(k)$, more precisely, the usual evaluation and coevaluation maps associated with a dual exist, but as maps with target or source some $\mathbb{Z}(d)[2d]$ rather than the unit. For a general proper modulus pair (X, D), this does not seem to be the case; one seems to need modulus pairs with an anti-effective Cartier divisor. Another way to say the same thing, if one looks for a proper modulus pair (X, D') such that $\mathrm{Hom}_{\mathrm{MDM}^{\mathrm{eff}}(k)}(M(X, D'), \mathbb{Z}(q)[p])$ looks at all like $\mathrm{CH}^q(X, D, 2q-p)$ for some given proper modulus pair (X, D), the defining inequalities in Corr_k suggest that D' could be $-D$. See the section "Perspectives" in [71, INTRODUCTION] for further details in this direction.

5.4. Logarithmic motives and reciprocity sheaves

Grothendieck motives for log schemes have been constructed in [66], where a version for mixed motives has also been constructed using systems of realizations. There the emphasis is on versions of motives for homological or numerical equivalence in the setting of log schemes. In this section we discuss a recent construction of a triangulated category of log motives, by Binda–Park–Østvær [19], that follows the Voevodsky program. We refer the reader to the lectures notes of Ogus [97] for the facts about log schemes.

Recall that a log scheme is a pair $(X, \alpha : \mathcal{M} \to (\mathcal{O}_X, \times))$ consisting of a scheme X and a homomorphism of sheaves of commutative monoids $\alpha : \mathcal{M} \to (\mathcal{O}_X, \times)$ such that $\alpha^{-1}(\mathcal{O}_X^\times) \to \mathcal{O}_X^\times$ is an isomorphism; without this last condition, the pair $(X, \alpha : \mathcal{M} \to (\mathcal{O}_X, \times))$ is called a *pre-log structure*. A pre-log structure $\alpha : \mathcal{M} \to (\mathcal{O}_X, \times)$ induces a log structure $\alpha^{\log} : \mathcal{M}^{\log} \to (\mathcal{O}_X, \times)$ by taking $\mathcal{M}^{\log}$ to be the push-out (in the category of sheaves of monoids) in

$$\begin{array}{ccc} \alpha^{-1}(\mathcal{O}_X^\times) & \longrightarrow & \mathcal{M} \\ \big\downarrow & & \\ \mathcal{O}_X^\times & & \end{array}$$

Given a modulus pair (X, D), there are a number of (in general distinct) induced log structures on X. For example, one can take the *compactifying* log structure, with $\mathcal{M} := \mathcal{O}_X \cap j_*\mathcal{O}_U^\times$, where $U = X \setminus D$ and $j : U \to X$ is the inclusion. There are other log structures, which in general depend on a choice of decomposition of D as a sum of effective Cartier divisors (for example, the Deligne–Faltings log structure, discussed in [97, III, DEFINITION 1.7.1]).

Replacing the category of smooth k-schemes is the category lSm_k of *fine, saturated, log smooth* and separated log schemes over the log scheme $\mathrm{Spec}\, k$ endowed with the trivial log structure. We refer the reader to [19] for details; one needs these technical conditions to construct the category of finite log correspondences. We call a separated, fine, saturated log scheme an fs log scheme.

We sketch the construction of the category of finite log correspondences, and describe how Binda–Park–Østvær follow Voevodsky's program to define the triangulated category $\mathrm{logDM}^{\mathrm{eff}}(k)$ of effective log motives over k.

For $X \in \mathrm{lSm}_k$, let $\underline{X}$ denote the underlying k-scheme. We let $X^\circ \subset \underline{X}$ denote the maximal open subscheme over which the log structure $\mathcal{M}_X \to \mathcal{O}_X$ is trivial, that is, $\mathcal{M}_{X|U} = \mathcal{O}_U^\times$, and let $\partial X = \underline{X} \setminus X^\circ$.

Definition 5.5. 1. For $X, Y \in \mathrm{lSm}_k$, the group $\mathrm{lCor}_k(X, Y)$ consisting of *finite log correspondences* from X to Y is the free abelian group on integral closed subschemes $\underline{Z} \subset \underline{X} \times \underline{Y}$ such that

(i) $\underline{Z} \to \underline{X}$ is finite and is surjective to a component of $\underline{X}$.

(ii) Let Z^N be the log scheme with underlying scheme the normalization $\nu : \underline{Z}^N \to X \times Y$ of $\underline{Z}$ and log structure $(\nu \circ p_1)^*_{\log}\mathcal{M}_X \to \mathcal{O}_{Z^N}$. Here $\mathcal{M}_X \to \mathcal{O}_X$ is the given log structure on X and $(\nu \circ p_1)^*_{\log}\mathcal{M}_X \to \mathcal{O}_Z$ is the log structure induced by the pre-log structure $(\nu \circ p_1)^{-1}\mathcal{M}_X \to (\nu \circ p_1)^{-1}\mathcal{O}_X \to \mathcal{O}_Z$. Then the map of schemes $p_2 \circ \nu : \underline{Z}^N \to \underline{Y}$ extends to a map of log schemes $Z^N \to Y$.

Remark 5.6. It follows from (i) and (ii) above that, for $\underline{Z} \in \mathrm{lCor}_k(X, Y)$, the restriction of $\underline{Z}$ to a cycle on the open subset $X^\circ \times Y^\circ$ of $\underline{X} \times \underline{Y}$ actually lands in $\mathrm{Cor}_k(X^\circ, Y^\circ)$. Moreover, by [19, LEMMA 2.3.1], if the extension in (ii) exists, it is unique, so there is no need to include this as part of the data. In particular, the restriction map $\mathrm{lCor}_k(X, Y) \to \mathrm{Cor}_k(X^\circ, Y^\circ)$ is injective ([19, LEMMA 2.3.2]).

The condition that there exists a map of log schemes $(Z^N, (p_1 \circ \nu)^*_{\log}\mathcal{M}_X) \to (Y, \mathcal{M}_Y)$ extending $p_2 \circ \nu : \underline{Z}^N \to \underline{Y}$ is analogous to the modulus condition

$$\nu^*(D \times Y) \geq \nu^*(X \times E)$$

for a subvariety $W \subset X \setminus D \times Y \setminus E$ to define a finite correspondence of modulus pairs from (X, D) to (Y, E).

For the composition law, the proof of [19, LEMMA 2.3.3] shows that, given elementary log correspondences $\underline{W} \in \mathrm{lCor}_k(X, Y)$, and $\underline{W}' \in \mathrm{lCor}_k(Y, Z)$, each integral component $\underline{R}$ of $\underline{W} \times \underline{Z} \cap \underline{X} \times \underline{W}'$ is the underlying scheme of a (unique!) elementary log correspondence $\underline{R} \in \mathrm{lCor}_k(X, Z)$. It is then easy to show that there is a unique composition law

$$\circ : \mathrm{lCor}_k(Y, Z) \times \mathrm{lCor}_k(X, Y) \to \mathrm{lCor}_k(X, Z)$$

that is compatible with the composition law in Cor_k via the respective restriction maps.

This defines the additive category of *finite log correspondences* lCor_k with the same objects as for lSm_k, giving the category of *presheaves with log transfers*, lPST_k, defined as the category of additive presheaves of abelian groups on lCor_k. For a log scheme $X \in \mathrm{lSm}_k$, let $\mathbb{Z}_{\mathrm{ltr}}(X)$ denote the representable presheaf

$$\mathbb{Z}_{\mathrm{ltr}}(X)(Y) := \mathrm{lCor}_k(Y, X).$$

The fiber product of log schemes induces a tensor product structure on lPST_k.

The next step is to define the log version of the Nisnevich topology.

A morphism of log schemes $f : (X, \mathcal{M}_X \to \mathcal{O}_X), (Y, \mathcal{M}_Y \to \mathcal{O}_Y)$ is *strict* if the map of log structures $f^*\mathcal{M}_Y \to \mathcal{M}_X$ is an isomorphism. An *elementary log Nisnevich square* is a cartesian square in the category of fs log schemes

$$\begin{array}{ccc} V & \longrightarrow & Y \\ \downarrow{\scriptstyle f'} & & \downarrow{\scriptstyle f} \\ U & \xrightarrow{g} & X \end{array} \tag{5.1}$$

where f is strict étale, g is an open immersion, and f induces an isomorphism on reduced schemes $\underline{Y} \setminus \underline{V} \to \underline{X} \setminus \underline{U}$.

A log modification is a generalization of the notion of a log blow-up, which in turn is a morphism of log schemes modeled on the birational morphism of toric varieties given by a subdivision of the fan defining the target. We refer the reader to [19, APPENDIX A] for details. The Grothendieck topology generated by the log modifications and strict Nisnevich elementary squares is called the *dividing Nisnevich topology* on fs log schemes. In a sense, this is a log version of the cdh topology, where all the modifications are taking place in the boundary.

With this topology in hand, we have the subcategory lNST_k of lPST_k of Nisnevich sheaves with log transfers, just as for $\mathrm{NST}_k \subset \mathrm{PST}_k$, by requiring that a presheaf with log transfers be a sheaf for the dividing Nisnevich topology when restricted to lSm_k.

Finally, we need a suitable interval object to define a good notion of homotopy invariance. This is just as for the category $\mathrm{MDM}^{\mathrm{eff}}(k)$, where we consider $\overline{\square}$ as the scheme $\mathbb{P}^1$ with compactifying log structure for $(\mathbb{P}^1, \{1\})$. The product log scheme $\overline{\square}^2$ also has the compactifying log structure for the divisor $1 \times \mathbb{P}^1 + \mathbb{P}^1 \times 1$. However, the closure $\bar{\Gamma}_m$ of the graph of the multiplication map $m : \overline{\square}^2 \to \overline{\square}$ is not a morphism $\tilde{m}$ in lCor_k, as the requirement that the map of $\bar{\Gamma}_m$ to $\overline{\square}^2$ be finite is not satisfied.

Another way to look at this is to note that the projection $\bar{\Gamma}_m \to \overline{\square}^2$ is a cover of $\overline{\square}^2$ in the dividing Nisnevich topology, and becomes an isomorphism after dNis-localization. In a sense, this allows one to consider the sheaf $a_{\mathrm{dNis}}\overline{\square}$ as a version of a cylinder object and allows many of the constructions of Morel–Voevodsky for a site with interval to go through, although there are occasional technical difficulties that arise.

Definition 5.7. The tensor triangulated category of *effective log motives over* k, $\mathrm{logDM}^{\mathrm{eff}}(k)$, is the Verdier localization of the derived category $D(\mathrm{lPST}_k)$ with respect to the localizing subcategory generated by:

(lMV) for an elementary log Nisnevich square

$$\begin{array}{ccc} V & \longrightarrow & Y \\ \downarrow{\scriptstyle f'} & & \downarrow{\scriptstyle f} \\ U & \xrightarrow{g} & X \end{array}$$

we have the complex

$$\mathbb{Z}_{\mathrm{ltr}}(V) \to \mathbb{Z}_{\mathrm{ltr}}(U) \oplus \mathbb{Z}_{\mathrm{ltr}}(Y) \to \mathbb{Z}_{\mathrm{ltr}}(X).$$

(lM) For a log modification $f : Y \to X$ in lSm_k, we have the complex

$$\mathbb{Z}_{\mathrm{ltr}}(Y) \to \mathbb{Z}_{\mathrm{ltr}}(X).$$

(lCI) For $X \in \mathrm{lSm}_k$, we have the complex

$$\mathbb{Z}_{\mathrm{ltr}}(X \times \underline{\square}) \to \mathbb{Z}_{\mathrm{ltr}}(X).$$

For each fs smooth log scheme $X \in \mathrm{lSm}_k$, the image of $\mathbb{Z}_{\mathrm{ltr}}(X)$ in $\mathrm{logDM}^{\mathrm{eff}}(k)$ is the *effective log motive* $\mathrm{lM}^{\mathrm{eff}}(X)$, giving the functor

$$\mathrm{lM}^{\mathrm{eff}} : \mathrm{lSm}_k \to \mathrm{logDM}^{\mathrm{eff}}(k).$$

The functor $\mathrm{lM}^{\mathrm{eff}}$ shares many of the formal properties of $M^{\mathrm{eff}} : \mathrm{Sm}_k \to \mathrm{DM}^{\mathrm{eff}}(k)$; we refer the reader to the [19, INTRODUCTION] for an overview.

Questions of representing known constructions such as the higher Chow groups with modulus in $\mathrm{logDM}^{\mathrm{eff}}(k)$, or finding direct connections of $\mathrm{logDM}^{\mathrm{eff}}(k)$ with the category $\mathrm{MDM}^{\mathrm{eff}}(k)$ are not discussed in [19]. However, for (X, D) a proper modulus pair, one has the log scheme $l(X, D)$, defined using the Deligne–Faltings log structure on X associated to the ideal sheaf $\mathcal{O}_X(-D)$. In general, this is not saturated. Still, there should be presheaves with log transfers $\mathbb{Z}_{\mathrm{ltr}}(X, D)$ and $\mathbb{Z}^c_{\mathrm{ltr}}(X, D)$ using finite and quasi-finite "log correspondences," with value on $Y \in \mathrm{lSm}_k$ the free abelian group of integral subschemes W of $\underline{Y} \times X$ that admit a map of log schemes $(W^N, (p_1 \circ \nu)^*(\mathcal{M}_Y)) \to l(X, D)$, as in the definition of $\mathrm{lCor}_k(-,-)$. One could also expect to have presheaves $lz(X, D, r)$ similarly defined, and corresponding to the presheaves with modulus transfers $z(X, D, r)$ constructed by Miyazaki. These could be used to give a map

$$H_p\big(z_r^{\mathrm{equi}}(X, D, *)\big) \to \mathrm{Hom}_{\mathrm{logDM}^{\mathrm{eff}}(k)}\big(\mathbb{Z}(0)[p], M\big(lz(X, D, r)\big)\big).$$

We have briefly mentioned *reciprocity sheaves* in our discussion of motives with modulus. There is a nice connection of $\mathrm{logDM}^{\mathrm{eff}}(k)$ with the theory of reciprocity sheaves, so we take the opportunity to say a few words about reciprocity sheaves before we describe the theorem of Shuji Saito, which gives the connection between these two theories.

The notion of a reciprocity sheaf and its relation to motives with modulus goes back to the theorem of Rosenlicht–Serre. In our discussion of reciprocity sheaves, we work over a fixed perfect field k.

Theorem 5.8 (Rosenlicht–Serre [189, III]). *Let k be a perfect field, let C be a smooth complete curve over k, let G be an smooth commutative algebraic group over k, and let $f : C \dashrightarrow G$ be a rational map over k. Let $S \subset C$ be a finite subset such that f is a morphism on $C \setminus S$. Then there is an effective divisor D supported in S such that, for g a rational function on C with $g \equiv 1 \bmod D$, one has*

$$\sum_{P \in C \setminus S} \mathrm{ord}_P(g) \cdot f(P) = 0$$

in G.

In [72], reciprocity functors and reciprocity sheaves are defined. We will just give a sketch. One first defines for F a presheaf with transfers (in the Voevodsky sense), and for a proper modulus pair (X, D) with a section $a \in F(X \setminus D)$, what it means for a to have modulus D. As an example, if $p : C \to X$ is a non-constant morphism of a smooth proper integral curve C over k to X with $p(C)$ not contained in D, and g is a rational function on C such that $g \equiv 1 \bmod p^*(D)$, then one is required to have

$$a\big(p_*\big(\mathrm{div}(g)\big)\big) = 0 \in F(\mathrm{Spec}\, k).$$

Here, for a 0-cycle $\sum_x n_x \cdot x$ on $X \setminus D$, $a(\sum_x n_x \cdot x) = \sum_x n_x \cdot p_{x*} i_x^*(a) \in F(\mathrm{Spec}\, k)$, where for a closed point x of $X \setminus D$, $i_x : x \to X \setminus D$ is the inclusion and $p_x : x \to \mathrm{Spec}\, k$ is the (finite) structure morphism. In general, one imposes a similar condition in $F(S)$ for a "relative curve" on $X \times S$ over some smooth base scheme S.

A presheaf with transfers F is a reciprocity sheaf if for each quasi-affine U and section $a \in F(U)$, there is a proper modulus pair (X, D) with $U = X \setminus D$ such that a has modulus D. Roughly speaking, one should think that each section of F has "bounded ramification," although the "ramification" for F itself may be unbounded.

This definition is not quite accurate, as a slightly different notion of "modulus pair" from what we have defined here is used in [72]. A more elegant definition of reciprocity sheaf is given in [73]. This new notion is a bit more restrictive than the old one, but by [73, THEOREM 2], the two notions agree on for $F \in \underline{\mathrm{M}}\mathrm{NST}_k$.

Using the definition of [73], the reciprocity sheaves define a the full subcategory $\mathbf{RST}_k$ of PST_k, strictly containing the subcategory $\mathbf{HI}_k \subset \mathrm{PST}_k$ of $\mathbb{A}^1$-homotopy invariant presheaves with transfer. There is also the subcategory $\mathbf{RST}_{\mathrm{Nis},k}$ of NST_k, consisting of those reciprocity presheaves that are Nisnevich sheaves.

Some examples of non-homotopy invariant sheaves in $\mathbf{RST}_{\mathrm{Nis},k}$ include the sheaf of n-forms over k, $X \mapsto \Omega^n_{X/k}$, the sheaf of absolute n-forms, $X \mapsto \Omega^n_X$, and for k of positive characteristic, the truncated de Rham–Witt sheaves, $X \mapsto W_m\Omega^n_X$. The representable sheaf of a commutative algebraic group G over k, $X \mapsto G(X)$, is in $\mathbf{RST}_{\mathrm{Nis},k}$, and for some G (e.g. $G = \mathbb{G}_m^n$) this is also $\mathbb{A}^1$-homotopy invariant. This is not the case for unipotent G (e.g. $G = \mathbb{G}_a^n$).

Here is the promised theorem of Saito. For a sheaf $G \in \mathrm{lNST}_k$, we say that G is *strictly $\overline{\square}$-invariant* if for all $X \in \mathrm{lSm}_k$, the map

$$H^*_{d\mathrm{Nis}}(X, G_{|X_{d\mathrm{Nis}}}) \to H^*_{d\mathrm{Nis}}(X \times \overline{\square}, G_{|X \times \overline{\square}_{d\mathrm{Nis}}})$$

induced by the projection $X \times \overline{\square} \to X$ is an isomorphism. Here $_{d\mathrm{Nis}}$ refers to the divided Nisnevich site.

Theorem 5.9 (Saito [105, THEOREM 0.2]). *There exists a fully faithful exact functor*

$$\log : \mathbf{RST}_{\mathrm{Nis},k} \to \mathrm{lNST}_k$$

such that $\log(F)$ *is strictly $\overline{\square}$-invariant for every $F \in \mathbf{RST}_{\mathrm{Nis},k}$. Moreover, for each $X \in \mathrm{Sm}_k$, there is a natural isomorphism*

$$H^i_{\mathrm{Nis}}(X, F_{|X}) \cong \mathrm{Hom}_{\mathrm{logDM}^{\mathrm{eff}}(k)}\big(\mathrm{lM}^{\mathrm{eff}}(X), \log(F)[i]\big).$$

6. p-ADIC ÉTALE MOTIVIC COHOMOLOGY AND p-ADIC HODGE THEORY

We discuss yet another theory of motivic cohomology that is not $\mathbb{A}^1$-homotopy invariant.

Working over a base-field k and for m prime to the characteristic, we have the isomorphism of the étale sheafification $\mathbb{Z}/m(r)_{\text{ét}}$ with the étale sheaf $\mu_m^{\otimes r}$. The étale sheaves $\mathbb{Z}/m(r)_{\text{ét}}$ can be considered as objects in a version of Voevodsky's DM constructed using the étale topology rather than the Nisnevich topology, and their categorical cohomology agrees with the usual étale cohomology [127, CHAP. 5, §3.3]. In particular, the complexes $\mathbb{Z}/m(r)_{\text{ét}}$ have $\mathbb{A}^1$-homotopy invariant étale cohomology.

On the other hand, if k has characteristic $p > 0$, we have the isomorphism [52] of the Nisnevich sheaves on Sm_k,

$$\mathbb{Z}/p^n(r) \cong W_n\Omega^r_{\log}[-r], \tag{6.1}$$

hence of étale sheaves

$$\mathbb{Z}/p^n(r)_{\text{ét}} \cong W_n\Omega^r_{\log}[-r].$$

Here something strange happens: the étale sheaf $\mathbb{Z}/p^n(r)_{\text{ét}}$ is no longer strictly homotopy invariant! In fact, the existence of the Artin–Schreyer étale cover $\mathbb{A}^1 \to \mathbb{A}^1$ of degree p implies that the étale version of $\mathrm{DM}^{\mathrm{eff}}(k)$ with coefficients modulo p^n is zero if k has characteristic $p > 0$ [127, CHAP. 5, PROPOSITION 3.3.3]. Thus $\mathbb{Z}/p^n(r)_{\text{ét}}$ leaves the world of Voevodsky's motives and motivic cohomology.

For $S = \operatorname{Spec}\Lambda$, with Λ a mixed characteristic $(0, p)$ dvr, the complex $\mathbb{Z}/p^n(r)_{\text{ét}}$ on $\mathrm{Sm}_{S,\text{ét}}$ yields an interesting gluing of $\mathbb{Z}/p^n(r)_{\text{ét}} = \mu_{p^n}^{\otimes r}$ over the characteristic zero quotient field of Λ and $\mathbb{Z}/p^n(r)_{\text{ét}} = W_n\Omega^r_{\log}[-r]$ over the characteristic p residue field. The positive characteristic part again says that we have left homotopy invariance behind.

The complexes $\mathbb{Z}/p^n(r)_{\text{ét}}$ have an interesting connection with a certain complex of sheaves arising in p-adic Hodge theory. A version of this complex first appears in the paper [40] of Fontaine–Messing, and plays an important role in the proof of their main result. Its construction was reinterpreted by Kurihara [82], relying on the work of Bloch–Kato [28] and Kato [79], and was generalized by Sato [106]. Geisser [49], following work of Schneider [108], established the connection of the Fontaine–Messing/Kurihara/Sato complex with $\mathbb{Z}/p^n(r)_{\text{ét}}$ in the case of a smooth degeneration, and this connection was partially extended by Zhong [128] to the semi-stable case.

In their recent work on integral p-adic Hodge theory, Bhatt–Morrow–Scholze [18] have introduced a "motivic filtration" on p-adic étale K-theory, relying on a Postnikov tower for topological cyclic homology, and the layers in this tower have been identified with the pro-system $\{\mathbb{Z}/p^n(r)_{\text{ét}}\}_n$ in a work-in-progress [16] by Bhargav Bhatt and Akhil Mathew.

Our goal in this section is to give some details of the story sketched above.

We first discuss the papers of Bloch–Kato, Fontaine–Messing, Kurihara and Sato without reference to all the advances in p-adic Hodge theory that followed these works; we wanted to give the reader just enough background to put the connections with motivic cohomology in context. We will then describe the works of Geisser and Zhong, as well as results of Geisser–Hesselholt that form some of the foundations for the work of Bhatt–Morrow–Scholze. We conclude with a description of the Bhatt–Morrow–Scholze motivic tower and its connection with the p-adic cycle complexes.

We refer the reader to [15] for background on crystalline cohomology.

6.1. A quick overview of some p-adic Hodge theory

We begin with a few comments on the paper of Bloch and Kato [28], which we have already mentioned in our discussion of the Beilinson–Lichtenbaum conjectures. They consider the spectrum S of a complete dvr Λ with generic point $\eta = \operatorname{Spec} K \hookrightarrow S$ and closed point $s = \operatorname{Spec} k \hookrightarrow S$, and a smooth and proper S-scheme $X \to S$ with generic fiber $V := X_\eta$ and special fiber $Y := X_s$. $\bar{V}$, $\bar{Y}$ denote V, Y over the respective algebraic closures $\bar{K}$ and $\bar{k}$ of K and k. Let $\bar{\Lambda}$ be the integral closure of Λ in $\bar{K}$, $\bar{S} := \operatorname{Spec} \bar{\Lambda}$, and $\bar{X} = X \times_S \bar{S}$. Let $G = \operatorname{Gal}(\bar{K}/K)$ and let C denote the completion of $\bar{K}$.

The closure $\bar{Y}$ has its crystalline cohomology $H^*_{\text{crys}}(\bar{Y}/W(\bar{k}))$ with action of Frobenius, giving the p^i-eigenspace

$$H^*_{\text{crys}}(\bar{Y}/W(\bar{k}))^{(i)} \subset H^*_{\text{crys}}(\bar{Y}/W(\bar{k}))_{\mathbb{Q}}.$$

We say $\bar{Y}$ is *ordinary* if

$$\dim_{W(\bar{k})_{\mathbb{Q}}} H^m_{\text{crys}}(\bar{Y}/W(\bar{k}))^{(i)} = \dim_{\bar{k}} H^{m-i}(\bar{Y}, \Omega^i_{\bar{Y}/\bar{k}}).$$

We have the inclusions $\bar{i} : \bar{Y} \to \bar{X}$, $\bar{j} : \bar{V} \to \bar{X}$ and the spectral sequence

$$E_2^{s,t} = H^s_{\text{ét}}(\bar{Y}, \bar{i}^* R^t \bar{j}_*(\mathbb{Z}/p^n\mathbb{Z})) \Rightarrow H^{s+t}_{\text{ét}}(\bar{V}, \mathbb{Z}/p^n\mathbb{Z}),$$

inducing a descending filtration $F^* H^*_{\text{ét}}(\bar{V}, \mathbb{Q}_p)$ on $H^*_{\text{ét}}(\bar{V}, \mathbb{Q}_p)$ with $F^0 H^q = H^q$ and $F^{q+1} H^q = 0$.

We have the de Rham–Witt sheaf $W\Omega^i$ on $\mathrm{Sm}_{\bar{k}}$ and the sheaf of differential forms $\Omega^i_{-/K}$ on Sm_K.

Theorem 6.1 (Bloch–Kato [28, THEOREM 8.7]). *Suppose that k is perfect and that Y is ordinary. Then there are natural G-equivariant isomorphisms*

$$\begin{aligned}
&\text{(i)} \quad \mathrm{gr}^{q-i} H^q_{\acute{e}t}(\bar{V}, \mathbb{Q}_p) \cong H^q_{\text{crys}}(\bar{Y}/W(\bar{k}))^{(i)}_{\mathbb{Q}}(-i),\\
&\text{(ii)} \quad \mathrm{gr}^{q-i} H^q_{\acute{e}t}(\bar{V}, \mathbb{Q}_p) \otimes_{\mathbb{Z}_p} W(\bar{k}) \cong H^q_{\text{crys}}(\bar{Y}, W\Omega^i)_{\mathbb{Q}}(-i),\\
&\text{(iii)} \quad \mathrm{gr}^{q-i} H^q_{\acute{e}t}(\bar{V}, \mathbb{Q}_p) \otimes_{\mathbb{Q}_p} C \cong H^q(V, \Omega^i_{V/K}) \otimes_K C(-i).
\end{aligned}$$

We will not give any details of the proof here, but do want to mention that what ties these different theories together is the sheaf of Milnor K-groups $\mathcal{K}^M_q$. This maps to étale cohomology by the Galois symbol

$$\theta_{q,m} : \mathcal{K}^M_q/m \to \mathcal{H}^q_{\text{ét}}(\mu_m^{\otimes q})$$

for m prime to the characteristic, to the de Rham–Witt sheaf by the $d\log$ map on Sm_k,

$$d\log^W_{q,p^n} : \mathcal{K}^M_q/p^n \to W_n\Omega^q,$$

and to the sheaf of differential forms, by the $d\log$ map on Sm_S,

$$d\log_{q,p^n} : \mathcal{K}_q^M/p^n \to \Omega^q_{-/S}/p^n.$$

The main structural results that underpin the proof of the Bloch–Kato theorem are two comparison isomorphisms on the sheaf level. For the first, let $W_n\Omega^q_{\log} \subset W_n\Omega^q$ be the étale subsheaf locally generated by the image $d\log(\mathcal{K}_q^M/p^n)$.

Theorem 6.2 ([28, COROLLARY 2.8]). *Let F be a field of characteristic $p > 0$. Then the map $d\log : K_q^M(F)/p^n \to W_n\Omega^q(F)$ defines an isomorphism of $K_q^M(F)/p^n$ with $W_n\Omega^q_{\log}(F)$.*

Note that the composition

$$\mathbb{Z}/p^n(q) \to \tau_{\geq q}\mathbb{Z}/p^n(q) \cong \mathcal{K}_q^M/p^n[-q] \xrightarrow{d\log} W_n\Omega^q_{\log}[-q]$$

is the map that defines the isomorphism (6.1).

The second result is a special case of the Bloch–Kato conjecture.

Theorem 6.3 (Bloch–Kato [28, THEOREM 5.12]). *Let F be a henselian discretely valued field of characteristic 0, with residue field of characteristic $p > 0$. Then the Galois symbol*

$$K_q^M(F)/p^n \to H^q_{ét}(F, \mu_{p^n}^{\otimes q})$$

is an isomorphism for all $n \geq 1$.[2]

Bloch and Kato use $\mathcal{K}_q^M$ to relate $i^*R^q j_*\mu_{p^n}^{\otimes q}$ to $\Omega^q_{-/K}/p^n$ and $W_n\Omega^{q-1}_{\log}$ via the respective $d\log$ maps. Relying on the isomorphisms of Theorem 6.2 and Theorem 6.3, these maps from Milnor K-theory tie de Rham cohomology, crystalline cohomology and étale cohomology together, and eventually lead to a proof of Theorem 6.1.

As part of the proof, they define a surjective map

$$\gamma : i^*R^q j_*\mu_{p^n}^{\otimes r} \to W_n\Omega^{q-1}_{\log} \tag{6.2}$$

on $Y_{\text{ét}}$ with the following property: Let $\theta : i^*j_*\mathcal{K}^M_{q,\text{ét}} \to i^*R^q j_*\mu_{p^n}^{\otimes r}$ be the Galois symbol map, let $u_2, \dots, u_q$ be units on X near some point y of Y with restrictions $\bar{u}_1, \dots, \bar{u}_q$ to Y and let π be a parameter in Λ. Then

$$\gamma \circ \theta\big(\{u_1, \dots, u_{q-1}, \pi\}\big) = d\log\big(\{\bar{u}_1, \dots, \bar{u}_{q-1}\}\big).$$

We highlight this because it will be used later on in a gluing construction that defines an object of central interest for this section.

The next paper I want to mention is by Fontaine–Messing [40]. They construct a comparison isomorphism between de Rham cohomology and étale cohomology for a smooth, proper scheme X over the ring of integers $\mathcal{O}_K$ for K a characteristic zero local field (under some additional assumptions). The de Rham cohomology $H^q_{\mathrm{dR}}(V/K)$ has its Hodge

---

2 In fact, at the beginning of §3 of [28], Bloch and Kato write, "The cohomological symbol defined by Tate [114] gives a map . . . , which one conjectures to be an isomorphism quite generally."

filtration and via the comparison isomorphism $H^q_{\mathrm{dR}}(V/K) \cong H^q_{\mathrm{crys}}(Y/W(k)) \otimes_{W(k)} K$ $H^q_{\mathrm{dR}}(V/K)$ acquires a Frobenius operator ϕ; call this object $H^q_{\mathrm{crys}}(X)$. Fontaine–Messing construct the p-adic period ring $B_{\mathrm{crys}} \supset K$ with a Galois action, a Frobenius and a filtration, and show there are isomorphisms

$$\mathrm{Fil}^0\big(B_{\mathrm{crys}} \otimes_K H^q_{\mathrm{crys}}(X)\big)^{\phi=\mathrm{Id}} \cong H^q_{\text{ét}}(V_{\bar{K}}, \mathbb{Q}_p)$$

and

$$\big(B_{\mathrm{crys}} \otimes_{\mathbb{Q}_p} H^q_{\text{ét}}(V_{\bar{K}}, \mathbb{Q}_p)\big)^G \cong H^q_{\mathrm{crys}}(X)$$

both compatible with the "remaining" structures.

To set this up, they consider the syntomic topology on $\mathrm{Sch}_{W_n(k)}$, where a cover is a surjective syntomic map (we described syntomic maps in Section 3.3). The crystalline structure sheaf $\mathcal{O}_n^{\mathrm{crys}}$ defines a sheaf for the syntomic topology with a surjection to the usual structure sheaf $\mathcal{O}_n$ on $\mathrm{Sch}_{W_n(k)}$. Letting J_n denote the kernel of $\mathcal{O}_n^{\mathrm{crys}} \to \mathcal{O}_n$, one has the rth divided power $J_n^{[r]}$; this gives us the sheaf $\tilde{S}_n^r$ defined as the kernel of $\phi - p^r : J_n^{[r]} \to \mathcal{O}_n^{\mathrm{crys}}$. Modifying this by taking the image S_n^r of the reduction map $\tilde{S}_{n+r}^r \to \tilde{S}_n^r$ gives the inverse system $\{S_n^r\}_n$ and the cohomology

$$H^*(\bar{X}, S^r_{\mathbb{Q}_p}) := \Big(\varprojlim_n H^*(\bar{X}, S_n^r)\Big) \otimes_{\mathbb{Z}_p} \mathbb{Q}_p.$$

The ring B^+_{crys} is defined as follows. The characteristic p ring $\mathcal{O}_{\bar{K}}/p$ forms an inverse system via the Frobenius endomorphism; let

$$\mathcal{O}^\flat = \varprojlim_{\mathrm{Frob}} \mathcal{O}_{\bar{K}}/p,$$

a perfect characteristic p ring. We have the ring of truncated Witt vectors $W_n(\mathcal{O}^\flat)$ and a surjection $\pi_n : W_n(\mathcal{O}^\flat) \to \mathcal{O}_{\bar{K}}/p^n$. Let $W_n^{\mathrm{DP}}(\mathcal{O}_{\bar{K}})$ be the divided power envelope of the kernel of π_n, forming the inverse system $\{W_n^{\mathrm{DP}}(\mathcal{O}_{\bar{K}})\}_{n\geq 0}$. Let

$$B^+_{\mathrm{crys}} := K \otimes_{W(k)} \varprojlim_n W_n^{\mathrm{DP}}(\mathcal{O}_{\bar{K}}).$$

The Frobenius on $W_n(\mathcal{O}^\flat)$ induces a Frobenius on B^+_{crys} and the filtration $J_n^{[*]}$ of $W_n^{\mathrm{DP}}(\mathcal{O}_{\bar{K}})$ induces a filtration $\mathrm{Fil}^* B^+_{\mathrm{crys}}$ on B^+_{crys}.

The derived push-forward of the complex $J_n^{[r]} \xrightarrow{\phi - p^r} \mathcal{O}_n^{\mathrm{crys}}$ is an analog of the Deligne complex, as expressed in the following theorem.

Theorem 6.4 ([**40, Corollary to Theorem 1.6, Lemma 3.1**]). *Suppose that X is admissible*[3] *and $\Lambda = W(k)$. Then for $m \leq r < p$ there is an exact sequence*

$$0 \to H^m(\bar{X}, S^r_{\mathbb{Q}_p}) \to \mathrm{Fil}^r\big(B^+_{\mathrm{crys}} \otimes_K H^m_{\mathrm{dR}}(V/K)\big) \xrightarrow{\phi - p^r} H_{\mathrm{dR}}(V/K) \to 0.$$

In other words,

$$H^m(\bar{X}, S^r_{\mathbb{Q}_p}) = \big(\mathrm{Fil}^r\big(B^+_{\mathrm{crys}} \otimes_K H^m_{\mathrm{dR}}(V/K)\big)\big)^{\phi = p^r}.$$

---

3 See [**40, §2.1**] for the definition of admissible X.

To involve étale cohomology in the picture, Fontaine–Messing introduce the syntomic–étale site on formal $\mathrm{Spf}(W(k))$-schemes, where a cover is a map $\{U_n\}_n \to \{V_n\}_n$ such that $U_n \to V_n$ is a syntomic cover for all n and is an étale cover on the rigid analytic generic fibers. This extends to the syntomic–étale site on $\bar{X}$, where an object is $U \to \bar{X}$, quasi-finite and syntomic, with $U_{\bar{K}} \to V_{\bar{K}}$ étale. Letting $\mathcal{Z}$ be the formal completion of $\bar{X}$, we have the diagram of sites

$$\mathcal{Z}_{\text{syn}-\text{ét}} \xrightarrow{i} \bar{X}_{\text{syn}-\text{ét}} \xleftarrow{j} V_{\bar{K},\text{ét}}.$$

Fontaine–Messing prove a patching result, that a sheaf on $\bar{X}_{\text{syn}-\text{ét}}$ is given by a triple $(\mathcal{F}, \mathcal{G}, \alpha)$, with $\mathcal{F}$ a sheaf on $\mathcal{Z}_{\text{syn}-\text{ét}}$, $\mathcal{G}$ a sheaf on $V_{\bar{K},\text{ét}}$, and $\alpha : \mathcal{F} \to i^* Rj_* \mathcal{G}$ a morphism. Using this description, they construct a sheaf on $\bar{X}_{\text{syn}-\text{ét}}$ by defining a certain morphism (see [40, §5.1])

$$\alpha : S_n^r \to i^* Rj_* \mu_{p^n}^{\otimes r}.$$

The resulting sheaf $\tilde{S}_n^r$ has

$$j^* \tilde{S}_n^r \cong \mu_{p^n}^{\otimes r}, \quad i^* \tilde{S}_n^r \cong S_n^r.$$

It follows from proper base-change (see the proof of [40, PROPOSITION 6.2]) that the restriction map $H^*(\bar{X}_{\text{syn}-\text{ét}}, \tilde{S}_n^r) \to H^*(\mathcal{Z}_{\text{syn}-\text{ét}}, S_n^r)$ is an isomorphism, and we also have

$$\Bigl(\varprojlim_n H^*(\mathcal{Z}_{\text{syn}-\text{ét}}, S_n^r)\Bigr) \otimes_{\mathbb{Z}_p} \mathbb{Q}_p \cong \bigl(\mathrm{Fil}^r\bigl(B_{\text{crys}}^+ \otimes_K H_{\text{dR}}^m(V/K)\bigr)\bigr)^{\phi=p^r}.$$

Moreover, the restriction map j^* gives

$$j^* : H^*(\bar{X}_{\text{syn}-\text{ét}}, \tilde{S}_n^r) \to H_{\text{ét}}^*(V_{\bar{K}}, \mu_{p^n}^{\otimes r}),$$

so passing to the limit, we have the map

$$\beta : \bigl(\mathrm{Fil}^r\bigl(B_{\text{crys}}^+ \otimes_K H_{\text{dR}}^m(V/K)\bigr)\bigr)^{\phi=p^r} \to H_{\text{ét}}^m\bigl(V_{\bar{K}}, \mathbb{Q}_p(r)\bigr),$$

which they show is an isomorphism.

This gives a twisted version of the result announced at the beginning of our discussion. To recover the untwisted version, they define a map

$$\mathbb{Q}_p(1)(\bar{K}) \to B_{\text{crys}}^+$$

by sending a p^n-root of unity ε in $\bar{K}$ to the logarithm of the Teichmüller lift of the mod p reduction of ε, and passing to the limit in n. Let $t \in \mathbb{Q}_p(1)(\bar{K})$ be a nonzero element and define $B_{\text{crys}} = B_{\text{crys}}^+[1/t]$, with induced filtration and Galois action. Twisting with respect to t translates the twisted version to the untwisted one.

The sheaf $\tilde{S}_n^r$ is only defined on $\bar{X}_{\text{syn}-\text{ét}}$ for X smooth over S and for $r < p$, and with base-ring Λ equal to $W(k)$, i.e., in the unramified case. Kato [79] studies the derived push-forward $\mathcal{S}_n(r)$ of the syntomic sheaf $\tilde{S}_n^r$ to $\mathrm{Sm}_{k,\text{ét}}$. Kurihara [82, §1, THEOREM] considers the ramified case and also clarifies the relation of $\mathcal{S}_n(r)$ with the sheaf of log forms $W_n\Omega_{\log}^{r-1}$.

Theorem 6.5 (Kurihara). *Suppose that $[k : k^p] < \infty$. Let $X \to S$ be smooth and projective and suppose that $r < p-1$. Then there is a distinguished triangle in $D(Y_{\text{ét}})$,*

$$W_n\Omega^{r-1}_{\log}[-r-1] \to \mathcal{S}_n(r) \to i^*Rj_*\mu_{p^n}^{\otimes r} \to W_n\Omega^{r-1}_{\log}[-r].$$

Schneider [108] extends the construction of $\mathcal{S}_n(r)$ to all $r \geq 0$ by using the Bloch–Kato symbol map γ of (6.2) to give a map $s : \tau_{\leq r}Rj_*\mu_{p^n}^{\otimes r} \to i_*W_n\Omega^{r-1}_{\log}$ with i^*s the composition

$$i^*\tau_{\leq r}Rj_*\mu_{p^n}^{\otimes r} \to i^*R^r j_*\mu_{p^n}^{\otimes r}[-r] \xrightarrow{\gamma} W_n\Omega^{r-1}_{\log}[-r].$$

Schneider then defines the sheaf $\mathcal{S}_n(r)$ as $\text{Cone}(s)[-1]$, giving the distinguished triangle

$$i_*W_n\Omega^{r-1}_{\log}[-r-1] \to \mathcal{S}_n(r) \to Rj_*\mu_{p^n}^{\otimes r} \xrightarrow{s} i_*W_n\Omega^{r-1}_{\log}[-r], \tag{6.3}$$

which recovers the one in Kurihara's theorem for $r < p-1$ by applying i^*.

Using a similar method, Schneider's construction was extended to the semi-stable case by Sato [106], who defines the object $\mathfrak{T}_n(r) \in D(X_{\text{ét}})$ with $\mathfrak{T}_n(r) \cong \mathcal{S}_n(r)$ in the smooth case.

6.2. Étale motivic cohomology

We return to algebraic cycles. As before, we consider a smooth separated finite type S-scheme $X \to S = \text{Spec}\,\Lambda$ with generic fiber $j : V \to X$ and special fiber $i : Y \to X$, and with Λ a mixed characteristic $(0, p)$ dvr with perfect residue field.

Geisser [49] considers the motivic complex $\mathbb{Z}(r)_X$ on a smooth S-scheme $X \to S$ as a sheaf of complexes on X_{Nis}. Here we use the reindexed Bloch cycle complex to define $\mathbb{Z}(r)^*_X(U)$ as

$$\mathbb{Z}(r)^*_X(U) := z^r(U, 2r - *),$$

and define the motivic complexes $\mathbb{Z}(r)_V$ and $\mathbb{Z}(r)_Y$ on V and Y similarly.

Let $\alpha : (-)_{\text{ét}} \to (-)_{\text{Nis}}$ be the change of topology map. Sheafifying for the étale topology gives complexes $\mathbb{Z}(r)_{\text{ét},X}$, $\mathbb{Z}(r)_{\text{ét},V}$, and $\mathbb{Z}(r)_{\text{ét},Y}$. Geisser shows that various known properties of $\mathbb{Z}(r)_X$, $\mathbb{Z}(r)_V$, and $\mathbb{Z}(r)_Y$, such as the purity isomorphism [84, THEOREM 1.7]

$$i^!\mathbb{Z}(r)_X \cong \mathbb{Z}(r-1)_Y[-2],$$

the theorem of Geisser–Levine [52]

$$\mathbb{Z}/p^n(r)_Y \cong W_n\Omega^r_{\log,Y}[-r],$$

the Suslin–Voevodsky isomorphism in $D^b(V_{\text{ét}})$ (Beilinson's axiom (iv)(a))

$$j^*\mathbb{Z}/p^n(r)_{\text{ét},X} \cong \mathbb{Z}/p^n(n)_{\text{ét},V} \cong \mu_{p^n}^{\otimes r},$$

and the Beilinson–Lichtenbaum conjectures (now a theorem)

$$\mathbb{Z}/p^n(r)_V \cong \tau_{\leq r}R\alpha_*\mu_{p^n}^{\otimes r}, \quad R^{r+1}\alpha_*\mathbb{Z}(r)_{\text{ét},V} = 0$$

have as consequence

Theorem 6.6 (Geisser [49, THEOREM 1.3]). *Let $X \to \operatorname{Spec}\Lambda$ be smooth and essentially of finite type, with Λ a complete discrete valuation ring of mixed characteristic $(0, p)$. Then there is a distinguished triangle in $D^b(X_{\acute{e}t})$,*

$$i_* W_n\Omega^{r-1}_{\log}[-r-1] \to \mathbb{Z}/p^n(r)_{\acute{e}t} \to \tau_{\leq r} Rj_*\mu_{p^n}^{\otimes r} \to i_* W_n\Omega^{r-1}_{\log}[-r],$$

and an isomorphism $\mathbb{Z}/p^n(r)_{\acute{e}t} \cong \mathcal{S}_n(r)$ in $D^b(X_{\acute{e}t})$ that transforms this triangle to Schneider's defining triangle (6.3).

Zhong has extended this to the semi-stable case, establishing an isomorphism with Sato's construction $\mathfrak{T}_n(r)$ after a truncation [128, PROPOSITION 4.5]:

$$\tau_{\leq r}\mathbb{Z}/p^n(r)_{\text{ét}} \cong \mathfrak{T}_n(r).$$

Assuming a "weak Gersten conjecture" for $\mathbb{Z}/p^n(r)_{\text{ét}}$, the truncation is removed [128, THEOREM 4.8].

6.3. The theorems of Geisser–Hesselholt

The construction of a motivic tower for integral p-adic Hodge theory by Bhatt–Morrow–Scholze relies on properties of p-completed topological cyclic homology, including the results of Geisser–Hesselholt identifying this with the p-completed étale K-theory. We give a brief résumé of these constructions. Fix as before our mixed characteristic dvr Λ with perfect residue field k.

Topological cyclic homology for a fixed prime p is a spectrum refined version of Connes' cyclic homology and is defined for a scheme X with a topology $\tau \in \{\text{ét}, \text{Nis}, \text{Zar}\}$; we use the étale topology throughout. There is an inverse system of spectra $\{\mathrm{TC}^m(X, p)\}_{m\in\mathbb{N}}$ defining $\mathrm{TC}(X; p)$ as the homotopy inverse limit

$$\mathrm{TC}(X; p) := \operatorname*{holim}_m \mathrm{TC}^m(X, p).$$

Let $\mathcal{TC}^{\cdot}_i$ denote the étale sheaf associated to the presheaf of the ith pro-homotopy groups $U \mapsto \pi_i\mathrm{TC}^{\cdot}(U; p)$. There is a descent spectral sequence

$$E_2^{s,t} = H^s_{\text{cont}}(X, \mathcal{TC}^{\cdot}_{-t}) \Rightarrow \mathrm{TC}_{-s-t}(X; p)$$

and a cyclotomic trace map

$$\mathrm{trc} : K(X) \to \mathrm{TC}(X; p).$$

Let k be a perfect field of characteristic $p > 0$. It follows from a result of Hesselholt [60, THEOREM B] that there is a isomorphism of pro-sheaves on $\mathrm{Sm}_{k\text{ét}}$

$$\mathcal{TC}^{\cdot}_r \cong W_{\cdot}\Omega^r_{\log}. \tag{6.4}$$

The map trc induces the map of pro-sheaves on $\mathrm{Sm}_{k\text{ét}}$

$$\mathrm{trc} : \mathcal{K}_i(\mathbb{Z}/p^{\cdot}) \to \mathcal{TC}^{\cdot}_i;$$

where $\mathcal{K}_i(\mathbb{Z}/p^{\cdot})$ is the pro-étale sheaf associated to the system of presheaves $U \mapsto \{K_i(U, \mathbb{Z}/p^{\nu})\}_{\nu}$. Relying on the main theorem of [52] and the isomorphism (6.4), Geisser and Hesselholt show

Theorem 6.7 (Geisser–Hesselholt [50, COROLLARY 4.2.5, THEOREM 4.2.6]).

1. *The trace map* $\mathrm{trc} : \mathcal{K}_i(\mathbb{Z}/p^{\cdot}) \to \mathcal{TC}_i^{\cdot}$ *is an isomorphism of pro-sheaves on* $\mathrm{Sm}_{k\,\acute{e}t}$.
2. *For* $Y \in \mathrm{Sm}_k$, $\mathrm{TC}(Y; p)$ *is weakly equivalent to the p-completed étale K-theory spectrum of* Y,
$$K^{\acute{e}t}(Y)^{\wedge_p} := \operatorname*{holim}_m K^{\acute{e}t}(Y, \mathbb{Z}/p^n) \cong \mathrm{TC}(Y; p),$$
and this weak equivalence arises from the weak equivalences at the finite level
$$\mathrm{trc} : K^{\acute{e}t}(Y, \mathbb{Z}/p^\nu) \xrightarrow{\sim} \mathrm{TC}(Y; p, \mathbb{Z}/p^\nu).$$

Now consider a smooth finite type scheme $X \to \operatorname{Spec}\Lambda$ with special fiber $i : Y \to X$ and generic fiber $j : V \to X$, as before.

Theorem 6.8 (Geisser–Hesselholt [51, THEOREMS A AND B]). *Suppose* Λ *is henselian.*

A. *Suppose* $X \to \operatorname{Spec}\Lambda$ *is smooth and proper. Then*
$$\mathrm{trc} : K_q^{\acute{e}t}(X, \mathbb{Z}/p^\nu) \to \mathrm{TC}_q(X; p, \mathbb{Z}/p^\nu)$$
is an isomorphism for all $q \in \mathbb{Z}$ *and* $\nu \geq 1$.

B. *Suppose that* $X \to \operatorname{Spec}\Lambda$ *is smooth and finite type. Then the map of pro-sheaves on* $Y_{\acute{e}t}$,
$$i^*\mathcal{K}_q(\mathbb{Z}/p^\nu) \to \{i^*\mathcal{TC}_q^m(p, \mathbb{Z}/p^\nu)\}_{m\in\mathbb{N}},$$
is an isomorphism for all $q \in \mathbb{Z}$ *and all* $\nu \geq 1$.

Remark 6.9. To pass from the isomorphism of Theorem 6.7 to that of Theorem 6.8(B), Geisser–Hesselholt rely on the theorem of McCarthy [88], stating that the cyclotomic trace map from relative K-theory to relative TC,
$$\mathrm{trc} : K_q(X/\pi^n, X/\pi^{n-r}, \mathbb{Z}/p^\nu) \to \mathrm{TC}_q(X/\pi^n, X/\pi^{n-r}, \mathbb{Z}/p^\nu),$$
is an isomorphism for affine X. Thus, the K-theory and topological cyclic homology of non-reduced schemes play a central role in the proof of Theorem 6.8.

6.4. Integral p-adic Hodge theory and the motivic filtration

Bhatt–Morrow–Scholze [17, 18] have constructed integral versions of p-adic Hodge theory. Here we discuss some aspects of the theory of [18] and its relation to p-adic étale motivic cohomology. This uses (p-completed) topological Hochschild homology $\mathrm{THH}(-, \mathbb{Z}_p)$, topological negative cyclic homology $\mathrm{TC}^-(-, \mathbb{Z}_p)$, and topological periodic cyclic homology $\mathrm{TP}(-, \mathbb{Z}_p)$. For a nice, quick overview of these theories, we refer the reader to [18, §1.2, §2.3], and to [18, THEOREM 1.12] for their relation to $\mathrm{TC}(-, \mathbb{Z}_p)$.

Let $\mathbb{C}_p$ be the completion of the algebraic closure of $\mathbb{Q}_p$, with ring of integers $\mathcal{O}_{\mathbb{C}_p}$. As in our review of the work of Fontaine–Messing, we have the $\mathbb{F}_p$-algebra $\mathcal{O}_{\mathbb{C}_p}/p$, its perfection $\mathcal{O}^\flat_{\mathbb{C}_p}$ and the ring of Witt vectors $A_{\mathrm{inf}}(\mathcal{O}_{\mathbb{C}_p}) := W(\mathcal{O}^\flat_{\mathbb{C}_p})$. Hesselholt has connected this with negative cyclic homology TC^-, constructing an isomorphism

$$\pi_0\mathrm{TC}^-(\mathcal{O}_{\mathbb{C}_p}, \mathbb{Z}_p) \cong A_{\mathrm{inf}}(\mathcal{O}_{\mathbb{C}_p}).$$

This has been generalized by Bhatt–Morrow–Scholze in the setting of *perfectoid* rings (see [17, DEFINITION 3.5]). For a perfectoid ring R, we have Scholze's ring $R^\flat$, defined as for $\mathcal{O}^\flat_{\mathbb{C}_p}$ by taking the perfection of R/p,

$$R^\flat := \lim_{\leftarrow \mathrm{Frob}} R/p.$$

This gives the ring of Witt vectors $A_{\mathrm{inf}}(R) := W(R^\flat)$ with Frobenius ϕ induced by the Frobenius on $R^\flat$.

Theorem 6.10 (Bhatt–Morrow–Scholze [18, THEOREM 1.6]). *Let R be a perfectoid ring. Then there is a canonical ϕ-equivariant isomorphism $\pi_0\mathrm{TC}^-(R, \mathbb{Z}_p) \cong A_{\mathrm{inf}}(R)$.*

Fix a discretely valued extension K of $\mathbb{Q}_p$, with ring of integers $\mathcal{O}_K$ having perfect residue field k. Let C be the completed algebraic closure of K, with ring of integers $\mathcal{O}_C$. Let $A_{\mathrm{inf}} := A_{\mathrm{inf}}(\mathcal{O}_C)$.

Let $\mathfrak{X}$ be a smooth formal scheme over $\mathcal{O}_C$. In [17], Bhatt–Morrow–Scholze construct a presheaf of complexes of A_{inf}-algebras on $\mathfrak{X}_{\mathrm{Zar}}$, $A\Omega_{\mathfrak{X}}$, whose Zariski hypercohomology specializes to crystalline cohomology, p-adic étale cohomology and de Rham cohomology via base-change with respect to suitable ring homomorphisms out of A_{inf}, replacing the ring homomorphism $A_{\mathrm{inf}}(\mathcal{O}_{\mathbb{C}_p}) \to B_{\mathrm{crys}}$ used in the Fontaine–Messing theory. In [18], they refine and reinterpret this theory using TC^-. They define the notion of a quasi-syntomic ring and the associated quasi-syntomic site [18, DEFINITION 1.7]; this gives the presheaf $\pi_0\mathrm{TC}^-(-;\mathbb{Z}_p)$ on the quasi-syntomic site $\mathrm{qsyn}_{\tilde{A}}$ over a quasi-syntomic ring $\tilde{A}$ and the associated derived global sections functor $R\Gamma_{\mathrm{syn}}(\tilde{A}, -)$.

Theorem 6.11 ([18, THEOREM 1.8]). *Let $\tilde{A}$ be an $\mathcal{O}_C$-algebra that can be written as the p-adic completion of a smooth $\mathcal{O}_C$-algebra. There is a functorial (in $\tilde{A}$) ϕ-equivariant isomorphism of E_∞-A_{inf}-algebras*

$$A\Omega_{\tilde{A}} \cong R\Gamma_{\mathrm{syn}}\big(\tilde{A}, \pi_0\mathrm{TC}^-(-;\mathbb{Z}_p)\big).$$

The Postnikov tower $\tau_{\geq *}\mathrm{TC}(-;\mathbb{Z}_p)$ for the presheaf of spectra $\mathrm{TC}(-;\mathbb{Z}_p)$ on $\tilde{A}_{\mathrm{qsyn}}$ induces the tower over $\mathrm{TC}(\tilde{A};\mathbb{Z}_p$:

$$\cdots \to \mathrm{Fil}^{n+1}\mathrm{TC}(\tilde{A};\mathbb{Z}_p) \to \mathrm{Fil}^n\mathrm{TC}(\tilde{A};\mathbb{Z}_p) \to \cdots \to \mathrm{TC}(\tilde{A};\mathbb{Z}_p, \tag{6.5}$$

by setting

$$\mathrm{Fil}^n\mathrm{TC}(\tilde{A};\mathbb{Z}_p) := R\Gamma_{\mathrm{syn}}\big(\tilde{A}, \tau_{\geq 2n}\mathrm{TC}(-;\mathbb{Z}_p)\big)$$

(see [18, §1.4]). Define the sheaves $\mathbb{Z}_p^{\mathrm{BMS}}(n)$ by sheafifying the presheaf

$$\tilde{A} \mapsto \mathbb{Z}_p^{\mathrm{BMS}}(n)(\tilde{A}) := \mathrm{gr}^n_{\mathrm{Fil}}\mathrm{TC}(\tilde{A};\mathbb{Z}_p)[-2n],$$

where $\mathrm{gr}^n_{\mathrm{Fil}}\mathrm{TC}(\tilde{A};\mathbb{Z}_p)$ is the homotopy cofiber of $\mathrm{Fil}^{n+1}\mathrm{TC}(\tilde{A};\mathbb{Z}_p) \to \mathrm{Fil}^n\mathrm{TC}(\tilde{A};\mathbb{Z}_p)$.

Theorem 6.12 ([18, THEOREM 1.15]). (1) *Let k be a perfect field of characteristic $p > 0$, let A be a smooth k-algebra and let $X = \mathrm{Spec}\, A$. Then there is an isomorphism in the derived category of sheaves on the pro-étale site of X,*

$$\mathbb{Z}_p^{\mathrm{BMS}}(r)_X \cong W\Omega^r_{X,\log}[-r].$$

(2) *Let C be an algebraically closed complete extension of $\mathbb{Q}_p$, let A be the completion of a smooth $\mathcal{O}_C$-algebra, and let $\mathfrak{X} = \mathrm{Spf} A$. Then there is an isomorphism in the derived category of sheaves on the pro-étale site of $\mathfrak{X}$,*

$$\mathbb{Z}_p^{\mathrm{BMS}}(r)_{\mathfrak{X}} \cong \tau_{\leq r} R\psi \mathbb{Z}_p(r)_{X,\acute{e}t}.$$

Here $\mathbb{Z}_p(r)_{X,\acute{e}t}$ denotes the pro-étale sheaf $\{\mu_{p^n}^{\otimes r}\}_n$ on the rigid analytic generic fiber X of $\mathfrak{X}$ and $R\psi$ is the nearby cycles functor.

The isomorphism in (1) above, combined with the main result of [52], gives us the identification of pro-objects

$$\mathbb{Z}_p^{\mathrm{BMS}}(r)_X \cong \{\mathbb{Z}/p^n(r)_{X,\text{ét}}\}_n$$

in the setting of (1).

Consider the case of a smooth $\mathcal{O}_K$-scheme X as before. Bhatt–Morrow–Scholze suggest in [18, REMARK 1.16] that $\mathbb{Z}_p^{\mathrm{BMS}}/p^n(r)_X$ should be Schneider's sheaf $\mathcal{S}_n(r)$, and by passage to the limit, there should be a distinguished triangle

$$i_*W\Omega^{r-1}_{\log}[-r-1] \to \mathbb{Z}_p^{\mathrm{BMS}}(r) \to \tau_{\leq r} Rj_*\big(\mathbb{Z}_p(r)_{V,\text{ét}}\big) \to i_*W\Omega^{r-1}_{\log}[-r]. \tag{6.6}$$

For X a smooth $\mathcal{O}_K$-scheme with associated formal scheme $\mathfrak{X}$ and special fiber $i : Y \to \mathfrak{X}$, this would give an isomorphism of $i^*\mathbb{Z}_p^{\mathrm{BMS}}/p^n(r)_{\mathfrak{X}}$ with the étale motivic complex $i^*\mathbb{Z}/p^n(r)_{\text{ét}}$ on $Y_{\text{ét}}$ considered by Geisser.

This has been proven in a work–in–progress by Bhargav Bhatt and Akhil Mathew [16]. They construct an isomorphism of a version of $\mathbb{Z}_p^{\mathrm{BMS}}/p^n(r)_X$ with Sato's sheaf $\mathfrak{T}_n(r)_X$ in the semi-stable case; using Zhong's extension of Geisser's results, this gives an isomorphism

$$i^*\mathbb{Z}_p^{\mathrm{BMS}}/p^n(r)_{\mathfrak{X}} \cong i^*\tau_{\leq r}\mathbb{Z}/p^n(r)_{\text{ét}}$$

in the semi-stable case.

One has the Geisser–Hesselholt isomorphism (Theorem 6.8) of étale K-theory and topological cyclic homology given by the cyclotomic trace map. Perhaps one can compare the localization pro-distinguished triangle

$$K(Y;\mathbb{Z}_p) \to K(X;\mathbb{Z}_p) \to K(X \setminus Y;\mathbb{Z}_p)$$

with the distinguished triangle (6.6). Assuming one does have the pro-isomorphism $\mathbb{Z}_p(r)_{\text{ét}} \cong \mathbb{Z}_p^{\mathrm{BMS}}(r)$ as suggested above, it would be interesting to see if the identification of the sheaves $\mathcal{S}_n(r)$ with the étale motivic complexes $\mathbb{Z}/p^n(r)_{\text{ét}}$ and the Atiyah–Hirzebruch spectral sequence from motivic cohomology to K-theory could yield a comparison with the spectral sequence corresponding to the motivic tower $\mathrm{Fil}^*\mathrm{TC}(\mathfrak{X};\mathbb{Z}_p)$ described above.

The sheaf $\mathbb{Z}_p^{\mathrm{BMS}}(r)$ is built from $\mathrm{TC}(-;\mathbb{Z}_p)$, which by the Geisser–Hesselholt theorem is p-completed étale K-theory. As we mentioned before, the Geisser–Hesselholt isomorphism arises at least in part from McCarthy's theorem identifying the relative K-theory and relative TC of the nilpotent thickenings $X/(\pi^n)$ of the special fiber Y. However, the motivic cohomology complexes do not detect the difference between $X/(\pi^n)$ and Y. Supposing again that one does have a pro-isomorphism $\mathbb{Z}_p(r)_{\text{ét}} \cong \mathbb{Z}_p^{\mathrm{BMS}}(r)$, this says that in mixed characteristic $(0, p)$, one can still see the K-theory of the thickened fibers $X/(\pi^n)$ reflected in the motivic complexes $\mathbb{Z}_p(r)_{\text{ét}}$.

I am not aware of a categorical framework for the tower $\mathrm{Fil}^n\mathrm{TC}(\tilde{A};\mathbb{Z}_p)$ and its layers $\mathbb{Z}_p^{\mathrm{BMS}}(r)$, analogous to the framework for Voevodsky's slice tower for K-theory given by $\mathrm{SH}(k)$. As $\mathbb{A}^1$-homotopy invariance fails for these theories, one would need a stable homotopy theory with a weaker invariance property, perhaps modeled on the one of the categories of motives with modulus discussed in the previous section, for these theories to find a home, in which the Bhatt–Morrow–Scholze tower (6.5) would be seen as a parallel to Voevodsky's slice tower.

FUNDING

The author gratefully acknowledges support from the DFG through the SPP 1786, and through a project that has received funding from the European Research Council (ERC) under the European Union's Horizon 2020 research and innovation programme (grant agreement No. 832833).

REFERENCES

[1] A. Ananyevskiy, SL-oriented cohomology theories. In *Motivic homotopy theory and refined enumerative geometry*, pp. 1–19, Contemp. Math. 745, Amer. Math. Soc., Providence, RI, 2020.

[2] A. Ananyevskiy, G. Garkusha, and I. Panin, Cancellation theorem for framed motives of algebraic varieties. *Adv. Math.* **383** (2021), 107681, 38 pp.

[3] A. Ananyevskiy, M. Levine, and I. Panin, Witt sheaves and the η-inverted sphere spectrum. *J. Topol.* **10** (2017), no. 2, 370–385.

[4] A. Ananyevskiy and A. Neshitov, Framed and MW-transfers for homotopy modules. *Selecta Math. (N.S.)* **25** (2019), 26.

[5] J. Ayoub, Les six opérations de Grothendieck et le formalisme des cycles évanescents dans le monde motivique. I. *Astérisque* **314** (2007), x+466 pp.

[6] J. Ayoub, Les six opérations de Grothendieck et le formalisme des cycles évanescents dans le monde motivique. II. *Astérisque* **315** (2007), vi+364 pp.

[7] T. Bachmann, The generalized slices of Hermitian K-theory. *J. Topol.* **10** (2017), no. 4, 1124–1144.

[8] T. Bachmann, B. Calmès, F. Déglise, J. Fasel, and P. A. Østvær, Milnor–Witt Motives. (2020), arXiv:2004.06634.

[9] T. Bachmann and E. Elmanto, Voevodsky's slice conjectures via Hilbert schemes. *Algebr. Geom.* **8** (2021), no. 5, 626–636.

[10] T. Bachmann and K. Wickelgren, $\mathbb{A}^1$–Euler classes: six functors formalisms, dualities, integrality and linear subspaces of complete intersections. 2021, arXiv:2002.01848.

[11] T. Bachmann and K. Wickelgren, On quadratically enriched excess and residual intersections. 2021, arXiv:2112.05960.

[12] J. Barge and F. Morel, Groupe de Chow des cycles orientés et classe d'Euler des fibrés vectoriels. *C. R. Acad. Sci. Paris Sér. I Math.* **330** (2000), no. 4, 287–290.

[13] A. A. Beilinson, Higher regulators and values of L-functions. *Itogi Nauki Tekh. Ser. Sovrem. Probl. Mat. Nov. Dostizh.* **24** (1984), no. 2, 181–238; *J. Sov. Math.* **30**, (1985), no. 2, 2036–2070.

[14] A. A. Beilinson, Height pairing between algebraic cycles. In *K-theory, arithmetic and geometry (Moscow, 1984–1986)*, pp. 1–25, Lecture Notes in Math. 1289, Springer, Berlin, 1987.

[15] P. Berthelot and A. Ogus, *Notes on crystalline cohomology*. Princeton University Press, Princeton, NJ; University of Tokyo Press, Tokyo, 1978.

[16] B. Bhatt and A. Mathew, p-adic étale twists and algebraic K-theory. Work in progress.

[17] B. Bhatt, M. Morrow, and P. Scholze, Integral p-adic Hodge theory. *Publ. Math. Inst. Hautes Études Sci.* **128** (2018), 219–397.

[18] B. Bhatt, M. Morrow, and P. Scholze, Topological Hochschild homology and integral p-adic Hodge theory. *Publ. Math. Inst. Hautes Études Sci.* **129** (2019), 199–310.

[19] F. Binda, D. Park, and R. A. Østvær, Triangulated categories of logarithmic motives over a field. 2020, arXiv:2004.12298.

[20] F. Binda and S. Saito, Relative cycles with moduli and regulator maps. *J. Inst. Math. Jussieu* **18** (2019), no. 6, 1233–1293.

[21] F. Binda et al., In *Motivic homotopy theory and refined enumerative geometry. Workshop on Motivic Homotopy Theory and Refined Enumerative Geometry, May 14–18, 2018, Universität Duisburg-Essen, Essen, Germany*, edited by F. Binda, M. Levine, M. T. Nguyen and O. Röndigs, Contemporary Mathematics, 745, American Mathematical Society, Providence, RI, 2020.

[22] S. Bloch, K_2 and algebraic cycles. *Ann. of Math. (2)* **99** (1974), 349–379.

[23] S. Bloch, K_2 of Artinian $\mathbb{Q}$-algebras, with application to algebraic cycles. *Comm. Algebra* **3** (1975), 405–428.

[24] S. Bloch, Algebraic cycles and higher K-theory. *Adv. Math.* **61** (1986), no. 3, 267–304.

[25] S. Bloch, The moving lemma for higher Chow groups. *J. Algebraic Geom.* **3** (1994), no. 3, 537–568.

[26] S. Bloch and H. Esnault, An additive version of higher Chow groups. *Ann. Sci. Éc. Norm. Supér. (4)* **36** (2003), no. 3, 463–477.

[27] S. Bloch and H. Esnault, The additive dilogarithm. Kazuya Kato's fiftieth birthday. *Doc. Math.* **Extra Vol.** (2003), 131–155.

[28] S. Bloch and K. Kato, p-adic étale cohomology. *Publ. Math. Inst. Hautes Études Sci.* **63** (1986), 107–152.

[29] S. Bloch and S. Lichtenbaum, A Spectral Sequence for Motivic Cohomology. Preprint, 1995, http://www.math.uiuc.edu/K-theory/0062/.

[30] S. Bloch and A. Ogus, Gersten's conjecture and the homology of schemes. *Ann. Sci. Éc. Norm. Supér. (4)* **7** (1974), 181–201.

[31] B. Calmès and J. Fasel, The category of finite Chow–Witt correspondences. Preprint, 2014, arXiv:1412.2989.

[32] J-L. Cathelineau, Remarques sur les différentielles des polylogarithmes uniformes. *Ann. Inst. Fourier (Grenoble)* **46** (1996), no. 5, 1327–1347.

[33] D-C. Cisinski and F. Déglise, *Triangulated categories of mixed motives*. Springer Monogr. Math., Springer, Cham, 2019.

[34] A. Dold and R. Thom, Quasifaserungen und unendliche symmetrische Produkte. *Ann. of Math. (2)* **67** (1958), 239–281.

[35] W. G. Dwyer and E. M. Friedlander, Algebraic and etale K-theory. *Trans. Amer. Math. Soc.* **292** (1985), no. 1, 247–280.

[36] E. Elmanto, M. Hoyois, A. A. Khan, V. Sosnilo, and M. Yakerson, Framed transfers and motivic fundamental classes. *J. Topol.* **13** (2020), no. 2, 460–500.

[37] E. Elmanto, M. Hoyois, A. A. Khan, V. Sosnilo, and M. Yakerson, Modules over algebraic cobordism. *Forum Math. Pi* **8** (2020), e14, 44 pp.

[38] E. Elmanto, M. Hoyois, A. A. Khan, V. Sosnilo, and M. Yakerson, Motivic infinite loop spaces. *Camb. J. Math.* **9** (2021), no. 2, 431–549.

[39] J. Fasel, *Groupes de Chow–Witt*. Mém. Soc. Math. Fr. (N.S.) 113, 2008.

[40] J-M. Fontaine and W. Messing, p-adic periods and p-adic étale cohomology. In *Current trends in arithmetical algebraic geometry (Arcata, Calif., 1985)*, pp. 179–207, Contemp. Math. 67, Amer. Math. Soc., Providence, RI, 1987.

[41] E. M. Friedlander, Étale K-theory. I. Connections with etale cohomology and algebraic vector bundles. *Invent. Math.* **60** (1980), no. 2, 105–134.

[42] E. M. Friedlander, Étale K-theory. II. Connections with algebraic K-theory. *Ann. Sci. Éc. Norm. Supér. (4)* **15** (1982), no. 2, 231–256.

[43] E. M. Friedlander and A. Suslin, The spectral sequence relating algebraic K-theory to motivic cohomology. *Ann. Sci. Éc. Norm. Supér. (4)* **35** (2002), no. 6, 773–875.

[44] G. Garkusha and I. Panin, On the motivic spectral sequence. *J. Inst. Math. Jussieu* **17** (2018), no. 1, 137–170.

[45] G. Garkusha and A. Neshitov, Fibrant resolutions for motivic Thom spectra. 2018, arXiv:1804.07621.

[46] G. Garkusha, A. Neshitov, and I. Panin, Framed motives of relative motivic spheres. *Trans. Amer. Math. Soc.* **374** (2021), no. 7, 5131–5161.

[47] G. Garkusha and I. Panin, Homotopy invariant presheaves with framed transfers. *Cambridge J. Math.* **8** (2020), no. 1, 1–94.

[48] G. Garkusha and I. Panin, Framed motives of algebraic varieties (after V. Voevodsky). *J. Amer. Math. Soc.* **34** (2021), no. 1, 261–313.

[49] T. Geisser, Motivic cohomology over Dedekind rings. *Math. Z.* **248** (2004), no. 4, 773–794.

[50] T. Geisser and L. Hesselholt, Topological cyclic homology of schemes. In *Algebraic K-theory (Seattle, WA, 1997)*, pp. 41–87, Proc. Sympos. Pure Math. 67, Amer. Math. Soc., Providence, RI, 1999.

[51] T. Geisser and L. Hesselholt, On the K-theory and topological cyclic homology of smooth schemes over a discrete valuation ring. *Trans. Amer. Math. Soc.* **358** (2006), no. 1, 131–145.

[52] T. Geisser and M. Levine, The K-theory of fields in characteristic p. *Invent. Math.* **139** (2000), no. 3, 459–493.

[53] H. Gillet, Riemann–Roch theorems for higher algebraic K-theory. *Adv. Math.* **40** (1981), no. 3, 203–289.

[54] D. R. Grayson, Weight filtrations via commuting automorphisms. *K-Theory* **9** (1995), no. 2, 139–172.

[55] J. J. Gutiérrez, C. Röndigs, M. Spitzweck, and P. A. Østvær, Motivic slices and coloured operads. *J. Topol.* **5** (2012), no. 3, 727–755.

[56] C. Haesemeyer and C. A. Weibel, *The norm residue theorem in motivic cohomology*. Ann. of Math. Stud. 200, Princeton University Press, Princeton, NJ, 2019.

[57] M. Hanamura, Mixed motives and algebraic cycles. I. *Math. Res. Lett.* **2** (1995), no. 6, 811–821.

[58] M. Hanamura, Mixed motives and algebraic cycles. III. *Math. Res. Lett.* **6** (1999), no. 1, 61–82.

[59] M. Hanamura, Mixed motives and algebraic cycles. II. *Invent. Math.* **158** (2004), no. 1, 105–179.

[60] L. Hesselholt, On the p-typical curves in Quillen's K-theory. *Acta Math.* **177** (1996), no. 1, 1–53.

[61] J. Hornbostel and M. Wendt, Chow-Witt rings of classifying spaces for symplectic and special linear groups. *J. Topol.* **12** (2019), no. 3, 916–966.

[62] M. Hoyois, The six operations in equivariant motivic homotopy theory. *Adv. Math.* **305** (2017), 197–279.

[63] M. Hoyois, The localization theorem for framed motivic spaces. *Compos. Math.* **157** (2021), no. 1, 1–11.

[64] A. Huber, *Mixed motives and their realization in derived categories*. Lecture Notes in Math. 1604, Springer, Berlin, 1995.

[65] A. Huber and S. Müller-Stach, *Periods and Nori motives. With contributions by Benjamin Friedrich and Jonas von Wangenheim*. Ergeb. Math. Grenzgeb. (3) 65, Springer, Cham, 2017.

[66] T. Ito, K. Kato, C. Nakayama, and S. Usui, On log motives. *Tunis. J. Math.* **2** (2020), no. 4, 733–789.

[67] R. Iwasa and W. Kai, Chern classes with modulus. *Nagoya Math. J.* **236** (2019), 84–133.

[68] U. Jannsen, Mixed motives, motivic cohomology, and Ext-groups. In *Proceedings of the International Congress of Mathematicians, Vol. 1, 2 (Zürich, 1994)*, pp. 667–679, Birkhäuser, Basel, 1995.

[69] B. Kahn, H. Miyazaki, S. Saito, and T. Yamazaki, Motives with modulus, I: Modulus sheaves with transfers for non-proper modulus pairs. *Épijournal Géom. Algébrique* **5** (2021), 1, 46 pp.

[70] B. Kahn, H. Miyazaki, S. Saito, and T. Yamazaki, Motives with modulus, II: modulus sheaves with transfers for proper modulus pairs. *Épijournal Géom. Algébrique* **5** (2021), 2, 31 pp.

[71] B. Kahn, H. Miyazaki, S. Saito, and T. Yamazaki, Motives with modulus, III: the categories of motives. 2021, arXiv:2011.11859.

[72] B. Kahn, S. Saito, and T. Yamazaki, Reciprocity sheaves. With two appendices by Kay Rülling. *Compos. Math.* **152** (2016), no. 9, 1851–1898.

[73] B. Kahn, S. Saito, and T. Yamazaki, Reciprocity sheaves, II. 2021, arXiv:1707.07398.

[74] W. Kai, A moving lemma for algebraic cycles with modulus and contravariance. *Int. Math. Res. Not. IMRN* **1** (2021), 475–522.

[75] W. Kai and H. Miyazaki, Suslin's moving lemma with modulus. *Ann. K-Theory* **3** (2018), no. 1, 55–70.

[76] J. L. Kass and K. Wickelgren, An arithmetic count of the lines on a smooth cubic surface. *Compos. Math.* **157** (2021), no. 4, 677–709.

[77] J. L. Kass and K. Wickelgren, The class of Eisenbud–Khimshiashvili–Levine is the local $\mathbb{A}^1$-Brouwer degree. *Duke Math. J.* **168** (2019), no. 3, 429–469.

[78] K. Kato, Milnor K-theory and the Chow group of zero cycles. In *Applications of algebraic K-theory to algebraic geometry and number theory, Part I, II (Boulder, CO, 1983)*, pp. 241–253, Contemp. Math. 55, Amer. Math. Soc., Providence, RI, 1986.

[79] K. Kato, On p-adic vanishing cycles (application of ideas of Fontaine–Messing). *Adv. Stud. Pure Math.* **10** (1987), 207–251.

[80] K. Kato and S. Saito, Global class field theory of arithmetic schemes. In *Applications of algebraic K-theory to algebraic geometry and number theory, Part I, II (Boulder, CO, 1983)*, pp. 255–331, Contemp. Math. 55, Amer. Math. Soc., Providence, RI, 1986.

[81] M. Kerz and S. Saito, Chow group of 0-cycles with modulus and higher-dimensional class field theory. *Duke Math. J.* **165** (2016), no. 15, 2811–2897.

[82] M. Kurihara, A note on p-adic étale cohomology. *Proc. Japan Acad. Ser. A Math. Sci.* **63** (1987), no. 7, 275–278.

[83] M. Levine, *Mixed motives*. Math. Surveys Monogr. 57, American Mathematical Society, Providence, RI, 1998.

[84] M. Levine, Techniques of localization in the theory of algebraic cycles. *J. Algebraic Geom.* **10** (2001), no. 2, 299–363.

[85] M. Levine, The homotopy coniveau tower. *J. Topol.* **1** (2008), no. 1, 217–267.

[86] M. Levine, Aspects of enumerative geometry with quadratic forms. *Doc. Math.* **25** (2020), 2179–2239.

[87] S. Lichtenbaum, Values of zeta-functions at nonnegative integers. In *Number theory (Noordwijkerhout, 1983)*, pp. 127–138, Lecture Notes in Math. 1068, Springer, Berlin, 1984.

[88] R. McCarthy, Relative algebraic K-theory and topological cyclic homology. *Acta Math.* **179** (1997), 197–222.

[89] A. S. Merkurjev and A. A. Suslin, K-cohomology of Severi–Brauer varieties and the norm residue homomorphism. *Izv. Akad. Nauk SSSR Ser. Mat.* **46** (1982), no. 5, 1011–1046, 1135–1136; *Math. USSR, Izv.* **21** (1983), no. 2, 307–340.

[90] J. S. Milne, Motivic cohomology and values of zeta functions. *Compos. Math.* **68** (1988), no. 1, 59–100, 101–102.

[91] H. Miyazaki, Private communication, Sept. 6, 2021.

[92] F. Morel, An introduction to $\mathbb{A}^1$-homotopy theory. In *Contemporary developments in algebraic K-theory*, pp. 357–441, ICTP Lect. Notes XV, Abdus Salam Int. Cent. Theoret. Phys., Trieste, 2004.

[93] F. Morel, *$\mathbb{A}^1$-algebraic topology over a field*. Lecture Notes in Math. 2052, Springer, Heidelberg, 2012.

[94] F. Morel and V. Voevodsky, $\mathbb{A}^1$-homotopy theory of schemes. *Publ. Math. Inst. Hautes Études Sci.* **90** (1999), 45–143.

[95] D. Mumford, Rational equivalence of 0-cycles on surfaces. *J. Math. Kyoto Univ.* **9** (1968), 195–204.

[96] Yu. P. Nesterenko and A. A. Suslin, Homology of the general linear group over a local ring, and Milnor's K-theory. *Izv. Akad. Nauk SSSR Ser. Mat.* **53** (1989), no. 1, 121–146; *Math. USSR, Izv.* **34** (1990), no. 1, 121–145.

[97] A. Ogus, *Lectures on logarithmic algebraic geometry*. Cambridge Stud. Adv. Math. 178, Cambridge University Press, Cambridge, 2018.

[98] J. Park, Regulators on additive higher Chow groups. *Amer. J. Math.* **131** (2009), no. 1, 257–276.

[99] P. Pelaez, Multiplicative properties of the slice filtration. *Astérisque* **335** (2011).

[100] P. Pelaez, On the functoriality of the slice filtration. *J. K-Theory* **11** (2013), no. 1, 55–71.

[101] D. Quillen, Higher algebraic K-theory, In *Proceedings of the International Congress of Mathematicians (Vancouver, BC, 1974), Vol. 1* pp. 171–176, Canad. Math. Congress, Montreal, QC, 1974.

[102] J. Riou, La conjecture de Bloch–Kato (d'aprés M. Rost et V. Voevodsky). *Astérisque* **361** (2014), Exp. No. 1073, x, 421–463.

[103] C. Röndigs and P. A. Østvær, Modules over motivic cohomology. *Adv. Math.* **219** (2008), no. 2, 689–727.

[104] K. Rülling, The generalized de Rham–Witt complex over a field is a complex of zero-cycles. *J. Algebraic Geom.* **16** (2007), no. 1, 109–169.

[105] S. Saito, Reciprocity sheaves and logarithmic motives. 2021, arXiv:2107.00381.

[106] K. Sato, p-adic étale Tate twists and arithmetic duality. With an appendix by Kei Hagihara. *Ann. Sci. Éc. Norm. Supér. (4)* **40** (2007), no. 4, 519–588.

[107] W. Scharlau, *Quadratic and Hermitian forms*. Grundlehren Math. Wiss. 270, Springer, Berlin, 1985.

[108] P. Schneider, p-adic points of motives. In *Motives (Seattle, WA, 1991)*, pp. 225–249, Proc. Sympos. Pure Math. 55, Amer. Math. Soc., Providence, RI, 1994.

[109] J.-P. Serre, *Groupes algébriques et corps de classes. 2nd edn.*, Publ. Inst. Math. Univ. Nancago 7, Actual. Sci. Ind., 1264, Hermann, Paris, 1984.

[110] M. Spitzweck, A commutative $\mathbb{P}^1$-spectrum representing motivic cohomology over Dedekind domains. *Mém. Soc. Math. Fr. (N.S.)* **157** (2018), 110 pp.

[111] A. Suslin, Lecture, K-theory conference at Luminy, 1988.

[112] A. Suslin and V. Voevodsky, Singular homology of abstract algebraic varieties. *Invent. Math.* **123** (1996), no. 1, 61–94.

[113] A. Suslin and V. Voevodsky, Bloch–Kato conjecture and motivic cohomology with finite coefficients. In *The arithmetic and geometry of algebraic cycles (Banff, AB, 1998)*, pp. 117–189, NATO Sci. Ser. C Math. Phys. Sci. 548, Kluwer Acad. Publ., Dordrecht, 2000.

[114] J. Tate, Relations between K_2 and Galois cohomology. *Invent. Math.* **36** (1976), 257–274.

[115] B. Totaro, Milnor K-theory is the simplest part of algebraic K-theory. *K-Theory* **6** (1992), no. 2, 177–189.

[116] V. Voevodsky, $\mathbb{A}^1$-homotopy theory. Proceedings of the International Congress of Mathematicians, Vol. I (Berlin, 1998). *Doc. Math.* **Extra Vol. I** (1998), 579–604.

[117] V. Voevodsky, Motivic cohomology groups are isomorphic to higher Chow groups in any characteristic. *Int. Math. Res. Not.* **7** (2002), 351–355.

[118] V. Voevodsky, Open problems in the motivic stable homotopy theory. I. Motives. In *polylogarithms and Hodge theory, Part I (Irvine, CA, 1998)*, pp. 3–34, Int. Press Lect. Ser. 3, Int. Press, Somerville, MA, 2002.

[119] V. Voevodsky, A possible new approach to the motivic spectral sequence for algebraic K-theory. In *Recent progress in homotopy theory (Baltimore, MD, 2000)*, pp. 371–379, Contemp. Math. 293, Amer. Math. Soc., Providence, RI, 2002.

[120] V. Voevodsky, Motivic cohomology with $\mathbb{Z}/2$-coefficients. *Publ. Math. Inst. Hautes Études Sci.* **98** (2003), 59–104.

[121] V. Voevodsky, Reduced power operations in motivic cohomology. *Publ. Math. Inst. Hautes Études Sci.* **98** (2003), 1–57.

[122] V. Voevodsky, On the zero slice of the sphere spectrum. *Tr. Mat. Inst. Steklova* **246** (2004); *Algebr. Geom. Metody, Svyazi Prilozh.* 106–115; reprinted in *Proc. Steklov Inst. Math.* **3(246)** (2004), 93–102.

[123] V. Voevodsky, Motivic Eilenberg–Maclane spaces. *Publ. Math. Inst. Hautes Études Sci.* **112** (2010), 1–99.

[124] V. Voevodsky, On motivic cohomology with $\mathbb{Z}/l$-coefficients. *Ann. of Math. (2)* **174** (2011), no. 1, 401–438.

[125] V. Voevodsky, Cancellation theorem. *Doc. Math.* **Extra vol.: Andrei A. Suslin sixtieth birthday** (2010), 671–685.

[126] V. Voevodsky, Notes on framed correspondences, https://www.uni-due.de/~bm0032/MotSemWS201617/VoevFramedCorr.pdf.

[127] V. Voevodsky, A. Suslin, and E. M. Friedlander, *Cycles, transfers, and motivic homology theories*. Ann. of Math. Stud. 143, Princeton University Press, Princeton, NJ, 2000.

[128] C. Zhong, Comparison of dualizing complexes. *J. Reine Angew. Math.* **695** (2014), 1–39.

MARC LEVINE

Universität Duisburg-Essen, Fakultät Mathematik, Campus Essen, 45117 Essen, Germany, marc.levine@uni-due.de

4. ALGEBRAIC AND COMPLEX GEOMETRY

HOMOLOGICAL KNOT INVARIANTS FROM MIRROR SYMMETRY

MINA AGANAGIC

ABSTRACT

In 1999, Khovanov showed that a link invariant known as the Jones polynomial is the Euler characteristic of a homology theory. The knot categorification problem is to find a general construction of knot homology groups, and to explain their meaning – what are they homologies of?

Homological mirror symmetry, formulated by Kontsevich in 1994, naturally produces hosts of homological invariants. Typically though, it leads to invariants which have no particular interest outside of the problem at hand.

I showed recently that there is a new family of mirror pairs of manifolds, for which homological mirror symmetry does lead to interesting invariants and solves the knot categorification problem. The resulting invariants are computable explicitly for any simple Lie algebra, and certain Lie superalgebras.

MATHEMATICS SUBJECT CLASSIFICATION 2020

Primary 57K18; Secondary 14J33, 53D37, 53D45

KEYWORDS

Homological mirror symmetry, knot homology theory, categorification

ICM INTERNATIONAL CONGRESS OF MATHEMATICIANS 2022 JULY 6–14

Proc. Int. Cong. Math. 2022, Vol. 3, pp. 2108–2144
DOI 10.4171/ICM2022/197

1. INTRODUCTION

There are many beautiful strands that connect mathematics and physics. Two of the most fruitful ones are knot theory and mirror symmetry. I will describe a new connection between them. We will find a solution to the knot categorification problem as a new application of homological mirror symmetry.

1.1. Quantum link invariants

In 1984, Jones constructed a polynomial invariant of links in $\mathbb{R}^3$ [42]. The Jones polynomial is defined by picking a projection of the link to a plane, the skein relation it satisfies

$$\mathfrak{q}^{n/2} P_{L_+} - \mathfrak{q}^{-n/2} P_{L_-} = (\mathfrak{q}^{1/2} - \mathfrak{q}^{-1/2}) P_{L_0}$$

where $n = 2$, and the value for the unknot. It has the same flavor as the Alexander polynomial, dating back to 1928 [8], which one gets by setting $n = 0$ instead.

The proper framework for these invariants was provided by Witten in 1988, who showed that they originate from three-dimensional Chern–Simons theory based on a Lie algebra ${}^L\mathfrak{g}$ [82]. In particular, the Jones polynomial comes from ${}^L\mathfrak{g} = \mathfrak{su}_2$ with links colored by the defining two-dimensional representation. The Alexander polynomial comes from the same setting by taking ${}^L\mathfrak{g}$ to be a Lie superalgebra $\mathfrak{gl}_{1|1}$. The resulting link invariants are known as the $U_{\mathfrak{q}}({}^L\mathfrak{g})$ quantum group invariants. The relation to quantum groups was discovered by Reshetikhin and Turaev [67].

1.2. The knot categorification problem

The quantum invariants of links are Laurant polynomials in $\mathfrak{q}^{1/2}$, with integer coefficients. In 1999, Khovanov showed [48,49] that one can associate to a projection of the link to a plane a bigraded complex of vector spaces

$$C^{*,j}(K) = \cdots C^{i-1,j}(K) \xrightarrow{\partial^{i-1}} C^{i,j}(K) \xrightarrow{\partial^{i}} \cdots,$$

whose cohomology $\mathcal{H}^{i,j}(K) = \ker \partial^i / \operatorname{im} \partial^{i-1}$ categorifies the Jones polynomial,

$$J_K(\mathfrak{q}) = \sum_{i,j} (-1)^i \mathfrak{q}^{j/2} \operatorname{rk} \mathcal{H}^{i,j}(K).$$

Moreover, the cohomology groups

$$\mathcal{H}^{*,*}(K) = \bigoplus_{i,j} \mathcal{H}^{i,j}(K)$$

are independent of the choice of projection; they are themselves link invariants.

1.2.1.

Khovanov's construction is part of the categorification program initiated by Crane and Frenkel [25], which aims to lift integers to vector spaces and vector spaces to categories.

A toy model of categorification comes from a Riemannian manifold M, whose Euler characteristic

$$\chi(M)=\sum_{k\in\mathbb{Z}}(-1)^k \dim \mathcal{H}^k(M)$$

is categorified by the cohomology $\mathcal{H}^k(M)=\ker d_k/\mathrm{im}\, d_{k-1}$ of the de Rham complex

$$C^* = \cdots C^{k-1} \xrightarrow{d_{k-1}} C^k \xrightarrow{d_k} \cdots.$$

The Euler characteristic is, from the physics perspective, the partition function of supersymmetric quantum mechanics with M as a target space $\chi(M)=\mathrm{Tr}(-1)^F e^{-\beta H}$, with Laplacian $H=dd^*+d^*d$ as the Hamiltonian, and $d=\sum_k d_k$ as the supersymmetry operator. If h is a Morse function on M, the complex can be replaced by a Morse–Smale–Witten complex C_h^* with the differential $d_h=e^h d e^{-h}$. The complex C_h^* is the space of perturbative ground states of a σ-model on M with potential h [81]. The action of the differential d_h is generated by solutions to flow equations, called instantons.

1.2.2.

Khovanov's remarkable categorification of the Jones polynomial is explicit and easily computable. It has generalizations of similar flavor for ${}^L\mathfrak{g}=\mathfrak{su}_n$, and links colored by its minuscule representations [51].

In 2013, Webster showed [78] that for any ${}^L\mathfrak{g}$, there exists an algebraic framework for categorification of $U_\mathfrak{q}({}^L\mathfrak{g})$ invariants of links in $\mathbb{R}^3$, based on a derived category of modules of an associative algebra. The KLRW algebra, defined in [78], generalizes the algebras of Khovanov and Lauda [50] and Rouquier [68]. Unlike Khovanov's construction, Webster's categorification is anything but explicit.

1.2.3.

Despite the successes of the program, one is missing a fundamental principle which explains why is categorification possible – the construction has no right to exist. Unlike in our toy example of categorification of the Euler characteristic of a Riemanniann manifold, Khovanov's construction and its generalizations did not come from either geometry or physics in any unified way. The problem Khovanov initiated is to find a general framework for link homology, that works uniformly for all Lie algebras, explains what link homology groups are, and why they exist.

1.3. Homological invariants from mirror symmetry

The solution to the problem comes from a new relation between mirror symmetry and representation theory.

Homological mirror symmetry relates pairs of categories of geometric origin [55]: a derived category of coherent sheaves and a version of the derived Fukaya category, in which complementary aspects of the theory are simple to understand. Occasionally, one can make mirror symmetry manifest, by showing that both categories are equivalent to a derived category of modules of a single algebra.

I will describe a new family of mirror pairs, in which homological mirror symmetry can be made manifest and leads to the solution to the knot categorification problem [1,2]. Many special features exist in this family, in part due to its deep connections to representation theory. As a result, the theory is solvable explicitly, as opposed to only formally [4,5].

1.4. The solution

We will get not one, but two solutions to the knot categorification problem. The first solution [1] is based on $\mathscr{D}_{\mathcal{X}}$, the derived category of equivariant coherent sheaves on a certain holomorphic symplectic manifold $\mathcal{X}$, which plays a role in the geometric Langlands correspondence. Recently, Webster proved that $\mathscr{D}_{\mathcal{X}}$ is equivalent to $\mathscr{D}_{\mathscr{A}}$, the derived category of modules of an algebra $\mathscr{A}$ which is a cylindrical version of the KLRW algebra from [79,80]. The generalization allows the theory to describe links in $\mathbb{R}^2 \times S^1$, as well as in $\mathbb{R}^3$.

The second solution [2] is based on $\mathscr{D}_Y$, the derived Fukaya–Seidel category of a certain manifold Y with potential W. The theory generalizes Heegard–Floer theory [63,64,66], which categorifies the Alexander polynomial, from ${}^L\mathfrak{g} = \mathfrak{gl}_{1|1}$, to arbitrary ${}^L\mathfrak{g}$.

The two solutions are related by equivariant homological mirror symmetry, which is not an equivalence of categories, but a correspondence of objects and morphisms coming from a pair of adjoint functors. In $\mathscr{D}_{\mathcal{X}}$, we will learn which question we need to ask to obtain $U_{\mathfrak{q}}({}^L\mathfrak{g})$ link homology. In $\mathscr{D}_Y$, we will learn how to answer it.

In [5], we give an explicit algorithm for computing homological link invariants from $\mathscr{D}_Y$, for any simple Lie algebra ${}^L\mathfrak{g}$ and links colored by its minuscule representations. It has an extension to Lie superalgebras ${}^L\mathfrak{g} = \mathfrak{gl}_{m|n}$ and $\mathfrak{sp}_{m|2n}$. In [4], we set the mathematical foundations of $\mathscr{D}_Y$ and prove (equivariant) homological mirror symmetry relating it to $\mathscr{D}_{\mathcal{X}}$.

2. KNOT INVARIANTS AND CONFORMAL FIELD THEORY

Most approaches to categorification of $U_{\mathfrak{q}}({}^L\mathfrak{g})$ link invariants start with quantum groups and their modules. We will start by recalling how quantum groups came into the story. The seeming detour will help us understand how $U_{\mathfrak{q}}({}^L\mathfrak{g})$ link invariants arise from geometry, and what categorifies them.

2.1. Knizhnik–Zamolodchikov equation and quantum groups

Chern–Simons theory associates to a punctured Riemann surface $\mathcal{A}$ a vector space, its Hilbert space. As Witten showed [82], the Hilbert space is finite dimensional, and spanned by vectors that have a name. They are known as conformal blocks of the affine Lie algebra $\widehat{{}^L\mathfrak{g}}_\kappa$. The effective level κ is an arbitrary complex number, related to $\mathfrak{q}$ by $\mathfrak{q} = e^{\frac{2\pi i}{\kappa}}$. While in principle arbitrary representations of ${}^L\mathfrak{g}$ can occur, in relating to geometry and categorification, we will take them to be minuscule.

To get invariants of knots in $\mathbb{R}^3$, one typically takes $\mathcal{A}$ to be a complex plane with punctures. It is equivalent, but for our purposes better, to take $\mathcal{A}$ to be an infinite complex cylinder. This way, we will be able to describe invariants of links in $\mathbb{R}^2 \times S^1$, as well.

2.1.1.

Every conformal block, and hence every state in the Hilbert space, can be obtained explicitly as a solution to a linear differential equation discovered by Knizhnik and Zamolodchikov in 1984 [53]. The KZ equation we get is of trigonometric type, schematically

$$\kappa\, \partial_i\, \mathcal{V} = \sum_{j \neq i} r_{ij}(a_i/a_j)\, \mathcal{V}, \tag{2.1}$$

since $\mathcal{A}$ is an infinite cylinder. Here, $\partial_i = a_i \frac{\partial}{\partial a_i}$, where a_i is any of the n punctures in the interior of $\mathcal{A}$, colored by a representation V_i of ${}^L\mathfrak{g}$. The right hand side of (2.1) is given in terms of classical r-matrices of ${}^L\mathfrak{g}$, and acts irreducibly on a subspace of $V_1 \otimes \cdots \otimes V_n$ of a fixed weight ν, where $\mathcal{V}$ takes values [32,33].

The KZ equations define a flat connection on a vector bundle over the configuration space of distinct points $a_1, \ldots, a_n \in \mathcal{A}$. The flatness of the connection is the integrability condition for the equation.

2.1.2.

The monodromy problem of the KZ equation, which is to describe analytic continuation of its fundamental solution along a path in the configuration space, has an explicit solution. Drinfeld [30] and Kazhdan and Lustig [47] proved that that the monodromy matrix $\mathfrak{B}$ of the KZ connection is a product of R-matrices of the $U_{\mathfrak{q}}({}^L\mathfrak{g})$ quantum group corresponding to ${}^L\mathfrak{g}$. The R-matrices describe reorderings of neighboring pairs of punctures.

The monodromy matrix $\mathfrak{B}$ is the Chern–Simons path integral on $\mathcal{A} \times [0,1]$ in presence of a colored braid. By composing braids, we get a representation of the affine braid group based on the $U_{\mathfrak{q}}({}^L\mathfrak{g})$ quantum group, acting on the space of solutions to the KZ equation. The braid group is affine, since $\mathcal{A}$ is a cylinder and not a plane.

2.1.3.

Any link can be represented as a plat closure of some braid. The Chern-Simons path integral together with the link computes a very specific matrix element of the braiding matrix $\mathfrak{B}$, picked out by a pair of states in the Hilbert space corresponding to the top and the bottom of Figure 1.

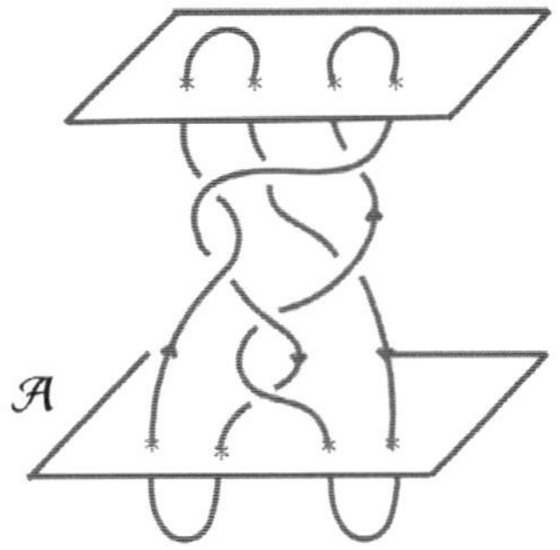

FIGURE 1
Every link arises as a plat closure of a braid.

These states, describing a collection of cups and caps, are very special solutions of the KZ equation in which pairs of punctures, colored by conjugate representations V_i and V_i^*, come together and fuse to disappear. In this way, both fusion and braiding enter the problem.

2.2. A categorification wishlist

To categorify $U_{\mathfrak{q}}({}^L\mathfrak{g})$ invariants of links in $\mathbb{R}^3$, we would like to associate, to the space of conformal blocks of $\widehat{{}^L\mathfrak{g}}$ on the Riemann surface $\mathcal{A}$, a bigraded category, which in addition to the cohomological grading has a grading associated to $\mathfrak{q}$. Additional $\mathrm{rk}({}^L\mathfrak{g})$ gradings are needed to categorify invariants of links in $\mathbb{R}^2 \times S^1$, as they depend on the choice of a flat ${}^L\mathfrak{g}$ connection around the S^1. To braids, we would like to associate functors between the categories corresponding to the top and bottom. To links, we would like to associate a vector space whose elements are morphisms between the objects of the categories associated to the top and bottom, up to the action of the braiding functor. Moreover, we would like to do this in a way that recovers quantum link invariants upon decategorification. One typically proceeds by coming up with a category, and then works to prove that decategorification gives the link invariants one set out to categorify. A virtue of the solutions in [1,2] is that the second step is automatic.

3. MIRROR SYMMETRY

Mirror symmetry is a string duality which relates σ-models on a pair of Calabi–Yau manifolds $\mathcal{X}$ and $\mathcal{Y}$. Its mathematical imprint are relations between very different problems in complex geometry of $\mathcal{X}$ ("B-type") and symplectic geometry of $\mathcal{Y}$ ("A-type"), and vice versa.

Mirror symmetry was discovered as a duality of σ-models on closed Riemann surfaces D. In string theory, one must allow Riemann surfaces with boundaries. This enriches the theory by introducing "branes," which are boundary conditions at ∂D and naturally objects of a category [9].

By asking how mirror symmetry acts on branes turned out to yield deep insights into mirror symmetry. One such insight is due to Strominger, Yau, and Zaslov [75], who showed that in order for every point-like brane on $\mathcal{X}$ to have a mirror on $\mathcal{Y}$, mirror manifolds have to be fibered by a pair of (special Lagrangian) dual tori T and $T^\vee$, over a common base.

3.1. Homological mirror symmetry

Kontsevich conjectured in his 1994 ICM address [55] that mirror symmetry should be understood as an equivalence of a pair of categories of branes, one associated to complex geometry of $\mathcal{X}$, the other to symplectic geometry of $\mathcal{Y}$.

The category of branes associated to complex geometry of $\mathcal{X}$ is the derived category of coherent sheaves,

$$\mathscr{D}_{\mathcal{X}} = D^b\mathrm{Coh}_T(\mathcal{X}).$$

Its objects are "B-type branes," supported on complex submanifolds of $\mathcal{X}$. The category of branes associated to symplectic geometry is the derived Fukaya category

$$\mathscr{D}_{\mathcal{Y}} = D\mathrm{Fuk}(\mathcal{Y}),$$

whose objects are "A-type branes," supported on Lagrangian submanifolds of $\mathcal{Y}$, together with a choice of orientation and a flat bundle. For example, mirror symmetry should map the structure sheaf of a point in $\mathcal{X}$ to a Lagrangian brane in $\mathcal{Y}$ supported on a $T^\vee$ fiber. The choice of a flat $U(1)$ connection is the position of the point in the dual fiber T.

Kontsevich' homological mirror symmetry is a conjecture that the category of B-branes on $\mathcal{X}$ and the category of A-branes on $\mathcal{Y}$ are equivalent,

$$\mathscr{D}_{\mathcal{X}} \cong \mathscr{D}_{\mathcal{Y}},$$

and that this equivalence characterizes what mirror symmetry is.

3.2. Quantum differential equation and its monodromy

Knizhnik–Zamolodchikov equation, which plays a central role in knot theory, has a geometric counterpart. In the world of mirror symmetry, there is an equally fundamental differential equation,

$$\partial_i \mathcal{V}_\alpha - (C_i)_\alpha^\beta \, \mathcal{V}_\beta = 0. \tag{3.1}$$

The equation is known as the quantum differential equation of $\mathcal{X}$. Both the equation and its monodromy problem featured prominently, starting with the first papers on mirror symmetry, see [37] for an early account. In (3.1), $(C_i)_\alpha^\beta = C_{\gamma_i \alpha\delta}\eta^{\delta\beta}$ is a connection on a vector bundle with fibers $H^{\text{even}}(\mathcal{X}) = \bigoplus_k H^k(\mathcal{X}, \wedge^k T^*_{\mathcal{X}})$ over the complexified Kahler moduli space. The derivative stands for $\partial_i = a_i \frac{\partial}{\partial a_i}$, so that $\partial_i a^d = (\gamma_i, d) a^d$ for a curve of degree $d \in H_2(\mathcal{X})$. The connection comes from quantum multiplication with classes $\gamma_i \in H^2(\mathcal{X})$. Given three de Rham cohomology classes on $\mathcal{X}$, their quantum product

$$C_{\alpha\beta\gamma} = \sum_{d\geq 0, d\in H_2(\mathcal{X})} (\alpha, \beta, \gamma)_d \; a^d \tag{3.2}$$

is a deformation of the classical cup product $(\alpha, \beta, \gamma)_0 = \int_{\mathcal{X}} \alpha \wedge \beta \wedge \gamma$ coming from Gromov–Witten theory of $\mathcal{X}$: $(\alpha, \beta, \gamma)_d$ is computed by an integral over the moduli space of degree d holomorphic maps from $D = \mathbb{P}^1$ to $\mathcal{X}$ whose image meets classes Poincaré dual of α, β and γ at points. The quantum product, together with the invariant inner product $\eta_{\alpha\beta} = \int_{\mathcal{X}} \alpha \wedge \beta$, gives rise to an associative algebra with structure constants $C_{\alpha\beta}{}^\delta = C_{\alpha\beta\gamma}\eta^{\gamma\delta}$. Flatness of the connection follows from the WDVV equations [27, 31, 83].

From the mirror perspective of $\mathcal{Y}$, the connection is the classical Gauss–Manin connection on the vector bundle over the moduli space of complex structures on $\mathcal{Y}$, with fibers the mid-dimensional cohomology $H^d(\mathcal{Y})$ as mirror symmetry identifies $H^k(\mathcal{X}, \wedge^k T^*_{\mathcal{X}})$ with $H^k(\mathcal{Y}, \wedge^{d-k} T^*_{\mathcal{Y}})$.

3.2.1.

Solutions to the equation live in a vector space, spanned by K-theory classes of branes [22,36,41,46]. These are B-type branes on $\mathcal{X}$, objects of $\mathscr{D}_{\mathcal{X}}$, and A-type branes on $\mathcal{Y}$, objects of $\mathscr{D}_{\mathcal{Y}}$. A characteristic feature is that the equation and its solutions mix the A- and B-type structures on the same manifold.

From the perspective of $\mathcal{X}$, the solutions of the quantum differential equation come from Gromov–Witten theory. They are obtained by counting holomorphic maps from a domain curve D to $\mathcal{X}$, where D is best thought of as an infinite cigar [39,40] together with insertions of a class in $\alpha \in H^*_{\text{even}}(\mathcal{X})$ at the origin, and $[\mathcal{F}] \in K(\mathcal{X})$ at infinity. The latter is the K-theory class of a B-type brane $\mathcal{F} \in \mathscr{D}_{\mathcal{X}}$ which serves as the boundary condition at the S^1 boundary at infinity of D. In the mirror $\mathcal{Y}$, the A- and B-type structures get exchanged. In the interior of D, supersymmetry is preserved by B-type twist, and at the boundary at infinity we place an A-type brane $\mathcal{L} \in \mathscr{D}_{\mathcal{Y}}$, whose K-theory class picks which solution of the equation we get.

3.2.2.

One of the key mirror symmetry predictions is that monodromy of the quantum differential equation gets categorified by the action of derived autoequivalences of $\mathscr{D}_{\mathcal{X}}$. It is related by mirror symmetry to the monodromy of the Gauss–Manin connection, computed by Picard–Lefshetz theory, whose categorification by $\mathscr{D}_{\mathcal{Y}}$ is developed by Seidel [71].

The flat section $\mathcal{V}$ of the connection in (3.1) has a close cousin. This is Douglas' [9,28,29] Π-stability central charge function $\mathcal{Z}^0 : K(\mathscr{D}) \to \mathbb{C}$, whose existence motivated Bridgeland's formulation of stability conditions [17]. The Π-stability central charge $\mathcal{Z}^0$ arises from the same setting as $\mathcal{V}$, except one places trivial insertions at the origin of D. This implies that monodromies of $\mathcal{V}$ and $\mathcal{Z}^0$ coincide [22]. In the context of the mirror $\mathcal{Y}$, given any brane $\mathcal{L} \in \mathscr{D}_{\mathcal{Y}}$, its central charge is simply $\mathcal{Z}^0[\mathcal{L}] = \int_{\mathcal{L}} \Omega$, where Ω is the top holomorphic form on $\mathcal{Y}$. The stable objects are special Lagrangians, on which the phase of Ω is constant. By mirror symmetry, monodromy of $\mathcal{Z}^0$ is expected to induce the action of monodromy on $\mathscr{D}_{\mathcal{X}}$. Examples of braid group actions on the derived categories include works of Khovanov and Seidel [52], Seidel and Thomas [74], and others [18,77].

4. HOMOLOGICAL LINK INVARIANTS FROM B-BRANES

The Knizhnik–Zamolodchikov equation not only has the same flavor as the quantum differential equation, but for some very special choices of $\mathcal{X}$, they coincide. For the time being, we will take ${}^L\mathfrak{g}$ to be simply laced, so it coincides with its Langlands dual $\mathfrak{g}$.

4.1. The geometry

The manifold $\mathcal{X}$ may be described as the moduli space of G-monopoles on

$$\mathbb{R}^3 = \mathbb{R} \times \mathbb{C}, \tag{4.1}$$

with prescribed singularities. The monopole group G is related to ${}^LG$, the Chern–Simons gauge group, by Langlands or electric–magnetic duality. In Chern–Simons theory, the knots are labeled by representations of ${}^LG$ and viewed as paths of heavy particles, charged electrically under ${}^LG$. In the geometric description, the same heavy particles appear as singular, Dirac-type monopoles of the Langlands dual group G. The fact the magnetic description is what is needed to understand categorification was anticipated by Witten [84–87].

4.1.1.

Place a singular G monopole for every finite puncture on $\mathcal{A} \cong \mathbb{R} \times S^1$, at the point on $\mathbb{R}$ obtained by forgetting the S^1. Singular monopole charges are elements of the cocharacter lattice of G, which Langlands duality identifies with the character lattice of ${}^LG$. Pick the charge of the monopole to be the highest weight μ_i of the ${}^LG$ representation V_i coloring the puncture. The relative positions of singular monopoles on $\mathbb{R}^3$ are the moduli of the metric on $\mathcal{X}$, so we will hold them fixed.

The smooth monopole charge is a positive root of ${}^LG$; choose it so that the total monopole charge is the weight ν of subspace of representation $\bigotimes_i V_i$, where the conformal blocks take values. For our current purpose, it suffices to assume

$$\nu = \sum_i \mu_i - \sum_{a=1}^{rk} d_a \, {}^Le_a, \tag{4.2}$$

is a dominant weight; ${}^Le_a$ are the simple positive roots of ${}^L\mathfrak{g}$. Provided μ_i are minuscule co-weights of G and no pairs of singular monopoles coincide, the monopole moduli space $\mathcal{X}$ is a smooth hyper-Kahler manifold of dimension

$$\dim_{\mathbb{C}}(\mathcal{X}) = 2\sum_a d_a.$$

It is parameterized, in part, by positions of smooth monopoles on $\mathbb{R}^3$.

4.1.2.

A choice of complex structure on $\mathcal{X}$ reflects a split of $\mathbb{R}^3$ as $\mathbb{R} \times \mathbb{C}$. The relative positions of singular monopoles on $\mathbb{C}$ become the complex structure moduli, and the relative positions of monopoles on $\mathbb{R}$ the Kahler moduli.

This identifies the complexified Kahler moduli space of $\mathcal{X}$ (where the Kahler form gets complexified by a periodic two-form) with the configuration space of n distinct punctures on $\mathcal{A} = \mathbb{R} \times S^1$, modulo overall translations, as in Figure 2.

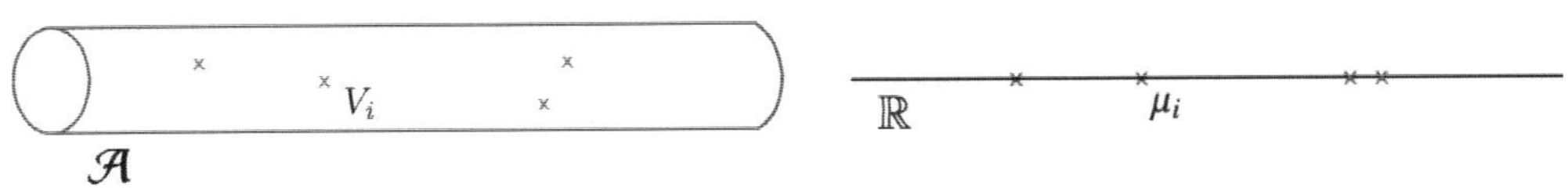

FIGURE 2
Punctures on $\mathcal{A}$ correspond to singular G-monopoles on $\mathbb{R} \in \mathbb{R} \times \mathbb{C}$.

4.1.3.

As a hyper-Kahler manifold, $\mathcal{X}$ has more symmetries than a typical Calabi–Yau. For its quantum cohomology to be nontrivial, and for the quantum differential equation to coincide with the KZ equation, we need to work equivariantly with respect to a torus action that scales its holomorphic symplectic form

$$\omega^{2,0} \to \mathfrak{q}\, \omega^{2,0}.$$

For this to be a symmetry, we will place all the singular monopoles at the origin of $\mathbb{C}$; $\mathcal{X}$ has a larger torus of symmetries

$$T = \Lambda \times \mathbb{C}^\times_{\mathfrak{q}},$$

where Λ preserves the holomorphic symplectic form, and comes from the Cartan torus of G. The equivariant parameters of the Λ-action correspond to the choice of a flat ${}^L G$ connection of Chern–Simons theory on $\mathbb{R}^2 \times S^1$.

4.1.4.

The same manifold $\mathcal{X}$ has appeared in mathematics before, as a resolution of a transversal slice in the affine Grassmannian $\mathrm{Gr}_G = G((z))/G[[z]]$ of G, often denoted by

$$\mathcal{X} = \mathrm{Gr}^{\vec{\mu}}{}_{\nu}. \tag{4.3}$$

The two are related by thinking of monopole moduli space $\mathcal{X}$ as obtained by a sequence of Hecke modifications of holomorphic G-bundles on $\mathbb{C}$ [45].

Manifold $\mathcal{X}$ is also the Coulomb branch of a 3d quiver gauge theory with $\mathcal{N} = 4$ supersymmetry, with quiver based on the Dynkin diagram of $\mathfrak{g}$, see e.g. [19]. The ranks of the flavor and gauge symmetry groups are determined from the weights μ and ν.

4.1.5.

The vector $\vec{\mu} = (\mu_1, \ldots, \mu_n)$ in (4.3) encodes singular monopole charges, and the order in which they appear on $\mathbb{R}$, and ν is the total monopole charge. The ordering of entries of $\vec{\mu}$ is a choice of a chamber in the Kahler moduli. We will suppress $\vec{\mu}$ for the most part, and denote all the corresponding distinct symplectic manifolds by $\mathcal{X}$.

4.1.6.

By a recent theorem of Danilenko [26], the Knizhnik–Zamolodchikov equation corresponding to the Riemann surface $\mathcal{A} = \mathbb{R} \times S^1$, with punctures colored by minuscule representations V_i of ${}^L\mathfrak{g}$, coincides with the quantum differential equation of the T-equivariant Gromov–Witten theory on $\mathcal{X} = \mathrm{Gr}^{\vec{\mu}}{}_{\nu}$.

This has many deep consequences.

4.2. Branes and braiding

Since the KZ equation is the quantum-differential equation of T-equivariant Gromov–Witten theory of $\mathcal{X}$, the space of its solutions gets identified with $K_T(\mathcal{X})$, the T-equivariant K-theory of $\mathcal{X}$.

This is the K-group of the category of its B-type branes, the derived category of T-equivariant coherent sheaves on $\mathcal{X}$,

$$\mathscr{D}_{\mathcal{X}} = D^b\mathrm{Coh}_T(\mathcal{X}).$$

This connection between the KZ equation and $\mathscr{D}_{\mathcal{X}}$ is the starting point for our first solution of the categorification problem.

4.2.1.

A colored braid with n strands in $\mathcal{A} \times [0, 1]$ has a geometric interpretation as a path in the complexified Kahler moduli of $\mathcal{X}$ that avoids singularities, as the order of punctures on $\mathcal{A}$ corresponds to a choice of chamber in the Kahler moduli of $\mathcal{X}$.

The monodromy of the quantum differential equation along this path acts on $K_T(\mathcal{X})$. Since the quantum differential equation coincides with the KZ equation, by the theorem of [26], $K_T(\mathcal{X})$ becomes a module for $U_{\mathfrak{q}}({}^L\mathfrak{g})$, corresponding to the weight ν subspace of representation $V_1 \otimes \cdots \otimes V_n$.

The fact that derived equivalences of $\mathscr{D}_{\mathcal{X}}$ categorify this action is not only an expectation, but also a theorem by Bezrukavnikov and Okounkov [14], whose proof makes use of quantization of $\mathcal{X}$ in characteristic p.

4.2.2.

From physics perspective, the reason derived equivalences of $\mathscr{D}_{\mathcal{X}}$ had to categorify the action of monodromy of the quantum differential equation on $K_T(\mathcal{X})$ is as follows.

Braid group acts, in the σ-model on the cigar D from Section 3.2.1, by letting the moduli of $\mathcal{X}$ vary according to the braid near the boundary at infinity. The Euclidean time, running along the cigar, is identified with the time along the braid. This leads to a Berry phase-type problem studied by Cecotti and Vafa [22]. It follows that the σ-model on the annulus, with moduli that vary according to the braid, computes the matrix element of the monodromy $\mathfrak{B}$, picked out by a pair of branes $\mathcal{F}_0$ and $\mathcal{F}_1$ at its boundaries.

The σ-model on the same Euclidian annulus, where we take the time to run around S^1 instead, computes the index of the supercharge Q preserved by the two branes. The cohomology of Q is computed by $\mathscr{D}_{\mathcal{X}}$ as its most basic ingredient, the space of morphisms

$$\mathrm{Hom}^{*,*}_{\mathscr{D}_{\mathcal{X}}}(\mathscr{B}\mathcal{F}_0, \mathcal{F}_1)$$

between the branes. This is the space of supersymmetric ground states of the σ-model on a strip, obtained by cutting the annulus open. We took here all the variations of moduli to happen near one boundary, at the expense of changing a boundary condition from $\mathcal{F}_0$ to $\mathscr{B}\mathcal{F}_0$. This does not affect the homology [1,35], for the very same reason the theory depends on the homotopy type of the braid only. Per construction, the graded Euler characteristic of the homology theory, computed by closing the strip back up to the annulus, is the braiding matrix element,

$$\chi(\mathscr{B}\,\mathcal{F}_0, \mathcal{F}_1) = (\mathfrak{B}\,\mathcal{V}_0, \mathcal{V}_1), \tag{4.4}$$

between the conformal blocks $\mathcal{V}_{0,1} = \mathcal{V}[\mathcal{F}_{0,1}]$ of the two branes.

Thus, by viewing the same Euclidian annulus in two different ways, we learn that the braid group action on the derived category

$$\mathscr{B} : \mathscr{D}_{\mathcal{X}_{\vec{\mu}}} \to \mathscr{D}_{\mathcal{X}_{\vec{\mu}'}}, \tag{4.5}$$

manifestly categorifies the monodromy matrix $\mathfrak{B} \in U_{\mathfrak{q}}({}^L\mathfrak{g})$ of the KZ equation.

4.3. Link invariants from perverse equivalences

The quantum $U_{\mathfrak{q}}({}^L\mathfrak{g})$ invariants of knots and links are matrix elements of the braiding matrix $\mathfrak{B}$, so they too will be categorified by $\mathscr{D}_{\mathcal{X}}$, provided we can identify objects $\mathcal{U} \in \mathscr{D}_{\mathcal{X}}$ which serve as cups and caps.

Conformal blocks corresponding to cups and caps are defined using fusion [62]. The geometric analogue of fusion, in terms of $\mathcal{X}$ and its category of branes, was shown in [1] to be the existence of certain perverse filtrations on $\mathscr{D}_{\mathcal{X}}$, defined by abstractly by Chuang and Rouqiuer [24]. The utility of perverse filtrations for understanding the action of braiding on $\mathscr{D}_{\mathcal{X}}$ parallels the utility of fusion in describing the action of braiding in conformal field theory. In particular, it leads to identification of the cup and cap branes $\mathcal{U}$ we need, and a simple proof that $\mathrm{Hom}^{*,*}_{\mathscr{D}_{\mathcal{X}}}(\mathscr{B}\mathcal{U}, \mathcal{U})$ are homological invariants of links [1].

4.3.1.

As we bring a pair of punctures at a_i and a_j on $\mathcal{A}$ together, we get a new natural basis of solutions to the KZ equation, called the fusion basis, whose virtue is that it diagonalizes braiding. The possible eigenvectors are labeled by the representations

$$V_i \otimes V_j = \bigotimes_{m=0}^{m_{\max}} V_{k_m}, \tag{4.6}$$

that occur in the tensor product of representations V_i and V_j labeling the punctures. Because V_i and V_j are minuscule representations, the nonzero multiplicities on the right-hand side are all equal to 1. The cap arises as a special case, obtained by starting with a pair of conjugate representations V_i and $V_i^{\star}$, and picking the trivial representation in their tensor product.

4.3.2.

From perspective of $\mathcal{X}$, a pair of singular monopoles of charges μ_i and μ_j are coming together on $\mathbb{R}$, as in Figure 2, and we approach a wall in Kahler moduli at which $\mathcal{X}$ develops a singularity. At the singularity, a collection of cycles vanishes. This is due to monopole bubbling phenomena described by Kapustin and Witten in [45].

The types of monopole bubbling that can occur are labeled by representations V_{k_m} that occur in the tensor product $V_i \otimes V_j$. The moduli space of monopoles whose positions we need to tune for the bubbling of type V_{k_m} to occur is $\mathrm{Gr}^{(\mu_i,\mu_j)}_{\mu_{k_m}} = T^*F_{k_m}$, where μ_{k_m} is the highest weight of V_{k_m}. This space is transverse to the locus where exactly $\mu_i + \mu_j - \mu_{k_m}$ monopoles have bubbled off [1]. It has a vanishing cycle F_{k_m}, corresponding to the representation V_{k_m}, as its zero section. (Viewing $\mathcal{X}$ as the Coulomb branch, monopole bubbling is related to partial Higgsing phenomena.)

4.3.3.

Conformal blocks which diagonalize the action of braiding do not in general come from actual objects of the derived category $\mathscr{D}_{\mathcal{X}}$. As is well known from Picard–Lefshetz theory, eigensheaves of braiding, branes on which the braiding acts only by degree shifts $\mathscr{B}\mathcal{E} = \mathcal{E}[D_{\mathcal{E}}]\{C_{\mathcal{E}}\}$, are very rare.

What one gets instead [1] is a filtration

$$\mathscr{D}_{k_0} \subset \mathscr{D}_{k_1} \subset \cdots \subset \mathscr{D}_{k_{\max}} = \mathscr{D}_{\mathcal{X}}, \tag{4.7}$$

by the order of vanishing of the Π-stability central charge $\mathcal{Z}^0 : K(\mathcal{X}) \to \mathbb{C}$. More precisely, one gets a pair of such filtrations, one on each side of the wall. Crossing the wall preserves the filtration, but has the effect of mixing up branes at a given order in the filtration, with those at lower orders, whose central charge vanishes faster. Because $\mathcal{X}$ is hyper-Kahler, the Π-stability central charge is given in terms of classical geometry (by Eq. (4.7) of [1]).

The existence of the filtration with the stated properties follows from the existence of the equivariant central charge function $\mathcal{Z}$,

$$\mathcal{Z} : K_T(\mathcal{X}) \to \mathbb{C}, \tag{4.8}$$

and the fact the action of braiding on $K_T(\mathcal{X})$ lifts to the action on $\mathscr{D}_{\mathcal{X}}$, by the theorem of [14]. The equivariant central charge $\mathcal{Z}$ is computed by the equivariant Gromov–Witten theory on $\mathcal{X}$ in a manner analogous to $\mathcal{V}$, starting with the σ-model on the cigar D except with no insertion at its tip. It reduces to the Π-stability central charge $\mathcal{Z}^0$ by turning the equivariant parameters off.

4.3.4.

While $\mathscr{B}$ has few eigensheaves in $\mathscr{D}_{\mathcal{X}}$, it acts by degree shifts

$$\mathscr{B} : \mathscr{D}_{k_m}/\mathscr{D}_{k_{m-1}} \to \mathscr{D}_{k_m}/\mathscr{D}_{k_{m-1}}[D_{k_m}]\{C_{k_m}\}, \tag{4.9}$$

on the quotient subcategories. The degree shifts may be read off from the eigenvectors of the action of braiding on the equivariant central charge function $\mathcal{Z}$. As the punctures at a_i and a_j come together, the eigenvector corresponding to the representation V_{k_m} in (4.6), vanishes as [1]

$$\mathcal{Z}_{k_m} = (a_i - a_j)^{D_{km} + C_{km}/\kappa} \times \text{finite}.$$

It follows that braiding a_i and a_j counterclockwise acts by

$$\mathcal{Z}_{k_m} \to (-1)^{D_{km}} \mathfrak{q}^{\frac{1}{2} C_{km}} \mathcal{Z}_{k_m}.$$

The cohomological degree shift $D_{k_m} = \dim_{\mathbb{C}} F_{k_m}$ is by the dimension of the vanishing cycle. The equivariant degree shift C_{k_m} is essentially the one familiar from the action of braiding on conformal blocks of $\widehat{^L\mathfrak{g}}$ in the fusion basis [1].

4.3.5.

The derived equivalences of this type are the perverse equivalences of Chuang and Rouquier [23, 24]. They envisioned them as a way to describe derived equivalences which come from variations of Bridgeland stability conditions, but with few examples from geometry.

Traditionally, braid group actions on derived categories of coherent sheaves, or B-branes, are fairly difficult to describe, see for example [20, 21]. Braid group actions on the categories of A-branes are much easier to understand, via Picard–Lefshetz theory and its categorical uplifts [71], see e.g. [52, 77]. The theory of variations of stability conditions, by Douglas and Bridgeland, was invented to bridge the two [9, 29].

4.3.6.

As a by-product, we learn that conformal blocks describing collections cups or caps colored by minuscule representations, come from branes in $\mathscr{D}_{\mathcal{X}}$ which have a simple geometric meaning [1].

Take $\mathcal{X} = \mathrm{Gr}_0^{(\mu_1, \mu_1^*, \ldots, \mu_d, \mu_d^*)}$ corresponding to $\mathcal{A}$ with $n = 2d$ punctures, colored by pairs of complex conjugate, minuscule representations V_i and V_i^*. We get a vanishing cycle U in $\mathcal{X}$ which is a product of d minuscule Grassmannians,

$$U = G/P_1 \times \cdots \times G/P_d,$$

where P_i is the maximal parabolic subgroup of G associated to representation V_i. This vanishing cycle embeds in $\mathcal{X}$ as a compact holomorphic Lagrangian, so in the neighborhood of U, we can model $\mathcal{X}$ as T^*U. The structure sheaf

$$\mathcal{U} = \mathcal{O}_U \in \mathscr{D}_{\mathcal{X}}$$

of U is the brane we are after. The Grassmannian G/P_i is the cycle that vanishes when a single pair of singular monopoles of charges μ_i and μ_i^* come together, as $\mathrm{Gr}_0^{(\mu_i, \mu_i^*)} = T^*G/P_i$.

The brane $\mathcal{U}$ lives at the very bottom of a d-fold filtration which $\mathscr{D}_{\mathcal{X}}$ develops at the intersection of d walls in the Kahler moduli of $\mathcal{X}$ corresponding to bringing punctures together pairwise. It follows $\mathcal{U}$ is the eigensheaf of braiding each pair of matched endpoints. It is extremely special, for the same reason the trivial representation is special.

4.3.7.

Just as fusion provides the right language to understand the action of braiding in conformal field theory, the perverse filtrations provide the right language to describe the action of braiding on derived categories. Using perverse filtrations and the very special properties of the vanishing cycle branes $\mathcal{U} \in \mathscr{D}_{\mathcal{X}}$, one gets the following theorem [1]:

Theorem 1. *For any simply laced Lie algebra* $^L\mathfrak{g}$, *the homology groups*

$$\mathrm{Hom}^{*,*}_{\mathscr{D}_{\mathcal{X}}}(\mathscr{B}\mathcal{U}, \mathcal{U}),$$

categorify $U_{\mathfrak{q}}(^L\mathfrak{g})$ *quantum link invariants, and are themselves link invariants.*

4.3.8.

As an illustration, proving that (the equivalent of) the pitchfork move in the figure below holds in $\mathscr{D}_{\mathcal{X}}$

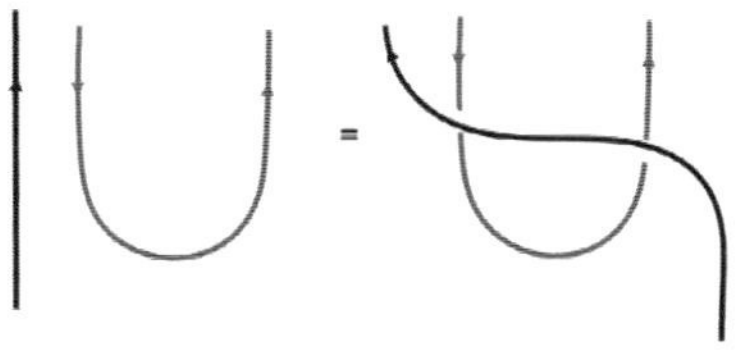

FIGURE 3
A move equivalent to the pitchfork move.

requires showing that we have a derived equivalence

$$\mathscr{B} \circ \mathscr{C}_i \cong \mathscr{C}_i'', \tag{4.10}$$

where $\mathscr{C}_i$ and $\mathscr{C}_i''$ are cup functors on the right and the left in Figure 3, respectively. They increase the number of strands by two and map

$$\mathscr{C}_i : \mathscr{D}_{\mathcal{X}_{n-2}} \rightarrow \mathscr{D}_{\mathcal{X}_n} \quad \text{and} \quad \mathscr{C}_i'' : \mathscr{D}_{\mathcal{X}_{n-2}} \rightarrow \mathscr{D}_{\mathcal{X}_n''},$$

where the subscript serves to indicate the number of strands. The functor $\mathscr{B}$ is the equivalence of categories from the theorem of **[14]**

$$\mathscr{B} : \mathscr{D}_{\mathcal{X}_n} \rightarrow \mathscr{D}_{\mathcal{X}_n''},$$

corresponding to braiding $V_k(a_k)$ with $V_i(a_i) \otimes V_i^*(a_j)$ where V_i and V_i^* color the red and V_k the black strand in Figure 3.

To prove the identity (4.10) note that

$$\mathscr{C}_i \mathscr{D}_{\mathcal{X}_{n-2}} \subset \mathscr{D}_{\mathcal{X}_n} \quad \text{and} \quad \mathscr{C}_i'' \mathscr{D}_{\mathcal{X}_{n-2}} \subset \mathscr{D}_{\mathcal{X}_n''}, \tag{4.11}$$

are the subcategories which are the bottom-most part of the double filtrations of $\mathscr{D}_{\mathcal{X}_n}$ and $\mathscr{D}_{\mathcal{X}_n''}$, corresponding to the intersection of walls at which the three punctures come together. By the definition of perverse filtrations, the functor $\mathscr{B}$ acts at a bottom part of a double filtration at most by degree shifts. The degree shifts are trivial too, since if they were not, the relation we are after would not hold even in conformal field theory, and we know it does. To complete the proof, one recalls that a perverse equivalence that acts by degree shifts that are trivial is an equivalence of categories **[24]**.

Proofs of invariance under the Reidermeister 0 and the framed Reidermeister I moves are similar. The invariance under Reidermeister II and III moves follows from the theorem of **[14]**. One should compare this to proofs of the same relations in **[20, 21]**, which are more technical and less general.

4.3.9.

An elementary consequence is a geometric explanation of mirror symmetry which relates the $U_{\mathfrak{q}}({}^L\mathfrak{g})$ invariants of a link K and its mirror reflection K^*.

It is a consequence of a basic property of $\mathscr{D}_{\mathcal{X}}$, Serre duality. Serre duality implies the isomorphism of homology groups on $\mathcal{X}$ which is a $2d$ complex-dimensional Calabi–Yau manifold,

$$\mathrm{Hom}_{\mathscr{D}_{\mathcal{X}}}\big(\mathscr{B}\,\mathcal{U}, \mathcal{U}[M]\{J_0, \vec{J}\}\big) = \mathrm{Hom}_{\mathscr{D}_{\mathcal{X}}}\big(\mathscr{B}\mathcal{U}, \mathcal{U}[2d - M]\{-d - J_0, -\vec{J}\}\big). \quad (4.12)$$

The equivariant degree shift comes from the fact the unique holomorphic section of the canonical bundle has weight d under the $\mathbb{C}_{\mathfrak{q}}^{\times} \subset \mathrm{T}$ action. Mirror symmetry follows by taking Euler characteristic of both sides [1].

4.4. Algebra from B-branes

Bezrukavnikov and Kaledin, using quantization in characteristic p, constructed a tilting vector bundle $\mathcal{T}$, on any $\mathcal{X}$ which is a symplectic resolution [12, 13, 43, 44]. Its endomorphism algebra

$$\mathscr{A} = \mathrm{Hom}^*_{\mathscr{D}_{\mathcal{X}}}(\mathcal{T}, \mathcal{T})$$

is an ordinary associative algebra, graded only by equivariant degrees. The derived category $\mathscr{D}_{\mathscr{A}}$ of its modules is equivalent to $\mathscr{D}_{\mathcal{X}}$,

$$\mathscr{D}_{\mathcal{X}} \cong \mathscr{D}_{\mathscr{A}},$$

essentially per definition.

Webster recently computed the algebra $\mathscr{A}$ for our $\mathcal{X}$ [80], and showed that it coincides with a cylindrical version of the KLRW algebra from [78]. Working with the cylindrical KLRW algebra, as opposed to the ordinary one, leads to invariants of links in $\mathbb{R}^2 \times S^1$ and not just in $\mathbb{R}^3$. The KLRW algebra generalizes the algebras of Khovanov and Lauda [50] and Rouquier [68]. The cylindrical version of the KLR algebra corresponds to $\mathcal{X}$ which is a Coulomb branch of a pure 3D gauge theory.

4.4.1.

The description of link homologies via $\mathscr{D}_{\mathcal{X}} = D\mathrm{Coh}_T(\mathcal{X})$ provides a geometric meaning of homological $U_{\mathfrak{q}}({}^L\mathfrak{g})$ link invariants. Even so, without further input, the description of link homologies either in terms of $\mathscr{D}_{\mathcal{X}}$ or $\mathscr{D}_{\mathscr{A}}$ is purely formal. With the help of (equivariant) homological mirror symmetry, we will give a description of link homology groups which is explicit and explicitly computable; in this sense, link homology groups come to life in the mirror.

5. MIRROR SYMMETRY FOR MONOPOLE MODULI SPACE

In the very best instances, homological mirror symmetry relating $\mathscr{D}_{\mathcal{Y}}$ and $\mathscr{D}_{\mathcal{X}}$ can be made manifest, by showing that each is equivalent to $\mathscr{D}_{\mathscr{A}}$, the derived category of modules of the same associative algebra $\mathscr{A}$,

$$\mathscr{D}_{\mathcal{X}} \cong \mathscr{D}_{\mathscr{A}} \cong \mathscr{D}_{\mathcal{Y}}. \tag{5.1}$$

The algebra

$$\mathscr{A} = \mathrm{Hom}^*_{\mathscr{D}}(\mathcal{T}, \mathcal{T})$$

is the endomorphism algebra of a set of branes $\mathcal{T} = \bigoplus_{\mathcal{C}} \mathcal{T}_{\mathcal{C}}$, which generate $\mathscr{D}_{\mathcal{X}}$ and $\mathscr{D}_{\mathcal{Y}}$. For economy, we will be denoting branes related by mirror symmetry by the same letter.

An elementary example [10] is mirror symmetry relating a pair of infinite cylinders, $\mathcal{X} = \mathbb{C}^\times$ and $\mathcal{Y} = \mathbb{R} \times S^1$, whose torus fibers are dual S^1's. Both $\mathscr{D}_{\mathcal{X}}$, the derived category of coherent sheaves on $\mathcal{X}$, and $\mathscr{D}_{\mathcal{Y}}$, based on the wrapped Fukaya category, are generated by a single object $\mathcal{T}$, a flat line bundle on $\mathcal{X}$ and a real-line Lagrangian on $\mathcal{Y}$. Their algebras of open strings are the same, equal to the algebra $\mathscr{A} = \mathbb{C}[x^{\pm 1}]$ of holomorphic functions on the cylinder.

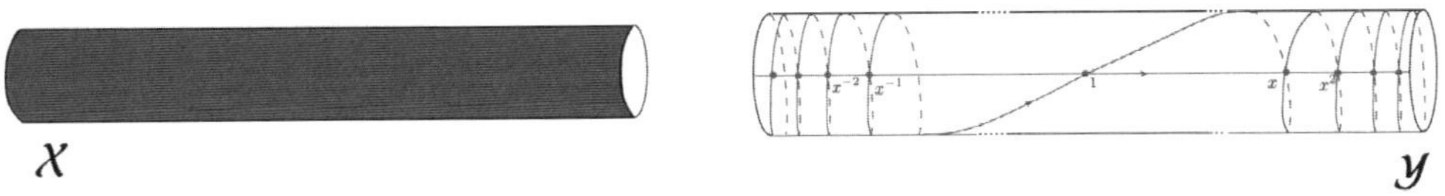

FIGURE 4
A simple example of manifest mirror symmetry.

5.1. The algebra for homological mirror symmetry

In our setting, the generator $\mathcal{T}$ of $\mathscr{D}_{\mathcal{X}}$ is the tilting generator of Bezrukavnikov and Kaledin from Section 4.4. Webster's proof of the equivalence of categorification of $U_{\mathfrak{q}}({}^L\mathfrak{g})$ link invariants and B-type branes on $\mathcal{X}$ and via the cKLRW algebra $\mathscr{A}$ should be understood as the first of the two equivalences in (5.1).

5.1.1.

The mirror $\mathcal{Y}$ of $\mathcal{X}$ is the moduli space of G monopoles, of the same charges as $\mathcal{X}$ except on $\mathbb{R}^2 \times S^1$ instead of on $\mathbb{R}^3$, with only complex and no Kahler moduli turned on, and equipped with a potential [2]. Without the potential, the mirror to $\mathcal{Y}$ would be another moduli space of G monopoles on $\mathbb{R}^2 \times S^1$.

The theory based on $\mathscr{D}_{\mathcal{Y}}$, the derived Fukaya–Seidel category of $\mathcal{Y}$, is in the same spirit as the work of Seidel and Smith [72]. They pioneered geometric approaches to link homology, but produced a only singly graded theory, known as symplectic Khovanov homology. The computation of $\mathscr{D}_{\mathcal{Y}}$, which makes mirror symmetry in (5.1) manifest, is given in the joint work with Danilenko, Li, and Zhou [4].

5.2. The core of the monopole moduli space

Working equivariantly with respect to a $\mathbb{C}^\times_{\mathfrak{q}}$-symmetry which scales the holomorphic symplectic form of $\mathcal{X}$, all the information about its geometry should be encoded in a core locus preserved by such actions.

The core X is a singular holomorphic Lagrangian in $\mathcal{X}$ which is the union of supports of all stable envelopes [7,61]. Equivalently, X is the union of all attracting sets of Λ-torus actions on $\mathcal{X}$, where we let Λ vary over all chambers. If we view $\mathcal{X}$ as the monopole moduli space, we can put this more simply: X is the locus where all the monopoles, singular or not, are at the origin of $\mathbb{C}$ in $\mathbb{R} \times \mathbb{C}$. Viewing it as a Coulomb branch, X is the locus at which the complex scalar fields in vector multiplets vanish.

We will define the equivariant mirror Y of $\mathcal{X}$ to be the ordinary mirror of its core, so we have

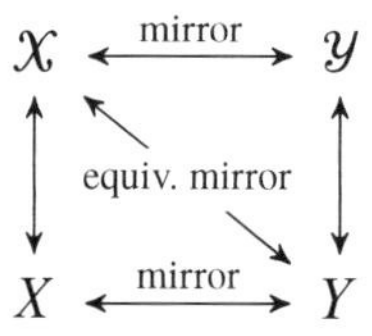

Working equivariantly with respect to the T-action on $\mathcal{X}$, the equivariant mirror gets a potential W, making the theory on Y into a Landau–Ginsburg model. While X embeds into $\mathcal{X}$ as a holomorphic Lagrangian of dimension d, $\mathcal{Y}$ fibers over Y with holomorphic Lagrangian $(\mathbb{C}^{\times})^d$ fibers.

5.2.1.

A model example is $\mathcal{X}$ which is the resolution of an A_{n-1} hypersurface singularity, $uv = z^n$; $\mathcal{X}$ is the moduli space of a single smooth $G = \mathrm{SU}(2)/\mathbb{Z}_2$ monopole, in the presence of n singular ones. The core X is a collection of $n-1$ $\mathbb{P}^1$'s with a pair of infinite discs attached, as in Figure 5.

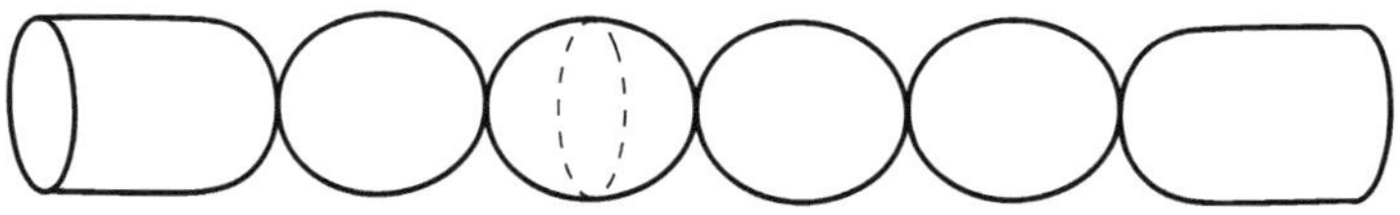

FIGURE 5
Core X of a resolution of the A_{n-1} singularity.

The ordinary mirror $\mathcal{Y}$ of $\mathcal{X}$ is the complex structure deformation of the "multiplicative" A_{n-1} surface singularity, with a potential which we will not need. $\mathcal{Y}$ is a $\mathbb{C}^{\times}$ fibration over Y which is itself an infinite cylinder, a copy of $\mathbb{C}^{\times}$ with n points deleted. At the marked points, the $\mathbb{C}^{\times}$ fibers degenerate. There are $n-1$ Lagrangian spheres in $\mathcal{Y}$, which are mirror to $n-1$ $\mathbb{P}^1$'s in $\mathcal{X}$. They project to Lagrangians in Y which begin and end at the punctures.

5.2.2.

The model example corresponds to ${}^L G = \mathrm{SU}(2)$ Chern–Simons theory on $\mathbb{R}^2 \times S^1$, and $\widehat{\mathfrak{su}_2}$ conformal blocks on $\mathcal{A} = \mathbb{R} \times S^1$. The n punctures on $\mathcal{A}$ are colored by the fundamental, two-dimensional representation $V_{1/2}$ of $\mathfrak{su}_2$, and we take the subspace of weight

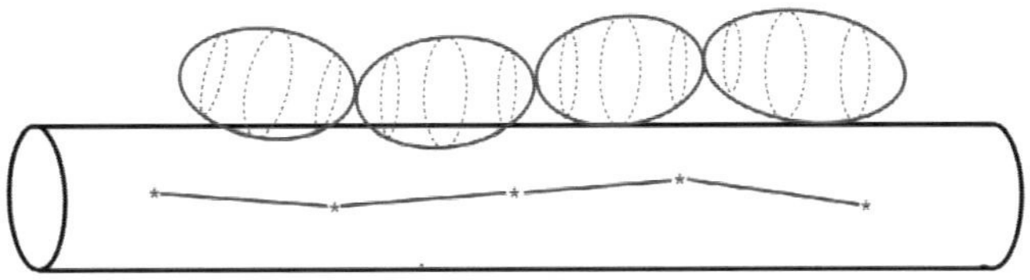

FIGURE 6
Lagrangian spheres in $\mathcal{Y}$ mirror the vanishing $\mathbb{P}^1$'s in $\mathcal{X}$.

one level below the highest. Note that Y coincides with the Riemann surface $\mathcal{A}$ where the conformal blocks live. This is not an accident.

In the model example, both X and Y are S^1 fibrations over $\mathbb{R}$ with n marked points. At the marked points, the S^1 fibers of X degenerate. In Y, this is mirrored by fibers that decompactify, due to points which are deleted.

5.2.3.

More generally, for $\mathcal{X} = \mathrm{Gr}^{\vec{\mu}}{}_{\nu}$ we have d_a smooth G-monopoles colored by simple roots ${}^L e_a$ and otherwise identical. It follows that the common base of SYZ fibrations of X and Y is the configuration space of the smooth monopoles on the real line $\mathbb{R}$ with n marked points. The marked points are labeled by the weights μ_i of ${}^L\mathfrak{g}$, which are the singular monopole charges.

An explicit description of Y, as well as its category of A-branes $\mathscr{D}_Y$, is given **[4]**. Here we will only describe some of its features. In an open set, Y coincides with

$$Y_0 = \bigotimes_{a=1}^{rk} \mathrm{Sym}^{d_a} \mathcal{A},$$

the configuration space of $d = \sum_{a=1}^{rk} d_a$ points on the punctured Riemann surface $\mathcal{A}$, "colored" by simple roots ${}^L e_a$ of ${}^L\mathfrak{g}$, but otherwise identical. The open set is the complement of the divisor of zeros and of poles of function f^0 in (5.5).

The top holomorphic form on Y is

$$\Omega = \bigwedge_{a=1}^{rk} \bigwedge_{\alpha=1}^{d_a} \frac{dy_{\alpha,a}}{y_{\alpha,a}}, \tag{5.2}$$

where $y_{\alpha,a}$ are coordinates on d copies of $\mathcal{A}$, viewed as the complex plane with 0 and ∞ deleted. While Ω itself is not globally well defined, so K_Y is not trivial, $\Omega^{\otimes 2}$ is well defined and

$$2c_1(K_Y) = 0. \tag{5.3}$$

This allows $\mathscr{D}_Y$ to have a $\mathbb{Z}$-valued cohomological grading. The symplectic form on Y is inherited from the symplectic form on $\mathcal{Y}$, by restricting it to the vanishing $(S^1)^d$ in each of its $(\mathbb{C}^\times)^d$ fibers over Y **[4]**. The precise choice of symplectic structure is the one compatible with mirror symmetry which we used to define Y, as the equivariant mirror of $\mathcal{X} = \mathrm{Gr}^{\vec{\mu}}{}_{\nu}$ and the ordinary mirror of its core.

Including the equivariant T-equivariant action on $\mathcal{X}$ and X corresponds to adding to the σ-model on Y a potential

$$W = \lambda_0 W^0 + \sum_{a=1}^{rk} \lambda_a W^a, \tag{5.4}$$

which is a multivalued holomorphic function on Y; λ_a are the equivariant parameters of the Λ-action on $\mathcal{X}$, and

$$\mathfrak{q} = e^{2\pi i \lambda_0}.$$

The potentials W^0 and W^a are given by

$$W^0 = \ln f^0, \quad W^a = \ln \prod_{\alpha=1}^{d_a} y_{a,\alpha},$$

where

$$f^0(y) = \prod_{a=1}^{rk} \prod_{\alpha=1}^{d_a} \frac{\prod_i (1 - a_i / y_{\alpha,a})^{\langle {}^L e_a, \mu_i \rangle}}{\prod_{(b,\beta)\neq(a,\alpha)} (1 - y_{\beta,b}/y_{\alpha,a})^{\langle {}^L e_a, {}^L e_b \rangle / 2}}. \tag{5.5}$$

The superpotential W breaks the conformal invariance of the σ-model to Y if $\lambda_0 \neq 0$, since only a quasihomogenous superpotential is compatible with it. This is mirror to breaking of conformal invariance on $\mathcal{X}$ by the $\mathbb{C}^\times_{\mathfrak{q}}$-action for $\mathfrak{q} \neq 1$.

Since W^0 and W^a are multivalued, Y is equipped with a collection of closed one-forms with integer periods

$$c^0 = dW^0/2\pi i, \quad c^a = dW^a/2\pi i \in H^1(Y, \mathbb{Z}),$$

which introduce additional gradings in the category of A-branes, as in [73].

5.2.4.

From the mirror perspective, the conformal blocks of $\widehat{{}^L\mathfrak{g}}$ come from the B-twisted Landau–Ginsburg model (Y, W) on D which is an infinitely long cigar, with A-type boundary condition at infinity corresponding to a Lagrangian $L \in Y$. The partition function of the theory has the following form:

$$\mathcal{V}_\alpha[L] = \int_L \Phi_\alpha \, \Omega \, e^{-W}, \tag{5.6}$$

where Φ_α are chiral ring operators, inserted at the tip of the cigar [22, 39, 40]. By placing the trivial insertion at the origin instead, we get the equivariant central charge function $\mathcal{Z}[L] = \int_L \Omega \, e^{-W}$; by further turning the equivariant parameters off, the potential W vanishes and the equivariant central charge becomes the ordinary brane central charge $\mathcal{Z}^0[L] = \int_L \Omega$.

We have (re)discovered, from mirror symmetry, an integral representation of the conformal blocks of $\widehat{{}^L\mathfrak{g}}$. This "free field representation" of conformal blocks, remarkable for its simplicity [32], goes back to the 1980s work of Kohno and Feigin and Frenkel [34, 54], and of Schechtman and Varchenko [69, 70].

5.2.5.

There is a reconstruction theory, due to Givental [38] and Teleman [76], which says that, starting with the solution of the quantum differential equation or its mirror counterpart, one gets to reconstruct all genus topological string amplitudes for any semisimple 2D field theory. The semisimplicity condition is satisfied in our case, as W has isolated critical points. It follows the B-twisted Landau–Ginsburg model on (Y, W) and A-twisted T-equivariant sigma model on $\mathcal{X}$ are equivalent to all genus [2]. Thus, equivariant mirror symmetry holds as an equivalence of topological string amplitudes.

5.3. Equivariant Fukaya–Seidel category

For every A-brane L at the boundary at infinity of the cigar D , we get a solution of the KZ equation. The brane is an object of

$$\mathscr{D}_Y = D\big(\mathcal{F}S(Y, W)\big),$$

the derived Fukaya–Seidel category of Y with potential W. The category should be thought of as a category of equivariant A-branes, due to the fact W in (5.4) is multivalued. Another novel aspect of $\mathscr{D}_Y$ is that it provides an example of Fukaya–Seidel category with coefficients in perverse schobers. This structure, inherited from equivariant mirror symmetry, was discovered in [4].

5.3.1.

Objects of $\mathscr{D}_Y$ are Lagrangians in Y, equipped with some extra data. A Lagrangian in Y is a product of d one-dimensional curves on $\mathcal{A}$ which are colored by simple roots and may be immersed; or a simplex obtained from an embedded curve, as a configuration space of d partially ordered colored points. The theory also includes more abstract branes, which are iterated mapping cones over morphisms between Lagrangians.

5.3.2.

The extra data includes a grading by Maslov and equivariant degrees. The equivariant grading of a brane in $\mathscr{D}_Y$ is defined by choosing a lift of the phase of e^{-W} to a real-valued function on the Lagrangian L. The equivariant degree shift operation,

$$L \to L\{\vec{d}\},$$

with $\vec{d} \in \mathbb{Z}^{rk+1}$, corresponds to changing the lift of W on L, now viewed as a graded Lagrangian, $W|_{L\{\vec{d}\}} = W|_L + 2\pi i \vec{\lambda} \cdot \vec{d}$. This is analogous to how a choice of a lift of the phase of $\Omega^{\otimes 2}$ defines the Maslov, or cohomological, grading of a Lagrangian. This restricts the Lagrangians that give rise to objects of $\mathscr{D}_Y$ to those for which such lifts can be defined.

More generally, branes in $\mathscr{D}_Y$ are graded Lagrangians L equipped with an extra structure of a local system Λ of modules of a certain algebra $\mathscr{B}$ we will describe shortly. For the time being, only branes for which Λ is trivial will play a role for us.

5.3.3.

The space of morphisms between a pair of Lagrangian branes in a derived Fukaya category

$$\mathrm{Hom}^{*,*}_{\mathscr{D}_Y}(L_0, L_1) = \ker Q / \mathrm{im}\, Q,$$

is defined by Floer theory, which itself is modeled after Morse theory approach to supersymmetric quantum mechanics, from the introduction. The role of the Morse complex is taken by the Floer complex.

For branes equipped with a trivial local system, the Floer complex

$$\mathrm{CF}^{*,*}(L_0, L_1) = \bigoplus_{\mathscr{P} \in L_0 \cap L_1} \mathbb{C}\mathscr{P} \tag{5.7}$$

is a graded vector space spanned by the intersection points of the two Lagrangians, together with the action of a differential Q. The complex is graded by the fermion number, which is the Maslov index, and the equivariant gradings, thanks to the fact W is multivalued.

The action of the differential on this space

$$Q : \mathrm{CF}^{*,*}(L_0, L_1) \to \mathrm{CF}^{*+1,*}(L_0, L_1)$$

is generated by instantons. In Floer theory, the coefficient of $\mathscr{P}'$ in $Q\mathscr{P}$ is obtained by "counting" holomorphic strips in Y with boundary on L_0 and L_1, interpolating from $\mathscr{P}$ to $\mathscr{P}'$, of Maslov index 1 and equivariant degree 0. The cohomology of the resulting complex is Floer cohomology.

5.3.4.

A simplification in the present case is that, just as branes have a description in terms of the Riemann surface, so do their intersection points, as well as the maps between them.

The theory that results is a generalization of Heegard–Floer theory, which is associated to ${}^L\mathfrak{g} = \mathfrak{gl}_{1|1}$ and categorifies the Alexander polynomial [63,64]. Heegard–Floer theory has target $Y_{\mathfrak{gl}_{1|1}} = \mathrm{Sym}^d(\mathcal{A})$, the symmetric product of d copies of $\mathcal{A}$. $Y_{\mathfrak{gl}_{1|1}}$ should be thought of as a configuration space of fermions on the Riemann surface, as opposed to anyons for $Y_{\mathfrak{su}_2}$; in particular, their top holomorphic forms differ.

While we so far assumed that ${}^L\mathfrak{g}$ is simply laced, the $\mathscr{D}_Y$ has an extension to non-simply-laced Lie algebras, as well as $\mathfrak{gl}_{m|n}$ and $\mathfrak{sp}_{m|2n}$ Lie superalgebras, described in [3,5].

5.4. Link invariants and equivariant mirror symmetry

Mirror symmetry helps us understand exactly which questions we need to ask to recover homological knot invariants from Y.

5.4.1.

Since Y is the ordinary mirror of X, we should start by understanding how to recover homological knot invariants from X, rather than $\mathcal{X}$. Every B-brane on $\mathcal{X}$ which is relevant for us comes from a B-brane on X via an exact functor

$$f_* : \mathscr{D}_X \to \mathscr{D}_{\mathcal{X}}, \tag{5.8}$$

which interprets a sheaf "downstairs" on X as a sheaf "upstairs" on $\mathcal{X}$. The functor f_* is more precisely the right-derived functor Rf_*. Its adjoint

$$f^* : \mathscr{D}_{\mathcal{X}} \to \mathscr{D}_X \tag{5.9}$$

is the left derived functor Lf^*, and corresponds to tensoring with the structure sheaf $\otimes \mathcal{O}_X$, and restricting. Adjointness implies that, given any pair of branes on $\mathcal{X}$ that come from X,

$$\mathcal{F} = f_* F, \quad \mathcal{G} = f_* G,$$

the Hom's upstairs, in $\mathscr{D}_{\mathcal{X}}$, agree with the Hom's downstairs, in $\mathscr{D}_X$,

$$\mathrm{Hom}_{\mathscr{D}_{\mathcal{X}}}(\mathcal{F}, \mathcal{G}) = \mathrm{Hom}_{\mathscr{D}_X}(f^* f_* F, G), \tag{5.10}$$

after replacing F with $f^* f_* F$. The functor $f^* f_*$ is not identity on $\mathscr{D}_X$.

5.4.2.

The equivariant homological mirror symmetry relating $\mathscr{D}_{\mathcal{X}}$ and $\mathscr{D}_Y$ is not an equivalence of categories, but a correspondence of branes and Hom's which come from a pair of adjoint functors h_* and h^*, inherited from f_* and f^* via the downstairs homological mirror symmetry:

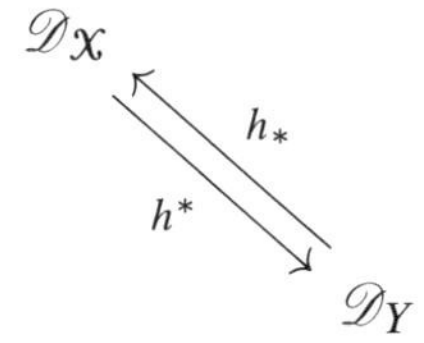

Alternatively, h^* and h_* come by composing the upstairs mirror symmetry with a pair of functors $k^* : \mathscr{D}_{\mathcal{Y}} \to \mathscr{D}_Y$ and $k_* : \mathscr{D}_Y \to \mathscr{D}_{\mathcal{Y}}$, which are mirror to f^* and f_*. The functors k^*, k_* come from Lagrangian correspondences; their construction is described in joint work with McBreen, Shende, and Zhou [6]. The functor k_* amounts to pairing a brane downstairs, with a vanishing torus fiber over it; this is how Figure 6 arises in our model example. The adjoint functors let us recover answers to all interesting questions about $\mathcal{X}$ from Y.

5.4.3.

For any simply laced Lie algebra ${}^L\mathfrak{g}$, the branes $\mathcal{U} \in \mathscr{D}_{\mathcal{X}}$ which serve as cups upstairs are the structure sheaves of (products of) minuscule Grassmannians, as described in Section 4.3.6. They come via the functor h_* from branes $I_{\mathcal{U}} \in \mathscr{D}_Y$ downstairs, on Y

$$\mathcal{U} = h_* I_{\mathcal{U}},$$

which are (products of) generalized intervals. A minuscule Grassmannian G/P_i is the h_*-image of a brane which is the configuration space of colored points on an interval ending on a pair of punctures on $\mathcal{A}$ corresponding to representations V_i and V_i^*. The points are colored by simple positive roots in $\mu_i + \mu_i^* = \sum_a d_{a,i}\, {}^L e_a$, and ordered in the sequence by which, to obtain the lowest weight μ_i^* in representation V_i, we subtract simple positive

FIGURE 7
The cup and cap A-branes corresponding to the defining representation of ${}^L\mathfrak{g} = \mathfrak{su}_4$, colored by its three simple roots; they are equivariant mirror to a B-brane supported on a $\mathbb{P}^4$ as its structure sheaf.

roots from the highest weight μ_i. Because V_i and V_i^* are minuscule representations, the ordering and hence the brane $I_{\mathcal{U}}$ is unique, up to equivalence and a choice of grading. The $\mathcal{U}$ branes project back down as generalized figure-eight branes; these are nested products of figure-eights, colored by simple roots

$$h^*\mathcal{U} = h^*h_*I_{\mathcal{U}} = E_{\mathcal{U}},$$

and ordered analogously, as in Figure 7. As objects of $\mathscr{D}_Y$, these branes are best described iterated cones over more elementary branes, mirror to stable basis branes **[5]**. The cup and cap branes all come with trivial local systems, for which the Floer complexes are the familiar ones, given by (5.7).

As an example, for ${}^L\mathfrak{g} = \mathfrak{su}_2$ the only minuscule representation is the defining representation $V_i = V_{\frac{1}{2}}$, which is self-conjugate. The cup brane $\mathcal{U}$ in $\mathcal{X}$ is a product of d non-intersecting $\mathbb{P}^1$'s. It comes, as the image of h_*, from a brane $I_{\mathcal{U}}$ in Y which is a product of d simple intervals, connecting pairs of punctures that come together. The $\mathcal{U}$-brane projects back down, via the h^* functor, as a product of d elementary figure-eight branes. The branes are graded by Maslov and equivariant gradings, as described in **[2]**.

5.4.4.

In the description based on Y, both the branes, and the action of braiding on them is geometric, so we can simply start with a link and a choice of projection to the surface $\mathcal{A} = \mathbb{R} \times S^1$. A link contained in a three ball in $\mathbb{R}^2 \times S^1$ is equivalent to the same link in $\mathbb{R}^3$, and projects to a contractible patch on $\mathcal{A}$.

To translate the link to a pair of A-branes, start by choosing bicoloring of the link projection, such that each of its components has an equal number of red and blue segments, and the red always underpass the blue. For a link component colored by a representation V_i of ${}^L\mathfrak{g}$, place a puncture colored by its highest weight μ_i where a blue segment begins and its conjugate μ_i^* where it ends; the orientation of the link component distinguishes the two. The mirror Lagrangians $I_{\mathcal{U}}$ and $\mathscr{B}E_{\mathcal{U}}$ are obtained by replacing all the blue segments by the interval branes, and all the red segments by figure-eight branes, related by equivariant mirror symmetry to minuscule Grassmannian branes. This data determines both Y and the branes on it we need. The variant of the second step, applicable for Lie superalgebras, is described in **[5]**.

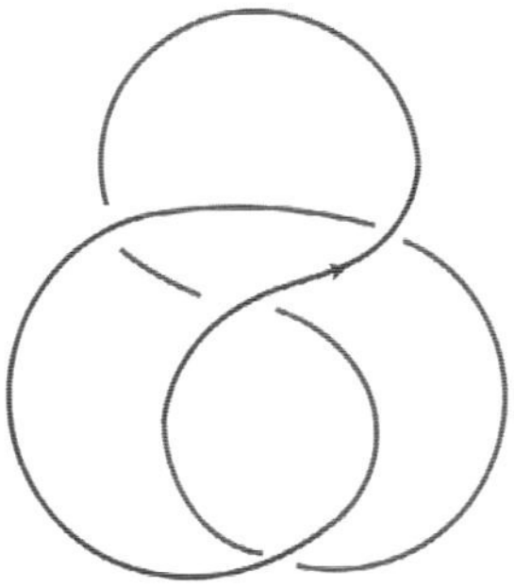

FIGURE 8
A bicoloring of the left-handed trefoil.

Equivariant mirror symmetry predicts that a homological link invariant is the space of morphisms

$$\mathrm{Hom}^{*,*}_{\mathscr{D}_Y}(\mathscr{B}E_{\mathcal{U}}, I_{\mathcal{U}}) = \bigoplus_{k\in\mathbb{Z},\vec{d}\in\mathbb{Z}^{rk+1}} \mathrm{Hom}_{\mathscr{D}_Y}\big(\mathscr{B}E_{\mathcal{U}}, I_{\mathcal{U}}[k]\{\vec{d}\}\big), \tag{5.11}$$

the cohomology of the Floer complex of the two branes. In what follows, will explain how to compute it.

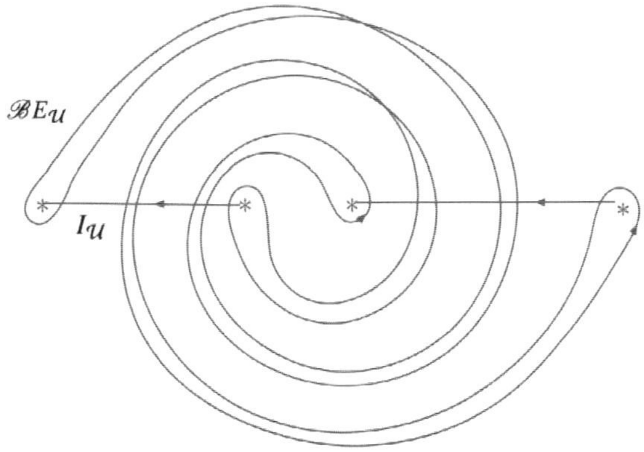

FIGURE 9
The branes corresponding to the left-handed trefoil in ${}^L\mathfrak{g} = \mathfrak{su}_2$. The knot was isotoped relative to Figure 8.

5.4.5.

To evaluate the Euler characteristic of the homology in (5.11), one simply counts intersection points of Lagrangians, keeping track of gradings. For links in $\mathbb{R}^3$, the equivariant grading in (5.11) collapses to a $\mathbb{Z}$-grading. The Euler characteristic becomes

$$\chi(\mathscr{B}E_{\mathcal{U}}, I_{\mathcal{U}}) = \bigoplus_{\mathscr{P}\in\mathscr{B}E_{\mathcal{U}}\cap I_{\mathcal{U}}} (-1)^{M(\mathscr{P})}\mathfrak{q}^{J(\mathscr{P})}, \tag{5.12}$$

where $M(\mathscr{P})$ and $J(\mathscr{P})$ are the Maslov and c^0-grading of the point $\mathscr{P}$; as in Heegard–Floer theory, there are purely combinatorial formulas for them **[3,5]**. Mirror symmetry implies that this is the $U_{\mathfrak{q}}({}^L\mathfrak{g})$ invariant of the link in $\mathbb{R}^3$.

The fact that for ${}^L\mathfrak{g} = \mathfrak{su}_2$ the graded count of intersection points in (5.12) reproduces the Jones polynomial is a theorem of Bigelow [15], building on the work of Lawrence [56–58]. Bigelow also proved the statement for ${}^L\mathfrak{g} = \mathfrak{su}_N$ with links colored by the defining representation [16]. The equivariant homological mirror symmetry explains the origin of Bigelow's peculiar construction, and generalizes it to other $U_{\mathfrak{q}}({}^L\mathfrak{g})$ link invariants.[1]

5.4.6.

The action of the differential Q on the Floer complex, defined by counting holomorphic maps from a disk D to Y with boundaries on the pair of Lagrangians, should have a reformulation [2] in terms of counting holomorphic curves embedded in $D \times \mathcal{A}$ with certain properties, generalizing the cylindrical formulation of Heegard–Floer theory due to Lipshitz [59]. The curve must have a projection to D as a $d = \sum_a d_a$-fold cover, with branching only between components of one color, and a projection to $\mathcal{A}$ as a domain with boundaries on one-dimensional Lagrangians of matching colors. In addition, the potential W must pull back to D as a regular holomorphic function. Computing the action of Q in this framework reduces to solving a sequence of well defined, but hard, problems in complex analysis in one dimension, which are applications of the Riemannian mapping theorem, similar to that in [63].

6. HOMOLOGICAL LINK INVARIANTS FROM A-BRANES

To compute the link homology groups

$$\mathrm{Hom}^{*,*}_{\mathscr{D}_Y}(\mathscr{B}E_{\mathcal{U}}, I_{\mathcal{U}}), \tag{6.1}$$

we will make use of mirror symmetry which relates X and Y and is the equivalence of categories

$$\mathscr{D}_X \cong \mathscr{D}_A \cong \mathscr{D}_Y, \tag{6.2}$$

proven in [4]. A basic virtue of mirror symmetry is that it sums up holomorphic curve counts. In our case, it solves all the disk-counting problems required to find the action of the differential Q on the Floer complex underlying (6.1).

6.1. The algebra of A-branes

As in the simplest examples of homological mirror symmetry, $\mathscr{D}_X$ and $\mathscr{D}_Y$ are both generated by a finite set of branes.

6.1.1.

From perspective of Y, the generating set of branes come from products of real line Lagrangians on $\mathcal{A}$, colored by $d = \sum_a d_a$ simple roots. The brane is unchanged by reordering a pair of its neighboring Lagrangian components, provided they are colored by roots

1 In [60], Bigelow's representation of the Jones polynomial, specialized to $\mathfrak{q} = 1$, was related to the Euler characteristic of symplectic Khovanov homology of [73].

which are not linked in the Dynkin diagram $\langle {}^L e_a, {}^L e_b\rangle = 0$. It is also unchanged by passing a component colored by ${}^L e_a$ across a puncture colored by a weight μ_i with $\langle {}^L e_a, \mu_i\rangle = 0$.

There is a generating brane

$$T_{\mathcal{C}} = T_{i_1} \times \cdots \times T_{i_d} \in \mathscr{D}_Y,$$

for every inequivalent ordering of d colored real lines on the cylinder. Their direct sum

$$T = \bigoplus_{\mathcal{C}} T_{\mathcal{C}} \in \mathscr{D}_Y,$$

is the generator of $\mathscr{D}_Y$ which is mirror to the tilting vector bundle on X, which generates $\mathscr{D}_X$. This generalizes the simplest example of mirror symmetry from Section 5.1. As before, we will be denoting branes on X and on Y related by homological mirror symmetry by the same letter.

6.1.2.

A well known phenomenon in mirror symmetry is that it may introduce Lagrangians with an extra structure of a local system, a nontrivial flat $U(1)$ bundle. The mirror of a structure sheaf of a generic point, in our model example of mirror symmetry from Section 5.1, is a Lagrangian of this sort.

Here, we find a generalization of this [4]. The pair of adjoint functors h_* and h^* that relate $\mathscr{D}_Y$ with its equivariant mirror $\mathscr{D}_\mathcal{X}$ equip each T-brane with a vector bundle or, more precisely, with a local system of modules for a graded algebra $\mathcal{B}$. The algebra is a product $\mathcal{B} = \bigotimes_{a=1}^{rk} \mathcal{B}_{d_a}$, where $\mathcal{B}_d$ has a representation as the quotient of the algebra of polynomials in d variables $z_1, \ldots, z_d$ which sets their symmetric functions to zero. The z's have equivariant $\mathfrak{q}$-degree equal to one.

As a consequence, the downstairs theory is not simply the Fukaya category of Y, but a Fukaya category with coefficients [4]: Floer complexes assign to each intersection point $\mathcal{P} \in L_0 \cap L_1$ a vector space $\hom_{\mathcal{B}}(\Lambda_0|_{\mathcal{P}}, \Lambda_1|_{\mathcal{P}})$ of homomorphisms of $\mathcal{B}$-modules $\Lambda_{0,1}$ which $L_{0,1}$ are equipped with. The cup and cap branes $E_{\mathcal{U}}$ and $I_{\mathcal{U}}$ come with trivial modules for $\mathcal{B}$. The $T_{\mathcal{C}}$ branes correspond to modules for $\mathcal{B}$ equal to $\mathcal{B}$ itself.

6.1.3.

Since the $T_{\mathcal{C}}$-branes are noncompact, defining the Hom's between them requires care. The Hom's

$$\mathrm{Hom}_{\mathscr{D}_Y}\big(T_{\mathcal{C}}, T'_{\mathcal{C}}[k]\{\vec{d}\}\big) = \mathrm{HF}\big(T^{\xi}_{\mathcal{C}}, T'_{\mathcal{C}}[k]\{\vec{d}\}\big)$$

are defined through a perturbation of $T_{\mathcal{C}}$ which induces wrapping near infinities of $\mathcal{A}$, as in Figure 4, and other examples of wrapped Fukaya categories.

The Floer cohomology groups HF are cohomology groups of the Floer complex whose generators are intersection points of the $T_{\mathcal{C}}$ branes, with coefficients in $\mathcal{B}$. The generators all have homological degree zero, so the Floer differential is trivial, and

$$\mathrm{Hom}_{\mathscr{D}_Y}\big(T_{\mathcal{C}}, T_{\mathcal{C}'}[k]\{\vec{d}\}\big) = 0, \quad \text{for all } k \neq 0 \text{ and all } \vec{d}\,. \tag{6.3}$$

The Floer product on $\mathscr{D}_Y$ makes

$$A = \mathrm{Hom}^*_{\mathscr{D}_Y}(T, T) = \bigoplus_{\mathcal{C},\mathcal{C}'} \bigoplus_{\vec{d}\in\mathbb{Z}^{rk+1}} \mathrm{Hom}_{\mathscr{D}_X}\left(T_{\mathcal{C}}, T_{\mathcal{C}'}\{\vec{d}\}\right)$$

into an algebra, which is an ordinary associative algebra, graded only by equivariant degrees.

6.1.4.

The vanishing in (6.3) mirrors the tilting property of T viewed as the generator of $\mathscr{D}_X$. The tilting vector bundle $T \in \mathscr{D}_X$ is inherited from the Bezrukavnikov–Kaledin tilting bundle $\mathcal{T}$ on $\mathcal{X}$,

$$\mathcal{T} = \bigoplus_{\mathcal{C}} \mathcal{T}_{\mathcal{C}} \in \mathscr{D}_{\mathcal{X}},$$

from Section 4.4, as the image of the f^* functor, which is tensoring with the structure sheaf of X and restriction, $f^*\mathcal{T} = T \in \mathscr{D}_X$. The endomorphism of the upstairs tilting generator $\mathcal{T}$,

$$\mathscr{A} = \mathrm{Hom}^*_{\mathscr{D}_{\mathcal{X}}}(\mathcal{T}, \mathcal{T}),$$

is the cylindrical KLRW algebra.

Since $\mathcal{T}$ is a vector bundle on $\mathcal{X}$, the center of $\mathscr{A}$ is the algebra of holomorphic functions on $\mathcal{X}$. The downstairs algebra is a quotient of the upstairs one, by a two-sided ideal

$$A = \mathscr{A}/\mathscr{I}. \tag{6.4}$$

The ideal $\mathscr{I}$ is generated by holomorphic functions that vanish on the core X.

6.1.5.

The cKLRW algebra $\mathscr{A}$ is defined as an algebra of colored strands on a cylinder, decorated with dots, where composition is represented by stacking cylinders and rescaling **[80]**. The local algebra relations are those of the ordinary KLRW algebra from **[78]**. Placing the theory on the cylinder, it gets additional gradings by the winding number of strands of a given color, corresponding to the equivariant Λ-action on $\mathcal{X}$.

The elements of the algebra $A = \mathscr{A}/\mathscr{I}$ have a geometric interpretation by recalling the Floer complex $\mathrm{CF}^*(T_{\mathcal{C}}, T_{\mathcal{C}'})$ is fundamentally generated by paths rather the intersection points. The S^1 of the algebra cylinder is the S^1 in the Riemann surface $\mathcal{A}$; its vertical direction parameterizes the path. The geometric intersection points of the T-branes on $\mathcal{A}$ correspond to strings in A. The flat bundle morphisms, a copy of $\mathcal{B}$ for each geometric intersection point, dress the strings by dots of the same color. The algebra $\mathcal{B}$ is the quotient, of the subalgebra of $\mathscr{A}$ generated by dots, by the ideal $\mathscr{I}$ of their symmetric functions.

6.2. The meaning of link homology

Since $T = \bigoplus_{\mathcal{C}} T_{\mathcal{C}}$ generates $\mathscr{D}_Y$, like every Lagrangian in $\mathscr{D}_Y$, the $\mathscr{B}E_{\mathcal{U}}$ brane has a description as a complex

$$\mathscr{B}E_{\mathcal{U}} \cong \cdots \xrightarrow{t_1} \mathscr{B}E_1(T) \xrightarrow{t_0} \mathscr{B}E_0(T), \tag{6.5}$$

every term of which is a direct sum of $T_{\mathcal{C}}$-branes. The complex is the projective resolution of the $\mathscr{B}E_{\mathcal{U}}$ brane. It describes how to get the $\mathscr{B}E_{\mathcal{U}} \in \mathscr{D}_Y$ brane by starting with the direct sum brane

$$\mathscr{B}E(T) = \bigoplus_k \mathscr{B}E_k(T)[k], \tag{6.6}$$

with a trivial differential, and taking iterated cones over elements $t_k \in A$. This deforms the differential to

$$Q_A = \sum_k t_k \in A, \tag{6.7}$$

which takes

$$Q_A : \mathscr{B}E(T) \to \mathscr{B}E(T)[1],$$

as a cohomological degree 1 and equivariant degree 0 operator, which squares to zero $Q_A^2 = 0$ in the algebra A.

6.2.1.

The category of A-branes $\mathscr{D}_Y$ has a second, Koszul dual set of generators, which are the vanishing cycle branes $I = \bigoplus_{\mathcal{C}} I_{\mathcal{C}}$ of **[2]**. The I- and the T-branes are dual in the sense that the only nonvanishing morphisms from the T- to the I-branes are

$$\mathrm{Hom}_{\mathscr{D}_Y}(T_{\mathcal{C}}, I_{\mathcal{C}'}) = \mathbb{C}\delta_{\mathcal{C},\mathcal{C}'}. \tag{6.8}$$

The $I_{\mathcal{C}}$-branes and the $T\mathcal{C}$-branes are, respectively, the simple and the projective modules of the algebra A.

6.2.2.

Among the I-branes, we find the branes $I_{\mathcal{U}} \in \mathscr{D}_Y$ which serve as cups. This is a wonderful simplification because it implies that from the complex in (6.5), we get for free a complex of vector spaces:

$$0 \to \mathrm{hom}_A\big(\mathscr{B}E_0(T), I_{\mathcal{U}}\{\vec{d}\}\big) \xrightarrow{t_0} \mathrm{hom}_A\big(\mathscr{B}E_1(T), I_{\mathcal{U}}\{\vec{d}\}\big) \xrightarrow{t_1} \cdots. \tag{6.9}$$

The maps in the complex (6.9) are induced from the complex in (6.5). The cohomologies of this complex are the link homologies we are after,

$$\mathrm{Hom}_{\mathscr{D}_Y}\big(\mathscr{B}E_{\mathcal{U}}, I_{\mathcal{U}}[k]\{\vec{d}\}\big) = H^k\big(\mathrm{hom}_A(\mathscr{B}E_{\mathcal{U}}, I_{\mathcal{U}})\big). \tag{6.10}$$

6.2.3.

We learn that link homology captures only a small part of the geometry of $\mathscr{B}E_{\mathcal{U}}$, the braided cup brane, or more precisely, of the complex that resolves it. Because the T-branes are dual to the I-branes by (6.8), almost all terms in the complex (6.9) vanish. The cohomology (6.10) of small complex that remains is the $U_{\mathfrak{q}}({}^L\mathfrak{g})$ link homology.

6.2.4.

The complex (6.9) itself has a geometric interpretation as the Floer complex,

$$\mathrm{CF}^{*,*}(\mathscr{B}E_{\mathcal{U}}, I_{\mathcal{U}}).$$

Namely, the vector space at the kth term of the complex

$$\hom_A\big(\mathscr{B}E_k(T), I_{\mathcal{U}}\{\vec{d}\}\big)$$

is identified, by construction described in section 6.3, with that spanned by the intersection points of the $\mathscr{B}E_{\mathcal{U}}$ brane and the $I_{\mathcal{U}}$ brane, of cohomological degree $[k]$ and equivariant degree $\{\vec{d}\}$.

The maps in the complex

$$\cdots \xrightarrow{t_{k-1}} \hom_A\big(\mathscr{B}E_k(T), I_{\mathcal{U}}\{\vec{d}\}\big) \xrightarrow{t_k} \hom_A\big(\mathscr{B}E_{k+1}(T), I_{\mathcal{U}}\{\vec{d}\}\big) \xrightarrow{t_{k+1}} \cdots$$

encode the action of the Floer differential. A priori, computing these requires counting holomorphic disk instantons. In our case, mirror symmetry (6.2) has summed them up.

6.3. Projective resolutions from geometry

The projective resolution in (6.5) encodes all the $U_{\mathfrak{q}}({}^L\mathfrak{g})$ link homology, and more. Finding the resolution requires solving two problems, both in general difficult. We will solve simultaneously [5].

6.3.1.

The first problem is to compute which module of the algebra A the brane $\mathscr{B}E_{\mathcal{U}}$ gets mapped to by the Yoneda functor

$$L \in \mathscr{D}_Y \to \mathrm{Hom}^{*,*}_{\mathscr{D}_Y}(T, L) \in \mathscr{D}_A.$$

This functor, which is the source of the equivalence $\mathscr{D}_Y \cong \mathscr{D}_A$, maps a brane L to a right module for A, on which the algebra acts as

$$\mathrm{Hom}^{*,*}_{\mathscr{D}_Y}(T, L) \otimes \mathrm{Hom}^{*}_{\mathscr{D}_Y}(T, T) \to \mathrm{Hom}^{*,*}_{\mathscr{D}_Y}(T, L).$$

Evaluating this action requires counting disk instantons.

6.3.2.

The second problem is to find the resolution of this module, as in (6.5). The Yoneda functor maps the $T_{\mathcal{C}}$ branes to projective modules of the algebra A, so the resolution in (6.5) is a projective resolution of the A module corresponding to the $\mathscr{B}E_{\mathcal{U}}$ brane. This problem is known to be solvable, however, only formally so, by infinite bar resolutions.

6.3.3.

In our setting, these two problems get solved together. Fortune smiles since the $\mathscr{B}E_{\mathcal{U}} \in \mathscr{D}_Y$ branes are products of d one-dimensional Lagrangians on $\mathcal{A}$, for which the complex resolving brane (6.5) can be deduced explicitly, from the geometry of the brane.

6.3.4.

Take a pair of branes L' and L'' on Y which are products of d one-dimensional Lagrangians on $\mathcal{A}$, chosen to coincide up to one of their factors. Up to permutation, we can take

$$L' = L_1 \times L_2 \times \cdots \times L_d, \quad L'' = L_1'' \times L_2 \times \cdots \times L_d.$$

If L_1' and L_1'' (which are necessarily of the same color) intersect over a point $p \in L_1' \cap L_1''$ of Maslov index zero, we get a new one dimensional Lagrangian L_1 which is a cone over p,

$$L_1 = \mathrm{Cone}(p) = L_1' \xrightarrow{p} L_1'',$$

as well as a new d-dimensional Lagrangian L on Y given by

$$L = L_1 \times L_2 \times \cdots \times L_d. \tag{6.11}$$

The Lagrangian is a cone over the intersection point $\mathcal{P}$ of L' and L'' which is of the form

$$\mathcal{P} = (p, \mathrm{id}_{L_2}, \ldots, \mathrm{id}_{L_d}) \in L' \cap L'', \tag{6.12}$$

and which also has Maslov index zero, $L = \mathrm{Cone}(\mathcal{P})$.

Conversely, any L brane which is of the product form in (6.11) can be written as a complex [11]

$$L \cong L' \xrightarrow{\mathcal{P}} L'' \tag{6.13}$$

with an explicit map $\mathcal{P}$ coming from a one-dimensional intersection point in one of its factors, as in (6.12).

6.3.5.

To find the projective resolution of the $\mathcal{B}E_{\mathcal{U}}$ brane in (6.5), start by isotoping the brane, by stretching it straight along the cylinder.

Let the brane break at the two infinities of $\mathcal{A}$, to get the direct sum brane $\mathcal{B}E(T)$ in (6.6), on which the complex is based. To find the maps in the complex, record how the brane breaks, iterating the above construction, one one-dimensional intersection point at the time. Every intersection point of the form (6.12) translates to a specific element of the algebra A and a specific map in the complex. The result is a product of d one-dimensional complexes, which describes factors of $\mathcal{B}E_{\mathcal{U}}$, and captures almost all the terms in the differential Q_A. The remaining ones follow, up to quasi-isomorphisms, by asking that the differential closes $Q_A^2 = 0$ in the algebra A. The fact that not all terms in Q_A are geometric is a general feature of $d > 1$ theories.

In practice, it is convenient to first break the brane one of the two infinities of $\mathcal{A}$, and only then on the other. The branes at the intermediate stage are images, under the h^* functor, of stable basis branes [7, 61] on $\mathcal{D}_{\mathcal{X}}$. The stable basis branes play a similar role to that of Verma modules in category $\mathcal{O}$. The detailed algorithm is given in [5].

6.3.6.

As an example, take the left-handed trefoil and ${}^L\mathfrak{g} = \mathfrak{su}_2$, which leads to the brane configuration from Figure 9. For simplicity, consider the reduced knot homology, where the unknot homology is set to be trivial. As in Heegard-Floer theory, this corresponds to erasing a component from the $\mathscr{B}E_{\mathcal{U}}$ and the $I_{\mathcal{U}}$ branes, and leads to Figure 10. This also brings us back to the setting of our running example, where Y is the equivariant mirror to $\mathcal{X}$, the resolution of the A_{n-1} surface singularity, with $n = 4$.

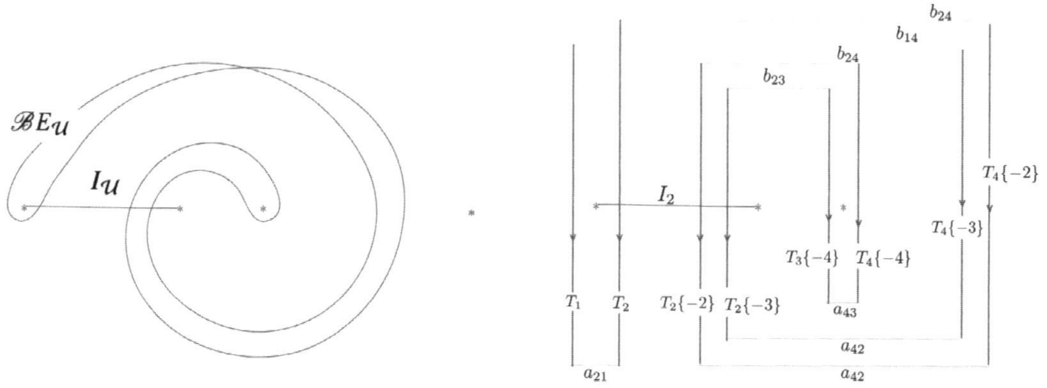

FIGURE 10
Resolution of the $\mathscr{B}E_{\mathcal{U}}$ brane corresponding to the reduced trefoil. The axis of the cylinder $\mathcal{A}$ is oriented vertically here; the branes do not wind around the S^1.

The corresponding algebra $A = \bigoplus_{i,j=0}^{n-1} \mathrm{Hom}^*_{\mathscr{D}_Y}(T_i, T_j)$ is the path algebra of an affine A_{n-1} quiver, whose nodes correspond to T_i branes. The arrows $a_{i+1,i} \in \mathrm{Hom}_{\mathscr{D}_Y}(T_i, T_{i+1})$ and $b_{i,i+1} \in \mathrm{Hom}_{\mathscr{D}_Y}(T_{i+1}, T_i\{1\})$ satisfy $a_{i,i-1}b_{i-1,i} = 0 = b_{i,i+1}a_{i+1,i}$, with i defined modulo n. The a's and b's correspond to intersections of T-branes, near one or the other infinity of $\mathcal{A}$; we have suppressed their Λ-equivariant degrees.

Isotope the $\mathscr{B}E_{\mathcal{U}}$ brane straight along the cylinder $\mathcal{A}$. Let it break into T-branes, as in Figure 10, while recording how the brane breaks, one connected sum at a time. Every connected sum of a pair of T-branes is a cone over their intersection point, at one of the two infinities of $\mathcal{A}$, and a specific element of the algebra A. This leads to the complex

$$\mathscr{B}E_{\mathcal{U}} \cong T_3\{-4\} \xrightarrow{\begin{pmatrix} a_{4,3} \\ b_{2,3}\end{pmatrix}} \begin{matrix} T_4\{-4\} \\ \oplus \\ T_2\{-3\}\end{matrix} \xrightarrow{\begin{pmatrix} b_{2,4} & 0 \\ 0 & a_{4,2}\end{pmatrix}} \begin{matrix} T_2\{-2\}\\ \oplus \\ T_4\{-3\}\end{matrix} \xrightarrow{\begin{pmatrix} a_{42} & 0\\ 0 & b_{14}\end{pmatrix}} \begin{matrix}T_4\{-2\}\\ \oplus \\ T_1\end{matrix} \xrightarrow{\begin{pmatrix} b_{24} & -a_{21}\end{pmatrix}} T_2$$

which closes by the A-algebra relations.

The reduced homology of the trefoil is the cohomology of the complex $\hom_A(\mathscr{B}E^\bullet, I_{\mathcal{U}}\{d\})$ in (6.9). The only non-zero contributions come from the T_2 brane, since the cup brane $I_{\mathcal{U}} = I_2$ is dual to it. All the maps evaluate to zero, as $I_{\mathcal{U}}$ brane is a simple module for A. As a consequence,

$$\mathrm{Hom}_{\mathscr{D}_Y}(\mathscr{B}E_{\mathcal{U}}, I_{\mathcal{U}}[k]\{d\}) = H^k(\hom_A(\mathscr{B}E^\bullet, I_2\{d\})),$$

equals to $\mathbb{Z}$ only for $(k, d) = (0,0), (2,-2), (3,-3)$, and vanishes otherwise. Here, $k = M$ is the Maslov or cohomological degree and $d = J$ the Jones grading. This is the reduced Khovanov homology of the left-handed trefoil, up to regrading: Khovanov's (i, j) gradings are related to (M, J) by $i = M + 2J + i_0$ and $j = 2J + j_0$ where $i_0 = 0$, $j_0 = d + n_+ - n_-$, where $n_+ = 0$, $n_- = 3$ are the numbers of positive and negative crossings, and $d = 1$ is the dimension of Y [2].

6.3.7.

The theory extends to non-simply-laced Lie algebras, and to Lie superalgebras $\mathfrak{gl}_{m|n}$ and $\mathfrak{sp}_{m|2n}$, as described in [5]. The algebra A corresponding to ${}^L\mathfrak{g}$ which is a Lie superalgebra, is not an ordinary associative algebra but a differential graded algebra; the projective resolutions are then in terms of twisted complexes. This section gives a method for solving the theory which is new even for ${}^L\mathfrak{g} = \mathfrak{gl}_{1|1}$, corresponding to Heegard-Floer theory. The solution differs from that in [65], in particular since our Heegard surface is $\mathcal{A} = \mathbb{R} \times S^1$, independent of the link.

ACKNOWLEDGMENTS

This work grew out of earlier collaborations with Andrei Okounkov, which were indispensable. It includes results obtained jointly with Ivan Danilenko, Elise LePage, Yixuan Li, Michael McBreen, Miroslav Rapcak, Vivek Shende, and Peng Zhou. I am grateful to all of them for collaboration. I was supported by the NSF foundation grant PHY1820912, by the Simons Investigator Award, and by the Berkeley Center for Theoretical Physics.

REFERENCES

[1] M. Aganagic, Knot categorification from mirror symmetry, Part I: coherent sheaves. 2020, arXiv:2004.14518.

[2] M. Aganagic, Knot categorification from mirror symmetry, Part II: lagrangians. 2021, arXiv:2105.06039.

[3] M. Aganagic, Knot categorification from mirror symmetry, Part III: string theory origin (to appear).

[4] M. Aganagic, I. Danilenko, Y. Li, and P. Zhou, Homological mirror symmetry for monopole moduli spaces (to appear).

[5] M. Aganagic, E. LePage, and M. Rapcak, Homological link invariants from mirror symmetry: computations (to appear).

[6] M. Aganagic, M. McBreen, V. Shende, and P. Zhou, Intertwining toric and hypertoric mirror symmetry (to appear).

[7] M. Aganagic and A. Okounkov, Elliptic stable envelope. 2020, arXiv:1604.00423v4.

[8] J. W. Alexander, Topological invariants of knots and links. *Trans. Amer. Math. Soc.* **30** (1923), 275–306.

[9] P. S. Aspinwall, et al., *Dirichlet branes and mirror symmetry*. Clay Math. Monogr. 4, AMS, Providence, RI, 2009, 681 pages.

[10] D. Auroux, A beginner's introduction to Fukaya categories. 2013, arXiv:1301.7056.

[11] D. Auroux, Fukaya categories of symmetric products and bordered Heegaard-Floer homology. arXiv: 1001.4323; D. Auroux,Fukaya categories and bordered Heegaard-Floer homology, arXiv:1003.2962.

[12] R. Bezrukavnikov, Noncommutative counterparts of the Springer resolution. 2006, arXiv:math/0604445.

[13] R. Bezrukavnikov and D. Kaledin, Fedosov quantization in positive characteristic. 2005, arXiv:math/0501247.

[14] R. Bezrukavnikov and A. Okounkov, Monodromy and derived equivalences, (to appear).

[15] S. Bigelow, A homological definition of the Jones polynomial. *Geom. Topol. Monogr.* **4** (2002), 29–41. arXiv:math.GT/0201221v2.

[16] S. Bigelow, A homological definition of the HOMFLY polynomial. *Algebr. Geom. Topol.* **7** (2007), no. 3, 1409–1440.

[17] T. Bridgeland, Stability conditions on triangulated categories. *Ann. Math.* **166** (2007), 317–345.

[18] T. Bridgeland, Stability conditions and Kleinian singularities. 2020, arXiv:math/0508257.

[19] M. Bullimore, T. Dimofte, and D. Gaiotto, The Coulomb branch of 3D $\mathcal{N} = 4$ theories. *Comm. Math. Phys.* **354** (2017), no. 2, 671. arXiv:1503.04817.

[20] S. Cautis and J. Kamnitzer, Knot homology via derived categories of coherent sheaves I, $sl(2)$ case. 2007, arXiv:math/0701194.

[21] S. Cautis and J. Kamnitzer, Knot homology via derived categories of coherent sheaves II, sl_m case. *Invent. Math.* **174** (2008), Article number 165.

[22] S. Cecotti and C. Vafa, Topological antitopological fusion. *Nuclear Phys. B* **367** (1991), 359.

[23] J. Chuang and R. Rouquier, Derived equivalences for symmetric groups and sl_2-categorification. 2005, arXiv:math/0407205.

[24] J. Chuang and R. Rouquier, Perverse equivalence, preprint, available at https://www.math.ucla.edu/~rouquier/papers/perverse.pdf.

[25] L. Crane and I. Frenkel, Four-dimensional topological field theory, Hopf categories, and the canonical bases. *J. Math. Phys.* **35** (1994), 5136–5154.

[26] I. Danilenko, *Slices of the affine Grassmannian and quantum cohomology*. PhD thesis, Columbia University, 2020.

[27] R. Dijkgraaf, H. L. Verlinde, and E. P. Verlinde, Topological strings in $d < 1$. *Nuclear Phys. B* **352** (1991), 59–86.

[28] M. R. Douglas, D-branes, categories and $N = 1$ supersymmetry. *J. Math. Phys.* **42** (2001), 2818.

[29] M. R. Douglas, B. Fiol, and C. Romelsberger, Stability and BPS branes. *J. High Energy Phys.* **0509** (2005), 006.
[30] V. G. Drinfeld, Quasi-Hopf algebras and Knizhnik–Zamolodchikov equations. In *Problems of modern quantum field theory*. Research Reports in Physics, edited by A. A. Belavin, A. U. Klimyk, and A. B. Zamolodchikov, Springer, Berlin, Heidelberg, 1989.
[31] B. Dubrovin, Geometry of 2d topological field theories. 1994, arXiv:hep-th/9407018.
[32] P. Etingof, I. Frenkel, and A. A. Jr. Kirillov, *Lectures on representation theory and Knizhnik–Zamolodchikov equations*. Math. Surveys Monogr. 58, Amer. Math. Soc., Providence, RI, 1998, 198 pp.
[33] P. Etingof and a. Geer, Monodromy of trigonometric KZ equations. 2006, arXiv:math/0611003.
[34] B. Feigin and E. Frenkel, A family of representations of affine Lie algebras. *Uspekhi Mat. Nauk* **43** (1988), no. 5, 227–228 [English translation: Russ. Math. Surv., **43** (1988), no. 5, 221–222].
[35] D. Gaiotto, G. W. Moore, and E. Witten, Algebra of the infrared: string field theoretic structures in massive $\mathcal{N} = (2, 2)$ field theory in two dimensions. 2015, arXiv:1506.04087.
[36] S. Galkin, V. Golyshev, and H. Iritani, Gamma classes and quantum cohomology of Fano manifolds: Gamma conjectures. 2014, arXiv:1404.6407.
[37] A. Givental, Homological geometry and mirror symmetry. In *Proceedings of the International Congress of Mathematicians*, edited by S. D. Chatterji, Birkhäuser, 1994.
[38] A. Givental, Semisimple Frobenius structures at higher genus. 2001, arXiv:math/0008067v4.
[39] K. Hori, A. Iqbal, and C. Vafa, D-branes and mirror symmetry. 2000, arXiv:hep-th/0005247.
[40] K. Hori, S. Katz, A. Klemm, R. Pandharipande, R. Thomas, C. Vafa, R. Vakil, and E. Zaslow, *Mirror symmetry*. 2006, 929 pp.
[41] H. Iritani, An integral structure in quantum cohomology and mirror symmetry for toric orbifolds. 2009, arXiv:0903.1463.
[42] V. F. R. Jones, A polynomial invariant for knots via von Neumann algebras. *Bull. Amer. Math. Soc.* **12** (1985), 103–112.
[43] D. Kaledin, Derived equivalences by quantization. 2006, arXiv:math/0504584.
[44] D. Kaledin, Geometry and topology of symplectic resolutions. 2008, arXiv:math/0608143.
[45] A. Kapustin and E. Witten, Electric-Magnetic duality and the geometric Langlands program. *Commun. Number Theory Phys.* **1** (2007), arXiv:hep-th/0604151.
[46] L. Katzarkov, M. Kontsevich, and T. Pantev, Hodge theoretic aspects of mirror symmetry. *Proc. Sympos. Pure Math.* **78** (2008), 87.

[47] D. Kazhdan and G. Lusztig, Tensor structures arising from affine Lie algebras. I–IV. *J. Amer. Math. Soc.* **6** (1993), 905–947, 949–1011; **7** (1994), 355–381, 383–453.

[48] M. Khovanov, A categorification of the Jones polynomial. *Duke Math. J.* **101** (2000), no. 3, 359–426.

[49] M. Khovanov, Link homology and categorification. 2006, arXiv:math/0605339.

[50] M. Khovanov and A. D. Lauda, A diagrammatic approach to categorification of quantum groups I. 2008, arXiv:0803.4121.

[51] M. Khovanov and L. Rozansky, Matrix factorizations and link homology. 2004, arXiv:math.QA/0401268.

[52] M. Khovanov and P. Seidel, Quivers, Floer cohomology, and braid group actions. *J. Amer. Math. Soc.* **15** (2002), no. 1, 203–271.

[53] V. G. Knizhnik and A. B. Zamolodchikov, Current algebra and Wess–Zumino model in two-dimensions. *Nuclear Phys. B* **247** (1984), 83.

[54] T. Kohno, Monodromy representations of braid groups and Yang–Baxter equations. *Ann. Inst. Fourier (Grenoble)* **37** (1987), 139–160.

[55] M. Kontsevich, Homological algebra of mirror symmetry. In *Proceedings of the International Congress of Mathematicians*, Birkhäuser, 1994.

[56] R. J. Lawrence, *Homology representations of braid groups*. D.Phil. Thesis, University of Oxford, June 1989.

[57] R. J. Lawrence, The homological approach applied to higher representations. 1990. Harvard preprint, available at http://www.ma.huji.ac.il/~ruthel/.

[58] R. J. Lawrence, A functorial approach to the one-variable Jones polynomial. *J. Differential Geom.* **37** (1993), 689–710.

[59] R. Lipshitz, A cylindrical reformulation of Heegaard Floer homology. *Geom. Topol.* **10** (2006), 955–1097.

[60] C. Manolescu, Nilpotent slices, Hilbert schemes, and the Jones polynomial, arXiv:math/0411015.

[61] D. Maulik and A. Okounkov, Quantum groups and quantum cohomology, 2018, arXiv:1211.1287.

[62] G. W. Moore and N. Seiberg, Lectures on RCFT, RU-89-32, YCTP-P13-89.

[63] P. S. Ozsvath and Z. Szabo, Holomorphic disks and three-manifold invariants, *Ann. of Math.* Vol. 159 (2004), Issue 3, p 1027-1158, p 1159-1245. .

[64] P. S. Ozsvath and Z. Szabo, Holomorphic disks, link invariants and the multivariable Alexander polynomial. *Algebr. Geom. Topol.* **8** (2008), 615–692.

[65] P. S. Ozsvath and Z. Szabo, Algebras with matchings and link Floer homology. 2020, arXiv:2004.07309.

[66] J. A. Rasmussen, *Floer homology and knot complements*. PhD thesis, Harvard University, 2003.

[67] N. Yu. Reshetikhin and V. G. Turaev, Ribbon graphs and their invariants derived from quantum groups. *Comm. Math. Phys.* **127** (1990), no. 1, 1–26.

[68] R. Rouquier, 2-Kac–Moody algebras. 2008, arXiv:0812.5023.

[69] V. Schechtman and A. Varchenko, Arrangements of hyperplanes and Lie algebra homology. *Invent. Math.*, **106** (1991), 139–194.

[70] V. Schechtman and A. Varchenko, Quantum groups and homology of local systems. In *Algebraic Geometry and Analytic Geometry, ICM-90 Satellite Conference Proceedings*, pp. 182–197, Springer, 1991.

[71] P. Seidel, *Fukaya categories and Picard–Lefschetz theory*. Zur. Lect. Adv. Math. 10, European Mathematical Society, 2007, 292 pp.

[72] P. Seidel and I. Smith, A link invariant from the symplectic geometry of nilpotent slices. *Duke Math. J.* **134** (2006), no. 3, 453–514.

[73] P. Seidel and J. P. Solomon, Symplectic cohomology and q-intersection numbers. 2012, arXiv:1005.5156.

[74] P. Seidel and R. P. Thomas, Braid group actions on derived categories of coherent sheaves. 2000, arXiv:math/0001043.

[75] A. Strominger, S. T. Yau, and E. Zaslow, Mirror symmetry is T duality. *Nuclear Phys. B* **479** (1996), 243.

[76] C. Teleman, The structure of semi-simple field theory at higher genus. 2012, arXiv:0712.0160.

[77] R. P. Thomas, Stability conditions and the braid group. 2006, arXiv:math/0212214v5.

[78] B. Webster, Knot invariants and higher representation theory. 2015, arXiv:1309.3796v3.

[79] B. Webster, Coherent sheaves and quantum Coulomb branches I: tilting bundles from integrable systems. 2019, arXiv:1905.04623v2.

[80] B. Webster, Coherent sheaves and quantum Coulomb branches II: quiver gauge theories and knot homology (to appear).

[81] E. Witten, Supersymmetry and Morse theory. *J. Differential Geom.* **17** (1982).

[82] E. Witten, Quantum field theory and the Jones polynomial. *Comm. Math. Phys.* **121** (1989), 351.

[83] E. Witten, On the structure of the topological phase of two-dimensional gravity. *Nuclear Phys. B* **340** (1990), 281–332.

[84] E. Witten, Fivebranes and knots. 2011, arXiv:1101.3216v2.

[85] E. Witten, Khovanov homology and gauge theory. 2012, arXiv:1108.3103v2.

[86] E. Witten, Two lectures on the Jones polynomial and Khovanov homology. 2014, arXiv:1401.6996.

[87] E. Witten, Two lectures on gauge theory and Khovanov homology. 2017, arXiv:1603.03854v2.

MINA AGANAGIC

Department of Mathematics, University of California, Berkeley, USA and Center for Theoretical Physics, University of California, Berkeley, USA, aganagic@berkeley.edu

This Page Intentionally Left Blank

VECTOR BUNDLES ON ALGEBRAIC VARIETIES

ARAVIND ASOK AND JEAN FASEL

ABSTRACT

We survey recent developments related to the problem of classifying vector bundles on algebraic varieties. We focus on the striking analogies between topology and algebraic geometry, and the way in which the Morel–Voevodsky motivic homotopy category can be used to exploit those analogies.

MATHEMATICS SUBJECT CLASSIFICATION 2020

Primary 14F42; Secondary 13C10, 18F25, 19A13, 55R45, 55S35

KEYWORDS

Vector bundles, motivic homotopy theory, obstruction theory

INTERNATIONAL CONGRESS OF MATHEMATICIANS 2022 JULY 6–14

Proc. Int. Cong. Math. 2022, Vol. 3, pp. 2146–2170
DOI 10.4171/ICM2022/98

1. INTRODUCTION

The celebrated Poincaré–Hopf theorem implies that the vanishing locus of a suitably generic vector field on a closed, smooth manifold M is topologically constrained: the number of points at which a generic vector field vanishes is equal to the Euler characteristic of M. More generally, one may ask: given a vector bundle E on a compact smooth manifold, what sorts of constraints are present on the topology of vanishing loci of generic sections? If M is a connected, closed, smooth manifold of dimension d and E is a rank r vector bundle on M, then by the corank of E we will mean the difference $d - r$. The classical work of Eilenberg, Stiefel, Steenrod, and Whitney laid down the foundations for results restricting the topology of vanishing loci of generic sections for bundles of a fixed corank; these results appear essentially in modern form in Steenrod's book [57]. For example, one knows that if the corank of E is negative, then E admits a nowhere vanishing section and if the corank of E is 0, then a generic section vanishes at a finite set of points, and the cardinality of that finite set is determined by purely cohomological data (the Euler class of E and the corresponding Euler number of E). The situation becomes more interesting when the corank of E is positive, to which we will return momentarily.

In the mid-1950s, Serre created a dictionary between the theory of vector bundles in topology and the theory of projective modules over a commutative ring [55, 56]. Echoing M. M. Postnikov's MathSciNet review of Serre's paper, J. F. Adams prosaically wrote in his review of H. Bass' paper [22]: "This leads to the following programme: take definitions, constructions and theorems from bundle-theory; express them as particular cases of definitions, constructions and statements about finitely-generated projective modules over a general ring; and finally, try to prove the statements under suitable assumptions". One of the results Serre presented to illustrate this dictionary was the algebro-geometric analog of existence of nowhere vanishing sections for negative corank projective modules, now frequently referred to as Serre's splitting theorem, which we recall in algebro-geometric formulation: if $\mathcal{E}$ is a rank r vector bundle over a Noetherian affine scheme X of dimension d, then when $r > d$, $\mathcal{E} \cong \mathcal{E}' \oplus \mathcal{O}_X$.

After the Pontryagin–Steenrod representability theorem, topological vector bundles on smooth manifolds (or spaces having the homotopy type of a CW complex) can be analyzed using homotopy theoretic techniques. Extending Serre's analogy further and using celebrated work of Bass, Quillen, Suslin, and Lindel, F. Morel showed that algebraic vector bundles on smooth affine varieties could be studied using an algebro-geometric homotopy theory: the Morel–Voevodsky motivic homotopy theory. In this note, we survey recent developments in the theory of algebraic vector bundles motivated by this circle of ideas, making sure to indicate the striking analogies between topology and algebraic geometry.

To give the reader a taste of the methods we will use, we mention two results here. First, we state an improvement of Serre's splitting theorem mentioned above (for the moment it suffices to know that $\mathbb{A}^1$-cohomological dimension is bounded above by Krull dimension, but can be strictly smaller). Second, we will discuss the splitting problem for projective modules in corank 1, which goes beyond any classical results.

Theorem 1.1. *If k is a field, and X is a smooth affine k-scheme of $\mathbb{A}^1$-cohomological dimension $\leq d$, then any rank $r > d$ bundle splits off a trivial rank 1 summand.*

Conjecture 1.2. *Assume k is an algebraically closed field, and $X = \operatorname{Spec} R$ is a smooth affine k-variety of dimension d. A rank $d-1$ vector bundle $\mathcal{E}$ on X splits off a free rank 1 summand if and only if $0 = c_{d-1}(\mathcal{E}) \in \mathrm{CH}^{d-1}(X)$.*

In Theorem 4.12 we verify Conjecture 1.2 in case $d = 3, 4$ (and k has characteristic not equal to 2). To motivate the techniques used to establish these results, we begin by analyzing topological variants of these conjectures. We close this note with a discussion of joint work with Mike Hopkins which addresses the difficult problem of constructing interesting low rank vector bundles on "simple" algebraic varieties. As with any survey, this one reflects the biases and knowledge of the authors. Limitations of space have prevented us from talking about a number of very exciting and closely related topics.

2. A FEW TOPOLOGICAL STORIES

In this section, we recall a few topological constructions that elucidate the approaches we use to analyze corresponding algebro-geometric questions studied later.

2.1. Moore–Postnikov factorizations

Suppose $f : E \to B$ is a morphism of pointed, connected topological spaces having the homotopy type of CW complexes that induces an isomorphism of fundamental groups (for simplicity of discussion). Write F for the "homotopy" fiber of f, so that there is a fiber sequence

$$F \to E \xrightarrow{f} B$$

yielding a long exact sequence relating the homotopy of F, E, and B.

A basic question that arises repeatedly is the following: given a map $M \to B$, when can it be lifted along f to a map $M \to E$? To approach this problem, one method is to factor f in such a way as to break the original lifting problem into simpler problems where existence of a lift can be checked by, say, cohomological means.

One systematic approach to analyzing this question was laid out in the work of Moore–Postnikov. In this case, one factors f so as to build E out of B by sequentially adding higher homotopy of f (keeping track of the induced action of $\pi_1(E) \cong \pi_1(B)$ on the fiber). In more detail, the Moore–Postnikov tower of f consists of a sequence of spaces $\tau_{\leq i} f$, $i \geq 0$ and morphisms fitting into the following diagram:

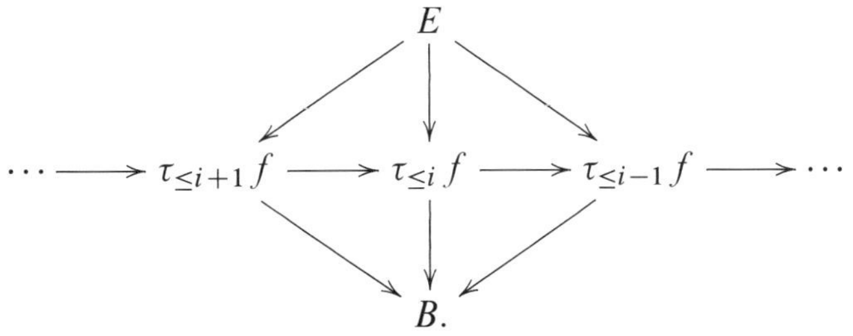

(2.1)

The key properties of this factorization are that (i) the composite maps $E \to \tau_{\leq i} f \to B$ all coincide with f, (ii) the maps $E \to \tau_{\leq i} f$ induce isomorphisms on homotopy groups in degrees $\leq i$, (iii) the maps $\tau_{\leq i} f \to B$ induce isomorphisms on homotopy in degrees $> i+1$, and (iv) there is a homotopy pullback diagram of the form

$$\begin{array}{ccc} \tau_{\leq i} f & \longrightarrow & \mathrm{B}\pi_1(E) \\ \downarrow & & \downarrow \\ \tau_{\leq i-1} f & \longrightarrow & K^{\pi_1(E)}(\pi_i(F), i+1). \end{array} \tag{2.2}$$

In particular, the morphism $\tau_{\leq i} f \to \tau_{\leq i-1} f$ is a twisted principal fibration, which means that a morphism $M \to \tau_{\leq i-1} f$ lifts along the tower if and only if the composite $M \to K^{\pi_1(E)}(\pi_i(F), i+1)$ lifts to $\mathrm{B}\pi_1(E)$. The latter map amounts to a cohomology class on M with coefficients in a local coefficient system; this cohomology class is pulled back from a "universal example" the k-invariant at the corresponding stage. If the obstruction vanishes, a lift exists. Lifts are not unique in general, but the ambiguity in choice of a lift can also be described.

2.2. The topological splitting problem

In this section, to motivate some of the algebro-geometric results we will describe later, we review the problem of deciding whether a bundle of corank 0 or 1 on a closed smooth manifold M of dimension $d+1$ has a nowhere vanishing section. We now phrase this problem as a lifting problem of the type described in the preceding section.

In this case, the relevant lifting problem is:

$$\begin{array}{ccc} & & \mathrm{B}O(d-1) \\ & \overset{\exists?}{\nearrow} & \downarrow f \\ M & \xrightarrow[\varphi]{} & \mathrm{B}O(d). \end{array}$$

To analyze the lifting problem, we describe the Moore–Postnikov factorization of f. The homotopy fiber of f coincides with the standard sphere $S^{d-1} \cong O(d)/O(d-1)$.

The stabilization map $O(d-1) \to O(d)$ is compatible with the determinant, and there are thus induced isomorphisms $\pi_1(\mathrm{B}O(d-1)) \to \pi_1(\mathrm{B}O(d)) \cong \mathbb{Z}/2$ compatible with f. Note, however, that the action of $\mathbb{Z}/2$ on the higher homotopy of $\mathrm{B}O(d)$ depends on the parity of d: when d is odd the action is trivial, while if d is even the action is nontrivial in general and even fails to be nilpotent. Of course, S^{d-1} is $(d-2)$-connected.

Remark 2.1. At this stage, the fact that bundles of negative corank on spaces have the homotopy type of a CW complex of dimension d follows immediately from obstruction theory granted the assertion that the sphere S^r is an $(r-1)$-connected space in conjunction with the fact that negative corank means $r > d$.

In order to write down obstructions, we need some information about the homotopy of spheres: the first nonvanishing homotopy group of S^{d-1} is $\pi_{d-1}(S^{d-1})$ which coincides with $\mathbb{Z}$ for all $d \geq 2$ (via the degree map). Likewise, $\pi_d(S^{d-1})$ is $\mathbb{Z}$ if $d = 3$ and $\mathbb{Z}/2$ if $d > 3$ and is generated by a suitable suspension of the classical Hopf map $\eta : S^3 \to S^2$.

Assume now X is a space having the homotopy type of a finite CW complex of dimension $d + 1$ for some fixed integer $d \geq 2$ (to eliminate some uninteresting cases) and $\xi : X \to \mathrm{B}O(d)$ classifies a rank d vector bundle on X. The first nonzero k-invariant for f yields a map $X \to K^{\mathbb{Z}/2}(\pi_{d-1}(S^{d-1}), d)$, i.e., an element

$$e(\xi) \in H^d\big(X, \mathbb{Z}[\sigma]\big)$$

called the (twisted) Euler class, where $\mathbb{Z}[\sigma]$ is $\mathbb{Z}$ twisted by the orientation character σ defined by applying π_1 to the morphism $X \to \mathrm{B}O(d) \to \mathrm{B}(\mathbb{Z}/2)$.

Assuming this primary obstruction vanishes, one may choose a lift to the next stage of the Postnikov tower. If we fix a lift, then there is a well-defined secondary obstruction to lifting, that comes from the next k-invariant: this obstruction is given by a map $X \to K^{\mathbb{Z}/2}(\pi_d(S^{d-1}), d + 1)$, i.e., a cohomology class in $H^{d+1}(X, \mathbb{Z}[\sigma])$ if $d = 3$ or $H^{d+1}(X, \mathbb{Z}/2)$ if $d \neq 3$; in the latter case the choice of orientation character no longer affects this cohomology group.

If one tracks the effect of choice of lift on the obstruction class described above, one obtains a map $K^{\mathbb{Z}/2}(\pi_{d-1}(S^{d-1}), d - 1) \to K^{\mathbb{Z}/2}(\pi_d(S^{d-1}), d + 1)$, which is a twisted cohomology operation. If $d = 3$, the map in question is a twisted version of the Pontryagin squaring operation, while if $d > 3$ the operation can be described as $\mathrm{Sq}^2 + w_2\cup$, where w_2 is the second Stiefel–Whitney class of the bundle. In that case, the secondary obstruction yields a well-defined coset in

$$o_2(\xi) \in H^{d+1}(X, \mathbb{Z}/2)/(\mathrm{Sq}^2 + w_2\cup)H^{d-1}\big(X, \mathbb{Z}[\sigma]\big)$$

This description of the primary and secondary obstructions was laid out carefully by the early 1950s by S. D. Liao [37].

Finally, the dimension assumption on X guarantees that a lift of ξ along f exists if and only if these two obstructions vanish. In principle, this kind of analysis can be continued, though the calculations become more involved as the indeterminacy created by successive choices of lifts becomes harder to control and information about higher unstable homotopy of spheres is also harder to obtain. For a thorough treatment of this and even more general situations, we refer the reader to [61].

Remark 2.2. The analysis of the obstructions can be improved by organizing the calculations differently. The Moore–Postnikov factorization has the effect of factoring a map $f : X \to Y$ as a tower of fibrations where the relevant fibers are Eilenberg–Mac Lane spaces. However, there are many other ways to produce factorizations of f with different constraints on the "cohomological" properties of pieces of the tower.

3. A QUICK REVIEW OF MOTIVIC HOMOTOPY THEORY

Motivic homotopy theory, introduced by F. Morel and V. Voevodsky [41], provides a homotopy theory for schemes over a base. While there are a number of different approaches to constructing the motivic homotopy category that work in great generality, we work in a very concrete situation. By an algebraic variety over a field k, we will mean a separated, finite type, reduced k-scheme. We write Sm_k for the category of smooth algebraic varieties; for later use, we will also write $\mathrm{Sm}_k^{\mathrm{aff}}$ for the full subcategory of Sm_k consisting of affine schemes.

The category Sm_k is "too small" to do homotopy theory, in the sense that various natural categorical constructions one would like to make (increasing unions, quotients by subspaces, etc.) can leave the category. As such, one first enlarges Sm_k to a suitable category Spc_k of "spaces"; one may take Spc_k to be the category of simplicial presheaves on Sm_k and the functor $\mathrm{Sm}_k \to \mathrm{Spc}_k$ is given by the Yoneda embedding followed by the functor viewing a presheaf on Sm_k as a constant simplicial presheaf.

3.1. Homotopical sheaf theory

Passing to Spc_k has the effect of destroying certain colimits that one would like to retain. To recover the colimits that have been lost, one localizes Spc_k and passes to a suitable "local" homotopy category of the sort first studied in detail by K. Brown–S. Gersten, A. Joyal, and J. F. Jardine: one fixes a Grothendieck topology τ on Sm_k and inverts the so-called τ-local weak equivalences on Spc_k; we refer the reader to [34] for a textbook treatment. We write $\mathrm{H}_\tau(k)$ for the resulting localization of Spc_k. If $\mathscr{X} \in \mathrm{Spc}_k$, then a base-point for $\mathscr{X}$ is a morphism $x : \operatorname{Spec} k \to \mathscr{X}$ splitting the structure morphism. There is an associated pointed homotopy category and these homotopy categories can be thought of as providing a convenient framework for "nonabelian" homological algebra.

Henceforth, we take τ to be the Nisnevich topology (which is finer than the Zariski topology, but coarser than the étale topology). For the purposes of this note, it suffices to observe that the Nisnevich cohomological dimension of a k-scheme is equal to its Krull dimension, like the Zariski topology.

In the category of pointed spaces, we can make sense of wedge sums and smash products, just as in ordinary topology. We also define spheres S^i, $i \geq 0$, as the constant simplicial presheaves corresponding to the simplicial sets S^i. For any pointed space $(\mathscr{X}, x)$, we define its homotopy sheaves $\boldsymbol{\pi}_i(\mathscr{X}, x)$ as the Nisnevich sheaves associated with the presheaves on Sm_k defined by

$$U \mapsto \hom_{\mathrm{H}_{\mathrm{Nis}}(k)}(S^i \wedge U_+, \mathscr{X}, x);$$

here the subscript $+$ means adjoint a disjoint base-point. These homotopy sheaves may be used to formulate a Whitehead theorem.

If $\mathbf{G}$ is a Nisnevich sheaf of groups on Sm_k, then there is a classifying space $\mathrm{B}\,\mathbf{G}$ such that for any smooth k-scheme X one has a functorial identification of pointed sets of the form

$$\hom_{\mathrm{H}_{\mathrm{Nis}}(k)}(X, \mathrm{B}\,\mathbf{G}) = \mathrm{H}^1_{\mathrm{Nis}}(X, \mathbf{G}).$$

For later use, we set

$$\mathrm{Vect}_n(X) := \mathrm{H}^1_{\mathrm{Zar}}(X, \mathrm{GL}_n) = \mathrm{H}^1_{\mathrm{Nis}}(X, \mathrm{GL}_n) = \hom_{\mathrm{H}_{\mathrm{Nis}}(k)}(X, \mathrm{BGL}_n);$$

where we as usual identify isomorphism classes of rank n vector bundles locally trivial with respect to the Zariski topology on X with GL_n-torsors (and the choice of topology does not matter).

If $\mathbf{A}$ is any Nisnevich sheaf of abelian groups on Sm_k, then for any integer $n \geq 0$ there are Eilenberg–Mac Lane spaces $\mathrm{K}(\mathbf{A}, n)$, i.e., spaces with exactly one nonvanishing homotopy sheaf, appearing in degree n, isomorphic to $\mathbf{A}$. For such spaces, $\hom_{\mathrm{H_{Nis}}(k)}(X, \mathrm{K}(\mathbf{A}, n))$ has a natural abelian group structure, and there are functorial isomorphisms of abelian groups

$$\hom_{\mathrm{H_{Nis}}(k)}\big(X, \mathrm{K}(\mathbf{A}, n)\big) = \mathrm{H}^n_{\mathrm{Nis}}(X, \mathbf{A}).$$

With this definition, for essentially formal reasons there is a suspension isomorphism for Nisnevich cohomology with respect to the suspension $S^1 \wedge (-)$.

3.2. The motivic homotopy category

The motivic homotopy category is obtained as a further localization of $\mathrm{H_{Nis}}(k)$: one localizes at the projection morphisms $\mathscr{X} \times \mathbb{A}^1 \to \mathscr{X}$. We write $\mathrm{H}(k)$ for the resulting homotopy category; isomorphisms in this category will be referred to as $\mathbb{A}^1$-weak equivalences. Following the notation in classical homotopy theory, we write

$$[\mathscr{X}, \mathscr{Y}]_{\mathbb{A}^1} := \hom_{\mathrm{H}(k)}(\mathscr{X}, \mathscr{Y})$$

and refer to this set as the set of $\mathbb{A}^1$-homotopy classes of maps from $\mathscr{X}$ to $\mathscr{Y}$.

If $\mathscr{X}$ is a space, we will write $\boldsymbol{\pi}_0^{\mathbb{A}^1}(\mathscr{X})$ for the Nisnevich sheaf associated with the presheaf $U \mapsto [U, \mathscr{X}]_{\mathbb{A}^1}$ on Sm_k; we refer to $\boldsymbol{\pi}_0^{\mathbb{A}^1}(\mathscr{X})$ as the sheaf of connected components, and we say that $\mathscr{X}$ is $\mathbb{A}^1$-connected if $\boldsymbol{\pi}_0^{\mathbb{A}^1}(\mathscr{X})$ is the sheaf $\mathrm{Spec}(k)$.

We consider $\mathbb{G}_m$ as a pointed space, with base point its identity section 1. In that case, we define motivic spheres

$$S^{i,j} := S^i \wedge \mathbb{G}_m^{\wedge j}.$$

We caution the reader that there are a number of different indexing conventions used for motivic spheres. One defines bigraded homotopy sheaves $\boldsymbol{\pi}_{i,j}^{\mathbb{A}^1}(\mathscr{X}, x)$ for any pointed space as the Nisnevich sheaves associated with the presheaves on Sm_k

$$U \mapsto \big[S^{i,j} \wedge U_+, \mathscr{X}\big]_{\mathbb{A}^1};$$

we write $\boldsymbol{\pi}_i^{\mathbb{A}^1}(\mathscr{X}, x)$ for $\boldsymbol{\pi}_{i,0}^{\mathbb{A}^1}(\mathscr{X})$. We will say that a pointed space $(\mathscr{X}, x)$ is $\mathbb{A}^1$-k-connected for some integer $k \geq 1$ if it is $\mathbb{A}^1$-connected and the sheaves $\boldsymbol{\pi}_i^{\mathbb{A}^1}(\mathscr{X}, x)$ are trivial for $1 \leq i \leq k$. Because of the form of the Whitehead theorem in the Nisnevich local homotopy category, the sheaves $\boldsymbol{\pi}_i^{\mathbb{A}^1}(-)$ detect $\mathbb{A}^1$-weak equivalences.

We write $\Delta_k^\bullet$ for the cosimplicial affine space with

$$\Delta_k^n := \mathrm{Spec}\, k[x_0, \dots, x_n]/\Big\langle \sum_i x_i = 1 \Big\rangle.$$

For any space $\mathscr{X}$, we write $\mathrm{Sing}^{\mathbb{A}^1} \mathscr{X}$ for the space $\mathrm{diag}\, \underline{\hom}(\Delta^\bullet, \mathscr{X})$. There is a canonical map $\mathscr{X} \to \mathrm{Sing}^{\mathbb{A}^1} \mathscr{X}$ and the space $\mathrm{Sing}^{\mathbb{A}^1} \mathscr{X}$ is called the singular construction on $\mathscr{X}$. For a smooth scheme U, the set of connected components $\pi_0(\mathrm{Sing}^{\mathbb{A}^1} \mathscr{X}(U))$ will be called the set of *naive* $\mathbb{A}^1$*-homotopy classes* of maps $U \to \mathscr{X}$ (by construction, it is the quotient of the set of morphisms $U \to \mathscr{X}$ by the equivalence relation generated by maps $U \times \mathbb{A}^1 \to \mathscr{X}$). Again, by definition there is a comparison morphism

$$\pi_0(\mathrm{Sing}^{\mathbb{A}^1} \mathscr{X}(U)) \to [U, \mathscr{X}]_{\mathbb{A}^1}. \tag{3.1}$$

Typically, the map (3.1) is far from being a bijection.

3.3. $\mathbb{A}^1$-weak equivalences

We now give a number of examples of $\mathbb{A}^1$-weak equivalences, highlighting some examples and constructions that will be important in the sequel.

Example 3.1. A smooth k-scheme X is called $\mathbb{A}^1$*-contractible* if the structure morphism $X \to \operatorname{Spec} k$ is an $\mathbb{A}^1$-weak equivalence. By construction, $\mathbb{A}^n$ is an $\mathbb{A}^1$-contractible smooth k-scheme. However, there are a plethora of $\mathbb{A}^1$-contractible smooth k-schemes that are nonisomorphic to $\mathbb{A}^n$. For instance, the Russell cubic threefold, defined by the hypersurface equation $x + x^2y + z^2 + t^3 = 0$ is known to be nonisomorphic to affine space and also $\mathbb{A}^1$-contractible [28]. See [16] for a survey of further examples.

Example 3.2. If $f : X \to Y$ is a Nisnevich locally trivial morphism with fibers that are $\mathbb{A}^1$-contractible smooth k-schemes, then f is an $\mathbb{A}^1$-weak equivalence. Thus, the projection morphism for a vector bundle is an $\mathbb{A}^1$-weak equivalence. A vector bundle E over a scheme X can be seen as a commutative algebraic X-group scheme, so we may speak of E-torsors; E-torsors are classified by the coherent cohomology group $\mathrm{H}^1(X, \mathscr{E})$ (in particular, vector bundle torsors over affine schemes may always be trivialized). Vector bundle torsors are Zariski locally trivial fiber bundles with fibers isomorphic to affine spaces, and the projection morphism for a vector bundle torsor is an $\mathbb{A}^1$-weak equivalence.

By an *affine vector bundle torsor* over a scheme X we will mean a torsor $\pi : Y \to X$ for some vector bundle E on X such that Y is an affine scheme. Jouanolou proved [35, LEMME 1.5] that any quasiprojective variety admits an affine vector bundle torsor. Thomason [63, PROPOSITION 4.4] generalized Jouanolou's observation, and the following result is a special case of his results.

Lemma 3.3 (Jouanolou–Thomason homotopy lemma). *If X is a smooth k-variety, then X admits an affine vector bundle torsor. In particular, any smooth k-variety is isomorphic in* $\mathrm{H}(k)$ *to a smooth affine variety.*

Definition 3.4. By a *Jouanolou device* for a smooth k-variety X we will mean a choice of an affine vector bundle torsor $p : Y \to X$.

Example 3.5. When $X = \mathbb{P}^n$ there is a very simple construction of a "standard" Jouanolou device $\tilde{\mathbb{P}}^n$. Geometrically, the standard Jouanolou device for $\mathbb{P}^n$ may be described as the complement of the incidence divisor in $\mathbb{P}^n \times \mathbb{P}^n$ where the second projective space is viewed as the dual of the first, with structure morphism the projection onto either factor.

Example 3.6. If X is a smooth projective variety of dimension d, then we may choose a finite morphism $\psi : X \to \mathbb{P}^d$. Pulling back the standard Jouanolou device for $\mathbb{P}^d$ along ψ, we see that X admits a Jouanolou device $\tilde{X}$ of dimension $2d$.

Example 3.7. For $n \in \mathbb{N}$, consider the smooth affine k-scheme Q_{2n-1} defined as the hypersurface in $\mathbb{A}^{2n}_k$ given by the equation $\sum_{i=1}^{n} x_i y_i = 1$. Projecting onto the first n-factors, we obtain a map $p : Q_{2n-1} \to \mathbb{A}^n \setminus 0$ which one may check is an affine vector bundle torsor. For any integer $n \geq 0$, $\mathbb{A}^n \setminus 0$ is $\mathbb{A}^1$-weakly equivalent to $S^{n-1,n}$ (see [41, §3.2, EXAMPLE 2.20]) and consequently Q_{2n-1} is $\mathbb{A}^1$-weakly equivalent to $S^{n-1,n}$ as well.

Example 3.8. For $n \in \mathbb{N}$, consider the smooth affine k-scheme Q_{2n} defined as the hypersurface in $\mathbb{A}^{2n+1}_k$ given by the equation

$$\sum_{i=1}^{n} x_i y_i = z(1-z).$$

The variety Q_2 is isomorphic to the standard Jouanolou device over $\mathbb{P}^1$. The variety $\mathbb{P}^1$ is $\mathbb{A}^1$-weakly equivalent to $S^{1,1}$ and therefore Q_2 is $\mathbb{A}^1$-weakly equivalent to $S^{1,1}$ as well. For $n \geq 2$, one knows that Q_{2n} is $\mathbb{A}^1$-weakly equivalent to $S^{n,n}$ [2, THEOREM 2].

3.4. Representability results

If $\mathscr{F}$ is a presheaf on Sm_k, we will say that $\mathscr{F}$ is $\mathbb{A}^1$-invariant (resp. $\mathbb{A}^1$-invariant on affines) if the pullback map $\mathscr{F}(X) \to \mathscr{F}(X \times \mathbb{A}^1)$ is an isomorphism for all $X \in \mathrm{Sm}_k$ (resp. $X \in \mathrm{Sm}_k^{\mathrm{aff}}$). A necessary condition for a cohomology theory on smooth schemes to be representable in $\mathrm{H}(k)$ is that it is $\mathbb{A}^1$-invariant and has a Mayer–Vietoris property with respect to the Nisnevich topology. One of the first functors that one encounters with these properties is that which assigns to a smooth k-scheme its Picard group. Morel and Voevodsky showed [41, §4 PROPOSITION 3.8] that if X is a smooth k-scheme, then the $\mathbb{A}^1$-weak equivalence $\mathbb{P}^\infty \to \mathrm{B}\mathbb{G}_m$ induces a bijection $[X, \mathbb{P}^\infty]_{\mathbb{A}^1} \cong \mathrm{Pic}(X)$.

If $\mathbf{A}$ is a sheaf of abelian groups on Sm_k, then the functors $\mathrm{H}^i_{\mathrm{Nis}}(-, \mathbf{A})$ frequently fail to be $\mathbb{A}^1$-invariant (taking $\mathbf{A} = \mathbb{G}_a$ gives a simple example) and therefore fail to be representable on Sm_k. The situation above where $\mathbf{A} = \mathbb{G}_m$ provides the prototypical example of a sheaf whose cohomology *is* $\mathbb{A}^1$-invariant (here the zeroth cohomology is the presheaf of units, which is even $\mathbb{A}^1$-invariant on reduced schemes). Following Morel and Voevodsky, we distinguish the cases where sheaf cohomology is $\mathbb{A}^1$-invariant.

Definition 3.9. A sheaf of groups $\mathbf{G}$ on Sm_k is called *strongly* $\mathbb{A}^1$*-invariant* if for $i = 0, 1$ the functors $\mathrm{H}^i_{\mathrm{Nis}}(-, \mathbf{G})$ on Sm_k are $\mathbb{A}^1$-invariant. A sheaf of abelian groups $\mathbf{A}$ on Sm_k is called *strictly* $\mathbb{A}^1$*-invariant* if for all $i \geq 0$ the functors $\mathrm{H}^i_{\mathrm{Nis}}(-, \mathbf{A})$ on Sm_k are $\mathbb{A}^1$-invariant.

The fundamental work of Morel, which we will review shortly, demonstrates the key role played by strongly and strictly $\mathbb{A}^1$-invariant sheaves. Nevertheless, various natural functors of geometric origin *fail* to be $\mathbb{A}^1$-invariant on smooth schemes.

Example 3.10. If $r \geq 2$, then the functor $\mathrm{H}^1_{\mathrm{Nis}}(-, \mathrm{GL}_r)$ fails to be $\mathbb{A}^1$-invariant on all schemes. For an explicit example, consider the simplest case. By a theorem of Dedekind–Weber frequently attributed to Grothendieck every rank n vector bundle on $\mathbb{P}^1$ is isomorphic to a unique line bundle of the form $\bigoplus_{i=1}^n \mathcal{O}(a_i)$ with the a_i weakly increasing. On the other hand, consider $\mathbb{P}^1 \times \mathbb{A}^1$ with coordinates t and x. The matrix

$$\begin{pmatrix} t & x \\ 0 & t^{-1} \end{pmatrix}$$

determines a rank 2 vector bundle on $\mathbb{P}^1 \times \mathbb{A}^1$ whose restriction to $\mathbb{P}^1 \times 0$ is $\mathcal{O}(1) \oplus \mathcal{O}(-1)$ and whose restriction to $\mathbb{P}^1 \times 1$ is $\mathcal{O} \oplus \mathcal{O}$. In contrast, Lindel's theorem affirming the Bass–Quillen conjecture in the geometric case shows that $\mathrm{H}^1_{\mathrm{Nis}}(-, \mathrm{GL}_r)$ is $\mathbb{A}^1$-invariant on affines. The next result generalizes this last observation.

Theorem 3.11 (Morel, Schlichting, Asok–Hoyois–Wendt). *If X is a smooth affine k-scheme, then for any $r \in \mathbb{N}$ there are functorial bijections of the form*

$$\pi_0\big(\mathrm{Sing}^{\mathbb{A}^1} \mathrm{Gr}_r(X)\big) \xrightarrow{\sim} [X, \mathrm{Gr}_r]_{\mathbb{A}^1} \xrightarrow{\sim} \mathrm{Vect}_r(X).$$

Remark 3.12. The above result was first established by F. Morel in [40] for $r \neq 2$ and k an infinite, perfect field, and his proof was partly simplified by M. Schlichting whose argument also established the case $r = 2$ [51]. The version above is stated in [13].

Remark 3.13. While the functor of isomorphism classes of vector bundles is $\mathbb{A}^1$-invariant on smooth affine k-schemes, even the latter can fail for G-torsors under more general group schemes, e.g., the special orthogonal group scheme SO_n (see [47] or [45]). Furthermore, while GL_n-torsors are always locally trivial with respect to the Nisnevich (and even the Zariski) topology, for an arbitrary smooth k-group scheme G, one only knows that G-torsors are locally trivial with respect to the étale topology.

In [14, 15], it is shown that if G is an isotropic reductive group scheme (see [14, DEFINITION 3.3.5] for the definition), then the functor assigning to $X \in \mathrm{Sm}_k^{\mathrm{aff}}$ the set $\mathrm{H}^1_{\mathrm{Nis}}(X, \mathrm{G})$ is representable by BG. This observation has a number of consequences, e.g., the following result about quadrics (see Examples 3.7 and 3.8).

Theorem 3.14 ([1, 14, 15]). *For any integer $i \geq 1$ and any $X \in \mathrm{Sm}_k^{\mathrm{aff}}$, the comparison map*

$$\pi_0\big(\mathrm{Sing}^{\mathbb{A}^1} Q_i(X)\big) \xrightarrow{\sim} [X, Q_i]_{\mathbb{A}^1}$$

is a bijection, contravariantly functorial in X.

3.5. Postnikov towers, connectedness and strictly $\mathbb{A}^1$-invariant sheaves

Recall from Definition 3.9 the notion of strongly or strictly $\mathbb{A}^1$-invariant sheaves of groups. F. Morel showed that such sheaves can be thought of as "building blocks" for the unstable $\mathbb{A}^1$-homotopy category. Morel's foundational works [39, 40] can be viewed as a careful analysis of strictly and strongly $\mathbb{A}^1$-invariant sheaves of groups and the relationship between the two notions. More precisely, Morel showed that working over a perfect field k, the $\mathbb{A}^1$-homotopy sheaves of a motivic space are *always* strongly $\mathbb{A}^1$-invariant, and that strongly $\mathbb{A}^1$-invariant sheaves of abelian groups are automatically strictly $\mathbb{A}^1$-invariant.

To check this, Morel showed that strongly (resp. strictly) $\mathbb{A}^1$-invariant sheaves of groups come equipped with a package of results/tools that are central to computations; this package of results is essentially an extension/amalgam/axiomatization of work of Bloch–Ogus and Gabber on étale cohomology exposed in [26] and Rost [50].

Example 3.15. Some examples of $\mathbb{A}^1$-invariant sheaves that will appear in the sequel are:

- unramified Milnor K-theory sheaves $\mathbf{K}_i^M$, $i \geq 0$ (see [50, COROLLARY 6.5, PROPOSITION 8.6] where, more generally, it is shown that any Rost cycle module gives rise to a strictly $\mathbb{A}^1$-invariant sheaf);
- the Witt sheaf $\mathbf{W}$ or unramified powers of the fundamental ideal in the Witt ring $\mathbf{I}^j$, $j \geq 0$ (this follows from [46]); and
- unramified Milnor–Witt K-theory sheaves $\mathbf{K}_i^{MW}$, $i \in \mathbb{Z}$ (see [40, CHAPTER 3] for this assertion, or [31, COROLLARY 8.5, PROPOSITION 9.1] where this observation is generalized to so-called Milnor–Witt cycle modules).

3.16 (Moore–Postnikov factorizations). There is an analog of the Moore–Postnikov factorization of a map $f : \mathscr{E} \to \mathscr{B}$ of spaces along the lines described in Section 2. For concreteness we discuss the case where $\mathscr{E}$ and $\mathscr{B}$ are $\mathbb{A}^1$-connected and f induces an isomorphism on $\mathbb{A}^1$-fundamental sheaves of groups for some choice of base-point in $\mathscr{E}$.

Given f as above, there are $\tau_{\leq i} f \in \mathrm{Spc}_k$ together with maps $\mathscr{E} \to \tau_{\leq i} f$, $\tau_{\leq i} f \to \mathscr{B}$ and $\tau_{\leq i} f \to \tau_{\leq i-1} f$ fitting into a diagram of exactly the same form as (2.1) (replacing E by $\mathscr{E}$ and B by $\mathscr{B}$). The relevant properties of this presentation are similar to those sketched before (replacing homotopy groups by homotopy sheaves), together with a homotopy pullback diagram of exactly the same form as (2.2). We refer to this tower as the $\mathbb{A}^1$-Moore–Postnikov tower of f and the reader may consult [40, APPENDIX B] or [5, §6] for a more detailed presentation.

If X is a smooth scheme, then a map $\psi : X \to \mathscr{B}$ lifts to $\tilde{\psi} : X \to \mathscr{E}$ if and only if lifts exist at each stage of the tower, i.e., if and only if a suitable obstruction vanishes. These obstructions are, by construction, valued in Nisnevich cohomology on X with values in a strictly $\mathbb{A}^1$-invariant sheaf (see [5, §6] for a more detailed explanation).

By analogy with the situation in topology, we will use the $\mathbb{A}^1$-Moore–Postnikov factorization to study lifting problems by means of obstruction theory. The relevant obstructions will lie in cohomology groups of a smooth scheme with coefficients in a strictly $\mathbb{A}^1$-invariant sheaf. This motivates the following definition.

Definition 3.17. Let X be a smooth k-scheme. We say that X has $\mathbb{A}^1$*-cohomological dimension* $\leq d$ if for any integer $i > d$ and any strictly $\mathbb{A}^1$-invariant sheaf $\mathbf{F}$, $\mathrm{H}^i_{\mathrm{Nis}}(X, \mathbf{F}) = 0$. In that case, we write $cd_{\mathbb{A}^1}(X) \leq d$.

Example 3.18. If X is a smooth k-scheme of dimension d, then X necessarily has $\mathbb{A}^1$-cohomological dimension $\leq d$ as well. Since $\mathbb{A}^n$ has $\mathbb{A}^1$-cohomological dimension ≤ 0, the $\mathbb{A}^1$-cohomological dimension can be strictly smaller than Krull dimension; Example 3.6 gives numerous other such examples.

3.6. Complex realization

Assume k is a field that admits an embedding $\iota_{\mathbb{C}} : k \hookrightarrow \mathbb{C}$. The functor that assigns to a smooth k-variety X the complex manifold $X(\mathbb{C})$ equipped with its classical topology extends to a complex realization functor

$$\mathfrak{R}_{\mathbb{C}} : \mathrm{H}(k) \to \mathrm{H}$$

where H is the usual homotopy category of topological spaces [41, §3.3]. By construction, complex realization preserves finite products and homotopy colimits. It follows that the complex realization of the motivic sphere $S^{p,q}$ is the ordinary sphere S^{p+q}, and consequently the complex realization functor induces group homomorphisms of the form

$$\pi_{i,j}^{\mathbb{A}^1}(X,x)(\mathbb{C}) \to \pi_{i+j}\big(X(\mathbb{C}),x\big)$$

for any pointed smooth k-scheme (X,x).

Suppose X is any k-scheme admitting a complex embedding and fix such an embedding. Write $\mathrm{Vect}_r^{\mathrm{top}}(X)$ for the set of isomorphism classes of complex topological vector bundles on X. There is a function

$$\mathrm{Vect}_r(X) \to \mathrm{Vect}_r^{\mathrm{top}}(X)$$

sending an algebraic vector bundle E over X to the topological vector bundle on $X(\mathbb{C})$ attached to the base change of E to $X_{\mathbb{C}}$. We will say that an algebraic vector bundle is *algebraizable* if it lies in the image of this map.

As rank r topological vector bundles are classified by the set $[X(\mathbb{C}), BU(r)]$ of homotopy classes of maps from $X(\mathbb{C})$ to the complex Grassmannian, it follows that the function of the preceding paragraph factors as

$$\mathrm{Vect}_r(X) \to [X, \mathrm{Gr}_r]_{\mathbb{A}^1} \to \mathrm{Vect}_r^{\mathrm{top}}(X).$$

Theorem 3.11 implies that the first map is a bijection if X is a smooth affine k-scheme (or, alternatively, if $r = 1$). More generally, combining Theorem 3.11 and Lemma 3.3 one knows that any element of $[X, \mathrm{Gr}_r]_{\mathbb{A}^1}$ may be represented by an actual rank r vector bundle on any Jouanolou device $\tilde{X}$ of X; this suggests the following definition.

Definition 3.19. If X is a smooth k-scheme, then by a *rank r motivic vector bundle* on X we mean an element of the set $[X, \mathrm{Gr}_r]_{\mathbb{A}^1}$.

Question 3.20. If X is a smooth complex algebraic variety, then which topological vector bundles are algebraizable (resp. motivic)?

4. OBSTRUCTION THEORY AND VECTOR BUNDLES

In order to apply the obstruction theory described in the previous sections to analyze algebraic vector bundles, we need more information about the structure of the classifying space BGL_n including information about its $\mathbb{A}^1$-homotopy sheaves, and the structure of the homotopy fiber of the stabilization map $\mathrm{BGL}_n \to \mathrm{BGL}_{n+1}$ induced by the map $\mathrm{GL}_n \to \mathrm{GL}_{n+1}$ sending an invertible matrix X to the block matrix $\mathrm{diag}(1, X)$.

4.1. The homotopy sheaves of the classifying space of $\mathbf{BGL}_n$

We observed earlier that $\mathrm{BGL}_1 = \mathrm{B}\mathbb{G}_m$ is an Eilenberg–Mac Lane space for the sheaf $\mathbb{G}_m$: it is $\mathbb{A}^1$-connected, and has exactly 1 nonvanishing $\mathbb{A}^1$-homotopy sheaf in degree 1, which is isomorphic to $\mathbb{G}_m$. For $n \geq 1$, the analysis of homotopy sheaves of BGL_n uses several ingredients. First, Morel–Voevodsky observed that $\mathrm{BGL} = \mathrm{colim}_n \mathrm{BGL}_n$ (for the inclusions described above) represents (reduced) algebraic K-theory after [41, §4 THEOREM 3.13]. Second, Morel observed that there is an $\mathbb{A}^1$-fiber sequence of the form

$$\mathbb{A}^{n+1} \setminus 0 \to \mathrm{BGL}_n \to \mathrm{BGL}_{n+1}, \tag{4.1}$$

and that $\mathbb{A}^{n+1} \setminus 0$ is $\mathbb{A}^1$-$(n-1)$-connected. Furthermore, Morel computed [40] the first nonvanishing $\mathbb{A}^1$-homotopy sheaf of $\mathbb{A}^{n+1} \setminus 0$ in terms of what he called Milnor–Witt K-theory sheaves (Example 3.15).

Putting these ingredients together, one deduces

$$\boldsymbol{\pi}_i^{\mathbb{A}^1}(\mathrm{BGL}_n) \cong \mathbf{K}_i^Q, \quad 1 \leq i \leq n-1,$$

where $\mathbf{K}_i^Q$ is the (Nisnevich) sheafification of the Quillen K-theory presheaf on Sm_k. Following terminology from topology, sheaves in this range are called stable, and the case $i = n$ is called the first unstable homotopy sheaf. In [3], we described the first unstable homotopy sheaf of BGL_n.

The group scheme GL_n maps to $\mathrm{GL}_n(\mathbb{C})$ under complex realization; the latter is homotopy equivalent to $U(n)$. For context, we recall some facts about homotopy of $U(n)$. A classical result of Bott, refining results of Borel–Hirzeburch [24, THEOREM 25.8] asserts that the image of $\pi_{2n}(BU(n))$ in $H_{2n}(BU(n))$ is divisible by precisely $(n-1)!$ [25]. This result implies the assertion that $\pi_{2n}(U(n)) = n!$.

Complex realization yields a map $\pi_{n,n}^{\mathbb{A}^1}(\mathrm{GL}_n) \to \pi_{2n}(U(n))$. One can view the celebrated "Suslin matrices" [59] as providing an algebro-geometric realization of the generator of $\pi_{2n}(U(n))$. Analyzing the fiber sequence of (4.1) and putting all of the ingredients above together, we obtain the following result (we refer the reader to Example 3.15 for notation).

Theorem 4.1 ([3, THEOREM 1.1]). *Assume k is a field that has characteristic not equal to 2. For any integer $n \geq 2$, there are strictly $\mathbb{A}^1$-invariant sheaves $\mathbf{S}_n$ fitting into exact sequences of the form:*

$$0 \to \mathbf{S}_{n+1} \to \boldsymbol{\pi}_n^{\mathbb{A}^1}(\mathrm{BGL}_n) \to \mathbf{K}_n^Q \to 0, \quad n \text{ odd};$$
$$0 \to \mathbf{S}_{n+1} \times_{\mathbf{K}_{n+1}^M/2} \mathbf{I}^{n+1} \to \boldsymbol{\pi}_n^{\mathbb{A}^1}(\mathrm{BGL}_n) \to \mathbf{K}_n^Q \to 0, \quad n \text{ even},$$

where

(1) *there is a canonical epimorphism* $\mathbf{K}_n^M/(n-1)! \to \mathbf{S}_n$ *which becomes an isomorphism after* $n-2$ *contractions (see* [3, §2.3] *for this terminology);*

(2) *there is a canonical epimorphism* $\mathbf{S}_n \to \mathbf{K}_n^M/2$ *such that the composite*

$$\mathbf{K}_n^M/(n-1)! \to \mathbf{S}_n \to \mathbf{K}_n^M/2$$

is reduction modulo 2;

(3) *the fiber product is taken over the epimorphism* $\mathbf{S}_{n+1} \to \mathbf{K}_{n+1}^M/2$ *and a sheafified version of Milnor's homomorphism* $\mathbf{I}^{n+1} \to \mathbf{K}_{n+1}^M/2$.

Moreover, if k *admits a complex embedding, then the map*

$$\boldsymbol{\pi}_{n,n+1}^{\mathbb{A}^1}(\mathrm{BGL}_n)(\mathbb{C}) \to \pi_{2n+1}\big(BU(n)\big) \cong \mathbb{Z}/n!$$

induced by complex realization is an isomorphism.

Bott's refinement of the theorem of Borel–Hirzebruch turns out to have an algebro-geometric interpretation. Indeed, in joint work with T. B. Williams [12] we showed that $\mathbf{S}_n$ can described using a "Hurewicz map" analyzed by Andrei Suslin [60]. Suslin's conjecture on the image of this map is equivalent to the following conjecture.

Conjecture 4.2 (Suslin's factorial conjecture). *The canonical epimorphism* $\mathbf{K}_n^M/(n-1)! \to \mathbf{S}_n$ *is an isomorphism.*

Remark 4.3. The conjecture holds tautologically for $n=2$. For $n=3$, Suslin observed the conjecture was equivalent to the Milnor conjecture on quadratic forms, which was resolved later independently by Merkurjev–Suslin and Rost. The conjecture was established for $n=5$ in "most" cases in [12] (see the latter for a precise statement); this work relies heavily on the computation by Østvær–Röndigs–Spitzweck of the motivic stable 1-stem [49].

4.2. Splitting bundles, Euler classes, and cohomotopy

Morel's computations around $\mathbb{A}^n \setminus 0$ in conjunction with the fiber sequences of (4.1) allow a significant improvement of Serre's celebrated splitting theorem for smooth affine varieties over a field that we stated in the introduction.

Proof of the motivic Serre Splitting Theorem 1.1. Suppose X is a smooth affine k-variety having $\mathbb{A}^1$-cohomological dimension $\leq d$, and suppose $\xi : X \to \mathrm{BGL}_r$ classifies a rank $r > d$ vector bundle on X. We proceed by analyzing the $\mathbb{A}^1$-Moore–Postnikov factorization of the stabilization map (4.1) with $n = r-1$. In that case, combining the fact that $\mathbb{A}^r \setminus 0$ is $\mathbb{A}^1$-$(r-2)$-connected and the $\mathbb{A}^1$-cohomological dimension assumption on X, one sees all obstructions to splitting vanish. ■

Remark 4.4. The proof of this result does not rely on the Serre splitting theorem. Since $\mathbb{A}^1$-cohomological dimension can be strictly smaller than Krull dimension (Example 3.18), this statement is strictly stronger than Serre splitting. Importantly, the improvement achieved here seems inaccessible to classical techniques.

The algebro-geometric splitting problem in corank 0 on smooth affine varieties of dimension d over a field k has been analyzed by many authors. When k is an algebraically closed field, M. P. Murthy proved that the top Chern class in Chow groups is the only obstruction to splitting [42]. When k is not algebraically closed, vanishing of the top Chern class is known to be insufficient to guarantee splitting, and Nori proposed some ideas to analyze this situation. His ideas led Bhatwadekar and Sridharan [23] to introduce what they called Euler class groups and to provide one explicit "generators and relations" answer to this question. At the same time, F. Morel proposed an approach to the splitting problem in corank 0, which we recall here.

Theorem 4.5 (Morel's splitting theorem [40, THEOREM 1.32]). *Assume k is a field and X is a smooth affine k-variety of $\mathbb{A}^1$-cohomological dimension $\leq d$. If $\mathcal{E}$ is a rank d vector bundle on X, then $\mathcal{E}$ splits off a free rank 1 summand if and only if an Euler class $e(\mathcal{E}) \in \mathrm{H}^d_{\mathrm{Nis}}(X, \mathbf{K}^{MW}_d(\det \mathcal{E}))$ vanishes.*

Remark 4.6. The Euler class of Theorem 4.5 is precisely the first nonvanishing obstruction class, as described in Paragraph 3.16. A related "cohomological" approach to the splitting problem in corank 0 was proposed by Barge–Morel [20] and analyzed in the thesis of the second author [29]. The cohomological approach was in most cases shown to be equivalent to the "obstruction-theoretic" approach in [6]. We also refer the reader to [51] for related results on the theory of Euler classes, extending also to singular varieties.

The next result shows that the relationship between Euler classes à la Bhatwadekar–Sridharan and Euler classes à la Morel is mediated by another topologically inspired notion: cohomotopy (at least for bundles of trivial determinant).

Theorem 4.7 ([8, THEOREM 1]). *Suppose k is a field, n and d are integers, $n \geq 2$, and X is a smooth affine k-scheme of dimension $d \leq 2n-2$. Write $\mathrm{E}^n(X)$ for the Bhatwadekar–Sridharan Euler class group.*

- *The set $[X, Q_{2n}]_{\mathbb{A}^1}$ carries a functorial abelian group structure;*
- *There are functorial homomorphisms:*

$$\mathrm{E}^n(X) \xrightarrow{s} [X, Q_{2n}]_{\mathbb{A}^1} \xrightarrow{h} \mathrm{H}^d_{\mathrm{Nis}}(X, \mathbf{K}^{MW}_n)$$

  where the "Segre class" homomorphism s is surjective and an isomorphism if k is infinite and $d \geq 2$, and the Hurewicz homomorphism h is an isomorphism if $d \leq n$.

Remark 4.8. The group structure on $[X, Q_{2n}]_{\mathbb{A}^1}$ is an algebro-geometric variant of Borsuk's group structure on cohomotopy. The second point of the statement includes the algebro-geometric analog of the Hopf classification theorem from topology.

4.3. The next nontrivial $\mathbb{A}^1$-homotopy sheaf of spheres

In Section 2.2 we described a cohomological approach to the splitting problem in corank 1 for smooth closed manifolds of dimension d; this approach relied on the computation of $\pi_d(S^{d-1})$. In order to analyze the algebro-geometric splitting problem in corank 1 using the $\mathbb{A}^1$-Moore–Postnikov factorization we will need as input further information about the homotopy sheaves of $\mathbb{A}^d \setminus 0$. We now describe known results in this direction. For technical reasons, we assume 2 is invertible in what follows.

4.3.1. The KO-degree map

In classical algebraic topology, all of the "low degree" elements in the homotopy of spheres can be realized by constructions of "linear algebraic" nature. The situation in algebraic geometry appears to be broadly similar. The first contribution to the "next" nontrivial homotopy sheaves of motivic spheres requires recalling the geometric formulation of Bott periodicity for Hermitian K-theory given by Schlichting–Tripathi.

We write O for the infinite orthogonal group. In topology, Bott periodicity identifies the 8-fold loop space of O with itself and identifies the intermediate loop spaces in concrete geometric terms. In algebraic geometry, Schlichting and Tripathi proved that the 4-fold $\mathbb{P}^1$-loop space $\Omega^4_{\mathbb{P}^1}\mathrm{O}$ also coincides with O and realized suitable intermediate loop spaces: $\Omega^n_{\mathbb{P}^1}\mathrm{O}$ is isomorphic to GL / O when $n = 1$, Sp when $n = 2$ and GL / Sp when $n = 3$, where Sp is the stable sympletic group, GL/O is the ind-variety of invertible symmetric matrices, and GL/Sp is the ind-variety of invertible skew-symmetric matrices [52, THEOREMS 8.2 AND 8.4].

A slight modification of the Suslin matrix construction [59, LEMMA 5.3] yields a map

$$u_n : Q_{2n-1} \to \Omega^{-n}_{\mathbb{P}^1} O$$

called the *(unstable)* KO-*degree map in weight n* that was analyzed in detail in [7]. The terminology stems from the fact that this map stabilizes to the "unit map from the sphere spectrum to the Hermitian K-theory spectrum" in an appropriate sense. The scheme Q_{2n-1} is $\mathbb{A}^1$-$(n-2)$-connected by combining the weak equivalence of Example 3.7 and Morel's connectivity results for $\mathbb{A}^n \setminus 0$. Thus, u_n factors through the $\mathbb{A}^1$-$(n-2)$-connected cover of $\Omega^{-n}_{\mathbb{P}^1} O$.

Taking homotopy sheaves on both sides, there are induced morphisms

$$\boldsymbol{\pi}_i^{\mathbb{A}^1}(u) : \boldsymbol{\pi}_i^{\mathbb{A}^1}(Q_{2n-1}) \to \boldsymbol{\pi}_i^{\mathbb{A}^1}(\Omega^{-n}_{\mathbb{P}^1} O).$$

This homomorphism is trivial if $i < n-1$ by connectivity estimates. If $i = n-1$, via Morel's calculations one obtains a morphism $\mathbf{K}^{\mathrm{MW}}_n \to \mathbf{GW}^n_n$ whose sections over finitely generated field extensions of k can be viewed as a quadratic enhancement of the "natural" map from Milnor K-theory to Quillen K-theory defined by symbols; we will refer to it as the natural homomorphism (the natural homomorphism is known to be an isomorphism if $n \leq 4$; the case $n \leq 2$ is essentially Suslin's, $n = 3$ is [7, THEOREM 4.3.1], and $n = 4$ is unpublished work of O. Röndigs).

When $i = n$, we obtain a morphism

$$\boldsymbol{\pi}_n^{\mathbb{A}^1}(\mathbb{A}^n \setminus 0) \cong \boldsymbol{\pi}_n^{\mathbb{A}^1}(Q_{2n-1}) \to \boldsymbol{\pi}_n^{\mathbb{A}^1}(\Omega^{-n}_{\mathbb{P}^1} O) \cong \mathbf{GW}^n_{n+1},$$

where the right-hand term is by definition a higher Grothendieck–Witt sheaf (obtained by sheafifying the corresponding higher Grothendieck–Witt presheaf on Sm_k). The above map is an epimorphism for $n = 2, 3$ and it follows from these observations that the morphism is an epimorphism after $(n-3)$ contractions [7, THEOREM 4.4.5].

4.3.2. The motivic J-homomorphism

The classical J-homomorphism has an algebro-geometric counterpart that yields the second contribution to the "next" homotopy sheaf of motivic spheres. The standard action of SL_n on $\mathbb{A}^n$ extends to an action on the one-point compactification $\mathbb{P}^n/\mathbb{P}^{n-1}$. The latter space is a motivic sphere $\mathbb{P}^{1^{\wedge n}}$ and thus one obtains a map

$$\Sigma^n_{\mathbb{P}^1} SL_n \to \mathbb{P}^{1^{\wedge n}}.$$

As SL_n is $\mathbb{A}^1$-connected, it follows that $\Sigma^n_{\mathbb{P}^1} \mathrm{SL}_n$ is $\mathbb{A}^1$-n-connected.

The first nonvanishing $\mathbb{A}^1$-homotopy sheaf appears in degree $n+1$; for $n = 2$, it is isomorphic to $\mathbf{K}^{MW}_4$, while for $n \geq 3$ it is isomorphic to $\mathbf{K}^M_{n+2}$; this follows from $\mathbb{A}^1$-Hurewicz theorem combined with [17, PROPOSITION 3.3.9] using the fact that $\pi_1^{\mathbb{A}^1}(\mathrm{SL}_n) = \mathbf{K}^M_2$ for $n \geq 3$ and properties of the $\mathbb{A}^1$-tensor product [17, LEMMA 5.1.8].

Combining the above discussion with that of the previous section, we see that for $n \geq 3$, we may consider the composite maps $\mathbf{K}^M_{n+2} \to \pi^{\mathbb{A}^1}_{n+1}(\mathbb{P}^{1^{\wedge n}}) \to \mathbf{GW}^n_{n+1}$; this composite is known to be zero, but the map induced by the J-homomorphism fails to be injective. Instead, it factors through a morphism

$$\mathbf{K}^M_{n+2}/24 \to \pi^{\mathbb{A}^1}_{n+1}(\mathbb{P}^{1^{\wedge n}}) \to \mathbf{GW}^n_{n+1}.$$

Furthermore, the map on the right fails to be surjective. The unstable description above is not present in the literature, but it is equivalent to the results stated in [12]. In [49], the stable motivic 1-stem was computed in the terms above: the above sequence is exact on the left stably. The next result compares the unstable group to the corresponding stable group.

Theorem 4.9. *For any integer $n \geq 3$, the kernel $\mathbf{U}_{n+1}$ of the stabilization map*

$$\pi^{\mathbb{A}^1}_{n+1}(\mathbb{P}^{1^{\wedge n}}) \to \pi^{\mathbb{A}^1}_{n+1}(\Omega^\infty_{\mathbb{P}^1} \Sigma^\infty_{\mathbb{P}^1} \mathbb{P}^{1^{\wedge n}})$$

is a direct summand; the stabilization map is an isomorphism if $n = 3$, i.e., $\mathbf{U}_4 = 0$.

Conjecture 4.10. *For $n \geq 4$, the sheaf $\mathbf{U}_{n+1}$ is zero.*

Remark 4.11. Conjecture 4.10 would follow from a suitable version of the Freudenthal suspension theorem for $\mathbb{P}^1$-suspension.

4.4. Splitting in corank 1

Using the results above, we can analyze the splitting problem for vector bundles in corank 1. The expected result was posed as a question by Murthy [43, P. 173] which we stated in the introduction as Conjecture 1.2. Murthy's conjecture is trivial if $d = 2$. In [4] and [5] we established the following result, which reduces Murthy's question to Conjecture 4.10.

Theorem 4.12. *Let X be a smooth affine scheme of dimension $d \geq 2$ over an algebraically closed field k. A rank $d-1$ vector bundle $\mathcal{E}$ on X splits off a trivial rank 1 summand if and* *only if* $c_{d-1}(\mathcal{E}) \in \mathrm{CH}^{d-1}(X)$ *is trivial and a secondary obstruction*

$$o_2(\mathcal{E}) \in \mathrm{H}^d_{\mathrm{Nis}}\big(X, \boldsymbol{\pi}^{\mathbb{A}^1}_{d-1}(\mathbb{A}^{d-1} \smallsetminus 0)\big)$$

vanishes. This secondary obstruction vanishes if $d = 3, 4$ *or if Conjecture* 4.10 *has a positive answer.*

To establish this result, one uses the assumptions that X is smooth affine of Krull dimension d and k is algebraically closed in a strong way. Indeed, these assertions can be leveraged to show that the primary obstruction, which is a priori an Euler class, actually coincides with the $(d-1)$st Chern class. The secondary obstruction can be described by Theorem 4.9 and the form of the secondary obstruction is extremely similar to Liao's description in Section 2.2: it is a coset in $\mathrm{Ch}^d(X)/(\mathrm{Sq}^2 + c_1(\mathcal{E})\cup)\mathrm{Ch}^{d-1}(X)$ where $\mathrm{Ch}^i(X) = \mathrm{CH}^i(X)/2$. Once more, the assumptions on X guarantee that $\mathrm{Ch}^d(X)$ is trivial and thus the secondary obstruction is so as well.

4.5. The enumeration problem

If a vector bundle E splits off a free rank 1 summand, then another natural question is to enumerate the possible E' that become isomorphic to E after adding a free rank 1 summand. This problem may also be analyzed in homotopy theoretic terms as it amounts to enumerating the number of distinct lifts. This kind of problem was studied in detail in topology by James and Thomas [33] and the same kind of analysis can be pursued in algebraic geometry.

The history of the enumeration problem in algebraic geometry goes back to early days of algebraic K-theory. Indeed, the Bass–Schanuel cancellation theorem [22] solves the enumeration problem for bundles of negative corank. Suslin's celebrated cancellation theorem [58] solved the enumeration problem in corank 0. In all of these statements, "cancellation" means that there is a unique lift. On the other hand, Mohan Kumar observed [38] that for bundles of corank 2, uniqueness was no longer true in general. Nevertheless, Suslin conjectured that the enumeration problem had a particularly nice solution in corank 1.

Conjecture 4.13 (Suslin's cancellation conjecture). *If k is an algebraically closed field, and X is a smooth affine k-scheme of dimension $d \geq 2$. If $\mathcal{E}$ and $\mathcal{E}'$ are corank* 1 *bundles that become isomorphic after addition of a trivial rank* 1 *summand, then $\mathcal{E}$ and $\mathcal{E}'$ are isomorphic.*

The above conjecture is trivial when $d = 2$. It was established for $\mathcal{E}$ the trivial bundle of rank $d-1$ in [30] and $d = \dim(X)$ under the condition that $(d-1)!$ is invertible in k. The above conjecture was also established for $d = 3$ in [4] (assuming 2 is invertible in k). Paralleling the results of James–Thomas in topology [33], P. Du was able to prove in [27] that Suslin's question has a positive answer for oriented vector bundles in case the cohomology group $\mathrm{H}^d_{\mathrm{Nis}}(X, \boldsymbol{\pi}^{\mathbb{A}^1}_d(\mathbb{A}^d \smallsetminus 0))$ vanishes. This vanishing statement would follow immediately from Conjecture 4.10.

5. VECTOR BUNDLES: NONAFFINE VARIETIES AND ALGEBRAIZABILITY

In this final section, we survey some joint work with M. J. Hopkins related to the classification of motivic vector bundles (see Definition 3.19), its relationship to the algebraizability question (see Question 3.20), and investigate the extent to which $\mathbb{A}^1$-homotopy theory can be used to analyze vector bundles on projective varieties.

5.1. Descent along a Jouanolou device

If X is a smooth algebraic k-variety, then there is always the map

$$\mathrm{Vect}_r(X) \to [X, \mathrm{BGL}_r]_{\mathbb{A}^1} \tag{5.1}$$

from rank r vector bundles to rank r motivic vector bundles. When X is affine, Theorem 3.11 guarantees that this map is a bijection, and examples show that the map fails to be an isomorphism outside of this case. Nevertheless, it is very interesting to try to quantify the failure of the above map to be a bijection.

If $\pi : \tilde{X} \to X$ is a Jouanolou device for X, then it follows from the definitions that the map (5.1) coincides with $\pi^* : \mathrm{Vect}_r(X) \to \mathrm{Vect}_r(\tilde{X})$ under the bijection of Theorem 3.11. The morphism π is faithfully flat by construction, and therefore, vector bundles on X are precisely vector bundles on $\tilde{X}$ equipped with a descent datum along π.

Since $\pi : \tilde{X} \to X$ is an affine morphism, it follows that $\tilde{X} \times_X \tilde{X}$ is itself an affine scheme, and the two projections $p_1, p_2 : \tilde{X} \times_X \tilde{X} \to \tilde{X}$ are $\mathbb{A}^1$-weak equivalences. Thus, pullbacks p_1^* and p_2^* are bijections on sets of isomorphism classes of vector bundles. In fact, since the relative diagonal map splits the two projections, the two pullbacks actually coincide on isomorphism classes. In descent-theoretic terms, these observations mean that any vector bundle $\mathscr{E}$ on $\tilde{X}$ can always be equipped with an isomorphism $p_1^*\mathscr{E} \xrightarrow{\sim} p_2^*\mathscr{E}$, i.e., a predescent datum. Thus, the only obstruction to descending a vector bundle along π is whether one may choose a predescent datum that actually satisfies the cocycle condition. With this observation in mind, it seems natural to analyze the question of whether *every* vector bundle can be equipped with a descent datum along π.

Question 5.1. If X is a smooth k-variety and $\pi : \tilde{X} \to X$ is a Jouanolou device for X, then is the pull-back map

$$p^* : \mathrm{Vect}_n(X) \to \mathrm{Vect}_n(\tilde{X})$$

surjective?

Theorem 5.2 (Asok, Fasel, Hopkins). *Suppose X is a smooth projective k-variety of dimension d. If either (i) $d \leq 2$ or (ii) k is algebraically closed and $d \leq 3$, then Question 5.1 admits a positive answer, i.e., every vector bundle on $\tilde{X}$ admits a descent datum relative to π.*

5.2. Algebraizability I: obstructions

If X is a smooth complex algebraic variety, then we considered the map

$$\mathrm{Vect}_r(X) \to \mathrm{Vect}_r^{\mathrm{top}}(X)$$

and posed the question of characterizing its image. We observed that this map factors through the set of motivic vector bundles, so one necessary condition for a topological vector bundle to be algebraizable is that it admits a motivic lift. In particular, this means that the Chern classes of the topological vector bundle in integral cohomology must lie in the image of the cycle class map. It is natural to ask if algebraizability of Chern classes is sufficient to guarantee that a vector bundle admits a motivic left.

In case X is projective, this question has been for instance studied in [53] where it is proved that any vector bundle with algebraic Chern classes is algebraizable if $\dim(X) = 2$. In case of projective threefolds, positive results are given by Atiyah–Rees and Bănică–Putinar respectively in [18] and [19]. If X is affine, the works of Swan–Murthy [44] and Murthy–Kumar [36] show that the answer to the question is positive if X is of dimension ≤ 3 as a consequence of the following statement: Given any pair $(\alpha_1, \alpha_2) \in \mathrm{CH}^1(X) \times \mathrm{CH}^2(X)$, there exists a vector bundle $\mathcal{E}$ on X with $c_i(\mathcal{E}) = \alpha_i$. However, in dimension 4, additional restrictions on Chern classes arise from the action of the motivic Steenrod algebra.

Theorem 5.3 ([10, THEOREM 2]). *If X is a smooth affine 4-fold, then a pair $(c_1, c_2) \in \mathrm{CH}^1(X) \times \mathrm{CH}^2(X)$ are Chern classes of a rank 2 bundle on X if and only if c_1, c_2 satisfy the additional condition $\mathrm{Sq}^2(c_2) + c_1c_2 = 0$, where*

$$\mathrm{Sq}^2 : \mathrm{CH}^2(X) \to \mathrm{CH}^3(X)/2$$

is the Steenrod squaring operation, and c_1c_2 is the reduction modulo 2 of the cup product.

Remark 5.4. This obstruction is sufficient to identify topological vector bundles on a smooth affine fourfold X having algebraic Chern classes which are not algebraizable [10, COROLLARY 3.1.5]. One example of such an X is provided by the open complement in $\mathbb{P}^1 \times \mathbb{P}^3$ of a suitable smooth hypersurface Z of bidegree $(3, 4)$.

5.3. Algebraizability II: building motivic vector bundles

The notion of a cellular space goes back to the work of Dror Farjoun. By a cellular motivic space, we will mean a space that can be built out of the motivic spheres $S^{p,q}$ by formation of homotopy colimits. It is straightforward to see inductively that $\mathbb{P}^n$ is cellular. In the presence of cellularity assumptions, many obstructions to producing a motivic lift of a vector bundle vanish and this motivates the following conjecture.

Conjecture 5.5. *If X is a smooth cellular $\mathbb{C}$-variety, then the map*

$$[X, \mathrm{Gr}_r]_{\mathbb{A}^1} \to \mathrm{Vect}^{\mathrm{top}}(X)$$

is surjective (resp. bijective).

Remark 5.6. The conjecture holds for $\mathbb{P}^n$ for $n \leq 3$ (this follows, for example, from the results of Schwarzenberger and Atiyah–Rees mentioned above); in these cases, bijectivity holds. For $\mathbb{P}^4$, the "surjective" formulation of Conjecture 5.5 is known, but the "bijective" formulation is not.

We now analyze Conjecture 5.5 for a class of "interesting" topological vector bundles on $\mathbb{P}^n$ introduced by E. Rees and L. Smith. We briefly recall the construction of these topological vector bundles here. By a classical result of Serre [54, PROPOSITION 11], we know that if p is a prime, then the p-primary component of $\pi_{4p-3}(S^3)$ is isomorphic to $\mathbb{Z}/p$, generated by the composite of a generator α_1 of the p-primary component of $\pi_{2p}(S^3)$ and the $(2p-3)$rd suspension of itself; we will write α_1^2 for this class.

The map $\mathbb{P}^n \to S^{2n}$ that collapses $\mathbb{P}^{n-1}$ to a point determines a function

$$[S^{2n-1}, S^3] \cong [S^{2n}, \mathrm{BSU}(2)] \to [\mathbb{P}^n, \mathrm{BSU}(2)]$$

Rees established that the class α_1^2 determines a nontrivial rank 2 vector bundle $\xi_p \in [\mathbb{P}^{2p-1}, \mathrm{BSU}(2)]$; we will refer to this bundle as a *Rees bundle* [48]. By construction, ξ_p is a nontrivial rank 2 bundle with trivial Chern classes.

The motivation for Rees' construction originated from results of Grauert–Schneider [32]. If the bundles ξ_p were algebraizable, then the fact that they have trivial Chern classes would imply they were necessarily unstable by Barth's results on Chern classes of stable vector bundles [21, COROLLARY 1 P. 127] (here, stability means slope stability in the sense of Mumford). Grauert and Schneider analyzed unstable rank 2 vector bundles on projective space and they aimed to prove that such vector bundles were necessarily direct sums of line bundles; this assertion is now sometimes known as the Grauert–Schneider conjecture. In view of the Grauert–Schneider conjecture, the bundles ξ_p should not be algebraizable. On the other hand, one of the motivations for Conjecture 5.5 is the following result.

Theorem 5.7 ([11, THEOREM 2.2.16]). *For every prime number p, the bundle ξ_p lifts to a class in $[\mathbb{P}^{2p-1}, \mathrm{Gr}_2]_{\mathbb{A}^1}$.*

Remark 5.8. This is established by constructing motivic homotopy classes lifting α_1 and α_1^2. In our situation, the collapse map takes the form

$$\mathbb{P}^n \to S^{n,n}$$

and the lift must come from an element of $[S^{n-1,n}, SL_2]_{\mathbb{A}^1}$. The class α_1 can be lifted using ideas related to those discussed in 4.1 in conjunction with a motivic version of Serre's classical p-local splitting of compact Lie groups [9, THEOREM 2], the resulting lift has the wrong weight to lift to a group as above. Since the class α_1^2 is torsion, we can employ a weight-shifting mechanism to fix this issue. In this direction, there are host of other vector bundles that are analogous to the Rees bundles that one might investigate from this point of view, e.g., bundles that can be built out of Toda's unstable α-family [62]. Likewise, even the surjectivity assertion in Conjecture 5.5 is unknown for $\mathbb{P}^5$.

5.9 (The Wilson Space Hypothesis). To close, we briefly sketch an approach to the resolution of Conjecture 5.5 building on Mike Hopkins' Wilson Space Hypothesis. The latter asserts that the Voevodsky motive of the $\mathbb{P}^1$-infinite loop spaces $\Omega^{\infty}_{\mathbb{P}^1}\Sigma^n_{\mathbb{P}^1}$ MGL arising from algebraic cobordism are pure Tate (the space is "homologically even"); this hypothesis is an algebro-geometric version of a result of Steve Wilson on the infinite loop spaces of the classical cobordism spectrum.

The motivic version of the unstable Adams–Novikov resolution for BGL_r yields a spectral sequence that, under the cellularity assumption on X should converge to a (completion of) the set of rank r motivic vector bundles on X. The resulting spectral sequence can be compared to its topological counterpart and Wilson Space Hypothesis combined with the cellularity assumption on X would imply that the two spectral sequences coincide.

ACKNOWLEDGMENTS

The authors would like to acknowledge the profound influence of M. Hopkins and F. Morel on their work. Many of the results presented here would not have been possible without their insights and vision. Both authors would also like to thank all their colleagues, coauthors, and friends for their help and support.

FUNDING

Aravind Asok was partially supported by NSF Awards DMS-1802060 and DMS-2101898.

REFERENCES

[1] A. Asok, Affine representability of quadrics revisited. 2021, arXiv:2104.08208.

[2] A. Asok, B. Doran, and J. Fasel, Smooth models of motivic spheres and the clutching construction. *Int. Math. Res. Not. IMRN* **6** (2016), no. 1, 1890–1925.

[3] A. Asok and J. Fasel, Algebraic vector bundles on spheres. *J. Topol.* **7** (2014), no. 3, 894–926.

[4] A. Asok and J. Fasel, A cohomological classification of vector bundles on smooth affine threefolds. *Duke Math. J.* **163** (2014), no. 14, 2561–2601.

[5] A. Asok and J. Fasel, Splitting vector bundles outside the stable range and $\mathbb{A}^1$-homotopy sheaves of punctured affine spaces. *J. Amer. Math. Soc.* **28** (2015), no. 4, 1031–1062.

[6] A. Asok and J. Fasel, Comparing Euler classes. *Quart. J. Math.* **67** (2016), 603–635.

[7] A. Asok and J. Fasel, An explicit KO-degree map and applications. *J. Topol.* **10** (2017), no. 1, 268–300.

[8] A. Asok and J. Fasel, Euler class groups and motivic stable cohomotopy (with an appendix by M. K. Das). *J. Eur. Math. Soc. (JEMS)* (accepted for publication).

[9] A. Asok, J. Fasel, and M. Hopkins, Localization and nilpotent spaces in $\mathbb{A}^1$-homotopy theory. 2020, arXiv:1909.05185.

[10] A. Asok, J. Fasel, and M. J. Hopkins, Obstructions to algebraizing topological vector bundles. *Forum Math. Sigma* **7** (2019), e6, 18 pp.

[11] A. Asok, J. Fasel, and M. J. Hopkins, Algebraic vector bundles and p-local $\mathbb{A}^1$-homotopy theory. 2020, arXiv:2008.03363.

[12] A. Asok, J. Fasel, and B. Williams, Motivic spheres and the image of the Suslin–Hurewicz map. *Invent. Math.* **219** (2020), no. 1, 39–73.

[13] A. Asok, M. Hoyois, and M. Wendt, Affine representability results in $\mathbb{A}^1$-homotopy theory, I: vector bundles. *Duke Math. J.* **166** (2017), no. 10, 1923–1953.

[14] A. Asok, M. Hoyois, and M. Wendt, Affine representability results in $\mathbb{A}^1$-homotopy theory, II: principal bundles and homogeneous spaces. *Geom. Topol.* **22** (2018), no. 2, 1181–1225.

[15] A. Asok, M. Hoyois, and M. Wendt, Affine representability results in $\mathbb{A}^1$-homotopy theory III: finite fields and complements. *Algebr. Geom.* **7** (2020), no. 5, 634–644.

[16] A. Asok and P. Østvær, $\mathbb{A}^1$-homotopy theory and contractible varieties: a survey. *Lecture Notes in Math.* (to appear).

[17] A. Asok, K. Wickelgren, and B. Williams, The simplicial suspension sequence in $\mathbb{A}^1$-homotopy. *Geom. Topol.* **21** (2017), no. 4, 2093–2160.

[18] M. F. Atiyah and E. Rees, Vector bundles on projective 3-space. *Invent. Math.* **35** (1976), 131–153.

[19] C. Bănică and M. Putinar, On complex vector bundles on projective threefolds. *Invent. Math.* **88** (1987), no. 2, 427–438.

[20] J. Barge and F. Morel, Groupe de Chow des cycles orientés et classe d'Euler des fibrés vectoriels. *C. R. Acad. Sci. Paris Sér. I Math.* **330** (2000), no. 4, 287–290.

[21] W. Barth, Some properties of stable rank-2 vector bundles on $\mathbf{P}_n$. *Math. Ann.* **226** (1977), no. 2, 125–150.

[22] H. Bass, K-theory and stable algebra. *Publ. Math. Inst. Hautes Études Sci.* **22** (1964), 5–60.

[23] S. M. Bhatwadekar and R. Sridharan, The Euler class group of a Noetherian ring. *Compos. Math.* **122** (2000), no. 2, 183–222.

[24] A. Borel and F. Hirzebruch, Characteristic classes and homogeneous spaces. II. *Amer. J. Math.* **81** (1959), 315–382.

[25] R. Bott, The space of loops on a Lie group. *Michigan Math. J.* **5** (1958), 35–61.

[26] J.-L. Colliot-Thélène, R. Hoobler, and B. Kahn, The Bloch–Ogus–Gabber theorem. In *Proceedings of the Great Lakes K-theory conference (Toronto 1996)*, edited by R. Jardine and V. Snaith, pp. 31–94, Fields Inst. Res. Math. Sci. Commun. Ser. 16, Amer. Math. Soc., Providence, RI, 1997.

[27] P. Du, Enumerating non-stable vector bundles. *Int. Math. Res. Not.* (to appear).

[28] A. Dubouloz and J. Fasel, Families of $\mathbb{A}^1$-contractible affine threefolds. *Algebr. Geom.* **5** (2018), no. 1, 1–14.

[29] J. Fasel. Groupes de Chow-Witt. *Mém. Soc. Math. Fr. (N.S.)* (2008), no. 113, viii+197 pp.

[30] J. Fasel, R. A. Rao, and R. G. Swan, On stably free modules over affinc algebras. *Publ. Math. Inst. Hautes Études Sci.* **116** (2012), no. 1, 223–243.

[31] N. Feld, Milnor–Witt Cycle Modules. *J. Pure Appl. Algebra* **224** (2020), no. 7.

[32] H. Grauert and M. Schneider, Komplexe Unterräume und holomorphe Vektorraumbündel vom Rang zwei. *Math. Ann.* **230** (1977), no. 1, 75–90.

[33] I. James and E. Thomas, An Approach to the Enumeration Problem for Non-Stable vector Bundles. *Indiana Univ. Math. J.* **14** (1965), no. 3, 485–506.

[34] J. F. Jardine, *Local homotopy theory*. Springer Monogr. Math., Springer, New York, 2015.

[35] J.-P. Jouanolou, Une suite exacte de Mayer–Vietoris en K-théorie algébrique. In *Higher K-theories*, edited by H. Bass, pp. 293–316, Lecture Notes in Math. 341, Springer, Berlin–Heidelberg–New York–Tokyo, 1973.

[36] N. M. Kumar and M. P. Murthy, Algebraic cycles and vector bundles over affine three-folds. *Ann. of Math. (2)* **116** (1982), no. 3, 579–591.

[37] S. D. Liao, On the theory of obstructions of fiber bundles. *Ann. of Math. (2)* **60** (1954), 146–191.

[38] N. Mohan Kumar, Stably free modules. *Amer. J. Math.* **107** (1985), no. 6, 1439–1444.

[39] F. Morel, The stable $\mathbb{A}^1$-connectivity theorems. *K-Theory* **35** (2005), no. 1–2, 1–68.

[40] F. Morel, *$\mathbb{A}^1$-algebraic topology over a field*. Lecture Notes in Math. 2052, Springer, New York, 2012.

[41] F. Morel and V. Voevodsky, $\mathbb{A}^1$-homotopy theory of schemes. *Publ. Math. Inst. Hautes Études Sci.* **90** (1999), 45–143.

[42] M. P. Murthy, Zero cycles and projective modules. *Ann. of Math. (2)* **140** (1994), no. 2, 405–434.

[43] M. P. Murthy, A survey of obstruction theory for projective modules of top rank. In *Algebra, K-theory, groups, and education. On the occasion of Hyman Bass's 65th birthday. Mainly the proceedings of the conference, Columbia University, New York, NY, November 6–7, 1997*, pp. 153–174, American Mathematical Society, Providence, RI, 1997.

[44] M. P. Murthy and R. G. Swan, Vector bundles over affine surfaces. *Invent. Math.* **36** (1976), 125–165.

[45] M. Ojanguren, Formes quadratiques sur les algèbres de polynômes. *C. R. Acad. Sci., Sér. A* **287** (1978), 695–698.

[46] M. Ojanguren and I. Panin, A purity theorem for the Witt group. *Ann. Sci. Éc. Norm. Supér. (4)* **32** (1999), no. 1, 71–86.

[47] S. Parimala, Failure of a quadratic analogue of Serre's conjecture. *Amer. J. Math.* **100** (1978), 913–924.

[48] E. Rees, Some rank two bundles on $P_n\mathbf{C}$, whose Chern classes vanish. In *Variétés analytiques compactes (Colloq., Nice, 1977)*, pp. 25–28, Lecture Notes in Math. 683, Springer, Berlin, 1978.

[49] O. Röndigs, M. Spitzweck, and P.-A. Østvær, The first stable homotopy groups of motivic spheres. *Ann. of Math. (2)* **189** (2019), no. 1, 1–74.

[50] M. Rost, Chow groups with coefficients. *Doc. Math.* **1** (1996), no. 16, 319–393 (electronic).

[51] M. Schlichting, Euler class groups and the homology of elementary and special linear groups. *Adv. Math.* **320** (2017), 1–81.

[52] M. Schlichting and G. Tripathi, Geometric models for higher Grothendieck–Witt groups in $\mathbb{A}^1$-homotopy theory. *Math. Ann.* **362** (2015), no. 3–4, 1143–1167.

[53] R. L. E. Schwarzenberger, Vector bundles on algebraic surfaces. *Proc. Lond. Math. Soc. (3)* **11** (1961), 601–622.

[54] J.-P. Serre, Groupes d'homotopie et classes de groupes abéliens. *Ann. of Math. (2)* **58** (1953), 258–294.

[55] J.-P. Serre, Faisceaux algébriques cohérents. *Ann. of Math. (2)* **61** (1955), 197–278.

[56] J.-P. Serre, Modules projectifs et espaces fibrés à fibre vectorielle. In *Séminaire P. Dubreil, M.-L. Dubreil-Jacotin et C. Pisot, 1957/58, Fasc. 2, Exposé 23*, p. 18, Secrétariat mathématique, Paris, 1957/1958.

[57] N. Steenrod, *The topology of fibre bundles*. Princeton Landmarks Math., Princeton University Press, Princeton, NJ, 1999.

[58] A. A. Suslin, A cancellation theorem for projective modules over algebras. *Dokl. Akad. Nauk SSSR* **236** (1977), no. 4, 808–811.

[59] A. A. Suslin, On stably free modules. *Mat. Sb. (N.S.)* **102(144)** (1977), no. 4, 537–550, 632.

[60] A. A. Suslin, Homology of GL_n, characteristic classes and Milnor K-theory. In *Algebraic K-theory, number theory, geometry and analysis (Bielefeld, 1982)*, pp. 357–375, Lecture Notes in Math. 1046, Springer, Berlin, 1984.

[61] E. Thomas, *Seminar on fiber spaces*. Lecture Notes in Math. 13, Springer, Berlin–New York, 1966.

[62] H. Toda, *Composition methods in homotopy groups of spheres*. Ann. of Math. Stud. 49, Princeton University Press, Princeton, NJ, 1962.

[63] C. Weibel, Homotopy Algebraic K-theory. In *Algebraic K-theory and algebraic number theory, Proc. Semin., Honolulu/Hawaii 1987*, pp. 461–488, Contemp. Math. 83, Amer. Math. Soc., 1989.

ARAVIND ASOK

Department of Mathematics, University of Southern California, 3620 S. Vermont Ave., Los Angeles, CA 90089-2532, USA, asok@usc.edu

JEAN FASEL

Institut Fourier, Université Grenoble-Alpes, CS 40700, 38058 Grenoble cedex 9, France, jean.fasel@univ-grenoble-alpes.fr

This Page Intentionally Left Blank

THE UNREASONABLE EFFECTIVENESS OF WALL-CROSSING IN ALGEBRAIC GEOMETRY

AREND BAYER AND EMANUELE MACRÌ

ABSTRACT

We survey applications of Bridgeland stability conditions in algebraic geometry and discuss open questions for future research.

MATHEMATICS SUBJECT CLASSIFICATION 2020

Primary 14F08; Secondary 14F17, 14H51, 14J28, 14J32, 14J42

KEYWORDS

Bridgeland stability conditions, K3 categories, hyperkähler varieties, Calabi–Yau threefolds, Brill–Noether theory, cubic fourfolds, Donaldson–Thomas theory

ICM INTERNATIONAL CONGRESS OF MATHEMATICIANS 2022 JULY 6–14

Proc. Int. Cong. Math. 2022, Vol. 3, pp. 2172–2195
DOI 10.4171/ICM2022/80

Published by EMS Press

1. INTRODUCTION

Bridgeland stability conditions and wall-crossing have provided answers to many questions in algebraic geometry a priori unrelated to derived categories, including hyperkähler varieties—their rational curves, their birational geometry, their automorphisms, and their moduli spaces—, Brill–Noether questions, Noether–Lefschetz loci, geometry of cubic fourfolds, or higher-rank Donaldson–Thomas theory. Our goal is to answer the question: why? What makes these techniques so effective, and what exactly do they add beyond, for example, classical vector bundle techniques?

The common underlying strategy can be roughly summarized as follows. For each stability condition on a derived category $D^b(X)$ of an algebraic variety X and each numerical class, moduli spaces of semistable objects in $D^b(X)$ exist as proper algebraic spaces. This formalism includes many previously studied moduli spaces: moduli spaces of Gieseker- or slope-stable sheaves, of stable pairs, or of certain equivalences classes of rational curves in X. The set of stability conditions on $D^b(X)$ has the structure of a complex manifold; when we vary the stability condition, stability of a given object only changes when we cross the walls of a well-defined wall and chamber structure.

The typical ingredients when approaching a problem with stability conditions are the following:

(large volume) There is a point in the space of stability conditions where stable objects have a "classical" interpretation, e.g. as Gieseker-stable sheaves.

(point of interest) There is a point in the space of stability conditions where stability has strong implications, e.g., vanishing properties, or even there is no semistable object of a given numerical class.

(wall-crossing) It is possible to analyze the finite set of walls between these two points, and how stability changes when crossing each wall.

In general, it is quite clear from the problem which are the points of interest, and the main difficulty consists in analyzing the walls. In the ideal situation, which leads to sharp exact results, these walls can be characterized purely numerically; there are only a few such ideal situations, K3 surfaces being one of them. Otherwise, the study of walls can get quite involved, even though there are now a number of more general results available, e.g., a wall-crossing formula for counting invariants arising from moduli spaces.

We illustrate the case of K3 surfaces, or more generally *K3 categories*, in Section 3, with applications to hyperkähler varieties, to Brill–Noether theory of curves, and to the geometry of special cubic fourfolds. The study of other surfaces or the higher-dimensional case becomes more technical, and the existence of Bridgeland stability conditions is not yet known in full generality. There are weaker notions of stability, which in the threefold case already lead to striking results. We give an overview of this circle of ideas in Section 4, along with three applications related to curves. We give a brief introduction to stability conditions in Section 2, and pose some questions for future research in Section 5.

Derived categories of coherent sheaves on varieties have been hugely influential in recent years; we refer to [17, 19, 46, 74] for an overview of the theory. Moduli spaces of sheaves on K3 surfaces have largely been influenced by [61]; we refer to [63] for an overview of applications of these techniques and to [39] for the higher-dimensional case of hyperkähler manifolds. For the original motivation from physics, we refer to [26, 41].

Our survey completely omits the quickly developing theory and applications of stability conditions on Kuznetsov components of Fano threefolds. We also will not touch the rich subject of extra structures on spaces of stability conditions, developed, for example, in the foundational papers [22, 23]; we also refer to [72] for a symplectic perspective.

2. STABILITY CONDITIONS ON DERIVED CATEGORIES

Recall slope-stability for sheaves on an integral projective curve C: we set

$$\mu(E) = \frac{\deg E}{\operatorname{rk} E} \in (-\infty, +\infty], \quad \text{visualized by } Z(E) = -\deg E + i \operatorname{rk} E,$$

and call a sheaf *slope-semistable* if every subsheaf $F \subset E$ satisfies $\mu(F) \leq \mu(F)$. The set of semistable sheaves of fixed rank and degree is *bounded* and can be parameterized by a projective moduli space. Moreover, semistable sheaves generate $\operatorname{Coh}(C)$, in the sense that every sheaf E admits a (unique and functorial) *Harder–Narasimhan (HN) filtration*

$$0 = E_0 \subset E_1 \subset E_2 \subset \cdots \subset E_m = E$$

with E_l/E_{l-1} semistable, and $\mu(E_1) > \mu(E_2/E_1) > \cdots > \mu(E_m/E_{m-1})$.

How to generalize this to a variety X of dimension $n \geq 2$? Given a polarization H, one can define the slope μ_H using $H^{n-1} \cdot \operatorname{ch}_1(E)$ as the degree. To distinguish, e.g., the slope of the structure sheaf $\mathcal{O}_X$ from that of an ideal sheaf $\mathcal{I}_x \subset \mathcal{O}_X$ for $x \in X$, we can further refine the notion of slope and use lower-degree terms of the Hilbert polynomial $p_E(m) = \chi(E(mH))$ as successive tie breakers; this yields Gieseker stability.

One of the key insights in Bridgeland's notion of stability conditions introduced in [20] is that instead we can, in fact, still use a notion of slope-stability, defined as the quotient of "degree" by "rank." The price we have to pay is to replace $\operatorname{Coh}(X)$ by another abelian subcategory $\mathcal{A}$ of the bounded derived category $\mathrm{D}^{\mathrm{b}}(X)$ of coherent sheaves on X, and to generalize the notions of "degree" and "rank" (combined into a central charge Z as above).

To motivate the definition, consider again slope-stability for a curve C. First, for $\phi \in (0, 1]$, let $\mathcal{P}(\phi) \subset \operatorname{Coh}(C) \subset \mathrm{D}^{\mathrm{b}}(C)$ be the category of slope-semistable sheaves E with $Z(E) \in \mathbb{R}_{>0} \cdot e^{i\pi\phi}$, i.e., of slope $\mu(E) = -\cot(\pi\phi)$, and let $\mathcal{P}(\phi + n) = \mathcal{P}(\phi)[n]$ for $n \in \mathbb{Z}$ be the set of *semistable objects of phase* $\phi + n$. Every complex $E \in \mathrm{D}^{\mathrm{b}}(C)$ has a filtration into its cohomology objects $\mathcal{H}^l(E)[-l]$. We can combine this with the classical HN filtration of $\mathcal{H}^l(E)[-l]$ for each l to obtain a fincr filtration for E where every filtration quotient is semistable, i.e., an object of $\mathcal{P}(\phi)$ for $\phi \in \mathbb{R}$. The properties of this structure are axiomatized by conditions (1)–(4) in Definition 2.1 below. But crucially it can always be obtained from slope-stability in an abelian category $\mathcal{A}$; we just have to generalize the setting $\mathcal{A} \subset \mathrm{D}^{\mathrm{b}}(\mathcal{A})$ to $\mathcal{A} \subset \mathcal{D}$ being the "heart of a bounded t-structure" in a triangulated category.

When combined with the remaining conditions in Definition 2.1, the main payoff are the strong deformation and wall-crossing properties of Bridgeland stability conditions. Given any small deformation of "rank" and "degree" (equivalently, of the central charge Z), we can accordingly adjust the abelian category $\mathcal{A}$ (or, equivalently, the set of semistable objects $\mathcal{P}$) and obtain a new stability condition. Along such a deformation, moduli spaces of semistable objects undergo very well-behaved wall-crossing transformations.

2.1. Bridgeland stability conditions

We now consider more generally an *admissible subcategory* $\mathcal{D}$ of $\mathrm{D^b}(X)$ for a smooth and proper variety X over a field k, namely a full triangulated subcategory whose inclusion $\mathcal{D} \hookrightarrow \mathrm{D^b}(X)$ admits both a left and a right adjoint. For instance, $\mathcal{D} = \mathrm{D^b}(X)$; otherwise we think of $\mathcal{D}$ as a smooth and proper *noncommutative variety*.

We fix a finite rank free abelian group Λ and a group homomorphism

$$v : K_0(\mathcal{D}) \to \Lambda$$

from the Grothendieck group $K_0(\mathcal{D})$ of $\mathcal{D}$ to Λ.

Definition 2.1. A *Bridgeland stability condition* on $\mathcal{D}$ with respect to (v, Λ) is a pair $\sigma = (Z, \mathcal{P})$ where

- $Z : \Lambda \to \mathbb{C}$ is a group homomorphism, called *central charge*, and
- $\mathcal{P} = (\mathcal{P}(\phi))_{\phi \in \mathbb{R}}$ is a collection of full additive subcategories $\mathcal{P}(\phi) \subset \mathcal{D}$

satisfying the following conditions:

(1) for all nonzero $E \in \mathcal{P}(\phi)$, we have $Z(v(E)) \in \mathbb{R}_{>0} \cdot e^{i\pi\phi}$;

(2) for all $\phi \in \mathbb{R}$, we have $\mathcal{P}(\phi + 1) = \mathcal{P}(\phi)[1]$;

(3) if $\phi_1 > \phi_2$ and $E_j \in \mathcal{P}(\phi_j)$, then $\mathrm{Hom}(E_1, E_2) = 0$;

(4) (Harder–Narasimhan filtrations) for all nonzero $E \in \mathcal{D}$, there exist real numbers $\phi_1 > \phi_2 > \cdots > \phi_m$ and a finite sequence of morphisms

$$0 = E_0 \xrightarrow{s_1} E_1 \xrightarrow{s_2} \cdots \xrightarrow{s_m} E_m = E$$

such that the cone of s_l is a non-zero object of $\mathcal{P}(\phi_l)$;

(5) (support property) there exists a quadratic form Q on $\Lambda_{\mathbb{R}} = \Lambda \otimes \mathbb{R}$ such that

- the kernel of Z is negative definite with respect to Q, and
- for all $E \in \mathcal{P}(\phi)$ for any ϕ we have $Q(v(E)) \geq 0$;

(6) (openness of stability) the property of being in $\mathcal{P}(\phi)$ is open in families of objects in $\mathcal{D}$ over any scheme;

(7) (boundedness) objects in $\mathcal{P}(\phi)$ with fixed class $v \in \Lambda$ are parameterized by a k-scheme of finite type.

An object of the subcategory $\mathcal{P}(\phi)$ is called σ*-semistable* of phase ϕ, and σ*-stable* if it admits no non-trivial subobject in $\mathcal{P}(\phi)$. The set of Bridgeland stability conditions on $\mathcal{D}$ is denoted by $\mathrm{Stab}(\mathcal{D})$, where we omit the dependence on (v, Λ) from the notation.

Conditions (1)–(4) form the original definition in [20] and ensure we have a notion of slope-stability. The support property is necessary to show that stability conditions vary continuously (see Theorem 2.2 below) and admit a well-behaved wall and chamber structure: fundamentally, this is due to the simple linear algebra consequence that given $C > 0$, there are only finitely many classes $w \in \Lambda$ of semistable objects with $|Z(w)| < C$. Conditions (6) and (7) were introduced in [73, 74], with similar versions appearing previously in [42, SECTION 3]; they guarantee the existence of moduli spaces of semistable objects.

Theorem 2.2 (Bridgeland deformation theorem). *The set* $\mathrm{Stab}(\mathcal{D})$ *has the structure of a complex manifold such that the natural map*

$$\mathcal{Z} : \mathrm{Stab}(\mathcal{D}) \to \mathrm{Hom}(\Lambda, \mathbb{C}), \quad (Z, \mathcal{P}) \mapsto Z$$

is a local isomorphism at every point.

For conditions (1)–(5), this is a reformulation of Bridgeland's main result [20, THEOREM 1.2]. It says that $\sigma = (Z, \mathcal{P})$ can be deformed uniquely given a small deformation of $Z \rightsquigarrow Z'$, roughly as long as $Z'(E) \neq 0$ remains true for all σ-semistable objects E. (More precisely, any path where Q remains negative definite on $\mathrm{Ker}\, Z'$ can be lifted uniquely to a path in $\mathrm{Stab}(\mathcal{D})$.) With the additional conditions (6) and (7), Theorem 2.2 was proved in [73, THEOREM 3.20] and [67, SECTION 4.4], where the most difficult aspect is to show that openness of stability is preserved under deformations.

The theory has been developed over an arbitrary base scheme in [8]. A stability condition over a base is the datum of a stability condition on each fiber, such that families of objects over the base have locally constant central charges, satisfy openness of stability, and a global notion of HN filtration after base change to a one-dimensional scheme; moreover, we impose a global version of the support property and of boundedness. An analogue of Theorem 2.2 holds; differently to the absolute case, assuming the support property is not enough and the proof requires the additional conditions (6) and (7).

The construction of Bridgeland stability conditions is discussed in Section 4; in particular, they exist on surfaces and certain threefolds.

2.2. Stability conditions as polarizations

It was first suggested in the arXiv version of [21] to think of σ as a polarization of the noncommutative variety $\mathcal{D}$. We now review some results partly justifying this analogy. A polarization of a variety X by an ample line bundle H gives projective moduli spaces of H-Gieseker-stable sheaves; the following two results provide an analogue.

Theorem 2.3 (Toda, Alper, Halpern-Leistner, Heinloth). *Given* $\sigma \in \mathrm{Stab}(\mathcal{D})$ *and* $v \in \Lambda$, *there is a finite type Artin stack* $\mathcal{M}_\sigma(v)$ *of* σ*-semistable objects* E *with* $v(E) = v$ *and fixed phase. In characteristic zero, it has a proper* good moduli space $M_\sigma(v)$ *in the sense of Alper.*

Proof. The existence as Artin stack is [73, THEOREM 3.20], while the existence of a good moduli space is proven in [2, THEOREM 7.25]; see also [8, THEOREM 21.24]. ■

Theorem 2.4. *The algebraic space $M_\sigma(v)$ admits a Cartier divisor ℓ_σ that has strictly positive degree on every curve. In characteristic zero, if $M_\sigma(v)$ is smooth, or more generally if it has $\mathbb{Q}$-factorial log-terminal singularities, then $M_\sigma(v)$ is projective.*

Proof. The existence of the Cartier divisor and its properties is the *Positivity Lemma* in [12], see also [8, THEOREM 21.25]. The projectivity follows from [76, COROLLARY 3.4]. ■

As studied extensively in Donaldson theory in the 1990s, the Gieseker-moduli spaces may change as H crosses walls in the ample cone.

Theorem 2.5. *Fix a vector $v \in \Lambda$. Then there exists a locally finite union $\mathcal{W}_v$ of real-codimension one submanifolds in* $\mathrm{Stab}(\mathcal{D})$*, called walls, such that on every connected component $\mathcal{C}$ of the complement* $\mathrm{Stab}(\mathcal{D}) \setminus \mathcal{W}_v$*, called a chamber, the moduli space $M_\sigma(v)$ is independent of the choice $\sigma \in \mathcal{C}$.*

Theorem 2.5 follows from the results in [21, SECTION 9]; see also [73, PROPOSITION 2.8] and [10, PROPOSITION 3.3]. The set $\mathcal{W}_v$ consists of stability conditions for which there exists an exact triangle $A \to E \to B$ of semistable objects of the same phase with $v(E) = v$, but $v(A)$ not proportional to v. Locally, the wall is defined by $Z(A)$ being proportional to $Z(E)$, and the objects E is unstable on the side where $\arg(Z(A)) > \arg(Z(E))$; often it is stable near the wall on the other side, e.g., when A and B are *stable* and the extension is nontrivial. The support property (5) is again crucial in the proof of Theorem 2.5: it constrains the classes $a = v(A)$, $b = v(B)$ involved in a wall, and locally that produces a finite set.

Sometimes, one can describe $\mathcal{W}_v$ completely, namely when we know which of the moduli spaces $M_\sigma(a)$ and $M_\sigma(b)$ are nonempty.

2.3. K3 categories

Such descriptions of $\mathcal{W}_v$ have been particularly powerful in the case of K3 categories; it has also been carried out completely for $\mathrm{D^b}(\mathbb{P}^2)$, where the answer is more involved [24, 54]. For this section, we work over the complex numbers and let $\mathcal{D}$ be

(1) the derived category $\mathcal{D} = \mathrm{D^b}(S)$ of a smooth projective K3 surface, or

(2) the *Kuznetsov component*

$$\mathcal{D} = \mathcal{K}u(Y) = \mathcal{O}_Y^\perp \cap \mathcal{O}_Y(1)^\perp \cap \mathcal{O}_Y(2)^\perp \subset \mathrm{D^b}(Y)$$

of the derived category of a smooth cubic fourfold Y, or

(3) the Kuznetsov component of a Gushel–Mukai fourfold defined in [47].

In (1) we can also allow a Brauer twist; one expects further examples of Kuznetsov components of Fano varieties where similar results hold. In all these cases, $\mathcal{D}$ is a Calabi–Yau-2 category: there is a functorial isomorphism $\mathrm{Hom}(E, F) = \mathrm{Hom}(F, E[2])^\vee$ for all $E, F \in \mathcal{D}$.

Moreover, it has an associated integral weight two Hodge structure $H^*(\mathcal{D}, \mathbb{Z})$ with an even pairing $(\_, \_)$; in the case of a K3 surface, $H^*(\mathrm{D^b}(S), \mathbb{Z}) = H^*(S, \mathbb{Z})$ with H^0 and H^4 considered to be $(1, 1)$-classes; in the other cases, the underlying lattice is the same, and after the initial indirect construction in [1] there is now an intrinsic construction based on the topological K-theory and Hochschild homology of $\mathcal{D}$ [64]. There is a Mukai vector $v: K_0(\mathcal{D}) \to H^{1,1}(\mathcal{D}, \mathbb{Z})$ satisfying $\chi(E, F) = -(v(E), v(F))$ for all $E, F \in \mathcal{D}$.

In all three cases, there is a main component $\mathrm{Stab}^\dagger(\mathcal{D}) \subset \mathrm{Stab}(\mathcal{D})$ with an effective version of Theorem 2.2 for $\Lambda = H^{1,1}(\mathcal{D}, \mathbb{Z})$: the map $\mathcal{Z}$ is a covering of an explicitly described open subset of $\mathrm{Hom}(\Lambda, \mathbb{C})$, see [21] for case (1), [8,9] for case (2), and [65] for case (3).

Now consider a family of such K3 categories, given by a family of K3 surfaces or Fano fourfolds over a base scheme, respectively. In this case, Mukai's classical deformation argument applies: every stable object E in a given fiber is simple, i.e., it satisfies $\mathrm{Hom}(E, E) = \mathbb{C}$, and so $\mathrm{Ext}^2(E, E) = \mathbb{C}$ by Serre duality; therefore the obvious obstruction to extending E across the family, namely that $v(E)$ remains a Hodge class, is the only one. Extending such deformation arguments to $\mathcal{D}$ was the original motivation for introducing stability conditions for families of noncommutative varieties, see [8, SECTION 31]. They allows us to deduce nonemptiness of moduli spaces from the previously known case of K3 surfaces (and simplify the previous classical argument for Gieseker stability on K3 surfaces by reduction to elliptically fibered K3s, see [18]), which leads to the following result.

Theorem 2.6 (Mukai, Huybrechts, O'Grady, Yoshioka, Toda [8,12,65]). *Let $v \in H^{1,1}(\mathcal{D}, \mathbb{Z})$ be primitive, and $\sigma \in \mathrm{Stab}^\dagger(\mathcal{D})$ be generic. Then $M_\sigma(v)$ is nonempty if and only if $v^2 := (v, v) \geq -2$; in this case, it is a smooth projective irreducible holomorphic symplectic (IHS) variety.*

More precisely, $M_\sigma(v)$ is of $\mathrm{K3}^{[n]}$-type, where $n = (v^2 + 2)/2$, i.e., it is deformation equivalent to the Hilbert scheme of n points on a K3 surface (see [25,34] for the basic theory of irreducible holomorphic symplectic varieties). If $v^2 \geq 2$, the Mukai morphism

$$\vartheta: H^2\big(M_\sigma(v), \mathbb{Z}\big) \to H^*(\mathcal{D}, \mathbb{Z})$$

induced by a (quasi)universal family gives an identification of $H^2(M_\sigma(v), \mathbb{Z})$ with $v^\perp$. If $v^2 = 0$, then $M_\sigma(v)$ is a K3 surface and $H^2(M_\sigma(v), \mathbb{Z})$ is identified with $v^\perp / v$.

Knowing exactly which semistable objects exist then allows us to describe exactly when we are on a wall. While a complete result as in [11, THEOREM 5.7] also needs to treat essential aspects of the wall-crossing behavior, the basic result is simple to state:

Theorem 2.7 ([11]). *Let $v \in H^{1,1}(\mathcal{D}, \mathbb{Z})$ be a primitive class. Then $\sigma = (Z, \mathcal{P}) \in \mathrm{Stab}^\dagger(\mathcal{D})$ lies on a wall for v if and only if there exists classes $a, b \in H^{1,1}(\mathcal{D}, \mathbb{Z})$ with $v = a + b$, $a^2, b^2 \geq -2$ and $Z(a), Z(b)$ are positive real multiples of $Z(v)$.*

And the fundamental reason is similarly simple to explain: by Theorem 2.6, this allows for the existence of extensions

$$0 \to A \to E \to B \to 0 \tag{2.1}$$

where $v(A) = a$, $v(E) = v$, $v(B) = b$, and A, E, B are all semistable of the same phase. For stronger results, we need to know when such E can become *stable* near the wall.

3. CONSTRUCTIONS BASED ON K3 CATEGORIES

In this section we present three applications of stability conditions on K3 categories, to irreducible holomorphic symplectic varieties, to curves, and to cubic fourfolds.

3.1. Curves in irreducible holomorphic symplectic manifolds

Let M be a smooth projective irreducible holomorphic symplectic (IHS) variety of $\mathrm{K3}^{[n]}$-type, with $n \geq 2$. We let q_M be the Beauville–Bogomolov–Fujiki quadratic form on $H^2(M, \mathbb{Z})$. By **[25, SECTION 3.7.1]**, there exists a canonical extension

$$\vartheta_M\colon (H^2(M,\mathbb{Z}), q_M) \hookrightarrow \widetilde{\Lambda}_M$$

of lattices and weight-2 Hodge structures, where the lattice $\widetilde{\Lambda}_M$ is isometric to the extended K3 lattice $U^{\oplus 4} \oplus E_8(-1)^{\oplus 2}$. Let us denote by $v \in \widetilde{\Lambda}_M$ a generator of $\vartheta(H^2(M,\mathbb{Z}))^{\perp}$: it is of type $(1,1)$ and square $v^2 = 2n-2$. The lattice $\widetilde{\Lambda}_M$ is called the *Markman–Mukai* lattice associated to M. If $M = M_\sigma(v)$, for a stability condition $\sigma \in \mathrm{Stab}^\dagger(\mathrm{D}^{\mathrm{b}}(S))$ on a K3 surface S, then $\widetilde{\Lambda}_M = H^*(S,\mathbb{Z})$ with the Mukai pairing, the notation for the vector v is coherent, and ϑ_M is the Mukai morphism mentioned after Theorem 2.6.

We let $\mathrm{Pos}(M)$ be the connected component of the positive cone of M containing an ample divisor class:

$$\mathrm{Pos}(M) := \left\{D \in H^2(M,\mathbb{R}) : q_M(D) > 0\right\}^+.$$

The following result rephrases and proves a conjecture by Hassett–Tschinkel and gives a complete description of the ample cone of M.

Theorem 3.1. *Let M be a smooth projective IHS variety of* $\mathrm{K3}^{[n]}$*-type. The ample cone of M is a connected component of*

$$\mathrm{Pos}(M) \setminus \bigcup_{\substack{a \in \widetilde{\Lambda}_M^{1,1} \text{ such that} \\ a^2 \geq -2 \text{ and} \\ 0 \leq (a,v) \leq v^2/2}} a^{\perp}.$$

Theorem 3.1 is proved in **[11]** for moduli spaces of stable sheaves/complexes on a K3 surface, and extended in **[7]** to all IHS of $\mathrm{K3}^{[n]}$-type, by using deformation theory of rational curves on IHS varieties.

The approach to Theorem 3.1 via wall-crossing is as follows. Let S be a K3 surface and $M = M_{\sigma_0}(v)$ be a moduli space of σ_0-stable objects in $\mathrm{D}^{\mathrm{b}}(S)$, where $v \in H^{1,1}(\mathrm{D}^{\mathrm{b}}(S), \mathbb{Z})$ is a primitive vector of square $v^2 \geq 2$. As σ varies in the chamber $\mathcal{C}$ containing σ_0, Theorem 2.4 gives a family of ample divisor classes ℓ_σ in $\mathrm{Pos}(M)$. When σ reaches a wall of $\mathcal{C}$, as given by Theorem 2.7, the class ℓ_σ remains nef. On the other hand, consider an object E that becomes strictly semistable on the wall, admitting an exact sequence as in (2.1). Varying the extension class in a line in $\mathbb{P}(\mathrm{Ext}^1(B, A))$ produces a $\mathbb{P}^1$ of such objects, and Theorem 2.4 shows that ℓ_σ has degree zero on this curve. We have found an extremal curve and, dually, a boundary wall of the ample cone.

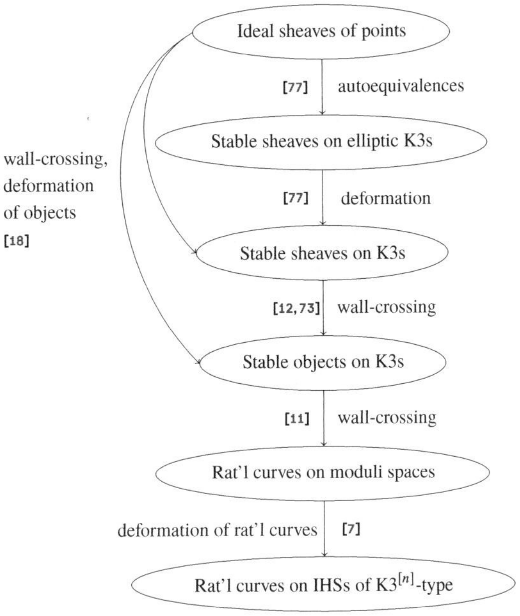

FIGURE 1
The approach to Theorem 3.1.

We summarize the history underlying Theorem 3.1 with the diagram in Figure 1. The analogue of Theorem 2.6 for Gieseker-stable sheaves involves a two-step argument, using autoequivalences and deformations. Wall-crossing techniques then imply the existence of Bridgeland stable objects on K3 surfaces, and thus Theorem 2.6. As discussed above, a finer wall-crossing analysis based on Theorem 2.7 then produces the extremal rational curves on moduli spaces that appear implicitly as extremal curves in Theorem 3.1. Finally, another deformation argument, involving rational curves, deduces Theorem 3.1 for all IHS manifolds of K3$^{[n]}$-type. Wall-crossing combined with stability conditions in families can also simplify the approach to Theorem 2.6, see **[18]**.

3.2. Curves

Consider a *Brill–Noether (BN) wall* in $\mathrm{Stab}(\mathrm{D^b}(X))$ for a variety X: the structure sheaf $\mathcal{O}_X$ is stable and of the same phase ϕ as objects E of a fixed class v. Then $\mathcal{O}_X$ is an object of the abelian category $\mathcal{P}(\phi)$ with no subobjects; hence the evaluation map $\mathcal{O}_X \otimes H^0(E) \to E$ must be injective, giving a short exact sequence

$$0 \to \mathcal{O}_X \otimes H^0(E) \to E \to Q \to 0 \tag{3.1}$$

where $Q \in \mathcal{P}(\phi)$ is also semistable. Applying known inequalities for Chern classes of semistable objects to $\mathrm{ch}(Q) = v - h^0(E)\,\mathrm{ch}(\mathcal{O}_X)$ can directly lead to bounds on $h^0(E)$.

This simple idea turns out to be powerful. For a K3 surface S, we can be more precise: applying Theorem 2.6 to the class of Q, we can *construct* all E with given $r = h^0$ as a Grassmannian bundle $\mathrm{Gr}(r, \mathrm{Ext}^1(Q, \mathcal{O}_S))$ of extensions over the moduli space of such Q.

Corollary 3.2. *Let S be a K3 surface, $v \in H^{1,1}(S, \mathbb{Z})$ primitive and σ be a stability condition near the Brill–Noether wall for v. If the lattice generated by v and $v(\mathcal{O}_S)$ is saturated, then the locus of objects $E \in M_\sigma(v)$ with $h^0(E) = r$ has expected dimension.*

In [5], this is applied, in the case where $\mathrm{Pic}(S) = \mathbb{Z} \cdot H$, to rank zero classes of the form $v = (0, H, s)$. In this case, there are no walls between the BN wall and the large volume limit; hence Corollary 3.2 applies in the large volume limit, and thus to zero-dimensional torsion sheaves supported on curves in the primitive linear system. This gives a variant of Lazarsfeld's proof [49] of the Brill–Noether theorem: *every* curve in the primitive linear system is Brill–Noether general.

This approach has been significantly strengthened in [27]: instead of requiring E to be semistable near the Brill–Noether wall, it is sufficient to control the classes occurring in its HN filtration. A bound on h^0 is obtained by applying Corollary 3.2 to all HN filtration factors. Thus we need to consider a point near the Brill–Noether walls for all HN factors, and which is the limit point where $Z(\mathcal{O}_X) \rightsquigarrow 0$.

Proposition 3.3 ([27, PROPOSITION 3.4]). *Let S be a K3 surface of Picard rank one. There exists a limit point $\overline{\sigma}$ of the space of stability conditions, with central charge $\overline{Z}$, and a constant C, such that for (most) objects in the heart, we have*

$$h^0(E) + h^1(E) \leq C \cdot \sum_l \left|\overline{Z}(E_l / E_{l-1})\right|,$$

where $E_0 \subset E_1 \subset \cdots \subset E_m$ is the HN filtration of E near $\overline{\sigma}$.

The following application completes a program originally proposed by Mukai [62]:

Theorem 3.4 ([27, 28]). *Let S be a polarised K3 surface with $\mathrm{Pic}(S) = \mathbb{Z} \cdot H$ and genus $g \geq 11$, and let $C \in |H|$. Then S is the unique K3 surface containing C, and can be reconstructed as a Fourier–Mukai partner of a Brill–Noether locus of stable vector bundles on C with prescribed number of sections.*

The structure of the argument is as follows. The numerics are chosen such that there is a two–dimensional moduli space $\hat{S}$, necessarily a K3 surface, of stable bundles E on S whose restriction $E|_C$ is automatically in the Brill–Noether locus. Conversely, given a stable bundle V on C, its push-forward i_*V along $i: C \hookrightarrow S$ is stable at the large volume limit. Standard wall-crossing arguments bound its HN filtration near the limit point $\overline{\sigma}$ in Proposition 3.3, which then gives a bound on $h^0(V)$. The argument also shows that equality—the Brill–Noether condition—only holds for the HN filtration $E \to E|_C = i_*V \to E(-H)[1]$, i.e., when V is the restriction of a vector bundle in $\hat{S}$. Thus $\hat{S}$ is a Brill–Noether locus on C, and S can be reconstructed as a Fourier-Mukai partner of $\hat{S}$.

3.3. Surfaces in cubic fourfolds

Let $Y \subset \mathbb{P}^5$ be a complex smooth cubic fourfold and let h be the class of a hyperplane section. Following [38], we say that Y is *special* of discriminant d, and write $Y \in \mathcal{C}_d$, if there exists a surface $\Sigma \subset Y$ such that h^2 and Σ span a saturated rank two lattice in $H^4(Y, \mathbb{Z})$ with

$$\det \begin{pmatrix} h^4 & h^2 \cdot \Sigma \\ h^2 \cdot \Sigma & \Sigma^2 \end{pmatrix} = d.$$

The locus $\mathcal{C}_d$ is non-empty if and only if $d \equiv 0, 2 \pmod 6$ and $d > 6$; in this case, $\mathcal{C}_d$ is an irreducible divisor in the moduli space of cubic fourfolds.

Given d it is not known in general which degree $h^2 \cdot \Sigma$ and self-intersection Σ^2 can be realised. The following gives an answer for an infinite series of d.

Theorem 3.5 ([6]). *Let $a \geq 1$ be an integer and let $d := 6a^2 + 6a + 2$. Let Y be a general cubic fourfold in $\mathcal{C}_d$. Then there exist surfaces $\Sigma \subset Y$ such that*

- $\deg(\Sigma) := h^2 \cdot \Sigma = 1 + \frac{3}{2}a(a+1)$ *and* $\Sigma^2 = \frac{d + \deg(\Sigma)^2}{3}$;
- $H^*(Y, \mathcal{I}_\Sigma(a - j)) = 0$, *for all* $j = 0, 1, 2$.

In fact, we obtain a family of such surfaces Σ. It is parameterized by (an open subset of) the K3 surface S of degree $d = 6a^2 + 6a + 2$ associated to every $Y \in \mathcal{C}_d$ by Hassett [38] via the Hodge structures of Y and S.

The proof of Theorem 3.5 is based on an enhancement of Hassett's Hodge-theoretic relation to the derived category: by [1,8], we have $\mathcal{K}u(Y) = \mathrm{D}^{\mathrm{b}}(S)$. By [53], the Lehn-Lehn-Sorger-van Straten IHS eightfold X associated to Y of [58] can be realised as a moduli space of stable objects in $\mathcal{K}u(Y)$. It admits a Lagrangian embedding $Y \hookrightarrow X$. Markman's Torelli Theorem [59] implies that X and the Hilbert scheme $S^{[4]}$ are birational; for $a \geq 2$, and $Y \in \mathcal{C}_d$ very general, we use Theorem 3.1 to show that the nef cone and the movable cone of $S^{[4]}$ are the same, and hence X and $S^{[4]}$ are isomorphic. Now for $s \in S$ general let $\Gamma_s \subset S^{[4]}$ be the locus of subschemes containing s; then $\Sigma = Y \cap \Gamma_s$ is the desired surface.

We are interested in this construction because it produces many rational morphisms from Y as follows. The rational map $S^{[4]} \to S^{[5]}$, $Z \mapsto Z \cup s$ is resolved by the blow-up $\mathrm{Bl}_{\Gamma_s} S^{[4]} \to S^{[5]}$. Restricting to the cubic, we obtain an embedding $\mathrm{Bl}_\Sigma Y \to S^{[5]}$. We then use wall-crossing for $S^{[5]}$, interpreted as a moduli space in $\mathcal{K}u(Y) = \mathrm{D}^{\mathrm{b}}(S)$, and restrict the resulting birational transformations to $\mathrm{Bl}_\Sigma Y$. In the case $a = 2$, this recovers completely the picture described in [69] and thus the rationality of all cubics in $\mathcal{C}_{38}$. Analogous constructions likely exist for arbitrary $d \equiv 2 \pmod 6$ if we replace Γ_s with a locus of sheaves not locally free at s inside a moduli space of stable sheaves.

4. THREEFOLDS

Stability conditions on a threefold X have been constructed in a three step process. Initially we consider slope-stability. In the second step, we reinterpret slope-stability by changing the abelian subcategory of $\mathrm{D}^{\mathrm{b}}(X)$ and rotating the central charge; the classical Bogomolov–Gieseker inequality then allows us to deform this to obtain *tilt-stability*, which behaves much like Bridgeland stability conditions on surfaces. Finally, a conjectural generalized Bogomolov–Gieseker type inequality, Conjecture 4.7, for tilt-stable objects allows one to repeat this procedure and produce actual stability conditions.

Already the second step, tilt-stability, has geometric consequences when combined with Conjecture 4.7. We present three applications: to a bound for the genus of a curve on a threefold [58], to higher rank Donaldson–Thomas theory on Calabi–Yau threefolds [30, 31], and to Clifford-type bounds for vector bundles on curves and quintic threefolds [51].

4.1. The generalized Bogomolov–Gieseker inequality

We will now describe the first and second step of this construction. For more details, we refer to [8, PART V], [67] and [13]. Throughout this section, we let X be a smooth projective variety of dimension n, over the complex numbers unless noted otherwise, and let H in $\mathrm{NS}(X)$ be the class of an ample divisor on X.

The twisted Chern character. Let

$$\gamma := e^{-B} \cdot (1, 0, -\Gamma, \gamma_3, \dots, \gamma_n) \in \bigoplus_{l=0}^{n} \mathrm{CH}^l_{\mathrm{num}}(X)_{\mathbb{Q}}, \tag{4.1}$$

with $B \in \mathrm{NS}(X)_{\mathbb{Q}}$, $\Gamma \in \mathrm{CH}^2_{\mathrm{num}}(X)_{\mathbb{Q}}$ such that $H^{n-2} \cdot \Gamma = 0$, and $\gamma_3, \dots, \gamma_n$ arbitrary. We let

$$\mathrm{ch}^{\gamma} := \gamma \cdot \mathrm{ch}\colon K_0(X) \to \bigoplus_{l=0}^{n} \mathrm{CH}^l_{\mathrm{num}}(X)_{\mathbb{Q}}$$

be the *Chern character twisted by* γ. If $\gamma = e^{-B}$ (e.g. in the case of surfaces) then ch^{γ} is usually denoted by ch^B. When X is a threefold of Picard rank 1, $\Gamma = 0$.

Let $\Lambda^{\gamma}_H \subset \mathbb{Q}^{n+1}$ be the image of the morphism $v^{\gamma}_H\colon K_0(\mathrm{D}^{\mathrm{b}}(X)) \to \mathbb{Q}^{n+1}$ given by

$$v^{\gamma}_H(E) := \big(H^n \cdot \mathrm{ch}^{\gamma}_0(E), H^{n-1} \cdot \mathrm{ch}^{\gamma}_1(E), \dots, H \cdot \mathrm{ch}^{\gamma}_{n-1}(E), \mathrm{ch}^{\gamma}_n(E)\big).$$

It is a free abelian group of rank $n+1$. Given $v \in \Lambda^{\gamma}_H$, we denote by v_l its l-th component. For $l = 0, \dots, n$, we denote by $\Lambda^{\gamma}_{H,\le l}$ the sublattice of rank $l+1$ generated by the first $l+1$ components and by $v_{\le l}$ the corresponding truncated vector.

Slope-stability (step 1). This is analogous to the curve case in Section 2. We define the slope of a coherent sheaf $E \in \mathrm{Coh}(X)$ as

$$\mu^{\gamma}_H(E) := \mu\big(v^{\gamma}_H(E)\big) = v_1(E)/v_0(E), \quad \text{with } \mu^{\gamma}_H(E) = +\infty \text{ if } v_0(E) = 0.$$

A sheaf E is μ^{γ}_H*-semistable* if for every non-zero subsheaf $F \hookrightarrow E$, we have $\mu^{\gamma}_H(F) \le \mu^{\gamma}_H(E/F)$. In particular, torsion sheaves are semistable of slope $+\infty$. Harder–Narasimhan (HN) filtrations for slope stability exist; if the sheaf has torsion, the first HN factor is the torsion part. As before, we visualize the slope with the (weak) stability function $Z : \Lambda^{\gamma}_H \to \mathbb{C}$ given by

$$Z(v) := -v_1 + i v_0.$$

The Bogomolov–Gieseker inequality. Let us assume $n \geq 2$ and define the quadratic form $\overline{\Delta}$ on $\Lambda^{\gamma}_{H,\leq 2}$ of signature $(2,1)$ by

$$\overline{\Delta}(v) := v_1^2 - 2v_0v_2.$$

The following result is a consequence of [16, 33, 68] and the Hodge index theorem:

Theorem 4.1 (Bogomolov–Gieseker inequality). *Let $E \in \mathrm{Coh}(X)$ be a μ^{γ}_H-semistable sheaf. Then*

$$\overline{\Delta}\big(v^{\gamma}_H(E)\big) \geq 0. \tag{4.2}$$

Remark 4.2. The theory works similarly in finite characteristic. Inequalities similar to (4.2) proved by Langer in [48] are not sufficient to construct stability conditions. Instead, by [43, THEOREM 1.3], there exists a constant $C_{X,H} \in \mathbb{R}_{\geq 0}$ such that inequality (4.2) holds if we add the term $C_{X,H} \cdot H^2/2$ to γ. This is sufficient for the construction tilt-stability below.

Tilt-stability (step 2). Constructing tilt-stability from slope stability needs two operations, see also Figure 2. First, let $\beta \in \mathbb{R}$. We rotate the central charge Z by setting $Z_\beta := v_0 + i(v_1 - \beta v_0)$, and modify the abelian category accordingly to obtain $\mathrm{Coh}^{\gamma}_{H,\beta}(X)$ as follows:

$$\begin{aligned}
\mathcal{T}^\beta &:= \big\{E \in \mathrm{Coh}(X) \colon \mu^{\gamma,-}_H(E) > \beta\big\}, \\
\mathcal{F}^\beta &:= \big\{E \in \mathrm{Coh}(X) \colon \mu^{\gamma,+}_H(E) \leq \beta\big\}, \\
\mathrm{Coh}^{\gamma}_{H,\beta}(X) &:= \big\{E \in \mathrm{D^b}(X) : \mathcal{H}^l(E) = 0, \text{ for } l \neq 0, -1, \mathcal{H}^{-1}(E) \in \mathcal{F}^\beta, \mathcal{H}^0(E) \in \mathcal{T}^\beta\big\},
\end{aligned} \tag{4.3}$$

where we denoted by $\mu^{\gamma,\pm}_H$ the first and last slope of the HN filtration with respect to μ^{γ}_H-stability. By tilting theory [36], $\mathrm{Coh}^{\gamma}_{H,\beta}(X)$, the extension-closure of $\mathcal{F}^\beta[1]$ and $\mathcal{T}^\beta$, is the heart of a bounded t-structure on $\mathrm{D^b}(X)$; in particular, it is an abelian category.

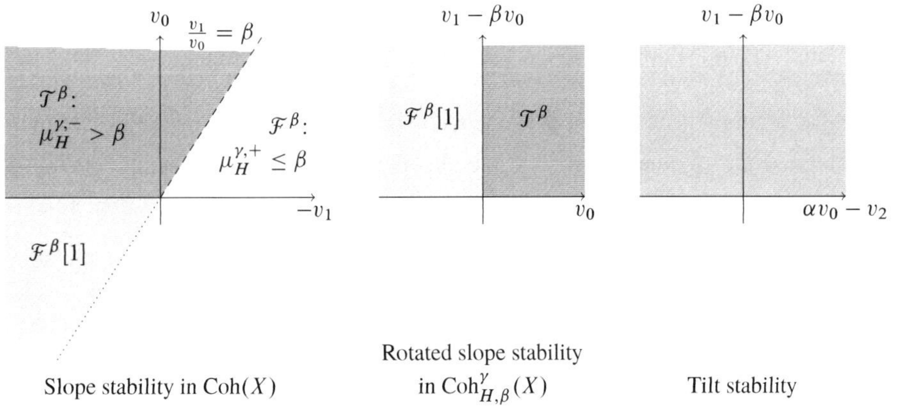

FIGURE 2
Rotating and deforming slope stability to obtain tilt stability.

The pair $(Z_\beta, \mathrm{Coh}^{\gamma}_{H,\beta}(X))$ admits HN filtrations. The difference to slope stability in $\mathrm{Coh}(X)$ is subtle: torsion sheaves supported codimension ≥ 2 are in both categories, considered to have slope $+\infty$, and thus now have bigger phase than objects in $\mathcal{F}^\beta[1]$.

For the second part, we deform Z_β while preserving the category $\mathrm{Coh}^{\gamma}_{H,\beta}(X)$. We follow the presentation in [51] and denote by U the open subset of $\mathbb{R}^2$ given by

$$U := \left\{ (\alpha, \beta) \in \mathbb{R}^2 : \alpha > \frac{\beta^2}{2} \right\}. \tag{4.4}$$

For $(\alpha, \beta) \in U$, we consider the slope $\nu^{\gamma}_{H,\alpha,\beta}$ induced on $\mathrm{Coh}^{\gamma}_{H,\beta}(X)$ by the central charge

$$Z_{\alpha,\beta}(v) := -(v_2 - \alpha v_0) + i(v_1 - \beta v_0). \tag{4.5}$$

Objects semistable with respect to $\nu^{\gamma}_{H,\alpha,\beta}$ are called *tilt-semistable*. HN filtrations exist and Theorem 4.1 applies equally to tilt-semistable objects, which ensures that tilt-stability has a wall-and-chamber structure analogous to Bridgeland stability as we deform $(\alpha, \beta) \in U$.

Remark 4.3. We can be more precise and sometimes obtain a larger set of tilt-stability conditions. Following [32], we define the *Le Potier function* $\Phi^{\gamma}_{X,H} : \mathbb{R} \to \mathbb{R}$ as

$$\Phi^{\gamma}_{X,H}(x) := \limsup_{\mu \to x} \left\{ \frac{v^{\gamma}_H(E)_2}{v^{\gamma}_H(E)_0} : E \in \mathrm{Coh}(X) \text{ is } \mu^{\gamma}_H\text{-semistable with } \mu^{\gamma}_H(E) = \mu \right\}.$$

It is upper semicontinuous, and by Theorem 4.1 it satisfies $\Phi^{\gamma}_{X,H}(x) \le x^2/2$. Tilt-stability is well defined for all $(\alpha, \beta) \in \mathbb{R}^2$ such that $\alpha > \Phi^{\gamma}_{X,H}(\beta)$.

Example 4.4. If X is a surface, then $(Z_{\alpha,\beta}, \mathrm{Coh}^{\gamma}_{H,\beta}(X))$ induces a Bridgeland stability condition on $\mathrm{D}^{\mathrm{b}}(X)$, as constructed in [3]. The support property is given by the quadratic form (4.2). If X has finite Albanese morphism, then all stability conditions on $\mathrm{D}^{\mathrm{b}}(X)$ with respect to $(v^{\gamma}_H, \Lambda^{\gamma}_H)$ are given by those constructed in Remark 4.3, up to linear action [32]. A stronger version of Theorem 4.1, involving $\mathrm{ch}_1(E)^2$ rather than just $(H.\mathrm{ch}_1(E))^2$, gives a support property with respect to the full numerical Grothendieck group of $\mathrm{D}^{\mathrm{b}}(X)$ and allows us to vary H; see [13, THEOREM 3.5]. In the case of a K3 surface S, the closure of the locus of such stability conditions and its translates under autoequivalences give the connected component $\mathrm{Stab}^{\dagger}(\mathrm{D}^{\mathrm{b}}(S))$ mentioned in Section 2.3.

The generalized Bogomolov–Gieseker inequality. Let us assume $n \ge 3$. The main open question is to find an inequality involving v_3 for tilt-semistable objects, generalizing (4.2). For $(\alpha, \beta) \in U$, let us define the quadratic form $\overline{Q}_{\alpha,\beta}$ on $\Lambda^{\gamma}_{H,\le 3}$ of signature $(2, 2)$ by

$$\overline{Q}_{\alpha,\beta}(v) := \alpha(v_1^2 - 2v_0v_2) + \beta(3v_0v_3 - v_1v_2) + (2v_2^2 - 3v_1v_3).$$

Definition 4.5. We say that (X, H) satisfies the *γ-generalized Bogomolov–Gieseker (BG) inequality* at $(\alpha, \beta) \in U$, if for all $E \in \mathrm{Coh}^{\gamma}_{H,\beta}(X)$ which are $\nu^{\gamma}_{H,\alpha,\beta}$-semistable, we have

$$\overline{Q}_{\alpha,\beta}\big(v^{\gamma}_H(E)\big) \ge 0. \tag{4.6}$$

Theorem 4.6 ([13, 67]). *A polarized threefold satisfying the γ-generalized BG inequality at a point $(\alpha, \beta) \in U$ admits Bridgeland stability conditions.*

The construction of these stability conditions from tilt stability is completely analogous to the construction of tilt stability from slope-stability discussed above. It was conjectured in [14, CONJECTURE 1.3.1] that all polarized threefolds satisfy the generalized BG inequality for all $(\alpha, \beta) \in U$ and $\gamma = e^{-B}$; this turned out to be too optimistic, see [71]. The following is a modification of the original conjecture, based on [51, 66]:

Conjecture 4.7. *Let (X, H) be a smooth complex projective polarized variety. There exists a class $\gamma = \gamma_{X,H}$ as in (4.1) and an upper semicontinuous function $f = f^{\gamma}_{X,H} : \mathbb{R} \to \mathbb{R}$ such that (X, H) satisfies the γ-generalized BG inequality for all $(\alpha, \beta) \in \mathbb{R}^2$ with $\alpha > f(\beta)$.*

Conjecture 4.7 has been established in a number of three-dimensional cases:

- prime Fano threefolds [52, 57, 70], with $\gamma = 1$ and $f^{\gamma}_{X,H}(x) = x^2/2$;
- abelian threefolds [13, 56], with $\gamma = e^{-B}$, for all $B \in \mathrm{NS}(X)_{\mathbb{R}}$, and $f^{\gamma}_{X,H}(x) = x^2/2$;
- the quintic threefold [51], with $\gamma = 1$ and $f^{\gamma}_{X,H}(x) = x^2/2 + (x - \lfloor x \rfloor)(\lfloor x \rfloor + 1 - x)/2$;
- the complete intersection of quadratic and quartic hypersurfaces in $\mathbb{P}^5$ [55], with $\gamma = 1$ and
$$f^{\gamma}_{X,H}(x) = x^2/2 + \big(x - \lfloor x \rfloor\big)\big(\lfloor x \rfloor + 1 - x\big)/2;$$
- the blow-up of $\mathbb{P}^3$ at a point [66], with $H = -K_X/2$, $\gamma = (1, 0, -\Gamma, 0)$,
$$\Gamma = \frac{1}{12}\left(c_2(X) - \frac{H \cdot c_2(X)}{H^3} H^2\right),$$
and $f^{\gamma}_{X,H}(x) = x^2/2 + (x - \lfloor x \rfloor + 1)^2$; and
- threefolds with nef tangent bundle [15, 45], with H any ample divisor, $\gamma = e^{-B}$, for all $B \in \mathrm{NS}(X)_{\mathbb{R}}$ (except in the case $X = \mathbb{P}(T_{\mathbb{P}^2})$, where $H = -K_X/2$ and $\gamma = 1$), and $f^{\gamma}_{X,H}(x) = x^2/2$.

Similar versions have been proved for all Fano threefolds [15, 66] and for Calabi–Yau double and triple solids [44]. It is also known in some cases in arbitrary characteristic, e.g., $\mathbb{P}^3$ (where $C_{\mathbb{P}^3,H} = 0$, $\gamma = 1$, $f^{\gamma}_{X,H}(x) = x^2/2$). There is no known counterexample to Conjecture 4.7 with $f^{\gamma}_{X,H}(x) = x^2/2$; a nontrivial choice of γ is instead necessary: the blow-up of $\mathbb{P}^3$ at a point with the anticanonical polarization does not satisfy the γ-generalized BG inequality if we take $\gamma = 1$ and $f^{\gamma}_{X,H}(x) = x^2/2$ [71].

4.2. Tilt-stability methods

We now describe three limit points of tilt-stability in the set U defined in (4.4), the *small and large volume limit* and the *Brill–Noether* point. The latter two are generalizations of limits discussed in Section 3.2 for Bridgeland stability conditions in the K3 surface case.

Throughout this section we fix a class $v \in \Lambda^{\gamma}_{H}$, and assume $v_0 \neq 0$ for simplicity.

Walls. We want to understand walls in U for tilt-stability $(Z_{\alpha,\beta}, \mathrm{Coh}_{H,\beta}(X))$, defined by (4.5) and (4.3), of objects of class v. We may assume $\bar{\Delta}(v) \geq 0$ and define
$$p(v) := \left(\frac{v_2}{v_0}, \frac{v_1}{v_0}\right) \in \mathbb{R}^2 \setminus U.$$

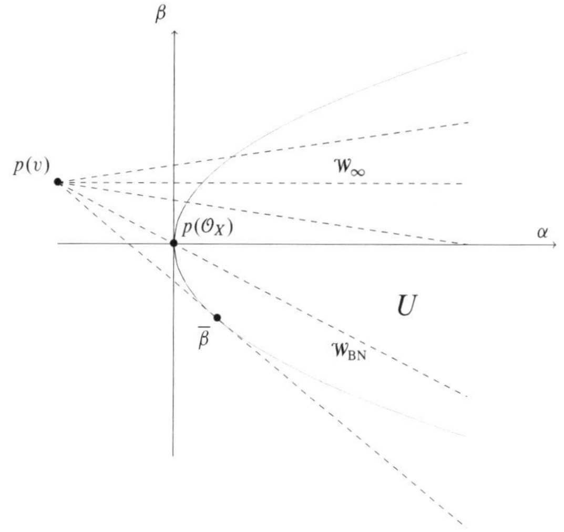

FIGURE 3
Walls for v.

Given a line L containing $p(v)$, we define the *potential wall* in U for v associated to L by

$$\mathcal{W}_L := \{(\alpha, \beta) \in U \cap L\}.$$

The main property of walls are the following, see also Figure 3. Two objects of $\mathrm{Coh}^{\gamma}_{H,\beta}(X)$ with classes v and w have the same slope with respect to $\nu^{\gamma}_{\alpha,\beta,H}$ if and only if (α, β) lies on the line passing through $p(v)$ and $p(w)$. If an object of class v is tilt-(semi)stable at one point of $\mathcal{W}_L$, then it is tilt-(semi)stable for all points in $\mathcal{W}_L$. Moreover, actual walls for v are a locally finite set of potential walls, and tilt-stability is unchanged except when crossing one of these walls.

The following elementary observation is often useful for induction arguments:

Remark 4.8. Given a wall $\mathcal{W}$ for v in U, let $w_1, \dots, w_m$ be the classes of the Jordan–Hölder factors of a tilt-semistable object with class v at a point of $\mathcal{W}$. Then the version of Theorem 4.1 for tilt-stable objects implies that $\overline{\Delta}(w_l) \le \overline{\Delta}(v)$, for all $l = 1, \dots, m$, with equality if and only if all $(w_l)_{\le 2}$ and $v_{\le 2}$ are proportional and $\overline{\Delta}(w_l) = \overline{\Delta}(v) = 0$. In particular, the structure sheaf $\mathcal{O}_X$ or its shift $\mathcal{O}_X[1]$ is tilt-stable everywhere in U.

Example 4.9. The equation $\overline{Q}_{\alpha,\beta}(v) = 0$ defines a line containing $p(v)$. If it intersects U, we call the associated potential wall the *BG wall*, which gives a bound on walls for v.

The small volume limit point. Assume that $\overline{\Delta}(v) > 0$. We define the two *small volume limit points* $\overline{\beta}(v)$ as the points $(\beta^2/2, \beta) \in \mathbb{R}^2$ where the tangent to the parabola contains $p(v)$.

By the local finiteness of walls, they can accumulate only at the small volume limit points, and only finitely many lie outside a neighborhood of them. Objects which are tilt-stable in a neighborhood of a small volume limit point are called *$\overline{\beta}$-stable* and have strong vanishing properties, which are useful both in proving cases of Conjecture 4.7 or in applications of it. For instance, if Conjecture 4.7 is true and the BG wall exists, there are no $\overline{\beta}$-stable objects for the corresponding small limit point. In the approach mentioned in the Introduction, the small limit point is a promising choice for a point of interest.

The large volume limit point. As in Figure 3, consider the horizontal line

$$L_\infty := \{\beta = v_1/v_0\}.$$

It is called the *large volume limit wall* for v and a point in it is called a *large volume limit point*. Object which are tilt-stable on a point in U nearby the large volume limit wall—equivalently, for $\alpha \rightsquigarrow +\infty$—essentially correspond to Gieseker-stable sheaves or derived duals of them, according to which side of the wall we are.

The Brill–Noether point. The last point we are interested in helps studying global sections of objects. In the K3 surface case, as discussed in Section 3.2, this is where applications to Brill–Noether theory come from. In tilt-stability, it is given by the potential wall $\mathcal{W}_{\mathrm{BN}}$ associated to the line passing through $p(v)$ and $(0, 0)$, see again Figure 3. We call the point $(0, 0)$ the *Brill–Noether (BN) point*.

4.3. Tilt-stability applications

In this section, we give an informal exposition of three applications of tilt-stability and Conjecture 4.7.

Curves on threefolds. Let $X = \mathbb{P}^3$. We want to study the following question, called the *Halphen problem*: what is the maximal genus g of an integral curve in $\mathbb{P}^3$ of degree d which is not contained in a surface of degree $< k$? While the question is open for smaller d, a celebrated theorem of Gruson–Peskine and Harris [35, 37] gives such maximal genus $G(d, k)$, when $d > k(k-1)$ We give an idea how to reprove this theorem by using tilt-stability. The approach works for any threefold which satisfies Conjecture 4.7 and a few extra assumptions, see [58, THEOREM 1.2] for the precise statement. For instance, it yields new results for principally polarized abelian threefolds of Picard rank 1.

Let C be a curve as above whose genus is larger than the expected bound $G(d, k)$. We look at the twisted ideal sheaf $\mathcal{I}_C(k-1)$ and we let $v = v(\mathcal{I}_C(k-1))$; here $\gamma = 1$ and the generalized BG inequality holds, for all $(\alpha, \beta) \in U$. The first step is a straightforward application of Conjecture 4.7, which shows that in a neighborhood of the small volume limit point, there are no tilt-semistable objects with class v. The second step is to use the information that C is not contained in a surface of degree $k-1$, which says that the BN wall does not give a wall for $\mathcal{I}_C(k-1)$. To summarize: $\mathcal{I}_C(k-1)$ is tilt-stable at the large volume limit, it must be destabilized at a certain wall, which cannot be on the BN wall.

To get a contradiction we need to analyze the finitely many walls for $\mathcal{I}_C(k-1)$. This is where the assumption $d > k(k-1)$ comes in. In fact, possible destabilizing subobjects for $\mathcal{I}_C(k-1)$ are always reflexive sheaves: in the range $d > k(k-1)$ they have rank either 1 or 2. The rank 1 case, namely invertible sheaves, can be dealt with by the fact that the BN wall is not an actual wall. For the rank 2 case, we use once more a similar strategy to get bounds on the third Chern character of such rank 2 sheaves and thus a contradiction, since we have control on the discriminant, by Remark 4.8.

Higher rank DT invariants on CY3s. Let (X, H) be a complex polarized Calabi–Yau threefold. In a recent sequence of papers [30, 31], Feyzbakhsh and Thomas proved the following theorem: if (X, H) satisfies the generalized BG inequality on U, then the higher rank Donaldson–Thomas (DT) theory is completely governed by the rank 1 theory, i.e., Hilbert schemes of curves. There is some flexibility on the assumption on the generalized BG inequality; in particular, their theorem holds for the examples of Calabi–Yau threefolds where Conjecture 4.7 has been proved, e.g. for the quintic threefold.

The fundamental idea is as follows. Fix a class v of rank r. The DT invariant $\mathrm{DT}_{\text{large-v}}(v)$ of tilt-semistable objects of class v near the large-volume limit is essentially the classical DT invariant. Now let $n \gg 0$. The role of the point of interest is first played by a variant of the Brill–Noether wall: the Joyce–Song wall $\mathcal{W}_{\mathrm{JS}}$ where v and $\mathcal{O}_X(-n)[1]$ have the same slope. This is a wall not for v (where it is contained in the large-volume chamber), but for the class $v^{(n)} = v - v(\mathcal{O}_X(-n))$ of rank $r - 1$: there are objects E of class $v^{(n)}$ destabilised by a short exact sequence $F \to E \to \mathcal{O}_X(-n)[1]$ with $F \in M_{\alpha,\beta}(v)$. They are generically parameterised by a projective bundle over the DT moduli space for v.

Now consider the DT invariant $\mathrm{DT}_{(\alpha,\beta)}(v^{(n)})$ as (α, β) moves on a path from the small-volume limit—our point of interest—for $v^{(n)}$ to its large-volume limit. Conjecture 4.7 shows $\mathrm{DT}_{\text{small-v}}(v^{(n)}) = 0$. Applying the generalised BG inequality again, the authors show that *except for the Joyce–Song wall*, all other walls are defined by sheaves of rank $\leq r - 1$. Thus, the Joyce–Song wall-crossing formula in [40] gives a relation of the form

$$\begin{aligned}\mathrm{DT}_{\text{large-v}}(v^{(n)}) &= \mathrm{DT}_{\text{small-v}}(v^{(n)}) + \text{Wall-crossing contributions}\\ &= 0 + \text{WallCr(lower rank DT invariants)} + \chi(\mathcal{O}_X(-n), v) \cdot \mathrm{DT}_{\text{large-v}}(v).\end{aligned}$$

This shows that $\mathrm{DT}_{\text{large-v}}(v)$ is determined by lower rank DT invariants.

The quintic threefold and Clifford-type bounds. In the proof of the generalized BG inequality for the quintic threefold in [51], the first idea is the following: if we know the generalized BG inequality at the BN point, we know it everywhere (for an appropriately chosen function f in Conjecture 4.7). Here, Li uses the same idea as in Section 3.2: a stronger *classical* Bogomolov–Gieseker type inequality for the quotient Q in (3.1) implies a bound for $h^0(E)$, and consequently for $\chi(E)$ and thus $\mathrm{ch}_3(E)$. The stronger bound for Q is deduced by a restriction theorem from stronger bounds on $(2, 5)$-complete intersection surfaces. Using the logic of Section 3.2 in reverse, this bound is reduced to Clifford-type bounds for stable vector bundles on $(2, 2, 5)$-complete intersection curves C. Now we consider the embedding $C \subset S_{2,2}$ of C into the $(2, 2)$-complete intersection del Pezzo surface $S_{2,2}$. Riemann-Roch directly implies a stronger Bogomolov–Gieseker inequality on $S_{2,2}$, and shifting the logic of Section 3.2 back to forward gear implies the desired Clifford bounds.

These Clifford bound arguments yield new results even for planar curves [29].

5. FURTHER RESEARCH DIRECTIONS

A proof of Conjecture 4.7, and thus the existence of Bridgeland stability conditions on threefolds, would evidently be tremendous progress. We present here some more specific open questions related to the topic of our survey.

The quintic threefold and Toda's conjecture. The picture for Bridgeland stability conditions is not yet complete, even for quintic threefolds. The expectation in [4] (see also [19, REMARK 3.9] for more details) is the following: there exists a closed embedding

$$I : \mathfrak{M}_K \to \left[\mathrm{Aut}\left(\mathrm{D^b}(X)\right)\backslash \mathrm{Stab}\left(\mathrm{D^b}(X)\right)/\mathbb{C}\right], \quad \text{where } \mathfrak{M}_K := \left[\left\{\psi \in \mathbb{C} : \psi^5 \neq 1\right\}/\mu_5\right]$$

and $I(\psi) = (Z_\psi, \mathcal{P}_\psi)$ where Z_ψ is a solution of the associated Picard–Fuchs equation (see [75, SECTION 3.2]). Li's Theorem in [51] describes only a neighborhood of the large volume limit. The global picture would follow from an appropriate answer to the following question.

Question 5.1. Is there a better bound for the Le Potier function for quintic threefolds (extending [75, CONJECTURE 1.2] for slope $\frac{1}{2}$ to arbitrary slope), and a version of Conjecture 4.7 with $f(x)$ approximating this Le Potier function?

Higher dimension. The first step towards stability conditions in higher dimension would be the following:

Question 5.2. Can we prove Conjecture 4.7 by induction on the dimension of X once it is known for (a suitable class of) threefolds?

Moduli spaces of polarized noncommutative varieties. A natural extension of the results in Section 2.3 would be to answer the following:

Question 5.3. Is any smooth projective polarized irreducible holomorphic symplectic variety (M, H) of $\mathrm{K3}^{[n]}$-type isomorphic to the moduli space $(M_\sigma(v), \ell_\sigma)$ of σ-stable objects for a stability condition σ on a K3 category $\mathcal{D}$?

A candidate for $\mathcal{D}$ has been constructed in [60], over an open subset of the moduli space. It is not known if it can be realized as an admissible subcategory. The question is closely related to the following, completing the analogy between stability conditions and polarizations:

Question 5.4. Is there an algebraic moduli space of polarised non-commutative K3 surfaces, parameterising pairs $(\mathcal{D}, \sigma)$ where $\mathcal{D}$ is a K3 category deformation–equivalent to $\mathrm{D^b}(S)$ for a projective K3 surface S, and σ is a stability condition on $\mathcal{D}$?

The theory of stability conditions in families developed in [8] provides a definition of morphisms to this moduli space.

ACKNOWLEDGMENTS

The work reviewed here benefited immensely from the input of our mentors, collaborators and friends. Aaron Bertram envisioned many of the topics surveyed here and persistently encouraged us to explore them. Many of our (current or former) mentees carried these topics much further than we imagined. We want to thank in particular Enrico Arbarello, Claudio Bartocci, Marcello Bernardara, Tom Bridgeland, Ugo Bruzzo, François Charles, İzzet Coşkun, Alastair Craw, Olivier Debarre, Gavril Farkas, Tommaso de Fernex, Soheyla Feyzbakhsh, Laure Flapan, Brendan Hassett, Jack Huizenga, Daniel Huybrechts, Qingyuang Jiang, Naoki Koseki, Alexander Kuznetsov, Martí Lahoz, Chunyi Li, Antony Maciocia, Yuri I. Manin, Eyal Markman, Sukhendu Mehrotra, Howard Nuer, Kieran O'Grady, Tony Pantev, Alexander Perry, Laura Pertusi, Giulia Saccà, Benjamin Schmidt, Paolo Stellari, Richard Thomas, Yukinobu Toda, Claire Voisin, Michael Wemyss, Kōta Yoshioka, Ziyu Zhang and Xiaolei Zhao for their invaluable constant support.

FUNDING

This work was partially supported by EPSRC grant EP/R034826/1, and the ERC Grants ERC-2018-CoG-819864-WallCrossAG and ERC-2020-SyG-854361-HyperK.

REFERENCES

[1] N. Addington and R. Thomas, Hodge theory and derived categories of cubic fourfolds. *Duke Math. J.* **163** (2014), no. 10, 1885–1927.

[2] J. Alper, D. Halpern-Leistner, and J. Heinloth, Existence of moduli spaces for algebraic stacks. 2019, arXiv:1812.01128.

[3] D. Arcara and A. Bertram, Bridgeland-stable moduli spaces for K-trivial surfaces. *J. Eur. Math. Soc. (JEMS)* **15** (2013), no. 1, 1–38.

[4] P. Aspinwall, D-branes on Calabi–Yau manifolds. In *Progress in string theory*, pp. 1–152, World Sci. Publ., Hackensack, NJ, 2005.

[5] A. Bayer, Wall-crossing implies Brill–Noether: applications of stability conditions on surfaces. In *Algebraic geometry: Salt Lake City 2015*, pp. 3–27, Proc. Sympos. Pure Math. 97.1, Amer. Math. Soc., Providence, RI, 2018.

[6] A. Bayer, A. Bertram, E. Macrì, and A. Perry, Surfaces on cubic fourfolds via stability conditions. In preparation.

[7] A. Bayer, B. Hassett, and Y. Tschinkel, Mori cones of holomorphic symplectic varieties of K3 type. *Ann. Sci. Éc. Norm. Supér. (4)* **48** (2015), no. 4, 941–950.

[8] A. Bayer, M. Lahoz, E. Macrì, H. Nuer, A. Perry, and P. Stellari, Stability conditions in families. *Publ. Math. Inst. Hautes Études Sci.* **133** (2021), 157–325.

[9] A. Bayer, M. Lahoz, E. Macrì, and P. Stellari, Stability conditions on Kuznetsov components. 2021, arXiv:1703.10839.

[10] A. Bayer and E. Macrì, The space of stability conditions on the local projective plane. *Duke Math. J.* **160** (2011), no. 2, 263–322.

[11] A. Bayer and E. Macrì, MMP for moduli of sheaves on K3s via wall-crossing: nef and movable cones, Lagrangian fibrations. *Invent. Math.* **198** (2014), no. 3, 505–590.

[12] A. Bayer and E. Macrì, Projectivity and birational geometry of Bridgeland moduli spaces. *J. Amer. Math. Soc.* **27** (2014), no. 3, 707–752.

[13] A. Bayer, E. Macrì, and P. Stellari, The space of stability conditions on abelian threefolds, and on some Calabi–Yau threefolds. *Invent. Math.* **206** (2016), no. 3, 869–933.

[14] A. Bayer, E. Macrì, and Y. Toda, Bridgeland stability conditions on threefolds I: Bogomolov–Gieseker type inequalities. *J. Algebraic Geom.* **23** (2014), no. 1, 117–163.

[15] M. Bernardara, E. Macrì, B. Schmidt, and X. Zhao, Bridgeland stability conditions on Fano threefolds. *Épij. Geom. Algébr.* **1** (2017) 2, 24 pp.

[16] F. A. Bogomolov, Holomorphic tensors and vector bundles on projective manifolds. *Izv. Akad. Nauk SSSR Ser. Mat.* **42** (1978), no. 6, 1227–1287, 1439.

[17] A. Bondal and D. Orlov, Derived categories of coherent sheaves. In *Proceedings of the international congress of mathematicians, Vol. II (Beijing, 2002)*, pp. 47–56, Higher Ed. Press, Beijing, 2002.

[18] A. Bottini, Stable sheaves on K3 surfaces via wall-crossing. 2021, arXiv:2103.09661.

[19] T. Bridgeland, Derived categories of coherent sheaves. In *International Congress of Mathematicians, Vol. II*, pp. 563–582, Eur. Math. Soc., Zürich, 2006.

[20] T. Bridgeland, Stability conditions on triangulated categories. *Ann. of Math. (2)* **166** (2007), no. 2, 317–345.

[21] T. Bridgeland, Stability conditions on $K3$ surfaces. *Duke Math. J.* **141** (2008), no. 2, 241–291.

[22] T. Bridgeland, Geometry from Donaldson–Thomas invariants. 2020, arXiv:1912.06504.

[23] T. Bridgeland and I. Smith, Quadratic differentials as stability conditions. *Publ. Math. Inst. Hautes Études Sci.* **121** (2015), 155–278.

[24] I. Coskun, J. Huizenga, and M. Woolf, The effective cone of the moduli space of sheaves on the plane. *J. Eur. Math. Soc. (JEMS)* **19** (2017), no. 5, 1421–1467.

[25] O. Debarre, Hyperkähler manifolds. 2020, arXiv:1810.02087.

[26] M. R. Douglas, Dirichlet branes, homological mirror symmetry, and stability. In *Proceedings of the international congress of mathematicians, Vol. III (Beijing, 2002)*, pp. 395–408, Higher Ed. Press, Beijing, 2002.

[27] S. Feyzbakhsh, Mukai's program (reconstructing a K3 surface from a curve) via wall-crossing. *J. Reine Angew. Math.* **765** (2020), 101–137.

[28] S. Feyzbakhsh, Mukai's program (reconstructing a K3 surface from a curve) via wall-crossing, II. 2020, arXiv:2006.08410.

[29] S. Feyzbakhsh and C. Li, Higher rank Clifford indices of curves on a K3 surface. *Selecta Math. (N.S.)* **27** (2021), no. 3, 48, 34 pp.

[30] S. Feyzbakhsh and R. Thomas, Rank r DT theory from rank 0. 2021, arXiv:2103.02915.

[31] S. Feyzbakhsh and R. Thomas, Rank r DT theory from rank 1. 2021, arXiv:2108.02828.

[32] L. Fu, C. Li, and X. Zhao, Stability manifolds of varieties with finite Albanese morphisms. 2021, arXiv:2103.07728.

[33] D. Gieseker, On a theorem of Bogomolov on Chern classes of stable bundles. *Amer. J. Math.* **101** (1979), no. 1, 77–85.

[34] M. Gross, D. Huybrechts, and D. Joyce, *Calabi–Yau manifolds and related geometries*. Universitext, Springer, Berlin, 2003.

[35] L. Gruson and C. Peskine, Genre des courbes de l'espace projectif. In *Algebraic geometry (Proc. Sympos., Univ. Tromsø, Tromsø, 1977)*, pp. 31–59, Lecture Notes in Math. 687, Springer, Berlin, 1978.

[36] D. Happel, I. Reiten, and S. O. Smalø, Tilting in abelian categories and quasitilted algebras. *Mem. Amer. Math. Soc.* **120** (1996), no. 575, viii+88.

[37] J. Harris, The genus of space curves. *Math. Ann.* **249** (1980), no. 3, 191–204.

[38] B. Hassett, Special cubic fourfolds. *Compos. Math.* **120** (2000), no. 1, 1–23.

[39] D. Huybrechts, Hyperkähler manifolds and sheaves. In *Proceedings of the International Congress of Mathematicians, Vol. II*, pp. 450–460, Hindustan Book Agency, New Delhi, 2010.

[40] D. Joyce and Y. Song, A theory of generalized Donaldson–Thomas invariants. *Mem. Amer. Math. Soc.* **217** (2012), no. 1020, iv+199 pp.

[41] M. Kontsevich, Homological algebra of mirror symmetry. In *Proceedings of the International Congress of Mathematicians, Vols. 1, 2 (Zürich, 1994)*, pp. 120–139, Birkhäuser, Basel, 1995.

[42] M. Kontsevich and Y. Soibelman, Stability structures, motivic Donaldson–Thomas invariants and cluster transformations. 2008, arXiv:0811.2435.

[43] N. Koseki, On the Bogomolov–Gieseker inequality in positive characteristic. 2020, arXiv:2008.09800.

[44] N. Koseki, Stability conditions on Calabi–Yau double/triple solids. 2020, arXiv:2007.00044.

[45] N. Koseki, Stability conditions on threefolds with nef tangent bundles. *Adv. Math.* **372** (2020), 107316, 29 pp.

[46] A. Kuznetsov, Semiorthogonal decompositions in algebraic geometry. In *Proceedings of the International Congress of Mathematicians—Seoul 2014, Vol. II*, pp. 635–660, Kyung Moon Sa, Seoul, 2014.

[47] A. Kuznetsov and A. Perry, Derived categories of Gushel–Mukai varieties. *Compos. Math.* **154** (2018), no. 7, 1362–1406.

[48] A. Langer, Semistable sheaves in positive characteristic. *Ann. of Math. (2)* **159** (2004), no. 1, 251–276.

[49] R. Lazarsfeld, Brill–Noether–Petri without degenerations. *J. Differential Geom.* **23** (1986), no. 3, 299–307.

[50] C. Lehn, M. Lehn, C. Sorger, and D. van Straten, Twisted cubics on cubic fourfolds. *J. Reine Angew. Math.* **731** (2017), 87–128.

[51] C. Li, On stability conditions for the quintic threefold. *Invent. Math.* **218** (2019), no. 1, 301–340.

[52] C. Li, Stability conditions on Fano threefolds of Picard number 1. *J. Eur. Math. Soc. (JEMS)* **21** (2019), no. 3, 709–726.

[53] C. Li, L. Pertusi, and X. Zhao, Twisted cubics on cubic fourfolds and stability conditions. 2020, arXiv:1802.01134.

[54] C. Li and X. Zhao, Birational models of moduli spaces of coherent sheaves on the projective plane. *Geom. Topol.* **23** (2019), no. 1, 347–426.

[55] S. Liu, Stability condition on Calabi–Yau threefold of complete intersection of quadratic and quartic hypersurfaces. 2021, arXiv:2108.08934.

[56] A. Maciocia and D. Piyaratne, Fourier–Mukai transforms and Bridgeland stability conditions on abelian threefolds II. *Internat. J. Math.* **27** (2016), no. 1, 1650007.

[57] E. Macrì, A generalized Bogomolov–Gieseker inequality for the three-dimensional projective space. *Algebra Number Theory* **8** (2014), no. 1, 173–190.

[58] E. Macrì and B. Schmidt, Derived categories and the genus of space curves. *Algebr. Geom.* **7** (2020), no. 2, 153–191.

[59] E. Markman, A survey of Torelli and monodromy results for holomorphic-symplectic varieties. In *Complex and differential geometry*, pp. 257–322, Springer Proc. Math. 8, Springer, Heidelberg, 2011.

[60] E. Markman and S. Mehrotra, Integral transforms and deformations of K3 surfaces. 2015, arXiv:1507.03108.

[61] S. Mukai, On the moduli space of bundles on K3 surfaces. I. In *Vector bundles on algebraic varieties (Bombay, 1984)*, pp. 341–413, Tata Inst. Fund. Res. Stud. Math. 11, Tata Inst. Fund. Res, Bombay, 1987.

[62] S. Mukai, Non-abelian Brill–Noether theory and Fano 3-folds. *Sugaku Expositions* **14** (2001), no. 2, 125–153.

[63] S. Mukai, Vector bundles on a K3 surface. In *Proceedings of the International Congress of Mathematicians, Vol. II (Beijing, 2002)*, pp. 495–502, Higher Ed. Press, Beijing, 2002.

[64] A. Perry, The integral Hodge conjecture for two-dimensional Calabi–Yau categories. 2020, arXiv:2004.03163.

[65] A. Perry, L. Pertusi, and X. Zhao, Stability conditions and moduli spaces for Kuznetsov components of Gushel–Mukai varieties. 2019, arXiv:1912.06935.

[66] D. Piyaratne, Stability conditions, Bogomolov–Gieseker type inequalities and Fano 3-folds. 2017, arXiv:1705.04011.

[67] D. Piyaratne and Y. Toda, Moduli of Bridgeland semistable objects on 3-folds and Donaldson–Thomas invariants. *J. Reine Angew. Math.* **747** (2019), 175–219.

[68] M. Reid, Bogomolov's theorem $c_1^2 \leq 4c_2$. In *Proceedings of the International Symposium on Algebraic Geometry (Kyoto Univ., Kyoto, 1977)*, pp. 623–642, Kinokuniya Book Store, Tokyo, 1978.

[69] F. Russo and G. Staglianò, Trisecant Flops, their associated K3 surfaces and the rationality of some Fano fourfolds. 2020, arXiv:1909.01263.

[70] B. Schmidt, A generalized Bogomolov–Gieseker inequality for the smooth quadric threefold. *Bull. Lond. Math. Soc.* **46** (2014), no. 5, 915–923.

[71] B. Schmidt, Counterexample to the generalized Bogomolov–Gieseker inequality for threefolds. *Int. Math. Res. Not. IMRN* **8** (2017), 2562–2566.

[72] I. Smith, Stability conditions in symplectic topology. In *Proceedings of the International Congress of Mathematicians—Rio de Janeiro 2018, Vol. II. Invited lectures*, pp. 969–991, World Sci. Publ., Hackensack, NJ, 2018.

[73] Y. Toda, Moduli stacks and invariants of semistable objects on $K3$ surfaces. *Adv. Math.* **217** (2008), no. 6, 2736–2781.

[74] Y. Toda, Derived category of coherent sheaves and counting invariants. In *Proceedings of the International Congress of Mathematicians—Seoul 2014, Vol. II*, pp. 745–767, Kyung Moon Sa, Seoul, 2014.

[75] Y. Toda, Gepner point and strong Bogomolov–Gieseker inequality for quintic 3-folds. In *Higher dimensional algebraic geometry-in honour of Professor Yujiro Kawamata's sixtieth birthday*, pp. 381–405, Adv. Stud. Pure Math. 74, Math. Soc. Japan, Tokyo, 2017.

[76] D. Villalobos-Paz, Moishezon spaces and projectivity criteria. 2001, arXiv:2105.14630.

[77] K. Yoshioka, Moduli spaces of stable sheaves on abelian surfaces. *Math. Ann.* **321** (2001), no. 4, 817–884.

AREND BAYER

School of Mathematics and Maxwell Institute, University of Edinburgh, Edinburgh, United Kingdom, arend.bayer@ed.ac.uk

EMANUELE MACRÌ

Université Paris-Saclay, CNRS, Laboratoire de Mathématiques d'Orsay, Rue Michel Magat, Bât. 307, 91405 Orsay, France, emanuele.macri@universite-paris-saclay.fr

COUNTING LATTICE POINTS IN MODULI SPACES OF QUADRATIC DIFFERENTIALS

VINCENT DELECROIX, ÉLISE GOUJARD, PETER ZOGRAF, AND ANTON ZORICH

ABSTRACT

We show how to count lattice points represented by square-tiled surfaces in the moduli spaces of meromorphic quadratic differentials with simple poles on complex algebraic curves. We demonstrate the versatility of the lattice point count on three different examples, including evaluation of Masur–Veech volumes of the moduli spaces of quadratic differentials, computation of asymptotic frequencies of geodesic multicurves on hyperbolic surfaces, and asymptotic enumeration of meanders with a fixed number of minimal arcs.

MATHEMATICS SUBJECT CLASSIFICATION 2020

Primary 30F30; Secondary 05A16

KEYWORDS

Quadratic differentials, square-tiled surfaces, Masur–Veech volumes, simple closed geodesics, meanders

INTERNATIONAL CONGRESS OF MATHEMATICIANS 2022 JULY 6–14

Proc. Int. Cong. Math. 2022, Vol. 3, pp. 2196–2211
DOI 10.4171/ICM2022/56

1. INTRODUCTION

Quadratic differentials on complex algebraic curves and their moduli spaces is an actively developing area of modern mathematics with close connections to algebraic geometry, dynamics, and mathematical physics (for example, this was one of the main topics at the MSRI semester program "Holomorphic Differentials in Mathematics and Physics" in the Fall 2019). In this paper we deal with the moduli spaces of meromorphic quadratic differentials with simple poles. These spaces are naturally equipped with linear coordinates (called period coordinates), volume form (called the Masur–Veech volume form), and integer lattice (whose points are represented by square-tiled surfaces). We show how to count lattice points in the moduli spaces of meromorphic quadratic differentials and apply this count to three seemingly different problems—namely, to the computation of Masur–Veech volumes of these moduli spaces, to the distribution of simple closed geodesics on hyperbolic surfaces, and to the enumeration of meanders.

A lattice point count relevant to moduli spaces $\mathcal{M}_{g,n}$ of genus g complex curves with n labeled marked points was first performed by P. Norbury in [22]. He considers a lattice in the Harer–Mumford combinatorial model of $\mathcal{M}_{g,n}$ whose points correspond to metric ribbon graphs with edges of integer length. Moreover, he shows that the top degree homogeneous part of his lattice point count polynomial is Kontsevich's volume polynomial [13] whose coefficients are given by intersection numbers of tautological ψ-classes on the Deligne–Mumford compactification $\overline{\mathcal{M}}_{g,n}$ (in fact, the top degree part of Norbury's polynomial coincides, up to the factor of 2^{2g-3+n}, with the top degree homogeneous part of Mirzakhani's Weil–Petersson volume polynomial for the moduli space of bordered hyperbolic surfaces [17, 18]). He then derives a simple recursion for the lattice count polynomials that yields an elementary proof of Witten's conjecture [24].

A few words about the structure of the paper. In Section 2 we collect basic facts about moduli spaces of meromorphic quadratic differentials, and in Section 3 we establish relationships between square-tiled surfaces, geodesic multicurves on hyperbolic surfaces, and stable graphs. Section 4 contains a formula for the Masur–Veech volume of the moduli space of quadratic differentials. Our formula expresses the Masur–Veech volumes in terms of the top degree parts of Norbury's counting polynomials with explicit rational coefficients. Note that for the moduli spaces of holomorphic quadratic differentials a similar formula for the Masur–Veech volumes was obtained (without explicit evaluation) by M. Mirzakhani [19] using a different approach. In Section 5 we use the interpretation of geodesic (multi)curves on hyperbolic surfaces as square-tiled surfaces to analyze their large genus asymptotic distribution. In particular, following the ideas of M. Mirzakhani [20], we show that the nonseparating simple closed geodesics become exponentially more frequent than the separating ones when the genus of the surface grows. Finally, in Section 6 we use the correspondence between meanders and square-tiled surfaces of special type to address the problem of meander enumeration.

2. MODULI SPACES OF QUADRATIC DIFFERENTIALS AND SQUARE-TILED SURFACES

The moduli space $\mathcal{Q}_{g,n}$ of (meromorphic) quadratic differentials is defined as the set of isomorphism classes of pairs (C, q), where C is a smooth genus g complex curve with $n \geq 0$ labeled distinct marked points, and q is a meromorphic quadratic differential on C with at most simple poles at the marked points and no other poles (throughout the paper we will assume that $2g - 2 + n > 0$). It is well known that $\mathcal{Q}_{g,n}$ is naturally isomorphic to $T^*\mathcal{M}_{g,n}$, the total space of the holomorphic cotangent bundle on the moduli space $\mathcal{M}_{g,n}$ of n-pointed genus g complex algebraic curves.

The moduli space $\mathcal{Q}_{g,n}$ is stratified according to the set $\mu = (m_1, \dots, m_k)$ of multiplicities of zeros of q (μ is a partition of $4g - 4 + n$). In what follows, we will mostly deal with the *principal stratum* $\mathcal{Q}(1^{4g-4+n}, -1^n)$ that consists of isomorphism classes of pairs (C, q), where C is a smooth curve and q has exactly $4g - 4 + n$ simple zeros and n simple poles. Moreover, we will always assume that the zeros of q are labeled. The natural map $\mathcal{Q}(1^{4g-4+n}, -1^n) \to \mathcal{Q}_{g,n}$ is a $(4g - 4 + n)!$-fold covering of its image that is open and dense in $\mathcal{Q}_{g,n}$, and its complement is closed and has positive codimension.

A nonzero differential q in $\mathcal{Q}_{g,n}$ defines a flat metric $|q|$ on the complex curve C. This metric has conical singularities at zeroes and poles of q. It defines the area function by the formula

$$A(C, q) = \int_C |q|, \tag{1}$$

so that $A(C, q)$ is finite and positive. The area function $A(C, q)$ may be viewed as a level function on the moduli space $\mathcal{Q}_{g,n}$.

To introduce the period coordinates on $\mathcal{Q}(1^{4g-4+n}, -1^n)$ consider the canonical cover of C defined by the meromorphic quadratic differential q with simple zeros and poles. More precisely, to each such pair (C, q) one associates a twofold covering $f : \hat{C} \to C$ and an abelian differential ω on $\hat{C}$, where $\hat{C} = \{(x, \omega(x)) \mid x \in C, \omega(x) \in T_x^*C, \omega(x)^2 = q(x)\}$. The curve $\hat{C}$ is smooth of genus $\hat{g} = 4g - 3 + n$, and the covering f is ramified precisely over zeros and poles of q. The differential ω is holomorphic on $\hat{C}$, has second order zeros at the preimages of zeros of q, and does not vanish at the preimages of poles of q

The map f is invariant under the canonical involution $\iota : \hat{C} \to \hat{C}$, $(x, \omega(x)) \mapsto (x, -\omega(x))$. The induced map $\iota_* : H_1(\hat{C}, \mathbb{C}) \to H_1(\hat{C}, \mathbb{C})$ allows decomposing $H_1(\hat{C}, \mathbb{C})$ into the direct sum $H_1^+(\hat{C}, \mathbb{C}) \oplus H_1^-(\hat{C}, \mathbb{C})$ of eigenspaces corresponding to the eigenvalues ± 1 of ι_*, where $\dim_{\mathbb{C}} H_1^+(\hat{C}, \mathbb{C}) = 2g$, $\dim_{\mathbb{C}} H_1^- = 6g - 6 + 2n$. Now put $H_1^-(\hat{C}, \mathbb{Z}) = H_1(\hat{C}, \mathbb{Z}) \cap H_1^-(\hat{C}, \mathbb{C})$ and consider the period map $P : H_1^-(\hat{C}, \mathbb{Z}) \to \mathbb{C}$ defined by $\alpha \mapsto P(\alpha) = \int_\alpha \omega$, $\alpha \in H_1^-(\hat{C}, \mathbb{Z})$. We can think of P as an element of $\mathrm{Hom}(H_1^-(\hat{C}, \mathbb{Z}), \mathbb{C}) = H_-^1(\hat{C}, \mathbb{C})$, the anti-invariant part of $H^1(\hat{C}, \mathbb{C})$ with respect to the involution ι^*. This gives us *period* (or *homological*) coordinates in a neighborhood of $(C, q) \in \mathcal{Q}(1^{4g-4+n}, -1^n)$.

Consider the lattice $L = \mathrm{Hom}(H_1^-(\hat{C}, \mathbb{Z}), \mathbb{Z} \oplus i\mathbb{Z})$, $i^2 = -1$, in the vector space $H_-^1(\hat{C}, \mathbb{C})$. By definition, the points of the lattice L are those elements of $H_-^1(\hat{C}, \mathbb{C})$ that take values in $\mathbb{Z} \oplus i\mathbb{Z}$ on $H_1^-(\hat{C}, \mathbb{Z})$. Note that L is a sublattice in the lattice $H_-^1(\hat{C}, \mathbb{Z} \oplus i\mathbb{Z}) \hookrightarrow H_-^1(\hat{C}, \mathbb{C})$ of index 4^{2g} induced by the inclusion $\mathbb{Z} \oplus i\mathbb{Z} \hookrightarrow \mathbb{C}$. The

Masur–Veech volume form dV on $\mathcal{Q}(1^{4g-4+n}, -1^n)$ is defined as the linear volume form in the vector space $H^1_-(\hat{C}, \mathbb{C})$ normalized in such a way that $\mathrm{Vol}(H^1_-(\hat{C}, \mathbb{C})/L) = 1$. The form dV induces a volume form on the level sets of the area function A, and we define

$$\mathrm{Vol}\,\mathcal{Q}(1^{4g-4+n}, -1^n) = \mathrm{Vol}'\,\mathcal{Q}^{A=1/2}(1^{4g-4+n}, -1^n) = 2d\,\mathrm{Vol}\,\mathcal{Q}^{A\le 1/2}(1^{4g-4+n}, -1^n). \tag{2}$$

Here

$$\mathcal{Q}^{A=1/2}_{g,n} = \{(C, q) \in \mathcal{Q}_{g,n} \mid A(C, q) = 1/2\}$$

is the level $A = 1/2$ hypersurface (sphere bundle on $\mathcal{M}_{g,n}$),

$$\mathcal{Q}^{A\le 1/2}_{g,n} = \{(C, q) \in \mathcal{Q}_{g,n} \mid A(C, q) \le 1/2\}$$

is the radius-1/2 disc bundle, and $d = 6g - 6 + 2n$ is the complex dimension of $\mathcal{Q}_{g,n}$. By a result of H. Masur [15] (cf. also W. Veech [23]), the volume of $\mathcal{Q}^{A\le 1/2}_{g,n}$ is finite. One of the main objectives of this paper is to provide a formula for this volume in terms of the lattice point count.

Remark 1. The moduli space $\mathcal{Q}_{g,n} = T^*\mathcal{M}_{g,n}$, as a cotangent bundle, carries a canonical symplectic form that we denote by Ω. The corresponding volume form $\frac{1}{d!}\Omega^d$ is invariant under the Teichmüller flow on $\mathcal{Q}_{g,n}$, see [16]. Previously it was shown in [15, 23] that the volume form dV is also preserved by the Teichmüller flow. Since $\frac{1}{d!}\Omega^d$ and dV belong to the same Lebesgue measure class, the ergodicity of the Teichmüller flow implies that these two volume forms coincide up to a constant proportionality coefficient. We postpone the evaluation of this constant factor to another occasion.

To give a geometric description of the lattice points in $\mathcal{Q}(1^{4g-4+n}, -1^n)$, we introduce a notion of square-tiled surfaces. It is possible to construct a discrete collection of meromorphic quadratic differentials with simple poles by assembling together identical flat squares in the following way. Take a finite number of oriented $1/2 \times 1/2$-squares with two opposite sides called horizontal and the other two sides called vertical. Identify pairs of sides of the squares by orientation reversing isometries in such a way that horizontal sides are glued to horizontal ones and vertical sides are glued to vertical ones. We get an oriented topological surface C without boundary, and we consider only those gluings for which C is connected. Quadratic differential dz^2 on each square is compatible with gluing and endows C with a complex structure and a nonzero quadratic differential q with at most simple poles. The total area $A(C, q)$ of C is 1/4 times the number of squares. We call such a pair (C, q) a *square-tiled surface*.

We denote by $ST(\mathcal{Q}(\mu, -1^n))$ the set of all square-tiled surfaces in the stratum $\mathcal{Q}(\mu, -1^n)$, and by $ST(\mathcal{Q}(\mu, -1^n), 2N)$ its subset consisting of surfaces made up of at most $2N$ squares.[1] Since each square-tiled surface (C, q) is glued from $1/2 \times 1/2$-squares,

1 For example, the square-tiled surface in Figure 1 is made up of 54 squares, has 3 conical points of angle 3π (corresponding to simple zeros of q), and 7 conical points of angle π (corresponding to simple poles of q). Therefore, it has genus 0 and belongs to the principal stratum $\mathcal{Q}(1^3, -1^7)$.

the periods of the abelian differential $\omega = \sqrt{q}$ on the double cover $\hat{C}$ belong to $\frac{1}{2}\mathbb{Z} \oplus \frac{i}{2}\mathbb{Z}$. In particular, this yields a natural inclusion $ST(\mathcal{Q}(1^{4g-4+n}, -1^n)) \hookrightarrow L$, where the lattice L was defined above. This inclusion is actually a bijection in each coordinate chart (in the case $g = 0$ a proof of the bijection between the set $ST(\mathcal{Q}(1^{n-4}, -1^n))$ and the lattice L can be found in **[3]**, see also **[5, APPENDIX A]**). Therefore, we have

$$\operatorname{Vol} \mathcal{Q}(1^{4g-4+n}, -1^n) = 2(6g - 6 + 2n) \lim_{N \to \infty} \frac{|ST(\mathcal{Q}(1^{4g-4+n}, -1^n), 2N)|}{N^{6g-6+2n}}, \tag{3}$$

where the vertical bars $|\,|$ denote the cardinality of a set. Formula (3) will serve as a starting point for our computation of $\operatorname{Vol} \mathcal{Q}_{g,n}$. Note that the volumes of $\mathcal{Q}_{g,n}$ and $\mathcal{Q}(1^{4g-4+n}, -1^n)$ are related by a simple formula

$$\operatorname{Vol} \mathcal{Q}_{g,n} = \frac{1}{(4g-4+n)!} \operatorname{Vol} \mathcal{Q}(1^{4g-4+n}, -1^n).$$

3. CYLINDER DECOMPOSITION, MULTICURVES, AND STABLE GRAPHS

A square-tiled surface admits a decomposition into maximal horizontal cylinders filled with isometric simple closed flat geodesics. Every such maximal horizontal cylinder has at least one conical singularity on each of its two boundary components. The square-tiled surface in Figure 1 has four maximal horizontal cylinders that are highlighted by different shades:

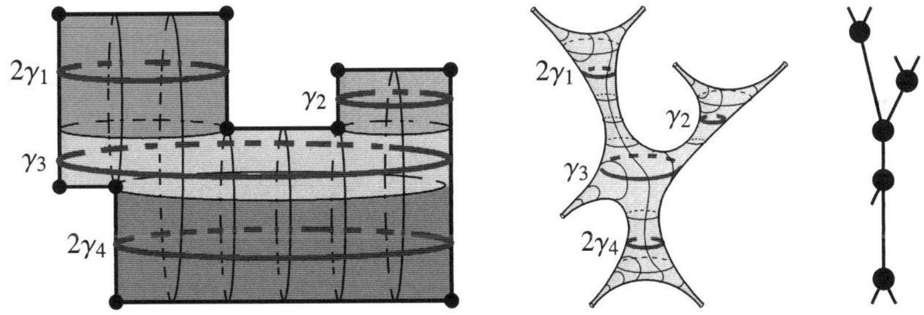

FIGURE 1
A square-tiled surface in $\mathcal{Q}(1^3, -1^7)$, and its associated multicurve and stable graph.

For a square-tiled surface (C, q), consider its decomposition into the set of k maximal horizontal cylinders. To each cylinder we associate the corresponding waist curve γ_i, $i = 1, \dots, k$, considered up to a free homotopy. The curves γ_i are nonperipheral (i.e., none of them bounds a disc containing a single pole) and pairwise nonhomotopic. We denote the number of circular horizontal bands of squares contained in the ith maximal horizontal cylinder by h_i. The formal linear combination $\gamma = \sum_{i=1}^k h_i \gamma_i$ is called a simple closed *multicurve* on C. (For example, the multicurve associated to the square-tiled surface in Figure 1 is $\gamma = 2\gamma_1 + \gamma_2 + \gamma_3 + 2\gamma_4$.)

To each multicurve $\gamma = \sum_{i=1}^k h_i \gamma_i$ as above we associate its *stable graph* $\Gamma(\gamma)$. A stable graph is a graph that is dual to the nodal curve obtained from C by pinching each

γ_i to a point. Stable graphs are used to describe the natural stratification of the Deligne–Mumford boundary of the moduli space $\mathcal{M}_{g,n}$. More precisely, $\Gamma(\gamma)$ is a decorated graph whose vertices represent the components of $S \setminus \{\gamma_1 \cup \cdots \cup \gamma_k\}$ and are labeled with the genus of the corresponding component. The edges of $\Gamma(\gamma)$ correspond to the curves γ_i and connect the vertices representing the components of $S \setminus \{\gamma_1 \cup \cdots \cup \gamma_k\}$ adjacent to γ_i (which may actually be the same). Finally, $\Gamma(\gamma)$ is endowed with n "legs" (or half-edges) labeled from 1 to n. The ith leg is attached to the vertex that represents the component that contains the ith marked point of C (i.e., the position of the ith pole of q). In addition to that, it is required that at each vertex v the stability condition $2g(v) - 2 + n(v) > 0$ is satisfied, where $g(v)$ is the genus assigned to v and $n(v)$ is the degree (or valency) of v. The right picture in Figure 1 shows the stable graph associated to the multicurve γ drawn in the middle picture.

For a pair of nonnegative integers g and n with $2g - 2 + n > 0$, denote by $\mathcal{G}_{g,n}$ the set of (isomorphism classes of) stable graphs of genus g with n legs (recall that the genus of a stable graph Γ is defined as $g = \sum_{v \in V(\Gamma)} g(v) + b_1(\Gamma)$, where $V(\Gamma)$ is the set of vertices and $b_1(\Gamma)$ is the first Betty number of the graph Γ). The number of such graphs $|\mathcal{G}_{g,n}|$ is finite, though grows rapidly with g and n.[2]

For a stable graph Γ in $\mathcal{G}_{g,n}$, consider the subset $ST_{\Gamma,h}(\mathcal{Q}(1^{4g-4+n}, -1^n), 2N)$ of square-tiled surfaces with at most $2N$ squares, having Γ as the associated stable graph and $h = (h_1, \ldots, h_k)$ as the set of heights of the cylinders (in the units of 1/2). Denote by $ST_\Gamma(\mathcal{Q}(1^{4g-4+n}, -1^n), 2N)$ the analogous subset without restriction on heights. Then the contributions to $\operatorname{Vol} \mathcal{Q}(1^{4g-4+n}, -1^n)$ in (2) from these subsets are given by

$$\operatorname{Vol}(\Gamma) = 2d \cdot \lim_{N \to \infty} \frac{|ST_\Gamma(\mathcal{Q}(1^{4g-4+n}, -1^n), 2N)|}{N^d}, \tag{4}$$

$$\operatorname{Vol}(\Gamma, h) = 2d \cdot \lim_{N \to \infty} \frac{|ST_{\Gamma,h}(\mathcal{Q}(1^{4g-4+n}, -1^n), 2N)|}{N^d}, \tag{5}$$

where $d = 6g - 6 + 2n$. The results in [5] imply that for any Γ in $\mathcal{G}_{g,n}$ the above limits exist, are strictly positive, and that

$$\operatorname{Vol} \mathcal{Q}(1^{4g-4+n}, -1^n) = \sum_{\Gamma \in \mathcal{G}_{g,n}} \operatorname{Vol}(\Gamma) = \sum_{\Gamma \in \mathcal{G}_{g,n}} \sum_{h \in \mathbb{N}^k} \operatorname{Vol}(\Gamma, h), \tag{6}$$

where k is the number of horizontal cylinders in C (or, equivalently, the number of edges of the stable graph Γ). Dividing both sides of (6) by $\operatorname{Vol} \mathcal{Q}(1^{4g-4+n}, -1^n)$, we see that the ratio $\operatorname{Vol}(\Gamma)/\operatorname{Vol} \mathcal{Q}(1^{4g-4+n}, -1^n)$ can be interpreted as the "asymptotic probability" that a random square-tiled surface has Γ as the stable graph associated to its horizontal cylinder decomposition.

4. FORMULA FOR THE MASUR–VEECH VOLUME

To evaluate the Masur–Veech volume of $\mathcal{Q}_{g,n}$, we introduce certain multivariate polynomials $N_{g,n}(b_1, \ldots, b_n)$ that have already appeared in the work on intersection theory

2 Table 1 below lists all stable graphs in $\mathcal{G}_{1,2}$.

of the moduli space of stable curves $\overline{\mathcal{M}}_{g,n}$, cf. [13,17,18,22]. Let $\boldsymbol{b} = (b_1, \ldots, b_n)$ be a set of variables, and let $(d_1, \ldots, d_n)$ be a set of nonnegative integers such that $d_1 + \cdots + d_n = 3g - 3 + n$. Define the homogeneous polynomial $N_{g,n}(b_1, \ldots, b_n)$ of degree $6g - 6 + 2n$ in the variables $b_1, \ldots, b_n$ by the formula

$$N_{g,n}(b_1, \ldots, b_n) = \frac{1}{2^{5g-6+2n}} \sum_{d_1,\ldots,d_n} \frac{b_1^{2d_1} \cdots b_n^{2d_n}}{d_1! \cdots d_n!} \int_{\overline{\mathcal{M}}_{g,n}} \psi_1^{d_1} \cdots \psi_n^{d_n}, \tag{7}$$

where $d_1 + \cdots + d_n = 3g - 3 + n$ and $\psi_1, \ldots, \psi_n$ are the tautological ψ-classes on the Deligne–Mumford compactification $\overline{\mathcal{M}}_{g,n}$ of the moduli space of curves (informally speaking, ψ_i is the class of the line bundle whose fiber over $(C, p_1, \ldots, p_n)$ is the cotangent line $T^*_{p_i}C$ to C at the ith marked point p_i). For small g and n, we have

$$\begin{aligned}
N_{0,3}(b_1, b_2, b_3) &= 1, \\
N_{0,4}(b_1, b_2, b_3, b_4) &= \frac{1}{4}(b_1^2 + b_2^2 + b_3^2 + b_4^2), \\
N_{1,1}(b_1) &= \frac{1}{48} b_1^2, \\
N_{1,2}(b_1, b_2) &= \frac{1}{384}(b_1^2 + b_2^2)(b_1^2 + b_2^2), \\
N_{2,1}(b_1) &= \frac{1}{1\,769\,472} b_1^8.
\end{aligned}$$

Following [3], we introduce the linear operators Y_h and Z acting on the space of polynomials in variables $b_1, \ldots, b_k$. The operator Y_h is defined on monomials as

$$Y_h : \prod_{i=1}^{k} b_i^{m_i} \mapsto \prod_{i=1}^{k} \frac{m_i!}{h_i^{m_i+1}}, \tag{8}$$

and extended to arbitrary polynomials by linearity. The operator Z is defined on monomials as

$$Z : \prod_{i=1}^{k} b_i^{m_i} \mapsto \prod_{i=1}^{k} m_i! \cdot \zeta(m_i + 1), \tag{9}$$

and extended to arbitrary polynomials by linearity. In the above formula, ζ is the Riemann zeta function

$$\zeta(s) = \sum_{n \geq 1} \frac{1}{n^s},$$

so that for any collection of positive integers $(m_1, \ldots, m_k)$ we have

$$Z\left(\prod_{i=1}^{k} b_i^{m_i}\right) = \sum_{h \in \mathbb{N}^k} Y_h\left(\prod_{i=1}^{k} b_i^{m_i}\right).$$

For a stable graph Γ with the vertex set $V(\Gamma)$ and the edge set $E(\Gamma)$, we associate a homogeneous polynomial P_Γ of degree $6g - 6 + 2n$ as follows. To every edge $e \in E(\Gamma)$ we assign a formal variable b_e. For a vertex $v \in V(\Gamma)$ of weight $g(v)$ and valency $n(v)$, we assign the variable b_e to each half of the edge e incident to v, and we assign 0 to each leg of Γ. We denote by $\boldsymbol{b}_v$ the resulting collection of $n(v)$ variables. More specifically, if an edge e is a loop joining v to itself, b_e enters $\boldsymbol{b}_v$ twice; if an edge e joins v to a distinct vertex, b_e enters $\boldsymbol{b}_v$ once; the remaining entries of $\boldsymbol{b}_v$ correspond to legs (they are represented by zeros). To each vertex $v \in V(\Gamma)$ we associate the polynomial $N_{g(v),n(v)}(\boldsymbol{b}_v)$, where $N_{g,n}$ is defined in (7). We define P_Γ by the formula

$$P_\Gamma = \frac{2^{6g-5+2n} \cdot (4g-4+n)!}{(6g-7+2n)!} \cdot \frac{1}{2^{|V(\Gamma)|-1}} \cdot \frac{1}{|\operatorname{Aut}(\Gamma)|} \cdot \prod_{e \in E(\Gamma)} b_e \cdot \prod_{v \in V(\Gamma)} N_{g(v),n(v)}(\boldsymbol{b}_v). \tag{10}$$

Lemma 1. *The contributions* $\operatorname{Vol}(\Gamma)$ *and* $\operatorname{Vol}(\Gamma, h)$ *to* $\operatorname{Vol}\mathcal{Q}(1^{4g-4+n}, -1^n)$*, given by formulas* (4) *and* (5)*, respectively, satisfy the relations*

$$\operatorname{Vol}(\Gamma) = Z(P_\Gamma) \quad \text{and} \quad \operatorname{Vol}(\Gamma, h) = Y_h(P_\Gamma). \tag{11}$$

Substituting these expressions into formula (6), we immediately obtain

Theorem 1. *The Masur–Veech volume of the principal stratum* $\mathcal{Q}(1^{4g-4+n}, -1^n)$ *is given by the formula*

$$\operatorname{Vol}\mathcal{Q}(1^{4g-4+n}, -1^n) = \sum_{\Gamma \in \mathcal{G}_{g,n}} Z(P_\Gamma) = \sum_{\Gamma \in \mathcal{G}_{g,n}} \sum_{h \in \mathbb{N}^{|E(\Gamma)|}} Y_h(P_\Gamma). \tag{12}$$

Table 1 illustrates the computation of the polynomials P_Γ, as well as the contributions $\operatorname{Vol}(\Gamma)$ to the Masur–Veech volume $\operatorname{Vol}\mathcal{Q}(1^{4g-4+n}, -1^n)$, in the simplest nontrivial case of $(g,n) = (1,2)$. To make the computations tractable, we follow the structure of formula (10). The first numerical factor $\frac{32}{3}$ in the first line of each calculation in the right column of Table 1 is $\frac{2^{6g-5+2n}(4g-4+n)!}{(6g-7+2n)!}$ evaluated at $(g,n) = (1,2)$; it is common for all stable graphs in $\mathcal{G}_{1,2}$. The second numerical factor is $\frac{1}{2^{|V(\Gamma)|-1}}$. The third numerical factor is $\frac{1}{|\operatorname{Aut}(\Gamma)|}$ (while the vertices and edges of Γ are not labeled, the automorphism group $\operatorname{Aut}(\Gamma)$ respects the decoration of the graph). The resulting value

$$\operatorname{Vol}\mathcal{Q}(1^2, -1^2) = \left(\frac{8}{45} + \frac{1}{135} + \frac{2}{27} + \frac{2}{27}\right) \cdot \pi^4 = \frac{\pi^4}{3}$$

matches that previously found in [11] by implementing a completely different algorithm of [10].

For $g = 0$, the formula for the Masur–Veech volumes simplifies considerably. As it was shown in [3], for all $n \geq 4$,

$$\operatorname{Vol}\mathcal{Q}(1^{n-4}, -1^n) = 4\left(\frac{\pi^2}{2}\right)^{n-3} \tag{13}$$

($\operatorname{Vol}\mathcal{Q}_{0,3} = 4$ and $\operatorname{Vol}\mathcal{Q}_{1,1} = \frac{2}{3}\pi^2$ by convention).

Table 2 below provides the volumes of principal strata $V_{g,n} = \operatorname{Vol}\mathcal{Q}(1^{4g-4+n}, -1^n)$ for small g and n (some additional data can be found in [7]).

Remark 2. An alternative formula for the Masur–Veech volume of the principal stratum $\mathcal{Q}(1^{4g-4+n}, -1^n)$ was obtained in [4]. The formula is based on the intersection theory and expresses $\operatorname{Vol}\mathcal{Q}(1^{4g-4+n}, -1^n)$ as a weighted sum of certain linear Hodge integrals over the

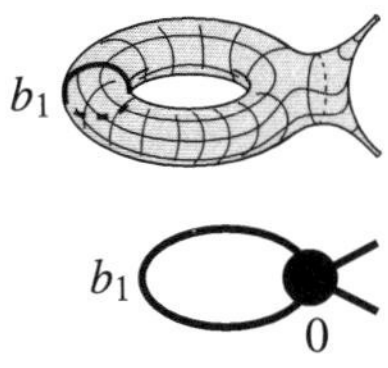	$\frac{32}{3}\cdot 1\cdot\frac{1}{2}\cdot b_1\cdot N_{0,4}(b_1,b_1,0,0)$ $=\frac{16}{3}\cdot b_1\cdot(\frac{1}{4}(2b_1^2))=\frac{8}{3}\cdot b_1^3$ $\overset{Z}{\mapsto}\ \frac{8}{3}\cdot 3!\cdot\zeta(4)=\frac{8}{45}\pi^4$
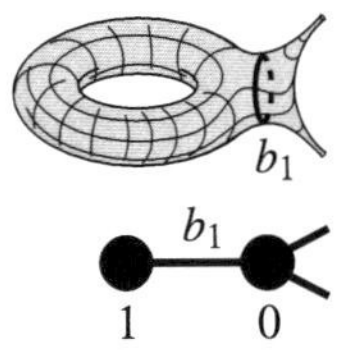	$\frac{32}{3}\cdot\frac{1}{2}\cdot 1\cdot b_1\cdot N_{1,1}(b_1)\cdot N_{0,3}(0,0,b_1)$ $=\frac{16}{3}\cdot b_1\cdot(\frac{1}{48}b_1^2)\cdot(1)=\frac{1}{9}\cdot b_1^3$ $\overset{Z}{\mapsto}\ \frac{1}{9}\cdot 3!\cdot\zeta(4)=\frac{1}{135}\pi^4$
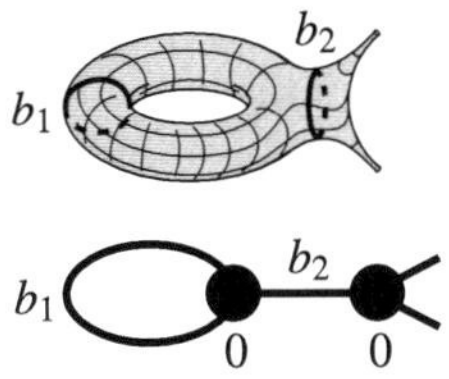	$\frac{32}{3}\cdot\frac{1}{2}\cdot\frac{1}{2}\cdot b_1b_2\cdot N_{0,3}(b_1,b_1,b_2)\cdot N_{0,3}(b_1,0,0)$ $=\frac{8}{3}\cdot b_1b_2\cdot(1)\cdot(1)=\frac{8}{3}\cdot b_1b_2$ $\overset{Z}{\mapsto}\ \frac{8}{3}\cdot(\zeta(2))^2=\frac{2}{27}\pi^4$
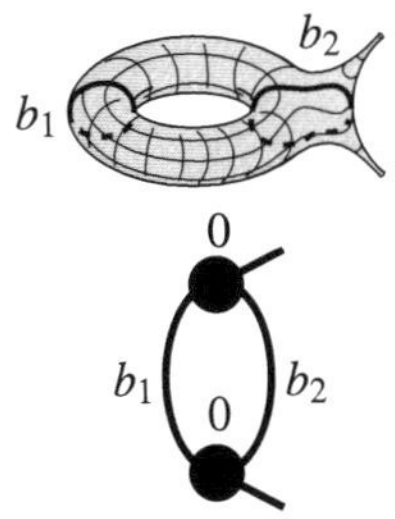	$\frac{32}{3}\cdot\frac{1}{2}\cdot\frac{1}{2}\cdot b_1b_2\cdot N_{0,3}(0,b_1,b_2)\cdot N_{0,3}(b_1,b_2,0)$ $=\frac{8}{3}\cdot b_1b_2\cdot(1)\cdot(1)=\frac{8}{3}\cdot b_1b_2$ $\overset{Z}{\mapsto}\ \frac{8}{3}\cdot(\zeta(2))^2=\frac{2}{27}\pi^4$

TABLE 1

Computation of $\operatorname{Vol}\mathcal{Q}(1^2,-1^2)=2!\cdot\operatorname{Vol}\mathcal{Q}_{1,2}$. The left column lists the multicurves and their associated stable graphs Γ, while the right column gives the polynomials P_Γ and the corresponding volume contributions $\operatorname{Vol}(\Gamma)=Z(P_\Gamma)$.

g	n	$V_{g,n}$
1	3	$\frac{11}{60}\pi^6$
1	4	$\frac{1}{10}\pi^8$
1	5	$\frac{163}{3\,024}\pi^{10}$
2	0	$\frac{1}{15}\pi^6$
2	1	$\frac{29}{840}\pi^8$
2	2	$\frac{337}{18\,144}\pi^{10}$
3	0	$\frac{115}{33\,264}\pi^{12}$
4	0	$\frac{2\,106\,241}{11\,548\,293\,120}\pi^{18}$

TABLE 2

Numerical values of Masur–Veech volumes of low-dimensional principal strata.

moduli space of stable curves $\overline{\mathcal{M}}_{g,n}$. The subsequent papers [12, 25] provided efficient algorithms for computing these Hodge integrals that allowed computing $\operatorname{Vol}\mathcal{Q}(1^{4g-4+n}, -1^n)$ for large enough values of g and n reasonably fast. In particular, this approach yields the same data as in Table 2.

Remark 3. A conjectural large genus asymptotic formula for the Masur–Veech volume of any stratum in the moduli space $\mathcal{Q}_{g,n}$ was proposed in [2]. For the principal stratum $\mathcal{Q}(1^{4g-4+n}, -1^n)$, it reads

$$\operatorname{Vol}\mathcal{Q}(1^{4g-4+n}, -1^n) \approx \frac{4}{\pi} \cdot 2^n \cdot \left(\frac{8}{3}\right)^{4g-4+n} \quad \text{as } g \to \infty. \tag{14}$$

Formula (14) was recently proven in [1] on the basis of (12) and a uniform large genus estimate for intersection numbers of ψ-classes on the moduli space $\overline{\mathcal{M}}_{g,n}$.

More details concerning the results of this section can be found in [8].

5. RANDOM SQUARE-TILED SURFACES AND RANDOM MULTICURVES

Here we discuss statistical geometry of simple closed hyperbolic multicurves. The research in this direction was inspired by the pioneering work of M. Mirzakhani [19–21]. The new tool that we bring in is a natural bijection between square-tiled surfaces and multicurves described in Section 3.

To give a flavor of these results, let us take a complex curve C of genus g with n punctures endowed with the compatible complete hyperbolic metric. Let $\gamma = \sum_{i=1}^{k} h_i \gamma_i$ be a multicurve on C consisting of pairwise disjoint primitive simple closed geodesics γ_i. Furthermore, denote by ℓ the hyperbolic length function, and put $L = \sum_{i=1}^{k} h_i \ell(\gamma_i)$ to be the total length of γ.

Denote by $\mathcal{ML}_{g,n}(\mathbb{Z})$ the lattice of integer points in the space of measured laminations on C (i.e., the set of all simple closed geodesic multicurves). We say that two multicurves have the same topological type if they belong to the same orbit of the mapping class group $\operatorname{Mod}_{g,n}$ in $\mathcal{ML}_{g,n}(\mathbb{Z})$. By definition, the asymptotic probability that a random multicurve belongs to the orbit $\operatorname{Mod}_{g,n} \cdot \gamma$ is

$$P_{g,n}(\gamma) = \lim_{L \to \infty} \frac{|\{\gamma' \in \operatorname{Mod}_{g,n} \cdot \gamma \mid \ell(\gamma') \le L\}|}{|\{\gamma' \in \mathcal{ML}_{g,n}(\mathbb{Z}) \mid \ell(\gamma') \le L\}|}. \tag{15}$$

M. Mirzakhani [20] expressed the probability $P_{g,n}(\gamma)$ in terms of the intersection numbers of tautological ψ-classes on the moduli spaces $\overline{\mathcal{M}}_{g,n}$. Her formula implies, in particular, that $P_{g,n}(\gamma)$ depends only on topology of the pair C, γ and is independent of the complex structure or hyperbolic metric on C. We have the following refinement of her result, cf. [8]:

Theorem 2. *The asymptotic probability $P_{g,n}(\gamma)$ is given in terms of the volume contribution $\operatorname{Vol}(\Gamma, h)$ by the formula*

$$P_{g,n}(\gamma) = \frac{\operatorname{Vol}(\Gamma, h)}{\operatorname{Vol}\mathcal{Q}(1^{4g-4+n}, -1^n)}, \tag{16}$$

where Γ is the stable graph corresponding to the multicurve γ.

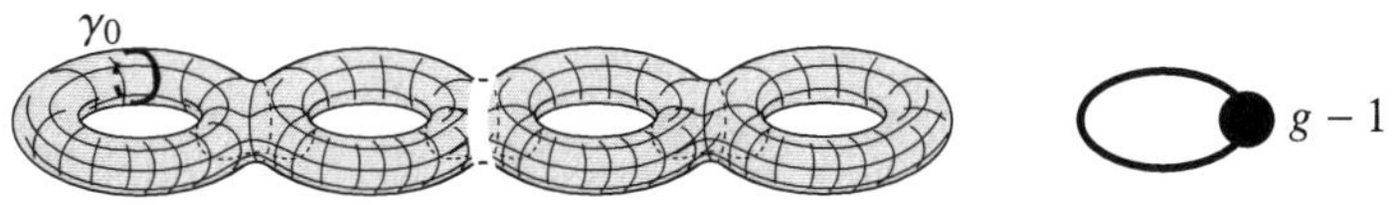

FIGURE 2
The non-separating curve γ_0 and the associated stable graph Γ_0.

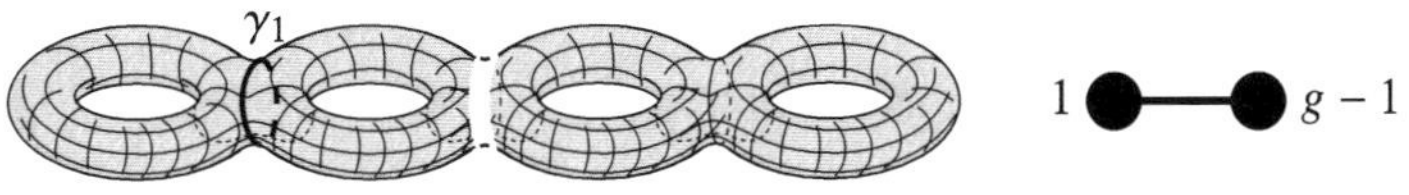

FIGURE 3
The separating curve γ_1 and the associated stable graph Γ_1.

Let us now present a consequence of Theorem 2 for $n = 0$, $g \geq 2$. In this case there is just one topological type of nonseparating simple closed geodesics γ_0 as in Figure 2, and $[g/2]$ topological types of separating closed geodesics $\gamma_1, \ldots, \gamma_{[g/2]}$ as in Figure 3, where γ_i splits the complex curve C into two parts of genera i and $g - i$, respectively. Then the following asymptotic formula holds:

$$\frac{\sum_{i=1}^{[g/2]} P_{g,0}(\gamma_i)}{P_{g,0}(\gamma_0)} \approx \sqrt{\frac{2}{3\pi g}} \cdot \frac{1}{2^{2g}} \quad \text{as } g \to \infty. \tag{17}$$

In plain words, on a compact hyperbolic surface of large genus, nonseparating simple closed curves are exponentially more frequent than separating ones. The proof of formula (17) comprises evaluation of individual contributions of particularly simple stable graphs displayed in Figures 2 and 3 and analysis of their large genus asymptotic behavior using an explicit closed-form expression for the intersection numbers $\int_{\overline{\mathcal{M}}_{g,2}} \psi_1^{d_1} \psi_2^{d_2}$ obtained in **[26]**, cf. **[8]** for details.

In order to go beyond the case of simple closed curves, a much more involved asymptotic analysis of intersection numbers of ψ-classes is needed that is performed in full generality in **[1, 6]**.

6. SQUARE-TILED SURFACES AND ENUMERATION OF MEANDERS

Here we apply the lattice point count to the enumeration problem of meanders. A meander is a configuration in the plane that consists of a straight line and a simple closed curve transversely intersecting it, considered up to isotopy (see Figure 4 for an example). Meanders naturally appear in various areas of mathematics and theoretical physics (for instance, they provide a model of polymer folding). A brief introduction to meanders can be found in **[14]**

Enumeration of meanders is a long-standing difficult combinatorial problem. Let $\mathcal{M}(N)$ be the number of meanders with $2N$ crossings. Then, conjecturally,

$$\mathcal{M}(N) \approx \text{const} \cdot R^N N^\alpha \quad \text{as } N \to \infty.$$

There is a plausible prediction that $\alpha = -\frac{29+\sqrt{145}}{12}$ coming from its interpretation as the critical exponent in a two dimensional conformal field theory with central charge $c = -4$, see [9]. However, even a hypothetical value of R is not known.

The situation becomes more tractable if we impose on meanders an additional topological restriction. We call an arc *minimal* if it connects two adjacent intersections. The maximal arc connecting the first and the last intersections, if present, is also treated as a minimal arc at infinity (for instance, the meander displayed on Figure 4 has 6 minimal arcs including the one at infinity).

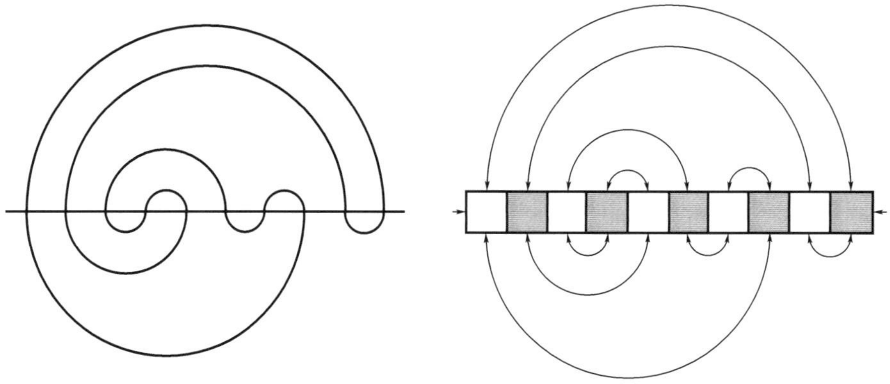

FIGURE 4
A meander with 10 crossings and 6 minimal arcs, and the corresponding square-tiled surface in $\mathcal{Q}(1^2, 0, -1^6)$ (pairs of sides connected with arrowed arcs are identified).

Denote by $\mathcal{M}_n(N)$ the number of meanders with not more than $2N$ crossings and exactly n minimal arcs (it is not hard to see that $n \geq 4$). Then we have

Theorem 3. *For any fixed $n \geq 4$, the number of meanders $\mathcal{M}_n(N)$ satisfies the following asymptotic formula:*

$$\mathcal{M}_n(N) = \frac{4}{n!(n-4)!}\left(\frac{2}{\pi^2}\right)^{n-3}\binom{2n-4}{n-2}^2 \frac{N^{2n-5}}{4n-10} + o(N^{2n-5}) \quad \text{as } N \to \infty. \tag{18}$$

Below we sketch a derivation of formula (18). To begin with, we establish a relationship between meanders and certain square-tiled surfaces. A meander with $2N$ crossings may be viewed as a 4-regular plane graph, and its dual graph is a quadrangulation of the sphere made up of $2N$ squares. We can think of these squares to be identical of size $1/2 \times 1/2$ as above. Since this quadrangulation respects the horizontal and vertical sides of the squares, it gives rise to a genus 0 square-tiled surface (C, q), where the quadratic differential q has exactly n simple poles. Besides the horizontal cylinder decomposition of C considered earlier, we define its vertical cylinder decomposition in the same way. Clearly, a square-tiled surface associated with a meander has exactly one horizontal and one vertical cylinders, both of maximal circumference $2N$ (in the units of $1/2$), see Figure 4. At last, we need to mark a point on C that corresponds to the infinity (by convention, we put it at the endpoint of a vertical side of one of the squares). Thus, we get a correspondence between the set of meanders with $2N$ crossings and n minimal arcs on one side and the set of genus 0 square-tiled surfaces glued from $2N$ squares with exactly n simple poles, one marked point, one horizontal and one vertical cylinders of maximal circumference $2N$ on the other (note that the latter carries no labeling of either poles or zeros).[3] This correspondence is generically two-to-one for large N since, for a regular vertex of a square-tiled surface, there are two vertical edges incident to it, and either of them can be chosen as a distinguished edge at infinity.

Now we want to realize these square-tiled surfaces as lattice points in a moduli space of quadratic differentials. Denote by $\mathcal{Q}_{g,n,1}$ the moduli space of pairs (C, q), where C is a genus g complex curve with $n+1$ labeled marked points and q is a quadratic differentials with at most simple poles at the first n marked points and regular at the last one (in fact, $\mathcal{Q}_{g,n,1} = p^*T^*\mathcal{M}_{g,n}$, where $p : \mathcal{M}_{g,n+1} \to \mathcal{M}_{g,n}$ is the forgetful map). Let $\mathcal{Q}(1^{4g-4+n}, 0, -1^n)$ be the (principal) stratum of quadratic differentials with $4g-4+n$ labeled simple zeros, so that the natural map $\mathcal{Q}(1^{4g-4+n}, 0, -1^n) \to \mathcal{Q}_{g,n,1}$ is a $(4g-4+n)!$-fold covering of its image. Period coordinates on $\mathcal{Q}(1^{4g-4+n}, 0, -1^n)$ are defined as in Section 2, the only difference is that there is an additional coordinate given by the integral of ω along a path connecting two preimages of the marked point. The Masur–Veech volume form is also well defined in this case.

Consider three nested sets of square-tiled surfaces defined as follows:

- the set $ST(\mathcal{Q}(1^{n-4}, 0, -1^n), 2N)$ of all square-tiled surfaces in $\mathcal{Q}(1^{n-4}, 0, -1^n)$ made up of at most $2N$ squares;
- the subset $ST_1(\mathcal{Q}(1^{n-4}, 0, -1^n), 2N) \subset ST(\mathcal{Q}(1^{n-4}, 0, -1^n), 2N)$ of surfaces with one horizontal cylinder of maximal circumference;
- the subset $ST_{1,1}(\mathcal{Q}(1^{n-4}, 0, -1^n), 2N) \subset ST_1(\mathcal{Q}(1^{n-4}, 0, -1^n), 2N)$ of surfaces with one horizontal and one vertical cylinders of maximal circumference.

Since square-tiled surfaces are uniformly distributed in $\mathcal{Q}(1^{n-4}, 0, -1^n)$ relative to the Masur–Veech volume form, we get the asymptotics

$$\left|ST\big(\mathcal{Q}(1^{n-4}, 0, -1^n), 2N\big)\right| = c(n)\frac{N^d}{2d} + o(N^d) \quad \text{as } N \to \infty, \tag{19}$$

where $d = 2n-5 = \dim_{\mathbb{C}} \mathcal{Q}(1^{n-4}, 0, -1^n)$ and $c(n) = \operatorname{Vol} \mathcal{Q}(1^{n-4}, 0, -1^n)$, cf. formula (3). By a more subtle argument, we obtain that

$$\left|ST_1\big(\mathcal{Q}(1^{n-4}, 0, -1^n), 2N\big)\right| = c_1(n)\frac{N^d}{2d} + o(N^d) \quad \text{as } N \to \infty, \tag{20}$$

$$\left|ST_{1,1}\big(\mathcal{Q}(1^{n-4}, 0, -1^n), 2N\big)\right| = c_{1,1}(n)\frac{N^d}{2d} + o(N^d) \quad \text{as } N \to \infty. \tag{21}$$

As it was shown in [5], the coefficients $c_1(n)$ and $c_{1,1}(n)$ also exist and satisfy the condition $c(n) > c_1(n) > c_{1,1}(n) > 0$. Though we cannot access $c_{1,1}(n)$ directly, we can use the fun-

3 Actually, such a square-tiled surface additionally admits a proper chessboard coloring, but we will not need it here.

damental fact proven in [5] that the numbers of horizontal and vertical cylinders of a random square-tiled surface are asymptotically uncorrelated when $N \to \infty$. In particular, this yields

$$\frac{c_{1,1}(n)}{c_1(n)} = \frac{c_1(n)}{c(n)}, \tag{22}$$

so that $c_{1,1}(n) = \frac{c_1(n)^2}{c(n)}$. The coefficients $c(n)$ and $c_1(n)$ can be computed explicitly. By (13), we have

$$c(n) = \operatorname{Vol} \mathcal{Q}(1^{n-4}, 0, -1^n) = 2 \operatorname{Vol} \mathcal{Q}(1^{n-4}, -1^n) = 8\left(\frac{\pi^2}{2}\right)^{n-3}, \tag{23}$$

and, by a direct combinatorial argument in [5],

$$c_1(n) = 4\binom{2n-4}{n-2}. \tag{24}$$

Note that the contribution to $\mathcal{M}_n(N)$ from meanders whose associated square-tiled surfaces lie in the principle stratum $\mathcal{Q}(1^{n-4}, 0, -1^n)$ becomes predominant as $N \to \infty$. This implies that for $N \to \infty$ we have

$$\begin{aligned} \mathcal{M}_n(N) &= \frac{2c_{1,1}(n)}{n!(n-4)!} \frac{N^d}{2d} + o(N^d) \\ &= \frac{4}{n!(n-4)!} \left(\frac{2}{\pi^2}\right)^{n-3} \binom{2n-4}{n-2}^2 \frac{N^{2n-5}}{4n-10} + o(N^{2n-5}) \end{aligned}$$

as claimed.

Remark 4. The techniques outlined in this section can be applied to asymptotic enumeration of pairs of multicurves on surfaces of arbitrary genus satisfying certain topological restrictions. The details will appear elsewhere.

ACKNOWLEDGMENTS

We are grateful to everybody with whom we discussed our work at its various stages and apologize for not being able to thank everyone personally. Figures 1–3 and images in Table 1 originally appeared in [8] (Copyright 2021, Duke University Press. All rights reserved.)

FUNDING

This work was supported by the Russian Science Foundation grant 19-71-30002.

REFERENCES

[1] A. Aggarwal, Large genus asymptotics for intersection numbers and principal strata volumes of quadratic differentials. *Invent. Math.* **226** (2021), no. 3, 897–1010.

[2] A. Aggarwal, V. Delecroix, É. Goujard, P. Zograf, and A. Zorich, Conjectural large genus asymptotics of Masur–Veech volumes and of area Siegel–Veech constants of strata of quadratic differentials. *Arnold Math. J.* **6** (2020), no. 2, 149–161.

[3] J. Athreya, A. Eskin, and A. Zorich, Right-angled billiards and volumes of moduli spaces of quadratic differentials on $\mathbb{C}P^1$. *Ann. Sci. Éc. Norm. Supér.* **49** (2016), 1307–1381.

[4] D. Chen, M. Möller, and A. Sauvaget, with an appendix by G. Borot, A. Giacchetto, and D. Lewanski, Masur–Veech volumes and intersection theory: the principal strata of quadratic differentials. 2019, arXiv:1912.02267.

[5] V. Delecroix, É. Goujard, P. Zograf, and A. Zorich, Enumeration of meanders and Masur–Veech volumes. *Forum Math. Pi* **8** (2020), e4, 80 pp.

[6] V. Delecroix, É. Goujard, P. Zograf, and A. Zorich, Large genus asymptotic geometry of random square-tiled surfaces and of random multicurves. 2020, arXiv:2007.04740.

[7] V. Delecroix, É. Goujard, P. Zograf, and A. Zorich, Tables of Masur–Veech volumes and of area Siegel–Veech constants of principal strata of quadratic differentials. University of Bordeaux, 2020, https://www.math.u-bordeaux.fr/~egoujard/tablevol2020.pdf.

[8] V. Delecroix, É. Goujard, P. Zograf, and A. Zorich, Masur–Veech volumes, frequencies of simple closed geodesics and intersection numbers of moduli spaces of curves. *Duke Math. J.* **170** (2021), no. 12, 2633–2718.

[9] P. Di Francesco, O. Golinelli, and E. Guitter, Meanders: exact asymptotics. *Nuclear Phys. B* **570** (2000), no. 3, 699–712.

[10] A. Eskin and A. Okounkov, Pillowcases and quasimodular forms. In *Algebraic geometry and number theory*, edited by V. Ginzburg, pp. 1–25, Progr. Math. 253, Birkhäuser, 2006.

[11] É. Goujard, Volumes of strata of moduli spaces of quadratic differentials: getting explicit values. *Ann. Inst. Fourier* **66** (2016), no. 6, 2203–2251.

[12] M. Kazarian, Recursion for Masur–Veech volumes of moduli spaces of quadratic differentials. *J. Inst. Math. Jussieu* (2021), 1–6.

[13] M. Kontsevich, Intersection theory on the moduli space of curves and the matrix Airy function. *Comm. Math. Phys.* **147** (1992), 1–23.

[14] S. Lando and A. Zvonkin, *Graphs on Surfaces and Their Applications*. Encyclopaedia Math. Sci. 141, Springer, 2004.

[15] H. Masur, Interval exchange transformations and measured foliations. *Ann. of Math.* **115** (1982), 169–200.

[16] H. Masur, The Teichmüller flow is Hamiltonian. *Proc. Amer. Math. Soc.* **123** (1995), no. 12, 3739–3747.

[17] M. Mirzakhani, Simple geodesics and Weil–Petersson volumes of moduli spaces of bordered Riemann surfaces. *Invent. Math.* **167** (2007), no. 1, 179–222.

[18] M. Mirzakhani, Weil–Petersson volumes and intersection theory on the moduli space of curves. *J. Amer. Math. Soc.* **20** (2007), no. 1, 1–23.

[19] M. Mirzakhani, Ergodic theory of the earthquake flow. *Int. Math. Res. Not.* **2008** (2008), 1–39.

[20] M. Mirzakhani, Growth of the number of simple closed geodesics on hyperbolic surfaces. *Ann. of Math.* **168** (2008), no. 1, 97–125.

[21] M. Mirzakhani, Counting mapping class group orbits on hyperbolic surfaces. 2016, arXiv:1601.03342.

[22] P. Norbury, Counting lattice points in the moduli space of curves. *Math. Res. Lett.* **17** (2010), no. 3, 467–481.

[23] W. Veech, Gauss measures for transformations on the space of interval exchange maps. *Ann. of Math.* **115** (1982), 201–242.

[24] E. Witten, Two-dimensional gravity and intersection theory on moduli space. *Surv. Differ. Geom.* **1** (1991), 243–310.

[25] D. Yang, D. Zagier, and Y. Zhang, Masur–Veech volumes of quadratic differentials and their asymptotics. *J. Geom. Phys.* **158** (2020), no. 103870, 1–12.

[26] P. Zograf, An explicit formula for Witten's 2-correlators. *J. Math. Sci.* **240** (2019), 535–538 (corrected version: arXiv:2012.03268).

VINCENT DELECROIX

Laboratoire Bordelais de Recherche en Informatique, Université de Bordeaux, Talence, France, vincent.delecroix@u-bordeaux.fr

ÉLISE GOUJARD

Institut de Mathématiques de Bordeaux, Université de Bordeaux, Talence, France, elise.goujard@gmail.com

PETER ZOGRAF

Euler International Mathematical Institute, St. Petersburg, Russia, zograf@pdmi.ras.ru

ANTON ZORICH

Center for Advanced Studies, Skoltech, Moscow, Russia, and Institut de Mathématiques de Jussieu–Paris Rive Gauche, Paris, France, anton.zorich@gmail.com

K-THEORY OF LARGE CATEGORIES

ALEXANDER I. EFIMOV

ABSTRACT

We give a short overview of a new notion of *continuous* K-theory, which is defined for a certain class of large (enhanced) triangulated categories. For compactly generated triangulated categories, this continuous K-theory gives the usual nonconnective K-theory of the category of compact objects. We formulate a general theorem about the computation of continuous K-theory for the category of sheaves (of modules) on a locally compact Hausdorff space. This result already gives a surprising answer for the category of sheaves on the real line.

MATHEMATICS SUBJECT CLASSIFICATION 2020

Primary 19D55; Secondary 18G80, 54B40, 14F08

KEYWORDS

Algebraic K-theory, sheaves on locally compact spaces, differential graded categories

INTERNATIONAL CONGRESS OF MATHEMATICIANS 2022 JULY 6–14

Proc. Int. Cong. Math. 2022, Vol. 3, pp. 2212–2226
DOI 10.4171/ICM2022/33

Published by EMS Press

1. INTRODUCTION

This paper is a very short introduction to a recent new notion of K-theory for a certain class of "large" (enhanced) triangulated categories. We call it "continuous K-theory." We explain the general idea and formulate some results, including the computation of K-theory for categories of sheaves on locally compact (Hausdorff) spaces. The detailed study with complete proofs will appear in [2].

It is well known that the usual Grothendieck group $K_0(\mathcal{A})$ of an additive category $\mathcal{A}$ vanishes when $\mathcal{A}$ has countable direct sums (Eilenberg swindle). More generally, for any enhanced triangulated category $\mathcal{C}$, its (nonconnective) K-theory spectrum $\mathbb{K}(\mathcal{C})$ is contractible. In particular, when $\mathcal{C} = D(R)$ is the unbounded derived category of modules over a ring, we get $\mathbb{K}_n(D(R)) = 0$ for $n \in \mathbb{Z}$.

This very observation has been used by M. Schlichting to define the negative K-theory. Namely, a short exact sequence of enhanced triangulated categories

$$0 \to \mathcal{A} \to \mathcal{B} \to \mathcal{C} \to 0$$

gives rise to the long exact sequence of K-groups. If $\mathcal{B}$ has countable direct sums, we have $\mathbb{K}_n(\mathcal{A}) \cong \mathbb{K}_{n+1}(\mathcal{C})$. In particular, ignoring the set-theoretic issues, we can take $\mathcal{B} = \mathrm{Ind}(\mathcal{A})$ and $\mathcal{C} = \mathrm{Calk}_{\mathcal{A}} := (\mathrm{Ind}(\mathcal{A})/\mathcal{A})^{\mathrm{Kar}}$, which is the Calkin category of $\mathcal{A}$ (an algebraic analogue of the usual Calkin algebra of a Hilbert space, namely bounded operators modulo compact operators).

One characterization of dualizable presentable categories is that they can be represented as a kernel of a localization, $\mathcal{C} = \ker(\mathrm{Ind}(\mathcal{A}) \xrightarrow{F} \mathrm{Ind}(\mathcal{B}))$, where the functor F commutes with direct sums and takes compact objects to compact objects. The idea is to define K-theory of such a category using the localization property for K-theory. Namely, $\mathbb{K}^{\mathrm{cont}}(\mathcal{C}) := \mathrm{Fiber}(\mathbb{K}(\mathcal{A}) \to \mathbb{K}(\mathcal{B}))$. However, such an approach would require checking the independence on the choice of the representation of $\mathcal{C}$ as such a kernel (and also the functoriality of $\mathbb{K}^{\mathrm{cont}}$ is not really immediate from such a definition).

An alternative characterization of dualizable categories is the following: these are presentable categories $\mathcal{C}$ such that the Yoneda embedding $Y_{\mathcal{C}} : \mathcal{C} \to \mathrm{Ind}(\mathcal{C})$ has a twice left adjoint. Using it, one can define the continuous Calkin category $\mathrm{Calk}_{\mathcal{C}}^{\mathrm{cont}}$, namely the "virtual quotient by compact objects." Then we simply put

$$\mathbb{K}^{\mathrm{cont}}(\mathcal{C}) := \Omega\mathbb{K}\big(\mathrm{Calk}_{\mathcal{C}}^{\mathrm{cont}}\big).$$

Continuous K-theory is functorial with respect to *strongly continuous* functors, that is, the functors whose right adjoint commutes with infinite direct sums.

We have the following computation of the continuous K-theory for categories of sheaves on locally compact Hausdorff spaces.

Theorem 1.1. *Let X be a locally compact Hausdorff space, and let $\underline{R}$ be a presheaf of DG rings on X. Then we have a natural isomorphism*

$$\mathbb{K}^{\mathrm{cont}}\big(\mathrm{Shv}(X, \mathrm{Mod}\text{-}\underline{R})\big) \cong \Gamma_c\big(X, \mathbb{K}(\underline{R})\big).$$

In particular, for any $n \in \mathbb{Z}_{\geq 0}$ and any DG ring A, we have

$$\mathbb{K}^{\mathrm{cont}}(\mathrm{Shv}(\mathbb{R}^n, \mathrm{Mod}\text{-}A)) \cong \Omega^n \mathbb{K}(A),$$

hence

$$\mathbb{K}_0^{\mathrm{cont}}(\mathrm{Shv}(\mathbb{R}^n, \mathrm{Mod}\text{-}A)) \cong \mathbb{K}_n(A).$$

Another interesting class of dualizable categories for which it would be very interesting to compute the continuous K-theory comes from the theory of condensed mathematics due to Clausen and Scholze [17,18]. These are the so-called categories of nuclear modules. We only briefly mention them in the case of affine formal schemes.

Although it would be natural to consider stable ∞-categories, in this paper we will restrict to DG categories to fix the ideas.

The paper is organized as follows. In Section 2 we recall the basic notions about DG categories. In Section 3 we recall presentable and dualizable DG categories, and give some examples.

Section 4 is devoted to the classical algebraic K-theory. In Section 5 we introduce the notion of continuous K-theory of dualizable categories.

Finally, in Section 6 we discuss in some detail Theorem 1.1, giving a sketch of its proof in the case of $X = \mathbb{R}$ and a constant presheaf of DG rings.

2. PRELIMINARIES ON DG CATEGORIES

We refer to [8,10] for a general introduction and overview of DG algebras, DG categories, and the derived categories of DG modules. We refer to [1] for the notion of an enhanced triangulated category. We refer to [20,21] for the model structures on the category of small DG categories.

For a small DG category $\mathcal{A}$ over a base ring k, we denote by $D(\mathcal{A})$ the derived category of right $\mathcal{A}$-modules. We will denote by Mod-$\mathcal{A}$ the "correct" DG category of $\mathcal{A}$-modules, for example, the DG category of semi-free $\mathcal{A}$-modules. Moreover, for a usual associative ring R (considered as a DG algebra over $\mathbb{Z}$ concentrated in degree zero), we write Mod-R for the (correct) category of DG R-modules, i.e., complexes of usual R-modules

In some cases we will also write Ind($\mathcal{A}$) (ind-objects) instead of Mod-$\mathcal{A}$, provided that $\mathcal{A}$ is pretriangulated. We will denote by $\mathrm{dgcat}_{\mathrm{k}}^{\mathrm{tr}}$ the ∞-category of pretriangulated Karoubi complete small DG categories.

We denote by Perf($\mathcal{A}$) $\subset$ Mod-$\mathcal{A}$ the full DG subcategory of perfect A-modules. We denote by PsPerf($\mathcal{A}$) $\subset$ Mod-$\mathcal{A}$ the full DG subcategory of *pseudoperfect* modules, i.e., A-modules which are perfect over k.

The tensor product of DG categories will always be derived over k. Given small DG categories $\mathcal{A}$ and $\mathcal{B}$, we denote by Fun($\mathcal{A}$, $\mathcal{B}$) the "correct" DG category of functors $\mathcal{A} \to \mathcal{B}$, i.e., the internal Hom in the symmetric monoidal homotopy category of DG categories (with inverted quasiequivalences).

Although in this paper we do not need the notions of smoothness and properness of DG algebras and DG categories, we recall them here for completeness.

Definition 2.1. A DG algebra A over k is called proper if A is perfect as a complex of k-modules.

Definition 2.2. A DG algebra A over k is called smooth if $A \in \mathrm{Perf}(A \otimes A^{op})$, i.e., the diagonal A-A-bimodule is perfect.

Note that for a proper (resp. smooth) DG algebra A, we have $\mathrm{Perf}(A) \subseteq \mathrm{PsPerf}(A)$ (resp. $\mathrm{PsPerf}(A) \subseteq \mathrm{Perf}(A)$). If X is a separated scheme of finite type over a field k, and $\mathrm{Perf}(X) \simeq \mathrm{Perf}(A)$, then X is smooth (resp. proper) if and only if such is A.

A morphism of DG algebras $f : A \to B$ is called a *homological epimorphism* if the map $B \overset{\mathbf{L}}{\otimes}_A B \to B$ is an isomorphism in $D(\mathrm{k})$. This is equivalent to f being a homotopy epimorphism, i.e., the map $B \sqcup^h_A B \to B$ is an isomorphism in $\mathrm{Ho}(\mathrm{dgalg}_{\mathrm{k}})$.

The notion of a homological epimorphism for a functor between small DG categories is defined similarly; it is also equivalent to the property of being a homotopy epimorphism (in the Morita model structure).

3. PRESENTABLE AND DUALIZABLE DG CATEGORIES

We fix some base commutative ring k. In what follows, we will mostly ignore set-theoretic issues. Recall that a DG category $\mathcal{C}$ is *presentable* if

1) $\mathcal{C}$ is pretriangulated;

2) the homotopy category $H^0(\mathcal{C})$ has small direct sums;

3) there exists a regular cardinal κ such that the triangulated category $H^0(\mathcal{C})$ is generated by κ-compact objects.

Recall that an object x of a triangulated category $\mathcal{T}$ is κ-compact if for any small family of objects $\{y_j\}_{j \in J}$, and for any morphism $x \xrightarrow{f} \bigoplus_{j \in J} y_j$, there exists a subset $I \subseteq J$ with $|I| < \kappa$ such that f factors through $\bigoplus_{i \in I} y_i$.

In other words, conditions 2) and 3) mean that the triangulated category $H^0(\mathcal{C})$ is well generated in the terminology of Neeman [14].

We will call a functor $F : \mathcal{C} \to \mathcal{D}$ between such DG categories *continuous* if it commutes with small direct sums. We denote by $\mathrm{dgcat}_{\mathrm{k}}^{\mathrm{cont}}$ the ∞-category of presentable DG categories and continuous functors.

There is a natural symmetric monoidal structure on $\mathrm{dgcat}_{\mathrm{k}}^{\mathrm{cont}}$—the Lurie tensor product. It is uniquely determined by the internal Hom, given by $\mathrm{Fun}^{\mathrm{cont}}(\mathcal{C}, \mathcal{D})$—the full subcategory of $\mathrm{Fun}(\mathcal{C}, \mathcal{D})$ formed by continuous functors. Denoting this tensor product by $-\widehat{\otimes}-$, we thus have for any $\mathcal{C}_1, \mathcal{C}_2, \mathcal{D} \in \mathrm{dgcat}_{\mathrm{k}}^{\mathrm{cont}}$ a full embedding

$$\mathrm{Fun}^{\mathrm{cont}}(\mathcal{C}_1 \widehat{\otimes}_{\mathrm{k}} \mathcal{C}_2, \mathcal{D}) \hookrightarrow \mathrm{Fun}(\mathcal{C}_1 \otimes_{\mathrm{k}} \mathcal{C}_2, \mathcal{D}),$$

and the essential image consists of bicontinuous bifunctors.

Definition 3.1. A presentable DG category $\mathcal{C}$ is dualizable if it is a dualizable object in the symmetric monoidal category $(\mathrm{dgcat}_{\mathrm{k}}^{\mathrm{cont}}, \widehat{\otimes}_{\mathrm{k}})$.

Remark 3.2. In fact, dualizability of a presentable DG category is independent of the base ring k.

The following is due to Lurie.

Theorem 3.3 ([12]). *Let $\mathcal{C}$ be a presentable DG category. The following are equivalent:*

(i) *$\mathcal{C}$ is dualizable.*

(ii) *There is a short exact sequence*

$$0 \to \mathcal{C} \to \text{Mod-}A \xrightarrow{-\otimes_A B} \text{Mod-}B \to 0,$$

where $A \to B$ is a homological epimorphism of DG algebras, i.e., $B \overset{\mathbf{L}}{\otimes}_A B \xrightarrow{\sim} B$.

(iii) *Same as* (ii) *with small DG categories instead of DG algebras.*

(iv) *$\mathcal{C}$ is a retract in* $\mathrm{dgcat}_{\mathrm{k}}^{\mathrm{cont}}$ *of a compactly generated category.*

(v) *The Yoneda embedding $Y_{\mathcal{C}} : \mathcal{C} \to \mathrm{Ind}(\mathcal{C})$ has a twice left adjoint.*

(vi) *Any continuous localization of presentable DG categories $\mathcal{D} \to \mathcal{C}$ has a continuous section (not necessarily fully faithful).*

We call a continuous functor $F : \mathcal{C} \to \mathcal{D}$ *strongly continuous* if its right adjoint is continuous. We denote by $\mathrm{dgcat}_{\mathrm{k}}^{\mathrm{dual}} \subset \mathrm{dgcat}_{\mathrm{k}}^{\mathrm{cont}}$ the (nonfull) subcategory formed by dualizable categories and strongly continuous functors. We have a fully faithful embedding

$$\mathrm{dgcat}_{\mathrm{k}}^{\mathrm{tr}} \hookrightarrow \mathrm{dgcat}_{\mathrm{k}}^{\mathrm{dual}}, \quad \mathcal{A} \mapsto \text{Mod-}\mathcal{A}.$$

Moreover, we have

$$(\text{Mod-}\mathcal{A})\widehat{\otimes}_{\mathrm{k}}(\text{Mod-}\mathcal{B}) \simeq \text{Mod-}(\mathcal{A} \otimes_{\mathrm{k}} \mathcal{B}).$$

For a homological epimorphism $A \to B$, we have

$$\ker(\text{Mod-}A \to \text{Mod-}B)^{\vee} \simeq \ker(A\text{-Mod} \to B\text{-Mod}).$$

Example: derived categories of almost modules. The first class of examples is given by (a not necessarily commutative version of) the basic setup for almost mathematics [3, 4]. Let R be an associative ring, and $I \subset R$ an ideal such that $I^2 = I$ and I is flat as a left or right R-module. Then $I \overset{\mathbf{L}}{\otimes}_R I \cong I$, hence $R \to R/I$ is a homological epimorphism. We get a dualizable category

$$\mathcal{C}(R, I) := \ker(\text{Mod-}R \to \text{Mod-}R/I).$$

If $1 + I \subset R^{\times}$, then $\mathcal{C}(R, I)$ has no nonzero compact objects. This observation is originally due to Keller [9].

Example: sheaves on exponentiable spaces. Recall that a topological space X is exponentiable if the functor $X \times - : \mathrm{Top} \to \mathrm{Top}$ commutes with colimits. This is equivalent to X being *core-compact*: for any point $x \in X$ and any open neighborhood U of x, there exists an open neighborhood V of x such that $V \ll U$. Here $V \ll U$ means that V is relatively compact in U, i.e., any open cover of U admits a finite subcover of V. In particular, any locally compact Hausdorff space is exponentiable, and such is any spectral space.

Let R be a DG ring and denote by $\mathrm{Shv}(X, \text{Mod-}R)$ the DG category of sheaves of R-modules. Then the category $\mathrm{Shv}(X, \text{Mod-}R)$ is dualizable [12]. Similarly, one can replace a single DG ring R with a presheaf of DG rings $\underline{\mathcal{R}}$.

Remark 3.4. We consider sheaves of R-modules which are not necessarily hypercomplete. In particular, even if X is sober (that is, any irreducible closed subset has a unique generic point), there might be some sheaves $\mathcal{F}$ which have zero stalks at all points of X. For example, this happens when $R = \mathbb{Z}$, $X = [0,1]^{\mathbb{N}}$ is the Hilbert cube, and $\mathcal{F}$ is the sheaf such that $\mathcal{F}(U)$ is the (complex computing the) Borel–Moore homology $H_{\bullet}^{\mathrm{BM}}(U, \mathbb{Z})$. See [11] for details.

Below we will formulate a result on the computation of the continuous K-theory of such categories when X is locally compact Hausdorff.

Example: sheaves with condition on the singular support. Let X be a (paracompact Hausdorff) C^1-manifold, and R a DG ring. Recall that the singular support $SS(\mathcal{F}) \subset T^*X$ of a sheaf $\mathcal{F} \in \mathrm{Shv}(X, \text{Mod-}R)$ is defined as follows (see [6] for details). For a point $(x_0, \xi_0) \in T^*X$, we have $(x_0, \xi_0) \notin SS(\mathcal{F})$ if and only if there exists an open neighborhood $(x_0, \xi_0) \in U \subset T^*X$ such that, for any point $(x_1, \xi_1) \in U$ and for any C^1-function $f : V \to \mathbb{R}$, $x_1 \in V \subset \pi(U)$ such that $df(x_1) = \xi_1$, we have $\Gamma_{\{f(x) \geq f(x_1)\}}(\mathcal{F})_{x_1} = 0$. Hence, $SS(\mathcal{F})$ is a conical closed subset.

Now, let $\Lambda \subset T^*X$ be any conical closed subset. Denote by $\mathrm{Shv}_{\Lambda}(X, \text{Mod-}R) \subset \mathrm{Shv}(X, \text{Mod-}R)$ the full subcategory of sheaves R such that $SS(\mathcal{F}) \subset \Lambda$. Then the inclusion functor $\mathrm{Shv}_{\Lambda}(X, \text{Mod-}R) \hookrightarrow \mathrm{Shv}(X, \text{Mod-}R)$ has a left adjoint. In particular, the category $\mathrm{Shv}_{\Lambda}(X, \text{Mod-}R)$ is dualizable.

Example: nuclear modules. The following class of examples comes from the theory of condensed mathematics due to Clausen and Scholze [17, 18]. These are the categories of nuclear modules on sufficiently nice analytic spaces.

We mention here the following characterization of the category of nuclear modules on an affine formal scheme (for simplicity assumed to be noetherian).

Theorem 3.5. *Let R be a commutative noetherian ring, and $I \subset R$ an ideal. Then the following hold:*

1) *The inverse limit $\varprojlim_n (\text{Mod-}R/I^n)$ exists in $\mathrm{dgcat}_{\mathbb{Z}}^{\mathrm{dual}}$ (and is totally different from the usual inverse limit).*

2) *The category of nuclear modules on $\mathrm{Spf}(R_{\hat{I}})$ defined in [17] embeds fully faithfully (and strongly continuously) into the above inverse limit.*

Nonexample. One of the characterizations of dualizable categories (condition (ii) of Theorem 3.3) is via taking kernels of extension of scalars for a homological epimorphism of DG algebras $A \to B$. However, for a general morphism $A \to B$ the kernel of the functor $- \otimes_A B$ is some "random" presentable DG category which does not have to be dualizable. For example, take $A = \mathrm{k}[x, y]$, $B = \mathrm{k}[t, t^{-1}]$ ($\deg(x) = \deg(y) = \deg(t) = 0$), and consider the morphism

$$f : A \to B, \quad f(x) = t, \; f(y) = 0.$$

Then one can show that the kernel $\ker(\text{Mod-}A \xrightarrow{-\otimes_A B} \text{Mod-}B)$ is not dualizable (a pleasant exercise).

4. K-THEORY OF RINGS, ABELIAN CATEGORIES, AND TRIANGULATED CATEGORIES

In this section we recall the notion of a Grothendieck group in various contexts, and very briefly recall the higher K-theory, as well as negative K-theory.

Recall that for an associative ring R, the Grothendieck group $K_0(R)$ is generated by the isomorphism classes $[P]$ of finitely generated (right) projective R-modules, subject to relation $[P \oplus Q] = [P] + [Q]$.

For an abelian category $\mathcal{A}$, the group $K_0(\mathcal{A})$ is defined similarly, but now we have a relation $[Y] = [X] \oplus [Z]$ for each short exact sequence

$$0 \to X \to Y \to Z \to 0.$$

For a triangulated category $\mathcal{T}$, we obtain the group $K_0(\mathcal{T})$ by replacing short exact sequences with exact triangles

$$0 \to X \to Y \to Z \to X[1].$$

For an abelian category $\mathcal{A}$, we have a natural isomorphism

$$K_0\big(D^b(\mathcal{A})\big) \xrightarrow{\sim} K_0(\mathcal{A}), \quad [X] \mapsto \sum_i (-1)^i \big[H^i(X)\big].$$

For a ring R, we have an isomorphism

$$K_0\big(\mathrm{Perf}(R)\big) \xrightarrow{\sim} K_0(R), \quad [P^\bullet] \mapsto \sum_i [P^i],$$

for a bounded complex $P^\bullet$ of finitely generated projective R-modules.

In particular, if R is right noetherian and has finite homological dimension, then

$$K_0(R) \cong K_0\big(\mathrm{Perf}(R)\big) \cong K_0(\mathrm{Mod}_{f.g.}\text{-}R).$$

For a scheme X (quasicompact, quasiseparated), one defines

$$K_0(X) := K_0(D_{\mathrm{perf}}(X)).$$

If X is noetherian, we have $G_0(X) = K_0(\mathrm{Coh}(X)) = K_0(D^b_{\mathrm{coh}}(X))$.

It is easy to check that for a noetherian scheme X and for a closed subset Z, we have an exact sequence

$$G_0(Z) \cong K_0(\mathrm{Coh}_Z(X)) \to G_0(X) \to G_0(X - Z) \to 0.$$

More generally, for an abelian category $\mathcal{A}$ and a Serre subcategory $\mathcal{B} \subset \mathcal{A}$ (that is, a full subcategory closed under subobjects, quotient objects, and extensions), we have an exact sequence

$$K_0(\mathcal{B}) \to K_0(\mathcal{A}) \to K_0(\mathcal{A}/\mathcal{B}) \to 0.$$

Similarly, for a triangulated category $\mathcal{T}$ and a full idempotent complete triangulated subcategory $\mathcal{S}$, we have an exact sequence

$$K_0(\mathcal{S}) \to K_0(\mathcal{T}) \to K_0(\mathcal{T}/\mathcal{S}) \to 0.$$

It is well known that the above definition of K_0 can be extended to higher K-groups $K_n(-)$: for associative rings (via Quillen's plus construction **[15]**), for abelian categories (via Quillen's Q-construction **[15]**) and for *enhanced* triangulated categories (via Waldhausen's S-construction **[23]**). These are again compatible with each other: $K_n(R) \cong K_n(\mathrm{Perf}(R))$, $K_n(\mathcal{A}) \cong K_n(D^b(\mathcal{A}))$.

One of the most important properties of higher algebraic K-theory is the localization sequence. Namely, for an enhanced triangulated category $\mathcal{T}$ and a full idempotent complete triangulated subcategory $\mathcal{S} \subset \mathcal{T}$, we have a long exact sequence

$$\cdots \to K_1(\mathcal{S}) \to K_1(\mathcal{T}) \to K_1(\mathcal{T}/\mathcal{S}) \to \\ \to K_0(\mathcal{S}) \to K_0(\mathcal{T}) \to K_0(\mathcal{T}/\mathcal{S}) \to 0.$$

Now, an important observation is that if a triangulated category $\mathcal{T}$ has at least countable direct sums, then $K_0(\mathcal{T}) = 0$. Indeed, for any object $X \in \mathcal{T}$, we have

$$[X] + [X^{(\mathbb{N})}] = [X \oplus X^{(\mathbb{N})}] = [X^{(\mathbb{N})}], \quad \text{hence } [X] = 0$$

(Eilenberg swindle).

Essentially the same argument (applied to the identity functor) shows that, for an enhanced triangulated category $\mathcal{T}$ with countable direct sums, we have $K_n(\mathcal{T}) = 0$, $n \geq 0$. This actually allows defining negative K-theory, which was studied by Schlichting **[16]**.

Namely, let $\mathcal{T} = \mathrm{Perf}(\mathcal{B})$ for a DG category $\mathcal{B}$. Let us define the Calkin category $\mathrm{Calk}_{\mathcal{B}}$ as the (Karoubi completion of) the quotient $(\text{Mod-}\mathcal{B}/\mathrm{Perf}(\mathcal{B}))^{\mathrm{Kar}}$. Then we have a short exact sequence

$$0 \to \mathrm{Perf}(\mathcal{B}) \to \text{Mod-}\mathcal{B} \to \mathrm{Calk}_{\mathcal{B}} \to 0,$$

which forces the definition $K_{-1}(\mathrm{Perf}(\mathcal{B})) := K_0(\mathrm{Calk}_{\mathcal{B}})$ (ignoring set theory for simplicity). Iterating, one gets the groups $K_n(\mathrm{Perf}(\mathcal{B}))$ for all negative integers n.

Let us remark that the terminology "Calkin category" comes from functional analysis. Namely, for a field k and a vector space V, we have $\mathrm{End}_{\mathrm{Calk}_{\mathrm{k}}}(V) = \mathrm{End}_{\mathrm{k}}(V)/V^* \otimes V$.

This is an algebraic version of the usual Calkin algebra of a Hilbert space $\mathcal{H}$, defined as $B(\mathcal{H})/C(\mathcal{H})$. Here $C(\mathcal{H})$ is the ideal of compact operators, and we have $C(\mathcal{H}) = \overline{\mathcal{H}^* \otimes \mathcal{H}}$.

Summarizing, for an enhanced triangulated category $\mathcal{T}$, one has K-groups $K_n(\mathcal{T})$, which are stable homotopy groups of the nonconnective K-theory spectrum $\mathbb{K}(\mathcal{T})$. For a short exact sequence of enhanced triangulated categories

$$0 \to \mathcal{T}_1 \to \mathcal{T}_2 \to \mathcal{T}_3 \to 0,$$

we have an exact triangle of spectra

$$\mathbb{K}(\mathcal{T}_1) \to \mathbb{K}(\mathcal{T}_2) \to \mathbb{K}(\mathcal{T}_3) \to \mathbb{K}(\mathcal{T}_1)[1].$$

5. CONTINUOUS K-THEORY

Condition (v) of Theorem 3.3 gives a canonical short exact sequence (strictly speaking, we have to choose a regular cardinal to get presentable compactly generated categories)

$$0 \to \mathcal{C} \xrightarrow{Y_{\mathcal{C}}^{LL}} \text{Mod-}\mathcal{C} \to \text{Mod-Calk}^{\text{cont}}(\mathcal{C}) \to 0.$$

Here $\text{Calk}^{\text{cont}}$ is the Karoubi closure of the image of $\text{Cone}(Y_{\mathcal{C}}^{LL} \to Y_{\mathcal{C}}) : \mathcal{C} \to \text{Mod-}\mathcal{C}$. This allows defining continuous K-theory for dualizable categories (and a continuous version of any localizing invariant):

$$\mathbb{K}^{\text{cont}} : \text{dgcat}_{\text{k}}^{\text{dual}} \to \text{Sp}, \quad \mathbb{K}^{\text{cont}}(\mathcal{C}) := \Omega\mathbb{K}\big(\text{Calk}^{\text{cont}}(\mathcal{C})\big).$$

Moreover, this is the only way to extend K-theory to dualizable categories such that

- For a small DG category $\mathcal{A}$, we have $\mathbb{K}(\text{Mod-}\mathcal{A}) \cong \mathbb{K}(\mathcal{A})$.
- $\mathbb{K}^{\text{cont}}$ is a localizing invariant.

In particular, if a dualizable category is represented as a kernel of extension of scalars for a homological epimorphism $A \to B$, then we have

$$\mathbb{K}^{\text{cont}}\big(\ker(\text{Mod-}A \to \text{Mod-}B)\big) \cong \text{Fiber}\big(\mathbb{K}(A) \to \mathbb{K}(B)\big).$$

The independence of this fiber of the choice of a homological epimorphism $A \to B$ is closely related with (and, in fact, reproves) the excision theorem of Tamme [22].

It is not hard to deduce the general properties of continuous K-theory from the corresponding properties of the usual nonconnective K-theory. In particular, continuous K-theory commutes with filtered colimits in $\text{dgcat}_{\mathbb{Z}}^{\text{dual}}$. Furthermore, one can deduce from the results of [7] that continuous K-theory commutes with the products in $\text{dgcat}_{\mathbb{Z}}^{\text{dual}}$ (which are quite different from the products in $\text{dgcat}_{\mathbb{Z}}^{\text{cont}}$).

6. CONTINUOUS K-THEORY OF CATEGORIES OF SHEAVES

This section is devoted to Theorem 1.1 from the introduction. We first recall its formulation. Let X be a locally compact Hausdorff space. Let $\underline{\mathcal{R}}$ be a presheaf of DG rings on X.

Theorem 6.1. *We have a natural isomorphism* $\mathbb{K}^{\mathrm{cont}}(\mathrm{Shv}(X, \mathrm{Mod}\text{-}\underline{R})) \cong \Gamma_c(X, \mathbb{K}(\underline{R}))$. *In particular, for any* $n \in \mathbb{Z}_{\geq 0}$ *and any DG ring A, we have*

$$\mathbb{K}^{\mathrm{cont}}\big(\mathrm{Shv}(\mathbb{R}^n, \mathrm{Mod}\text{-}A)\big) \cong \Omega^n \mathbb{K}(A),$$

hence

$$\mathbb{K}_0^{\mathrm{cont}}\big(\mathrm{Shv}(\mathbb{R}^n, \mathrm{Mod}\text{-}A)\big) \cong \mathbb{K}_n(A).$$

6.1. The case of the real line

Below we sketch the proof of Theorem 6.1 in the special case when $X = \mathbb{R}$ and $\underline{R}$ is the constant presheaf of DG rings with value A. We consider the subcategories $\mathrm{Shv}_{\mathbb{R}\times\mathbb{R}_{\geq 0}}(\mathbb{R}, \mathrm{Mod}\text{-}A)$, $\mathrm{Shv}_{\mathbb{R}\times\mathbb{R}_{\leq 0}}(\mathbb{R}, \mathrm{Mod}\text{-}A)$, and $\mathrm{Shv}_{\mathbb{R}\times\{0\}}(\mathbb{R}, \mathrm{Mod}\text{-}A) \simeq \mathrm{Mod}\text{-}A$. Here, of course, we identify $T^*\mathbb{R}$ with $\mathbb{R} \times \mathbb{R}$. The following assertion is standard.

Proposition 6.2. *For a sheaf* $\mathcal{F} \in \mathrm{Shv}(\mathbb{R}, \mathrm{Mod}\text{-}A)$, *the following are equivalent:*

(i) $SS(\mathcal{F}) \subset \mathbb{R} \times \mathbb{R}_{\geq 0}$;

(ii) *for any real numbers* $a < b$, *the natural map* $\Gamma((-\infty, b), \mathcal{F}) \to \Gamma((a, b), \mathcal{F})$ *is an isomorphism.*

It is not hard to check the following "gluing" statement.

Proposition 6.3. *We have a homotopy Cartesian square of DG categories*

$$\begin{array}{ccc} \mathrm{Shv}(\mathbb{R}, \mathrm{Mod}\text{-}A) & \longrightarrow & \mathrm{Shv}_{\mathbb{R}\times\mathbb{R}_{\geq 0}}(\mathbb{R}, \mathrm{Mod}\text{-}A) \\ \downarrow & & \downarrow \\ \mathrm{Shv}_{\mathbb{R}\times\mathbb{R}_{\leq 0}}(\mathbb{R}, \mathrm{Mod}\text{-}A) & \longrightarrow & \mathrm{Shv}_{\mathbb{R}\times\{0\}}(\mathbb{R}, \mathrm{Mod}\text{-}A). \end{array} \tag{6.1}$$

Here each of the functors is left adjoint to the inclusion.

Since all the functors in the diagram (6.1) are localizations, we obtain the homotopy Cartesian square of (continuous) K-theory spectra

$$\begin{array}{ccc} \mathbb{K}^{\mathrm{cont}}(\mathrm{Shv}(\mathbb{R}, \mathrm{Mod}\text{-}A)) & \longrightarrow & \mathbb{K}^{\mathrm{cont}}(\mathrm{Shv}_{\mathbb{R}\times\mathbb{R}_{\geq 0}}(\mathbb{R}, \mathrm{Mod}\text{-}A)) \\ \downarrow & & \downarrow \\ \mathbb{K}^{\mathrm{cont}}(\mathrm{Shv}_{\mathbb{R}\times\mathbb{R}_{\leq 0}}(\mathbb{R}, \mathrm{Mod}\text{-}A)) & \longrightarrow & \mathbb{K}^{\mathrm{cont}}(\mathrm{Shv}_{\mathbb{R}\times\{0\}}(\mathbb{R}, \mathrm{Mod}\text{-}A)) = \mathbb{K}(A). \end{array} \tag{6.2}$$

The special case of Theorem 6.1 states that $\mathbb{K}^{\mathrm{cont}}(\mathrm{Shv}(\mathbb{R}, \mathrm{Mod}\text{-}A)) \cong \Omega\mathbb{K}(A)$. Using the commutative diagram (6.2), we observe that this assertion reduces to the vanishing of $\mathbb{K}^{\mathrm{cont}}(\mathrm{Shv}_{\mathbb{R}\times\mathbb{R}_{\geq 0}}(\mathbb{R}, \mathrm{Mod}\text{-}A))$ (note that, by symmetry, this implies the vanishing of $\mathbb{K}^{\mathrm{cont}}(\mathrm{Shv}_{\mathbb{R}\times\mathbb{R}_{\leq 0}}(\mathbb{R}, \mathrm{Mod}\text{-}A))$).

Sketch of the proof of the vanishing of $\mathbb{K}^{\mathrm{cont}}(\mathrm{Shv}_{\mathbb{R}\times\mathbb{R}_{\geq 0}}(\mathbb{R}, \text{Mod-}A))$. We define two auxiliary DG categories $\mathcal{B}$ and $\mathcal{C}$ as follows. We have $Ob(\mathcal{B}) = Ob(\mathcal{C}) = \mathbb{R}$, and the morphisms are given by

$$\mathcal{B}(a,b) = \begin{cases} A, & \text{for } a \leq b, \\ 0, & \text{otherwise,} \end{cases} \qquad \mathcal{C}(a,b) = \begin{cases} A, & \text{for } a = b, \\ 0, & \text{otherwise.} \end{cases}$$

The compositions are induced by the multiplication in A. Denote by $F : \mathcal{B} \to \mathcal{C}$ the obvious functor given by $F(a) = a$. We claim that this functor is a homological epimorphism. This is straightforward to check. The only property of $\mathbb{R}$ we need here is that $\mathbb{R}$ is a dense linearly ordered set, i.e., for any $a < b$ there exists c such that $a < c < b$.

The above description of sheaves with singular support in $\mathbb{R} \times \mathbb{R}_{\geq 0}$ (Proposition 6.2) implies a short exact sequence in $\mathrm{dgcat}_{\mathbb{Z}}^{\mathrm{dual}}$, namely

$$0 \to \mathrm{Shv}_{\mathbb{R}\times\mathbb{R}_{\geq 0}}(\mathbb{R}, \text{Mod-}A) \to \text{Mod-}\mathcal{B} \xrightarrow{F^*} \text{Mod-}\mathcal{C} \to 0.$$

Indeed, condition (ii) of Proposition 6.2 means that a sheaf $\mathcal{F} \in \mathrm{Shv}_{\mathbb{R}\times\mathbb{R}_{\geq 0}}(\mathbb{R}, \text{Mod-}A)$ is determined by its sections on the negative rays $(-\infty, a)$, together with compatible restriction maps $\Gamma((-\infty, b), \mathcal{F}) \to \Gamma((-\infty, a), \mathcal{F})$ subject to the following condition: for any $a \in \mathbb{R}$, the natural map

$$\Gamma\big((-\infty, a), \mathcal{F}\big) \to \varprojlim_{b<a} \Gamma\big((-\infty, b), \mathcal{F}\big)$$

is an isomorphism.

This exactly identifies $\mathrm{Shv}_{\mathbb{R}\times\mathbb{R}_{\geq 0}}(\mathbb{R}, \text{Mod-}A)$ with the right orthogonal complement $F_*(\text{Mod-}\mathcal{C})^{\perp} \subset \text{Mod-}\mathcal{B}$. But the full subcategory $F_*(\text{Mod-}\mathcal{C}) \subset \text{Mod-}\mathcal{B}$ is both left and right admissible, hence its right orthogonal complement is equivalent to the left orthogonal complement, which exactly equals the kernel $\ker(\text{Mod-}\mathcal{B} \to \text{Mod-}\mathcal{C})$.

Finally, we obtain an exact triangle of spectra

$$\mathbb{K}^{\mathrm{cont}}\big(\mathrm{Shv}_{\mathbb{R}\times\mathbb{R}_{\geq 0}}(\mathbb{R}, \text{Mod-}A)\big) \to \mathbb{K}(\mathcal{B}) \to \mathbb{K}(\mathcal{C}),$$

so we are reduced to showing that the map $\mathbb{K}(\mathcal{B}) \to \mathbb{K}(\mathcal{C})$ is an isomorphism. This is immediate from the two standard properties of K-theory:

1) it is additive with respect to finite semiorthogonal decompositions;

2) it commutes with filtered colimits.

This finishes the (sketch of the) proof. ■

6.2. Reduction to the hypercomplete case

We recall the definition of a hypercomplete space in a slightly different form to avoid going into details.

Definition 6.4. A sober topological space X is hypercomplete if for any nonzero (homotopy) sheaf of spectra $\mathcal{F}$ on X there is a point $x \in X$ such that the stalk $\mathcal{F}_x$ is nonzero.

A paracompact space of finite covering dimension is hypercomplete. One can prove Theorem 6.1 in the hypercomplete case indirectly by "sheafifying" the assertion and reducing to stalks.

The general case can be reduced to the hypercomplete case using Urysohn's lemma. Namely, each compact Hausdorff space can be embedded into a product of an infinite number of copies of the closed unit interval $[0, 1]$ (for example, one can take the product over all continuous functions $X \to [0, 1]$). This allows representing X as a cofiltered limit of spaces X_i which are closed subsets of finite-dimensional cubes, hence are hypercomplete. This allows reducing Theorem 6.1 to the hypercomplete case.

6.3. Continuous presheaves and continuous partially ordered sets

We now sketch another more conceptual approach to computing the K-theory of the category of sheaves, and also recall the relevant notion of a continuous poset and a continuous category.

The notion of a continuous (abstract, discrete) category $\mathcal{C}$ is due to Johnstone and Joyal **[5]**. This means that the Yoneda embedding functor from $\mathcal{C}$ the the category $\operatorname{Ind}(\mathcal{C})$ of ind-objects has a twice left adjoint (as in condition (v) of Theorem 3.3). Note that having a single left adjoint means exactly that the category $\mathcal{C}$ has small filtered colimits. It is proved in **[5]** that a category $\mathcal{C}$ with small filtered colimits is continuous if and only if there is some category $\mathcal{D}$ and a pair of functors $F : \mathcal{C} \to \operatorname{Ind}(\mathcal{D})$, $G : \operatorname{Ind}(\mathcal{D}) \to \mathcal{C}$ such that $G \circ F \cong \mathrm{id}_{\mathcal{C}}$, and both F and G commute with filtered colimits (as in condition (iv) of Theorem 3.3).

If one considers a partially ordered set P as a category, then we recover the notion of a continuous poset **[13]**, which generalizes the notion of a continuous lattice due to Scott **[19]**. Namely, recall that in a partially ordered set an element x is *way below* y, written as $x \ll y$, if for any directed family of elements $\{z_i \in P\}_{i \in I}$ such that the supremum (join) $\sup\{z_i, i \in I\}$ exists and $\sup\{z_i, i \in I\} \geq y$, there exists some $i \in I$ such that $z_i \geq x$.

Definition 6.5. A partially ordered set P is continuous if the following conditions hold:

1) any directed subset of elements of P has a supremum;
2) for any element $x \in P$, the set $\{y \in P : y \ll x\}$ is directed and $\sup\{y \in P : y \ll x\} = x$.

It is proved in **[5]** that a poset P is continuous if and only if the associated category is continuous. The collection of open subsets of a locally compact Hausdorff space X is a continuous poset. Here we have $U \ll V$ if and only if $\bar{U} \subset V$ and $\bar{U}$ is compact. More generally, a topological space is exponentiable (= core-compact) if and only if the collection of its open subsets is a continuous poset.

Let now $\underline{\mathcal{R}}$ be a presheaf of DG rings on a continuous poset P. Let us call a presheaf $\mathcal{F}$ of $\underline{\mathcal{R}}$-modules a sheaf if for any $x \in P$ we have an isomorphism

$$\mathcal{F}(x) \xrightarrow{\sim} \varprojlim_{y \ll x} \mathcal{F}(y).$$

Denote by $\mathrm{Shv}(P, \mathrm{Mod}\text{-}\underline{\mathcal{R}})$ the DG category of sheaves (they are actually the sheaves with respect to a suitable Grothendieck topology). We have the following result.

Proposition 6.6.

1) *The category* $\mathrm{Shv}(P, \mathrm{Mod}\text{-}\underline{\mathcal{R}})$ *is dualizable.*

2) *We have a natural isomorphism*

$$\mathbb{K}^{\mathrm{cont}}(\mathrm{Shv}(P, \mathrm{Mod}\text{-}\underline{\mathcal{R}})) \cong \bigoplus_{x \ll x} \mathbb{K}(\mathcal{R}(x)).$$

Recall that an element $x \in P$ such that $x \ll x$ is called *compact*, and this is the special case of the usual notion of a compact object of a category with small filtered colimits.

Now, let us return to a locally compact Hausdorff space X with a presheaf of DG rings $\underline{\mathcal{R}}$. For a presheaf $\mathcal{F}$ of $\underline{\mathcal{R}}$-modules, the sheafiness condition can be separated into the following two conditions:

(i) for any open $U \subset X$, we have an isomorphism

$$\mathcal{F}(U) \xrightarrow{\sim} \varprojlim_{V \ll U} \mathcal{F}(V),$$

(ii) for any pair of open subsets $U_1, U_2 \subset X$, we have a Cartesian square

$$\begin{array}{ccc} \mathcal{F}(U_1 \cup U_2) & \longrightarrow & \mathcal{F}(U_1) \\ \downarrow & & \downarrow \\ \mathcal{F}(U_2) & \longrightarrow & \mathcal{F}(U_1 \cap U_2). \end{array}$$

Let us call a presheaf $\mathcal{F}$ of $\underline{\mathcal{R}}$-modules *continuous* if it satisfies condition (i) (but not necessarily condition (ii)). We denote by $\mathrm{PSh}^{\mathrm{cont}}(X, \mathrm{Mod}\text{-}\underline{\mathcal{R}})$ the DG category of continuous presheaves. We get the following corollary of Proposition 6.6.

Corollary 6.7.

1) *The category* $\mathrm{PSh}^{\mathrm{cont}}(X, \mathrm{Mod}\text{-}\underline{\mathcal{R}})$ *is dualizable.*

2) *We have*

$$\mathbb{K}^{\mathrm{cont}}(\mathrm{PSh}^{\mathrm{cont}}(X, \mathrm{Mod}\text{-}\underline{\mathcal{R}})) \cong \bigoplus_{U \ll U} \mathbb{K}(\mathcal{R}(U)).$$

Here the summation is taken over open-compact subsets $U \subset X$.

This computation leads to another way of computing the continuous K-theory of sheaves of $\underline{\mathcal{R}}$-modules. Namely, assuming X is compact for simplicity, one can "approximate" the category $\mathrm{Shv}(X, \mathrm{Mod}\text{-}\underline{\mathcal{R}})$ by certain finite limits of categories of continuous presheaves on various closed subsets of X, and eventually reduce the statement of Theorem 6.1 to Corollary 6.7.

ACKNOWLEDGMENTS

I am grateful to Alexander Beilinson, Dennis Gaitsgory, Marc Hoyois, Dmitry Kaledin, Maxim Kontsevich, Dmitri Orlov, Vivek Shende, and Bertrand Toën for useful discussions.

A. I. EFIMOV

FUNDING

The author was partially supported by the HSE University Basic Research Program.

REFERENCES

[1] A. I. Bondal and M. M. Kapranov, Framed triangulated categories. *Mat. Sb.* **181** (1990), no. 5, 669–683 (Russian), translation in Math. USSR-Sb. **70** (1991), no. 1, 93–107.

[2] A. Efimov, K-theory and localizing invariants of large categories, in preparation.

[3] G. Faltings, Almost étale extensions. In *Cohomologies p-adiques et applications arithmétiques, II*, edited by P. Berthelot, J. M. Fontaine, L. Illusie, K. Kato, and M. Rapoport, pp. 185–270, Astérisque 279, 2002.

[4] O. Gabber and L. Ramero, *Almost ring theory*. Lecture Notes in Math. 1800, Springer, Berlin, 2003.

[5] P. Johnstone and A. Joyal, Continuous categories and exponentiable toposes. *J. Pure Appl. Algebra* **25** (1982), no. 3, 255–296.

[6] M. Kashiwara and P. Schapira, *Sheaves on manifolds*. Grundlehren Math. Wiss. 292, Springer, 1994.

[7] D. Kasprowski and C. Winges, Algebraic K-theory of stable ∞-categories via binary complexes. *J. Topol.* **12** (2019), no. 2, 442–462.

[8] B. Keller, Deriving DG categories. *Ann. Sci. Éc. Norm. Supér. (4)* **27** (1994), no. 1, 63–102.

[9] B. Keller, A remark on the generalized smashing conjecture. *Manuscripta Math.* **84** (1994), no. 2, 193–198.

[10] B. Keller, On differential graded categories. In *International Congress of Mathematicians*, edited by M. Sanz-Solé, J. Soria, J. L. Varona, and J. Verdera, pp. 151–190, II, Eur. Math. Soc., Zürich, 2006.

[11] J. Lurie, *Higher topos theory*. Ann. of Math. Stud. 170, Princeton University Press, Princeton, NJ, 2009.

[12] J. Lurie, Spectral algebraic geometry. 2018. https://www.math.ias.edu/~lurie/papers/SAG-rootfile.pdf

[13] G. Markowsky, A motivation and generalization of Scott's notion of a continuous lattice. In *Continuous lattices*, edited by B. Banaschewski and R.-E. Hoffmann, pp. 298–307, Lecture Notes in Math. 871, Springer, Berlin–New York, 1981.

[14] A. Neeman, *Triangulated categories*. Ann. of Math. Stud. 148, Princeton University Press, Princeton, NJ, 2001.

[15] D. Quillen, Higher algebraic K-theory: I. In *Higher K-theories (Proc. Seattle 1972)*, edited by H. Bass, pp. 85–147, Lecture Notes in Math. 341, Springer, 1973.

[16] M. Schlichting, Negative K-theory of derived categories. *Math. Z.* **253** (2006), no. 1, 97–134.

[17] P. Scholze, Lectures on analytic geometry. Lecture notes for a course taught at. the University of Bonn. Winter term 2019. https://www.math.uni-bonn.de/people/scholze/Analytic.pdf

[18] P. Scholze, Lectures on condensed mathematics. Lecture notes for a course taught at the University of Bonn. Summer term 2019. https://www.math.uni-bonn.de/people/scholze/Condensed.pdf

[19] D. S. Scott, Continuous lattices. In *Toposes, algebraic geometry and logic*, edited by F. W. Lawvere, pp. 97–136, Lecture Notes in Math. 274, Springer, Berlin–New York, 1972.

[20] G. Tabuada, Une structure de catégorie de modéles de Quillen sur la catégorie des dg-catégories. *C. R. Acad. Sci. Paris Ser. I Math.* **340** (2005), no. 1, 15–19.

[21] G. Tabuada, *Théorie homotopique des DG-catégories*, Thése de L'Université Paris Diderot – Paris 7.

[22] G. Tamme, Excision in algebraic K-theory revisited. *Compos. Math.* **154** (2018), no. 9, 1801–1814.

[23] F. Waldhausen, Algebraic K-theory of spaces. In *Algebraic and geometric topology*, edited by A. Ranicki, N. Levitt, and F. Quinn, pp. 318–419, Lecture Notes in Math. 1126, Springer, 1985.

ALEXANDER I. EFIMOV

Steklov Mathematical Institute of RAS, Gubkin str. 8, GSP-1, Moscow 119991, Russia, and National Research University Higher School of Economics, Russian Federation, efimov@mi-ras.ru

This Page Intentionally Left Blank

ENHANCED MIRROR SYMMETRY FOR LANGLANDS DUAL HITCHIN SYSTEMS

TAMÁS HAUSEL

ABSTRACT

The first part of this paper is a survey of mathematical results on mirror symmetry phenomena between Hitchin systems for Langlands dual groups. The second part introduces and discusses multiplicity algebras of the Hitchin system on Lagrangians, and considers corresponding conjectural structures on their mirror.

MATHEMATICS SUBJECT CLASSIFICATION 2020

Primary 14D21; Secondary 14C17, 14J33, 14D24, 20G05, 81T13

KEYWORDS

Hitchin system, mirror symmetry, multiplicity algebra, geometric Langlands

INTERNATIONAL CONGRESS OF MATHEMATICIANS 2022 JULY 6–14

Proc. Int. Cong. Math. 2022, Vol. 3, pp. 2228–2249
DOI 10.4171/ICM2022/164

1. INTRODUCTION

Considering the 2-dimensional reduction of the Yang–Mills equations in 4 dimensions, Hitchin [34] in 1987 introduced and studied the moduli space of solutions to certain self-duality equations on a Riemann surface. The moduli space turns out to have an "extremely rich geometric structure."

The moduli space $\mathcal{M}$ of solutions for a complex reductive structure group G (Hitchin first considered $G = SL_2$), carries a canonical hyperkähler metric g with complex structures I, J, and K, and corresponding Käher forms ω_I, ω_J, and ω_K. In complex structure I, it agrees with the moduli space $\mathcal{M}_{\mathrm{Dol}}$ of Higgs bundles—or Hitchin pairs—(E, Φ), where E is a G-bundle and the Higgs field $\Phi \in H^0(C; \mathrm{ad}(E) \otimes K)$ is a section of the adjoint bundle twisted by the canonical bundle on a complex curve C:

$$(\mathcal{M}, I) \cong \mathcal{M}_{\mathrm{Dol}}.$$

Under an isomorphism induced by multiplying the Higgs field with i, the Kähler manifolds $(\mathcal{M}, J, \omega_J)$ and $(\mathcal{M}, K, \omega_K)$ are isomorphic. In turn, they are both isomorphic with the moduli space $\mathcal{M}_{\mathrm{DR}}$ of flat G-connections on the curve C:

$$(\mathcal{M}, J) \cong (\mathcal{M}, K) \cong \mathcal{M}_{\mathrm{DR}}. \tag{1.1}$$

The notation $\mathcal{M}_{\mathrm{Dol}}$ for Dolbeault and $\mathcal{M}_{\mathrm{DR}}$ for de Rham nonabelian cohomologies follows [48] who introduced the viewpoint of nonabelian Hodge theory in the study of $\mathcal{M}$.

In turn, Hitchin [35] introduced the Hitchin map

$$h : \mathcal{M} \to \mathcal{A}.$$

In the $G = GL_n$ case, this is just the characteristic polynomial of the Higgs field

$$h(E, \Phi) = \det(x - \Phi) \in \mathcal{A} := \bigoplus_{i=1}^{n} H^0(C; K^i).$$

For general G, one needs to consider invariant polynomials on $\mathfrak{g} := \mathrm{Lie}(G)$ and compute them on the Higgs field. He proved in [35] that h is an algebraically completely integrable Hamiltonian system with respect to the I-holomorphic symplectic form $\omega_{\mathbb{C}} := \omega_J + i\omega_K$. Thus h is sometimes referred to as the *Hitchin system*. This means that $\dim(\mathcal{A}) = \dim(\mathcal{M})/2$ and that the component functions of h are independent and Poisson commute. Additionally, the Hitchin map is proper, which was proved by Hitchin for SL_2 in [35], for GL_n by Nitsure [46] and Simpson [48], and by Faltings [16] for general G. The complete integrability and properness of the Hitchin map together imply that its generic fiber is a torsor over an abelian variety. In particular, topologically they are isomorphic to compact tori.

Due to the flexibility of their constructions (choice of curves and structure groups—but also various types of ramification data) Hitchin systems have been related to most of the known integrable systems [11, 12]. They thus play a central role in the field of integrable systems.

Our main interest in this survey will be how mirror symmetry and Langlands duality relate to the Hitchin system. In 2003 the paper [33] mathematically related the Hitchin system for Langlands dual groups to mirror symmetry. In particular, it formulated a topological mirror symmetry conjecture for certain SL_n and PGL_n Hitchin systems.

In 2007 Kapustin–Witten [38] placed the Hitchin system in the framework of a certain supersymmetric 4-dimensional Yang–Mills theory reduced to 2 dimensions. It also offered a detailed understanding of mirror symmetry and the geometrical Langlands program as a reduction of S-duality in 4 dimensions. This led to many papers, such as [2, 8, 29, 31, 36] discussing pairs of mirror branes in Langlands dual Hitchin systems. The last two papers emphasized a further structure on $\mathcal{M}$, namely a canonical $\mathbb{T}$-action given by $(E, \Phi) \mapsto (E, \lambda\Phi)$.

In 2010 Ngô [45] proved the Fundamental Lemma in the Langlands program, via a detailed understanding of the cohomology of certain singular fibers of the Hitchin map. In [28] similarities between the topological mirror symmetry conjecture of [33] and such cohomological results of Ngô were discussed and were conjecturally related. In [24] Gröchenig–Wyss–Ziegler proved the topological mirror symmetry conjecture of [33] using an arithmetic p-adic integration technique. In turn, in [23] the same authors managed to reprove Ngô's cohomological results with these new p-adic techniques. More recently Maulik–Shen [42] managed to complete some of the suggestions of [28] and derived a proof of the topological mirror symmetry conjecture from Ngô's results.

First we will discuss some of the background to these developments, and then in Section 3 we will explain some unpublished results about enhanced mirror symmetry for Langlands dual groups at the tip of the nilpotent cone.

2. BACKGROUND

2.1. Mirror symmetry

Three aspects of mirror symmetry will be relevant for us: topological and homological mirror symmetry and Strominger–Yau–Zaslow mirror symmetry.

Mirror symmetry in a nutshell relates the complex geometry of a Calabi–Yau X with complex structure I_X and Kähler, in particular symplectic, 2-form ω_X to the symplectic geometry of a same dimensional mirror Calabi–Yau (Y, I_Y, ω_Y). Originally [22] 3-dimensional examples of such a correspondence appeared in string theory and mirror symmetry as the statement that the physics of a certain 2-dimensional type A nonlinear sigma model with target (X, ω_X) matches that of a 2-dimensional type B nonlinear sigma model with target (Y, I_Y).

The study of mathematical aspects of this mirror symmetry has been one of the central subjects in modern symplectic/ complex algebraic geometry. The first mathematical aspect of the mirror relationship is the agreement of Hodge numbers $h^{n-p,q}(X) = h^{p,q}(Y)$, which we call *topological mirror symmetry*.

The formulation of Kontsevich [39] in 1994 of *homological mirror symmetry*

$$\mathrm{Fuk}(X, \omega_X) \cong D^b(Y, I_Y)$$

— the agreement of the Fukaya category of X and the derived category of coherent sheaves on Y—gave a profound mathematical conjecture for what mirror symmetry should mean.

The early 1990s saw several constructions of conjectured mirror pairs in [3,5,9]. In 1996 Strominger–Yau–Zaslow [49] suggested a way to construct the mirror of a Calabi–Yau 3-fold X out of the geometry of X. They argued that there should be fibrations of X and Y over the same base $B \cong S^3$:

$$\begin{array}{ccc} X & & Y \\ & \searrow \quad \swarrow & \\ & B & \end{array} \tag{2.1}$$

so that the generic fibers are dual special Lagrangian 3-tori. Here $L \subset X$ special Lagrangian means that L is Lagrangian $\omega_X|_L = 0$ and additionally $\mathrm{Im}(\Omega_X)|_L = 0$ the imaginary part of the Calabi–Yau volume form vanishes on L. In turn, then Y should be thought of as the moduli space of certain objects in $\mathrm{Fuk}(X, \omega_X)$, generically defined by special Lagrangian 3-tori equipped with a $U(1)$-local system. A mathematical formulation of Strominger–Yau–Zaslow was pursued by Gross–Siebert [26], with many accomplishments and recent breakthroughs [25,27].

Between mirror Calabi–Yau 3-folds, a complete construction of the dual special Lagrangian fibrations (2.1) is still missing. The Higgs bundle moduli spaces for Langlands dual groups, where the Hitchin systems will automatically give us such dual special Lagrangian fibrations, is a natural example, albeit in a geometrically different scenario from the original [49].

2.2. Geometric Langlands correspondence

In the works of [6,14,40], a geometric version of the Langlands correspondence has been proposed. Recall that the Langlands program in number theory (see [18,19] for some introductory ideas) for a reductive group G over a number field relates automorphic data for G (like modular forms for SL_2) with spectral data (like a Galois representation on the cohomology of an elliptic curve for SL_2) for a Langlands dual G^L group. The conjectures can be formulated over the other kind of global field as well: the function field of a curve over a finite field. The conjectures become more tractable in this case as the algebraic geometry of curves can be efficiently used. Here we will consider the even more geometric version of this program for function fields of a curve over the complex numbers.

Over $\mathbb{C}$ the Langlands dual $\mathrm{G}^\vee := \mathrm{G}^L$ of a complex reductive group G is simple to construct. The classification of complex reductive groups is via their root datum $(X, \Phi, X^\vee, \Phi^\vee)$—consisting of a rank n lattice X, a root system and coroot system $\Phi \subset X$ and $\Phi^\vee \subset X^\vee$ in the dual lattice $X^\vee$, satisfying certain properties—attached to all rank n complex reductive groups G. The Langlands dual of G then is the reductive group $\mathrm{G}^\vee$ whose root datum is the dual root datum $(X^\vee, \Phi^\vee, X, \Phi)$. For example, $\mathrm{GL}_n^\vee \cong \mathrm{GL}_n$ and $\mathrm{SL}_n^\vee \cong \mathrm{PGL}_n$.

The geometric Langlands correspondence of Beilinson–Drinfeld [6] for a smooth projective curve C proposes to construct from a G-local system on C (a geometric analogue of a Galois representation) a holonomic D-module on the moduli stack of bundles $\mathrm{Bun}_{\mathrm{G}^\vee}$

(a geometric analogue of an automorphic form). The main property of this construction is that the holonomic D-module must be an eigensheaf of certain Hecke operators. Beilinson–Drinfeld [6] succeed in this construction for a certain set of G-local systems on C, the so-called opers.

We will see below how Beilinson–Drinfeld's picture can be understood also as enhanced mirror symmetry between the hyperkähler, and thus in particular Calabi–Yau, moduli space of flat G connections on $\mathcal{M}_{\mathrm{DR}}$ and the moduli space of flat $\mathrm{G}^{\vee}$ connections $\mathcal{M}_{\mathrm{DR}}^{\vee}$.

2.3. SYZ mirror symmetry for Langlands dual Hitchin systems

The starting point of [33] was the observation that for $\mathrm{G} = \mathrm{SL}_n$ and $\mathrm{G}^{\vee} = \mathrm{PGL}_n$ the two Hitchin systems $h : \mathcal{M}_{\mathrm{Dol}} \to \mathcal{A}$ for G-Higgs bundles and $\mathcal{M}_{\mathrm{Dol}}^{\vee} \to \mathcal{A}^{\vee}$ for $\mathrm{G}^{\vee}$ Higgs bundles have the same base $\mathcal{A} \cong \mathcal{A}^{\vee}$ and in the diagram

$$\begin{array}{ccc} \mathcal{M}_{\mathrm{Dol}} & & \mathcal{M}_{\mathrm{Dol}}^{\vee} \\ & h\searrow \quad \swarrow h^{\vee} & \\ & \mathcal{A} & \end{array} \tag{2.2}$$

the generic fibers are dual abelian varieties. The fibers are holomorphic Lagrangian with respect to the holomorphic symplectic forms $\omega_{\mathbb{C}} := \omega_J + i\omega_K$, i.e., $\omega_{\mathbb{C}} = 0$, thus both $\omega_J = \omega_K = 0$, along the fibers because of the complete integrability of the Hitchin systems.

The same maps in complex structure J then yield

$$\begin{array}{ccc} \mathcal{M}_{\mathrm{DR}} & & \mathcal{M}_{\mathrm{DR}}^{\vee} \\ & h\searrow \quad \swarrow h^{\vee} & \\ & \mathcal{A} & \end{array} \tag{2.3}$$

where the generic fibers are special Lagrangian fibrations. This means that ω_J vanishes on the fibers, thus they are Lagrangian. Additionally, ω_K and the imaginary part of the J-holomorphic Calabi–Yau form $(\omega_K + i\omega_I)^{2d}$ also vanish along the fibers.

To find a mathematically testable form of mirror symmetry, the paper [33] formulated *topological mirror symmetry* between the mirror Calabi–Yau's, $\mathcal{M}_{\mathrm{DR}}$ and $\mathcal{M}_{\mathrm{DR}}^{\vee}$.

2.4. Topological mirror symmetry for Langlands dual Hitchin systems

To formulate this version of mirror symmetry from [33], we will be a bit more precise about our moduli spaces. For SL_n, we consider $\mathcal{M} := \mathcal{M}_{\mathrm{Dol}}$ the moduli space of stable Hitchin pairs of rank n fixed bundles E on C of fixed determinant line bundle of degree 1 with trace-free Higgs fields $\Phi \in H^0(C; \mathrm{End}_0(E) \otimes K)$. For PGL_n, we consider the action of $\Gamma := \mathrm{Jac}_C[n]$, the group of order n line bundles on C, on the SL_n moduli space $\mathcal{M}$ and define $\mathcal{M}^{\vee} := \mathcal{M}/\Gamma$. Then $\mathcal{M}$ will be a smooth quasiprojective variety, while $\mathcal{M}^{\vee}$ is a quasiprojective orbifold. For our considerations, we will need extra twisting structure on the moduli spaces in the form of a gerbe

$$\alpha \in H^2\big(\mathcal{M}^{\vee}, \mathrm{U}(1)\big) \cong H^2\big(\mathcal{M}, \mathrm{U}(1)\big)^{\Gamma}$$

which can be constructed using the universal bundle on $\mathcal{M}$.

One can then define certain mixed Hodge numbers

$$h^{p,q}(\mathcal{M}) = h^{p,q}\big(H_c^{p+q}(\mathcal{M})\big)$$

for the smooth $\mathcal{M}$ and α-twisted stringy Hodge numbers for the orbifold $\mathcal{M}^\vee = \mathcal{M}/\Gamma$:

$$h_{st,\alpha}^{p,q}(\mathcal{M}^\vee) := \sum_{\gamma\in\Gamma} h^{p-F(\gamma);q-F(\gamma)}\big(H^{p+q-2F(\gamma)}(\mathcal{M}^\gamma; L_\alpha)^\Gamma\big),$$

where $F(\gamma)$ is the fermionic shift, defined from the action of γ on the tangent space of $\mathcal{M}$ at a γ-fixed point.

With these we can formulate our topological mirror symmetry conjecture:

Conjecture 2.1 ([33]). *We have an agreement of Hodge numbers* $h^{p,q}(\mathcal{M}) = h_{st,\alpha}^{p,q}(\mathcal{M}^\vee)$.

Note that, as it stands, the conjecture is about the Hodge numbers of the Higgs moduli spaces $\mathcal{M} = \mathcal{M}_{\mathrm{Dol}}$. However, it was proved in [33] that the Hodge numbers of $\mathcal{M}_{\mathrm{Dol}}$ and $\mathcal{M}_{\mathrm{DR}}$ agree, and so (2.1) is also about the agreement of (mixed) Hodge numbers of the proposed mirrors $\mathcal{M}_{\mathrm{DR}}$ and $\mathcal{M}_{\mathrm{DR}}^\vee$. This way we can interpret Conjecture 2.1 as topological mirror symmetry for our SYZ mirror pair $\mathcal{M}_{\mathrm{DR}}$ and $\mathcal{M}_{\mathrm{DR}}^\vee$.

Conjecture 2.1 was proved for SL_2 and SL_3 in [33]. More recently, in 2020, the general case of SL_n was settled by Gröchenig–Wyss–Ziegler [24]. They used an arithmetic p-adic integration technique, pioneered by Denef–Loeser [10] and used by Batyrev [4] to check some topological mirror symmetry conjectures in the usual mirror symmetry.

In [28] we observed a curious similarity. Namely, Ngô in his proof [45] of the Fundamental Lemma in the Langlands program reduced the Fundamental Lemma to the agreement of the number of points of certain singular Hitchin fibers over finite fields and in turn to the agreement of certain Hodge numbers of singular fibers of the Hitchin fibration. In [28, §5.4] we argued that Ngô's cohomological result is a relative version of the topological mirror symmetry conjecture along the Hitchin fibration. A strategy was also proposed to deduce topological mirror symmetry using Ngô's techniques. This proposal has been recently completed by Maulik–Shen in [42] in 2020, giving a new proof of the topological mirror symmetry Conjecture 2.1 using Ngô's techniques.

Finally, in 2020, Gröchenig–Wyss–Ziegler in [23] managed to extend their p-adic integration techniques from [24] for Higgs moduli spaces of general reductive groups G and in turn they found a new proof of Ngô's cohomological result.

2.5. Geometric Langlands as enhanced homological mirror symmetry

In 2007 Kapustin–Witten [38] put forward a detailed circle of ideas amounting to a physics derivation of the geometric Langlands Correspondence as an enhanced mirror symmetry. They argued that a well-studied S-duality (or electro-magnetic or Montonen–Olive duality [43]) in a certain four-dimensional $N = 4$ supersymmetric Yang–Mills theory, when reduced to two dimensions, yields an enhanced mirror symmetry, which in turn recovers the geometric Langlands correspondence as formulated by [6].

In this two-dimensional reduction, Montonen–Olive duality becomes an equivalence of a type B sigma model with target the moduli space $\mathcal{M}_{\mathrm{DR}}$ of flat G-connections on a complex curve C and a type A sigma model with target $\mathcal{M}_{\mathrm{DR}}^{\vee}$, the moduli space of flat $\mathrm{G}^{\vee}$-connections on C. As a consequence, the category of boundary conditions in the two theories should be equivalent,

$$\mathcal{S} : D^b(\mathcal{M}_{\mathrm{DR}}) \simeq \mathrm{Fuk}(\mathcal{M}_{\mathrm{DR}}^{\vee}), \tag{2.4}$$

which can be interpreted as Kontsevich's homological mirror symmetry conjecture applied to the mirror pair $\mathcal{M}_{\mathrm{DR}}$ and $\mathcal{M}_{\mathrm{DR}}^{\vee}$.

Kapustin–Witten [38] explained that this equivalence of categories has more structure due to the hyperkähler targets, and more symmetries due to their origin in 4-dimensions than the usual homological mirror symmetry, which arises from an equivalence of two 2-dimensional sigma models. They use these additional ideas to construct from a flat connection α in $\mathcal{M}_{\mathrm{DR}}$, considered by its skycraper sheaf $\mathcal{O}_\alpha \in D^b(\mathcal{M}_{\mathrm{DR}})$, an element of $\mathrm{Fuk}(\mathcal{M}_{\mathrm{DR}}^{\vee})$ which they interpret as a D-module on the moduli space of $\mathrm{G}^{\vee}$-bundles on our curve C. The Hecke eigensheaf property then in turn is deduced from the extra symmetry stemming from the 4-dimensional origin.

First, due to the hyperkähler targets, Kapustin–Witten talk about more structured branes (also known as boundary conditions) by proposing that a brane should be either type A or B with respect to all the three complex structures I, J, and K. This way they consider type (B, A, A), (A, B, A), (A, A, B) and type (B, B, B) branes on hyperkähler manifolds. For example, a type (B, A, A) brane could be an I-holomorphic $\omega_{\mathbb{C}} = \omega_J + i\omega_K$ Lagrangian subvariety together with a local system. Or a type (B, B, B) brane should be a hyperkähler submanifold together with a hyperholomorphic connection on a bundle over it.

In their framework, Kapustin–Witten [38] argue that the mirror (S-dual) of an (B, A, A) brane on the hyperkähler $\mathcal{M}$ should be a (B, B, B) brane on $\mathcal{M}^{\vee}$. In particular, if we just concentrate on complex structure J, that of $\mathcal{M}_{\mathrm{DR}}$, we see that the mirror of an A brane should be a B brane. The mirror relationships are slightly more subtle [38, TABLE 2, P. 74] in that the A-model in complex structure J should be mirror to the B-model in complex structure K, which in turn, by (1.1), yields our version. This more refined version of mirror symmetry matches a type B brane in complex structure I to another type B brane in complex structure I on the mirror.

This latter correspondence was also formulated by Donagi–Pantev [13] as a classical limit—a first approximation of (2.4)—as

$$\mathcal{S} : D^b(\mathcal{M}_{\mathrm{Dol}}) \simeq D^b(\mathcal{M}_{\mathrm{Dol}}^{\vee}) \tag{2.5}$$

of the homological mirror symmetry (2.4). First, [13] checks that (2.2) generically gives dual abelian varieties as fibers, for every reductive group G. Second, they check that generically the Fourier–Mukai transform [44] relative to the Hitchin base gives an equivalence like (2.5) which satisfies the additional intertwining of Hecke–Wilson symmetries discussed below. It is expected that (2.5) will have to be modified when extended over certain singular points of the moduli spaces.

Another direction of research—motivated by [38]'s consideration of hyperkähler branes—lead to new understandings of Lagrangian subvarieties in $\mathcal{M}_{\mathrm{Dol}}$ and $\mathcal{M}_{\mathrm{DR}}$ and hyperholomorphic sheaves on $\mathcal{M}$. For example, [2, 8, 36] studied various constructions of such hyperkähler branes in all four different types, and contemplated what their mirror should be. In particular, Hitchin [36] proposed pairs of a (B, A, A) brane on $\mathcal{M}$ and (B, B, B) brane on $\mathcal{M}^\vee$ for which he could show that generically over the Hitchin base they are Fourier–Mukai dual. The $\mathrm{G} = \mathrm{GL}_2$ case of Hitchin's suggestion was the starting point of [31], where an additional structure, the $\mathbb{T}$-action, played an important role. This point of view will be explained below in more detail.

Second, due to the 4-dimensional origin of their derivation of (2.4), Kapustin–Witten considered extra symmetries on these categories, arising from line operators in the 4-dimensional theory. Namely, Wilson operators $\mathcal{W}^\mu$ attached to representations μ of G act on $D^b(\mathcal{M}_{\mathrm{DR}})$ via tensoring with the vector bundle in the representation μ of the G-bundle underlying the universal G flat connection. On the other hand, Hecke operators (or t'Hooft operators for the physicists) $\mathcal{H}^\mu$ attached to irreducible representations μ of G act on the moduli space of $\mathrm{G}^\vee$ bundles and in turn on D-modules on them. The extra symmetry observation of [38] is that these operators should intertwine the mirror symmetry of (2.4). We will spell out these operators in a more detailed way in the more symmetric classical limit (2.5) of [13] in Section 3.4 below.

The homological mirror symmetry (2.4) with these two additional structures: matching of hyperkähler branes under mirror symmetry, and the Wilson–Hecke symmetry is what we call *enhanced mirror symmetry*. These go beyond the usual homological mirror symmetry of Kontsevich and stem from the supersymmetric and 4-dimensional origins of S-duality.

3. ENHANCED MIRROR SYMMETRY AT THE TIP OF THE NILPOTENT CONE

The original motivation for the considerations below is to find a way to test the conjectured mirror pairs of (B, A, A) and (B, B, B) branes put forward in [2, 8, 36]. The only tests so far—which were often carried out in [2, 8, 36]—are to check if the proposed mirror pairs are indeed Fourier–Mukai dual relative to the Hitchin maps. This can only be checked generically over the Hitchin base. We would like to see more global checks, in particular ones which can verify mirror symmetry proposals over the 0-fibers of the Hitchin maps, the global nilpotent cones.

We introduced a technique in [29, 31] which can verify mirror symmetry proposals over the nilpotent cone by considering the effect of mirror symmetry on morphisms in the corresponding categories. The difficulty to consider the morphisms in our categories arises from the noncompactness of our moduli spaces. For example, the vector spaces of morphisms in the derived category $D(\mathcal{M}_{\mathrm{Dol}})$ are typically infinite dimensional. To measure their size, we will be looking at the $\mathbb{T}$-equivariant structure on them. Recall that the multiplicative group $\mathbb{T} := \mathbb{C}^\times$ of the complex numbers acts on $\mathcal{M}$ by $\lambda : (E, \Phi) \mapsto (E, \lambda\Phi)$, scalar multiplication of the Higgs field. Here we will be interested in a $\mathbb{T}$-equivariant extension of the classical limit (2.5) of the geometric Langlands correspondence, which should be as a first approximation an equivalence $\mathcal{S} : D_{\mathbb{T}}(\mathcal{M}_{\mathrm{Dol}}) \sim D_{\mathbb{T}}(\mathcal{M}_{\mathrm{Dol}}^\vee)$ between the $\mathbb{T}$-equivariant derived categories of $\mathcal{M}_{\mathrm{Dol}}$ and $\mathcal{M}_{\mathrm{Dol}}^\vee$. The morphisms between two objects $\mathcal{F}_1$ and $\mathcal{F}_2$ in

$D_{\mathbb{T}}(\mathcal{M}_{\mathrm{Dol}})$ can be identified with the graded vector space $\mathrm{Ext}^*(X;\mathcal{F}_1,\mathcal{F}_2)$. To measure this graded vector space, we note that $\mathbb{T}$ acts on it, and assuming that the weight spaces are finite dimensional and vanish for large enough weights (which we expect for semiprojective varieties) we can define the *equivariant Euler form* as

$$\begin{aligned}\chi_{\mathbb{T}}(X;\mathcal{F}_1,\mathcal{F}_2) &= \sum_{k,l} \dim\bigl(H^k\bigl(R\mathcal{H}om(\mathcal{F}_1,\mathcal{F}_2)\bigr)^l\bigr)(-1)^k t^{-l} \\ &= \sum_{k,l} \dim\bigl(\mathrm{Hom}_{\mathbf{D}_{\mathrm{coh}}(X)}\bigl(\mathcal{F}_1,\mathcal{F}_2[k]\bigr)^l\bigr)(-1)^k t^{-l} \\ &= \sum_{k,l} \dim\bigl(\mathrm{Ext}^k(X;\mathcal{F}_1,\mathcal{F}_2)^l\bigr)(-1)^k t^{-l} \in \mathbb{C}\bigl((t)\bigr).\end{aligned}$$

With this we expect that $\mathcal{S}$ is an isometry,

$$\chi_{\mathbb{T}}\bigl(\mathcal{S}(\mathcal{F}),\mathcal{S}(\mathcal{G})\bigr) = \chi_{\mathbb{T}}(\mathcal{F},\mathcal{G}).$$

In [31] we managed to check this isometry for several pairs of conjectured mirror branes from [36], while in [29] we checked this isometry for the conjectured mirror pairs relevant here; see below. In the second part of this paper, we will recall some results of [29] about the mirror of very stable upward flows, introduce and study the multiplicity algebras of the Lagrangian upward flows following [30], and finally we will consider what the multiplicity algebra should correspond to on the mirror.

3.1. Very stable Higgs bundles and mirror symmetry

The starting point is the recent paper [29]. First we recall its formalism.

3.1.1. Białynicki-Birula decomposition of semiprojective varieties

Let the multiplicative group $\mathbb{T} := \mathbb{C}^\times$ of the complex numbers act on a (possibly reducible) variety (a reduced separated scheme of finite type over $\mathbb{C}$). We say that the action is *semiprojective* [32, §1] if the following three conditions hold:

(1) the action is *linear*, i.e., there is a locally closed $\mathbb{T}$-equivariant embedding of X into $\mathbb{P}^N$ with a linear action of $\mathbb{T}$ (for example when X is normal and quasi-projective);

(2) the fixed point subvariety $X^{\mathbb{T}}$ is proper and thus projective;

(3) $\lim_{\lambda\to 0}\lambda\cdot x$ exists for all $x \in X$.

For $\alpha \in X^{\mathbb{T}}$, we define $W_\alpha^+ := \{x \in X \mid \lim_{\lambda\to 0}\lambda\cdot x = \alpha\}$, the *upward flow* from α, and $W_\alpha^- := \{x \in X \mid \lim_{\lambda\to\infty}\lambda\cdot x = \alpha\}$, the *downward flow* from α. Then we have

$$X = \coprod_{\alpha\in X^{\mathbb{T}}} W_\alpha^+,$$

the Białyinicki-Birula partition of X, and we define the projective variety

$$\mathcal{C} := \coprod_{\alpha\in X^{\mathbb{T}}} W_\alpha^-$$

to be the *core* of X. We then have the following

Theorem 3.1 ([7], [29, PROPOSITIONS 2.1 AND 2.10]). *When $\alpha \in (X^s)^{\mathbb{T}}$, a $\mathbb{T}$-fixed point on the smooth locus, then $W_\alpha^\pm \subset X$ are locally closed subvarieties and $W_\alpha^\pm \cong T_\alpha^\pm X$ as $\mathbb{T}$-varieties. Moreover, when $\omega \in \Omega^2(X^s)$ is a homogeneity 1 symplectic form, $W_\alpha^+ \subset X$ and $\mathcal{C} \subset X$ are Lagrangian subvarieties.*

We call $\alpha \in (X^s)^{\mathbb{T}}$ and W_α^+ *very stable* when $W_\alpha^+ \cap \mathcal{C} = \{\alpha\}$. Equivalently, $\alpha \in (X^s)^{\mathbb{T}}$ is very stable if and only if $W_\alpha^+ \cap W_\beta^- \neq \emptyset$ implies $\beta = \alpha$. More generally, one can show that the relation $\alpha \leq \beta$ when $W_\alpha^+ \cap W_\beta^- \neq \emptyset$ induces a partial ordering. Then $\alpha \in (X^s)^{\mathbb{T}}$ is very stable if it is maximal with respect to this ordering. We then have

Theorem 3.2 ([29, PROPOSITION 2.14]). *$\alpha \in (X^s)^{\mathbb{T}}$ is very stable if and only if $W_\alpha^+ \subset X$ is closed.*

3.1.2. Białynicki-Birula partition for Higgs bundles

We will work with $\mathrm{G} = \mathrm{GL}_n$. We will denote by $\mathcal{M}$ the moduli space of semistable rank n degree d Higgs bundles (E, Φ) on a smooth projective curve of genus g. Here E is a rank n vector bundle of degree d on C and the Higgs field $\Phi \in H^0(C; \mathrm{End}(E) \otimes K)$.

We have the *Hitchin map* [35] given by the characteristic polynomial of the Higgs field:

$$h : \mathcal{M} \to \mathcal{A} := \times_{i=1}^n H^0(C; K^i),$$
$$(E, \Phi) \mapsto \det(x - \Phi).$$

Then h is a proper map [34, 46, 48] and a completely integrable Hamiltonian system [16, 35] with respect to a natural holomorphic symplectic form on $\mathcal{M}$. In particular, the generic fibers are Lagrangian abelian varieties, Jacobians of certain spectral curves.

The $\mathbb{T}$-action on $\mathcal{M}$ is given by $(E, \Phi) \mapsto (E, \lambda\Phi)$. This makes the Hitchin map $\mathbb{T}$-equivariant if we let $\mathbb{T}$ act on $H^0(C; K^i)$ with weight i. As these weights are all positive on $\mathcal{A} \cong \times_{i=1}^n H^0(C; K^i)$, it is semiprojective, and as the Hitchin map is proper and $\mathbb{T}$-equivariant we get that our $\mathbb{T}$-action on $\mathcal{M}$ is also semiprojective. Additionally, we get that the core of $\mathcal{A}$, i.e., the origin $0 \in \mathcal{A}$ pulls back to the core of $\mathcal{M}$. For $\mathcal{M}$, the core agrees with the nilpotent cone $h^{-1}(0)_{\mathrm{red}} = \mathcal{C}$.

Generalizing the notion of very stable bundle of Drinfeld and Laumon [15, 41], we can thus define a *very stable Higgs bundle* as a stable $\mathbb{T}$-fixed Higgs bundle $\mathcal{E} \in \mathcal{M}^{s\mathbb{T}}$ for which $W_\mathcal{E}^+ \cap h^{-1}(0) = \{\mathcal{E}\}$ the only nilpotent Higgs bundle in its upward flow is itself. Thus by Theorem 3.2, we know $\mathcal{E} \in \mathcal{M}^{s\mathbb{T}}$ is very stable exactly when its upward flow is closed. To reformulate in terms of the Hitchin map, we have an alternate version of Theorem 3.2.

Theorem 3.3 ([30], cf. [51]). *$\mathcal{E} \in \mathcal{M}^{s\mathbb{T}}$ is very stable if and only if $h^{-1}(0) \cap W_\mathcal{E}^+ = \{\mathcal{E}\}$ if and only $h_\mathcal{E} := W_\mathcal{E}^+ \to \mathcal{A}$ is proper if and only if it is finite.*

Definition 3.4. For $\mathcal{E} \in \mathcal{M}^{s\mathbb{T}}$, define the rational function

$$m_\mathcal{E}(t) := \frac{\chi_\mathbb{T}(\mathrm{Sym}(T_\mathcal{E}^{+*}))}{\chi_\mathbb{T}(\mathrm{Sym}(\mathcal{A}^*))} \in \mathbb{Z}(t).$$

We call it the *equivariant multiplicity* of $W_\mathcal{E}^+$.

We have the following

Theorem 3.5 ([29, COROLLARY 5.4]). *When $\mathcal{E} \in \mathcal{M}^{s\mathbb{T}}$ is very stable, $m_{\mathcal{E}}(t)$ is a polynomial*

- *with nonnegative coefficients, which is*
- *palindromic and*
- *monic, such that*
- *$m_{\mathcal{E}}(1) = m_{F_{\mathcal{E}}}$ is the multiplicity of the component $N_{F_{\mathcal{E}}} \subset N$ in the nilpotent cone.*

Let $\ell \in \mathbb{Z}$, $m_i \in \mathbb{Z}_{\geq 0}$ and

$$\delta := (\delta_0, \delta_1, \ldots, \delta_{n-1}) \in \mathrm{Jac}_\ell(C) \times C^{[m_1]} \times \cdots \times C^{[m_{n-1}]}$$

be a vector of representative divisors on C. To this we can construct a type $(1, \ldots, 1)$ $\mathbb{T}$-fixed Higgs bundle $\mathcal{E}_\delta = (E_\delta, \Phi_\delta)$ where

$$\mathcal{E}_\delta = M_0 \oplus \cdots \oplus M_{n-1}$$

is a rank n vector bundle

$$M_i := \mathcal{O}(\delta_0 + \cdots + \delta_i)K^{-i}$$

and

$$\Phi_\delta|_{M_i} : M_i \to M_{i+1}K \subset E_\delta K$$

is given by the defining section of

$$H^0(C; M_i^{-1}M_{i+1}K) \cong H^0\big(C; \mathcal{O}(\delta_i)\big).$$

The following classifies all very stable $\mathbb{T}$-fixed type $(1, \ldots, 1)$ Higgs bundles and gives their equivariant multiplicity.

Theorem 3.6 ([29, THEOREM 4.16, (5.18)]). *Let δ be as above and suppose that $\mathcal{E}_\delta$ is a stable Higgs bundle. Then $\mathcal{E}_\delta$ is very stable if and only if the effective divisor $\delta_1 + \cdots + \delta_{n-1}$ is reduced. Its equivariant multiplicity is given by*

$$m_{\mathcal{E}_\delta}(t) = \prod_{i=1}^{n-1} \begin{bmatrix} n \\ i \end{bmatrix}_t^{m_i},$$

product of t-binomial coefficients.

3.2. Multiplicity algebra and explicit Hitchin system on Lagrangians

The main idea is to study, for $\mathcal{E} \in \mathcal{M}^{s\mathbb{T}}$, the restricted Hitchin map

$$h_{\mathcal{E}} := h|_{W^+_{\mathcal{E}}} : W^+_{\mathcal{E}} \to \mathcal{A}$$

in the framework of the Arnold school [1]. It is a $\mathbb{T}$-equivariant *Lagrangian map* between semiprojective vector spaces (i.e., only positive $\mathbb{T}$-weights) of the same dimension. Such maps are called *quasihomogeneous* in [1, §12.3].

We recall from Theorem 3.3 that $\mathcal{E}$ is very stable if and only if $h_{\mathcal{E}}^{-1}(0) = \{\mathcal{E}\}$ if and only if $h_{\mathcal{E}}$ is proper. For such maps—called *nondegenerate*—[1, §4,5,12] introduces and studies its local multiplicity algebra.

Definition 3.7. When $\mathcal{E} \in \mathcal{M}^{s\mathbb{T}}$ is very stable, define

$$Q_{\mathcal{E}} := Q_{h_{\mathcal{E}}} := \mathbb{C}[W_{\mathcal{E}}^+]/(h_{\mathcal{E}}^{-1}(\mathfrak{m}_0)) = \mathbb{C}[W_{\mathcal{E}}^+]/(h_1, \ldots, h_N),$$

the *local multiplicity algebra* of $h_{\mathcal{E}}$ at $\mathcal{E}$. Here $\mathfrak{m}_0 \subset \mathbb{C}[\mathcal{A}]$ is the maximal ideal at $0 \in \mathcal{A}$ and

$$h_{\mathcal{E}} = (h_1, \ldots, h_N) : \mathbb{C}^N \cong W_{\mathcal{E}}^+ \to \mathbb{C}^N \cong \mathcal{A}$$

in some homogeneous coordinates.

Scheme-theoretically, $Q_{\mathcal{E}}$ is just the coordinate ring of the scheme-theoretical fiber of $h_{\mathcal{E}}$ over 0 or the scheme-theoretical intersection of $W_{\mathcal{E}}^+ \cap h^{-1}(0)$ of the upward flow with the nilpotent cone. Because $h_{\mathcal{E}}$ is $\mathbb{T}$-equivariant, we will get a $\mathbb{T}$-action, and thus a grading on $Q_{\mathcal{E}}$. Because of this, sometimes we call $Q_{\mathcal{E}}$ the *equivariant multiplicity algebra* of $h_{\mathcal{E}}$ at $\mathcal{E}$.

Remark 3.8. Note that determining the algebra $Q_{\mathcal{E}}$ explicitly by N generators and N relations gives us coordinates on $W_{\mathcal{E}}^+$ such that the Hitchin map is given *explicitly* by the relations.

Using results of [1, §4,5,12], we have the following

Theorem 3.9 ([30]). *Let $\mathcal{E} \in \mathcal{M}^s$ be very stable. Then its local multiplicity algebra is*

(1) *finite dimensional,*

(2) *graded $Q_{\mathcal{E}} := \bigoplus_{k=0}^m Q_{\mathcal{E}}^k$ such that $Q_{\mathcal{E}}^0 \cong \mathbb{C}$,*

(3) *Gorenstein, with socle $Q_{\mathcal{E}}^m \cong \mathbb{C}J_{h_{\mathcal{E}}}$, which is one-dimensional and spanned by the Jacobian $J_{h_{\mathcal{E}}}$ of $h_{\mathcal{E}} = (h_1, \ldots, h_N)$ and*

(4) *a Poincaré duality ring, that is, it has a natural bilinear pairing $(\cdot, \cdot) : Q_{\mathcal{E}} \times Q_{\mathcal{E}} \to \mathbb{C}$ inducing a perfect pairing $Q_{\mathcal{E}}^k \times Q_{\mathcal{E}}^{m-k} \to \mathbb{C}$ for all k.*

(5) *Finally, its Poincaré polynomial $\sum_{k=0}^m \dim(Q_{\mathcal{E}}^k)t^k = m_{\mathcal{E}}(t)$ agrees with the equivariant multiplicity.*

This result gives a satisfactory explanation of all the properties of the equivariant multiplicity polynomial we observed in Theorem 3.5. The following then gives an explanation for the appearance of quantum binomial coefficients—which is well known to be the Poincaré polynomial of a Grassmannian—for the equivariant multiplicity.

Theorem 3.10 ([30]). *Let $\delta = (\delta_0, \delta_1, \ldots, \delta_{n-1}) \in J_\ell(C) \times C^{[m_1]} \times \cdots \times C^{[m_{n-1}]}$ be a vector of divisors. Assume that $\delta_1 + \cdots + \delta_{n-1}$ is reduced, then $\mathcal{E}_\delta$ is very stable from Theorem 3.6* *and its equivariant multiplicity algebra is*

$$Q_\delta \cong \prod_{i=1}^{n-1} H^*\big(\mathrm{Gr}(i,n);\mathbb{C}\big)^{m_i},$$

the product of the complex cohomology rings of Grassmanians $\mathrm{Gr}(i,n)$ *of* i*-dimensional subspaces in* $\mathbb{C}^n$.

Our attack on Theorem 3.10 starts with understanding the map $h_\delta : W_\delta^+ \to \mathcal{A}$ using Hecke transformations. We explain it here in the first nontrivial case when

$$\delta_c^k := (\delta_0, \dots, \delta_{n-1})$$

where $\delta_i = 0$ unless $i = k$ and the divisor $\delta_k = c$ is one point. For this, we consider the kth fundamental Hecke correspondence over the Hitchin section $W_0^+ := W_{\mathcal{E}_0}^+$ which coincides with the upward flow of the *canonical uniformizing Higgs bundle* $\mathcal{E}_0 = (E_0, \Phi_0)$. Recall **[29, (3.16)]** that it is given in chain notation as

$$(E_0, \Phi_0) = (\mathcal{O} \oplus K^{-1} \oplus \cdots \oplus K^{1-n}, \Phi_0) = \mathcal{O} \xrightarrow{1} K^{-1} \xrightarrow{1} \cdots \xrightarrow{1} K^{1-n},$$

where morphisms on the arrows are twisted by K. The canonical uniformizing Higgs bundle is at the top of the nilpotent cone. We will use Hecke transformations to generate all other type $(1, \dots, 1)$ upward flows from the Hitchin section W_0^+.

We let

$$\mathcal{H}_k := \big\{(\mathcal{E}_a, V) \in W_0^+ \times \mathrm{Gr}(k, E_0|_c) | \Phi_a|_c(V) \subset V\big\} \subset W_0^+ \times \mathrm{Gr}(k, E_0|_c). \tag{3.1}$$

It can be constructed étale locally over $\mathcal{A}$ as the fixed point scheme of the self-map of $\mathrm{Gr}(k, E_0|_c)$ induced by the invertible $\Phi_a|_c - \lambda I : E_0 \to E_0$ where $\lambda \in \mathbb{C}$ is not an eigenvalue of Φ_a. By performing a Hecke transformation of $\mathcal{E}_a$ at V for a point $(\mathcal{E}_a, V) \in \mathcal{H}_k$, we get that $\mathcal{H}_k \cong W_{\delta_c^k}^+$ and, moreover, have the commutative diagram

$$\begin{array}{ccc} \mathcal{H}_k & \xrightarrow{\cong} & W_{\delta_c^k}^+ \\ \pi_k \downarrow & & \downarrow h_{\delta_c^k} \\ W_0^+ & \xrightarrow{h_0} & \mathcal{A}. \end{array}$$

Hence $Q_{\delta_c^k} \cong Q_{\pi_k}$. In turn, the computation of Q_{π_k} can be done in the Grassmannian $\mathrm{Gr}(k, E_0|_c)$ and will yield in two different ways the two isomorphisms

$$\begin{aligned} &H^*\big(\mathrm{Gr}(k, E_0|_c); \mathbb{C}\big) \\ &\cong Q_{\pi_k} \\ &\cong \frac{\mathbb{C}[p_1, \dots, p_k, q_1, \dots, q_{n-k}]}{((p_k + \cdots + p_1 x^{k-1} + x^k)(q_{n-k} + \cdots + q_1 x^{n-k-1} + x^{n-k}) = x^n)}. \end{aligned} \tag{3.2}$$

Of course, the isomorphism of the first and third rings gives the well-known presentation of the cohomology ring of the Grassmannian.

In effect, we can think of the determination of the multiplicity algebra in (3.2) to give coordinates on the upward flow $W_{\delta_c^k}^+$ so that the Hitchin map $h_{\delta_c^k}$ becomes explicit—basically given by the relations in the cohomology ring of the Grassmannian.

3.3. Explicit Hitchin system for wobbly Lagrangians

Here we show how one can generalize the technique above to understand multiplicity algebras of wobbly, i.e., not very stable, upward flows using the affine Grassmannian [21,52]. We start with a generalized notion of the kth fundamental Hecke correspondence over the Hitchin section. Recall the *affine Grassmannian*

$$\mathrm{Gr} := \mathrm{GL}_n((z))/\mathrm{GL}_n[[z]],$$

where $\mathrm{GL}_n((z)) := \mathrm{GL}_n(\mathbb{C}((z)))$ and $\mathrm{GL}_n[[z]] := \mathrm{GL}_n(\mathbb{C}[[z]])$. It is a projective ind scheme, in particular its reduced is a nested union of projective varieties. It parametrizes higher Hecke transformations of a vector bundle at a point on a curve.

Let

$$\mu = (\mu_1 \geq \mu_2 \geq \cdots \geq \mu_n) \in P^+ \subset P \cong \mathbb{Z}^n$$

be a dominant weight and

$$z^\mu := \begin{pmatrix} z^{\mu_1} & 0 & \dots & 0 \\ 0 & z^{\mu_2} & \dots & 0 \\ \vdots & \vdots & \ddots & \vdots \\ 0 & 0 & \dots & z^{\mu_n} \end{pmatrix} \in \mathrm{GL}_n\big(\mathbb{C}((z))\big).$$

We note that $\mathrm{GL}_n[[z]]$ acts from the left on Gr with orbit decomposition

$$\mathrm{Gr} = \coprod_{\mu \in P^+} \mathrm{Gr}^\mu = \coprod_{\mu \in P^+} \mathrm{GL}_n[[z]]\big[z^\mu\big],$$

where Gr^μ are labeled by dominant weights $\mu \in P^+$ as they are the orbits of $[z^\mu]$. We have a natural map

$$\mathrm{Gr}^\mu \to \mathrm{GL}_n\big[z^\mu\big] \cong \mathrm{GL}_n/P_\mu \tag{3.3}$$

given by setting $z = 0$. We note that $\mathrm{GL}_n[z^\mu] \cong \mathrm{GL}_n/P_\mu$ is a partial flag variety and the map (3.3) is a finite-rank vector bundle on GL_n/P_μ. We denote by $\overline{\mathrm{Gr}}^\mu$ the reduced of the closure of Gr^μ in Gr. We then have

$$\overline{\mathrm{Gr}}^\mu = \coprod_{\mu \geq \lambda \in P^+} \mathrm{Gr}^\lambda, \tag{3.4}$$

where $\mu \geq \lambda$ is meant in the *dominance order* on P^+, i.e., when $\mu - \lambda$ is some sum of positive roots (possibly with multiplicity).

One important example is when $\mu = \omega_k = (\underbrace{1, \dots, 1}_{k}, 0, \dots, 0)$ is the kth fundamental weight. Then ω_k is minuscule (minimal in dominance order) and

$$\overline{\mathrm{Gr}}^{\omega_k} = \mathrm{Gr}^{\omega_k} \cong \mathrm{Gr}(k, n),$$

the classical Grassmannian.

For a dominant μ, we can now define the Hecke correspondence of type μ over the Hitchin section as

$$\mathcal{H}^\mu := \big\{(\mathcal{E}_a, V) \in W_0^+ \times \mathrm{Gr}^\mu \text{ such that } \Phi_a|_{\Delta_c}(V) \subset V\big\} \subset W_0^+ \times \mathrm{Gr}^\mu, \tag{3.5}$$

where $\Delta_c \to C$ is the formal neighborhood of c in C. In particular, $\Delta_c \cong \operatorname{Spec}(\mathbb{C}[[z]])$. We fix a trivialization $E_0|_{\Delta_c} \cong \mathcal{O}^n_{\Delta_c}$ and a trivialization $K|_{\Delta_c} \cong \mathcal{O}_{\Delta_c}$, then we can think of $\gamma_a := \Phi_a|_{\Delta_c} \in \mathfrak{gl}_n[[z]]$. Then if $V = [g] \in \mathrm{Gr}^\mu$ is represented by $g \in \mathrm{GL}_n((z))$ the condition $\Phi_a|_{\Delta_c} \in \mathfrak{gl}_n[[z]]$ translates as

$$\gamma_a g^{-1} \in \mathfrak{gl}_n[[z]]. \tag{3.6}$$

For a fixed $a \in \mathcal{A}$, this defines the *affine Springer fiber* of γ_a.

By performing a type μ Hecke transformation of $\mathcal{E}_0$ at $V \in \mathrm{Gr}^\mu$, this will yield a new vector bundle together with a Higgs field, thanks to the invariance condition (3.6). Provided some stability conditions are satisfied, the new Higgs bundle will be stable and on the upward flow of $\mathcal{E}_{\delta_c^\mu}$, where

$$\delta_c^\mu = (\alpha_n c, \alpha_{n-1} c, \ldots, \alpha_1 c), \tag{3.7}$$

by writing $\mu = \sum_i \alpha_i \omega_i$ in terms of the fundamental weights and $\alpha_i \in \mathbb{Z}_{\geq 0}$ for $1 \leq i \leq n-1$ and $\alpha_n \in \mathbb{Z}$. We claim that such a Hecke transformation will induce an isomorphism, and so we get the following diagram:

$$\begin{array}{ccc} \mathcal{H}^\mu & \overset{\cong}{\longrightarrow} & W^+_{\delta_c^\mu} \\ \pi_\mu \downarrow & & \downarrow h_{\delta_c^\mu} \\ W_0^+ & \overset{h_0}{\longrightarrow} & \mathcal{A}. \end{array} \tag{3.8}$$

This way we get $Q_{\pi_\mu} \cong Q_{h_{\delta_c^\mu}}$. Thus we reduced the computation of the equivariant multiplicity algebra for a computation inside Gr^μ by studying the equations describing $\mathcal{H}^\mu \subset W_0^+ \times \mathrm{Gr}^\mu$. It turns out that this reduces to a relatively simple linear algebra computation. We have the following results and conjectures. In order to formulate them, we will need to introduce the notion of dominant upward flows and their multiplicity.

Definition 3.11. Let $\mathcal{E} \in \mathcal{M}^{s\mathbb{T}}$. We call the upward flow $W_\mathcal{E}^+$ *dominant* if the Hitchin map $h_\mathcal{E} : W_\mathcal{E}^+ \to \mathcal{A}$ is dominant. In this case the induced map on algebra of functions $h_\mathcal{E}^* : \mathbb{C}[\mathcal{A}] \to \mathbb{C}[W_\mathcal{E}^+]$ is injective and thus we get a extension $\mathbb{C}(\mathcal{A}) \subset \mathbb{C}(W_\mathcal{E}^+)$ of function fields. We define its degree to be the *multiplicity* of $W_\mathcal{E}^+$,

$$m_\mathcal{E} := \left[\mathbb{C}(W_\mathcal{E}^+) : \mathbb{C}(\mathcal{A})\right].$$

We note that $m_\mathcal{E}$ also agrees with the rank

$$m_\mathcal{E} = \dim_{\mathbb{C}(\mathcal{A})}\left(M_\mathcal{E} \otimes_{\mathbb{C}[\mathcal{A}]} \mathbb{C}(\mathcal{A})\right)$$

of the $\mathbb{C}[\mathcal{A}]$-module $M_\mathcal{E}$ given by $h_\mathcal{E}^*$ and also with the cardinality of the generic fiber

$$m_\mathcal{E} = \#\left(h_\mathcal{E}^{-1}(a) \cap W_\mathcal{E}^+\right)$$

for generic $a \in \mathcal{A}$. Notice that the latter two definitions make sense for all upward flows. Dominance in turn then will be equivalent to nonzero multiplicity.

We have the following results:

Theorem 3.12 ([30]). *Let $\mu = (d+1)\omega_k \in P^+$ for $d \in \mathbb{Z}_{>0}$ and $1 \le k \le n-1$ and $c \in C$. When $\mathcal{E}_{\delta_c^\mu} \in \mathcal{M}^{s\mathbb{T}}$, the following hold:*

(1) *Hecke modification of type μ of W_0^+ induces an isomorphism $\mathcal{H}^\mu \to W^+_{\delta_c^\mu}$;*

(2) *$W^+_{\mathcal{E}_{\delta_c^\mu}}$ is dominant;*

(3) *$m_{\mathcal{E}} = |W \cdot \mu|$ is the order of the Weyl orbit of μ;*

(4) *we have*

$$Q_\mu \cong \mathbb{C}\big[J_d(Q_{\omega_k})\big] \cong \mathbb{C}\big[J_d\big(\operatorname{Spec}\big(H^*(\operatorname{Gr}(k,n),\mathbb{C})\big)\big)\big],$$

where for a scheme X we denote by $J_d(X)$ the $(d-1)$th jet scheme of Spec(R). *In particular, $J_d(X)(\mathbb{C}) = \operatorname{Hom}(\operatorname{Spec}(\mathbb{C}[z]/(z^d)), X)$ is the set of $d-1$ jets in X.*

Remark 3.13. We conjecture that (1), (2) and (3) hold for any $\mu \in P^+$.

Remark 3.14. As examples in the $n = 2$ case, let us give the multiplicity algebra for $d = 1, 2,$ and 3. First we have

$$Q_{\delta_c^{\omega_1}} \cong \mathbb{C}[a_0]/(a_0^2) \cong H^*(\mathbb{P}^1, \mathbb{C}). \tag{3.9}$$

Then we have

$$Q_{\delta_c^{2\omega_1}} \cong \mathbb{C}[a_0, a_1]/(a_0^2, a_0 a_1). \tag{3.10}$$

Note that $(a_0^2, a_0a_1) = (a_0, a_1)^2 \cap (a_0)$, thus the scheme-theoretical intersection $\operatorname{Spec}(Q_{\delta_c^{2\omega_1}})$ of the upward flow $W^+_{\delta_c^{2\omega_1}}$ and the nilpotent cone $h^{-1}(0)$ is the line (a_0) with a double embedded point at the origin. Note that this upward flow was studied in [29, §8.2].

For $d = 3$, we have the multiplicity-2 algebra

$$Q_{\delta_c^{3\omega_1}} \cong \mathbb{C}[a_0, a_1, a_2]/(a_0^2, a_0a_1, a_0a_2 + a_1^2). \tag{3.11}$$

Both (3.10) and (3.11) follow from Conjecture 3.12.4, and both can be proved by direct computation in Gr^μ as explained above.

Remark 3.15. It is surprising how complex $J_d(\operatorname{Spec}(H^*(\operatorname{Gr}(k,n),\mathbb{C})))$ can be. In particular, in the $k = 1$ case (i.e., jet schemes of the cohomology ring of projective space) there is only a conjecture about its multiplicity in [50, CONJECTURE III.21].

Remark 3.16. Finally, we remark that already for type (2) we have new phenomena. As discussed in [37, §5.4], there are multiplicity algebras depending on continuous parameters, in particular they cannot be isomorphic to cohomology rings, because cohomology rings are integral.

3.3.1. Lagrangian closure of $W_{\mathcal{E}}^+$

Definition 3.17. Let $\mathcal{E} \in \mathcal{M}^{s\mathbb{T}}$. The *Lagrangian closure* $\overline{\overline{W_{\mathcal{E}}^+}}$ of $W_{\mathcal{E}}^+$ is the smallest closed union of upward flows containing $W_{\mathcal{E}}^+$. In other words, the Lagrangian closure is the closure in the quotient space by the BB partition.

Using (3.8) and (3.4), we can deduce the following

Theorem 3.18 ([38]). *Let $\mu \in P^+$ and $c \in C$. Recall δ_c^μ from (3.7). Assume $\mathcal{E}_{\delta_c^\mu} \in \mathcal{M}^{s\mathbb{T}}$. Then*

$$\overline{\overline{W_{\mathcal{E}}^+}} = \coprod_{\mu \geq \lambda \in P^+} W_{\delta_c^\lambda}^+,$$

i.e., the upward flows correspond to dominant weights λ less than or equal to μ in dominance order.

3.4. Towards a classical limit of the geometric Satake correspondence

Finally, we will formulate some conjectures which were the original motivation of much of the previous ideas. In particular, they hint at a new construction of the irreducible representations of $\mathrm{GL}_n(\mathbb{C})$, and more generally of any complex reductive group G.

The general setup comes from the classical limit (2.5) of the geometric Langlands program, as formulated in [13]. Here we sketch some of the expectations of this classical limit in a schematic (not completely well defined) manner. It should be an equivalence of some sort of derived categories of coherent sheaves

$$\mathscr{S} : D^b(\mathcal{M}_{\mathrm{Dol}}) \to D^b(\mathcal{M}_{\mathrm{Dol}}^\vee).$$

Several properties of this equivalence were proposed and some established in [13]. In particular, $\mathscr{S}$ should be a relative Fourier–Mukai transform along the generic locus in $\mathcal{A}_{\mathrm{G}} \cong \mathcal{A}_{\mathrm{G}^\vee}$. Another crucial property [38], which we called enhanced mirror symmetry in Section 2.5 above, is that $\mathscr{S}$ should intertwine the actions of certain Hecke operators on $D^b(\mathcal{M}_{\mathrm{Dol}})$ and the Wilson operators on $D^b(\mathcal{M}_{\mathrm{Dol}})$. Let $\mu \in X_+(\mathrm{G}^\vee) = X^+(\mathrm{G})$ be a dominant character of $\mathrm{G}^\vee$. We denote by

$$\mathcal{H}^\mu := \{(E, \Phi) \in \mathcal{M}_{\mathrm{Dol}}, [g] \in \mathrm{Gr}^\mu \mid g^{-1}\Phi_c g \in \mathrm{G}[[z]]\} \subset \mathcal{M}_{\mathrm{Dol}} \times \mathrm{Gr}^\mu$$

some space of Hecke correspondences at a point $c \in C$. Indeed, this gives us

$$\begin{array}{ccc} & \mathcal{H}^\mu & \\ \pi_\mu \swarrow & & \searrow f^\mu \\ \mathcal{M}_{\mathrm{Dol}} & & \mathcal{M}_{\mathrm{Dol}} \end{array}$$

two maps to $\mathcal{M}_{\mathrm{Dol}}$, first the projection π_μ to the first factor, and second f^μ, the Hecke transformation[1] of (E, Φ) by the compatible Hecke transform $[g] \in \mathrm{Gr}^\mu$, which is expected to induce

$$\mathscr{H}^\mu := f_*^\mu \circ \pi_\mu^* : D^b(\mathcal{M}_{\mathrm{Dol}}) \to D^b(\mathcal{M}_{\mathrm{Dol}})$$

a Hecke (or the physicists' *t'Hooft*) operator.

On the other hand, we can consider the so-called *Wilson* operators

$$\mathscr{W}^\mu : D^b(\mathcal{M}_{\mathrm{Dol}}^\vee) \to D^b(\mathcal{M}_{\mathrm{Dol}}^\vee),$$
$$\mathcal{F} \mapsto \mathcal{F} \otimes \rho_\mu(\mathbb{E}|_{\mathcal{M}_{\mathrm{Dol}}^\vee \times \{c\}})$$

given by tensoring with the universal $\mathrm{G}^\vee$ bundle $\mathbb{E}$ in the representation $\rho_\mu : \mathrm{G}^\vee \to \mathrm{GL}(V_{\rho_\mu})$.

1 Here we ignore stability issues.

We then expect [13, 38] that

$$\mathscr{W}^{\mu} \circ \mathscr{S} = \mathscr{S} \circ \mathscr{H}^{\mu}. \tag{3.12}$$

There are two more expectations for the classical limit $\mathscr{S}$, both are motivated from Fourier–Mukai transform where the analogous statements hold. First, we expect that for any $\mathcal{F} \in D^b(\mathcal{M}_{\mathrm{Dol}})$ we should have

$$(h_{\mathrm{G}})_*(\mathcal{F}) \cong \mathscr{S}(\mathcal{F})|_{W_0^+}. \tag{3.13}$$

Second, the structure sheaf of the Hitchin section should be mirror to the structure sheaf of the mirror Higgs moduli space,

$$\mathscr{S}(\mathcal{O}_{W_0^+}) \cong \mathcal{O}_{\mathcal{M}_{\mathrm{Dol}}^{\vee}}. \tag{3.14}$$

Combining (3.12) with (3.14), we can deduce that

$$\mathscr{S}\big(\mathscr{H}^{\mu}(\mathcal{O}_{W_0^+})\big) = \mathscr{W}^{\mu}\big(\mathscr{S}(\mathcal{O}_{W_0^+})\big) = \mathscr{W}^{\mu}(\mathcal{O}_{\mathcal{M}_{\mathrm{Dol}}^{\vee}}).$$

On the one hand, we should have

$$\operatorname{supp}\big(\mathscr{H}^{\mu}(\mathcal{O}_{W_0^+})\big) = \overline{\overline{W_{\mu}^+}},$$

where W_{μ}^+ is the upward flow from a certain $\mathcal{E}_{\mu}$ maximally split G-Higgs bundle of type μ at $c \in C$. On the other hand,

$$\mathscr{W}^{\mu}(\mathcal{O}_{\mathcal{M}_{\mathrm{Dol}}^{\vee}}) = \rho_{\mu}(\mathbb{E})_c =: \Lambda_{\mu},$$

the vector bundle associated to the principal bundle $\mathbb{E}_c$ in the representation ρ_{μ}.

Thus Kapustin–Witten's (3.12) implies

$$\mathscr{S}\big(\mathscr{H}^{\mu}(\mathcal{O}_{W_0^+})\big) = \Lambda_{\mu}.$$

We can test this by (3.13) as

$$\Lambda_{\mu}|_{{}^L W_0^+} = \mathscr{S}\big(\mathscr{H}^{\mu}(\mathcal{O}_{W_0^+})\big)|_{{}^L W_0^+} = (h_{\mathrm{G}})_*\big(\mathscr{H}^{\mu}(\mathcal{O}_{W_0^+})\big).$$

In [29] we have argued that the mirror of the structure sheaf of a very stable type $(1, \dots, 1)$ upward flow W_{δ} is

$$\Lambda_{\delta} := \bigotimes_{i=1}^{n-1} \bigotimes_{j=1}^{m_i} \Lambda^i \mathbb{E}_{c_{ij}},$$

where $(\mathbb{E}, \Phi)$ is a universal Higgs bundle on $\mathcal{M} \times C$ and

$$\delta_i = c_{i1} + c_{i2} + \cdots \in C^{[m_i]}.$$

In particular, one expectation of mirror symmetry is that

$$h_*(\mathcal{O}_{W_{\delta}^+}) \cong \Lambda_{\delta}|_{W_0^+}.$$

This follows from Theorem 3.6 and a direct computation for $\chi_{\mathbb{T}}(\mathbb{E}_c)$.

In [29, §8.2] we proposed that for $n=2$ the mirror of $\mathrm{Sym}^2(\mathbb{E}_c)$ should be the structure sheaf of the Lagrangian closure $\overline{W^+_{\delta^2_c}}$ where $\delta^2_c=(0,2c)$. We can generalize this as follows.

Conjecture 3.19. *Let $c\in C$ and* G *a reductive group. Then we have the following conjectures:*

(1) *For any $\mu\in X_+(\mathrm{G}^\vee)$, the support of the mirror of $\rho_\mu(\mathbb{E}_c)$ is $\overline{W^+_{\delta^\mu_c}}$.*

(2) *Let $\mu\in X_+(\mathrm{G}^\vee)$ such that the corresponding irreducible $\mathrm{G}^\vee$ representation ρ_μ is multiplicity free. Then the mirror of $\rho_\mu(\mathbb{E}_c)$ is $\mathcal{O}_{\overline{W^+_{\delta^\mu_c}}}$.*

(3) *In the latter case, the multiplicity algebra of the restricted Hitchin map $h_{\mathrm{G}}:\overline{W^+_{\delta^\mu_c}}\to\mathcal{A}$ is isomorphic with the cohomology ring of $\overline{\mathrm{Gr}}^\mu$.*

Remark 3.20. In [17], studying opers in the geometric Langlands program, the authors construct a canonical Poincaré duality ring structure on the underlying vector space V_μ of all irreducible representation ρ_μ of $\mathrm{G}^\vee$. In the case when ρ_μ is multiplicity-free, this ring is isomorphic with the cohomology ring $H^*(\overline{\mathrm{Gr}}^\mu)$. Note that, according to [20, THEOREM 1.5], these are precisely the cases when

$$H^*(\overline{\mathrm{Gr}}^\mu)\cong IH^*(\overline{\mathrm{Gr}}^\mu),$$

when the cohomology ring satisfies Poincaré duality. In this case, this ring was more carefully studied in [47].

ACKNOWLEDGMENTS

The author thanks Nigel Hitchin for introducing him to Higgs bundles during 1995–1998, suggesting the SYZ picture for Langlands dual Hitchin systems in 1996, and for the more recent collaborations [29, 30]. He also thanks David Ben-Zvi, Pierre-Henri Chaudouard, Pierre Deligne, Ron Donagi, Sergei Gukov, Jochen Heinloth, Vadim Kaloshin, Joel Kamnitzer, Gérard Laumon, Anton Mellit, David Nadler, Andy Neitzke, Ngô Bao Châu, Michael Thaddeus, Tony Pantev, Du Pei, Richárd Rimányi, Leonid Rybnikov, Vivek Shende, Balázs Szendrői, András Szenes, Fernando Rodriguez-Villegas, Edward Witten, and Zhiwei Yun for discussions about the subjects in this paper over the years. Thanks are also due to Hülya Argüz, Jakub Löwit, Balázs Szendrői, and Nigel Hitchin for the careful reading of the paper.

REFERENCES

[1] V. I. Arnold, S. M. Gusein-Zade, and A. N. Varchenko, *Singularities of differentiable maps. Vol. I.* Monogr. Math. 82, Birkhäuser Boston, Inc., Boston, MA, 1985.

[2] D. Baraglia and L. Schaposnik, Real structures on moduli spaces of Higgs bundles. *Adv. Theor. Math. Phys.* **20** (2016), no. 3, 525–551.

[3] V. Batyrev, Dual polyhedra and mirror symmetry for Calabi–Yau hypersurfaces in toric varieties. *J. Algebraic Geom.* **3** (1994), no. 3, 493–535.

[4] V. Batyrev, Birational Calabi–Yau n-folds have equal Betti numbers. In *New trends in algebraic geometry (Warwick, 1996)*, pp. 1–11, London Math. Soc. Lecture Note Ser. 264, Cambridge Univ. Press, Cambridge, 1999.

[5] V. Batyrev and L. Borisov, Mirror duality and string-theoretic Hodge numbers. *Invent. Math.* **126** (1996), no. 1, 183–203.

[6] A. Beilinson and V. Drinfeld, Quantization of Hitchin's integrable system and Hecke eigensheaves. Preprint, 1995. https://math.uchicago.edu/~drinfeld/langlands/QuantizationHitchin.pdf.

[7] A. Białynicki-Birula, Some theorems on actions of algebraic groups. *Ann. of Math. (2)* **98** (1973), 480–497.

[8] I. Biswas, O. García-Prada, and J. Hurtubise, Higgs bundles, branes and Langlands duality. *Comm. Math. Phys.* **365** (2019), no. 3, 1005–1018.

[9] P. Candelas, M. Lynker, and R. Schimmrigk, Calabi–Yau manifolds in weighted $\mathbb{P}^4$. *Nuclear Phys. B* **341** (1990), no. 2, 383–402.

[10] J. Denef and F. Loeser, Caractéristiques d'Euler–Poincaré, fonctions zêta locales et modifications analytiques. *J. Amer. Math. Soc.* **5** (1992), no. 4, 705–720.

[11] R. Donagi, Seiberg–Witten integrable systems. In *Surveys in differential geometry: integral systems*, pp. 83–129, Surv. Differ. Geom. 4, Int. Press, Boston, MA, 1998.

[12] R. Donagi and E. Markman, Spectral covers, algebraically completely integrable, Hamiltonian systems, and moduli of bundles. In *Integrable systems and quantum groups (Montecatini Terme, 1993)*, pp. 1–119, Lecture Notes in Math. 1620, Springer, Berlin, 1996.

[13] R. Donagi and T. Pantev, Langlands duality for Hitchin systems. *Invent. Math.* **189** (2006), no. 3, 653–735. arXiv:math/0604617.

[14] V. G. Drinfeld, Letter to P. Deligne, June 1981.

[15] V. G. Drinfeld, Two dimensional ℓ-adic representations of the fundamental group of a curve over a finite field and automorphic forms on GL(2). *Amer. J. Math.* **105** (1983), 85–114.

[16] G. Faltings, Stable G-bundles and projective connections. *J. Algebraic Geom.* **2** (1993), 507–568.

[17] B. Feigin, E. Frenkel, and L. Rybnikov, Opers with irregular singularity and spectra of the shift of argument subalgebra. *Duke Math. J.* **155** (2010), no. 2, 337–363.

[18] E. Frenkel, Lectures on the Langlands program and conformal field theory. In *Frontiers in number theory, physics, and geometry. II*, pp. 387–533, Springer, Berlin, 2007.

[19] S. Gelbart, An elementary introduction to the Langlands program. *Bull. Amer. Math. Soc. (N.S.)* **10**, no. 2, 177–219.

[20] V. Ginzburg, Loop Grassmannian cohomology, the principal nilpotent and Kostant theorem. 1998, arXiv:math/9803141.

[21] U. Görtz, Affine Springer fibers and affine Deligne–Lusztig varieties. In *Affine flag manifolds and principal bundles*, pp. 1–50, Trends Math., Springer, Basel AG, Basel, 2010.

[22] B. R. Greene and M. R. Plesser, Duality in Calabi–Yau moduli space. *Nuclear Phys. B* **338** (1990), no. 1, 15–37.

[23] M. Gröchenig, D. Wyss, and P. Ziegler, Geometric stabilisation via p-adic integration. *J. Amer. Math. Soc.* **33** (2020), no. 3, 807–873.

[24] M. Gröchenig, D. Wyss, and P. Ziegler, Mirror symmetry for moduli spaces of Higgs bundles via p-adic integration. *Invent. Math.* **221** (2020), no. 2, 505–596.

[25] M. Gross, P. Hacking, S. Keel, and M. Kontsevich, Canonical bases for cluster algebras. *J. Amer. Math. Soc.* **31** (2018), no. 2, 497–608.

[26] M. Gross and B. Siebert, Affine manifolds, log structures, and mirror symmetry. *Turkish J. Math.* **27** (2003), no. 1, 33–60.

[27] M. Gross and B. Siebert, From real affine geometry to complex geometry. *Ann. of Math.* (2011), 1301–1428.

[28] T. Hausel, Global Topology of the Hitchin system, to appear in Handbook of moduli. In *dedicated to David Mumford*, edited by G. Farkas and I. Morrison, pp. 29–69, Adv. Lectures Math. 25, International Press, 2013, arXiv:1102.1717.

[29] T. Hausel and N. Hitchin, Very stable Higgs bundles, equivariant multiplicity and mirror symmetry. 2021, arXiv:2101.08583. To appear in *Invent. Math.*

[30] T. Hausel and N. Hitchin, Multiplicity algebras in integrable systems, ongoing project.

[31] T. Hausel, A. Mellit, and D. Pei, Mirror symmetry with branes by equivariant Verlinde formulas. In *Geometry and physics. Vol. I*, edited by J. E. Andersen et al., pp. 189–218, Oxford Univ. Press, Oxford, 2018.

[32] T. Hausel and F. Rodriguez-Villegas, Cohomology of large semiprojective hyperkähler varieties. *Astérisque* **370** (2015), 113–156.

[33] T. Hausel and M. Thaddeus, Mirror symmetry, Langlands duality and Hitchin systems. *Invent. Math.* **153** (2003), no. 1, 197–229.

[34] N. Hitchin, The self-duality equations on a Riemann surface. *Proc. Lond. Math. Soc. (3)* **55** (1987), 59–126.

[35] N. Hitchin, Stable bundles and integrable systems. *Duke Math. J.* **54** (1987), no. 1, 91–114.

[36] N. Hitchin, Higgs bundles and characteristic classes. In *Arbeitstagung Bonn 2013; In Memory of Friedrich Hirzebruch*, pp. 247–264, Progr. Math., Birkhäuser, 2016, arXiv:1308.4603.

[37] N. Hitchin, Spinors, twistors and classical geometry. *SIGMA Symmetry Integrability Geom. Methods Appl.* **17** (2021), 090, 9 pp.

[38] K. Kapustin and E. Witten, Electric-magnetic duality and the geometric Langlands program. *Commun. Number Theory Phys.* **1** (2007), no. 1, 1–236.

[39] M. Kontsevich, Homological algebra of mirror symmetry. In *Proceedings of the International Congress of Mathematicians (Zürich, 1994)*, pp. 120–139, Birkhäuser, Basel, 1995.

[40] G. Laumon, Correspondance de Langlands géométrique pour les corps de fonctions. *Duke Math. J.* **54** (1987), 309–359.

[41] G. Laumon, Un analogue du cone nilpotent. *Duke Math. J. (2)* **57** (1988), 647–671.

[42] D. Maulik and J. Shen, Endoscopic decompositions and the Hausel–Thaddeus conjecture. *Forum Math. Pi* **9** (2021), e8, 49 pp.

[43] C. Montonen and C. Olive, Magnetic monopoles as gauge particles? *Phys. Lett. B* **72** (1977), 117–120.

[44] S. Mukai, Duality between $D(X)$ and $D(\hat{X})$ with its application to Picard sheaves. *Nagoya Math. J.* **81** (1981), 153–175.

[45] B. C. Ngô, Le lemme fondamentale pour les algèbres de Lie. *Publ. Math. Inst. Hautes Études Sci.* **111** (2010), 1–169.

[46] N. Nitsure, Moduli space of semistable pairs on a curve. *Proc. Lond. Math. Soc. (3)* **62** (1991), 275–300.

[47] D. Panyushev, Weight multiplicity free representations, $\mathfrak{g}$-endomorphism algebras, and Dynkin polynomials. *J. Lond. Math. Soc. (2)* **69** (2004), no. 2, 273–290.

[48] C. Simpson, Nonabelian Hodge theory, In *Proceedings of the International Congress of Mathematicians (Kyoto, 1990)*, pp. 747–756, Math. Soc. Japan, 1991.

[49] A. Strominger, S-T. Yau, and E. Zaslow, Mirror symmetry is T-duality. *Nucl. Phys. B* **479** (1996), 243–259.

[50] C. Yuen, *Jet Schemes and Truncated Wedge Schemes*. Ph.D. thesis, University of Michigan, 2006. http://www.math.lsa.umich.edu/~kesmith/YuenThesis.pdf.

[51] H. Zelaci, On very stablity of principal G-bundles. *Geom. Dedicata* **204** (2020), 165–173.

[52] X. Zhu, Introduction to affine Grassmannians and the geometric Satake equivalence. In *Geometry of moduli spaces and representation theory*, pp. 59–154, IAS/Park City Math. Ser. 24, Amer. Math. Soc., Providence, RI, 2017.

TAMÁS HAUSEL

Institute of Science and Technology Austria, Am Campus 1, Klosterneuburg 3400, Austria, tamas.hausel@ist.ac.at

HODGE THEORY, BETWEEN ALGEBRAICITY AND TRANSCENDENCE

BRUNO KLINGLER

ABSTRACT

The Hodge theory of complex algebraic varieties is at heart a transcendental comparison of two algebraic structures. We survey the recent advances bounding this transcendence, mainly due to the introduction of o-minimal geometry as a natural framework for Hodge theory.

MATHEMATICS SUBJECT CLASSIFICATION 2020

Primary 14D; Secondary 14D07, 14C30, 32G20

KEYWORDS

Hodge theory, variations of Hodge structures, periods

INTERNATIONAL CONGRESS OF MATHEMATICIANS 2022 JULY 6–14

Proc. Int. Cong. Math. 2022, Vol. 3, pp. 2250–2284
DOI 10.4171/ICM2022/158

Published by EMS Press

1. INTRODUCTION

Let X be a smooth connected projective variety over $\mathbb{C}$, and X^{an} its associated compact complex manifold. Classical Hodge theory [52] states that the Betti (i.e., singular) cohomology group $H^k_{\mathrm{B}}(X^{\mathrm{an}}, \mathbb{Z})$ is a *polarizable* $\mathbb{Z}$*-Hodge structure of weight* k: there exists a canonical decomposition (called the Hodge decomposition) of complex vector spaces

$$H^k_{\mathrm{B}}(X^{\mathrm{an}}, \mathbb{Z}) \otimes_{\mathbb{Z}} \mathbb{C} = \bigoplus_{p+q=k} H^{p,q}(X^{\mathrm{an}}) \quad \text{satisfying } \overline{H^{p,q}(X^{\mathrm{an}})} = H^{q,p}(X^{\mathrm{an}})$$

and a $(-1)^k$-symmetric bilinear pairing $q_k : H^k_{\mathrm{B}}(X^{\mathrm{an}}, \mathbb{Z}) \times H^k_{\mathrm{B}}(X^{\mathrm{an}}, \mathbb{Z}) \to \mathbb{Z}$ whose complexification makes the above decomposition orthogonal, and satisfies the positivity condition (the signs are complicated but are imposed to us by geometry)

$$\mathrm{i}^{p-q} q_{k,\mathbb{C}}(\alpha, \overline{\alpha}) > 0 \quad \text{for any nonzero } \alpha \in H^{p,q}(X^{\mathrm{an}}).$$

Deligne [29] vastly generalized Hodge's result, showing that the cohomology $H^k_{\mathrm{B}}(X^{\mathrm{an}}, \mathbb{Z})$ of *any* complex algebraic variety X is functorially endowed with a slightly more general *graded polarizable mixed* $\mathbb{Z}$*-Hodge structure*, that makes, after tensoring with $\mathbb{Q}$, $H^k_{\mathrm{B}}(X^{\mathrm{an}}, \mathbb{Q})$ a successive extension of polarizable $\mathbb{Q}$-Hodge structures, with weights between 0 and $2k$. As mixed $\mathbb{Q}$-Hodge structures form a Tannakian category $\mathrm{MHS}_{\mathbb{Q}}$, one can conveniently (although rather abstractly) summarize the Hodge–Deligne theory as functorially assigning to any complex algebraic variety X a $\mathbb{Q}$-algebraic group: *the Mumford–Tate group* $\mathbf{MT}_X$ *of* X, defined as the Tannaka group of the Tannakian subcategory $\langle H^\bullet_{\mathrm{B}}(X^{\mathrm{an}}, \mathbb{Q})\rangle$ of $\mathrm{MHS}_{\mathbb{Q}}$ generated by $H^\bullet_{\mathrm{B}}(X^{\mathrm{an}}, \mathbb{Q})$. The knowledge of the group $\mathbf{MT}_X$ is equivalent to the knowledge of all *Hodge tensors* for the Hodge structure $H^\bullet_{\mathrm{B}}(X^{\mathrm{an}}, \mathbb{Q})$.

These apparently rather innocuous semilinear algebra statements are anything but trivial. They have become the main tool for analyzing the topology, geometry and arithmetic of complex algebraic varieties. Let us illustrate what we mean with regard to topology, which we will not go into later. The existence of the Hodge decomposition for smooth projective complex varieties, which holds more generally for compact Kähler manifolds, imposes many constraints on the cohomology of such spaces, the most obvious being that their odd Betti numbers have to be even. Such constraints are not satisfied even by compact complex manifolds as simple as the Hopf surfaces, quotients of $\mathbb{C}^2 \setminus \{0\}$ by the action of $\mathbb{Z}$ given by multiplication by $\lambda \neq 0$, $|\lambda| \neq 1$, whose first Betti number is one. Characterizing the homotopy types of compact Kähler manifolds is an essentially open question, which we will not discuss here.

The mystery of the Hodge–Deligne theory lies in the fact that it is at heart *not* an algebraic theory, but rather the transcendental comparison of two algebraic structures. For simplicity, let X be a smooth connected projective variety over $\mathbb{C}$. The Betti cohomology $H^\bullet_{\mathrm{B}}(X^{\mathrm{an}}, \mathbb{Q})$ defines a $\mathbb{Q}$-structure on the complex vector space of the algebraic de Rham cohomology $H^\bullet_{\mathrm{dR}}(X/\mathbb{C}) := H^\bullet(X, \Omega^\bullet_{X/\mathbb{C}})$ under the transcendental comparison isomorphism:

$$\varpi : H^\bullet_{\mathrm{dR}}(X/\mathbb{C}) \xrightarrow{\sim} H^\bullet\big(X^{\mathrm{an}}, \Omega^\bullet_{X^{\mathrm{an}}}\big) =: H^\bullet_{\mathrm{dR}}(X^{\mathrm{an}}, \mathbb{C}) \xrightarrow{\sim} H^\bullet_{\mathrm{B}}(X^{\mathrm{an}}, \mathbb{Q}) \otimes_{\mathbb{Q}} \mathbb{C}, \quad (1.1)$$

where the first canonical isomorphism is the comparison between algebraic and analytic de Rham cohomology provided by GAGA, and the second one is provided by integrating complex C^∞ differential forms over cycles (de Rham's theorem). The Hodge filtration F^p on $H^\bullet_B(X^{an}, \mathbb{Q}) \otimes_\mathbb{Q} \mathbb{C}$ is the image under ϖ of the algebraic filtration $F^p = \mathrm{Im}(H^\bullet(X, \Omega^{\bullet \geq p}_{X/\mathbb{C}}) \to H^\bullet_{dR}(X/\mathbb{C}))$ on the left-hand side.

The surprising power of the Hodge–Deligne theory lies in the fact that, although the comparison between the two algebraic structures is transcendental, this transcendence should be severely constrained, as predicted, for instance, by the Hodge conjecture and the Grothendieck period conjecture:

- For X smooth projective, it is well known that the cycle class $[Z]$ of any codimension k algebraic cycle on X with $\mathbb{Q}$ coefficients is a Hodge class in the Hodge structure $H^{2k}(X^{an}, \mathbb{Q})(k)$. Hodge **[52]** famously conjectured that the converse holds true: any Hodge class in $H^{2k}(X, \mathbb{Q})(k)$ should be such a cycle class.

- For X smooth and defined over a number field $K \subset \mathbb{C}$, its *periods* are the coefficients of the matrix of Grothendieck's isomorphism (generalizing (1.1))

$$\varpi : H^\bullet_{dR}(X/K) \otimes_K \mathbb{C} \xrightarrow{\sim} H^\bullet_B(X^{an}, \mathbb{Q}) \otimes_\mathbb{Q} \mathbb{C}$$

  with respect to bases of $H^\bullet_{dR}(X/K)$ and $H^\bullet_B(X^{an}, \mathbb{Q})$. The Grothendieck period conjecture (combined with the Hodge conjecture) predicts that the transcendence degree of the field $k_X \subset \mathbb{C}$ generated by the periods of X coincides with the dimension of $\mathbf{MT}_X$.

This tension between algebraicity and transcendence is perhaps best revealed when considering Hodge theory *in families*, as developed by Griffiths **[43]**. Let $f : X \to S$ be a smooth projective morphism of smooth connected quasiprojective varieties over $\mathbb{C}$. Its complex analytic fibers X^{an}_s, $s \in S^{an}$, are diffeomorphic, hence their cohomologies $\mathbb{V}_{\mathbb{Z},s} := H^\bullet_B(X^{an}_s, \mathbb{Z})$, $s \in S^{an}$ are all isomorphic to a fixed abelian group $V_\mathbb{Z}$ and glue together into a locally constant sheaf $\mathbb{V}_\mathbb{Z} := R^\bullet {f^{an}}_* \mathbb{Z}$ on S^{an}. However, the complex algebraic structure on X_s, hence also the Hodge structure on $\mathbb{V}_{\mathbb{Z},s}$, varies with s, making $R^\bullet f^{an}_* \mathbb{Z}$ a variation of $\mathbb{Z}$-Hodge structures ($\mathbb{Z}$VHS) $\mathbb{V}$ on S^{an}, which can be naturally polarized. One easily checks that the Mumford–Tate group $\mathbf{G}_s := \mathbf{MT}_{X_s}$, $s \in S^{an}$, is locally constant equal to the so-called *generic Mumford–Tate group* $\mathbf{G}$, outside of a meagre set $\mathrm{HL}(S, f) \subset S^{an}$, the *Hodge locus of the morphism* f, where it shrinks as exceptional Hodge tensors appear in $H^\bullet_B(X^{an}_s, \mathbb{Z})$. The variation $\mathbb{V}$ is completely described by its *period map*

$$\Phi : S^{an} \to \Gamma \backslash D.$$

Here the period domain D classifies all possible $\mathbb{Z}$-Hodge structure on the abelian group $V_\mathbb{Z}$, with a fixed polarization and Mumford–Tate group contained in $\mathbf{G}$; and Φ maps a point $s \in S^{an}$ to the point of D parameterizing the polarized $\mathbb{Z}$-Hodge structure on $V_\mathbb{Z}$ defined by $\mathbb{V}_{\mathbb{Z},s}$ (well defined up to the action of the arithmetic group $\Gamma := G \cap \mathbf{GL}(V_\mathbb{Z})$).

The transcendence of the comparison isomorphism (1.1) for each fiber X_s is embodied in the fact that the Hodge variety $\Gamma \backslash D$ is, in general, a mere complex analytic variety not admitting any algebraic structure; and that the period map Φ is a mere complex analytic map. On the other hand this transcendence is sufficiently constrained so that the following corollary of the Hodge conjecture [96] holds true, as proven by Cattani–Deligne–Kaplan [22]: the Hodge locus $\mathrm{HL}(S, f)$ is a countable union of *algebraic* subvarieties of S. Remarkably, their result is in fact valid for any polarized $\mathbb{Z}$VHS $\mathbb{V}$ on S^{an}, not necessarily coming from geometry: the Hodge locus $\mathrm{HL}(S, \mathbb{V}^{\otimes})$ is a countable union of algebraic subvarieties of S.

In this paper we report on recent advances in the understanding of this interplay between algebraicity and transcendence in Hodge theory, our main object of interest being period maps $\Phi : S^{\mathrm{an}} \to \Gamma\backslash D$. The paper is written for nonexperts: we present the mathematical objects involved, the questions, and the results but give only vague ideas of proofs, if any. It is organized as follows. After Section 2 presenting the objects of Hodge theory (which the advanced reader will skip to refer to on occasion), we present in Section 3 the main driving force behind the recent advances: although period maps are very rarely complex algebraic, their geometry is tame and does not suffer from any of the many possible pathologies of a general holomorphic map. In model-theoretic terms, period maps are definable in the *o-minimal structure* $\mathbb{R}_{\mathrm{an,exp}}$. In Section 4, we introduce the general format of *bialgebraic structures* for comparing the algebraic structure on S and that on (the compact dual $\check{D}$ of) the period domain D. The heuristic provided by this format, combined with o-minimal geometry, leads to a powerful functional transcendence result: the Ax–Schanuel theorem for polarized $\mathbb{Z}$VHS. It also suggests to interpret variational Hodge theory as a special case of an *atypical intersection* problem. In Section 5 we describe how this viewpoint leads to a stunning improvement of the result of Cattani, Deligne, and Kaplan: in most cases $\mathrm{HL}(S, \mathbb{V}^{\otimes})$ is not only a countable union of algebraic varieties, but is actually algebraic on the nose (at least if we restrict to its components of positive period dimension). Finally, in Section 6 we turn briefly to some arithmetic aspects of the theory.

For the sake of simplicity, we focus on the case of pure Hodge structures, only mentioning the references dealing with the mixed case.

2. VARIATIONS OF HODGE STRUCTURES AND PERIOD MAPS

2.1. Polarizable Hodge structures

Let $n \in \mathbb{Z}$. Let $R = \mathbb{Z}, \mathbb{Q}$, or $\mathbb{R}$. An *R-Hodge structure V of weight n* is a finitely generated R-module V_R together with one of the following equivalent data: a bigrading $V_{\mathbb{C}} (:= V_R \otimes_R \mathbb{C}) = \bigoplus_{p+q=n} V^{p,q}$, called the Hodge decomposition, such that $\overline{V^{p,q}} = V^{q,p}$ (the numbers $(\dim V^{p,q})_{p+q=n}$ are called the Hodge numbers of V); or a decreasing filtration $F^{\bullet}$ of $V_{\mathbb{C}}$, called the Hodge filtration, satisfying $F^p \oplus \overline{F^{n+1-p}} = V_{\mathbb{C}}$. One goes from one to the other through $F^p = \bigoplus_{r \geq p} V^{r,n-r}$ and $V^{p,q} = F^p \cap \overline{F^q}$. The following group-theoretic description will be most useful to us: a Hodge structure is an R-module V_R and a real algebraic representation $\varphi : \mathbf{S} \to \mathbf{GL}(V_{\mathbb{R}})$ whose restriction to $\mathbf{G}_{m,\mathbb{R}}$ is defined over $\mathbb{Q}$. Here the Deligne torus $\mathbf{S}$ denotes the real algebraic group $\mathbb{C}^*$ of invertible matrices of the forms $\left(\begin{smallmatrix} a & -b \\ b & a \end{smallmatrix}\right)$, which contains the diagonal subgroup $\mathbf{G}_{m,\mathbb{R}}$. Being of weight n is the requirement that $\varphi_{|\mathbf{G}_{m,\mathbb{R}}}$ acts via the character $z \mapsto z^{-n}$. The space $V^{p,q}$ is recovered as the eigenspace for the character $z \mapsto z^{-p}\bar{z}^{-q}$ of $\mathbf{S}(\mathbb{R}) \simeq \mathbb{C}^*$. A *morphism of Hodge structures* is a morphism of R-modules compatible with the bigrading (equivalently, with the Hodge filtration or the $\mathbf{S}$-action).

Example 2.1. We write $R(n)$ for the unique R-Hodge structure of weight $-2n$, called the Tate–Hodge structure of weight $-2n$, on the rank-one free R-module $(2\pi \mathrm{i})^n R \subset \mathbb{C}$.

One easily checks that the category of R-Hodge structures is an abelian category (where the kernels and cokernels coincide with the usual kernels and cokernels in the category of R-modules, with the induced Hodge filtrations on their complexifications), with natural tensor products $V \otimes W$ and internal homs $\hom(V, W)$ (in particular, duals $V^\vee := \hom(V, R(0))$). For $R = \mathbb{Q}$, or $\mathbb{R}$, we obtain a Tannakian category, with an obvious exact faithful R-linear tensor functor $\omega : (V_R, \varphi) \mapsto V_R$. In particular, $R(n) = R(1)^{\otimes n}$. If V is an R-Hodge structure, we write $V(n) := V \otimes R(n)$ its nth Tate twist.

If $V = (V_R, \varphi)$ is an R-Hodge structure of weight n, a *polarization for* V is a morphism of R-Hodge structures $q : V^{\otimes 2} \to R(-n)$ such that $(2\pi \mathrm{i})^n q(x, \varphi(\mathrm{i})y)$ is a positive-definite bilinear form on $V_{\mathbb{R}}$, called the *Hodge form* associated with the polarization. If there exists a polarization for V then V is said *polarizable*. One easily checks that the category of polarizable $\mathbb{Q}$-Hodge structures is semisimple.

Example 2.2. Let M be a compact complex manifold. If M admits a Kähler metric, the singular cohomology $H^n_{\mathrm{B}}(M, \mathbb{Z})$ is naturally a $\mathbb{Z}$-Hodge structure of weight n, see [52], [94, CHAP. 6]:

$$H^n_{\mathrm{B}}(M, \mathbb{Z}) \otimes_{\mathbb{Z}} \mathbb{C} = H^n_{dR}(M, \mathbb{C}) = \bigoplus_{p+q=n} H^{p,q}(M),$$

where $H^\bullet_{dR}(M, \mathbb{C})$ denotes the de Rham cohomology of the complex $(A^\bullet(M, \mathbb{C}), d)$ of C^∞ differential forms on M, the first equality is the canonical isomorphism obtained by integrating forms on cycles (de Rham theorem), and the complex vector subspace $H^{p,q}(M)$ of $H^n_{\mathrm{dR}}(M, \mathbb{C})$ is generated by the d-closed forms of type (p, q), and thus satisfies automatically $\overline{H^{p,q}(M)} = H^{q,p}(M)$. Although the second equality depends only on the complex structure on M, its proof relies on the choice of a Kähler form ω on M through the following sequence of isomorphisms:

$$H^n_{dR}(M, \mathbb{C}) \xrightarrow{\sim} \mathcal{H}^n_{\Delta_\omega}(M) = \bigoplus_{p+q=n} \mathcal{H}^{p,q}_{\Delta_\omega}(M) \xrightarrow{\sim} \bigoplus_{p+q=n} H^{p,q}(M),$$

where $\mathcal{H}^n_{\Delta_\omega}(M)$ denotes the vector space of Δ_ω-harmonic differential forms on M and $\mathcal{H}^{p,q}_{\Delta_\omega}(M)$ its subspace of Δ_ω-harmonic (p, q)-forms. The heart of Hodge theory is thus reduced to the statement that the Laplacian Δ_ω of a Kähler metric preserves the type of forms. The choice of a Kähler form ω on M also defines, through the hard Lefschetz theorem [94, THEOREM 6.25], a polarization *of the* $\mathbb{R}$*-Hodge structure* $H^n(M, \mathbb{R})$, see [94, THEOREM 6.32]. If $f : M \to N$ is any holomorphic map between compact complex manifolds admitting Kähler metrics then both $f^* : H^n_{\mathrm{B}}(N, \mathbb{Z}) \to H^n_{\mathrm{B}}(M, \mathbb{Z})$ and the Gysin morphism $f_* : H^n_{\mathrm{B}}(M, \mathbb{Z}) \to H^{n-2r}_{\mathrm{B}}(N, \mathbb{Z})(-r)$ are morphism of $\mathbb{Z}$-Hodge structures, where $r = \dim M - \dim N$.

Example 2.3. Suppose moreover that $M = X^{\mathrm{an}}$ is the compact complex manifold analytification of a smooth projective variety X over $\mathbb{C}$. In that case, $H^n_{\mathrm{B}}(X, \mathbb{Z})$ is a *polarizable $\mathbb{Z}$-Hodge structure*. Indeed, the Kähler class $[\omega]$ can be chosen as the first Chern class of an ample line bundle on X, giving rise to a rational Lefschetz decomposition and (after clearing denominators by multiplying by a sufficiently large integer) to an integral polarization. Moreover, the Hodge filtration $F^\bullet$ on $H^n_{\mathrm{B}}(X^{\mathrm{an}}, \mathbb{C})$ can be defined algebraically: upon identifying $H^n_{\mathrm{B}}(X^{\mathrm{an}}, \mathbb{C})$ with the algebraic de Rham cohomology $H^n_{\mathrm{dR}}(X/\mathbb{C}) := H^n(X, \Omega^\bullet_{X/\mathbb{C}})$, the Hodge filtration is given by $F^p = \mathrm{Im}(H^n(X, \Omega^{\bullet \geq p}_{X/\mathbb{C}}) \to H^n_{\mathrm{B}}(X^{\mathrm{an}}, \mathbb{C}))$. It follows that if X is defined over a subfield K of $\mathbb{C}$, then the Hodge filtration $F^\bullet$ on $H^n_{\mathrm{B}}(X^{\mathrm{an}}, \mathbb{C}) = H^n_{\mathrm{dR}}(X/K) \otimes_K \mathbb{C}$ is defined over K.

Example 2.4. The functor which assigns to a complex abelian variety A its $H^1_{\mathrm{B}}(A^{\mathrm{an}}, \mathbb{Z})$ defines an equivalence of categories between abelian varieties and polarizable $\mathbb{Z}$-Hodge structures of weight 1 and type $(1, 0)$ and $(0, 1)$.

2.2. Hodge classes and Mumford–Tate group

Let $R = \mathbb{Z}$ or $\mathbb{Q}$ and let V be an R-Hodge structure. A *Hodge class for* V is a vector in $V^{0,0} \cap V_{\mathbb{Q}} = F^0 V_{\mathbb{C}} \cap V_{\mathbb{Q}}$. For instance, any morphism of R-Hodge structures $f : V \to W$ defines a Hodge class in the internal $\mathrm{hom}(V, W)$. Let $T^{m,n} V_{\mathbb{Q}}$ denote the $\mathbb{Q}$-Hodge structure $V_{\mathbb{Q}}^{\otimes m} \otimes \mathrm{hom}(V, R(0))_{\mathbb{Q}}^{\otimes n}$. A *Hodge tensor for* V is a Hodge class in some $T^{m,n} V_{\mathbb{Q}}$.

The main invariant of an R-Hodge structure is its *Mumford–Tate group*. For any R-Hodge structure V we denote by $\langle V \rangle$ the Tannakian subcategory of the category of $\mathbb{Q}$-Hodge structures generated by $V_{\mathbb{Q}}$; in other words, $\langle V \rangle$ is the smallest full subcategory containing V, $\mathbb{Q}(0)$ and stable under $\oplus$, $\otimes$, and taking subquotients. If ω_V denotes the restriction of the tensor functor ω to $\langle V \rangle$, the functor $\mathrm{Aut}^{\otimes}(\omega_V)$ is representable by some closed $\mathbb{Q}$-algebraic subgroup $\mathbf{G}_V \subset \mathbf{GL}(V_{\mathbb{Q}})$, called the Mumford–Tate group of V, and ω_V defines an equivalence of categories $\langle V \rangle \simeq \mathrm{Rep}_{\mathbb{Q}} \mathbf{G}_V$. See [33, II, 2.11].

The Mumford–Tate group $\mathbf{G}_V$ can also be characterized as the fixator in $\mathbf{GL}(V_{\mathbb{Q}})$ of the Hodge tensors for V, or equivalently, writing $V = (V_R, \varphi)$, as the smallest $\mathbb{Q}$-algebraic subgroup of $\mathbf{GL}(V_{\mathbb{Q}})$ whose base change to $\mathbb{R}$ contains the image $\mathrm{Im}\, \varphi$. In particular φ factorizes as $\varphi : \mathbf{S} \to \mathbf{G}_{V,\mathbb{R}}$. The group $\mathbf{G}_V$ is thus connected, and reductive if V is polarizable. See [2, LEMMA 2].

Example 2.5. $\mathbf{G}_{\mathbb{Z}(n)} = \mathbf{G}_m$ if $n \neq 0$ and $\mathbf{G}_{\mathbb{Z}(0)} = \{1\}$.

Example 2.6. Let A be a complex abelian variety and let $V := H^1_{\mathrm{B}}(A^{\mathrm{an}}, \mathbb{Z})$ be the associated $\mathbb{Z}$-Hodge structure of weight 1. We write $\mathbf{G}_A := \mathbf{G}_V$. The choice of an ample line bundle on A defines a polarization q on V. On the one hand, the endomorphism algebra $D := \mathrm{End}^0(A)(:= \mathrm{End}(A) \otimes_{\mathbb{Z}} \mathbb{Q})$ is a finite-dimensional semisimple $\mathbb{Q}$-algebra which, in view of Example 2.4, identifies with $\mathrm{End}(V_{\mathbb{Q}})^{\mathbf{G}_A}$. Thus $\mathbf{G}_A \subset \mathbf{GL}_D(V_{\mathbb{Q}})$. On the other hand, the polarization q defines a Hodge class in $\mathrm{hom}(V_{\mathbb{Q}}^{\otimes 2}, \mathbb{Q}(-1))$ thus $\mathbf{G}_A$ has to be contained in the group $\mathbf{GSp}(V_{\mathbb{Q}}, q)$ of symplectic similitudes of $V_{\mathbb{Q}}$ with respect to the symplectic form q. Finally, $\mathbf{G}_A \subset \mathbf{GL}_D(V_{\mathbb{Q}}) \cap \mathbf{GSp}(V_{\mathbb{Q}}, q)$.

If $A = E$ is an elliptic curve, it follows readily that either $D = \mathbb{Q}$ and $\mathbf{G}_E = \mathbf{GL}_2$, or D is an imaginary quadratic field (E has complex multiplication) and $\mathbf{G}_E = \mathbf{T}_D$, the $\mathbb{Q}$-torus defined by $\mathbf{T}_D(S) = (D \otimes_{\mathbb{Q}} S)^*$ for any $\mathbb{Q}$-algebra S.

2.3. Period domains and Hodge data

Let $V_{\mathbb{Z}}$ be a finitely generated abelian group $V_{\mathbb{Z}}$ of rank r. Fix a positive integer n, a $(-1)^n$-symmetric bilinear form $q_{\mathbb{Z}}$ on $V_{\mathbb{Z}}$ and a collection of nonnegative integers $(h^{p,q})$ $(p, q \geq 0,\ p+q = n)$ such that $h^{p,q} = h^{q,p}$ and $\sum h^{p,q} = r$. Associated with $(n, q_{\mathbb{Z}}, (h^{p,q}))$ we want to define a *period domain* D classifying $\mathbb{Z}$-Hodge structures of weight n on $V_{\mathbb{Z}}$, polarized by $q_{\mathbb{Z}}$, and with Hodge numbers $h^{p,q}$. Setting $f^p = \sum_{r \geq p} h^{r,n-r}$, we first define the *compact dual* $\check{D}$ parametrizing the finite decreasing filtrations $F^\bullet$ on $V_{\mathbb{C}}$ satisfying $(F^p)^{\perp_{q_{\mathbb{Z}}}} = F^{n+1-p}$ and $\dim F^p = f^p$. This is a closed algebraic subvariety of the product of Grassmannians $\prod_p \mathrm{Gr}(f^p, V_{\mathbb{C}})$. The period domain $D \subset \check{D}^{\mathrm{an}}$ is the open subset where the Hodge form is positive definite. If $\mathbf{G} := \mathbf{GAut}(V_{\mathbb{Q}}, q_{\mathbb{Q}})$ denotes the group of similitudes of $q_{\mathbb{Q}}$, one easily checks that $\mathbf{G}(\mathbb{C})$ acts transitively on $\check{D}^{\mathrm{an}}$, which is thus a flag variety for $\mathbf{G}_{\mathbb{C}}$; and that the connected component $G := \mathbf{G}^{\mathrm{der}}(\mathbb{R})^+$ of the identity in the derived group $\mathbf{G}^{\mathrm{der}}(\mathbb{R})$ acts transitively on D, which identifies with an open G-orbit in $\check{D}$. If we fix a base point $o \in D$ and denote by P and M its stabilizer in $\mathbf{G}(\mathbb{C})$ and G, respectively, the period domain D is thus the homogeneous space

$$D = G/M \hookrightarrow \check{D}^{\mathrm{an}} = \mathbf{G}(\mathbb{C})/P.$$

The group P is a parabolic subgroup of $\mathbf{G}(\mathbb{C})$. Its subgroup $M = P \cap G$, consisting of real elements, not only fixes the filtration $F_o^\bullet$ but also the Hodge decomposition, hence the Hodge form, at o. It is thus a compact subgroup of G and D is an open elliptic orbit of G in $\check{D}$.

Example 2.7. Let $n = 1$, suppose that the only nonzero Hodge numbers are $h^{1,0} = h^{0,1} = g$, $q_{\mathbb{Z}}$ is a symplectic form and D is the subset of $\mathrm{Gr}(g, V_{\mathbb{C}})$ consisting of $q_{\mathbb{C}}$-Lagrangian subspaces F^1 on which $\mathrm{i}q_{\mathbb{C}}(u, \bar{u})$ is positive definite. In this case $\mathbf{G} = \mathbf{GSp}_{2g}$, $G = \mathbf{Sp}_{2g}(\mathbb{R})$, $M = \mathbf{SO}_{2g}(\mathbb{R})$ is a maximal compact subgroup of the connected Lie group G, and $D = G/M$ is a bounded symmetric domain naturally biholomorphic to Siegel's upper half-space $\mathfrak{H}_g$ of $g \times g$-complex symmetric matrices $Z = X + \mathrm{i}Y$ with Y positive definite. When $g = 1$, D is the Poincaré disk, biholomorphic to the Poincaré upper half-space $\mathfrak{H}$.

More generally, let $\mathbf{G}$ be a connected reductive $\mathbb{Q}$-algebraic group and let $\varphi : \mathbf{S} \to \mathbf{G}_{\mathbb{R}}$ be a real algebraic morphism such that $\varphi_{|\mathbf{G}_{m,\mathbb{R}}}$ is defined over $\mathbb{Q}$. We assume that $\mathbf{G}$ is the Mumford–Tate group of φ. The *period domain* (or Hodge domain) D associated with $\varphi : \mathbf{S} \to \mathbf{G}_{\mathbb{R}}$ is the connected component of the $\mathbf{G}(\mathbb{R})$-conjugacy class of $\varphi : \mathbf{S} \to \mathbf{G}_{\mathbb{R}}$ in $\mathrm{Hom}(\mathbf{S}, \mathbf{G}_{\mathbb{R}})$. Again, one easily checks that D is an open elliptic orbit of $G := \mathbf{G}^{\mathrm{der}}(\mathbb{R})^+$ in the compact dual flag variety $\check{D}^{\mathrm{an}}$, the $\mathbf{G}(\mathbb{C})$-conjugacy class of $\varphi_{\mathbb{C}} \circ \mu : \mathbf{G}_{\mathrm{m},\mathbb{C}} \to \mathbf{G}_{\mathbb{C}}$, where $\mu : \mathbf{G}_{\mathrm{m},\mathbb{C}} \to \mathbf{S}_{\mathbb{C}} = \mathbf{G}_{\mathrm{m},\mathbb{C}} \times \mathbf{G}_{\mathrm{m},\mathbb{C}}$ is the cocharacter $z \mapsto (z, 1)$. See [41] for details. The pair $(\mathbf{G}, D)$ is called a (connected) *Hodge datum*. A morphism of Hodge data $(\mathbf{G}, D) \to$

$(\mathbf{G}', D')$ is a morphism $\rho : \mathbf{G} \to \mathbf{G}'$ sending D to D'. Any linear representation $\lambda : \mathbf{G} \to \mathbf{GL}(V_{\mathbb{Q}})$ defines a $\mathbf{G}(\mathbb{Q})$-equivariant local system $\check{\mathbb{V}}_\lambda$ on $\check{D}^{\mathrm{an}}$. Moreover, each point $x \in D$, seen as a morphism $\varphi_x : \mathbf{S} \to \mathbf{G}_{\mathbb{R}}$, defines a $\mathbb{Q}$-Hodge structure $V_x := (V_{\mathbb{Q}}, \lambda \circ \varphi_x)$. The $\mathbf{G}(\mathbb{C})$-equivariant filtration $F^\bullet \check{\mathcal{V}}_\lambda := \mathbf{G}^{\mathrm{ad}}(\mathbb{C}) \times_{P,\lambda} F^\bullet V_{o,\mathbb{C}}$ of the holomorphic vector bundle $\check{\mathcal{V}}_\lambda := \mathbf{G}^{\mathrm{ad}}(\mathbb{C}) \times_{P,\lambda} V_{o,\mathbb{C}}$ on $\check{D}^{\mathrm{an}}$ induces the Hodge filtration on V_x for each $x \in D$. The Mumford–Tate group of V_x is $\mathbf{G}$ precisely when $x \in D \setminus \bigcup_\tau \tau(D')$, where τ ranges through the countable set of morphisms of Hodge data $\tau : (\mathbf{G}', D') \to (\mathbf{G}, D)$. The complex analytic subvarieties $\tau(D')$ of D are called the *special subvarieties* of D.

The following geometric feature of $\check{D}$ will be crucial for us. The algebraic tangent bundle $T\check{D}$ naturally identifies, as a $\mathbf{G}_{\mathbb{C}}$-equivariant bundle, with the quotient vector bundle $\check{\mathcal{V}}_{\mathrm{Ad}}/F^0\check{\mathcal{V}}_{\mathrm{Ad}}$, where $\mathrm{Ad} : \mathbf{G} \to \mathbf{GL}(\mathfrak{g})$ is the adjoint representation on the Lie algebra $\mathfrak{g}$ of $\mathbf{G}$. In particular, it is naturally filtered by the $F^i T\check{D} := F^i\check{\mathcal{V}}_{\mathrm{Ad}}/F^0\check{\mathcal{V}}_{\mathrm{Ad}}$, $i \leq -1$. The subbundle $F^{-1}T\check{D}$ is called the *horizontal tangent bundle* of $\check{D}$.

2.4. Hodge varieties

Let $(\mathbf{G}, D)$ be a Hodge datum as in Section 2.3. A *Hodge variety* is the quotient $\Gamma\backslash D$ of D by an arithmetic lattice Γ of $\mathbf{G}(\mathbb{Q})^+ := \mathbf{G}(\mathbb{Q}) \cap G$. It is thus naturally a complex analytic variety, which is smooth if Γ is torsion-free. The *special subvarieties* of $\Gamma\backslash D$ are the images of the special subvarieties of D under the projection $\pi : D \to \Gamma\backslash D$ (one easily checks these are closed complex analytic subvarieties of $\Gamma\backslash D$). For any algebraic representation $\lambda : \mathbf{G} \to \mathbf{GL}(V_{\mathbb{Q}})$, the $\mathbf{G}(\mathbb{Q})$-equivariant local system $\check{\mathbb{V}}_\lambda$ as well as the filtered holomorphic vector bundle $(\check{\mathcal{V}}_\lambda, F^\bullet)$ on $\check{D}$ are G-equivariant when restricted to D, hence descend to a triple $(\mathbb{V}_\lambda, (\mathcal{V}_\lambda, F^\bullet), \nabla)$ on $\Gamma\backslash D$. Similarly, the horizontal tangent bundle of $\check{D}$ defines the *horizontal tangent bundle* $T_h(\Gamma\backslash D) \subset T(\Gamma\backslash D)$ of the Hodge variety $\Gamma\backslash D$.

2.5. Polarized $\mathbb{Z}$-variations of Hodge structures

Hodge theory as recalled in Section 2.1 can be considered as the particular case over a point of Hodge theory over an arbitrary base. Again, the motivation comes from geometry. Let $f : Y \to B$ be a proper surjective complex analytic submersion from a connected Kähler manifold Y to a complex manifold B. It defines a locally constant sheaf $\mathbb{V}_{\mathbb{Z}} := R^\bullet f_*\mathbb{Z}$ of finitely generated abelian groups on B, gathering the cohomologies $H^\bullet_{\mathrm{B}}(Y_b, \mathbb{Z})$, $b \in B$. Upon choosing a base point $b_0 \in B$, the datum of $\mathbb{V}_{\mathbb{Z}}$ is equivalent to the datum of a *monodromy representation* $\rho : \pi_1(B, b_0) \to \mathbf{GL}(\mathbb{V}_{\mathbb{Z},b_0})$. On the other hand, the de Rham incarnation of the cohomology of the fibers of f is the holomorphic flat vector bundle $(\mathcal{V} := \mathbb{V}_{\mathbb{Z}} \otimes_{\mathbb{Z}_B} \mathcal{O}_B \simeq R^\bullet f_*\Omega^\bullet_{Y/B}, \nabla)$, where $\mathcal{O}_B$ is the sheaf of holomorphic functions on B, $\Omega^\bullet_{Y/B}$ is the relative holomorphic de Rham complex and ∇ is the Gauss–Manin connection. The Hodge filtration on each $H^\bullet_{\mathrm{B}}(Y_b, \mathbb{C})$ is induced by the holomorphic subbundles $F^p := R^\bullet f_*\Omega^{\bullet \geq p}_{Y/B}$ of $\mathcal{V}$. The Hodge filtration is usually not preserved by the connection, but Griffiths [42] crucially observed that it satisfies the *transversality constraint* $\nabla F^p \subset \Omega^1_B \otimes_{\mathcal{O}_B} F^{p-1}$. More generally, *a variation of $\mathbb{Z}$-Hodge structures ($\mathbb{Z}$VHS)* on a connected complex manifold $(B, \mathcal{O}_B)$ is a pair $\mathbb{V} := (\mathbb{V}_{\mathbb{Z}}, F^\bullet)$, consisting of a locally constant sheaf of finitely generated abelian groups $\mathbb{V}_{\mathbb{Z}}$ on B and a (decreasing) filtration $F^\bullet$ of the holomorphic vector bundle $\mathcal{V} := \mathbb{V}_{\mathbb{Z}} \otimes_{\mathbb{Z}_B} \mathcal{O}_B$ by holomorphic subbundles, called the Hodge filtration, satisfying the following conditions: for each $b \in B$, the pair $(\mathbb{V}_b, F_b^\bullet)$ is a $\mathbb{Z}$-Hodge structure; and the flat connection ∇ on $\mathcal{V}$ defined by $\mathbb{V}_{\mathbb{C}}$ satisfies Griffiths' transversality,

$$\nabla F^\bullet \subset \Omega_B^1 \otimes_{\mathcal{O}_B} F^{\bullet-1}. \tag{2.1}$$

A *morphism* $\mathbb{V} \to \mathbb{V}'$ *of* $\mathbb{Z}$*VHSs* on B is a morphism $f : \mathbb{V}_{\mathbb{Z}} \to \mathbb{V}'_{\mathbb{Z}}$ of local systems such that the associated morphism of vector bundles $f : \mathcal{V} \to \mathcal{V}'$ is compatible with the Hodge filtrations. If $\mathbb{V}$ has weight k, a *polarization of* $\mathbb{V}$ is a morphism $\mathrm{q} : \mathbb{V} \otimes \mathbb{V} \to \mathbb{Z}_B(-k)$ inducing a polarization on each $\mathbb{Z}$-Hodge structure $\mathbb{V}_b$, $b \in B$. In the geometric situation, such a polarization exists if there exists an element $\eta \in H^2(Y, \mathbb{Z})$ whose restriction to each fiber Y_b defines a Kähler class, for instance if f is the analytification of a smooth projective morphism of smooth connected algebraic varieties over $\mathbb{C}$.

2.6. Generic Hodge datum and period map

Let S be a smooth connected quasiprojective variety over $\mathbb{C}$ and let $\mathbb{V}$ be a polarized $\mathbb{Z}$VHS on S^{an}. Fix a base point $o \in S^{\mathrm{an}}$, let $p : \widetilde{S^{\mathrm{an}}} \to S^{\mathrm{an}}$ be the corresponding universal cover and write $V_{\mathbb{Z}} := \mathbb{V}_{\mathbb{Z},o}$, $q_{\mathbb{Z}} := \mathrm{q}_{\mathbb{Z},o}$. The pulled-back polarized $\mathbb{Z}$VHS $p^*\mathbb{V}$ is canonically trivialized as $(\widetilde{S^{\mathrm{an}}} \times V_{\mathbb{Z}}, (\widetilde{S^{\mathrm{an}}} \times V_{\mathbb{C}}, F^\bullet), \nabla = d, q_{\mathbb{Z}})$. In **[31, 7.5]**, Deligne proved that there exists a reductive $\mathbb{Q}$-algebraic subgroup $\iota : \mathbf{G} \hookrightarrow \mathbf{GL}(V_{\mathbb{Q}})$, called the *generic Mumford–Tate group* of $\mathbb{V}$, such that, for all points $\tilde{s} \in \widetilde{S^{\mathrm{an}}}$, the Mumford–Tate group $\mathbf{G}_{(V_{\mathbb{Z}}, F_{\tilde{s}}^\bullet)}$ is contained in $\mathbf{G}$, and is equal to $\mathbf{G}$ outside of a meagre set of $\widetilde{S^{\mathrm{an}}}$ (such points $\tilde{s}$ are said *Hodge generic* for $\mathbb{V}$). A closed irreducible subvariety $Y \subset S$ is said Hodge generic for $\mathbb{V}$ if it contains a Hodge generic point. The setup of Section 2.3 is thus in force. Without loss of generality, we can assume that the point $\tilde{o}$ is Hodge generic. Let $(\mathbf{G}, D)$ be the Hodge datum (called the *generic Hodge datum* of S^{an} for $\mathbb{V}$) associated with the polarized Hodge structure $(V_{\mathbb{Z}}, F_{\tilde{o}}^\bullet)$. The $\mathbb{Z}$VHS $p^*\mathbb{V}$ is completely described by a holomorphic map $\widetilde{\Phi} : \widetilde{S^{\mathrm{an}}} \to D$, which is naturally equivariant under the monodromy representation $\rho : \pi_1(S^{\mathrm{an}}, o) \to \Gamma := G \cap \mathbf{GL}(V_{\mathbb{Z}})$, hence descends to a holomorphic map $\Phi : S^{\mathrm{an}} \to \Gamma \backslash D$, called the *period map* of S for $\mathbb{V}$. We thus obtain the following commutative diagram in the category of complex analytic spaces:

$$\begin{array}{ccccc} \widetilde{S^{\mathrm{an}}} & \xrightarrow{\widetilde{\Phi}} & D & \hookrightarrow & \check{D}^{\mathrm{an}} \\ {\scriptstyle p}\downarrow & & \downarrow{\scriptstyle \pi} & & \\ S^{\mathrm{an}} & \xrightarrow{\Phi} & \Gamma\backslash D. & & \end{array} \tag{2.2}$$

Notice that the pair $(\mathbb{V}_{\mathbb{Q}}, (\mathcal{V}, F^\bullet))$ is the pullback under Φ of the pair $(\mathbb{V}_\iota, (\mathcal{V}_\iota, F^\bullet))$ on the Hodge variety $\Gamma\backslash D$ defined by the inclusion $\iota : \mathbf{G} \hookrightarrow \mathbf{GL}(V_{\mathbb{Q}})$. Griffiths' transversality condition is equivalent to the statement that Φ is *horizontal*, $d\Phi(TS^{\mathrm{an}}) \subset T_h(\Gamma\backslash D)$. By extension we call *period map* any holomorphic, horizontal, locally liftable map from S^{an} to a Hodge variety $\Gamma\backslash D$.

The *Hodge locus* $\mathrm{HL}(S, \mathbb{V}^{\otimes})$ of S for $\mathbb{V}$ is the subset of points $s \in S^{\mathrm{an}}$ for which the Mumford–Tate group $\mathbf{G}_s$ is a strict subgroup of $\mathbf{G}$, or equivalently for which the Hodge structure $\mathbb{V}_s$ admits more Hodge tensors than the very general fiber $\mathbb{V}_{s'}$. Thus

$$\mathrm{HL}(S, \mathbb{V}^{\otimes}) = \bigcup_{(\mathbf{G}', D') \hookrightarrow (\mathbf{G}, D)} \Phi^{-1}(\Gamma' \backslash D'), \tag{2.3}$$

where the union is over all strict Hodge subdata and $\Gamma' \backslash D'$ is a slight abuse of notation for denoting the projection of $D' \subset D$ to $\Gamma \backslash D$.

Let $Y \subset S$ be a closed irreducible algebraic subvariety $i : Y \hookrightarrow S$. Let $(\mathbf{G}_Y, D_Y)$ be the generic Hodge datum of the $\mathbb{Z}$VHS $\mathbb{V}$ restricted to the smooth locus of Y. The algebraic monodromy group $\mathbf{H}_Y$ of Y for $\mathbb{V}$ is the identity component of the Zariski-closure in $\mathbf{GL}(V_{\mathbb{Q}})$ of the monodromy of the restriction to Y of the local system $\mathbb{V}_{\mathbb{Z}}$. It follows from Deligne's (in the geometric case) and Schmid's (in general) "Theorem of the fixed part" and "Semisimplicity Theorem" that $\mathbf{H}_Y$ is a normal subgroup of the derived group $\mathbf{G}_Y^{\mathrm{der}}$, see [2, THEOREM 1].

3. HODGE THEORY AND TAME GEOMETRY

3.1. Variational Hodge theory between algebraicity and transcendence

Let S be a smooth connected quasi-projective variety over $\mathbb{C}$ and let $\mathbb{V} = (\mathbb{V}_{\mathbb{Z}}, F^{\bullet})$ be a polarized $\mathbb{Z}$VHS on S^{an}. Let $(\mathbf{G}, D)$ be the generic Hodge datum of S for $\mathbb{V}$ and let $\Phi : S^{\mathrm{an}} \to \Gamma \backslash D$ be the period map defined by $\mathbb{V}$.

The fact that Hodge theory is a transcendental theory is reflected in the following facts:

- First, the triplets $(\mathbb{V}_{\lambda}, (\mathcal{V}_{\lambda}, F^{\bullet}), \nabla)$ on $\Gamma \backslash D$ (for $\lambda : \mathbf{G} \to \mathbf{GL}(V_{\mathbb{Q}})$ an algebraic representation) do not in general satisfy Griffiths' transversality, hence do not define a $\mathbb{Z}$VHS on $\Gamma \backslash D$. They do if and only if $\mathbb{V}$ is *of Shimura type*, i.e., $(\mathbf{G}, D)$ is a (connected) *Shimura datum* (meaning that the weight zero Hodge structures on the fibers of $\mathbb{V}_{\mathrm{Ad}}$ are of type $\{(-1, 1), (0, 0), (1, -1)\}$; or equivalently, if the horizontal tangent bundle $T_h D$ coincides with TD. In other words, Hodge varieties are in general not classifying spaces for polarized $\mathbb{Z}$VHS.

- Second, and more importantly, the complex analytic Hodge variety $\Gamma \backslash D$ is in general not algebraizable (i.e., it is not the analytification of a complex quasiprojective variety). More precisely, let us write $D = G/M$ as in Section 2.3. A classical property of elliptic orbits like D is that there exists a unique maximal compact subgroup K of G containing M [46]. Supposing for simplicity that G is a real simple Lie group G, then $\Gamma \backslash D$ is algebraizable only if G/K is a hermitian symmetric domain and the projection $D \to G/K$ is holomorphic or antiholomorphic, see [45].

On the other hand, this transcendence is severely constrained, as shown by the following algebraicity results:

- If $(\mathbf{G}, D)$ is *of Shimura type*, then $\Gamma\backslash D = \mathrm{Sh}^{\mathrm{an}}$ is the analytification of an algebraic variety, called a Shimura variety Sh [8,30,32]. In that case Borel [17, THEOREM 3.10] proved that the complex analytic period map $\Phi : S^{\mathrm{an}} \to \mathrm{Sh}^{\mathrm{an}}$ is the analytification of an algebraic map.

- Let $S \subset \overline{S}$ be a log-smooth compactification of S by a simple normal crossing divisor Z. Following Deligne [28], the flat holomorphic connection ∇ on $\mathcal{V}$ defines a canonical extension $\overline{\mathcal{V}}$ of $\mathcal{V}$ to $\overline{S}$. Using GAGA for $\overline{S}$, this defines an algebraic structure on $(\mathcal{V}, \nabla)$, for which the connection ∇ is regular. Around any point of Z, the complex manifold S^{an} is locally isomorphic to a product $(\Delta^*)^k \times \Delta^l$ of punctured polydisks. Borel showed that the monodromy representation $\rho : \pi_1(S^{\mathrm{an}}, s_o) \to \Gamma \subset \mathbf{G}(\mathbb{Q})$ of $\mathbb{V}$ is "tame at infinity," that is, its restriction to $\mathbb{Z}^k = \pi_1((\Delta^*)^k \times \Delta^l)$ is quasiunipotent, see [82, (4.5)]. Using this result, Schmid showed that the Hodge filtration $F^\bullet$ extends holomorphically to the Deligne extension $\overline{\mathcal{V}}$. This is the celebrated Nilpotent Orbit theorem [82, (4.12)]. It follows, as noticed by Griffiths [82, (4.13)], that the Hodge filtration on $\mathcal{V}$ comes from an algebraic filtration on the underlying algebraic bundle, whether $\mathbb{V}$ is of geometric origin or not.

- More recently, an even stronger evidence came from the study of Hodge loci. Cattani, Deligne, and Kaplan proved the following celebrated result (generalized to the mixed case in [18–21]):

Theorem 3.1 ([22]). *Let S be a smooth connected quasiprojective variety over $\mathbb{C}$ and $\mathbb{V}$ be a polarized $\mathbb{Z}$VHS over S. Then $\mathrm{HL}(S, \mathbb{V}^{\otimes})$ is a countable union of closed irreducible algebraic subvarieties of S.*

In view of this tension between algebraicity and transcendence, it is natural to ask if there is a framework, less strict than complex algebraic geometry but more constraining than complex analytic geometry, where to analyze period maps and explain its remarkable properties.

3.2. O-minimal geometry

Such a framework was in fact envisioned by Grothendieck in [47, §5] under the name "tame topology," as a way out of the pathologies of general topological spaces. Examples of pathologies are Cantor sets, space-filling curves but also much simpler objects like the graph $\Gamma := \{(x, \sin\frac{1}{x}),\ 0 < x \leq 1\} \subset \mathbb{R}^2$: its closure $\overline{\Gamma} := \Gamma \amalg \mathrm{I}$, where $\mathrm{I} := \{0\} \times [-1, 1] \subset \mathbb{R}^2$ is connected but not arc-connected; $\dim(\overline{\Gamma} \setminus \Gamma) = \dim \Gamma$, which prevents any reasonable stratification theory; and $\Gamma \cap \mathbb{R}$ is not "of finite type." Tame geometry has been developed by model theorists as o-minimal geometry, which studies structures where every definable set has a finite geometric complexity. Its prototype is real semialgebraic geometry, but it is much richer. We refer to [34] for a nice survey.

Definition 3.2. A structure $\mathcal{S}$ expanding the real field is a collection $\mathcal{S} = (S_n)_{n\in\mathbb{N}}$, where S_n is a set of subsets of $\mathbb{R}^n$ such that for every $n \in \mathbb{N}$:

(1) all algebraic subsets of $\mathbb{R}^n$ are in S_n.

(2) S_n is a boolean subalgebra of the power set of $\mathbb{R}^n$ (i.e., S_n is stable by finite union, intersection, and complement).

(3) If $A \in S_n$ and $B \in S_m$ then $A \times B \in S_{n+m}$.

(4) Let $p : \mathbb{R}^{n+1} \to \mathbb{R}^n$ be a linear projection. If $A \in S_{n+1}$ then $p(A) \in S_n$.

The elements of S_n are called the $\mathcal{S}$-definable sets of $\mathbb{R}^n$. A map $f : A \to B$ between $\mathcal{S}$-definable sets is said to be $\mathcal{S}$-definable if its graph is $\mathcal{S}$-definable.

A dual point of view starts from the functions, namely considers sets definable in a first-order structure $\mathcal{S} = \langle\mathbb{R}, +, \times, <, (f_i)_{i\in I}\rangle$ where I is a set and the $f_i : \mathbb{R}^{n_i} \to \mathbb{R}, i \in I$, are functions. A subset $Z \subset \mathbb{R}^n$ is $\mathcal{S}$-definable if it can be defined by a formula

$$Z := \{(x_1, \dots, x_n) \in \mathbb{R}^n \mid \phi(x_1, \dots, x_n) \text{ is true}\},$$

where ϕ is a first-order formula that can be written using only the quantifiers $\forall$ and $\exists$ applied to real variables; logical connectors; algebraic expressions written with the f_i; the order symbol $<$; and fixed parameters $\lambda_i \in \mathbb{R}$. When the set I is empty the $\mathcal{S}$-definable subsets are the semialgebraic sets. Semialgebraic subsets are thus always $\mathcal{S}$-definable.

One easily checks that the composite of $\mathcal{S}$-definable functions is $\mathcal{S}$-definable, as are the images and the preimages of $\mathcal{S}$-definable sets under $\mathcal{S}$-definable maps. Using that the euclidean distance is a real-algebraic function, one shows easily that the closure and interior of an $\mathcal{S}$-definable set are again $\mathcal{S}$-definable.

The following o-minimal axiom for a structure $\mathcal{S}$ guarantees the possibility of doing geometry using $\mathcal{S}$-definable sets as basic blocks.

Definition 3.3. A structure $\mathcal{S}$ is said to be o-minimal if S_1 consists precisely of the finite unions of points and intervals (i.e., the semialgebraic subsets of $\mathbb{R}$).

Example 3.4. The structure $\mathbb{R}_{\sin} := \langle\mathbb{R}, +, \times, <, \sin\rangle$ is not o-minimal. Indeed, the infinite union of points $\pi\mathbb{Z} = \{x \in \mathbb{R} \mid \sin x = 0\}$ is a definable subset of $\mathbb{R}$ in this structure.

Any o-minimal structure $\mathcal{S}$ has the following main tameness property: given finitely many $\mathcal{S}$-definable sets $U_1, \dots, U_k \subset \mathbb{R}^n$, there exists a definable cylindrical cellular decomposition of $\mathbb{R}^n$ such that each U_i is a finite union of cells. Such a decomposition is defined inductively on n. For $n = 1$, this is a finite partition of $\mathbb{R}$ into cells which are points or open intervals. For $n > 1$, it is obtained from a definable cylindrical cellular decomposition of $\mathbb{R}^{n-1}$ by fixing, for any cell $C \subset \mathbb{R}^{n-1}$, finitely many definable functions $f_{C,i} : C \to \mathbb{R}$, $1 \leq i \leq k_C$, with $f_{C,0} := -\infty < f_{C,1} < \dots < f_{C,k_C} < f_{C,k_C+1} := +\infty$, and defining the cells of $\mathbb{R}^n$ as the graphs $\{(x, f_{C,i}(x)), x \in C\}$, $1 \leq i \leq k_C$, and the bands $\{(x, f_{C,i}(x) < y < f_{C,i+1}(x)), x \in C, y \in \mathbb{R}\}$, $0 \leq i \leq k_C$, for all cells C of $\mathbb{R}^{n-1}$.

The simplest o-minimal structure is the structure $\mathbb{R}_{\text{alg}}$ consisting of semialgebraic sets. It is too close to algebraic geometry to be used for studying transcendence phenomena. Luckily much richer o-minimal geometries do exist. A fundamental result of Wilkie, building on the result of Khovanskii [54] that any exponential set $\{(x_1, \ldots, x_n) \in \mathbb{R}^n \mid P(x_1, \ldots, x_n, \exp(x_1), \ldots, \exp(x_n)) = 0\}$ (where $P \in \mathbb{R}[X_1, \ldots, X_n, Y_1, \ldots, Y_n]$) has finitely many connected components, states:

Theorem 3.5 ([97]). *The structure* $\mathbb{R}_{\exp} := \langle \mathbb{R}, +, \times, <, \exp : \mathbb{R} \to \mathbb{R} \rangle$ *is o-minimal.*

In another direction, let us define

$$\mathbb{R}_{\text{an}} := \big\langle \mathbb{R}, +, \times, <, \{f\} \text{ for } f \text{ restricted real analytic function} \big\rangle,$$

where a function $f : \mathbb{R}^n \to \mathbb{R}$ is a restricted real analytic function if it is zero outside $[0, 1]^n$ and if there exists a real analytic function g on a neighborhood of $[0, 1]^n$ such that f and g are equal on $[0, 1]^n$. Gabrielov's result [37] that the difference of two subanalytic sets is subanalytic implies rather easily that the structure $\mathbb{R}_{\text{an}}$ is o-minimal. The structure generated by two o-minimal structures is not o-minimal in general, but Van den Dries and Miller [35] proved that the structure $\mathbb{R}_{\text{an,exp}}$ generated by $\mathbb{R}_{\text{an}}$ and $\mathbb{R}_{\exp}$ is o-minimal. This is the o-minimal structure which will be mainly used in the rest of this text.

Let us now globalize the notion of definable set using charts:

Definition 3.6. A definable topological space $\mathcal{X}$ is the data of a Hausdorff topological space $\mathcal{X}$, a *finite* open covering $(U_i)_{1 \leq i \leq k}$ of $\mathcal{X}$, and homeomorphisms $\psi_i : U_i \to V_i \subset \mathbb{R}^n$ such that all V_i, $V_{ij} := \psi_i(U_i \cap U_j)$ and $\psi_i \circ \psi_j^{-1} : V_{ij} \to V_{ji}$ are definable. As usual the pairs (U_i, ψ_i) are called charts. A morphism of definable topological spaces is a continuous map which is definable when read in the charts. The definable site $\underline{\mathcal{X}}$ of a definable topological space $\mathcal{X}$ has for objects definable open subsets $U \subset X$ and admissible coverings are the finite ones.

Example 3.7. Let X be an algebraic variety over $\mathbb{R}$. Then $X(\mathbb{R})$ equipped with the euclidean topology carries a natural $\mathbb{R}_{\text{alg}}$-definable structure (up to isomorphism): one covers X by finitely many (Zariski) open affine subvarieties X_i and take $U_i := X_i(\mathbb{R})$ which is naturally a semialgebraic set. One easily check that any two finite open affine covers define isomorphic $\mathbb{R}_{\text{alg}}$-structures on $X(\mathbb{R})$. If X is an algebraic variety over $\mathbb{C}$ then $X(\mathbb{C}) = (\text{Res}_{\mathbb{C}/\mathbb{R}} X)(\mathbb{R})$ carries thus a natural $\mathbb{R}_{\text{alg}}$-structure. We call this the $\mathbb{R}_{\text{alg}}$-*definabilization* of X and denote it by $X^{\mathbb{R}_{\text{alg}}}$.

In the rest of this section, we fix an o-minimal structure $\mathcal{S}$ and write "definable" for $\mathcal{S}$-definable. Given a complex algebraic variety X we write X^{def} for the $\mathcal{S}$-definabilization $X^{\mathcal{S}}$.

3.3. O-minimal geometry and algebraization

Why should an algebraic geometer care about o-minimal geometry? Because o-minimal geometry provides strong algebraization results.

3.3.1. Diophantine criterion

The first algebraization result is the celebrated Pila–Wilkie theorem:

Theorem 3.8 ([77]). *Let $Z \subset \mathbb{R}^n$ be a definable set. We define Z^{alg} as the union of all connected* positive-dimensional *semialgebraic subsets of Z. Then, denoting by $H : \mathbb{Q}^n \to \mathbb{R}$ the standard height function:*

$$\forall\, \varepsilon > 0, \quad \exists\, C_\varepsilon > 0, \quad \forall\, T > 0, \quad \left|\left\{x \in (Z \setminus Z^{\mathrm{alg}}) \cap \mathbb{Q}^n,\ H(x) \leq T\right\}\right| < C_\varepsilon T^\varepsilon.$$

In words, if a definable set contains at least polynomially many rational points (with respect to their height), then it contains a positive dimensional semialgebraic set! For instance, if $f : \mathbb{R} \to \mathbb{R}$ is a real analytic function such that its graph $\Gamma_f \cap [0,1] \times [0,1]$ contains at least polynomially many rational points (with respect to their height), then the function f is real algebraic [15]. This algebraization result is a crucial ingredient in the proof of functional transcendence results for period maps, see Section 4.

3.3.2. Definable Chow and definable GAGA

In another direction, algebraicity follows from the meeting of o-minimal geometry with complex geometry. The motto is that o-minimal geometry is incompatible with the many pathologies of complex analysis. As a simple illustration, let $f : \Delta^* \to \mathbb{C}$ be a holomorphic function, and assume that f is definable (where we identify $\mathbb{C}$ with $\mathbb{R}^2$ and $\Delta^* \subset \mathbb{R}^2$ is semialgebraic). Then f does not have any essential singularity at 0 (i.e., f is meromorphic). Otherwise, by the Big Picard theorem, the boundary $\overline{\Gamma_f} \setminus \Gamma_f$ of its graph would contain $\{0\} \times \mathbb{C}$, hence would have the same real dimension (two) as Γ_f, contradicting the fact that Γ_f is definable.

Let us first define a good notion of a definable topological space "endowed with a complex analytic structure." We identify $\mathbb{C}^n$ with $\mathbb{R}^{2n}$ by taking real and imaginary parts. Given $U \subset \mathbb{C}^n$ a definable open subset, let $\mathcal{O}_{\mathbb{C}^n}(U)$ denote the $\mathbb{C}$-algebra of holomorphic definable functions $U \to \mathbb{C}$. The assignment $U \rightsquigarrow \mathcal{O}_{\mathbb{C}^n}(U)$ defines a sheaf $\mathcal{O}_{\mathbb{C}^n}$ on $\underline{\mathbb{C}^n}$ whose stalks are local rings. Given a finitely generated ideal $I \subset \mathcal{O}_{\mathbb{C}^n}(U)$, its zero locus $V(I) \subset U$ is definable and the restriction $\mathcal{O}_{V(I)} := (\mathcal{O}_U / I\mathcal{O}_U)_{|\underline{V(I)}}$ define a sheaf of local rings on $\underline{V(I)}$.

Definition 3.9. A definable complex analytic space is a pair $(\mathcal{X}, \mathcal{O}_\mathcal{X})$ consisting of a definable topological space $\mathcal{X}$ and a sheaf $\mathcal{O}_\mathcal{X}$ on $\underline{\mathcal{X}}$ such that there exists a finite covering of $\mathcal{X}$ by definable open subsets $\mathcal{X}_i$ on which $(\mathcal{X}, \mathcal{O}_\mathcal{X})_{|\mathcal{X}_i}$ is isomorphic to some $(V(I), \mathcal{O}_{V(I)})$.

Bakker et al. [10, THEOREM 2.16] show that this is a reasonable definition: the sheaf $\mathcal{O}_\mathcal{X}$, in analogy with the classical Oka's theorem, is a coherent sheaf of rings. Moreover, one has a natural definabilization functor $(X, \mathcal{O}_X) \rightsquigarrow (X^{\mathrm{def}}, \mathcal{O}_{X^{\mathrm{def}}})$ from the category of separated schemes (or algebraic spaces) of finite type over $\mathbb{C}$ to the category of definable complex analytic spaces, which induces a morphism $g : (\underline{X^{\mathrm{def}}}, \mathcal{O}_{X^{\mathrm{def}}}) \to (\underline{X}, \mathcal{O}_X)$ of locally ringed sites.

Let us now describe the promised algebraization results. The classical Chow's theorem states that a closed complex analytic subset Z of X^{an} for X smooth projective over $\mathbb{C}$ is in fact algebraic. This fails dramatically if X is only quasiprojective, as shown by the graph of the complex exponential in $(\mathbb{A}^2)^{\mathrm{an}}$. However, Peterzil and Starchenko, generalizing [36] in the $\mathbb{R}_{\mathrm{alg}}$ case, have shown the following:

Theorem 3.10 ([69,70]). *Let X be a complex quasiprojective variety and let $Z \subset X^{\mathrm{an}}$ be a closed analytic subvariety. If Z is definable in X^{def} then Z is complex algebraic in X.*

Chow's theorem, which deals only with spaces, was extended to sheaves by Serre [83]: when X is proper, the analytification functor $(\cdot)^{\mathrm{an}} : \mathrm{Coh}(X) \to \mathrm{Coh}(X^{\mathrm{an}})$ defines an equivalence of categories between the categories of coherent sheaves $\mathrm{Coh}(X)$ and $\mathrm{Coh}(X^{\mathrm{an}})$. In the definable world, let X be a separated scheme (or algebraic space) of finite type over $\mathbb{C}$. Associating with a coherent sheaf F on X the coherent sheaf $F^{\mathrm{def}} := F \otimes_{g^{-1}\mathcal{O}_X} \mathcal{O}_{X^{\mathrm{def}}}$ on the $\mathcal{S}$-definabilization X^{def} of X, one obtains a definabilization functor $(\cdot)^{\mathrm{def}} : \mathrm{Coh}(X) \to \mathrm{Coh}(X^{\mathrm{def}})$. Similarly there is an analytification functor $\mathcal{X} \rightsquigarrow \mathcal{X}^{\mathrm{an}}$ from complex definable analytic spaces to complex analytic spaces, that induces a functor $(\cdot)^{\mathrm{an}} : \mathrm{Coh}(\mathcal{X}) \to \mathrm{Coh}(\mathcal{X}^{\mathrm{an}})$.

Theorem 3.11 ([10]). *For every separated algebraic space of finite type X, the definabilization functor $(\cdot)^{\mathrm{def}} : \mathrm{Coh}(X) \to \mathrm{Coh}(X^{\mathrm{def}})$ is exact and fully faithful (but it is not necessarily essentially surjective). Its essential image is stable under subobjects and subquotients.*

Using Theorem 3.11 and Artin's algebraization theorem for formal modification [4], one obtains the following useful algebraization result for definable images of algebraic spaces, which will be used in Section 3.6.2:

Theorem 3.12 ([10]). *Let X be a separated algebraic space of finite type and let $\mathcal{E}$ be a definable analytic space. Any proper definable analytic map $\Phi : X^{\mathrm{def}} \to \mathcal{E}$ factors uniquely as $\iota \circ f^{\mathrm{def}}$, where $f : X \to Y$ is a proper morphism of separated algebraic spaces (of finite type) such that $\mathcal{O}_Y \to f_*\mathcal{O}_X$ is injective, and $\iota : Y^{\mathrm{def}} \hookrightarrow \mathcal{E}$ is a closed immersion of definable analytic spaces.*

3.4. Definability of Hodge varieties

Let us now describe the first result establishing that o-minimal geometry is potentially interesting for Hodge theory.

Theorem 3.13 ([11]). *Any Hodge variety $\Gamma\backslash D$ can be naturally endowed with a functorial structure $(\Gamma\backslash D)^{\mathbb{R}_{\mathrm{alg}}}$ of $\mathbb{R}_{\mathrm{alg}}$-definable complex analytic space.*

Here "functorial" means that that any morphism $(\mathbf{G}', D') \to (\mathbf{G}, D)$ of Hodge data induces a definable map $(\Gamma'\backslash D')^{\mathbb{R}_{\mathrm{alg}}} \to (\Gamma\backslash D)^{\mathbb{R}_{\mathrm{alg}}}$ of Hodge varieties. Let us sketch the construction of $(\Gamma\backslash D)^{\mathbb{R}_{\mathrm{alg}}}$. Without loss of generality (replacing $\mathbf{G}$ by its adjoint group if necessary), we can assume that $\mathbf{G}$ is semisimple, $G = \mathbf{G}(\mathbb{R})^+$. For simplicity, let us assume that the arithmetic lattice Γ is torsion free. We choose a base point in $D = G/M$. Notice that

G and $G/M \subset \check{D}^{\mathbb{R}_{\text{alg}}}$ are naturally endowed with a G-equivariant semialgebraic structure, making the projection $G \to G/M$ semialgebraic. To define an $\mathbb{R}_{\text{alg}}$-structure on $\Gamma\backslash(G/M)$, it is thus enough to find a semialgebraic open fundamental set $F \subset G/M$ for the action of Γ and to write $\Gamma\backslash G/M = \Gamma\backslash F$, where the right-hand side is the quotient of F by the closed étale semialgebraic equivalence relation induced by the action of Γ on D. Here by fundamental set we mean that the set of $\gamma \in \Gamma$ such that $\gamma F \cap F \neq \emptyset$ is finite. We construct the fundamental set F using the reduction theory of arithmetic groups, namely the theory of Siegel sets. Let K be the unique maximal compact subgroup of G containing M. For any $\mathbb{Q}$-parabolic $\mathbf{P}$ of $\mathbf{G}$ with unipotent radical $\mathbf{N}$, the maximal compact subgroup K of G determines a real Levi $L \subset G$ which decomposes as $L = AQ$ where A is the center and Q is semisimple. A semialgebraic Siegel set of G associated to $\mathbf{P}$ and K is then a set of the form $\mathfrak{S} = U(aA_{>0})W$ where $U \subset \mathbf{N}(\mathbb{R})$, $W \subset QK$ are bounded semialgebraic subsets, $a \in A$, and $A_{>0}$ is the cone corresponding to the positive root chamber. By a Siegel set of G associated to K we mean a semialgebraic Siegel set associated to $\mathbf{P}$ and K for some $\mathbb{Q}$-parabolic $\mathbf{P}$ of $\mathbf{G}$. Suppose now that $\Gamma \subset G$ is an arithmetic group. A fundamental result of Borel [16] states that there exists finitely many Siegel sets $\mathfrak{S}_i \subset G$, $1 \leq i \leq s$, associated with K, whose images in $\Gamma\backslash G/K$ cover the whole space; and such that for any $1 \leq i \neq j \leq s$, the set of $\gamma \in \Gamma$ such that $\gamma\mathfrak{S}_i \cap \mathfrak{S}_j \neq \emptyset$ is finite. We call the images $\mathfrak{S}_{i,D} := \mathfrak{S}_i/M$ *Siegel sets for* D. Noticing that these Siegel sets for D are semialgebraic in D, we can take $F = \coprod_{i=1}^{s} \mathfrak{S}_{i,D}$. It is not difficult to show that the $\mathbb{R}_{\text{alg}}$-structure thus constructed is independent of the choice of the base point $eM \in G/M$. The functoriality follows from a nontrivial property of Siegel sets with respect to morphisms of algebraic groups, due to Orr [68].

3.5. Definability of period maps

Once Theorem 3.13 is in place, the following result shows that o-minimal geometry is a natural framework for Hodge theory:

Theorem 3.14 ([11]). *Let S be a smooth connected complex quasiprojective variety. Any period map $\Phi : S^{\text{an}} \to \Gamma\backslash D$ is the analytification of a morphism $\Phi : S^{\mathbb{R}_{\text{an,exp}}} \to (\Gamma\backslash D)^{\mathbb{R}_{\text{an,exp}}}$ of $\mathbb{R}_{\text{an,exp}}$-definable complex analytic spaces, where the $\mathbb{R}_{\text{an,exp}}$-structures on $S(\mathbb{C})$ and $\Gamma\backslash D$ extend their natural $\mathbb{R}_{\text{alg}}$-structures defined in Example 3.7 and Theorem 3.13, respectively.*

In down-to-earth terms, this means that we can cover S by finitely many open affine charts S_i such that Φ restricted to $(\text{Res}_{\mathbb{C}/\mathbb{R}} S_i)(\mathbb{R}) = S_i(\mathbb{C})$ and read in a chart of $\Gamma\backslash D$ defined by a Siegel set of D, can be written using only real polynomials, the real exponential function, and restricted real analytic functions! This statement is already nontrivial when $S = \text{Sh}$ is a Shimura variety and $\Phi^{\text{an}} : S^{\text{an}} \to \Gamma\backslash D$ is the identity map coming from the uniformization $\pi : D \to S^{\text{an}}$ of S^{an} by the hermitian symmetric domain $D = G/K$. In that case the $\mathbb{R}_{\text{alg}}$-definable varieties $\text{Sh}^{\mathbb{R}_{\text{alg}}}$ and $(\Gamma\backslash D)^{\mathbb{R}_{\text{alg}}}$ are not isomorphic, but Theorem 3.14 claims that their $\mathbb{R}_{\text{an,exp}}$-extensions $\text{Sh}^{\mathbb{R}_{\text{an,exp}}}$ and $(\Gamma\backslash D)^{\mathbb{R}_{\text{an,exp}}}$ are. This is equivalent to showing that the restriction $\pi_{|\mathfrak{S}_D} : \mathfrak{S}_D \to S^{\mathbb{R}_{\text{an,exp}}}$ to a Siegel set for D can be written using only real polynomials, the real exponential function, and restricted real analytic functions. This is a nice exercise on the j-function when Sh is a modular curve, was done in [71] and [76] for $\mathrm{Sh} = \mathcal{A}_g$, and [58] in general.

Let us sketch the proof of Theorem 3.14. We choose a log-smooth compactification of S, hence providing us with a definable cover of $S^{\mathbb{R}_{\mathrm{an}}}$ by punctured polydisks $(\Delta^*)^k \times \Delta^l$. We are reduced to showing that the restriction of Φ to such a punctured polydisk is $\mathbb{R}_{\mathrm{an,exp}}$-definable. This is clear if $k = 0$, as in this case $\varphi : \Delta^{k+l} \to \Gamma\backslash D$ is even $\mathbb{R}_{\mathrm{an}}$-definable. For $k > 0$, let $\mathrm{e} : \exp(2\pi \mathrm{i}\cdot) : \mathfrak{H} \to \Delta^*$ be the universal covering map. Its restriction to a sufficiently large bounded vertical strip $V := [a, b] \times]0, +\infty[\subset \mathfrak{H} = \{x + \mathrm{i}y,\ y > 0\}$ is $\mathbb{R}_{\mathrm{an,exp}}$-definable. Considering the following commutative diagram:

$$\begin{array}{ccccc} V^k \times \Delta^l & & \xrightarrow{\quad\widetilde{\Phi}\quad} & D & \hookleftarrow F \\ \Big\downarrow \mathrm{e} & & & \Big\downarrow \pi & \\ (\Delta^*)^k \times \Delta^l & \hookrightarrow & S^{\mathrm{an}} \xrightarrow[\Phi]{} & \Gamma\backslash D, & \end{array}$$

it is thus enough to show that $\pi \circ \widetilde{\Phi} : V^k \times \Delta^l \to \Gamma\backslash D$ is $\mathbb{R}_{\mathrm{an,exp}}$-definable.

Let the coordinates of $(\Delta^*)^k \times \Delta^l$ be t_i, $1 \leq i \leq k + l$, those of $\mathfrak{H}^k$ be z_i, $1 \leq i \leq k$, so that $\mathrm{e}(z_i) = t_i$. Let T_i be the monodromy at infinity of Φ around the hyperplane $(z_i = 0)$, boundary component of $\overline{S} \setminus S$. By Borel's theorem T_i is quasiunipotent. Replacing S by a finite étale cover, we can without loss of generality assume that each $T_i = \exp(N_i)$, with $N_i \in \mathfrak{g}$ nilpotent. The Nilpotent Orbit Theorem of Schmid is equivalent to saying that $\widetilde{\Phi} : V^k \times \Delta^l \to D$ can be written as $\widetilde{\Phi}(z_1, \ldots, z_k, t_{k+1}, \ldots, t_{k+l}) = \exp(\sum_{i=1}^k z_i N_i) \cdot \Psi(t_1, \ldots, t_{k+l})$ for $\Psi : \Delta^k \times \Delta^l \to \check{D}^{\mathrm{an}}$ a holomorphic map. On the one hand, Ψ is $\mathbb{R}_{\mathrm{an}}$-definable as a function of the variables t_i, hence $\mathbb{R}_{\mathrm{an,exp}}$-definable as a function of the variables z_i, $1 \leq i \leq k$, and the variables t_j, $k + 1 \leq j \leq k + l$. On the other hand, $\exp(\sum_{i=1}^k z_i N_i) \in \mathbf{G}(\mathbb{C})$ is polynomial in the variables z_i, as the monodromies N_i are nilpotent and commute pairwise. As the action of $\mathbf{G}(\mathbb{C})$ on $\check{D}$ is algebraic, it follows that $\widetilde{\Phi} : V^k \times \Delta^l \to D$ is $\mathbb{R}_{\mathrm{an,exp}}$-definable. The proof of Theorem 3.14 is thus reduced to the following, proven by Schmid when $k = 1$, $l = 0$ [82, 5.29]:

Theorem 3.15 ([11]). *The image $\widetilde{\Phi}(V^k \times \Delta^l)$ lies in a finite union of Siegel sets of D.*

This can be interpreted as showing that, possibly after passing to a definable cover of $V^k \times \Delta^l$, the Hodge form of $\widetilde{\Phi}$ is Minkowski reduced with respect to a flat frame. This is done using the hard analytic theory of Hodge forms estimates for degenerations of variations of Hodge structure, as in [53, THEOREMS 3.4.1 AND 3.4.2] and [23, THEOREM 5.21].

Remark 3.16. Theorems 3.13 and 3.14 have been extended to the mixed case in [9].

3.6. Applications

3.6.1. About the Cattani–Deligne–Kaplan theorem

As a corollary of Theorems 3.14 and 3.10 one obtains the following, which, in view of (2.3), implies immediately Theorem 3.1:

Theorem 3.17 ([11]). *Let S be a smooth quasiprojective complex variety. Let $\mathbb{V}$ be a polarized $\mathbb{Z}$VHS on S^{an} with period map $\Phi : S^{\mathrm{an}} \to \Gamma\backslash D$. For any special subvariety $\Gamma'\backslash D' \subset \Gamma\backslash D$, its preimage $\Phi^{-1}(\Gamma'\backslash D')$ is a finite union of irreducible algebraic subvarieties of S.*

Indeed, it follows from Theorem 3.13 that $\Gamma'\backslash D'$ is definable in $(\Gamma\backslash D)^{\mathbb{R}_{\mathrm{alg}}}$. By Theorem 3.14, its preimage $\Phi^{-1}(\Gamma'\backslash D')$ is definable in $S^{\mathbb{R}_{\mathrm{an,exp}}}$. As Φ is holomorphic and $\Gamma'\backslash D' \subset \Gamma\backslash D$ is a closed complex analytic subvariety, $\Phi^{-1}(\Gamma'\backslash D')$ is also a closed complex analytic subvariety of S^{an}. By Theorem 3.10, it is thus algebraic in S.

Remark 3.18. Theorem 3.17 has been extended to the mixed case in [9], thus recovering [18–21].

Let $Y \subset S$ be a closed irreducible algebraic subvariety. Let $(\mathbf{G}_Y, D_Y) \subset (\mathbf{G}, D)$ be the generic Hodge datum of $\mathbb{V}$ restricted to the smooth locus of of Y. There exist a smallest Hodge subvariety $\Gamma_Y\backslash D_Y$ of $\Gamma\backslash D$ containing $\Phi(Y^{\mathrm{an}})$. The following terminology will be convenient:

Definition 3.19. Let S be a smooth quasiprojective complex variety. Let $\mathbb{V}$ be a polarized $\mathbb{Z}$VHS on S^{an} with period map $\Phi : S^{\mathrm{an}} \to \Gamma\backslash D$. A closed irreducible subvariety $Y \subset S$ is called a *special subvariety* of S for $\mathbb{V}$ if it coincides with an irreducible component of the preimage $\Phi^{-1}(\Gamma_Y\backslash D_Y)$.

Equivalently, a special subvariety of S for $\mathbb{V}$ is a closed irreducible algebraic subvariety $Y \subset S$ maximal among the closed irreducible algebraic subvarieties Z of S such that the generic Mumford–Tate group $\mathbf{G}_Z$ of $\mathbb{V}_{|Z}$ equals $\mathbf{G}_Y$.

3.6.2. A conjecture of Griffiths

Combining Theorem 3.14 this time with Theorem 3.12 leads to a proof of an old conjecture of Griffiths [44], claiming that the image of any period map has a natural structure of quasiprojective variety (Griffiths proved it when the target Hodge variety is compact):

Theorem 3.20 ([10]). *Let S be a smooth connected quasiprojective complex variety and let $\Phi : S^{\mathrm{an}} \to \Gamma\backslash D$ be a period map. There exists a unique dominant morphism of complex algebraic varieties $f : S \to T$, with T quasiprojective, and a closed complex analytic immersion $\iota : T^{\mathrm{an}} \hookrightarrow \Gamma\backslash D$ such that $\Phi = \iota \circ f^{\mathrm{an}}$.*

Let us sketch the proof. As before, let $S \subset \overline{S}$ be a log-smooth compactification by a simple normal crossing divisor Z. It follows from a result of Griffiths [43, PROP. 9.11I)] that Φ extends to a *proper* period map over the components of Z around which the monodromy is finite. Hence, without loss of generality, we can assume that Φ is proper. The existence of f in the category of algebraic spaces then follows immediately from Theorems 3.14 and 3.12 (for $\mathcal{S} = \mathbb{R}_{\mathrm{an,exp}}$). The proof that T is in fact quasiprojective exploits a crucial observation of Griffiths that $\Gamma\backslash D$ carries a positively curved $\mathbb{Q}$-line bundle $\mathcal{L} := \bigotimes_p \det(F^p)$. This line bundle is naturally definable on $(\Gamma\backslash D)^{\mathrm{def}}$. Using the definable GAGA Theorem 3.11, one shows that its restriction to T^{def} comes from an algebraic $\mathbb{Q}$-line bundle L_T on T, which one manages to show to be ample.

4. FUNCTIONAL TRANSCENDENCE

4.1. Bialgebraic geometry

As we saw, Hodge theory, which compares the Hodge filtration on $H^\bullet_{\mathrm{dR}}(X/\mathbb{C})$ with the rational structure on $H^\bullet_{\mathrm{B}}(X^{\mathrm{an}}, \mathbb{C})$, gives rise to variational Hodge theory, whose fundamental diagram (2.2) compares the algebraic structure of S with the algebraic structure on the dual period domain $\check{D}$. As such, it is a partial answer to one of the most classical problem of complex algebraic geometry: the transcendental nature of the topological universal cover of complex algebraic varieties. If S is a connected complex algebraic variety, the universal cover $\widetilde{S^{\mathrm{an}}}$ has usually no algebraic structure as soon as the topological fundamental group $\pi_1(S^{\mathrm{an}})$ is infinite. As an aside, let us mention an interesting conjecture of Kóllar and Pardon **[60]**, predicting that if X is a normal projective irreducible complex variety whose universal cover $\widetilde{X^{\mathrm{an}}}$ is biholomorphic to a semialgebraic open subset of an algebraic variety then $\widetilde{X^{\mathrm{an}}}$ is biholomorphic to $\mathbb{C}^n \times D \times F^{\mathrm{an}}$, where D is a bounded symmetric domain and F is a normal, projective, irreducible, topologically simply connected, complex algebraic variety. We want to think of variational Hodge theory as an attempt to provide a partial *algebraic uniformization*: the period map emulates an algebraic structure on $\widetilde{S^{\mathrm{an}}}$, modeled on the flag variety $\check{D}$. The remaining task is then to describe the transcendence properties of the complex analytic uniformization map $p : \widetilde{S^{\mathrm{an}}} \to S^{\mathrm{an}}$ with respects to the emulated algebraic structure on $\widetilde{S^{\mathrm{an}}}$ and the algebraic structure S on S^{an}. A few years ago, the author **[55]**, together with Ullmo and Yafaev **[59]**, introduced a convenient format for studying such questions, which encompasses many classical transcendence problems and provides a powerful heuristic.

Definition 4.1. A bialgebraic structure on a connected quasiprojective variety S over $\mathbb{C}$ is a pair

$$\big(f : \widetilde{S^{\mathrm{an}}} \to Z^{\mathrm{an}},\ \rho : \pi_1(S^{\mathrm{an}}) \to \mathrm{Aut}(Z)\big)$$

where Z denotes an algebraic variety (called the *algebraic model* of $\widetilde{S^{\mathrm{an}}}$), $\mathrm{Aut}(Z)$ is its group of algebraic automorphisms, ρ is a group morphism (called the monodromy representation) and f is a ρ-equivariant holomorphic map (called the developing map).

An irreducible analytic subvariety $Y \subset \widetilde{S^{\mathrm{an}}}$ is said to be an *algebraic subvariety of* $\widetilde{S^{\mathrm{an}}}$ for the bialgebraic structure (f, ρ) if Y is an analytic irreducible component of $f^{-1}(\overline{f(Y)}^{\mathrm{Zar}})$ (where $\overline{f(Y)}^{\mathrm{Zar}}$ denotes the Zariski-closure of $f(Y)$ in Z). An irreducible algebraic subvariety $Y \subset \widetilde{S^{\mathrm{an}}}$, resp. $W \subset S$, is said to be *bialgebraic* if $p(Y)$ is an algebraic subvariety of S, resp. any (equivalently one) analytic irreducible component of $p^{-1}(W)$ is an irreducible algebraic subvariety of $\widetilde{S^{\mathrm{an}}}$. The bialgebraic subvarieties of S are precisely the ones where the emulated algebraic structure on $\widetilde{S^{\mathrm{an}}}$ and the one on S interact nontrivially.

Example 4.2. (a) *tori*, $S = (\mathbb{C}^*)^n$. The uniformization map is the multiexponential

$$p := \big(\exp(2\pi i\cdot), \dots, \exp(2\pi i\cdot)\big) : \mathbb{C}^n \to (\mathbb{C}^*)^n,$$

and f is the identity morphism of $\mathbb{C}^n$. An irreducible algebraic subvariety $Y \subset \mathbb{C}^n$ (resp. $W \subset (\mathbb{C}^*)^n$) is bialgebraic if and only if Y is a translate of a rational linear subspace of $\mathbb{C}^n = \mathbb{Q}^n \otimes_{\mathbb{Q}} \mathbb{C}$ (resp. W is a translate of a subtorus of $(\mathbb{C}^*)^n$).

(b) *abelian varieties*, $S = A$ is a complex abelian variety of dimension n. Let $p : \mathrm{Lie}\, A \simeq \mathbb{C}^n \to A$ be the uniformizing map of a complex abelian variety A of dimension n. Once more $\widetilde{S^{\mathrm{an}}} = \mathbb{C}^n$ and f is the identity morphism. One checks easily that an irreducible algebraic subvariety $W \subset A$ is bialgebraic if and only if W is the translate of an abelian subvariety of A.

(c) *Shimura varieties*, $(\mathbf{G}, D)$ is a Shimura datum. The quotient $S^{\mathrm{an}} = \Gamma \backslash D$ (for $\Gamma \subset G := \mathbf{G}^{\mathrm{der}}(\mathbb{R})^+$ a congruence torsion-free lattice) is the complex analytification of a (connected) Shimura variety Sh, defined over a number field (a finite extension of the reflex field of $(\mathbf{G}, D)$). And f is the open embedding $D \hookrightarrow \check{D}^{\mathrm{an}}$.

Let us come back to the case of the bialgebraic structure on S

$$\big(\widetilde{\Phi} : \widetilde{S^{\mathrm{an}}} \to \check{D}^{\mathrm{an}},\ \rho : \pi_1(S^{\mathrm{an}}) \to \Gamma \subset \mathbf{G}(\mathbb{Q})\big)$$

defined by a polarized $\mathbb{Z}$VHS $\mathbb{V}$ and its period map $\Phi : S^{\mathrm{an}} \to \Gamma \backslash D$ with monodromy $\rho : \pi_1(S^{\mathrm{an}}) \to \Gamma \subset \mathbf{G}(\mathbb{Q})$ (in fact, all the examples above are of this form if we consider more generally graded-polarized variations of mixed $\mathbb{Z}$-Hodge structures). What are its bialgebraic subvarieties? To answer this question, we need to define the *weakly special* subvarieties of $\Gamma \backslash D$, as either a special subvariety or a subvariety of the form

$$\Gamma_{\mathbf{H}} \backslash D_H \times \{t\} \subset \Gamma_{\mathbf{H}} \backslash D_H \times \Gamma_{\mathbf{L}} \backslash D_L \subset \Gamma \backslash D,$$

where $(\mathbf{H} \times \mathbf{L}, D_H \times D_L)$ is a Hodge subdatum of $(\mathbf{G}^{\mathrm{ad}}, D)$ and $\{t\}$ is a Hodge generic point in $\Gamma_{\mathbf{L}} \backslash D_L$. Generalizing Theorem 3.17, the preimage under Φ of any weakly special subvariety of $\Gamma \backslash D$ is an algebraic subvariety of S [56]. An irreducible component of such a preimage is called a *weakly special* subvariety of S for $\mathbb{V}$ (or Φ).

Theorem 4.3 ([56]). *Let $\Phi : S^{\mathrm{an}} \to \Gamma \backslash D$ be a period map. The bialgebraic subvarieties of S for the bialgebraic structure defined by Φ are precisely the weakly special subvarieties of S for Φ. In analogy with Definition 3.19, they are also the closed irreducible algebraic subvarieties $Y \subset S$ maximal among the closed irreducible algebraic subvarieties Z of S whose algebraic monodromy group $\mathbf{H}_Z$ equals $\mathbf{H}_Y$.*

When $S = \mathrm{Sh}$ is a Shimura variety, these results are due to Moonen [65] and [91]. In that case the weakly special subvarieties are also the irreducible algebraic subvarieties of Sh whose smooth locus is totally geodesic in $\mathrm{Sh}^{\mathrm{an}}$ for the canonical Kähler–Einstein metric on $\mathrm{Sh}^{\mathrm{an}} = \Gamma \backslash D$ coming from the Bergman metric on D, see [65].

To study not only functional transcendence but also arithmetic transcendence, we enrich bialgebraic structures over $\overline{\mathbb{Q}}$. A $\overline{\mathbb{Q}}$-bialgebraic structure on a quasi-projective variety S defined over $\overline{\mathbb{Q}}$ is a bialgebraic structure $(f : \widetilde{S^{\mathrm{an}}} \to Z^{\mathrm{an}}, h : \pi_1(S^{\mathrm{an}}) \to \mathrm{Aut}(Z))$ such that Z is defined over $\overline{\mathbb{Q}}$ and the homomorphism h takes values in $\mathrm{Aut}_{\overline{\mathbb{Q}}} Z$. An algebraic subvariety $Y \subset \widetilde{S^{\mathrm{an}}}$ is said to be defined over $\overline{\mathbb{Q}}$ if its model $\overline{f(Y)}^{\mathrm{Zar}} \subset Z$ is. A $\overline{\mathbb{Q}}$-bialgebraic subvariety $W \subset S$ is an algebraic subvariety of S defined over $\overline{\mathbb{Q}}$ and such that any (equivalently one) of the analytic irreducible components of $p^{-1}(W)$ is an algebraic subvariety of

$\widetilde{S^{\mathrm{an}}}$ defined over $\overline{\mathbb{Q}}$. A $\overline{\mathbb{Q}}$-bialgebraic point $s \in S(\overline{\mathbb{Q}})$ is also called an *arithmetic point*. Example 4.2a) is naturally defined over $\overline{\mathbb{Q}}$, with arithmetic points the torsion points of $(\mathbb{C}^*)^n$. In Example 4.2b) the bialgebraic structure can be defined over $\overline{\mathbb{Q}}$ if the abelian variety A has CM, and its arithmetic points are its torsion points, see [90]. Example 4.2c) is naturally a $\overline{\mathbb{Q}}$-bialgebraic structure, with arithmetic points the *special points* of the Shimura variety (namely the special subvarieties of dimension zero), at least when the pure part of the Shimura variety is of Abelian type, see [84]. In all these cases it is interesting to notice that the $\overline{\mathbb{Q}}$-bialgebraic subvarieties are the bialgebraic subvarieties containing one arithmetic point (in Example 4.2c) these are the special subvarieties of the Shimura variety).

The bi-algebraic structure associated with a period map $\Phi : S^{\mathrm{an}} \to \Gamma\backslash D$ is defined over $\overline{\mathbb{Q}}$ as soon as S is. In this case, we expect the $\overline{\mathbb{Q}}$-bi-algebraic subvarieties to be precisely the special subvarieties, see [55, 2.6 AND 3.4].

4.2. The Ax–Schanuel theorem for period maps

The geometry of bialgebraic structures is controlled by the following functional transcendence heuristic, whose idea was introduced by Pila in the case of Shimura varieties, see [73, 74]:

Ax–Schanuel principle. Let S be an irreducible algebraic variety endowed with a non-trivial bialgebraic structure. Let $U \subset \widetilde{S^{\mathrm{an}}} \times S^{\mathrm{an}}$ be an algebraic subvariety (for the product bialgebraic structure) and let W be an analytic irreducible component of $U \cap \Delta$, where Δ denotes the graph of $p : \widetilde{S^{\mathrm{an}}} \to S^{\mathrm{an}}$. Then $\operatorname{codim}_U W \geq \dim \overline{W}^{\mathrm{bi}}$, where $\overline{W}^{\mathrm{bi}}$ denotes the smallest bialgebraic subvariety of S containing $p(W)$.

When applied to a subvariety $U \subset \widetilde{S^{\mathrm{an}}} \times S^{\mathrm{an}}$ of the form $Y \times \overline{p(Y)}^{\mathrm{Zar}}$ for $Y \subset \widetilde{S^{\mathrm{an}}}$ algebraic, the Ax–Schanuel principle specializes to the following:

Ax–Lindemann principle. Let S be an irreducible algebraic variety endowed with a nontrivial bialgebraic structure. Let $Y \subset \widetilde{S^{\mathrm{an}}}$ be an algebraic subvariety. Then $\overline{p(Y)}^{\mathrm{Zar}}$ is a bialgebraic subvariety of S.

Ax [5, 6] showed that the abstract Ax–Schanuel principle holds true for Example 4.2a) and Example 4.2b) above, using differential algebra. Notice that the Ax–Lindemann principle in Example 4.2a) is the functional analog of the classical Lindemann theorem stating that if $\alpha_1, \dots, \alpha_n$ are $\mathbb{Q}$-linearly independent algebraic numbers then $e^{\alpha_1}, \dots, e^{\alpha_n}$ are algebraically independent over $\mathbb{Q}$. This explains the terminology. The Ax–Lindemann principle in Example 4.2c) was proven by Pila [72] when S is a product $Y(1)^n \times (\mathbb{C}^*)^k$, by Ullmo–Yafaev [92] for projective Shimura varieties, by Pila–Tsimerman [76] for $\mathcal{A}_g$, and by Klingler–Ullmo–Yafaev [58] for any pure Shimura variety. The full Ax–Schanuel principle was proven by Mok–Pila–Tsimerman for pure Shimura varieties [64].

We conjectured in [55, CONJ. 7.5] that the Ax-Schanuel principle holds true for the bi-algebraic structure associated to a (graded-)polarized variation of (mixed) $\mathbb{Z}$HS on an arbitrary quasiprojective variety S. Bakker and Tsimerman proved this conjecture in the pure case:

Theorem 4.4 (Ax–Schanuel for $\mathbb{Z}$VHS, [12]). *Let $\Phi : S^{an} \to \Gamma\backslash D$ be a period map. Let $V \subset S \times \check{D}$ be an algebraic subvariety. Let U be an irreducible complex analytic component of $W \cap (S \times_{\Gamma\backslash D} D)$ such that*

$$\mathrm{codim}_{S\times D} U < \mathrm{codim}_{S\times\check{D}} W + \mathrm{codim}_{S\times D}(S \times_{\Gamma\backslash D} D). \tag{4.1}$$

Then the projection of U to S is contained in a strict weakly special subvariety of S for Φ.

Remark 4.5. The results of [64] were extended by Gao [39] to mixed Shimura varieties of Kuga type. Recently the full Ax–Schanuel [55, CONJ. 7.5] for variations of mixed Hodge structures has been fully proven independently in [40] and [26].

The proof of Theorem 4.4 follows a strategy started in [58] and fully developed in [64] in the Shimura case, see [88] for an introduction. It does not use Theorem 3.14, but only a weak version equivalent to the Nilpotent Orbit Theorem, and relies crucially on the definable Chow Theorem 3.10, the Pila–Wilkie Theorem 3.8, and the proof that the volume (for the natural metric on $\Gamma\backslash D$) of the intersection of a ball of radius R in $\Gamma\backslash D$ with the horizontal complex analytic subvariety $\Phi(S^{an})$ grows exponentially with R (a negative curvature property of the horizontal tangent bundle).

4.3. On the distribution of the Hodge locus

Theorem 4.4 is most useful, even in its simplest version of the Ax–Lindemann theorem. After Theorem 3.1 one would like to understand the distribution in S of the special subvarieties for $\mathbb{V}$. For instance, are there any geometric constraints on the Zariski closure of $\mathrm{HL}(S, \mathbb{V}^{\otimes})$? To approach this question, let us decompose the adjoint group $\mathbf{G}^{ad}$ into a product $\mathbf{G}_1 \times \cdots \times \mathbf{G}_r$ of its simple factors. It gives rise (after passing to a finite étale covering if necessary) to a decomposition of the Hodge variety $\Gamma\backslash D$ into a product of Hodge varieties $\Gamma_1\backslash D_1 \times \cdots \times \Gamma_r\backslash D_r$. A special subvariety Z of S for $\mathbb{V}$ is said of *positive period dimension* if $\dim_{\mathbb{C}} \Phi(Z^{an}) > 0$; and of *factorwise positive period dimension* if, moreover, the projection of $\Phi(Z^{an})$ on each factor $\Gamma_i\backslash D_i$ has positive dimension. The *Hodge locus of factorwise positive period dimension* $\mathrm{HL}(S, \mathbb{V}^{\otimes})_{\mathrm{fpos}}$ is the union of the strict special subvarieties of positive period dimension, it is contained in the *Hodge locus of positive period dimension* $\mathrm{HL}(S, \mathbb{V}^{\otimes})_{\mathrm{pos}}$ union of the strict special subvarieties of positive period dimension, and the two coincide if $\mathbf{G}^{ad}$ is simple.

Using the Ax–Lindemann theorem special case of Theorem 4.4 and a global algebraicity result in the total bundle of $\mathcal{V}$, Otwinowska and the author proved the following:

Theorem 4.6 ([56]). *Let $\mathbb{V}$ be a polarized $\mathbb{Z}$VHS on a smooth connected complex quasi-projective variety S. Then either $\mathrm{HL}(S, \mathbb{V}^{\otimes})_{\mathrm{fpos}}$ is Zariski-dense in S; or it is an algebraic subvariety of S (i.e., the set of strict special subvarieties of S for $\mathbb{V}$ of factorwise positive period dimension has only finitely many maximal elements for the inclusion).*

Example 4.7. The simplest example of Theorem 4.6 is the following. Let $S \subset \mathcal{A}_g$ be a Hodge-generic closed irreducible subvariety. Either the set of positive-dimensional closed irreducible subvarieties of S which are not Hodge generic has finitely many maximal elements (for the inclusion), or their union is Zariski-dense in S.

Example 4.8. Let $B \subset \mathbb{P}H^0(\mathbb{P}^3_{\mathbb{C}}, \mathcal{O}(d))$ be the open subvariety parametrizing the smooth surfaces of degree d in $\mathbb{P}^3_{\mathbb{C}}$. Suppose $d > 3$. The classical Noether theorem states that any surface $Y \subset \mathbb{P}^3_{\mathbb{C}}$ corresponding to a very general point $[Y] \in B$ has Picard group $\mathbb{Z}$: every curve on Y is a complete intersection of Y with another surface in $\mathbb{P}^3_{\mathbb{C}}$. The countable union $\mathrm{NL}(B)$ of closed algebraic subvarieties of B corresponding to surfaces with bigger Picard group is called the Noether–Lefschetz locus of B. Let $\mathbb{V} \to B$ be the $\mathbb{Z}$VHS $R^2 f_* \mathbb{Z}_{\mathrm{prim}}$, where $f : \mathcal{Y} \to B$ denotes the universal family of surfaces of degree d. Clearly $\mathrm{NL}(B) \subset \mathrm{HL}(B, \mathbb{V}^{\otimes})$. Green (see [94, PROP. 5.20]) proved that $\mathrm{NL}(B)$, hence also $\mathrm{HL}(B, \mathbb{V}^{\otimes})$, is analytically dense in B. Now Theorem 4.6 implies the following: Let $S \subset B$ be a Hodge-generic closed irreducible subvariety. Either $S \cap \mathrm{HL}(B, \mathbb{V}^{\otimes})_{\mathrm{fpos}}$ contains only finitely many maximal positive-dimensional closed irreducible subvarieties of S, or the union of such subvarieties is Zariski-dense in S.

5. TYPICAL AND ATYPICAL INTERSECTIONS: THE ZILBER–PINK CONJECTURE FOR PERIOD MAPS

5.1. The Zilber–Pink conjecture for $\mathbb{Z}$VHS: Conjectures

In the same way that the Ax–Schanuel principle controls the geometry of bialgebraic structures, the diophantine geometry of $\overline{\mathbb{Q}}$-bialgebraic structures is controlled by the following heuristic:

Atypical intersection principle. Let S be an irreducible algebraic $\overline{\mathbb{Q}}$-variety endowed with a $\overline{\mathbb{Q}}$-bialgebraic structure. Then the union S_{atyp} of atypical $\overline{\mathbb{Q}}$-bialgebraic subvarieties of S is an algebraic subvariety of S (i.e., it contains only finitely many atypical $\overline{\mathbb{Q}}$-bialgebraic subvarieties maximal for the inclusion).

Here a $\overline{\mathbb{Q}}$-bialgebraic subvariety $Y \subset S$ is said to be *atypical* for the given bialgebraic structure on S if it is obtained as an excess intersection of $f(\widetilde{S^{\mathrm{an}}})$ with its model $\overline{f(\tilde{Y})}^{\mathrm{Zar}} \subset Z$; and S_{atyp} denotes the union of all atypical subvarieties of S. As a particular case of the atypical intersection principle:

Sparsity of arithmetic points principle. Let S be an irreducible algebraic $\overline{\mathbb{Q}}$-variety endowed with a $\overline{\mathbb{Q}}$-bialgebraic structure. Then any irreducible algebraic subvariety of S containing a Zariski-dense set of atypical arithmetic points is a $\overline{\mathbb{Q}}$-bialgebraic subvariety.

This principle that arithmetic points are sparse is a theorem of Mann [63] in Example 4.2a). For abelian varieties over $\overline{\mathbb{Q}}$ (Example 4.2b)), this is the Manin–Mumford conjecture proven first by Raynaud [80], saying that an irreducible subvariety of an abelian variety over $\overline{\mathbb{Q}}$ containing a Zariski-dense set of torsion point is the translate of an abelian subvariety by a torsion point. For Shimura varieties of abelian type (Example 4.2c)), this is the classical André–Oort conjecture [1, 67] stating that an irreducible subvariety of a Shimura variety containing a Zariski-dense set of special points is special. It has been proven in this case using tame geometry and following the strategy proposed by Pila–Zannier [78] (let us mention [3, 58, 72, 76, 87, 89, 98]; and [38] in the mixed case; see [59] for a survey). Recently the André–Oort conjecture in full generality has been obtained in [75], reducing to the case of abelian type using ingredients from p-adic Hodge theory. We refer to [99] for many examples of atypical intersection problems.

In the case of Shimura varieties (Example 4.2c)) the general atypical intersection principle is the Zilber–Pink conjecture [51, 79, 100]. Only very few instances of the Zilber–Pink conjecture are known outside of the André–Oort conjecture, see [27, 49, 50], for example.

For a general polarized $\mathbb{Z}$VHS $\mathbb{V}$ with period map $\Phi : S^{an} \to \Gamma\backslash D$, which we can assume to be proper without loss of generality, we already mentioned that even the geometric characterization of the $\overline{\mathbb{Q}}$-bialgebraic subvarieties as the special subvarieties is unknown. Replacing the $\overline{\mathbb{Q}}$-bialgebraic subvarieties of S by the special ones, we define:

Definition 5.1. A special subvariety $Z = \Phi^{-1}(\Gamma_Z\backslash D_Z)^0 \subset S$ is said *atypical* if either Z is *singular for* $\mathbb{V}$ (meaning that $\Phi(Z^{an})$ is contained in the singular locus of the complex analytic variety $\Phi(S^{an})$), or if $\Phi(S^{an})$ and $\Gamma_Z\backslash D_Z$ do not intersect generically along $\Phi(Z)$:

$$\operatorname{codim}_{\Gamma\backslash D} \Phi(Z^{an}) < \operatorname{codim}_{\Gamma\backslash D} \Phi(S^{an}) + \operatorname{codim}_{\Gamma\backslash D} \Gamma_Z\backslash D_Z.$$

Otherwise, it is said to be *typical*.

Defining the *atypical Hodge locus* $\mathrm{HL}(S, \mathbb{V}^{\otimes})_{atyp} \subset \mathrm{HL}(S, \mathbb{V}^{\otimes})$ as the union of the atypical special subvarieties of S for $\mathbb{V}$, we obtain the following precise atypical intersection principle for $\mathbb{Z}$VHS, first proposed in [55] in a more restrictive form:

Conjecture 5.2 (Zilber–Pink conjecture for $\mathbb{Z}$VHS, [13, 55]). *Let $\mathbb{V}$ be a polarizable $\mathbb{Z}$VHS on an irreducible smooth quasiprojective variety S. The atypical Hodge locus $\mathrm{HL}(S, \mathbb{V}^{\otimes})_{atyp}$ is a finite union of atypical special subvarieties of S for $\mathbb{V}$. Equivalently, the set of atypical special subvarieties of S for $\mathbb{V}$ has finitely many maximal elements for the inclusion.*

Notice that this conjecture is in some sense more general than the above atypical intersection principle, as we do not assume that S is defined over $\overline{\mathbb{Q}}$; this has to be compared to the fact that the Manin–Mumford conjecture holds true for every complex abelian variety, not necessarily defined over $\overline{\mathbb{Q}}$.

Example 5.3. Recently Baldi and Ullmo [14] proved a special case of Conjecture 5.2 of much interest. Margulis' arithmeticity theorem states that any lattice in a simple real Lie group G of real rank at least 2 is arithmetic: it is commensurable with a group $\mathbf{G}(\mathbb{Z})$, for $\mathbf{G}$ a $\mathbb{Q}$-algebraic group such that $\mathbf{G}(\mathbb{R}) = G$ up to a compact factor. On the other hand, the structure of lattices in a simple real Lie group of rank 1, like the group $\mathrm{PU}(n, 1)$ of holomorphic isometries of the complex unit ball $\mathbf{B}^n_{\mathbb{C}}$ endowed with its Bergman metric, is an essentially open question. In particular, there exist nonarithmetic lattices in $\mathrm{PU}(n, 1)$, $n = 2, 3$. Let $\iota : \Lambda \hookrightarrow \mathrm{PU}(n, 1)$ be a lattice. The ball quotient $S^{an} := \Lambda\backslash\mathbf{B}^n_{\mathbb{C}}$ is the analytification of a complex algebraic variety S. By results of Simpson and Esnault-Groechenig, there exists a $\mathbb{Z}$VHS $\Phi : S^{an} \to \Gamma\backslash(\mathbf{B}^n_{\mathbb{C}} \times D')$ with monodromy representation $\rho : \Lambda \to \mathrm{PU}(n, 1) \times G'$

whose first factor $\Lambda \to \mathrm{PU}(n,1)$ is the rigid representation ι. The special subvarieties of S for $\mathbb{V}$ are the totally geodesic complex subvarieties of S^{an}. When Λ is nonarithmetic, they are automatically atypical. In accordance with Conjecture 5.2 in this case, Baldi and Ullmo prove that if Λ is nonarithmetic, then S^{an} contains only finitely many maximal totally geodesic subvarieties. This result has been proved independently by Bader, Fisher, Miller, and Stover [7], using completely different methods from homogeneous dynamics.

Among the special points for a $\mathbb{Z}$VHS $\mathbb{V}$, the CM-points (i.e., those for which the Mumford–Tate group is a torus) are always atypical except if the generic Hodge datum $(\mathbf{G}, D)$ is of Shimura type and the period map Φ is dominant. Hence, as explained in [55, SECTION 5.2], Conjecture 5.2 implies the following:

Conjecture 5.4 (André–Oort conjecture for $\mathbb{Z}$VHS, [55]). *Let $\mathbb{V}$ be a polarizable $\mathbb{Z}$VHS on an irreducible smooth quasiprojective variety S. If S contains a Zariski-dense set of CM-points then the generic Hodge datum $(\mathbf{G}, D)$ of $\mathbb{V}$ is a Shimura datum, and the period map $\Phi : S^{\mathrm{an}} \to \Gamma\backslash D$ is an algebraic map, dominant on the Shimura variety $\Gamma\backslash D$.*

Example 5.5. Consider the Calabi–Yau Hodge structure V of weight 3 with Hodge numbers $h^{3,0} = h^{2,1} = 1$ given by the mirror dual quintic. Its universal deformation space S is the projective line minus 3 points, which carries a $\mathbb{Z}$VHS $\mathbb{V}$ of the same type. This gives a nontrivial period map $\Phi : S^{\mathrm{an}} \to \Gamma\backslash D$, where $D = \mathbf{Sp}(4,\mathbb{R})/U(1) \times U(1)$ is a 4-dimensional period domain. This period map is known not to factorize through a Shimura subvariety (its algebraic monodromy group is $\mathbf{Sp}_4$). Conjecture 5.4 in that case predicts that S contains only finitely many points CM-points s. A version of this prediction already appears in [48]. The more general Conjecture 5.2 also predicts that S contains only finitely many points s where $\mathbb{V}_s$ splits as a direct sum of two (Tate twisted) weight one Hodge structures $(\mathbb{V}_s^{2,1} \oplus \mathbb{V}_s^{1,2})$ and its orthogonal for the Hodge metric $(\mathbb{V}_s^{3,0} \oplus \mathbb{V}_s^{0,3})$ (the so-called "rank two attractors" points, see [66]).

Conjecture 5.2 about the atypical Hodge locus takes all its meaning if we compare it to the expected behavior of its complement, the *typical Hodge locus* $\mathrm{HL}(S, \mathbb{V}^{\otimes})_{\mathrm{typ}} := \mathrm{HL}(S, \mathbb{V}^{\otimes}) \setminus \mathrm{HL}(S, \mathbb{V}^{\otimes}_{\mathrm{atyp}})$:

Conjecture 5.6 (Density of the typical Hodge locus, [13]). *If $\mathrm{HL}(S, \mathbb{V}^{\otimes})_{\mathrm{typ}}$ is not empty then it is dense (for the analytic topology) in S^{an}.*

Conjectures 5.2 and 5.6 imply immediately the following, which clarifies the possible alternatives in Theorem 4.6:

Conjecture 5.7 ([13]). *Let $\mathbb{V}$ be a polarizable $\mathbb{Z}$VHS on an irreducible smooth quasiprojective variety S. If $\mathrm{HL}(S, \mathbb{V}^{\otimes})_{\mathrm{typ}}$ is empty then $\mathrm{HL}(S, \mathbb{V}^{\otimes})$ is algebraic; otherwise, $\mathrm{HL}(S, \mathbb{V}^{\otimes})$ is analytically dense in S^{an}.*

5.2. The Zilber–Pink conjecture for $\mathbb{Z}$VHS: Results

In [13] Baldi, Ullmo, and I establish the *geometric part* of Conjecture 5.2: the maximal atypical special subvarieties *of positive period dimension* arise in a finite number of families whose geometry is well understood. We cannot say anything on the atypical locus of zero period dimension (for which different ideas are certainly needed):

Theorem 5.8 (Geometric Zilber–Pink, [13]). *Let $\mathbb{V}$ be a polarizable $\mathbb{Z}$VHS on a smooth connected complex quasiprojective variety S. Let Z be an irreducible component of the Zariski closure of* $\mathrm{HL}(S, \mathbb{V}^{\otimes})_{\mathrm{pos,atyp}} := \mathrm{HL}(S, \mathbb{V}^{\otimes})_{\mathrm{pos}} \cap \mathrm{HL}(S, \mathbb{V}^{\otimes})_{\mathrm{atyp}}$ *in S. Then:*

(a) *Either Z is a maximal atypical special subvariety;*

(b) *Or the generic adjoint Hodge datum $(\mathbf{G}_Z^{\mathrm{ad}}, D_{G_Z})$ decomposes as a nontrivial product $(\mathbf{G}', D') \times (\mathbf{G}'', D'')$, inducing (after replacing S by a finite étale cover if necessary)*

$$\Phi_{|Z^{\mathrm{an}}} = (\Phi', \Phi'') : Z^{\mathrm{an}} \to \Gamma_{\mathbf{G}_Z} \backslash D_{G_Z} = \Gamma' \backslash D' \times \Gamma'' \backslash D'' \subset \Gamma \backslash D,$$

such that Z contains a Zariski-dense set of atypical special subvarieties for Φ'' of zero period dimension. Moreover, Z is Hodge generic in the special subvariety $\Phi^{-1}(\Gamma_{\mathbf{G}_Z} \backslash D_{G_Z})^0$ of S for Φ, which is typical.

Conjecture 5.2, which also takes into account the atypical special subvarieties of zero period dimension, predicts that the branch (b) of the alternative in the conclusion of Theorem 5.8 never occurs. Theorem 5.8 is proven using properties of definable sets and the Ax–Schanuel Theorem 4.4, following an idea originating in [89].

As an application of Theorem 5.8, let us consider the *Shimura locus* of S for $\mathbb{V}$, namely the union of the special subvarieties of S for $\mathbb{V}$ which are of Shimura type (but not necessarily with dominant period maps). In [55], I asked (generalizing the André–Oort conjecture for $\mathbb{Z}$VHS) whether a polarizable $\mathbb{Z}$VHS $\mathbb{V}$ on S whose Shimura locus in Zariski-dense in S is necessarily of Shimura type. As a corollary of Theorem 5.8 we obtain:

Theorem 5.9 ([13]). *Let $\mathbb{V}$ be a polarizable $\mathbb{Z}$VHS on a smooth irreducible complex quasi-projective variety S, with generic Hodge datum $(\mathbf{G}, D)$. Suppose that the Shimura locus of S for $\mathbb{V}$* of positive period dimension *is Zariski-dense in S. If $\mathbf{G}^{\mathrm{ad}}$ is simple then $\mathbb{V}$ is of Shimura type.*

5.3. On the algebraicity of the Hodge locus

In view of Conjecture 5.7, it is natural to ask if there a simple combinatorial criterion on $(\mathbf{G}, D)$ for deciding whether $\mathrm{HL}(S, \mathbb{V})_{\mathrm{typ}}$ is empty. Intuitively, one expects that the more "complicated" the Hodge structure is, the smaller the typical Hodge locus should be, due to the constraint imposed by Griffiths' transversality. Let us measure the complexity of $\mathbb{V}$ by its *level*: when $\mathbf{G}^{\mathrm{ad}}$ is simple, it is the greatest integer k such that $\mathfrak{g}^{k,-k} \neq 0$ in the Hodge decomposition of the Lie algebra $\mathfrak{g}$ of $\mathbf{G}$; in general one takes the minimum of these integers obtained for each simple $\mathbb{Q}$-factor of $\mathbf{G}^{\mathrm{ad}}$. While strict typical special subvarieties usually abound for $\mathbb{Z}$VHSs of level one (e.g., families of abelian varieties, see Example 4.7; or families of K3 surfaces) and can occur in level two (see Example 4.8), they do not exist in level at least three!

Theorem 5.10 ([13]). *Let $\mathbb{V}$ be a polarizable $\mathbb{Z}$VHS on a smooth connected complex quasi-projective variety S. If $\mathbb{V}$ is of level at least 3 then $\mathrm{HL}(S, \mathbb{V}^{\otimes})_{\mathrm{typ}} = \emptyset$ (and thus $\mathrm{HL}(S, \mathbb{V}^{\otimes}) = \mathrm{HL}(S, \mathbb{V}^{\otimes})_{\mathrm{atyp}}$).*

The proof of Theorem 5.10 is purely Lie-theoretic. Let $(\mathbf{G}, D)$ be the generic Hodge datum of $\mathbb{V}$ and $\Phi : S^{\mathrm{an}} \to \Gamma\backslash D$ its period map. Suppose that $Y \subset S$ is a typical special subvariety, with generic Hodge datum $(\mathbf{G}_Y, D_Y)$. The typicality condition and the horizontality of the period map Φ imply that $\mathfrak{g}_Y^{-i,i} = \mathfrak{g}^{-i,i}$ for all $i \geq 2$ (for the Hodge structures on the Lie algebras $\mathfrak{g}_Y$ and $\mathfrak{g}$ defined by some point of D_Y). Under the assumption that $\mathbb{V}$ has level at least 3, we show that this is enough to ensure that $\mathfrak{g}_Y = \mathfrak{g}$, hence $Y = S$. Hence there are no strict typical special subvariety.

Notice that Conjecture 5.2 and Theorem 5.10 imply:

Conjecture 5.11 (Algebraicity of the Hodge locus in level at least 3, [13]). *Let $\mathbb{V}$ be a polarizable $\mathbb{Z}$VHS on a smooth connected complex quasiprojective variety S. If $\mathbb{V}$ is of level at least 3 then $\mathrm{HL}(S, \mathbb{V}^{\otimes})$ is algebraic.*

The main result of [13], which follows immediately from Theorems 5.8 and 5.10, is the following stunning geometric reinforcement of Theorems 3.1 and 4.6:

Theorem 5.12 ([13]). *If $\mathbb{V}$ is of level at least 3 then $\mathrm{HL}(S, \mathbb{V}^{\otimes})_{\mathrm{fpos}}$ is algebraic.*

As a simple geometric illustration of Theorem 5.12, we prove the following, to be contrasted with the $n = 2$ case (see Example 4.8):

Corollary 5.13. *Let $\mathbf{P}_{\mathbb{C}}^{N(n,d)}$ be the projective space parametrizing the hypersurfaces X of $\mathbf{P}_{\mathbb{C}}^{n+1}$ of degree d (where $N(n,d) = \binom{n+d+1}{d} - 1$). Let $U_{n,d} \subset \mathbf{P}_{\mathbb{C}}^{N(n,d)}$ be the Zariski-open subset parametrizing the smooth hypersurfaces X and let $\mathbb{V} \to U_{n,d}$ be the $\mathbb{Z}$VHS corresponding to the primitive cohomology $H^n(X, \mathbb{Z})_{\mathrm{prim}}$. If $n \geq 3$ and $d > 5$, then $\mathrm{HL}(U_{n,d}, \mathbb{V}^{\otimes})_{\mathrm{pos}} \subset U_{n,d}$ is algebraic.*

5.4. On the typical Hodge locus in level one and two

In the direction of Conjecture 5.6, we obtain:

Theorem 5.14 (Density of the typical locus, [13]). *Let $\mathbb{V}$ be a polarized $\mathbb{Z}$VHS on a smooth connected complex quasiprojective variety S. If the typical Hodge locus $\mathrm{HL}(S, \mathbb{V}^{\otimes})_{\mathrm{typ}}$ is nonempty (hence the level of $\mathbb{V}$ is one or two by Theorem 5.10) then $\mathrm{HL}(S, \mathbb{V}^{\otimes})$ is analytically (hence Zariski) dense in S.*

Notice that, in Theorem 5.14, we also treat the typical Hodge locus of zero period dimension. Theorem 5.14 is new even for S a subvariety of a Shimura variety. Its proof is inspired by the arguments of Chai [24] in that case.

It remains to find a criterion for deciding whether, in level one or two, the typical Hodge locus $\mathrm{HL}(S, \mathbb{V}^{\otimes})_{\mathrm{typ}}$ is empty or not. We refer to [57, THEOREM 2.15] and [85, 86] for results in this direction.

6. ARITHMETIC ASPECTS

We turn briefly to some arithmetic aspects of period maps.

6.1. Field of definition of special subvarieties

Once more the geometric case provides us with a motivation and a heuristic. Let $f : X \to S$ be a smooth projective morphism of connected algebraic varieties defined over a number field $L \subset \mathbb{C}$ and let $\mathbb{V}$ be the natural polarizable $\mathbb{Z}$VHS on S^{an} with underlying local system $R^\bullet f_*^{\mathrm{an}}\mathbb{Z}$. In that case, the Hodge conjecture implies that each special subvariety Y of S for $\mathbb{V}$ is defined over $\overline{\mathbb{Q}}$ and that each of the $\mathrm{Gal}(\overline{\mathbb{Q}}/L)$-conjugates of Y is again a special subvariety of S for $\mathbb{V}$. More generally, let us say that a polarized $\mathbb{Z}$VHS $\mathbb{V} = (\mathbb{V}_{\mathbb{Z}}, (\mathcal{V}, F^\bullet, \nabla), \mathrm{q})$ on S^{an} is *defined over a number field* $L \subset \mathbb{C}$ if S, $\mathcal{V}$, $F^\bullet$ and ∇ are defined over L (with the obvious compatibilities).

Conjecture 6.1. *Let $\mathbb{V}$ be a $\mathbb{Z}$VHS defined over a number field $L \subset \mathbb{C}$. Then any special subvariety of S for $\mathbb{V}$ is defined over $\overline{\mathbb{Q}}$, and any of its finitely many* $\mathrm{Gal}(\overline{\mathbb{Q}}/L)$*-conjugates is a special subvariety of S for $\mathbb{V}$.*

There are only few results in that direction: see [95, THEOREM 0.6] for a proof under a strong geometric assumption; and [81], where it is shown that when S (not necessarily $\mathbb{V}$) is defined over $\overline{\mathbb{Q}}$, then a special subvariety of S for $\mathbb{V}$ is defined over $\overline{\mathbb{Q}}$ if and only if it contains a $\overline{\mathbb{Q}}$-point of S. In [57] Otwinowska, Urbanik, and I provide a simple geometric criterion for a special subvariety of S for $\mathbb{V}$ to satisfy Conjecture 6.1. In particular we obtain:

Theorem 6.2 ([57]). *Let $\mathbb{V}$ be a polarized $\mathbb{Z}$VHS on a smooth connected complex quasi-projective variety S. Suppose that the adjoint generic Mumford–Tate group $\mathbf{G}^{\mathrm{ad}}$ of $\mathbb{V}$ is simple. If S is defined over a number field L, then any maximal (strict) special subvariety $Y \subset S$ of positive period dimension is defined over $\overline{\mathbb{Q}}$. If, moreover, $\mathbb{V}$ is defined over L then the finitely many* $\mathrm{Gal}(\overline{\mathbb{Q}}/L)$*-translates of Y are special subvarieties of S for $\mathbb{V}$.*

As a corollary of Theorems 5.12 and 6.2, one obtains the following, which applies for instance in the situation of Corollary 5.13.

Corollary 6.3. *Let $\mathbb{V}$ be a polarized variation of $\mathbb{Z}$-Hodge structure on a smooth connected quasiprojective variety S. Suppose that $\mathbb{V}$ is of level at least 3, and that it is defined over $\overline{\mathbb{Q}}$. Then* $\mathrm{HL}(S, \mathbb{V}^{\otimes})_{\mathrm{fpos}}$ *is an algebraic subvariety of S, defined over $\overline{\mathbb{Q}}$.*

It is interesting to notice that Conjecture 5.11, which is stronger than Theorem 5.12, predicts the existence of a Hodge generic $\overline{\mathbb{Q}}$-point in S for $\mathbb{V}$ in the situation of Corollary 6.3.

As the criterion given in [57] is purely geometric, it says nothing about fields of definitions of special points. It is, however, strong enough to reduce the first part of Conjecture 6.1 to this particular case:

Theorem 6.4. *Special subvarieties for $\mathbb{Z}$VHSs defined over $\overline{\mathbb{Q}}$ are defined over $\overline{\mathbb{Q}}$ if and only if it holds true for special points.*

6.2. Absolute Hodge locus

Interestingly, Conjecture 6.1 in the geometric case follows from an *a priori* much weaker conjecture than the Hodge conjecture. Let $f : X \to S$ be a smooth projective morphism of smooth connected complex quasiprojective varieties. For any automorphism $\sigma \in \mathrm{Aut}(\mathbb{C}/\mathbb{Q})$, we can consider the algebraic family $f^\sigma : X^\sigma \to S^\sigma$, where $\sigma^{-1} : S^\sigma = S \times_{\mathbb{C},\sigma} \mathbb{C} \xrightarrow{\sim} S$ is the natural isomorphism of abstract schemes; and the attached polarizable $\mathbb{Z}$VSH $\mathbb{V}^\sigma = (\mathbb{V}^\sigma_{\mathbb{Z}}, \mathcal{V}^\sigma, F^{\bullet\sigma}, \nabla^\sigma)$ with underlying local system $\mathbb{V}^\sigma_{\mathbb{Z}} = R f^{\sigma\,\mathrm{an}}_{*}\mathbb{Z}$ on $(S^\sigma)^{\mathrm{an}}$. The algebraic construction of the algebraic de Rham cohomology provides compatible canonical comparison isomorphisms $\iota^\sigma : (\mathcal{V}^\sigma, F^{\bullet\sigma}, \nabla^\sigma) \xrightarrow{\sim} \sigma^{-1*}(\mathcal{V}, F^\bullet, \nabla)$ of the associated algebraic filtered vector bundles with connection. More generally, a collection of $\mathbb{Z}$VHS $(\mathbb{V}^\sigma)_\sigma$ with such compatible comparison isomorphisms is called a *(de Rham) motivic variation of Hodge structures* on S, in which case we write $\mathbb{V} := \mathbb{V}^{\mathrm{Id}}$. Following Deligne (see [25] for a nice exposition), an *absolute Hodge tensor* for such a collection is a Hodge tensor α for $\mathbb{V}_s$ such that the conjugates $\sigma^{-1*}\alpha_{\mathrm{dR}}$ of the de Rham component of α defines a Hodge tensor in $\mathbb{V}^\sigma_{\sigma(s)}$ for all σ. The *generic absolute Mumford–Tate group* for $(\mathbb{V}^\sigma)_\sigma$ is defined in terms of the absolute Hodge tensors as the generic Mumford–Tate group is defined in terms of the Hodge tensors. Thus $\mathbf{G} \subset \mathbf{G}^{\mathrm{AH}}$. In view of Definition 3.19 the following is natural:

Definition 6.5. Let $(\mathbb{V}^\sigma)_\sigma$ be a (de Rham) motivic variation of Hodge structure on a smooth connected complex quasiprojective variety S. A closed irreducible algebraic subvariety Y of S is called *absolutely special* if it is maximal among the closed irreducible algebraic subvarieties Z of S satisfying $\mathbf{G}^{\mathrm{AH}}_Z = \mathbf{G}^{\mathrm{AH}}_Y$.

In the geometric case, the Hodge conjecture implies, since any automorphism $\sigma \in \mathrm{Aut}(\mathbb{C}/\mathbb{Q})$ maps algebraic cycles in X to algebraic cycles on X^σ, the following conjecture of Deligne:

Conjecture 6.6 ([33]). *Let $(\mathbb{V}^\sigma)_\sigma$ be a (de Rham) motivic variation of Hodge structure on S. Then all Hodge tensors are absolute Hodge tensors, i.e., $\mathbf{G} = \mathbf{G}^{\mathrm{AH}}$.*

This conjecture immediately implies:

Conjecture 6.7. *Let $(\mathbb{V}^\sigma)_\sigma$ be a (de Rham) motivic variation of Hodge structure on S. Then any special subvariety of S for $\mathbb{V}$ is absolutely special for $(\mathbb{V}^\sigma)_\sigma$.*

Let us say that a (de Rham) motivic variation $(\mathbb{V}^\sigma)_\sigma$ is defined over $\overline{\mathbb{Q}}$ if $\mathbb{V}^\sigma = \mathbb{V}$ for all $\sigma \in \mathrm{Aut}(\mathbb{C}/\overline{\mathbb{Q}})$. In the geometric case, any morphism $f : X \to S$ defined over $\overline{\mathbb{Q}}$ defines such a (de Rham) motivic variation $(\mathbb{V}^\sigma)_\sigma$ over $\overline{\mathbb{Q}}$. Notice that the absolutely special subvarieties of S for $(\mathbb{V}^\sigma)_\sigma$ are then by their very definition defined over $\overline{\mathbb{Q}}$, and their Galois conjugates are also special. In particular, Conjecture 6.7 implies Conjecture 6.1 in the geometric case. As proven in [95], Deligne's conjecture is actually equivalent to a much stronger version of Conjecture 6.1, where one replaces the special subvarieties of S (components of the Hodge locus) with the special subvarieties in the total bundle of $\mathcal{V}^\otimes$ (components of the locus of Hodge tensors).

Recently T. Kreutz, using the same geometric argument as in [57], justified Theorem 6.2 by proving:

Theorem 6.8 ([62]). *Let $(\mathbb{V}^\sigma)_\sigma$ be a (de Rham) motivic variation of Hodge structure on S. Suppose that the adjoint generic Mumford–Tate group $\mathbf{G}^{\mathrm{ad}}$ is simple. Then any strict maximal special subvariety $Y \subset S$ of positive period dimension for $\mathbb{V}$ is absolutely special.*

We refer the reader to [61], as well as [93], for other arithmetic aspects of Hodge loci taking into account not only the de Rham incarnation of absolute Hodge classes but also their ℓ-adic components.

ACKNOWLEDGMENTS

I would like to thank Gregorio Baldi, Benjamin Bakker, Yohan Brunebarbe, Jeremy Daniel, Philippe Eyssidieux, Ania Otwinowska, Carlos Simpson, Emmanuel Ullmo, Claire Voisin, and Andrei Yafaev for many interesting discussions on Hodge theory. I also thank Gregorio Baldi, Tobias Kreutz, and Leonardo Lerer for their comments on this text.

FUNDING

This work was partially supported by the ERC Advanced Grant 101020009 "TameHodge."

REFERENCES

[1] Y. André, *G-functions and geometry*. Aspects Math. E13, 1989.

[2] Y. André, Mumford–Tate groups of mixed Hodge structures and the theorem of the fixed part. *Compos. Math.* **82** (1992), 1–24.

[3] F. Andreatta, E. Goren, B. Howard, and K. Madapusi-Pera, Faltings heights of abelian varieties with complex multiplication. *Ann. of Math. (2)* **187** (2018), no. 2, 391–531.

[4] M. Artin, Algebraization of formal moduli. II. Existence of modifications. *Ann. of Math.* **91** (1970), 88–135.

[5] J. Ax, On Schanuel's conjecture. *Ann. of Math.* **93** (1971), 1–24.

[6] J. Ax, Some topics in differential algebraic geometry. I. Analytic subgroups of algebraic groups. *Amer. J. Math.* **94** (1972), 1195–1204.

[7] U. Bader, D. Fisher, N. Miller, and M. Stover, Arithmeticity, superrigidity and totally geodesic submanifolds of complex hyperbolic manifolds. 2020, arXiv:2006.03008.

[8] W. L. Baily and A. Borel, Compactification of arithmetic quotients of bounded symmetric domains. *Ann. of Math.* **84** (1966), 442–528.

[9] B. Bakker, Y. Brunebarbe, B. Klingler, and J. Tsimerman, Definability of mixed period maps. *J. Eur. Math. Soc.* (to appear), arXiv:2006.12403.

[10] B. Bakker, Y. Brunebarbe, and J. Tsimerman, o-minimal GAGA and a conjecture of Griffiths. 2019, arXiv:1811.12230.

[11] B. Bakker, B. Klingler, and J. Tsimerman, Tame topology of arithmetic quotients and algebraicity of Hodge loci. *J. Amer. Math. Soc.* **33** (2020), 917–939.

[12] B. Bakker and J. Tsimerman, The Ax–Schanuel conjecture for variations of Hodge structures. *Invent. Math.* **217** (2019), no. 1, 77–94.

[13] G. Baldi, B. Klingler, and E. Ullmo, On the distribution of the Hodge locus. 2021, arXiv:2107.08838.

[14] G. Baldi and E. Ullmo, Special subvarieties of non-arithmetic ball quotients and Hodge theory. 2021, arXiv:2005.03524.

[15] E. Bombieri and J. Pila, The number of integral points on arcs and ovals. *Duke Math. J.* **59** (1989), no. 2, 337–357.

[16] A. Borel, *Introduction aux groupes arithmétiques*. Publications de l'Institut de Mathématique de l'Université de Strasbourg, XV. Actualités Scientifiques et Industrielles 1341, Hermann, Paris, 1969.

[17] A. Borel, Some metric properties of arithmetic quotients of symmetric spaces and an extension theorem. *J. Differential Geom.* **6** (1972), 543–560.

[18] P. Brosnan and G. Pearlstein, Zero loci of admissible normal functions with torsion singularities. *Duke Math. J.* **150** (2009), no. 1, 77–100.

[19] P. Brosnan and G. Pearlstein, The zero locus of an admissible normal function. *Ann. of Math. (2)* **170** (2009), no. 2, 883–897.

[20] P. Brosnan and G. Pearlstein, On the algebraicity of the zero locus of an admissible normal function. *Compos. Math.* **149** (2013), no. 11, 1913–1962.

[21] P. Brosnan, G. Pearlstein, and C. Schnell, The locus of Hodge classes in an admissible variation of mixed Hodge structure. *C. R. Math. Acad. Sci. Paris* **348** (2010), no. 11–12, 657–660.

[22] E. Cattani, P. Deligne, and A. Kaplan, On the locus of Hodge classes. *J. Amer. Math. Soc.* **8** (1995), 483–506.

[23] E. Cattani, A. Kaplan, and W. Schmid, Degeneration of Hodge structures. *Ann. of Math. (2)* **123** (1986), no. 3, 457–535.

[24] C. L. Chai, Density of members with extra Hodge cycles in a family of Hodge structures. *Asian J. Math.* **2** (1998), no. 3, 405–418.

[25] F. Charles and C. Schnell, Notes on absolute Hodge classes, Hodge theory. *Math. Notes* **49** (2014), 469–530.

[26] K. C. T. Chiu, Ax–Schanuel for variations of mixed Hodge structures. 2021, arXiv:2101.10968.

[27] C. Daw and J. Ren, Applications of the hyperbolic Ax–Schanuel conjecture. *Compos. Math.* **154** (2018), no. 9, 1843–1888.

[28] P. Deligne, *Equations différentielles à points singuliers réguliers*. Lecture Notes in Math. 163, Springer, 1970.

[29] P. Deligne, Théorie de Hodge II. *Publ. Math. Inst. Hautes Études Sci.* **40** (1971), 5–57.

[30] P. Deligne, Travaux de Shimura. *In Séminaire Bourbaki Exposé 389, Lecture Notes in Math.* **244** (1971), 123–165.

[31] P. Deligne, La conjecture de Weil pour les surfaces $K3$. *Invent. Math.* **15** (1972), 206–226.

[32] P. Deligne, Variétés de Shimura: interprétation modulaire et techniques de construction de modèles canoniques. In *Automorphic Forms, Representations, and L-functions part. 2*, pp. 247–290, Proc. Sympos. Pure Math. 33, American Mathematical Society, 1979.

[33] P. Deligne, J. Milne, A. Ogus, and K.-Y. Shih, *Hodge cycles, motives, and Shimura varieties*. Lecture Notes in Math. 900, Springer, 1982.

[34] L. van den Dries, *Tame Topology and o-minimal structures*. London Math. Soc. Lecture Note Ser. 248, Cambridge University Press, 1998.

[35] L. van den Dries and C. Miller, On the real exponential field with restricted analytic functions. *Israel J. Math.* **85** (1994), 19–56.

[36] E. Fortuna and S. Łojasiewicz, Sur l'algébricité des ensembles analytiques complexes. *J. Reine Angew. Math.* **329** (1981), 215–220.

[37] A. M. Gabrielov, Projections of semi-analytic sets. *Funktsional. Anal. i Prilozhen.* **2** (1968), 18–30.

[38] Z. Gao, Towards the André–Oort conjecture for mixed Shimura varieties: the Ax–Lindemann theorem and lower bounds for Galois orbits of special points. *J. Reine Angew. Math.* **732** (2017), 85–146.

[39] Z. Gao, Mixed Ax–Schanuel for the universal abelian varieties and some applications. *Compos. Math.* **156** (2020), no. 11, 2263–2297.

[40] Z. Gao and B. Klingler, The Ax–Schanuel conjecture for variations of mixed Hodge structures. 2021, arXiv:2101.10938.

[41] M. Green, P. Griffiths, and M. Kerr, *Mumford–Tate groups and domains. Their geometry and arithmetic*. Ann. of Math. Stud. 183, Princeton University Press, 2012.

[42] P. Griffiths, Period of integrals on algebraic manifolds, I. *Amer. J. Math.* **90** (1968), 568–626.

[43] P. Griffiths, Period of integrals on algebraic manifolds, III. *Publ. Math. Inst. Hautes Études Sci.* **38** (1970), 125–180.

[44] P. Griffiths, Periods of integrals on algebraic manifolds: Summary of main results and discussion of open problems. *Bull. Amer. Math. Soc.* **76** (1970), 228–296.

[45] P. Griffiths, C. Robles, and D. Toledo, Quotients of non-classical flag domains are not algebraic. *Algebr. Geom.* **1** (2014), no. 1, 1–13.

[46] P. Griffiths and W. Schmid, Locally homogeneous complex manifolds. *Acta Math.* **123** (1969), 253–302.

[47] A. Grothendieck, Esquisse d'un programme in Geometric Galois Actions vol. I, London Math. Soc. Lecture Note Ser. 242, Cambridge University Press, 1997.

[48] S. Gukov and C. Vafa, Rational conformal field theories and complex multiplication. *Comm. Math. Phys.* **246** (2004), no. 1, 181–210.

[49] P. Habegger and J. Pila, Some unlikely intersections beyond André–Oort. *Compos. Math.* **148** (2012), no. 1, 1–27.

[50] P. Habegger and J. Pila, O-minimality and certain atypical intersections. *Ann. Sci. Éc. Norm. Supér. (4)* **49** (2016), no. 4, 813–858.

[51] P. Habegger, G. Rémond, T. Scanlon, E. Ullmo, and A. Yafaev, *Around the Zilber–Pink conjecture*. Panor. Synthèses 52, Société Mathématique de France, 2017.

[52] W. Hodge, Differential forms on a Kähler manifold. *Proc. Camb. Philos. Soc.* **47** (1951), 504–517.

[53] M. Kashiwara, The asymptotic behavior of a variation of polarized Hodge structure. *Publ. Res. Inst. Math. Sci.* **21** (1985), no. 4, 853–875.

[54] A. Khovanskii, On a class of systems of transcendental equations. *Sov. Math., Dokl.* **22** (1980), 762–765.

[55] B. Klingler, Hodge loci and atypical intersections: conjectures. 2017, arXiv:1711.09387, accepted for publication in the book *Motives and complex multiplication*, Birkhaüser.

[56] B. Klingler and A. Otwinowska, On the closure of the positive dimensional Hodge locus. *Invent. Math.* **225** (2021), no. 3, 857–883.

[57] B. Klingler, A. Otwinowska, and D. Urbanik, On the fields of definition of Hodge loci. *Ann. Éc. Norm. Sup.* (to appear), arXiv:2010.03359.

[58] B. Klingler, E. Ullmo, and A. Yafaev, The hyperbolic Ax–Lindemann–Weierstraß conjecture. *Publ. Math. Inst. Hautes Études Sci.* **123** (2016), no. 1, 333–360.

[59] B. Klingler, E. Ullmo, and A. Yafaev, Bi-algebraic geometry and the André–Oort conjecture: a survey. In *Proceedings of 2015 AMS Summer Institute in Algebraic Geometry, PSPMS 97-2*, pp. 319–360, AMS, 2018.

[60] J. Kóllar and J. Pardon, Algebraic varieties with semialgebraic universal cover. *J. Topol.* **5** (2012), no. 1, 199–212.

[61] T. Kreutz, ℓ-Galois special subvarieties and the Mumford–Tate conjecture. 2021, arXiv:2111.01126.

[62] T. Kreutz, Absolutely special subvarieties and absolute Hodge cycles. 2021, arXiv:2111.00216.

[63] H. B. Mann, On linear relations between roots of unity. *Mathematika* **12** (1965).

[64] N. Mok, J. Pila, and J. Tsimerman, Ax–Schanuel for Shimura varieties. *Ann. of Math. (2)* **189** (2019), no. 3, 945–978.

[65] B. Moonen, Linearity properties of Shimura varieties. I. *J. Algebraic Geom.* **7** (1998), 539–567.

[66] G. Moore, Arithmetic and Attractors. 2007, arXiv:hep-th/9807087.

[67] F. Oort, Canonical liftings and dense sets of CM-points. In *Arithmetic Geometry (Cortona, 1994)*, pp. 228–234, Symp. Math. XXXVII, Cambridge Univ. Press, Cambridge, 1997.

[68] M. Orr, Height bounds and the Siegel property. *Algebra Number Theory* **12** (2018), no. 2, 455–478.

[69] Y. Peterzil and S. Starchenko, Complex analytic geometry and analytic-geometric categories. *J. Reine Angew. Math.* **626** (2009), 39–74.

[70] Y. Peterzil and S. Starchenko, Tame complex analysis and o-minimality. In *Proceedings of the ICM*, Hyderabad, 2010. Available on first author's webpage.

[71] Y. Peterzil and S. Starchenko, Definability of restricted theta functions and families of abelian varieties. *Duke Math. J.* **162** (2013), 731–765.

[72] J. Pila, O-minimality and the André–Oort conjecture for $\mathbb{C}^n$. *Ann. of Math.* **173** (2011), 1779–1840.

[73] J. Pila, O-minimality and diophantine geometry. In *Proceedings of the ICM – Seoul*, 2014.

[74] J. Pila, Functional transcendence via o-minimality. In *O-minimality and diophantine geometry*, pp. 66–99, London Math. Soc. Lecture Note Ser. 421, Cambridge Univ. Press, Cambridge, 2015.

[75] J. Pila, A. Shankar, and J. Tsimerman, with an appendix by H. Esnault and M. Groechenig, Canonical heights on Shimura varieties and the André–Oort conjecture. 2021, arXiv:2109.08788.

[76] J. Pila and J. Tsimerman, Ax–Lindemann for $\mathcal{A}_g$. *Ann. of Math.* **179** (2014), 659–681.

[77] J. Pila and A. Wilkie, The rational points on a definable set. *Duke Math. J.* **133** (2006), 591–616.

[78] J. Pila and U. Zannier, Rational points in periodic analytic sets and the Manin–Mumford conjecture. *Atti Accad. Naz. Lincei Cl. Sci. Fis. Mat. Natur. Rend. Lincei (9) Mat. Appl.* **19** (2008), no. 2, 149–162.

[79] R. Pink, A combination of the conjectures of Mordell–Lang and André–Oort. In *Geometric Methods in Algebra and Number Theory*, pp. 251–282, Progr. Math. 253, Birkhaüser, 2005.

[80] M. Raynaud, Sous-variétés d'une variété abélienne et points de torsion. In *Arithmetic and Geometry, Vol. I*, Progr. Math. 35, Birkhäuser, 1988.

[81] M. Saito and C. Schnell, Fields of definition of Hodge loci. In *Recent advances in Hodge theory*, pp. 275–291, London Math. Soc. Lecture Note Ser. 427, Cambridge Univ. Press, 2016.

[82] W. Schmid, Variation of Hodge structure: the singularities of the period mapping. *Invent. Math.* **22** (1973), 211–319.

[83] J. P. Serre, Géométrie algébrique et géométrie analytique. *Ann. Inst. Fourier (Grenoble)* **6** (1955–1956), 1–42.

[84] H. Shiga and J. Wolfart, Criteria for complex multiplication and transcendence properties of automorphic functions. *J. Reine Angew. Math.* **463** (1995), 1–25.

[85] S. Tayou, On the equidistribution of some Hodge loci. *J. Reine Angew. Math.* **762** (2020), 167–194.

[86] S. Tayou and N. Tholozan, Equidistribution of Hodge loci II. 2021, arXiv:2103.15717.

[87] J. Tsimerman, A proof of the André–Oort conjecture for $\mathbb{A}_g$. *Ann. of Math. (2)* **187** (2018), no. 2, 379–390.

[88] J. Tsimerman, Functional transcendence and arithmetic applications. In *Proc. Int. Cong. of Math.* 2, pp. 453–472, 2018.

[89] E. Ullmo, Applications du théorème d'Ax–Lindemann hyperbolique. *Compos. Math.* **150** (2014), 175–190.

[90] E. Ullmo, Structures spéciales et problème de Zilber–Pink. In *Around the Zilber–Pink conjecture/Autour de la conjecture de Zilber–Pink*, pp. 1–30, Panor. Synthèses 52, Soc. Math. France, Paris, 2017.

[91] E. Ullmo and A. Yafaev, A characterization of special subvarieties. *Mathematika* **57** (2011), 263–273.

[92] E. Ullmo and A. Yafaev, Hyperbolic Ax–Lindemann theorem in the cocompact case. *Duke Math. J.* **163** (2014), 433–463.

[93] D. Urbanik, Absolute Hodge and ℓ-adic monodromy. 2020, arXiv:2011.10703.

[94] C. Voisin, *Hodge theory and complex algebraic geometry I*. Cambridge Stud. Adv. Math. 76, Cambridge University Press, 2002.

[95] C. Voisin, Hodge loci and absolute Hodge classes. *Compos. Math.* **143** (2007), 945–958.

[96] A. Weil, Abelian varieties and the Hodge ring. In *Collected papers III*, pp. 421–429, Springer, 1979.

[97] J. Wilkie, Model completeness results for expansions of the ordered field or real numbers by restricted Pfaffian functions and the exponential function. *J. Amer. Math. Soc.* **9** (1996), no. 4, 1051–1094.

[98] X. Yuan and S. Zhang, On the averaged Colmez conjecture. *Ann. of Math. (2)* **187** (2018), no. 2, 533–638.

[99] U. Zannier, *Some problems of unlikely intersections in arithmetic and geometry*, with appendices by David Masser. Ann. of Math. Stud. 181, Princeton University Press, 2012.

[100] B. Zilber, Exponential sums equations and the Schanuel conjecture. *J. Lond. Math. Soc. (2)* **65** (2002), no. 1, 27–44.

BRUNO KLINGLER

Humboldt-Universität zu Berlin, Unter den Linden 6, 10099 Berlin, Germany, bruno.klingler@hu-berlin.de

This Page Intentionally Left Blank

CANONICAL KÄHLER METRICS AND STABILITY OF ALGEBRAIC VARIETIES

CHI LI

ABSTRACT

We survey some recent developments in the study of canonical Kähler metrics on algebraic varieties and their relation with stability in algebraic geometry.

MATHEMATICS SUBJECT CLASSIFICATION 2020

Primary 53C55; Secondary 32Q26, 14J45, 32Q20, 58E11, 32P05

KEYWORDS

Constant scalar curvature Kähler metrics, Kähler–Einstein metrics, K-stability, valuative criterion, non-Archimedean geometry, Yau–Tian–Donaldson conjecture

ICM INTERNATIONAL CONGRESS OF MATHEMATICIANS 2022 JULY 6–14

Proc. Int. Cong. Math. 2022, Vol. 3, pp. 2286–2310
DOI 10.4171/ICM2022/37

The study of canonical Kähler metrics on algebraic varieties is a very active program in complex geometry. It is a common playground of several fields: differential geometry, partial differential equations, pluripotential theory, birational algebraic geometry, and non-Archimedean analysis. We will try to give the reader a tour of this vast program, emphasizing recent developments and highlighting interactions of different concepts and techniques. This article consists of three parts. In the *first* part, we discuss important classes of canonical Kähler metrics, and explain a well-established variational formalism for studying their existence. In the *second* part, we discuss algebraic aspects by reviewing recent developments in the study of K-stability with the help of deep tools from algebraic geometry and non-Archimedean analysis. In the *third* part, we discuss how the previous two parts are connected with each other. In particular, we will discuss the Yau–Tian–Donaldson (YTD) conjecture for canonical Kähler metrics in the first part.

1. CANONICAL KÄHLER METRICS ON ALGEBRAIC VARIETIES

1.1. Constant scalar curvature Kähler metrics

Let X be an n-dimensional projective manifold equipped with an ample line bundle L. By Kodaira's theorem, we have an embedding $\iota_m : X \to \mathbb{P}^N$ by using a complete linear system $|mL|$ for $m \gg 1$. If we denote by h_{FS} the standard Fubini–Study metric on the hyperplane bundle over $\mathbb{P}^N$ with Chern curvature $\omega_{\mathrm{FS}} = -\mathrm{dd^c} \log h_{\mathrm{FS}}$, then $h_0 = \iota_m^* h_{\mathrm{FS}}^{1/m}$ is a smooth Hermitian metric on L whose Chern curvature $\omega_0 = \frac{1}{m}\iota_m^*\omega_{\mathrm{FS}} = -\mathrm{dd^c}\log h_0$ is a Kähler form in $c_1(L) \in H^2(X,\mathbb{R})$. In this paper we will use the convention $\mathrm{dd^c} = \frac{\sqrt{-1}}{2\pi}\partial\bar{\partial}$.

We will also use singular Hermitian metrics. An upper-semicontinuous function $\varphi \in L^1(\omega^n)$ is called an ω_0-psh potential if $\psi + \varphi$ is a plurisubharmonic function for any local potential ψ of ω_0 (i.e., $\omega_0 = \mathrm{dd^c}\psi$ locally); $h_\varphi := h_0 e^{-\varphi}$ is then called a psh Hermitian metric on L. Denote by $\mathrm{PSH}(\omega_0)$ the space of ω_0-psh functions. By a $\partial\bar{\partial}$-lemma, any closed positive $(1,1)$-current in $c_1(L)$ is of the form $\omega_\varphi := \omega_0 + \mathrm{dd^c}\varphi = -\mathrm{dd^c}\log h_\varphi$ with $\varphi \in \mathrm{PSH}(\omega_0)$. Moreover, $\omega_{\varphi_2} = \omega_{\varphi_1}$ if and only if $\varphi_2 - \varphi_1$ is a constant. Define the space of smooth strictly ω_0-psh potentials (also called Kähler potentials) by

$$\mathcal{H} := \mathcal{H}(\omega_0) = \{\varphi \in C^\infty(X) : \omega_\varphi = \omega_0 + \mathrm{dd^c}\varphi > 0\}. \tag{1.1}$$

Fix any $\varphi \in \mathcal{H}$. If $\omega_\varphi = \sqrt{-1}\sum_{i,j}(\omega_\varphi)_{i\bar{j}} dz_i \wedge d\bar{z}_j$ under a holomorphic coordinate chart, then its *Ricci curvature* form $\mathrm{Ric}(\omega_\varphi) = \frac{\sqrt{-1}}{2\pi}\sum_{i,j} R_{i\bar{j}} dz_i \wedge d\bar{z}_j$ is given by

$$R_{i\bar{j}} := \mathrm{Ric}(\omega_\varphi)_{i\bar{j}} = -\frac{\partial^2 \log\det((\omega_\varphi)_{k\bar{l}})}{\partial z_i \partial\bar{z}_j}.$$

Then $\mathrm{Ric}(\omega_\varphi)$ is a real closed $(1,1)$-form which represents the cohomology class $c_1(-K_X) =: c_1(X)$. Here $-K_X = \wedge^n T^{(1,0)}X$ is the anticanonical line bundle of X. The *scalar curvature* of ω_φ is given by the contraction

$$S(\omega_\varphi) = \omega_\varphi^{i\bar{j}}\big(\mathrm{Ric}(\omega_\varphi)\big)_{i\bar{j}} = \frac{n\cdot \mathrm{Ric}(\omega_\varphi)\wedge\omega_\varphi^{n-1}}{\omega_\varphi^n}.$$

Further, ω_φ is called a *constant scalar curvature Kähler* (cscK) metric if $S(\omega_\varphi)$ is the constant $\underline{S}$ which is the average scalar curvature and is determined by cohomology classes:

$$\underline{S} = \frac{n\langle c_1(X) \cdot c_1(L)^{\cdot n-1}, [X]\rangle}{\mathbf{V}} \quad \text{with } \mathbf{V} = \langle c_1(L)^{\cdot n}, [X]\rangle. \tag{1.2}$$

The Kähler potential of a cscK metric is a solution to a 4th order nonlinear PDE. In general, there are obstructions to the existence of cscK metrics. For example, the Matsushima–Lichnerowicz theorem states that if (X, L) admits a cscK metric then the automorphism group $\mathrm{Aut}(X, L)$ must be reductive. Our goal is to discuss the Yau-Tian-Donaldson conjecture which would provide a sufficient and necessary algebraic criterion for the existence of cscK metrics.

1.2. Kähler–Einstein metrics and weighted Kähler–Ricci soliton

Kähler–Einstein metrics form an important class of cscK metrics. A Kähler form ω_φ is called Kähler–Einstein (KE) if $\mathrm{Ric}(\omega_\varphi) = \lambda\omega_\varphi$ for a real constant λ. A necessary condition for the existence of KE metrics is that the cohomology class $c_1(X) \in H^2(X, \mathbb{R})$ is either negative, numerically trivial, or positive. The existence for the first two cases was understood in 1970s: there always exists a Kähler–Einstein metric if $c_1(X)$ is negative (by the work of Aubin and Yau), or if $c_1(X)$ is numerically trivial (by the work of Yau).

Now we assume that X is a Fano manifold. In other words, $-K_X$ is an ample line bundle, and we set $L = -K_X$. Any $\varphi \in \mathcal{H}$ corresponds to a volume form

$$\Omega_\varphi := |s^*|^2_{h_\varphi}(\sqrt{-1})^{n^2} s \wedge \bar{s} = \Omega_0 e^{-\varphi} \quad \text{with } s = dz_1 \wedge \cdots \wedge dz_n, \ s^* = \partial_{z_1} \wedge \cdots \wedge \partial_{z_n}.$$

The KE equation in this case is reduced to a complex Monge–Ampère equation for φ, namely

$$(\omega + \mathrm{dd^c}\varphi)^n = e^{-\varphi}\Omega_0.$$

We also consider an interesting generalization of Kähler–Einstein metrics on Fano manifolds with torus actions. Assume that $\mathbb{T} \cong (\mathbb{C}^*)^r$ is an algebraic torus and $T \cong (S^1)^r \subset \mathbb{T}$ is a compact real subtorus. We will use the following notation:

$$N_\mathbb{Z} = \mathrm{Hom}_{\mathrm{alg}}(\mathbb{C}^*, \mathbb{T}), \quad N_\mathbb{Q} = N_\mathbb{Z} \otimes_\mathbb{Z} \mathbb{Q}, \quad N_\mathbb{R} = N_\mathbb{Z} \otimes_\mathbb{Z} \mathbb{R}. \tag{1.3}$$

Assume that $\mathbb{T}$ acts faithfully on X. Then there is an induced $\mathbb{T}$-action on $-K_X$. Each $\xi \in N_\mathbb{R}$ corresponds to a holomorphic vector field V_ξ on X. Denote by $\mathcal{H}^T$ the set of T-invariant Kähler potentials. For any $\varphi \in \mathcal{H}^T$, the T-action becomes Hamiltonian with respect to ω_φ. Denote by $\mathbf{m}_\varphi : X \to N^*_\mathbb{R} \cong \mathbb{R}^r$ the corresponding moment map, and let P be the image of $\mathbf{m}_\varphi$. By a theorem of Atiyah–Guillemin–Sternberg, P is a convex polytope which depends only on the Kähler class $c_1(L)$. Let $g : P \to \mathbb{R}$ be a smooth *positive* function. The following equation will be called the g-weighted soliton (or just g-soliton) equation for $\varphi \in \mathcal{H}(-K_X)^T$.

$$g(\mathbf{m}_\varphi)(\omega_0 + \mathrm{dd^c}\varphi)^n = e^{-\varphi}\Omega_0.$$

An equivalent tensorial equation is given by $\mathrm{Ric}(\omega_\varphi) = \omega_\varphi + \mathrm{dd^c} \log g(\mathbf{m}_\varphi)$.

Example 1.1. If $g(y) = e^{-\langle y, \xi\rangle}$, then the above equation becomes the standard Kähler–Ricci soliton equation $\mathrm{Ric}(\omega_\varphi) = \omega_\varphi + \mathscr{L}_{V_\xi}\omega_\varphi$ where $\mathscr{L}$ denotes the Lie derivative.

1.3. Kähler–Einstein metrics on log Fano pairs

Singular algebraic varieties and log pairs are important objects in algebraic geometry, and appear naturally for studying limits of smooth varieties. It is thus natural to study canonical Kähler metric on general log pairs. We recall a definition from birational algebraic geometry. Let X be a normal projective variety and D be a $\mathbb{Q}$-Weil divisor. Assume that $K_X + D$ is $\mathbb{Q}$-Cartier. Let $\mu : Y \to X$ be a resolution of singularities of (X, D) with simple normal crossing exceptional divisors $\sum_i E_i$. We then have an identity

$$K_Y = \mu^*(K_X + D) + \sum_i a_i E_i. \tag{1.4}$$

Here $A_{(X,D)}(E_i) := a_i + 1$ is called the log discrepancy of E_i. The pair (X, D) has klt singularities if $A_{(X,D)}(E_i) > 0$ for any E_i. We will always assume that (X, D) has klt singularities.

If $K_X + D$ is ample or numerically trivial, Yau and Aubin's existence result had been generalized to the singular and log case in [32], partly based on Kołodziej's pluripotential estimates. There were many related works by Yau, Tian, H. Tsuji, Z. Zhang, and many others.

Now we assume that $-(K_X + D)$ is ample and call (X, D) a log Fano pair. Then one can consider Kähler–Einstein equation or, more generally, g-soliton equation on (X, D). Note that there is a globally defined volume form as in the smooth case: choose a local trivializing section s of $m(K_X + D)$ with the dual s^* and define $\Omega_0 = |s^*|_{h_0}^{2/m}(\sqrt{-1}^{mn^2} s \wedge \bar{s})^{1/m}$. Assume that $\mathbb{T}$ acts on X faithfully and preserves the divisor D. With the notation from before, we say that φ is the potential for a g-weighted soliton on (X, D) if φ is a bounded ω_0-psh function that satisfies the equation

$$g(\mathbf{m}_\varphi)(\omega + dd^c\varphi)^n = e^{-\varphi}\Omega_0. \tag{1.5}$$

For any bounded $\varphi \in \mathrm{PSH}(\omega_0)$, the g-weighted Monge–Ampère measure on the left-hand side of (1.5) is well defined by the work of Berman–Witt–Nyström [10] and also by Han–Li [38], generalizing the definition of Bedford–Taylor (when $g = 1$). It is known that any bounded solution φ, if it exists, is orbifold smooth over the orbifold locus of (X, D). Moreover, if p is a regular point of supp(D) such that $D = (1 - \beta)\{z_1 = 0\}$ locally for a holomorphic function z_1 (with $\beta \in (0, 1]$), then the associated Kähler metric is modeled by $\mathbb{C}_\beta \times \mathbb{C}^{n-1}$ where $\mathbb{C}_\beta = (\mathbb{C}, dr^2 + \beta^2 r^2 d\theta^2)$ is the 2-dimensional flat cone with cone angle $2\pi\beta$.

1.4. Ricci-flat Kähler cone metrics

The class of Ricci-flat Kähler cone metrics is closely related to KE/g-soliton metrics, and is interesting in both complex geometry and mathematical physics (see [57]).

Let $Y = \mathbf{Spec}(R)$ be an $(n + 1)$-dimensional normal affine variety with a singularity $o \in Y$. Assume that an algebraic torus $\hat{\mathbb{T}} \cong (\mathbb{C}^*)^{r+1}$ acts faithfully on Y, with o being the only fixed point. Define $\hat{N}_{\mathbb{Q}}$, $\hat{N}_{\mathbb{R}}$ similar to (1.3). The $\hat{\mathbb{T}}$-action corresponds to a weight decomposition of the coordinate ring $R = \bigoplus_{\alpha \in \mathbb{Z}^{r+1}} R_\alpha$. The Reeb cone can be defined as

$$\hat{N}_{\mathbb{R}}^+ = \{\xi \in \hat{N}_{\mathbb{R}} : \langle \alpha, \xi \rangle > 0 \text{ for all } \alpha \in \mathbb{Z}^{r+1} \setminus \{0\} \text{ with } R_\alpha \neq 0\}.$$

Any $\hat{\xi} \in \hat{N}_{\mathbb{R}}^{+}$ is called a Reeb vector and corresponds to an expanding holomorphic vector field $V_{\hat{\xi}}$. Assume, furthermore, that Y is $\mathbb{Q}$-Gorenstein and there is a $\hat{\mathbb{T}}$-equivariant nonvanishing section $s \in |mK_Y|$, which induces a $\hat{\mathbb{T}}$-equivariant volume form $dV_Y = (\sqrt{-1}^{m(n+1)^2} s \wedge \bar{s})^{1/m}$ on Y. We call the data $(Y, \hat{\xi})$ with $\hat{\xi} \in \hat{N}_{\mathbb{R}}^{+}$ a polarized Fano cone.

Let $\hat{T} \cong (S^1)^{r+1}$ be a compact real subtorus of $\hat{\mathbb{T}}$. A $\hat{T}$-invariant function $r : Y \to \mathbb{R}_{\geq 0}$ is called a radius function for $\hat{\xi} \in \hat{N}_{\mathbb{R}}^{+}$ if $\widehat{\omega} = \mathrm{dd^c} r^2$ is a Kähler cone metric on $Y^* = Y \setminus \{o\}$ and $\frac{1}{2}(r\partial_r - \sqrt{-1}J(r\partial_r)) = V_{\hat{\xi}}$. Here J is a complex structure on Y^* and $\widehat{\omega}$ is called a Kähler cone metric if $G := \frac{1}{2}\widehat{\omega}(\cdot, J\cdot)$ on Y^* is isometric to $dr^2 + r^2 G_S$ where $S = \{r = 1\}$ and $G_S = G|_S$. In the literature of CR geometry, the induced structure on the link S by a Kähler cone metric is called a Sasaki structure. Also $\widehat{\omega} = \mathrm{dd^c} r^2$ is called *Ricci-flat* if $\mathrm{Ric}(\widehat{\omega}) = 0$. In this case, the radius function satisfies the equation (up to rescaling)

$$(\mathrm{dd^c} r^2)^{n+1} = dV_Y.$$

If $\hat{\xi} \in \hat{N}_{\mathbb{Q}}$, then $\widehat{\omega}$ is called quasiregular, and $V_{\hat{\xi}}$ generates a $\mathbb{C}^*$-subgroup $\langle \hat{\xi} \rangle$ of $\hat{\mathbb{T}}$. The GIT quotient $X = Y /\!/ \langle \hat{\xi} \rangle$ admits an orbifold structure encoded by a log Fano pair (X, D). A straightforward calculation shows that a quasiregular $(Y, \hat{\xi})$ admits a Ricci-flat Kähler cone metric if and only if (X, D) admits a Kähler–Einstein metric.

In general, there are many irregular Ricci-flat Kähler cone metrics, i.e., with $\hat{\xi} \in \hat{N}_{\mathbb{R}} \setminus \hat{N}_{\mathbb{Q}}$. Recent works by Apostolov–Calderbank–Jubert–Lahdili establish an equivalence between Ricci-flat Kähler cone metrics and special g-soliton metrics. More precisely, fix any $\hat{\chi} \in \hat{N}_{\mathbb{Q}}^{+}$ and consider the quotient $(X, D) = Y /\!/ \langle \hat{\chi} \rangle$ as above. It is shown in [2] (see also [47]) that the Ricci-flat Kähler cone metric on $(Y, \hat{\xi})$ is equivalent to the g-soliton metric on (X, D) with $g(y) = (n + 1 + \langle y, \xi \rangle)^{-n-2}$ where ξ (equivalently, V_ξ) is induced by $\hat{\xi}$ on X.

1.5. Analytic criteria for the existence

We now review a well-understood criterion for the existence of above canonical Kähler metrics. The general idea is to view corresponding equations as Euler–Lagrange equations of appropriate energy functionals and then use a variational approach to prove that the existence of solutions is equivalent to the coercivity of the energy functionals. First we have the following functionals defined for any $\varphi \in \mathcal{H}$ (see (1.1)):

$$\mathbf{E}(\varphi) = \frac{1}{(n+1)\mathbf{V}} \sum_{k=0}^{n} \int_X \varphi \omega_\varphi^k \wedge \omega_0^{n-k}, \quad \mathbf{\Lambda}(\varphi) = \frac{1}{\mathbf{V}} \int_X \varphi \omega_0^n, \tag{1.6}$$

$$\mathbf{J}(\varphi) = \mathbf{\Lambda}(\varphi) - \mathbf{E}(\varphi), \quad \mathbf{E}^\chi(\varphi) = \frac{1}{\mathbf{V}} \sum_{k=0}^{n-1} \int_X \varphi \chi \wedge \omega_\varphi^k \wedge \omega^{n-1-k}. \tag{1.7}$$

Here $\mathbf{V}$ is defined in (1.2) and χ is any closed real $(1,1)$-form.

The following functionals are important for studying the cscK problem:

$$\mathbf{H}(\varphi) = \frac{1}{\mathbf{V}} \int_X \log \frac{\omega_\varphi^n}{\Omega_0} \omega_\varphi^n, \quad \mathbf{M}(\varphi) = \mathbf{H}(\varphi) + \mathbf{E}^{-\mathrm{Ric}(\omega_0)}(\varphi) + \underline{S} \cdot \mathbf{E}(\varphi). \tag{1.8}$$

The above $\mathbf{H}(\varphi)$ is usually called the entropy of the measure ω_φ^n. One can verify that any critical point of $\mathbf{M}$ is the potential of a cscK metric.

For Kähler–Einstein (KE) metrics on Fano manifolds, we have more functionals:

$$\mathbf{L}(\varphi) = -\log\left(\frac{1}{\mathbf{V}}\int_X e^{-\varphi}\Omega_0\right), \quad \mathbf{D}(\varphi) = -\mathbf{E}(\varphi) + \mathbf{L}(\varphi). \tag{1.9}$$

A critical point of $\mathbf{D}$ is also a KE potential. These functionals can be generalized to the settings of g-weighted solitons and Ricci-flat Kähler cone metrics (see [47] for references).

To apply the variational approach, one first needs a "completion" of $\mathcal{H}$. Such a completion was defined by Guedj–Zeriahi extending the local study of Cegrell. Following [7], one way to introduce this is to first define the $\mathbf{E}$ functional for any $\varphi \in \mathrm{PSH}(\omega_0)$ by

$$\mathbf{E}(\varphi) = \inf\{\mathbf{E}(\tilde{\varphi}) : \tilde{\varphi} \geq \varphi,\ \tilde{\varphi} \in \mathcal{H}(\omega_0)\}, \tag{1.10}$$

Then define the set of finite energy potentials as

$$\mathcal{E}^1 := \mathcal{E}^1(\omega_0) = \{\varphi \in \mathrm{PSH}(\omega_0) : \mathbf{E}(\varphi) > -\infty\}. \tag{1.11}$$

After the work [6], $\mathcal{E}^1$ can be endowed with a strong topology which is the coarsest refinement of the weak topology (i.e., the L^1-topology) that makes $\mathbf{E}$ continuous. The above energy functionals can be extended to $\mathcal{E}^1$, and they satisfy important regularization properties:

Theorem 1.2 (see [6,8]). *For any $\varphi \in \mathcal{E}^1$, there exists $\{\varphi_k\}_{k\in\mathbb{N}} \subset \mathcal{H}$ such that $\mathbf{F}(\varphi_k) \to \mathbf{F}(\varphi)$ for $\mathbf{F} \in \{\mathbf{E}, \mathbf{\Lambda}, \mathbf{E}^{-\mathrm{Ric}}, \mathbf{H}\}$.*

We would like to emphasize the result for $\mathbf{F} = \mathbf{H}$, which was proved in [8]. The idea of proof there is to first regularize the measure ω_φ^n with converging entropy and then use Yau's solution to complex Monge–Ampère equations with prescribed volume forms. Later we will encounter the same idea in the non-Archimedean setting.

Another key concept is the geodesic between two finite energy potentials. For $\varphi_i \in \mathcal{E}^1$, $i = 0, 1$, the *geodesic* connecting them is the following $p_1^*\omega_0$-psh function on $X \times [0,1] \times S^1$ where p_1 is the projection to the first factor (see [7,26]):

$$\Phi = \sup\left\{\Psi : \Psi \text{ is } S^1\text{-invariant and } p_1^*\omega_0\text{-psh}, \lim_{s\to i}\Psi(\cdot, s) \leq \varphi(i), i = 0, 1\right\}. \tag{1.12}$$

The concept of geodesics originates from Mabuchi's L^2-Riemannian metric on $\mathcal{H}$. According to the work of Semmes and Donaldson, if $\varphi_i \in \mathcal{H}$, $i = 0, 1$, then the geodesic Φ is a solution to the Dirichlet problem of homogeneous complex Monge–Ampère equation

$$\left(p_1^*\omega_0 + \mathrm{dd^c}\Phi\right)^{n+1} = 0, \quad \Phi(\cdot, i) = \varphi_i, i = 0, 1. \tag{1.13}$$

Since Φ is S^1-invariant, we can consider Φ as a family of ω_0-psh functions $\{\varphi(s)\}_{s\in[0,1]}$.

Theorem 1.3 ([5,8]). *Let $\Phi = \{\varphi(s)\}_{s\in[0,1]}$ be a geodesic segment in $\mathcal{E}^1$. Then* (1) *$s \mapsto \mathbf{E}(\varphi(s))$ is affine;* (2) *$s \mapsto \mathbf{M}(\varphi(s))$ is convex.*

Results in Theorem 1.3 are important in the variational approach. If a geodesic is smooth, the statements follow from straightforward calculations. However, there are examples (first due to Lempert–Vivas) showing that the solution to (1.13) in general does not have sufficient regularity. So the proofs of the above results are more involved.

In this paper $\tilde{\mathbb{T}}$ will always denote a maximal torus of the linear algebraic group $\mathrm{Aut}(X, L)$ and $\tilde{T}$ is a maximal real subtorus of $\tilde{\mathbb{T}}$. In the following result, we use the translation invariance $\mathbf{F}(\varphi + c) = \mathbf{F}(\varphi)$ for $\mathbf{F} \in \{\mathbf{M}, \mathbf{J}\}$ and hence $\mathbf{F}(\omega_\varphi) := \mathbf{F}(\varphi)$ is well defined.

Theorem 1.4 ([9, 23, 27]). *There exists a $\tilde{T}$-invariant cscK metric in $c_1(L)$ if and only if* $\mathbf{M}$ *is reduced coercive, which means that there exist $\gamma, C > 0$ such that for any $\varphi \in \mathcal{H}^{\tilde{T}}$,*

$$\mathbf{M}(\omega_\varphi) \geq \gamma \cdot \inf_{\sigma \in \tilde{\mathbb{T}}} \mathbf{J}(\sigma^* \omega_\varphi) - C. \tag{1.14}$$

This type of result goes back to Tian's pioneering work in [64] which proves that if X is a Fano manifold with a discrete automorphism group, then the existence of Kähler–Einstein metric is equivalent to the properness of the $\mathbf{M}$-functional, and is also equivalent to the properness of the $\mathbf{D}$ functional. Tian's work has since been refined and generalized for other canonical metrics. For the necessity direction (from existence to reduced coercivity), there is now a general principle due to Darvas–Rubinstein ([27]) that can be applied for all previously-mentioned canonical Kähler metrics. The sufficient direction (from reduced coercivity to existence) for Kähler–Einstein metrics is reproved in [6] using pluripotential theory, which works equally well in the setting of log Fano pairs. See [10, 38] for the extension to the g-soliton case. The existence result for smooth cscK metrics is accomplished recently by Chen–Cheng's new estimates [23]. The use of maximal torus appears in [44, 45], refining an earlier formulation of Hisamoto [39]. There is also an existence criterion when $\tilde{\mathbb{T}}$ is replaced by any connected reductive subgroup of $\mathrm{Aut}(X, L)$ that contains a maximal torus.

2. STABILITY OF ALGEBRAIC VARIETIES AND NON-ARCHIMEDEAN GEOMETRY

2.1. K-stability and non-Archimedean geometry

The concept of K-stability, as first introduced by Tian [64] and Donaldson [30], is motivated by results from geometric analysis. On the other hand, the recent development shows that various tools from algebraic geometry are crucial in unlocking many of its mysteries.

Definition 2.1. A test configuration for a polarized manifold (X, L) consists of $(\mathcal{X}, \mathcal{L})$ that satisfies: (i) $\pi : \mathcal{X} \to \mathbb{C}$ is a flat projective morphism from a normal variety $\mathcal{X}$, and $\mathcal{L}$ is a π-semiample $\mathbb{Q}$-line bundle; (ii) There is a $\mathbb{C}^*$-action on $(\mathcal{X}, \mathcal{L})$ such that π is $\mathbb{C}^*$-equivariant; (iii) There is a $\mathbb{C}^*$-equivariant isomorphism $(\mathcal{X}, \mathcal{L}) \times_{\mathbb{C}} \mathbb{C}^* \cong (X \times \mathbb{C}^*, p_1^* L)$.

Configuration $(\mathcal{X}, \mathcal{L})$ is called a product test configuration if there is a $\mathbb{C}^*$-equivariant isomorphism $(\mathcal{X}, \mathcal{L}) \cong (X \times \mathbb{C}, p_1^* L)$ where the $\mathbb{C}^*$-action on the right-hand side is the product action of a $\mathbb{C}^*$-action on (X, L) with the standard multiplication on $\mathbb{C}$.

Two test configurations $(\mathcal{X}_i, \mathcal{L}_i)$, $i = 1, 2$ are called equivalent if there exists a test configuration $(\mathcal{X}', \mathcal{L}')$ with $\mathbb{C}^*$-equivariant birational morphisms $\rho_i : \mathcal{X}' \to \mathcal{X}_i$ satisfying $\rho_1^* \mathcal{L}_1 = \mathcal{L}' = \rho_2^* \mathcal{L}_2$. For any test configuration $(\mathcal{X}, \mathcal{L})$, by taking fiber product, one can always find an equivalent test configuration $(\mathcal{X}', \mathcal{L}')$ such that $\mathcal{X}'$ dominates $X \times \mathbb{C}$.

Given any test configuration $(\mathcal{X}, \mathcal{L})$, there is a canonical compactification over $\mathbb{P}^1$ denoted by $(\overline{\mathcal{X}}, \overline{\mathcal{L}})$ which is obtained by adding a trivial fiber over $\{\infty\} = \mathbb{P}^1 \setminus \mathbb{C}$.

The notion of a test configuration is a way to formulate the degeneration of (X, L). In fact, any test configuration is induced by a one-parameter subgroup of $\mathrm{PGL}(N+1, \mathbb{C})$ for a Kodaira embedding $X \to \mathbb{P}^N$.

We will continue our discussion in a framework of non-Archimedean geometry as proposed by Boucksom–Jonsson. Let X^{NA} denote the Berkovich analytification of X with respect to the trivial absolute value on $\mathbb{C}$ (see [18] for references). X^{NA} is a topological space consisting of real valuations on subvarieties of X, and contains a dense subset $X^{\mathrm{div}}_{\mathbb{Q}}$ consisting of divisorial valuations on X. Any test configuration $(\mathcal{X}, \mathcal{L})$ defines a function on X^{NA} in the following way. First, up to equivalence, we can assume that there is a birational morphism $\rho : \mathcal{X} \to X_{\mathbb{C}} := X \times \mathbb{C}$. Write $\mathcal{L} = \rho^* p_1^* L + E$ where E is a $\mathbb{Q}$-divisor supported on $\mathcal{X}_0$. For any $v \in X^{\mathrm{NA}}$, denote by $G(v)$ the $\mathbb{C}^*$-invariant semivaluation on $X_{\mathbb{C}}$ that satisfies $G(v)|_{\mathbb{C}(X)} = v$ and $G(v)(t) = 1$ where t is the coordinate of $\mathbb{C}$. One then defines

$$\phi_{(\mathcal{X},\mathcal{L})}(v) = G(v)(E), \quad \text{for any } v \in X^{\mathrm{NA}}. \tag{2.1}$$

The set of such functions on X^{NA} obtained from test configurations is denoted by $\mathcal{H}^{\mathrm{NA}}$ which is considered as the set of smooth non-Archimedean psh potentials on the analytification of L. The following functionals, defined on the space of test configurations, correspond to the Archimedean (i.e., complex-analytic) functionals in (1.6)–(1.7):

$$\mathbf{E}^{\mathrm{NA}}(\mathcal{X}, \mathcal{L}) = \frac{\overline{\mathcal{L}}^{\cdot n+1}}{(n+1)\mathbf{V}}, \quad \mathbf{\Lambda}^{\mathrm{NA}}(\mathcal{X}, \mathcal{L}) = \frac{1}{\mathbf{V}} \overline{\mathcal{L}}^{\cdot n} \cdot \rho^* L_{\mathbb{P}^1}, \tag{2.2}$$

$$\mathbf{J}^{\mathrm{NA}}(\mathcal{X}, \mathcal{L}) = \mathbf{\Lambda}^{\mathrm{NA}}(\mathcal{X}, \mathcal{L}) - \mathbf{E}^{\mathrm{NA}}(\mathcal{X}, \mathcal{L}), \quad (\mathbf{E}^{K_X})^{\mathrm{NA}}(\mathcal{X}, \mathcal{L}) = \frac{1}{\mathbf{V}} K_X \cdot \overline{\mathcal{L}}^{\cdot n}, \tag{2.3}$$

$$\mathbf{H}^{\mathrm{NA}}(\mathcal{X}, \mathcal{L}) = \frac{1}{\mathbf{V}} K^{\log}_{\overline{\mathcal{X}}/X_{\mathbb{P}^1}} \cdot \mathcal{L}^{\cdot n}, \quad \mathbf{M}^{\mathrm{NA}}(\mathcal{X}, \mathcal{L}) = \mathbf{H}^{\mathrm{NA}} + (\mathbf{E}^{K_X})^{\mathrm{NA}} + \underline{S} \cdot \mathbf{E}^{\mathrm{NA}}, \tag{2.4}$$

where we assume that $\overline{\mathcal{X}}$ dominates $X_{\mathbb{P}^1} = X \times \mathbb{P}^1$ by ρ, and $L_{\mathbb{P}^1} = p_1^* L$, $K^{\log}_{\overline{\mathcal{X}}/X_{\mathbb{P}^1}} = K_{\overline{\mathcal{X}}} + \mathcal{X}_0^{\mathrm{red}} - (\rho^*(K_{X \times \mathbb{P}^1} + X \times \{0\}))$. These functionals were defined before the introduction of the non-Archimedean framework. For example, the $\mathbf{E}^{\mathrm{NA}}$ functional appeared in Mumford's study of Chow stability of projective varieties.

Assume that $\mathcal{X}_0 = \sum_i b_i F_i$ where F_i are irreducible components. Set $v_i = b_i^{-1} \mathrm{ord}_{F_i} \circ p_1^* \in X^{\mathrm{div}}_{\mathbb{Q}}$ and let δ_{v_i} be the Dirac measure supported at $\{v_i\}$. Chambert-Loir defined the following non-Archimedean Monge–Ampère measure using the intersection theory:

$$\mathrm{MA}^{\mathrm{NA}}(\phi_{(\mathcal{X},\mathcal{L})}) = \sum_i b_i \big(\mathcal{L}^{\cdot n} \cdot F_i\big) \delta_{v_i}. \tag{2.5}$$

Mixed non-Archimedean Monge–Ampère measures are similarly defined. It then turns out that the functionals from (2.2)–(2.3) can be obtained by using the same formula as in (1.6)–(1.7) but with the ordinary integrals replaced by corresponding non-Archimedean ones, while the $\mathbf{H}^{\mathrm{NA}}$ functional has the following expression after [19]:

$$\mathbf{H}^{\mathrm{NA}}(\mathcal{X}, \mathcal{L}) = \frac{1}{\mathbf{V}} \int_{X^{\mathrm{NA}}} A_X(v) \mathrm{MA}^{\mathrm{NA}}(\phi_{(\mathcal{X},\mathcal{L})})(v). \tag{2.6}$$

Here A_X is a functional defined on X^{NA} that generalizes the log discrepancy functional on $X^{\mathrm{div}}_{\mathbb{Q}}$ (see [41]). We can now recall the notion of K-stability:

Definition 2.2. A polarized manifold (X, L) is K-semistable, K-stable or K-polystable if any nontrivial test configuration $(\mathcal{X}, \mathcal{L})$ for (X, L) satisfies $\mathbf{M}^{\mathrm{NA}}(\mathcal{X}, \mathcal{L}) \geq 0$, $\mathbf{M}^{\mathrm{NA}}(\mathcal{X}, \mathcal{L}) > 0$, or $\mathbf{M}^{\mathrm{NA}}(\mathcal{X}, \mathcal{L}) \geq 0$ and $= 0$ only if $(\mathcal{X}, \mathcal{L})$ is a product test configuration, respectively.

This is like a Hilbert–Mumford's numerical criterion in the Geometric Invariant Theory.[1] The recent development of K-stability involves a strengthened notion called reduced uniform K-stability, which matches the reduced coercivity in (1.14) (see [19, 29, 39]). Recall that $\tilde{\mathbb{T}}$ denotes a maximal torus of $\mathrm{Aut}(X, L)$, and $\tilde{N}_{\mathbb{Q}}$ is defined similar to (1.3).

Definition 2.3. A polarized manifold (X, L) is uniformly K-stable (resp. reduced uniformly K-stable) if there exists $\gamma > 0$ such that any test configuration $(\mathcal{X}, \mathcal{L})$ satisfies $\mathbf{M}^{\mathrm{NA}}(\mathcal{X}, \mathcal{L}) \geq \gamma \cdot \mathbf{J}^{\mathrm{NA}}(\mathcal{X}, \mathcal{L})$ (resp. $\mathbf{M}^{\mathrm{NA}}(\mathcal{X}, \mathcal{L}) \geq \gamma \cdot \inf_{\xi \in \tilde{N}_{\mathbb{Q}}} \mathbf{J}^{\mathrm{NA}}(\mathcal{X}_\xi, \mathcal{L}_\xi)$).

Here the twist $(\mathcal{X}_\xi, \mathcal{L}_\xi)$ is introduced by Hisamoto [39]. One way to define it as a test configuration is by resolving the composition of birational morphisms $(\mathcal{X}, \mathcal{L}) \dashrightarrow (X_{\mathbb{C}} = X \times \mathbb{C}, L_{\mathbb{C}} = p_1^* L) \xrightarrow{\sigma_\xi} (X_{\mathbb{C}}, L_{\mathbb{C}})$ where σ_ξ is the $\mathbb{C}^*$-action generated by ξ. Alternatively, it can be defined in a more general setting of filtrations (see Example 2.8).

2.2. Non-Archimedean pluripotential theory

We discuss how non-Archimedean pluripotential theory as developed by Boucksom–Jonsson can be applied to study K-stability. Corresponding to a regularization result in the complex analytic case, an u.s.c. function $\phi : X^{\mathrm{NA}} \to \mathbb{R} \cup \{+\infty\}$ is called a non-Archimedean psh potential if it is a decreasing limit of a sequence from $\mathcal{H}^{\mathrm{NA}}$. Denote the space of such functions by $\mathrm{PSH}^{\mathrm{NA}}$. Boucksom–Jonsson introduced the following non-Archimedean version of the finite energy space. First corresponding to (1.10), for any $\phi \in \mathrm{PSH}^{\mathrm{NA}}$, define

$$\mathbf{E}^{\mathrm{NA}}(\phi) = \inf\{\mathbf{E}^{\mathrm{NA}}(\tilde{\phi}) : \tilde{\phi} \geq \phi, \tilde{\phi} \in \mathcal{H}^{\mathrm{NA}}\}.$$

Then, corresponding to (1.11), define the space of non-Archimedean finite energy potentials by

$$(\mathcal{E}^1)^{\mathrm{NA}} = \{\phi \in \mathrm{PSH}^{\mathrm{NA}} : \mathbf{E}^{\mathrm{NA}}(\phi) > -\infty\}.$$

This space is again equipped with a strong topology which makes $\mathbf{E}^{\mathrm{NA}}$ continuous. Boucksom–Jonsson showed in [22] that the non-Archimedean Monge–Ampère measure $\mathrm{MA}^{\mathrm{NA}}(\phi)$ is well defined for any $\phi \in (\mathcal{E}^1)^{\mathrm{NA}}$ such that if $\{\phi_k\}_{k \in \mathbb{N}} \subset \mathcal{H}^{\mathrm{NA}}$ converges to ϕ strongly, then $\mathrm{MA}^{\mathrm{NA}}(\phi_k)$ converges to $\mathrm{MA}^{\mathrm{NA}}(\phi)$ weakly.

A large class of potentials come from filtrations (see [19]). Set $R_m = H^0(X, mL)$.

Definition 2.4. A filtration is the data $\mathcal{F} = \{\mathcal{F}^\lambda R_m \subseteq R_m; \lambda \in \mathbb{R}, m \in \mathbb{N}\}$ that satisfies the following four conditions:

(i) $\mathcal{F}^\lambda R_m \subseteq \mathcal{F}^{\lambda'} R_m$, if $\lambda \geq \lambda'$;

[1] In the classical formulation, Tian's CM weight or, equivalently, the Donaldson–Futaki invariant is used to define the K-stability. However, to fit our discussion in the non-Archimedean framework, we use the equivalent formulation via the $\mathbf{M}^{\mathrm{NA}}$ functional.

(ii) $\mathcal{F}^{\lambda} R_m = \bigcap_{\lambda' < \lambda} \mathcal{F}^{\lambda'} R_m$;

(iii) $\mathcal{F}^{\lambda} R_m \cdot \mathcal{F}^{\lambda'} R_{m'} \subseteq \mathcal{F}^{\lambda+\lambda'} R_{m+m'}$, for $\lambda, \lambda' \in \mathbb{R}$ and $m, m' \in \mathbb{N}$;

(iv) There exist $e_-, e_+ \in \mathbb{Z}$ such that $\mathcal{F}^{me_-} R_m = R_m$ and $\mathcal{F}^{me_+} R_m = 0$ for $m \in \mathbb{Z}_{\geq 0}$.

Filtration $\mathcal{F}$ is finitely generated if its extended Rees algebra $\mathcal{R}(\mathcal{F})$ is finitely generated where

$$\mathcal{R}(\mathcal{F}) = \bigoplus_{\lambda \in \mathbb{R}} \bigoplus_{m \in \mathbb{N}} t^{-\lambda} \mathcal{F}^{\lambda} R_m.$$

In this case $\mathcal{F}$ induces a degeneration of X into $\mathcal{X}_0 = \mathrm{Proj}(\bigoplus_{m,\lambda} \mathcal{F}^{\lambda} R_m / \mathcal{F}^{>\lambda} R_m)$.

For a general $\mathcal{F}$, $\{\mathcal{F}^{\lambda} R_\ell; \lambda \in \mathbb{R}\}$ generates a filtration $\check{\mathcal{F}}^{(\ell)}$ on $R^{(\ell)} := \bigoplus_{m \in \mathbb{N}} R_{m\ell}$, which induces a non-Archimedean psh potential $\check{\phi}^{(\ell)} \in \mathcal{H}^{\mathrm{NA}}$. Define

$$\phi_{\mathcal{F}} = \Big(\limsup_{\ell \to +\infty} \check{\phi}^{(\ell)}\Big)^*$$

where $(\cdot)^*$ denotes the upper-semicontinuous regularization.

Example 2.5. Filtration $\mathcal{F}$ is a $\mathbb{Z}$-filtration if $\mathcal{F}^{\lambda} R_m = \mathcal{F}^{\lceil \lambda \rceil} R_m$. By [19, 63, 69], there is a one-to-one correspondence between test configurations equipped with relatively ample $\mathbb{Q}$-polarizations and finitely generated $\mathbb{Z}$-filtrations. Any test configuration $(\mathcal{X}, \mathcal{L})$ defines such a filtration by

$$\mathcal{F}^{\lambda} R_m = \{s \in R_m : t^{-\lceil \lambda \rceil} s \in H^0(\mathcal{X}, m\mathcal{L})\}. \tag{2.7}$$

Conversely, if $\mathcal{F}$ is a finitely generated $\mathbb{Z}$-filtration, then $(\mathcal{X} := \mathrm{Proj}_{\mathbb{C}[t]}(\mathcal{R}(\check{\mathcal{F}}^{(\ell)})), \frac{1}{\ell}\mathcal{O}_{\mathcal{X}}(1))$ is a test configuration for ℓ sufficiently divisible.

Example 2.6. In Definition 2.1 of test configurations, if we do not require $\mathcal{L}$ to be π-semiample, then we call $(\mathcal{X}, \mathcal{L})$ a model (of $(X \times \mathbb{C}, p_1^* L)$). The same definition in (2.7) defines a filtration also denoted by $\mathcal{F}_{(\mathcal{X},\mathcal{L})}$. However, in general the filtration is not finitely generated anymore. Fix any model $(\mathcal{X}, \mathcal{L})$ such that $\overline{\mathcal{L}}$ is big over $\overline{\mathcal{X}}$ (we call such $(\mathcal{X}, \mathcal{L})$ a big model for (X, L)). In [46] we obtained the following formula for the non-Archimedean Monge–Ampère measure of $\phi = \phi_{(\mathcal{X},\mathcal{L})} := \phi_{\mathcal{F}_{(\mathcal{X},\mathcal{L})}}$ which generalizes (2.5):

$$\mathrm{MA}^{\mathrm{NA}}(\phi) = \sum_i b_i \big(\langle \overline{\mathcal{L}}^{\cdot n} \rangle \cdot F_i\big) \delta_{v_i}. \tag{2.8}$$

Here for any divisor D, we use the notion of a positive intersection product introduced in [17]:

$$\langle \overline{\mathcal{L}}^{\cdot n+1} \rangle := \mathrm{vol}(\overline{\mathcal{L}}) = \lim_{m \to +\infty} \frac{h^0(\overline{\mathcal{X}}, m\overline{\mathcal{L}})}{\frac{m^{n+1}}{(n+1)!}}, \quad \langle \overline{\mathcal{L}}^{\cdot n} \rangle \cdot D := \frac{1}{n+1} \frac{d}{dt}\Big|_{t=0} \mathrm{vol}(\overline{\mathcal{L}} + tD).$$

Example 2.7. Any $v \in X^{\mathrm{div}}_{\mathbb{Q}}$ defines a filtration: for any $\lambda \in \mathbb{R}$ and $m \in \mathbb{Z}_{\geq 0}$, define

$$\mathcal{F}_v^{\lambda} R_m = \{s \in R_m : v(s) \geq \lambda\}. \tag{2.9}$$

Boucksom–Jonsson proved in [21] that $\mathrm{MA}^{\mathrm{NA}}(\phi_{\mathcal{F}_v}) = \mathbf{V} \cdot \delta_v$.

Example 2.8. Assume a torus $\tilde{\mathbb{T}} \cong (\mathbb{C}^*)^r$-acts on (X, L). Then we have a weight decomposition $R_m = \bigoplus_{\alpha \in \mathbb{Z}^r} R_{m,\alpha}$. For any $\xi \in \tilde{N}_{\mathbb{R}}$, we can define the ξ-twist of a given filtration: $\mathcal{F}_\xi^\lambda R_m = \mathcal{F}^{\lambda - \langle \alpha, \xi \rangle} R_{m,\alpha}$. On the other hand, there is an induced $\tilde{N}_{\mathbb{R}}$-action on $(X^{\mathrm{NA}})^{\tilde{\mathbb{T}}}$ which sends (ξ, v) to $v_\xi \in (X^{\mathrm{NA}})^{\tilde{\mathbb{T}}}$ determined by the following condition: if $f \in \mathbb{C}(X)_\alpha$ which means $f \circ \mathrm{t}^{-1} = \mathrm{t}^\alpha \cdot f$ for any $\mathrm{t} \in \tilde{\mathbb{T}}$, then $v_\xi(f) = \langle \alpha, \xi \rangle + v(f)$. We then have the following formula: $\mathrm{MA}^{\mathrm{NA}}(\phi_{\mathcal{F}_\xi}) = (-\xi)_* \mathrm{MA}^{\mathrm{NA}}(\phi_{\mathcal{F}})$ (see **[44, 45]**).

Generalizing the case of test configurations, Boucksom–Jonsson showed that the non-Archimedean functionals from (2.2)–(2.4) are well defined for all $\phi \in (\mathcal{E}^1)^{\mathrm{NA}}$ by using integrals over X^{NA} mentioned before (for example, for $\mathbf{H}^{\mathrm{NA}}$ use (2.6)).

Example 2.9. For any filtration $\mathcal{F}$, it is known that $\phi_{\mathcal{F}} \in (\mathcal{E}^1)^{\mathrm{NA}}$. Following **[19]**, define

$$\mathrm{vol}(\mathcal{F}^{(t)}) = \lim_{m \to +\infty} \frac{\dim_{\mathbb{C}} \mathcal{F}^{mt} R_m}{m^n/n!}.$$

Then $\mathbf{E}^{\mathrm{NA}}$ is the following "expected vanishing order" with respect to $\mathcal{F}$ (see **[21]**).

$$\mathbf{E}^{\mathrm{NA}}(\phi_{\mathcal{F}}) = \frac{1}{\mathbf{V}} \int_{\mathbb{R}} t\big(-d\,\mathrm{vol}(\mathcal{F}^{(t)})\big). \tag{2.10}$$

Similar to Theorem 1.2, we also have important regularization properties:

Theorem 2.10 (**[22]**). *For any $\phi \in (\mathcal{E}^1)^{\mathrm{NA}}$, there exists $\{\phi_k\}_{k \in \mathbb{N}} \subset \mathcal{H}^{\mathrm{NA}}$ (i.e., $\phi_k = \phi_{(\mathcal{X}_k, \mathcal{L}_k)}$ for a test configuration $(\mathcal{X}_k, \mathcal{L}_k)$) such that $\phi_k \to \phi$ in the strong topology and $\mathbf{F}^{\mathrm{NA}}(\phi_k) \to \mathbf{F}^{\mathrm{NA}}(\phi)$ for $\mathbf{F} \in \{\mathbf{E}, \mathbf{\Lambda}, \mathbf{E}^{K_X}\}$.*

Boucksom–Jonsson conjectured that the same conclusion should also hold for $\mathbf{H}^{\mathrm{NA}}$. This conjecture is still open in general and it is important in the non-Archimedean approach to the YTD conjecture. We have made progress in this direction.

Theorem 2.11 (**[45, 46]**). (1) *For any $\phi \in (\mathcal{E}^1)^{\mathrm{NA}}$, there exist models $\{(\mathcal{X}_k, \mathcal{L}_k)\}_{k \in \mathbb{N}}$ such that $\phi_k = \phi_{(\mathcal{X}_k, \mathcal{L}_k)} \to \phi$ in the strong topology and $\mathbf{H}^{\mathrm{NA}}(\phi_k) \to \mathbf{H}^{\mathrm{NA}}(\phi)$.*

(2) *For any big model $(\mathcal{X}, \mathcal{L})$, we have the following formula that generalizes (2.4):*

$$\mathbf{M}^{\mathrm{NA}}(\mathcal{X}, \mathcal{L}) = \frac{1}{\mathbf{V}} \langle \overline{\mathcal{L}}^{\cdot n} \rangle \cdot K_{\overline{\mathcal{X}}/\mathbb{P}^1} + \frac{\underline{S}}{(n+1)\mathbf{V}} \langle \overline{\mathcal{L}}^{\cdot n+1} \rangle.$$

The idea for proving the first statement is similar to the Archimedean setting in **[8]**. First we regularize the measure $\mathrm{MA}^{\mathrm{NA}}(\phi)$ with converging entropy. In fact, we find a way to regularize it by using measures supported at finitely many points in $X^{\mathrm{div}}_{\mathbb{Q}}$. Then we use the solution of non-Archimedean Monge–Ampère equations obtained in **[18]** to get the wanted potentials which are known to be associated to models. However, in the non-Archimedean case, there is not yet a characterization of measures associated to test configurations which prevents us from regularizing via test configurations. The second statement in Theorem 2.11 follows from the formula (2.8), and it prompts us to propose the following algebro-geometric conjecture which would strengthen the classical Fujita approximation theorem.

Conjecture 2.12. *Let $\overline{\mathcal{X}}$ be a smooth $(n+1)$-dimensional projective variety. Let $\overline{\mathcal{L}}$ be a big line bundle over $\overline{\mathcal{X}}$. Then there exist birational morphisms $\mu_k : \overline{\mathcal{X}}_k \to \overline{\mathcal{X}}$ and decompositions $\mu_k^* \overline{\mathcal{L}} = \overline{\mathcal{L}}_k + E_k$ in $N^1(\overline{\mathcal{X}})_{\mathbb{Q}}$ with $\overline{\mathcal{L}}_k$ semiample and E_k effective such that*

$$\lim_{k\to+\infty} \overline{\mathcal{L}}_k^{\cdot n+1} = \mathrm{vol}(\overline{\mathcal{L}}), \quad \lim_{k\to+\infty} \overline{\mathcal{L}}_k^{\cdot n} \cdot K_{\overline{\mathcal{X}}_k} = \frac{1}{n+1}\frac{d}{dt}\mathrm{vol}(\overline{\mathcal{L}} + tK_{\overline{\mathcal{X}}})\Big|_{t=0} = \langle \overline{\mathcal{L}}^{\cdot n} \rangle \cdot K_{\overline{\mathcal{X}}}.$$

It is easy to show that this conjecture is true if $\overline{\mathcal{L}}$ admits a birational Zariski decomposition. The author verified this conjecture for certain examples of big line bundles due to Nakamaya which do not admit such decompositions (see [46]). Y. Odaka observed that when $(\mathcal{X}, \mathcal{L})$ is a big model for a polarized spherical manifold (for example, a polarized toric manifold), $\overline{\mathcal{X}}$ is a Mori dream space which implies that $\overline{\mathcal{L}}$ admits a Zariski decomposition and hence the above conjecture holds true.

2.3. Stability of Fano varieties

In this section, we assume that X is a $\mathbb{Q}$-Fano variety (i.e., $-K_X$ is an ample $\mathbb{Q}$-line bundle and X has at worst klt singularities). Corresponding to (1.9), we have a non-Archimedean **D** functional. For general test configurations, it first appeared in Berman's work [4] and was reformulated in [19] using non-Archimedean potentials:

$$\mathbf{L}^{\mathrm{NA}}(\mathcal{X}, \mathcal{L}) = \inf_{v\in X_{\mathbb{Q}}^{\mathrm{div}}} \big(A_X(v) + \phi_{(\mathcal{X},\mathcal{L})}(v)\big), \quad \mathbf{D}^{\mathrm{NA}}(\mathcal{X}, \mathcal{L}) = -\mathbf{E}^{\mathrm{NA}}(\mathcal{X}, \mathcal{L}) + \mathbf{L}^{\mathrm{NA}}(\mathcal{X}, \mathcal{L}).$$

The notions of Ding-stability and uniform Ding-stability are defined if $\mathbf{M}^{\mathrm{NA}}$ is replaced by $\mathbf{D}^{\mathrm{NA}}$ in Definitions 2.2 and 2.3. In general, we have the inequality $\mathbf{M}^{\mathrm{NA}}(\mathcal{X}, \mathcal{L}) \geq \mathbf{D}^{\mathrm{NA}}(\mathcal{X}, \mathcal{L})$. For Fano varieties, special test configurations play important roles. A test configuration $(\mathcal{X}, \mathcal{L})$ is called special if the central fiber $\mathcal{X}_0$ is a $\mathbb{Q}$-Fano variety and $\mathcal{L} = -K_{\mathcal{X}/\mathbb{P}^1}$. For special test configurations, we have $\mathbf{D}^{\mathrm{NA}} = \mathbf{M}^{\mathrm{NA}} = -\mathbf{E}^{\mathrm{NA}} =: \mathrm{Fut}_{\mathcal{X}_0}(\xi)$, the last quantity being the Futaki invariant on $\mathcal{X}_0$ for the holomorphic vector field ξ that generates the $\mathbb{C}^*$-action. The importance of special test configurations was first pointed out in Tian's work [64] motivated by compactness results from metric geometry. The following results show their importance from the point of view of algebraic geometry:

Theorem 2.13 ([35,44,52], see also [7,21]). *For any $\mathbb{Q}$-Fano variety, K-stability is equivalent to Ding-stability, and they are equivalent to K-stability or Ding-stability over special test configurations. Moreover, the same conclusion holds true if stability is replaced by semistability, polystability, or reduced uniform stability.*

The proofs of these results depend on a careful process of Minimal Model Program first used in [52] to transform any given test configuration into a special one. Moreover, crucial calculations show that the relevant invariants such as $\mathbf{M}^{\mathrm{NA}}$ or $\mathbf{D}^{\mathrm{NA}}$ decrease along the MMP process. Theorem 2.13 leads directly to a valuative criterion for K-stability. To state it, first define for any $v \in X_{\mathbb{Q}}^{\mathrm{div}}$ an invariant (see Example 2.9):

$$S_L(v) := \frac{1}{\mathbf{V}}\int_0^{+\infty} \mathrm{vol}\big(\mathcal{F}_v^{(t)}\big)dt = \frac{1}{\mathbf{V}}\int_{\mathbb{R}} t\big(-d\,\mathrm{vol}\big(\mathcal{F}_v^{(t)}\big)\big) = \mathbf{E}^{\mathrm{NA}}(\phi_{\mathcal{F}_v}). \tag{2.11}$$

Let $\tilde{\mathbb{T}}$ be a maximal torus of $\mathrm{Aut}(X)$ and $(X_{\mathbb{Q}}^{\mathrm{div}})^{\tilde{\mathbb{T}}}$ be the set of $\tilde{\mathbb{T}}$-invariant divisorial valuations. Define the following invariant ($(\xi, v) \mapsto v_\xi$ is the action appeared in Example 2.8):

$$\delta(X) = \inf_{v \in X_{\mathbb{Q}}^{\mathrm{div}}} \frac{A_X(v)}{S_X(v)}, \quad \delta_{\tilde{\mathbb{T}}}(X) = \inf_{v \in (X_{\mathbb{Q}}^{\mathrm{div}})^{\tilde{\mathbb{T}}}} \sup_{\xi \in \tilde{N}_{\mathbb{R}}} \frac{A_X(v_\xi)}{S_X(v_\xi)}.$$

Here we use the convention that $A_X(v_{\mathrm{triv}})/S_X(v_{\mathrm{triv}}) = +\infty$ for the trivial valuation v_{triv}.

Theorem 2.14. *The following statements are true.*

(1) ([35,42]) *X is K-semistable if $\delta(X) \geq 1$.*

(2) ([34,35]) *X is uniformly K-stable if and only if $\delta(X) > 1$.*

(3) ([15,35,42]) *X is K-stable if and only if $A_X(v) > S(v)$ for any nontrivial $v \in X_{\mathbb{Q}}^{\mathrm{div}}$.*

(4) ([44]) *X is reduced uniformly K-stable if and only if $\delta_{\tilde{\mathbb{T}}}(X) > 1$.*

To get these, we first use the fact as pointed out in [19] that for a special test configuration $(\mathcal{X}, \mathcal{L})$, the valuation $\mathrm{ord}_{\mathcal{X}_0}$ of the function field $\mathbb{C}(X \times \mathbb{C})$ restricts to become a divisorial valuation $v \in X_{\mathbb{Q}}^{\mathrm{div}}$. A crucial observation is then made in [42]: $\mathcal{F}^{\lambda}_{(\mathcal{X},\mathcal{L})} R_m = \mathcal{F}_v^{\lambda + m A_X(v)} R_m$ (see (2.9)). This implies $\mathrm{vol}(\mathcal{F}^{(t)}_{(\mathcal{X},\mathcal{L})}) = \mathrm{vol}(\mathcal{F}_v^{(t + A_X(v))})$, which, together with (2.10), leads to[2]

$$\mathbf{M}^{\mathrm{NA}}(\mathcal{X}, \mathcal{L}) = \mathbf{D}^{\mathrm{NA}}(\mathcal{X}, \mathcal{L}) = A_X(v) - \mathbf{E}^{\mathrm{NA}}(\mathcal{F}_v) = A_X(v) - S(v). \tag{2.12}$$

This, together with Theorem 2.13, already gives the sufficient condition for the K-(semi)stability. The criterion for uniform K-stability follows from a similar argument and K. Fujita's inequality, $\frac{1}{n} S(v) \leq \mathbf{J}^{\mathrm{NA}}(\mathcal{F}_v) \leq n S(v)$ [34]. For reduced uniform stability, another identity $A_X(v_\xi) - S(v_\xi) = A_X(v) - S(v) + \mathrm{Fut}_X(\xi)$ proved in [44] is needed.

As we will see in Section 3.2, a main reason for introducing the (reduced) uniform K-stability is that it is much easier to use in making connection with the (reduced) coercivity in the complex analytic setting. On the other hand, we now have the following fundamental result:

Theorem 2.15 ([56]). *Let X be a $\mathbb{Q}$-Fano variety. Then X is K-stable if and only if X is uniformly K-stable. More generally, X is reduced uniformly stable if and only if X is K-polystable. Moreover, these statements hold true for any log Fano pair.*

This is achieved by several works. First, according to a work of Blum–Liu–Xu [13], divisorial valuations on X associated to special test configurations are log canonical places of complements. By deep boundedness of Birkar and Haccon–McKernan–Xu, it was also shown that there exists a quasimonomial valuation (i.e., a monomial valuation on a smooth birational model) that achieves the infimum defining $\delta(X)$ (and a similar result holds more generally for $\delta_{\tilde{\mathbb{T}}}(X)$). Then the main problem becomes proving a finite generation property for the minimizing valuation, which is achieved by using deep techniques from birational

2 The original argument in [42] also explicitly relates the filtration $\mathcal{F}_{(\mathcal{X},\mathcal{L})}$ to a filtration of the section ring of $\mathcal{X}_0$ induced by the $\mathbb{C}^*$-action.

algebraic geometry in [56]. In fact, in the past several years, the algebraic study of K-stability for Fano varieties has flourished, and there are many important results which answer fundamental questions in this subject. We highlight two such achievements:

(1) Algebraic construction of projective moduli space of K-polystable Fano varieties. This is achieved in a collection of works, settling different issues in the construction including boundedness, separatedness, properness, and projectivity. Moreover, concrete examples of compact moduli spaces have been identified. We refer to [56,70] for extensive discussions on related topics.

(2) Fujita–Odaka [36] introduced quantizations of the $\delta(X)$ invariant: for each $m \in \mathbb{N}$,

$$\delta_m(X) = \inf\{\mathrm{lct}(X, D) : D \text{ is of } m\text{-basis type}\}$$

where D is of m-basis type if $D = \frac{1}{mN_m}\sum_{i=1}^{N_m}\{s_i = 0\}$ where $\{s_i\}$ is a basis of $H^0(X, mL)$. Blum–Jonsson [12] proved $\lim_{m\to+\infty}\delta_m(X) = \delta(X)$. This provides a practical tool to verify uniform stability of Fano varieties. Ahmadinezhad–Zhuang [1] further introduced new techniques for estimating the δ_m and δ invariant which lead to many new examples of K-stable Fano varieties. All of these culminate in the recent determination of deformation types of smooth Fano threefolds that contain K-polystable ones (see [3]).

In another direction, Han–Li [37] establishes a valuative criterion for g-weighted stability, corresponding to the study of g-solitons. A key idea in such an extension is using a fibration technique for a polynomial weights (as motivated by the theory of equivariant de Rham cohomology) and then using the Stone–Weierstrass approximation to deal with the general g. Moreover, there is a notion of stability for Fano cones introduced earlier by Collins–Székelyhidi associated to Ricci-flat Kähler cone metrics. It is shown recently that this stability of Fano cones is, in fact, equivalent to a particular g-weighted stability of log Fano quotients (see [2,47]).

The techniques developed in the study of (weighted) K-stability of Fano varieties have also been applied to treat an optimal degeneration problem that is motivated by the Hamilton–Tian conjecture in differential geometry (see [74] for background of this conjecture). This is formulated as a minimization problem for valuations in [38] which defines the following functional (cf. (2.12) and (2.11)), for any valuation $v \in X^{\mathrm{NA}}$,

$$\tilde{\beta}(v) = A_X(v) + \log\left(\frac{1}{\mathbf{V}}\int_0^{+\infty} e^{-t}\left(-d\,\mathrm{vol}\left(\mathcal{F}_v^{(t)}\right)\right)\right).$$

Very roughly speaking, the $\tilde{\beta}$ functional is an antiderivative of a certain weighted Futaki invariant. This functional is a variant of invariants that appeared in previous works of Tian–Zhang–Zhang–Zhu, Dervan–Székelyhidi, and Hisamoto (see [74] for more details). The results from [14, 37, 51] together prove the following algebraic version of Hamilton–Tian conjecture:

Theorem 2.16. *For any $\mathbb{Q}$-Fano variety, there exists a unique quasimonomial valuation v_* that minimizes $\tilde{\beta}$, whose associated filtration $\mathcal{F}_{v_*}$ is finitely generated and induces a degener-*

ation of X to a $\mathbb{Q}$-Fano variety $\mathcal{X}_0$ together with a vector field V_ξ. Moreover, $\mathcal{X}_0$ degenerates uniquely to an $e^{-\langle \cdot,\xi\rangle}$-weighted polystable $\mathbb{Q}$-Fano variety (cf. Example 1.1).

Combined with previous works, the uniqueness part in particular confirms a conjecture of Chen–Sun–Wang about the algebraic uniqueness of limits under normalized Kähler–Ricci flows on Fano manifolds (see **[62]**).

2.4. Normalized volume and local stability theory of klt singularities

A similar minimization problem for valuations was actually studied even earlier in the local setting, which motivates the formulation and the proof of Theorem 2.16. Let (X, x) be a klt singularity. Denote by $\mathrm{Val}_{X,x}$ the space of real valuations that have center x. The following *normalized volume* functional was introduced in **[43]**: for any $v \in \mathrm{Val}_{X,x}$,

$$\widehat{\mathrm{vol}}(v) := \begin{cases} A_X(v)^n \cdot \mathrm{vol}(v), & \text{if } A_X(v) < +\infty, \\ +\infty, & \text{otherwise.} \end{cases} \tag{2.13}$$

Here $A_X(v)$ is again the log discrepancy functional and $\mathrm{vol}(v)$ is defined as

$$\mathrm{vol}(v) = \lim_{p\to+\infty} \frac{\dim_{\mathbb{C}}(\mathcal{O}_{X,x}/\mathfrak{a}_p(v))}{p^n/n!} \quad \text{where } \mathfrak{a}_p(v) = \{f \in \mathcal{O}_{X,x}; v(f) \geq p\}.$$

The expression in (2.13) is inspired by the work of Martelli–Sparks–Yau **[57]** on a volume minimization property of Reeb vector fields associated to Ricci-flat Kähler cone metrics. In **[43]** we started to consider the minimization of $\widehat{\mathrm{vol}}$ over $\mathrm{Val}_{X,x}$ and define the invariant $\widehat{\mathrm{vol}}(X, x) = \inf_{v\in \mathrm{Val}_{X,x}} \widehat{\mathrm{vol}}(v)$. We proved that the invariant $\widehat{\mathrm{vol}}(X, x)$ is strictly positive and further conjectured the existence, uniqueness of minimizing valuations which should have finite generated associated graded rings. For a concrete example, it was shown by the author and Y. Liu that for an isolated quotient singularity $X = \mathbb{C}^n/\Gamma$, $\widehat{\mathrm{vol}}(\mathbb{C}^n/\Gamma, 0) = \frac{n^n}{|\Gamma|}$ and the exceptional divisor of the standard blowup obtains the infimum.

This minimization problem was proposed by the author to attack a conjecture of Donaldson–Sun, which states that the metric tangent cone at any point on a Gromov–Hausdorff limit of Kähler–Einstein manifolds depends only on the algebraic structure (see **[62]**). This conjecture has been confirmed in a series of following-up papers **[51, 53, 54]**. Algebraically, we have the following results regarding this minimization problem.

Theorem 2.17. (1) *There exists a valuation that achieves the infimum in defining $\widehat{\mathrm{vol}}(X, x)$. Moreover, this minimizing valuation is quasimonomial and unique up to rescaling.*

(2) *A divisorial valuation v_* is the minimizer if and only if it is the exceptional divisor of a plt blowup and also the associated log Fano pair is K-semistable.*

The first statement is a combination of works by Harold Blum, Chenyang Xu, and Ziquan Zhuang **[11,71,72]** partly based on calculations from **[42,43]**. The second statement was proved in Li–Xu **[54]** (see also **[11]**) by extending the global argument from **[52]** to the local case, and it shows a close relationship between the local and global theory. In fact, it is in proving the affine cone case of this statement when valuative criterion for K-(semi)stability was first discovered in [42]. A similar statement is true for more general quasimonomial minimizing valuations [53]. However, the finite generation conjecture from [43] is still open in general, and seems to require deeper boundedness property of Fano varieties.

We also like to mention that Yuchen Liu obtained a surprising local-to-global comparison inequality by generalizing an estimate of Kento Fujita:

Theorem 2.18 ([55]). *For any closed point x on a K-semistable $\mathbb{Q}$-Fano variety X, we have*

$$(-K_X)^{\cdot n} \le \frac{(n+1)^n}{n^n}\widehat{\mathrm{vol}}(X,x). \tag{2.14}$$

For example, if $x \in X$ is a regular point, (2.14) recovers Fujita's beautiful inequality, namely $(-K_X)^{\cdot n} \le (n+1)^n$ for any K-semistable X [33]. Inequality (2.14) has applications in controlling singularities on the varieties that correspond to boundary points of moduli spaces. In order for this to be effective, good estimates of $\widehat{\mathrm{vol}}(X,x)$ for klt singularities need to be developed. In particular, it is still interesting to understand better the $\widehat{\mathrm{vol}}$ invariants and associated minimizers for 3-dimensional klt singularities. For more discussion on related topics, we refer to the survey [48].

3. ARCHIMEDEAN (COMPLEX ANALYTIC) THEORY VS. NON-ARCHIMEDEAN THEORY

3.1. Correspondence between Archimedean and non-Archimedean objects

In this section, we discuss results showing a general philosophy that non-Archimedean objects encode the information of corresponding Archimedean objects "at infinity."

Let $(\mathcal{X},\mathcal{L})$ be a test configuration and $\tilde{h}$ be a smooth psh metric on $\mathcal{L}$. Via the isomorphism $(\mathcal{X},\mathcal{L})\times_{\mathbb{C}}\mathbb{C}^* \cong X\times\mathbb{C}^*$, we get a path $\widetilde{\Phi} = \{\tilde{\varphi}(s)\}_{s\in\mathbb{R}}$ of smooth ω_0-psh potentials where $s = -\log|t|^2$. With these notation, we have the following slope formula:

Theorem 3.1 ([20,59,64,67]). *The slope at infinity of a functional $\mathbf{F} \in \{\mathbf{E},\mathbf{\Lambda},\mathbf{I},\mathbf{J},\mathbf{M}\}$ is given by the corresponding non-Archimedean functional*

$$\mathbf{F}'^{\infty}(\widetilde{\Phi}) := \lim_{s\to+\infty}\frac{\mathbf{F}(\tilde{\varphi}(s))}{s} = \mathbf{F}^{\mathrm{NA}}(\mathcal{X},\mathcal{L}) = \mathbf{F}^{\mathrm{NA}}(\phi_{(\mathcal{X},\mathcal{L})}).$$

There is a more canonical analytic object associated to a test configuration. Recall from section 1.5 that by a geodesic ray $\Phi = \{\varphi(s)\}_{s\in[0,+\infty)}$ in $\mathcal{E}^1$ we mean that $\Phi|_{[s_1,s_2]}$ is a geodesic connecting $\varphi(s_1), \varphi(s_2)$ for any $s_1, s_2 \in [0,\infty)$ (see (1.12)).

Theorem 3.2 ([60]). *Given any test configuration $(\mathcal{X},\mathcal{L})$ for (X,L), there exists a geodesic ray $\Phi_{(\mathcal{X},\mathcal{L})}$ emanating from any given smooth potential φ_0.*

On the other hand, recall from section 2.1 that there is a non-Archimedean potential associated to $(\mathcal{X},\mathcal{L})$ (see (2.1)). Berman–Boucksom–Jonsson proved that there is a direct relation between geodesic rays and non-Archimedean potentials. First they showed that any geodesic ray Φ defines a non-Archimedean potential (cf. (2.1))

$$\Phi_{\mathrm{NA}}(v) := -G(v)(\Phi), \quad \text{for any } v \in X^{\mathrm{div}}_{\mathbb{Q}},$$

where $G(v)(\Phi)$ is the generic Lelong number of Φ considered as a singular quasi-psh potential on a birational model where the center of the valuation $G(v)$ is a prime divisor.

Theorem 3.3 ([7]). *The following statements are true:*

(1) *The map $\Phi \mapsto \Phi_{\mathrm{NA}}$ has the image contained in $(\mathcal{E}^1)^{\mathrm{NA}}$. Conversely, for any $\phi \in (\mathcal{E}^1)^{\mathrm{NA}}$, there exists a geodesic ray denoted by $\gamma(\phi)$ that satisfies $\gamma(\phi)_{\mathrm{NA}} = \phi$.*

(2) *For any geodesic ray Φ, $\widehat{\Phi} = \gamma(\Phi_{\mathrm{NA}})$ satisfies $\widehat{\Phi}_{\mathrm{NA}} = \Phi_{\mathrm{NA}} \in (\mathcal{E}^1)^{\mathrm{NA}}$ and $\widehat{\Phi} \geq \Phi$.*

(3) *For $\Phi = \gamma(\phi)$ with $\phi \in (\mathcal{E}^1)^{\mathrm{NA}}$, $\mathbf{E}'^{\infty}(\Phi) = \mathbf{E}^{\mathrm{NA}}(\phi)$, and there exists a sequence of test configurations $(\mathcal{X}_m, \mathcal{L}_m)$ such that Φ is the decreasing limit of $\Phi_{(\mathcal{X}_m, \mathcal{L}_m)}$ (see Theorem 3.2).*

Berman–Boucksom–Jonsson proved $\Phi_{\mathrm{NA}} \in \mathrm{PSH}^{\mathrm{NA}}$ by blowing up multiplier ideal sheaves $\{\mathcal{J}(m\Phi)\}_{m\in\mathbb{N}}$ and using their global generation properties to construct test configurations $\{(\mathcal{X}_m, \mathcal{L}_m)\}$ such that $\phi_{(\mathcal{X}_m, \mathcal{L}_m)}$ decreases to Φ_{NA}. Because of the second statement, any geodesic ray $\gamma(\phi)$ with $\phi \in (\mathcal{E}^1)^{\mathrm{NA}}$ is called *maximal* in [7]. By the last statement, maximal geodesic rays can be approximated by (geodesic rays associated to) test configurations. Moreover, when $\phi = \phi_{(\mathcal{X},\mathcal{L})} \in \mathcal{H}^{\mathrm{NA}}$, $\gamma(\phi)$ coincides with the geodesic ray from Theorem 3.2. Further useful properties of maximal geodesic rays are known (cf. Theorem 3.1):

Theorem 3.4 ([45]). *Let Φ be a maximal geodesic ray.*

(1) *We have the identity $(\mathbf{E}^{-\mathrm{Ric}(\omega_0)})'^{\infty}(\Phi) = (\mathbf{E}^{K_X})^{\mathrm{NA}}(\Phi_{\mathrm{NA}})$.*

(2) $\mathbf{H}'^{\infty}(\Phi) \geq \mathbf{H}^{\mathrm{NA}}(\Phi_{\mathrm{NA}})$. *Moreover, if $\Phi = \Phi_{(\mathcal{X},\mathcal{L})}$ is associated to a test configuration, then $\mathbf{H}'^{\infty}(\Phi) = \mathbf{H}^{\mathrm{NA}}(\Phi_{\mathrm{NA}})$.*

It is natural to conjecture that $\mathbf{H}'^{\infty}(\Phi) = \mathbf{H}^{\mathrm{NA}}(\Phi_{\mathrm{NA}})$ always holds for any maximal geodesic ray Φ. This is implied by the algebraic Conjecture 2.12, according to [45, 46].

As pointed out in [7], by a construction of Darvas, there are abundant nonmaximal geodesic rays. In fact, analogous local examples have been used by the author to disprove a conjecture of Demailly on Monge–Ampère mass of psh singularities. It is thus a surprising fact that maximal geodesic rays are the only ones of interest in the cscK problem.

Theorem 3.5 ([45]). *If a geodesic ray Φ satisfies $\mathbf{M}'^{\infty}(\Phi) < +\infty$, then Φ is maximal.*

Note that $\mathbf{M}'^{\infty}(\Phi) = \lim_{s\to+\infty} \frac{\mathbf{M}(\varphi(s))}{s}$ exists by Theorem 1.3. This result resolves a difficulty raised in Boucksom's ICM talk [16], and implies that destabilizing geodesic rays can always be approximated by test configurations, thus giving a very strong evidence for the validity of Yau–Tian–Donaldson Conjecture 3.6. The proof of Theorem 3.5 starts with an equisingularity property $\int_{X\times\{|t|<1\}} e^{-\alpha(\widehat{\Phi}-\Phi)} < +\infty$ for any $\alpha > 0$, and then uses Jensen's inequality, together with a comparison principle, for the $\mathbf{E}$ functional to get a contradiction with the finite slope assumption if $\widehat{\Phi} = \gamma(\Phi_{\mathrm{NA}}) \neq \Phi$.

3.2. Yau–Tian–Donaldson conjecture for general polarized manifolds

The Yau–Tian–Donaldson (YTD) conjecture says that the existence of canonical Kähler metrics on projective manifolds should be equivalent to a certain K-stability condition. For a general polarization, it is believed that one needs to use a strengthened definition of K-stability such as Definition 2.3. In particular, we have the following version.

Conjecture 3.6 (uniform YTD conjecture). *A polarized manifold (X, L) admits a cscK metric if and only if (X, L) is reduced uniformly K-stable.*

The implication from existence to stability is known, and follows from Theorems 1.4 and 3.1. The other direction is still open in general. However, based on the results discussed thus far, we can explain the proof of a weak version.

Theorem 3.7 ([45]). *If (X, L) is uniformly stable over models (i.e., there exists $\gamma > 0$ such that $\mathbf{M}^{\mathrm{NA}}(\mathcal{X}, \mathcal{L}) \geq \gamma \cdot \mathbf{J}^{\mathrm{NA}}(\mathcal{X}, \mathcal{L})$ for any model $(\mathcal{X}, \mathcal{L})$), then it admits a cscK metric.*

Summary of proof. Step 1. By Theorem 1.4, we need to show that $\mathbf{M}$ is coercive. Assume that the coercivity fails. Then there exists a geodesic ray $\Phi = \{\varphi(s)\}_{s\in[0,\infty)}$ satisfying

$$\mathbf{M}'^{\infty}(\Phi) \leq 0, \quad \mathbf{J}'^{\infty}(\Phi) = 1, \quad \sup\big(\varphi(s)\big) = 0.$$

Such a destabilizing geodesic ray Φ was constructed in [7,27] from a destabilizing sequence. In this construction, both the convexity of $\mathbf{M}$ from Theorem 1.3 and a compactness result for potentials with uniform entropy bounds from [6] play crucial roles.

Step 2. By Theorem 3.5, Φ is maximal. Set $\phi = \Phi_{\mathrm{NA}}$. By using Theorem 3.3(3) and Theorem 3.4(1), we derive the identities

$$\mathbf{E}'^{\infty}(\Phi) = \mathbf{E}^{\mathrm{NA}}(\phi), \quad (\mathbf{E}^{-\mathrm{Ric}(\omega_0)})'^{\infty}(\Phi) = (\mathbf{E}^{K_X})^{\mathrm{NA}}(\phi), \quad \mathbf{J}'^{\infty}(\Phi) = \mathbf{J}^{\mathrm{NA}}(\phi).$$

Moreover, by Theorem 3.4(2), $\mathbf{H}'^{\infty}(\Phi) \geq \mathbf{H}^{\mathrm{NA}}(\phi)$ so that $\mathbf{M}'^{\infty}(\Phi) \geq \mathbf{M}^{\mathrm{NA}}(\phi)$.

Step 3. By Theorem 2.11, there exist models $(\mathcal{X}_m, \mathcal{L}_m)$ such that $\phi_m = \phi_{(\mathcal{X}_m, \mathcal{L}_m)}$ converges to ϕ in the strong topology and

$$\lim_{m\to+\infty} \mathbf{M}^{\mathrm{NA}}(\phi_m) = \mathbf{M}^{\mathrm{NA}}(\phi), \qquad \lim_{m\to+\infty} \mathbf{J}^{\mathrm{NA}}(\phi_m) = \mathbf{J}^{\mathrm{NA}}(\phi).$$

Step 4. We can complete the proof by getting a contradiction to the stability assumption:

$$0 \geq \mathbf{M}'^{\infty}(\Phi) \geq \mathbf{M}^{\mathrm{NA}}(\phi) = \lim_{m\to+\infty} \mathbf{M}^{\mathrm{NA}}(\phi_m) \geq_{\text{stability}} \lim_{m\to+\infty} \mathbf{J}^{\mathrm{NA}}(\phi_m) = \mathbf{J}^{\mathrm{NA}}(\phi) = 1. \ \blacksquare$$

There is a version of Theorem 3.7 in [45] when $\mathrm{Aut}(X, L)$ is continuous. Moreover, it is shown in [46] that Conjecture 2.12 implies Conjecture 3.6. As mentioned earlier, if (X, L) is any polarized spherical manifold, Conjecture 2.12 is true and hence in this case the YTD Conjecture 3.6 is proved. Based on this fact, Delcroix [28] obtained further refined existence results in this case.

We should mention that Sean Paul (see [58]) has works that give a beautiful interpretation of the coercivity of $\mathbf{M}$-functional using a new notion of stability for pairs. However, it is not clear how K-stability discussed here can directly imply his stability notion.

3.3. YTD conjecture for Fano varieties

3.3.1. Non-Archimedean approach

Our proof of Theorem 3.7 is in fact modeled on a non-Archimedean approach to the uniform YTD conjecture proposed by Berman–Boucksom–Jonsson in [7]. They carried it out sucessfully for smooth Fano manifolds with discrete automorphism groups. The main advantage in the Fano case is that $\mathbf{D}^{\mathrm{NA}}$ satisfies a regularization property and can be used in place of $\mathbf{M}^{\mathrm{NA}}$ to complete the argument. Recently their work has been extended to the most general setting of log Fano pairs.

Theorem 3.8 ([44,49,50]). *A log Fano pair (X, D) admits a Kähler–Einstein metric if and only if it is reduced uniformly stable for all special test configurations.*

Note that this combined with Theorem 2.15 also proves the K-polystable version of the YTD conjecture. Theorem 3.8 can be used to get examples of Kähler–Einstein metrics on Fano varieties with large symmetry groups (see, for example, [40]). The proof of Theorem 3.8 is much more technical than [7] because we need to overcome the difficulties caused by singularities. The first key idea is to use an approximation approach initiated in [49]. Consider the log resolution $\mu : X' \to X$ as in Section 1.3 and reorganize (1.4) as

$$-K_{X'} - D_\varepsilon = \frac{1}{1+\varepsilon}\big(\mu^*(-K_X - D) + \varepsilon H\big) =: L_\varepsilon,$$

where $H = \mu^*(-K_X - D) - \sum_k \theta_k E_k$ is ample by choosing appropriate $\{\theta_k\}$ and $D_\varepsilon = \sum_k(-a_k + \frac{\varepsilon}{1+\varepsilon})\theta_k E_k$ with $0 \le \varepsilon \ll 1$. In [49] we considered the simple case when $a_k \in (-1, 0]$ for all k. In this case for $0 < \varepsilon \ll 1$, (X', D_ε) is a smooth log Fano pair. A crucial calculation using the valuative criterion from Theorem 2.14 shows that (semi)stability of (X, D) implies the uniform stability of (X', D_ε) for $\varepsilon > 0$. Moreover, we can prove a version of YTD conjecture for (X', D_ε) and deduce that it admits a Kähler–Einstein metric. Next we take a limit as $\varepsilon \to 0$ to get a Kähler–Einstein metric on (X, D) itself. The proof of this convergence depends on technical uniform pluripotential and geometric estimates.

In [50], we dealt with the general case when D_ε is not necessarily effective. A key difficulty for the argument in [7] to work on singular varieties is that it is not clear how to use multiplier ideal sheaves to approximate a destabilizing geodesic ray Φ when X is singular. To circumvent this difficulty, we first need to perturb Φ to become a singular quasi-psh potential Φ_ε on $(X' \times \mathbb{C}, p_1'^* L_\varepsilon)$. Since X' is smooth, we know how to approximate Φ_ε by test configurations for (X', L_ε) thanks to [7]. However, due to the ineffectiveness of D_ε, the remaining arguments depend more heavily on non-Archimedean analysis and some key observation on convergence of slopes. In [45] we further derived the valuative criterion for reduced uniform stability and understood how the torus action induces an action on the space of non-Archimedean potentials in order to incorporate group actions in the argument. Note that the non-Archimedean approach a priori does not prove the statement in Theorem 3.8 involving special test configurations. Fortunately, Theorem 2.13 fills this gap.

By using the fibration and approximation techniques mentioned earlier, Theorem 3.8 has been extended to the case of g-soliton on log Fano pairs in [38]. As explained in [2,47], this can be used prove the YTD conjecture for Ricci-flat Kähler cone metrics thanks to its equivalence to particular g-solitons (see Section 1.4). This generalizes the result of Collins–Székelyhidi on YTD conjecture for Fano cones with isolated singularities [25].

3.3.2. Other approaches

For completeness, we briefly mention other approaches to the YTD conjecture on Fano manifolds. The classical way to solve the Kähler–Einstein equation is through various continuity methods. Traditionally, one uses Aubin's continuity method involving twisted KE metrics. A more recent continuity method uses KE metrics with edge cone singularities as proposed by Donaldson. Finally, there is a Kähler–Ricci flow approach. Tian's early works showed that the most difficult part in proving the YTD conjecture by continuity methods is to establish the algebraicity of limit objects in the Gromov–Hausdorff topology, and he had essentially reduced this difficulty to proving some partial C^0-estimates. The partial C^0-estimates were later proved in different settings, starting with Donaldson–Sun's work in the Kähler–Einstein case, which led to the solution of the YTD conjecture for smooth Fano manifolds in [24,65]. Moreover, the partial C^0-estimates has applications in constructing moduli spaces of smoothable Kähler–Einstein varieties and proving quasiprojectivity of the moduli spaces of KE manifolds, and these applications preclude the algebraic approach mentioned earlier (see [68]). We also refer to [31,66] for surveys on related topics in this approach.

Very recently, yet another quantization approach is carried out by Kewei Zhang based partly on an earlier work of Rubinstein–Tian–Zhang. Zhang considered an analytic invariant of Moser–Trudinger type, namely

$$\delta^A(X) = \sup\left\{c : \sup_{\varphi\in\mathcal{H}} \int_X e^{-c(\varphi-\mathbf{E}(\varphi))} < +\infty\right\}.$$

It is easy to show that the coercivity of $\mathbf{D}$-functional is equivalent to $\delta^A(X) > 1$. The authors of [61] introduced a quantization $\delta^A_m(X)$ by using a quantization of $\mathbf{E}$ on the space of Bergman metrics, and further proved $\delta^A_m(X) = \delta_m(X)$. Using some deep results in complex geometry including Tian's work on Bergman kernels and Berndtsson's subharmonicity theorem, it is proved in [73] that $\lim_{m\to+\infty} \delta^A_m(X) = \delta^A(X)$. Combining these discussions with the algebraic convergence result of Blum–Jonsson and the valuative criterion of uniform stability of Fujita discussed earlier, Zhang gets $\delta^A(X) = \delta(X)$ and completes the proof of uniform version of YTD conjecture for smooth Fano manifolds. It would be interesting to extend this approach to the more general case (i.e., Fano varieties with continuous automorphism groups).

We finish by remarking that it is of interest to apply the ideas and methods from the above two approaches to study the YTD conjecture for general polarizations. For the approach involving partial C^0-estimates, the geometry is complicated by collapsing phenomenon in the Gromov–Hausdorff convergence with only scalar curvature bounds, which is very difficult to study with current techniques. For the quantization approach, there were some attempts by Mabuchi in several works. But the precise picture seems again unclear.

ACKNOWLEDGMENTS

The author thanks Gang Tian, Chenyang Xu, Feng Wang, Xiaowei Wang, Jiyuan Han, and Yuchen Liu for collaborations on works discussed in this survey. He also thanks many mathematicians for various discussions related to our works: Vestislav Apostolov, Robert Berman, Sébastien Boucksom, Tamás Darvas, Simon Donaldson, Kento Fujita, Hans-Joachim Hein, Mattias Jonsson, László Lempert, Mircea Mustaţă, Yuji Odaka, Yanir Rubinstein, Jian Song, Song Sun, Jun Yu, and many others. A large part of the author's works discussed here were carried out at the Department of Mathematics, Purdue University, and he is grateful to his colleagues and friends there for creating a great environment of research.

FUNDING

This work is partially supported by NSF grant (DMS-181086) and an Alfred Sloan Fellowship.

REFERENCES

[1] H. Ahmadinezhad and Z. Zhuang, K-stability of Fano varieties via admissible flags. 2020, arXiv:2003.13788.

[2] V. Apostolov, S. Jubert, and A. Lahdili, Weighted K-stability and coercivity with applications to extremal Kähler and Sasaki metrics. 2021, arXiv:2104.09709.

[3] C. Araujo, A.-M. Castravet, I. Cheltsov, K. Fujita, A.-S. Kaloghiros, J. Martinez-Garcia, C. Shramov, H. Süss, and N. Viswanathan, The Calabi problem for Fano threefolds. 2021, MPIM preprint.

[4] R. Berman, K-polystability of $\mathbb{Q}$-Fano varieties admitting Kähler–Einstein metrics. *Invent. Math.* **203** (2015), no. 3, 973–1025.

[5] R. Berman and R. Berndtsson, Convexity of the K-energy on the space of Kähler metrics. *J. Amer. Math. Soc.* **30** (2017), 1165–1196.

[6] R. Berman, S. Boucksom, P. Eyssidieux, V. Guedj, and A. Zeriahi, Kähler–Einstein metrics and the Kähler–Ricci flow on log Fano varieties. *J. Reine Angew. Math.* **751** (2019), 27–89.

[7] R. Berman, S. Boucksom, and M. Jonsson, A variational approach to the Yau–Tian–Donaldson conjecture. *J. Amer. Math. Soc.* **34** (2020), no. 3, 605–652.

[8] R. Berman, T. Darvas, and C. H. Lu, Convexity of the extended K-energy and the large time behaviour of the weak Calabi flow. *Geom. Topol.* **21** (2017), 2945–2988.

[9] R. Berman, T. Darvas, and C. H. Lu, Regularity of weak minimizers of the K-energy and applications to properness and K-stability. *Ann. Sci. Éc. Norm. Supér. (4)* **53** (2020), no. 2, 267–289.

[10] R. Berman and D. Witt Nyström, Complex optimal transport and the pluripotential theory of Kähler–Ricci solitons, Preprint, 2014, arXiv:1401.8264.

[11] H. Blum, Existence of valuations with smallest normalized volume. *Compos. Math.* **154** (2018), no. 4, 820–849.

[12] H. Blum and M. Jonsson, Thresholds, valuations, and K-stability. *Adv. Math.* **365** (2020), 107062.

[13] H. Blum, Y. Liu, and C. Xu, Openness of K-semistability for Fano varieties. 2019, arXiv:1907.02408.

[14] H. Blum, Y. Liu, C. Xu, and Z. Zhuang, The existence of the Kähler–Ricci soliton degeneration. 2021, arXiv:2103.15278.

[15] H. Blum and C. Xu, Uniqueness of K-polystable degenerations of Fano varieties. *Ann. of Math. (2)* **190** (2019), no. 2, 609–656.

[16] S. Boucksom, Variational and non-Archimedean aspects of the Yau–Tian–Donaldson conjecture. In *Proceedings of the International Congress of Mathematicians – Rio de Janeiro 2018. Vol. II. Invited lectures*, pp. 591–617, World Sci. Publ., Hackensack, NJ, 2018.

[17] S. Boucksom, C. Favre, and M. Jonsson, Differentiability of volumes of divisors and a problem of Teissier. *J. Algebraic Geom.* **18** (2009), no. 2, 279–308.

[18] S. Boucksom, C. Favre, and M. Jonsson, Solution to a non-Archimedean Monge–Ampère equation. *J. Amer. Math. Soc.* **28** (2015), 617–667.

[19] S. Boucksom, T. Hisamoto, and M. Jonsson, Uniform K-stability, Duistermaat–Heckman measures and singularities of pairs. *Ann. Inst. Fourier (Grenoble)* **67** (2017), 87–139.

[20] S. Boucksom, T. Hisamoto, and M. Jonsson, Uniform K-stability and asymptotics of energy functionals in Kähler geometry. *J. Eur. Math. Soc. (JEMS)* **21** (2019), no. 9, 2905–2944.

[21] S. Boucksom and M. Jonsson, A non-Archimedean approach to K-stability. 2018, arXiv:1805.11160v1.

[22] S. Boucksom and M. Jonsson, Global pluripotential theory over a trivially valued field. 2021, arXiv:1801.08229v2.

[23] X. Chen and J. Cheng, On the constant scalar curvature Kähler metrics. *J. Amer. Math. Soc.* **34** (2021), no. 4, 909–1009.

[24] X. Chen, S. K. Donaldson, and S. Sun, Kähler–Einstein metrics on Fano manifolds, I–III. *J. Amer. Math. Soc.* **28** (2015), 183–197, 199–234, 235–278.

[25] T. Collins and G. Székelyhidi, Sasaki–Einstein metrics and K-stability. *Geom. Topol.* **23** (2019), no. 3, 1339–1413.

[26] T. Darvas, The Mabuchi geometry of finite energy classes. *Adv. Math.* **285** (2015), 182–219.

[27] T. Darvas and Y. Rubinstein, Tian's properness conjectures and Finsler geometry of the space of Kähler metrics. *J. Amer. Math. Soc.* **30** (2017), 347–387.

[28] T. Delcroix, Uniform K-stability of polarized spherical varieties. 2020, arXiv:2009.06463.

[29] R. Dervan, Uniform stability of twisted constant scalar curvature Kähler metrics. *Int. Math. Res. Not.* **2016** (2016), no. 15, 4728–4783.

[30] S. Donaldson, Scalar curvature and stability of toric varieties. *J. Differential Geom.* **62** (2002), no. 2, 289–349.

[31] S. Donaldson, Kähler–Einstein metrics and algebraic geometry. In *Current developments in mathematics 2015*, pp. 1–25, International Press, Somerville, MA, 2016.

[32] P. Eyssidieux, V. Guedj, and A. Zeriahi, Singular Kähler–Einstein metrics. *J. Amer. Math. Soc.* **22** (2009), 607–639.

[33] K. Fujita, Optimal bounds for the volumes of Kähler–Einstein Fano manifolds. *Amer. J. Math.* **140** (2018), no. 2, 391–414.

[34] K. Fujita, Uniform K-stability and plt blowups of log Fano pairs. *Kyoto J. Math.* **59** (2019), no. 2, 399–418.

[35] K. Fujita, A valuative criterion for uniform K-stability of $\mathbb{Q}$-Fano varieties. *J. Reine Angew. Math.* **751** (2019), 309–338.

[36] K. Fujita and Y. Odaka, On the K-stability of Fano varieties and anticanonical divisors. *Tohoku Math. J.* **70** (2018), no. 4, 511–521.

[37] J. Han and C. Li, Algebraic uniqueness of Kähler–Ricci flow limits and optimal degenerations of Fano varieties. 2009, arXiv:2009.01010.

[38] J. Han and C. Li, On the Yau–Tian–Donaldson conjecture for g-weighted Kähler–Ricci soliton equations. *Comm. Pure Appl. Math.* (to appear), arXiv:2006.00903.

[39] T. Hisamoto, Stability and coercivity for toric polarizations. 2016, arXiv:1610.07998.

[40] N. Ilten and H. Süß, K-stability for Fano manifolds with torus action of complexity 1. *Duke Math. J.* **166** (2017), no. 1, 177–204.

[41] M. Jonsson and M. Mustaţă, Valuations and asymptotic invariants for sequences of ideals. *Ann. Inst. Fourier* **62** (2012), no. 6, 2145–2209.

[42] C. Li, K-semistability is equivariant volume minimization. *Duke Math. J.* **166** (2017), no. 16, 3147–3218.

[43] C. Li, Minimizing normalized volume of valuations. *Math. Z.* **289** (2018), no. 1–2, 491–513.

[44] C. Li, G-uniform stability and Kähler–Einstein metrics on Fano varieties. *Invent. Math.* (to appear), arXiv:1907.09399v3.

[45] C. Li, Geodesic rays and stability in the cscK problem. *Ann. Sci. Éc. Norm. Supér.* (to appear), arXiv:2001.01366.

[46] C. Li, K-stability and Fujita approximation. 2021, arXiv:2102.09457.

[47] C. Li, Notes on weighted Kähler–Ricci solitons and application to Ricci-flat Kähler cone metrics. 2021, arXiv:2107.02088.

[48] C. Li, Y. Liu, and C. Xu, A guided tour to normalized volume. In *Geometric analysis*, pp. 167–219, Progr. Math. 333, Springer, Cham, 2020.

[49] C. Li, G. Tian, and F. Wang, On the Yau–Tian–Donaldson conjecture for singular Fano varieties. *Comm. Pure Appl. Math.* **74** (2020), no. 8, 1748–1800.

[50] C. Li, G. Tian, and F. Wang, The uniform version of Yau–Tian–Donaldson conjecture for singular Fano varieties. *Peking Math. J.* (to appear), arXiv:1903.01215.

[51] C. Li, X. Wang, and C. Xu, Algebraicity of metric tangent cones and equivariant K-stability. *J. Amer. Math. Soc.* **34** (2021), no. 4, 1175–1214.

[52] C. Li and C. Xu, Special test configurations and K-stability of Fano varieties. *Ann. of Math. (2)* **180** (2014), no. 1, 197–232.

[53] C. Li and C. Xu, Stability of valuations: Higher rational rank. *Peking Math. J.* **1** (2018), no. 1, 1–79.

[54] C. Li and C. Xu, Stability of valuations and Kollár components. *J. Eur. Math. Soc. (JEMS)* **22** (2020), no. 8, 2573–2627.

[55] Y. Liu, The volume of singular Kähler–Einstein Fano varieties. *Compos. Math.* **154** (2018), no. 6, 1131–1158.

[56] Y. Liu, C. Xu, and Z. Zhuang, Finite generation for valuations computing stability thresholds and applications to K-stability. 2021, arXiv:2012.09405.

[57] D. Martelli, J. Sparks, and S.-T. Yau, Sasaki–Einstein manifolds and volume minimisation. *Comm. Math. Phys.* **280** (2008), no. 3, 611–673.

[58] S. T. Paul, Hyperdiscriminant polytopes, Chow polytopes, and Mabuchi energy asymptotics. *Ann. of Math.* **175** (2012), no. 1, 255–296.

[59] D. H. Phong, J. Ross, and J. Sturm, Deligne pairings and the Knudsen–Mumford expansion. *J. Differential Geom.* **78** (2008), 475–496.

[60] D. H. Phong and J. Sturm, Test configurations for K-stability and geodesic rays. *J. Symplectic Geom.* **5** (2007), no. 2, 221–247.

[61] Y. A. Rubinstein, G. Tian, and K. Zhang, Basis divisors and balanced metrics. *J. Reine Angew. Math.*, **778** (2021), 171–218.

[62] S. Sun, Degenerations and moduli spaces in Kähler geometry. In *Proceedings of the International Congress of Mathematicians – Rio de Janeiro 2018, Vol. II. Invited lectures*, pp. 993–1012, World Sci. Publ., Hackensack, NJ, 2018.

[63] G. Székelyhidi, Filtrations and test configurations, with an appendix by S. Boucksom. *Math. Ann.* **362** (2015), no. 1–2, 451–484.

[64] G. Tian, Kähler–Einstein metrics with positive scalar curvature. *Invent. Math.* **130** (1997), 239–265.

[65] G. Tian, K-stability and Kähler–Einstein metrics. *Comm. Pure Appl. Math.* **68** (2015), no. 7, 1085–1156.

[66] G. Tian, Kähler–Einstein metrics on Fano manifolds. *Jpn. J. Math.* **10** (2015), no. 1, 1–41.

[67] G. Tian, K-stability implies CM-stability. In *Geometry, analysis and probability*, pp. 245–261, Progr. Math. 310, Springer, Cham, 2017.

[68] X. Wang, GIT stability, K-stability and the moduli space of Fano varieties. In *Moduli of K-stable varieties*, pp. 153–181, Springer INdAM Series 31, Springer, Cham, 2019.

[69] D. Witt Nyström, Test configuration and Okounkov bodies. *Compos. Math.* **148** (2012), no. 6, 1736–1756.

[70] C. Xu, K-stability of Fano varieties: an algebro-geometric approach. *EMS Surv. Math. Sci.* **8** (2021), no. 1-2, 265–354.

[71] C. Xu, A minimizing valuation is quasi-monomial. *Ann. of Math.* **191** (2020), no. 3, 1003–1030.

[72] C. Xu and Z. Zhuang, Uniqueness of the minimizer of the normalized volume function. *Camb. J. Math.* **9** (2021), no. 1, 149–176.

[73] K. Zhang, A quantization proof of the uniform Yau–Tian–Donaldson conjecture. 2021, arXiv:2102.02438v2.

[74] X. Zhu, Kähler–Ricci flow on Fano manifolds. In *Proceedings of the International Congress of Mathematicians 2022, Vol. 4*, pp. 2718–2737, EMS Press, 2022.

CHI LI

Department of Mathematics, Rutgers University, 110 Frelinghuysen Rd. Piscataway, NJ 08854-8019, USA, chi.li@rutgers.edu

This Page Intentionally Left Blank

THE DOUBLE RAMIFICATION CYCLE FORMULA

AARON PIXTON

ABSTRACT

The double ramification cycle $\mathrm{DR}_g(A) = \mathrm{DR}_g(\mu, \nu)$ is a cycle in the moduli space of stable curves parametrizing genus g curves admitting a map to $\mathbb{P}^1$ with specified ramification profiles μ, ν over two points. In 2016, Janda, Pandharipande, Zvonkine, and the author proved a formula expressing the double ramification cycle in terms of basic tautological classes, answering a question of Eliashberg from 2001. This formula has an intricate combinatorial shape involving an unusual way to sum divergent series using polynomial interpolation. Here we give some motivation for where this formula came from, relating it both to an older partial formula of Hain and to Givental's R-matrix action on cohomological field theories.

MATHEMATICS SUBJECT CLASSIFICATION 2020

Primary 14N35; Secondary 14H10, 14C17

KEYWORDS

Moduli of curves, double ramification cycle

INTERNATIONAL CONGRESS OF MATHEMATICIANS 2022 JULY 6–14

Proc. Int. Cong. Math. 2022, Vol. 3, pp. 2312–2322
DOI 10.4171/ICM2022/146

Published by EMS Press

1. INTRODUCTION

Let g, n be nonnegative integers satisfying $2g - 2 + n > 0$, so that the moduli space $\overline{\mathcal{M}}_{g,n}$ of stable curves of genus g with n markings is nonempty. Let $A = (a_1, \ldots, a_n) \in \mathbb{Z}^n$ be a vector of n integers satisfying $a_1 + \cdots + a_n = 0$. In this paper we will be interested in a Chow cycle class

$$\mathrm{DR}_g(A) \in A^g(\overline{\mathcal{M}}_{g,n})$$

that depends on this data.

There are two main perspectives on how to think about and define $\mathrm{DR}_g(A)$, the *double ramification cycle*. The first is the source of its name; we think of it as parametrizing the genus g curves C that admit a finite map $C \to \mathbb{P}^1$ with specified ramification profiles μ, ν over two points (say 0 and ∞). These two ramification profiles are encoded in the vector A: we can take the positive and negative components of A separately to get two partitions of equal size. The marked points with nonzero a_i should then be the points in C lying above 0 and ∞, while the marked points with $a_i = 0$ are unconstrained. Ramification above points other than 0 and ∞ is unconstrained.

The above description defines a *double ramification locus* inside the moduli space of smooth curves $\mathcal{M}_{g,n}$ that is usually (but not always) of pure codimension g. To extend this to a codimension g class on $\overline{\mathcal{M}}_{g,n}$, we can use the virtual class in relative Gromov–Witten theory. There is a moduli space of stable (rubber) maps to $\mathbb{P}^1$ with given marked ramification over two points, $\overline{\mathcal{M}}_{g,n}(\mathbb{P}^1/\{0,\infty\}, \mu, \nu)^\sim$, equipped with a forgetful map $p : \overline{\mathcal{M}}_{g,n}(\mathbb{P}^1/\{0,\infty\}, \mu, \nu)^\sim \to \overline{\mathcal{M}}_{g,n}$, and the double ramification cycle can be taken to be the pushforward under this map of the virtual class,

$$\mathrm{DR}_g(A) = p_*\big[\overline{\mathcal{M}}_{g,n}\big(\mathbb{P}^1/\{0,\infty\}, \mu, \nu\big)^\sim\big]^{\mathrm{vir}} \in A^g(\overline{\mathcal{M}}_{g,n}).$$

The second perspective on $\mathrm{DR}_g(A)$ is via Abel–Jacobi maps. Let $\mathcal{X}_g \to \mathcal{A}_g$ be the universal abelian variety of dimension g. Then the data in the vector A can be used to define a morphism $j_A : \mathcal{M}_{g,n} \to \mathcal{X}_g$ by

$$(C, p_1, \ldots, p_n) \mapsto \big(\mathrm{Jac}(C), \mathcal{O}_C(a_1 p_1 + \cdots + a_n p_n)\big).$$

The double ramification locus is then the inverse image under this map of the zero section $\mathcal{Z}_g$ of $\mathcal{X}_g$, since C admits a map to $\mathbb{P}^1$ with the given ramification profiles if and only if $\mathcal{O}_C(a_1 p_1 + \cdots + a_n p_n)$ is trivial.

This Abel–Jacobi map extends easily to $\mathcal{M}^{\mathrm{ct}}_{g,n}$, the moduli space of curves of compact type (those with compact Jacobians), but using this perspective to define the double ramification cycle on all of $\overline{\mathcal{M}}_{g,n}$ requires more work. It also is not obvious that constructing $\mathrm{DR}_g(A)$ in this way will give the same class as that given by relative Gromov–Witten theory, even after restriction to $\mathcal{M}^{\mathrm{ct}}_{g,n}$. For one approach to these questions using logarithmic and tropical geometry, see the work of Marcus and Wise [13].

Eliashberg proposed the problem of giving a formula for the double ramification cycle in 2001, in the context of symplectic field theory. This problem was solved by Janda, Pandharipande, Zvonkine, and the author in 2016 [11], giving an explicit combinatorial formula for the double ramification cycle. This formula has an unexpected form—an additional integer parameter $r > 0$ is introduced, then an expression is written down that becomes polynomial in r for r sufficiently large, and finally this polynomial is specialized to $r = 0$. Subsequent papers extending or generalizing the double ramification cycle formula in various ways (e.g., [2, 5, 12]) have left the basic combinatorial structure of the formula virtually unchanged. The purpose of this paper is to discuss this structure and give some motivation for where it comes from.

In Section 2, we review the tautological classes in the Chow ring of the moduli space of stable curves. In Section 3, we discuss results leading up to the formula of [11], most notably Hain's formula for the compact type double ramification cycle. Section 4 is the heart of the paper and consists of an extended discussion motivating the shape of the double ramification cycle formula. We conclude in Section 5 by stating the formula and briefly explaining how its proof in [11] is related to some of the motivation in Section 4.

2. TAUTOLOGICAL CLASSES

2.1. Preliminaries

In this section we review the language in which the double ramification cycle formula is written. This is the language of the tautological ring, a subring $R^*(\overline{\mathcal{M}}_{g,n}) \subseteq A^*(\overline{\mathcal{M}}_{g,n})$ containing most classes that arise naturally in geometry.

Following Faber and Pandharipande [6], the tautological rings $R^*(\overline{\mathcal{M}}_{g,n})$ can be defined simultaneously for all $g, n \geq 0$ satisfying $2g - 2 + n > 0$ as the smallest subrings of the Chow rings $A^*(\overline{\mathcal{M}}_{g,n})$ closed under pushforward by forgetful maps $\overline{\mathcal{M}}_{g,n+1} \to \overline{\mathcal{M}}_{g,n}$ and gluing maps $\overline{\mathcal{M}}_{g,n+2} \to \overline{\mathcal{M}}_{g+1,n}$ or $\overline{\mathcal{M}}_{g_1,n_1+1} \times \overline{\mathcal{M}}_{g_2,n_2+1} \to \overline{\mathcal{M}}_{g_1+g_2,n_1+n_2}$. Our discussions of tautological classes will use a more explicit description of them. Graber and Pandharipande [8, APPENDIX A] gave a set of additive generators and a multiplication law satisfied by these generators.

These additive generators are formed from three ingredients: psi classes, kappa classes, and generalized gluing maps corresponding to stable graphs. The psi classes $\psi_i \in A^1(\overline{\mathcal{M}}_{g,n})$, $i = 1, \dots, n$ correspond to the n marked points and are defined as the first Chern classes of the cotangent line bundles to the curves at those points. The Arbarello–Cornalba [1] kappa classes are then the pushforwards of powers of psi classes,

$$\kappa_a := \pi_*\big(\psi_{n+1}^{a+1}\big) \in A^a(\overline{\mathcal{M}}_{g,n}),$$

where $\pi : \overline{\mathcal{M}}_{g,n+1} \to \overline{\mathcal{M}}_{g,n}$ forgets the last marking. The kappa classes will not appear in any of the formulas in this paper.

The tautological ring of the moduli space of smooth marked curves, $R^*(\mathcal{M}_{g,n})$, is the ring generated by these ψ_i and κ_a. To extend this to $R^*(\overline{\mathcal{M}}_{g,n})$ we need classes supported on boundary strata.

2.2. Stable graphs

A *stable graph* Γ is the combinatorial data of a boundary stratum in $\overline{\mathcal{M}}_{g,n}$. It consists of the following:

(1) a set of vertices $V(\Gamma)$;

(2) a genus $g_v \geq 0$ at each vertex $v \in V(\Gamma)$;

(3) a set of half-edges $H(\Gamma)$;

(4) an incidence map $H(\Gamma) \to V(\Gamma)$;

(5) a partition of $H(\Gamma)$ into sets of size 2 (called edges, the set of which is denoted $E(\Gamma)$) and sets of size 1 (called legs);

(6) a bijection between the set of legs and $\{1, \dots, n\}$.

The underlying graph is required to be connected. The stability constraint is that

$$2g_v - 2 + n_v > 0$$

at each vertex v, where n_v is the number of half-edges incident to v. The genera are constrained by the identity

$$2g - 2 + n = \sum_{v \in V(\Gamma)} (2g_v - 2 + n_v),$$

or equivalently that $g - \sum_v g_v = h^1(\Gamma)$, the first Betti number of the graph. Such a stable graph Γ corresponds to a generalized gluing map

$$\xi_\Gamma : \prod_{v \in V(\Gamma)} \overline{\mathcal{M}}_{g_v, n_v} \to \overline{\mathcal{M}}_{g,n}.$$

We can then consider classes

$$\xi_{\Gamma *}(\alpha) \in A^*(\overline{\mathcal{M}}_{g,n}),$$

where Γ is a stable graph and α is a monomial in the psi and kappa classes on the $\overline{\mathcal{M}}_{g_v,n_v}$ factors. These are the additive generators for the tautological ring considered in [8].

2.3. Compact type

The moduli space of curves of compact type, denoted $\mathcal{M}^{\mathrm{ct}}_{g,n}$, is the open subscheme of $\overline{\mathcal{M}}_{g,n}$ consisting of those curves whose dual graph is a tree. Its tautological ring $R^*(\mathcal{M}^{\mathrm{ct}}_{g,n})$ is the image of $R^*(\overline{\mathcal{M}}_{g,n})$ under restriction, so it is additively generated by classes $\xi_{\Gamma *}(\alpha)$ as above where Γ is a tree.

It will be convenient for us to have notation for the compact type boundary divisor classes when stating Hain's formula below, (3.2). If Γ is a stable graph with 2 vertices and 1 edge and one of the vertices is genus h and has those legs with markings in a set $S \subseteq 1, 2, \dots, n$, let $\delta_{h,S} = \xi_{\Gamma *}(1)$ be the corresponding boundary divisor class.

3. PREVIOUS FORMULAS AND RESULTS

The first progress towards a formula for the double ramification cycle was when Faber and Pandharipande [7] proved that the double ramification cycle lies in the tautological ring, and thus in theory must be expressible in terms of the generators described in the previous section. Their proof, although in principle constructive, involves a complicated recursion and does not seem to yield a practical formula.

The first progress towards an explicit formula came when Hain [10] computed the double ramification cycle when restricted to the compact type locus $\mathcal{M}^{\mathrm{ct}}_{g,n}$. On this locus the double ramification cycle is the pullback along an Abel–Jacobi map $j_A : \mathcal{M}^{\mathrm{ct}}_{g,n} \to \mathcal{X}_{g,n}$ of the class of the zero section $\mathcal{Z}_{g,n}$ of the universal abelian variety $\mathcal{X}_{g,n} \to \mathcal{A}_{g,n}$. Hain showed that the class of this zero section is $[\mathcal{Z}_{g,n}] = \Theta^g/g!$ and computed the pullback of the theta divisor Θ as an explicit divisor on $\mathcal{M}^{\mathrm{ct}}_{g,n}$,

$$j_A^*\Theta = \sum_{i=1}^{n} \frac{a_i^2}{2}\psi_i - \sum_{h,S} \frac{a_S^2}{4}\delta_{h,S}, \tag{3.1}$$

where $a_S = \sum_{i\in S} a_i$ and the second sum runs over all h, S defining boundary divisor classes.

Hain's formula for the compact type double ramification cycle is then

$$\mathrm{DR}_g^{\mathrm{ct}}(A) = \frac{1}{g!}(j_A^*\Theta)^g = \frac{1}{g!}\left(\sum_{i=1}^{n} \frac{a_i^2}{2}\psi_i - \sum_{h,S} \frac{a_S^2}{4}\delta_{h,S}\right)^g. \tag{3.2}$$

The divisor formula (3.1) is a homogeneous polynomial of degree 2 in A, so Hain's DR formula (3.2) is a homogeneous polynomial of degree $2g$ in A.

Grushevsky and Zakharov [9] extended Hain's computation slightly, expanding from $\mathcal{M}^{\mathrm{ct}}_{g,n}$ to a slightly larger open subscheme of $\overline{\mathcal{M}}_{g,n}$ by adding the locus of curves whose dual graph is a tree with a single loop added at one vertex. If Γ is the stable graph with a single vertex and single loop, then their correction term is the codimension g part of

$$\xi_{\Gamma *}\left(-\prod_{i=1}^{n} \exp\left(\frac{1}{2}a_i^2\psi_i\right)\sum_{k=1}^{\infty} \frac{B_{2k}}{2^k k!}(\psi+\psi')^{k-1}\right), \tag{3.3}$$

where $\psi_1, \ldots, \psi_n$ are the psi classes on the legs, ψ, ψ' are the psi classes on the two half-edges of the loop, and B_{2k} is a Bernoulli number.

In particular, the double ramification cycle is no longer a homogeneous polynomial in A when computed beyond compact type. This was also seen in work of Buryak, Shadrin, Spitz, and Zvonkine [3], who showed that the top degree intersections of double ramification cycles with monomials in the psi classes are inhomogeneous polynomials of degree $2g$ in A.

4. MOTIVATION FOR THE FORMULA

In this section we discuss various observations and ideas that come about when one tries to extend Hain's formula (3.2) to $\overline{\mathcal{M}}_{g,n}$ to obtain a full double ramification cycle formula.

4.1. Expanding Hain's formula

Exponentiating a boundary divisor class can be done using the multiplication laws for tautological classes [8, APPENDIX A]. Multiplying out Hain's formula (3.2) in this way gives a nice sum over trees: $\mathrm{DR}_g^{\mathrm{ct}}(A)$ is the codimension g part of

$$\sum_{T \text{ stable tree}} \frac{1}{|\mathrm{Aut}(T)|}(\xi_T)_*\left[\prod_{i=1}^{n} \exp\left(\frac{1}{2}a_i^2\psi_{h_i}\right) \cdot \prod_{e=\{h,h'\}\in E(T)} \frac{1-\exp(-\frac{1}{2}w(h)w(h')(\psi_h+\psi_{h'}))}{\psi_h+\psi_{h'}}\right], \quad (4.1)$$

where the function $w : H(T) \to \mathbb{Z}$ is defined here by contracting all the edges in the tree T other than the one containing h and then letting $w(h)$ be the sum of the a_i for the legs i on the same vertex as the half-edge h.

Extending this formula to $\overline{\mathcal{M}}_{g,n}$ requires us to provide a polynomial (or power series) in the psi classes for every stable graph Γ, not just every stable tree. The $w(h)$ definition above does not naturally extend to non-separating edges, so it is not immediately clear how to do this. Moreover, we know that this power series needs to be (3.3) for the single-loop graph, so something quite new is going on even there.

4.2. Cohomological field theory axioms

A cohomological field theory (CohFT) is a collection of classes $\Omega_{g,n}(\gamma_1,\dots,\gamma_n)$ on $\overline{\mathcal{M}}_{g,n}$ for all g and n, where the inputs γ_i belong to some finite set S (a basis for the state space of the CohFT). These classes must satisfy certain compatibility axioms relating them to each other under pullback by natural maps between the $\overline{\mathcal{M}}_{g,n}$. For one basic treatment of CohFTs and Givental's R-matrix action, see **[14]**. The double ramification cycle is not quite a CohFT, but it satisfies some subset of the properties of one. For example, if $j : \overline{\mathcal{M}}_{g_1,n_1+1} \times \overline{\mathcal{M}}_{g_2,n_2+1} \to \overline{\mathcal{M}}_{g,n}$ is a separating gluing map where the marked points split into sets I_1, I_2 with $|I_i| = n_i$, then we have

$$j^*\,\mathrm{DR}_g(a_1,\dots,a_n) = \mathrm{DR}_{g_1}\big(\{a_i \mid i \in I_1\}, t\big) \otimes \mathrm{DR}_{g_2}\big(\{a_i \mid i \in I_2\}, -t\big),$$

where $t \in \mathbb{Z}$ is the unique insertion that makes the parameters sum to 0 in each DR term on the right.

If the double ramification cycle were a CohFT, we would want a similar formula for the pullback along the nonseparating gluing map $k : \overline{\mathcal{M}}_{g-1,n+2} \to \overline{\mathcal{M}}_{g,n}$: the natural thing to write down would be

$$k^*\,\mathrm{DR}_g(a_1,\dots,a_n) = \sum_{t\in\mathbb{Z}} \mathrm{DR}_{g-1}(a_1,\dots,a_n,t,-t),$$

but it is not clear how one might make sense of this infinite sum—it will not converge in any standard sense. What is going wrong here is that CohFTs are supposed to depend multilinearly on parameters from a finite-dimensional state space, but double ramification cycles take inputs in $\mathbb{Z}$ so the state space appears to be infinite-dimensional.

So the double ramification cycle behaves like a CohFT as far as separating nodes are concerned, but the state space would have to be infinite-dimensional and this makes it unclear what to do at nonseparating nodes.

4.3. Givental's R-matrix action

Teleman [16] proved that semisimple CohFTs all have a very particular graph sum form, given by applying Givental's R-matrix action to a CohFT that lives fully in codimension zero. The rough shape of the resulting formula for a semisimple CohFT is

$$\Omega_{g,n}(\gamma_1,\ldots,\gamma_n) = \sum_{\Gamma \text{ stable graph}} \sum_{w:H(\Gamma)\to S} \frac{1}{|\mathrm{Aut}(\Gamma)|}(\xi_\Gamma)_* \Bigg[\prod_{v\in V(\Gamma)} (\textit{vertex factor}) \prod_{i=1}^{n} (\textit{leg factor}) \prod_{e=\{h,h'\}\in E(\Gamma)} (\textit{edge factor}) \Bigg],$$

where the second sum is over functions w on the half-edges of the graph taking values in some set S (a basis for the state space of the CohFT) and the values of w on the legs $h_1,\ldots,h_n$ are given as $w(h_i) = \gamma_i$. The various factors are then power series (that depend on w) in the corresponding kappa and psi classes. The expanded version of Hain's compact type formula (4.1) is of this shape: we take $S = \mathbb{Z}$, the vertex factor is 0 unless all of the incident $w(h)$ sum to zero, and the edge factor is 0 unless the two $w(h)$ along the edge sum to zero. These vanishings effectively place the following constraints on w (to get a nonzero contribution to $\mathrm{DR}_g^{\mathrm{ct}}(A)$):

(1) $w(h_i) = a_i$ for $i = 1,2,\ldots,n$, where h_i is the ith leg;

(2) $w(h) + w(h') = 0$ if $\{h,h'\}$ is an edge;

(3) $\sum_{h\to v} w(h) = 0$ for each vertex v.

We say w is *balanced* (with respect to A) if it satisfies these constraints. In other words, w is a flow on Γ with sources/sinks at the legs (with specified values given there by A). When Γ is a tree, there is a unique such balanced w and we recover the $w(h)$ used in (4.1).

From this perspective it is natural to just try to take (4.1) and extend it to be a Givental-type sum over arbitrary graphs (not just trees), but then there will be infinitely many choices of w and the resulting infinite sums will be nonconvergent. Moreover, careful comparison with the exact form of Givental's R-matrix action suggests that the vertex factor should contribute a total factor of something like "$|\mathbb{Z}|^{-h^1(\Gamma)}$." Note that the set of balanced w is a torsor over $H_1(\Gamma;\mathbb{Z}) \cong \mathbb{Z}^{h^1(\Gamma)}$, so this factor feels like some sort of infinite averaging procedure.

4.4. Divergent averages

Returning to the simplest non-tree case, the graph with one vertex and one loop, matching things up with (3.3) would then require making sense of the "infinite average" identity

$$\frac{1}{|\mathbb{Z}|}\sum_{c\in\mathbb{Z}} c^{2k} = B_{2k}. \tag{4.2}$$

This is reminiscent of the zeta regularization sum

$$\sum_{c\geq 1} c^{2k-1} = \zeta(1-2k) = -\frac{B_{2k}}{2k},$$

but there is no obvious way to make sense of this similarity. Moreover, more complicated graphs require much more complicated divergent sums; for example, a graph with two vertices, a double edge between them, and one leg on each vertex gives rise to infinite sums like

$$\frac{1}{|\mathbb{Z}|}\sum_{c+d=a} c^{2k}d^{2l} \tag{4.3}$$

which must be interpreted.

4.5. Interpolating finite rank CohFTs

The problem with writing down a double ramification cycle formula of this type is clearly that the state space is infinite-dimensional. If we replace $\mathbb{Z}$ with $\mathbb{Z}/r\mathbb{Z}$ everywhere then there is no difficulty with writing down a similar-looking finite rank CohFT. The result might be something like the following (the case of a diagonal R-matrix—for an example of a more complicated CohFT of this general type, see **[15]**):

$$\sum_{\Gamma \text{ stable graph}} \sum_{\substack{w:H(\Gamma)\to\mathbb{Z}/r\mathbb{Z} \\ \text{balanced}}} \frac{1}{|\mathrm{Aut}(\Gamma)|}(\xi_\Gamma)_*\Bigg[\frac{1}{r^{h^1(\Gamma)}}\prod_{i=1}^{n}\exp\big(F_{w(h_i)}(\psi_{h_i})\big)$$
$$\cdot \prod_{e=\{h,h'\}\in E(\Gamma)} \frac{1-\exp(F_{w(h)}(\psi_h)+F_{w(h')}(\psi_{h'}))}{\psi_h+\psi_{h'}}\Bigg],$$

for power series $F_a(Z)$ for $a \in \mathbb{Z}/r\mathbb{Z}$ with $F_0(Z) = 0$ and $F_{-a}(-Z) = -F_a(Z)$.

If we take $F_a(Z) = \frac{1}{2}a^2 Z$ for $-\frac{r}{2} < a \le \frac{r}{2}$ then this CohFT starts to look very much like the expanded version of Hain's formula, (4.1). In fact, if Γ is a tree then the Γ-term in this sum agrees with that in Hain's formula for all sufficiently large r. So it is tempting to try to take the limit as $r \to \infty$ of these CohFTs. But this is not quite right: the r-version of the left side of (4.2) is then

$$\frac{1}{r}\sum_{-\frac{r}{2}<c\le\frac{r}{2}} c^{2k}.$$

This certainly does not converge as $r \to \infty$. However, if we restrict to even r then it is polynomial in r, and if we examine the coefficients of this polynomial then we see that B_{2k}, the desired value, is the constant coefficient in r.

This suggests a potential interpretation even of more complicated sums like (4.3):

$$\frac{1}{|\mathbb{Z}|}\sum_{c+d=a} c^{2k}d^{2l} = \frac{1}{|\mathbb{Z}/0\mathbb{Z}|}\sum_{\substack{c,d\in\mathbb{Z}/0\mathbb{Z} \\ c+d=a \pmod 0}} c^{2k}d^{2l}$$
$$:= \Bigg[\frac{1}{|\mathbb{Z}/r\mathbb{Z}|}\sum_{\substack{c,d\in\mathbb{Z}/r\mathbb{Z} \\ c+d=a \pmod r}} c^{2k}d^{2l}\Bigg]_{r=0},$$

where c and d must be interpreted inside $c^{2k}d^{2l}$ as elements of $\mathbb{Z}$ via some choice of mod r representatives (we used $-r/2+1,\dots,r/2$ before but $0,\dots,r-1$ will give the same final answer) and setting $r = 0$ at the end is done by polynomial interpolation.

4.6. Geometric interpretation from (k/r)-spin structures

An (k/r)-spin structure on a smooth curve C with marked points p_i and weights a_i is a choice of line bundle L on C such that $L^{\otimes r} \equiv \omega_C^{\otimes k}(a_1 p_1 + \cdots + a_n p_n)$. If we take $k = 0$ and assume the weights a_i sum to zero then for any positive r any smooth curve will have such "rth root structures." But we can also interpret this construction as meaningful when $r = 0$, when we get that a curve only admits a $(0/0)$-spin structure if it is in the double ramification locus. This observation gives a vague geometric idea for what it might mean to think of the double ramification cycle as given by specializing some parameter r to 0.

5. THE DOUBLE RAMIFICATION CYCLE FORMULA

We can now state the main result of [11], the double ramification cycle formula:

Theorem 1 ([11]). $\mathrm{DR}_g(A)$ *is the codimension* g *part of*

$$\sum_{\Gamma \text{ stable graph}} \frac{1}{|\mathrm{Aut}(\Gamma)|}(\xi_\Gamma)_* \Bigg[\frac{1}{|\mathbb{Z}|^{h^1(\Gamma)}} \sum_{\substack{w:H(\Gamma)\to\mathbb{Z} \\ \text{balanced}}} \prod_{i=1}^{n} \exp\left(\frac{1}{2}a_i^2 \psi_{h_i}\right) \cdot \prod_{e=\{h,h'\}\in E(\Gamma)} \frac{1-\exp(-\frac{1}{2}w(h)w(h')(\psi_h+\psi_{h'}))}{\psi_h+\psi_{h'}} \Bigg],$$

where formal expressions of the form

$$\frac{1}{|\mathbb{Z}|^{h^1(\Gamma)}} \sum_{\substack{w:H(\Gamma)\to\mathbb{Z} \\ \text{balanced}}} P(\{w(h)\})$$

(for P *a polynomial) are evaluated by setting* $r = 0$ *in the corresponding* r*-polynomial*

$$\frac{1}{r^{h^1(\Gamma)}} \sum_{\substack{w:H(\Gamma)\to\{0,1,\ldots,r-1\} \\ \text{balanced mod } r}} P(\{w(h)\}).$$

The combinatorial result (necessary for this theorem statement to make sense) that the expression in the final line is in fact a polynomial in r (for r sufficiently large) was proved in [11, APPENDIX A].

The proof of Theorem 1 in [11] follows some of the motivation in Section 4. We first explain the meaning of the additional r parameter. For each $r > 0$, let $\mathbb{P}^1[r]$ denote the projective line with a $B\mathbb{Z}_r$ orbifold point at 0. One can then use $\mathbb{C}^*$-localization on the moduli space of relative stable maps to $\mathbb{P}^1[r]/\{\infty\}$ to obtain complicated relations that entangle double ramification cycles, classes coming from moduli of $(0/r)$-spin curves (discussed briefly in the case of smooth curves in Section 4.6), and other basic tautological classes. The relevant $(0/r)$-spin classes were previously computed by Chiodo [4] using Grothendieck–Riemann–Roch.

These localization relations are too difficult to study effectively for specific values of r, but it turns out that they have polynomial dependence on r. Taking the constant term in r simplifies them greatly: most of the terms vanish, and the only remaining terms are a single double ramification cycle and the $r = 0$ interpolation of certain classes written in terms of the Chern characters of the pushforward of the universal rth root line bundle on the moduli space of $(0/r)$-spin curves. Chiodo's formula [4] for these Chern characters gives that these classes are CohFTs with formulas of the type described in Section 4.5. The power series in psi classes appearing in these formulas do not look exactly like those appearing in Theorem 1, but they have the same $r = 0$ interpolation. (In the language of Section 4.5, the power series $F_a(Z)$ will be congruent to $\frac{1}{2}a^2Z$ mod r.) The result is a proof of Theorem 1.

ACKNOWLEDGMENTS

The author is grateful to R. Cavalieri, D. Holmes, F. Janda, S. Grushevsky, S. Molcho, G. Oberdieck, R. Pandharipande, J. Schmitt, D. Zakharov, and D. Zvonkine for many discussions about double ramification cycles and related topics over the years.

FUNDING

The author was partially supported by National Science Foundation Grant No. 1807079.

REFERENCES

[1] E. Arbarello and M. Cornalba, Combinatorial and algebro-geometric cohomology classes on the moduli spaces of curves. *J. Algebraic Geom.* **5** (1996), no. 4, 705–749.

[2] Y. Bae, D. Holmes, R. Pandharipande, J. Schmitt, and R. Schwarz, Pixton's formula and Abel–Jacobi theory on the Picard stack. 2021, arXiv:2004.08676v2.

[3] A. Buryak, S. Shadrin, L. Spitz, and D. Zvonkine, Integrals of ψ-classes over double ramification cycles. *Amer. J. Math.* **137** (2015), no. 3, 699–737.

[4] A. Chiodo, Towards an enumerative geometry of the moduli space of twisted curves and rth roots. *Compos. Math.* **144** (2008), no. 6, 1461–1496.

[5] E. Clader and F. Janda, Pixton's double ramification cycle relations. *Geom. Topol.* **22** (2018), no. 2, 1069–1108.

[6] C. Faber and R. Pandharipande, Logarithmic series and Hodge integrals in the tautological ring. *Michigan Math. J.* **48** (2000), 215–252.

[7] C. Faber and R. Pandharipande, Relative maps and tautological classes. *J. Eur. Math. Soc. (JEMS)* **7** (2005), no. 1, 13–49.

[8] T. Graber and R. Pandharipande, Constructions of nontautological classes on moduli spaces of curves. *Michigan Math. J.* **51** (2003), no. 1, 93–109.

[9] S. Grushevsky and D. Zakharov, The zero section of the universal semiabelian variety and the double ramification cycle. *Duke Math. J.* **163** (2014), no. 5, 953–982.

[10] R. Hain, Normal functions and the geometry of moduli spaces of curves. In *Handbook of moduli. Vol. I*, pp. 527–578, Adv. Lect. Math. (ALM) 24, Int. Press, Somerville, MA, 2013.

[11] F. Janda, R. Pandharipande, A. Pixton, and D. Zvonkine, Double ramification cycles on the moduli spaces of curves. *Publ. Math. Inst. Hautes Études Sci.* **125** (2017), no. 1, 221–266.

[12] F. Janda, R. Pandharipande, A. Pixton, and D. Zvonkine, Double ramification cycles with target varieties. *J. Topol.* **13** (2020), no. 4, 1725–1766.

[13] S. Marcus and J. Wise, Logarithmic compactification of the Abel–Jacobi section. *Proc. Lond. Math. Soc. (3)* **121** (2020), no. 5, 1207–1250.

[14] R. Pandharipande, A. Pixton, and D. Zvonkine, Relations on $\overline{M}_{g,n}$ via 3-spin structures. *J. Amer. Math. Soc.* **28** (2015), no. 1, 279–309.

[15] R. Pandharipande, A. Pixton, and D. Zvonkine, Tautological relations via r-spin structures. *J. Algebraic Geom.* **28** (2019), no. 3, 439–496.

[16] C. Teleman, The structure of 2D semi-simple field theories. *Invent. Math.* **188** (2012), no. 3, 525–588.

AARON PIXTON

Department of Mathematics, University of Michigan, Ann Arbor, MI 48109, USA, pixton@umich.edu

This Page Intentionally Left Blank

EFFECTIVE RESULTS IN THE THREE-DIMENSIONAL MINIMAL MODEL PROGRAM

YURI PROKHOROV

ABSTRACT

We give a brief review on recent developments in the three-dimensional minimal model program.

MATHEMATICS SUBJECT CLASSIFICATION 2020

Primary 14E30; Secondary 14J30, 14B05, 14E05

KEYWORDS

Minimal model, Mori contraction, terminal (canonical) singularity, flip, extremal ray

ICM INTERNATIONAL CONGRESS OF MATHEMATICIANS 2022 JULY 6–14

Proc. Int. Cong. Math. 2022, Vol. 3, pp. 2324–2345
DOI 10.4171/ICM2022/30

In this note we give a brief review on recent developments in the three-dimensional minimal model program (MMP for short). Certainly, this is not a complete survey of all advances in this area. For example, we do not discuss the minimal models of varieties of nonnegative Kodaira dimension, as well as applications to birational geometry and moduli spaces.

The aim of the MMP is to find a good representative in a fixed birational equivalence class of algebraic varieties. Starting with an arbitrary smooth projective variety, one can perform a finite number of certain elementary transformations, called divisorial contractions and flips, and at the end obtain a variety which is simpler in some sense. Most parts of the MMP are completed in arbitrary dimension. One of the basic remaining problems is the following:

Describe all the intermediate steps and the outcome of the MMP.

The MMP makes sense only in dimensions ≥ 2, and for surfaces it is classical and well known. So the first nontrivial case is the three-dimensional one. It turns out that to proceed with the MMP in dimension ≥ 3, one has to work with varieties admitting certain types of very mild, the so-called terminal, singularities. On the other hand, dimension 3 is the last dimension where one can expect effective results: in higher dimensions, classification results become very complicated and unreasonably long.

We will work over the field $\mathbb{C}$ of complex numbers throughout. A variety is either an algebraic variety or a reduced complex space.

1. SINGULARITIES

Recall that a Weil divisor D on a normal variety is said to be $\mathbb{Q}$-*Cartier* if its multiple nD, for some n, is a Cartier divisor. For any morphism $f : Y \to X$, the pull-back f^*D of a $\mathbb{Q}$-Cartier divisor D is well defined as a divisor with rational coefficients ($\mathbb{Q}$-divisor). A variety X has $\mathbb{Q}$-*factorial singularities* if any Weil divisor on X is $\mathbb{Q}$-Cartier.

Definition 1.1. A normal algebraic variety (or an analytic space) X is said to have *terminal* (resp. *canonical*, *log terminal*, *log canonical*) singularities if the canonical Weil divisor K_X is $\mathbb{Q}$-Cartier and, for any birational morphism $f : Y \to X$, one can write

$$K_Y = f^*K_X + \sum a_i E_i, \qquad (1.1.1)$$

where E_i are all the exceptional divisors and $a_i > 0$ (resp. $a_i \geq 0$, $a_i > -1$, $a_i \geq -1$) for all i. The smallest positive m such that mK_X is Cartier is called the *Gorenstein index* of X. Canonical singularities of index 1 are rational Gorenstein.

The class of terminal $\mathbb{Q}$-factorial singularities is the smallest class that is closed under the MMP. Canonical singularities are important because they appear in the canonical models of varieties of general type. A crucial observation is that terminal singularities lie in codimension ≥ 3. In particular, terminal surface singularities are smooth and terminal threefold singularities are isolated. Canonical singularities of surfaces are called *Du Val* or rational double points. Any two-dimensional log terminal singularity is a quotient of a smooth surface germ by a finite group [32]. Terminal threefolds singularities were classified by M. Reid [65] and S. Mori [43].

Example. Let $X \subset \mathbb{C}^4$ be a hypersurface given by the equation

$$\phi(x_1, x_2, x_3) + x_4 \psi(x_1, \dots, x_4) = 0,$$

where $\phi = 0$ is an equation of a Du Val (ADE) singularity. Then the singularity of X at 0 is canonical Gorenstein. It is terminal if and only if it is isolated. Singularities of this type are called cDV.

According to [65], any three-dimensional terminal singularity of index $m > 1$ is a quotient of an isolated cDV-singularity by the cyclic group $\boldsymbol{\mu}_m$ of order m. More precisely, we have the following

Theorem 1.2 ([65]). *Let $(X \ni P)$ be an analytic germ of a three-dimensional terminal singularity of index $m \geq 1$. Then there exist an isolated* cDV*-singularity $(X^\sharp \ni P^\sharp)$ and a cyclic $\boldsymbol{\mu}_m$-cover*

$$\pi : (X^\sharp \ni P^\sharp) \longrightarrow (X \ni P)$$

which is étale outside P.

The morphism π in the above theorem is called the *index-one cover*. A detailed classification of all possibilities for the equations of $X^\sharp \subset \mathbb{C}^4$ and the actions of $\boldsymbol{\mu}_m$ was obtained in [43] (see also [66]).

Example. Let the cyclic group $\boldsymbol{\mu}_m$ act on $\mathbb{C}^n$ diagonally via

$$(x_1, \dots, x_n) \mapsto (\zeta^{a_1} x_1, \dots, \zeta^{a_n} x_n), \quad \zeta = \zeta_m = \exp(2\pi \mathrm{i}/m).$$

Then we say that $(a_1, \dots, a_n)$ is the collection of *weights* of the action. Assume that the action is free in codimension 1. Then the quotient singularity $\mathbb{C}^n/\boldsymbol{\mu}_m \ni 0$ is said to be of type $\frac{1}{m}(a_1, \dots, a_n)$. According to the criterion (see [66, THEOREM 4.11]), this singularity is terminal if and only if

$$\sum_{i=1}^{n} \overline{k a_i} > m \quad \text{for } k = 1, \dots, m-1,$$

where $\overline{}$ is the smallest residue mod m. In the threefold case this criterion has a very simple form: a quotient singularity $\mathbb{C}^m/\boldsymbol{\mu}_m$ is terminal if and only if it is of type $\frac{1}{m}(1, -1, a)$, where $\gcd(m, a) = 1$. This is a *cyclic quotient terminal singularity*.

Example ([43,66]). Let the cyclic group $\boldsymbol{\mu}_m$ act on $\mathbb{C}^4$ diagonally with weights $(1, -1, a, 0)$, where $\gcd(m, a) = 1$. Then for a polynomial $\phi(u, v)$, the singularity at 0 of the quotient

$$\{x_1 x_2 + \phi(x_3^m, x_4) = 0\}/\boldsymbol{\mu}_m$$

is terminal whenever it is isolated. The index of this singularity equals m.

As a consequence of the classification, we obtain that the local fundamental group of the (analytic) germ of a three-dimensional terminal singularity of index m is cyclic of order m:

$$\pi_1\big(X \setminus \{P\}\big) \simeq \mathbb{Z}/m\mathbb{Z}. \tag{1.2.1}$$

In particular, for any Weil $\mathbb{Q}$-Cartier divisor D on X, its mth multiple mD is Cartier [32, LEMMA 5.1].

The class of canonical threefold singularities is much larger than the class of terminal ones. However, there are certain boundedness results. For example, it is known that the index of a strictly canonical isolated singularity is at most 6 [31].

The modern higher-dimensional MMP often works with pairs, and one needs to extend Definition 1.1 to a wider class of objects.

Definition. Let X be a normal variety and let B be an effective $\mathbb{Q}$-divisor on X. The pair (X, B) is said to be *plt* (resp. *lc*) if $K_X + B$ is $\mathbb{Q}$-Cartier and, for any birational morphism $f : Y \to X$, one can write

$$K_Y + B_Y = f^*(K_X + B) + \sum a_i E_i,$$

where B_Y is the proper transform of B, E_i are all the exceptional divisors and $a_i > -1$ (resp. $a_i \geq -1$) for all i. The pair (X, B) is said to be *klt* if it is plt and $\lfloor B \rfloor = 0$.

2. MINIMAL MODEL PROGRAM

Basic elementary operations in the MMP are Mori contractions.

A *contraction* is a proper surjective morphism $f : X \to Z$ of normal varieties with connected fibers. The *exceptional locus* of a contraction f is the subset $\mathrm{Exc}(f) \subset X$ of points at which f is not an isomorphism. A *Mori contraction* is a contraction $f : X \to Z$ such that the variety X has at worst terminal $\mathbb{Q}$-factorial singularities, the anticanonical class $-K_X$ is f-ample, and the relative Picard number $\rho(X/Z)$ equals 1. A Mori contraction is said to be *divisorial* (resp. *flipping*) if it is birational and the locus $\mathrm{Exc}(f)$ has codimension 1 (resp. ≥ 2). For a divisorial contraction, the exceptional locus $\mathrm{Exc}(f)$ is a prime divisor. A Mori contraction whose target is a lower-dimensional variety is called a *Mori fiber space*. Then the general fiber is a Fano variety with at worst terminal singularities. In the particular cases where the relative dimension of X/Z equals 1 (resp. 2), the Mori fiber space $f : X \to Z$ is called a $\mathbb{Q}$-*conic bundle* (resp. $\mathbb{Q}$-*del Pezzo fibration*). If Z is a point, then X is a Fano variety with at worst terminal $\mathbb{Q}$-factorial singularities and $\mathrm{Pic}(X) \simeq \mathbb{Z}$. For short, we call such varieties $\mathbb{Q}$-*Fano*.

The MMP procedure is a sequence of elementary transformations which are constructed inductively [35,39]. Let X be a projective algebraic variety with terminal $\mathbb{Q}$-factorial singularities. If the canonical divisor K_X is not nef, then there exists a Mori contraction $f : X \to Z$. If f is divisorial, then Z is again a variety with terminal $\mathbb{Q}$-factorial singularities and, in this situation, we can proceed with the MMP replacing X with Z. In contrast, a flipping contraction takes us out the category of terminal $\mathbb{Q}$-factorial varieties. To proceed, one has to perform a surgery operation as follows:

$$\begin{array}{ccc} X & \dashrightarrow & X^+ \\ & {\scriptstyle f}\searrow \quad \swarrow{\scriptstyle f^+} & \\ & Z & \end{array}$$

where f^+ is a contraction whose exceptional locus has codimension ≥ 2 and the divisor K_{X^+} is $\mathbb{Q}$-Cartier and f^+-ample. Then the variety X^+ again has terminal $\mathbb{Q}$-factorial singularities, and we can proceed by replacing X with X^+.

The process described above should terminate, and at the end we obtain a variety $\bar{X}$ such that either $\bar{X}$ has a Mori fiber space structure $\bar{X} \to \bar{Z}$ or $K_{\bar{X}}$ is nef. One of the remaining open problems is the termination of the program, to be more precise, termination of a sequence of flips. The problem was solved affirmatively in dimension ≤ 4 **[35, 69]**, for varieties of general type, for uniruled varieties **[5]**, and in some other special cases. We refer to **[3]** for more comprehensive survey of the higher-dimensional MMP.

The MMP has a huge number of applications in algebraic geometry. The most impressive consequence of the MMP is the finite generation of the canonical ring

$$\mathrm{R}(X, K_X) := \bigoplus_{n \geq 0} H^0(X, \mathscr{O}_X(mK_X))$$

of a smooth projective variety X **[5, 15]**. Another application of the MMP is the so-called Sarkisov program which allows decomposing birational maps between Mori fiber spaces into composition of elementary transformations, called Sarkisov links **[9, 16, 68]**. Also the MMP can be applied to varieties with finite group actions and to varieties over nonclosed fields (see **[63]**).

As was explained above, the Mori contractions are fundamental building blocks in the MMP. To apply the MMP effectively, one needs to understand the structure of its steps in details. For a Mori contraction $f : X \to Z$ of a three-dimensional variety X, there are only the following possibilities:

- f is divisorial and the image of the (prime) exceptional divisor $E := \mathrm{Exc}(f)$ is either a point or an irreducible curve,
- f is flipping and the exceptional locus $\mathrm{Exc}(f)$ is a union of a finite number of irreducible curves,
- Z is a surface and f is a $\mathbb{Q}$-conic bundle,
- Z is a curve and f is a $\mathbb{Q}$-del Pezzo fibration,
- Z is a point and X is a $\mathbb{Q}$-Fano threefold.

Mori contractions of smooth threefolds to varieties of positive dimension where classified in the pioneering work of S. Mori **[42]**. S. Cutkosky **[12]** extended this classification to the case of Gorenstein terminal varieties. Smooth Fano threefolds of Picard number 1 where classified by Iskovskikh **[22, 23]** (see also **[25]**). Fano threefolds with Gorenstein terminal singularities are degenerated smooth ones **[57]**. Below we are going to discuss Mori contractions of threefolds. We are interested only in the biregular structure of a contraction $f : X \to Z$ near a fixed fiber $f^{-1}(o)$, $o \in Z$. Typically, we do not consider the simple case where X is Gorenstein.

3. GENERAL ELEPHANT

A natural way to study higher-dimensional varieties is the inductive one. Typically, to apply this method, we need to find a certain subvariety of dimension one less (divisor) which is sufficiently good is the sense of singularities.

Conjecture 3.1. *Let $f : X \to (Z \ni o)$ be a threefold Mori contraction, where $(Z \ni o)$ is a small neighborhood. Then the general member $D \in |-K_X|$ is a normal surface with Du Val singularities.*

The conjecture was proposed by M. Reid who called a good member of $|-K_X|$ "elephant." We follow this language and call Conjecture 3.1 the General Elephant Conjecture. The importance of the existence of a good member in $|-K_X|$ is motivated by many reasons:

- The general elephant passes through all the non-Gorenstein points of X and so it encodes the information about their types and configuration (cf. Proposition 3.2 below).
- For flipping contractions, Conjecture 3.1 is a sufficient condition for the existence of threefold flips [32].
- For a divisorial contraction $f : X \to Z$ whose fibers have dimension ≤ 1, the image $D_Z := f(D)$ of a Du Val elephant $D \in |-K_X|$ must be again Du Val and the image $\Gamma := f(E)$ of the exceptional divisor is a curve on D_Z. Then one can reconstruct f starting from the triple $(Z \supset D_Z \supset \Gamma)$ by using a certain birational procedure. Such an approach was successfully worked out in many cases by N. Tziolas [71–74].
- If $f : X \to (Z \ni o)$ is a $\mathbb{Q}$-del Pezzo fibration such that general $D \in |-K_X|$ is Du Val, then, composing the projection $D \to Z$ with minimal resolution $\tilde{D} \to D$, we obtain a relatively minimal elliptic fibration whose singular fibers are classified by Kodaira [36]. Then one can get a bound of multiplicities of fibers and describe the configuration of non-Gorenstein singularities.
- For a $\mathbb{Q}$-Fano threefold X, a Du Val general elephant is a (singular) K3 surface. In the case where the linear system $|-K_X|$ is "sufficiently big," this implies the existence of a good Gorenstein model [1].

Shokurov [70] generalized Conjecture 3.1 and introduced a new notion which is very efficient in the study of pluri-anticanonical linear systems. Omitting technicalities, we reproduce a weak form of Shokurov's definition.

Definition. An *n-complement* of the canonical class K_X is a member $D \in |-nK_X|$ such that the pair $(X, \frac{1}{n}D)$ is lc. An n-complement is said to be klt (resp. plt) if such is the pair $(X, \frac{1}{n}D)$.

According to the inversion of adjunction [70], the existence of a Du Val general elephant $D \in |-K_X|$ is equivalent to the existence of a plt 1-complement. Shokurov developed a powerful theory that works in arbitrary dimension and allows constructing complements inductively (see [64, 70] and references therein).

Note that Reid's general elephant fails for Fano threefolds. For example, in [6, 21] one can find examples of $\mathbb{Q}$-Fano threefolds with an empty anticanonical linear system. Because of this, the statement of Conjecture 3.1 sometimes is called a "principle." Nonetheless, there are only a few examples of such Fano threefolds. In the case $\dim(Z) > 0$, Conjecture 3.1 is expected to be true without exceptions. The following should be considered as the local version of Conjecture 3.1.

Proposition 3.2 (Reid [66]). *Let $(X \ni P)$ be the analytic germ of a threefold terminal singularity of index $m > 1$. Then the general member $D \in |-K_X|$ is a Du Val singularity. Furthermore, let $\pi : X' \to X$ be the index-one cover and let $D' := \pi^{-1}(D)$. Then the cover $D' \to D$ belongs to one of the following six types:*

$(X \ni P)$	$D' \to D$	$(X \ni P)$	$D' \to D$
cA/m	$A_{k-1} \xrightarrow{m:1} A_{km-1}$	cAx/2	$A_{2k-1} \xrightarrow{2:1} D_{k+2}$
cAx/4	$A_{2k-2} \xrightarrow{4:1} D_{2k+1}$	cD/2	$D_{k+1} \xrightarrow{2:1} D_{2k}$
cD/3	$D_4 \xrightarrow{3:1} E_6$	cE/2	$E_6 \xrightarrow{2:1} E_7$

4. DIVISORIAL CONTRACTIONS TO A POINT

In this section we treat divisorial Mori contractions of a divisor to a point. Such contractions are studied very well due to works of Y. Kawamata [34], A. Corti [10], M. Kawakita [26–30], T. Hayakawa [18–20], and others. In this case, General Elephant Conjecture 3.1 has been verified:

Theorem 4.1 (Kawakita [28, 29]). *Let $f : X \to (Z \ni o)$ be a divisorial Mori contraction that contracts a divisor to a point. Then the general member $D \in |-K_X|$ is Du Val.*

One of the main tools in the proofs is the orbifold Riemann–Roch formula [66]: if X is a three-dimensional projective variety with terminal singularities and D is a Weil $\mathbb{Q}$-Cartier divisor on X, then for the sheaf $\mathscr{L} = \mathscr{O}_X(D)$ there is a formula of the form

$$\chi(\mathscr{L}) = \chi(\mathscr{O}_X) + \frac{1}{12} D \cdot (D - K_X) \cdot (2D - K_X) + \frac{1}{12} D \cdot c_2 + \sum_P c_P(D), \quad (4.1.1)$$

where the sum rungs over all the virtual quotient singularities of X, i.e., over the actual singularities of X that are replaced with their small deformations [66], and $c_P(D)$ is a local contribution due to singularity at P, depending only on the local analytic type of D at P. There is an explicit formula for the computation of $c_P(D)$.

Except for a few hard cases, the classification of divisorial Mori contractions of a divisor to a point has been completed. A typical result here is to show that a contraction is a weighted blowup with some explicit collection of weights:

Theorem 4.2 (Y. Kawamata [34]). *Let $f : X \to (Z \ni o)$ be a divisorial Mori contraction that contracts a divisor to a point. Assume that $o \in Z$ is a cyclic quotient singularity of type $\frac{1}{r}(a, -a, 1)$. Then f is the weighted blowup with weights $(a/r, 1 - a/r, 1/r)$.*

Theorem 4.3 (M. Kawakita [26]). *Let $f : X \to (Z \ni o)$ be a divisorial Mori contraction that contracts a divisor to a smooth point. Then f is the weighted blowup with weights $(1, a, b)$, where* $\gcd(a, b) = 1$.

These results are intensively used in the three-dimensional birational geometry, for example, in the proof of birational rigidity of index-1 Fano threefold weighted hypersurfaces [11].

5. DEL PEZZO FIBRATIONS

Much less is known about the local structure of $\mathbb{Q}$-del Pezzo fibrations. As was explained in Section 3, the existence of a Du Val general elephant would be very helpful in the study such contractions. However, in this case Conjecture 3.1 is established only in some special situations.

An important question that can be asked in the Del Pezzo fibration case is the presence of multiple fibers.

Theorem 5.1 ([49]). *Let $f : X \to Z$ be a $\mathbb{Q}$-del Pezzo fibration and let $f^*(o) = m_o F_o$ be a special fiber of multiplicity m_o. Then $m_o \le 6$ and all the cases $1 \le m_o \le 6$ occur. Moreover, the possibilities for the local behavior of F_o near singular points are described.*

The main idea of the proof is to apply the orbifold Riemann–Roch formula (4.1.1) to the divisor F_o and its multiples.

Example. Suppose that the cyclic group $\boldsymbol{\mu}_4$ acts on $\mathbb{P}^1_x \times \mathbb{P}^1_y \times \mathbb{C}_t$ via

$$(x, y; t) \longmapsto (y, -x, \sqrt{-1}t).$$

Then the quotient

$$X = (\mathbb{P}^1 \times \mathbb{P}^1 \times \mathbb{C})/\boldsymbol{\mu}_4 \longrightarrow Z = \mathbb{C}/\boldsymbol{\mu}_4$$

is the germ of a $\mathbb{Q}$-del Pezzo fibration with central fiber of multiplicity 4.

Another type of $\mathbb{Q}$-del Pezzo fibrations which are investigated relatively well are those whose central fiber $F := f^{-1}(o)$ is reduced, normal, and has "good" singularities. Then X can be viewed as a one-parameter *smoothing* of F. The total space of this smoothing must be $\mathbb{Q}$-*Gorenstein* and F can be viewed as a *degeneration* of a general fiber (smooth del Pezzo surface) in a $\mathbb{Q}$-Gorenstein family. The most studied class of singularities admitting $\mathbb{Q}$-Gorenstein smoothings is the class of singularities of type T.

Definition (Kollár, Shepherd-Barron [40]). A two-dimensional quotient singularity is said to be of *type* T if it admits a smoothing in a one-parameter $\mathbb{Q}$-Gorenstein family $X \to B$.

In this case, by the inversion of adjunction [70], the pair (X, F) is plt and the total family X is terminal. Conversely, if $X \ni P$ is a $\mathbb{Q}$-Gorenstein point and F is an effective Cartier divisor at P such that the pair (X, F) is plt, then $F \ni P$ is a T-singularity and the point $X \ni P$ is terminal. Singularities of type T and their deformations were studied by Kollár and Shepherd-Barron [40]. In particular, they proved that any T-singularity is either a Du Val point or a cyclic quotient of type $\frac{1}{m}(q_1, q_2)$ with

$$\gcd(m, q_1) = \gcd(m, q_2) = 1, \quad (q_1 + q_2)^2 \equiv 0 \bmod m.$$

Minimal resolutions of these singularities are also described [40, § 3].

Thus to study $\mathbb{Q}$-del Pezzo fibrations whose central fiber has only quotient singularities, one has to consider $\mathbb{Q}$-Gorenstein smoothings of del Pezzo surfaces with singularities of type T. The important auxiliary fact here is the unobstructedness of deformations:

Proposition 5.2 ([13, 41]). *Let F be a projective surface with log canonical singularities such that $-K_F$ is big. Then there are no local-to-global obstructions to deformations of F. In particular, if F has* T*-singularities, then F admits a $\mathbb{Q}$-Gorenstein smoothing.*

Theorem 5.3 (Hacking–Prokhorov [13]). *Let F be a projective surface with quotient singularities such that $-K_F$ is ample, $\rho(F) = 1$, and F admits a $\mathbb{Q}$-Gorenstein smoothing. Then F belongs to one of the following:*

- 14 *infinite sequences of toric surfaces (see below);*
- *partial smoothing of a toric surface as above;*
- 18 *sporadic families of surfaces of index* ≤ 2 [2].

Toric surfaces appearing in the above theorem are determined by a Markov-type equation. More precisely, for $K_F^2 \geq 5$ these surfaces are weighted projective spaces given by the following table:

K_F^2	F	Markov-type equation
9	$\mathbb{P}(a^2, b^2, c^2)$	$a^2 + b^2 + c^2 = 3abc$
8	$\mathbb{P}(a^2, b^2, 2c^2)$	$a^2 + b^2 + 2c^2 = 4abc$
6	$\mathbb{P}(a^2, 2b^2, 3c^2)$	$a^2 + 2b^2 + 3c^2 = 6abc$
5	$\mathbb{P}(a^2, b^2, 5c^2)$	$a^2 + b^2 + 5c^2 = 5abc$

and for $K^2 \leq 4$ they are certain abelian quotients of the weighted projective spaces as above. Note, however, that in general we cannot assert that, for central fiber F of a $\mathbb{Q}$-del Pezzo fibration, the condition $\rho(F) = 1$ holds. Some partial results in the case $\rho(F) > 1$ where obtained in [60]. In particular, [60] establishes the existence of Du Val general elephant for $\mathbb{Q}$-del Pezzo fibrations with "good" fibers:

Theorem 5.4. *Let $f : X \to (Z \ni o)$ be a $\mathbb{Q}$-del Pezzo fibration over a curve germ $Z \ni o$. Assume that the fiber $f^{-1}(o)$ is reduced, normal, and has only log terminal singularities. Then the general elephant $D \in |-K_X|$ is Du Val.*

Theorem 5.3 gives a complete answer to the question posed by M. Manetti **[41]**:

Corollary 5.5 (**[13]**). *Let X be a projective surface with quotient singularities which admits a smoothing to $\mathbb{P}^2$. Then X is a $\mathbb{Q}$-Gorenstein deformation of a weighted projective plane $\mathbb{P}(a^2, b^2, c^2)$, where the triple (a, b, c) is a solution of the Markov equation*

$$a^2 + b^2 + c^2 = 3abc.$$

Results similar to Theorem 5.3 were obtained for $\mathbb{Q}$-del Pezzo fibrations whose central fiber is log canonical **[62]**. However, in this case the classification is not complete.

6. EXTREMAL CURVE GERMS

To study Mori contractions with fibers of dimension ≤ 1, it is convenient to work with analytic threefolds and to localize to situation near a curve contained in a fiber.

Definition 6.1. Let $(X \supset C)$ be the *analytic* germ of a threefold with terminal singularities along a reduced connected complete curve. Then $(X \supset C)$ is called an *extremal curve germ* if there exists a contraction

$$f : (X \supset C) \longrightarrow (Z \ni o)$$

such that $C = f^{-1}(o)_{\mathrm{red}}$ and $-K_X$ is f-ample. The curve C is called the *central fiber* of the germ and $Z \ni o$ is called the *target variety* or the *base* of $(X \supset C)$. An extremal curve germ is said to be *irreducible* if such is its central fiber.

In the definition above, we do not assume that X is $\mathbb{Q}$-factorial or $\rho(X/Z) = 1$. This is because $\mathbb{Q}$-factoriality typically is not a local condition in the analytic category (see **[32, § 1]**). There are three types of extremal curve germs:

- *flipping* if f is birational and does not contract divisors;
- *divisorial* if the exceptional locus is two-dimensional;
- $\mathbb{Q}$-*conic bundle germ* if the target variety Z is a surface.

If a divisorial curve germ is irreducible, then the exceptional locus of the corresponding contraction is a $\mathbb{Q}$-Cartier divisor and the target variety Z has terminal singularities **[51, §3]**. In general, this is not true. It may happen that the exceptional locus is a union of a divisor and some curves.

As an example, we consider the case where X has singularities of indices 1 and 2.

Theorem 6.2 (**[47]**). *Let $(X \supset C)$ be a $\mathbb{Q}$-conic bundle germ over a smooth base. Assume that X is not Gorenstein and $2K_X$ is Cartier. Then X can be embedded to $\mathbb{P}(1, 1, 1, 2) \times \mathbb{C}^2$*

and given there by two quadratic equations. In particular, the point $P \in X$ of index 2 is unique, the curve C has at most 4 components, all of them pass through P.

Theorem 6.3 ([38]). *Let $(X \supset C)$ be a flipping extremal curve germ and let*

$$\begin{array}{ccc} (X \supset C) & \dashrightarrow & (X^+ \supset C^+) \\ & {\scriptstyle f} \searrow \quad \swarrow {\scriptstyle f^+} & \\ & (Z \ni o) & \end{array}$$

be the corresponding flip. Assume that $2K_X$ is Cartier. Then $(Z \ni o)$ is the quotient of the isolated hypersurface singularity

$$\{x_1x_3 + x_2\phi(x_2^2, x_4) = 0\} \ni 0$$

by the $\boldsymbol{\mu}_2$-action given by the weights $(1,1,0,0)$. The contraction f (resp. f^+) is the quotient of the blowup of the plane $\{x_2 = x_3 = 0\}$ (resp. $\{x_1 = x_2 = 0\}$) by $\boldsymbol{\mu}_2$. In particular, X contains a unique point of index 2 and the central fiber C is irreducible. The variety X^+ is Gorenstein.

A similar description is known for divisorial extremal curve germs of index 2 [38, § 4].

First properties. Let $(X \supset C)$ be an extremal curve germ and let $f : (X \supset C) \to (Z \ni o)$ be the corresponding contraction. For any connected subcurve $C' \subset C$, the germ $(X \supset C')$ is again an extremal curve germ. If, moreover, $C' \subsetneqq C$, then $(X \supset C')$ is birational. By the Kawamata–Viehweg vanishing theorem,

$$R^1 f_* \mathscr{O}_X = 0 \tag{6.3.1}$$

(see, e.g., [35]). As a consequence, one has $\mathrm{p_a}(C') \leq 0$ for any subcurve $C' \subset C$. In particular, $C = \bigcup C_i$ is a “tree” of smooth rational curves. Furthermore,

$$\mathrm{Pic}(X) \simeq H^2(X, \mathbb{Z}) \simeq \mathbb{Z}^{\oplus n}, \tag{6.3.2}$$

where n is the number of irreducible components of C. For more delicate properties of extremal curve germs, one needs to know the cohomology of the dualizing sheaf, see [44,47]:

$$R^1 f_* \omega_X = \begin{cases} 0, & \text{if } f \text{ is birational,} \\ \omega_Z, & \text{if } f \text{ is } \mathbb{Q}\text{-conic bundle and } Z \text{ is smooth.} \end{cases} \tag{6.3.3}$$

Definition. An irreducible extremal curve germ $(X \supset C)$ is (locally) *imprimitive* at a point P if the inverse image of C under the index-one cover $(X^\sharp \ni P^\sharp) \to (X \ni P)$ splits.

Theorem 6.4 ([44, 47]). *Let $(X \supset C)$ be an extremal curve germ and let $C_1, \dots, C_n$ be irreducible components of C.*

- *Each C_i contains at most 3 singular points of X.*
- *Each C_i contains at most 2 non-Gorenstein points of X and at most 1 point which is imprimitive for $(X \supset C_i)$.*

- *If X is Gorenstein at the intersection point $P = C_i \cap C_j$, $C_i \neq C_j$, then X is smooth outside P and $(X \supset C)$ is a $\mathbb{Q}$-conic bundle germ over a smooth base.*

To prove the first assertion, one needs to analyze the conormal sheaf I_C/I_C^2 and use the vanishing $H^1(\mathscr{O}_X/J) = 0$ for any $J \subset \mathscr{O}_X$ with $\mathrm{Supp}(\mathscr{O}_X/J) = C$ (see [44, 55]). For the second assertion, one can use topological arguments based on (1.2.1) (see [55]). For the last assertion, we refer to [44, 1.15], [37, 4.2], and [55, 4.7.6]

The techniques applied in the proof of the above proposition allow obtaining much stronger results. In particular, they alow classifying all the possibilities for the local behavior of an irreducible germ $(X \supset C)$ near a singular point P [44]. Thus, according to [44] and [47], the triple $(X \supset C \ni P)$ belongs to one of the following types:

$$\text{(IA)}, \text{(IC)}, \text{(IIA)}, \text{(IIB)}, (\text{IA}^\vee), (\text{II}^\vee), (\text{ID}^\vee), (\text{IE}^\vee), \text{(III)}.$$

Here the symbol $^\vee$ means that $(X \supset C \ni P)$ is locally imprimitive, the symbol II means that $(X \ni P)$ is a terminal point of exceptional type cAx/4 (see Proposition 3.2), and III means that $(X \ni P)$ is an (isolated) cDV-point.

For example, a triple $(X \supset C \ni P)$ is of type (IC) if there are analytic isomorphisms

$$(X \ni P) \simeq \mathbb{C}^3_{y_1,y_2,y_4}/\boldsymbol{\mu}_m(2, m-2, 1), \quad C \simeq \{y_1^{m-2} - y_2^2 = y_4 = 0\}/\boldsymbol{\mu}_m,$$

where m is odd and $m \geq 5$. For definitions other types, we refer to [44] and [47].

6.1. Construction of germs by deformations

Let $(X \supset C)$ be an extremal curve germ and let $f : X \to (Z \ni o)$ be the corresponding contraction. Denote by $|\mathscr{O}_Z|$ the infinite-dimensional linear system of hyperplane sections passing through o and let $|\mathscr{O}_X| := f^*|\mathscr{O}_Z|$. The *general hyperplane section* of $(X \supset C)$ is the general member $H \in |\mathscr{O}_X|$. The divisor H contains much more information on the total space than a general elephant $D \in |-K_X|$. However, the singularities of H typically are more complicated, in particular, H can be nonnormal.

The variety X (resp. Z) can be viewed as the total space of a one-parameter deformation of H (resp. $H_Z := f(H)$). We are going to reverse this consideration.

Construction (see [38, § 11], [44, § 1B]). Suppose we are given a normal surface germ $(H \supset C)$ along a proper curve C and a contraction $f_H : H \to H_Z$ such that C is a fiber and $-K_H$ is f_H-ample. Let $P_1, \ldots, P_r \in H$ be all the singular points. Assume also that near each P_i there exists a small one-parameter deformation $\mathfrak{H}_i$ of a neighborhood H_i of P_i in H such that the total space $\mathfrak{H}_i$ has a terminal singularity at P_i. The obstruction to globalize deformations

$$\Psi : \mathrm{Def}(H) \longrightarrow \prod_{P_i \in \mathrm{Sing}(H)} \mathrm{Def}(H, P_i)$$

lies in $R^2 f_* \mathscr{T}_H$, where $\mathscr{T}_H = \mathscr{H}om(\Omega_H, \mathscr{O}_H)$ is the tangent sheaf of H. Since $R^2 f_* \mathscr{T}_H = 0$ due to dimension reasons, the morphism Ψ is smooth, and so there exists a global one-parameter deformation $\mathfrak{H}$ of H inducing a local deformation of $\mathfrak{H}_i$ near P_i.

Then we have a threefold $X := \mathfrak{H} \supset C$ with $H \in |\mathscr{O}_X|$ such that locally near P_i it has the desired structure and one can extend f_H to a contraction $f : X \to Z$ which is birational (resp. a $\mathbb{Q}$-conic bundle) if H_Z is a surface (resp. a curve).

Example. Consider a rational curve fibration $f_{\tilde{H}} : \tilde{H} \to H_Z$ over a smooth curve germ $H_Z \ni o$, where $\tilde{H}$ is a smooth surface such that the fiber over o has the following weighted dual graph:

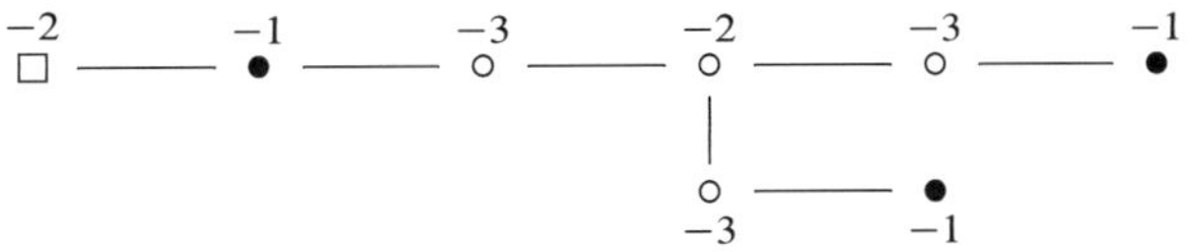

Contracting the curves corresponding to the white vertices □ and ∘, we obtain a singular surface H and a K_H-negative contraction $f_H : H \to H_Z$ whose fiber over o is a curve $C \subset H$ having three irreducible components that correspond to the black vertices •. The singular locus of H consists of a Du Val point $P_0 \in H$ of type A_1 and a log canonical singularity $P \in H$ whose dual graph is formed by the white circle vertices ∘. Both P_0 and P have 1-parameter $\mathbb{Q}$-Gorenstein smoothings [38, COMPUTATION 6.7.1]. Applying the above construction to $H \supset C$, we obtain an example of a $\mathbb{Q}$-conic bundle contraction $f : (X \supset C) \to (Z \ni o)$ with a unique non-Gorenstein point which is of type cD/3. If we remove the (-2)-curve corresponding to □ on the left-hand side of the graph, we get a birational contraction of surfaces $f'_H : H' \to H'_Z$. Applying the same construction to $H' \supset C$, we obtain an example of a divisorial contraction. Similarly, removing further one of the (-1)-curves, we get a flip.

7. EXTREMAL CURVE GERMS: GENERAL ELEPHANT

Theorem 7.1 (Mori [44], Kollár–Mori [38], Mori–Prokhorov [50]). *Let $(X \supset C)$ be an irreducible extremal curve germ. Then the general member $D \in |-K_X|$ has only Du Val singularities.*

The existence of a Du Val elephant for extremal curve germs with reducible central fiber is not known at the moment. See Theorem 9.2 below for partial results in this direction.

Comment on the proof. Essentially, there are three methods to find a good elephant $D \in |-K_X|$. We outline them below.

7.1. Local method

As in Proposition 3.2, near each non-Gorenstein point $P_i \in X$ take a local general elephant $D_i \in |-K_{(X \ni P_i)}|$. Since D_i is general, we have $D_i \cap C = \{P_i\}$. Then we can regard $D := \sum D_i$ as a Weil divisor on X. By the construction, $K_X + D$ is a Cartier divisor near each P_i, hence it is Cartier everywhere. In some cases it is possible to compute the intersection numbers $D_i \cdot C$ and show that $D \cdot C < 1$. Then we may assume that $K_X + D \sim 0$ by (6.3.2) and so D is a Du Val anticanonical divisor. For example, this method works for extremal curve germs described in Theorems 6.2 and 6.3, as well as in Example 7.3 below.

7.2. Extension from $S \in |-2K_X|$

In some cases, the above approach does not work, but it allows showing the existence of a klt 2-complement $S \in |-2K_X|$ such that $\dim(D \cap C) = 0$. Then one can try to extend a good element from the surface S. The crucial fact here is that the natural map

$$\tau : H^0\big(X, \mathscr{O}_X(-K_X)\big) \longrightarrow H^0\big(S, \mathscr{O}_S(-K_X)\big) = \omega_S$$

is surjective if (X, C) is birational and surjective modulo Ω_S^2 if (X, C) is a $\mathbb{Q}$-conic bundle. This immediately follows from (6.3.3). Details can be found in **[38, § 2]** and **[50]**.

7.3. Global method

Finally, in the most complicated cases, none of the above methods work. Then one needs more subtle techniques which require detailed analysis of singularities and infinitesimal structure of X along C **[44, §§ 8–9]**. Then, roughly speaking, the good section $D \in |-K_X|$ is recovered as the formal Weil divisor $\varprojlim C_n$ of the completion $X^\wedge$ of X along C, where C_n are subschemes with support C constructed by using certain inductive procedure **[44, § 9]**.

As a consequence of Theorem 7.1, in the $\mathbb{Q}$-conic bundle case, one obtains the following fact which proves Iskovskikh's conjecture **[24]**.

Corollary 7.2. *Let $(X \supset C)$ be a $\mathbb{Q}$-conic bundle germ over $(Z \ni o)$, where C can be reducible. Then $(Z \ni o)$ is a Du Val singularity of type $\mathrm{A_n}$ (or smooth).*

This result is very useful for applications to the rationality problem of three-dimensional varieties having conic bundle structure **[24, 61]** and some problems of biregular geometry **[58, 59]**.

It turns out that the structure of $\mathbb{Q}$-conic bundle germs over a singular base $(Z \ni o)$ is much simpler and shorter than others. In fact, these germs can be exhibited as certain quotients of $\mathbb{Q}$-conic bundles of index ≤ 2 (see Theorem 6.2). A complete classification of such germs was obtained in **[47, 48]**. Here is a typical example.

Example 7.3. Let the group $\boldsymbol{\mu}_n$ act on $\mathbb{C}^2_{u,v}$ and $\mathbb{P}^1_{x,y} \times \mathbb{C}^2_{u,v}$ via

$$(x : y; u, v) \longmapsto (x : \zeta^a y; \zeta u, \zeta^{-1} v),$$

where $\zeta = \zeta_n = \exp(2\pi i/n)$ and $\gcd(n, a) = 1$. Then the projection

$$f : X = (\mathbb{P}^1 \times \mathbb{C}^2)/\boldsymbol{\mu}_n \longrightarrow Z = \mathbb{C}^2/\boldsymbol{\mu}_n$$

is a $\mathbb{Q}$-conic bundle. The variety X has exactly two singular points which are terminal cyclic quotients of type $\frac{1}{n}(1, -1, \pm a)$. The surface Z has at 0 a Du Val point of type $\mathrm{A_{n-1}}$.

McKernan proposed a natural higher-dimensional analogue of Corollary 7.2:

Conjecture 7.4. *Let $f : X \to Z$ be a K-negative contraction such that $\rho(X/Z) = 1$ and X is ε-lc, that is, all the coefficients in (1.1.1) satisfy $a_i \geq -1 + \varepsilon$. Then Z is δ-lc, where δ depends on ε and the dimension.*

A deeper version of this conjecture which generalizes Theorem 5.1 and uses the notion was proposed by Shokurov. He also suggested that the optimal value of δ, in the case where singularities of X are canonical and f has one-dimensional fibers, equals $1/2$. Recently, this was proved by J. Han, C. Jiang, and Y. Luo [17].

Once we have a Du Val general elephants, all extremal curve germs can be divided into two large classes which will be discussed separately in the next two sections.

Definition 7.5. Let $(X \supset C)$ be an extremal curve germ and let $f : X \to (Z \ni o)$ be the corresponding contraction. Assume that the general member $D \in |-K_X|$ is Du Val. Consider the Stein factorization:

$$f_D : D \longrightarrow D' \longrightarrow f(D) \qquad \text{(put } D' = f(D) \text{ if } f \text{ is birational).}$$

Then the germ $(X \supset C)$ is said to be *semistable* if D' has only (Du Val) singularities of type $\mathrm{A_n}$. Otherwise, $(X \supset C)$ is called *exceptional.*

8. SEMISTABLE GERMS

Let $(X \supset C)$ be an irreducible extremal curve germ. By Theorem 7.1, the general member $D \in |-K_X|$ is Du Val. In this section we assume that $(X \supset C)$ is semistable. Excluding simple cases, we assume also that X is not Gorenstein [12] and $(X \supset C)$ is not a $\mathbb{Q}$-conic bundle germ over a singular base [47, 48]. According to Theorem 6.4, the threefold X has at most two non-Gorenstein points. Thus the following case division is natural:

Case (k1A): the set of non-Gorenstein points consists of a single point P;

Case (k2A): the set of non-Gorenstein points consists of exactly two points P_1, P_2.

Proposition 8.1. *In the above hypothesis, for the general member $H \in |\mathscr{O}_X|$, the pair $(X, H + D)$ is lc. If, moreover, $D \supset C$, then H is normal and has only cyclic quotient singularities. In this case the singularities of H are of type* T.

The proof uses the inversion of adjunction [70] to extend a general hyperplane section from D to X (see [51, PROPOSITION 2.6]).

For an extremal curve germ of type (k2A), any member $D \in |-K_X|$ contains C [38]. Hence the general hyperplane section $H \in |\mathscr{O}_X|$ has only T-singularities and X can be restored as a one-parameter deformation space of H. In this case X has no singularities other than P_1, P_2. Moreover, $(X \supset C)$ cannot be a $\mathbb{Q}$-conic bundle germ [47, 50]. The birational germs of type (k2A) were completely described by Mori [46]. He gave an explicit algorithm for computing divisorial contractions and flips in this case.

The structure of extremal curve germs of type (k1A) is more complicated. They were studied in [51]. In particular, the general hyperplane section $H \in |\mathscr{O}_X|$ was computed. However, [51] does not provide a good description of the infinitesimal structure of X along C or an algorithm similar to [46]. This was done only in a special situation in [14]. Note that in the case (k1A) a general member $H \in |\mathscr{O}_X|$ can be nonnormal.

Examples. Similar to the example in Section 6.1, consider a surface germ $H \supset C \simeq \mathbb{P}^1$ whose dual graph has the following graph of the minimal resolution:

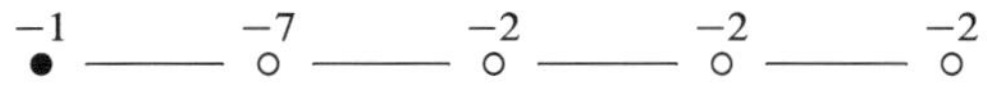

where • is a (-1)-curve. The chain formed by white circle vertices ∘ corresponds to a T-singularity of type $\frac{1}{25}(1,4)$. The whole configuration can be contracted to a cyclic quotient singularity $H_Z \ni o$ of type $\frac{1}{21}(1,16)$. Since this is not a T-singularity, the induced threefold contraction must be flipping.

9. EXCEPTIONAL CURVE GERMS

In this section we assume that $(X \supset C)$ is an exceptional irreducible extremal curve germ. As in the previous section we also assume that X is not Gorenstein and $(X \supset C)$ is not a $\mathbb{Q}$-conic bundle germ over a singular base. According to the classification **[38, 44, 50]**, the germ $(X \supset C)$ belongs to one of following types:

- X has a unique non-Gorenstein point P which is of type cD/2, cAx/2, cE/2, or cD/3 and $(X \supset C)$ is of type (IA) at P;
- X has a unique non-Gorenstein point P which is of exceptional type cAx/4 and $(X \supset C)$ is of type (IIA), (II$^\vee$), or (IIB) at P;
- X has a unique singular point P which is a cyclic quotient singularity of index $m \geq 5$ (odd) and $(X \supset C)$ is of type (IC) at P;
- X has two singular points of indices $m \geq 3$ (odd) and 2, then $(X \supset C)$ is said to be of type (kAD);
- X has three singular points of indices $m \geq 3$ (odd), 2 and 1, then $(X \supset C)$ is said to be of type (k3A).

In each case the general elephant is completely described in terms of its minimal resolution:

Theorem 9.1 (**[38, 50]**). *In the above hypothesis assume that the general element $D \in |-K_X|$ contains C. Then the dual graph of $(D \supset C)$ is one of the following, where white vertices ∘ denote (-2)-curves on the minimal resolution of D and the black vertex • corresponds to the proper transform of C:*

(IC)

(IIB)

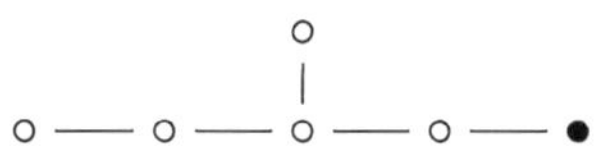

(kAD)

(k3A)

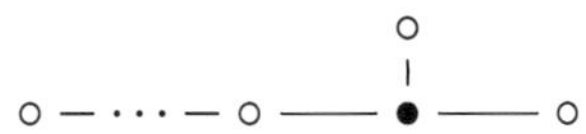

Exceptional irreducible extremal curve germs are well studied (see [38,55], and references therein). For flipping ones, the general hyperplane section $H \in |\mathscr{O}_X|$ is normal and has only rational singularities. It is computed in [38] and the flip is reconstructed as a one-parameter deformation space of H. For divisorial and $\mathbb{Q}$-conic bundle germs, the situation is more complicated. Then the general hyperplane section H can be nonnormal (see, e.g., [54]). Nevertheless, in almost all cases, except for types (kAD)and (k3A), there is a description of $H \in |\mathscr{O}_X|$ and infinitesimal structure of these germs. For convenience of reference, in the table below we collect the known information on the exceptional irreducible extremal curve germs.

Type	(X, C)	References
index 2 germs	divisorial, $\mathbb{Q}$-conic bundle	[38, § 4], [47, § 12], [51, § 7]
cD/3	flip, divisorial	[38, § 6], [51, § 4]
(IC)	flip, $\mathbb{Q}$-conic bundle (only for $m = 5$)	[38, § 8], [52]
(IIA)	flip, divisorial, $\mathbb{Q}$-conic bundle	[38, § 7], [53,54]
(IIB)	divisorial, $\mathbb{Q}$-conic bundle	[52]
$(\mathrm{II}^{\vee})$	divisorial, $\mathbb{Q}$-conic bundle	[38, 4.11.2], [47]
(kAD)	flip, divisorial, $\mathbb{Q}$-conic bundle	[38, § 9], [45,47,50]
(k3A)	divisorial, $\mathbb{Q}$-conic bundle	[38, § 5], [47,50]

Detailed analysis of the local structure of exceptional extremal curve germs allows extending the result of Theorem 7.1 to the case of reducible central fiber containing an exceptional component:

Theorem 9.2 (Mori–Prokhorov [56]). *Let $(X \supset C)$ be an extremal curve germ such that C is reducible and satisfies the following condition:*

(*) *each component $C_i \subset C$ contains at most one point of index > 2.*

Then the general member $D \in |-K_X|$ has only Du Val singularities. Moreover, for each irreducible component $C_i \subset C$ with two non-Gorenstein points or of types (IC) *or* (IIB), *the dual graph of (D, C_i) has the same form as the irreducible extremal curve germ $(X \supset C_i)$.*

The proof uses the extension techniques of sections of $|-K_X|$ from a good member $S \in |-2K_X|$ (see Section 7.2).

10. $\mathbb{Q}$-FANO THREEFOLDS

In arbitrary dimension, $\mathbb{Q}$-Fano threefolds are bounded, i.e., they are contained in fibers of a morphism of schemes of finite type. This is a consequence of the much more general fact [4]. In dimension 3, there are effective results based on the orbifold Riemann–Roch formula (4.1.1) and Bogomolov–Miyaoka inequality applied to the restriction of the reflexive sheaf $(\Omega^1_X)^{\vee\vee}$ to a sufficiently general hyperplane section [33]. In particular, combining (4.1.1) with Serre duality, we obtain

$$\chi(\mathscr{O}_X) = \frac{1}{24}\left(-K_X \cdot c_2(X) + \sum_P \left(m_P - \frac{1}{m_p}\right)\right)$$

where m_P is the index of a virtual quotient singularity of X [66]. Since X is $\mathbb{Q}$-Fano, by Kawamata–Viehweg vanishing theorem [35], one has $\chi(\mathscr{O}_X) = 1$. Arguments based on Bogomolov–Miyaoka inequality show that $-K_X \cdot c_2(X)$ is positive (see [33]). This gives an effective bound of the indices of singularities of X. Similarly, one can get an upper bound of the anticanonical degree $-K_X^3$. Moreover, analyzing the methods of [33], it is possible to enumerate Hilbert series of all $\mathbb{Q}$-Fano threefolds. This information is collected in [6] in a form of a huge computer database of possible "candidates." It was extensively explored by many authors, basically to obtain lists of examples representing $\mathbb{Q}$-Fano threefolds as subvarieties of small codimension in a weighted projective space (see, e.g., [7, 21] and references therein).

Examples.

- There are 130 (resp. 125) families of $\mathbb{Q}$-Fano threefolds that are representable as hypersurfaces (resp. codimension 2 complete intersections) in weighted projective spaces [6, 21].
- Toric $\mathbb{Q}$-Fano threefolds are exactly weighted projective spaces $\mathbb{P}(3,4,5,7)$, $\mathbb{P}(2,3,5,7)$, $\mathbb{P}(1,3,4,5)$, $\mathbb{P}(1,2,3,5)$, $\mathbb{P}(1,1,2,3)$, $\mathbb{P}(1,1,1,2)$, $\mathbb{P}^3 = \mathbb{P}(1,1,1,1)$, and the quotient of $\mathbb{P}^3$ by $\boldsymbol{\mu}_5$ that acts diagonally with weights $(1,2,3,4)$ [6].

Although the classification is very far from completion, there are several systematic results. For example, the optimal upper bound of the degree $-K_X^3$ of $\mathbb{Q}$-Fano threefolds was obtained in [58]. If X is singular, it is equal to $125/2$ and achieved for the weighted projective space $\mathbb{P}(1,1,1,2)$. The lower bound of the degree equals $1/330$ [8] and is achieved for a hypersurface of degree 66 in $\mathbb{P}(1,5,6,22,33)$. It is known that, under certain conditions, General Elephant Conjecture 3.1 holds for $\mathbb{Q}$-Fano threefolds modulo deformations [67].

ACKNOWLEDGMENTS

The author would like to thank Professors Shigefumi Mori and Vyacheslav Shokurov for helpful comments on the original version of this paper.

FUNDING

This work was performed at the Steklov International Mathematical Center and supported by the Ministry of Science and Higher Education of the Russian Federation (agreement no. 075-15-2019-1614).

REFERENCES

[1] V. Alexeev, General elephants of **Q**-Fano 3-folds. *Compos. Math.* **91** (1994), no. 1, 91–116.

[2] V. Alexeev and V. V. Nikulin, *Del Pezzo and K3 surfaces*. MSJ Mem. 15, Mathematical Society of Japan, Tokyo, 2006.

[3] C. Birkar, Birational geometry of algebraic varieties. In *Proceedings of the International Congress of Mathematicians—Rio de Janeiro 2018. Vol. II. Invited lectures*, pp. 565–590, World Sci. Publ., Hackensack, NJ, 2018.

[4] C. Birkar, Singularities of linear systems and boundedness of Fano varieties. *Ann. of Math. (2)* **193** (2021), no. 2, 347–405.

[5] C. Birkar, P. Cascini, C. D. Hacon, and J. McKernan, Existence of minimal models for varieties of log general type. *J. Amer. Math. Soc.* **23** (2010), no. 2, 405–468.

[6] G. Brown, et al., Graded ring database, http://www.grdb.co.uk/.

[7] G. Brown, M. Kerber, and M. Reid, Fano 3-folds in codimension 4, Tom and Jerry. Part I. *Compos. Math.* **148** (2012), no. 4, 1171–1194.

[8] J. A. Chen and M. Chen, An optimal boundedness on weak **Q**-Fano 3-folds. *Adv. Math.* **219** (2008), no. 6, 2086–2104.

[9] A. Corti, Factoring birational maps of threefolds after Sarkisov. *J. Algebraic Geom.* **4** (1995), no. 2, 223–254.

[10] A. Corti, Singularities of linear systems and 3-fold birational geometry. In *Explicit birational geometry of 3-folds*, pp. 259–312, London Math. Soc. Lecture Note Ser. 281, Cambridge Univ. Press, Cambridge, 2000.

[11] A. Corti, A. Pukhlikov, and M. Reid, Fano 3-fold hypersurfaces. In *Explicit birational geometry of 3-folds*, pp. 175–258, London Math. Soc. Lecture Note Ser. 281, Cambridge Univ. Press, Cambridge, 2000.

[12] S. Cutkosky, Elementary contractions of Gorenstein threefolds. *Math. Ann.* **280** (1988), no. 3, 521–525.

[13] P. Hacking and Y. Prokhorov, Smoothable del Pezzo surfaces with quotient singularities. *Compos. Math.* **146** (2010), no. 1, 169–192.

[14] P. Hacking, J. Tevelev, and G. Urzúa, Flipping surfaces. *J. Algebraic Geom.* **26** (2017), no. 2, 279–345.

[15] C. D. Hacon and J. McKernan, Existence of minimal models for varieties of log general type. II. *J. Amer. Math. Soc.* **23** (2010), no. 2, 469–490.

[16] C. D. Hacon and J. McKernan, The Sarkisov program. *J. Algebraic Geom.* **22** (2013), no. 2, 389–405.

[17] J. Han, C. Jiang, and Y. Luo, Shokurov's conjecture on conic bundles with canonical singularities. 2021, arXiv:2104.15072.

[18] T. Hayakawa, Blowing ups of 3-dimensional terminal singularities. *Publ. Res. Inst. Math. Sci.* **35** (1999), no. 3, 515–570.

[19] T. Hayakawa, Blowing ups of 3-dimensional terminal singularities. II. *Publ. Res. Inst. Math. Sci.* **36** (2000), no. 3, 423–456.

[20] T. Hayakawa, Divisorial contractions to 3-dimensional terminal singularities with discrepancy one. *J. Math. Soc. Japan* **57** (2005), no. 3, 651–668.

[21] A. R. Iano-Fletcher, Working with weighted complete intersections. In *Explicit birational geometry of 3-folds*, pp. 101–173, London Math. Soc. Lecture Note Ser. 281, Cambridge Univ. Press, Cambridge, 2000.

[22] V. A. Iskovskih, Fano threefolds. I. *Izv. Ross. Akad. Nauk Ser. Mat.* **41** (1977), no. 3, 516–562, 717.

[23] V. A. Iskovskih, Fano threefolds. II. *Izv. Ross. Akad. Nauk Ser. Mat.* **42** (1978), no. 3, 506–549.

[24] V. A. Iskovskikh, On a rationality criterion for conic bundles. *Sb. Math.* **187** (1996), no. 7, 1021–1038.

[25] V. A. Iskovskikh and Y. Prokhorov, *Fano varieties. Algebraic geometry V*. Encyclopaedia Math. Sci. 47, Springer, Berlin, 1999.

[26] M. Kawakita, Divisorial contractions in dimension three which contract divisors to smooth points. *Invent. Math.* **145** (2001), no. 1, 105–119.

[27] M. Kawakita, Divisorial contractions in dimension three which contract divisors to compound A_1 points. *Compos. Math.* **133** (2002), no. 1, 95–116.

[28] M. Kawakita, General elephants of three-fold divisorial contractions. *J. Amer. Math. Soc.* **16** (2003), no. 2, 331–362.

[29] M. Kawakita, Three-fold divisorial contractions to singularities of higher indices. *Duke Math. J.* **130** (2005), no. 1, 57–126.

[30] M. Kawakita, Supplement to classification of threefold divisorial contractions. *Nagoya Math. J.* **206** (2012), 67–73.

[31] M. Kawakita, The index of a threefold canonical singularity. *Amer. J. Math.* **137** (2015), no. 1, 271–280.

[32] Y. Kawamata, Crepant blowing-up of 3-dimensional canonical singularities and its application to degenerations of surfaces. *Ann. of Math. (2)* **127** (1988), no. 1, 93–163.

[33] Y. Kawamata, Boundedness of **Q**-Fano threefolds. In *Proceedings of the International Conference on Algebra, Part 3 (Novosibirsk, 1989)*, pp. 439–445, Contemp. Math. 131, Amer. Math. Soc., Providence, RI, 1992.

[34] Y. Kawamata, Divisorial contractions to 3-dimensional terminal quotient singularities. In *Higher-dimensional complex varieties (Trento, 1994)*, pp. 241–246, de Gruyter, Berlin, 1996.

[35] Y. Kawamata, K. Matsuda, and K. Matsuki, Introduction to the minimal model problem. In *Algebraic geometry (Sendai, 1985)*, pp. 283–360, Adv. Stud. Pure Math. 10, North-Holland, Amsterdam, 1987.

[36] K. Kodaira, On compact analytic surfaces. II, III. *Ann. of Math. (2)* **77** (1963), 563–626; ibid. **78** (1963), 1–40

[37] J. Kollár, Real algebraic threefolds. III. Conic bundles. *J. Math. Sci. (N. Y.)* **94** (1999), no. 1, 996–1020.

[38] J. Kollár and S. Mori, Classification of three-dimensional flips. *J. Amer. Math. Soc.* **5** (1992), no. 3, 533–703.

[39] J. Kollár and S. Mori, *Birational geometry of algebraic varieties*. Cambridge Tracts in Math. 134, Cambridge University Press, Cambridge, 1998.

[40] J. Kollár and N. I. Shepherd-Barron, Threefolds and deformations of surface singularities. *Invent. Math.* **91** (1988), no. 2, 299–338.

[41] M. Manetti, Normal degenerations of the complex projective plane. *J. Reine Angew. Math.* **419** (1991), 89–118.

[42] S. Mori, Threefolds whose canonical bundles are not numerically effective. *Ann. of Math. (2)* **116** (1982), 133–176.

[43] S. Mori, On 3-dimensional terminal singularities. *Nagoya Math. J.* **98** (1985), 43–66.

[44] S. Mori, Flip theorem and the existence of minimal models for 3-folds. *J. Amer. Math. Soc.* **1** (1988), no. 1, 117–253.

[45] S. Mori, Errata to: "Classification of three-dimensional flips". [*J. Amer. Math. Soc.* **5** (1992), no. 3, 533–703; mr1149195] by J. Kollár and S. Mori, *J. Amer. Math. Soc.* **20** (2007), no. 1, 269–271.

[46] S. Mori, On semistable extremal neighborhoods. In *Higher dimensional birational geometry (Kyoto, 1997)*, pp. 157–184, Adv. Stud. Pure Math. 35, Math. Soc. Japan, Tokyo, 2002.

[47] S. Mori and Y. Prokhorov, On **Q**-conic bundles. *Publ. Res. Inst. Math. Sci.* **44** (2008), no. 2, 315–369.

[48] S. Mori and Y. Prokhorov, On **Q**-conic bundles. II. *Publ. Res. Inst. Math. Sci.* **44** (2008), no. 3, 955–971.

[49] S. Mori and Y. Prokhorov, Multiple fibers of del Pezzo fibrations. *Proc. Steklov Inst. Math.* **264** (2009), no. 1, 131–145.

[50] S. Mori and Y. Prokhorov, On **Q**-conic bundles, III. *Publ. Res. Inst. Math. Sci.* **45** (2009), no. 3, 787–810.

[51] S. Mori and Y. Prokhorov, Threefold extremal contractions of type IA. *Kyoto J. Math.* **51** (2011), no. 2, 393–438.

[52] S. Mori and Y. Prokhorov, Threefold extremal contractions of types (IC) and (IIB). *Proc. Edinb. Math. Soc.* **57** (2014), no. 1, 231–252.

[53] S. Mori and Y. Prokhorov, Threefold extremal contractions of type (IIA), I. *Izv. Math.* **80** (2016), no. 5, 884–909.

[54] S. Mori and Y. Prokhorov, Threefold extremal contractions of type (IIA), II. In *Geometry and physics: a festschrift in honour of Nigel Hitchin: volume 2*, edited by A. Dancer, J. E. Andersen, and O. Garcia-Prada, Oxford University Press, 2018.

[55] S. Mori and Y. Prokhorov, Threefold extremal curve germs with one non-Gorenstein point. *Izv. Math.* **83** (2019), no. 3, 565–612.

[56] S. Mori and Y. G. Prokhorov, General elephants for threefold extremal contractions with one-dimensional fibres: exceptional case. *Mat. Sb.* **212** (2021), no. 3, 88–111.

[57] Y. Namikawa, Smoothing Fano 3-folds. *J. Algebraic Geom.* **6** (1997), no. 2, 307–324.

[58] Y. Prokhorov, The degree of **Q**-Fano threefolds. *Sb. Math.* **198** (2007), no. 11, 153–174.

[59] Y. Prokhorov, **Q**-Fano threefolds of large Fano index, I. *Doc. Math.* **15** (2010), 843–872.

[60] Y. Prokhorov, A note on degenerations of del Pezzo surfaces. *Ann. Inst. Fourier* **65** (2015), no. 1, 369–388.

[61] Y. Prokhorov, The rationality problem for conic bundles. *Russian Math. Surveys* **73** (2018), no. 3, 375–456.

[62] Y. Prokhorov, Log canonical degenerations of del Pezzo surfaces in Q-Gorenstein families. *Kyoto J. Math.* **59** (2019), no. 4, 1041–1073.

[63] Y. Prokhorov, Equivariant minimal model program. *Russian Math. Surveys* **76** (2021), no. 3, 461–542.

[64] Y. G. Prokhorov, *Lectures on complements on log surfaces*, MSJ Mem. 10, Mathematical Society of Japan, Tokyo, 2001.

[65] M. Reid, Minimal models of canonical 3-folds. In *Algebraic varieties and analytic varieties (Tokyo, 1981)*, pp. 131–180, Adv. Stud. Pure Math. 1, 1981, North-Holland, Amsterdam, 1983.

[66] M. Reid, Young person's guide to canonical singularities. In *Algebraic geometry, Bowdoin, 1985 (Brunswick, Maine, 1985)*, pp. 345–414, Proc. Sympos. Pure Math. 46, Amer. Math. Soc., Providence, RI, 1987.

[67] T. Sano, Deforming elephants of $\mathbb{Q}$-Fano 3-folds. *J. Lond. Math. Soc. (2)* **95** (2017), no. 1, 23–51.

[68] V. Shokurov and S. R. Choi, Geography of log models: theory and applications. *Cent. Eur. J. Math.* **9** (2011), no. 3, 489–534.

[69] V. V. Shokurov, A nonvanishing theorem. *Math. USSR, Izv.* **26** (1986), no. 3, 591–604.

[70] V. V. Shokurov, 3-fold log flips. *Izv. Math.* **40** (1993), no. 1, 95–202.

[71] N. Tziolas, Terminal 3-fold divisorial contractions of a surface to a curve. I. *Compos. Math.* **139** (2003), no. 3, 239–261.

[72] N. Tziolas, Families of D-minimal models and applications to 3-fold divisorial contractions. *Proc. Lond. Math. Soc. (3)* **90** (2005), no. 2, 345–370.

[73] N. Tziolas, Three dimensional divisorial extremal neighborhoods. *Math. Ann.* **333** (2005), no. 2, 315–354.

[74] N. Tziolas, Three-fold divisorial extremal neighborhoods over cE_7 and cE_6 compound DuVal singularities. *Internat. J. Math.* **21** (2010), no. 1, 1–23.

YURI PROKHOROV

Steklov Mathematical Institute, 8 Gubkina street, Moscow 119991, Russia, prokhoro@mi-ras.ru

SOME ASPECTS OF RATIONAL POINTS AND RATIONAL CURVES

OLIVIER WITTENBERG

ABSTRACT

Various methods have been used to construct rational points and rational curves on rationally connected algebraic varieties. We survey recent advances in two of them, the descent and the fibration method, in a number-theoretical context (rational points over number fields) and in an algebro-geometric one (rational curves on real varieties), and discuss the question of rational points over function fields of p-adic curves.

MATHEMATICS SUBJECT CLASSIFICATION 2020

Primary 11G35; Secondary 14G05, 14C25, 14P99

KEYWORDS

Rational points, rational curves, inverse Galois problem, tight approximation, p-adic function fields

ICM INTERNATIONAL CONGRESS OF MATHEMATICIANS 2022 JULY 6–14

Proc. Int. Cong. Math. 2022, Vol. 3, pp. 2346–2368
DOI 10.4171/ICM2022/133

1. INTRODUCTION

Let X be an algebraic variety over a field k and $X(k)$ the set of its rational points.

The search for explicit descriptions of the set $X(k)$ when k is a number field is one of the oldest themes of number theory. A modern point of view on this problem consists in embedding $X(k)$ diagonally into the topological space $X(\mathbf{A}_k)$ of adelic points of X and attempting to identify its topological closure. By general principles that were formulated by Lang after the works of Mordell, Weil, and Siegel, the answer is expected to depend in a crucial manner on the geometry of X. For instance, assuming that X is smooth and projective and that an embedding $k \hookrightarrow \mathbf{C}$ is given, the set $X(k)$ is conjectured to be finite if the complex variety $X_{\mathbf{C}}$ is *hyperbolic* (see [71]). One may then seek to count, list, or bound its elements. At the other end of the spectrum, if $X_{\mathbf{C}}$ is a *rationally connected* smooth projective variety in the sense of Campana [12] and Kollár–Miyaoka–Mori [69], then one expects that the set $X(k)$ is Zariski dense in X whenever it is nonempty; more precisely, by a conjecture of Colliot-Thélène, the closure of $X(k)$ in $X(\mathbf{A}_k)$ should coincide in this case with the Brauer–Manin set $X(\mathbf{A}_k)^{\mathrm{Br}(X)}$ defined by Manin [76]. This far-reaching conjecture encompasses in particular the inverse Galois problem, and its refinement the Grunwald problem (see [28, 33, 36], [95, §3.5]).

Criteria for the existence of rational points on X are also of relevance outside of number theory, when k is no longer assumed to be a number field. For instance, the Graber–Harris–Starr theorem [34], a central result in the theory of rational curves on complex algebraic varieties, is equivalent to the statement that $X(k) \neq \varnothing$ if k is the function field of a complex curve and X is a rationally connected variety. (We say that X is *rationally connected* to mean that for any algebraically closed field extension K of k, the variety X_K over K is rationally connected in the sense of [12, 69].) As another example, if X is a real algebraic variety with no real point and k denotes the function field of the real conic given by $x^2 + y^2 = -1$, the existence of a geometrically rational curve on X—a property conjectured by Kollár to hold whenever X is a positive-dimensional rationally connected variety—is equivalent to the statement that $X(k) \neq \varnothing$.

The results we discuss in this expository article concern the existence of rational points in two very distinct contexts, leading to the following two concrete theorems, obtained in collaboration with Yonatan Harpaz and with Olivier Benoist, respectively. As we shall see, their proofs roughly follow, perhaps somewhat surprisingly, a common general strategy.

Theorem A (see [58]). *Let G be a finite nilpotent group. Let k be a number field.*

(1) *There exist Galois extensions K/k with Galois group G.*

(2) *If $v_1, \dots, v_n$ are pairwise distinct places of k none of which is a finite place dividing the order of G, and $w_1, \dots, w_n$ are places of K above $v_1, \dots, v_n$, then in* (1)*, one can require that the extensions K_{w_i}/k_{v_i} be isomorphic to any prescribed collection of Galois extensions of $k_{v_1}, \dots, k_{v_n}$ whose Galois groups are subgroups of G.*

Theorem B (see [6]). *Let X be a smooth, proper variety over $\mathbf{R}$. Let $\varepsilon : \mathbf{S}^1 \to X(\mathbf{R})$ be a continuous map. Assume that X is birationally equivalent to a homogeneous space of a linear algebraic group over $\mathbf{R}$. Then there exist morphisms of algebraic varieties $\mathbf{P}^1_{\mathbf{R}} \to X$ that induce maps $\mathbf{P}^1(\mathbf{R}) = \mathbf{S}^1 \to X(\mathbf{R})$ arbitrarily close to ε in the compact-open topology.*

Theorem A (1) was first proved by Shafarevich in his seminal work on the inverse Galois problem for solvable groups (see [82, CHAPTER IX, § 6]; it should be noted that nilpotent groups form the most difficult case in his proof); the proof given in [50] is independent from his and has a geometric flavour. Theorem A (2), on the other hand, was new in [50] and was not accessible with Shafarevich's methods.

As far as we know, Theorem B might hold under the sole assumption that X is rationally connected. This is a question we raise in [6]. Theorem B provides the first examples of a positive answer to it for varieties that are not $\mathbf{R}$-rational (indeed, not even $\mathbf{C}$-rational). For $\mathbf{R}$-rational varieties, the conclusion of Theorem B was previously shown, by Bochnak and Kucharz [8], to follow from the Stone–Weierstrass theorem.

The first step in the proofs of Theorems A and B consists in strengthening and reformulating the desired conclusion in terms of the existence of suitable rational points on suitable varieties over suitable fields. In the case of Theorem A, the varieties in question are homogeneous spaces of SL_n over number fields; for the proof, though not for the statement, it is crucial to *not* restrict to homogeneous spaces that have rational points (i.e., to homogeneous spaces of the form SL_n/G). In the case of Theorem B, the varieties in question are homogeneous spaces of linear algebraic groups, over the rational function field $\mathbf{R}(t)$; for the proof, though not for the statement, it is crucial to *not* restrict to homogeneous spaces or algebraic groups that are defined over $\mathbf{R}$. In the remainder of the proofs of Theorems A and B, one establishes the validity of these strengthened formulations by combining geometric *dévissages* of the underlying algebraic varieties with two general tools: the descent method and the fibration method. The fibration method, whose first instance can be found in the work of Hasse on the local–global principle for quadratic forms, consists in reducing the desired property for a variety V endowed with a morphism $p : V \to B$ with geometrically irreducible generic fibre to the same property for B and for a collection of smooth fibres of p. The descent method, which goes back to Fermat, attempts to reduce the desired property for a variety V endowed with a torsor $p : W \to V$ under a (possibly disconnected) linear algebraic group over k to the same property for W and for all of its twists. It was developed in the context of elliptic curves, for torsors under finite abelian groups, by Mordell, Cassels, and Tate, and the setup was later extended to torsors under positive-dimensional linear algebraic groups by Colliot-Thélène and Sansuc, Skorobogatov, Harari.

We take Theorems A and B as excuses leading us to the general study of rational points on rationally connected varieties defined over number fields or over function fields of real curves. We discuss recent advances in the fibration and descent methods in these two contexts in Sections 2 and 3, stating along the way the main open questions that surround Theorems A and B and their proofs. We then turn, in Section 4, to function fields of p-adic curves, and speculate about the existence of a p-adic analogue of the "tight approximation" property discussed in Section 3 that would enable one to exploit fibration and descent methods in the study of rational curves over p-adic fields and more generally of rational points over function fields of p-adic curves.

2. SOLVABLE GROUPS AND THE GRUNWALD PROBLEM IN INVERSE GALOIS THEORY

2.1. Homogeneous spaces

It is the following general theorem about the arithmetic of homogeneous spaces of linear algebraic groups that underlies Theorem A.

Theorem 2.1. *Let V be a homogeneous space of a connected linear algebraic group L over a number field k. Let X be a smooth compactification of V. Let $\bar{v} \in V(\bar{k})$. Assume that the group of connected components G of the stabiliser of $\bar{v}$ is supersolvable, in the sense that it possesses a normal series $1 = G_0 \lhd \cdots \lhd G_m = G$ such that the quotients G_{i+1}/G_i are cyclic while the subgroups G_i are normal in G and are stable under the natural outer action of $\mathrm{Gal}(\bar{k}/k)$ on G. Then the subset $X(k)$ is dense in $X(\mathbf{A}_k)^{\mathrm{Br}(X)}$.*

Here and elsewhere, by "compactification of V," we mean a proper variety over k that contains V as a dense open subset; we do not require that the algebraic group L act on the compactification. Examples of supersolvable groups with respect to the trivial outer action of $\mathrm{Gal}(\bar{k}/k)$ include finite nilpotent groups and dihedral groups. With a nontrivial outer action of $\mathrm{Gal}(\bar{k}/k)$, however, even abelian groups need not be supersolvable. Previous work of Borovoi [10] nevertheless establishes the conclusion of Theorem 2.1 in many cases where the stabiliser of $\bar{v}$ is abelian but not necessarily supersolvable.

Theorem 2.1 can be found in [50, THÉORÈME B] in the particular case where L is semi-simple simply connected and the stabiliser of $\bar{v}$ is finite, and in [51, COROLLARY 4.5] in general. To deduce Theorem A from it, embed G into $\mathrm{SL}_n(k)$ for some n, take $L = \mathrm{SL}_n$ and $V = \mathrm{SL}_n/G$ and let H denote the set of points of V above which the fibre of the étale cover $\pi : L \to V$ is irreducible. The function field of the fibre of π above any rational point contained in H is a Galois extension of k with Galois group G. On the other hand, by a theorem of Ekedahl [33], the density of $X(k)$ in $X(\mathbf{A}_k)^{\mathrm{Br}(X)}$ implies that of $X(k) \cap H$ in $X(\mathbf{A}_k)^{\mathrm{Br}(X)}$. Thus, Theorem 2.1 ensures the existence of Galois extensions K/k with Galois group G having a local behaviour prescribed by any element of the Brauer–Manin set $X(\mathbf{A}_k)^{\mathrm{Br}(X)}$; that is, one may freely prescribe the completions of K at any finite set of places of k, as long as these prescriptions satisfy a certain global reciprocity condition determined by $\mathrm{Br}(X)$. By a theorem of Lucchini Arteche [75, § 6], this reciprocity condition imposes, in fact, no restriction at the places indicated in Theorem A (2).

2.2. Geometry

In the special case where $L = \mathrm{SL}_n$ and the stabiliser of $\bar{v}$ is a finite group G, the geometry behind the proof of Theorem 2.1 can be summarised with the following assertion:

there exist an algebraic torus T over k and a torsor $\bar{Y} \to X_{\bar{k}}$ under $T_{\bar{k}}$ whose isomorphism class is invariant under $\mathrm{Gal}(\bar{k}/k)$, such that for any torsor $Y \to X$ under T whose base change to $X_{\bar{k}}$ is isomorphic to $\bar{Y}$, there exist a dense open subset $W \subseteq Y$ and a smooth morphism $p : W \to Q$ to a quasitrivial torus Q (i.e., a torus of the form $R_{E/k}\mathbf{G}_{\mathrm{m}}$ for a nonzero étale k-algebra E) whose fibres are homogeneous spaces of SL_n with geometric stabiliser isomorphic to G_{m-1}. In addition, the morphism p admits a rational section over $\bar{k}$.

This geometry is the key to a proof of Theorem 2.1 by an induction on m, at each step of which one applies the descent method and the fibration method, in the form of Theorems 2.2 and 2.3 below. It should be noted that even if G is embedded into $\mathrm{SL}_n(k)$ and $V = \mathrm{SL}_n/G$, the homogeneous spaces of SL_n that arise as fibres of p need not possess rational points. Thus, for the induction to be possible, one cannot restrict the statement of Theorem 2.1 to homogeneous spaces of the form SL_n/G, even though only homogeneous spaces of this form are relevant for Theorem A.

2.3. Descent

The following theorem, which was established in [50] and can also be deduced from [13], is the definitive statement of descent theory in the case of smooth and proper rationally connected varieties over number fields. For geometrically rational X, this theorem is due to Colliot-Thélène and Sansuc [25]. The homogeneous spaces of Theorem 2.1 are not geometrically rational in general (Saltman, Bogomolov; see [26]).

Theorem 2.2. *Let X be a smooth and proper rationally connected variety over a number field k. Let T be a torus over k and $\bar{Y} \to X_{\bar{k}}$ a torsor under $T_{\bar{k}}$ whose isomorphism class is invariant under* $\mathrm{Gal}(\bar{k}/k)$. *Then*

$$X(\mathbf{A}_k)^{\mathrm{Br}(X)} = \bigcup_{f:Y\to X} f'\big(Y'(\mathbf{A}_k)^{\mathrm{Br}(Y')}\big),$$

where the union ranges over the torsors $f : Y \to X$ under T whose base change to $X_{\bar{k}}$ is isomorphic to $\bar{Y}$, and Y' denotes a smooth compactification of Y such that f extends to a morphism $f' : Y' \to X$. In particular, if $Y'(k)$ is dense in $Y'(\mathbf{A}_k)^{\mathrm{Br}(Y')}$ for every such f, then $X(k)$ is dense in $X(\mathbf{A}_k)^{\mathrm{Br}(X)}$.

(To bridge the gap between Theorem 2.2 and [50, THÉORÈME 2.1], one needs to know that $X(\mathbf{A}_k)^{\mathrm{Br}(X)} \neq \varnothing$ implies the existence of at least one f. This goes back to [25] and follows from [102, THEOREM 3.3.1], [25, PROPOSITION 2.2.5], [103, (3.3)].)

2.4. Fibration

The following fibration theorem suffices for the proof of Theorem 2.1. It results from combining a descent with the work of Harari [35] on the fibration method.

Theorem 2.3. *Let $p : Z \to B$ be a dominant morphism between irreducible, smooth, and proper varieties over a number field k, with rationally connected generic fibre. Assume that*

(1) *there exist dense open subsets $W \subset Z$ and $Q \subset B$ such that Q is a quasitrivial torus over k and p induces a smooth morphism $W \to Q$ with geometrically irreducible fibres;*

(2) *the morphism p admits a rational section over $\bar{k}$;*

(3) *for all $b \in B(k)$ in a dense open subset of B, the set $Z_b(k)$ is dense in $Z_b(\mathbf{A}_k)^{\mathrm{Br}(Z_b)}$.*

Then $Z(k)$ is dense in $Z(\mathbf{A}_k)^{\mathrm{Br}(Z)}$.

The assumptions of Theorem 2.3 imply that B is k-rational. Under the condition that B is k-rational, the first two assumptions of Theorem 2.3 are expected to be superfluous (even under weaker hypotheses on the generic fibre of p than rational connectedness, see [48, COROLLARY 9.23 (1)–(2)]), but removing them altogether is a wide-open problem, well connected with analytic number theory (see [48, § 9], [47]). Removing (2) while keeping (1) might be within reach, though:

Question 2.4. In the statement of Theorem 2.3, can one dispense with the assumption that p admit a rational section over $\bar{k}$?

This would allow one to replace "supersolvable" with "solvable" in the statement of Theorem 2.1. Indeed, in Section 2.2, the cyclicity of the quotient G_m/G_{m-1} plays a rôle only to ensure the existence of a rational section of p over $\bar{k}$ (see [50, PROPOSITION 3.3 (II)]).

2.5. An application to Massey products

Theorem 2.1 has concrete applications, over number fields, beyond the inverse Galois problem: for the homogeneous spaces that appear in its statement, it turns the problem of deciding the existence of a rational point into the much more approachable question of deciding the non-vacuity of the Brauer–Manin set. In this way, Theorem 2.1 can be used to confirm, in the case of number fields, the conjecture of Mináč and Tân on the vanishing of Massey products in Galois cohomology (see [49]). Indeed, this conjecture—which posits that for any field k, any prime number p, any integer $m \geq 3$ and any classes $a_1, \dots, a_m \in H^1(k, \mathbf{Z}/p\mathbf{Z})$, the m-fold Massey product of $a_1, \dots, a_m$ vanishes if it is defined (see [78, 79])—can be reinterpreted, according to Pál and Schlank [83], in terms of the existence of rational points on appropriate homogeneous spaces of SL_n over k (with $n \gg 0$), and it so happens that the geometric stabilisers of these homogeneous spaces are finite and supersolvable.

3. RATIONAL CURVES ON REAL ALGEBRAIC VARIETIES

3.1. A few questions

Let X be a smooth variety over $\mathbf{R}$. The interplay between the topology of the $\mathscr{C}^\infty$ manifold $X(\mathbf{R})$ and the geometry of the algebraic variety X lies at the core of classical real algebraic geometry. One of the fundamental problems in this area consists in investigating which submanifolds of $X(\mathbf{R})$ can be approximated, in the Euclidean topology, by Zariski closed submanifolds. Even for 1-dimensional submanifolds, i.e., disjoint unions of $\mathscr{C}^\infty$ loops, various phenomena—of a topological, Hodge-theoretic, or yet more subtle nature—can obstruct the existence of algebraic approximations (see [4, § 4]). In the case of 1-dimensional submanifolds, however, all known obstructions vanish when X is rationally connected. One can thus raise the following questions, in which $H_1^{\mathrm{alg}}(X(\mathbf{R}), \mathbf{Z}/2\mathbf{Z})$ denotes the image of the cycle class map $\mathrm{CH}_1(X) \to H_1(X(\mathbf{R}), \mathbf{Z}/2\mathbf{Z})$ defined by Borel and Haefliger [9].

Questions 3.1. Let X be a smooth, proper, rationally connected variety, over $\mathbf{R}$.

(1) Can all $\mathscr{C}^\infty$ loops in $X(\mathbf{R})$ be approximated, in the Euclidean topology, by real loci of algebraic curves? or even by real loci of rational algebraic curves?

(2) Is $H_1(X(\mathbf{R}), \mathbf{Z}/2\mathbf{Z}) = H_1^{\mathrm{alg}}(X(\mathbf{R}), \mathbf{Z}/2\mathbf{Z})$? Is $H_1(X(\mathbf{R}), \mathbf{Z}/2\mathbf{Z})$ generated by classes of rational algebraic curves on X?

The first parts of Questions 3.1 (1) and (2) are in fact equivalent to each other, by the work of Akbulut and King (see [5, THEOREM 6.8]), and were studied in a systematic fashion in [4, 5]. The second part of Question 3.1 (1) is, however, as far as we know, genuinely stronger than the second part of Question 3.1 (2). We note that in order to formulate the second part of Question 3.1 (1) precisely, it is better to work with possibly noninjective $\mathscr{C}^\infty$ maps $\mathbf{P}^1(\mathbf{R}) \to X(\mathbf{R})$ rather than with submanifolds of $X(\mathbf{R})$. Indeed, there are examples of $\mathbf{R}$-rational surfaces X and of $\mathscr{C}^\infty$ loops in $X(\mathbf{R})$ such that the desired rational algebraic curves necessarily have singular real points (see [68, THEOREM 3]).

A specific motivation for Question 3.1 (2) is its analogy with the following questions in complex geometry raised by Voisin [101] and by Kollár [67]:

Questions 3.2. Let X be a smooth, proper, rationally connected variety, over $\mathbf{C}$. Is the group $H_2(X(\mathbf{C}), \mathbf{Z})$ generated by homology classes of algebraic curves? Is it generated by homology classes of rational algebraic curves?

The two parts of Questions 3.2 are in fact equivalent: Tian and Zong [100] have shown that the homology class of any algebraic curve on a rationally connected variety over $\mathbf{C}$ is a linear combination of homology classes of rational curves. The real analogue of their result remains unknown in general. Its validity is an interesting open problem.

The first parts of Questions 3.1 (2) and of Questions 3.2 are in fact related by more than an analogy: if X is a smooth, proper, rationally connected variety over $\mathbf{R}$ such that $X(\mathbf{R}) \neq \varnothing$ and such that Questions 3.2 admit a positive answer for $X_{\mathbf{C}}$, then the equality $H_1(X(\mathbf{R}), \mathbf{Z}/2\mathbf{Z}) = H_1^{\mathrm{alg}}(X(\mathbf{R}), \mathbf{Z}/2\mathbf{Z})$ is equivalent to the *real integral Hodge conjecture* for 1-cycles on X, a property formulated and studied in [4, 5].

In a different line of investigation around the abundance of rational curves on rationally connected varieties, many authors have considered the problem of finding rational curves through a prescribed set of points, or more generally through a prescribed curvilinear

0-dimensional subscheme, on any smooth, proper, rationally connected variety X. Over the complex numbers, such curves exist unconditionally (Kollár, Miyaoka, Mori, see [63, CHAPTER IV.3]). Over the real numbers, such curves exist under the necessary condition that all the prescribed points that are real belong to the same connected component of $X(\mathbf{R})$ (Kollár, see [64,66]). This problem can be generalised to one-parameter families: given a morphism $f : \mathscr{X} \to B$ with rationally connected generic fibre between smooth and proper varieties, where B is a curve, one looks for sections of f whose restriction to a given 0-dimensional subscheme of B is prescribed, thus leading to Questions 3.3 below. For simplicity of notation, in the statement of Question 3.3 (2), this 0-dimensional subscheme of B is assumed to be reduced; there is, however, no loss of generality in doing this, since jets of sections can be prescribed at any higher order by replacing $\mathscr{X}$ with a suitable iterated blow-up (see [52, PROPOSITION 1.4]).

Questions 3.3. Let B be a smooth, proper, connected curve over a field k_0. Let $\mathscr{X}$ be a smooth, proper variety over k_0, endowed with a flat morphism $f : \mathscr{X} \to B$ with rationally connected generic fibre. Let $P \subset B$ be a reduced 0-dimensional subscheme. Let $s : P \to \mathscr{X}$ be a section of f over P.

(1) If $k_0 = \mathbf{C}$, can s be extended to a section of f?

(2) If $k_0 = \mathbf{R}$ and the map $s|_{P(\mathbf{R})} : P(\mathbf{R}) \to \mathscr{X}(\mathbf{R})$ can be extended to a $\mathscr{C}^\infty$ section of $f|_{\mathscr{X}(\mathbf{R})} : \mathscr{X}(\mathbf{R}) \to B(\mathbf{R})$, can then s be extended to a section of f?

Let X be the generic fibre of f and k the function field of B. The existence of sections extending any given s as above is equivalent to the density of $X(k)$ in the topological space $X(\mathbf{A}_k) = \prod_b X(k_b)$ of adelic points of X, where the product runs over the closed points b of B and k_b denotes the completion of k at b. This is the *weak approximation* property.

The Graber–Harris–Starr theorem [34] provides a positive answer to Question 3.3 (1) when $P = \varnothing$ and it is a conjecture of Hassett and Tschinkel that the answer to this question is in the affirmative in general (see [17,52,53,99] for known results). Particular cases of Question 3.3 (2) were first studied by Colliot-Thélène [14], who conjectured the validity of weak approximation (i.e., a positive answer to Question 3.3 (2) even without assuming that $s|_{P(\mathbf{R})}$ can be extended to a $\mathscr{C}^\infty$ section of $f|_{\mathscr{X}(\mathbf{R})}$) when X is birationally equivalent to a homogeneous space of a connected linear algebraic group over k, and proved his conjecture when the geometric stabilisers are trivial. Scheiderer [94] then proved the same conjecture when the geometric stabilisers are connected. Ducros [30,31] stated Question 3.3 (2) in these exact terms, and gave a positive answer when X is a conic bundle surface, or more generally when there exists a dominant map $X \to \mathbf{P}^1_k$ whose generic fibre is a Severi–Brauer variety.

3.2. Tight approximation

The main insight behind the proof of Theorem B is the observation that formulating a suitable common strengthening of Questions 3.1 and 3.3, through the notion of *tight approximation*, can render all of these questions fully amenable to both the descent method and the fibration method. We note that Questions 3.1 and Questions 3.3 are somewhat orthogonal in spirit, insofar as the former consider global constraints on curves lying on X, while the latter are aimed at local constraints.

The idea of establishing a descent method (resp. fibration method) for Question 3.3 (2) already appeared in [31] (resp. [84]), though in [31] and [84] the implementations are subject to miscellaneous restrictions. The possibility of a descent method and a fibration method for studying Questions 3.1, however, is new and turns out to require a shift in perspective from single rationally connected varieties to one-parameter families of such.

Let us illustrate how Questions 3.1 need to be strengthened for a fibration argument to go through. We start with a dominant morphism $p : X \to Y$ with rationally connected generic fibre between smooth, proper, rationally connected varieties, over $\mathbf{R}$, and a $\mathscr{C}^\infty$ loop $\gamma : \mathbf{S}^1 \to X(\mathbf{R})$ that we want to approximate, in the Euclidean topology, by a Zariski closed submanifold of X, assuming that we can solve the same problem on Y as well as on the fibres of p. By assumption, we can approximate $p \circ \gamma : \mathbf{S}^1 \to Y(\mathbf{R})$ by a $\mathscr{C}^\infty$ map $\xi : \mathbf{S}^1 \to Y(\mathbf{R})$ with Zariski closed image. The best we can hope to find, then, is a $\mathscr{C}^\infty$ loop $\tilde{\gamma} : \mathbf{S}^1 \to X(\mathbf{R})$ arbitrarily close to γ and such that $p \circ \tilde{\gamma} = \xi$. We draw two conclusions:

(1) If such a $\tilde{\gamma}$ exists, the next and final step is not finding an algebraic approximation for a $\mathscr{C}^\infty$ loop in a fibre of p, but, rather, considering the algebraic curve B underlying $\xi(\mathbf{S}^1)$, viewing $\tilde{\gamma}$ as a $\mathscr{C}^\infty$ section of the projection

$$(X \times_Y B)(\mathbf{R}) \to B(\mathbf{R}),$$

and looking for an algebraic section of $X \times_Y B \to B$ approximating $\tilde{\gamma}$. Thus, even when we start with just two real varieties X and Y, we need to consider one-parameter algebraic families of fibres of p, rather than single fibres.

(2) Consider the example where p is the blow-up of a surface Y at a real point b and γ meets $p^{-1}(b)(\mathbf{R})$, transversally. Then for any $\tilde{\gamma}$ sufficiently close to γ in the Euclidean topology, the loop $p \circ \tilde{\gamma}$ has to go through b. Hence ξ has to be required to go through b for a loop $\tilde{\gamma}$ as above to exist. Thus, a condition of weak approximation type must be considered in conjunction with Questions 3.1 (as was already noted by Bochnak and Kucharz [8]).

Let us now similarly contemplate a fibration argument in the context of Question 3.3 (2). We assume that $\mathscr{X} \xrightarrow{f} B$ can be factored as $\mathscr{X} \xrightarrow{p} \mathscr{Y} \xrightarrow{g} B$, where the variety $\mathscr{Y}$ is smooth and proper over $\mathbf{R}$, the morphism p is dominant with rationally connected generic fibre, and g is flat. Starting from a section $s : P \to \mathscr{X}$ of f over P such that $s|_{P(\mathbf{R})}$ can be extended to a $\mathscr{C}^\infty$ section s' of $f|_{\mathscr{X}(\mathbf{R})}$, a positive answer to Question 3.3 (2) for g produces for us a section τ of g that extends $p \circ s$. Let $\mathscr{Z} = p^{-1}(\tau(B))$ and let $h : \mathscr{Z} \to B$ denote the restriction of f. At this point, one would like to apply a positive answer to Question 3.3 (2) for h to obtain a section of h extending s, thus completing the argument, as $\mathscr{Z} \subseteq \mathscr{X}$. In order to do so, one needs to know that $s|_{P(\mathbf{R})} : P(\mathbf{R}) \to \mathscr{Z}(\mathbf{R})$ can be extended to a $\mathscr{C}^\infty$ section of $h|_{\mathscr{Z}(\mathbf{R})} : \mathscr{Z}(\mathbf{R}) \to B(\mathbf{R})$. However, the map $h|_{\mathscr{Z}(\mathbf{R})}$ in general even fails to be surjective. To correct this problem, one should require, at the very least, that $\tau(B(\mathbf{R}))$ approximate, in the Euclidean topology, the image of $p \circ s' : B(\mathbf{R}) \to \mathscr{Y}(\mathbf{R})$. Thus, all in all, an approximation condition in the Euclidean topology has to be considered in conjunction with Question 3.3 (2).

The above discussion leads to the following definition. (This definition slightly differs from that given in [6], which considers the more general question of approximating holomorphic maps by algebraic ones, à la Runge, and which, as a consequence, is useful also for studying complex curves on complex varieties, without reference to the reals; however, all of the statements we make below are true with respect to either of the definitions.)

Definition 3.4. Let B be a smooth, proper, connected curve over $\mathbf{R}$. A variety X over $k = \mathbf{R}(B)$ satisfies the *tight approximation* property if for any proper model $f : \mathscr{X} \to B$ of X over B with $\mathscr{X}$ smooth over $\mathbf{R}$, any reduced 0-dimensional subscheme $P \subset B$, any section $s' : P \to \mathscr{X}$ of f over P and any $\mathscr{C}^\infty$ section $s : B(\mathbf{R}) \to \mathscr{X}(\mathbf{R})$ of $f|_{\mathscr{X}(\mathbf{R})}$ such that $s|_{P(\mathbf{R})} = s'|_{P(\mathbf{R})}$, there exists a section $\sigma : B \to \mathscr{X}$ of f such that $\sigma|_P = s'|_P$ and such that $\sigma|_{B(\mathbf{R})}$ lies arbitrarily close to s in the compact-open topology.

Given a smooth, proper, rationally connected variety X over $\mathbf{R}$, the validity of the tight approximation property for the variety obtained from X by extension of scalars from $\mathbf{R}$ to $\mathbf{R}(t)$ implies positive answers to Questions 3.1 for X.

The tight approximation property is (tautologically) a birational invariant, and it holds for $\mathbf{P}^n_k$ by a theorem of Bochnak and Kucharz [8]. (In [8], weak approximation conditions at complex points are ignored, but they create no additional difficulty.) The next two results provide more examples of varieties satisfying tight approximation.

3.3. Descent

The following theorem implements the descent method for the tight approximation property, in full generality (including non-abelian descent, as formalised by Harari and Skorobogatov). Its proof, given in [6], builds on the work of Scheiderer [94] and, in the case where G is finite, on an argument of Colliot-Thélène and Gille [17].

Theorem 3.5. *Let k be the function field of a real curve. Let X be a smooth variety over k. Let G be a linear algebraic group over k. Let $f : Y \to X$ be a left torsor under G. Consider twists $f' : Y' \to X$ of f by right torsors under G, over k. If every such Y' satisfies the tight approximation property, then so does X.*

3.4. Fibration

The next theorem implements the fibration method for the tight approximation property, in full generality. Its proof, contained in [6], makes essential use of the weak toroidalisation theorem of Abramovich, Denef, and Karu [1] to establish a version of the Néron smoothening process (as in [11, 3.1/3]) for higher-dimensional bases—the point being that in the discussion at the beginning of Section 3.2, the loop $\tilde{\gamma}$ is easily seen to exist once the morphism p is smooth along γ (see [5, LEMMA 6.11]).

Theorem 3.6. *Let k be the function field of a real curve. Let $p : Z \to B$ be a dominant morphism between smooth varieties over k. If B and the fibres of p above the rational points of a dense open subset of B satisfy the tight approximation property, then so does Z.*

3.5. Homogeneous spaces

We are now in a position to sketch the proof of the following theorem, which in the "constant case," i.e., when the algebraic group and the homogeneous space are both defined over **R**, immediately implies Theorem B.

Theorem 3.7. *Homogeneous spaces of connected linear algebraic groups over the function field of a real curve satisfy the tight approximation property.*

The proof of Theorem 3.7 starts by noting that quasitrivial tori over k are k-rational, hence satisfy the tight approximation property (since so does $\mathbf{P}^n_k$). Any torus T can be inserted into an exact sequence $1 \to S \to Q \to T \to 1$ where S is a torus and Q is a quasitrivial torus. As any twist of Q as a torsor remains isomorphic to Q (Hilbert's Theorem 90) and hence satisfies the tight approximation property, we deduce, by the descent method (Theorem 3.5), that all tori over k satisfy the tight approximation property. Next, as every connected linear algebraic group over k is birationally equivalent to a relative torus over a k-rational variety (namely over the variety of maximal tori, when the algebraic group is reductive), we deduce, by the fibration method (Theorem 3.6), that connected linear algebraic groups over k satisfy the tight approximation property. By descent (Theorem 3.5 again), it follows that homogeneous spaces of connected linear algebraic groups over k satisfy the tight approximation property when they have a rational point. Finally, it is a theorem of Scheiderer that homogeneous spaces of connected linear algebraic groups over k satisfy the Hasse principle with respect to the real closures of k, so that if X denotes such a homogeneous space, then $X(k) \neq \varnothing$ whenever a $\mathscr{C}^\infty$ section $s : B(\mathbf{R}) \to \mathscr{X}(\mathbf{R})$ as in Definition 3.4 exists. This completes the proof of Theorem 3.7.

3.6. Further comments

Theorem 3.7 implies that homogeneous spaces of connected linear algebraic groups over the function field of a real curve satisfy weak approximation, as conjectured by Colliot-Thélène. Indeed, in the notation of Definition 3.4, if X is such a homogeneous space and P contains the locus of singular fibres of f, Scheiderer's work implies that $f^{-1}(b)(\mathbf{R})$ is nonempty and connected for all $b \in B(\mathbf{R}) \setminus P(\mathbf{R})$, so that a $\mathscr{C}^\infty$ section $s : B(\mathbf{R}) \to \mathscr{X}(\mathbf{R})$ with $s|_{P(\mathbf{R})} = s'|_{P(\mathbf{R})}$ always exists.

The main open problem surrounding the notion of tight approximation is the following.

Question 3.8. Let k be the function field of a real curve. Do all rationally connected varieties over k satisfy the tight approximation property?

Building on Theorems 3.5 and 3.6, the tight approximation property is shown in [6] to hold for various classes of rationally connected varieties beyond homogeneous spaces of connected linear algebraic groups. For instance, it holds for smooth cubic hypersurfaces of dimension ≥ 2 that are defined over $\mathbf{R}$, thus yielding, for such hypersurfaces, a positive answer to (the second part of) Question 3.1 (1).

Question 3.8 is open for cubic surfaces over k. Even Question 3.3 (2) is open when X is a cubic surface, although Question 3.3 (1) has an affirmative answer in this case, by a theorem of Tian [99].

In another direction, Question 3.8 is open for surfaces defined over $\mathbf{R}$, and so is (the second part of) Question 3.1 (1). By inspecting the birational classification of geometrically rational surfaces and using the fibration method (Theorem 3.6), one can see that a positive answer to these questions for surfaces defined over $\mathbf{R}$ would follow from a positive answer for del Pezzo surfaces of degree 1 or 2 defined over $\mathbf{R}$. In these cases, it would suffice, by an application of the descent method (Theorem 3.5), to know that for any real del Pezzo surface X of degree 1 or 2, the universal torsors of X, in the sense of Colliot-Thélène and Sansuc [25], are $\mathbf{R}$-rational whenever they have a real point. This last question, unfortunately, is very much open—even the unirationality of real del Pezzo surfaces of degree 1 is unknown. In fact, not a single example of a minimal real del Pezzo surface of degree 1 is known to be unirational. For a description of these surfaces, see [93, § 5].

Naturally, one hopes for the answer to Question 3.8 to be in the affirmative in general. This conjecture would have a host of interesting consequences, among which: a version of the Graber–Harris–Starr theorem over the reals (i.e., a positive answer to Question 3.3 (2) when $P = \varnothing$); Lang's widely open conjecture from [70] that the function field of a real curve with no real point is C_1 (see [55, COROLLARY 1.5] for the implication); and the existence of a geometrically rational curve on any smooth, proper, rationally connected variety of dimension ≥ 1 over $\mathbf{R}$.

This last consequence is a conjecture of Kollár, who showed the existence of rational curves on those real rationally connected varieties of dimension ≥ 1 that have real points (see [2, REMARKS 20]). For real rationally connected varieties with no real point, it is interesting to consider a weaker property: the existence of a geometrically irreducible curve of even geometric genus. The latter can be reinterpreted in terms of the real integral Hodge conjecture (see [4]). Using Hodge theory and a real adaptation of Green's infinitesimal criterion for the density of Noether–Lefschetz loci, such curves of even genus can be shown to exist on all real Fano threefolds (see [5]). However, even on smooth quartic hypersurfaces in $\mathbf{P}^4_{\mathbf{R}}$, the existence of geometrically rational curves remains a challenge, as well as the mere existence of an absolute bound, independent of the chosen quartic hypersurface, on the minimal geometric genus of a geometrically irreducible curve of even geometric genus lying on such a hypersurface.

4. FUNCTION FIELDS OF CURVES OVER p-ADIC FIELDS

4.1. Some motivation: rational curves over number fields

Even though the main questions about rational points of rationally connected varieties over number fields and over function fields of real curves are still wide open, the

Brauer–Manin obstruction and the tight approximation property at least provide rather satisfactory conjectural answers. It would be highly desirable to obtain a similar conjectural picture for rational points over other fields, for significant classes of varieties—including, at a minimum, concrete criteria for the existence of rational points.

Over the field $\mathbf{Q}(t)$, this would encompass questions about rational curves on rationally connected varieties over $\mathbf{Q}$, about which very little is known. For example, it is unknown whether any rationally connected variety of dimension ≥ 1 over $\mathbf{Q}$ that possesses a rational point also contains a rational curve defined over $\mathbf{Q}$. Much more ambitiously, it is unknown whether any such variety contains enough rational curves to imply the finiteness of the set of R-equivalence classes of rational points, a question asked in [16, QUESTION 10.12]. (Known results on this problem are listed after Question 10.12 in [16].) As another example, the regular inverse Galois problem over $\mathbf{Q}$, which asks for the construction of a regular Galois extension of $\mathbf{Q}(t)$ with specified Galois group, and which can be reinterpreted as a problem about the existence of appropriate rational curves on the homogeneous space SL_n/G over $\mathbf{Q}$, is open even for finite nilpotent groups G. All of these problems are currently out of reach.

As a first step towards these questions, let us replace $\mathbf{Q}$ with its completions and turn to rational points over the field $\mathbf{Q}_p(t)$ or over its finite extensions.

4.2. Rational curves on varieties over p-adic fields

In the constant case (that is, for varieties obtained by scalar extension from varieties defined over a p-adic field, i.e., a finite extension of $\mathbf{Q}_p$), various existence results are known:

(1) the regular inverse Galois problem over $\mathbf{Q}_p$ has a positive solution (first proved by Harbater [39], by "formal patching"; reproved and generalised in different directions by Pop [91] and by Colliot-Thélène [15]; see also [65, 74, 80]);

(2) for any smooth, proper, rationally connected variety X over a p-adic field k, Kollár [64, 66] has shown that the rational points of X fall into finitely many R-equivalence classes, and that there exist rational curves on X, defined over k, passing through any finite set of rational points of X that belong to the same R-equivalence class (with prescribed jets of any given order at these points).

This last statement concerns conditions of weak approximation type that can be imposed on rational curves on rationally connected varieties over p-adic fields. It would be interesting to formulate an analogue, in this p-adic context, of the surjectivity of the Borel–Haefliger cycle class map $\mathrm{CH}_1(X) \to H_1(X(\mathbf{R}), \mathbf{Z}/2\mathbf{Z})$ (i.e., of Questions 3.1 (2)).

We saw in Section 3 that in order to answer questions about homology classes of rational curves on real varieties, it can be useful to consider more generally the tight approximation property, for nonconstant varieties over the function field of a real curve. By analogy, this gives incentive to investigate the possibility of a p-adic analogue of the tight approximation property for nonconstant varieties over the function field of a curve over a p-adic field, the validity of which would have consequences for a likely easier to formulate *p-adic integral Hodge conjecture* for 1-cycles on varieties over p-adic fields.

4.3. Quadrics and other homogeneous spaces

In the nonconstant case, even the simplest varieties over $\mathbf{Q}_p(t)$ lead to difficult problems when it comes to their rational points. For instance, it is only a relatively recent theorem of Parimala and Suresh [86], for $p \neq 2$, and of Leep [72], based on work of Heath-Brown [54], for arbitrary p, that every projective quadric of dimension ≥ 7 over $\mathbf{Q}_p(t)$ possesses a rational point. (In the language of quadratic forms, "the u-invariant of $\mathbf{Q}_p(t)$ is equal to 8.") Many other articles have been devoted to local–global principles for varieties over function fields of curves over p-adic fields (e.g., [19–21, 23, 24, 37, 38, 40–46, 56–58, 77, 85, 87, 88, 92, 98]).

A patching technique was developed by Harbater, Hartmann and Krashen ("patching over fields," a successor to formal patching), and was applied to study rational points of homogeneous spaces over such fields. It was used, in [42], to give another proof of the aforementioned theorem of Parimala and Suresh, and, in [23], to establish, more generally, the local–global principle for the existence of rational points on smooth projective quadrics of dimension ≥ 1 over $\mathbf{Q}_p(t)$ (or over a finite extension of $\mathbf{Q}_p(t)$), with respect to all discrete valuations on this field, when p is odd.

4.4. Reciprocity obstructions

Let k be a finite extension of $\mathbf{Q}_p(t)$. Let Ω denote the set of equivalence classes of discrete valuations (of rank 1) on k and, for $v \in \Omega$, let k_v denote the completion of k at v. Let X be an irreducible, smooth and proper variety over k. We embed $X(k)$ diagonally into the product topological space $\prod_{v\in\Omega} X(k_v)$, which we shall also denote $X(\mathbf{A}_k)$ (recall that X is proper).

We now explain how, building on the work of Bloch–Ogus and of Kato, an analogue of the Brauer–Manin obstruction can be set up in this context. These ideas, which are due to Colliot-Thélène, appear in print, and are put to use, in [24, § 2.3], in a very slightly different (equicharacteristic) situation. We refer the reader to [24, § 2.3] for more details. (The "reciprocity obstructions" of [37, § 4] are weaker than those we discuss here.)

Our goal is thus to define, in complete generality, a closed subset $X(\mathbf{A}_k)^{\mathrm{rec}} \subseteq X(\mathbf{A}_k)$ containing $X(k)$, using on the one hand a reciprocity law coming from k and on the other hand an analogue of the Brauer group of X.

Grothendieck's purity theorem for the Brauer group equates $\mathrm{Br}(X)$ with the unramified cohomology group $H^2_{\mathrm{nr}}(X/k, \mathbf{Q}/\mathbf{Z}(1))$. We recall the definition of unramified cohomology: for any irreducible smooth variety V over a field K of characteristic 0 and any torsion Galois module M over K, the group $H^q_{\mathrm{nr}}(V/K, M)$ is the subgroup of the Galois cohomology group $H^q(K(V), M)$ consisting of those classes whose residues along all codimension 1 points of V vanish. It is the unramified cohomology group $H^3_{\mathrm{nr}}(X/k, \mathbf{Q}/\mathbf{Z}(2))$ that will serve as a substitute for $\mathrm{Br}(X)$ here. (The shift in degree is explained by the fact that the field k has cohomological dimension 3 while number fields have virtual cohomological dimension 2.) For any field extension K/k, Bloch–Ogus theory provides an evaluation map $H^3_{\mathrm{nr}}(X/k, \mathbf{Q}/\mathbf{Z}(2)) \to H^3(K, \mathbf{Q}/\mathbf{Z}(2))$, $\alpha \mapsto \alpha(x)$ along any K-point x of X (see [7]).

Let $\mathscr{B}$ denote an irreducible normal proper scheme over $\mathbf{Z}_p$ with function field k. In contrast with what happens over number fields, here it is not one reciprocity law that will play a rôle, but infinitely many of them: one for each closed point of $\mathscr{B}$, for each such $\mathscr{B}$. Namely, given any closed point $b \in \mathscr{B}$, Kato [61, § 1] has constructed a complex

$$H^3(k, \mathbf{Q}/\mathbf{Z}(2)) \to \bigoplus_{\xi \in \mathscr{B}_{1,b}} \mathrm{Br}(\kappa(\xi)) \to \mathbf{Q}/\mathbf{Z}, \tag{4.1}$$

where ξ ranges over the set $\mathscr{B}_{1,b}$ of 1-dimensional irreducible closed subsets of $\mathscr{B}$ that contain b, and where $\kappa(\xi)$ denotes the function field of ξ (which is either a global field of characteristic p or a local field of characteristic 0). The second arrow in (4.1) is the sum of the invariant maps from local class field theory at the finitely many places of $\kappa(\xi)$ that lie over b. The first arrow of (4.1) is induced by residue maps $\partial_v : H^3(k_v, \mathbf{Q}/\mathbf{Z}(2)) \to \mathrm{Br}(\kappa(\xi))$ constructed by Kato in [61], where v denotes the discrete valuation of k defined by ξ.

For any $\alpha \in H^3_{\mathrm{nr}}(X/k, \mathbf{Q}/\mathbf{Z}(2))$, there are only finitely many 1-dimensional irreducible closed subsets ξ of $\mathscr{B}$ such that the map $X(k_v) \to \mathrm{Br}(\kappa(\xi))$, $x \mapsto \partial_v(\alpha(x))$ does not identically vanish, if we denote by v the discrete valuation of k defined by ξ (see [24, PROPOSITION 2.7 (II)] and note that for the proof given there, it is enough to assume that a dense open subset of $\mathscr{B}$, rather than $\mathscr{B}$ itself, is a scheme over a field—an assumption satisfied here). As a consequence, it makes sense to define $X(\mathbf{A}_k)^{\mathrm{rec}}$ to be the set of $(x_v)_{v\in\Omega} \in X(\mathbf{A}_k)$ such that for any irreducible normal proper scheme $\mathscr{B}$ over $\mathbf{Z}_p$ with function field k, for any closed point $b \in \mathscr{B}$, and for any $\alpha \in H^3_{\mathrm{nr}}(X/k, \mathbf{Q}/\mathbf{Z}(2))$, the family $(\partial_v(\alpha(x_v)))_{\xi\in\mathscr{B}_{1,b}} \in \bigoplus_{\xi\in\mathscr{B}_{1,b}} \mathrm{Br}(\kappa(\xi))$ belongs to the kernel of the second arrow of (4.1). The fact that (4.1) is a complex immediately implies that $X(k) \subseteq X(\mathbf{A}_k)^{\mathrm{rec}}$.

4.5. Sufficiency of the reciprocity obstruction

Although evidence is scarce, the answer to the following question might always be in the affirmative, as far as one knows:

Question 4.1. Let k be a finite extension of $\mathbf{Q}_p(t)$. Let X be a smooth, proper, rationally connected variety over k. If $X(\mathbf{A}_k)^{\mathrm{rec}} \neq \varnothing$, does it follow that $X(k) \neq \varnothing$?

Question 4.1 has a positive answer when X is a quadric and $p \neq 2$. Indeed, we recall from Section 4.3 that even $X(\mathbf{A}_k) \neq \varnothing$ then implies $X(k) \neq \varnothing$ (see [23]). It also has a positive answer when X is birationally equivalent to a torsor under a torus over k. This follows from the work of Harari, Scheiderer, Szamuely, Tian [38, THEOREM 5.1], [97, § 0.3.1] (modulo the comparison between the reciprocity obstruction defined here and the reciprocity obstruction considered in these articles; the latter is weaker, but turns out to suffice to detect rational points on torsors under tori). We note that there are examples of torsors under tori over k whose smooth compactifications X satisfy $X(\mathbf{A}_k)^{\mathrm{rec}} = \varnothing$ while $X(\mathbf{A}_k) \neq \varnothing$ (see [24, REMARQUE 5.10]). Positive answers to Question 4.1 are known in various other cases in which X is birationally equivalent to a homogeneous space of a connected linear algebraic group over k. For specific statements, we refer the reader to the articles quoted in Section 4.3. Question 4.1 remains open in general for smooth compactifications of torsors under connected linear algebraic groups over k, for smooth compactifications of homogeneous spaces of SL_n with finite stabilisers, and for conic bundle surfaces over $\mathbf{P}^1_k$.

Question 4.1 focuses on the existence of rational points rather than on the density of $X(k)$ in $X(\mathbf{A}_k)^{\mathrm{rec}}$ as the latter property is only known for projective space (see [3, THEOREM 1]) and hence for varieties that are rational as soon as they possess a rational point, such as quadrics. For smooth compactifications of tori, the density of $X(k)$ in $X(\mathbf{A}_k)^{\mathrm{rec}}$ is known to hold off the set of discrete valuations of k whose residue field has characteristic p (see [37, THEOREM 5.2]; for the meaning of "off" here, see [103, DEFINITION 2.9]).

To obtain more positive answers to Question 4.1, it is natural to wish for flexible tools such as general descent theorems and fibration theorems. In the same way that introducing the tight approximation property and replacing Question 3.3 (2) with Question 3.8 was a key step to obtain a problem that behaves well with respect to fibrations into rationally connected varieties (see the discussion in Section 3.2), it is likely that in order to obtain compatibility with descent and fibrations, one will have to strengthen Question 4.1 by incorporating into it a p-adic analogue of the approximation condition in the Euclidean topology that appears in Definition 3.4. The main challenge, here, is to provide the correct formulation for such a p-adic tight approximation property.

We note that in any case, a general fibration theorem has to lie deep, as it would presumably give a direct route to the local–global principle for the existence of rational points on smooth projective quadrics over k (so far unknown when $p = 2$) and hence to the computation of the u-invariant of k (equal to 8; see Section 4.3). Indeed, in the case of conics over k, this local–global principle follows from Tate–Lichtenbaum duality [73]; applying a fibration theorem to a general pencil of hyperplane sections of a fixed smooth projective quadric of dimension $n \geq 2$ would allow one to deduce the general case by induction on n.

4.6. Further questions

A good understanding of rational points of rationally connected varieties over function fields of curves over p-adic fields, be it via Question 4.1 or otherwise, should shed light on concrete test questions such as the following:

Questions 4.2. Let p be a prime number and k be a finite extension of $\mathbf{Q}_p(t)$.

(1) Does the conjecture of Mináč and Tân on the vanishing of Massey products in Galois cohomology hold for k? (See Section 2.5 and [78,79].)

(2) Is there an algorithm that takes as input a smooth, projective, rationally connected variety X over k and decides whether X has a rational point?

One might approach the first of these questions by trying to mimic [49] over k, which would require making progress on the arithmetic, over k, of homogeneous spaces of SL_n with finite supersolvable geometric stabilisers.

To put the second question in perspective, let us recall what is known about algorithms for deciding the existence of rational points on arbitrary varieties ("Hilbert's tenth problem") over various fields of interest. Over $\mathbf{Q}$ or $\mathbf{C}(t)$, the existence of such an algorithm is an outstanding open problem. Denef [29] showed that over $\mathbf{R}(t)$, such an algorithm does not exist. His method was extended to prove that there is no such algorithm over $\mathbf{Q}_p(t)$ (Kim and Roush [62], completed by Degroote and Demeyer [27]), over any finite extension of $\mathbf{R}(t)$ that possesses a real place (Moret-Bailly [81]), or, when $p \neq 2$, over any finite extension of $\mathbf{Q}_p(t)$ (Eisenträger [32], Moret-Bailly [81]). In addition, over number fields, it is known that restricting from arbitrary varieties to smooth projective varieties makes no difference (see [96, § II.7], [98, THEOREM 1.1 (I)]). Restricting to smooth, projective, rationally connected varieties, however, does make a drastic difference: Question 4.2 (2) might well have an affirmative answer for all of the fields just mentioned. Over $\mathbf{C}(t)$, this is trivially so, by the Graber–Harris–Starr theorem. Over $\mathbf{R}(t)$, a positive answer to Question 4.2 (2) would follow from a positive answer to Question 3.8. Indeed, in the notation of Definition 3.4, if X satisfies the tight approximation property, then X has a rational point if and only if $f|_{\mathscr{X}(\mathbf{R})}$ admits a $\mathscr{C}^\infty$ section, a property that can be decided algorithmically. Over number fields, as was observed by Poonen [89, REMARK 5.3], a positive answer to Question 4.2 (2) would follow from the conjecture that rational points are always dense in the Brauer–Manin set. It seems likely that a positive answer to Question 4.1 would similarly imply a positive answer to Question 4.2 (2). To mimic Poonen's argument, one runs into the difficulty that the elements of $H^3_{\mathrm{nr}}(X/k, \mathbf{Q}/\mathbf{Z}(2))$ are harder to describe than those of $H^2_{\mathrm{nr}}(X/k, \mathbf{Q}/\mathbf{Z}(1)) = \mathrm{Br}(X)$, whose interpretation in terms of Azumaya algebras is a key point in [89, REMARK 5.3]; however, this can be remedied by viewing $H^3_{\mathrm{nr}}(X/k, \mathbf{Q}/\mathbf{Z}(2))$, using Bloch–Ogus theory, as the group of global sections of the Zariski sheaf associated with the presheaf $U \mapsto H^3_{\text{ét}}(U, \mathbf{Q}/\mathbf{Z}(2))$, and describing $H^3_{\text{ét}}(U, \mathbf{Q}/\mathbf{Z}(2))$ via Čech cohomology.

4.7. Other fields

There are a number of other fields over which a better understanding of rational points of rationally connected varieties would be valuable. One of the simplest example is the fraction field $k = \mathbf{C}((x, y))$ of the ring of formal power series $\mathbf{C}[[x, y]]$, which can be seen as a first step before considering function fields of complex surfaces. This field presents both local and global features, and a reciprocity obstruction can again be defined (in terms of the unramified Brauer group—recall that k has cohomological dimension 2). This obstruction was used in [24] to produce the first example of a torsor Y under a torus, over k, such that $Y(k) = \varnothing$ but $Y(k_v) \neq \varnothing$ for every discrete valuation v on k. The analogues of Question 4.1 and of Questions 4.2 can be asked over this field, too. It is not known, however, whether the reciprocity obstruction explains the absence of rational points on smooth proper varieties that are birationally equivalent to torsors under tori over k (though see [59, COROLLAIRE 4.4] for a closely related result involving possibly ramified Brauer classes). We refer the interested reader to [18, 22, 59, 60] for the state of the art.

ACKNOWLEDGMENTS

I am grateful to Jean-Louis Colliot-Thélène, Antoine Ducros, Hélène Esnault, János Kollár, and Ján Mináč for their comments on a first version of the text, and to Olivier

Benoist and Yonatan Harpaz for the pleasant collaborations that have led to the results reported on in this article.

REFERENCES

[1] D. Abramovich, J. Denef, and K. Karu, Weak toroidalization over non-closed fields. *Manuscripta math.* **142** (2013), no. 1–2, 257–271.

[2] C. Araujo and J. Kollár, Rational curves on varieties. In *Higher dimensional varieties and rational points (Budapest, 2001)*, pp. 13–68, Bolyai Soc. Math. Stud. 12, Springer, Berlin, 2003.

[3] E. Artin and G. Whaples, Axiomatic characterization of fields by the product formula for valuations. *Bull. Amer. Math. Soc.* **51** (1945), 469–492.

[4] O. Benoist and O. Wittenberg, On the integral Hodge conjecture for real varieties, I. *Invent. math.* **222** (2020), no. 1, 1–77.

[5] O. Benoist and O. Wittenberg, On the integral Hodge conjecture for real varieties, II. *J. Éc. Polytech. Math.* **7** (2020), 373–429.

[6] O. Benoist and O. Wittenberg, The tight approximation property. *J. reine angew. Math.* **776** (2021), 151–200.

[7] S. Bloch and A. Ogus, Gersten's conjecture and the homology of schemes. *Ann. Sci. Éc. Norm. Supér. (4)* **7** (1974), 181–201.

[8] J. Bochnak and W. Kucharz, The Weierstrass approximation theorem for maps between real algebraic varieties. *Math. Ann.* **314** (1999), no. 4, 601–612.

[9] A. Borel and A. Haefliger, La classe d'homologie fondamentale d'un espace analytique. *Bull. Soc. Math. France* **89** (1961), 461–513.

[10] M. Borovoi, The Brauer–Manin obstructions for homogeneous spaces with connected or abelian stabilizer. *J. reine angew. Math.* **473** (1996), 181–194.

[11] S. Bosch, W. Lütkebohmert, and M. Raynaud, *Néron models*. Ergeb. Math. Grenzgeb. (3) 21, Springer, Berlin, 1990.

[12] F. Campana, Connexité rationnelle des variétés de Fano. *Ann. Sci. Éc. Norm. Supér. (4)* **25** (1992), no. 5, 539–545.

[13] Y. Cao, Approximation forte pour les variétés avec une action d'un groupe linéaire. *Compos. Math.* **154** (2018), no. 4, 773–819.

[14] J.-L. Colliot-Thélène, Groupes linéaires sur les corps de fonctions de courbes réelles. *J. reine angew. Math.* **474** (1996), 139–167.

[15] J.-L. Colliot-Thélène, Rational connectedness and Galois covers of the projective line. *Ann. of Math. (2)* **151** (2000), no. 1, 359–373.

[16] J.-L. Colliot-Thélène, Variétés presque rationnelles, leurs points rationnels et leurs dégénérescences. In *Arithmetic geometry*, pp. 1–44, Lecture Notes in Math. 2009, Springer, Berlin, 2011.

[17] J.-L. Colliot-Thélène and P. Gille, Remarques sur l'approximation faible sur un corps de fonctions d'une variable. In *Arithmetic of higher-dimensional algebraic varieties (Palo Alto, CA, 2002)*, pp. 121–134, Progr. Math. 226, Birkhäuser Boston, Boston, MA, 2004.

[18] J.-L. Colliot-Thélène, P. Gille, and R. Parimala, Arithmetic of linear algebraic groups over 2-dimensional geometric fields. *Duke Math. J.* **121** (2004), no. 2, 285–341.

[19] J.-L. Colliot-Thélène, D. Harbater, J. Hartmann, D. Krashen, R. Parimala, and V. Suresh, Local-global principles for zero-cycles on homogeneous spaces over arithmetic function fields. *Trans. Amer. Math. Soc.* **372** (2019), no. 8, 5263–5286.

[20] J.-L. Colliot-Thélène, D. Harbater, J. Hartmann, D. Krashen, R. Parimala, and V. Suresh, Local-global principles for tori over arithmetic curves. *Algebr. Geom.* **7** (2020), no. 5, 607–633.

[21] J.-L. Colliot-Thélène, D. Harbater, J. Hartmann, D. Krashen, R. Parimala, and V. Suresh, Local-global principles for constant reductive groups over semi-global fields. 2021, arXiv:2108.12349.

[22] J.-L. Colliot-Thélène, M. Ojanguren, and R. Parimala, Quadratic forms over fraction fields of two-dimensional Henselian rings and Brauer groups of related schemes. In *Algebra, arithmetic and geometry, Part I, II (Mumbai, 2000)*, pp. 185–217, Tata Inst. Fund. Res. Stud. Math. 16, Tata Inst. Fund. Res, Bombay, 2002.

[23] J.-L. Colliot-Thélène, R. Parimala, and V. Suresh, Patching and local-global principles for homogeneous spaces over function fields of p-adic curves. *Comment. Math. Helv.* **87** (2012), no. 4, 1011–1033.

[24] J.-L. Colliot-Thélène, R. Parimala, and V. Suresh, Lois de réciprocité supérieures et points rationnels. *Trans. Amer. Math. Soc.* **368** (2016), no. 6, 4219–4255.

[25] J.-L. Colliot-Thélène and J.-J. Sansuc, La descente sur les variétés rationnelles. II. *Duke Math. J.* **54** (1987), no. 2, 375–492.

[26] J.-L. Colliot-Thélène and J.-J. Sansuc, The rationality problem for fields of invariants under linear algebraic groups (with special regards to the Brauer group). In *Algebraic groups and homogeneous spaces*, pp. 113–186, Tata Inst. Fund. Res. Stud. Math. 19, Tata Inst. Fund. Res, Mumbai, 2007.

[27] C. Degroote and J. Demeyer, Hilbert's tenth problem for rational function fields over p-adic fields. *J. Algebra* **361** (2012), 172–187.

[28] C. Demarche, G. Lucchini Arteche, and D. Neftin, The Grunwald problem and approximation properties for homogeneous spaces. *Ann. Inst. Fourier (Grenoble)* **67** (2017), no. 3, 1009–1033.

[29] J. Denef, The Diophantine problem for polynomial rings and fields of rational functions. *Trans. Amer. Math. Soc.* **242** (1978), 391–399.

[30] A. Ducros, Fibrations en variétés de Severi–Brauer au-dessus de la droite projective sur le corps des fonctions d'une courbe réelle. *C. R. Acad. Sci. Paris Sér. I Math.* **327** (1998), no. 1, 71–75.

[31] A. Ducros, L'obstruction de réciprocité à l'existence de points rationnels pour certaines variétés sur le corps des fonctions d'une courbe réelle. *J. reine angew. Math.* **504** (1998), 73–114.

[32] K. Eisenträger, Hilbert's tenth problem for function fields of varieties over number fields and p-adic fields. *J. Algebra* **310** (2007), no. 2, 775–792.

[33] T. Ekedahl, An effective version of Hilbert's irreducibility theorem. In *Séminaire de Théorie des Nombres, Paris 1988–1989*, pp. 241–249, Progr. Math. 91, Birkhäuser Boston, Boston, MA, 1990.

[34] T. Graber, J. Harris, and J. Starr, Families of rationally connected varieties. *J. Amer. Math. Soc.* **16** (2003), no. 1, 57–67.

[35] D. Harari, Méthode des fibrations et obstruction de Manin. *Duke Math. J.* **75** (1994), no. 1, 221–260.

[36] D. Harari, Quelques propriétés d'approximation reliées à la cohomologie galoisienne d'un groupe algébrique fini. *Bull. Soc. Math. France* **135** (2007), no. 4, 549–564.

[37] D. Harari, C. Scheiderer, and T. Szamuely, Weak approximation for tori over p-adic function fields. *Int. Math. Res. Not. IMRN* **10** (2015), 2751–2783.

[38] D. Harari and T. Szamuely, Local-global questions for tori over p-adic function fields. *J. Algebraic Geom.* **25** (2016), no. 3, 571–605.

[39] D. Harbater, Galois coverings of the arithmetic line. In *Number theory*, pp. 165–195, Lecture Notes in Math. 1240, Springer, Berlin, 1987.

[40] D. Harbater and J. Hartmann, Patching over fields. *Israel J. Math.* **176** (2010), 61–107.

[41] D. Harbater, J. Hartmann, V. Karemaker, and F. Pop, A comparison between obstructions to local-global principles over semiglobal fields. 2019, arXiv:1903.08007.

[42] D. Harbater, J. Hartmann, and D. Krashen, Applications of patching to quadratic forms and central simple algebras. *Invent. math.* **178** (2009), no. 2, 231–263.

[43] D. Harbater, J. Hartmann, and D. Krashen, Local-global principles for Galois cohomology. *Comment. Math. Helv.* **89** (2014), no. 1, 215–253.

[44] D. Harbater, J. Hartmann, and D. Krashen, Local-global principles for torsors over arithmetic curves. *Amer. J. Math.* **137** (2015), no. 6, 1559–1612.

[45] D. Harbater, J. Hartmann, D. Krashen, R. Parimala, and V. Suresh, Local-global Galois theory of arithmetic function fields. *Israel J. Math.* **232** (2019), no. 2, 849–882.

[46] D. Harbater, D. Krashen, and A. Pirutka, Local-global principles for curves over semi-global fields. *Bull. Lond. Math. Soc.* **53** (2021), no. 1, 177–193.

[47] Y. Harpaz, D. Wei, and O. Wittenberg, Rational points on fibrations with few non-split fibres. 2021, arXiv:2109.03547.

[48] Y. Harpaz and O. Wittenberg, On the fibration method for zero-cycles and rational points. *Ann. of Math. (2)* **183** (2016), no. 1, 229–295.

[49] Y. Harpaz and O. Wittenberg, The Massey vanishing conjecture for number fields. 2019, arXiv:1904.06512, to appear, *Duke Math. J.*

[50] Y. Harpaz and O. Wittenberg, Zéro-cycles sur les espaces homogènes et problème de Galois inverse. *J. Amer. Math. Soc.* **33** (2020), no. 3, 775–805.

[51] Y. Harpaz and O. Wittenberg, Supersolvable descent for rational points. Preprint, 2021.

[52] B. Hassett, Weak approximation and rationally connected varieties over function fields of curves. In *Variétés rationnellement connexes: aspects géométriques et arithmétiques*, pp. 115–153, Panor. Synthèses 31, Soc. Math. France, Paris, 2010.

[53] B. Hassett and Y. Tschinkel, Weak approximation over function fields. *Invent. math.* **163** (2006), no. 1, 171–190.

[54] D. R. Heath-Brown, Zeros of systems of $\mathfrak{p}$-adic quadratic forms. *Compos. Math.* **146** (2010), no. 2, 271–287.

[55] A. Hogadi and C. Xu, Degenerations of rationally connected varieties. *Trans. Amer. Math. Soc.* **361** (2009), no. 7, 3931–3949.

[56] Y. Hu, Weak approximation over function fields of curves over large or finite fields. *Math. Ann.* **348** (2010), no. 2, 357–377.

[57] Y. Hu, Hasse principle for simply connected groups over function fields of surfaces. *J. Ramanujan Math. Soc.* **29** (2014), no. 2, 155–199.

[58] Y. Hu, A cohomological Hasse principle over two-dimensional local rings. *Int. Math. Res. Not. IMRN* **14** (2017), 4369–4397.

[59] D. Izquierdo, Dualité et principe local-global pour les anneaux locaux henséliens de dimension 2. *Algebr. Geom.* **6** (2019), no. 2, 148–176.

[60] D. Izquierdo and G. Lucchini Arteche, Local-global principles for homogeneous spaces over some two-dimensional geometric global fields. 2021, arXiv:2101.08245, to appear, *J. reine angew. Math.*

[61] K. Kato, A Hasse principle for two-dimensional global fields. *J. reine angew. Math.* **366** (1986), 142–183.

[62] K. H. Kim and F. W. Roush, Diophantine unsolvability over p-adic function fields. *J. Algebra* **176** (1995), no. 1, 83–110.

[63] J. Kollár, *Rational curves on algebraic varieties*. Ergeb. Math. Grenzgeb. (3) 32, Springer, Berlin, 1996.

[64] J. Kollár, Rationally connected varieties over local fields. *Ann. of Math. (2)* **150** (1999), no. 1, 357–367.

[65] J. Kollár, Rationally connected varieties and fundamental groups. In *Higher dimensional varieties and rational points (Budapest, 2001)*, pp. 69–92, Bolyai Soc. Math. Stud. 12, Springer, Berlin, 2003.

[66] J. Kollár, Specialization of zero cycles. *Publ. Res. Inst. Math. Sci.* **40** (2004), no. 3, 689–708.

[67] J. Kollár, Holomorphic and pseudo-holomorphic curves on rationally connected varieties. *Port. Math.* **67** (2010), no. 2, 155–179.

[68] J. Kollár and F. Mangolte, Approximating curves on real rational surfaces. *J. Algebraic Geom.* **25** (2016), no. 3, 549–570.

[69] J. Kollár, Y. Miyaoka, and S. Mori, Rationally connected varieties. *J. Algebraic Geom.* **1** (1992), no. 3, 429–448.

[70] S. Lang, The theory of real places. *Ann. of Math. (2)* **57** (1953), no. 2, 378–391.

[71] S. Lang, Higher dimensional diophantine problems. *Bull. Amer. Math. Soc.* **80** (1974), 779–787.

[72] D. B. Leep, The u-invariant of p-adic function fields. *J. reine angew. Math.* **679** (2013), 65–73.

[73] S. Lichtenbaum, Duality theorems for curves over p-adic fields. *Invent. math.* **7** (1969), 120–136.

[74] Q. Liu, Tout groupe fini est un groupe de Galois sur $\mathbf{Q}_p(T)$, d'après Harbater. In *Recent developments in the inverse Galois problem*, pp. 261–265, Contemp. Math. 186, Amer. Math. Soc., Providence, RI, 1995.

[75] G. Lucchini Arteche, The unramified Brauer group of homogeneous spaces with finite stabilizer. *Trans. Amer. Math. Soc.* **372** (2019), no. 8, 5393–5408.

[76] Yu. I. Manin, Le groupe de Brauer–Grothendieck en géométrie diophantienne. In *Actes du Congrès International des Mathématiciens (Nice, 1970), Tome 1*, pp. 401–411, Gauthier-Villars, Paris, 1970.

[77] V. Mehmeti, Patching over Berkovich curves and quadratic forms. *Compos. Math.* **155** (2019), no. 12, 2399–2438.

[78] J. Mináč and N. D. Tân, Triple Massey products vanish over all fields. *J. Lond. Math. Soc. (2)* **94** (2016), no. 3, 909–932.

[79] J. Mináč and N. D. Tân, Triple Massey products and Galois theory. *J. Eur. Math. Soc. (JEMS)* **19** (2017), no. 1, 255–284.

[80] L. Moret-Bailly, Construction de revêtements de courbes pointées. *J. Algebra* **240** (2001), no. 2, 505–534.

[81] L. Moret-Bailly, Elliptic curves and Hilbert's tenth problem for algebraic function fields over real and p-adic fields. *J. reine angew. Math.* **587** (2005), 77–143.

[82] J. Neukirch, A. Schmidt, and K. Wingberg, *Cohomology of number fields. 2nd edn.* Grundlehren Math. Wiss. 323, Springer, Berlin, 2008.

[83] A. Pál and T. Schlank, The Brauer–Manin obstruction to the local-global principle for the embedding problem. 2016, arXiv:1602.04998.

[84] A. Pál and E. Szabó, The fibration method over real function fields. *Math. Ann.* **378** (2020), no. 3–4, 993–1019.

[85] R. Parimala, R. Preeti, and V. Suresh, Local-global principle for reduced norms over function fields of p-adic curves. *Compos. Math.* **154** (2018), no. 2, 410–458.

[86] R. Parimala and V. Suresh, The u-invariant of the function fields of p-adic curves. *Ann. of Math. (2)* **172** (2010), no. 2, 1391–1405.

[87] R. Parimala and V. Suresh, Period-index and u-invariant questions for function fields over complete discretely valued fields. *Invent. math.* **197** (2014), no. 1, 215–235.

[88] R. Parimala and V. Suresh, Local-global principle for unitary groups over function fields of p-adic curves. 2020, arXiv:2004.10357.

[89] B. Poonen, Heuristics for the Brauer–Manin obstruction for curves. *Exp. Math.* **15** (2006), no. 4, 415–420.

[90] B. Poonen, Existence of rational points on smooth projective varieties. *J. Eur. Math. Soc. (JEMS)* **11** (2009), no. 3, 529–543.

[91] F. Pop, Embedding problems over large fields. *Ann. of Math. (2)* **144** (1996), no. 1, 1–34.

[92] R. Preeti, Classification theorems for Hermitian forms, the Rost kernel and Hasse principle over fields with $cd_2(k) \leq 3$. *J. Algebra* **385** (2013), 294–313.

[93] F. Russo, The antibirational involutions of the plane and the classification of real del Pezzo surfaces. In *Algebraic geometry*, pp. 289–312, de Gruyter, Berlin, 2002.

[94] C. Scheiderer, Hasse principles and approximation theorems for homogeneous spaces over fields of virtual cohomological dimension one. *Invent. math.* **125** (1996), no. 2, 307–365.

[95] J.-P. Serre, *Topics in Galois theory. 2nd edn.* Res. Notes Math. 1, A K Peters, Ltd., Wellesley, MA, 2008.

[96] C. Smoryński, *Logical number theory. I.* Universitext. Springer, Berlin, 1991.

[97] Y. Tian, *Arithmétique des groupes algébriques au-dessus du corps des fonctions d'une courbe sur un corps p-adique*. Ph.D. thesis, Université Paris-Sud, 2020. https://tel.archives-ouvertes.fr/tel-02985977.

[98] Y. Tian, Obstructions to weak approximation for reductive groups over p-adic function fields. *J. Number Theory* **220** (2021), 128–162.

[99] Z. Tian, Weak approximation for cubic hypersurfaces. *Duke Math. J.* **164** (2015), no. 7, 1401–1435.

[100] Z. Tian and H. R. Zong, One-cycles on rationally connected varieties. *Compos. Math.* **150** (2014), no. 3, 396–408.

[101] C. Voisin, Some aspects of the Hodge conjecture. *Jpn. J. Math.* **2** (2007), no. 2, 261–296.

[102] O. Wittenberg, On Albanese torsors and the elementary obstruction. *Math. Ann.* **340** (2008), no. 4, 805–838.

[103] O. Wittenberg, Rational points and zero-cycles on rationally connected varieties over number fields. In *Algebraic geometry: Salt Lake City 2015*, pp. 597–635, Proc. Sympos. Pure Math. 97, Amer. Math. Soc., Providence, RI, 2018.

OLIVIER WITTENBERG

Institut Galilée, Université Sorbonne Paris Nord, 99 avenue Jean-Baptiste Clément, 93430 Villetaneuse, France, wittenberg@math.univ-paris13.fr

LIST OF CONTRIBUTORS